Heilongjiang
Greater Khingan Range
Manchurian Plain
Jilin
N
NGOLIA
Liaoning
NORTH KOREA
JAPAN
Beijing
Tianjin
Gobi
SOUTH KOREA
Inner Mongolia
Hebei
Shanxi
Shandong
Yellow Sea
Ningxia
Yellow River
Jiangsu
Henan
Gansu
Anhui
Shanghai
Shaanxi
ow River
Hubei
Zhejiang
East China Sea
Yangtze River
Sichuan
Chongqing
Jiangxi
Sichuan Plain
Hunan
Fujian
TAIWAN
Guizhou
Guangdong
Guangxi
Yunnan
VIETNAM
South China Sea
PHILIPPINES
Salween
Mekong
LAOS
Hainan

CHINESE MEDICINAL PLANTS, HERBAL DRUGS AND SUBSTITUTES

an identification guide

Dedication

To my parents, Zena and Peter, for always being there

and

To Professor Harold and Joan Lambert, without whose support this book would not be a reality

CHINESE MEDICINAL PLANTS, HERBAL DRUGS AND SUBSTITUTES

an identification guide

Christine Leon and Lin Yu-Lin

Kew Publishing
Royal Botanic Gardens, Kew

Royal Botanic Gardens Kew

IMPLAD 药用植物研究所

First published in 2017 by the Royal Botanic Gardens, Kew, Richmond, Surrey, TW9 3AB, UK
www.kew.org

ISBN 978 1 84246 387 1
e-ISBN 978 1 84246 630 8

Distributed on behalf of the Royal Botanic Gardens, Kew in North America by the University of Chicago Press, 1427 East 60th St, Chicago, IL 60637, USA.

British Library Cataloguing in Publication Data
A catalogue record for this book is available from the British Library.

Production Management: Georgina Hills and Andrew Illes
Copyediting: Ruth Linklater and Sharon Whitehead
Design and page layout: Christine Beard and Nicola Thompson

For information or to purchase all Kew titles please visit shop.kew.org/kewbooksonline or email publishing@kew.org
Kew's mission is to be the global resource in plant and fungal knowledge, and the world's leading botanic garden.
Kew receives about half of its running costs from Government through the Department for Environment, Food and Rural Affairs (Defra). All other funding needed to support Kew's vital work comes from members, foundations, donors and commercial activities, including book sales.

DISCLAIMER: This book does not constitute any form of medical advice, and is not intended to be a substitute for professional medical opinion, diagnosis or treatment. Always seek the advice of your doctor. The *Safety* section includes brief information about the safety profile of individual herbal drugs in their crude or processed forms, when used within the practice of Traditional Chinese Medicine (see also *How to use this book*). This safety information is derived entirely from existing literature, as cited, and the reader is encouraged to consult the source text in the event of any queries. This book is not intended to endorse or recommend the use of the stated plants or herbal drugs, and the authors and publishers accept no liability for any harm or adverse effect arising from any use of the plants or herbal drugs described.

FSC www.fsc.org MIX Paper from responsible sources FSC® C015829

Printed and bound in Italy by Printer Trento srl.
Holder of the following quality, printing and environmental certifications:ISO 12647-2:2004, Fogra® PSO (Process Standard Offset), CERTIprint®, ISO 14001:2004, FSC®(C015829), Carbon Trust Standard.

Contents

Nelumbo nucifera

Foreword

Plants are the source of most Chinese *materia medica*, a crucial component of China's national healthcare system. In recent years they have become part of a vast trade which has become a global phenomenon. Increasing demand and ever-lengthening supply chains have brought into sharp focus the need to ensure that these herbal drugs are sourced from the correct species, and devoid of substitutions and adulterants. While their accurate identification is essential in ensuring their quality and safe use, it is also key to the development of conservation plans for the plant species themselves, to enable harvesting in a way that both protects plant populations and also supports sustainable livelihoods.

This book will be an essential tool for a wide range of users to check plant and herbal drug identities and is designed with a number of distinctive features that make it especially helpful to non-botanical audiences. Bridging the gap between plants in the field and herbal drugs in the dispensary or laboratory is vitally important, and this book takes a major step in that direction. Great effort has gone into creating reference materials of herbal drugs in the areas where they are harvested, with associated herbarium vouchers that ensure accurate identity. This reference material has, in turn, formed the basis of most of the descriptions and photographs in the book.

This landmark work of scholarship also reflects the strength of the collaboration between the Institute of Medicinal Plant Development (IMPLAD), in Beijing, and the Royal Botanic Gardens, Kew, in London. Over a period of 19 years there have been 14 joint field expeditions and a series of academic staff exchanges that have been mutually enriching. These have involved many people beyond the authors of the book who have benefitted from opportunities for travel and training in new skills. This is a time of rapid development of research into Chinese *materia medica*, and I look forward to future results as this exciting collaboration develops yet further.

Xiao Peigen

Academician Xiao Pei-Gen
Honorary Director of Institute of Medicinal Plant Development,
Chinese Academy of Medical Sciences, Beijing

Acknowledgements

Our heartfelt thanks go to Professor Harold and Joan Lambert, Kew Patrons, whose immensely generous support of Kew's work in complementary herbal medicine has enabled this publication to become a reality. In their capacity as private sponsors of this work we especially thank them for their patience and unwavering support throughout its long gestation.

This book was born out of many years' collaboration by the Royal Botanic Gardens, Kew, (RBG, Kew) UK and the Institute of Medicinal Plant Development, (IMPLAD), Beijing (Chinese Academy of Medical Sciences, Peking Union Medical College). We thank both institutes for allowing us to make use of their research facilities and especially their extensive plant and *materia medica* collections. To the many staff of both institutes, too many to list in full here, we express our sincere thanks for sharing with us your time and invaluable expertise.

Few publishers would have agreed to tackle such a large and complex publication so we are truly appreciative to the entire team of Kew Publishing, led by Gina Fullerlove, for their professionalism in managing all aspects of its production. Special thanks go to Christine Beard (design) for going that extra mile to ensure style as well as accuracy, especially for the annotated drug photos and for her unending patience during the text's many iterations, to Ruth Linklater (copy editing) for her unflappable efficiency and attention to detail, to Lydia White (sales and marketing) for her positivity and creativity, Georgie Smith (production manager) for steering the project it in its early days and to Andrew Illes (later production manager) who expertly guided it over the last hurdles. Thanks also to Nicola Thompson (freelance designer) for her speedy, efficient and ever-cheerful editing, Sharon Whitehead for stop-gap editorial support and John Stone for design of the Kew-IMPLAD expeditions map.

We single out the following for their particular contributions:

Chen Shi-Lin (Director of the Institute of Chinese Materia Medica, China Academy of Chinese Medical Sciences, Beijing; former Director of IMPLAD) for facilitating and supporting for many years our collaborative research at IMPLAD and its branch institutes across China and our extensive fieldwork.

Michael Gilbert (European Co-ordinator and taxonomist for the *Flora of China* project, Missouri Botanic Garden, USA, based at RBG, Kew) for sharing so readily his taxonomic knowledge of China's plants and his skill and stoicism in checking the accuracy of the plant descriptions.

Elizabeth Dauncey of RBG, Kew's Medicinal Plant Names Services *(Kew-MPNS)* team for her expert resolution of challenging scientific names and taxonomic and nomenclatural guidance.

Mark Nesbitt (RBG, Kew) for his invaluable guidance and encouragement during the final stages of getting this work to press and for his meticulous checking of the references.

Li Bao-Li (IMPLAD, Beijing) for her cheerful efficiency in helping to manage our post-harvest reference materials, for field guidance in Anhui and Henan provinces and keeping the 'wheels oiled' on so many other fronts.

Sven Landrein, formerly of RBG, Kew, for his meticulous verification of our herbal drug herbarium voucher specimens.

Zhang Bengang (IMPLAD, Beijing) for expert advice throughout, especially his leadership in the field and advice on GAP-certified herbs.

Hu Bing-Yi (IMPLAD, Beijing) for sharing his knowledge on sources of high quality TCM drugs and their assessment, and being our 'king of logistics'.

Yu Hong-Wen, formerly of East-West Herbs Ltd, UK, for helping to foster the Kew-IMPLAD collaboration in its formative years including our first field expedition to south-west China and for sharing so readily her TCM pharmacognosy skills.

Debbie Shaw, formerly Head of the Chinese Medicine Advisory Service, Guy's & St Thomas' Hospital Trust, London for her toxicological expertise, fieldwork support in Guangxi and Hainan, China and her wise counsel throughout.

Rebecca O'Cleirigh, for so generously sharing her time and expertise as a TCM practitioner and Council Member of the UK Register of Chinese Herbal Medicine (RCHM) to review and ensure consistency in the TCM Properties and Uses presented for each herbal drug.

Xie Chen and Zhong Zhong, formerly of IMPLAD, Beijing, for their expert advice on many aspects of herbal TCM *materia medica* and wider logistical support during our work in China.

Qin Hai-Nin, Institute of Botany, China Academy of Sciences, Beijing, for advice on China's threatened plants and collaborative links with the Institute of Botany.

Kew Foundation for obtaining sponsorship (see *Sponsors*) for the development of the Kew-IMPLAD Chinese Medicinal Plant Authentication Centre and which provided the context for the writing of this book.

Special thanks also to the following staff and former staff of:

IMPLAD

Directors: Sun Xiao-Bo, Li Biao, Academician Xiao Pei-Gen (Beijing), Yang Shi-Lin, Wei Jian-He and Peng Yong. Also Chen De-Rong, Li Xue-Lan, Miao Jian-Hua, Peng Chao-Zhong, Qi Yao-Dong, Song Jing-Yuan, Yao Hui, Zhang Xin-Guo, Zhang Zhao and Zou Zhong-Mei.

RBG, Kew

Hannah Banks, David Barnes, Nicholas Black, Angela Bond, Madeleine Bond, Michelle Cleave, Chris Clennett, Dave Cook, Phillip Cribb, Tracy Dawkins, Sally Dawson, Steve Davis, Rogier de Kok, Sarah Edwards,

Joanne Evans, Peter Gasson, Madeleine Groves, Nicholas Hind, Nick Johnson, Tonia Friedrich, Tony Kirkham, Geoffrey Kite, Sonny Larsson, Gwil Lewis, Paul Little, Carlos Magdalena, Brian Mathew, Noel McGough, Eimear Nic Lughadha, Jonas Mueller, Alan Paton, Lauren Phelan, Sam Phillips, Sylvia Phillips, Gerhard Prenner, Katie Price, Hugh Pritchard, Chrissie Prychid, Diane Scott, Monique Simmonds, David Simpson, Rose Simpson, Kit Strange, Wolfgang Stuppy, Marcia Tolfts, Gemma Toothill, James Wearn, China Williams, Richard Wilford, Paul Wilkin, Elizabeth Woodgyer and Sue Zmarzty.

Especially: Bob Allkin (Leader, Medicinal Plant Names Services, Kew-MPNS) for advice about handling plant names; Henk Beentje and Juliet Williamson for permission to reproduce some definitions and drawings from *The Kew Plant Glossary*, Frances Cook for commenting on early drafts, curatorial support and constant encouragement, Martin Cheek and Steven Bachman for advice on handling conservation categories, Rui Fang for accessing obscure Chinese texts and providing meticulous translations, Rafaël Govaerts (Kew's Senior Content Editor, Plant and Fungal Names) for his expert resolution of challenging scientific names, Melanie Howes for phyto-pharmacy and herbal safety advice, Anne Marshall and Craig Brough for efficient library support, Jill Turner for curatorial support of our extensive *materia medica* reference specimens and Tim Utteridge for advice on early designs and terminology.

Field-work in China

Our extensive travels in China from 1998 onwards to visit the field localities for gathering fresh plants with which to make the TCM reference drugs illustrated here, would not have been possible without the expertise and goodwill of a myriad of local institutions and individuals from growers, to harvesters, processing centres, wholesalers, medical teams, botanists and field assistants. Their generosity to guide and transport as well as to feed and water us (and on one occasion provide us with superb emergency medical care) has often been overwhelming. To them all we extend our deep and heartfelt thanks.

Special thanks go to: Liu Qi-Fu and Wang Chao-Xing for guiding our extensive fieldwork expeditions in Sichuan, Gansu and Guizhou, Kaisa Sulaiman for his leadership in Xinjiang and liaison with local ethnic communities, Yi Tu-Ming for skilfully driving us vast distances in Sichuan, Gansu and Guizhou, Zhou Kang-You for guiding (and driving) us on our fieldwork in Anhui and Zhejiang provinces, Aaron Davis for plant collecting advice during our first joint Kew-IMPLAD field expedition back in 1998, Sasha Barrow and Georgina Pearman for their field support, humour and resourcefulness on expeditions to Jiangxi, Jilin, Guangxi, Sichuan and Zhejiang, the staff of Wen Xian Hospital, Gansu, especially its Director Zhang Hai-Kuan.

And to: Chen Qing-Ying, Huang Bi-Kui, Li Hong-Jun, Hu Shilin, Liao Fang-Hua, Su Zhi-Zhang, Tian Ye-Hua, Wang Xiu-Yu, Yang Huai, Yang Qiu-Xiang, Ye Shu-Qing, Zhang Jian, Zhao Chao-Qian and Zhao Chun-Feng.

To our many other long-standing collaborators in China we extend our thanks, notably Zhao Zhong-Zhen and Eric Brand (School of Chinese Medicine, Hong Kong Baptist University, Hong Kong), Wang Zheng-Tao (Shanghai University of Traditional Chinese Medicine, Shanghai), Chang Yuan-Shiun and Chang Wen-Te (China Medical University, Taichung, Taiwan) and to our many other taxonomic colleagues, notably Ma Ke-Ping at the Institute of Botany, China Academy of Science, Beijing.

Outside China we are indebted to the many traders, practitioners and researchers who have shared their knowledge as well as provided medicinal trade samples in support of our background research.

We thank especially: Annie Bligh (University of Westminster), Michael McIntyre (EHTPA), staff and practitioner members of *the* Register for Chinese Herbal Medicine (RCHM in the UK*)* especially Mazin Al-Khafaji, Tony Booker, Tony Harrison, Emma Farrant, Andrew Flower and Nick Lampert, Patrick Chen (Herbprime), Peter Houghton (King's College, London), George He, Dan Jiang, Etelka Leadley, Vincent Mak, Robert Miller, Uwe Schippmann, Zhao Kai-Cun (Association of Traditional Chinese Medicine and Acupuncture, UK), and staff of the Royal Botanic Gardens Edinburgh, notably Mark Watson, Janette Latta and Jin-Hyub Paik.

Sponsors for the Kew-IMPLAD Chinese Medicine Authentication Centre

We gratefully acknowledge generous funding from the Garfield Weston Foundation, the Maurice Laing Foundation (MLF) and the Register for Chinese Herbal Medicine (RCHM), UK.

Photographs

We extend our thanks to those individuals kindly allowing us to reproduce their photographs. These are credited individually in the captions of their respective photos. Photographs taken by staff and former staff of the RBG Kew we thank as follows: Hannah Banks (pollen grains of *Pinus* and *Typha* species) and Peter Gasson (anatomical sections of *Aquilaria* wood and *Paulownia tomentosa*). Special thanks also go to Andrew McRobb, formerly scientific photographer of RBG, Kew, for photographs taken on two Kew-IMPLAD field expeditions and for close-up photography of some of our reference drugs. All other photographs, by far the majority, have been taken by the authors, mainly Lin Yu-Lin.

To you all, a warm and heart-felt thank you.

All errors are ours and ours alone.

Christine Leon and Lin Yu-Lin

Prunella vulgaris

Introduction

THE NEED FOR THIS BOOK

Herbal drug ingredients (*materia medica*) are pivotal to the practice of herbal Traditional Chinese Medicine (TCM), an integral component of China's national healthcare system. Equally, recent decades have seen the development of a globalised market for Chinese herbal drugs, both for TCM practised outside China and to supply many other sectors, notably natural products and pharmaceuticals (Cheung, 2011; Dang *et al.*, 2016; Williamson *et al.*, 2013). This book is a response to this global phenomenon, reflecting rising demand for guidance on the identification of these *materia medica* and the plants from which they are sourced; many have become critically endangered in the wild.

Ingredient accuracy and quality are central to the safety and efficacy of any medical system and Chinese herbal drug ingredients are no exception. Increasing concern, however, about the lack of rigorous quality control systems, especially in international markets, to prevent the use of unofficial substitutes and adulterants has become urgent. Confidence in the identity of these ingredients underpins many aspects of their supply:

Safe and efficacious practice of herbal TCM: TCM follows established principles of diagnosis specific to a medical paradigm developed over a period of 2,000 years. It is also characterised by combining *materia medica* (up to about 15 in a single formula) using traditional knowledge of their intended interactions either in decoction or manufactured form. Such specialised formulae, founded on the use of the correct herbs, has long been recognised as a central tenet of herbal TCM practice (Li & Luo, 2003). With each herb known to play a specific role in each formula, use of an incorrectly identified herb can at best lead to a formula being ineffective or, at worst, result in an adverse reaction (Shaw, 2010).

Unambiguous nomenclature: the complexity of TCM nomenclature is well-known (Farah *et al.*, 2006; Wiseman & Feng, 1988). It is designed to reflect not only herbal identity, but herbal provenance, plant part, processing and even clinical quality. To the uninitiated, however, it is a minefield and a source of much confusion. This is compounded by the lack of a standardised nomenclature for herbs common to different national pharmacopoeia (notably those of China, Korea, Japan, Europe, UK and USA) (Zhao *et al.*, 2006a). Mis-interpretation of ingredient names has led to serious adverse reactions, most notoriously those associated with the genus *Aristolochia* (Debelle *et al.*, 2008). Herbal authentication resources, such as those presented in this book, that link TCM names to verified *materia medica* and their source plants, aim to provide identification clarity and help avoid future mis-interpretations.

Botanical rigour in Quality Assurance monographs: good practice guidelines for the development of pharmacopoeia (notably European and British) and other QA monographs for herbals are increasingly recognising the need to authenticate herbal materials macroscopically ahead of developing analytical tests methods for inclusion in QA monographs (Bauer & Franz, 2010).

Evidence-based research and clinical trial design: an essential pre-requisite for the design of scientifically valid ethnopharmacological research and clinical trials for herbal medicines is proof of identity of herbal starting materials; recently published guidelines about the design of clinical trials around TCM plants highlight this specific need (Flower *et al.*, 2014; Buriani *et al.*, 2012; Chan *et al.*, 2012).

Pharmacovigilance: identification of herbs implicated in adverse reactions plays a vital role in toxicological case investigation and in the prevention of future adverse reactions (Farah *et al.*, 2006, Shaw *et al.*, 1997).

Sustainability and conservation: sustainable methods of growing and harvesting plants leads to a more reliable supply of raw materials as well as the conservation of their wild populations, their habitats, and associated livelihoods (Timoshyna *et al.*, 2015; Zhang *et al.*, 2010). Wild populations, for example, of *Glehnia littoralis* (source of Bei Sha Shan) are Critically Endangered. Despite the 'sustainable sourcing' of this species from medicinal cultivation, what is the long-term security of such crops if their narrow genepools are insufficiently robust to withstand the rigours of climate change, disease, etc? Reliance on healthy wild populations for breeding purposes therefore becomes vital, hence the urgency to conserve the wild species.

Erosion of knowledge that links *materia medica* with their source plants can have far-reaching impacts as outlined above. This book therefore gives equal weighting to the identification of *materia medica* and their source plants.

SCOPE OF THIS BOOK

This book is a guide to the macroscopic identification of 226 Chinese herbal drugs (in their crude and processed forms) and to the morphological identification of their source plants. These represent almost half of the herbal drugs documented in the Chinese Pharmacopoeia (CP2015), so-called *official* drugs. The book also includes detailed comparative identifications of the plants and trade items of over 150 unofficial substitute and adulterant species.

Main features of the book include:

- reference specimens purpose-made from fresh plants, with verified herbarium vouchers (see *Creation of Reference Materials*, page 3)
- photographs of living plants and drugs, annotated with key features
- up-to-date taxonomy and nomenclature for all Latin scientific names
- comparative identification of official (i.e. Pharmacopoeia) and unofficial species of living plants, and their herbal drug counterparts in international trade
- guidance on robustness of identifications
- conservation categories for all species
- safety summaries

SELECTION OF HERBAL DRUGS

A book of this kind can never claim to be comprehensive. Several criteria, however, have been selected to satisfy a wide readership: TCM traders, herbal dispensaries and practitioners, health regulators, conservationists including CITES enforcement officers, academia and the natural product industries as well as those involved in the cultivation and sustainable supply of the plants themselves. These criteria include their popular use in Traditional Chinese Medicine (TCM), their occurrence in international trade, their inclusion in the Core Curriculum of the European Herbal & Traditional Medicine Practitioners Association (EHTPA), and, most importantly, where there is evidence of their substitution or adulteration.

PHARMACOPOEIA SPECIES AND UNOFFICIAL SUBSTITUTES

A key feature of this book is the provision of identification characters to help differentiate official (CP2015) species and their drugs from unofficial substitutes. An *official* or *pharmacopoeia* species is one that is named in the CP2015 as the source of a herbal drug. The term *unofficial substitute* is used to refer to a plant species or trade item different to that named in the CP2015 for a given drug (see below).

As an actively practised medical system (used in over 100 countries) TCM is a dynamic knowledge system continuously evolving hence clinical preferences for different herbs evolve too. Such healthy growth in knowledge is reflected in the Chinese Pharmacopoeia which is revised every 5 years.

A feature of Chinese medicine is that a single drug can be sourced from a number of official plant species that have similar clinical properties. For example the herbal drug Bai Bu — Stemonae Radix can *officially* be sourced from the tuberous roots of *Stemona japonica*, *S. sessilifolia* or *S. tuberosa*. Such situations often reflect regional availability of different species.

In contrast, the use of unofficial substitutes can arise for a number of reasons. These include other clinical preferences, price and availability, as well as inadvertent use arising from identification confusion (Zhao *et al.*, 2006b). High value herbal drugs such as Chuan Bei Mu (Fritillaria Cirrhosae Bulbus) page 78 are especially prone to intentional substitution with 'less potent' look-alikes. Such information is given in a *Substitution Comment* box for each substitute described. It is beyond the scope of this book, however, to comment on whether substitutions described are clinically appropriate.

Unofficial substitutes known only from the literature and likely to be of local or historical interest in China only are listed separately within each account as 'Lesser Known Unofficial Substitutes'.

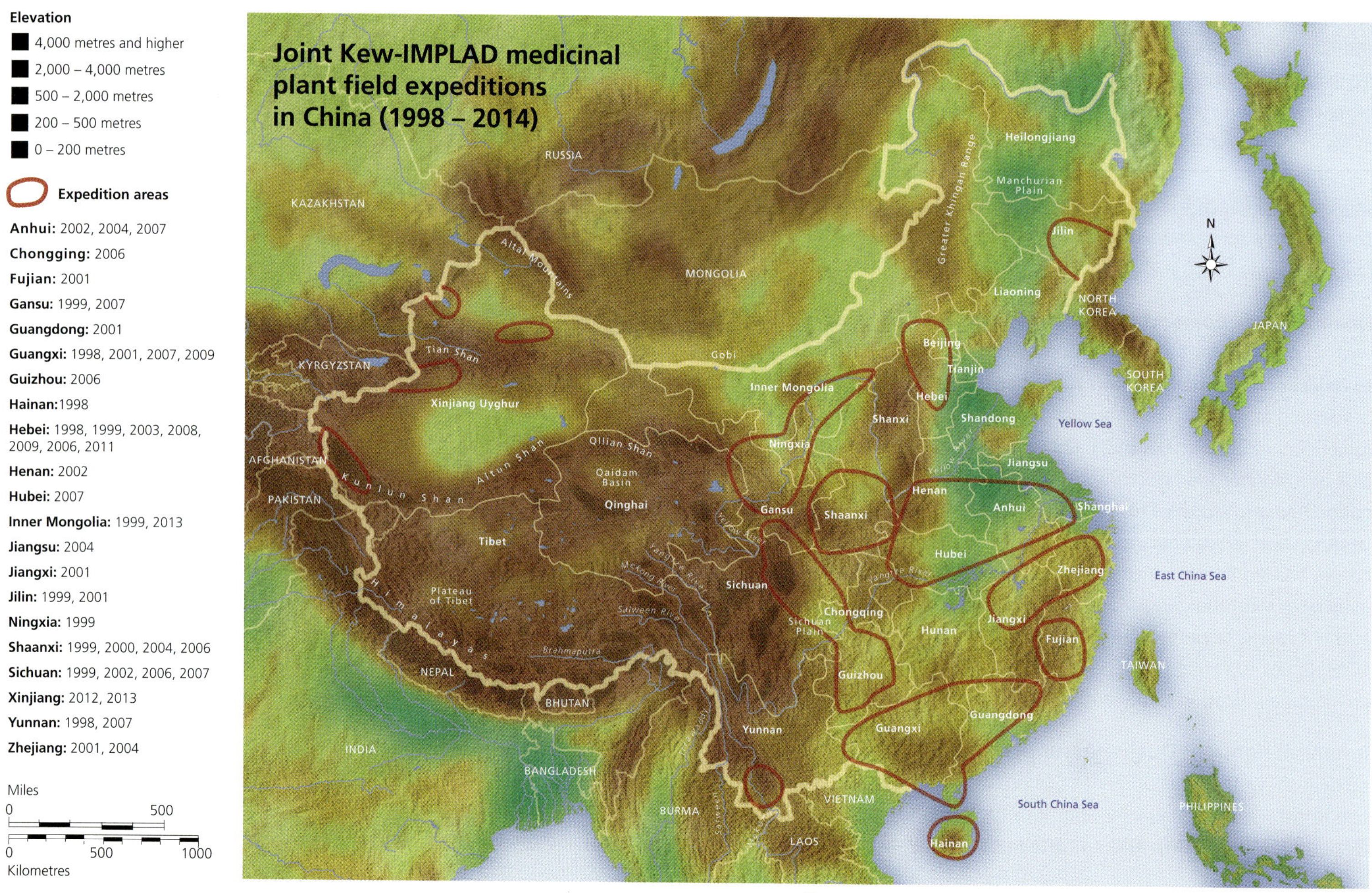

CREATION OF REFERENCE MATERIALS ILLUSTRATED IN THIS BOOK

Special attention has been given to ensuring all herbal drugs illustrated are sourced from the correct plant species. To this end, the authors have undertaken extensive field collecting of fresh plants across China during the period 1998–2014 (see map opposite and endpapers) to create typical TCM reference drugs prepared according to the traditional methods described in the CP. Herbarium voucher specimens were verified using floristic works such as FoC, Fu (1999) and Grey-Wilson & Cribb (2011). The resulting 4,500 physical reference specimens (drugs and cross-referenced herbarium vouchers) are housed in the Economic Botany Collection at the Royal Botanic Gardens, Kew (www.kew.org/science-conservation/collections/economic-botany) and at the Institute of Medicinal Plant Development (IMPLAD) in Beijing.

These purpose-made reference drugs have been sourced from wild-harvested plants or medicinal crop plants grown in accordance with current TCM production practices. With the help of local experts, special attention has been devoted to their preparation as crude and processed drugs and their decocting counterparts (yin pian) so as to be compliant with the macroscopic specifications detailed in the Chinese Pharmacopoeia. As such they have been prepared so as to be typical of good quality TCM *materia medica* trade items. In an effort to represent not just accurate species sourcing but also their traditional chemotypes/genotypes (notably those of *Di Dao*-quality, see Box 3) many duplicate reference drugs were created for a single species from different geographical localities.

CLOCKWISE FROM TOP LEFT

Making herbarium specimens in the field, Zhejiang province during a Kew-IMPLAD collecting trip;

Examining the quality of TCM herbal drugs for sale in a herbal market, Zhejiang province;

A herbarium-vouchered TCM reference drug, one of 4,500 housed at the Royal Botanic Gardens, Kew;

Seeds removed from freshly harvested fruits of *Trichosanthes kirilowii* to make a reference drug of Gua Lou Zi (Trichosanthis Semen);

Lin Yu-Lin (co-author) photographing *Coptis deltoidea* in Sichuan province to illustrate the drug account for Huang Lian (Coptidis Rhizoma).

How to use this book

ORDER OF MEDICINAL DRUG ACCOUNTS

The drug entries are arranged in the following order by colour-coded groups according to their plant parts:

- rhizomes, roots, tubers and bulbs
- aerial parts and whole plants
- stems and woods
- stem-barks and root-barks
- leaves
- flowers and flower parts
- fruits, seeds and other fruit parts

Within each of the above, drug accounts are arranged in alphabetical order of their Pin Yin names (as used in the CP2015).

GUIDE TO PAGE-LAYOUTS

Three page-layouts are used.

Layout A

A Pharmacopoeia drug sourced from a single species. Unofficial substitute species not known to occur regularly in international trade.

Layout B

A Pharmacopoeia drug sourced from more than one species. Comparative text and photographs highlight the key characters to distinguish them macroscopically. Unofficial substitute species not known to occur regularly in international trade.

Layout C

A Pharmacopoeia drug sourced from one or more species ***and*** where unofficial substitute species are known to be traded on the international market.

Layout A

Layout B

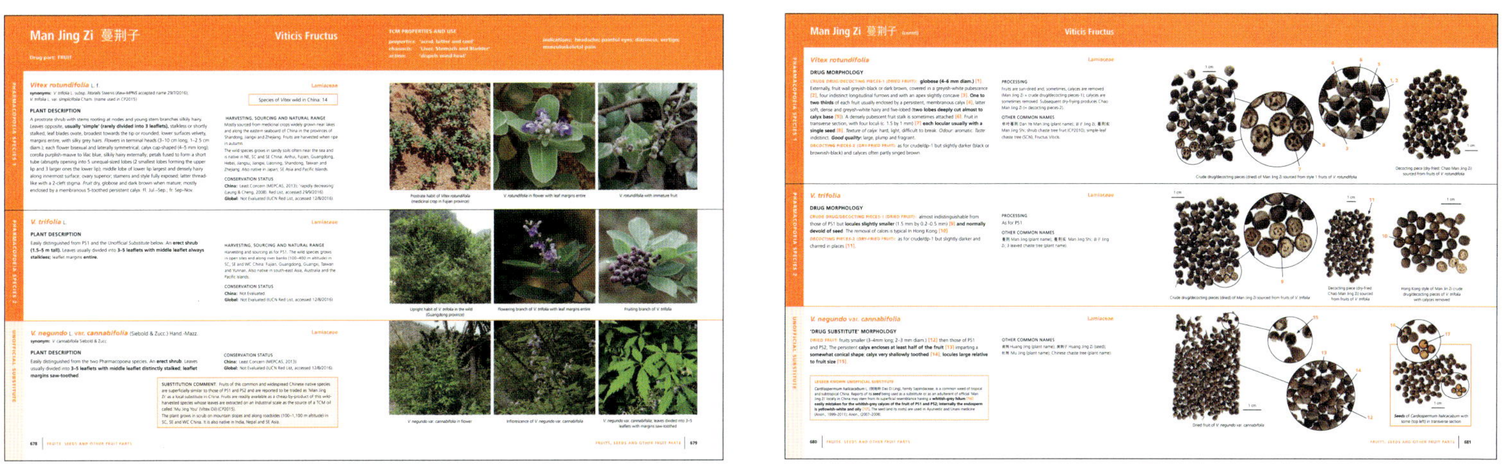

Layout C plants and drugs pages

OPPOSITE Kew's reference collection of vouchered Chinese herbal drugs

TOP BANNER

On the left-hand page the Pin Yin name (e.g. Di Gu Pi) is followed by its Chinese name (e.g. 地骨皮) pharmaceutical name (e.g. Lycii Cortex) then drug part (e.g. root-bark). These names follow those used in the CP2015. When processing of the herbal drug affects its safety, the words 'processed' or 'unprocessed' may follow the plant part.

TCM properties and use

Within TCM theory individual herbal drugs are traditionally classified according to their *Properties*, *Channels* and *Actions*. Those given in this book follow the *Chinese Pharmacopoeia* (English ed. CP2010). The CP2015 has yet to be translated.

These are for use only by those trained in herbal TCM. Note that the capitalisation of 'Kidney', 'Spleen', 'Lung', etc. follows TCM convention with each of these corresponding to a specific energetic system in the body; they do not equate to the specific organs as defined in Western medicine.

> **INDICATIONS** **BOX 1**
>
> These are included for the benefit of readers not trained in herbal TCM. TCM medical concepts such as Properties, Channels and Actions do not translate directly into those of Western medicine, thus the indications given are an approximation of the medical conditions treated. Typically TCM prescriptions are multi-ingredient, with 5–15 ingredients. Indications, and indeed safety, will therefore vary according to how they are combined. The prescribing of single herbs is uncommon in TCM.
>
> **IMPORTANT:** Indications given relate **only** to their use within the practice of TCM; they are not therefore indicative of much wider pharmacological activity that may have been reported by researchers from other domains.
>
> See *Glossary of Medical Terms*, page 12.

EHTPA

This denotes that the herbal drug is listed in the Chinese Herbal Medicine section of the European Herbal & Traditional Medicine Practitioners Association's *Core Curriculum* (EHTPA, 2014).

GAP

The abbreviation GAP indicates that a Good Agricultural Practice protocol has been published by the China Food and Drug Administration (CFDA) for this herbal drug and named growers in China have been licensed to grow and market it as a GAP-certified herb. GAP certification (Zhang *et al.*, 2010) provides an assurance of cultivated origin that the herbal drug meets strict standards of identity, purity and chemical quality, as well as traceability and compliance with heavy metal limits set by the Chinese Pharmacopoeia. Given that such licences are issued for a limited 5-year period only, readers should consult the CFDA website for the most up-to-date situation (China State Food and Drug Administration). Note that most GAP-certified herbs are destined for overseas markets. While they represent good quality they do not necessarily reflect top quality (see *Di Dao Herbal Drugs*, Box 3)

PLANT SPECIES PAGES

Taxonomy, nomenclature and synonyms

Each Latin scientific name follows the standard format of genus, species and author e.g. '*Lycium chinense* Mill.'. Where infraspecific varieties and subspecies are included they are represented as 'var.' and 'subsp.' respectively. See Box 2 for *Taxonomic Sources of Latin Scientific Names.*

Synonyms: the scientific synonyms (alternative names) provided after each Latin scientific name include the basionym (i.e. the original published name on which the current scientific name is based), together with synonyms in current use in the medicinal literature. The latter are mostly sourced from Kew's Medicinal Plant Names Services (*Kew-MPNS*).

Hybrids are indicated by the use of the symbol '×' before a species epithet e.g. *Citrus* × *wilsonii* Tanaka. Cultivar names are given in single quotes after the scientific name e.g. *Citrus maxima* (Burm.) Merr. 'Tomentosa'.

> **TAXONOMIC SOURCES OF LATIN SCIENTIFIC PLANT NAMES** **BOX 2**
>
> All Latin scientific plant names in the CP2015, and included in the present book, have been brought into line with accepted scientific names published in the *Flora of China* (*FoC*, 1994–2013) itself a revision of the earlier *Flora Reipublicae Popularis Sinicae* (*FRPS*, 1959–2004). Where a *FoC* accepted name differs from that in Kew's Medicinal Plant Names Services (*Kew-MPNS*), the MPNS accepted name is also added in the *Taxonomic/nomenclatural note* or *Synonyms* for that entry. For plant species not included in the *FoC*, Latin scientific names follow those accepted in *Kew-MPNS*. Where taxa are not included in either, then scientific names follow those accepted in the *World Checklist of Selected Plant Families* (*WCS*) or *The Plant List*.
>
> Any taxonomic or nomenclatural discrepancies with the CP2015 are clearly indicated under Taxonomic/nomenclatural notes or Synonyms for each species entry.
>
> The rationale for referencing Latin scientific names to both *FoC* and *Kew-MPNS* reflects the differing merits of the two taxonomic approaches. While the *FoC* offers a coherent and widely used taxonomy specific to China's native flora, it is taxonomically static and, like all published floras, is unable to keep abreast of subsequent changes in plant taxonomy or nomenclature. In contrast, the *Kew-MPNS* resource is global in coverage and regularly refreshed electronically to reflect current taxonomic opinion.

To provide a floristic context for each Pharmacopoeia species described, the number of species of that genus native in China is given in a box e.g. 'Species of *Prunella* wild in China: 4' with numbers taken from the *FoC*. This provides a useful point of reference when identifying plants in the wild.

Plant family names follow those adopted in the molecular-based Angiosperm Phylogeny Group (APG III) Classification (Stevens, 2001–). Family names with common alternatives are Apiaceae (= Umbelliferae), Asteraceae (= Compositae), Brassicaceae (=Cruciferae), Lamiaceae (= Labiatae) and Fabaceae (= Leguminosae).

Plant Description

Descriptions have been pitched for readers with minimal botanical knowledge although the use of some technical terms has been unavoidable (see the *Glossary of Plant Terms* page 12). Unless stated otherwise plant descriptions have been based on technical descriptions in the *FoC*. To aid identification when comparing species within a single drug account, key characters are **emboldened**.

Harvesting, Sourcing and Natural Range

All information in this section relates to the sourcing of the fresh plant in China, unless otherwise stated. Information includes whether the plant is wild-harvested or sourced from medicinal cultivation or both; season of harvest is indicated following that given in the CP2015; habitat, altitudinal range, endemism and natural range worldwide follows that in the *FoC* . Within China, wild distributions are indicated at the province level (see map, page 2) with especially wide distributions described in more general terms such as 'NW and C China'; 'Inner Mongolia' is used for 'Nei Mongol' and 'Tibet' for 'Xizang'.

Areas of commercial medicinal crop production are listed by province and derived from the recent TCM literature and the authors' field expeditions (see map and figs on pages 2 and 3). Except for slow-growing and perennial crops such as *Paeonia* spp. (Mu Dan Pi), *Eucommia ulmoides* (Du Zhong) and *Crataegus pinnatifida* (Shan Zha), the vagaries of market demand from one year to the next have a major influence on the choice of crops grown hence areas under cultivation are in a continual state of flux. For general patterns of wild and cultivated medicinal plant production, readers are referred to the comprehensive four volume publication: *National Survey of Chinese Medicine Resources* (Wang, 2014).

An important and long-established designation specific to the sourcing of medicinal plants in China and indicative of high quality is the concept of *Di Dao* (or Dao Di) (see Box 3).

DI DAO/DAO DI HERBAL DRUGS **BOX 3**

The *Di Dao* concept (variously translated as *geo-* or *geo-authentic* herbs) dates back many centuries in China to herbal drugs that denote 'best quality' and are well documented in China's classic Ben Cao (herbal medicine texts). Their recognition derives from the subjective clinical performance of these herbs made by TCM practitioners over many years compared to the same species sourced from other localities within China. The factors that contribute to *Di Dao* status are generally recognised as a combination of environmental factors (local climate, soil, drainage, terrain) coupled with traditional cultivation and processing methods (Zhao *et al.*, 2012). It is likely that *Di Dao* drugs represent chemotypes and/or genotypes but may or may not be phenotypically distinct. (The *Di Dao* concept is probably analogous with that of 'terroir', a concept used worldwide in the production of wine.)

Di Dao drugs are highly sought after and command the highest market prices due to their perceived clinical superiority compared to non-*Di Dao* drugs of the same species sourced from elsewhere. Not all Chinese herbal drugs, however, contain such *Di Dao* components. Where they do exist, their trade names often reflect this with prefixes commonplace, at least in local markets e.g. 'Dang Gui' (*Angelica sinensis* root) when traded as 'Min Dang Gui' denotes its *Di Dao* status, with 'Min' indicating its provenance as the Min Shan mountain region in southern Gansu province. Such nomenclatures can become highly complex and require a good knowledge of China's historical geography (many place names have changed) plus a good grasp of herbal processing nomenclature. Despite the considerable information that can be gleaned from these names, herbal drug names used in international trade are much simplified.

The popularity of *Di Dao*-quality drugs, especially from overseas markets, has led not only to high prices but reports of falsification. This has led to more systematic efforts to identify the unique character traits of these drugs using laboratory-based methods. These will support the identification and traceability of commercial trade items and also guide the selection of new production sites (Xiao *et al.*, 2009; Zhao *et al.*, 2012). Climate change is already, however, influencing the location of present and potential *Di Dao* production sites.

In the USA, where there is extensive cultivation of some TCM herbal drugs, researchers are investigating the local production of high quality TCM herbs, ideally of *Di Dao*-quality, to meet the growing demands of their own domestic market.

Conservation Status

Conservation status in the wild: IUCN Red List Categories *(see Box)* for individual species denote the conservation status of their wild populations in China and at a global level. 'Global status' assessments are derived from *The IUCN Red List of Threatened Species* (IUCN, 2012c).

Assessments at the 'China' level (e.g. China: Least Concern) have been derived from other sources, notably the IUCN China Plants Specialist Group published online (MEPCAS, 2013). Where no national assessment is available the authors provide a provisional assessment based on their own *ad-hoc* field experiences in China (e.g. 'Least Concern (authors' pers. obs. 2011)'). Unless otherwise stated *IUCN Red List Categories and Criteria Version 3.1* have been followed (IUCN, 2012b).

IUCN categories in abbreviated form (IUCN, 2012b)

EXTINCT (EX) A taxon is Extinct when there is no reasonable doubt that the last individual has died.

EXTINCT IN THE WILD (EW) A taxon is Extinct in the Wild when it is known only to survive in cultivation, in captivity or as a naturalised population (or populations) well outside the past range.

CRITICALLY ENDANGERED (CR) A taxon is Critically Endangered when the best available evidence indicates it is facing an extremely high risk of extinction in the wild.

ENDANGERED (EN) A taxon is Endangered when the best available evidence indicates that it is facing a very high risk of extinction in the wild.

VULNERABLE (VU) A taxon is Vulnerable when the best available evidence indicates that it is facing a high risk of extinction in the wild.

NEAR THREATENED (NT) A taxon is Near Threatened when it has been evaluated against the criteria but does not qualify for Critically Endangered, Endangered or Vulnerable now, but is close to qualifying for or is likely to qualify for a threatened category in the near future.

LEAST CONCERN (LC) A taxon is Least Concern when it has been evaluated against the criteria and does not qualify for Critically Endangered, Endangered, Vulnerable or Near Threatened. Widespread and abundant taxa are included in this category.

DATA DEFICIENT (DD) A taxon is Data Deficient when there is inadequate information to make a direct, or indirect, assessment of its risk of extinction based on its distribution and/or population status. A taxon in this category may be well studied, and its biology well known, but appropriate data on abundance and/or distribution are lacking. Data Deficient is therefore not a category of threat.

NOT EVALUATED (NE) A taxon is Not Evaluated when it has not yet been evaluated against the criteria.

> **CITES** **BOX 4**
>
> The aim of CITES is to ensure that international trade in specimens of wild animals and plants does not threaten their survival. CITES regulates international trade in over 30,000 plant species via their listing on 3 Appendices. Only 15 CITES-listed species are included in the CP2015 and, of these, 14 are included in this book (Table 1). The latter provides details of the precise plant parts and derivatives that are covered for each species or genus.
>
> For guidance on CITES and how to obtain permits for medicinal plants and their products contact the relevant national CITES Management Authority; contact details are given on the CITES website: cites.org/eng/cms/index.php/component/cp. It is important that permit applications are made well in advance (1–2 months) of exporting or importing CITES-listed material.
>
> Note that the European Union implements CITES using the EU Wildlife Trade Regulations (CITES-EU) and these are stricter. The most important difference between the basic CITES controls and EU Wildlife Trade Regulations is that the EU requires import permits for all CITES-listed material in addition to CITES export permits. Species listed on these Regulations are indicated as 'EU – Annexes A, B' etc. in Table 1.
>
> Reproduced from the CITES website: www.CITES.org

Protected Status

Where species are protected in China they appear on the 1999 List of Wild Plant Species Protected in China (SEPA, 1999). These are indicated within each herbal drug account under *Protected Status* as 'China: Cat. I, II or III (SEPA, 1999). An updated Version 2, still provisional, is available at: www.plant.csdb.cn/protectlist.

CITES (Convention on International Trade in Endangered Species of Wild Fauna and Flora): for CITES-listed species the relevant Appendix (or Annex, in the case of the European Union) is given. e.g. 'CITES Appendix I'; or 'EU Annex A'. See Box 4 and Table 1.

DRUG PAGES

Drug Morphology

All photographs of crude and processed drugs are of purpose-made reference specimens created by Kew and IMPLAD as a direct result of field-work in China (see *Creation of Reference Materials Illustrated in this Book*, page 3).

Macroscopic descriptions and annotated photographs are provided for crude and processed drugs and their decocting pieces (*Yin Pian*) based on the specifications given in the CP2010 (English ed.).

Where more than one species is included in an account (be it a Pharmacopoeia species or Unofficial Substitute species, key identification characters are **emboldened**. When macroscopic characters cannot reliably distinguish between species, this is stated, and laboratory-based authentication methods are recommended instead. Characters indicating 'good quality' are included when known.

The term *'good quality'* reflects characters used traditionally in TCM to denote good therapeutic quality based purely on organoleptic characters (physical appearance, taste, odour and fracture). Knowledge of these characters has been acquired and refined over the years in China as part of an informal iterative dialogue between harvesters, growers, traders and clinicians and adopted widely in TCM texts. In recent years such superior 'therapeutic' characters have often been validated by ethnopharmacological research (Zhao *et al.*, 2012; Xie & Leung, 2009).

The identification approach of herbal drugs presented in this book relies solely on macroscopic examination involving the use of the naked eye, x10 hand lens and stereomicroscope. Characters assessed include shape, size, colour, surface ornamentation, texture, fracture, longitudinal, transverse and oblique sections as well as the more traditional characters of smell and taste typical of the broader TCM organoleptic approach.

Many standard TCM *materia medica* identification texts have been used (notably Applequist, 2006; Bensky *et al.*, Bisset & Wichtl, 1984; 2004; Chang & But, 1986; Chang & Ho, 2013; Chen *et al.*, 2010a, Chen & Chen, 2004; Chen & Lin, 2008, 2013; Chen *et al.*, 2014; Department of Health, Hong Kong, 2005–, Ho *et al.*, 2006; Hsu *et al.*, 1986; Hu, 1989; Huang, 1999; Huang *et al.*, 2009b; Li & Luo, 2003; Lin, 1990–1996; Lin *et al.*, 2003; Lou, 1995–96; NICPBP, 1997–2002; Tang & Eisenbrand, 2011; Wang *et al.*, 2013; Wright, 2004; Xiao *et al.*, 2001–2002; Xu & Xu, 1997; Yen, 1992; Zhang, 1995–1996; Zhang, 2008; Zhao, 2004; Zhao *et al.*, 2006b; Zhao & Chen, 2014; Zhao & Li, 2007; Zhao & Xiao, 2009-10; Zhu, 1998; Zou & Qin, 1996). Herbarium and herbal drug collections have been consulted at the Royal Botanic Gardens Kew, IMPLAD (CAMS, Beijing) and the Institute of Botany (CAS, Beijing).

Taste: Within TCM it is traditional practice when authenticating or assessing the quality of a herbal drug to 'taste' the drug in question. This enables the drug to be assessed using its TCM attributes notably pungent, sweet, salty, etc. Such attributes are included in this book for completeness and to honour and reflect this tradition, but do not amount to a suggestion or recommendation to taste such herbal drugs (see Disclaimer, page 9).

> **CRUDE DRUGS AND DECOCTING PIECES** **BOX 5**
>
> A crude drug is one that has undergone one or more initial stages of preparation (usually washing, drying and sometimes cutting into short lengths as often required for large and bulky items such as certain rootstocks and barks). Such preparation optimises shelf-life, conserves medicinal attributes, facilitates transport once in trade, and enables size and quality grading. Crude drugs are unlikely to be seen in international trade unless they are identical to their decocting pieces.
>
> A decocting piece or 'yin pian' is a herbal drug ready for use in a multi-ingredient TCM decoction (the most traditional mode of administering a TCM herbal medicine). The preparation usually involves simple mechanical cutting or slicing to create sliced stems, roots, rhizomes and leaves, or hulled, de-seeded or halved fruits. The purpose is to optimise the release of active principles once in a decoction, while also to remove unwanted plant parts.
>
> The examples described in this book aim to correspond with those described in the CP2010 (English ed.); decocting pieces in the latter are only briefly described. These are expanded considerably in this book. The macroscopic appearance of some drugs will vary according to their production locality within mainland China. Wherever possible, some alternative forms of decocting piece (especially those typical of Hong Kong) are included. Decocting pieces often vary in the preparation of their underground parts, such as roots and rhizomes, where slicing may be longitudinal, transverse or oblique. See also *Processing*.

Processing

Processing methods are pharmaceutical techniques collectively known as 'Pao Zhi' and have evolved down the centuries in line with TCM theory and documented in detail in TCM *materia medica* texts 'Ben Cao' with the first monographic text on the topic appearing around 500 AD in the classic Lei Gong Pao Zhi Lun (Shang, 1991). Knowledge of their use and preparation is integral in TCM practice and their primary purpose is to modify therapeutic effects and reduce toxicity (Shaw, 2010; Sionneau, 1995; Zhao *et al.*, 2010).

Most herbal drugs undergo simple mechanical processing such as cutting or slicing (see Box 5) but for others processing is considerably more complex and results in chemical changes (e.g. 'stir-baking', 'boiling', 'carbonising' or 'calcining' with or without adjuvants such as with honey, wine or vinegar). Essentially 'processing' refers to the physical or chemical alteration of a crude drug into a herbal ingredient with the CP describing 15 such processing methods. Macroscopic examination can sometimes be used to distinguish processed from unprocessed drugs. Processing that affects the toxicity of a herbal drug is especially important to recognise (for reasons of safety) e.g. Ban Xia (Pinellia Rhizoma) pages 50–56.

Lack of processing standardisation across China, however, means that processing styles (physical and chemical) vary considerably between manufacturers for a single herb such that there is huge variability in the market place and no one text can claim to illustrate all. This book aims to illustrate only those in common use and known to be present in international trade.

Other Common Names

These have been obtained from diverse sources, including Bensky *et al.* (2004), the *Flora of China* (for plants), and other world pharmacopoeias covered by *Kew-MPNS*. Standard Common Names (SCN) are from McGuffin *et al.* (2000). Names in single quotes '.....' are a translation from the Chinese.

Safety

The following main texts have been consulted in order to provide brief information on the safety of each herbal drug: CP2015; Anon (2002a, b); Anon (2002b); Bensky *et al.*, (2004); Bone & Mills, 2013; FDA; Hong Kong Chinese Medicine Ordinance 'Schedules 1 and 2' (Hongkong, 1999a); Huai *et al.*, 2010 and TOXNET. Information from Gardner & McGuffin (2013) is presented using their classification of safety and interactions (see Box 6). Not included are allergic reactions or adverse effects arising due to over-dose. References to China's official healthfood and functional food legislation provide a wider context since many TCM herbal drugs are also officially marketed as healthfoods or functional foods.

Processing can have a profound affect on the toxicity of a herbal drug. See *Processing* for further details.

Lesser Known Unofficial Substitutes/Unofficial Substitutes

For some drugs a box gives brief details of those unofficial substitute species unlikely to be traded internationally. These are based on literature reports. See *Pharmacopoeia Species and Unofficial Substitutes*, page 2.

SAFETY AND INTERACTION CLASSES according to Gardner & McGuffin (2013) **BOX 6**

SAFETY

Class 1: herbs that can be safely consumed when used appropriately.

Class 2: herbs for which the following use restrictions apply, unless otherwise directed by an expert qualified in the use of the described substance:

- 2a: for external use only;
- 2b: not to be used during pregnancy;
- 2c: not to be used while nursing;
- 2d: other specific use restrictions as noted.

Class 3: herbs to be used only under the supervision of a qualified expert. Specific labelling is recommended for Class 3 herbs.

HERB/DRUG AND HERB/HERB INTERACTION

Class A: herbs for which no clinically relevant interactions are expected.

Class B: herbs for which clinically relevant interactions are plausible.

Class C: herbs for which clinically relevant interactions are known to occur.

Table 1

CITES-listed species included in the Chinese Pharmacopoeia 2015

Herbal Drug name (page in this book)	Latin scientific name of official source species (for taxonomy see Introduction)	Plant family (as on CITES)	CITES[1] Appendix (name on CITES plus annotations, accessed 12/11/2016)		EU Wildlife Trade Regulations[2] (name on Regulation accessed 12/11/2016)	
			Appendix I*	Appendix II**	Annex A	Annex B
Bai Ji 白及 Bletillae Rhizoma (page 26)	*Bletilla striata* (Thunb.) Rchb.f.	Orchidaceae		FAMILY listing Orchidaceae spp. #4		FAMILY listing Orchidaceae spp.
Chen Xiang 沉香 Aquilariae Lignum Resinatum (page 426)	*Aquilaria sinensis* (Lour.) Spreng.	Thymelaeaceae		GENUS listing *Aquilaria* spp. #14		GENUS listing *Aquilaria* spp.
Gan Song 甘松 Nardostachyos Radix et Rhizoma (*not included in this book*)	*Nardostachys jatamansi* (D. Don) DC.; syns: *N. chinensis* Batalin; *N. grandiflora* DC.	Caprifoliaceae (Valerianaceae)		*Nardostachys grandiflora* #2		*Nardostachys grandiflora* DC.
Gou Ji 狗脊 Cibotii Rhizoma (page 150)	*Cibotium barometz* (L.) J. Sm.	Cibotiaceae (Dicksoniaceae)		*Cibotium barometz* #4		*Cibotium barometz* (L.) Smith
Hong Shen 红参 Ginseng Radix et Rhizoma Rubra (page 244)	*Panax ginseng* C. A. Mey.	Araliaceae		*Panax ginseng* #3. Only the population of the Russian Federation; no other population is included in the Appendices.		*Panax ginseng* C. Meyer. (Only the population of the Russian Federation; no other population is included in the Annexes to this Regulation.)
Mu Xiang 木香 Aucklandiae Radix (page 210)	*Aucklandia costus* Falc.; syns: *A. lappa* Decne., *Saussurea costus* (Falc.) Lipsch.; *S. lappa* (Decne.) Sch. Bip.	Asteraceae	*Saussurea costus*		*Saussurea costus* (Falc.) Lipsch.(Also known as *S. lappa*, *Auklandia lappa* or *A. costus)*	
Ren Shen 人参 Ginseng Radix et Rhizoma (page 240)	*Panax ginseng* C. A. Mey.	Araliaceae		*Panax ginseng*#3. Only the population of the Russian Federation; no other population is included in the Appendices.		*Panax ginseng* C. Meyer. (Only the population of the Russian Federation; no other population is included in the Annexes to this Regulation.)
Rou Cong Rong 肉苁蓉 Cistanches Herba (page 468)	*Cistanche deserticola* Ma	Orobanchaceae		*Cistanche deserticola* #4		*Cistanche deserticola* Ma
Shan Ci Gu 山 慈菇 Cremastrae Pseudobulbus; Pleiones Pseudobulbus (page 252)	*Cremastra appendiculata* (D. Don) Makino	Orchidaceae		Orchidaceae spp. #4		FAMILY listing Orchidaceae spp.
Shan Ci Gu 山 慈菇 Cremastrae Pseudobulbus; Pleiones Pseudobulbus (page 252)	*Pleione bulbocodioides* (Franch.) Rolfe	Orchidaceae		Orchidaceae spp. #4		FAMILY listing Orchidaceae spp.
Shan Ci Gu 山 慈菇 Cremastrae Pseudobulbus; Pleiones Pseudobulbus (page 254)	*Pleione yunnanensis* (Rolfe) Rolfe	Orchidaceae		Orchidaceae spp. #4		FAMILY listing Orchidaceae spp.
Shi Hu 石斛 Dendrobii Caulis (page 478)	*Dendrobium fimbriatum* Hook.	Orchidaceae		Orchidaceae spp. #4 & '*Dendrobium* note'		FAMILY listing Orchidaceae spp.
Shi Hu 石斛 Dendrobii Caulis (page 478)	*Dendrobium nobile* Lindl.	Orchidaceae		Orchidaceae spp. #4 & '*Dendrobium* note'		FAMILY listing Orchidaceae spp.
Tie Pi Shi Hu 铁皮石斛 Dendrobii Officinalis Caulis (page 486)	*Dendrobium catenatum* Lindl. syn: *D. officinale* Kimura & Migo	Orchidaceae		Orchidaceae spp. #4 & '*Dendrobium* note'		FAMILY listing Orchidaceae spp.
Xi Yang Shen 西洋参 Panacis Quinquefolii Radix (page 310)	*Panax quinquefolius* L.	Araliaceae		*Panax quinquefolius* #3		*Panax quinquefolius* L.
Xiao Ye Lian 小叶莲 (Sinopodophyll Fructus) (roots reported as an adulterant of Wei Ling Xian) (page 292).	*Sinopodophyllum hexandrum* (Royle) T. S. Ying; syns: *Podophyllum hexandrum* Royle; *Sinopodophyllum emodi* (Wall. ex Honigberger) T. S. Ying	Berberidaceae		*Podophyllum hexandrum* #2		*Podophyllum hexandrum* Royle
stems of *Flickingeria* spp. reported as an adulterant of Shi Hu 石斛 Dendrobii Caulis (page 480).	*Flickingeria* spp. (treated by WCS and Kew-MPNS as part of *Dendrobium*).	Orchidaceae		Orchidaceae spp. #4		FAMILY listing Orchidaceae spp.

Notes on Table 1

1: CITES: Convention on International Trade in Endangered Species of Wild Fauna and Flora. see www.cites.org

2: EU WildlifeTrade Regulations (Council Regulation (EC) No 338/97): see ec.europa.eu/environment/cites/legislation_en.htm

* For Appendix 1 and EU Annex A species, international trade in specimens of this species is permitted only in exceptional circumstances, e.g. if "artificially propagated" or for "genuine conservation research" see www.cites.org and ec.europa.eu/environment/cites/legislation_en.htm

** For Appendix II and EU Annex B species, international trade is permitted under licence, of both wild and artificially propagated material, provided an appropriate CITES permit is obtained in advance. Permit exemptions exist for trade between CITES-registered scientific institutions and as indicated, where relevant, in the numbered notes below. See also www.cites.org and ec.europa.eu/environment/cites/legislation_en.htm

#2: All parts and derivatives except:
a) seeds and pollen; and
b) finished products packaged and ready for retail trade.

#3 Whole and sliced roots and parts of roots, excluding manufactured parts or derivatives, such as powders, pills, extracts, tonics, teas and confectionery.

#4 All parts and derivatives, except:
a) seeds (including seedpods of Orchidaceae), spores and pollen (including pollinia).
b) seedling or tissue cultures obtained in vitro, in solid or liquid media, transported in sterile containers;
c) cut flowers of artificially propagated plants;

#14: All parts and derivatives except:
a) seeds and pollen;
b) seedling or tissue cultures obtained in vitro, in solid or liquid media, transported in sterile containers;
c) fruits;
d) leaves;
e) exhausted agarwood powder, including compressed powder in all shapes; and
f) finished products packaged and ready for retail trade, this exemption does not apply to wood chips, beads, prayer beads and carvings.

***Dendrobium* note:** 'artificially propagated hybrids' of the genus *Dendrobium* are exempt from CITES permitting requirements if certain conditions are met (see www.cites.org/eng/app/appendices.php).

Dendrobium fimbriatum PHOTO: ANDRE SCHUITEMAN

Glossary of Medical Terms

adaptogen: describes a herbal drug considered to help the body adapt to stress by stimulating the body's own defensive mechanism

amenorrhoea: absence of menstruation

dysmenorrhoea: painful menstruation

dyspnoea: difficult breathing

dysuria: painful urination

enuresis: urinary incontinence

epistaxis: nosebleed

haematemesis: vomiting blood

haematochezia: blood in the stools

haematuria: blood in the urine

haemoptysis: coughing blood

hypertension: high blood pressure

hypochondriac pain: pain in the upper part of the abdomen on either side, inferior to the thorax in the area of the lower ribs

leucorrhoea: white or yellow mucosal vaginal discharge

mastititis: inflammation of breast tissue

metrorrhagia: abnormal bleeding from the uterus

musculoskeletal: disorders of the joints, muscles, ligaments and tendons

postpartum stasis: halt in flow of normal body substances after childbirth

spermatorrhoea: involuntary discharge of semen

stranguria: difficulty in passing urine

T: topical use

Glossary of Plant Terms

Based on *The Kew Plant Glossary* (Beentje, 2016) unless otherwise stated.

achene: a small dry thin-walled fruit, not splitting when ripe, and containing a single seed

alternate: (of leaves) inserted at different levels of the axis

amplexicaul: embracing the stem (e.g. the leaf base or *stipules* extending to the side of the stem opposite to the main blade)

annulate: consisting of, or marked with, one or more rings or ring-like segments; may be a swelling or a ring of hairs, or even a different colour)

anther: the part of the *stamen* containing the pollen (see *stamen* for illustration)

bract: a modified and specialised leaf in the *inflorescence*

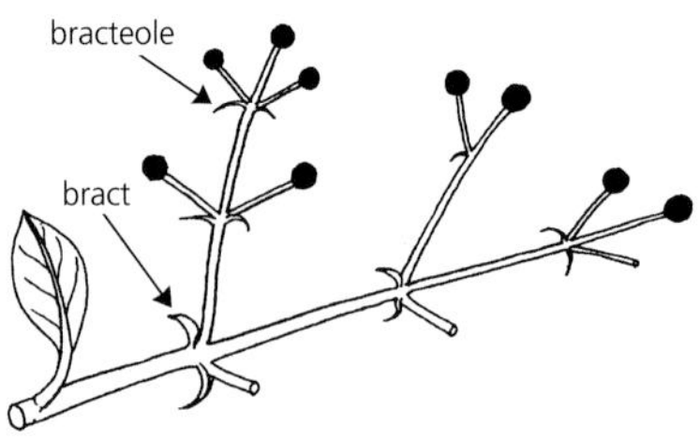

bracteole: a secondary *bract*, usually smaller than the *bracts* and always borne above them; a small modified leaf (or pair of modified leaves) borne just below the flower, or anywhere along a *pedicel* above the *bract*; often defined specifically as the *bract* at or near the base of the *pedicel*

callus: (plural calli), a thickening, for example, on one of the lips of some species in the Orchidaceae family (orchids)

calyx: (plural calyces), the outermost whorl of floral organs, often divided into *sepals* (see *sepal* for illustration)

campanulate: bell-shaped; with a tube about as long as wide, flared at its mouth

capitulum: (plural capitula), a compact cluster of more or less stalkless flowers; the capitulum may be surrounded by specialised *bracts*, the *involucre* (see *phyllary* for illustration)

carpel: the basic unit of the female sexual organ

chalaza: part of an ovule where the body joins the envelope

corolla: the second whorl of floral organs, inside or above the *calyx* and outside the *stamens*, consisting of free petals or of a joined tube with petal lobes

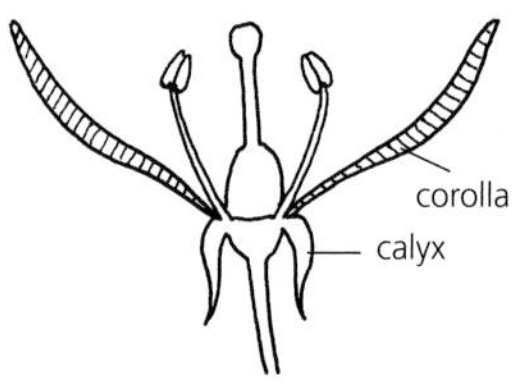

corymb: a more or less flat-topped, *racemose inflorescence* in which the branches or the *pedicels* (flower stalks) start from different points but all reach to about the same level

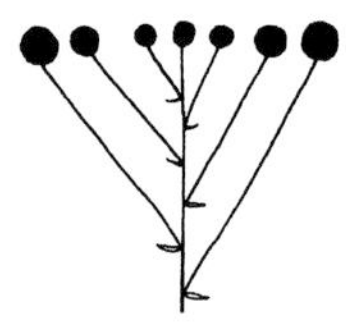

commissural surface: (of fruits in Apiaceae family) the place of joining (e.g. the faces of joining *carpels*)

compound: (of leaves) divided into leaflets

compound umbel: (of an *inflorescence*) an *umbel* of umbels

decussate: used of opposite organs (e.g. leaves), when alternate pairs are at right angles to each other

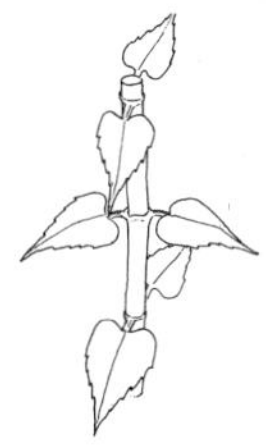

disc floret: (in the plant family Asteraceae) the florets in the centre of a flower-head (*capitulum*) (see also ray florets for illustration)

drupe: a stone fruit (e.g. plum, cherry), a fleshy fruit with the seed(s) enclosed in a stony *endocarp*

endocarp: (in a fruit) the innermost layer of a multi-layered fruit wall (e.g. the stone)

entire: (of leaf margin) smooth, unbroken by serrations, teeth or other irregularities

epithet: the second part of a Latin scientific name (e.g. in the name *Pteridium aquilinum*, '*aquilinum*' is the specific epithet)

filament: (of *stamens*) a stalk that bears an *anther* (see *stamen* for iillustration)

floral tube: (in flowers) a tube usually formed by the fusion of the lower parts of the petals of a flower (e.g. in a daffodil)

floret: 1. (in the plant family Asteraceae) a single flower; 2. (in Poaceae), the flower plus its *bracts* (*lemma* and *palea*)

fusiform: (of a 3-dimensional structure) straight, tapering from a wider middle towards both ends; = *spindle-shaped*

glume: (in the plant family Poaceae — grasses) the outermost of two *bracts* of a grass flower; see also *palea*

honey: (in the processing of herbal drugs) refined honey (as defined in Appendix II D of the Chinese Pharmacopoeia, 2015)

hypanthium: cup-shaped extension of the basal part of the flower sometimes imprecisely called a *floral tube*

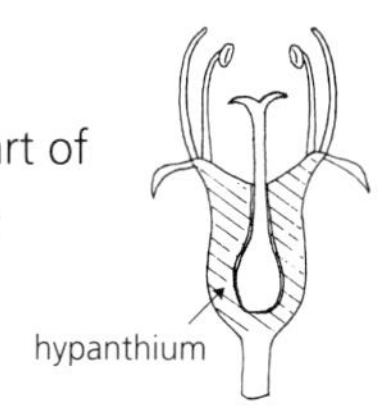

inferior: refers to an *ovary* that has the *calyx* above it

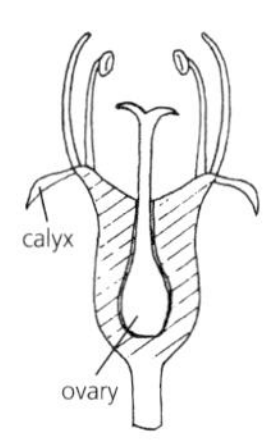

inflorescence: the part of the plant that bears the flowers, including all its *bracts*, branches and flowers

internode: the area between two *nodes*

involucre: group of *bracts* (the *phyllaries*), usually close together and below or around a compact head of flowers (*capitulum*) (in the plant family Asteraceae) (see *phyllary* for illustration)

lanceolate: narrowly *ovate* and tapering to a point at the apex

lateral: (in *leaflets*, flowers or *inflorescences*) on or at the side or margin; for example, lateral *leaflets* are those flanking the central *leaflet*

leaf midrib: the main support structure of a simple leaf-blade of *leaflet* running the full length of the leaf

leaflet: a part of a *compound* leaf (often mis-identified as a leaf by non-botanists) (see also *palmate*)

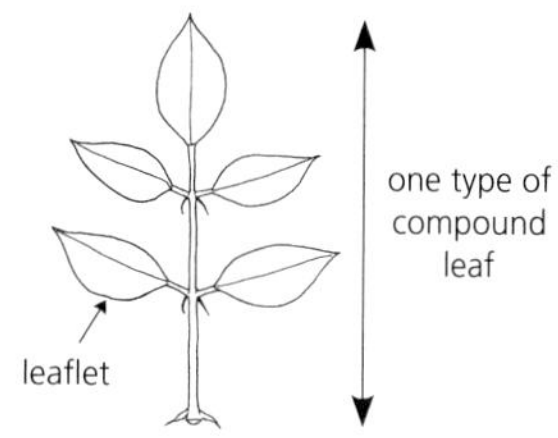

lemma: the outermost of two *bracts* of a grass flower; see also *palea*

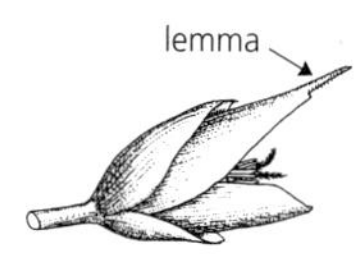

lenticels: corky eruptions on bark that allow gas exchange

ligule: 1. (in the plant family Poaceae — grasses) a projection of the leaf sheath at its tip; 2. (in the plant family Asteraceae), the 5-toothed strap-shaped kind of *floret*

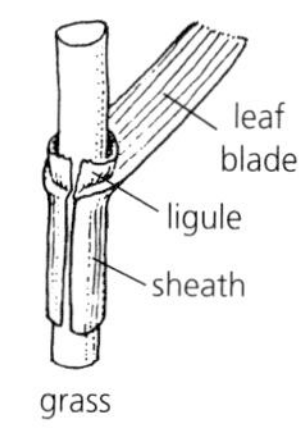

lyre-shaped: pinnately-lobed at base and with a large terminal lobe

node: the area of a stem where a leaf is attached or used to be attached; see also *internode*

obovate: egg-shaped (2-dimensional) with the broadest part near the tip

opposite: (of leaves) when two are borne on the same node but on diametrically opposed sides of the stem

organoleptic: characters identifiable using sight (macroscopic), smell, touch or taste

ovary: the (usually enlarged) part of the *pistil* that contains the immature seed and eventually becomes the fruit (see *inferior* for illustration)

ovate: egg-shaped (2-dimensional) with the wider part below the middle

palea: (plural paleae), (in the plant family Poaceae — grasses) the inner of two *bracts* of a grass flower; see also *glume*

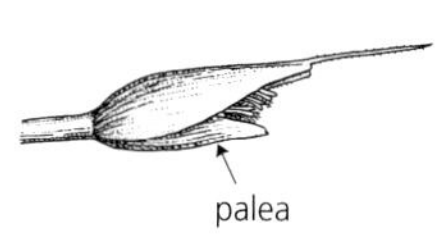

palmate: (in lobed or *compound leaves*) when all lobes or *leaflets* originate from one central point (as fingers originate from the palm of the hand)

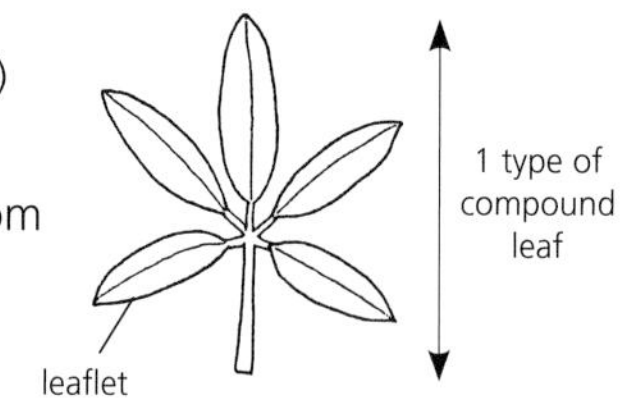

panicle: an *inflorescence* in which the main axis has several lateral branches, each of which is branched

pappus: a series of bristles, hairs or scales around the base of the *corolla* and later around the tip of a fruit (in the plant family Asteraceae)

pedicel: the stalk of an individual flower in an *inflorescence*

perianth: collective term for the *calyx* and *corolla*

phyllary: (plural phyllaries), (in the plant family Asteraceae) one or more of the *bracts* surrounding the *capitulum* and forming the *involucre*

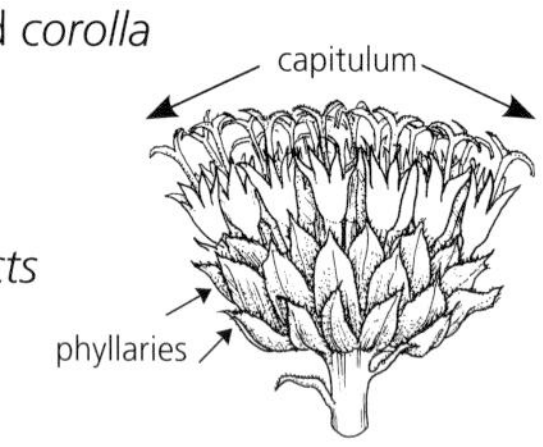

pinnate: (in *compound* leaves) divided into a central axis and several lateral *leaflets* (like a feather)

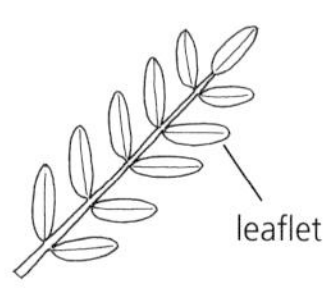

pistil: the female organ of a flower, consisting when complete of *ovary*, *style* and *stigma*

quilled: (in barks) see Box 6

raceme/racemose inflorescence: an *inflorescence* in which the flowers are borne (either on stalks or stalkless) along a central axis, with the *terminal* flowers being the youngest and last to open. There are many different types of racemes; the *spike* and *spadix* are racemes in which the flowers are stalkless

ray: one of the radiating branches of an *umbel* (in flower- heads of some species in plant family Apiaceae)

ray floret: (in the plant family Asteraceae) the floret at the margin of a flower-head (*capitulum*) (see also disc floret)

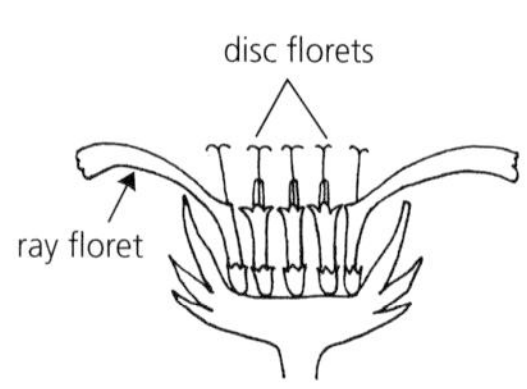

rhizome: underground stem, distinguished from root by its nodes, buds or scale-like leaves

sepal: a single part of the outermost whorl of floral organs, the *calyx*; usually green

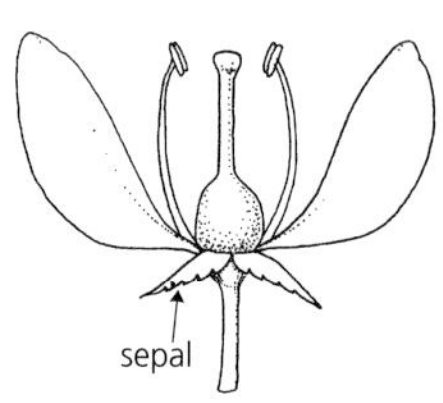

serrate: (usually of leaf margins) toothed like a saw, with regular acute and angled teeth pointing towards the leaf tip

serrulate: (usually of leaf margins) minutely *serrate*

simple: (of leaves) not divided into *leaflets*. (opposite to *compound*)

spathe: a large sheathing *bract*, surrounding the *inflorescence* or *spadix*

spadix: unbranched *inflorescence* with fleshy or thickened axis in which the flowers are (partly) sunken

spike: a *racemose inflorescence* with the flowers *alternate* and stalkless along a common unbranched axis

spikelet: (in the plant families Cyperaceae and Poaceae [grasses]) a typical flower structure in these families consisting of a group of *florets* (each consisting of *lemma*, *palea* and non-petal-like flower)

spindle-shaped: (of a 3-dimensional structure) straight, tapering from a wider middle towards both ends; = *fusiform*

spinule: a very small spine

spur: a tapering projection, usually short and curved

stamen: the male organ of a flower consisting of a stalk (*filament*) and *anthers* (the latter bearing the pollen)

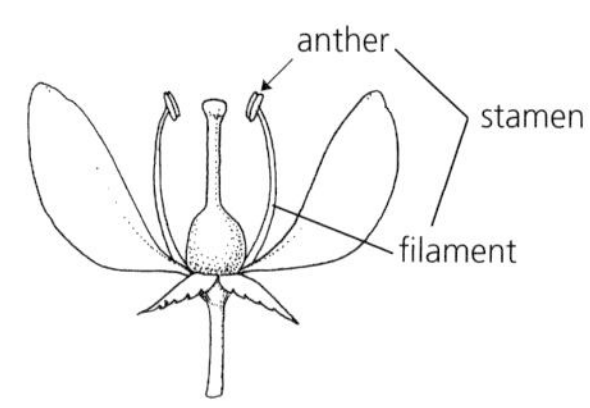

staminode: sterile *stamen*, usually much smaller than a true *stamen* and not bearing pollen

stele: (in anatomy) the part of a plant axis made up of the primary vascular system and its associated ground tissue

stigma: the pollen receptor in a flower (above the *style* and *ovary*)

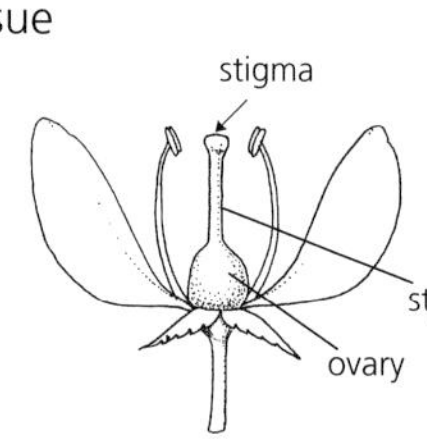

stipel: *stipule*-like outgrowth occurring (usually in pairs) at the base of a *leaflet* or a pair of *leaflets* in some *compound* leaves

stipule: leaf-like, spine-like or scale-like appendage of the leaf, usually in pairs at the base of the leaf-stalk

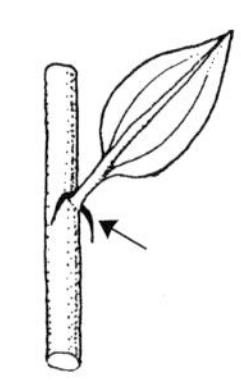

stolon: 1. vegetative shoot spreading along the surface of the ground and rooting at the nodes, where it may give rise to new plantlets; 2. (in the plant family Cyperaceae) a thin underground branch arising from the *rhizome* with each stolon terminating in an aerial shoot

style: the part of the flower between the *ovary* and the *stigma*, often slender and sometimes absent (see *stigma* for illustration)

superior: refers to an *ovary* that has the *calyx* below it

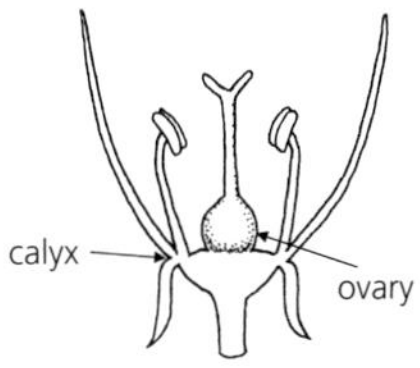

taxon: (plural taxa) a general term denoting a named group of any rank (e.g. variety, species or plant kingdom)

terminal: (of flowers or *inflorescences*) at the top of the plant

tortuous: twisted In different directions

trifoliate: a type of *compound* leaf with 3 *leaflets*

umbel: a type of *inflorescence* with branches arising from more or less the same point (in a simple *umbel*, each *ray* terminates in a flower; in a compound *umbel*, each *ray* itself bears an *umbel*)

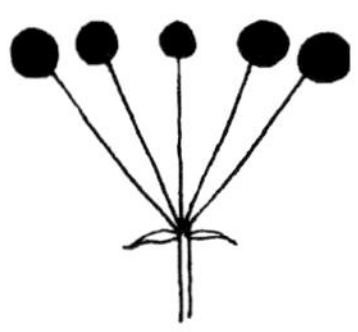

umbellet: a secondary *umbel* within a *compound umbel*

vinegar: (in the processing of herbal drugs) rice vinegar (as defined in Appendix II D of the Chinese Pharmacopoeia, 2015)

wine: (in the processing of herbal drugs) yellow rice wine (as defined in Appendix II D of the Chinese Pharmacopoeia, 2015)

BARKS (tree and root) **BOX 6**
based on Khatoon & Mehrotra, 2009

flat: (as in bark) often dried under pressure to become flat

curved: (as in bark) slightly concave on the inner side

re-curved: (as in bark) slightly concave on the outer side

channelled: (as in bark) when the curvature on the inside is deep enough to form a deep trough or channel

quill: (as in bark) strong curvature where one edge overlaps the other

double quill: (as in bark) when each edge is rolled independently into a *quill*

compound quill: (as in bark) when *quills* are packed one inside the other

Abbreviations

App. I, II and III:	Appendices I, II and III of CITES (see *Introduction and Bibliography*)
dbh.	diameter at breast height
dp	decocting pieces (yin pian), e.g. 'dp-1', 'dp-2' in Drug Morphology sections
CITES	Convention on International Trade in Endangered Species of Wild Fauna and Flora (see *Introduction* and *Bibliography)*
cm	centimetre
CP	Chinese Pharmacopoeia *(see Bibliography)*
EHTPA	European Herbal & Traditional Medicine Practitioners Association (see *Introduction* and *Bibliography)* EU Annex A, B, C, D (see *Introduction*)
FNA	*Flora of North America* (see *Bibliography*)
hl	×10 hand-lens
HPTLC	high performance thin layer chromatography
FoC	*Flora of China* (see *Bibliography*)
fl.	flowering
fr.	fruiting
GAP	Good Agricultural Practice (see *Introduction: How to use this book*)
IUCN	International Union for Conservation of Nature (see *Bibliography*)
Kew-MPNS	Kew Medicinal Plant Names Services (see Bibliography)
m	metre
mm	millimetre
NMR	nuclear magnetic resonance
nom. illeg.	illegitimate name
pers. comm.	personal communication
pers. obs.	personal observation
PS1, 2 etc.	Pharmacopoeia Species 1, 2 etc
RBG Kew	Royal Botanic Gardens, Kew, UK
sp.	species (singular)
spp.	species (plural)
SCN	Standard Common Names from Mc Guffin *et al.*, 2000
subsp.	subspecies
TCM	Traditional Chinese Medicine
WCS	World Checklist of Selected Plant Families
×	(in a Latin scientific name) before a specific epithet indicates hybrid origin e.g. *Citrus wilsonii* Tanaka
'...'	words in single quotes, usually under *Plant Description* (e.g. 'petals'), indicate that a popular but botanically slightly inaccurate term has been used in preference to a botanically correct one. Notably, the flowers of many species in families such as Liliaceae and Iridaceae (lilies, irises, crocuses) typically have tepals (i.e. sepals that look like petals) but since they resemble petals they are placed in quotes to help readers unfamiliar with botanical terms.

Strobilanthes cusia

rhizomes, roots, tubers and bulbs

Bai Bu 百部

Stemonae Radix

Drug part: ROOT TUBER

Stemona japonica (Blume) Miq.

Stemonaceae

synonym: *Roxburghia japonica* Blume

Species of *Stemona* wild in China: 7

PLANT DESCRIPTION

A **climbing**, slender and hairless perennial with clustered root tubers narrowing at both ends (6–13 cm long; 1–1.5 cm wide). *Stems* woody at base. *Leaves* in whorls, simple, **stalked** (1–4 cm long), leaf blade ovate or ovate to lance-shaped (usually 4–9 cm long; 1.5–4.5 cm wide), **veins 5–9**, margin slightly undulate and tip abruptly pointed. *Flowers* somewhat cryptic, at intervals up the stems; solitary or in groups and **emerging from base of leaf blade**; each flower bell-shaped, 'petals' 4 green, spreading and pointed; stamens 4 and purple. *Fruit* a 2-valved, oblong capsule (1–1.4 cm long; 0.4–0.8 cm wide); 2–3 seeded. Fl. May–Jul.

HARVESTING, SOURCING AND NATURAL RANGE

Tubers are harvested in summer or autumn from wild plants which are endemic to SE China (Anhui, Fujian, Hubei, Jiangsu, Jiangxi and Zhejiang) and where they grow in grassland (300–400 m altitude). Possibly also native to Japan.

CONSERVATION STATUS

Endemic to China: Data Deficient (MEPCAS, 2013); Not Evaluated (IUCN Red List, accessed 11/1/2016).

S. sessilifolia (Miq.) Miq.

Stemonaceae

synonyms: *Roxburghia sessilifolia* Miq.; *S. erecta* C. H. Wright

PLANT DESCRIPTION

Similar to PS1 but **shrub-like (30–70 cm high)** and **not climbing**. *Stems* unbranched. *Leaves* **shortly stalked or stalkless** with flowers **emerging from the base of the stems (opening at ground level only)**. ***Fruits* ovoid and smaller (0.7–0.9 cm long; 0.4–0.6 cm diam.)** than those of PS1. 1- or 2-seeded. Fl. Mar–May, fr. Jun–Jul.

HARVESTING, SOURCING AND NATURAL RANGE

Harvesting as for PS1. The species is endemic to SE China (Anhui, Fujian, Hubei, Jiangsu, Jiangxi and Zhejiang) where it grows on mountain slopes. Possibly also native to Japan.

CONSERVATION STATUS

Endemic to China: Data Deficient (MEPCAS, 2013); Not Evaluated (IUCN Red List, accessed 11/1/2016).

S. tuberosa Lour.

Stemonaceae

synonym: *S. acuta* C. H. Wright

PLANT DESCRIPTION

Similar to PS1 but **root tubers larger** (9–30 cm long; 1–2 cm diam.). *Leaves* opposite or whorled, leaf blades **7–13 veined**, flower stalks arise from the **base of the leaf stalks** not the leaf blades, 'petals' are striped green with purple veining and when mature are reflexed. ***Fruits*** ovoid-oblong, **much larger (2.5–6 cm long; 1–3 cm diam.).** *Seeds* several. Fl. Apr–Jul.

HARVESTING, SOURCING AND NATURAL RANGE

Tubers are harvested from wild plants which grow on mountain slopes, forest margins and thickets in parts of C, E and S China (Fujian, Guangdong, Guangxi, Guizhou, Hainan, Hubei, Hunan, Jiangxi, Sichuan, Taiwan and Yunnan); 300–2,300 m altitude. The species is also native to Bangladesh, Cambodia, India, Laos, Myanmar, Philippines, Thailand and Vietnam.

CONSERVATION STATUS

China: Least Concern (MEPCAS, 2013).
Global: Not Evaluated (IUCN Red List, accessed 19/1/2016).

TCM PROPERTIES AND USE

properties: 'sweet, bitter and neutral'
channels: 'Lung'
actions: 'relieves cough, moistens dryness'

indications: cough, dyspnoea, parasites; (T)

Stemona japonica in flower; leaf veins 5–9

S. japonica flower growing from leaf blade base

S. japonica tubers of freshly uprooted plant

Stemona sessilifolia; leaves virtually stalkless

Flowers of *S. sessilifolia* emerging at soil level

S. sessilifolia; small fruits at ground level

Stemona tuberosa, leaves 7–13-veined

Flowers of *S. tuberosa*

Large aerial fruits of *S. tuberosa*

Stemona japonica

Stemonaceae

DRUG MORPHOLOGY

CRUDE DRUG (DRIED TUBER): tapering at both ends **[1]**, often tortuous with a shrunken appearance **[2]** **(5–12 cm long; 0.5–1 cm wide)**. Externally yellowish-white or brownish-yellow, longitudinal furrows obvious, deep and irregular **[3]**; transverse furrows occasional and less obvious **[4]**. *Fracture*: even, yellowish-brown or yellowish-white and horny **[5]** with paler and denser central stele **[6]**. *Texture*: pliable, easily broken. *Odour*: indistinct. *Taste*: sweet with strongly bitter after-taste. **Good quality:** large, pliable and greyish-white.

DECOCTING PIECES-1 (SLICED TUBER): irregularly-shaped, often rhomboidal **[7]**, transverse slices (2–12 mm diam.); uncut edges deeply furrowed longitudinally **[3]**. *Texture*: pliable.

DECOCTING PIECES-2 (SLICED AND HONEY-PROCESSED TUBER): as above but all external surfaces brownish-yellow or brown, occasionally with scorch marks and slightly sticky. *Taste*: sweet.

PROCESSING

Tubers are steamed or blanched then sun-dried (= crude drug) before being obliquely cut into slices >2 mm thick (= decocting pieces-1). Further optional processing involves steeping in honey then dry-frying to produce 'Mi Bai Bu' (= decocting pieces-2).

OTHER COMMON NAMES

小百部 Xiao Bai Bu 'small bai bu' (refers to the size of tuber); 药虱药 Yao Shi Yao; Radix Stemonae; stemona (SCN); stemona root (CP2010).

SAFETY

Untreated tubers (i.e. neither steamed nor blanched) have been reported to cause respiratory problems (Bensky *et al.*, 2004).

S. sessilifolia

Stemonaceae

DRUG MORPHOLOGY

CRUDE DRUG (DRIED TUBER): tuber 5–12 cm long; 0.5–1 cm diam. Much smaller than drug of PS3 but macroscopically indistinguishable from that of PS1.

DECOCTING PIECES-1 AND -2 (SLICED AND HONEY-PROCESSED TUBER): Diameter usually narrower (up to 1 cm) than PS2 [7]. Macroscopically indistinguishable from their PS1 counterparts.

UNOFFICIAL SUBSTITUTES

Root tubers of other species have been reported as local and/or historical unofficial substitutes or adulterants (Bensky *et al.*, 2004; Xiao *et al.*, 2001–2002) but their occurrence as such in international trade is uncertain:

1. ***Stemona mairei*** (H. Lév.) K. Krause (云南百部 Yun Nan Bai Bu) family Stemonaceae. Endemic: Sichuan & Yunnan.
2. ***Asparagus acicularis*** F. T. Wang & S. C. Chen (山文竹 Shan Wen Zhu) family Asparagaceae. Surface of dried tuber greyish-brown or brown. Endemic to S and SE China.
3. ***A. filicinus*** D. Don (羊齿天门冬 Yang Chi Tian Men Dong) family Asparagaceae. Wild in C and W China in forests, thickets and shady moist places along valleys (1,200–3,000 m).

PROCESSING

As for PS1.

OTHER COMMON NAMES

小百部 Xiao Bai Bu 'small Bai Bu' (refers to the size of tuber); 药虱药 Yao Shi Yao; Man Sheng Bai Bu; Radix Stemonae; stemona (SCN); stemona root (CP2010).

SAFETY

As for PS1.

S. tuberosa

Stemonaceae

DRUG MORPHOLOGY

CRUDE DRUG (DRIED TUBER): as for PS1 and PS2 but **much larger (8–24 cm long [8]; 0.8–2 cm diam.)** and externally yellowish-brown to greyish-brown. *Fracture*: yellowish-white or dark brown, with the **diameter of the central stele larger**.

DECOCTING PIECES-1 (SLICED TUBER): transverse slices broader (c. 2.5–4.2 cm) than PS1 and PS2; cut surface yellowish-white or dark brown; **central stele usually broader than PS1 and 2 [9].**

DECOCTING PIECES-2 (HONEY-PROCESSED TUBER): as for sliced tuber above but darker. **Broader (c. 0.8–2 cm) than PS1 and PS2.**

PROCESSING

As for PS1.

OTHER COMMON NAMES

大百部 Da Bai Bu 'big bai bu' (refers to the size of the tuber); Dui Ye Ba Bu; Radix Stemonae; stemona (SCN); stemona root (CP2010).

SAFETY

As for PS1.

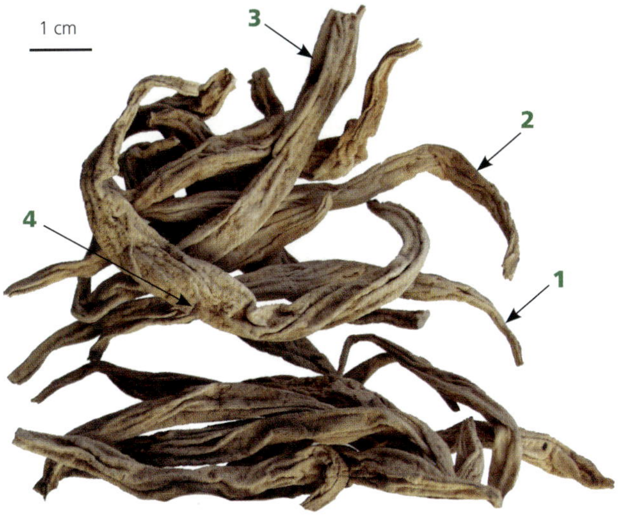

Crude drug (*S. japonica*)

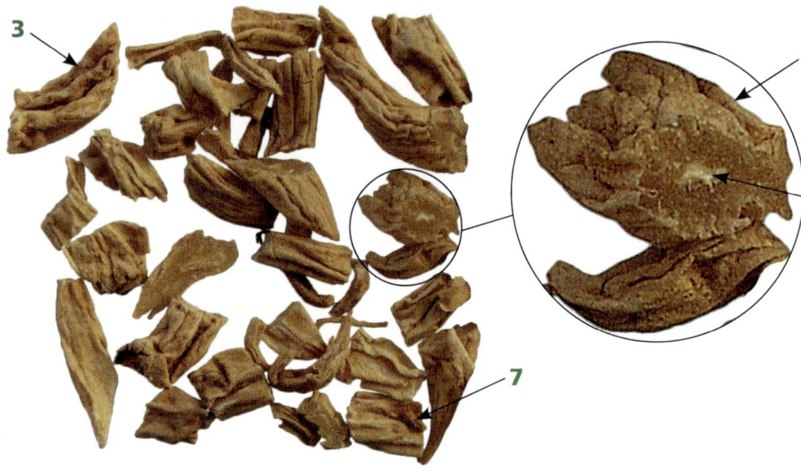

Decocting pieces-1 (*S. japonica*)

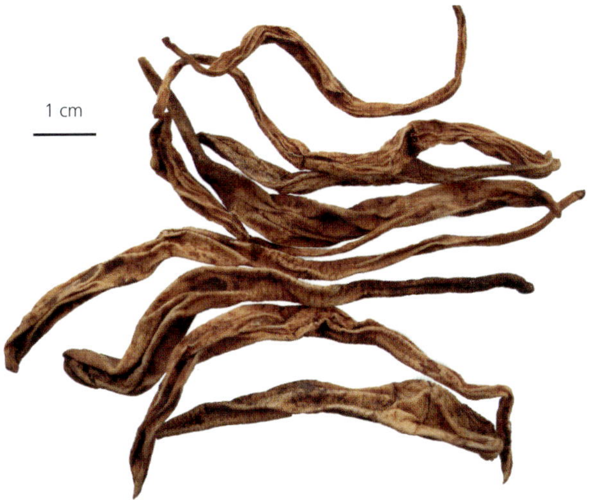

Crude drug (*S. sessilifolia*)

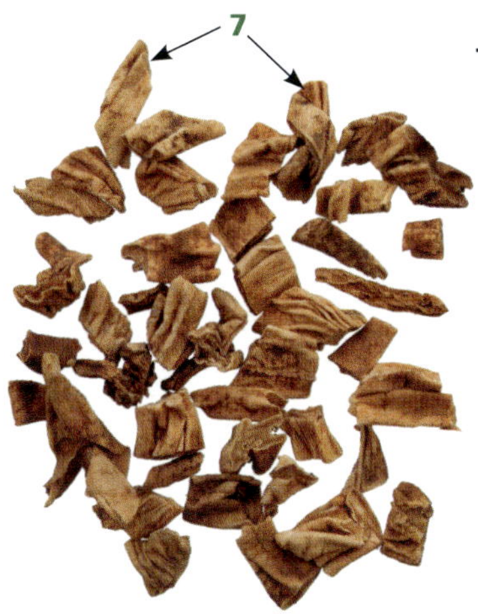

Decocting pieces-1 (*S. sessilifolia*)

Decocting pieces-2 'honey-processed' (*S. sessilifolia*)

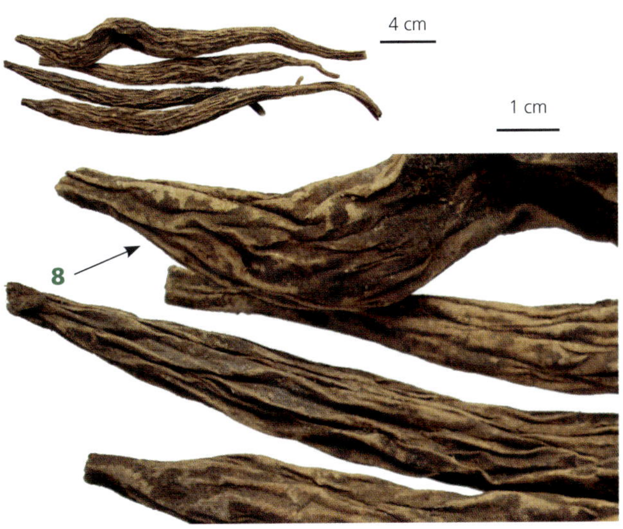

Crude drug (*S. tuberosa*)

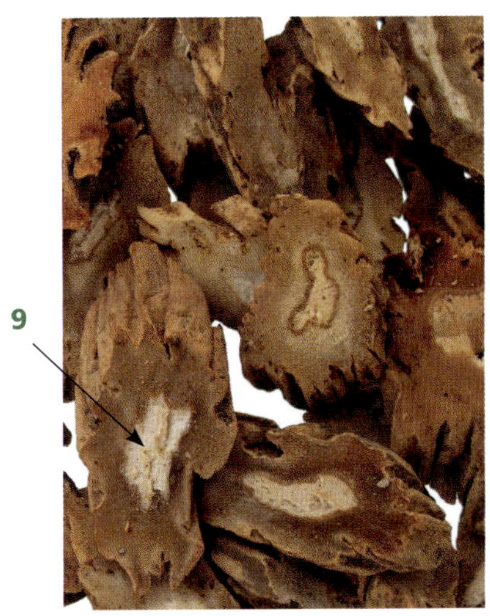

Decocting pieces-1 (*S. tuberosa*)

Decocting pieces-2 'honey-processed' (*S. tuberosa*)

Bai Fu Zi 白附子 or Zhi Bai Fu Zi 制白附子

Drug part: PROCESSED STEM TUBER (RHIZOME)

Typhonii Rhizoma

PHARMACOPOEIA SPECIES

Sauromatum giganteum (Engl.) Cusimano & Hett.

Araceae

synonym: *Typhonium giganteum* Engl. (name used in CP2015).

Species of *Sauromatum* wild in China: 7

PLANT DESCRIPTION

A herbaceous perennial (up to c. 45 cm tall) with an ovoid rhizome (5–8 cm long; 2–5 cm diam.) often producing many spindle-shaped offsets. *Leaves* all basal and stalked; stalks sometimes purple-spotted; leaf blade simple, large (15–45 cm long; 9–25 cm wide) and heart-shaped or with basal lobes pointing outwards; tip abruptly pointed; margins entire. *Inflorescence* emerging at same time as foliage with an erect cylindrical spadix of tightly packed flowers surrounded in part by a sail-like spathe with a long and often deep purplish limb (up to 15 cm long); latter with a sinuous margin and pointed tip. Spadix slightly shorter than the outer spathe with a zone (1–1.5 cm long) of crowded female flowers below a 2–3 cm sterile zone which separates the female flowers from the creamish male flowers above; sterile zone with scattered thread-like or stumpy 'staminodes' (sterile stamens); *male flower zone* (c. 2 cm long) with stalkess male flowers; a long (usually black) cylindrical appendix (up to 5 cm long) extends above the male zone. *Fruits* in a densely packed elongate group along the spadix; each fruit violet-purple; 1-seeded. *Fl.* Jul.

HARVESTING, SOURCING AND NATURAL RANGE

Rhizomes are harvested in autumn, mostly from wild populations which grow in grasslands, field margins and streamsides (up to 1,500 m altitude). The species is endemic to N, W and C China: Anhui, Gansu, Hebei, Henan, Jilin, Liaoning, Shandong, Shanxi, Sichuan and Tibet with the bulk of harvesting from Gansu, Henan and Hubei.

CONSERVATION STATUS

Endemic to China: Not Evaluated (IUCN Red List, accessed 11/11/2016).

UNOFFICIAL SUBSTITUTE

Aconitum carmichaelii Debeaux

Ranunculaceae

PLANT DESCRIPTION

The living plant is not confused with that of *S. giganteum* (the Pharmacopoeia species above). *Main differences*: an erect and branching plant (c. 1.5 m tall) with **leaves alternate up the stem** (not basal); leaf blades much smaller (6–11 cm long, 9–15 cm wide), pentagonal in outline with central stems leaves deeply divided palmately. **Flowers** large, conspicuous and **purplish-blue** in crowded heads **at the top of the plant;** each flower with 5 petal-like sepals (upper one hood-shaped) and 2 petals. Fruit a cluster of 3–5 dry carpels.

CONSERVATION STATUS

China: Least Concern (listed as 4 varieties, MEPCAS , 2013).
Global: Least Concern (IUCN Red List, accessed 11/11/2016).

SUBSTITUTION COMMENT. *Ad-hoc* sampling of 'Bai Fu Zi' on the UK TCM market in recent years suggests some level of substitution with materials identified macroscopically by the authors as 'Fu Zi' 附子 Aconiti Lateralis Radix Praeparata (*see photo of trade example on page 25*). 'Fu Zi' is a separate TCM drug (not included in this book) sourced from the processed lateral root tubers of *Aconitum carmichaelii*. Different degrees of processing 'Fu Zi' produce different end points; that called 'Bai Fu Pian' most closely resembles the UK material (see 'Drug Substitute' morphology).

The medicinal use of *Aconitum* species is restricted in many countries (e.g. UK, MHRA, 2014) (see *Safety*).

In the interests of patient safety and legal practice, these findings justify validation with laboratory testing. The cause of such substitution is unclear but given the similarity between the trade names 'Bai Fu Zi' 白附子 (official drug described here) and 'Fu Zi' 附子 sourced from *Aconitum*, trade name confusion may, in part, be responsible. The extent of such substitution in other markets is not known.

TCM PROPERTIES AND USE

properties: 'pungent and warm'
channels: 'Stomach and Liver'
actions: 'dries damp, dispels wind, resolves phlegm, eliminates toxins'

indications: muscle spasms; convulsions; stroke; nodules; itchy skin; (T)

Leaves of *Sauromatum giganteum*

S. giganteum with inflorescence

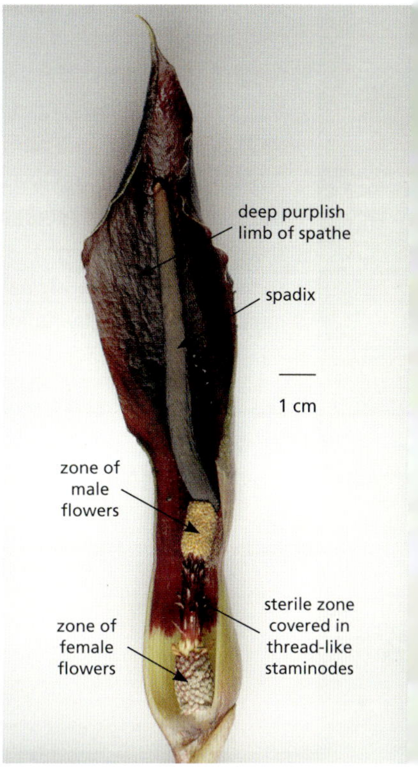

Dissected inflorescence of *S. giganteum*

Medicinal crop of *Aconitum carmichaelii*

Inflorescence of *A. carmichaelii*

Sauromatum giganteum

Araceae

DRUG MORPHOLOGY

CRUDE DRUG (DRIED AND PEELED RHIZOME): almost egg-shaped **[1]** or ellipsoid **[2]** (2–5 cm long; 1–3 cm diam.). Externally **yellowish-white or whitish**, surface slightly rough with indistinct annulations **[3]** and scattered root scars **[4]**; apex with stem and bud scars **[5]**. *Texture*: hard and powdery. *Fracture*: white. *Odour*: indistinct. *Taste*: **toxic** (NOT ADVISED) tongue-numbing and pungent. ***Good quality*:** large, plump, white and powdery, with 'skin' removed.

DECOCTING PIECES (SLICED AND PROCESSED RHIZOME): irregularly ovate or elliptical slices **[6]** (2–5 cm long; 1–3 cm diam.; 2–4 mm thick), uncut edges pale brown **[7]**. Cut surfaces homogeneously pale creamish-yellow and horny **[8]**. *Odour*: indistinct. *Taste*: indistinct; very slightly numbing and pungent.

PROCESSING

After removal of the fibrous roots and outer skin the rhizomes are sun-dried (= crude unprocessed drug; this is **toxic**); Rhizomes are then size-graded before undergoing a long processing procedure over several days to reduce their toxicity. This involves soaking in water, then boiling with fresh ginger and alumen powder before partially drying, slicing then fully drying (= decocting pieces 'Bai Fu Zi' or more accurately 'Zhi Bai Fu Zi' (i.e. processed Bai Fu Zi).

OTHER COMMON NAMES

独角莲 Du Jiao Lian (plant name); Yu Bai Fu; giant Typhonium tuber (CP2010); Rhizoma Typhonii; typhonium (SCN).

SAFETY

Processed and unprocessed rhizomes are **toxic** (Bensky *et al.*, 2004; CP2015). **Accidental ingestion or overdose may cause serious adverse reactions; as little as 30 g can be fatal** (Bensky *et al.*, 2004); other authorities (Chen & Chen, 2004) advise only topical use.

Aconitum carmichaelii

Ranunculaceae

'DRUG SUBSTITUTE' MORPHOLOGY

DECOCTING PIECES (PROCESSED, LONGITUDINALLY SLICED ROOT TUBER — BAI FU PIAN): in outline, **slices somewhat conical with apex rounded [9], truncate [10] or slightly indented [11] and base often tapering to a rounded tip [12]** (1.7–5 cm long; 0.9–3 cm diam.; 0.3 cm thick); edge of slice **occasionally with lateral, nodular-like root-bud [13]**. Uncut edges yellowish-white, relatively smooth, with root peel remnants occasional **[14]**; **cut edges sometimes distinctly raised and slightly incurved [15]**; cut surface brownish-yellow to yellowish-white, **cambium ring faintly discernible** 3–4 mm inside the perimeter **[16]**. *Texture*: hard and brittle. *Fracture*: translucent and horny. *Odour*: indistinct. *Taste*: indistinct.

PROCESSING

Fu Zi is processed into 3 main commercial forms: 'Yan Fu Zi', 'Hei Shun Pian' and 'Bai Fu Pian' with processing procedures often complex and lengthy and designed to reduce toxicity (see CP2015 or other TCM texts for more detailed descriptions).

OTHER COMMON NAMES

附片 Fu Pian; 附子 Fu Zi; 卦片 Gua Pian; 黑附子 Hei Fu Zi; 黄附片 Huang Fu Pian; 生附子 Sheng Fu Zi; aconite; monkshood (plant name); Aconiti Lateralis Radix Praeparata; Radix Aconiti Lateralis Praeparata; Sichuan aconite (SCN).

SAFETY

Toxic (CP2015). Safety Class: 3 *'to be used only under the supervision of a qualified expert'*; Interaction Class: A (Gardner & McGuffin, 2013). The *unprocessed* root tuber of this drug (Chuan Wu 川乌) should never be taken internally owing to its high toxicity resulting from the presence of aconitine alkaloids (Bensky *et al.*, 2004). Traditional processing to produce 'Fu Zi' (processed root tuber) greatly reduces the content of aconitine (Gardner & McGuffin, 2013).

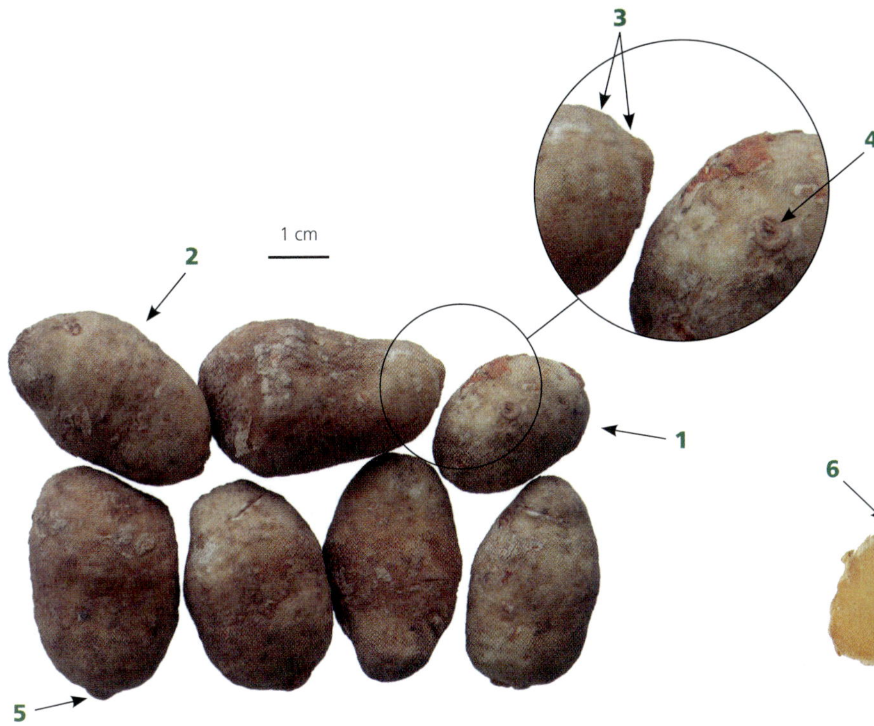

Crude drug of Bai Fu Zi (peeled rhizomes of *Sauromatum giganteum*)

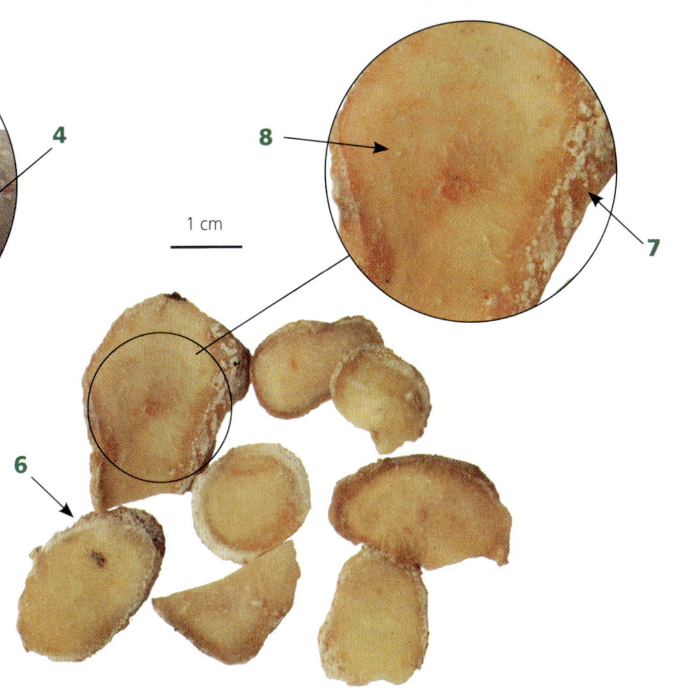

Decocting pieces of Bai Fu Zi (processed *S. giganteum* tuber)

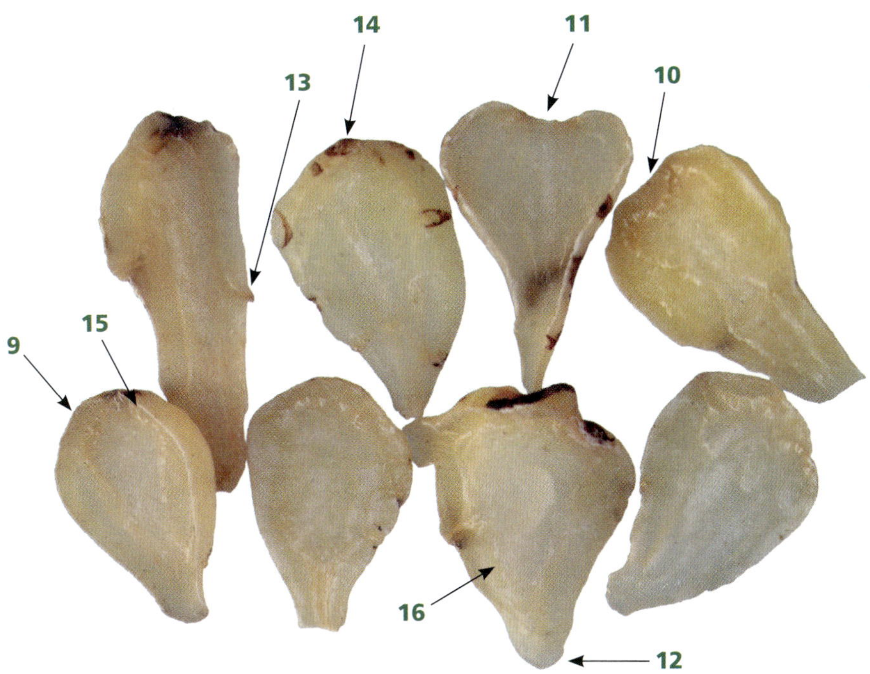

Reference sample of Bai Fu Pian sourced from *Aconitum carmichaelii*

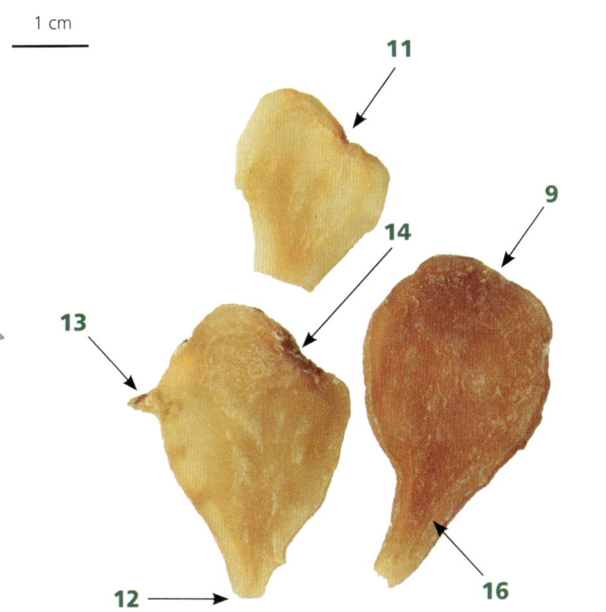

UK market sample sold as 'Bai Fu Zi' iin a TCM clinic, London

Bai Ji 白及

Bletillae Rhizoma

Drug part: RHIZOME (stem tuber)

Bletilla striata (Thunb.) Rchb. f.

Orchidaceae

synonyms: *Limodorum striatum* Thunb.; *Bletia striata* (Thunb.) Druce

Species of *Bletilla* wild in China: 4

PLANT DESCRIPTION

A perennial terrestrial orchid (18–60 cm tall) with an irregularly shaped and tuberous rhizome (1–5 cm diam.) with many long fibrous roots. *Leaves* 4–6, bases often forming a false stem; leaf blade linear to lance-shaped (8–29 cm long; 1.5–4 cm wide) with longitudinal parallel folds, margin entire and tip narrowing to a long point. *Flowering stem* erect, solitary and leafless (3–25 cm tall) emerging from between the leaves; flowers in a loose, few-flowered, occasionally branched, head at the top of the stem. *Each flower* large, showy and **purplish-red or pink**; 3 sepals and 2 of the petals similar, oblong and spreading; lip (the third and lowermost petal) larger, indistinctly 3-lobed (23–28 mm long), white with purplish-red blotching and ornamented with raised, strongly frilled longitudinal ridges. Stamen and style forming parts of the column which is rather petal-like, curved, slender and winged; anther sub-terminal with pollen fused into 2 groups of 4; ovary inferior. *Fruit* a dry, cylindrical capsule, narrowing at both ends; seeds many and dust-like. Fl. Apr–May.

HARVESTING, SOURCING AND NATURAL RANGE

Rhizomes are harvested from May to October exclusively from medicinally farmed plants (the harvesting of wild plants today is no longer economic due to their rarity).Traditional *Di Dao*-quality material is sourced from locations in Guizhou and Sichuan provinces.

The wild species is native to C, S and E China (provinces of Anhui, Fujian, SE Gansu, Guangdong, Guangxi, Guizhou, Hubei, Hunan, Jiangsu, Jiangxi, S Shaanxi, Sichuan and Zhejiang) where it grows in evergreen forests and grassy places (100–3,200 m altitude). The species is also native to Japan, the Korean Peninsula and Myanmar and is grown as an ornamental in China, Europe and North America.

CONSERVATION STATUS

China: Endangered (MEPCAS, 2013).

PROTECTED STATUS

CITES: Appendix II; EU Annex B (see Table 1 page 10)

Bletilla striata wild in north Sichuan PHOTO: PHILLIP CRIBB

Young fruiting head of *B. striata*

B. striata in full flower

TCM PROPERTIES AND USE

***properties*: 'bitter, sweet, sour and cool'**
***channels*: 'Lung, Stomach and Liver'**
***actions*: 'stops bleeding, promotes healing, astringent'**

***indications*: haemoptysis, haematemesis; boils and ulcers; (T)**

DRUG MORPHOLOGY

CRUDE DRUG (DRIED, PEELED RHIZOME): irregularly squat with 2 (occasionally 3) stout, sometimes claw-like, limbs **[1]** (1.5–5 cm long; 0.5–1.5 cm thick); apex of main body with a central raised stem remnant or scar **[2]** surrounded by concentric annual rings (up to 2 cm diam.) **[3]** and fibrous adventitious root remnants **[4]**. Externally greyish-white or yellowish-white, variously wrinkled and furrowed **[5]** and with occasional raised root scars **[6]**. *Texture*: hard and difficult to break. *Fracture*: creamish-white. *Odour*: indistinct. *Taste:* bitter; viscous when chewed.

Good quality: large, plump and translucent.

DECOCTING PIECES (DRIED, SLICED RHIZOME): thin slices of variable size and shape (up to 5 cm long but usually much shorter); occasionally, intact claw-like limbs discernible **[7]**. Cut surfaces creamish-white, somewhat translucent with vascular bundles scattered **[8]** or in dotted parallel lines **[9]**. Uncut surfaces greyish-white or yellowish-white, wrinkled and furrowed **[10]**; remnants of a raised stem scar at apex sometimes visible **[11]** along with fibrous root remains **[12]** and occasional raised root scars **[13]**.

PROCESSING

After removal of the roots, rhizomes are boiled or steamed thoroughly, partially sun-dried, then peeled and sun-dried again (= crude drug). Rhizomes are later sliced (usually longitudinally) for use in decoctions.

OTHER COMMON NAMES

bletilla (SCN); common bletilla tuber (CP2010); Rhizoma Bletillae.

SAFETY

An official Chinese healthfood (Anon., 2002a,b).

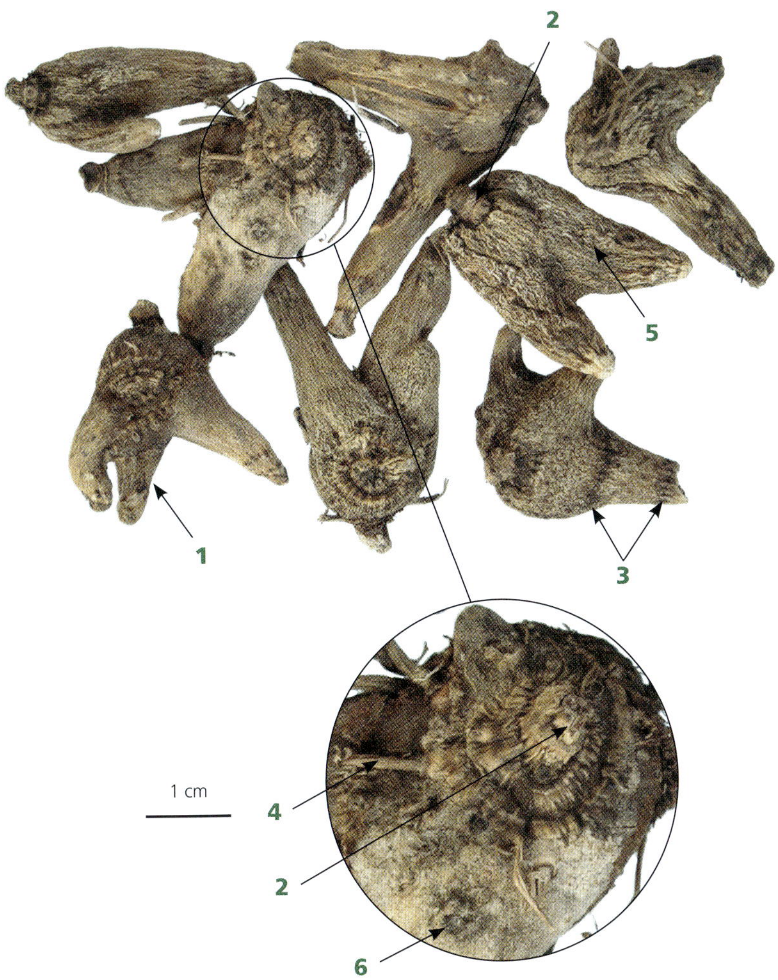

Crude drug of Bai Ji (*Bletilla striata*)

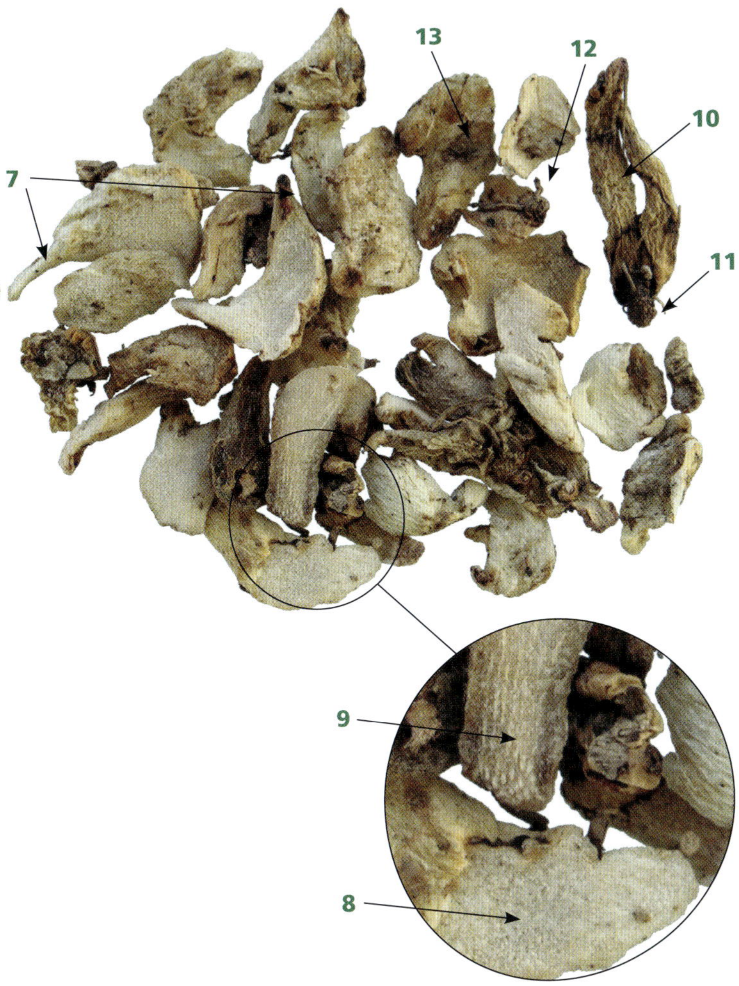

Decocting pieces of Bai Ji (*B. striata*)

Bai Qian 白前 Cynanchi Stauntonii Rhizoma et Radix

Drug parts: RHIZOME AND ROOT

Cynanchum stauntonii (Decne.) Schltr. ex H. Lév.

Asclepiadaceae

synonyms: *Pentasachme stauntonii* Decne.; *Vincetoxicum stauntonii* (Decne.) C. Y. Wu & D. Z. Li (*Kew-MPNS* accepted name, 20/8/2016).

Species of *Cynanchum* wild in China: 57

PLANT DESCRIPTION

An upright perennial herb (up to 1 m tall), hairless except for parts of the flower and with a fibrous and fascicled rootstock (medicinal part). *Stems* mostly unbranched. *Leaves* opposite, simple and short-stalked; leaf blades narrowly lance-shaped (6–13 cm long; 3–9 (rarely up to 17) mm wide), papery, tapering at both ends and lateral veins in c. 6 pairs. *Flowers* inconspicuous and purple in slender few-flowered heads arising from leaf axils; flower-head stalk (peduncle) up to 1.7 cm long. Each flower on a short (3–9 mm long) stalk (pedicel), bisexual and radially symmetrical; sepals 5, tiny, ovate to oblong; petals 5, purple (sometimes greenish-yellow), fused basally; lobes linear to oblong, spreading widely (3–5 (rarely up to 8) mm long; 0.6–1 mm wide) **minutely but conspicuously hairy at base**; stamens fused to the style to form a complex structure (gynostegium) which is surrounded by a set of 5 fleshy lobes (the corona); ovary superior. *Fruit* a dry follicle, linear to lance-shaped in outline (9–12 cm long; 3–6 mm diam.); seeds many with a conspicuous tuft of hairs (c. 2.5 cm long) at one end. Fl. May–Aug, fr. Sep–Dec.

HARVESTING, SOURCING AND NATURAL RANGE

Most 'Bai Qian' in trade is thought to be sourced from this species rather than *C. glaucescens* (PS2). Rootstocks are harvested in autumn from wild populations which grow in sunny, places in moist woodland or in water-logged places at low to medium elevations. The species is endemic to China and occurs in Anhui, Fujian, Gansu, Guangdong, Guangxi, Guizhou, Hunan, Jiangsu, Jiangxi, Yunnan and Zhejiang provinces.

CONSERVATION STATUS

Endemic to China: Least Concern (MEPCAS, 2013); Not Evaluated (IUCN Red List, accessed 28/8/2016).

C. glaucescens (Decne.) Hand.-Mazz.

Asclepiadaceae

synonyms: *Pentasachme glaucescens* Decne.; *Vincetoxicum glaucescens* (Decne.) C. Y. Wu & D. Z. Li (*Kew-MPNS* accepted name, 20/8/2016).

PLANT DESCRIPTION

Main differences from C. stauntonii (PS1): often shorter (up to 60 cm) with 2 rows of fine hairs up the stem. *Leaves* virtually stalkless; **leaf blades shorter and often wider** (1–7 cm long; 7–12 mm wide), leathery, bases wedge-shaped or rounded and lateral veins (indistinct) in 3–5 pairs. *Flowers* conspicuous and **pale yellow** in long-stalked clusters in leaf axils; petal lobes ovate to oblong, fleshy and bending inwards; **hairless.** *Fruit* **shorter and wider and narrowing at both ends** (4.5–6 cm long; 6–10 mm diam.). Fl. May–Nov, fr. Jul–Dec.

HARVESTING, SOURCING AND NATURAL RANGE

Harvesting as for *C. stauntonii* (PS1). The species is endemic to China and grows along rivers in mountains (100–800 m altitude) in the provinces of Fujian, Guangdong, Guangxi, Hunan, Jiangsu, Jiangxi, Sichuan, Zhejiang.

CONSERVATION STATUS

Endemic to China: Least Concern (MEPCAS, 2013); Not Evaluated (IUCN Red List, accessed 28/8/2016).

TCM PROPERTIES AND USE

properties: 'pungent, bitter and warm'
channel: 'Lung'
actions: 'resolves phlegm, descends Lung qi, relieves cough'
indications: coughs with phlegm

Cynanchum stauntonii in the wild (foreground and right)

Fresh rootstock of *C. stauntonii*

C. stauntonii in flower

Flowering plants of *C. glaucescens*

Cynanchum stauntonii

Asclepiadaceae

DRUG MORPHOLOGY

CRUDE DRUG (DRIED ROOT AND RHIZOME): a dense rootstock predominantly of rhizomes **[1]** interspersed with much thinner and fibrous roots **[2]**. Rhizomes externally yellowish-white or yellowish-brown, somewhat cylindrical, slightly curved (4–15 cm long; 1.5–4 mm diam.), occasionally branched, nodes slightly enlarged **[3]** bearing roots **[4]** or pale and prominent root scars **[5]**; internodes 1.5–4.5 cm long **[6]**; occasional membranous stem remnants at apex **[7]**. *Rhizome texture*: brittle; *fracture*: pith membranous and centre hollow **[8]**. Roots (up to 10 cm long; <1 mm diam.), curved, branched, fascicled at rhizome nodes and coiled in masses. *Odour*: indistinct. *Taste*: sweetish.

DECOCTING PIECES-1 (DRIED LENGTHS OF ROOT AND RHIZOME): a mix of rhizome lengths **[9]** interspersed with much narrower, fibrous roots **[10]** (<1 mm diam.). Rhizomes externally yellowish-white or yellowish-brown, somewhat cylindrical, slightly curved (1.5–4 mm diam.) and occasionally branched **[11]** with nodes slightly enlarged and scattered with pale, raised root scars **[5]** or roots still attached. *Texture*: brittle. *Fracture of rhizome*: hollow **[8]**. *Odour*: indistinct; *taste:* sweetish.

DECOCTING PIECES-2 (HONEYED LENGTHS OF ROOT AND RHIZOME — MI BAI QIAN): as for decocting pieces-1 above but all surfaces darker. *Odour*: honey-scented. *Taste*: sweet.

PROCESSING

Roots and rhizomes are shade-dried (= crude drug), then briefly washed before cutting into decocting lengths. The latter may later be dry-fried with honey to produce Mi Bai Qian (蜜白前).

OTHER COMMON NAMES

柳叶白前 Liu Ye Bai Qian; 沙消 Sha Xiao; Cynanchum stauntonii (SCN); Rhizoma et Radix Cynanchi Stauntonii; willowleaf swallowwort rhizome (CP2010).

SAFETY

To be avoided by patients with gastric or bleeding disorders since this drug irritates the gastric mucosa (Bensky *et al.*, 2004).

C. glaucescens

Asclepiadaceae

DRUG MORPHOLOGY

CRUDE DRUG (DRIED ROOT AND RHIZOME): very similar to those of *C. stauntonii* (PS1). *Main differences:* rhizomes slightly narrower and **externally greyish-green or greyish-yellow [12], nodes more nodular and swollen [13]** and internodes 1–2 cm long **[14]**. ***Rhizome texture:* relatively hard.** Roots slightly broader (c. 1 mm diam.).

DECOCTING PIECES-1 (DRIED LENGTHS OF ROOT AND RHIZOME): as for decocting pieces-1 of *C. stauntonii* (PS1). *Main differences*: **rhizomes greyish-green or greyish-yellow [12]**, nodes more nodular and swollen **[13]**. *Rhizome texture*: relatively hard. Roots slightly broader (c. 1 mm diam.). Laboratory-based methods are recommended to reliably differentiate decocting pieces from those of PS1.

DECOCTING PIECES-2 (HONEYED LENGTHS OF ROOT AND RHIZOME — MI BAI QIAN): as for decocting pieces-1 above but all surfaces darker. *Odour:* honey-scented. *Taste:* sweet.

PROCESSING

As for PS1.

OTHER COMMON NAMES

鹅白前 E Bai Qian; 草白前 Cao Bai Qian; Yuan Hua Ye Bai Qian; Rhizoma et Radix Cynanchi Stauntonii; willowleaf swallowwort rhizome (CP2010).

SAFETY

As for PS1.

Crude drug of Bai Qian
(sourced from *Cynanchum stauntonii*)

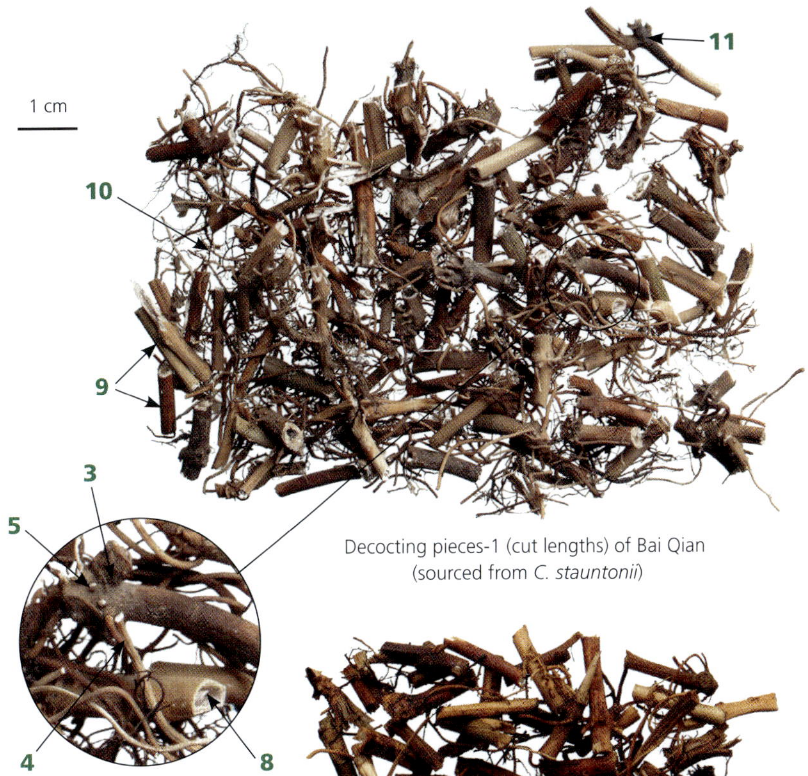

Decocting pieces-1 (cut lengths) of Bai Qian
(sourced from *C. stauntonii*)

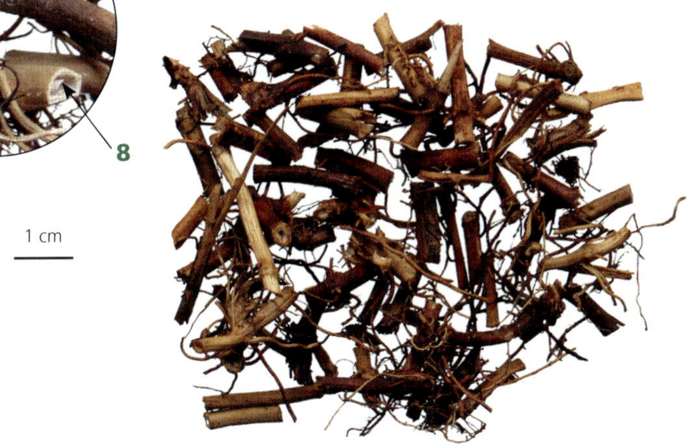

Honey-processed decocting pieces-2 of Bai Qian [Mi Bai Qian]
(sourced from *C. stauntonii*)

Crude drug of Bai Qian (sourced from *C. glaucescens*)

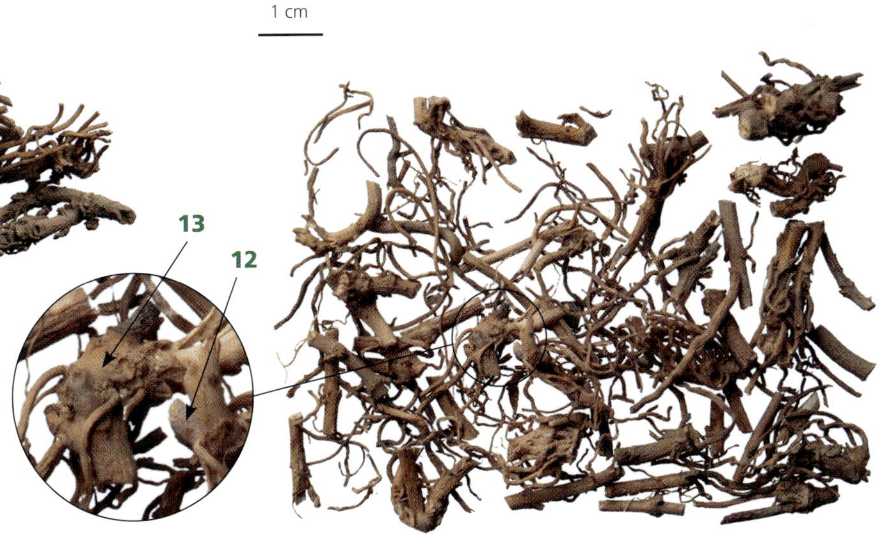

Decocting pieces-1 (cut lengths) of Bai Qian
(sourced from *C. glaucescens*)

Bai Shao 白芍

Paeoniae Radix Alba

Drug part: DE-BARKED ROOT

Paeonia lactiflora Pall.

Paeoniaceae

PLANT DESCRIPTION

A many-stemmed perennial herb (up to 70 cm tall) with a thick, sometimes branching rootstock. *Leaves* alternate, divided into not more than 15 long-ovate leaflets; blade of leaflets extending into leaflet stems almost as wings; margins thin, white and cartilaginous. *Bracts* below each flower, 4 or 5, green, lance-shaped and of different lengths. *Flowers* large and conspicuous (c. 8–13 cm diam.), usually several per stem; sepals 3 or 4; petals 9–13, variously coloured (mostly white or pink); stamens many and yellow. *Fruit* a group of 2–5 oblong-ellipsoid pods (follicles) (2.5–3 cm long; 1.2–1.5 cm wide), green or purple, hairless or sometimes covered in soft brown hairs. *Seeds* many, ovoid and black. Fl. May–Jun, fr. Aug. *(description based on FoC and Hong De-Yuan, 2010).* **Note:** flowers are seldom seen in medicinal crops of this species since, in order to improve root quality, flower-buds are usually removed.

Wild-harvested plants of this species, with root-bark intact, is the source of the separate TCM drug 'Chi Shao' (Paeoniae Radix Rubra) page 74.

Species of *Paeonia* wild in China: 15 (*FoC*; Hong, 2010)

HARVESTING, SOURCING AND NATURAL RANGE

Sourced entirely from medicinal crop plants grown using long-established cultivation methods dating back some 1,500 years. Roots are harvested from 4–5 year old plants likely to be medicinal cultivars developed over the years by specialist growers. *Di Dao*-quality material is sourced from the Chinese provinces of Anhui, Sichuan and Zhejiang (see also *Other Common Names*).

In the wild, the species grows in temperate forests, forest margins and grassland (400–2,300 m altitude) across northern China (Gansu, Hebei, Heilongjiang, Inner Mongolia, Jilin, Liaoning, Ningxia, Shaanxi and Shanxi) and its natural range extends into Japan, the Korean Peninsula, Mongolia and Russia (Far East, Siberia).

CONSERVATION STATUS

China: Least Concern (MEPCAS, 2013); Near Threatened (Wang & Xie, 2004).
Global: Not Evaluated (IUCN Red List, accessed 18/8/2016).

Medicinal crop of *Paeonia lactiflora* (cv.), Anhui province

P. lactiflora in flower

Freshly harvested rootstock of medicinally cultivated *P. lactiflora*

UNOFFICIAL SUBSTITUTE

Paeonia delavayi Franch. (滇牡丹 Dian Mu Dan). Roots reported as an historical or local substitute in China but unlikely to be seen in international trade. A **woody shrub** (0.2–1.8 m tall); leaves divided into **25–100 lobes**; flowers 1–3 per shoot (6–10 cm diam.); bracts 1–5; petals usually 7–11. Wild plants restricted to W Sichuan, SE Tibet, C and N Yunnan.

TCM PROPERTIES AND USE

properties: **'bitter, sour and cool'**
channels: **'Liver and Spleen'**
actions: **'tonifies Blood and yin, tonifies Liver, pacifies yang, relieves pain'**

indications: **dysmenorrhoea, metrorrhagia; night sweats; abdominal pain; musculoskeletal pain, spasms; diarrhoea**

DRUG MORPHOLOGY

CRUDE DRUG (DRIED DE-BARKED ROOT): solid cylindrical or slightly curved (5–18 cm long, 1–2.5 cm diam.); ends truncate. Externally pale brownish-red or white, sometimes lustrous; rootlet scars roundish and scattered **[1]**; occasional fine, longitudinal striations **[2]** and remnants of outer cortex (bark) **[3]** sometimes present. *Texture*: hard, dense and difficult to break. *Fracture*: relatively even, pinkish-white or creamish-white **[4]** with distinct paler cambium ring **[5]**; radiating vascular bundles of lines or dots **[6]**. *Odour*: indistinct. *Taste*: somewhat bitter and sour. ***Good quality***: large and heavy, internally white and devoid of cavities.

DECOCTING PIECES (OBLIQUELY SLICED DE-BARKED ROOT): elliptical or sub-rounded slices (1–2 mm thick); external uncut surfaces smooth, pale brownish-red or whitish **[7]**; cut surfaces pinkish-white or creamish-white **[4]**; cambium ring distinct **[5]**, white or paler than surrounding tissues with vascular bundles as interrupted radiating lines or dots **[6]**. *Texture, fracture, odour* and *taste* as for crude drug.

PROCESSING

After removal of rhizomes, rootlets and outer bark, roots are boiled then sun-dried (= crude drug). Sometimes the outer bark is removed after boiling. Roots are later cut into thin decocting slices (1–2 mm thick) and may be subsequently dry-fried to produce 炒白芍 Chao Bai Shao or steeped in yellow rice wine then dry-fried to produce 酒白芍 Jiu Bai Shao. See also 'Chi Shao' (Paeoniae Radix Rubra) page 74.

OTHER COMMON NAMES

亳白芍 Bo Bai Shao (sourced from Anhui province); 川白芍 Chuan Bai Shao 'Sichuan white peony'; 杭白芍 Hang Bai Shao/东白芍 Dong Bai Shao 'Hangzhou/Dongyang white peony' respectively, from Zhejiang province; 湖南白芍 Hu Nan Bai Shao 'Hunan white peony'; Chinese peony (SCN); Radix Paeoniae Alba; white peony root (CP2010).

SAFETY

An official healthfood in China (Anon., 2002a,b). Safety Class: 1; Interaction Class: A (Gardner & McGuffin, 2013).

Crude drug of Bai Shao (*P. lactiflora*)

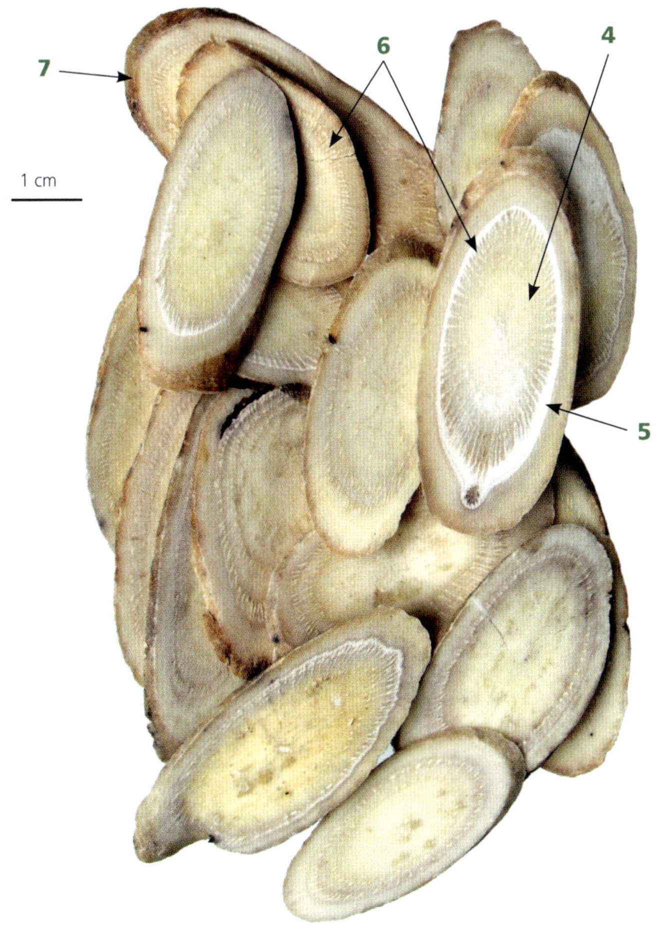

Decocting slices of Bai Shao (*P. lactiflora*)

Bai Wei 白薇

Cynanchi Atrati Radix et Rhizoma

Drug part: ROOT AND RHIZOME

Cynanchum atratum Bunge

Asclepiadaceae

synonym: *Vincetoxicum atratum* (Bunge) C. Morren & Decne. (*Kew-MPNS* accepted name, 28/8/2016).

Species of *Cynanchum* wild in China: 57

PLANT DESCRIPTION

An upright, perennial herb (up to 50 cm tall) with a fibrous and fragrant rootstock (medicinal part). *Stems* mostly unbranched, softly hairy and sometimes somewhat inflated. *Leaves* opposite, simple and short-stalked (stalks c. 5 mm long); leaf blades ovate to ovate-oblong (5–8 (rarely up to 12) cm long; 3–4 (rarely up to 7) cm wide); base rounded to somewhat heart-shaped; tip shortly pointed or tapering to a long point; softly hairy on both surfaces; lateral veins in 6 or 7 pairs. *Flowers* conspicuous and deep purple, in 8–10-flowered heads arranged in leaf axils up the stem; **flower stalks (pedicels) about 15 mm long**. Each flower unstalked, bisexual and radially symmetrical (c. 1–1.2 cm diam.); sepals 5, tiny (c. 3 mm long) and lance-shaped with fringed margins and 5 basal glands; petals 5 with lobes deep purple, ovate to triangular and spreading to give a flat star-like appearance; stamens fused to the style to form a complex structure (gynostegium) which is surrounded by a series of small, fleshy lobes (corona); ovary superior. *Fruit* a long, dry follicle usually tapering at both ends (5.5–11 cm long; 0.5–1.5 cm diam.); seeds many with a conspicuous tuft of hairs (3–4.5 cm long) at one end. Fl. Apr–Aug, fr. Jun–Oct.

HARVESTING, SOURCING AND NATURAL RANGE

Underground parts are harvested in spring or autumn from wild populations which grow in grassland, riversides and drylands (100–2000 m altitiude). The species is very local but widely distributed in parts of N, C and S China (Fujian, Guangdong, Guangxi, Guizhou, Hebei, Heilongjiang, Henan, Hunan, Inner Mongolia, Jiangsu, Jiangxi, Jilin, Liaoning, Shaanxi, Shandong, Shanxi, Sichuan and Yunnan); also native to Japan, the Korean Peninsula and E Russia.

CONSERVATION STATUS

China: Least Concern (MEPCAS, 2013).
Global: Not Evaluated (IUCN Red List, accessed 25/8/2016).

C. versicolor Bunge

Asclepiadaceae

synonym: *Vincetoxicum versicolor* (Bunge) Decne. (*Kew-MPNS* accepted name, 28/8/2016).

PLANT DESCRIPTION

Easily distinguished from C. atratum (PS1): a taller plant (to 2 m tall) with **upper part of stem twining**; leaves in lower parts broadly ovate or elliptic (7–10 cm long; 3–6 cm wide) and margins fringed with fine hairs. Flowers in heads of up to 12 flowers, **flower stalks (pedicels) 3–5 mm long**; petal lobes yellowish-white to dark purple. **Fruit much smaller** (4–5 cm long; 8–10 mm diam.) and broadly lance-shaped in outline; tuft of hairs at one end of the seed shorter (c. 2 cm). Fl. May–Aug, fr. Jul–Nov.

HARVESTING, SOURCING AND NATURAL RANGE

Underground parts are harvested in spring or autumn from wild populations which grow in thickets and along riversides (0–800 m altitude). The species is endemic to C, E and NE China, occurring in scattered populations in the provinces of Hebei, Henan, Hubei, Hunan, Jiangsu, Jilin, Liaoning, Shandong, Sichuan and Zhejiang,

CONSERVATION STATUS

Endemic to China: Least Concern (MEPCAS, 2013); Not Evaluated (IUCN Red List, accessed 25/8/2016).

UNOFFICIAL SUBSTITUTES

The rootstocks of many of other species in the genus *Cynanchum* (57 of which grow wild in China) have been reported as local, historical or accidental substitutes for 'Bai Wei'. Some of these are the source of closely related TCM drugs (see 'Xu Chang Qing' (Cynanchi Paniculati Radix et Rhizoma) page 316 and 'Bai Qian' (Cynanchi Stauntonii Rhizoma et Radix) page 28. Reliable identification of their dried rootstocks (as they appear in trade) requires laboratory-based authentication methods.

TCM PROPERTIES AND USE

properties: 'bitter, salty and cold'
channels: 'Liver, Lung and Kidney'
actions: 'clears heat, deficient heat, cools blood, eliminates toxins'

indications: fevers; postpartum conditions; phlegm cough; dysuria

Fresh rootstock of *Cynanchum atratum*

C. atratum with developing pod-like fruit

C. atratum in full flower

Fresh rootstocks *C. versicolor*

Young plant of *C. versicolor*

C. versicolor, cultivated plant in flower

Cynanchum atratum

Asclepiadaceae

DRUG MORPHOLOGY

CRUDE DRUG (DRIED ROOT AND RHIZOME): a **dense rootstock** predominantly of many slender, cylindrical roots **[1]** attached apically to short, thick rhizomes **[2]**, the latter often tortuous, and covered in rounded **stem scars (c. 5 mm diam.) [3]**. Roots sometimes growing in fascicles from the rhizomes. Roots slender and curved (10–25 cm long, 1–2 mm diam.), externally brownish-yellow with fine longitudinal wrinkles. Fibrous rootlets **[4]** occasional. *Root texture:* brittle. *Root fracture:* bark layer yellowish-white; central xylem paler. *Odour:* indistinct. *Taste:* slightly bitter.

DECOCTING PIECES (DRIED LENGTHS OF ROOT AND RHIZOME): predominantly of slender and slightly curved root lengths **[5]** (c. 1–3 cm long), externally brownish-yellow with fine longitudinal wrinkles. Denser irregularly-shaped rhizome pieces occasional **[6]** with pale yellowish-white cortex **[7]**; uncut surfaces densely covered with root bases of varying length **[8]**. Fibrous rootlets **[4]** thinner, unattached and scattered throughout. *Root texture:* brittle. *Root fracture:* bark layer yellowish-white with a paler central xylem. *Odour:* indistinct. *Taste:* slightly bitter.

PROCESSING

Roots and rhizomes are washed clean, then dried (= crude drug); decocting pieces are made by thoroughly softening the crude drug before cutting into short lengths.

OTHER COMMON NAMES

白尾 Bai Wei; blackend swallowwort root (CP2010); Radix et Rhizoma Cynanchi Atrati; Cynanchum atratum (SCN).

SAFETY

Safety Class: 1; Interaction Class: A (Gardner & McGuffin, 2013).

C. versicolor

Asclepiadaceae

DRUG MORPHOLOGY

CRUDE DRUG (DRIED ROOT AND RHIZOME): very similar to that of PS1 but with **fewer roots [9] and slightly smaller stem scars [10] (always less <5 mm diam.)** Laboratory-based authentication methods are recommended for reliable discrimination from the crude drug of PS1 above.

DECOCTING PIECES (DRIED LENGTHS OF ROOT AND RHIZOME): very similar to that of PS1 and cannot be distinguished reliably without laboratory-based authentication methods.

PROCESSING

As for PS1.

OTHER COMMON NAMES

蔓生白薇 Man Sheng Bai Wei; 变色白前 Bian Se Bai Qian (plant name); blackend swallowwort root (CP2010); Radix et Rhizoma Cynanchi Atrati.

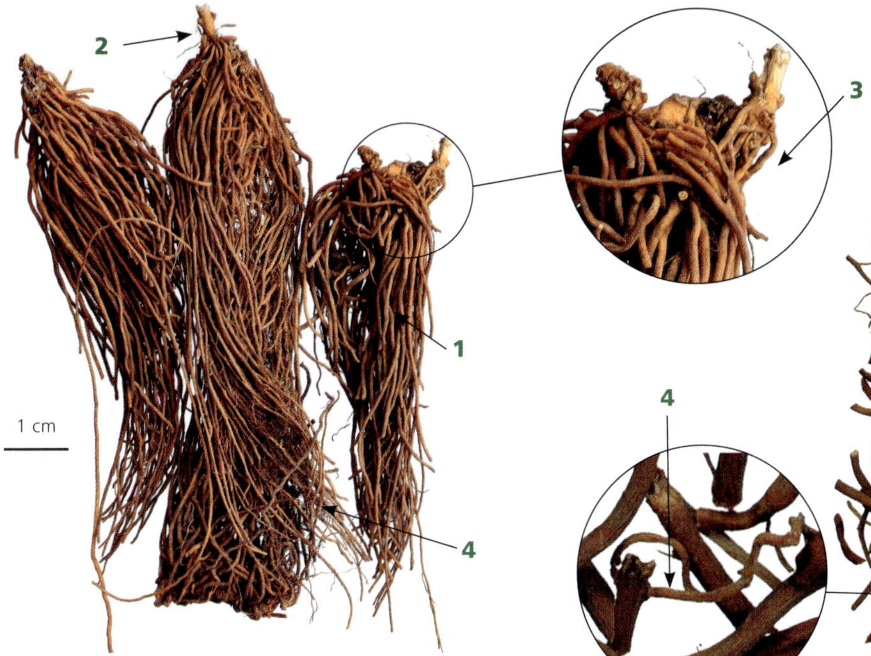

Crude drug of Bai Wei
(sourced from *C. atratum*)

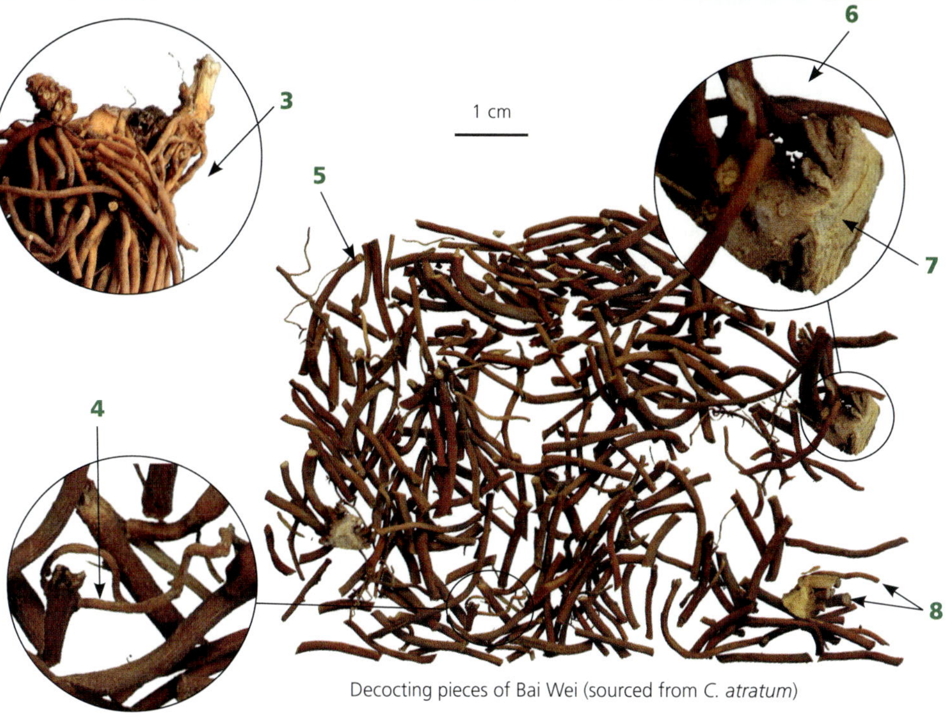

Decocting pieces of Bai Wei (sourced from *C. atratum*)

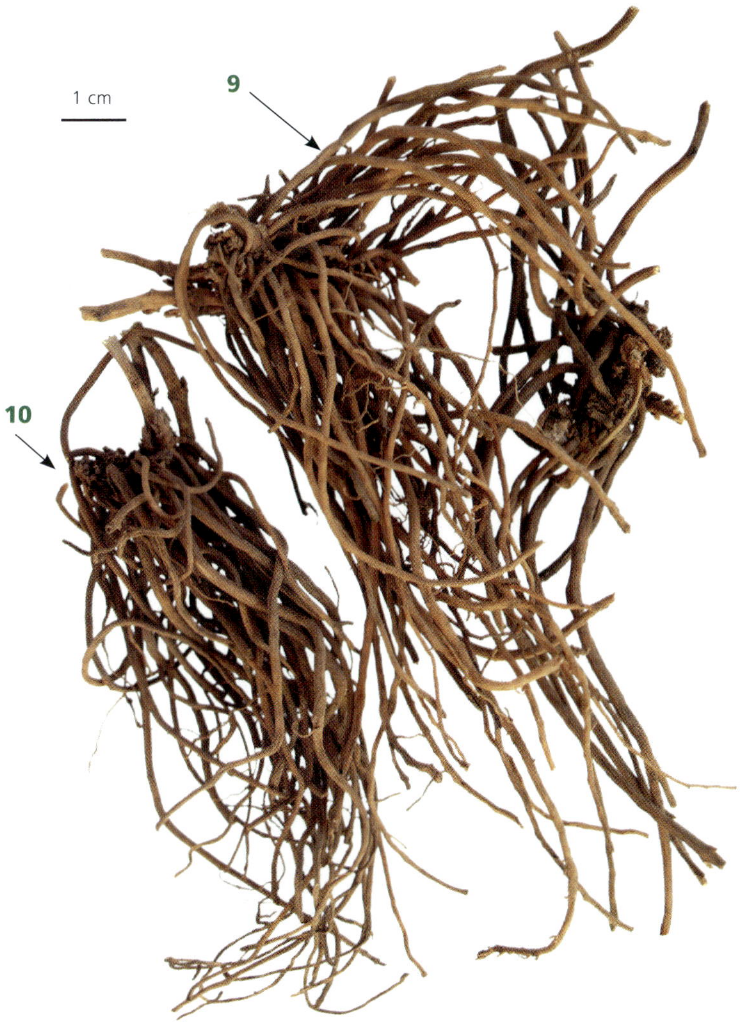

Crude drug of Bai Wei (sourced from *C. versicolor*)

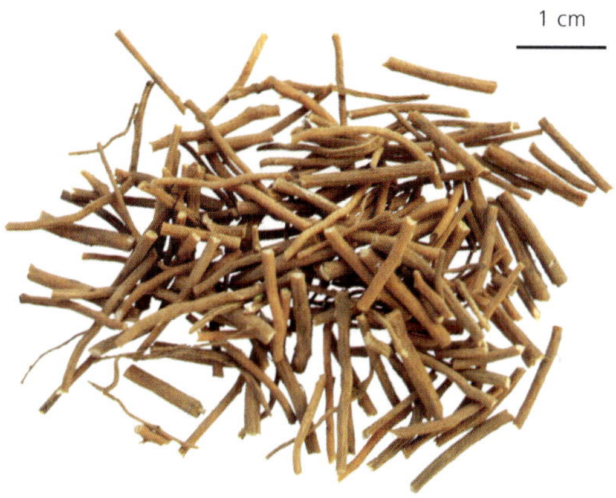

Decocting pieces of Bai Wei (sourced from *C. versicolor*)

Bai Zhi 白芷

Angelicae Dahuricae Radix

Drug part: ROOT

Angelica dahurica (Hoffm.) Benth. & Hook. f. ex Franch. & Sav.

Apiaceae

synonym: *Callisace dahurica* Hoffm.

taxonomic/nomenclatural note: two varieties are recognised, var. *formosana* (H. Boissieu) Yen and the type variety var. *dahurica* (in CP2015 as *A. dahurica* (Fisch. ex Hoffm.) Benth. et Hook. f.), the CP2015 states that 'Bai Zhi' can be sourced from either; see also *Harvesting Sourcing* etc....).

Species of *Angelica* wild in China: 45

PLANT DESCRIPTION

A stout, perennial herb (1–2.5 m tall) with a strongly aromatic tap root (3–5 cm diam.). *Stems* branched, ribbed and sometimes purplish-green. *Leaves* alternate; basal and lower leaves with inflated sheaths; leaf blades mostly large (30–50 cm long; 25–40 cm wide) and divided into unstalked leaflets (2–3-ternate-pinnate), margins coarsely toothed and white-cartilaginous; upper leaves reduced. *Inflorescences* many, in terminal or lateral compound umbels (10–30 cm diam.); main umbels with bracts 1, 2 or absent and resembling uppermost leaves and rays 18–40 (sometimes up to 70); umbellets with many thin, dry and linear to lance-shaped bracteoles. *Flowers* tiny; sepals absent; petals 5, white, equal-sized, broader at tips and notched; ovary inferior; styles 2. *Fruits* in umbellate clusters; **each fruit dry and sub-orbicular (4–7 mm long; 4–6 mm diam.)** (hairy in var. *formosana*) splitting into 2 one-seeded carpels (mericarps) with dorsal ribs thick and prominent and lateral ones with wings broader than dorsal; 1 oil gland in each furrow (between the ribs) and 2 on the flat internal (commissural) surfaces. Fl. Jul–Aug, fr. Aug–Sep.

HARVESTING, SOURCING AND NATURAL RANGE

Roots are harvested in summer or autumn when leaves have yellowed; they are sourced entirely from medicinal crop plants with those grown in mainland China being the type variety (var. *dahurica*). Variety *formosana* is confined to Taiwan. Mainly grown in Hebei, Henan, Sichuan and Zhejiang provinces. Two cultivars are commonly grown: 'Hang Bai Zhi' (primarily produced in Zhejiang) and 'Qi Bai Zhi' (primarily produced in Hebei province). In the wild the species grows in grassland along forest margins, valleys and stream-sides (500–1,000 m altitude) in north-eastern China (Hebei, Heilongjiang, Jilin, Liaoning and Shaanxi provinces); [also wild in Japan, Korean Peninsula and Russia (Siberia); with var. *formosana* confined to Taiwan].

CONSERVATION STATUS

China: Least Concern (MEPCAS, 2013).

Global: Not Evaluated (IUCN Red List accessed, 21/4/2015).

Medicinal crop of a cultivar of *A. dahurica* var. *dahurica* coming into flower, Sichuan province

Angelica dahurica var. *dahurica*, developing flower-head

A. dahurica var. *dahurica* with immature hairless fruits

TCM PROPERTIES AND USE

***properties*: 'pungent and warm'**
***channels*: 'Lung, Stomach and Spleen'**
***actions*: 'releases wind cold'**

***indications*: respiratory tract infection, allergic rhinitis; headache, toothache; sores, ulcers; vaginal discharge**

DRUG MORPHOLOGY

CRUDE DRUG (DRIED ROOT): robust, tapering and slightly curved **[1]** (10–25 cm long; 1.5–2.5 cm diam.), somewhat rounded or quadrangular in transverse outline **[2]** with occasional lateral roots strongly curving or with many large protuberant root scars **[3]**; apex (widest end) with sunken scars of stem and leaf bases **[4]**. Externally greyish- or yellowish-brown, with shallow longitudinal wrinkles and many, large and prominently raised transverse lenticels **[5]**, often aligned in close parallel. *Texture*: dense. *Fracture*: white or greyish-white; starchy. *Odour*: aromatic. *Taste*: pungent and somewhat bitter. ***Good quality*:** large and strongly aromatic, internally white and starchy with central xylem devoid of black spots and oil channels.

DECOCTING PIECES-1 (TRANSVERSELY SLICED ROOT): rounded, sub-rounded or squarish (1.5–2.5 cm diam.; 2–4 mm thick); remnants of external bark greyish- or yellowish-brown **[6]**; cut surfaces white or greyish-white with a thin cambium ring **[7]** (often distinctly oblong or sub-rounded) separating the very broad central wood **[8]** from the outer cortex; latter frequently mottled with brown oil glands **[9]**. *Texture* etc. as for crude drug.

DECOCTING PIECES-2 (LONGITUDINALLY SLICED ROOT — as sold in Hong Kong): long and tongue-shaped (c. 6–10 cm long; up to 2.5 cm wide; 2–4 mm thick) with a sharply truncate apex; remnants of external bark greyish- or yellowish-brown; cut surface white or greyish-white with a thin cambium ring separating the central wood from the outer cortex; latter frequently mottled with brown oil glands. *Texture* etc. as for crude drug.

PROCESSING

Rootlets are removed and discarded, then roots dried in the sun or at low temperature (= crude drug); later these are softened thoroughly and cut transversely into thick (2–4 mm) slices and dried again (= decocting pieces).

OTHER COMMON NAMES

杭白芷 Hang Bai Zhi (sourced from Zhejiang province); 祁白芷 Qi Bai Zhi (sourced from Anguo in Hebei province; 禹白芷 Yu Bai Zhi (sourced from Henan province); Dahurian angelica root (CP2010); fragrant angelica (SCN); Radix Angelicae Dahuricae.

SAFETY

Safety Class: 1; Interaction Class: A; (Gardner & McGuffin, 2013). An official Chinese food and healthfood (Anon., 2002a,b). Due to the presence of furanocoumarins, handling of the fresh plant may cause phototoxic skin reactions (Bensky *et al.*, 2004).

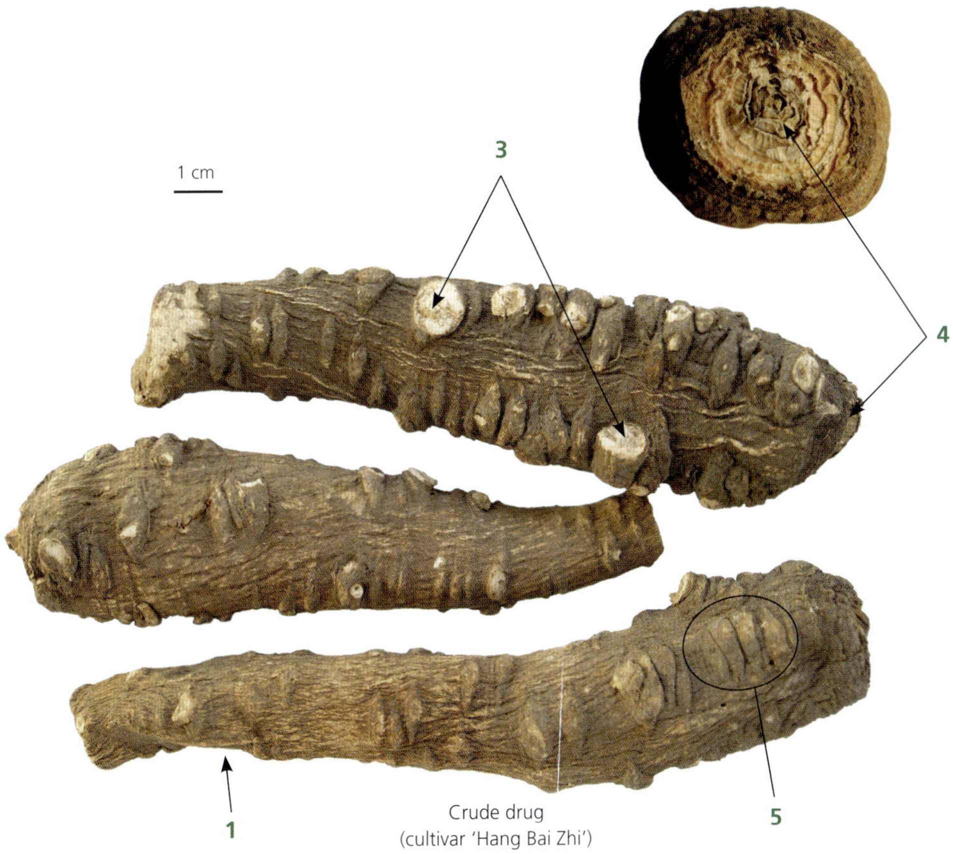

Crude drug
(cultivar 'Hang Bai Zhi')

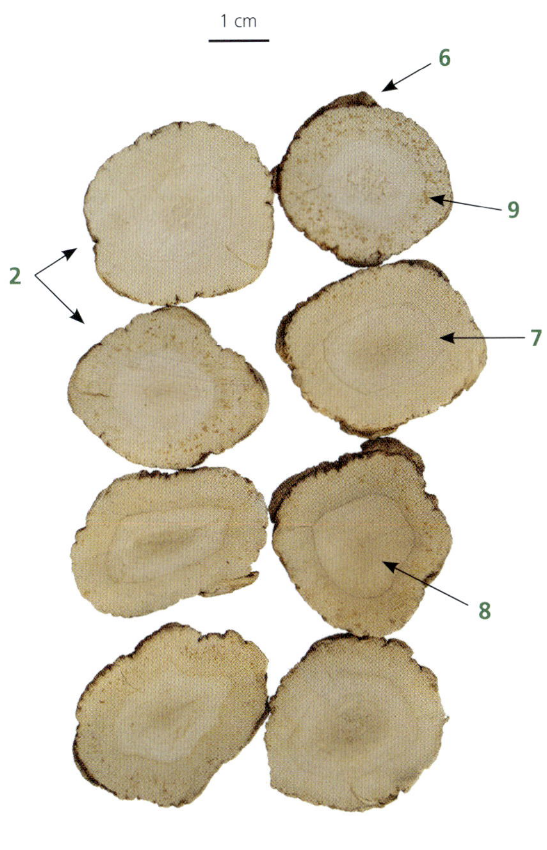

Decocting pieces-1 (transversely sliced)
(cultivar 'Hang Bai Zhi')

Bai Zhu 白术

Atractylodis Macrocephalae Rhizoma

Drug part: RHIZOME

Atractylodes macrocephala Koidz.

Asteraceae

PLANT DESCRIPTION

A perennial herb (20–60 cm tall) with a thick rhizome. *Stems* branching from base. *Leaves* alternate, simple and hairless. Middle stem leaves stalked, blades divided almost to base into 3–5 lance-shaped to narrowly elliptic segments (4.5–7 cm long; 1.5–2 cm wide); terminal segment largest. Upper stem leaves similar but smaller. *Flowers* densely packed into 6–10 thistle-like heads (3–4 cm diam.) surrounded by 2 whorls of green leaf-like bracts; outer bracts elliptic or narrowly elliptic with spiny margins, inner bracts deeply lobed. Overlapping green scale-like phyllaries form a goblet-shaped surround for the tiny purple florets within. *Fruits* single-seeded achenes (c. 7.5 mm long) with white hairs and a dirty white tuft of apical hairs (pappus) (c. 1.7 cm long) in a single row. Fl. and fr. Aug–Oct.

Species of *Atractylodes* wild in China: 4

HARVESTING, SOURCING AND NATURAL RANGE

Rhizomes are harvested in winter after the lower leaves have yellowed. Harvesting is exclusively from medicinal crops widely grown throughout China with the traditional *Di Dao* locality in Zhejiang province.

The wild species is endemic to China and confined to grassland and forest habitats between 600–2800 m altitude in Anhui, Chongqing, Fujian, Guizhou, Hubei, Hunan, Jiangxi and Zhejiang provinces.

CONSERVATION STATUS

Endemic to China: Least Concern (MEPCAS, 2013); Not Evaluated (IUCN Red List; accessed 10/3/2015).

Atractylodes macrocephala, medicinal crop in fruit, Zheijiang province, ready to harvest

A. macrocephala, flowering crop plants

A. macrocephala, single flowerhead

TCM PROPERTIES AND USE

***properties*: 'bitter, sweet and warm'**
***channels*: 'Spleen and Stomach'**
***actions*: 'tonifies qi, dries damp, resolves phlegm'**

***indications*: abdominal distension, poor appetite, diarrhoea, fatigue; leucorrhoea; oedema; threatened miscarriage**

DRUG MORPHOLOGY

CRUDE DRUG (DRIED RHIZOME): an irregular-shaped mass, often described as fist-shaped **[1]** or with 'ox-horns' (Ru Yi Tou) **[2]** (3–13 cm long, 1.5–7 cm diam.). Externally dull greyish-brown or greyish-yellow, rough and shallowly wrinkled longitudinally with many nodular masses and root scars **[3]**; remains of stem bases **[4]** and buds **[5]** at apical end. *Texture*: hard and difficult to break. *Fracture*: uneven, yellowish-white to brownish, and scattered with distinctive brownish-yellow oil glands **[6]**. *Odour*: aromatic. *Taste*: sweet and somewhat pungent; viscous when chewed.
Good quality: large, heavy and intensely aromatic.

DECOCTING PIECES-1 (SLICED RHIZOME): irregularly shaped slices (2–4 mm thick) with an undulating, somewhat nodular, outline. Rhizome skin dull greyish-brown or greyish-yellow and rough. Cut surface yellowish-white to pale brown dotted with brownish-yellow oil glands **[6]**; often with many clefts **[8]**. When oven-dried, cut surface somewhat horny in texture, with clefts more obvious and colour darker. *Odour*: faintly aromatic. *Taste*: somewhat viscous when chewed.

DECOCTING PIECES-2 (BRAN-PROCESSED RHIZOME SLICES — CHAO BAI ZHU): as for dp-1 but all surfaces yellowish-brown with occasional charred areas **[9]**. *Odour*: slightly burned.

PROCESSING

After removal of fibrous roots, rhizomes are sun-dried or oven-dried (= crude drug); rhizomes are subsequently softened thoroughly before slicing into thick decocting pieces. The latter may be further processed by frying with honeyed bran until the slices smoke and produce a burnt aroma (= bran-processed Bai Zhu or Chao Bai Zhu).

OTHER COMMON NAMES

冬术 Dong Zhu; 于术 Yu Zhu; bai-zhu atractylodes (SCN); large-head Atractylodes rhizome (CP2010); Rhizoma Atractylodis Macrocephalae.

SAFETY

Safety Class: 1; Interaction Class: A (Gardner & McGuffin, 2013). An official healthfood in China (Anon., 2002a,b).

Traditional processing method to create Chao Bai Zhu, Zhejiang province

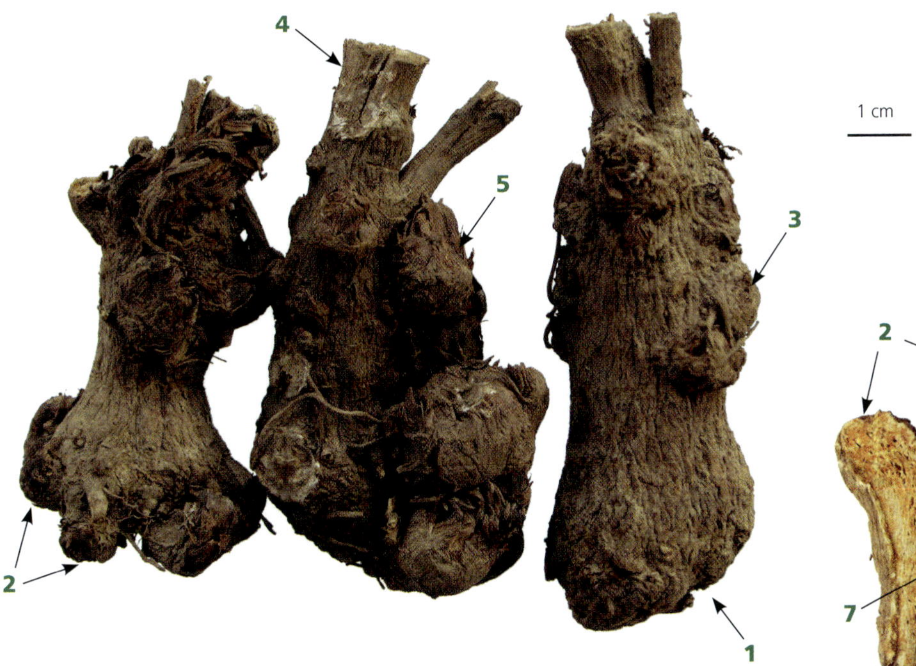

Crude drug of Bai Zhu (dried rhizome of *Atractylodes macrocephala*)

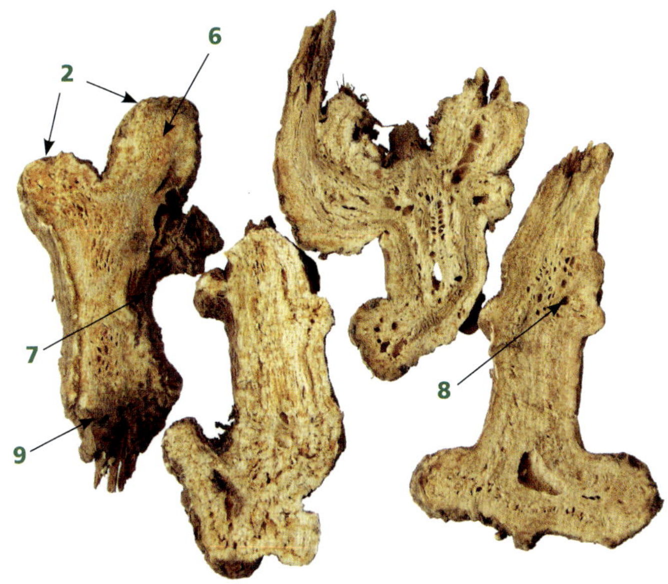

Bran-processed Bai Zhu (Chao Bai Zhu) (*A. macrocephala* rhizome)

Ba Ji Tian 巴戟天 Morindae Officinalis Radix

Drug part: ROOT

Morinda officinalis F. C. How

Rubiaceae

synonym: *Gynochthodes officinalis* (F. C. How) Razafim. & B. Bremer (*Kew-MPNS* accepted name, 25/8/2016).

Species of *Morinda* wild in China: 27

PLANT DESCRIPTION

A woody perennial climber. *Roots* distinctively fleshy and softly contorted, revealing pinkish-white inner cortex when broken and drying to purplish-red or purplish-blue. *Stems* turning brown or bluish-black when dried. *Leaves* opposite, small, membranous tube (3–5 mm long) of fused stipules present at base of opposite leaf bases; leaf blades ovate to oblong or elliptic (6–13 cm long; 3–6 cm wide), upper surface sparsely and minutely hairy, lower surface hairless or sparsely hairy on some veins and with small cavities (domatia) in the vein axils; margin entire. (Softly hairy plants from Hainan have been placed in var. *hirsuta* F. C. How). *Flowers* small and inconspicuous, in 1–3- (var. *uniflora*) or 4–10-flowered (var. *officinalis*) compact heads; flowers all joined at base into single mass; sepals 2–4, joined halfway, with a limb 1–2.5 mm long; stamens 3–4; corolla white with tube (3–4 mm long) and 3 or 4 triangular petal lobes as long as tube; ovary inferior, stigmas 2. *Fruit* an aggregated head of fused ovaries, fleshy, sub-globose (0.5–1.1 cm diam.), bright red when mature. Fl. May–Jul, fr. Oct–Nov.

HARVESTING, SOURCING AND NATURAL RANGE

Roots are harvested year round from medicinal crop plants which have reached 6 years old (i.e. when the fleshy medicinal cortex of the roots has developed). Main production areas occur throughout the natural range of the species in S and SE China: Fujian, Guangdong, Guangxi and Hainan with *Di Dao* material traditionally sourced from Guangdong (Gaoyao, Deqing and Yu'nan districts).

The wild species is endemic to China and grows in sub-tropical and tropical forests and thickets on mountainsides (100–500 m altitude).

CONSERVATION STATUS

The wild species began to decline many years ago due to over-harvesting of its medicinal roots.

Endemic to China: Vulnerable (MEPCAS, 2013, both wild varieties); Endangered (Fu, 2011); Vulnerable (Wang & Xie, 2004); Vulnerable (Fu, 1992); Not Evaluated (IUCN Red List, accessed 11/8/2016).

Young medicinal crop of *Morinda officinalis*, Guangdong province, S China

TCM PROPERTIES AND USE

properties: 'pungent, sweet and slightly warm'
channels: 'Kidney'
actions: 'tonifies Kidney, warms yang'

indications: impotence, spermatorrhoea, incontinence; infertility, dysmenorrhoea; musculoskeletal pain

Morinda officinalis, young plants

M. officinalis in full flower

M. officinalis with immature fruits

Freshly harvested roots of *M. officinalis* showing pinkish-white internal flesh

Morinda officinalis

Rubiaceae

DRUG MORPHOLOGY

CRUDE DRUG (FLATTENED ROOT): roots laterally compressed, tortuous, curved and partially broken **[1]** (length variable; 0.5–2 cm diam.). Externally greyish-yellow or dark grey, longitudinally coarsely wrinkled **[2]**, and with distinctive transverse divisions (breaks) **[3]** revealing the string-like woody core (xylem) (1–5 mm diam.) **[4]**. *Overall texture*: tough but pliable. *Fracture*: fleshy outer cortex purplish-red, purplish-blue or pale purple, easily separated from the inner yellowish-brown or yellowish-white xylem. *Odour*: mild. *Taste*: somewhat sweet and slightly astringent. ***Good quality*:** flattened roots with thick and purplish flesh.

DECOCTING PIECES-1 (STEAMED ROOT): flattened, somewhat tubular pieces with truncated ends or irregularly shaped **[5]** (0.5–2 cm diam.). Externally dull, greyish-yellow or dark grey, transversely strongly wrinkled **[6]**, with occasional transverse cracks revealing the darker cortex below **[7]**. Cut surface, fleshy purplish-red, purplish-blue or pale purple, with a hollow centre. *Overall texture*: tough but pliable. *Odour*: indistinct. *Taste*: somewhat sweet and slightly astringent.

DECOCTING PIECES-2 (SALT-STEAMED ROOT): as for dp-1 above but *odour* mild and *taste* sweet, salty and slightly astringent.

DECOCTING PIECES-3 (LIQUORICE-PROCESSED ROOT): as for dp-1 above but slightly liquorice-scented.

PROCESSING

Rootlets are removed from the rootstocks and discarded, then roots are partially dried (60–70%) in the sun before being mechanically flattened or beaten gently to break the fleshy purplish cortex and to enable subsequent thorough drying (= crude drug). The hard, woody core (xylem) is then removed (and discarded) from the fleshy cortex by steaming; the cortex is then cut into lengths and dried to form decocting pieces called Ba Ji Rou (巴戟肉); alternative processing either involves steaming with salt water (= Yan Ba Ji Tian 盐巴戟天 'salt-processed Ba Ji Tian') or boiling with liquorice root (Gan Cao — Glycyrrhizae Radix, not included in this book) to produce Zhi Ba Ji Tian 制巴戟天 'prepared Ba Ji Tian'. Both processes enable removal of the woody xylem before cutting and drying again.

OTHER COMMON NAMES

morinda (SCN); morinda root (CP2010); Radix Morindae Officinalis.

SAFETY

An official healthfood in China (Anon, 2002a,b).

UNOFFICIAL SUBSTITUTES

The roots of other species in the same genus (*Morinda*) along with those in other genera and families have been reported as local or historical unofficial substitutes; their occurrence as such, however, in international trade is unlikely:

1. ***Morinda cochinchinensis*** DC. Native to S China (Fujian, Guangdong, Guangxi, Hainan). **Leaves minutely hairy on lower surfaces only**; **lobes of flower tube longer** than the tube (tube 1–2 mm; lobes 3–4.5 mm); **fruit 1–2 cm diam.**
2. ***M. parvifolia*** Bartl. ex DC. Native to S China (Fujian, Guangdong, Guangxi, Hainan, Jiangxi, Taiwan). **Leaves smaller than the Pharmacopoeia species** (2–5 (–7) cm long, 0.3–3 cm wide); **fruits orange**.
3. ***M. shuanghuaensis*** C. Y. Chen & M. S. Huang. Endemic to S China (Fujian, Guangdong). **Calyx limb poorly developed**; flower tubes and lobes about equal in length (tube c. 3 mm, lobes c. 3 mm).
4. ***M. umbellata*** L. Native to S and C China (Anhui, Fujian, Guangdong, Guangxi, Hainan, Hunan, Jiangsu, Jiangxi, Taiwan, Zhejiang). Stems weakly angled, often channelled; **leaves narrower than the Pharmacopoeia species** (6–9 cm long; 2–3.5 cm wide); calyx limb **poorly developed** (< 1 mm).
5. ***Damnacanthus giganteus*** (Makino) Nakai (as *D. macrophyllus* var. *giganteus* (Makino) Koidz. in NICPBP, 1997–2002). Family Rubiaceae. Endemic to China (Guangxi, Guizhou, Yunnan). As for *D. indicus* but a **shrub or small tree** (0.5–2 (–7.5) m tall); **stems sometimes spiny**; **flower tube longer than Pharmacopoeia species** (15–18 mm long).
6. ***D. indicus*** C. F. Gaertn. Native to C, E, S and W China. A **shrub** (0.3–1.5 m tall); **roots** fleshy, **interrupted and bead-like**; **stems spiny**; flowers tightly packed but **not fused**; flower tube 7–9 mm long; fruit develops from 1 flower only.
7. ***D. officinarum*** C. C. Huang. Endemic to China (Hubei, Hunan, Sichuan). As for *D. indicus* but **1–2.5 m tall**, stems spineless.
8. ***Akebia trifoliata*** (Thunb.) Koidz. (family Lardizabalaceae). Endemic to C, N and S, W China (15 provinces). A woody climber; **leaves alternate** or clustered on short branches, divided into **3–5 oval leaflets, flowers pale purple, purple, purplish brown, dark purple, or purplish black**; fruit fleshy, oblong, straight or slightly curved; seeds many, in rows.
9. ***Kadsura coccinea*** (Lem.) A. C. Sm. (family Schisandraceae). Native in C, S and W China (8 provinces). Woody climber, **hairless; leaves alternate or clustered**; 'flowers' white, red, purplish red, or occasionally yellowish; flowers single-sexed; 'flower' bases **not fused** together; fruit red to purplish red.
10. ***K. longipedunculata*** Finet & Gagnep. Endemic to C and S China (9 provinces). As for *K. coccinea,* but **'flowers' pale yellow, yellow, or occasionally reddish**.
11. ***Schisandra propinqua*** (Wall.) Baill. subsp. ***sinensis*** (Oliv.) R. M. K. Saunders (family Schisandraceae). Endemic to C, S and W China (12 provinces). Woody climber; leaves alternate; **flowers stalked, of various colours (cream, yellow, orange, pink, or purplish)**; **fruits clustered in long, pendant spikes**.

Ba Ji Tian pre-crude drug (dried roots of *M. officinalis*)

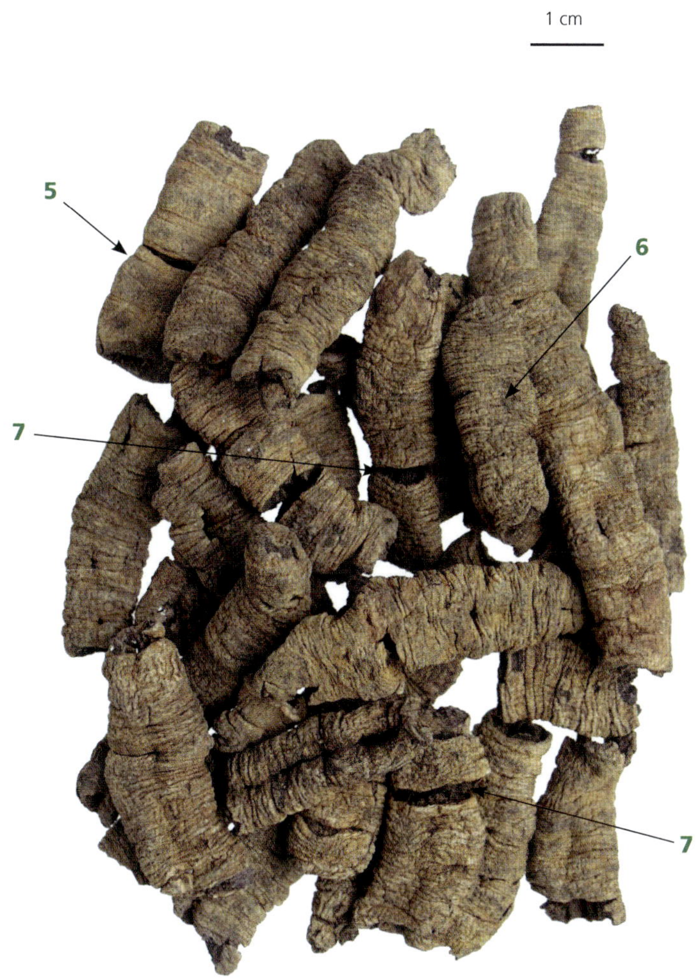

Liquorice-processed Ba Ji Tian (Zhi Ba Ji Tian) of *M. officinalis*

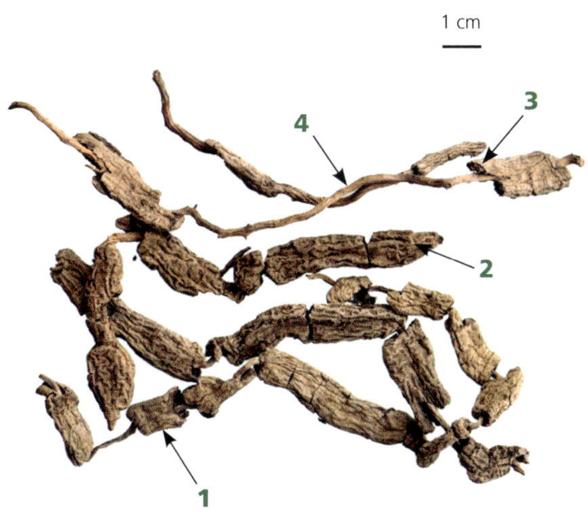

Crude drug (flattened and dried roots) of *M. officinalis*

Ban Lan Gen 板蓝根

Isatidis Radix

dyer's woad root

Drug part: ROOT

PHARMACOPOEIA SPECIES

Isatis tinctoria L.

Brassicaceae

Species of *Isatis* wild in China: 4

synonym: *I. indigotica* Fortune ex Lindl. (name used in CP2015 as '*I. indigotica* Fort.').

taxonomic notes: many TCM texts treat *I. indigotica* as a separate species rather than a synonym but the *FoC* recognises a huge variability in *I. tinctoria* and (like *Kew-MPNS*, 17/6/2016) includes *I. indigotica* within it.

PLANT DESCRIPTION

A highly variable species. A biennial herb (40–100 (rarely 30–150) cm tall). *Stem* branching above, often bluish-green, hairy or hairless. *Leaves* simple and **alternate**. Basal leaves (if present) often in a rosette, stalked, blades oblong or broadest above the middle (usually 5–15 cm long; 1.5–3.5 cm wide), bases tapering, margins wavy or sometimes toothed; tips broadly rounded; occasionally hairy. Middle stem leaves unstalked, becoming smaller up the stem, blades lance-shaped (usually 3–7 cm long; 0.8–2.5 cm wide), sometimes with 2 auricles (backward-pointing 'ears') or clasping the stem, margins entire, tips pointed. *Flowers* (appearing in second year) in many-branched sub-rounded heads; latter axillary and terminal. Each flower with 4, ovate or oblong sepals; petals 4, bright yellow, equal, wedge-shaped and alternating with sepals; stamens 6 (in 2 whorls); style 1. *Fruit* a dry, 2-valved capsule, flattened, pendulous and variously shaped (slightly broader towards the tip), black or purplish-brown; fruit-stalk (pedicel) swollen below the fruit. Fl. Apr–Jun, fr. May–Jul.

HARVESTING, SOURCING AND NATURAL RANGE

Roots are harvested in autumn mostly from cultivated plants grown on medicinal plant farms in northern China with the largest production being from Hebei, but also Jiangsu and Anhui provinces (Zhu, 1998). The species has been long cultivated in China and many forms or cultivars are thought to have evolved; it is now naturalised in fields, along roadsides and other disturbed places (600–2,800 m altitude). The species is native to Europe (introduced and naturalised in North America).

CONSERVATION STATUS

China: introduced.

Global: Not Evaluated (IUCN Red List, accessed 17/6/2015). Widely naturalised.

COMMENT For other TCM herbal drugs sourced from this species see 'Da Qing Ye' (Isatidis Folium) page 540.

UNOFFICIAL SUBSTITUTE

Strobilanthes cusia (Nees) Kuntze

Acanthaceae

synonyms: *Baphicacanthus cusia* (Nees) Bremek. (name used in CP2015); *Goldfussia cusia* Nees

PLANT DESCRIPTION

An upright perennial herb (0.5–1.5 m tall) **often woody at base.** *Leaves* **opposite**, blades **elliptic to ovate** (4–20 cm long; 2–9 cm wide) and **tips pointed**. ***Flowers* much larger (3.5–5 cm long), blue, conspicuous,** corolla **funnel-shaped** with 5 notched equal-sized lobes. Flowers intermingled with leaf-like bracts. *Fruit* a dry, elongate, 4-seeded capsule. Fl. Jul–Feb, fr. Dec–Feb.

LESSER KNOWN UNOFFICIAL SUBSTITUTES

The name 'Ban Lan Gen' has been widely used in China for many other herbal drugs causing potential confusion in trade (Bensky *et al.*, 2004). Examples include:

1. ***Clerodendrum cyrtophyllum*** Turcz. (大青) (Lamiaceae); wild in south China.
2. ***Strobilanthes penstemonoides*** (Nees) T. Anderson (球花马蓝) (Acanthaceae); wild in south-east Tibet and and Yunnan provinces.

TCM PROPERTIES AND USE

properties: **'bitter and cold'**
channels: **'Heart and Stomach'**
actions: **'clears heat, eliminate toxins, cools Blood'**

indications: **fevers; sore throat; rash**

Flowering crop of *Isatis tinctoria*, Gansu province

Close-up of flower-heads of *I. tinctoria*

I. tinctoria in full flower

I. tinctoria, middle stem leaf bases clasping or 'eared'

I. tinctoria with purplish-brown fruits

Strobilanthes cusia in flower

Isatis tinctoria

Brassicaceae

DRUG MORPHOLOGY

CRUDE DRUG (DRIED ROOT): roots long, cylindrical, **tapering [1]**, slightly curved and twisted (10–20 cm long; 0.5–1 cm diam.). External root surface yellowish-white, greyish-yellow or brownish-yellow, longitudinally wrinkled and scattered with faint and slightly raised transverse lenticels. Root apex with a dense and broader tubercular root head **[2]** bearing dense whorls of old dark green or dark brown fibrous stem and leaf bases. *Texture*: somewhat soft, compact or lax. *Fracture*: fairly even and powdery; cortex yellowish-white; central xylem pale yellow. *Odour*: faint. *Taste*: sweet, then bitter and astringent. ***Good quality*:** roots yellowish-white; powdery within.

DECOCTING PIECES (DRIED SLICED ROOT): sub-circular slices (0.5–1 cm diam.; 2–4 mm thick). Outer bark somewhat soft, yellowish-white, greyish-yellow or brownish-yellow with deep longitudinal wrinkles; transverse narrowly ovate lenticels sometimes present. Cut surface even, compact or lax with a broad yellowish-whitish cortex **[3]** up to half or 2/3 of root diameter (sometimes with radial cavities) and clearly separated from the pale yellow central xylem (wood) **[4]** by a thin dark brown or grey cambial ring **[5]**; rays sometimes present. *Texture*: somewhat powdery. *Odour*: indistinct. *Taste*: sweet, then bitter and astringent.

PROCESSING

Roots are sun-dried (= crude drug) then subsequently softened, cut into decocting slices (2–4 mm thick) and dried again.

OTHER COMMON NAMES

北板蓝根 Bei Ban Lan Gen (northern Ban Lan Gen); 菘蓝 Song Lan (plant name); dyer's woad (SCN) (for *I. tinctoria*); indigo blue woad; isatis (SCN) (for *I. indigotica*); isatis root (CP2010); Radix Isatidis; woad.

SAFETY

Safety Class: 1; Interaction Class: A (Gardner & McGuffin, 2013).

Strobilanthes cusia

Acanthaceae

'DRUG SUBSTITUTE' MORPHOLOGY

DRIED ROOTSTOCKS: rhizome and root somewhat **tortuous [6]** (10–30 cm long; width highly variable: 0.1–1.3 cm); occasionally branching; **rhizome nodes swollen [7]** and oval stem scars occasional. External surface brown, with outer cork bark **often exfoliating to reveal characteristic bluish-grey inner bark [8]** below; wood pale yellowish-brown to greyish-blue **[9]**. *Texture*: hard but brittle. *Fracture:* fibrous and uneven. *Odour and taste*: indistinct.

SLICED ROOTSTOCKS: sub-rounded slices of rhizome and root of varying width (0.1–1.3 cm). Thin, brown outer cork bark intact or exfoliating; easily removed completely to reveal slightly thicker **bluish-grey** or pale yellowish-brown **inner bark**; latter also easily removed. Cut surface predominantly with pale yellowish-brown or greyish-blue wood; **rhizome slices with a central white pith [10]**. *Texture*: hard but brittle. *Odour* and *taste*: indistinct.

OTHER COMMON NAMES

板蓝 Ban Lan (plant name); 板蓝根 Ban Lan Gen; 广东大青叶 Guangdong Da Qing Ye; 马蓝 Ma Lan; 马蓝根 Ma Lan Gen; 南板蓝根 Nan Ban Lan Gen (southern Ban Lan Gen); Baphicacanthis Cusiae Rhizoma et Radix (CP2015); Rhizoma et Radix Baphicacanthis Cusiae; strobilanthes cusia (SCN).

SUBSTITUTION COMMENT. Rootstocks are the source of the separate TCM drug 'Nan Ban Lan Gen' (page 220) from *Strobilanthes cusia*. This drug is commonly substituted for official 'Ban Lan Gen' in southern China (notably Guangdong) where it is traded either as 'Ban Lan Gen' or 'Nan (southern) Ban Lan Gen' (Lin, 1990–1996; Zhao & Li 2007). Bensky *et al.* (2004) state it "is used very similarly to Isatidis Radix .... but is considered to be of inferior quality". The respective source plants are morphologically quite different but their roots bear some superficial similarity and, to the inexperienced eye, may be confused.

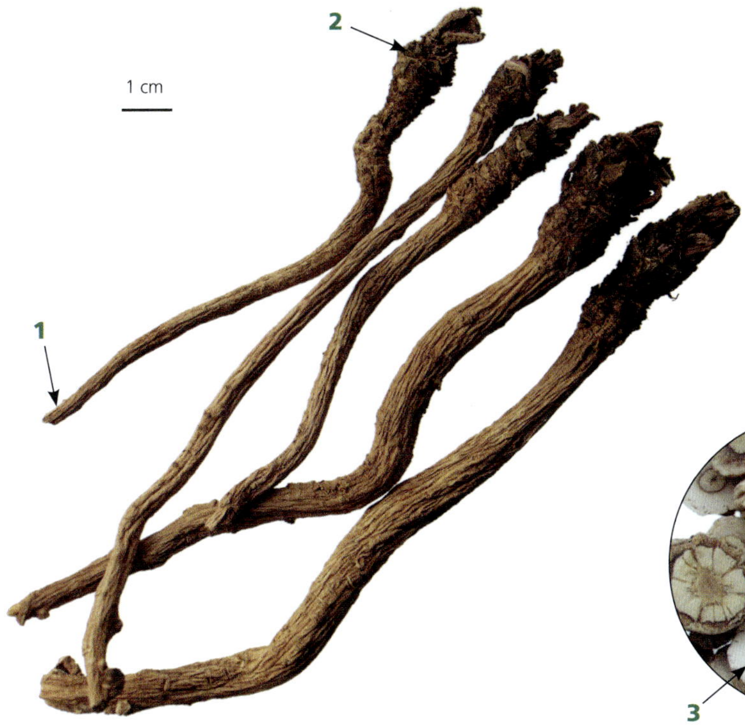

Ban Lan Gen crude drug (dried roots of *I. tinctoria*)

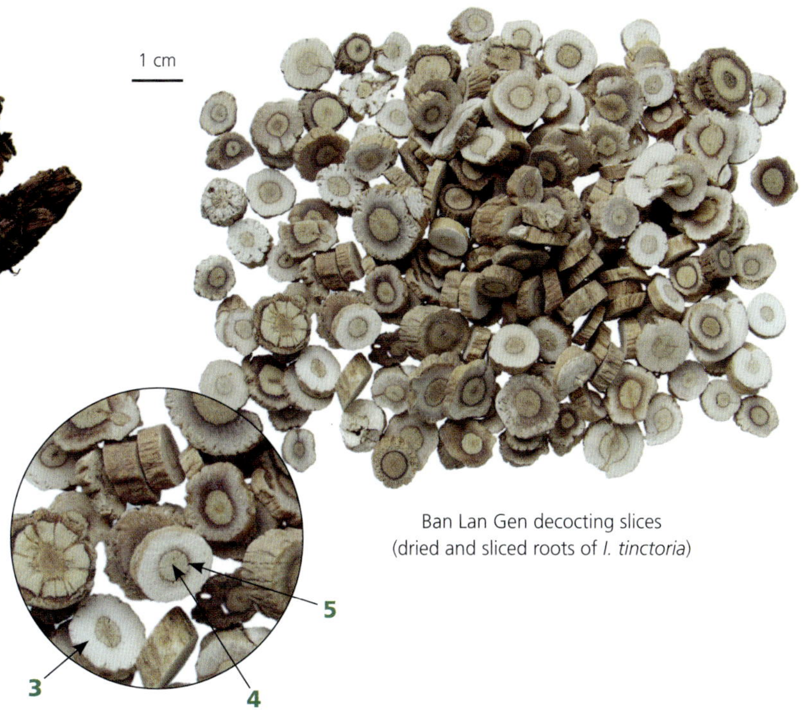

Ban Lan Gen decocting slices
(dried and sliced roots of *I. tinctoria*)

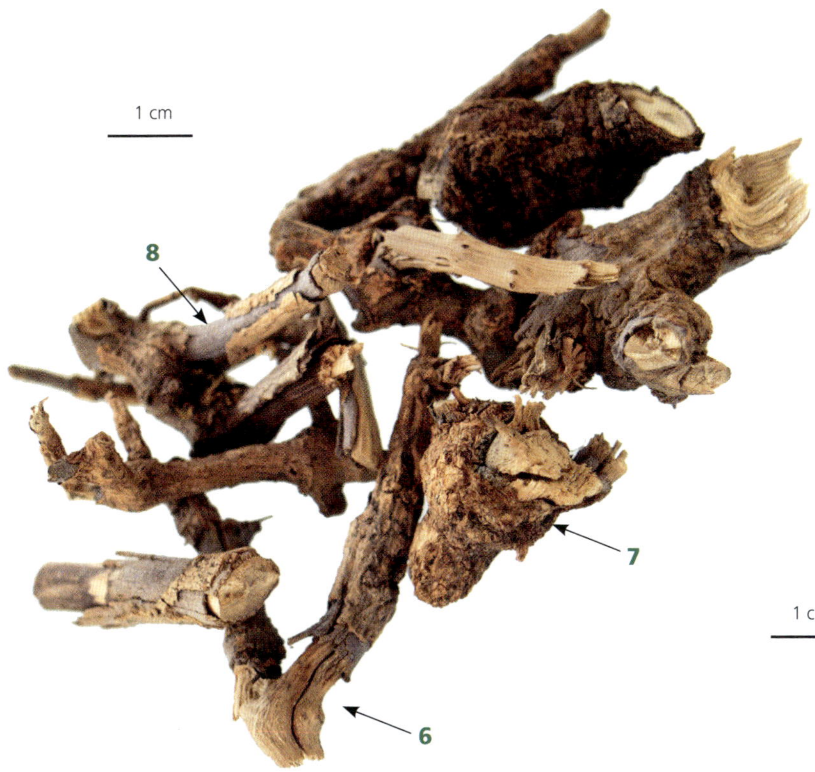

Dried roots and rhizomes of *Strobilanthes cusia*
(crude drug of Nan Ban Lan Gen)

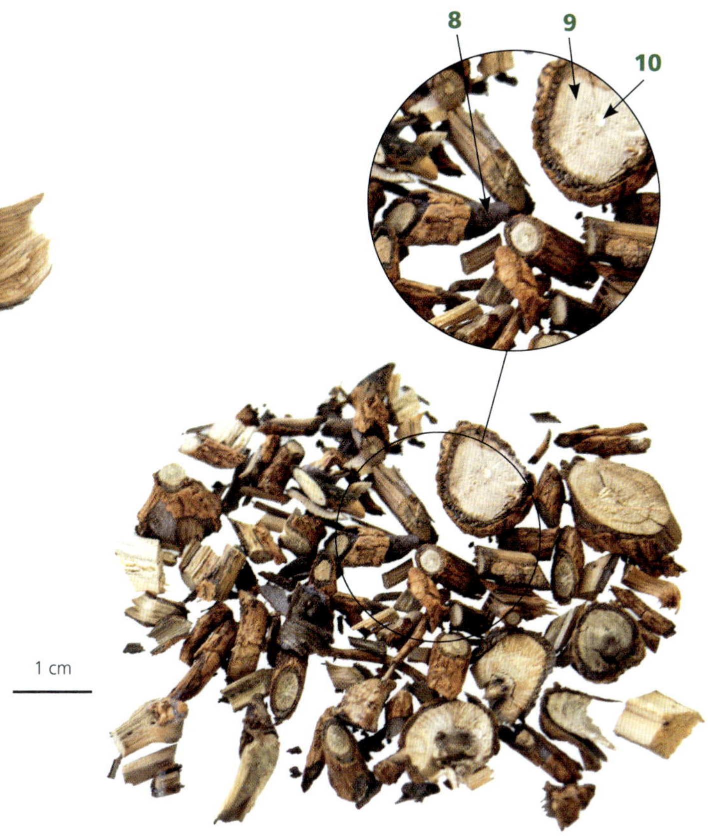

Dried and sliced roots and rhizomes of *S. cusia*
(decocting pieces of Nan Ban Lan Gen)

Ban Xia 半夏

Pinelliae Rhizoma

incl. Pinelliae Praeparatum Rhizoma

Drug parts: UNPROCESSED AND PROCESSED STEM TUBER (RHIZOME)

Pinellia ternata (Thunb.) Makino

Araceae

synonym: *Arum ternatum* Thunb.

taxonomic/nomenclatural note: in CP2015 as '*P. ternata* (Thunb.) Breit.'

Species of *Pinellia* wild in China: 9

PLANT DESCRIPTION

A highly variable species. A perennial herb (up to 20 cm tall) with sub-globose corm-like rhizomes (1–2 cm diam.). *Leaves* simple, 2–5, apparently basal, with sheathing stalks (15–20 cm long); bulbils present at varying positions along the leaf stalk; leaf blades divided into 3 (sometimes 5) oblong-elliptic or lance-shaped lobes, base wedge-shaped; tip pointed. *Inflorescence* appears at same time as foliage; with a spadix of tightly packed flowers surrounded in part by a **whitish-green spathe** (4–51.5 cm long), tubular and **closed at its throat,** extending into a long and often pinkish-purple-edged limb, obtuse or acute at its tip. Male and female flowers in separate zones within each inflorescence. **Female flowers** clustered in a 2 cm long zone, **fused to the spathe** and separated from the male flowers above by a 3 mm sterile zone; zone of male zone 5–7 mm wide; **uppermost part of male zone extends above the height of the spathe.** Terminal part of the spadix sigma-shaped or greenish-violet; much longer than the spathe. Sepals and petals absent. Male flowers with 2 fused stamens. *Fruits* ovoid, yellowish-green to whitish berries with persistent stigma and style.1-seeded. Fl. May–Jul, fr. Jul–Sep.

HARVESTING, SOURCING AND NATURAL RANGE

Rhizomes are harvested in summer and autumn mainly from medicinal crop plants grown extensively throughout the natural range of the species in China (i.e most provinces except Mongolia, Qinghai, Tibet and Xinjiang). The habitat of the wild species is grassland, secondary forest and disturbed ground up to 2,500 m altitude. Also native in Japan and the Korea Peninsula and naturalised in Europe and North America.

CONSERVATION STATUS

China: Least Concern (MEPCAS, 2013).
Global: Not Evaluated (IUCN Red List, accessed 9/10/2016).

TCM PROPERTIES AND USE (processed form)

properties: 'pungent and warm'
channels: 'Spleen, Stomach and Lung'
actions: 'dries damp, descends qi, resolves phlegm, disperses nodules'

indications (processed form): cough, phlegm; dizziness; vomiting; headache, sores; (T)

Medicinal crop of *Pinellia ternata*, Shaanxi province

Freshly harvested rhizomes of *P. ternata*

P. ternata in flower

UNOFFICIAL SUBSTITUTE 1

Typhonium flagelliforme (G. Lodd.) Blume

Araceae

synonym: *Arum flagelliforme* G. Lodd.

PLANT DESCRIPTION

Closely related and easily confused with the Pharmacopoeia species. *Leaves* simple, leaf blades 5–25 cm long; 0.5–18 cm wide, extremely variable in shape (linear, lance-shaped, elliptic and/or with 2 outward pointing basal triangular lobes**, spadix shorter than spathe,** throat of **spathe tube is open** and the appendix (arm) of the spadix typically **recurved**. Fruits pale greenish, each *fruit* 2 or 3-seeded. Fl. Apr–May.

SUBSTITUTION COMMENT. When supplies of *Pinellia ternata* rhizomes (official Ban Xia) were in short supply in China in the 1980s, rhizomes of *Typhonium flagelliforme*, a close relative, were used as a cheap substitute (Zhao & Li, 2007), a practice which continues today on the domestic and international markets. Rhizomes are used in local (non-TCM) Chinese medicine where they are known as Shui Ban Xia 水半夏 but their frequent abbreviation to Ban Xia 半夏 may create confusion.

T. flagelliforme is common and widespread in the wild in S and SW China (Guangdong, Guangxi and SE Yunnan provinces) where it grows in damp meadows, stream-sides and rice-paddies, up to 350 m altitude. It is cultivated as a local medicine in Anhui, Guangxi, Jiangxi and Zhejiang provinces and, once established, can become locally invasive.

Elsewhere the species is native in Bangladesh, Bhutan, Cambodia, India, Indonesia, Laos, Malaysia, Myanmar, Philippines, Singapore, Sri Lanka, Thailand and Australia.

CONSERVATION STATUS

China: Least Concern (MEPCAS, 2013).

Global Least Concern (IUCN Red List, accessed 9/10/2016).

UNOFFICIAL SUBSTITUTE 2

Pinellia pedatisecta Schott

Araceae

PLANT DESCRIPTION

Closely related and easily confused with the Pharmacopoeia species. **Leaf blade divided into 6–11** lance-shaped and long pointed leaflets; central leaflet 15–18 cm long, outermost lealfets 4–5 cm long. Spathe green outside, whitish inside. Fl. Jun–Jul.

SUBSTITUTION COMMENT. Easily confused with the official drug, the rhizome of this species is widely sold as a cheaper and more readily available substitute; it undergoes the same processing as the official drug in order to reduce its toxicity.

P. pedatisecta is endemic to China and is common and widespread in shady places in valleys and forests up to 1,000 m altitude; it grows throughout the country except in the northern provinces of Inner Mongolia, Qinghai, Tibet and Xinjiang. It is cultivated in Hebei, Henan and Shaanxi.

CONSERVATION STATUS

Endemic to China: Least Concern (authors' pers. obs. 2005); Not Evaluated (IUCN Red List, accessed 9/10/2016)

Crop of *Typhonium flagelliforme*, Anhui province

T. flagelliforme in flower

Medicinal crop of *Pinellia pedatisecta* near Anguo (Hebei province)

P. pedatisecta in flower

Pinellia ternata

Araceae

UNPROCESSED DRUG MORPHOLOGY

UNPROCESSED CRUDE DRUG (DRIED AND PEELED RHIZOME): globose to squat globose **[1]** (1–1.5 cm diam.) occasionally sub-ovoid or irregular. Externally white or pale yellow **[2]**, apex often depressed **[3]**, with scar of stem sometimes present; depression often encircled by a zone of sunken and dot-like root scars **[4]**; basal end rounded and somewhat smooth **[5]**. *Texture:* hard and uneasily broken. *Fracture*: white-yellowish and starchy. *Odour*: indistinct. *Taste* (*not advised*): strongly numbing, irritant and pungent.

PROCESSED DRUG MORPHOLOGY

FA BAN XIA (STANDARD OR LIQUORICE-PROCESSED RHIZOME): *cannot be reliably distinguished macroscopically from unprocessed Ban Xia above but the latter is unlikely to be encountered in trade. Minor differences*: *texture:* hard, crumbly or fragile; slightly powdery. *Fracture*: yellow or greyish-yellow. *Taste*: slightly sweet and slightly tongue-numbing. ***Good quality***: rhizomes large, sub-spherical, creamish-white and slightly powdery.

QING BAN XIA (SLICED, ALUM-PROCESSED RHIZOME): *as for Fa Ban Xia except*: root scars usually obscured by processing. *Texture*: easy to break; fragile. *Fracture*: slightly horny. *Taste*: slightly astringent and tongue-numbing. (*not illustrated*)

JIANG BAN XIA (WHOLE OR SLICED, GINGER-PROCESSED RHIZOME): *as for Fa Ban Xia except:* external surface pale to dark brown **[6]**. *Texture*: hard but brittle. *Fracture*: pale yellowish-brown, often horny and lustrous **[7]**. *Taste*: indistinct, slightly tongue-numbing and, on chewing, viscous.

Processed Ban Xia (Jiang Ban Xia) — sliced ginger-processed *P. ternata* PHOTO: ERIC BRAND

PROCESSING

(Sheng Ban Xia – unprocessed rhizomes)

After removal of outer skin and fibrous roots, rhizomes are sun-dried (= unprocessed crude drug 'Sheng Ban Xia') then broken into pieces for decocting, but given that rhizomes are **toxic** in this unprocessed state (see Safety) only processed Ban Xia (Zhi Ban Xia — Pinelliae Praeparatum Rhizoma) is normally traded; see below.

PROCESSING

(Ban Xia or Zhi Ban Xia – processed rhizomes)

Sheng Ban Xia (unprocessed rhizomes, above) are processed in one of the following ways until, according to traditional practice, rhizomes are 'only slightly tongue-numbing'. Each type of processed rhizome is described as a separate drug in the CP2015. Summary processing procedures only are given below (see CP2015 for full details).

1. **Fa Ban Xia (liquorice-processed)**: after size-grading, dried rhizomes are soaked in water before they undergo many days of processing in a solution of liquorice root (Gan Cao) and quicklime (calcium oxide) until yellow throughout, the fracture horny (hard and semi-transparent) and, rhizomes are then washed and dried in darkness or in an oven.
2. **Qing Ban Xia (alum-processed)**: after size-grading, dried rhizomes are soaked in a solution of water and alum, then dried.
3. **Jiang Ban Xia (ginger-processed)**: after size-grading, dried rhizomes are soaked in water then boiled in a decoction of ginger with alum, then dried.

OTHER COMMON NAMES

Plant names: 半夏属 Ban Xia Shu; 三步跳 San Bu Tiao; 三叶半夏 San Ye Ban Xia.

Drug names: the name Ban Xia is the commonest trade name likely to be encountered and is generally used to refer to any type of processed *P. ternata* rhizome. *Other drug names*: 法夏 Fa Xia 法半夏 Fa Ban Xia (processed *Pinellia* — usually with liquorice); Jiang Ban Xia (ginger-processed *Pinellia*); Qing Ban Xia (alum-processed *Pinellia*); Qing Shui Ban Xia (clear water- and alum-processed *Pinellia*); pinellia (SCN); pinellia rhizome (CP2010); Rhizoma Pinelliae; Sheng Ban Xia (unprocessed *Pinellia*).

SAFETY

Unprocessed rhizome (Sheng Ban Xia): toxic in fresh and dried states and inappropriate for internal use; topical use only (Bensky *et al.* 2004; CP2015; Gardner & McGuffin, 2013; Hong Kong Schedule 1, 1999a). Unlikely to be encountered in trade owing to its toxicity.

Processed rhizome (Ban Xia and other processed rhizome names above): Safety Class: 3; Interaction Class: A (Gardner & McGuffin, 2013). Officially forbidden in China both as food and healthfood (Anon., 2002a,b).

Unprocessed Ban Xia (Sheng Ban Xia) sourced from *Pinellia ternata*

Processed Ban Xia (Fa Ban Xia — liquorice-processed rhizomes) sourced from *P. ternata*

Processed Ban Xia (Jiang Ban Xia — ginger-processed rhizpomes) sourced from *P. ternata*

Typhonium flagelliforme

Araceae

'DRUG SUBSTITUTE' MORPHOLOGY

Rhizomes of this unofficial substitute may be processed in the same way as those for the official drug. They differ from their counterparts as follows: **usually tapering (sometimes acutely) at one or both ends [8] (apical depression absent)** and **root scars scattered over entire surface [9]** rather than localised around the apex (root scars usually obscured in alum-processed rhizomes).

OTHER COMMON NAMES

水半夏 Shui Ban Xia (water Pinellia).

SAFETY

The unprocessed rhizome is **toxic** and the effects of processing to minimise toxicity are thought to be the same as those performed for the official drug.

Pinellia pedatisecta

Araceae

'DRUG SUBSTITUTE' MORPHOLOGY

Rhizomes of this unofficial substitute may be processed in the same way as those for the official drug. Only **mature rhizomes** of this species can sometimes be differentiated macroscopically from their counterparts because they are **larger (up to 4 cm diam.)** (*compared to 1.5 cm diam. in the official drug*) and **one or more smaller nodular 'daughter' rhizomes [10]** often develop around the perimeter. Immature rhizomes **[11]** can be easily be mistaken for the official drug and cannot be reliably distinguished macroscopically. Laboratory-based authentication methods are needed for the latter.

OTHER COMMON NAMES

虎掌 Hu Zhang; 掌叶半夏 Zhang Ye Ban Xia (palm-leaved Pinellia).

SAFETY

The unprocessed rhizome is **toxic** and the effects of processing to minimise toxicity are thought to be the same as those performed for the official drug.

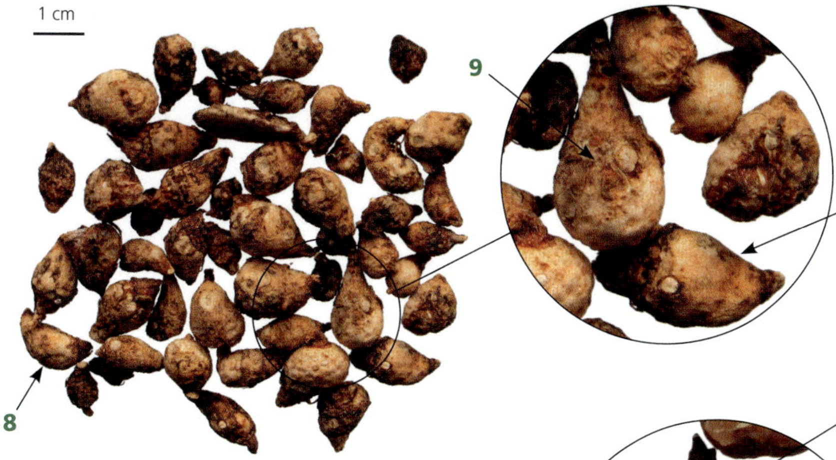

Unprocessed rhizomes of *T. flagelliforme*

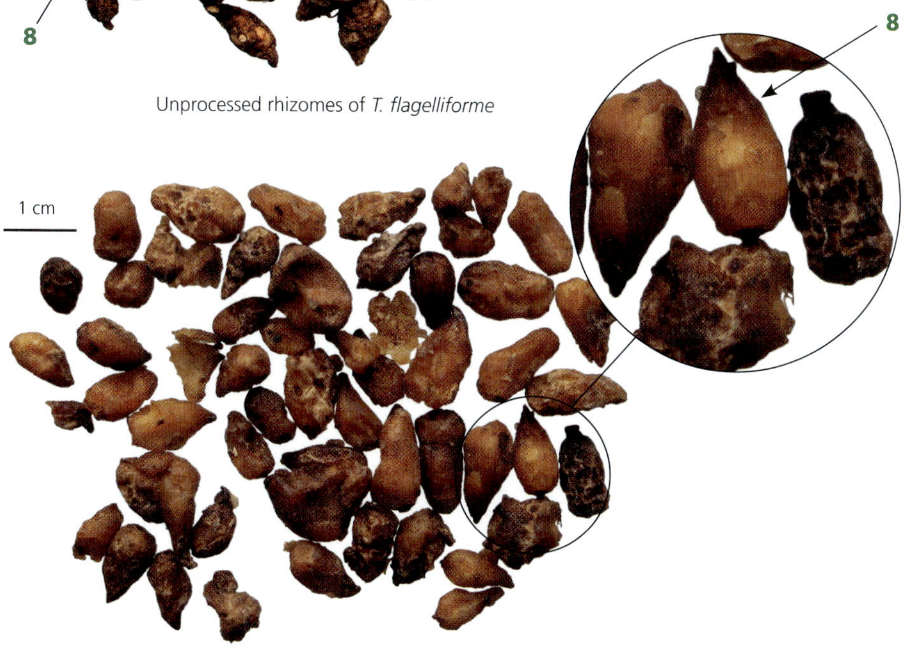

Qing Ban Xia (alum-processed rhizomes of *T. flagelliforme*)

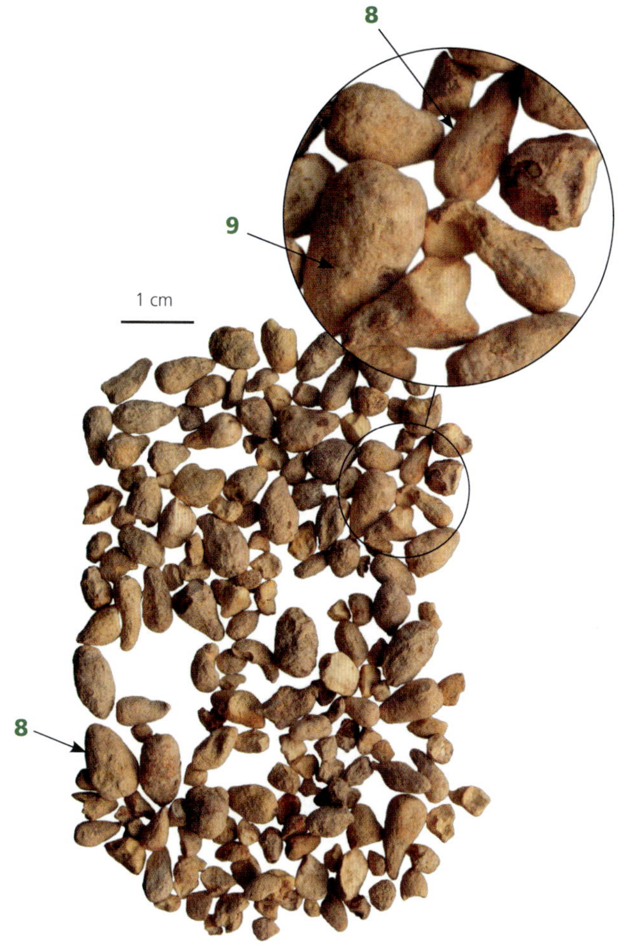

Fa Shui Ban Xia (liquorice-processed rhizomes of *Typhonium flagelliforme*)

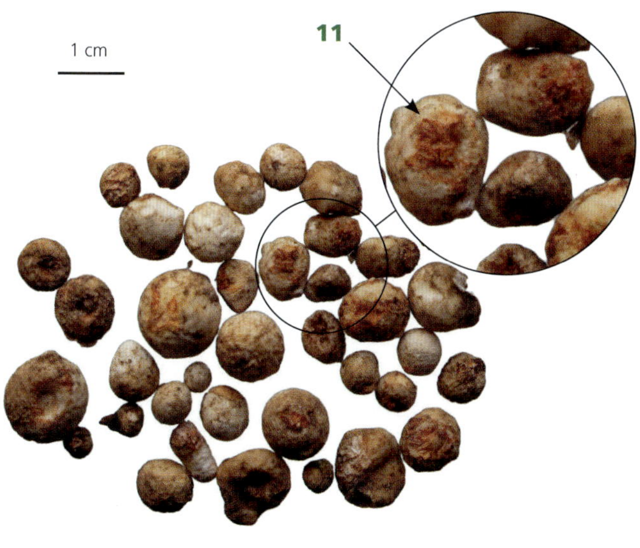

Processed immature rhizomes of *P. pedatisecta* (unofficial substitute)

Processed mature rhizomes of *P. pedatisecta*

Ba Qia 菝葜

Smilacis Chinae Rhizoma

Drug part: RHIZOME

Smilax china L.

Smilacaceae

Species of *Smilax* wild in China: 79

PLANT DESCRIPTION

A sparsely spiny, woody, perennial vine (1–5 m) usually with tendrils and a stout, fleshy, branching and densely nodular rhizome. *Stem* branched. *Leaves* alternate and simple; leaf stalk 0.5–1.5 cm long and narrowly winged along part of length; leaf blade somewhat leathery, elliptic to orbicular (3–10 cm long; 1.5–6 (rarely 10) cm wide); main veins 3–7, margins entire. *Male and female flowers* on separate plants. *Flowers* (both sexes) in a solitary 10–15-flowered cluster (umbel) on a 1–2 cm long stalk in leaf axils; each flower tiny, inconspicuous, 6-'petalled' and yellowish-green. Male flowers (3.5–4.5 mm long; 1.5–2.5 mm wide), with 6 stamens; female flowers with superior ovary, very short style and 3 stigmas. *Fruits* in spherical heads; each fruit a spherical red berry (6–15 mm diam.), minutely powdery. 1- or 2-seeded. Fl. Feb–May, fr. Sep–Nov.

HARVESTING, SOURCING AND NATURAL RANGE

Rhizomes are harvested year round, mostly from wild plants which grow in forests and shaded streamsides or open hillsides in grassland or scrub (up to 2000 m altitude). The species is distributed in S, C and W China: Anhui, Fujian, Guangdong, Guangxi, Guizhou, Henan, Hubei, Hunan, Jiangsu, Jiangxi, ?Liaoning, Shandong, Sichuan, Taiwan, Yunnan and Zhejiang [also wild in Myanmar, Philippines, Thailand and Vietnam].

CONSERVATION STATUS

China: Least Concern (authors' pers. obs.)
Global: Not Evaluated (IUCN Red List, accessed 17/8/2016).

Female inflorescence of *Smilax china*

S. china with ripe fruit

Fruiting plant of *S. china* and freshly uprooted rhizome, Anhui province

TCM PROPERTIES AND USE

***properties*: 'sweet, mildly bitter, sour and neutral'**
***channels*: 'Liver and Kidney'**
***actions*: 'releases the exterior, dispels wind damp, relieves pain'**

***indications*: leucorrhoea; dysuria; sores and abscesses**

DRUG MORPHOLOGY

CRUDE DRUG (DRIED RHIZOME): long, branching and tortuous, sometimes slightly compressed longitudinally (10–20 cm long; 2–4 cm diam.). External surface yellowish-brown or purplish-brown with many short conical protuberances ending in short fibrous root remnants **[1]**; stem scars visible as sunken and irregularly shaped depressions **[2]**. *Texture*: hard and difficult to break. *Fracture*: starchy. *Taste:* slightly bitter and astringent. *Odour:* absent.

DECOCTING PIECES (DRIED AND SLICED RHIZOME): irregularly shaped slices (0.3–1 cm thick), often with a convoluted outline **[3]**. Bark usually present, thin, yellowish-brown or purplish-brown. Cut surface pale brownish-yellow or reddish-brown **[4]**, vascular bundles starchy; in longitudinal slices vascular bundles are visible as fine lines **[5]** and in transverse section as circular dots **[6]**. *Texture*: hard. *Fracture*: starchy. *Taste*: slightly bitter and astringent. *Odour*: absent.

PROCESSING

After removal of aerial parts and fibrous roots, rhizomes are washed and either sun-dried (= crude drug) then cut into decocting slices or sliced fresh then dried.

OTHER COMMON NAMES

金刚头 Jin Gang Tou; Chinaroot greenbrier rhizome (CP2010); Chinese smilax (SCN); Rhizoma Smilacis Chinae.

Crude drug of Ba Qia (dried rhizome of *Smilax china*)

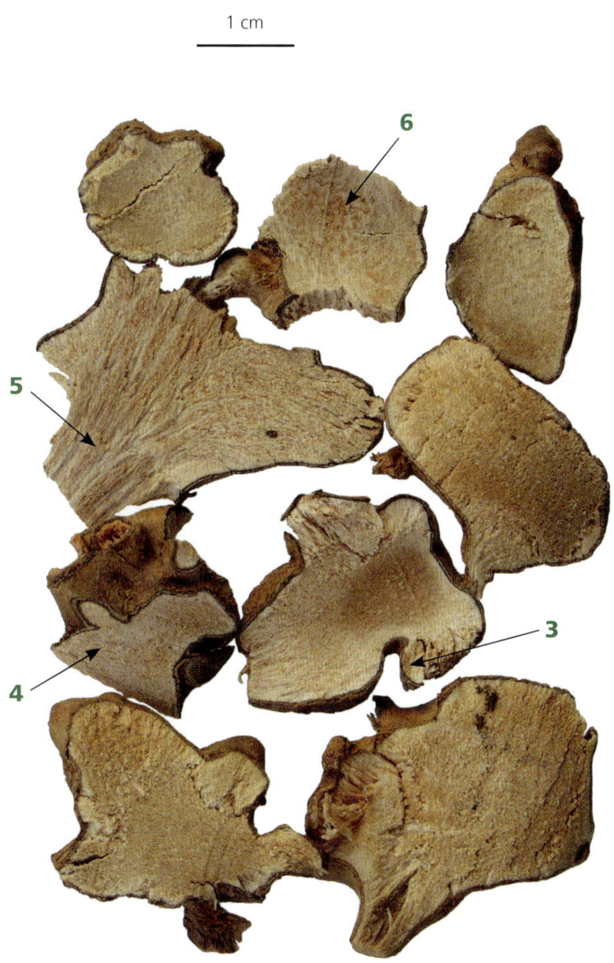

Decocting slices of Ba Qia (dried rhizome of *S. china*)

Bei Dou Gen 北豆根

Menispermi Rhizoma

Drug parts: RHIZOME

Menispermum dauricum DC.

Menispermaceae

taxonomic/nomenclatural note: the species epithet '*dauricum*' is often misspelt as '*dahuricum*'.

Species of *Menispermum* wild in China: 1

PLANT DESCRIPTION

A trailing, spineless, **herbaceous vine** with rhizomes. *Leaf stalks* slender, slightly swollen at both ends, often reddish and attached off-centre to the underside of each leaf blade, not the margin (peltate). *Leaves* alternate and simple; leaf blades almost as long as wide (3–12 cm) and angular to shallowly lobed; leaf veins radiating from a central point. *Male and female flowers* small, inconspicuous and on separate plants. Inflorescences in leaf axils, stalked and few-flowered. Male and female flowers minute (1.5–2.5 mm long) [hl]; petals 6–8 or 9–12, pale greenish-yellow, somewhat fleshy and margins curled inwards [hl]. Male flowers with 4–10, somewhat concave, sepals; **stamens free, 12–18,** usually longer than sepals. Female flowers with 6–12 sterile staminodes (c. 1 mm long). *Fruit* a spherical purplish-black and fleshy drupe (c. 1 cm diam.). Fl. Jun–Jul, fr. Aug–Sep.

HARVESTING, SOURCING AND NATURAL RANGE

Rhizomes are harvested in spring or autumn from wild plants which grow in open forest, scrub and along roadsides (up to 800 m altitude). The species is native in parts of N, C and E China (Anhui, Gansu, Guizhou, Hebei, Heilongjiang, Hubei, Hunan, Inner Mongolia, Jiangsu, Jiangxi, Jilin, Liaoning, Ningxia, Shaanxi, Shandong, Shanxi and Zhejiang) [also native in Japan, the Korean Peninsula and Russia (Siberia)].

CONSERVATION STATUS

China: Least Concern (MEPCAS, 2013).
Global: Not Evaluated (IUCN Red List, accessed 20/8/2016).

Typical trailing growth of *Menispermum dauricum* with flower-heads in bud (centre)

Fully open male flowers of *M. dauricum*

UNOFFICIAL SUBSTITUTES

The rhizomes of many other species have been reported in the TCM literature as unofficial substitutes or adulterants but their occurrence as such in international TCM trade is uncertain. Laboratory-based authentication methods are advised for reliable identification of the official Bei Dou Gen species. Substitute species reported in the literature include:

Ardisia crispa (Thunb.) A. DC. (family Primulaceae)
Beesia calthifolia Ulbr. (family Ranunculaceae)
Indigofera spp. (family Fabaceae)
Lespedeza bicolor Turcz. (family Fabaceae)
Rhynchosia volubilis Lour. (family Fabaceae)

TCM PROPERTIES AND USE

***properties*: 'bitter and cold'**
***channels*: 'Lung, Stomach and Large Intestine'**
***actions*: 'clears heat, eliminates toxins, dispels damp heat'**
***indications*: sore throat, cough; diarrhoea; musculoskeletal pain**

DRUG MORPHOLOGY

CRUDE DRUG (DRIED RHIZOME): long, slender and cylindrical **[1]** (up to 50 cm long; 3–8 mm diam.), occasionally contorted or branched **[2]**. Externally yellowish-brown to brown; finely wrinkled longitudinally **[3]** with raised and scattered branch root scars **[4]** and smaller fibrous rootlets **[5]** or their oval crater-like scars **[6]**; outer cortex easily removed. *Texture*: pliable and difficult to break. *Fracture*: uneven, with a distinctive white central pith **[7]** surrounded by fibrous and radiating yellow xylem (wood) **[8]**. *Odour*: indistinct. *Taste*: bitter. ***Good quality:*** long and thick with a bitter taste and few fibrous roots.

DECOCTING PIECES (DRIED SLICED RHIZOME): round or sub-rounded rhizome slices **[9]** (3–8 mm diam.). External uncut surfaces yellowish-brown to brown, finely wrinkled longitudinally **[3]**; raised root scars sometimes visible. Outer cortex easily removed **[10]**. Cut surfaces with a distinctive white central pith **[7]** surrounded by fibrous and radiating yellow xylem (wood) **[8]**. *Odour*: indistinct. *Taste*: bitter.

PROCESSING

After removal of the fibrous roots, rhizomes are washed and dried (= crude drug); they are subsequently cut into thick decocting slices (2–4 mm wide).

OTHER COMMON NAMES

蝙蝠葛 Bian Fu Ge; 山豆根 Shan Dou Gen; Asiatic moonseed rhizome (CP2010); Rhizoma Menispermae.

SAFETY

Slightly toxic (CP2015).

COMMENT. Bei Dou Gen is sometimes used as an unofficial substitute or adulterant of Shan Dou Gen (Sophorae Tonkinensis Radix et Rhizoma) see page 260 and Fang Ji (Stephaniae Tetrandrae Radix) see page 124.

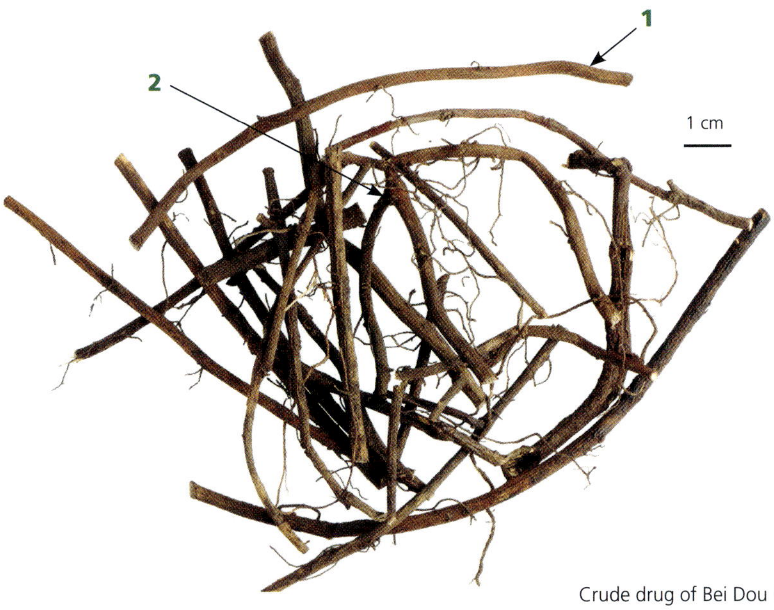

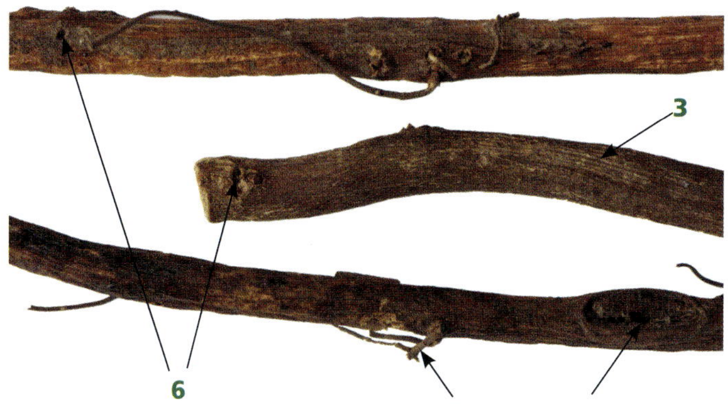

Crude drug of Bei Dou Gen (*M. dauricum*)

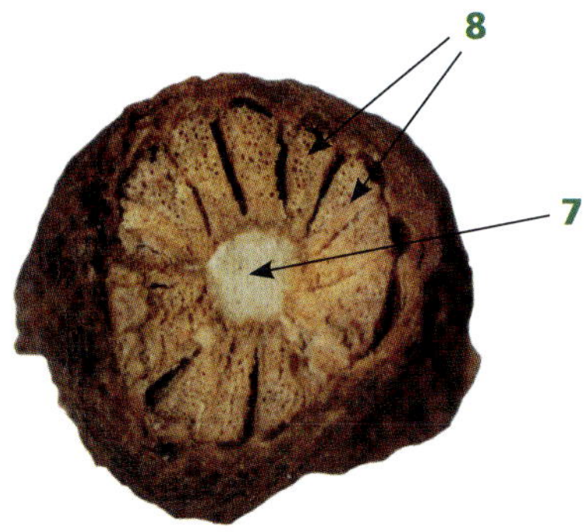

Decocting pieces of Bei Dou Gen

Bei Sha Shen 北沙参 Glehniae Radix

Drug part: ROOT

PHARMACOPOEIA SPECIES

Glehnia littoralis (A. Gray) F. Schmidt ex Miq.

Apiaceae

synonym: *Cymopterus littoralis* A. Gray

taxonomic/nomenclatural note: in CP2015 as *'G. littoralis* Fr. Schmidt ex Miq.'

Species of *Glehnia* wild in China: 1

PLANT DESCRIPTION

A perennial much branched herb (20–70 cm) **covered throughout with soft white hairs** and with a yellowish-white elongate, cylindrical or fusiform tap root (20–70 cm long; 0.5–1.5 cm diam.).Stem strongly shortened and branched. ***Basal and lower leaves* long-stalked** (5–15 cm) and sheathing at base; **leaf blades** broadly-ovate in outline, **divided once or twice into 3-parted segments**; margins saw-toothed and white-cartilaginous. ***Inflorescences*** crowded, terminal and lateral **compound umbels** (3–6 cm diam.) on thick flowering stems (2–6 cm long); rays 8–16 and unequal length (1–3 cm long); bracts absent; bracteoles several, linear to lance-shaped. ***Flowers*** tiny, petals 5, **creamish-white or purplish-red**, unequal-sized with tips bending inwards. *Fruit* dry, somewhat globose (6–13 mm long; 6–10 mm. diam.), densely hairy and splitting into 2 one-seeded carpels (mericarps); all ribs corky; 1–3 oil ducts in each furrow; 2–6 on internal (commissural) surfaces. Fl. and fr. Jun–Aug.

HARVESTING, SOURCING AND NATURAL RANGE

With a long history of medicinal use in China, spanning almost 600 years, the species is harvested today exclusively from medicinal crop plants. Harvesting takes place in summer and autumn. The main production areas are the N and NE Chinese provinces of Hebei, Inner Mongolia and Liaoning. Traditional *Di Dao*-quality drug is sourced from the Laiyang district of Shandong province. The wild species grows on sandy shores, 50–100 m altitude, along China's eastern seaboard: Fujian, Guangdong, Hebei, Jiangsu, Liaoning, Shandong, Taiwan and Zhejiang [also native in Japan, Korean Peninsula and Russia]. Wild populations have become seriously depleted due to habitat loss and over-harvesting for medicinal use (Wang *et al.* 2016).

CONSERVATION STATUS

China: Critically Endangered (MEPCAS, 2013).
Global: Not Evaluated (IUCN Red List, accessed 9/11/2016).

PROTECTED STATUS

China: Cat. II [(SEPA, 1999).

UNOFFICIAL SUBSTITUTES

Adenophora tetraphylla (Thunb.) Fisch. and *A. stricta* Miq.

Campanulaceae

PLANT DESCRIPTION

Easily distinguished from the Pharmacopoeia species. *Main differences*: sometimes much taller (*A. tetraphylla* up to 1.5 m); tuber cylindrical, conical or, more often, bulbous; **leaves simple (undivided) in whorls of 3–6 along the erect flowering stem**; ***inflorescence* not umbellate** but sparse with flowers on lateral branches at intervals up the stem; **flowers much larger, narrowly or broadly bell-shaped, blue or purplish-blue**, with 5 corolla lobes at mouth; fruit smaller (5–7 mm long; 4–5 mm diam.), a dry, ovoid-conical capsule.

SUBSTITUTION COMMENT. Ancient Chinese medicine texts describe the roots of *Adenophora* species as the original source of 'Sha Shen 沙参' (Bensky *et al.*, 2004; Brand, 2016; Tu *et al.*, 1991). Since then clinical preference has switched to the use of *Glehnia* roots but for many years both botanical sources were known as 'Sha Shen', creating much confusion. It was not until the mid-17th century that prefixes were formally introduced into the *materia medica* texts of the day in an attempt to resolve such confusion: 'Bei' (northern) for *G. littoralis* and 'Nan' (southern) for *Adenophora* roots (Tu *et al.*, 1994). Today 'Nan Sha Shen' (sourced from two *Adenophora* species described here) is recognised in the Pharmacopoeia as a separate TCM drug (see Adenophorae Radix page 222) (CP2015). Despite this formal distinction, confusion persists in the trade owing to inconsistent use of their Chinese names and the growing popularity of the less specific name 'Sha Shen'.

TCM PROPERTIES AND USE

properties: **'sweet, slightly bitter and cool'**
channels: **'Lung and Stomach'**
actions: **'tonifies yin, clears Lungs, generates fluids, tonifies Liver and Kidney yin'**

indications: **dry cough, phlegm; dry throat, thirst, constipation; hypochondriac pain, reflux, constipation**

Medicinal crop of *Glehnia littoralis* in full flower, Gansu province

Adenophora stricta

Glehnia littoralis — Apiaceae

DRUG MORPHOLOGY

CRUDE DRUG (WHOLE ROOT): slender and cylindrical **[1]** (15–45 cm long; 0.4–1.2 cm diam.), sometimes branching **[2]** and lower parts often slightly tapering **[3]**. Peeled roots pale yellowish-white **[4]** with occasional remnants of dark yellow peel **[5]**, finely wrinkled longitudinally and with **distinctly raised and scattered transverse lenticels [6]** creating a **rough surface**; raised or sunken rootlet scars **[7]**, often brownish-yellow, scattered. Apex often truncate where rootstock crown removed **[8]** or, when present, with shortly cylindrical dark brown rhizome **[9]** with paler, fibrous stem and leaf bases persistent at apex **[10]**. ***Texture*: hard and brittle**. *Fracture*: uneven, concentric rings of yellow vessels, palest at outer edge (see also cut surfaces of *Decocting Pieces* below). *Odour*: distinctive. *Taste*: sweet and somewhat viscous. ***Good quality*:** long, slender and solid.

DECOCTING PIECES (TRANSVERSE ROOT SLICES): round or ovate transverse slices **[11]** (0.4–1.2 cm diam.), peeled outer edges often rough **[12]**; cut surface with tissues in concentric rings: cortex (outermost ring) yellowish-white **[13]** and clearly demarcated by an inner slightly translucent broad cambium ring **[14]**; central xylem yellowish **[15]**. *Odour, Texture,* and *Taste* as for crude drug.

PROCESSING After washing and removal of rootlets, roots are partially air-dried, then blanched to enable removal of the outer skin before being thoroughly dried (= crude drug). Roots are then sliced transversely or, in Hong Kong, longitudinally, into decocting pieces. Alternatively fresh roots may be washed and dried without peeling.

OTHER COMMON NAMES

莱阳参 Lai Yang Shen; 沙参 Sha Shen, 珊瑚菜 Shan Hu Cai (plant name); coastal Glehnia root (CP2010); glehnia (SCN); Radix Glehniae.

SAFETY

An official healthfood in China (Anon., 2002a,b). Safety Class: 1; Interaction Class: A (Gardner & McGuffin, 2013).

Adenophora tetraphylla and *A. stricta* — Campanulaceae

'DRUG SUBSTITUTE' MORPHOLOGY

Despite the variability of rootstocks within and between species in the genus *Adenophora*, useful macroscopic characters exist to distinguish these from the official drug (*G. littoralis*) above.

DRIED ROOT (NAN SHA SHEN): size and shape highly variable but usually cylindrical or conical **[16]** (7–27 cm long; 0.8–3 cm diam.), slightly curved and often bulbous **[17]** towards the apex. Latter with 1 or 2 rhizome remnants and, depending on age of the root, extending to form a narrow rhizome 'neck' **[18]**. Externally, peeled roots yellowish-white or pale brownish-yellow with occasional dark brown peel remnants **[19]**. Lower (narrower) part of root longitudinally wrinkled. ***Texture:* lax** and easily broken. *Fracture*: uneven, **porous-looking with clefts of varying size [20]**. *Odour:* indistinct. *Taste*: somewhat sweet. **Good quality:** thick, plump, yellowish-white roots, devoid of dark outer rootbark.

TRANSVERSE ROOT SLICES (NAN SHA SHEN): irregularly ovate root slices of varying diameter **[21]** (0.8–3 cm); usually >2 mm thick; peeled outer edges often smooth with occasional dark brown peel remnants; **cut surfaces** yellowish-white and **not differentiated but lax and porous throughout with scattered clefts of varying size [20].** *Texture*: lax. *Fracture, Odour* and *Taste* as for *Dried Root* above.

OTHER COMMON NAMES

南沙参 Nan Sha Shen; 沙参 Sha Shen; adenophora (SCN); fourleaf ladybell root (CP2010); Radix Adenophorae; southern Sha Shen; Xing Ye Sha Shen.

SAFETY

Safety Class: 1; Interaction Class: A (Gardner & McGuffin, 2013).

OTHER UNOFFICIAL SUBSTITUTES

High market demand for this herbal drug has resulted in widespread fraudulent adulteration and substitution (Zhu *et al.*, 2015). Species whose rootstocks have been most frequently traded as Bei Sha Shen, because of their superficial macroscopic resemblance include:

1. Other species of *Adenophora* (notably *A. polyantha* Nakai) (family Campanulaceae). Shi Sha Shen.
2. *Changium smyrnioides* H. Wolff. (family Apiaceae). (Ming Dang Shen). Safety Class: 2b; Interaction Class: A (Gardner & McGuffin, 2013).
3. *Chuanminshen violaceum* M. L. Shen & R. H. Shan (family Apiaceae). Chuan Ming Shen.
4. *Silene tatarinowii* Regel (family Caryophyllaceae). Shi Sheng Ying Zi Cao.
5. *Sphallerocarpus gracilis* (Bess. ex Trev.) Koso-Pol. (family Apiaceae). Mi Guo Qin.

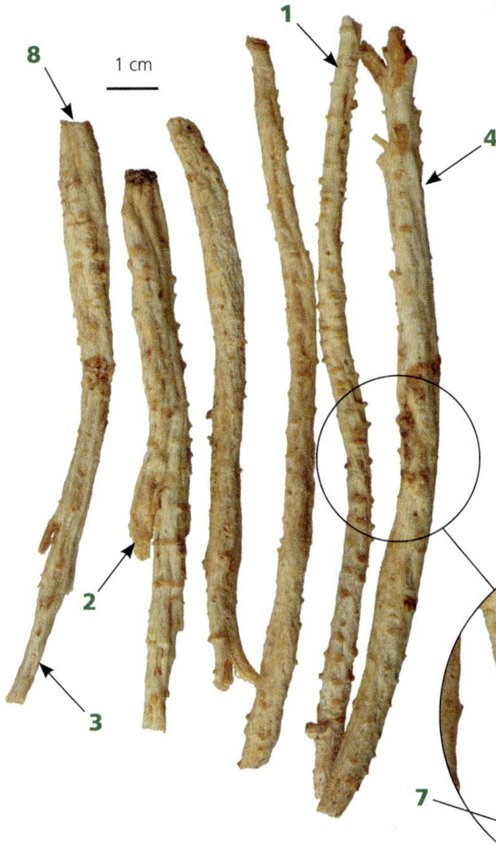
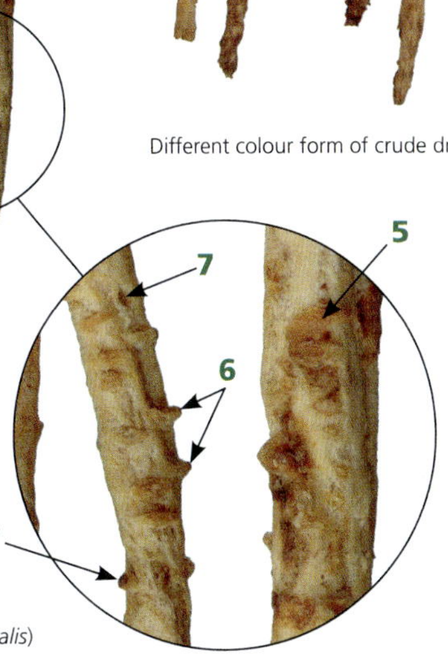

Peeled crude drug Bei Sha Shen (roots of *Glehnia littoralis*)

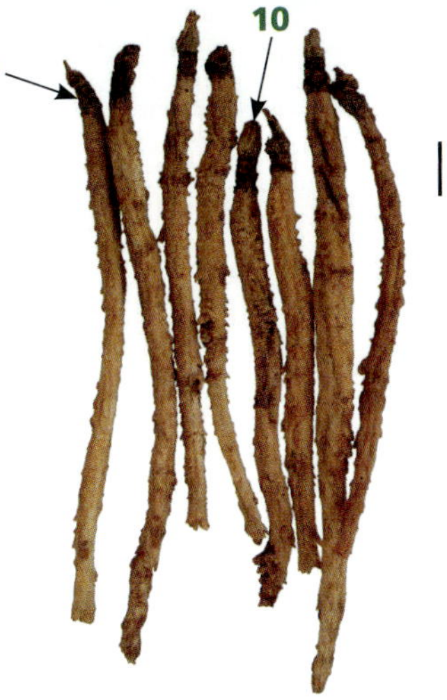

Different colour form of crude drug

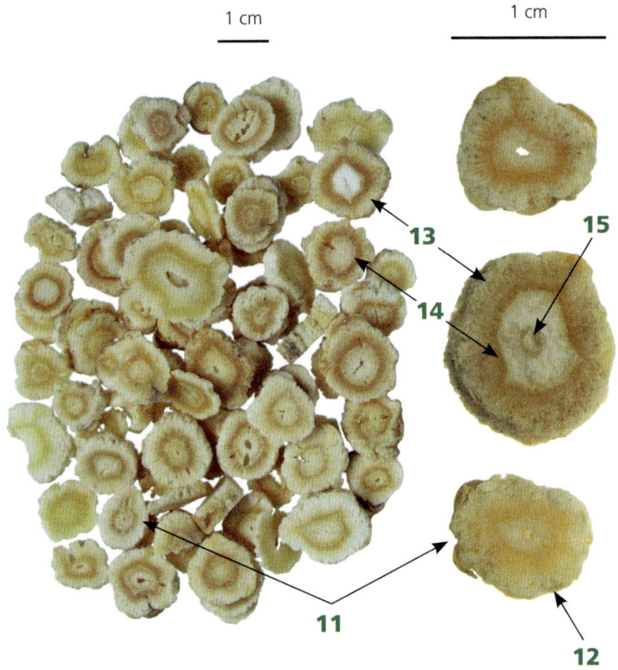

Decocting pieces of *G. littoralis* transversely sliced

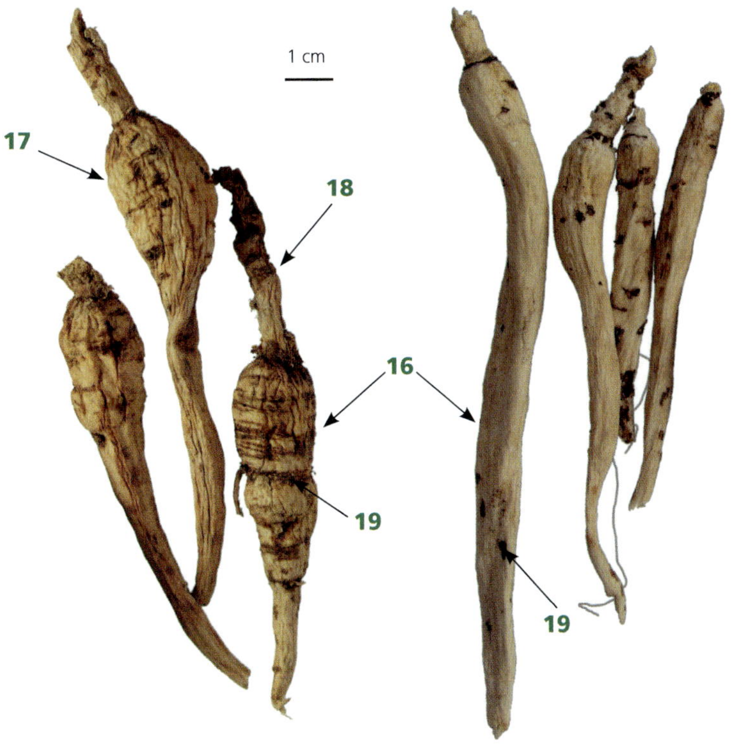

Crude drugs of Nan Sha Shen sourced from *A. tetraphylla* showing typical variation in root shape

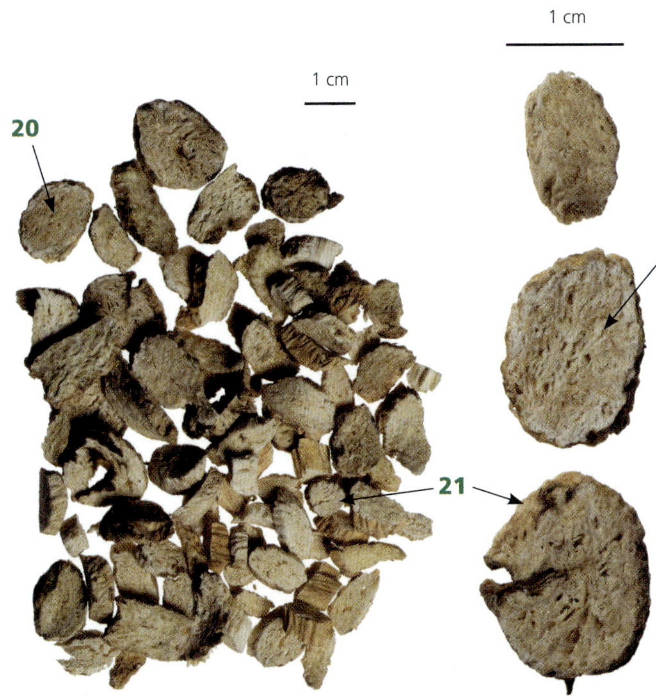

Decocting pieces of transversely sliced roots of *Adenophora stricta*

Cang Zhu 苍术

Atractylodis Rhizoma

Drug part: RHIZOME

Atractylodes lancea (Thunb.) DC.

Asteraceae

synonyms: *Atractylis lancea* Thunb.; *A. chinensis* (Bunge) DC.; *Atractylodes chinensis* (Bunge) Koidz.; *A. japonica* Koidz. ex Kitam.; *Acarna chinensis* Bunge.

taxonomic notes: 1. The CP2015 also includes '*A. chinensis* (DC.) Koidz.' as an alternative source of Cang Zhu (locally known as 'Bei (northern) Cang Zhu' owing to its production in the northern Chinese provinces of Hebei, Shanxi and Shaanxi). However its Latin scientific name is not a valid name (because the first author 'DC.' is incorrect) and its identity is therefore in doubt. The closest name match is *Atractylodes chinensis* (Bunge) Koidz. but this is a synonym of *A. lancea* (Thunb.) DC. (i.e. the Pharmacopoeia species). Further research is needed to determine the botanical identity of this entity; it is likely to be a cultivar of *A. lancea*. **2.** Some TCM texts list *Atractylodes japonica* Koidz. ex Kitam. as an unofficial substitute species used in NE China and called 'Guan Cang Zhu' (Bensky *et al.*, 2004; Zhao & Chen, 2014) but the *FoC* and *Kew-MPNS* (12/3/2015) treat this as a synonym of the official species (see synonyms above).

Species of *Atractylodes* wild in China: 4

PLANT DESCRIPTION

A perennial herb (30–100 cm (rarely 15 cm) tall with a thick rhizome. *Stems* solitary or tufted, unbranched or few-branched. *Leaves* alternate, simple, with lower surface hairless or sparsely cobwebby. Lower and middle stem leaves stalked, lower ones sometimes almost stalkless; leaf blades 8–12 cm long; 5–8 cm wide, undivided or divided almost to base into 3–5 (rarely 9) pinnately arranged segments some with spiny lobes. Upper stem leaves similar but smaller. *Flowers* densely packed into a thistle-like head (1–1.5cm diam.), each head surrounded by 2 whorls of bracts; outer bracts few and leaf-like; inner bracts many and resembling pinnate spines. Phyllaries many, overlapping, with cobwebby margins; outer phyllaries ovate, middle phyllaries becoming longer, inner phyllaries elliptic to linear. Each flower (floret) tiny, white or yellowish. *Fruits* single-seeded somewhat ovoid achenes (c. 5 mm long) with a dirty white to brown tuft of apical hairs (pappus) c. 7–9 mm long in a single row. Fl. and fr. Jun–Oct.

HARVESTING, SOURCING AND NATURAL RANGE

Rhizomes are harvested in spring and autumn mostly from cultivated plants with the traditional *Di Dao* source being the Jurong district of Jiangsu province. In the wild, the species can be found in grassland, thickets and forests, often in rock crevices, between 200–2,500 m altitude and is widespread in many parts of China: Anhui, Chongqing, Gansu, Hebei, Heilongjiang, Henan, Hubei, Hunan, Inner Mongolia, Jiangsu, Jiangxi, Jilin, Liaoning, Shaanxi, Shandong, Shanxi and Zhejiang provinces [also wild in Japan, Korean Peninsula and Russia].

CONSERVATION STATUS

China: 'rapidly decreasing' (Leung & Cheng, 2008).
Global: Not Evaluated (IUCN Red List; accessed 10//3/2015).

Atracylodes lancea in flower

TCM PROPERTIES AND USE

***properties*: 'pungent, bitter and warm'**
***channels*: 'Spleen and Stomach'**
***actions*: 'transforms damp, tonifies Spleen, dispels wind damp'**

***indications*: abdominal distension, diarrhoea; fever, chills; musculoskeletal pain**

DRUG MORPHOLOGY

CRUDE DRUG (DRIED RHIZOME): irregular-shaped rhizome masses, often nodular to cylindrical or constricted at intervals **[1]** to create a bead-like (moniliform) appearance (3–10 cm long, 1–4 cm diam.); sometimes branched. Externally dull greyish-brown or blackish-brown, rough, with twisted appearance, often with root remains **[2]** and, at apex, stem base remains **[3]** or scars **[4]**. *Texture*: compact or lax. *Fracture*: even and easily broken; yellowish-white to greyish-white and scattered with distinctive orangish-yellow, yellow or brownish-red oil glands **[5]** sometimes becoming crystallised. *Odour*: distinctly or weakly aromatic. *Taste*: bitter and pungent, sometimes somewhat sweet. ***Good quality***: heavy with many oil glands (ideally with white needle-like crystals) and intensely aromatic.

DECOCTING PIECES-1 (SLICED RHIZOME): sub-rounded or slat-shaped slices (c. 1–2.5 cm diam.; 2–4 mm thick). Rhizome skin dull greyish-brown or blackish-brown and rough **[6]**; root remains **[2]** or scars occasional. Cut surface yellowish-white or greyish-white dotted with many orangish-yellow, yellow or brownish-red oil glands **[5]** the latter sometimes crystallised with age forming thin, white needle-shaped crystals (the terms 'Qi Shuang' (frosting) or 'Bai Mao' (white hairs) are often coined to denote these crystals, a traditional measure of high quality). *Texture*: compact or lax. *Odour*: aromatic. *Taste*: pungent and bitter and sometimes sweet.

DECOCTING PIECES-2 (BRAN-PROCESSED RHIZOME SLICES — FU CHAO CANG ZHU): as for dried decocting slices but externally all surfaces dark yellow and with scattered oil cavities. *Odour*: aromatic and slightly burned.

PROCESSING

Rhizomes are dried then fibrous roots removed (= crude drug). Decocting pieces are produced by softening the rhizomes to facilitate slicing (= decocting pieces-1); further processing by frying with bran (then removing the bran by sifting) produces Fu Chao Cang Zhu (= decocting pieces-2).

OTHER COMMON NAMES

北苍术 Bei Cang Zhu (see opposite page) ; 北茅术 Bei Mao Zhu; 茅苍术 Mao Cang Zhu; 茅术 Mao Zhu; 南苍术 Nan Cang Zhu; cang-zhu atractylodes (SCN); Rhizoma Atractylodis.

SAFETY

Safety Class: 1; Interaction Class: A (Gardner & McGuffin, 2013). An official healthfood in China (Anon., 2002a,b).

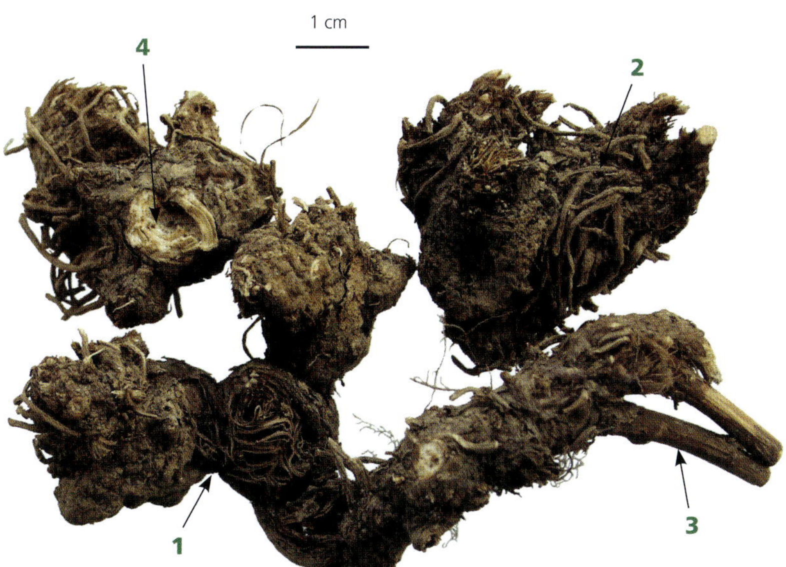

Pre-crude drug of Cang Zhu
(dried rhizome of *A. lancea*) before fibrous roots removed

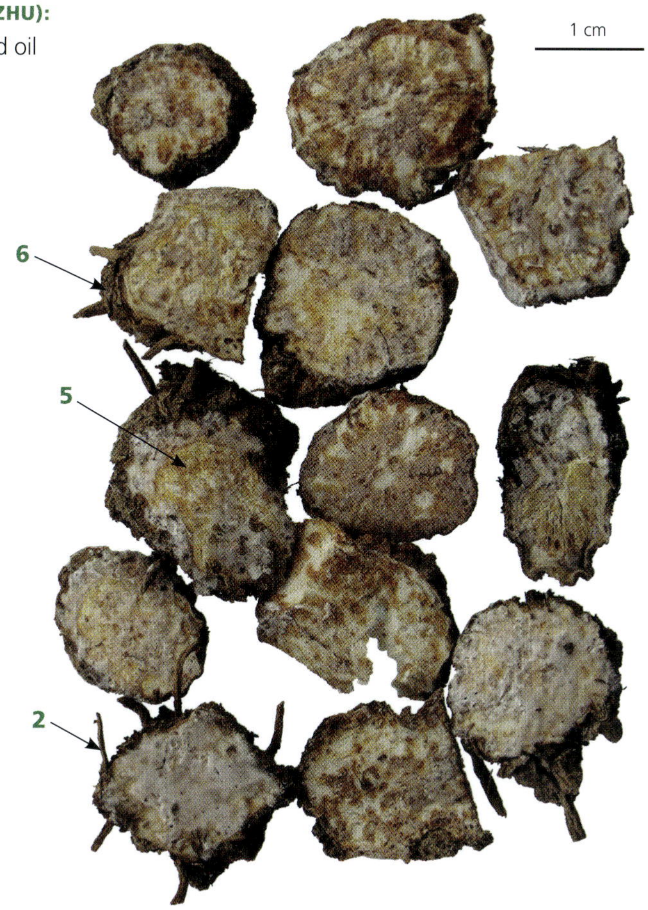

Decocting slices of Cang Zhu
(dried and sliced rhizome of *A. lancea*)

Chai Hu 柴胡 Bupleuri Radix

Drug part: ROOT

PHARMACOPOEIA SPECIES 1

Bupleurum chinense DC.

Apiaceae

synonym: *B. chinense* Franch., nom. illeg.

Species of *Bupleurum* wild in China: 42

PLANT DESCRIPTION

An erect, perennial herb **(50–85 cm tall)**, hairless throughout, with a long, woody, **brown taproot** (= medicinal part 'Chai Hu'). *Stems* 1 or several, much-branched, **bases without fibrous remnant sheaths**. *Leaves* alternate and simple, changing shape up the stem; bases tapering into leaf stalks; margins entire; basal leaves lance-shaped and wider towards their tips (4–7 cm long; **0.6–0.8 cm wide**), tips narrowing into a point; middle leaves broadly linear to lance-shaped, 7–9-nerved, lower surface of blade blue-green, tip abruptly ending in a point; upper leaves tiny. *Inflorescences* of large, loose and widely spreading terminal or lateral compound umbels (2–6 cm diam.); rays 3–8, very slender, unequal (1–3 cm); rays subtended by 0 or 2–3 tiny leaf-like bracts (1–5 mm long; 0.5–1 mm wide); umbellets 5–10-flowered (4–6 mm diam.) and subtended by 5 leaf-like and lance-shaped bracteoles of equal-length or longer than each umbellet. Each flower tiny; petals 5, bright yellow; ovary inferior, styles 2. *Fruits* in umbellate clusters; each fruit dry, **brown,** oblong (c. 3 mm long; 2 mm diam.) and splitting into 2 one-seeded carpels (mericarps) each with prominent ribs, **narrow pale brown wings** and 3 (rarely 4) oil ducts in each furrow and 4 on the flat internal (commissural) surface. Fl. and fr. Sep–Oct.

HARVESTING, SOURCING AND NATURAL RANGE

In recent years *B. chinense* has been extensively cultivated in China as a medicinal crop in Hebei, Henan and Shaanxi provinces and provides the bulk of 'Chai Hu' in trade. Roots are harvested in spring or autumn.

Endemic to China, the wild species grows in grassland and disturbed places at 100–2,700 m altitude in the provinces of Anhui, Gansu, Hebei, Heilongjiang, Henan, Hubei, Hunan, Inner Mongolia, Jiangsu, Jiangxi, Jilin, Liaoning, Shaanxi, Shandong, Shanxi and Zhejiang.

CONSERVATION STATUS

Endemic to China: Least Concern (MEPCAS, 2013); Not Evaluated (IUCN Red List accessed 25/6/2016).

PHARMACOPOEIA SPECIES 2

B. scorzonerifolium Willd.

Apiaceae

synonym: *B. falcatum* L. var. *scorzonerifolium* (Willd.) Ledeb.

PLANT DESCRIPTION

Similar to PS1. *Main differences*: taproot dark reddish-brown; **stem bases with fibrous remnant sheaths**; basal leaves linear (6–16 cm long; 0.2–0.7 cm wide); fruits dark brown, **not winged**, with 5–6 oil ducts in each furrow and 4–6 on the flat internal (commissural) surface of each mericarp. Fl. and fr. Jul–Sep.

HARVESTING, SOURCING AND NATURAL RANGE

Unlike *B. chinense* (PS1), *B. scorzonerifolium* is mostly wild harvested. It grows in dry grassland, forest and shrub margins (100–2,300 m altitude) in C, N and NE China (Anhui, Gansu, Guangxi, Hebei, Heilongjiang, Inner Mongolia, Jiangsu, Jilin, Liaoning, Shaanxi, Shandong and Shanxi) and roots are harvested in spring or autumn. The species is also native to Japan, Korean Peninsula, Mongolia and Russia.

CONSERVATION STATUS

China: Least Concern (MEPCAS, 2013, as the type variety '*scorzonerifolium*').

Global: Not Evaluated (IUCN Red List, accessed 25/8/2016).

UNOFFICIAL SUBSTITUTE

Bupleurum longiradiatum Turcz.

Apiaceae

PLANT DESCRIPTION

FoC recognises 2 varieties: the common, widespread and tall *type* variety (var. *longiradiatum*) and the shorter and less widespread var. *breviradiatum* F. Schmidt ex Maxim.

Main differences from PS1: usually **much taller** (80–150 cm = var. *longiradiatum*; rarely 50–80 cm = var. *breviradiatum*); basal leaves larger (8–17 cm long; 2.5–5 (rarely up to 8 cm) wide); **leaf bases usually clasping the stem**; fruits oblong, **not winged**, slightly larger (4–7 mm long; 2–2.5 mm diam.) and **bluish-green**.

Main differences from PS2: **usually much taller** (see above); **base without fibrous remnant sheaths**; basal leaves much broader (8–17 cm long; 2.5–5 (rarely up to 8 cm) wide); **base of upper stem leaves encircling the stem**; umbels often broader (3–10 cm wide); petals yellow or purple; fruits bluish-green; **3–4 oil ducts** in each furrow, 4–6 on the flat internal (commissural) surface.

SUBSTITUTION COMMENT. This species is closely related to the 2 Pharmacopoeia species and superficially similar as living plants. Previously used as a local 'Chai Hu' substitute in parts of China but today its use is prohibited owing to the toxicity of its roots and cause of many cases of acute adverse reaction (see *Safety*) (Bensky *et al.*, 2004; CP, 2015). Today, 'Chai Hu' is sourced predominantly from crop-grown *B. chinense* (PS1) which reduces the risk of accidental contamination from *B. longiradiatum*.

CONSERVATION STATUS

China: Least Concern (MEPCAS, 2013, as the type variety '*longiradiatum*' and var. *breviradiatum*).

Global: Not Evaluated (IUCN Red List, accessed 25/6/2016).

TCM PROPERTIES AND USE

properties: 'bitter, pungent and cool'
channels: 'Liver, Gall Bladder and Lung'
actions: 'releases wind heat, regulates qi'

indications: alternating chills and fever, chest and menstrual pain; prolapse

Flowering medicinal crop plants of *Bupleurum chinense*, Shaanxi province

Ripening fruits of *B. chinense*

B. chinense: fresh root

B. scorzonerifolium in flower

Umbellets of *B. scorzonerifolium*

B. scorzonerifolium: fresh root

Inflorescence of *Bupleurum longiradiatum* var. *longiradiatum* growing wild in China

Freshly uprooted wild plant of *B. longiradiatum* var. *longiradiatum* (note that this variety is twice the height of the 2 official species)

Base of upper stem leaves of *B. longiradiatum* encircling the stem

Bupleurum chinense

Apiaceae

DRUG MORPHOLOGY

CRUDE DRUG (DRIED ROOT): long, cylindrical and tapering (6–15 cm long; 0.3–0.8 cm diam.) **[1]**, occasionally branching. Apical part broadest **[2]** with remains of 3–15 stem bases **[3]**; **fibrous leaf bases and annulations largely absent**. External surface **blackish-brown or pale brown [4]** with fine longitudinal wrinkles; rootlets occasional or, more commonly, raised paler rootlet scars **[5]** and oval lenticels [hl]. *Texture*: **hard and difficult to break**. *Fracture*: **fibrous**, pale yellowish-cream and uneven. *Odour*: **slightly aromatic**. *Taste*: slightly bitter. ***Good quality*:** long, dry main root with some lateral ones.

DECOCTING PIECES-1 (SLICED ROOT): sub-ovate slices (0.3–1.3 cm diam.) **[6]**. **Bark blackish-brown or pale brown [7]** with fine longitudinal wrinkles; raised pale rootlet scars **[5]** and ovate lenticels [hl] very occasional. Cut surface with very narrow grey cambium layer **[8]** and very broad pale yellowish-cream wood with many faint medullary rays **[9]** sometimes fissured centrally **[10]**. *Texture*: **fibrous**. *Odour*: **slightly aromatic**. *Taste*: slightly bitter.

DECOCTING PIECES-2 (VINEGAR-PROCESSED): as for decocting slices above but all surfaces pale brownish-yellow. *Odour*: slightly vinegar-tainted.

PROCESSING

After removal of aerial parts, roots are dried (= crude drug) then later softened thoroughly with water, sliced thickly transversely or longitudinally and dried again (= decocting pieces-1). Further processing is sometimes performed by dry-frying with yellow rice-wine vinegar to produce 醋柴胡 Cu Chai Hu 'vinegared Chai Hu'.

OTHER COMMON NAMES

Popularly traded as 北柴胡 Bei Chai Hu 'northern Chai Hu'. Other names: bupleurum (SCN); Chinese thorowax root (CP2010); northern Chinese thorowax; Radix Bupleuri.

SAFETY

Safety Class: 1; Interaction Class: A (Gardner & McGuffin, 2013).

Crude drug of Chai Hu sourced from *B. chinense*

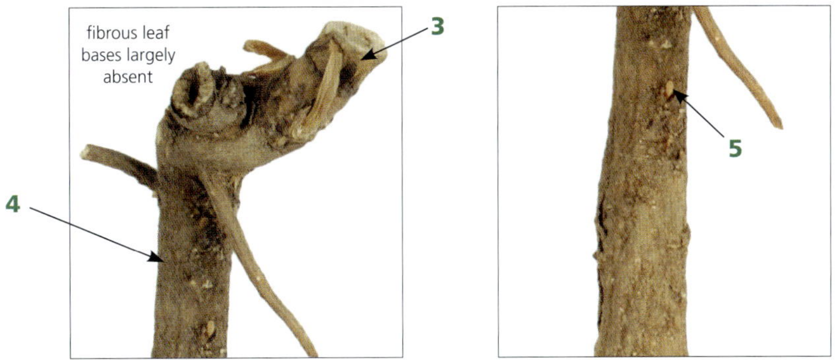

Chai Hu crude drug (*B. chinense*): root apex (left); middle part of root (right)

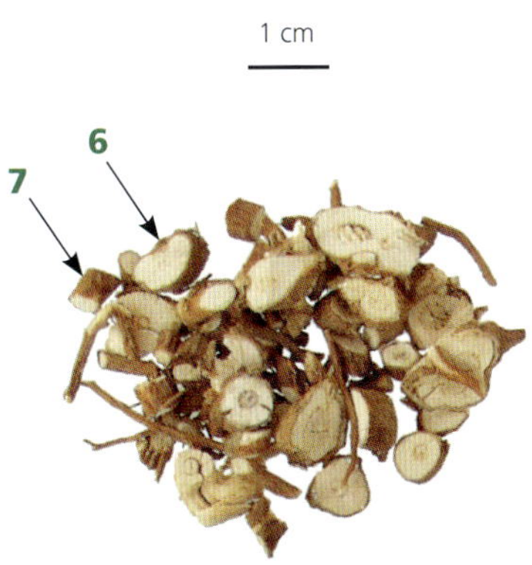

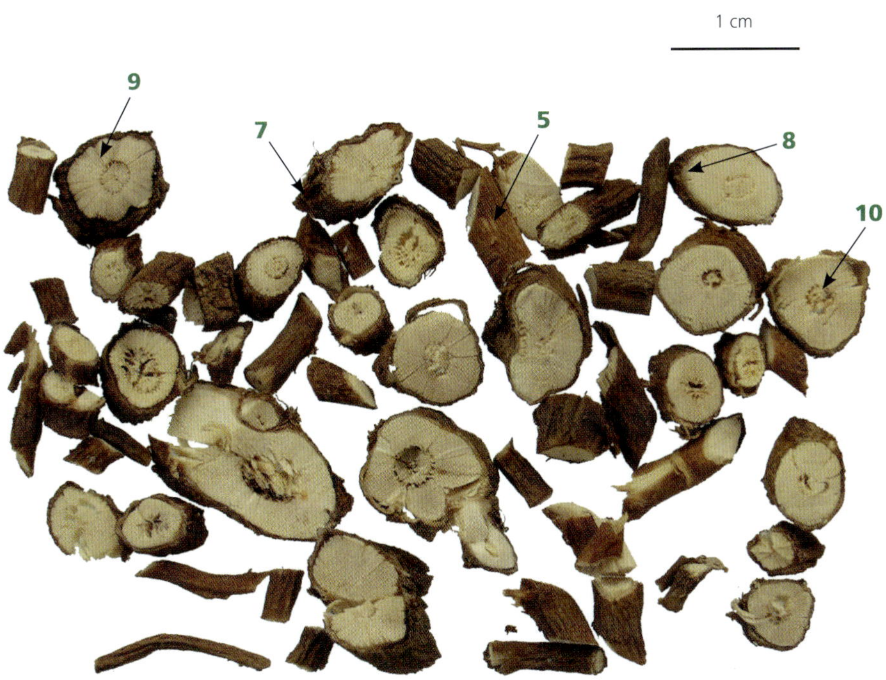

Decocting pieces-1 (transversely sliced) of Chai Hu sourced from *B. chinense*; pale brown form (left); blackish-brown form (right)

B. scorzonerifolium

Apiaceae

DRUG MORPHOLOGY

Crude drug and decocting pieces easily distinguished from those of PS1. Decocting pieces easily confused with those of *B. longiradiatum* and laboratory-based authentication methods are advised but see *Substitution Comment* for this unofficial substitute.

CRUDE DRUG (DRIED ROOT): *main differences from PS1*: more conical and rarely branched **[11]**. Apical part **densely covered in many bristle-like leaf-bases [12]** with an indistinct zone of dense annulations below **[13]**. External surface **reddish-brown or blackish-brown [14].** *Texture*: **slightly soft and easily broken**. *Fracture:* yellowish-white, **not fibrous**, slightly uneven. *Odour*: **rancid or creosote-like**. For a comparison with *B. longiradiatum* (toxic unofficial substitute) roots see *Drug Morphology* for that species.

DECOCTING PIECES-1 (DRIED, SLICED ROOT): sub-ovate or irregularly-shaped slices (0.3–0.8 cm diam.) **[15]**. Bark reddish-brown or blackish-brown **[14]** sometimes **bristle-like leaf bases and/or dense annulations** present **[12]**. Cut surface yellowish-white **[16]**. Fracture: **not fibrous**. *Taste and odour*: rancid or creosote-like.

DECOCTING PIECES-2 (VINEGAR-PROCESSED ROOT): as for decocting pieces-1 above but all surfaces pale brownish-yellow. *Odour*: slightly vinegar-tainted.

PROCESSING

As for PS1.

OTHER COMMON NAMES

Popularly traded as 南柴胡 Nan Chai Hu 'southern Chai Hu'. Other names: 狭叶柴胡 Xia Ye Chai Hu 'narrow-leaved Chai Hu'; bupleurum (SCN); Chinese thorowax root (CP2010); Radix Bupleuri.

SAFETY

Safety Class: 1; Interaction Class: A (Gardner & McGuffin, 2013).

Bupleurum longiradiatum

Apiaceae

'DRUG SUBSTITUTE' MORPHOLOGY

DRIED ROOT: main differences from the roots of *B. chinense* (PS1) and *B. scorzonerifolium* (PS2): a more cylindrical rootstock **[17]**; apical part of root-head not obviously enlarged and **without fibrous leaf bases [18]** but with a **long and distinct zone (1–4 cm) of widely spaced annulations** immediately below the root-head **[19]**. External surface brown to yellowish-brown. *Texture*: **pliable but difficult to break**. *Odour*: **somewhat pungent and creosote-like**. *Taste*: slightly bitter.

SLICED ROOT: details unavailable.

Given the toxicity of this species (see *Safety*), laboratory-based methods are recommended for reliable identification.

OTHER COMMON NAMES

大叶柴胡 Da Ye Chai Hu 'big-leaved Chai Hu'.

SAFETY

Toxic and prohibited as an official source of 'Chai Hu' (CP2015). Acute adverse reactions, including a fatality in 1973, have been reported following the use of the roots of this species as a local alternative to the Pharmacopoeia 'Chai Hu' species. Cytotoxic polyacetylenes, notably bupleurotoxin and acetylbupleurotoxin hsave been reported as the main toxins (Huang *et al.*, 2009a; Wright, 2004).

LESSER KNOWN UNOFFICIAL SUBSTITUTES

The roots of many other species of *Bupleurum* have been traded as 'Chai Hu' (and considerable expertise is need for their reliable identification, preferably using laboratory-based methods. Today, their occurrence as 'Chai Hu' substitutes is unlikely especially on the international market (see *Substitution Comment* page 68):

1. ***Bupleurum angustissimum*** (Franch.) Kitag. (线叶柴胡 Xian Ye Chai Hu).
2. ***B. bicaule*** Helm (锥叶柴胡 Zhui Ye Chai Hu).
3. ***B. chaishoui*** R. H. Shan & M. L. Sheh (柴首 Chai Shou, meaning 'large-headed Chai Hu' and refers to the large-headed rootstock). Endemic to China (north-west Sichuan).
4. ***B. falcatum* L. The official source of 'Bupleuri Radix' in Kampo medicine in Japan** (JP, 2012).
5. ***B. longicaule*** DC. var. ***giraldii*** (长茎柴胡 Chang Jing Chai Hu). N and W China (9 provinces).
6. ***B. marginatum*** Wall. (竹叶柴胡 Zhu Ye Chai Hu). Gansu, Guizhou, Hubei, Qinghai, Sichuan, Tibet and Yunnan provinces.
7. ***B. sibiricum*** Vest ex Sprengel var. ***jeholense*** (Nakai) Y. C. Chu ex R. H. Shan & Y. Li (雾灵柴胡 Wu Ling Chai Hu). Endemic to China (Hebei province) but wild populations are 'Endangered' (MEPCAS, 2013).
8. ***B. smithii*** H. Wolff var. ***parvifolium*** R. H. Shan & Yin Li (小叶黑柴胡 Xiao Ye Hei Chai Hu). A common but **toxic** endemic of N China (Gansu, Inner Mongolia, Ningxia and Qinghai (Wright, 2004); root dark brown and devoid of thick, fibrous leaf sheaths; bases of middle and upper leaves dilated, heart-shaped and clasping; bracteoles usually 6–9.
9. ***B. yinchowense*** R. H. Shan & Yin Li (银州柴胡 Yin Zhou Chai Hu).

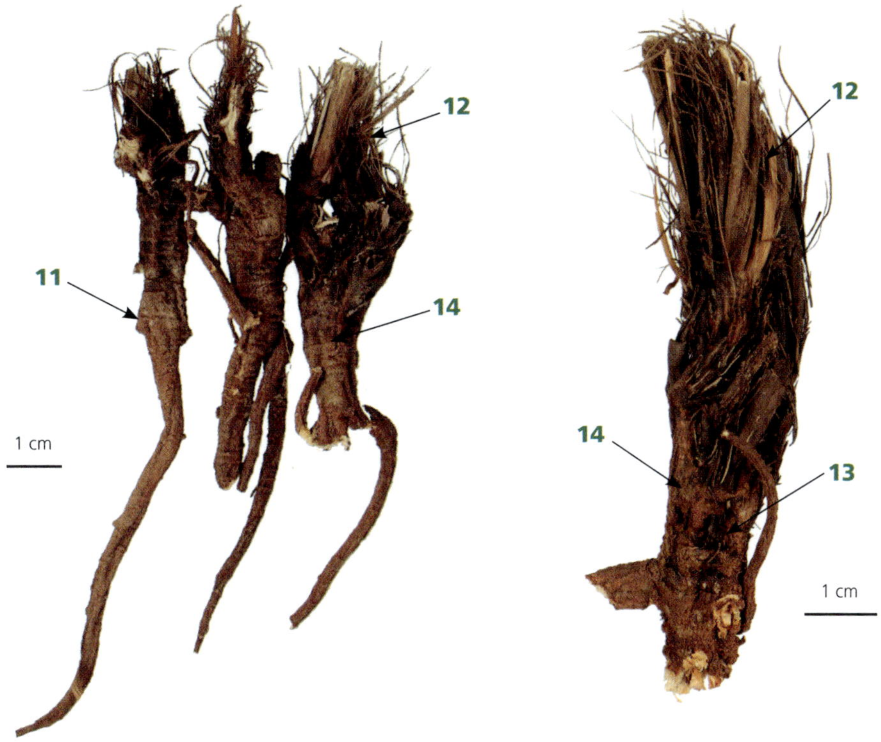

Crude drug of Chai Hu sourced from *B. scorzonerifolium*

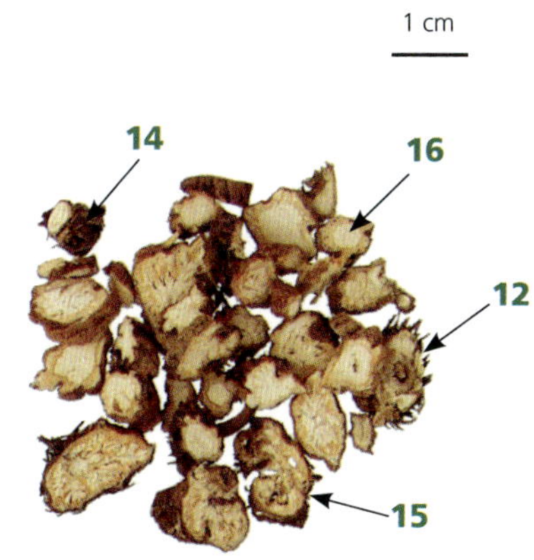

Decocting pieces-1 (sliced) of Chai Hu sourced from B. *scorzonerifolium*

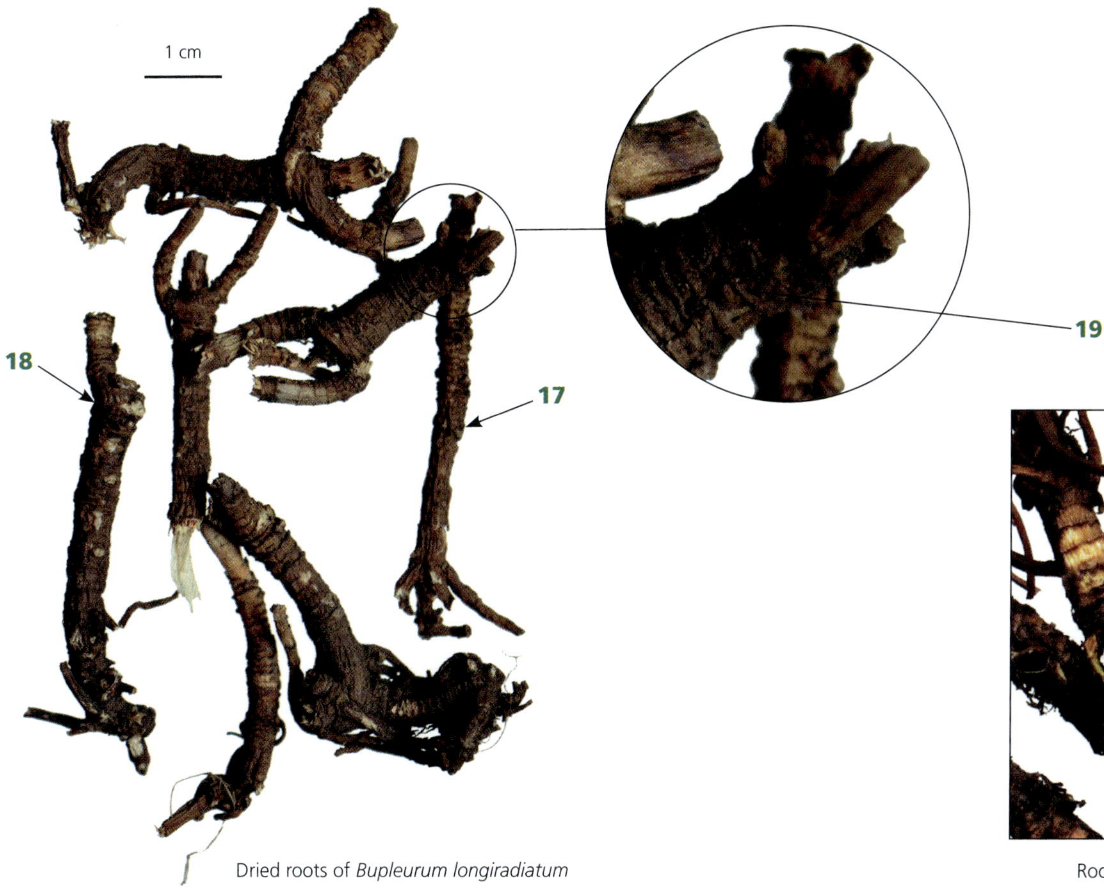

Dried roots of *Bupleurum longiradiatum*

Root-head of *B. longiradiatum*

Chi Shao 赤芍

Paeoniae Radix Rubra

Drug part: ROOT

PHARMACOPOEIA SPECIES 1

Paeonia lactiflora Pall.

Paeoniaceae

PLANT DESCRIPTION

A many-stemmed perennial herb (up to 70 cm tall) with a thick, sometimes branching rootstock. *Leaves* alternate, divided into not more than 15 long-ovate leaflets; blade of leaflets extending into leaflet stems almost as wings; margins thin, white and cartilaginous. *Bracts* below each flower, 4 or 5, green, lance-shaped and of different lengths. *Flowers* large and conspicuous (c. 8–13 cm diam.), usually several per stem; sepals 3 or 4; petals 9–13, variously coloured (mostly white or pink); stamens many and yellow. *Fruit* a group of 2–5 oblong-ellipsoid pods (follicles) (2.5–3 cm long; 1.2–1.5 cm wide), green or purple, hairless or sometimes covered in soft brown hairs. *Seeds* many, ovoid and black. Fl. May–Jun, fr. Aug. (*description based on FoC and Hong De-Yuan, 2010*).

De-barked roots from medicinal crop plants of this species are the source of the separate TCM drug 'Bai Shao' (Paeoniae Radix Alba) page 32.

Species of *Paeonia* wild in China: 15 (*FoC*; Hong, 2010)

HARVESTING, SOURCING AND NATURAL RANGE

Usually wild-harvested in spring or autumn from plants growing in temperate forest, forest margins and grassland (400–2,300 m altitude) with *Di Dao* material sourced from sites in Inner Mongolia. In recent years, due to high demand, cultivated plants have also begun to be used despite wild-sourced material still considered to be superior (note that the species is already well established in medicinal cultivation as the source of 'Bai Shao' (Paeoniae Radix Alba) page 32).

The species grows wild across the northern Chinese provinces of Gansu, Hebei, Heilongjiang, Inner Mongolia, Jilin, Liaoning, Ningxia, Shaanxi and Shanxi) and is also native in Japan, the Korean Peninsula, Mongolia and Russia (Far East, Siberia).

CONSERVATION STATUS

China: Least Concern (MEPCAS, 2013); Near Threatened (Wang & Xie, 2004).

Global: Not Evaluated (IUCN Red List, accessed 18/8/2016).

PHARMACOPOEIA SPECIES 2

P. anomala L. subsp. *veitchii* (Lynch) D. Y. Hong & K. Y. Pan

Paeoniaceae

synonym: *P. veitchii* Lynch (name used in CP2015).

PLANT DESCRIPTION

Similar to *P. lactiflora* but slightly taller (up to 1 m), **leaves divided into pinnae** (i.e. leaflets with a central axis) and not lobed; **blade of leaflet bases occasionally** extending down into the leaflet stalks. Bracts 2–5. **Flowers solitary or 2–4 per stem**, sometimes also with 1–3 undeveloped (sterile) flowerbuds in leaf axils; **petals 6–9**, oblong, rose to red, rarely almost white.

HARVESTING, SOURCING AND NATURAL RANGE

Roots are wild-harvested in spring or autumn from plants growing in moist places along forest margins, scrub and alpine meadows (1,800–3,900 m altitude). The species is endemic to N and NW China and grows in the provinces Gansu, Ningxia, Qinghai, Shaanxi, Shanxi, Sichuan, Tibet and Yunnan. It is also cultivated worldwide as an ornamental.

CONSERVATION STATUS

Endemic to China: Least Concern (MEPCAS, 2013); Vulnerable A2c (Wang & Xie, 2004); Not Evaluated (IUCN Red List, accessed 18/8/2016).

TCM PROPERTIES AND USE

properties: 'bitter and cool'
channel: 'Liver'
actions: 'clears heat, cools the Blood, alleviates pain'

indications: haematemesis; epistaxis; haematuria; amenorrhoea, dysmenorrhoea; sores and abscesses

Medicinal crop of flowering *Paeonia lactiflora*, Anhui province

P. lactiflora in flower

P. lactiflora with ripe fruit

Wild plant of *P. anomala* subsp. *veitchii*

Wild plant of *P. anomala* subsp. *veitchii*, Sichuan province

Fresh root stock of *P. anomala* subsp. *veitchii*

Paeonia lactiflora Paeoniaceae

DRUG MORPHOLOGY

CRUDE DRUG (DRIED ROOT): slightly curved and cylindrical, not tapering (5–40 cm long; 0.5–3 cm diam.). Externally root-bark dull brown, longitudinally wrinkled and grooved **[1]**, with raised transverse lenticels **[2]** and rootlet scars **[3]**; often exfoliating **[4]**. *Texture*: hard but brittle. *Fracture*: bark thin; cortex dense, chalky-white or pinkish-white **[5]**, wood (occupying most of the fractured surface) yellowish-white with many distinct rays **[6]**, often with clefts **[7]**. *Taste*: somewhat bitter, sour and astringent. ***Good quality***: wild-sourced roots; long, thick and powdery; internally pinkish or yellowish-white.

DECOCTING PIECES (TRANSVERSELY SLICED ROOT): sub-rounded slices (2–4 mm thick); externally with thin outer bark often flaking **[4]**; cut surface with cortex dense, chalky-white or pinkish-white **[5]**; wood (widest zone) yellowish-white with many distinct rays **[6]**, often with clefts **[7]**. *Taste* and *texture* as for crude drug. Material sold in Hong Kong is often pressed flat and sliced longitudinally.

PROCESSING

After removal of the rhizomes and rootlets the roots are sun-dried (= crude drug) and subsequently cut into decocting slices (2–4 mm thick).

OTHER COMMON NAMES

赤芍药 Chi Shao Yao; 芍药 Shao Yao; 西赤芍 Xi Chi Shao; Chinese peony (SCN); Radix Paeoniae Rubra; red peony root (CP2010); western red peony root.

SAFETY

An official healthfood in China (Anon., 2002a,b). Safety Class: 1; Interaction Class: A (Gardner & McGuffin, 2013).

P. anomala subsp. *veitchii* Paeoniaceae

DRUG MORPHOLOGY

CRUDE DRUG (DRIED ROOT-BARK): similar to that of PS1 but roots tapering **[8]**.

DECOCTING PIECES (TRANSVERSELY SLICED ROOT: cannot be reliably distinguished from that of PS1 using macroscopic characters; laboratory-based authentication methods are needed. Any colour differences visible in the photographs of the drugs of PS1 and PS2 opposite are due to natural variation since both drugs are wild-sourced. This variation is not helpful in differentiating the drugs of these two species.

PROCESSING

As for PS1.

OTHER COMMON NAMES

赤芍药 Chi Shao Yao; 川赤芍 Chuan Chi Shao; 山芍药 Shan Shao Yao; Chinese peony (SCN) (as '*P. veitchii* Lynch'); Radix Paeoniae Rubra; red peony root (CP2010).

SAFETY

An official healthfood in China (Anon., 2002a,b, as '*P. veitchii* Lynch').

UNOFFICIAL SUBSTITUTES/ADULTERANTS

The roots of the below have been reported in the TCM literature as accidental or intentional substitutes or adulterants; their occurrence as such in international trade is uncertain.

1. ***Paeonia anomala*** L. subsp. ***anomala*** (新疆芍药 Xinjiang Shao Yao). Flowers solitary and terminal; rarely with 1 or 2 under-developed flower-buds in axils of other leaves. In China, wild plants restricted to N Xinjiang province.
2. ***P. mairei*** H. Lév. (美丽芍药 Mei Li Shao Yao). Up to 1 m tall; leaflets and segments up to 19; flowers solitary; bracts 1–3; petals 7–9, pink to red.
3. ***P. obovata*** Maxim. (草芍药 Cao Shao Yao). Leaflets and leaf segments up to 9; flowers solitary; bracts 1 or 2, petals 4–7, white, rose, pink-red, red, purple-red, or rarely white with a pinkish base or margin.
4. ***Sanguisorba officinalis*** L. Family Rosaceae. Roots are the source of the separate TCM drug 'Di Yu' (Sanguisorbae Radix) see page 114.

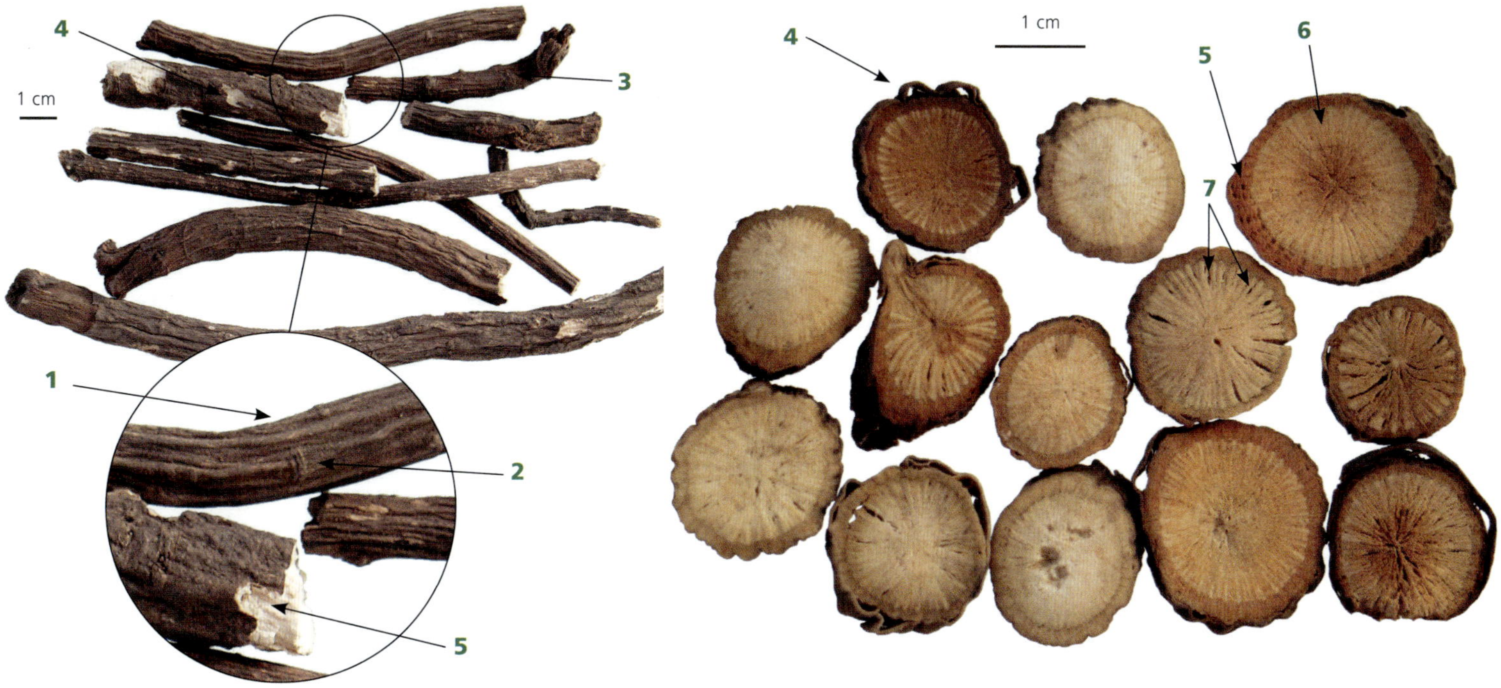

Crude drug and Chi Shao (sourced from *Paeonia lactiflora* growing wild in Inner Mongolia — *Di Dao*-quality)

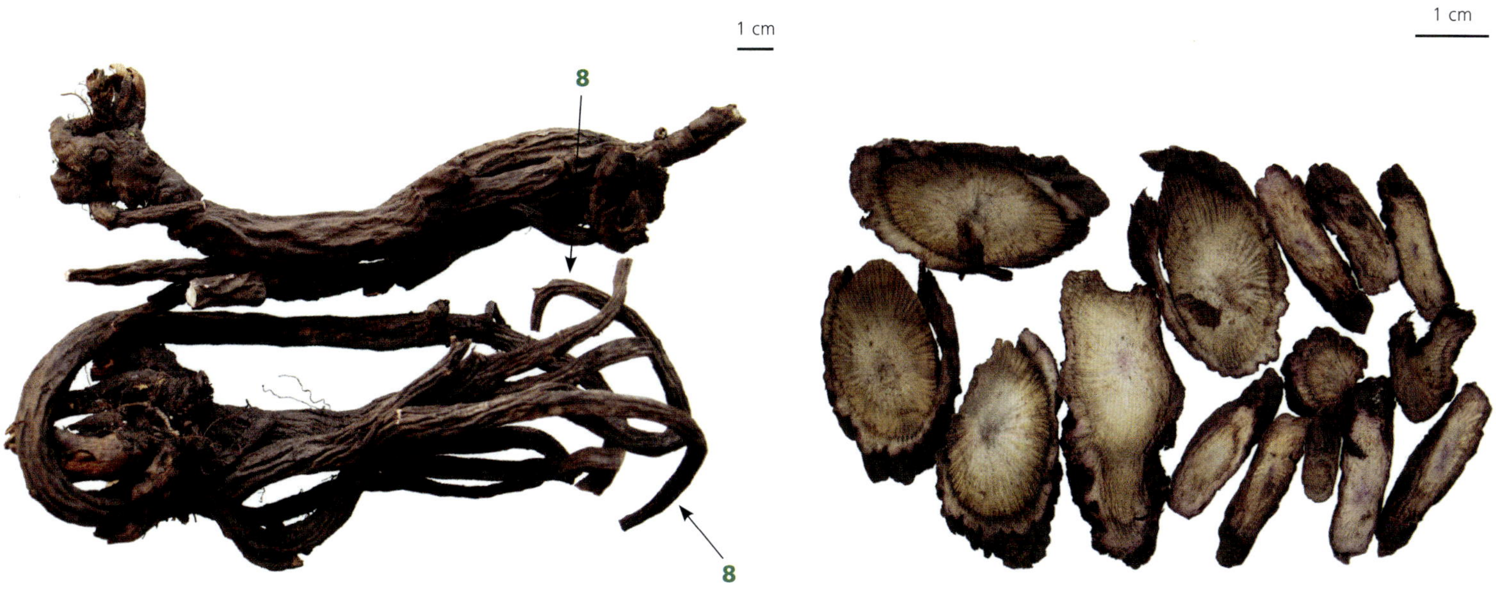

Crude drug and Chi Shao (sourced from *P. anomala* subsp. *veitchii* growing wild in Sichuan)

Chuan Bei Mu 川贝母 Fritillariae Cirrhosae Bulbus

Drug part: BULB

OVERVIEW

Liliaceae

Species of *Fritillaria* wild in China: **24**

Pharmacopoeia species in the 'Chuan Bei Mu' group may be sourced from the bulbs of six *Fritillaria* species (CP2015). These have long been the main ingredient in China's popular proprietary cough remedies with bulbs fetching at least $250/kg, used to fuel an industry in China worth an estimated US $400 million annually (Li *et al.,* 2012b). The largest volume of bulbs in recent decades have been wild-harvested leading to five of the official species becoming threatened due to unsustainable harvesting practices. This has resulted in widespread substitution and adulteration in domestic and overseas markets mostly using look-alike bulbs of other species in the genus. Four of the most widely reported substitutes are illustrated here.

Differentiating *Fritillaria* bulbs using macroscopic characters alone is seldom reliable such that laboratory-based authentication methods (notably using DNA) are advised. A sizeable analytical literature has emerged in recent years to address this need.

An understanding of the long-established bulb grading system of 'Chuan Bei Mu' may serve as a useful guide when directing authentication efforts; for example, top quality bulbs called 松贝 Songbei are especially prone to substitution or adulteration since they command the highest prices. These are sourced from *F. cirrhosa* (PS1), *F. przewalskii* (PS3) or *F. unibracteata* (PS5) with bulbs typically small, tightly closed, sourced from 1–2-year old pre-flowering plants and whose provenance is the Songpan district of Sichuan. Medium quality bulbs (青贝 Qingbei), also from these 3 species, are larger and typically open at their apex to reveal buds and old stem bases. 'Qingbei' are sourced from wild plants that have already flowered and that grow outside the Songpan district. The lower quality bulbs known as 炉贝 Lubei (sourced from *F. delavayi*, PS4) and 栽培品 Zai Pei Pin (sourced from *F. taipaiensis* PS4 or *F. unibracteata* var. *wabuensis* PS6) may be less prone to adulteration but this is dependent on market prices.

In an effort to halt the decline of wild *Fritillaria* species in the 'Chuan Bei Mu' group and satisfy continuing high market demand, a major sustainable harvesting initiative is now well-established for *F. cirrhosa* on the Tibetan plateau (see PS1), while *F. taipaiensis* (PS4) and *F. unibracteata* var. *wabuensis* (PS6) have been brought into commercial medicinal cultivation in Chongqing and Sichuan provinces respectively.

Fritillaria cirrhosa, growing under the 'natural fostering' conservation programme in Sichuan province PHOTO: MARTYN RIX

Fritillaria cirrhosa D. Don

PLANT DESCRIPTION

An erect, delicate-looking herbaceous perennial (15–60 cm tall) with a bulb (1–2 cm diam.) of 2 fleshy scales. *Stem* unbranched and bluish-green. *Leaves* 7–11, opposite, alternate or 3–4-whorled, leaf blades bluish-green, linear (4–12 cm long; c. 3–5 mm wide) with tips often curled and margins entire. *Inflorescence* 1-3-flowered; bracts 3, leaf-like and slightly curled. *Flowers* bisexual, corolla bell-shaped and nodding (3–5 cm long; 1.2–1.8 cm diam.); 'petals' 6, equal size, yellow or yellowish-green or often densely chequered purple; stamens 6; ovary superior; style lobes 3.5 mm long. *Fruit* a dry, 3-valved capsule with 3 narrow wings (wings 1–1.5 mm wide). *Seeds* many, flat and brown. Fl. May–Jul.; fr. Aug.–Oct.

HARVESTING, SOURCING AND NATURAL RANGE

A high altitude species (3,200–4,600 m) of the Tibetan plateau found in low-growing alpine scrub, meadows, forests and moist places in W and NW China (provinces of Gansu, Qinghai, Sichuan, Tibet and Yunnan). Although wild harvesting continues in places, in the Kangding district of Sichuan wild populations have long been the focus of an extensive sustainable harvesting initiative called 'natural fostering' (Li *et al.*, 2012b; Chen *et al.*, 2003; Xiao & Chen, 2003). Bulbs are harvested (under a quota system in Kangding) in autumn or in summer, after snows have melted with *Di Dao*-quality material traditionally sourced from Sichuan. The species is also native in Bhutan, India, Nepal and Sikkim.

CONSERVATION STATUS

China: Near Threatened (MEPCAS, 2013).
Global: Not Evaluated (IUCN Red List, accessed 16/8/2016).

TCM PROPERTIES AND USE

***properties*: 'bitter, sweet and cool'**
***channels*: 'Lung and Heart'**
***actions*: 'resolves phlegm, relieves cough, clears heat, disperses nodules'**

***indications*: cough, phlegm; sores and abscesses**

F. cirrhosa flower and freshly uprooted bulbs

F. cirrhosa medicinal crop, near Kangding, Sichuan province

DRUG MORPHOLOGY

CRUDE DRUG/DECOCTING PIECES (INTACT BULBS): bulbs of 'Songbei' and 'Qingbei' quality (see above) typically **bright white** with a **smooth surface.**
Songbei **[1]**: somewhat **conical or sub-spherical** (0.3–0.8 cm tall; 0.3–0.9 cm diam.) with 2 fleshy scales **[2]**, the larger tightly clasping the typically **crescent-shaped** inner and smaller inner one. **Bulb apex closed [3]**, obtuse or somewhat acute; base **[4]** even, slightly concave, with a greyish-brown central disc and occasional remains of fibrous roots. Internally 1 or 2 small scales **[5]** and slightly tapering buds **[6]**. *Texture*: hard and brittle. *Fracture*: white and powdery. *Odour*: indistinct. *Taste*: slightly bitter.
Qingbei **[7]**: somewhat broader than tall (0.4–1.4 cm tall; 0.4–1.6 cm diam.) with 2 clasping fleshy scales, almost equal in size. Bulb apex **open [8], revealing buds, 2–3 small scales and a shrunken stem base.**

PROCESSING

After removal of fibrous roots and coarse outer tunics, bulbs are dried (= crude drug/decocting pieces).

OTHER COMMON NAMES

卷叶贝母 Juan Ye Bei Mu; Bulbus Fritillariae Cirrhosae; tendrilleaf fritillary bulb (CP2010); Sichuan fritillary (SCN).

SAFETY

An official healthfood in China (Anon., 2002a,b). Safety Class: 1; Interaction Class: A (Gardner & McGuffin, 2013).

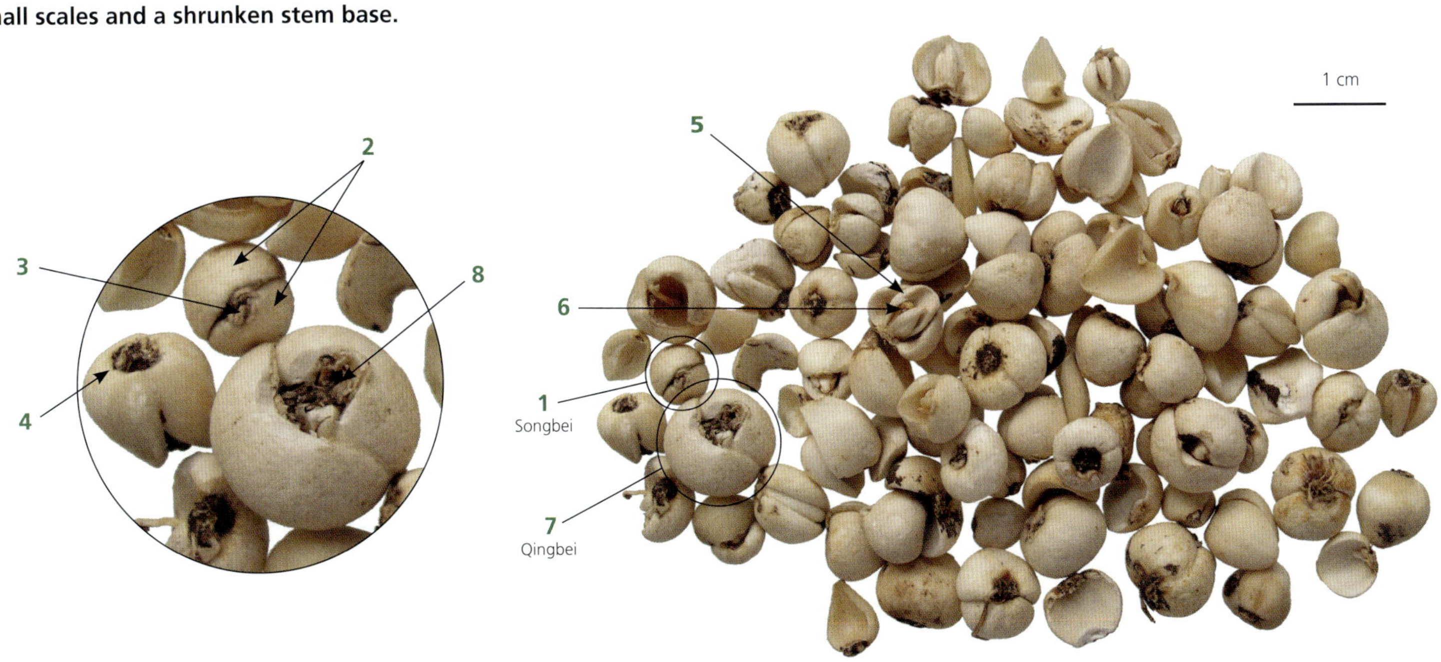

Chuan Bei Mu: a mixed batch of Qingbei and Songbei quality bulbs sourced from *Fritillaria cirrhosa*

Fritillaria delavayi Franch.

Liliaceae

PLANT DESCRIPTION

Easily distinguished from PS1. *Main differences:* **stems** usually **shorter** (15–35 cm long) and **bulbs much larger** (see *Drug Morphology*); **leaves 3–5, broader** (2–7 cm long; 1–3 cm wide), **often purplish**; **flowers solitary;** 'petals' **yellow, spotted or chequered reddish-brown** or completely **reddish-brown;** style lobes 0.5–4 mm long; fruit capsule somewhat enclosed by persistent 'petals'. Fl. Jun–Jul; fr. Aug–Sep.

HARVESTING, SOURCING AND NATURAL RANGE

Bulbs are harvested from wild populations which grow in W and NW China (Qinghai, Sichuan, Tibet and Yunnan provinces) on loose shaley rocks, or sandy places near flood plains (3,400–5,600 m altitude). Traditional *Di Dao*-quality material is sourced from Qinghai province. The species is also wild in Bhutan and Sikkim.

CONSERVATION STATUS

China: Vulnerable (MEPCAS, 2013).
Global: Not Evaluated (IUCN Red List; accessed 18/8/2016).

F. delavayi wild plant, Sichuan province; purple leaved form

Alpine habitat c. 4,000 m, of *Fritillaria delavayi*, Kangding district, Sichuan province

F. przewalskii Maxim. ex Batalin

Liliaceae

taxonomic/nomenclatural note: in CP2015 as '*F. przewalskii* Maxim.'

PLANT DESCRIPTION

Similar to PS1. *Main differences:* **leaves 4–7; bracts 1**; flowers solitary (rarely 2); **'petals' yellow,** sometimes speckled blackish-purple; nectaries inconspicuous; **style scarcely lobed** [hl] (each lobe < 1 mm long. Fl. Jun–Jul; fr. Aug.

HARVESTING, SOURCING AND NATURAL RANGE

Bulbs are harvested from wild populations which are endemic to W and NW China (Gansu, Qinghai and Sichuan provinces). The plant grows wild in scrub and grassland (2,800–4,400 m. altitude) with traditional *Di Dao*-quality drug sourced from all three provinces.

CONSERVATION STATUS

Endemic to China: Vulnerable (MEPCAS, 2013); Not Evaluated (IUCN Red List; accessed 18/8/2016).

F. przewalskii in the wild PHOTO: SJAAK DE GROOT

DRUG MORPHOLOGY

CRUDE DRUG/DECOCTING PIECES (INTACT BULBS): easily distinguished from those of PS1, PS5 and PS3 by their **larger size** (0.7–2.5 cm tall; 0.5–2.5 cm diam.) and **more ovoid, almost conical shape.** Externally **whitish or pale brownish-yellow, often speckled brown-maculate**. Bulb scales 2, almost equal sized with apex open and slightly tapering; base slightly acute or somewhat obtuse.

Crude drug/decocting pieces of Chuan Bei Mu (Lubei form), sourced from *F. delavayi*

PROCESSING

As for PS1.

OTHER COMMON NAMES

炉贝 Lu Bei; Bulbus Fritillariae Cirrhosae; Sichuan fritillary (SCN); tendrilleaf fritillary bulb (CP2010).

SAFETY

As for PS1.

DRUG MORPHOLOGY

CRUDE DRUG/DECOCTING PIECES (INTACT BULBS): as for PS1 (both 'Songbei' and 'Qingbei' quality). Bulbs of these 2 species cannot be reliably distinguished macroscopically.

Crude drug/decocting pieces of Chuan Bei Mu (left: Qingbei quality; right: Songbei quality; sourced from *F. przewalskii*

PROCESSING

As for PS1.

OTHER COMMON NAMES

As for PS1 plus 甘肃贝母 Gansu Bei Mu; Bulbus Fritillariae Cirrhosae; tendrilleaf fritillary bulb (CP2010); Sichuan fritillary (SCN).

SAFETY

As for PS1.

Fritillaria taipaiensis P. Y. Li

Liliaceae

PLANT DESCRIPTION

Similar to PS1. *Main differences*: **bracts not curled**; inflorescence 1(or 2)-flowered; 'petals' yellowish-green with irregularly-shaped purple spots, often merging to create larger mottling; nectaries slightly bulbous. Fl. May–Jun, fr. Jun–Jul.

HARVESTING, SOURCING AND NATURAL RANGE

Bulbs are harvested mostly from cultivation (Chongqing province) but also from wild populations in Shaanxi province (Ta Bei Shan). Endemic to W and NC China (Gansu, Hubei, Shaanxi and Sichuan provinces) where the species grows wild in scrub and on grassy slopes (2,000–3,200 m altitude).

CONSERVATION STATUS

Endemic to China: Endangered (MEPCAS, 2013); Not Evaluated (IUCN Red List, accessed 18/8/2016).

F. unibracteata P. K. Hsiao & K. C. Hsia

Liliaceae

(includes **var. *unibracteata*** and **var. *longinectarea*** S. Y. Tang & S. C. Yueh)

taxonomic/nomenclatural note: the above varieties are explicitly named here to distinguish this source of 'Chuan Bei Mu' from PS6 (var. *wabuensis*) described on page 84 as a separate official source of 'Chuan Bei Mu' (CP2015).

PLANT DESCRIPTION

Very similar and easily confused with PS1. *Main differences*: **leaves 5–7,** opposite or alternate **(never in whorls); bracts 1**; 'petals' blackish-purple, chequered yellowish-brown, sometimes with a coloured stripe near the tips; nectaries inconspicuous; **style scarcely or shortly lobed** [hl] (each lobe 0.5–2 mm long); wings of fruit capsule narrow (c. 1 mm). Fl. May–Jun.; Fr. Aug.

HARVESTING, SOURCING AND NATURAL RANGE

Bulbs are wild harvested. The species is endemic to W and NW China (Gansu, Qinghai and Sichuan provinces) where the plant grows in alpine scrub and meadows (3,200–4,700 m. altitude). Traditional *Di Dao*-quality material is sourced from Sichuan.

CONSERVATION STATUS

Endemic to China: Critically Endangered (MEPCAS, 2013, as '*F. unibracteata* var. *longinectarea*'); Endangered (MEPCAS, 2013, as '*F. unibracteata* var. *unibracteata*'); Not Evaluated (IUCN Red List, accessed 18/8/2016).

Wild flowering plant of *F. unibracteata* var. *longinectarea*

F. unibracteata var. *longinectarea*, freshly uprooted plant(s) and one dissected flower

DRUG MORPHOLOGY

CRUDE DRUG/DECOCTING PIECES (INTACT BULBS): similar to those of 'Songbei' (described under PS1 and PS5). *Main differences:* **somewhat squat (oblate) or shortly cylindrical** (0.5–2 cm tall; 1–2.5 cm wide). Externally **whitish or pale brownish-yellow, somewhat rough and slightly spotted yellow**; 2 outer scale leaves almost of equal size, usually open and obtuse.

PROCESSING

As for PS1.

OTHER COMMON NAMES

太白贝母 Tai Bai Bei Mu; Zai Pei Pin; tendrilleaf fritillary bulb (CP2010).

DRUG MORPHOLOGY

CRUDE DRUG/DECOCTING PIECES (INTACT BULBS): bulbs of 'Songbei' and 'Qingbei' quality cannot be distinguished macroscopically from those of PS1.

PROCESSING

As for PS1.

OTHER COMMON NAMES

暗紫贝母 An Zi Bei Mu (*F. unibracteata* var. *unibracteata*); 长腺贝母 Chang Xian Bei Mu (*F. unibracteata* var. *longinectarea*); Bulbus Fritillariae Cirrhosae; Sichuan fritillary (SCN); tendrilleaf fritillary bulb (CP2010).

SAFETY

As for PS1.

Chuan Bei Mu: mixed batch of Qingbei and Songbei quality bulbs sourced from *F. unibracteata*

F. unibracteata var. *wabuensis* (S. Y. Tang & S. C. Yueh) Z. D. Liu, Shu Wang & S. C. Chen

Liliaceae

synonym: *F. wabuensis* S. Y. Tang & S. C. Yueh

taxonomic/nomenclatural note: *WCS* and *Kew-MPNS* accepted name, 28/10/2016. *FoC* considers this taxon to be a synonym of *F. crassicaulis* S. C. Chen.

PLANT DESCRIPTION

Similar to PS1. *Main differences*: **sometimes taller** (30–80 cm tall); **leaves** more numerous **(10–18)**; **basal leaves broader** (1.5–3 cm wide); nectaries sometimes conspicuous (2–3 mm long); **style lobes 2–3 mm long**; bracts leaf-like but not curled. Fl. May–Jun, fr. Jul–Aug.

HARVESTING, SOURCING AND NATURAL RANGE

Bulbs are harvested from cultivated plants grown in Sichuan. Endemic to W China (Sichuan and Yunnan provinces) where the species grows wild in alpine grasslands and forest (2,500–3,400 m altitude).

CONSERVATION STATUS

Endemic to China: Vulnerable (IUCN (2012a), 31/3/2012 unpub.); Not Evaluated (IUCN Red List, accessed 18/8/2016).

F. anhuiensis S. C. Chen & S. F. Yin

Liliaceae

PLANT DESCRIPTION

Similar to PS1. *Main differences:* **bulbs** with 2 or 3 fleshy scales (1–2 cm diam.) and **with smaller bulbils** of varying shape and size; **leaves** usually more numerous **(6–18)**, **whorled** and **broader** (0.5–2 (rarely 3.5 cm) wide); fruit capsule with broad wings (wings 5–10 mm wide). Fl. Mar–Apr, fr. May–Jun.

SUBSTITUTION COMMENT. Bulbs have long been cultivated for use in local (not TCM) medicine in the eastern Chinese provinces of Anhui and Henan and where some 30–40 tons are produced annually. The wild species is endemic to Anhui and Henan where it grows in forests, scrub and grasslands between 600–900 m altitude.

CONSERVATION STATUS

Endemic to China: Vulnerable (MEPCAS, 2013); Not Evaluated (IUCN Red List, accessed 10/8/2016).

Basal leaves of *Fritillaria anhuiensis*

Fritillaria anhuiensis in the wild (Xinxian district, Henan province) PHOTO: ZHOU CHONGIAN

DRUG MORPHOLOGY

CRUDE DRUG/DECOCTING PIECES (INTACT BULB): cannot be reliably distinguished macroscopically from those of PS5.

PROCESSING

As for PS1.

OTHER COMMON NAMES

粗茎贝母 Cu Jing Bei Mu; tendrilleaf fritillary bulb (CP2010).

SAFETY

As for PS1.

'DRUG SUBSTITUTE' MORPHOLOGY

Bulbs with 2 or 3 fleshy scales (1–2 cm diam.) and **many smaller bulbils** within (often broadly conical). Since bulbils closely resemble the bulbs of PS1, PS3 or PS5 and cannot be reliably distinguished macroscopically, **laboratory-based authentication methods are recommended.**

OTHER COMMON NAMES

安徽贝母 Anhui Bei Mu; 皖贝 Wan Bei.

F. anhuiensis, fresh bulbs showing numerous bulbils

Bulbs of *F. anhuiensis*

OTHER UNOFFICIAL SUBSTITUTES

Bubs of other, cheaper and more readily available *Fritillaria* species are reported as frequent and intentional unofficial substitutes or adulterants of 'Chuan Bei Mu' in local and international trade. Common among these are the bulbs of other TCM drugs sourced from the *Fritillaria* species below and for which laboratory-based authentication methods are recommended for their discrimination. They are widely cultivated as medicinal crops and described elsewhere in this book:

1. *F. pallidiflora* Schrenk; source of 'Yi Bei Mu' (Fritillariae Pallidiflorae Bulbus) see page 320.
2. *F. thunbergii* Miq. var. *chekiangensis* Hsiao & K.C.Hsia.; one of the sources of 'Zhe Bei Mu' (Fritillariae Thunbergii Bulbus) see page 332.
3. *F. usuriensis* Maxim.; source of 'Ping Bei Mu' (Fritillariae Ussuriensis Bulbus) see page 232.

Chuan Niu Xi 川牛膝

(literal: Sichuan ox knee)

Drug part: ROOT

Cyathulae Radix

Cyathula officinalis K. C. Kuan

Amaranthaceae

PLANT DESCRIPTION

An upright and branching, herbaceous perennial (50–100 cm tall). *Stems* with distinctive red and swollen nodes (hence its Chinese name). *Leaves* opposite, softly hairy, elliptic or narrowly elliptic (3–10 cm long; 1.5–5.5 cm wide); margins entire. *Flowers* in dense globose heads (1–1.5 cm diam.), in leaf axils or in terminal groups, pale green to whitish. Individual flowers in groups of 1–3, some sterile, with 'petals' thin, dry, membranous and often spiny; stamens 5, alternating with rectangular lobes (0.3–0.4 mm long). *Fruits* dry, pale yellow and flask-shaped. *Seeds* many and glossy. Fl. Jun–Jul, fr. Aug–Sep.

Species of *Cyathula* wild in China: 4

HARVESTING, SOURCING AND NATURAL RANGE

Roots are harvested in autumn and winter mostly from medicinal crops grown on small-holdings or large-scale farms. The traditional *Di Dao* source occurs in the Tianquan district of Sichuan province. The wild species grows in disturbed places above 1,500 m in N, SC and SE China: Guizhou, Hebei, Shanxi, Sichuan, Yunnan and Zhejiang provinces [also wild in Nepal].

CONSERVATION STATUS

China: Not Evaluated.
Global: Not Evaluated (IUCN Red List; accessed 12/2/2015).

Young plants of *Cyathula officinalis* (left) are dwarfed by the larger *C. capitata* (right, surrounded by *C. officinalis*)

C. officinalis in flower

LESSER KNOWN UNOFFICIAL SUBSTITUTES

1. *Cyathula capitata* Moq. (麻牛膝 Ma Niu Xi) — roots. A common and similar-looking species, possibly the wild ancestor of *C. officinalis*; local sources in Sichuan claim it hybridises freely with the official species and contaminates cultivated plots (Lin Yu-Lin, pers. obs.). Leaves larger (5–14 cm long; 3–7 cm wide), broadly ovate or obovate-oblong; flowers greyish-black when dried and with larger lobes (0.6–1 mm long) visible [hl] between the bases of the fertile stamens (see photo, above). Roots usually **shorter (15–30 cm long) and thinner**; externally **dark brown** and twisted; *taste:* **bitter, astringent and numbing**.
2. Roots and rhizomes of a species of *Strobilanthes* (Acanthaceae): (see photo, right) roots/rhizomes easily distinguished from those of the offical species but frequently traded in error in mainland China, Taiwan and the UK probably owing to its very similar Chinese name 牛膝 Niu Xi (as well as 味牛膝 Wei Niu Xi). The TCM literature (Bensky *et al.*, 2004; Zhao & Li, 2007) states the botanical source to be *Strobilanthes forrestii* or *S. nemorosus*, however both names are under taxonomic review and the identity of this substitute remains uncertain.

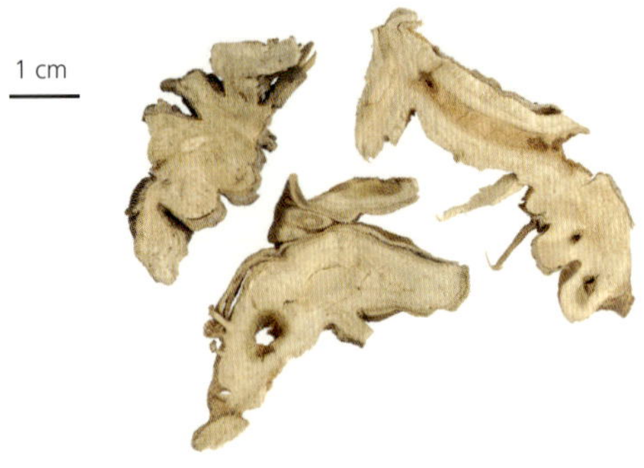

Sliced root and rhizome of a common but unidentified substitute, possibly *Strobilanthes* sp.

TCM PROPERTIES AND USE

***properties*: 'bitter, sweet and neutral'**
***channels*: 'Liver and Kidney'**
***actions*: 'invigorates blood, dispel stasis, promotes urination, tonifies Liver and Kidney'**

***indications*: gynaecological disorders; musculoskeletal pain; dysuria; toothache; mouth ulcers**

DRUG MORPHOLOGY

CRUDE DRUG (DRIED ROOT): roots somewhat cylindrical but strongly twisted and sometimes branched **[1]** (30–60 cm long; 0.5–3 cm diam.). Externally greyish-brown with distinct longitudinal striations, scattered rootlets or rootlet scars and slightly raised transverse, oval lenticels **[2]**. *Texture*: strong, pliable and difficult to break. *Fracture:* pale yellow or brownish-yellow with several concentric rings of pale yellow punctiform, vascular bundles **[3]**. *Odour*: indistinct. *Taste*: sweet. ***Good quality*:** roots dry, oily and yellowish-brown.

DECOCTING PIECES-1 (ROOT SLICES): sub-rounded or elliptical slices (0.5–3 cm diam., 1–2 mm thick). Root surface greyish-brown with distinct longitudinal striations **[4]**, rootlet scars or oval lenticels possibly present. Cut surfaces pale yellow or brownish-yellow often with several concentric rings of punctiform vascular bundles **[3]**. *Odour*: indistinct. *Taste*: sweet.

DECOCTING PEICES-2 (WINE-PROCESSED ROOT SLICES — JIU CHUAN NIU XI): as for dried decocting slices but externally brownish-black. *Odour:* faintly of wine. *Taste:* sweet.

PROCESSING

After removal of root-heads and rootlets, roots are partially sun- or oven-dried, then piled into heaps where they soften before drying is completed (= crude drug). The crude drug is later cut into thin transverse or oblique slices (= decocting pieces-1) and may undergo further processing by moistening with yellow rice wine then dry-frying (= decocting pieces-2) 酒川牛膝 Jiu Chuan Niu Xi).

OTHER COMMON NAMES

天全牛膝 Tian Quan Niu Xi; 甜牛膝 Tian Niu Xi (the two previous names are derived from the name of its *Di Dao* locality, 'Tianquan' in Sichuan); cyathula (SCN); Radix Cyathulae.

SAFETY

Safety Class: 2b and 2d (Gardner & McGuffin, 2013). An official Chinese food and healthfood (Anon., 2002a,b).

Crude drug sourced from *Cyathula officinalis*

Decocting pieces-1 (dried *C. officinalis* root)

Decocting pieces-2 (wine-processed *C. officinalis* root)

Chuan Xiong 川芎

Chuanxiong Rhizoma

Drug part: RHIZOME

Ligusticum chuanxiong Hort. ex S. H. Qiu *et al.*

Apiaceae

taxonomic/nomenclatural note: for many years the source plant of Chuan Xiong had not been named scientifically owing to it always being sterile (without flowers). One solution was to adopt its trade name (Chuan Xiong) as a cultivar (i.e. *L. sinense* Oliv. 'Chuanxiong') by which it has been known for some years. A chance flowering, however, has since enabled it to be formally (scientifically) named (i.e. to genus and species) as indicated above (Qiu *et al.*, 1979).

Species of *Ligusticum* wild in China: 40

PLANT DESCRIPTION

As for *Ligusticum sinense* (source plant of Gao Ben (Ligustici Rhizoma et Radix) see page 142) but plants never flower nor set seed and the outermost leaf segments are pinnately (and shallowly) lobed (i.e. pinnatifid).

Medicinal crop of *L. chuanxiong*, Sichuan province

Crop plant with swollen, purplish nodes

HARVESTING, SOURCING AND NATURAL RANGE

Mature rootstocks are harvested in summer when rhizome nodes become swollen and purplish. Sourcing is entirely from medicinal crops grown in China predominantly in Sichuan and Inner Mongolia but also in Gansu, Guizhou, Henan, Hubei, Hunan, Jiangxi, Shaanxi and Yunnan (since this cultivar is not known to flower or set seed it is propagated by vegetative division). Quality varies considerably between provinces with 'best quality' sourced from Sichuan (more specifically, the traditional *Di Dao* quality originates from the district of Guanxian). Young plants are first grown in the mountains on shaded nursery slopes (where they are Shan Chuanxiong — mountain Chuanxiong) before being transplanted to the fertile soils of the Sichuan Plain (the main production area) for an additional 9 months of unshaded growth until they are fully mature and of good quality.

CONSERVATION STATUS

China/Global: known only from cultivated plants.

Freshly harvested rootstock with stem bases and roots still attached

UNOFFICIAL SUBSTITUTES

When prices dictate Shan Chuanxiong (i.e. immature rhizomes) are sold directly into local markets, as a clinically inferior substitute (see *Harvesting, sourcing and natural range* above, *Drug Morphology* (and photos) opposite).

Additionally, some TCM texts (Bensky *et al.*, 2004, Xiao *et al.*, 2001–2002) mention occasional adulterants or unofficial substitutes sourced from rhizomes of related species but their botanical identity remains uncertain as is their presence in international trade.

TCM PROPERTIES AND USE

properties: **'pungent and warm'**
channels: **'Liver, Gall Bladder and Pericardium'**
actions: **'invigorates Blood and qi, dispels stasis, dispels wind, relieves pain'**

indications: **menstrual, postpartum disorders; abdominal pain, musculoskeletal pain; chronic sores; headaches**

DRUG MORPHOLOGY

CRUDE DRUG (DRIED RHIZOME): an irregular, fist-shaped mass **[1]** (2–7 cm diam.). Externally dull and yellowish-brown; uppermost surface with many protuberant, sub-rounded and swollen crater-like stem base scars **[2]**; lower surface with less pronounced wart-like root scars **[3]**; remainder of surface deeply and irregularly furrowed **[4]**. *Texture*: firm and difficult to break. *Fracture*: yellowish- or greyish-white with a markedly wavy or strongly undulating cambium ring **[5]** and many dark and scattered oil glands **[6]**. *Odour*: strongly aromatic. *Taste*: bitter and mildly numbing with a slightly sweet aftertaste. *Note*: inferior 'Shan Chuanxiong' are smaller and lighter (see also *Harvesting, sourcing and natural range*). ***Good quality***: large, solid, heavy and intensely aromatic.

DECOCTING PIECES (SLICED RHIZOME): strongly irregular in outline (2–7 cm diam.; 1–2 mm thick) with typically deep, round-edged indentations **[7]** (colloquially called 'Hu Die Pian' 蝴蝶片 (butterfly pieces) owing to their shape resembling butterfly wings). Outer bark dull, yellowish-brown and irregularly furrowed **[8]**; fragments of crater-like stem scars **[9]** sometimes visible. Cut surface often lustrous, predominantly yellowish- or greyish-white (central xylem and pith) **[10]** separated from the narrower and darker bark layers by a distinctly wavy or strongly undulating cambium ring **[11]**; many yellowish-brown oil glands scattered throughout **[6]**. *Texture*: hard. *Odour*: strongly aromatic. *Taste*: bitter, pungent and slightly sweet. *Note*: slices sourced from Shan Chuanxiong (see *Harvesting, sourcing and natural range*) are easily distinguished because they are lighter and smaller.

PROCESSING

Rootstocks are sun-dried slowly until thoroughly dry, then roots removed and discarded, leaving the rhizome (= crude drug). Subsequent size grading and softening is performed before rhizomes are thinly sliced (1-2 mm thick), usually longitudinally, into decocting pieces.

OTHER COMMON NAMES

Rhizoma Chuanxiong; Sichuan lovage (SCN); Sichuan lovage root (CP2010).

SAFETY

An official healthfood in China (Anon., 2002a,b). Safety Class: 2b;
Interaction Class: A (Gardner & McGuffin, 2013). Adverse events have been reported following ingestion (Bensky *et al.*, 2004).

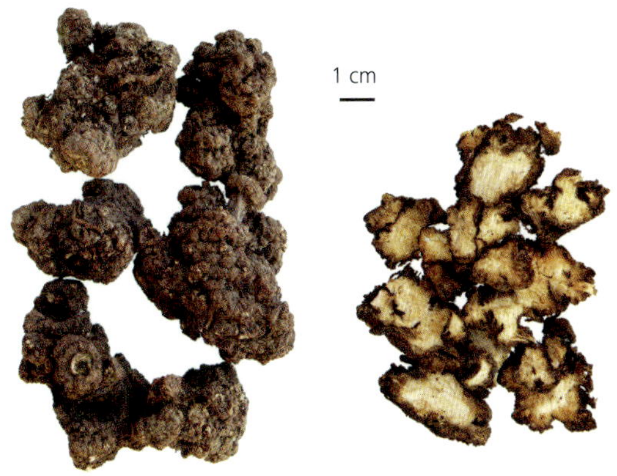

Shan Chuanxiong (mountain Chuanxiong)
(clinically "inferior")

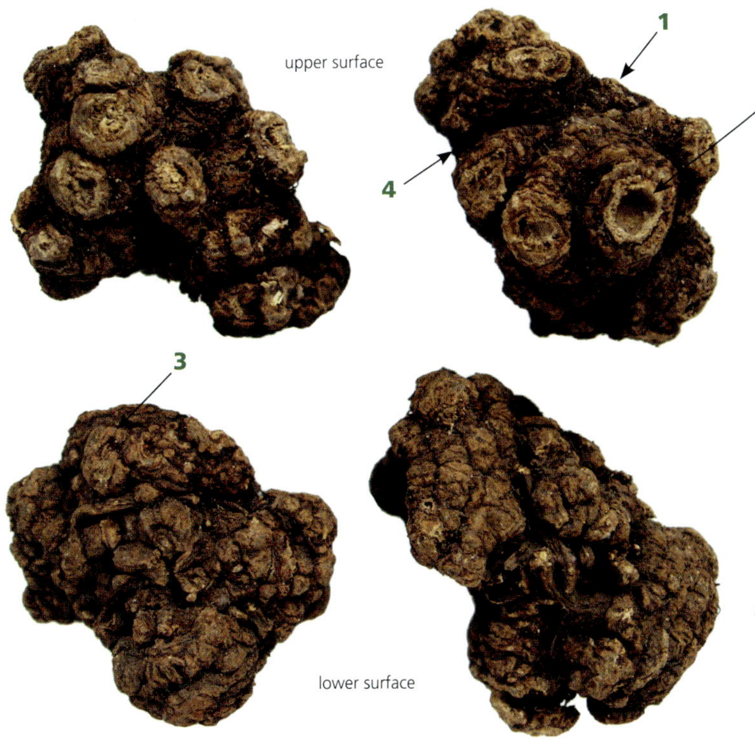

Crude drug

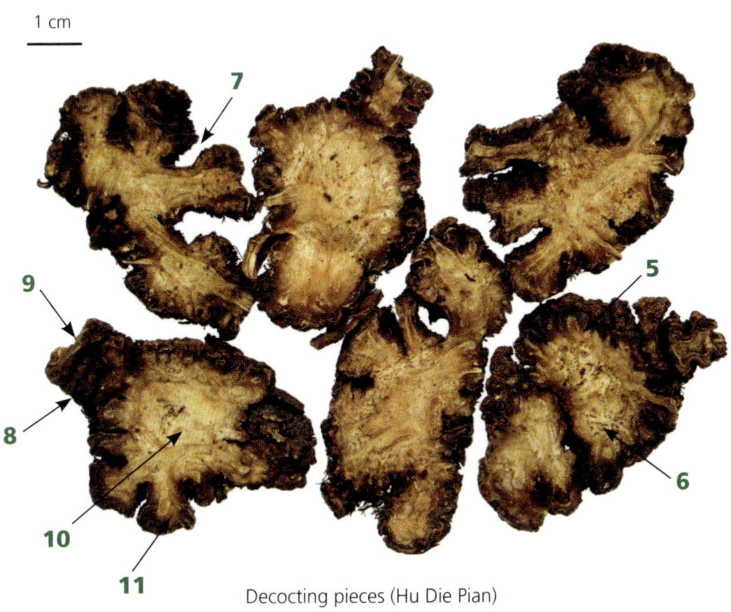

Decocting pieces (Hu Die Pian)

Da Huang 大黄

Rhei Radix et Rhizoma

Chinese rhubarb

Drug parts: ROOT AND RHIZOME

Hybridisation of *Rheum* species has occurred since its early cultivation thousands of years ago producing many intermediates and making their field identification seldom straightforward (Foster & Yue, 1992). A widely held traditional view in Chinese medicine is that only those *Rheum* plants with palmately lobed leaves are appropriate for medicinal use; leaves of the 3 CP species described below fit this broad tenet.

PHARMACOPOEIA SPECIES 1

Rheum palmatum L.

Polygonaceae

synonyms: *Rhabarbarum palmatum* (L.) Moench; *Rheum potaninii* Losinsk.

Species of *Rheum* wild in China: **38**

PLANT DESCRIPTION

A stout perennial herb (1.5–2 m tall) with a stout rootstock. *Stems* erect, hollow and grooved; each stem node with a large membranous sheath (ocrea) (up to 15 cm long) with an entire margin. *Leaves* large, simple and palmately lobed, alternate and stalked. Basal leaves about as long as wide (c. 40–60 cm), **usually lobed almost half-way to mid-vein** (pinnatisect), base heart-shaped, tip **sharply or narrowly pointed.** Stem towards top of plant. *Flowers* small and **tightly packed** in long lateral branching, upright cylindrical heads (panicles). Flowers bisexual and radially symmetrical, 'petals' 6, **purplish-red or rarely yellowish-white**, outer 3 smaller, elliptic to orbicular (1–1.5 mm long); stamens usually 9, not longer than 'petals'; ovary superior; styles 3, short and horizontal; stigma inflated and recurved. *Fruiting heads* erect, branched and **tightly closed;** each fruit a dry, brown, 3-sided achene with scale-like membranous wings; *seed* brownish-black, broadly ovoid. Fl. Jun, fr. Aug.

HARVESTING, SOURCING AND NATURAL RANGE

'Da Huang' destined for international markets is sourced from wild-harvested and medicinal crop plants with the bulk of production from the Chinese provinces of Gansu and Qinghai, with some from Shaanxi and Sichuan. It is known in TCM commerce (along with that from *R. tanguticum* (PS2) below) as 'Xi Bei Da Huang' (north-western rhubarb) to distinguish it from that of *R. officinale* (PS3 below) and generally traded as 'Nan Da Huang' (southern rhubarb).

Rootstocks are harvested in late autumn when stems and leaves have withered or in spring before new buds appear. Traditional *Di Dao*-quality material is produced in Gansu and Qinghai provinces. The species is also cultivated in Russia.

The wild plant grows in high mountains (1,500–4,400 m altitude) in cool, dry, occasionally damp soils and is endemic to N and W China (Gansu, Hubei, Inner Mongolia, Qinghai, Shaanxi, Sichuan, Tibet and Yunnan).

CONSERVATION STATUS

Endemic to China: Least Concern (MEPCAS, 2013); common and widespread (authors' pers. comm; 2010); Not Evaluated (IUCN Red List, accessed 10/8//2016).

PHARMACOPOEIA SPECIES 2

R. tanguticum (Maxim. ex Regel) Balf.

Polygonaceae

synonyms: *R. palmatum* L. var. *tanguticum* Maxim. ex Regel; *R. palmatum* subsp. *tanguticum* (Maxim. ex Regel) Stapf

taxonomic/nomenclatural note: in CP2015 as '*Rheum tanguticum* Maxim. ex Balf.'

PLANT DESCRIPTION

Distinguished from PS1 only by its basal and stem leaves which are **more deeply lobed** (palmately 5-lobed, with the middle 3 lobes **divided as far as the mid-vein**) and all lobes **narrowly triangular**; these features are most obvious in younger plants). Fl. Jun, fr. Jul–Aug.

HARVESTING, SOURCING AND NATURAL RANGE

In contrast to PS1 the sourcing of this species is totally from cultivated stock owing to the economic exhaustion of its wild populations back in the 1970s (Wang & Ren, 2009; Yu *et al.*, 2010). Crop-growing areas are located where the species once grew wild: Gansu, Qinghai, Shaanxi and Tibet (Foster & Yue, 1992) and, like PS1 above, it is often traded under the name 'Xi Bei Da Huang' (north-western rhubarb) to distinguish it from 'Nan Da Huang' (southern rhubarb) sourced from *R. officinale* (PS3 below). *Di Dao*-quality material is sourced from Qinghai. Rootstocks are harvested in late autumn when stems and leaves have withered or in spring before new buds appear.

The natural habitat and altitude (1,600–3,000 m) of the wild species is similar to that of PS1.

CONSERVATION STATUS

Endemic to China: Vulnerable (MEPCAS, 2013, as '*Rheum tanguticum* var. *tanguticum*'); Not Evaluated (IUCN Red List, accessed 10/8//2016).

Global: Not Evaluated (IUCN Red List, accessed 21/7/2015).

TCM PROPERTIES AND USE

properties: **'bitter and cold'**
channels: **'Spleen, Stomach, Large Intestine, Liver and Pericardium'**
actions: **'purgative, clears heat, eliminates toxins, invigorates blood, dispels damp heat'**

indications: **constipation; headaches; fever; thirst, abdominal distension; dysuria; amenorrhoea, skin lesions (charred: stops bleeding)**

Flowering plant of *Rheum palmatum*

Harvesting cultivated *R. palmatum*, Gansu province

Medicinal crop of *R. tanguticum* (with brown erect fruiting heads in distance), Qinghai province

Fruiting head of *R. tanguticum*

R. tanguticum in the wild

Rheum officinale Baill.

Polygonaceae

PLANT DESCRIPTION

Very similar to PS1 and PS2. ***Leaves*** palmately divided into **broad** (not narrowly triangular) **lobes divided almost to the mid-vein** (pinnatisect); flowers purplish-red, **often white;** fruiting branches **typically spreading** (divaricate), not tightly closed. Fl. Jun, fr. Aug.

HARVESTING, SOURCING AND NATURAL RANGE

Rootstocks are predominantly harvested from medicinal crops. The bulk of production is from Sichuan province (other sources are Hubei and Shaanxi provinces). This more southerly distribution (compared to PS1 and 2) is the origin of its common trade name 'Nan Da Huang' (southern rhubarb) while its more specific name 'Ya Huang' refers to material of *Di Dao*-quality sourced from the district of Ya'an in Sichuan.

The wild species, also endemic to China, has a more southerly natural range compared to that of PS1 and 2: Fujian (possibly), Guizhou, SW Henan, W Hubei, Shaanxi, Sichuan and Yunnan. It grows on forested hillsides at 1,200–4,000 m altitude.

CONSERVATION STATUS

Endemic to China: Least Concern (MEPCAS, 2013); 'rapidly decreasing' (Leung & Cheng, 2008);

Global: Not Evaluated (IUCN Red List, accessed 10/8/2016).

Cultivated plants of *R. officinale* with immature flower-head

Medicinal crop of *R. officinale*, white-flowered form

UNOFFICIAL SUBSTITUTES/ADULTERANTS

Chinese-sourced 'Da Huang' for overseas markets is unlikely to contain unofficial substitutes owing to large volumes of rootstocks produced from cultivated 'Da Huang' and because the smaller rootstocks of non-Pharmacopoeia species (i.e. substitute species) tend to be too small to make their harvesting economic for export (author Lin Y.L, pers. comm. 11/2016). Chinese markets are more prone to substitution (accidental or not) given the continuing preference in China for wild-harvested materials, reputedly of 'superior quality'. Identification of the three Pharmacopoeia species requires considerable botanical expertise, especially when wild-harvested since they need to be distinguished from the other 35 *Rheum* species that are native in China. Below are examples of local substitutes or adulterants most widely reported in the TCM literature (Bensky *et al*., 2004; Foster & Yue, 1992; Yang *et al*., 2004):

1. ***Rheum* species: *R. acuminatum*** Hook. f. & Thomson (心叶大黄 Xin Ye Da Huang; Himalayan rhubarb (SCN)). Roots **slender**; basal leaf **undersides purplish-red, broadly heart-shaped**; **stem sheaths 2 cm;** stem leaves ovate to heart-shaped; flowerhead sparse; flowers purplish-red. S Gansu, Sichuan, Tibet and Yunnan. 2,800–4,000 m altitude.
2. ***R. australe*** D. Don (syn. ***R. emodi*** Wall.) (藏边大黄 Zang Bian Da Huang; Himalayan or Indian rhubarb). Rootstock stout; basal leaves **broadly ovate with heart-shaped bases**; stem leaves ovate; flowers purplish-red. Native in Tibet [also native in India, Kashmir, Nepal and Pakistan] 3,400–4,300 m altitude. The species is widely cultivated in India and used as a 'Da Huang' substitute in Sichuan, Yunnan and Tibet (Foster & Yue,1992).
3. ***R. coreanum*** Nakai. The Japanese Pharmacopoeia (16th ed.) includes this species (along with the three CP2015 species described above) as an official source of 'Rhei Rhizoma'.
4. ***R. moorcroftianum*** Royle (卵果大黄 Luan Guo Da Huang). **Dwarf, leaning herb;** basal leaf **undersides purplish-red,** blades **much smaller** (6–12 cm long; 4–8.5 cm wide); flowers **yellowish-white, some red.** C and W Tibet. 4,500–5,300 m altitude.
5. ***R. nobile*** Hook. f. & Thomson (塔黄 Ta Huang). Rootstock stout; **basal leaves in a rosette;** stem leaves dense with **conspicuously raised veins** on underside; flower-head **rarely branched**; 'petals' **yellowish-green'.** S Tibet. 4,000–4,800 m altitude.
6. ***R. rhabarbarum*** L. (syn. ***R. franzenbachii*** Muent.) (波叶大黄 Bo Ye Da Huang 'wavy-leaved rhubarb'). Stem stout; leaf margins **curled, ruffled or wavy;** flower-head large, flowers **whitish-green or yellowish-white.** Hebei, Heilongjiang, Hubei, Inner Mogolia, Jilin and Shanxi. 1,000–1,600 m altitude.
7. ***R. webbianum*** Royle (须弥 大黄 Xu Mi Da Huang or 山大黄 Shan Da Huang). Basal leaves **kidney- to heart-shaped**; stem leaves **ovate**; flowers **yellowish-white.** SW Tibet (also India, Kashmir, Nepal and Pakstan). 3,500–3,600 m altitude.
8. ***R. wittrockii*** C. E. Lundstr. (天山大黄 Tian Shan Da Huang). Basal leaves **ovate to triangular-ovate**; membranous sheath (ocrea) **4–8 cm long**; 'petals' **whitish-green**; fruits with **broad wings 4–5 mm wide**. Tibet. 1,200–2,600 m altitude.

Plus:

9. ***Rumex chalepensis*** Mill. (family Polygonaceae). Rootstock blackish-brown; stems **branching**; basal leaves **much smaller and oblong (5–20 cm long; 3–8 cm wide),** mid-vein prominent, base rounded; ocrea deciduous; flowerheads lax; stamens 6; when fruiting 'petals' persistent **becoming hard, enlarged and enclosing the fruit**. Anhui, Gansu, Hebei, Henan, Hubei, Jiangsu, Shaanxi, Shandong, Shanxi, Xinjiang and Zhejiang. 100–1,500 m altitude.

Freshly cut rhizome (*R. palmatum*)

Freshly cut root (*R. tanguticum*)

Rheum palmatum, *R. tanguticum* and *R. officinale*

Polygonaceae

(PS1, 2 and 3 respectively) are described together below since they are not distinguishable macroscopically.

DRUG MORPHOLOGY

CRUDE DRUG (DRIED ROOT AND RHIZOME): Often difficult to discriminate root **[1]** from rhizome **[2]** especially in irregularly cut pieces **[3]**. Overall shape highly variable owing to irregular cutting of large rootstocks (3–17 cm long, 3–10 cm diam.); sometimes distinctly cylindrical, sub-ovoid or rhomboid; angular or sometimes with smooth rounded edges **[4]** where outer cortex evenly removed. Roughly peeled outer surface yellowish-brown to reddish-brown **[5]** with remnants of coarse outer cortex **[6]**; rootlet scars occasional; smooth-peeled surfaces with an almost reticulated or marbled appearance where creamish parenchyma tissues **[7]** alternate with reddish-brown wood rays **[8]**; scattered star-shaped vascular bundles ('star-spots') **[9]** occasional; a small man-made hole sometimes visible **[10]** where pieces have been hung to dry. Cut surfaces yellowish-brown to reddish-brown. **Pith or rhizome broad [11], with 'star-spots' scattered throughout [12] or forming an interrupted ring [13].** Wood (xylem) well-developed **[14]** with a distinct cambium ring **[15]**, clear rays **[16]**, star-spots absent; core sometimes porous and soft **[17]**. *Texture*: usually compact. *Odour:* aromatic. *Taste*: bitter and someti mes astringent; sticky when chewed. ***Good quality***: dense, heavy and oily; externally yellowish-brown; with a fresh aroma; bitter and slightly astringent.

DECOCTING PIECES-1 (UNPROCESSED, DRIED AND CUT REGULARLY SIZED PIECES OF ROOT AND RHIZOME): sharply-angled pieces displaying all or some of the features of the crude drug above (*see annotated photo).*

DECOCTING PIECES-2 (STEAM-PROCESSED DP-1): as for DP-1 but darker throughout **[18]**.

DECOCTING PIECES-3 (WINE-PROCESSED DP-1): as for DP-1 but darker throughout and wine-scented.

DECOCTING PIECES-4 (CARBONISED DP-1): as for DP-1 but most characters obscured; outer surface charred **[19]**; inner surface dark brown; often swollen **[20]**.

Yang *et al.*, (2004) claim that the presence of 'star-spots' is influenced by a range of environmental factors and is not a reliable character. Today, it is generally held that reliable differentiation of Pharmacopoeia Da Huang from unofficial substitutes requires laboratory-based methods; many (histological, chemical and DNA-based) have been published in recent years.

PROCESSING for PS1, 2 and 3

After removal of the rough outer cortex, rootstocks are cut into segments (usually transversely) or cubed Lou *et al.*, 1988. Traditionally they are hung on strings for very slow air-drying (= crude drug); then subsequently sliced into smaller pieces (= decocting pieces-1). Further processing may be performed by dry-frying with yellow rice wine to produce Jiu Da Huang 酒大黄 or stewed or steamed with yellow rice wine to produce Shou Da Huang 熟大黄 (= decocting pieces-2) or carbonised Da Huang Tan 大黄炭 (= decocting pieces-3). See *Safety* below.

OTHER COMMON NAMES

Names shared by PS1, 2 and 3: 川军 Chuan Jun; 将军 Jiang Jun; 锦纹大黄 Jin Wen Da Huang; Chinese rhubarb (SCN); Radix et Rhizoma Rhei; turkey rhubarb.

Additional names:

R. palmatum **(PS1):** 西北大黄 Xi Bei Da Huang (north-western rhubarb); 西大黄 Xi Da Huang; Ta Huang;

R. tanguticum **(PS2):** 西北大黄 Xi Bei Da Huang (north-western rhubarb); 西大黄 Xi Da Huang; 西宁Xi Ning Da Huang (from Xining district of Qinghai province).

R. officinale **(PS3):** 南大黄 Nan Da Huang (southern rhubarb); 雅黄 Ya Huang (from Ya'an district of Sichuan); Ta-Huang.

Many other trade names exist for the rootstocks of these species reflecting their wider international use (see *Kew-MPNS*).

SAFETY (PS1, 2 and 3)

Roots and rhizomes **toxic** when fresh (i.e. unprocessed); The processing methods described above are designed to reduce the concentration of toxins, notably those with purgative effects (Zhao & Xiao, 2009–2010). Processed products are recognised as official healthfoods in China (Anon., 2002a,b). Safety Classes 2b, 2c, 2d; Interaction Class: A. (Gardner & McGuffin, 2013).

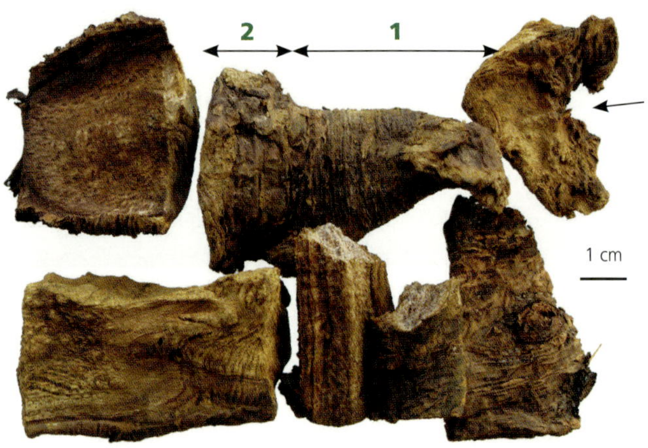

Crude drug of Da Huang showing variation in shape and size (partially peeled); source *R. officinale*

Crude drug of Da Huang (peeled rhizomes) sourced from *R. tanguticum*

Dried sliced rhizome of *R. tanguticum*

Rhizome (top) and root (below) of Da Huang, sourced from *R. officinale*

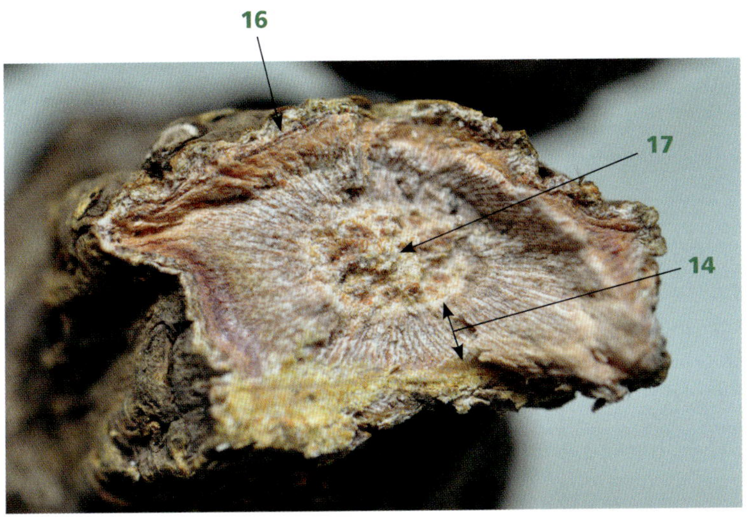

P. palmatum transverse root section

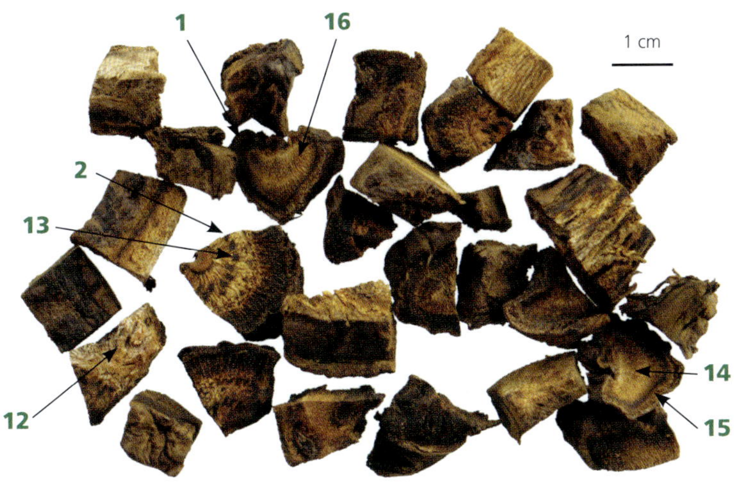

Decocting pieces-1 (unprocessed) of Da Huang, sourced from *P. palmatum*

Decocting pieces-2 (steam processed) of Da Huang sourced from *P. palmatum*

Decocting pieces-4 (carbonised) of Da Huang sourced from *R. palmatum*

Dang Gui 当归 Angelicae Sinensis Radix

Drug part: ROOT

PHARMACOPOEIA SPECIES

Angelica sinensis (Oliv.) Diels

Apiaceae (Umbelliferae)

synonym: *A. polymorpha* Maxim. var. *sinensis* Oliv.

Species of *Angelica* wild in China: **45**

PLANT DESCRIPTION

Long established as a medicinal crop in China with many cultivars grown (Zhang et al., 2012a). Their descriptions may depart slightly from the below of the wild species.
An erect perennial herb (40–100 cm tall) with a stout, many-branched and aromatic taproot (= medicinal part 'Dang Gui'). ***Stem* single,** branched and ribbed. *Leaves* alternate; **basal and lower leaves with inflated sheaths**; lowermost leaves 2–3-ternate-pinnate (10–30 cm long; 12–25 cm wide); pinnae 3–4 pairs; leaflets ovate to ovate-lance-shaped, widest below middle **(2–3.5 cm long; 0.8–2.5 cm wide)**, margins irregularly and sharply toothed; upper leaves 1-pinnate. *Inflorescences* many, in terminal or lateral compound umbels; main umbel with 10–30 rays; **bracts absent or 2,** linear; each umbellet 13–36-flowered; **bracteoles 2–4**, linear (3–5 mm long). *Each flower:* calyx absent; **petals** 5, **white, sometimes pink-purplish**, equal; styles 2. *Fruit* a dry, ellipsoid or suborbicular schizocarp (4–6 mm long; 3–4 mm diam.) splitting into 2 one-seeded carpels (mericarps); dorsal ribs slender and prominent; lateral ribs as wide as or wider than the fruit body, thin; oil ducts 1 in each furrow and 2–4 on the internal (commissural) surface. Fl. Jun–Jul, fr. Jul–Sep.

HARVESTING, SOURCING AND NATURAL RANGE

Rootstocks are harvested in late autumn entirely from medicinal crops grown on terraced mountain slopes (2,500–3,000 m altitude) mainly in Gansu province (N China) in the districts of Minxian, Wudu, Zhangxian, Chenxian and Wenxian (Zhao & Xiao, 2009–2010); Minxian is a source of *Di Dao* 'Dang Gui'. The species is also grown as a medicinal crop in Hubei, Shaanxi, Sichuan and Yunnan provinces. The plant is an endemic species of China but wild populations are likely to be acutely endangered or may longer exist (see *Conservation* below).

CONSERVATION STATUS

Endemic to China: 'endangered' (Zhang *et al.,* 2012a); 'wild populations of Dang Gui are thought to have disappeared' (Upton 2003); Not Evaluated (IUCN Red List, accessed 28/8/2016).

UNOFFICIAL SUBSTITUTE 1

Levisticum officinale W. D. J. Koch

Apiaceae

PLANT DESCRIPTION

Easily distinguished from the Pharmacopoeia species. *Main differences: Rootstock* with main root-body smaller; lateral roots stouter and longer. ***Stems* several**; **foliage and stems yellowish-green throughout.** *Leaves* with pinnae rhombic-ovate, widest at or above middle and much larger (4–11 cm long; 2–7 cm wide), margins with a few large teeth. *Inflorescence* with bracts many (7–11) of unequal length and reflexed; rays 12–20; bracteoles 8–11; **petals yellowish-green to yellow. *Fruit* longer** (5–7 cm) with lateral ribs thick-winged.

SUBSTITUTION COMMENT. Owing to its ease of cultivation and higher yields, *L. officinale* was extensively cultivated in China in the 1950s as a government approved substitute of 'Dang Gui' when the latter was in short supply. 'Dang Gui' production had recovered by the 1980s such that *L. officinale* as a substitute was banned in 1984 (Zhao & Xiao, 2009–10); it continues to be widely grown in mainland China today for medicated teas (rootstocks) and as a vegetable (leaves and shoots).

CONSERVATION STATUS

China: not native.
Global: Not Evaluated (IUCN Red List, accessed 22/8/2016).

UNOFFICIAL SUBSTITUTE 2

Angelica acutiloba (Siebold & Zucc.) Kitag.

Apiaceae

synonym: *Ligusticum acutilobum* Siebold & Zucc.

PLANT DESCRIPTION

Easily distinguished from the Pharmacopoeia species. *Main differences:* rootstock crown many-stemmed with pliable lateral roots more numerous creating a larger and more densely interwoven root mass. ***Stems* several.** *Leaves* 1–2 pinnate with leaflets usually much longer (2–9 cm long, 1–3 cm wide) and tips long-pointed, hairless, margins closely toothed, teeth sharply triangular. *Inflorescence* with **bracteoles 5–8** (linear and longer [5–15 mm]); petals white. ***Fruit* narrower** (4–5 mm long; 1–1.5 mm diam.) with narrowly winged lateral ribs.

SUBSTITUTION COMMENT. With clinical properties similar to those of the Pharmacopoeia species (Zhao & Xiao, 2009–10), roots of *A. acutiloba* are used as a local 'Dang Gui' substitute in Jilin province (NE China) and where long cultivated for this purpose. In Japan and Korea its roots are the official source of 'Riben Dang Gui' (Japanese Angelica Root) and 'Dang Gui' in their respective Pharmacopoeias (*JP*16th; *KP*, 2007). The occurrence of this species as a substitute in Western markets is unlikely.

CONSERVATION STATUS

China: not native (cultivated only.)
Global: Not Evaluated (IUCN Red List, accessed 22/8/2016).

EHTPA 4/14

TCM PROPERTIES AND USE

properties: **'sweet, pungent and warm'**
channels: **'Heart, Liver and Spleen'**
actions: **'tonifies Blood, invigorates Blood, relieves pain, moistens the intestines'**

indications: **menstrual, menopause, pregnancy, postpartum disorders; traumatic injury, sores; constipation; coughs**

Typical leaves of *A. sinensis* cultivated in Ghansu province

White-flowering form of *A. sinensis*, Gansu province

Fresh rootstock of *A. sinensis*

Levisticum officinale in cultivation, Gansu province, China

Fresh rootstock of *L. officinale*

Umbellet of *Angelica acutiloba*

Flowering plant of *A. acutiloba*

Fresh uprooted plant of *A. acutiloba*

Angelica sinensis

Apiaceae

DRUG MORPHOLOGY

CRUDE DRUG-1 (DRIED WHOLE ROOT — DANG GUI or QUAN DANG GUI): a many-branched taproot **[1]**, twisted and cylindrical (15–25 cm long) with 3–5 (sometimes more) lateral roots **[2]** and many long, pliable rootlets ('tails') **[3]**. **Externally pale brownish-yellow to dark brown** with longitudinal, often twisted wrinkles and scattered, **transverse, raised nodular lenticels [4].** Apex (root crown) usually broad (c. 1.5–5 cm diam.), blunt and rounded **[5]**, with **basal remnant or scar of a single stem** along with those of leaf bases **[6]**; transverse annulations scattered throughout (often indistinct) **[7]**. Lateral roots 10–15 cm long; tapering from 1.5 cm diam. in upper parts to 0.3 cm at tips **[8]**; externally strongly wrinkled and with occasional rootlet scars **[9]** and lenticels.

CRUDE DRUG-2 (DRIED ROOT-HEADS — GUI TOU or DANG GUI TOU): upper part of rootstock (1.5–5 cm diam.) **[5]** — see description of 'apex (root crown)' of crude drug above. Today, this trade item tends to include the root-body **[10]** too (see Dried Root-Body below); it is sometimes peeled and sulphured **[11]**.

CRUDE DRUG-3 (DRIED ROOT-BODY— GUI SHEN or DANG GUI SHEN): main root body (1.5–4 cm long; 3–5 cm diam.), irregularly shaped, typically peeled **[11]** leaving a somewhat smooth whitish to light brown surface; often with a gnarled appearance and frequently sulphured. Today, this trade item tends to include the root-head too (see Dried Root-Heads above).

CRUDE DRUG-4 (DRIED LATERAL ROOTS — GUI WEI or DANG GUI WEI): unsliced narrow branch roots ('tails') long and tapering **[12]** (10–15 cm long; (0.1)0.3–1.5 cm diam.), often heavily twisted; rootlet scars occasional **[13]**.

Crude drugs (1–4 above): *Texture*: **lateral roots pliable** and oily. ***Fracture*: mainly yellowish-white or yellowish-brown; outer root-bark yellowish-brown to brown [14]** with occasional clefts and dotted with many dark brown oil glands; cambium ring yellowish-brown **[15]**, wood paler **[16]**. *Odour*: highly aromatic. *Taste*: sweet, pungent and slightly bitter. ***Good quality*:** main roots large and thick, **externally golden- or yellowish-brown, internally yellowish-white, pliable** and strongly aromatic.

DECOCTING PIECES-1 (DRIED WHOLE ROOT SLICES — DANG GUI or QUAN DANG GUI PI): cut or pressed slices of varying size and shape (1–2 mm thick; transverse slices 1.5–5 cm diam.; longitudinal slices 10–15 cm long). Outer root-bark thick, brown or orangish-brown **[14]** with occasional clefts and dotted with dark brown oil glands; cambium ring yellowish-brown **[15]** with paler central wood **[16]**. *Odour:* highly aromatic. *Taste:* sweet, pungent and slightly bitter.

DECOCTING PIECES-2 (DRIED ROOT-HEAD SLICES — DANG GUI TOU or GUI TOU PI): (*not illustrated*) sub-rounded pieces (up to 5 cm diam.). Cut surfaces and odour etc. as for dp-1 above.

DECOCTING PIECES-3 (DRIED L ATERAL ROOT SLICES — DANG GUI WEI or GUI WEI PI): sub-rounded pieces ((0.1)0.3–1.5 cm diam.). Cut surfaces and odour etc. as for dp-1 above.

DECOCTING PIECES-4 (WINE-PROCESSED ROOT SLICES — JIU ZHI DANG GUI): slices of varying size and shape according to whether whole root, root-head or lateral roots used (see other dp-1, 2 and 3 above). Cut surfaces dark yellow or pale brownish-yellow **[17]**, with occasional darker scalded areas. *Odour:* highly aromatic, often wine-scented.

PROCESSING

Traditional processing methods of Dang Gui are varied and sometimes complex and likely to have evolved to minimise loss of therapeutic principles. Recent research confirms one of these to be ligustilide, present in the volatile oils and easily destroyed by sub-optimal processing and storage (Upton, 2003).

After removal of aerial parts and rootlets, drying takes place at low-temperatures either by i) shade-drying followed by a short period in the sun; ii) smoke-drying over a 1-3 month period (often with sulphur to prevent mould and discolouration) or iii) over-wintering (also often with sulphur). Sliced roots may undergo further processing by steeping in wine then dry-frying until completely dry.

The different rootstock parts ('head' [Tou 头], 'body' [Shen身] and 'tails' [Gui Wei 归尾, narrow branches and lower roots] are reputed to have different therapeutic profiles and are processed separately to meet the needs of different industry sectors: thinly sliced and pressed roots 1– 2 mm thick with or without 'tails' (Dang Gui or Quan Dang Gui Pi) are mainly used in TCM decoctions; sun-dried, whole, unprocessed roots often with 'tails' removed (Dang Gui or Quan Dang Gui 全当归) are used by the Chinese healthfood industry, while scraped and sulphured root-heads and -bodies (Dang Gui Tou 当归头 and Dang Gui Shen 当归身 respectively) are mainly supplied to Chinese food and Western healthfood industries (Upton, 2003).

OTHER COMMON NAMES

秦归 Qin Gui; 云归 Yun Gui; Chinese Angelica; Dong Quai (SCN); Tang-Kuei (Taiwan trade name); Radix Angelicae Sinensis; Tanggwi; Toki.

SAFETY

An official healthfood in China (Anon., 2002a,b) Safety Class: 1; Interaction Class: C (Gardner & McGuffin, 2013).

Landscape of cultivated *Angelica sinensis* extensively grown on terraces, Gansu province

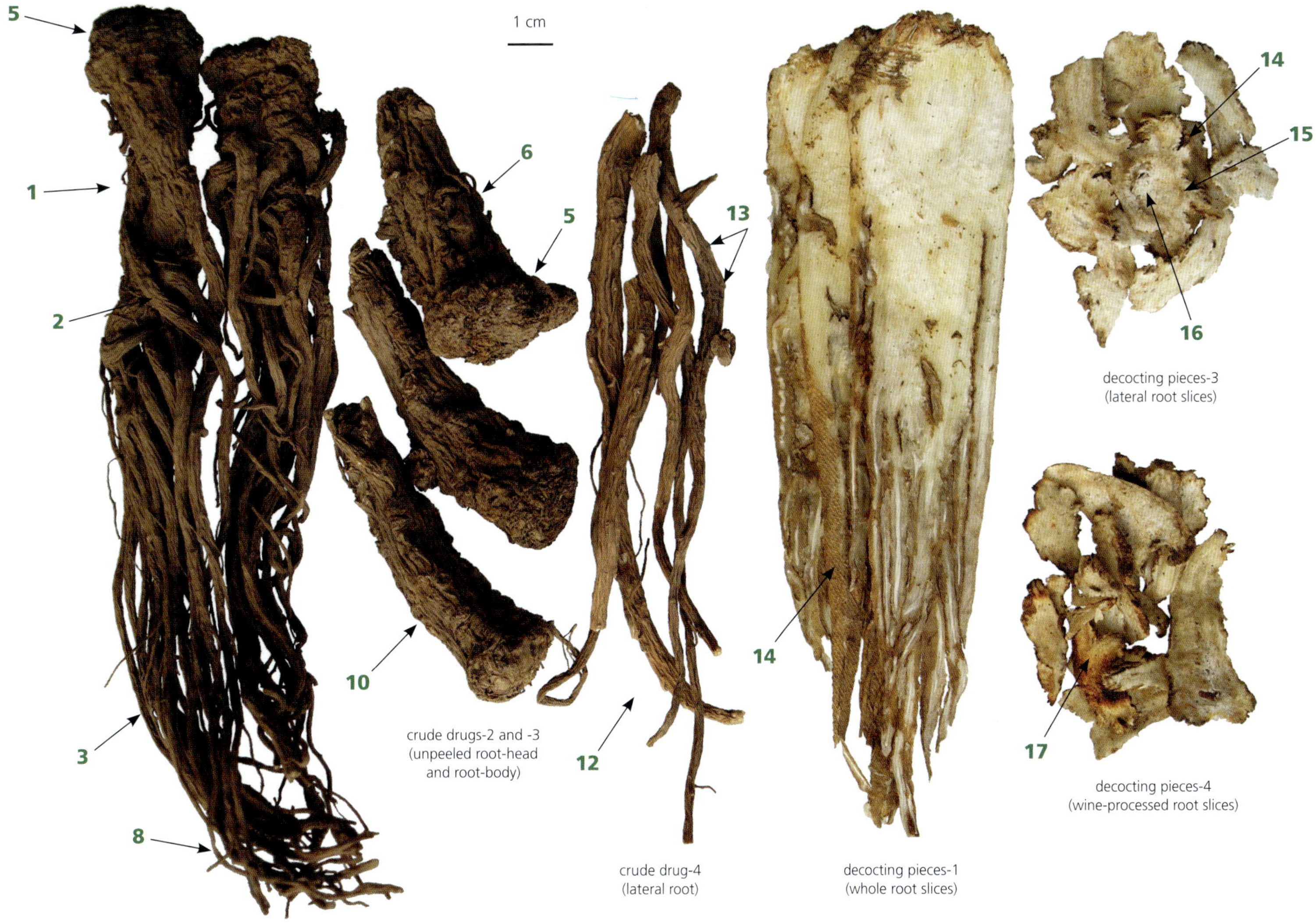

crude drug-1 (whole root)

crude drugs-2 and -3 (unpeeled root-head and root-body)

crude drug-4 (lateral root)

decocting pieces-1 (whole root slices)

decocting pieces-3 (lateral root slices)

decocting pieces-4 (wine-processed root slices)

Typical trade forms of Dang Gui sourced from *Angelica sinensis*

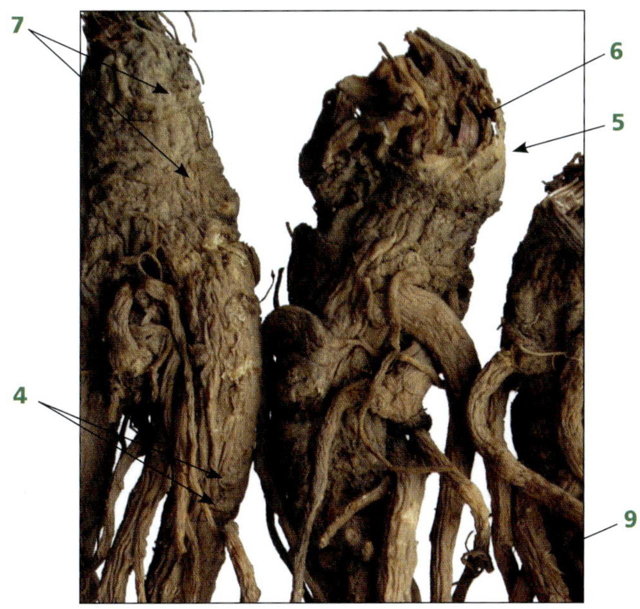

Crude drug-1 (apical parts only illustrated)

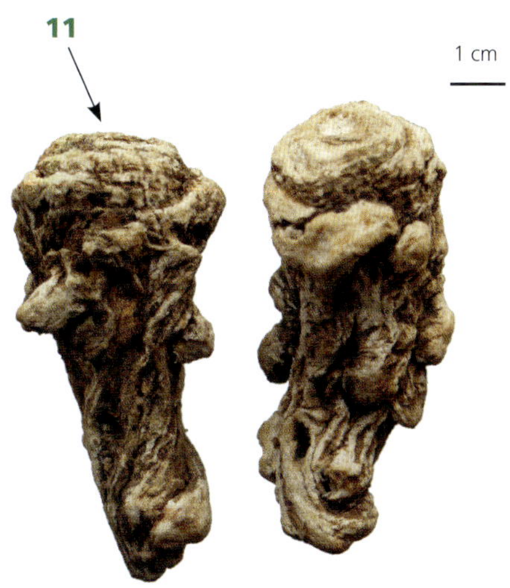

Crude drugs-2 and -3 (peeled root-head and root-body)

Levisticum officinale

Apiaceae

'DRUG SUBSTITUTE' MORPHOLOGY

Whole roots of *L. officinale* (Ou Dang Gui/European Dang Gui) are relatively easily distinguished from those of Dang Gui (*A. sinensis*); decocting pieces require laboratory-based techniques to distinguish them reliably (see Upton, 2003). *Main differences:*

WHOLE ROOTSTOCK: tends to be larger (especially longer) than that of *A. sinensis* but the main root-body smaller in relation to the size (especially length) of the lateral roots **[18]**; intertwining and pliable lateral roots **absent**. Root-heads less obviously club-shaped **[19]**. **External surface greyish-brown [20], conspicuously furrowed [21], and transverse lenticels absent**. *Fracture*: cortex often orangish-brown **[22]**, **wood (xylem) deep or pale yellow [23]**. *Taste*: **more bitter, less sweet and less oily.**

DECOCTING SLICES (OBLIQUELY-SLICED ROOT): elliptical to irregularly ovate, perimeter sometimes undulant **[24]**. Bark greyish-brown and conspicuously furrowed **[21]**. Cut surface: outer bark thin; inner bark (cortex) **[22]** creamish, lax with many irregular cavities and orange-brown oil glands, sometimes so dense as to colour the cortex orange-brown **[25]**; cambium ring sometimes distinct **[26]**; wood deep or pale yellow **[23]** with cavities and oil-glands.

OTHER COMMON NAMES

欧当归 Ou Dang Gui; European Dang Gui; Levistici Radix; lovage (SCN). Many other trade names exist for this drug reflecting its medicinal and other uses in other countries (*Kew-MPNS*).

SAFETY

Long-term oral use may increase the risk of phototoxic reactions in UV light.

Angelica acutiloba

Apiaceae

'DRUG SUBSTITUTE' MORPHOLOGY

Whole roots of *A. acutiloba* (Riben Dang Gui) are relatively easily distinguished from those of Dang Gui (*A. sinensis*); decocting pieces may require laboratory-based techniques to distinguish them reliably.

Main differences: crude drug (whole rootstock)10–25 cm long; **rootstock crown many-stemmed [27]** with **pliable lateral roots more numerous** creating a larger and **more densely interwoven** root mass **[28]**. **Externally dark brown to reddish-brown [29] and covered in numerous rootlet scars [30]**. *Fracture*: **dark brown to yellowish-brown [31]**. *Taste*: slightly sweet, with a pungent after-taste (based on *JP*16th).

OTHER COMMON NAMES:

东当归 Dong Dang Gui;日本当归 Riben Dang Gui 'Japanese Dang Gui'; Angelicae Radix; Japanese angelica root.

LESSER KNOWN UNOFFICIAL SUBSTITUTES

Rootstocks of the following species in the family Apiaceae (mostly species of *Angelica*) have also been reported to be traded locally in China as 'Dang Gui' (Bensky *et al.* 2004); Upton (2003). Their occurrence as substitutes in international trade is unlikely.

1. ***Angelica gigas*** Nakai (朝鲜当归 Chao Xian Dang Gui; Korean angelica). Native in NE China (Heilongjiang, Jilin, and Liaoning), Japan and Korean Peninsula. Leaflet bases decurrent; leaflet 'stems' obviously winged; leaflets 4–15 cm long, 1.5–5 cm wide; bracts 2, pouch-shaped and dark purple; petals dark purplish-red.
2. ***A. megaphylla*** Diels (大叶当归 A Ye Dang Gui). Endemic to Sichuan province. Leaflets 5–15 cm long; 2–10 cm wide; bracteoles absent.
3. ***A. polymorpha*** Maxim. (syn. *A. uchiyamae* Y. Yabe); C, E and NE China (also Japan and Korean Peninsula). Leaflets 3–5 cm long; 2.5–3.5 cm wide; flower stems (peduncles) and rays densely hairy; bracts absent or 1–3; rays 10–20; bracteoles 7–10 and purplish and ciliate; fruit 6–7 mm long; 3–5 mm wide.
4. ***A. valida*** Diels (金山当归 Jin Shan Dang Gui). Endemic to Chongqing province. Leaf sheaths spiny; leaves hairy along nerves; bracts 1–3 cm diam.; rays 25–55; bracteoles 8–10, 5–7 mm long; umbellules 30–40-flowered.
5. ***Ligusticum glaucescens*** Franch. ('Ye Dang Gui'). Native to Yunnan province.
6. ***Notopterygium franchetii*** H. Boissieu (syn. *Angelica tsinglingensis* K. T. Fu); Qinling Dang Gui; N and W China. Bracts 1–3; rays 10–17(–23), bracteoles 4 or 5; petals pale yellow or yellowish-green.

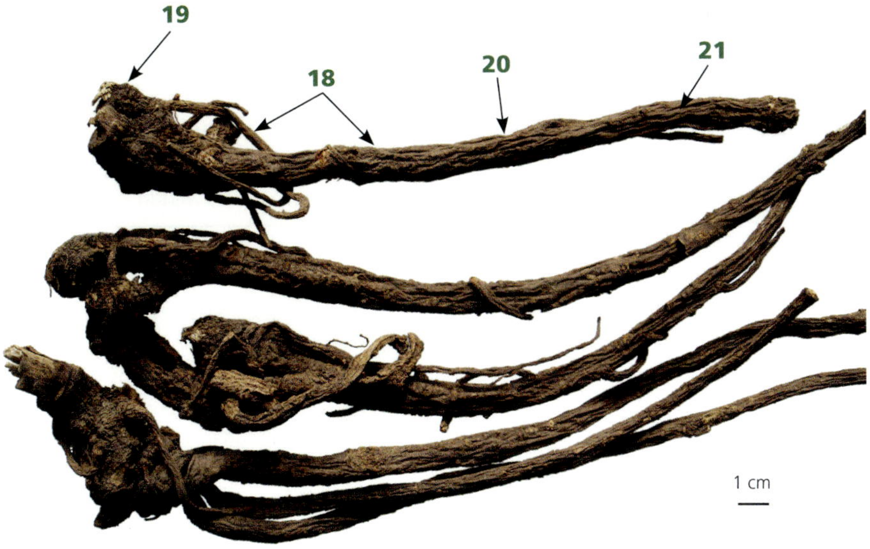

European Dang Gui (dried whole rootstocks of substitute *L. officinale*)

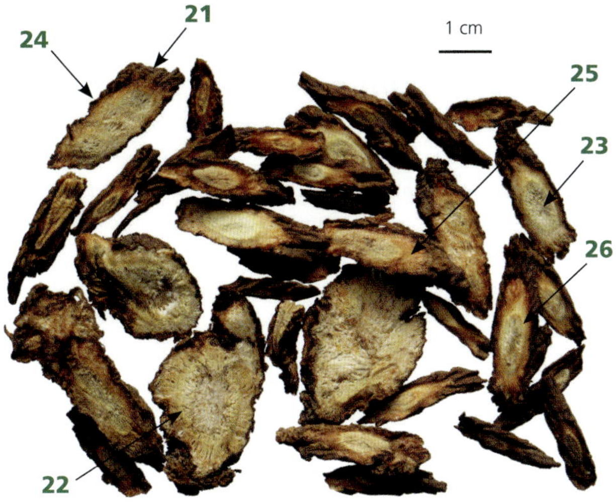

European Dang Gui (dried and sliced roots of substitute *L. officinale*)

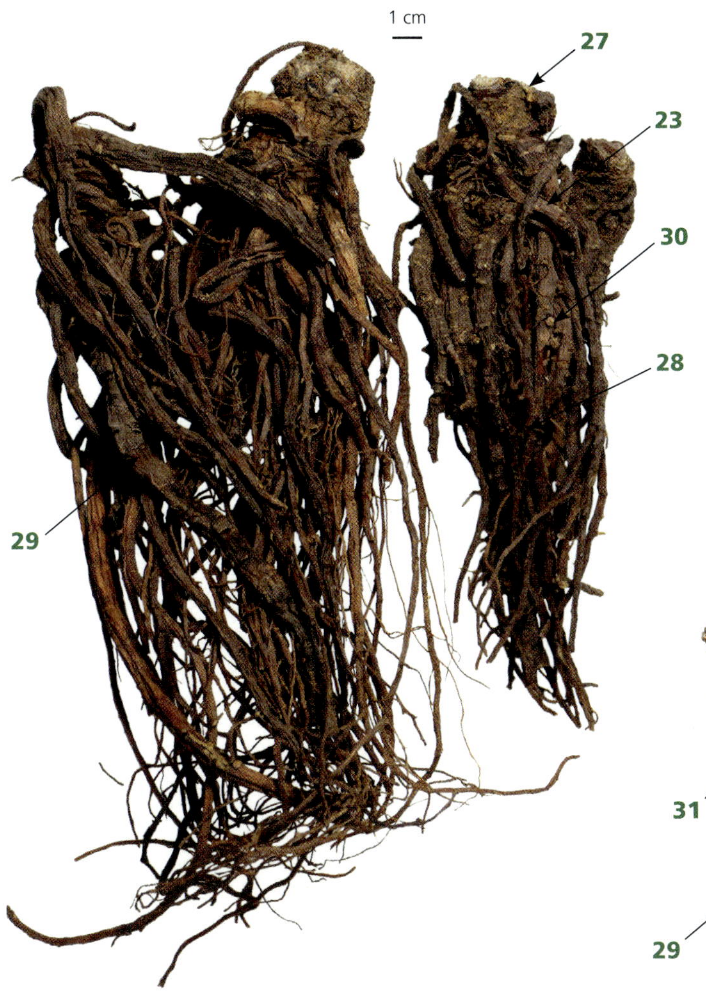

Japanese Dang Gui (whole dried rootstocks of *Angelica acutiloba*) — crude drug

Japanese Dang Gui (root-heads and root-body of *A. acutiloba*) — crude drug

Japanese Dang Gui (longitudinally sliced crude drug of root body of *A. acutiloba*)

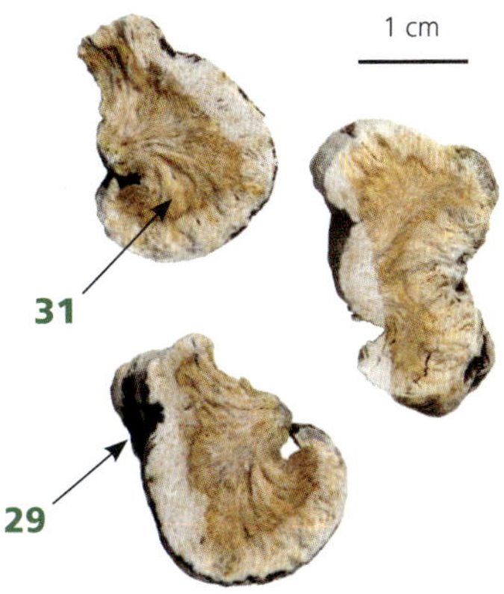

Japanese Dang Gui (transverse decocting slices of *A. acutiloba*)

Japanese Dang Gui (typical decocting pieces in Japan of *A. acutiloba*)

Dang Shen 党参

Codonopsis Radix

Drug part: ROOT

Codonopsis pilosula (Franch.) Nannf. **subsp. *pilosula***

Campanulaceae

synonym: *C. pilosula* var. *modesta* (Nannf.) L. D. Shen (CP2015 recognises this as a separate source of this drug).

taxonomic/nomenclatural note: in CP2015 as *'Codonopsis pilosula* (Franch.) Nannf.'; the type subspecies name (i.e. subsp. *pilosula*) is added above to differentiate this entity from *C. pilosula* subsp. *tangshen* below (PS2).

Species of *Codonopsis* wild in China: 40

PLANT DESCRIPTION

A trailing, perennial herb (1–2 m long) with tuberous, cylindrical and sometimes branching roots (15–35 cm long; 0.4–2.5 cm diam.). *Stems* slender, hairless, branched and twining; stipules absent. *Leaves* simple, those on main stem alternate, those on branchlets opposite; leaf stalks 0.5–2.5cm long; leaf blade ovate or narrowly ovate (1–7.3 cm long; 0.8–5 cm wide), hairless or minutely covered in stiff hairs, **leaf base rounded, straight or slightly heart-shaped**, margin minutely scalloped, tip broadly pointed or rounded. *Flowers* stalked, at end of branches, cryptic, solitary, broadly bell-shaped (1.7–2.3 cm long; 1.8–2.5 cm diam.), often nodding; sepals 5, large, green and spreading (lobes 1.4–1.8 cm long) with the **lower part of each sepal obviously fused to the base of the flower (ovary) forming a smooth green base**; petals 5, fused to form the broad flower tube, opening as 5 broadly pointed petal lobes at mouth, pale yellow or yellowish-green often purple-veined and inside often purple-spotted; stamens 5, ovary half-inferior to calyx; style solitary and 3-parted, stigma with white bristles. *Fruit* a squat dry capsule, hemispherical at base and becoming shortly conical at apex (1–2.4 cm long), **pale greenish-yellow when ripe**; sepals persistent; *seeds* many and tiny (c. 1 mm long). Fl. and fr. Jul–Oct.

HARVESTING, SOURCING AND NATURAL RANGE

Roots are harvested in autumn mostly from plants extensively and long-cultivated as medicinal crops in N, W and NE China (provinces of Gansu, Heilongjiang, Inner Mongolia, Jilin, Liaoning, Shaanxi, Shanxi and Sichuan). Many cultivars have been developed, whose names reflect their district of cultivation and hence provenance, a useful indicator of quality (see *Other Common Names*). *Di Dao*-quality 'Dang Shen' is traditionally sourced from the provinces of Gansu and Shanxi.

The wild species grows on mountain sides usually in thickets and along forest margins (900–2,900 m altitude) and has a wide distribution across temperate China (Gansu, Hebei, Heilongjiang, Henan, Inner Mongolia, Jilin, Liaoning, Ningxia, Qinghai, Shaanxi, Shandong, Shanxi and Sichuan). The species is also native to the Korean Peninsula, Mongolia and Russia (Far East).

CONSERVATION STATUS

China: Least Concern (MEPCAS, 2013).
Global: Not Evaluated (IUCN Red List, accessed 21/7/2015).

C. pilosula subsp. ***tangshen*** (Oliv.) D. Y. Hong

Campanulaceae

synonym: *C. tangshen* Oliv. (name used in CP2015).

PLANT DESCRIPTION

As for PS1 but **leaf base truncate** and sepal bases very narrowly fused to flower/fruit base, i.e. **smooth green base virtually absent**.

HARVESTING, SOURCING AND NATURAL RANGE

Harvesting is mostly from cultivated plants extensively farmed in Chongqing and Hubei provinces. The wild species is endemic to China and grows in scrub along forest margins (900–2,300 m altitude) in W and C China (Chongqing, Guizhou, Hubei, Hunan, Shaanxi and Sichuan).

CONSERVATION STATUS

Endemic to China: Least Concern (MEPCAS, 2013); Not Evaluated (IUCN Red List, accessed 21/7/2015).

UNOFFICIAL SUBSTITUTES

Many local and/or historical substitutes or adulterants have been reported; these include the tuberous roots of many other species in the genus *Codonopsis* as well as those in other genera in the same family together with members of the Apiaceae and Caryophyllaceae families. Their occurrence as substitutes in international trade, however, is unlikely since exported 'Dang Shen' is mostly sourced from large-scale medicinal crops..

TCM PROPERTIES AND USE

properties: **'sweet and neutral'**
channels: **'Spleen and Lung'**
actions: **'tonifies qi, tonifies Blood, generates fluids'**

indications: **fatigue; poor digestion; dyspnoea, chronic cough; organ prolapse**

Terraced hillsides extensively cultivated with *Codonopsis pilosula* subsp. *pilosula* in the Min Shan mountains, southern Gansu province (3,000 m)

Fresh Dang Shen hung on strings to dry, Gansu province

Underside of flower of *C. pilosula* subsp. *pilosula*

Flowering medicinal crop plants of *C. pilosula* subsp. *pilosula*, Gansu province

Unripe fruit of *C. pilosula* subsp. *pilosula*

Flowering plant of *C. pilosula* subsp. *tangshen*

Flower of *C. pilosula* subsp. *tangshen*

Almost ripe fruit of *C. pilosula* subsp. *tangshen*

C. pilosula subsp. *pilosula*

Campanulaceae

DRUG MORPHOLOGY

CRUDE DRUG (DRIED ROOT WITH ROOT-HEAD): root body sub-cylindrical **[1]** (10–35 cm long; 0.4–2.5 cm diam.), slightly curved, sometimes tapering and occasionally branching. Outer bark yellowish-brown to greyish-brown; longitudinal furrows distinct, somewhat irregular **[2]**; adventitious root scars slightly raised; lenticels large oval, transverse and scattered **[3]** (see photo of PS2); blackish resin-like patches occasional **[4]**. Distinct band of dense annulations **[5]** below the swollen root-head and often extending one-third to half of the entire root length *(absent in some cultivars)*. Root-head a slightly swollen, nodular mass of sunken stem scars and stem buds (colloquially called 'Shi Zi Pan Tou' — lion's head) **[6]**; a man-made hole through the base of the root-head sometimes visible where the root was strung for drying. *Texture*: relatively hard but pliable. *Fracture*: mostly even; cortex **[7]** greyish-white to pale brown with irregular cavities; wood (xylem) **[8]** much denser than cortex, pale creamish-yellow and occupying half the root diameter. *Odour*: distinctively aromatic. *Taste*: sweetish. ***Good quality***: thick, pliable and sweet-tasting.

DECOCTING PIECES (SLICED ROOT): mostly short cylindrical lengths or slices (0.4–2.5 cm diam.; 2–4 mm thick) **[9]**. Outer bark yellowish-brown to greyish-brown, longitudinal furrows distinct **[10]**, slightly raised adventitious root scars and large oval transverse lenticels occasionally visible **[3]** (see photo of PS2) along with narrowly spaced *transverse* annulations. Transverse outline undulating **[11]**. Cut surface with very thin outer bark closely adhering to the greyish-white to pale brown inner bark (cortex) below; cortex **[7]** often lax with many cavities; central wood **[8]** denser than cortex, pale creamish-yellow with very faint but characteristic medullary rays **[12]** (with the colloquial name 'Ju Hua Xin' — chrysanthemum centre). *Odour:* distinctively aromatic. *Taste*: sweetish.

DECOCTING PIECES (SLICED ROOT DRY-FRIED WITH RICE): as for the decocting pieces above but externally darker yellow with occasional charred spots.

PROCESSING

Roots are harvested in autumn then sun-dried (often hung on strings to dry evenly) (= crude drug); decocting pieces are made by thoroughly softening the roots with water in order to cut them into slices (> 2 mm thick) then drying again (decocting pieces-1). Further processing by dry-frying with rice may be undertaken (decocting pieces-1).

OTHER COMMON NAMES

东党 Dong Dang or 东党参 Dong Dang Shen 'eastern Codonopsis' from Heilongjiang, Jilin and Liaoning; 璐党参 Lu Dang Shen (mostly from Shanxi and Henan); 西党 Xi Dang 'western Codonopsis' from Gansu, Sichuan and Shaanxi; codonopsis (SCN); Radix Codonopsis; tangshen.

SAFETY

An official healthfood in China (Anon., 2002a,b). Safety Class: 1; Interaction Class: A (Gardner & McGuffin, 2013).

Cultivated roots of *C. pilosula* subsp. *pilosula* drying after being size-graded, Gansu province

C. pilosula subsp. *tangshen*

Campanulaceae

DRUG MORPHOLOGY

CRUDE DRUG (DRIED ROOT): almost indistinguishable from the roots of PS1. *Main differences*: Roots **often longer [13] and narrower (10–45 cm long; 0.5–2 cm wide)** and bark greyish-yellow to yellowish-brown **[14]**. *Texture*: slightly soft. ***Fracture*: fewer cavities [15]**. For the reliable differentiation from the roots of PS1 laboratory-based authentication methods are advised.

DECOCTING PIECES (SLICED ROOT): not reliably distinguished from those of PS1 using macroscopic characters alone; laboratory-based methods are advised.

PROCESSING

As for PS1.

OTHER COMMON NAMES

川参 Chuan Shen 'Sichuan-sourced Dang Shen'; 川党参 Chuan Dang Shen 'Sichuan-sourced Dang Shen'; 条党参 Tiao Dang Shen 'twiggy Codonopsis'; codonopsis (SCN); Radix Codonopsis; tangshen.

SAFETY

An official healthfood in China (Anon., 2002a,b). Safety Class: 1; Interaction Class: A (Gardner & McGuffin, 2013).

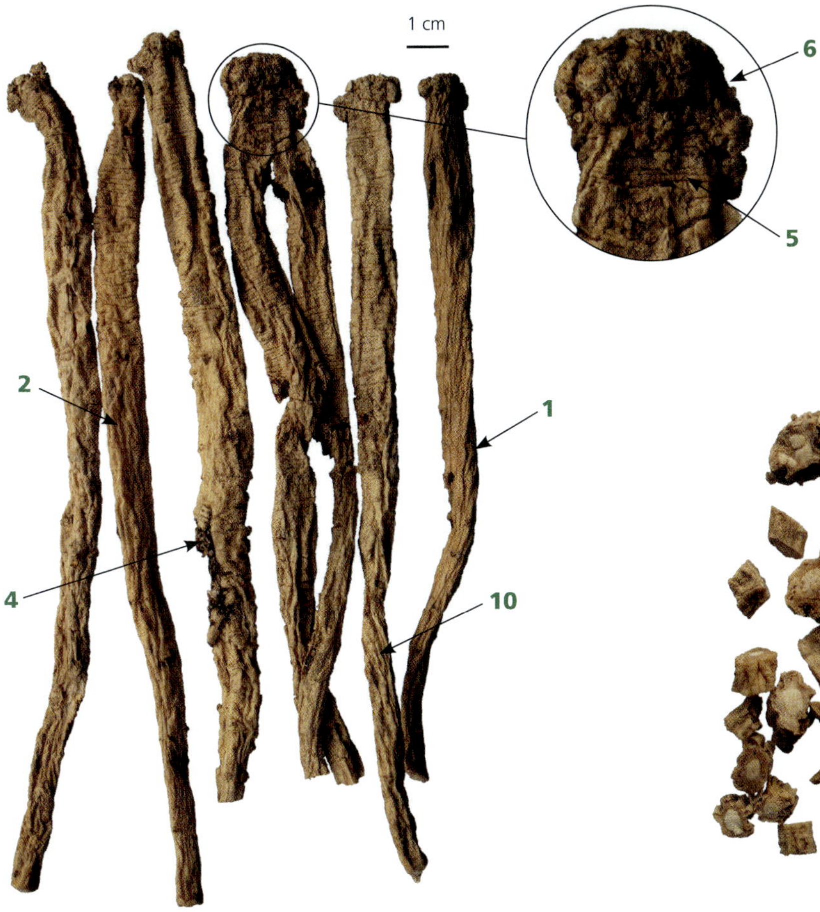

Crude drug of Dang Shen sourced from *C. pilosula* subsp. *pilosula*

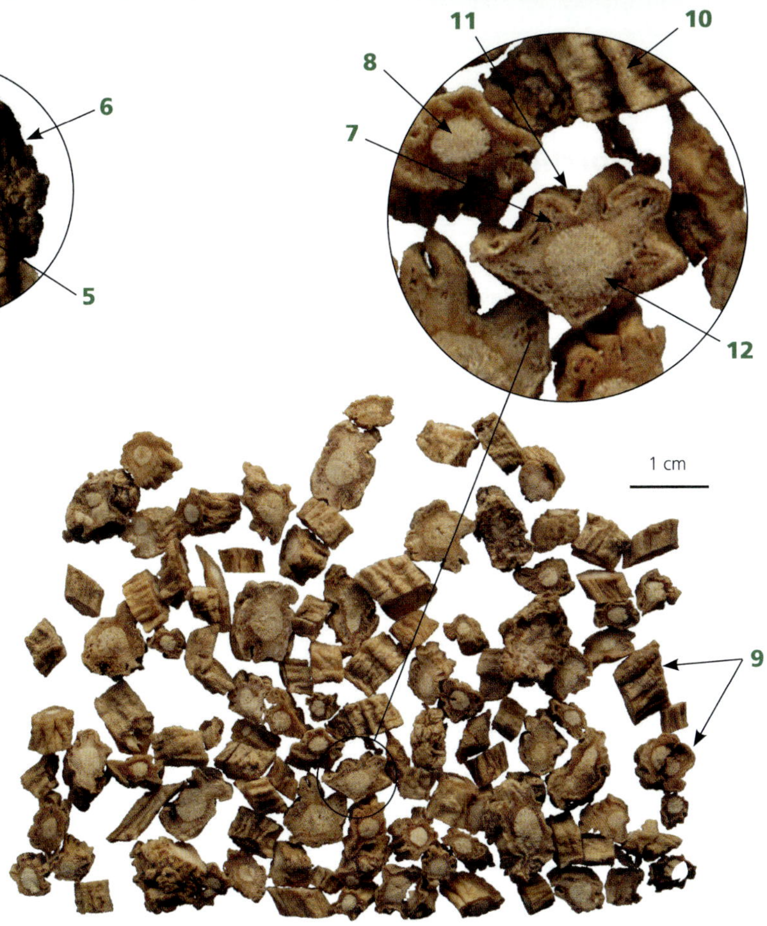

Decocting slices of Dang Shen sourced from *C. pilosula* subsp. *pilosula*

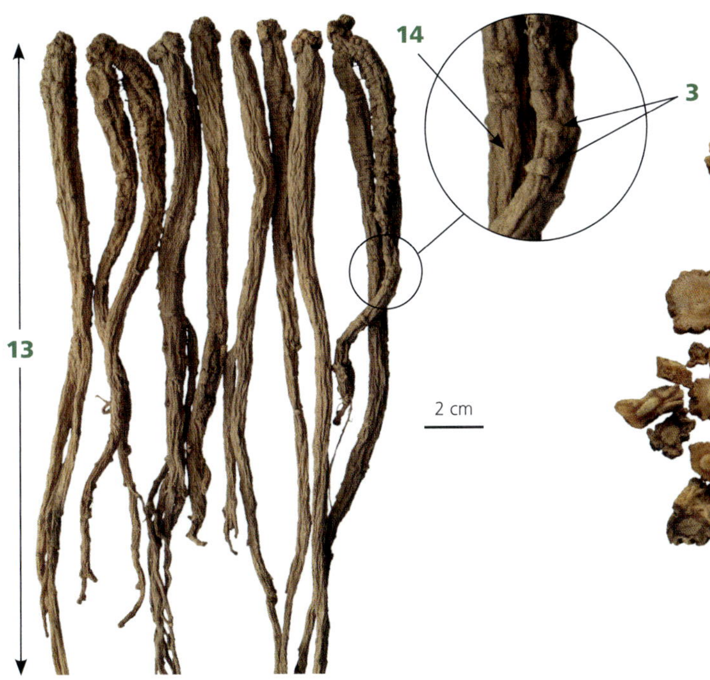

Crude drug of Dang Shen sourced from *C. pilosula* subsp. *tangshen*

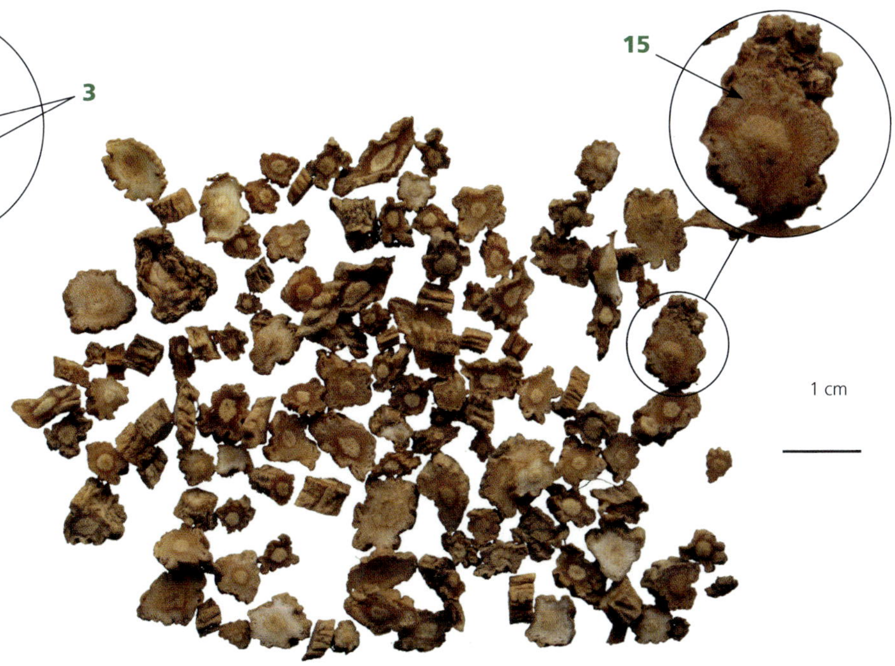

Decocting slices of Dang Shen sourced from *C. pilosula* subsp. *tangshen*

Dan Shen 丹参 Salviae Miltiorrhizae Radix et Rhizoma

Drug parts: ROOT AND RHIZOME

PHARMACOPOEIA SPECIES

Salvia miltiorrhiza Bunge

Lamiaceae

PLANT DESCRIPTION

An erect herbaceous perennial (40–80 cm tall) with a thick, red tap root. *Stems* 4-sided, much branched. ***Leaves*** **simple or more usually divided into 3–5 (rarely 7) leaflets; leaflet blades** circular to broadly lance-shaped (1.5–8 cm long; 1–4 cm wide) **with short thin hairs, especially dense on lower surfaces**, margins shallowly scalloped, tips narrowly pointed. *Flowers* in **6- to many-flowered whorls** (verticillasters) along an axillary or terminal **spike (4.5–17 cm long)**, flowers apically crowded; *bracts* lance-shaped. Each flower bisexual and laterally symmetrical; calyx bell-shaped (c. 1.1.cm long), purplish and glandular; corolla purplish-blue or white (2–2.7 cm long) and opening into a 2-lipped mouth; **upper lip distinctly hooded into a sickle-shape** and densely glandular (1.2–1.5 cm long); lower lip 3-lobed with middle lobe largest; stamens 2 (with short filaments); ovary superior; style extending beyond upper lip. *Fruit* with 4 dry, ellipsoid nutlets (c. 3.2 mm long; 1.5 mm diam.). Fl. Apr–Aug, fr. Sep–Oct.

Species of *Salvia* wild in China: 84

HARVESTING, SOURCING AND NATURAL RANGE

Mostly sourced from medicinal crop plants grown in the Chinese provinces of Anhui, Hebei, Shaanxi, Shandong, Shanxi and Sichuan, with some sites GAP-certified (recently in Shaanxi province). With 'Dan Shen' one of the earliest and most commonly used drugs in Chinese herbal medicine, its use dates back to the 1st century (Li *et al.*, 2008); many cultivars have developed during its long cultivation and traditional *Di Dao* sites are located in Sichuan province. Underground parts are harvested in spring or autumn.

Wild populations grow on wooded hillsides in China, often by streams (100–1,300 m altitude) in Anhui, Hebei, Henan, Hubei, Hunan, Jiangsu, Shaanxi, Shandong, Shanxi and Zhejiang [also native in Japan].

CONSERVATION STATUS

China: Least Concern MEPCAS, 2013 as var. *charbonnelii* (H. Lév.) C. Y. Wu and var. *miltiorrhiza*.
Global: Not Evaluated (IUCN Red List, accessed 19/8/2016).

UNOFFICIAL SUBSTITUTE 1

S. bowleyana Dunn

Lamiaceae

PLANT DESCRIPTION

Closely related and easily confured with the Pharmacopoeia species. *Main differnces*: Stems stout, growing slightly taller (up 1 m). Leaves with 5 (rarely 7) leaflets; **hairless except along the veins of the leaflet undersides; flower spikes longer (14–30 cm); calyx** 8–10 mm long; flower tube only slightly extending beyond the calyx; upper corolla lip slightly curved.

SUBSTITUTION COMMENT. A widely cultivated local substitute grown in south China, hence its common name 'Nan Dan Shen' (southern Dan Shen), Li Min-Hui *et al.*, 2008. The wild species is endemic to the S and SE China provinces of Fujian, Guangdong, Guangxi, Hunan, Jiangxi and Zhejiang.

CONSERVATION STATUS

Endemic to China: Least Concern (MEPCAS, 2013, as varieties *bowleyana* and *subpinnata*; Not Evaluated (IUCN Red List, accessed 19/8/2016).

UNOFFICIAL SUBSTITUTE 2

S. przewalskii Maxim.

Lamiaceae

PLANT DESCRIPTION

Easily distinguished from the Pharmacopoeia species. *Main differnces*: **Leaves simple (with triangular or rounded basal lobes)** (5–11 cm long; 3–7 cm wide); flowers purplish-red to reddish-brown or white in **2-4-flowered** widely spaced whorls along a flowering spike (8–20 cm long); calyx reddish-brown; **corolla tube mostly straight** and extending well beyond calyx with the 2 lips hardly dilated and **upper lip only very shallowly hooded**.

SUBSTITUTION COMMENT. A widely cultivated local substitute grown in NW China.
The species is endemic to this region, the provinces of Gansu, Hubei, Sichuan, Tibet and Yunnan, the home of the wild species which grows in margins of mountain forests, in thickets and along streams (1,100–4,000 m altitude).

CONSERVATION STATUS

Endemic to China: Least Concern (MEPCAS, 2013, as the varieties *alba, glabrescens, mandarinorum* and *przewalskii*); Not Evaluated (IUCN Red List, accessed 19/8/2016).

EHTPA 4/14
GAP

TCM PROPERTIES AND USE

properties: **'bitter and cool'**
channels: **'Heart and Liver'**
actions: **'invigorates Blood, dispels stasis, cools Blood, tonifies Blood, calms the shen'**

indications: **heart disease, stroke; poor circulation; menstrual conditions; irritability, insomnia, palpitations; sores and abscesses**

Flowering plant of *Salvia miltiorrhiza*

S. miltiorrhiza with 3–5 (rarely 7) parted leaves

Whorls of *S. miltiorrhiza* spike

S. bowleyana, Jiangxi province, 5–7 parted leaves

S. bowleyana leaf

Flowering spike of *S. bowleyana*

Medicinally cultivated plants of *S. przewalskii*

Salvia miltiorrhiza — Lamiaceae

DRUG MORPHOLOGY

CRUDE DRUG (DRIED ROOT): cylindrical, slightly curved and somewhat twisted **[1]** (10–20 cm long; 0.3–1 cm diam.). Apically, many roots often still attached to the darker rhizome **[2]**, a hard, short dense mass with remnants of multiple stem bases **[3]**. *Root*: outer bark thin, rough, brownish-red or dark brownish-red **[4]** (becoming purplish-brown with age) and irregularly furrowed longitudinally **[5]**; not exfoliating easily (unless old). *Root texture*: hard and brittle. *Root fracture*: either lax with clefts **[6]** or dense and solid **[7]**; cortex brownish-red **[8]**; wood (xylem) greyish-yellow or purplish-brown with yellowish-white radial vessels **[9]**. *Odour*: indistinct. *Taste*: slightly bitter and astringent.

Long established cultivars: usually stouter and broader (0.5–1.5 cm diam.); externally reddish-brown, longitudinally furrowed, thin outer bark adheres closely to tissues below and is difficult to remove. *Texture*: dense. *Fracture*: somewhat even and slightly horn-like. ***Good quality*:** unbroken roots, large, purplish-red without rootstock heads (rhizomes).

DECOCTING PIECES (ROOT SLICES OR LENGTHS): as for the crude drug but cut into short pieces of varying length and diameter or cut into sub-rounded (transverse) or oblique (diagonal) slices **[10]** (2–4 mm thick) or longitudinal ones of varying length. Cut surfaces somewhat even and solid **[7]** or with clefts **[6]** and sometimes horn-like (slightly translucent); cortex brownish-red **[8]** wood (xylem) greyish-yellow or purplish-brown with yellowish-white radiating vessels **[9]**. *Odour*: indistinct. *Taste*: slightly bitter and astringent.

PROCESSING

Rootstocks are dried (= crude drug), stem remnants removed and discarded, then rootstocks softened briefly before being cut into transversel sections (2–4 mm thick) or sometimes as in S China longitudinal sectrions (= decocting pieces).

OTHER COMMON NAMES

Chinese sage; Chinese salvia (SCN); danshen root (CP2010); Radix et Rhizoma Salviae Miltiorrhizae; red-root sage.

SAFETY

An official healthfood in China (Anon., 2002a,b). Safety Class: 2b; **Interaction Class: C**; a number of precautions, side effects and pharmacological considerations reported (Lou, 1995–1996; Bensky *et al*., 2004; Gardner & McGuffin, 2013).

S. bowleyana — Lamiaceae

'DRUG SUBSTITUTE' MORPHOLOGY

Cannot be reliably distinguished from the drug of the Pharmacopoeia species using macroscopic characters. Laboratory-based authentication methods are required and for which there is extensive literature (e.g. Kum *et al*., 2016).

OTHER COMMON NAMES

丹参 Dan Shen; 南丹参 Nan Dan Shen.

S. przewalskii — Lamiaceae

'DRUG SUBSTITUTE' MORPHOLOGY

Usually wider than the drug of the Pharmacopoeia species and root surface is purplish; For robust differentiation, however, laboratory-based authentication are recommended (see above).

OTHER COMMON NAMES

丹参 Dan Shen; 甘西鼠尾草 Gan Xi Shu Wei Cao.

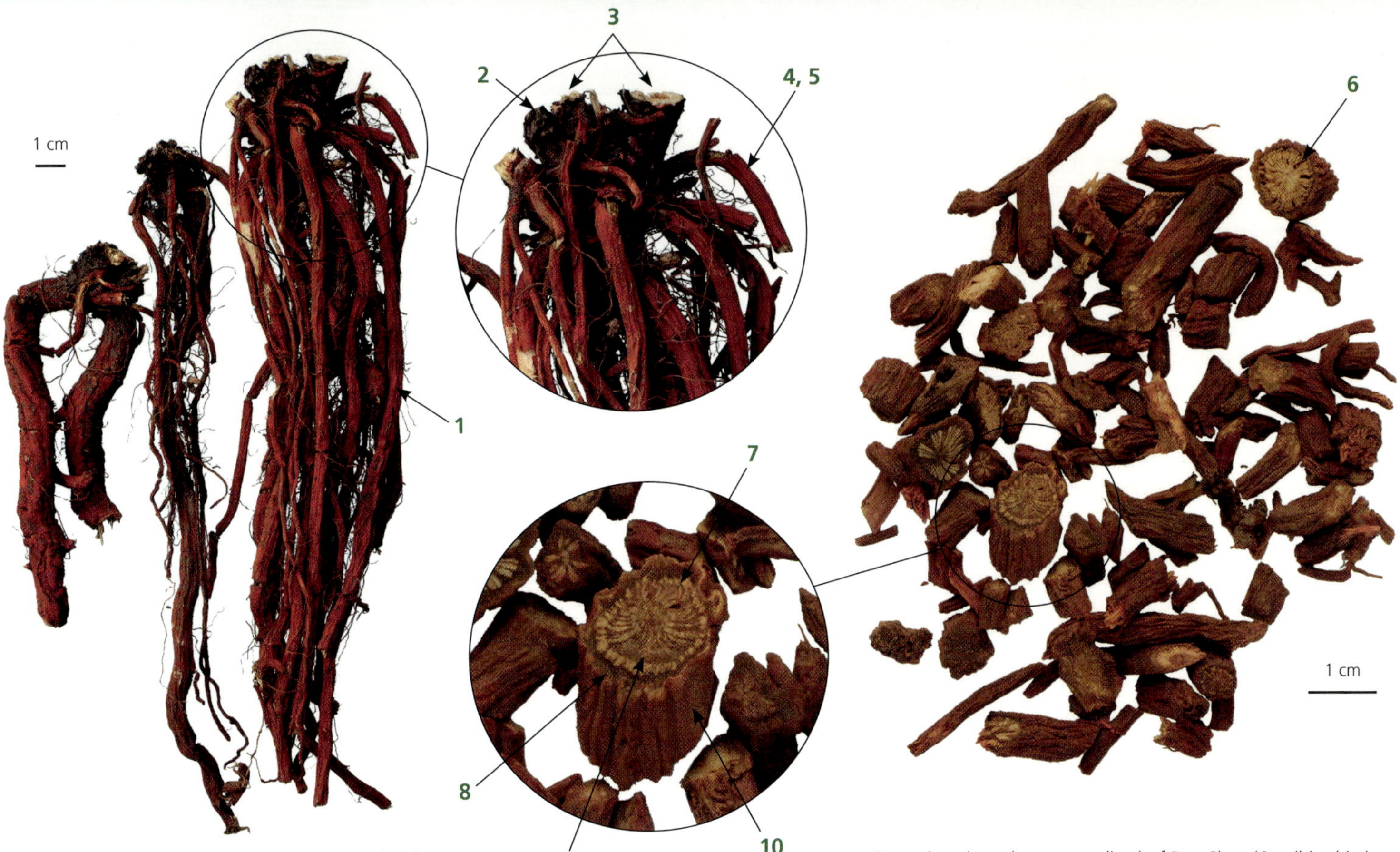

Crude drug of Dan Shen (*Salvia miltiorrhiza*)

Decocting pieces (transverse slices) of Dan Shen (*S. miltiorrhiza*)

OTHER UNOFFICIAL SUBSTITUTES

Dan Shen's long history as one of China's most popular medicinal herbs (see *Harvesting, Sourcing .....)* has seen the roots of many other *Salvia* species used and often cultivated as local or regional unofficial substitutes with Li *et al.* (2008) list over 18 of China's 84 native *Salvia* species. Clinical preference, yield, local availabilty and, for some, mis-identification are some of the reasons outlined (Kasimu *et al.*, 1998; Li *et al.*, 2008; Xiao *et al.*, 1997). The more widely documented substitutes are listed below; their occurrence in international markets, however, is unclear despite the huge volume of official 'Dan Shen' under cultivation in China. Note that the identification of living plants, even when flowers are available, requires considerable botanical expertise while dried roots (when traded as commercial 'Dan Shen') require laboratory-based methods for their robust identification (e.g. Kum *et al.*, 2016).

1. ***Salvia deserta*** Schangin (新疆鼠尾草 Xinjiang Shu Wei Cao). Native in Xinjiang provinces (NW China) where it is used to bulk up the official drug and where the latter is in short supply. **Leaves simple** (not divided into leaflets), ovate to lance-shaped (4–9 cm long; 1.5–5 cm); flowers in 4–6-flowered whorls along elongated spikes with **distinctive purplish-red bracts** (4–6 mm long); flowers bluish-purple to purple with yellowish-brown glandular hairs and **flower tube not extending beyond calyx.**
2. ***S. digitaloides*** Diels (毛地黄鼠尾草 Mao Di Huang Shu Wei Cao). Endemic to Guizhou, Sichuan and Yunnan provinces. **Leaves simple** (not divided into leaflets) and usually basal, oblong-elliptic (3.5–11 cm long; 2–5 cm wide), **tips broad or rounded** (not pointed); **flowers yellow** in 4–6-flowered whorls.
3. ***S. paramiltiorrhiza*** H. W. Li & X. L. Huang (拟丹参 Ni Dan Shen). Endemic to Anhui and Hebei provinces. Leaves usually divided into 5 leaflets; flowers in 4–6-flowered whorls and widely spaced; calyx tubular with a densely white hairy throat; flowers yellow to purplish-red, flower tube slightly extending beyond calyx, upper lip densely glandular and folded into a sickle-shape.
4. ***S. plectranthoides*** Griff. (长冠鼠尾草 Chang Guan Shu Wei Cao). Native in Guangxi, Guizhou, Hubei, Shaanxi, Sichuan and Yunnan provinces. Stems 1 to few; leaves divided into 3, 5 or 7 leaflets, **undersides purplish**; calyx bell-shaped to tubular and covered in yellowish hairy glands; flowers red to purplish or purplish-blue, rarely white and usually in 5–7-flowered whorls along elongated flowering spikes; **flower tube 2–3 times longer than calyx, upper lip straight, lower lip somewhat longer than upper lip.**
5. ***S. sinica*** Migo (拟丹参 Zhe Wan Dan Shen). Endemic to Anhui and Zhejiang provinces. 50–100 cm tall, with a prostrate rhizome; leaves divided into 7 (rarely 3 or 5) **narrowly ovate leaflets**; flowers in 5–12-flowered whorls; upper lip sickle-shaped and tawny, purplish or purple on upper lip; style extending far beyond flower mouth.
6. ***S. trijuga*** Diels (三叶鼠尾草 San Ye Shu Wei Cao). Endemic to Sichuan, Tibet and Yunnan provinces. Leaves simple or divided into 3 (rarely 5) leaflets; flowers bluish-purple **with yellow spots in 2-flowered whorls;** flower tube abruptly dilated with upper lip sickle-shaped.
7. ***S. yunnanensis*** C. H. Wright (云南鼠尾草 Yun Nan Shu Wei Cao). Endemic to Guizhou, Sichuan and Yunnan provinces. Roots tuberous, 2 or 3, bright red externally; leaves mostly basal, simple or divided into 3 or more **leaflets with broad or rounded tips;** flowers bluish-purple **in 4–6-flowered whorls**; upper lip sickle-shaped and middle lobe of lower lip with undulating margin.

Di Huang 地黄

Rehmanniae Radix

Drug part: FRESH or DRIED UNPROCESSED

Rehmannia glutinosa (Gaertn.) DC.

Orobanchaceae

synonyms: *Digitalis glutinosa* Gaertn.; *R. chinensis* Libosch. ex Fisch. & C. A. Mey.
taxonomic/nomenclatural note: in CP2015 as '*Rehmannia glutinosa* Libosch.'

Species of *Rehmannia* wild in China: 6

PLANT DESCRIPTION

A short perennial herb (10–30 cm tall) densely covered in short glandular hairs. Underground parts a cluster of fleshy rhizomes (8–24 cm long; 2–9 cm diam.). *Basal leaves* forming a rosette with leaf blades ovate to narrowly elliptic (2–13 cm long; 1–6 cm wide), bases tapering, margins with rounded- or bluntly acute teeth; stem leaves alternate and stalked, gradually or abruptly reducing in size up stem. *Inflorescence* terminal and erect, on a slender stem arising from centre of basal rosette. Flowers stalked and laterally symmetrical; calyx cup-shaped (1–1.5 cm long) with 5 apical teeth and hairy throughout; corolla tubular (3–4.5 cm long), dull reddish or yellowish-purple often with darker veining, mouth with 5 spreading lobes with tips broad and often minutely notched (lobes c. 5–7 mm long, 4–10 mm wide); stamens 4; ovary superior; style with 2-parted stigma. *Fruit* a dry capsule with persistent calyx. *Seeds* minute. Fl. and fr. Apr–Jul.

HARVESTING, SOURCING AND NATURAL RANGE

Tubers are harvested in autumn from medicinal crops grown in Hunan, Shaanxi, Shanxi and Shandong provinces. The species has a long history of medicinal cultivation and the cultivars 'Jin Zhuang Yuan' and 'Xiao Hei Ying' have been bred for their superior agronomic traits (Zhang *et al.*, 2010). Henan province is the source of traditional *Di Dao*-quality drug.

The wild species is endemic to China and grows in mountain grasslands and disturbed places (0–1,100 m altitude) in the provinces of Gansu, Hebei, Henan, Hubei, Inner Mongolia, Jiangsu, Liaoning, Shaanxi, Shandong and Shanxi.

CONSERVATION STATUS

Endemic to China: Least Concern (authors' pers. obs. 2009); Not Evaluated (IUCN Red List, accessed 16/8/2016).

Flowering plant of *Rehmannia glutinosa* in the wild

Flowering plant of *R. glutinosa* grown as a medicinal crop

Whole plant of *R. glutinosa* with fresh rhizomes

TCM PROPERTIES AND USE

properties: **'cold', 'sweet' and 'bitter'**
channels: **'Heart', 'Liver' and 'Kidney'**
actions: **'nourishes Blood, tonifies Liver and Kidney, tonifies jing, relieves cough'**
indications: **fever, dry mouth, delirium; haematemesis, haematuria, epistaxis, metrorrhagia; itchy skin, rash; constipation; dysuria**

DRUG MORPHOLOGY

CRUDE DRUG-1 (FRESH RHIZOME 'XIAN DI HUANG'): irregularly swollen **[1]** and often tapering at both ends **[2]** (8–24 cm long; 2–9 cm diam.). Outer skin pale pinkish-yellow with distinctive transverse furrows **[3]**, short transverse lenticels **[4]** and bud or root scars occasional **[5]**. *Texture*: hard, fleshy and easily broken. *Fracture*: cortex pale yellowish-white **[6]** scattered with orangish-red cells **[7]**; central pith white **[8]**. *Odour*: indistinct. *Taste*: bitter and slightly sweet.

CRUDE DRUG-2 (DRIED UNPREPARED ROOT TUBER 'SHENG DI HUANG'): irregular masses or oblong pieces, often centrally swollen **[9]** and sometimes tapering at both ends (6–12 cm long; 3–6 cm diam.); twisted or somewhat small and compressed. Externally greyish-brown or blackish-brown and shrunken and with pale transverse wrinkles scattered **[10]** and bud or root scars occasional **[11]**. *Texture:* heavy, soft and pliable but difficult to break. *Fracture:* jet-black, brownish-black or greyish-black **[12]**. *Odour*: indistinct. *Taste*: slightly sweet. ***Good quality*:** large, thick, heavy and pliable; interior black and lustrous.

DECOCTING PIECES-1 AND -2 (FRESH OR DRIED SLICED RHIZOME) ***Fresh***: irregularly shaped or sub-rounded slices (2–9 cm diam.; 2–4 mm thick); externally pale pinkish-yellow with thin outer skin; cut surface with pale yellowish-white cortex, radial vessels and scattered orangish-red oil cells; central pith white. *Odour*: indistinct. *Taste:* bitter and somewhat sweet. ***Dried***: irregularly shaped or sub-rounded slices (2–4 mm thick) **[13]**; externally brownish-black or brownish-grey with irregular folds **[14]**; cut surfaces jet black, brownish-black or greyish-black **[15]**. *Odour*: indistinct. *Taste*: somewhat sweet.

PROCESSING

Three main trade forms exist: fresh rhizomes (Xian Di Huang 鲜地黄) with roots and shrivelled aerial parts removed (= fresh crude drug-1); dried rhizomes (Sheng Di Huang 生地黄, Sheng Di 生地 or Gan Di Huang 干地黄 'dried Di Huang') are baked slowly until almost dry (= dried crude drug-2) and prepared rhizomes (Shu Di Huang 熟地黄); for latter see separate drug account page 112.

The crude drugs are cut into 2–4 mm thick slices (= decocting pieces-1 and -2).

OTHER COMMON NAMES

Additional names to those under *Processing* above: Radix Rehmanniae; Chinese foxglove root; rehmannia root (CP2010); rehmannia (SCN); unprocessed Di Huang.

SAFETY

An official Chinese healthfood in China (Anon., 2002a,b). Safety Class: 2d and Interaction Class: A (refers to fresh and dried unprepared trade forms; Gardner & McGuffin, 2013).

Crude drug-1: fresh rhizomes of Di Huang (Xian Di Huang) sourced from *Rehmannia glutinosa*

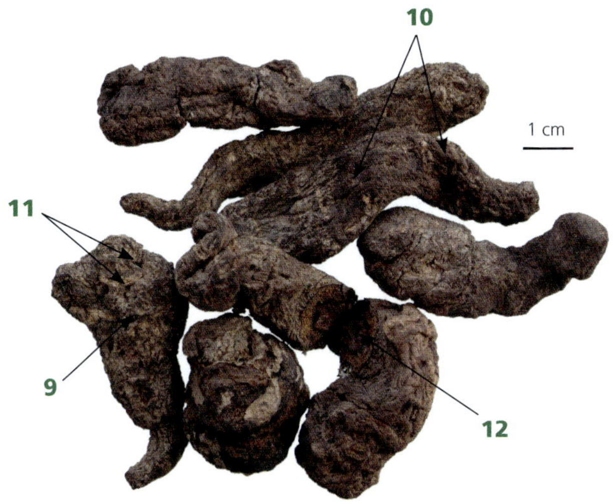

Crude drug-2: dried rhizomes of Di Huang (Sheng Di Huang) sourced from *R. glutinosa*

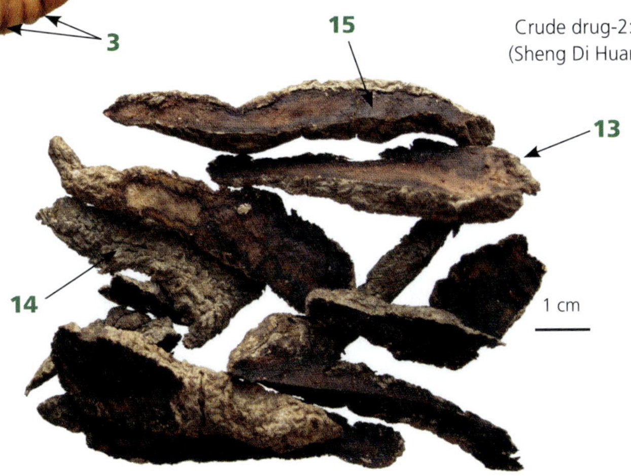

Decocting pieces-2 of Sheng Di Huang (dried rhizomes of *R. glutinosa*)

Di Huang – Shu Di Huang 熟地黄 Rehmanniae Radix Praeparata

Drug part: PROCESSED RHIZOMES

Rehmannia glutinosa (Gaertn.) DC.

Orobanchaceae

synonyms: *Digitalis glutinosa* Gaertn.; *R. chinensis* Libosch. ex Fisch. & C. A. Mey.
taxonomic/nomenclatural note: in CP2015 as '*Rehmannia glutinosa* Libosch.'

Species of *Rehmannia* wild in China: 6

PLANT DESCRIPTION

See 'Di Huang' (Rehmanniae Radix) page 110.

HARVESTING, SOURCING AND NATURAL RANGE

See 'Di Huang' (Rehmanniae Radix) page 110.

CONSERVATION STATUS

See 'Di Huang' (Rehmanniae Radix) page 110.

Medicinal crop of *Rehmannia glutinosa*

Medicinally cultivated of *R. glutinosa*

Freshly harvested root tubers of *R. glutinosa*

Top of inflorescence of wild plant of *R. glutinosa*

EHTPA 4/14
GAP

TCM PROPERTIES AND USE

***properties*: 'sweet and slightly warm'**
***channels*: 'Heart, Liver and Kidney'**
***actions*: 'tonifies Blood, tonifies Liver and Kidneys, tonifies jing, relieves cough'**

***indications*: dizziness, palpitations, insomnia, fatigue; menstrual disorders, infertility; weakness of back and knees; night sweats; cough**

DRUG MORPHOLOGY

PROCESSED RHIZOMES (WHOLE OR SLICED 'SHU DI HUANG'): shrivelled rhizomes of varying size and shape **[1]** (sometimes sliced **[2]**). Externally jet-black lustrous and slightly sticky **[3]**, internally less sticky. Jet-black and lustrous throughout; internally less sticky. *Texture:* soft and pliable but difficult to break. *Fracture:* jet-black and lustrous. *Odour:* indistinct. *Taste:* sweet. ***Good quality*:** rhizomes thick, soft and sweet; black throughout.

PROCESSING

The source material of Shu Di Huang is Sheng Di Huang (dried rhizomes of *R. glutinosa*), see Di Huang (page 110). Shu Di Huang can be prepared in two ways, processing with alcohol or steaming. For the former, Sheng Di Huang is stewed or boiled with yellow rice wine until totally absorbed, then dried until the outer skin is totally dry (= processed drug). It is then cut into thick pieces and dried again (= decocting pieces-1). Alternatively Sheng Di Huang is steamed until black and lustrous, then sun-dried until semi-dry and sliced into thick pieces and dried again (= decocting pieces-2).

OTHER COMMON NAMES

熟地 Shu Di; Chinese foxglove root; prepared rehmannia root (CP2010); Radix Rehmanniae Praeparata; rehmannia (SCN).

SAFETY

An official healthfood in China (Anon., 2002a,b). Safety Class: 1 and Interaction Class: A (Gardner & McGuffin, 2013).

Processed drug of 'Shu Di Huang' (alcohol-processed) sourced from *R. glutinosa*

1 cm

Decocting pieces of 'Shu Di Huang'

Di Yu 地榆

Sanguisorbae Radix

Drug part: ROOT

Sanguisorba officinalis L. var. ***officinalis***

Rosaceae

taxonomic/nomenclatural note: in CP2015 as *Sanguisorba officinalis* L.; the type variety (i.e. var. *officinalis*) is used here to differentiate this entity from *S. officinalis* var. *longifolia* below (PS2).

Species of *Sanguisorba* wild in China: 7

PLANT DESCRIPTION

A perennial herb (30–120 cm tall). *Roots* stout, brown or purplish-brown, usually fusiform, rarely cylindrical, with a yellowish-white or purple cross section. *Stems* erect and angular; base of leaf stems often sparsely glandular. *Leaves* alternate and pinnate. Basal leaves: stipules brown, membranous and sheathing; blade with 4–6 pairs of stalked leaflets (1–7 cm long; 0.5–3 cm wide overall); leaflet blades **ovate, oblong-ovate** and hairless with **somewhat heart-shaped bases,** margins coarsely or sharply toothed. Stem leaflets shortly stalked or stalkless; stipel (at base of each leaflet stalk) large and leaflet-like; blades oblong to oblong-lance-shaped; margin sharply toothed. *Flowers* in dense, terminal and erect, ovate or cylindrical heads usually 1–6 × 0.5–1 cm, with bracts and bracteoles; flowers developing from tip to base. Flowers tiny [hl]; petal-like sepals 4, purple, red or tinged purple, often extending beyond sepals; petals absent; style feathery; stamens 4. *Fruiting head* (hypanthium) hard, ovoid and longitudinally 4-ribbed, containing many achenes (single-seeded fruits). Fl. and fr. Jul–Oct.

HARVESTING, SOURCING AND NATURAL RANGE

Roots are harvested from wild populations in spring or autumn. The species grows in forest margins, scrub and grassland (up to 3,000 m altitude) in the Chinese provinces of Anhui, Gansu, Guangxi, Guizhou, Hebei, Heilongjiang, Henan, Hubei, Hunan, Inner Mongolia, Jiangsu, Jiangxi, Jilin, Liaoning, Qinghai, Shaanxi, Shandong, Shanxi, Sichuan, Tibet, Xinjiang, Yunnan and Zhejiang provinces [also wild in other parts of Asia and Europe].

CONSERVATION STATUS

China: Least Concern (MEPCAS, 2013).
Global: Not Evaluated (IUCN Red List, accessed 19/5/2015).

S. officinalis var. ***longifolia*** (Bertol.) T. T. Yu & C. L. Li

Rosaceae

synonym: *S. longifolia* Bertol.

PLANT DESCRIPTION

Similar to PS1 but l**eaflets of basal leaves much narrower and oblong to lance-shaped**; leaflet bases **somewhat heart-shaped or rounded to broadly wedge-shaped.** Fl. and fr. Aug–Nov.

HARVESTING, SOURCING AND NATURAL RANGE

Roots are harvested from wild populations in spring or autumn. The species grows in sparse forests, scrub, damp meadows and stream banks (100–3,000 m. altitude) in the Chinese provinces of Anhui, Gansu, Guangdong, Guangxi, Guizhou, Hebei, Heilongjiang, Henan, Hubei, Hunan, Jiangsu, Jiangxi, Liaoning, Shandong, Shanxi, Sichuan, Taiwan, Yunnan and Zhejiang [also wild in India, the Korean Peninsula, Mongolia and Russia].

CONSERVATION STATUS

China: Least Concern (MEPCAS, 2013).
Global: Not Evaluated (IUCN Red List, accessed 19/5/2015).

TCM PROPERTIES AND USE

properties: 'bitter, sour and cool'
channels: 'Liver, Large Intestine and Stomach'
actions: 'cools Blood, stops bleeding, promotes healing'

indications: metrorrhagia; haematochezia, haemorrhoids; haematemesis, epistaxis; burns, ulcers; (T)

Sanguisorba officinalis var. *officinalis* flowering in the wild, Shaanxi province

Flower-heads of *S. officinalis* var. *officinalis*

Freshly uprooted plants of *S. officinalis* var. *officinalis* (left) and var. *longifolia* (right)

Flowerheads of *S. officinalis* var. *longifolia*

Non-flowering plant showing typical basal leaves

Sanguisorba officinalis var. *officinalis*

Rosaceae

DRUG MORPHOLOGY

CRUDE DRUG (DRIED ROOT): roots **irregularly fusiform or cylindrical**, sometimes branching, slightly curved or twisted (5–25 cm long, 0.5–2 cm diam.) with occasional fine adventitious roots. Externally **greyish-brown to dark brown [1]**, rough, with distinct longitudinal furrows. *Texture*: hard and **brittle [2]**. *Fracture*: pale yellow, yellowish-brown or pinkish, **relatively even.** *Odour*: faint. *Taste*: somewhat bitter and astringent. ***Good quality*:** hard roots; pinkish-white internally.

DECOCTING PIECES-1 (SLICED ROOTS): sub-rounded transverse or oblique slices (0.5–2 cm diam.; 2–4 mm thick). Outer bark very thin and greyish-brown; cut surface pale yellow, yellowish-brown or pinkish; vascular tissues surrounding the dense central xylem often radiating; cambium tissue forms a distinct outer ring. *Texture*: mostly dense and granular **[3]**, sometimes fissured where inner bark splits away from the cambium ring.

DECOCTING PIECES-2 (CARBONISED ROOT): as for sliced roots above but charred black externally and brown internally. *Odour*: charred.

PROCESSING

After removal of rootlets, roots are dried (= crude drug) and subsequently softened before cutting transversely or obliquely into thick (2–4 mm) decocting slices or sliced while fresh. Carbonisation may be undertaken as an optional further processing stage (= 地榆炭 Di Yu Tan — carbonised Di Yu).

OTHER COMMON NAMES

黄瓜香 Huang Gua Xiang; 棉地榆 Mian Di Yu; 地榆 Di Yu (plant and drug name); garden burnet root (CP2010); Radix Sanguisorbae; sanguisorba (SCN).

S. officinalis var. *longifolia*

Rosaceae

DRUG MORPHOLOGY

CRUDE DRUG (DRIED ROOT): roots **long and cylindrical**, occasionally branching, slightly curved or twisted with occasional fine adventitious roots; sometimes adjoined to persistent rhizomes. Externally **reddish-brown or purplish-brown [4]**, rough, with longitudinal furrows; **bark with many yellowish-brown or yellowish-white wool-like fibres [5]**. *Texture*: hard and **fibrous [6]**; difficult to break. *Fracture*: reddish-brown or yellowish-brown, uneven. *Odour*: none. *Taste:* somewhat bitter and astringent.

DECOCTING PIECES (SLICED ROOTS): as for PS1 but with a **fibrous texture.**

PROCESSING

As for PS1.

OTHER COMMON NAMES

黄瓜香 Huang Gua Xiang; 棉地榆 Mian Di Yu; 长叶地榆 Chang Ye Di Yu (long leaved Di Yu — plant name); garden burnet root (CP2010); Radix Sanguisorbae; sanguisorba (SCN).

UNOFFICIAL SUBSTITUTES

1. ***Polygonum bistorta*** L. (syn. *Bistorta officinalis* Delarbre); (family Polygonaceae). Rhizomes are the source of the separate TCM drug 拳参 Quan Shen (Bistortae Rhizoma).
2. ***Reynoutria japonica*** Houtt. (family Polygonaceae). Root and rootbark are the source of the separate TCM drug 虎杖 Hu Zhang (Polygoni Cuspidati Rhizoma et Radix).

Crude drug sourced from roots of *Sanguisorba officinalis* var. *officinalis*

Transverse decocting slices showing example with yellowish-brown internal tissues; sourced from roots of *S. officinalis* var. *officinalis*

Crude drug sourced from roots of *S. officinalis* var. *longifolia*

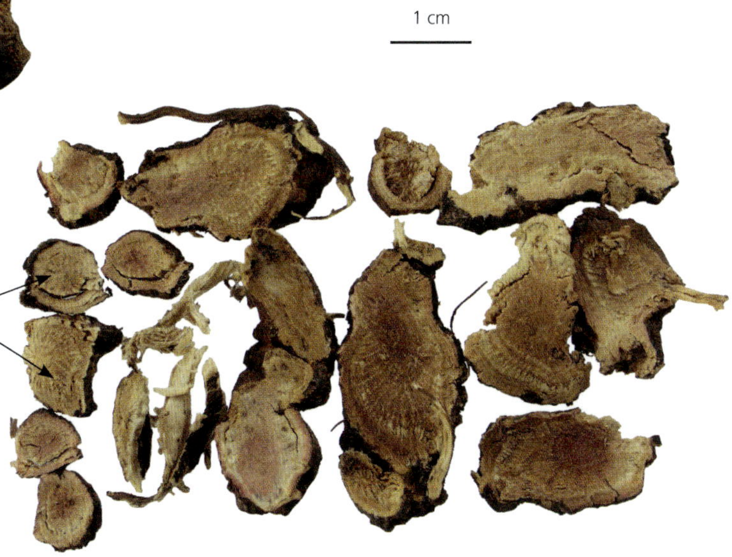

Transverse decocting slices sourced from roots of *S. officinalis* var. *longifolia*

Du Huo 独活

Angelicae Pubescentis Radix

Drug part: ROOT

Angelica biserrata (R. H. Shan & C. Q. Yuan) C. Q. Yuan & R. H. Shan

Apiaceae

synonym: *A. pubescens* Maxim. f. *biserrata* R. H. Shan & C. Q. Yuan (name used in CP2015).

Species of *Angelica* wild in China: 45

PLANT DESCRIPTION

A stout perennial herb (1–2 m tall) with an **aromatic tap root** (up to 15 cm long; 1–2.5 cm diam.) (= medicinal part 'Du Huo'). *Stem* purplish-green (up to 1.5 cm diam.), thinly ribbed and bristly above. *Leaves* alternate, with basal and lower ones long-stalked (30–50 cm) and with conspicuously inflated sheaths; blade broad to ovate in outline (20–30 (rarely 40) cm long; 5–25 cm wide), divided into leaflets (1–3 ternate-pinnate) with leaflet bases often extending narrowly along 'leaf stalks'; **leaf margins with alternating teeth of 2 different sizes**; tip tapering to a long point. *Inflorescences* many, in terminal or lateral compound umbels; lowest stalk of terminal compound umbel (peduncle) densely bristly with 1, needle-like and deciduous bract and rays 10–25; umbellets 17–28 (rarely 36)-flowered, unequal, with bracteoles 5–10 and broadly lance-shaped. *Flowers* tiny; calyx absent; petals 5, white, equal-sized and wedge-shaped; styles 2 and conspicuously long. *Fruits* in umbellate clusters; each fruit dry and **ellipsoid (6–8 mm long; 3–5 mm diam.),** splitting into 2 one-seeded carpels (mericarps) with **dorsal rib of each carpel thick and prominent**, **lateral ribs forming wings, broader than dorsal ribs; 2 or 3 oil ducts in each furrow** (between the ribs) and usually 2–4 on the flat internal (commissural) surface. Fl. Aug–Sep, fr. Sep–Oct.

HARVESTING, SOURCING AND NATURAL RANGE

Roots are harvested entirely from plants cultivated as medicinal crops in early spring before the leaves have appeared or at the end of autumn when the stems and leaves have withered. The main cultivation area in China is in Hubei province, from where the traditional *Di Dao* material is sourced and where it grows in thickets on damp slopes (1,000–1,700 m altitude). The distribution of the wild species is C and W China (Anhui, Hubei, Jiangxi, Sichuan and Zhejiang provinces).

CONSERVATION STATUS

China: Least Concern (MEPCAS, 2013).
Global: Not Evaluated (IUCN Red List, accessed 22/4/2015).

Medicinal crop of *Angelica biserrata* in Hubei province, China

Inflorescence of *A. biserrata*

Young plant of *A. biserrata*

UNOFFICIAL SUBSTITUTES

The roots of many other species, mostly in the Apiaceae family, have been reported as local or historical substitutes or adulterants. Since the official species of 'Du Huo', however, is farmed today on an extensive scale it is unlikely that these substitutes will enter international markets:

1. Wild-sourced ***Angelica dahurica*** (Hoffm.) Benth. & Hook. F. ex Franch. & Sav. var. ***dahurica***. (东北独活 Dong Bei Du Huo). Family Apiaceae. The cultivated roots of this species are the source of 'Bai Zhi' (Angelicae Dahuricae Radix, page 38 but only the wild roots of this species, which resemble 'Bai Zhi', are used as adulterants.
2. ***Levisticum officinale*** W. D. J. Koch (欧当归 Ou Dang Gui; European Dang Gui; lovage). Family Apiaceae. Cultivated widely in recent years in the Chinese province of Gansu where it is sold locally as 'Du Huo'. Whole dried roots often with >1 stem base at apex; tapering lateral branches virtually absent; xylem deep yellow; oil glands less frequent in transverse slices.
3. ***Heracleum hemsleyanum*** Diels. Family Apiaceae.
4. ***Aralia cordata*** Thunb. (食用土当归 Shi Yong Tu Dang Gui). Family Araliaceae.
5. ***A. atropurpurea*** Franch. (浓紫龙眼独活 Nong Zi Long Yan Du Huo). Family Araliaceae.

Levisticum officinale dried roots

L. officinale, transversely sliced roots

TCM PROPERTIES AND USE

properties: **'pungent, bitter and warm'**
channels: **'Liver, Kidney and Bladder'**
actions: **'dispels wind damp, relieves pain, releases exterior'**

indications: **acute, chronic musculoskeletal pains; headache, toothache**

DRUG MORPHOLOGY

CRUDE DRUG (DRIED ROOT): root with large, irregularly-shaped head dividing into curved and tapering branch roots (10–30 cm long; 1.5–3 cm diam.); apex sunken and surrounded by shrivelled remnants of leaves and stems **[1]**. Externally greyish-brown or brown, longitudinally wrinkled and irregularly covered with raised transverse lenticels **[2]** and slightly protuberant rootlet scars **[3]**. *Texture*: quite hard but soft when moistened. *Fracture*: cortex with densely scattered brown oil cells **[4]** and distinctly demarcated from central xylem by brown cambium ring **[5]**; xylem greyish-yellow to yellowish-brown **[6]**. *Odour*: distinctly aromatic. *Taste*: bitter, acrid and slightly tongue-numbing. ***Good quality***: large, solid and intensely aromatic.

DECOCTING PIECES (TRANSVERSELY OR OBLIQUELY SLICED ROOT): irregularly ovate slices (1.5–3.5 cm diam.; 1–2 mm thick). External bark rough and greyish-brown or brown, occasionally with raised transverse lenticels **[2]** and protuberant rootlet scars. Cut surface with clearly demarcated concentric rings of tissue: cortex (inner bark) greyish-white to yellowish and with many brown dot-like oil cells; cambium ring brown **[5]** and central xylem greyish-yellow to yellowish-brown **[6]**. *Odour* and *taste* as for crude drug.

Note: in Hong Kong, decocting pieces are typically traded as longitudinal slices.

PROCESSING

Roots are partially dried, then stored for 2–3 days to soften in large piles before being dried thoroughly (= crude drug); roots are then sliced thinly (1–2 mm) into transverse, oblique or longitudinal decocting pieces.

OTHER COMMON NAMES

重齿当归 Chong Chi Dang Gui (plant name); 重齿毛当归 Chong Chi Mao Dang Gui; 川独活 Chuan Du Huo; 恩施独活 En Shi Du Huo; 浙独活 Zhe Du Huo; 资丘独活 Zi Qiu Du Huo; doubleteeth pubescent angelica root (CP2010); pubescent angelica; Radix Angelicae Pubescentis.

SAFETY

Safety Class: 2d; Interaction Class: A (Gardner & McGuffin, 2013). **Side effects have been reported (dizziness, headache, GI upset etc.)** (Bensky *et al.*, 2004).

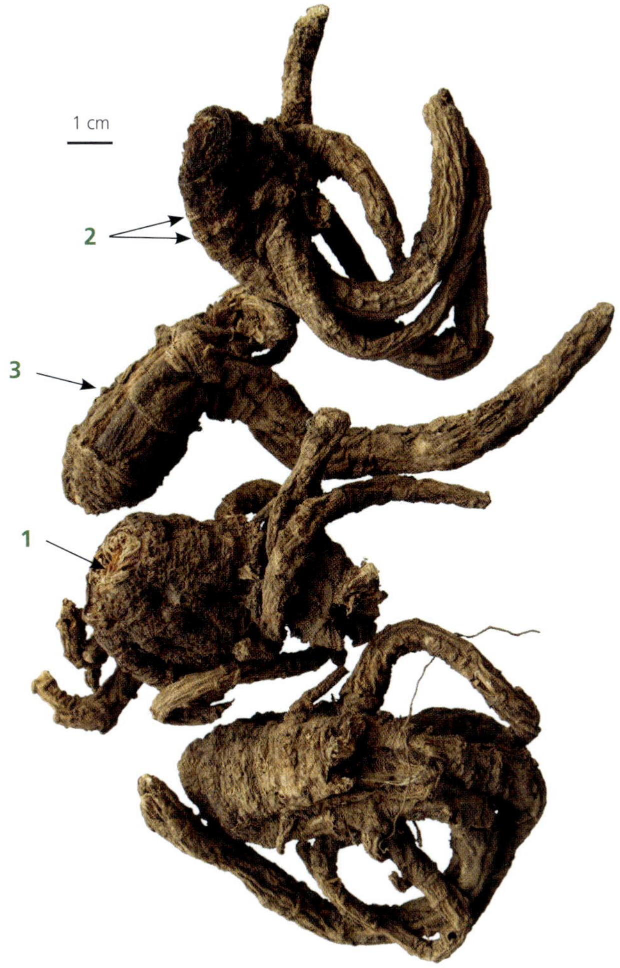

Crude drug of Du Huo (dried roots of *A. biserrata*)

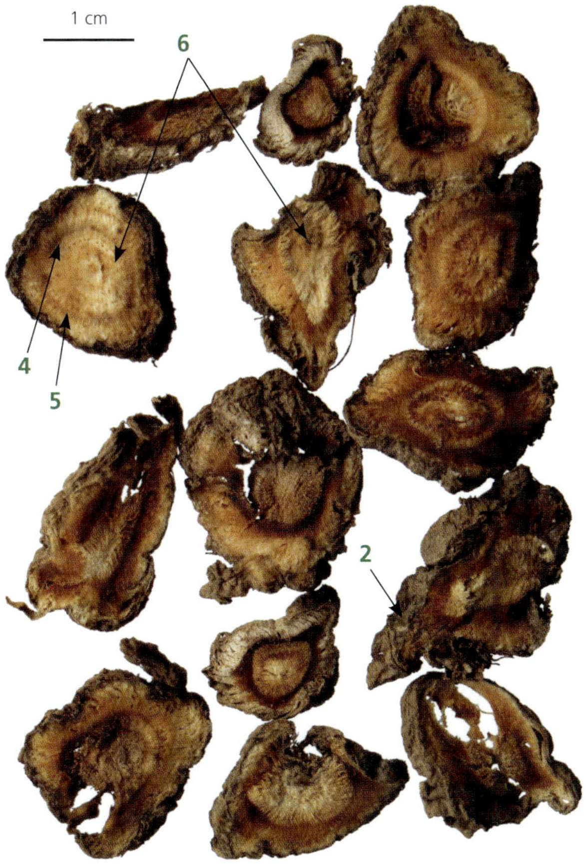

Transversely sliced decocting pieces of Du Huo (dried and sliced roots of *A. biserrata*)

E Zhu 莪术 Curcumae Rhizoma

Drug part: RHIZOME

PHARMACOPOEIA SPECIES 1

Curcuma phaeocaulis Valeton

Zingiberaceae

identification note: *FoC* states that this species "has been misidentified [in the Chinese literature] as *C. aeruginosa* Roxb., *C. caesia* Roxb. or *C. zedoaria* (Christm.) Roscoe". These names should not be used as the source of this drug.

Species of *Curcuma* wild in China: ??

PLANT DESCRIPTION

An erect perennial herb (c. 1 m tall). **Flesh of fresh rhizome** (source of 'E Zhu') **pale blue, pale green, yellowish-green or yellow.** *Root tubers* ellipsoid (source of 'Yu Jin', page 326). *Leaves* basal with leaf sheaths dark brown and leaf stalks longer than the leaf blades; blades large and oblong to lance-shaped (25–60 cm long; 10–15 cm wide); upper surface hairless **with a purple band along the green mid-vein**; **lower surface with dense mat of short, soft hairs**, margin entire, tip tapering to a narrow point. *Flowers* in tightly packed conical heads (10–18 cm long; 5–8 cm diam.) on separate shoots shorter than the leaf stems and arising directly from the rhizomes. Lower bracts (surrounding the flowers) whitish-green with a pink tip; upper petal-like bracts large and conspicuous, sterile (without flowers), white with a deep red tip. Each flower funnel-shaped and white (c. 3 cm long) with red lobes (c. 1.5 cm long) and a yellowish-white lip with a deeper yellow central band. *Fruit* an ellipsoid 3-valved capsule. Fl. Apr–Jun.

HARVESTING, SOURCING AND NATURAL RANGE

Rhizomes are harvested in winter, mostly from medicinal crop plants growing in S and W China (Fujian, Guangdong, Guangxi and Sichuan). The wild species grows in Yunnan [also native to Indonesia and Vietnam]. Root tubers, after stems and leaves have withered, are the source of the separate TCM drug 'Yu Jin' (Curcumae Radix) see page 326, and are harvested at the same time.

CONSERVATION STATUS

China: Not Evaluated.
Global: Not Evaluated (IUCN Red List, accessed 12/2/2016).

See also Yu Jin (Curcumae Radix) page 326 which is sourced from the roots of these three species plus one other.

PHARMACOPOEIA SPECIES 2

C. kwangsiensis S. G. Lee & C. F. Liang

Zingiberaceae

synonym: *C. chuanyujin* C. K. Hsieh & Hao Zhang

PLANT DESCRIPTION

Similar to the other 2 Pharmacopoeia species. *Main differences*: **usually smaller. Flesh of fresh rhizome milky-white or cream.** Leaf blades narrower (14–40 cm long, 4.5–9.5 cm wide); **hairy on both surfaces.** Lower bracts pale green (c. 4 cm long); upper (sterile) petal-like bracts pale red or pink. Flowers lobes red; lip (labellum) yellowish-white. Fl. May–Jul.

HARVESTING, SOURCING AND NATURAL RANGE

As for PS1 with rhizomes harvested from medicinal crops grown in S and SW China (Guangdong, Guangxi, Sichuan and Yunnan). The species is endemic to China where it grows wild in shrubby and grassy places in the same provinces where it is cultivated. Wild plants are seldom harvested since their rapid decline in recent years has made it uneconomic to do so (authors' pers. obs., 2006). *Di Dao*-quality material is sourced from Guangxi province.

CONSERVATION STATUS

Endemic to China: Near Threatened (authors' pers. obs., 2006); Not Evaluated (IUCN Red List, accessed 20/8/2016).

PHARMACOPOEIA SPECIES 3

Curcuma wenyujin Y. H. Chen & C. Ling

Zingiberaceae

taxonomic/nomenclatural note: *Kew-MPNS* accepted name,n10/10/2016, *C. aromatica* Salisb. [which *FoC* treat as a separate species].

PLANT DESCRIPTION

Similar to the other 2 Pharmacopoeia species. *Main differences:* hairless throughout and 0.8–1.6 m tall. **Flesh of fresh rhizome (source of E Zhu) yellow or whitish. Leaf blades large (35–75 cm long; 14–22 cm wide), hairless on both surfaces.** Lower bracts green; upper (sterile) petal-like bracts rose-tinged at tips and sterile (flowerless). Flowers white; the lip (labellum) reflexed, yellow with a deeper coloured centre. Fl. May–Jun.

HARVESTING, SOURCING AND NATURAL RANGE

As for PS1 but rhizomes harvested exclusively from cultivated crops grown in the Chinese provinces of Guangdong, Guangxi and Zhejiang. *Di Dao*-quality material is sourced from localities in Zhejiang. The species is endemic to China and grows wild in the same provinces as those in which it is cultivated.

CONSERVATION STATUS

Endemic to China: Not Evaluated (IUCN Red List, accessed 20/8/2016).

TCM PROPERTIES AND USE

properties: **'bitter', 'pungent' and 'warm'**
channels: **'Spleen and Liver'**
actions: **'breaks Blood, dispels stasis, regulates qi, relieves pain'**

indications: **abdominal pain, palpable masses; amenorrhoea; indigestion**

Crop of *Curcuma phaeocaulis* (inset: with purple banding along mid-vein)

C. phaeocaulis in flower (medicinal crop)

Fresh rootstock of *C. phaeocaulis*

Medicinal crop of *C. kwangsiensis*, Guangdong province

C. kwangsiensis in flower

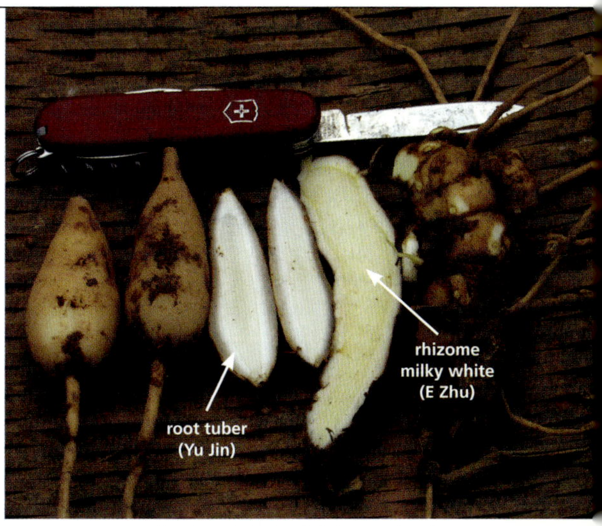

Fresh rootstock of *C. kwangsiensis*

Medicinal crop of *C. wenyujin*, Zhejiang province

Hairless leaf of *C. wenyujin*

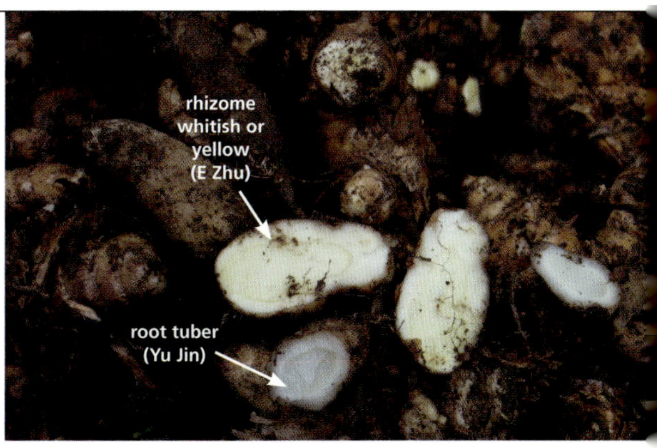

C. wenyujin showing rhizome flesh

Curcuma phaeocaulis

Zingiberaceae

DRUG MORPHOLOGY

CRUDE DRUG (DRIED RHIZOME): conical, ovoid or elongate-ovoid **[1]** (2–8 cm long, 1.5–4 cm diam.), broadly rounded at base **[2]** and sometimes at apex. Externally greyish-yellow to greyish-brown, with conspicuous raised-annulations **[3]** and, along one side, scattered basal remains of fibrous roots **[4]** or slightly sunken root scars; some annulations with a row of concave bud scars **[5]** and large and occasional knife cuts where lateral rhizomes have been removed **[6]**. *Texture:* heavy and hard. *Fracture:* greyish-brown to bluish-brown and somewhat wax-like, often with a greyish- or yellowish-brown powdery covering **[7]**; outer bark easily splitting away at the dark **brown endodermal ring [8]**. *Odour:* slightly aromatic. *Taste:* slightly bitter and pungent. ***Good quality*:** large, heavy and fragrant.

DECOCTING PIECES-1 (STEAMED RHIZOME SLICES): sub-rounded, elliptical or pear-shaped **[9]** slices (2–8 cm long, 1.5–4 cm diam., 2–4 mm thick). Uncut edges greyish-yellow or greyish-brown **[10]**; raised annulations, root remnants or root scars sometimes visible **[11]**. Cut surface yellowish-green, yellowish-brown or brown **[12]** with a distinct dark brown endodermal ring often visible **[13]**; vascular tissues scattered. *Odour:* slightly aromatic. *Taste:* somewhat bitter and pungent.

DECOCTING PIECES-2 (VINEGAR-PROCESSED SLICED RHIZOME): as for decocting pieces-1 above but darker throughout, with a horny texture and a slight scent of vinegar.

PROCESSING

Rhizomes are washed, steamed or boiled thoroughly, dried in the sun or at low temperature, then fibrous roots removed (= crude drug). Rhizomes are subsequently soaked for a short period, steamed until soft, then cut into 2–4 mm thick slices before being dried (= decocting pieces-1). Alternatively whole rhizomes are boiled in rice vinegar before being sliced to produce Cu E Zhu 醋莪术 (vinegar-processed E Zhu) (decocting pieces-2).

OTHER COMMON NAMES

黑心姜 Hei Xin Jiang; 蓝心姜 Lan Xin Jiang; 蓬莪术 Peng E Zhu (popular trade name in China); Curcuma phaeocaulis (SCN); ezhu; Indian saffron; tumeric; turmeric; Rhizoma Curcumae; zedoaria; zedoary rhizome (as 'zedoray' in CP2010). Many other trade names exist for this drug reflecting its medicinal use in other countries (see *Kew-MPNS*).

SAFETY

Some side effects reported; over-dose and long-term use should be avoided (Bensky *et al.*, 2004).

C. kwangsiensis

Zingiberaceae

DRUG MORPHOLOGY

As for PS1. In the authors' opinion, there are insufficient macroscopic characters to reliably differentiate between the drugs of the three Pharmacopoeia species of 'E Zhu'. Any variation shown in the photographs of the vouchered and verified material opposite is due to natural variation within rather than between each drug. Laboratory-based authentication methods are recommended.

PROCESSING

As for PS1.

OTHER COMMON NAMES

桂莪术 Gui E Zhu (popular trade name in China); 广西莪术 Guang Xi E Zhu (plant name) (denoting source from Guangxi); 毛莪术 Mao E Zhu 'hairy E Zhu'; common tumeric; Guangxi turmeric (SCN); Rhizoma Curcumae; zedoary rhizome (as 'zedoray' in CP2010); tumeric; turmeric. Many other trade names exist for this drug (see *Kew-MPNS*).

SAFETY

As for PS1.

Curcuma wenyujin

Zingiberaceae

DRUG MORPHOLOGY

As for PS1. See note under *Drug Morphology* for PS2.

PROCESSING

As for PS1.

OTHER COMMON NAMES

温莪术 Wen E Zhu (popular trade name in China); 温郁金 Wen Yu Jin (plant name); Curcuma wenyujin (SCN); Rhizoma Curcumae; wen curcuma; tumeric; turmeric; zedoary rhizome (as 'zedoray'in CP2010). Many other trade names exist for this drug (see *Kew-MPNS*).

SAFETY

As for PS1.

Crude drug of E Zhu sourced from *C. phaeocaulis*

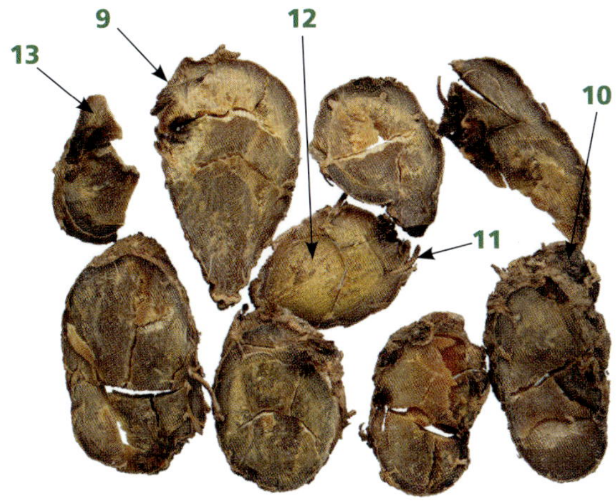

Steamed decocting slices of E Zhu sourced from *C. phaeocaulis* (decocting pieces-1)

Crude drug of E Zhu sourced from *C. kwangsiensis*

Steamed decocting slices of E Zhu sourced from *C. kwangsiensis* (= decocting pieces-1)

Vinegar-processed slices of E Zhu sourced from *C. kwangsiensis* (= decocting pieces-2)

Crude drug of E Zhu sourced from *C. wenyujin*

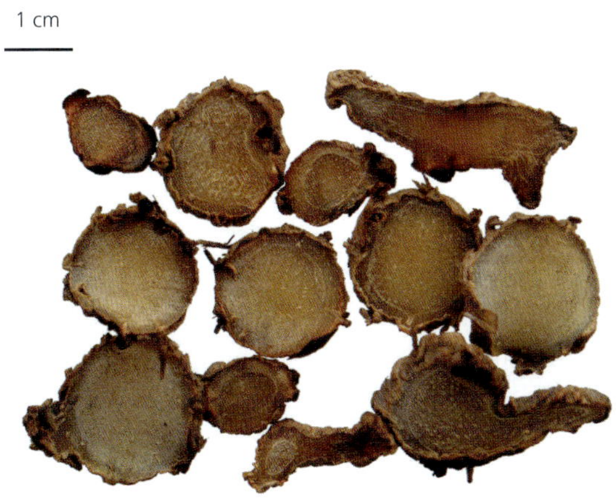

Steamed decocting slices of E Zhu sourced from *C. wenyujin* (= decocting pieces-1)

Vinegar-processed slices of E Zhu sourced from *C. wenyujin* (= decocting pieces-2)

Fang Feng 防风

Saposhnikoviae Radix

Drug part: ROOT

Saposhnikovia divaricata (Turcz.) Schischk.

Apiaceae

synonyms: *Ledebouriella divaricata* (Turcz.) M. Hiroe; *L. seseloides* (Hoffm.) H. Wolff; *Siler divaricatum* (Turcz.) Benth. & Hook.f.; *Stenocoelium divaricatum* Turcz.

Species of *Saposhnikovia* wild in China: 1

PLANT DESCRIPTION

A perennial herb (30–80 cm tall), hairless throughout; with a taproot c. 2 cm thick (= medicinal part 'Fang Feng') sometimes branched and with a distinctive crown surrounded by fibrous remnant leaf sheaths. *Stems* many-branched from base and thinly ribbed. *Leaves* alternate, mostly basal with stalks flattened and sheathing; leaf blades 2-pinnate and deeply divided (pinnatisect) into narrow, often branching linear to lance-shaped or wedge-shaped lobes, outermost lobe 3-parted; leaves becoming smaller up the plant with upper leaves unstalked. *Inflorescences* many, in terminal or lateral compound umbels (c. 6 cm diam.), bracts absent; rays (umbel stalks) 5–7 (3–5 cm long); umbellets 4- or 5-flowered with bracteoles 4–6 (c. 3 mm long). *Flowers* tiny; calyx with 5 short triangular teeth; petals 5 (c. 1.5 mm long), white with notched tips; styles 2. *Fruits* in umbellate clusters. Each fruit dry, oblong or ellipsoid (4–5 mm long; 2–3 mm diam.), covered in nodular swellings when young, smooth when mature, and splitting into 2 one-seeded carpels (mericarps); dorsal ribs slightly raised, lateral ribs narrowly winged; 1 oil duct in each furrow and 2 on the flat internal (commissural) surface; styles persistent, short and curving backwards. Fl. Aug–Sep, fr. Sep–Oct.

HARVESTING, SOURCING AND NATURAL RANGE

With wild populations becoming depleted due to over-harvesting, the species is now grown extensively as a medicinal crop. The province of Heilongjiang (NE China) is the traditional source of *Di Dao* material. Roots are harvested in spring, before the flowering stems appear, or in autumn. The wild species grows in scrub and grassland often on stony slopes (400–800 m altitude) in N China (Gansu, Hebei, Heilongjiang, Inner Mongolia, Jilin, Liaoning, Ningxia, Shaanxi, Shandong and Shanxi). [also wild in the Korean Peninsula, Mongolia and Russia (Siberia)].

CONSERVATION STATUS

China: Least Concern (MEPCAS, 2013); 'rapidly decreasing' (Leung & Cheng, 2008); Vulnerable (authors' pers. obs.: declining due to over-harvesting).
Global: Not Evaluated (IUCN Red List, accessed 21/4/2015).

Saposhnikovia divaricata growing wild (flowering in foreground), Inner Mongolia

S. divaricata

Young fruiting heads of *S. divaricata*

Uprooted immature plant of *S. divaricata*

TCM PROPERTIES AND USE

***properties*:** **'pungent, sweet and slightly warm'**
***channels*:** **'Bladder, Liver and Spleen'**
***actions*:** **'releases the exterior, dispels wind'**
***indications*:** **headache, respiratory tract infection; itching skin; musculoskeletal pain, cramps, diarrhoea**

DRUG MORPHOLOGY

CRUDE DRUG (DRIED ROOT): conical or cylindrical (15–30 cm long; 0.5–2 cm diam.), somewhat twisted and slightly wider towards apex. Externally greyish-brown, rugged, apex surrounded by fibrous remnant leaf sheaths **[1]**; often with zone of dense annulations below **[2]**; lower parts longitudinally wrinkled with raised elongated transverse lenticels **[3]** and dotted rootlet scars. Roots of cultivated plants often thinner and branched. *Texture:* light, hard and easily broken. *Fracture:* uneven, bark brown or reddish-brown, central wood (xylem) yellowish and sometimes withered or hollowed. *Odour:* distinctive. *Taste:* somewhat sweet. ***Good quality*:** long, thick, unbranched roots with closely spaced annular rings beneath the apex **[2]**.

DECOCTING PIECES (TRANSVERSELY SLICED ROOT): rounded or elliptical woody root slices (0.5–2 cm diam.; 2–4 mm thick). Outer bark greyish-brown, thin, rough and wrinkled longitudinally; friable and peeling easily **[4]**; a bluish-green coloration suggests poor processing; transverse annular rings and fibrous remnant leaf sheaths sometimes visible **[5]**. Cut surface with thin outer bark surrounding a brown and much broader cambium layer (1–3 mm wide) **[6]** with a loose, easily broken texture. This in turn surrounds a central pale yellow wood (xylem) **[7]** whose centre is sometimes brown and withered **[8]**. Patterning of the cut surface is classically described as *'resembling the shape and colour of a chrysanthemum flower'* [see Ju Hua / Chrysanthemi Flos page 574]. Occasionally sliced longitudinally.

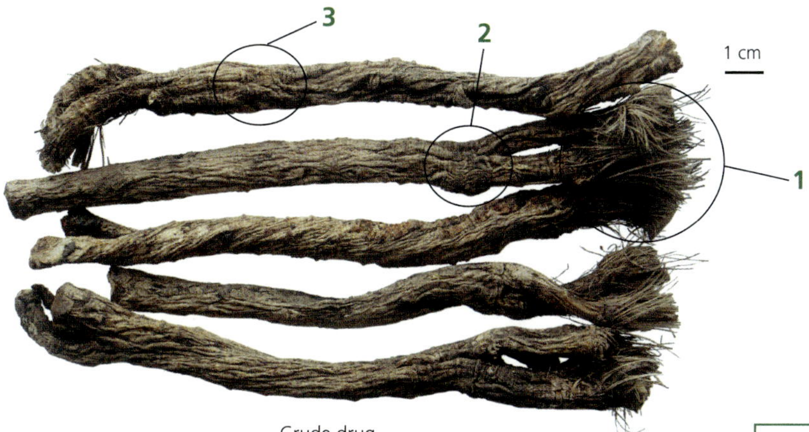

Crude drug

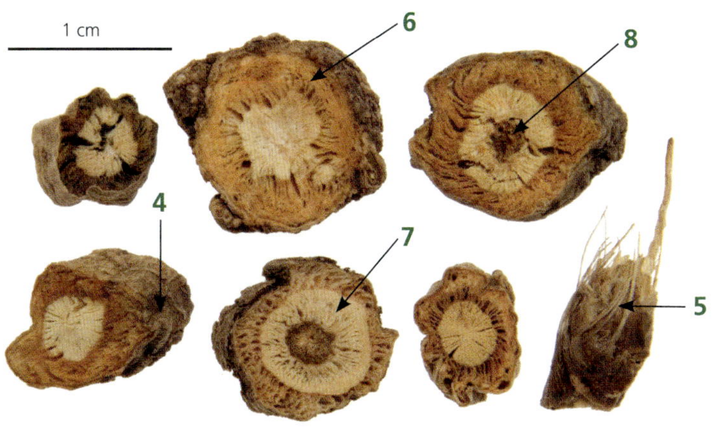

Transverse decocting slices

PROCESSING

After removal of stems, leaves and rootlets, the roots are sun-dried (= crude drug); later these are softened thoroughly and cut transversely into 2–4 mm thick slices and dried (= decocting pieces).

OTHER COMMON NAMES

关防风 Guan Fang Feng (sourced from Heilongjiang, Jilin or Liaoning provinces); ledebouriella root; Radix Saposhnikoviae; Siler (SCN); divaricate Saposhnikovia root.

SAFETY

Safety Class: 1; Interaction Class: A (Gardner & McGuffin, 2013).

Transverse decocting slices

UNOFFICIAL SUBSTITUTES

Due to plant mis-identification, especially when this drug is wild-harvested, roots of many other species of closely related genera have inadvertently traded as official 'Fang Feng' on Chinese and overseas markets (Zou 1995–1996). Laboratory-based authentication methods are advised to reliably differentiate official from unofficial trade items; macroscopic characters alone are considered insufficient.

1. ***Carum*** species: biennial or short-lived perennial herb. Stem base with or without papery remnant leaf sheaths. Calyx teeth rarely present. Petals white, rarely pinkish or purplish, with mid-vein yellow or yellowish-green. Fruit slightly laterally compressed, ribs 5, all similar, slender and prominent.
2. ***Peucedanum*** species: perennial herb. Base usually bearing remnant leaf sheaths. Calyx teeth very short or absent. Petals usually uniformly white, occasionally pinkish or purplish, rarely pale yellow. Fruit dorsally compressed; dorsal ribs slender, slightly prominent, lateral ribs thickened and narrowly winged, wings closely appressed to one another.
3. ***Libanotis*** species: perennial herb, rarely a subshrub, stout, sometimes small. Stem base densely clothed with fibrous leaf remains. Calyx teeth conspicuous. Petals uniformly white, rarely pinkish. Fruit slightly to moderately dorsally compressed; dorsal ribs slender, low or prominent, acute-ridged, lateral ribs sometimes slightly broader.

Fang Ji 防己

Stephaniae Tetrandrae Radix

Drug part: ROOT

Stephania tetrandra S. Moore

Menispermaceae

Species of *Stephania* wild in China: 37

PLANT DESCRIPTION

A deciduous, **non-woody climber** (1–3 m high). Root large, cylindrical, often intestine-shaped and tuberous. *Stems* round and striate *Stipules* absent. *Leaves* simple and alternate; leaf-stalk (3–7 cm) with **stalk attached to the leaf blade surface rather than leaf base** [peltate] with both ends of stalk slightly swollen; leaf blade triangular, deltoid or roundish (4–7 cm long; 5–8.5 cm wide) and palmately veined (with 9–10 spreading veins); margin entire; papery. *Flowers* tiny and inconspicuous in many-flowered, stalked umbel-like heads along a central axis arising from the leaf axils; the central flower of each 'umbel' opening first. Male and female flowers on separate plants**; male flowers with 4 (rarely 5) sepals**, elliptic and pale greenish-yellow; **petals 5**, fleshy and minute (c. 0.6 mm long); stamens 2–6; female flowers with sepals and petals as for those of male flowers. *Fruit* red (when mature), spherical (up to 6 mm diam.) and fleshy; in pendulous clusters like a grape vine. *Seed* horseshoe-shaped. Fl. summer, fr. autumn.

HARVESTING, SOURCING AND NATURAL RANGE

Roots are wild-harvested in autumn from plants growing in scrub and disturbed places. The species is endemic to parts of C and S China (Anhui, Fujian, Guangdong, Guangxi, Hainan, Hubei, Hunan, Jiangxi, Taiwan and Zhejiang).

CONSERVATION STATUS

Endemic to China: Least Concern (MEPCAS, 2013); Not Evaluated (IUCN Red List, accessed 26/4/2016).

Aristolochia fangchi Y. C. Wu ex L. D. Chow & S. M. Hwang

Aristolochiaceae

PLANT DESCRIPTION

Easily distinguished from the Pharmacopoeia species. *Main differences*: a deciduous, **woody climber**; stems round or **with cork-like wings;** leaf stalk attached to leaf base (*not* leaf blade surface); leaf blade oblong or lanceolate, **pinnately veined, tip tapering to a point** and leathery; flowers pendant in 2–4 flowered heads (4–6 cm long); **each flower much larger, with petals fused to form a distinctive horse-shaped tube** (utricle) the basal part **(4–5 cm long, 0.8–1 cm diam.)**, densely hairy on lower side of tube, purple with yellow blotches, mouth of tube (4–6 cm diam.) with 3 shallow lobes; **fruit brown,** ovoid, **much larger** (5-–10 cm long; 3–5 cm diam.) and **dry**; seeds many, stacked in layers.

SUBSTITUTION COMMENT. A **toxic** species (see *Safety*) whose roots were recognised as the source of the TCM drug 'Guang Fang Ji' (Radix Aristolochiae Fangchi) in pre-2005 editions of the Chinese Pharmacopoeia. Its roots (crude and sliced) are easily confused with those of 'Fang Ji' (*S. tetrandra*) and have been traded under this name, mistakenly or not (Upton & Graff, 2006). Its proven association with serious adverse reactions led to the removal of the 'Guang Fang Ji' monograph from the CP in 2003 (SFDA, 2003).
The species is endemic to S China (provinces of Guangdong, Guangxi and Guizhou) where it grows in dense forest and thickets (500–1,000 m altitude). Populations have declined due to medicinal over-harvesting and habitat loss.

CONSERVATION STATUS

Endemic to China: Least Concern (MEPCAS, 2013); Not Evaluated (IUCN Red List, accessed 26/4/2016); 'scarce in the wild' (Lin Yu-Lin, pers. obs., 2011).

Cocculus orbiculatus (L.) DC.

Menispermaceae

synonyms: *C. trilobus* (Thunb.) DC.; *Menispermum orbiculatum* L.

PLANT DESCRIPTION

Easily confused with the Pharmacopoeia species especially when plants are devoid of mature fruit. *Main differences:* a **woody climber**; **leaf stalk shorter** (1–3 cm) **and attached to leaf base** (*not* leaf blade surface); leaf blade variable in shape, sometimes 3–5-lobed (usually 3–8 cm long, variable in width) and palmately veined (with 3(or 5) veins); male flowers with **6 (rarely 9) sepals and 6 petals;** fruits red, **purple or purplish-black** and slightly larger (7–8 mm diam).

SUBSTITUTION COMMENT. The root of this species has been traded as 'Fang Ji' in China and elsewhere owing to its similar clinical properties (Bensky *et al.*, 2004).

CONSERVATION STATUS

China: Least Concern (authors' pers. obs. 2011).
Global: Not Evaluated (IUCN Red List, accessed 27/4/2016).

TCM PROPERTIES AND USE

properties: **'bitter and cold'**
channels: **'Bladder and Lung'**
action: **'dispels wind damp, relieves pain'**

indications: **musculoskeletal pain; oedema**

Leaves and developing inflorescences of *Stephania tetrandra*

S. tetrandra laden with young fruit, Zhejiang province

S. tetrandra with typical vine-like fruit-heads, red when mature

Woody stems and ovate leathery leaves of *Aristolochia fangchi* used to make an aerial part herbarium voucher for the root, an unofficial substitute of *S. tetrandra*

A. fangchi stem often winged and corky

A. fangchi seedling in the wild, Guangxi province

Male plant of *Cocculus orbiculatus* in flower

Female plant of *C. orbiculatus* with purple (mature) fruit

Stephania tetrandra

Menispermaceae

DRUG MORPHOLOGY

CRUDE DRUG (ROOT LENGTHS): sub-cylindrical and tuberous **[1]** (5–10 cm long; 1–5 cm diam.) , sometimes strongly contorted (intestine-like) with deep constrictions **[2]** or narrowing abruptly to form 'shoulders' **[3]**. Unpeeled root externally dull greyish-brown or greyish-yellow **[4]**; finely wrinkled longitudinally; pale and distinctive transverse scars, scattered and slightly raised **[5]**. Peeled root externally white or creamish-white **[6]** with occasional brown outer bark remnants **[7]** (especially in the constricted folds). *Texture*: hard and compact; **powder soft and smooth.** *Odour*: indistinct. *Taste*: bitter. *Fracture*: see cut surface of *Decocting Pieces* below.

DECOCTING PIECES (ROOT SLICES): ***Transverse***: sub-rounded (1–4 cm diam.; 2–4 mm thick); unpeeled root slices with outer bark thin and greyish-brown or greyish-yellow **[4]**; peeled root slices white or creamish-white (inner root-bark) **[8]** with remnants of outer bark. Cut surfaces hard and compact **[9]**; with a fine brown cambium line **[10]** 2 or 3 mm inside the perimeter; broad areas of creamish- or greyish-white parenchyma **[11]** alternate with a few narrow, radiating and irregularly spaced wood (xylem) tissues **[12]**, the latter brown or greyish-cream, some converging centrally; **minute orange dot-like glands absent.** *Longitudinal*: slices somewhat oblong or contorted **[13]** (5–10 cm long; c. 3–5 cm diam.) with or without greyish-brown or greyish-yellow coarse outer bark **[4]**. Cut surface as for transverse slices but wood tissues indistinct and irregular **[14]**. ***Texture*: powder white, soft and smooth.** *Taste*: bitter.

PROCESSING

After removal of any coarse outer bark, roots are part-dried in the sun, then cut into lengths (large diameter roots are halved longitudinally (= crude drug); decocting pieces are made by softening the roots thoroughly before cutting into 2–4 mm thick slices (transverse or longitudinal) and drying again.

OTHER COMMON NAMES

粉防己 Fen Fang Ji 'powdery Fang Ji'; 汉防己 Han Fang Ji; fourstamen Stephania root (CP2010); Radix Stephaniae Tetrandrae; stephania (SCN).

SAFETY

Safety Class: 1; Interaction Class: A (Gardner & McGuffin, 2013). In the interests of public safety, identification confusion between the root of this species and that of the toxic *A. fangchi* (Unofficial Substitute 1) has seen *S. tetrandra*, through association, become subject to legal restrictions regarding its medicinal supply and use in some countries (UK Government, 2001).

Aristolochia fangchi

Aristolochiaceae

'DRUG SUBSTITUTE' MORPHOLOGY

Roots of this species and the Pharmacopoeia species are 'very difficult' to distinguish reliably using macroscopic characters alone (Upton & Graff, 2006). The latter state that the "texture of the [root] powder may provide some guidance [to macroscopic differentiation] but should not be considered definitive" (see *Texture* for both species). A wealth of authentication literature is available to differentiate these species, using laboratory-based methods (e.g. Blatter and Reich, 2004a, 2004b; Flurer *et* al., 2001; Joshi *et al.*, 2008; Kite *et al.*, 2002; Koh *et al.*, 2006; Sorenson and Sullivan, 2007; Upton & Graff, 2006).

Given the above, the macroscopic characters below are indicative only; laboratory-based methods are strongly advised for robust identification of this root.

ROOT LENGTHS: cylindrical or sub-cylindrical, somewhat curved but **not obviously contorted [15]** (6–18 cm long; 1.5–4.5 cm diam.); sometimes halved longitudinally. Unpeeled root externally rough, pale greyish-brown with longitudinal wrinkles and some transverse fissures **[16]**. **Raised transverse scars absent.** Peeled root externally pale yellow **[17]**. *Texture:* compact, heavy and difficult to break; **powder gritty and fibrous**. *Odour*: indistinct or absent. *Taste*: mildly acrid and bitter. *Fracture*: **slightly fibrous xylem tissue (tracheids) remain intact when root broken.** See also cut surface of *Transverse root slices* below.

TRANSVERSE ROOT SLICES: sub-rounded (1.5–4.5 cm diam.; 2–4 mm thick) with remnants of thin and pale greyish-brown outer bark usually present **[18]**. Cut surfaces hard; cambium line **[19]** up to 6 mm inside the bark; **parenchyma** creamish-yellow **often narrow [20]** and alternating with narrow wood (xylem) tissues **[21]** often producing a dense radial effect **[22]**. **Minute orange dot-like secretory cells often visible [hl] in the xylem and outer bark [23].** *Texture*: **powder gritty and fibrous**. *Taste*: mildly acrid and bitter.

OTHER COMMON NAMES

防己 Fang Ji; 广防己 Guang Fang Ji; Aristolochia fangchi (SCN); Radix Aristolochiae Fangchi; southern Fang Ji.

SAFETY

Toxic. Over the last 20 years the medicinal use of this species has been proven to be the cause of many serious cases of adverse reaction (some fatal), including over 100 cases of nephropathy (Heinrich *et al.*, 2009). The genus contains aristolochic acids which are nephrotoxic and carcinogenic (Debelle *et al.*, 2008; Upton & Graff, 2006). Prohibition orders covering the sale and medicinal supply of this species (and others in the genus *Aristolochia*) have been issued in many countries including China (Lord *et al.*, 1999; Nortier *et al.*, 2003; SFDA, 2003; UK Government, 2001; Upton *et al.*, 2006; USFDA, 2000, 2001a, 2001b; Vanherweghem *et al.*, 1993). See also *Substitution Comment* page 124.

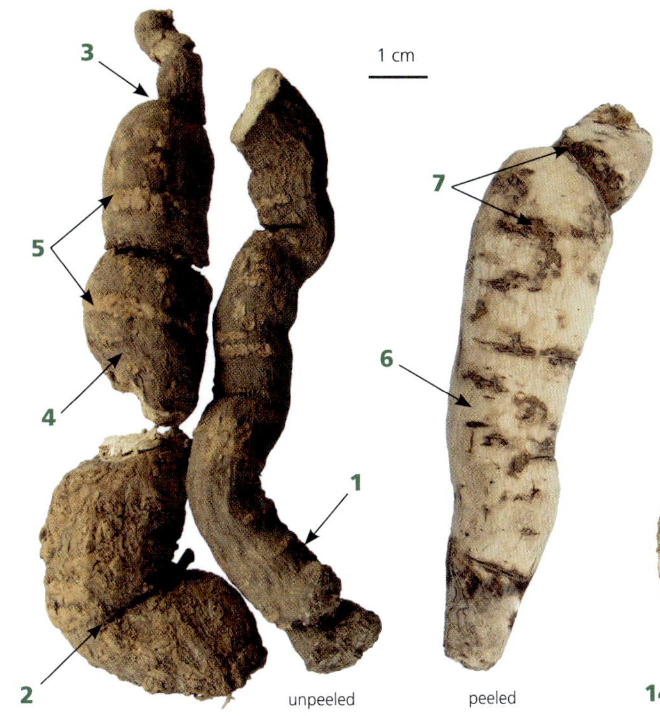

Crude drug of Fang Ji (*Stephania tetrandra* root)

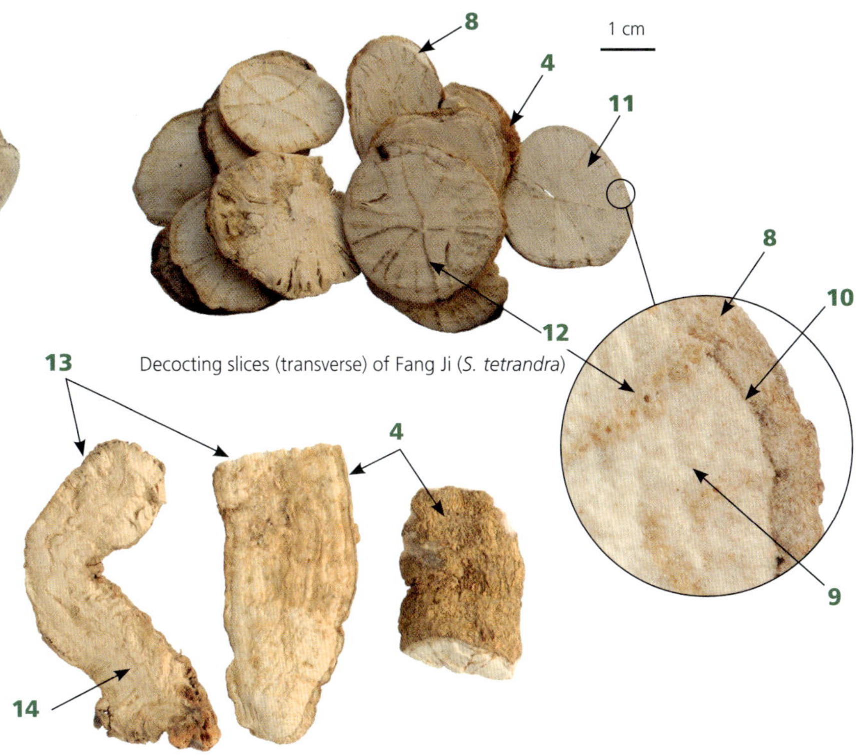

Decocting slices (transverse) of Fang Ji (*S. tetrandra*)

Decocting slices (longitudinal) of Fang Ji (*S. tetrandra*) root

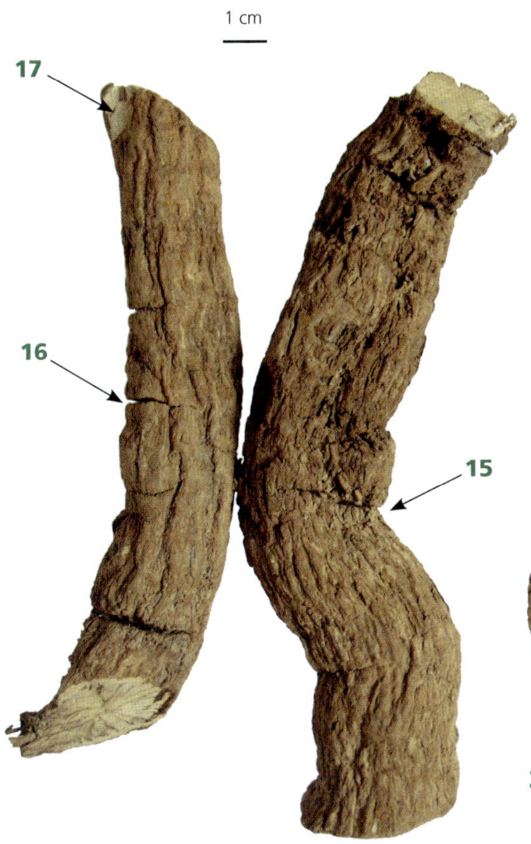

Dried unpeeled roots of *Aristolochia fangchi*

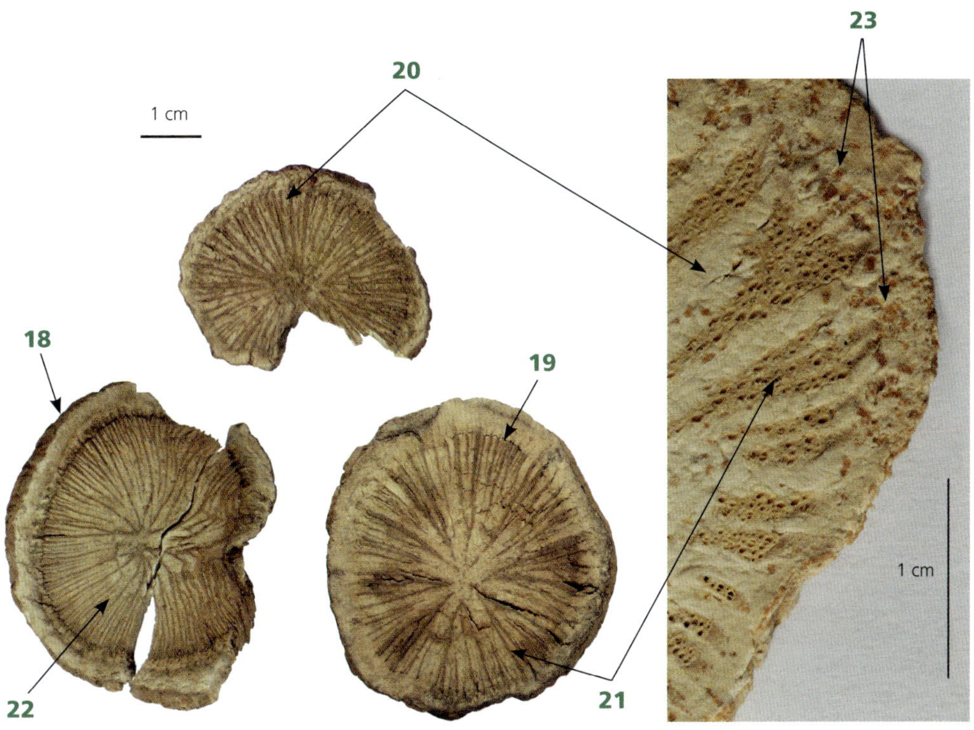

Transverse root slices of *A. fangchi*

Cocculus orbiculatus

Menispermaceae

'DRUG SUBSTITUTE' MORPHOLOGY

Easily confused with the official drug especially in the form of root slices (see below).

DRIED ROOT LENGTHS: irregularly cylindrical, somewhat contorted and shallowly constricted **[24]** but without obvious 'shoulders'. **Root diameter smaller (1–2.5 cm) [25]**. Outer bark thinner than that of *Aristolochia fangchi*. *Fracture*: **woody [26]**. *Odour*: indistinct. *Taste*: bitter.

TRANSVERSE ROOT SLICES: sub-rounded; **root diameter smaller** (**1–2.5 cm diam.**; 2–4 mm thick) **[25]** than that of *S. tetrandra*; bark pale greyish-brown; **cut surface woody** (not powdery) **[26]** (typical of *S. tetrandra*), cambium line indistinct. Compared to *S. tetrandra*, the radiating lines of xylem tissues are slightly broader and **with large roundish vessels [27]**. Minute orange dot-like glands **absent**.

OTHER COMMON NAMES

木防己 Mu Fang Ji 'woody Fang Ji'.

SAFETY

Identification confusion between official 'Fang Ji' and the toxic *A. fangchi* (Unofficial Substitute 1) has meant that *C. orbiculatus*, through association, has become subject to medicinal use restrictions in some countries (UK Government, 2001). See also *Substitution comment* (*C. orbiculatus*) and *Safety* (*A. fangchi*) — Unofficial Substitute 1.

LESSER KNOWN UNOFFICIAL SUBSTITUTES

The name Fang Ji 防己 has been used for many other Chinese plant species spanning several families (see below) (Upton, 2006 & Graff; Zhao & Xiao, 2009–2010) often creating confusion about the species traded under this name especially where markets are poorly policed. Greatest concerns arise when the Pharmacopoeia species (*S. tetrandra*) is substituted (mistakenly or not) by species of toxic *Aristolochia* (see *Unofficial Substitute 1* and the below). While fresh plants of *Aristolochia* are easily differentiated from those of *S. tetrandra* when in leaf and flower, reliable differentiation of their trade items (usually stems and roots) is best performed by combining macroscopic examination with laboratory-based authentication methods. The below (whose parts have been traded as 'Fang Ji') are likely to be of local concern in China only.

Aristolochiaceae family:

1. ***Aristolochia kaempferi*** Willd. (syn. *A. heterophylla* Hemsl.) (异叶马兜铃 Yi Ye Ma Dou Ling). Woody, sometimes non-woody climber; stems with long soft hairs; leaves ovate, oblong or linear (3–18 cm long; 0.3–8 cm wide); flower a yellowish-green, horseshoe-shaped tube with purple veins with 3 broadly ovate lobes; fruit a dry, ovoid or cylindrical many-seeded capsule; root curved but not deeply constricted; fracture woody. **Toxic**.
2. ***A. kwangsiensis*** W. Y. Chun & F. C. How (syn. *A. austroszechuanica* S. S. Chien & C. Y. Cheng) (广西马兜铃 Guang Xi Ma Dou Ling). Woody climber; covered in dense dull yellow or brownish hairs; leaves heart- or orb-shaped (11–13 cm long; 9–32 cm wide); flower a dark purple, spiny, horseshoe-shaped petal tube with protuberances, a yellow throat and flower lobes broadly deltoid; fruit a dry, many-seeded, cylindrical capsule. **Toxic**.
3. ***A. moupinensis*** Franch. (淮通 Huai Tong). Woody climber; leaves ovate to ovate-heart-shaped with lower surfaces densely covered in greyish-yellow long, soft hairs; flower a yellowish, bent petal tube with purple veins and 3 shallow, broadly ovate, recurved lobes; fruit a dry, many-seeded, cylindrical capsule. **Toxic**.
4. ***A. tagala*** Cham. (耳叶马兜铃 Er Ye Ma Dou Ling). Non-woody climber; leaves ovate to heart-shaped or oblong-ovate (8–12 cm long; 4–14 cm wide); leaf bases deeply heart-shaped; flower a pale yellowish or greenish, slightly curved petal tube (5–10 mm long; 2–3 mm diam.); fruit a dry, many-seeded, sub-globose to cylindrical capsule. **Toxic**.
5. ***A. westlandii*** Hemsl. (香港马兜铃 Xiang Gang Ma Dou Ling). Woody, densely hairy climber; leaves lance-shaped to narrowly oblong (12–20 cm long; 2–5 cm wide); flower a solitary, yellow bent petal tube blotched and veined purple (8–13 cm diam.). **Toxic**.

Lardizabalaceae family:

6. ***Akebia trifoliata*** (Thunb.) Koidz. Stems are the source of the TCM drug Mu Tong 木通 (Akebiae Caulis) page 458.

Menispermaceae family:

7. ***Diploclisia affinis*** (Oliv.) Diels (秤钩风 Cheng Gou Feng). Woody climber; leaves not or only slightly peltate, triangular, rhombic or ovate (3.5–9 cm long); fruit red and ovoid (8–10 mm long; c. 7 mm diam.).
8. ***Menispermum dauricum***. Rhizomes are the source of the TCM drug Bei Dou Gen 北豆根 (Menispermi Rhizoma) page 60.
9. ***Sinomenium acutum*** (Thunb.) Rehder & E. H. Wilson. Stems are the source of the TCM drug Qing Feng Teng 青风藤 (*not included in this book*).

Icacinaceae family:

10. ***Iodes seguinii*** (H. Lév.) Rehder (瘤枝微花藤 Liu Zhi Wei Hua Teng). As for *I. vitiginea* but branchlets with many tumour-like lenticels becoming prominent with age.
11. ***I. vitiginea*** (Hance) Hemsl. (syn. *I. ovalis* Blume var. *vitiginea* (Hance) Gagnep. (小果微花藤 Xiao Guo Wei Hua Teng). Woody climber, leaves opposite, narrowly ovate or ovate (6–17 cm long; 3–11 cm wide), densely white or light yellow on lower surfaces; sexes on separate plants; flowers small, yellowish-green in clusters; fruit red, ovoid (1.3–2.2 cm long; 1.2–1.6 cm diam.) covered in yellow hairs.

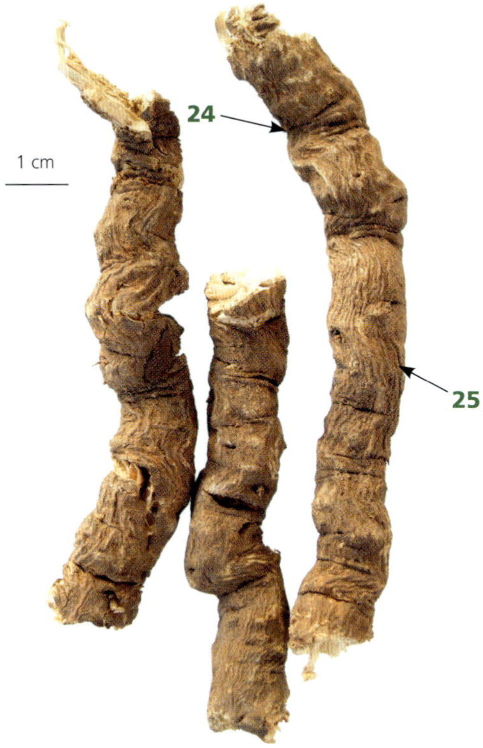

Dried roots of *Cocculus orbiculatus*

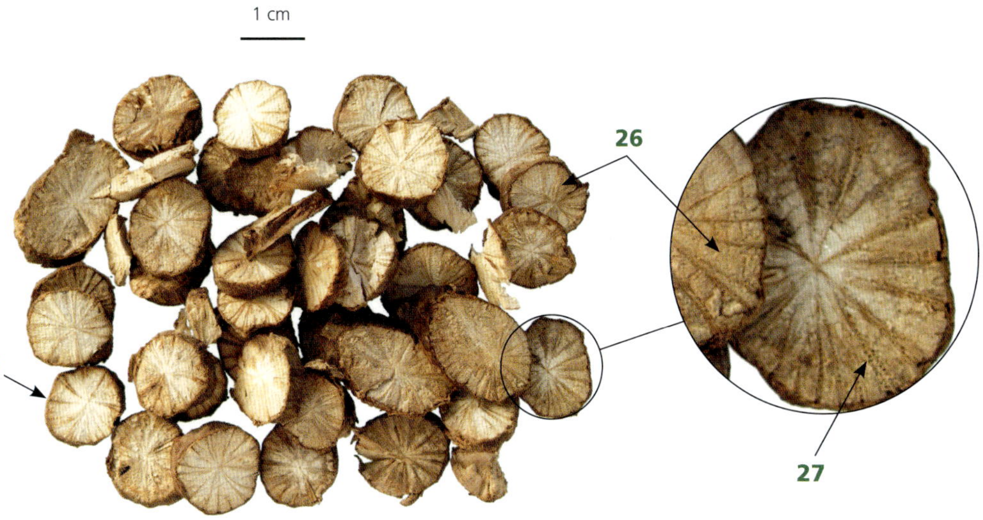

Dried, transversely sliced roots of *C. orbiculatus*

Freshly harvested roots of the Fang Ji unofficial substitute *C. orbiculatus* (top) and those of official Fang Ji sourced from *S. tetrandra* (bottom)

Fen Bi Xie 粉萆薢 Dioscoreae Hypoglaucae Rhizoma

Drug part: RHIZOME

PHARMACOPOEIA SPECIES

Dioscorea collettii Hook.f. var. ***hypoglauca*** (Palib.) S. J. Pei & C. T. Ting

Dioscoreaceae

synonym: *D. hypoglauca* Palib. (name used in CP2015).

Species of *Dioscorea* wild in China: 50

PLANT DESCRIPTION

A twining herb with a fleshy, horizontal, ginger-like rhizome; latter c. 2 cm in diameter, yellow in section and **powdery**. *Stems* twine to the left. *Leaves* alternate; stalks 4–7 cm long; **leaf blades simple, triangular or ovate, not leathery**, undersides sometimes minutely hairy, especially along veins; margins entire and transparent (membranous); base of leaf blade heart-shaped or straight; blades becoming blackish when dry. *Flowers* tiny in linear spikes; male and female flowers on separate plants. Male spike solitary or 2–3 together; flowers stalkless, saucer-shaped and **yellow** (drying blackish); stamens 3. Female flower-spike solitary (up to 5 cm long). *Fruit* a dry, 3-winged, **shiny**, ellipsoid capsule (16–21 mm diam.) with wings (8–11 mm wide. *Seeds* winged. Fl. May–Aug, fr. Jun–Oct.

HARVESTING, SOURCING AND NATURAL RANGE

Rhizomes are harvested in autumn and winter from wild populations which grow on slopes in mixed forests (200–1,300 m altitude). The species is endemic to China (S Anhui, Fujian, N Guangdong, NE Guangxi, S Henan, Hubei, Hunan, Jiangxi, N Taiwan and Zhejiang).

CONSERVATION STATUS

Endemic to China: Least Concern (MEPCAS, 2013); Not Evaluated (IUCN Red list, accessed 20/8/2016).

UNOFFICIAL SUBSTITUTE 1

D. spongiosa J. Q. Xi, M. Mizuno & W. L. Zhao

Dioscoreaceae

PLANT DESCRIPTION

Easily confused with the Pharmacopoeia species. *Main characters*: rhizome interior **spongy** (not powdery); **basal leaves** occasionally with **5–9 palmate lobes**; leaf stalks longer (12 cm); **leaves drying greenish;** leaf margins not transparent; **male flowers orange;** flower stalks very short (0.5–1 mm long); male flowers with 6 stamens; **female flower-spike longer** (12–30 cm) and occasionally branched. Fruits slightly narrower (13–16 mm diam.).

CONSERVATION STATUS

Endemic to China: Least Concern (MEPCAS, 2013); Not Evaluated (IUCN Red list, accessed 20/8/2016).

SUBSTITUTION COMMENT. *D. spongiosa* and *D. futschauensis* are close relatives of *D. collettii* var. *hypoglauca* (source of 'Fen Bi Xie' described above) and the living plants and rhizomes of these 3 species are easily confused. They occupy similar habitats and share similar distributions across China and require considerable expertise to distinguish them reliably. Note that the rhizomes of *D. spongiosa* and *D. futschauensis* are also the source of the separate TCM drug 'Mian Bi Xie' (Dioscoreae Spongiosae Rhizoma) (see page 206). These 2 drugs are reported to be used interchangeably in TCM clinical situations Bensky *et al*. (2004) owing to their similar actions.

UNOFFICIAL SUBSTITUTE 2

D. futschauensis Uline ex R. Knuth

PLANT DESCRIPTION

Easily confused with the Pharmacopoeia species. *Main characters*: rhizome **irregularly cylindrical** (1–3.5 cm diam.); leaf blades slightly leathery, drying dark grey; **basal leaves 7-lobed**; leaf margins not transparent; male flowers solitary, occasionally paired, **yellow, turning orange**, drying black; flowers with short stalks (1–1.5 mm long); stamens 3 or 6; fruits dark brown, **not shiny**, ovoid and widest above the middle (15–18 mm diam.).

CONSERVATION STATUS

Endemic to China: Near Threatened (MEPCAS, 2013); Not Evaluated (IUCN Red list, accessed 20/8/2016).

TCM PROPERTIES AND USE

***properties*:** **'bitter and neutral'**
***channels*:** **'Kidney and Stomach'**
***action*:** **'separates clear from turbid, dispels wind damp'**

***indications*:** **urinary infections; acute leucorrhoea; musculoskeletal pain**

Mature fruiting stem of *D. collettii* var. *hypoglauca*

D. collettii var. *hypoglauca* with male flowers

Fresh rhizomes of *D. collettii* var. *hypoglauca*

D. spongiosa (wild) with mature capsules

Fresh wild-sourced rhizomes of *D. spongiosa*

D. futschauensis with orange male flowers

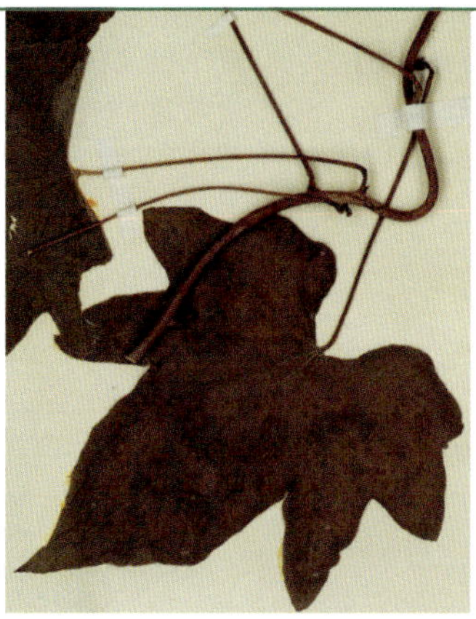

D. futschauensis with palmately 7-lobed basal leaves

Lower leaf surface of *D. futschauensis* clothed with dense mat of minute hairs [hl]

Fresh rhizomes of *D. futschauensis*

PHARMACOPOEIA SPECIES

Dioscorea collettii var. *hypoglauca*

Dioscoreaceae

DRUG MORPHOLOGY

CRUDE DRUG/DECOCTING PIECES (DRIED AND SLICED RHIZOME): thin and irregular in outline **[1]** (**c. 2 cm diam.**; c. 0.5 mm thick). Outer skin, when present, thin, **brownish-black or greyish-brown [2]**, rough, with scattered raised root-scars **[3]**; cut surface yellowish-white or pale greyish-white **[4]**, rough, with scattered vascular bundles. *Texture:* loose, slightly pliable and elastic; easily broken. *Fracture:* **slightly powdery** (hence the use of 粉 'fen' meaning powdery in the drug name). *Odour:* indistinct. *Taste:* acrid and slightly bitter. ***Good quality*:** thin, yellowish-white and somewhat elastic with a powdery fracture.

PROCESSING

After removal of fibrous roots, rhizomes are sliced and sun-dried (= crude drug/ decocting pieces).

OTHER COMMON NAMES

萆薢 Bi Xie; 粉背薯蓣 Fen Bei Shu Yu (plant name); 黄萆薢 Huang Bi Xie; hypoglaucous collett yam rhizome (CP2010); yellow tokoro yam; Rhizoma Dioscoreae Hypoglaucae; tokora; tokoro yam (SCN).

UNOFFICIAL SUBSTITUTE 1

D. spongiosa

Dioscoreaceae

'DRUG SUBSTITUTE' MORPHOLOGY

Similar to the Pharmacopoeia drug above. *Main differences:* **rhizome wider [5]** (**3–6 cm. diam**.; c. 2–5 mm thick). **Outer skin yellowish-brown [6].** Cut surface white to pale greyish-brown and rough **[7]**. ***Texture:* slightly elastic.** *Fracture:* spongy. *Taste:* slightly bitter. *For more robust differentiation from 'Fen Bi Xia' above, laboratory-based authentication methods are recommended.*

OTHER COMMON NAMES

萆薢 Bi Xie; 绵萆薢 Mian Bi Xie (also the plant name); 山薯 Shan Shu; sevenlobed yam rhizome (CP2010).

UNOFFICIAL SUBSTITUTE 2

D. futschauensis

Dioscoreaceae

'DRUG SUBSTITUTE' MORPHOLOGY

Similar to the official species but slices 1–3 cm diam.; **outer skin yellowish-brown [8]** and internal tissues. ***Fracture*: starchy [9]** (not powdery). Taste: slightly bitter. *For more robust differentiation from 'Fen Bi Xie' above, laboratory-based authentication methods are advised.*

OTHER COMMON NAMES

萆薢 Bi Xie; 福州薯蓣 Fu Zhou Shu Yu (plant name); sevenlobed yam rhizome (CP2010).

LESSER KNOWN UNOFFICIAL SUBSTITUTES

The genus *Dioscorea* is a difficult group to identify to species level with 52 species growing wild in China. Many species have been reported in the literature as accidental or intentional substitutes of 'Fen Bi Xie' along with closely related species from other genera in the same family. Their occurrence in international trade is uncertain:

1. ***D. gracillima*** Miq. (纤细薯蓣 Xian Xi Shu Yu) — plant name.
2. ***D. nipponica*** Makino (穿龙薯蓣 Chuan Long Shu Yu) — plant name.
3. ***D. tokoro*** Makino ex Miyabe (山萆薢 Shan Bei Xie) — plant name.

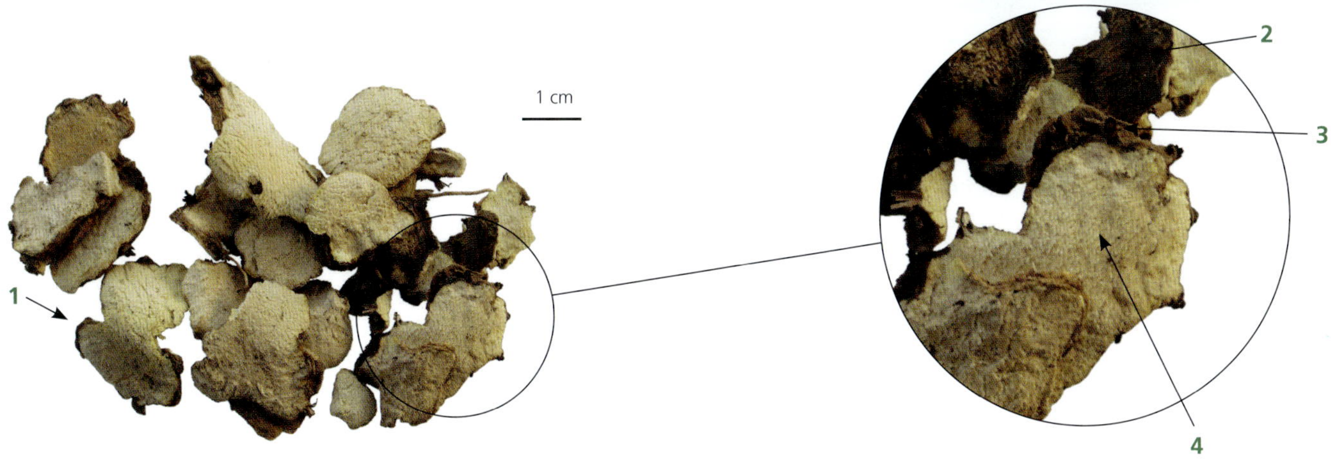

Crude drug/decocting slices of Fen Bi Xie (*D. spongiosa* var. *hypoglauca*)

Crude drug (decocting slices of Mian Bi Xie (*D. spongiosa*)

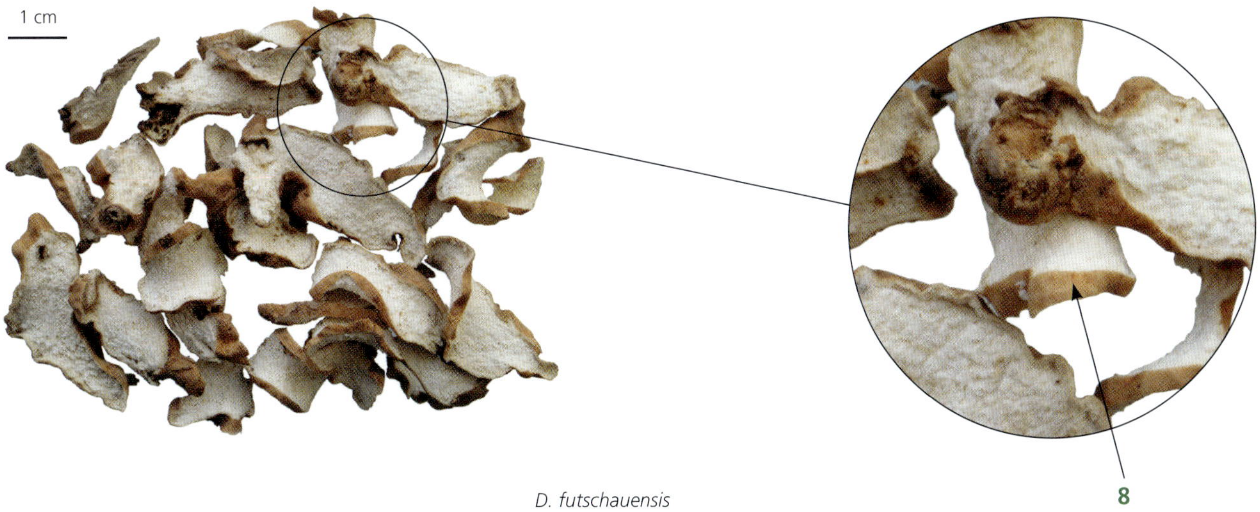

D. futschauensis

Fen Ge 粉葛

(literal: powdery Pueraria)

Drug part: ROOT TUBER

Puerariae Thomsonii Radix

Pueraria montana (Lour.) Merr. **var. *thomsonii*** (Benth.) M. R. Almeida

Fabaceae

synonym: *P. thomsonii* Benth. (name used in CP2015).

Species of *Pueraria* wild in China: **10**

PLANT DESCRIPTION

A robust, climbing perennial (up to 8 m) with starchy and powdery roots. Stem bases woody; covered throughout in yellowish hairs. Stipules and stipels present. *Leaves* alternate; leaf blades large and divided into 3 leaflets: terminal leaflet broadly ovate (usually 7–15 cm long; 5–12 cm wide); margin entire, tip pointed; lateral leaflets smaller, obliquely ovate, with yellow hairs (denser on lower surface). *Flowers* in a spike (raceme) along a single axis (15–30 cm long) with terminal flowers opening last; fragrant. Each flower short-stalked and pea-flower shaped and laterally symmetrical; calyx 5-lobed (up to 2 cm long) with lowermost lobe distinctly longer than lateral ones; petals 5, unequal, deep to pale purple, large posterior ('standard') petal suborbicular, 16–18 mm long; 2 lateral ('wing') petals sickle-shaped and slightly shorter than 'keel' (2 fused lower petals); 'keel' sickle-shaped to oblong. *Fruit* a long, flattened, elliptical pod (10–14 cm long; 10–13 mm diam.), covered in brown hairs. Fl. Sep., fr. Nov.

HARVESTING, SOURCING AND NATURAL RANGE

Root tubers are harvested in autumn and winter mostly from medicinal crop plants grown in the S and C Chinese provinces of Guangxi, Guangdong, Jiangxi and Chongqing. This variety is also commonly grown as a vegetable since it is a rich source of food starch.

The wild species grows in thickets and open forests mostly in S, C and W China (Chongqing, Guangdong, Guangxi, Hainan, Jiangxi, Taiwan, Tibet and Yunnan). [Also native to Bhutan, India, Laos, Myanmar, Philippines, Thailand and Vietnam].

CONSERVATION STATUS

China: Least Concern (MEPCAS, 2013).

Global: Not Evaluated (IUCN Red List, accessed 11/8/2016).

Flowering plant of *Pueraria montana* var. *thomsonii*

Freshly harvested crop plant of *P. montana* var. *thomsonii* with mature root tubers

TCM PROPERTIES AND USE

properties: **'sweet, pungent and cool'**
channels: **'Spleen and Stomach'**
action: **'dispels wind, releases exterior, clears heat, generates fluids**

indications: **headaches, fevers, stiffness in neck; diarrhoea; stroke**

DRUG MORPHOLOGY

CRUDE DRUG (WHOLE DRIED TUBER) — seldom traded: intact tubers semi-cylindrical or narrowing at both ends **[1]** (12–15 cm long, 4–8 cm diam.). Externally greyish-brown if root-bark present, or if removed, pale brown or creamish-white **[2]**). *Texture*: starchy, powdery and heavy. *Odour*: indistinct. *Taste*: slightly sweet. ***Good quality*:** white, solid and powdery.

DECOCTING PIECES (SLICED OR CUBED TUBER): size and shape varies according to style of slicing: transverse, longitudinal, oblique or, more usually, cubed **[3]**. Cut surfaces white or yellowish-white and powdery **[4]**; transverse sections with concentric fibre rings; longitudinal sections with fibres running lengthwise. *Texture*: starchy and powdery. *Odour*: indistinct. *Taste*: slightly sweet.

PROCESSING

Root tubers are cut, usually when fresh, into thick slices or cubes then sun-dried (decocting pieces); alternatively roots are dried first (= crude drug — rarely seen), then softened later for cutting and drying again (= decocting pieces). Sulphuring is sometimes undertaken to prolong shelf-life and enhance appearance.

OTHER COMMON NAMES

甘葛藤 Gang Ge Teng; 葛根 Ge Gen; kudzu; powdery kudzu; Radix Puerariae Thomsonii; Thomson kudzu vine root (CP2010, as *P. thomsonii* Benth.).

SAFETY

Safety Class: 1; Interaction Class: A (Gardner & McGuffin, 2013 as *P. montana*). Avoid use with anti-coagulant or anti-platelet medication (Bensky *et al.*, 2004, Chen & Chen, 2004). An official food and healthfood in China (Anon., 2002a,b).

Fen Ge crude drug (peeled and dried root-tuber)

Fen Ge decocting pieces
(peeled, dried and cubed root tuber)

UNOFFICIAL SUBSTITUTE

For many years (until the CP2005 edition) root tubers of this Pharmacopoeia variety were officially used interchangeably with those of variety *lobata* and under the shared name Ge Gen 葛根. Owing to differing clinical properties the two varieties have since been recognised as two official drugs in the CP (for an identification comparison see Ge Gen (Puerariae Lobatae Radix) page 146)

Gan Jiang 干姜

Zingiberis Rhizoma

dried ginger

Drug part: DRIED RHIZOME

Zingiber officinale Roscoe

Zingiberaceae

PLANT DESCRIPTION

An erect perennial herb with **branched, tuberous, yellow-fleshed and strongly aromatic rhizomes**. 'Stems' are pseudostems (overlapping leaf bases) up to 1 m tall. *Leaves* alternate, in two rows on opposite sides of the 'stem', virtually parallel to each other; leaf stalks swollen; ligule (at base of leaf stalk) membranous (2–4 mm long); leaf blades simple, lance-shaped to linear-lance-shaped (15–30 cm long; 2–2.5 cm wide), margins entire. *Flowers* bisexual, in erect, conspicuous and somewhat ovoid spikes (4–5 cm long; c.1.5 cm diam.) on a flower stalk (up to 25 cm long) **which arises directly from the rhizomes** at base of the plant. *Bracts* (encircling a single flower) conspicuous and overlapping, waxy, pale green (sometimes yellowish at the margins), ovate and with a short, rigid tip; *bracteoles* same size as bracts. Each flower with a tubular calyx split along one side; petals yellowish-green and fused into a slender 3-lobed tube (2–2.5 cm long) with lobes of different sizes; the petal-like lip (labellum) c. 6 mm long (shorter than petal lobes) with a purple stripe and cream blotches; stamens dark purple; ovary inferior. *Fruit* a dry, many-seeded capsule. Fl. Oct.

Species of *Zingiber* wild in China: 41

HARVESTING, SOURCING AND NATURAL RANGE

Not native to China but widely cultivated in S and C China as a medicine and a culinary spice in the provinces of Anhui, Fujian, Guangdong, Guangxi, Guizhou, Hainan, Henan, Hubei, Hunan, Jiangxi, Shaanxi, Shandong, Sichuan, Taiwan, Yunnan and Zhejiang. Rhizomes are harvested in autumn with *Di Dao*-quality traditionally sourced from localities in Sichuan province. Native origin unknown. Widely cultivated worldwide in the tropics and subtropics.

CONSERVATION STATUS

China: Introduced.

Global: Not Evaluated (IUCN Red List, accessed 20/8/2016).

Crop of *Zingiber officinale*, Yunnan province

Inflorescence in bud

Freshly unearthed rhizomes

TCM PROPERTIES AND USE

***properties*: 'pungent and hot'**
***channels*: 'Heart, Lung, Spleen and Stomach'**
***actions*: 'warms interior, resolves phlegm, dispels cold damp '**

***indications*: abdominal distension, vomiting, diarrhoea; cough, phlegm; metrorrhagia; haematemesis**

DRUG MORPHOLOGY

CRUDE DRUG (DRIED RHIZOME): irregularly-shaped, solid, somewhat flattened and rounded (3–7 cm long; 1–2 cm diam.); sometimes with short finger-like branches with slightly bulbous ends (stem-buds) **[1]**. Externally greyish-yellow, matt, with raised and irregularly-shaped wrinkles throughout **[2]**; one end sometimes truncated by a node **[3]** (sometimes annular rings visible); sunken crater-like stem and root scars occasional **[4]**. *Texture*: dense and solid. *Fracture:* yellowish-white or greyish-white, fibrous and somewhat starchy, with endodermis ring and oil drops visible. *Odour*: intensely and distinctly aromatic. *Taste*: acrid. ***Good quality*:** white, heavy and powdery; intensely aromatic.

DECOCTING PIECES-1 (DRIED SLICED RHIZOME 'GAN JIANG PIAN'): irregularly-shaped oblique or longitudinal slices (1–6 cm long; 1–2 cm wide; 0.2–0.4 cm thick), sometimes branching. Cut surfaces greyish-yellow or greyish-white, slightly starchy with traces of hair-like fibres **[5]**. Outer edges rough and irregularly wrinkled, greyish-yellow or pale yellowish-brown **[6]**. *Texture*, *Odour* etc as for crude drug.

DECOCTING PIECES-2 (CARBONISED RHIZOME 'JIANG TAN'): as for decocting pieces-1 but externally blackish-brown, internally brown **[7]**. *Texture*: light and brittle. *Taste*: slightly bitter and slightly pungent.

PROCESSING

After removal of fibrous roots, rhizomes are dried at low temperature (= crude drug, Gan Jiang) then sliced, or sliced fresh then dried = 'Gan Jiang Pian' (decocting pieces-1). Further processing may be undertaken by carbonising the rhizomes to produce 'Jiang Tan' (decocting pieces-2).

OTHER COMMON NAMES

白姜 Bai Jiang 'white ginger'; 川姜 Chuan Jiang (Sichuan ginger); 犍干姜 Jian Gan Jiang; ginger root; Rhizoma Zingiberis. Many other trade names exist for this entity reflecting its medicinal and culinary use in other countries (see *Kew-MPNS*).

SAFETY

Safety Class: 1; Interaction Class: B. Gardner & McGuffin, 2013.
An official food and health food in China (Anon, 2002a).

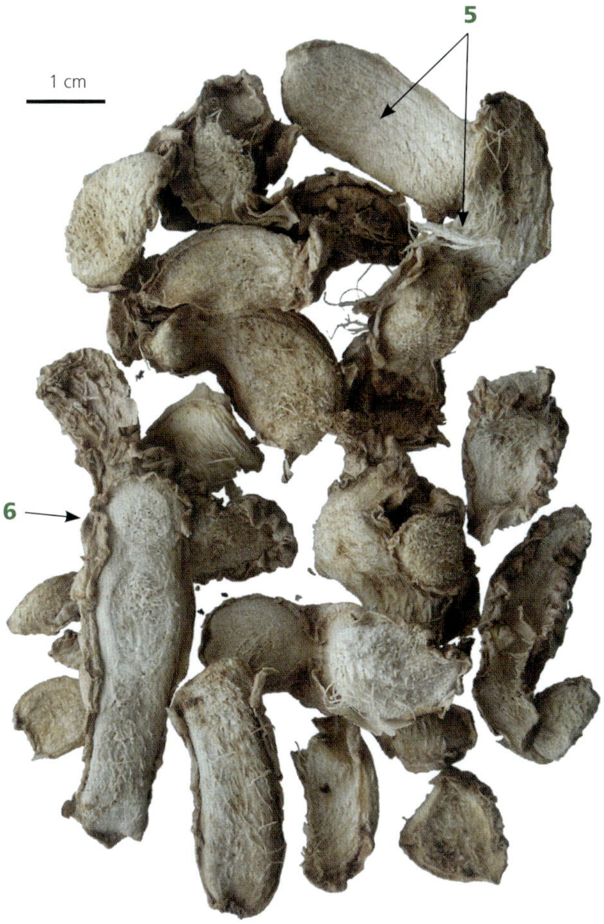

Decocting pieces-1 Gan Jiang Pian
(sliced then dried rhizomes)

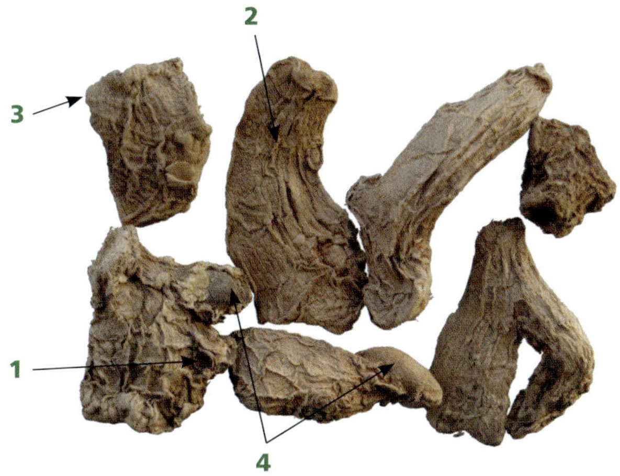

Crude drug of Gan Jiang (dried rhizome)

Decocting pieces-2 Jiang Tan
(carbonised rhizomes)

Gan Jiang – Pao Jiang 炮姜 Zingiberis Rhizoma Praeparatum

Drug part: PROCESSED DRIED RHIZOME

Zingiber officinale Roscoe **Zingiberaceae**

For PLANT DESCRIPTION, HARVESTING etc — see 'Gan Jiang' (Zingiberis Rhizoma) page 138.

PROCESSING

Clean sand is heated in a wok until very hot, then the crude drug of 'Gan Jiang' (page 138) is added, stirred until inflated then removed and sifted to remove any remaining sand (= Pao Jiang).

Gan Jiang – Sheng Jiang 生姜 Zingiberis Rhizoma Recens

fresh ginger

Drug part: FRESH RHIZOME

Zingiber officinale Roscoe **Zingiberaceae**

For PLANT DESCRIPTION, HARVESTING etc — see 'Gan Jiang' (Zingiberis Rhizoma) page 138.

PROCESSING

After removal of fibrous roots, rhizomes are washed (= crude drug). Thickly cut slices are called 'Sheng Jian Pian' (= decocting pieces-1). Finely sliced peel (with flesh discarded) is called 'Jiang Pi '(= decocting pieces-2).

Sheng Jiang crude drug (fresh rhizomes of *Zingiber officinale*)

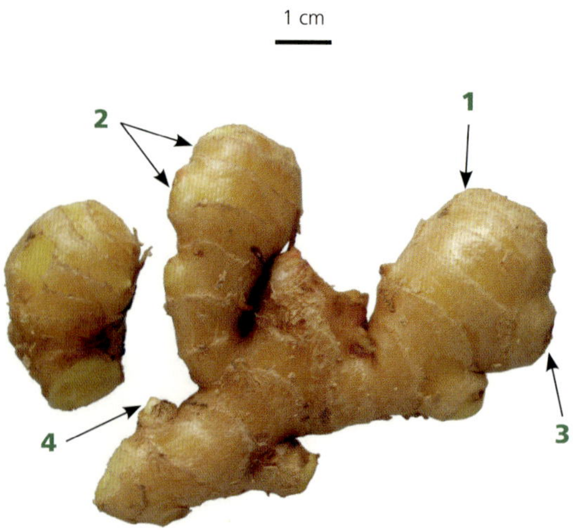

Crude drug of Sheng Jiang

TCM PROPERTIES AND USE

***properties*: 'pungent and hot'**
***channels*: 'Spleen, Stomach and Kidney'**
***actions*: 'warms interior, stops bleeding, astringes'**

***indications*: abdominal pain; vomiting, diarrhoea; metrorrhagia; haematemesis**

DRUG MORPHOLOGY

DECOCTING PIECES 'PAO JIANG': irregularly-shaped **[1]**, inflated and somewhat nodular **[2]**. Externally brownish-black or yellowish-brown, much smoother than 'Gan Jiang' (page 138). *Texture*: light and open. *Fracture*: brownish-yellow with scattered fine, vascular bundles. *Odour*: distinctly aromatic. *Taste*: faintly pungent.

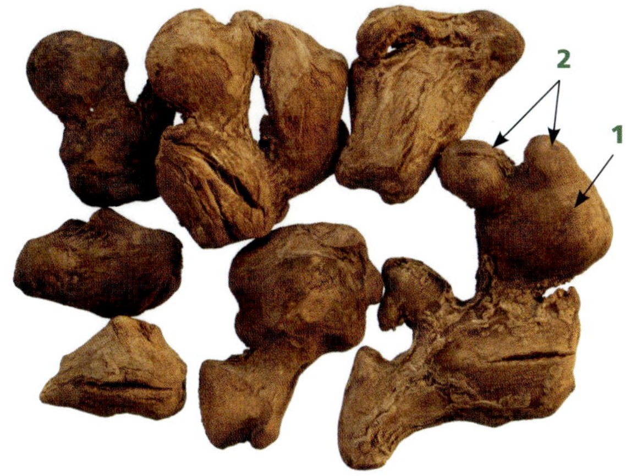

Pao Jiang (sand-scalded dried rhizome)

SAFETY

As for 'Gan Jiang' (page 138).

TCM PROPERTIES AND USE

***properties*: 'pungent and slightly warm'**
***channels*: 'Lung, Spleen and Stomach'**
***actions*: 'releases wind cold, alleviates nausea'**

***indications*: respiratory tract infection, cough; nausea, vomiting; seafood poisoning; reduces toxicity of other herbs**

DRUG MORPHOLOGY

CRUDE DRUG (FRESH RHIZOME): irregularly-shaped and slightly compressed (4–18 cm long, 1–3 cm thick), **often branched with tips of branches usually swollen [1].** Externally golden- or yellowish-brown to pinkish-white, with fine and widely spaced annular growth rings **[2]**; surfaces with scattered stem scars **[3]** and buds **[4]**. *Texture*: fragile and easily broken. *Fracture*: pale yellow; endodermis ring distinct; vascular bundles scattered. *Odour*: **distinctively aromatic.** *Taste*: **pungent**. ***Good quality*:** plump and pungent.

DECOCTING PIECES-1 'SHENG JIAN PIAN': long-ovate slices **[5]** (c. 3–4 cm long; c. 1–3 cm diam.; 2–3 mm thick). Uncut edges with remains of thin rhizome peel, golden- or yellowish-brown to pinkish-white **[6]**; widely spaced annular growth rings often visible as dark markings **[7]**. Cut surface pale yellow with a distinct endodermal ring **[8]** and scattered vascular bundles **[9]**. *Texture*: somewhat translucent and moist; easily broken. *Odour*: distinctively aromatic; *Taste*: pungent.

DECOCTING PIECES-2 'JIAN PI': irregularly-shaped slices of peel **[10]** (c. 3–4 cm long; c. 1–3 cm diam.; c. 1 mm thick). Surfaces golden- or yellowish-brown to pinkish-white; widely spaced annular growth rings often visible as slightly darker, regularly spaced lines **[11]**; stem **[12]** and root scars occasional. *Texture*: translucent and moist; fragile. *Odour*: distinctively aromatic; *Taste*: pungent.

Sheng Jiang Pian (decocting pieces-1)

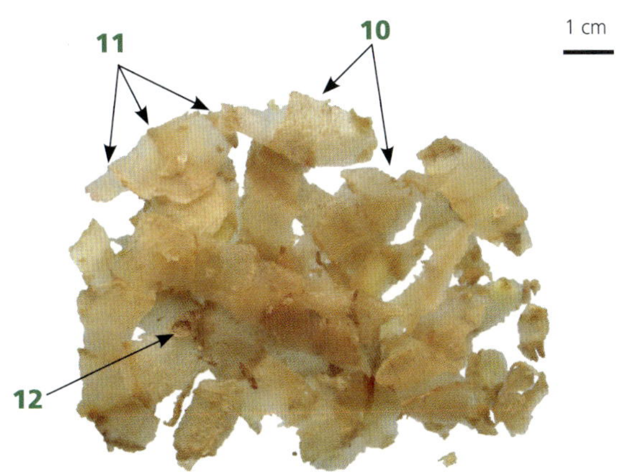

Jiang Pi (decocting pieces-2)

SAFETY

As for 'Gan Jiang' (page 138).

Gao Ben 藁本 Ligustici Rhizoma et Radix

Drug parts: RHIZOME AND ROOT

Ligusticum sinense Oliv.

Apiaceae

synonyms: *Carum anthriscoides* H. Boiss.; *Conioselinum anthriscoides* (H. Boiss.) Pimenov & Kljuykov (*Kew-MPNS* accepted name, 22/9/2016).

Species of *Ligusticum* wild in China: 40

PLANT DESCRIPTION

A robust, perennial herb (30–60 cm tall, rarely to 1 m). Rootstock thick and twisted (= medicinal part 'Gao Ben'); rhizomes highly aromatic with internodes short and **nodes swollen or tubercular;** roots curved and somewhat tapering. *Stem* single and branching. *Leaves* alternate; basal and lower ones stalked (10–20 cm) and with sheathing bases; leaf blades finely divided (1- to 2-ternate/pinnate); stem leaves unstalked, smaller than basal ones and 1-pinnate; leaflets ovate or oblong-ovate (2–3 cm long, 1–2 cm wide); margins irregularly saw-toothed; teeth sharply pointed (acute), *not* ending abruptly in a point. *Inflorescences* terminal or lateral compound umbels (6–8 cm diam. when in fruit); **main umbels with bracts 5 or 6 (rarely 10),** linear; rays (umbel stalks) 15–30, unequal (3–5 cm long); **umbellets with 5–8 bracteoles, shorter than flowers,** linear and reflexed. *Flowers* tiny; calyx absent; petals 5, white, wedge-shaped with notched tips; styles 2, reflexed. *Fruit* in umbellate clusters; each fruit dry, oblong-ovoid (**2–3 mm long; 1.5–2 mm diam.**), dorsally compressed and splitting into 2 one-seeded carpels (mericarps); dorsal ribs slender and prominent; lateral ribs narrowly winged; **persistent styles as long as the fruit,** reflexed; 1–3 (rarely 4) oil ducts in each furrow with 4–6 on the flat internal (commissural) surface. Fl. Jul–Aug; ft. Sep–Oct.

HARVESTING, SOURCING AND NATURAL RANGE

Rootstocks are harvested mostly from wild plants either in autumn after stems and leaves have withered or in spring when new growth has appeared. The species grows wild in forests, meadows and beside streams (500–2,700 m altitude) and is endemic to China (Gansu, Guizhou, Henan, Hubei, Jiangxi, Inner Mongolia, Shaanxi, Sichuan and Yunnan). It is occasionally grown commercially on medicinal plant farms.

CONSERVATION STATUS

China: Least Concern (MEPCAS, 2013, as the type variety '*sinense*').

Global: Not Evaluated (IUCN Red List, accessed 19/8/2016).

L. jeholense (Nakai & Kitag.) Nakai & Kitag.

Apiaceae

synonyms: *Cnidium jeholense* Nakai. & Kitag.; *Conioselinum smithii* (H. Wolff) Pimenov & Kljuykov (*Kew-MPNS* accepted name, 22/9/2016); *Ligusticum smithii* H. Wolff.

PLANT DESCRIPTION

Similar to *L. sinense* (PS1) but **rootstock not swollen at nodes**, roots tapering at both ends; stem purplish; leaflet margins 3–5-lobed; **main umbels with bracts 2; rays 8–16** (2–3 cm long), **umbellets with 8–10 bracteoles, longer than flowers**; fruit slightly larger (3–4 mm long; 2.2.5 mm diam.); **persistent styles c. half as long as the fruit**, reflexed; 2–4 oil ducts on the flat internal (commissural) surface.

HARVESTING, SOURCING AND NATURAL RANGE

Harvesting as for *L. sinense* (PS1). Endemic to north-east China (Hebei, Jilin, Liaoning, Shandong and Shanxi) where it grows between 1,200–2,500 m altitude. Occasionally cultivated for medicinal use.

CONSERVATION STATUS

China: Least Concern (MEPCAS, 2013).

Global: Not Evaluated (IUCN Red List, accessed 19/8/2016).

UNOFFICIAL SUBSTITUTES

Conioselinum vaginatum (Spreng.) Thell. 新疆藁本 Xinjiang Gao Ben (Apiaceae family). A close relative of the Pharmacopoeia species used as a local substitute of 'Gao Ben' in NW China (Xinjiang province) where it is native and, in recent years, has been successfully established as a medicinal crop. Recent pharmacological research of its rootstocks demonstrates their potential as a viable clinical alternative to 'Gao Ben' (Li *et al.*, 2012a and 2013) and may help to divert pressure from wild-harvested populations of this and the Pharmacopoeia species. A taller plant (60–120 cm); *stem* bases **without fibrous remnant sheaths; stem leaves stalked (6–9 cm long);** outermost leaflet lobes narrowly ovate to lanceolate (1.5–2 cm long, 0.5–0.8 cm wide). *Inflorescences:* **bracts absent; rays 10–14** (2–4 cm long); **bracteoles 5–8.** Fl. and fr. Jul–Sep.

Lesser Known Unofficial Substitutes:

Over the years, rootstocks of other closely related species in the Apiaceae family have been reported as regional substitutes or adulterants of official 'Gao Ben' in China (Bensky *et al.*, 2004; *FoC*; Wright, 2004); their occurrence as substitutes in international trade is unclear:

1. *Angelica cartilaginomarginata* (Makino ex Y. Yabe) Nakai. 长鞘当归 Chang Qiao Dang Gui 'long-sheathed Angelica' or 山藁本 Shan Gao Ben 'mountain Gao Ben'. Root short, conical and branched;
2. *Ligusticum acuminatum* Franch. 尖叶藁本 Jian Ye Gao Ben 'pointed leaf Gao Ben'. A regional substitute in west Sichuan province;
3. *L. pteridophyllum* Franch. 蕨叶藁本 Jue Ye Gao Ben 'fern-leaved Gao Ben'. A regional substitute in Yunnan province. Rootstock small and globose with slender rhizome internodes;
4. *L. chuanxiong* Hort. ex S. H. Qiu *et al.* 川芎 Chuan Xiong. Rootstock an irregular, fist-shaped mass (2–7 cm diam.) and the source of the separate TCM drug 'Chuan Xiong' (Chuanxiong Rhizoma) see page 88.
5. *L. tenuissimum* (Nakai) Kitag. 细叶藁本 Xi Ye Gao Ben or 'Han Gao Ben'. A regional substitute in Liaoning province;
6. *Meeboldia yunnanensis* (H. Wolff) Constance & F. T. Pu 滇芹Dian Qin. A regional substitute in Yunnan province. Rootstock a spindle-shaped taproot with prominent annual scars;
7. *Nothosmyrnium japonicum* Miq. 白苞芹 Bai Bao Qin. Rootstock a stout taproot with branching rootlets in fibrous clusters.

TCM PROPERTIES AND USE

properties: 'pungent and warm'
channel: 'Bladder'
actions: 'releases the exterior, disperses wind cold damp, relieves pain'

indications: headaches; musculoskeletal pain; abdominal pain, diarrhoea

Ligusticum sinense with young flower-heads

Compound umbel of *L. sinense*

L. jeholense

Compound umbel of *L. jeholense*

Flowering plant of *L. jeholense*

Developing fruit-head of *L. jeholense*

Fresh rootstock of *L. jeholense*

Ligusticum sinense

Apiaceae

DRUG MORPHOLOGY

CRUDE DRUG (DRIED RHIZOME AND ROOT): a contorted mass, predominantly of occasionally branched and longitudinally striated **[1]** with nodes swollen **[2]** or tubercular **[3]** rhizomes (3–10 cm long; 1–2 cm diam.); outer rhizome surface brown to dark brown. Apical parts of rhizome with **stem remnants > 5 mm diam. [4]** or with sunken, orbicular **stem scars (> 5 mm diam.) [5]**; lower and lateral parts dotted with prominent root scars **[6]**; some curved and tapering roots still attached **[7]**. *Texture:* light, hard and brittle. *Fracture:* light brownish-yellow, fibrous. *Odour:* aromatic. *Taste:* bitter and somewhat numbing. ***Good quality*:** large and intensely aromatic.

DECOCTING PIECES (SLICED RHIZOME AND ROOT): irregularly-shaped slices of rhizome (c. 0.5–4 cm long; 2–4 mm thick) with outer uncut rhizome surface dull brown to blackish-brown and coarse; when sliced longitudinally, with an undulating outline; swollen stem nodes, orbicular stem scars or root scars occasionally visible. Cut surface pale yellowish-cream or pale yellowish-brown **[8]** with a heterogeneous mix of tissues granular or fibrous, often with cavities **[9]**; yellowish-brown oil glands scattered throughout **[10]**. *Texture:* light, hard and easily broken. *Odour:* strongly aromatic. *Taste:* bitter, pungent and slightly numbing.

PROCESSING

Rootstocks are dried (= crude drug) then, later, softened thoroughly and cut into thick (2–4 mm) decocting slices before drying again in the sun.

OTHER COMMON NAMES

香藁本 Xiang Gao Ben 'aromatic Gao Ben'; 西芎 Xi Xiong; 西芎藁本 Xi Xiong Gao Ben; Chinese lovage (CP2010; SCN); Rhizoma et Radix Ligustici.

SAFETY

Safety Class: 1; Interaction Class: A (Gardner & McGuffin, 2013).

L. jeholense

Apiaceae

DRUG MORPHOLOGY

CRUDE DRUG (DRIED RHIZOME AND ROOT): similar to that of PS1 but **rhizomes less tubercular [11]** and smaller and **nodes less obviously swollen [12]**. Stem remnants and stem scars **< 5 mm diam. [13]. Roots more numerous and often distinctly curved [14].**

SLICED RHIZOMES AND ROOTS (DECOCTING PIECES): Similar to those of *L. sinense* (PS1) but **stem remnants < 5 mm diam. [14]. roots more numerous and curved [15].**

PROCESSING

As for PS1.

OTHER COMMON NAMES

北藁本 Bei Gao Ben 'northern Gao Ben'; 辽藁本 Liao Gao Ben; 香藁本 Xiang Gao Ben 'aromatic Gao Ben'; Chinese lovage (CP2010; SCN); Rhizoma et Radix Ligustici.

SAFETY

Safety Class: 1; Interaction Class: A (Gardner & McGuffin, 2013).

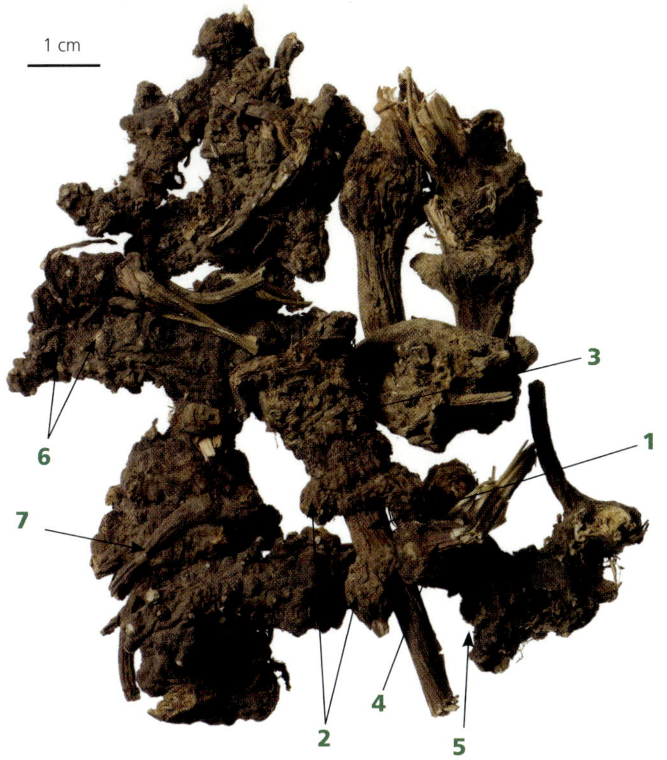

Crude drug of Gao Ben (sourced from wild-harvested *L. sinense*)

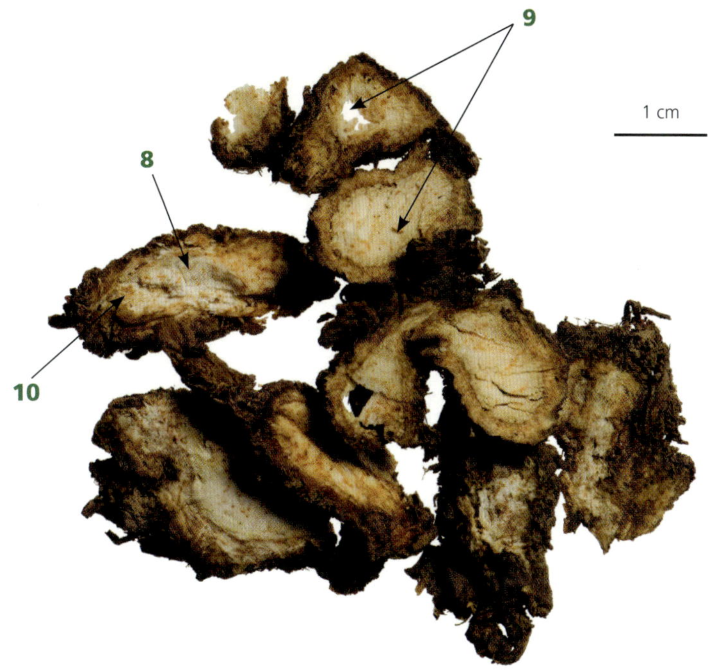

Decocting pieces of Gao Ben transversely sliced (sourced from *L. sinense*)

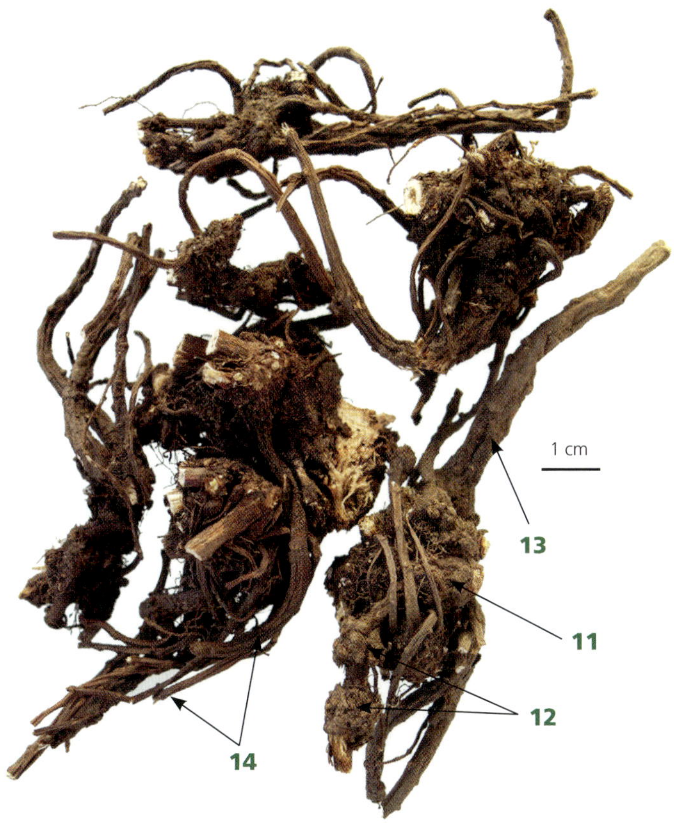

Crude drug of Gao Ben (sourced from *L. jeholense*)

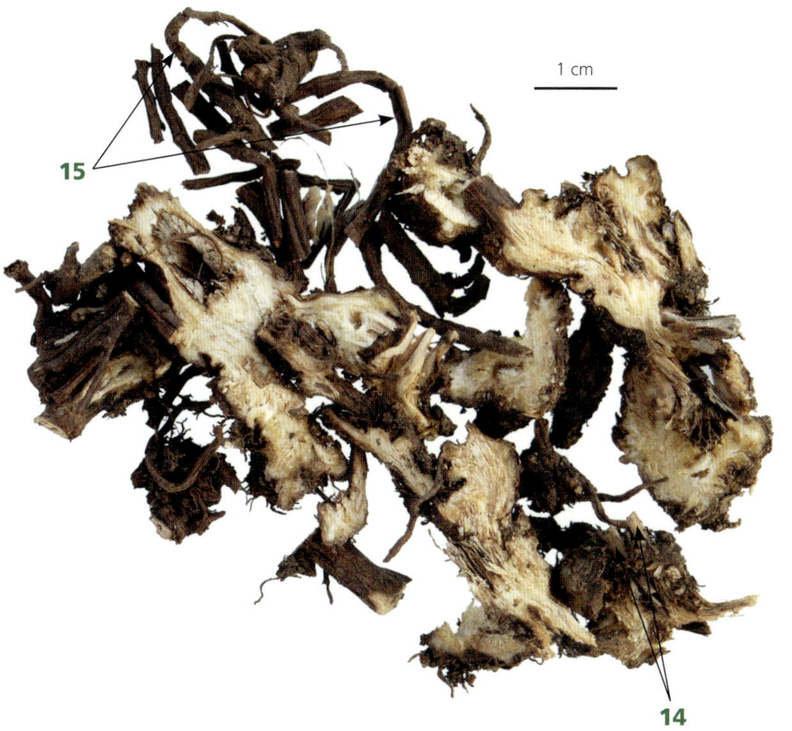

Decocting pieces of Gao Ben (sourced from *L. jeholense*)

Ge Gen 葛根 — Pueraria Lobatae Radix

Drug part: ROOT TUBER

PHARMACOPOEIA SPECIES

Pueraria montana (Lour.) Merr. **var.** ***lobata*** (Willd.) Maesen & S. M. Almeida ex Sanjappa & Predeep

Fabaceae

synonyms: *P. lobata* (Willd.) Ohwi (name used in CP2015); *Dolichos lobata* Willd.

Species of *Pueraria* wild in China: 10

PLANT DESCRIPTION

A vigorous, climbing perennial (up to 8 m) with fibrous and somewhat pliable root tubers and which often grow to a huge size. *Stem* bases woody; covered throughout in yellowish hairs. Stipules and stipels present. *Leaves* alternate, divided into 3 leaflets: terminal leaflet broadly ovate (c. 7–15 cm long; c. 5–12 cm wide); margin entire, tip pointed; lateral leaflets smaller, obliquely ovate, with yellow hairs (denser on lower surface). *Flowers* in a spike (raceme) along a single axis (15–30 cm long) with terminal flowers opening last; fragrant. Bracts deciduous; bracteoles <2 mm long. Each flower short-stalked and pea-flower shaped; **calyx** 5-lobed **(8–10 mm long)** with long, yellowish-brown hairs and lowermost lobe distinctly longer than lateral ones; petals unequal, deep to pale purple (large posterior 'standard' petal obovate, **10–12 mm long;** 2 lateral ('wing') petals sickle-shaped and similar length to the 'keel' (2 fused lower petals); keel sickle-shaped to oblong). *Fruit* a long, flattened, elliptical pod (**5–9 cm long**; 8–11 mm diam.) covered in brown hairs. Fl. Jul–Oct, fr. Oct–Dec.

HARVESTING, SOURCING AND NATURAL RANGE

Root tubers are harvested in autumn and winter from wild plants which grow in mountain forests, scrub and along roadsides throughout China (except Qinghai, Tibet and Xinjiang). Also cultivated locally for the medicinal use of its other plant parts (flower, stem and seed) (Zhao & Xiao, 2009) as well as a cover crop and for fodder. The fibrous tubers are always sliced before being dried. The species has a large natural range from SE Asia to Australia and has been introduced in Africa, America and Europe.

CONSERVATION STATUS

China: Least Concern (MEPCAS, 2013). Often an invasive weed *(*authors' pers. obs.,1999).
Global: Not Evaluated (IUCN Red List, accessed 11/8/2016).

UNOFFICIAL SUBSTITUTE

P. montana var. ***thomsonii*** (Benth.) M. R. Almeida

Fabaceae

synonym: *P. thomsonii* Benth. (name used in CP2015).

PLANT DESCRIPTION

Very similar to the Pharmacopoeia species above but **roots powdery and less fibrous, calyx much longer (up to 20 mm)**; 'standard' petal sub-orbicular and **larger (16–18 mm long)**, 'wing' petals slightly shorter than 'keel' and **fruits often slightly longer** and wider (**10–14 cm long**; 10–13 mm diam.).

HARVESTING, SOURCING AND NATURAL RANGE

Harvested mostly from cultivated crop plants (see Fen Ge (Puerariae Thomsonii Radix) page 136).

CONSERVATION STATUS

China: Least Concern (MEPCAS, 2013).
Global: Not Evaluated (IUCN Red List, accessed 11/8/2016).

LESSER KNOWN UNOFFICIAL SUBSTITUTES

Roots of other species of *Pueraria* have also been reported (below) in the TCM literature as local or historical substitutes; their occurrence as substitutes in international trade is unlikely:

1. *Pueraria stricta* Kurz; wild in China (Yunnan). A shrub (rarely climbing) up to 2.5 m tall. Flowers white, pink, purple, blue, or yellow; with shorter 'standard' petal (5–8 mm long).
2. *P. alopecuroides* Craib; wild in China (Yunnan). Leaf margins notched with regularly rounded teeth. Flowerheads with 10 cm long branches. Flower with white 'standard' petal with yellow basal spot; 'wings' oblong, longer than 'keel' and with long appendage.
3. *P. calycina* Franch.; endemic to Yunnan province, China. Leaflet margins irregularly 5–7-lobed. Flowers purplish; 'standard' petal apically notched.
4. *P. edulis* Pamp.; Sichuan province, Yunnan. Flower spikes up to 30 cm; flowers purple of pink; 'standard' petal suborbicular (14–17 mm long); apically notched.
5. *P. phaseoloides* (Roxb.) Benth.; wild in China (Guangdong, Guangxi, Taiwan, Yunnan, Zhejiang). Flower spikes 8–15 cm long; flowers bluish or lilac. Fruits 5–8 cm long; c. 4 mm diam.
6. *P. peduncularis* (Benth.) Benth. (苦葛根 Ku Ge Gen): wild in China (Guangxi, Guizhou, Sichuan, Tibet, Yunnan). **Toxic** (Bensky *et al.*, 2004). Distinguished from the official species by **smaller leaflets** (5–12 cm long; 3–8 cm wide) **usually densely hairy on both surfaces**, flower spikes 20–40 cm long; **petals white**, roots dense, difficult to break and **bitter**, hence the name 苦葛根 Ku Gé Gēn (**bitter kudzu vine**).
7. *Wisteria sinensis* (Sims) Sweet; C, S and E China.

TCM PROPERTIES AND USE

properties: 'sweet, pungent and cool'
channels: 'Spleen, Stomach and Lung'
action: 'dispels wind, releases exterior, clears heat'

indications: headaches, fevers, stiffness in neck; skin eruptions

Flowering plants of *Pueraria montana* var. *lobata*

Fresh root tuber densely fibrous throughout

Flowering plant of *Pueraria montana* var. *thomsonii*

Freshly harvested crop plant of *P. montana* var. *thomsonii* with mature root tubers of Fen Ge (c. 2 years old)

Fresh root tubers with white powdery flesh

Pueraria montana var. *lobata*

Fabaceae

DRUG MORPHOLOGY

CRUDE DRUG (FRESH ROOT TUBER, SLICED OR UNSLICED) – not traded in this form: *whole tubers* long and cylindrical **[1]**. Externally pale brown or pale reddish, somewhat mottled and rough **[2]**; very finely wrinkled longitudinally. *Fracture*: concentrically arranged tissues, yellowish-white and pale to dark brown **[3]**. *Texture*: **pliable and densely fibrous** throughout **[4]**. *Odour*: indistinct. *Taste*: sweetish. *Sliced tubers* rectangular, of varying size; concentric tissues of fracture (visible in whole tuber) no longer obvious.

DECOCTING PIECES (SLICED ROOT TUBER) — commonly traded: thick rectangular (5–35 cm long, 0.5–1 cm wide) or smaller cubed pieces **[5]**. **Cut surfaces creamish-white, pale or dark brown with dense longitudinal fibres throughout [6].** *Texture*: **densely fibrous** and easily broken. *Odour*: indistinct. *Taste*: slightly sweet.

Fresh Ge Gen drying by the road, Dangfeng, Shaanxi province

PROCESSING

Root tubers are usually cut when fresh, into thick slices or cubes then sun-dried; less commonly roots are dried first (= crude drug) then softened later for cutting and drying again (= decocting pieces).

OTHER COMMON NAMES

野葛 Ye Ge (wild kudzu vine); 野葛根 Ye Ge Gen (wild kudzu vine root); kudzu (SCN); kudzu vine root (as *P. lobata* (Willd.) Ohwi); Pueraria Radix; Radix Puerariae Lobatae.

SAFETY

Safety Class: 1; Interaction Class: A (Gardner & McGuffin, 2013). Avoid use with anti-coagulant or anti-platelet medication (Bensky *et al.*, 2004, Chen & Chen, 2004). An official food and healthfood in China (Anon., 2002a,b).

P. montana var. *thomsonii*

Fabaceae

'DRUG SUBSTITUTE' MORPHOLOGY

CRUDE DRUG (WHOLE DRIED TUBER) – seldom traded: intact tubers semi-cylindrical or narrowing at both ends **[7]** (12–15 cm long, 4–8 cm diam.). Externally greyish-brown if root-bark present **[8]** or, if removed, pale brown or creamish-white **[9]**). *Texture*: **starchy, powdery and heavy**. *Odour*: indistinct. *Taste*: slightly sweet. ***Good quality***: white, solid and powdery .

DECOCTING PIECES (SLICED OR CUBED TUBER): size and shape vary according to style of slicing: transverse, longitudinal, oblique or, more usually, cubed. Cut surfaces white or yellowish-white and powdery **[11]**; transverse sections with concentric fibre rings **[12]**; longitudinal sections with fibres running lengthwise. *Texture*: starchy and powdery. *Odour*: indistinct. *Taste*: slightly sweet.

SUBSTITUTION COMMENT. Although recent editions of the CP characterise the tubers of this variety as a separate TCM drug (see 'Fen Ge' (Puerariae Thomsonii Radix) page 136) they have long been used interchangeably in clinical settings and there is much confusion about which drug is intended when the trade name 'Ge Gen' is used. Note that decocting pieces of official Ge Gen and Fen Ge (as defined in CP2015) are easily distinguished macroscopically. Recent research (Wong *et al.*, 2015) describes these two TCM drugs as having differing chemical profiles (and possibly pharmacological activity) providing potential additional authentication (and quality control) markers.

OTHER COMMON NAMES

粉葛 Fen Ge; 甘葛藤 Gang Ge Teng; kudzu (SCN); powdery kudzu; Radix Puerariae Thomsonii; Thomson kudzu vine.

SAFETY

As for the Pharmacopoeia species above.

Crude drug of Ge Gen (*P. montana* var. *lobata*)
(fresh tubers, sliced and unsliced)

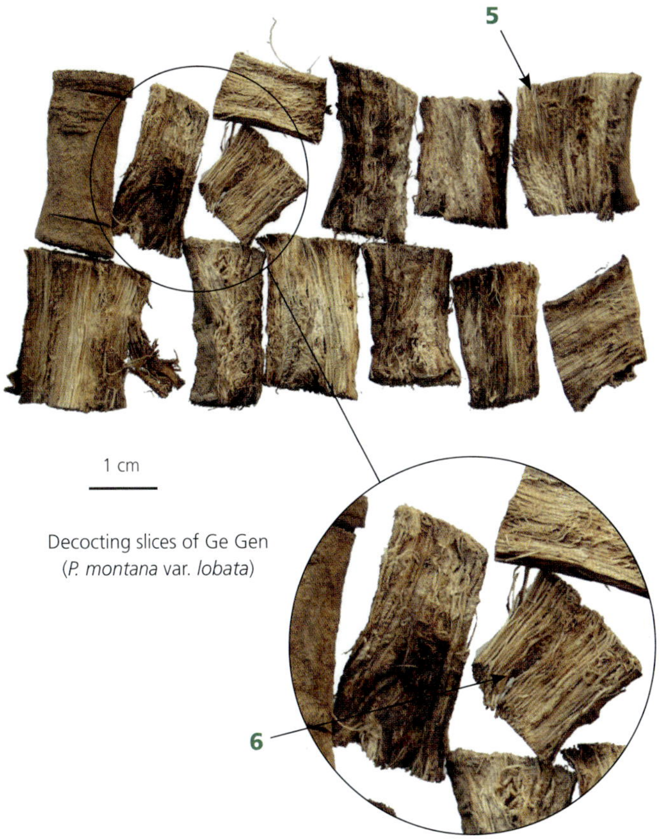

Decocting slices of Ge Gen
(*P. montana* var. *lobata*)

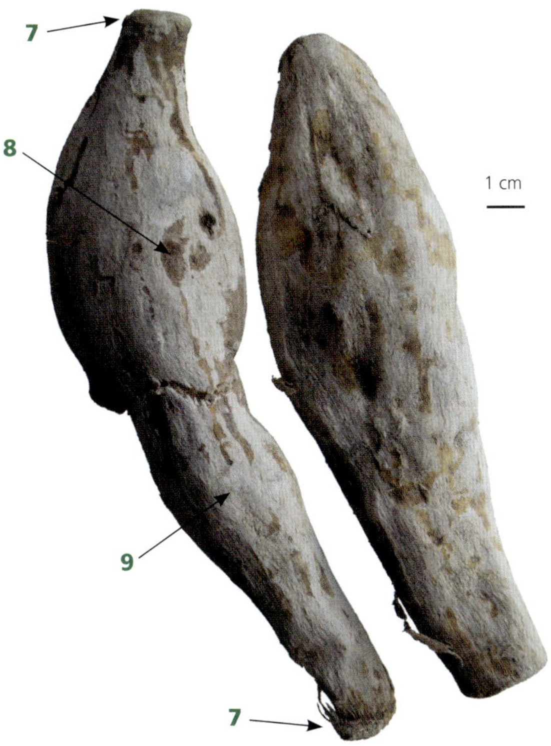

Crude drug of Fen Ge (seldom traded as whole tubers)
(*P. montana* var. *thomsonii*)

Decocting pieces (cubed) of Fen Ge (*P. montana* var. *thomsonii*)

Gou Ji 狗脊 Cibotii Rhizoma

Drug part: RHIZOME

PHARMACOPOEIA SPECIES

Cibotium barometz (L.) J. Sm. — Cibotiaceae

synonym: *Polypodium barometz* L.

Species of *Cibotium* wild in China: 2

PLANT DESCRIPTION

A large fern (up to 3 m tall) with a large, stout and prostrate **rhizome densely covered with long (1–1.5 cm), soft and shiny golden hairs**. *Fronds stalks* long and stout (usually up to 1 m long), triangular in cross-section and densely covered in hairs. *Frond blades* 1.5–3 m long and 2-pinnate; pinnules slightly curved, with a short sharp tip, **lower surface bluish-green**; both surfaces hairless; margins shallowly scalloped to minutely saw-toothed. *Sori* (containing reproductive spores) rhomboidal, usually 1–5, **along lateral veins** on the underside of lower pairs of pinnules; spores pale yellowish.

Mature plant of *Cibotium barometz* (centre background above man)

HARVESTING, SOURCING AND NATURAL RANGE

Rhizomes are harvested exclusively from wild populations all year round but usually from late autumn to early spring. This widespread species grows in acid soils in warm, humid and open places along streams, in forests, forest margins and valleys. It is distributed mainly in S China: Chongqing, Fujian, Guangdong, Guangxi, Guizhou, Hainan, Hubei, Hunan, Jiangxi, Sichuan, Taiwan, Tibet, Yunnan and Zhejiang [also native to NE India, Indonesia, Japan, Malaysia, Myanmar, Thailand and Vietnam].

CONSERVATION STATUS

China: Least Concern (MEPCAS, 2013).
Global: Not Evaluated (IUCN Red List, accessed 28/8/2016)

PROTECTED STATUS

CITES: Appendix II; Eu Annex B (see Table 1 page 10).

UNOFFICIAL SUBSTITUTE

Woodwardia unigemmata (Makino) Nakai — Blechnaceae

synonym: *W. radicans* (L.) Sm. var. *unigemmata* Makino

PLANT DESCRIPTION

Superficially very similar to the Pharmacopoeia species (*C. barometz*). *Main differences*: **rhizomes with long (2.2–2.7 cm) brown scales.** Each frond with (towards the tip on the lower surface) one or more **sub-spherical roundish scaly bulbils. Sori (containing reproductive spores) arranged along the central veins** (not lateral veins) **on the underside of each pinnule**, each covered with a pair of flaps (indusia).

SUBSTITUTION COMMENT. *W. unigemmata* is a common and widespread fern in the wild in China. One of its Chinese names (狗脊 Gou Ji) is identical to that of *Cibotium barometz* (the Pharmacopoeia species) suggesting this may be a cause of confusion. Its rhizomes are the source of a separate TCM drug known as 狗脊贯众 Gou Ji Guan Zhong (not included in the CP to date, 2015) and which has been described as '**slightly toxic**' owing to reports of adverse reaction (Chen & Chen, 2004). Sliced rhizomes consistent with reference specimens of *W. unigemmata* have frequently been seen in TCM herbal dispensaries in the UK labelled '狗脊 Gou Ji' (C. Leon pers. obs., 2011, 2014) while the listing of official 'Gou Ji' on CITES Appendix II (see above) may also be discouraging trade in *C. barometz* rhizomes because of the need to obtain CITES permits. Further research to better understand the frequency, cause and clinical implications of this substitution seems justified.

HARVESTING, SOURCING AND NATURAL RANGE

In China, wild populations of this species grow in forests, scrub and along roadsides between 400–3,000 m altitude in Fujian, Gansu, Guangdong, Guangxi, Guizhou, Hubei, Hunan, Jiangxi, Shanxi, Sichuan, Taiwan, Tibet and Yunnan; it is also native in Bhutan, India, Japan, Kashmir, Myanmar, Nepal, Pakistan, Philippines and Vietnam.

CONSERVATION STATUS

China: Not Evaluated.
Global: Not Evaluated (IUCN Red List, accessed 28/8/2016).

TCM PROPERTIES AND USE

properties: 'sweet, bitter and warm'
channels: 'Liver and Kidney'
actions: 'tonifies Liver and Kidney, tonifies yang, dispels wind damp, consolidates jing'

indications: musculoskeletal pain; incontinence, spermatorrhoea, leucorrhoea

Cibotium barometz in the wild, Guangxi province

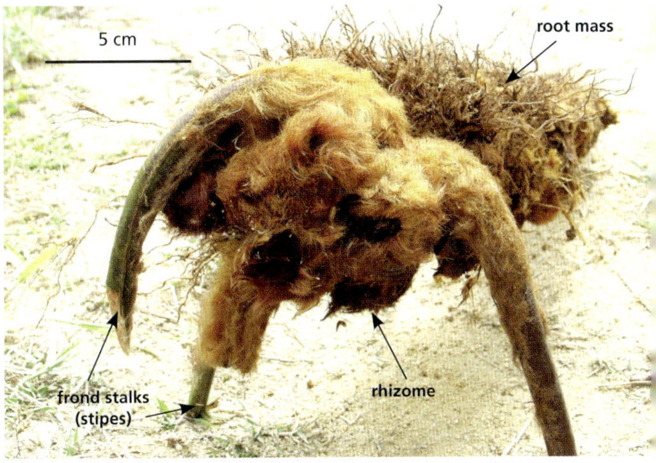

TOP RIGHT
Lines of rhomboidal reproductive sori on the lower surface of the pinnules of *C. barometz*

RIGHT
Fresh rhizome of *C. barometz* before removal of dense mass of long, soft golden hairs

Woodwardia unigemmata in the wild

Brown scaly bulbil on lower side of frond tip

Arrangement of sori on lower surface of frond

Fresh rhizomes of *W. unigemmata*

Cibotium barometz

Cibotiaceae

DRUG MORPHOLOGY

CRUDE DRUG (DRIED RHIZOME): irregularly cylindrical and somewhat stout (10–30 cm long; 2–10 cm diam.) **[1]**. Externally dull yellowish-brown, with occasional remains of soft, golden hairs in tufts **[2]**; dorsal surface sometimes with several reddish-brown and woody frond bases; ventral surface with remnants of black fibrous roots. *Texture*: hard and difficult to break. *Odour*: indistinct. *Taste*: weak and slightly astringent.

***Good quality*:** rhizomes long, hard, solid and densely clothed in golden-yellow hairs.

DECOCTING PIECES-1 (Sheng Gou Ji Pian): irregularly shaped slices usually with rounded **[3]**, or undulate margins **[4]** (5–20 cm long, 1.5–5 mm thick). Uncut rhizome edges dark brown, smooth and sometimes with tufts of golden hairs **[5]**. Transversely cut surfaces with an often distinctive and narrow, pale brown xylem ring **[6]** (1–4 mm from rhizome edge), sometimes raised. *Texture*: fragile, easily broken and starchy.

DECOCTING PIECES-2 (Shu Gou Ji Pian): as for 'Sheng Gou Ji Pian' above but blackish-brown **[7]** and texture hard.

DECOCTING PIECES-3 (Tang Gou Ji Pian): as for 'Sheng Gou Ji Pian' above but darker **[8]**, colour of xylem ring black and less distinctive **[9]**; cut surfaces somewhat inflated, and usually devoid of golden hairs. *Taste*: astringent.

PROCESSING

After removal of the root mass and fronds (= crude drug), rhizomes are either sliced thickly then sun-dried (= Sheng Gou Ji Pian 'uncooked Gou Ji slices' decocting pieces-1) or steamed, then partially dried and cut into thick slices before drying (= Shu Gou Ji Pian 'cooked Gou Ji slices' decocting pieces-2). Decocting slices may undergo further processing by scalding with sand (= Tang Gou Ji Pian decocting pieces-3).

OTHER COMMON NAMES

狗脊条 Gou Ji Tiao 'unprocessed cibotium'; 金毛狗 Jin Mao Gou 'golden-haired dog'; 金毛狗蕨 Jin Mao Gou Jue; cibot rhizome; golden moss; Rhizoma Cibotii; Scythian lamb (SCN); Tartarian lamb.

SAFETY

Safety Class: 1; Interaction Class: A (Gardner & McGuffin, 2013).

Woodwardia unigemmata

Blechnaceae

'DRUG SUBSTITUTE' MORPHOLOGY

Easily distinguished from rhizomes of the Pharmacopoeia species. *Main differences*: **densely covered in long (up to 2.7 cm), reddish-brown to blackish-brown, stout, convex and somewhat woody scales [10]** and occasional raised hollow frond bases **[11]**. Scales interspersed (especially towards apex) with **reddish-brown hairs [12] less dense and coarser than those of *C. barometz*,** together with many curved root remains **[13]**.

SLICED DRIED/STEAMED RHIZOME-1: uncut external surfaces reddish-brown or blackish-brown **[14]**. Outline of slices with **indentations more angular and often parallel [15]** (indicating the angle of the woody scale bases), interspersed with occasional **reddish-brown hairs [16]** and curved root remains **[17]**. Cut surfaces with **xylem interrupted, sometimes almost crescent-shaped [18]** (not forming a continuous ring). *Texture*: hard and woody.

SLICED DRIED AND SCALDED RHIZOMES-2: as for sliced rhizome above but slightly darker and hairs mostly destroyed.

OTHER COMMON NAMES

单芽狗脊 Dan Ya Gou Ji; 顶芽狗脊 Ding Ya Gou Ji; 狗脊 Gou Ji; 狗脊贯众 Gou Ji Guan Zhong.

SAFETY

Slightly toxic: adverse reactions have been reported (Chen & Chen, 2004).

LESSER KNOWN UNOFFICIAL SUBSTITUTES

The rhizomes of other fern species in the genus *Woodwardia* and other genera (e.g. *Pteris, Dryopteris* and *Athyrium*) have been reported in the literature as unofficial substitutes of 'Gou Ji' (Bensky *et al.,* 2004; Xiao *et al.*, 2001–2) but their occurrence in trade is likely to be confined to China.

Crude drug of 'Gou Ji' (in background) with decocting pieces 'Sheng Gou Ji Pian' (tied in bundles)

Decocting pieces -1 'Sheng Gou Ji Pian'

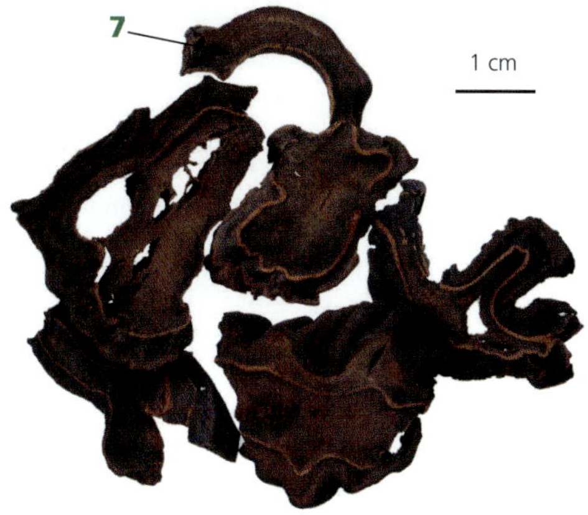

Decocting pieces-2 'Shu Gou Ji Pian'

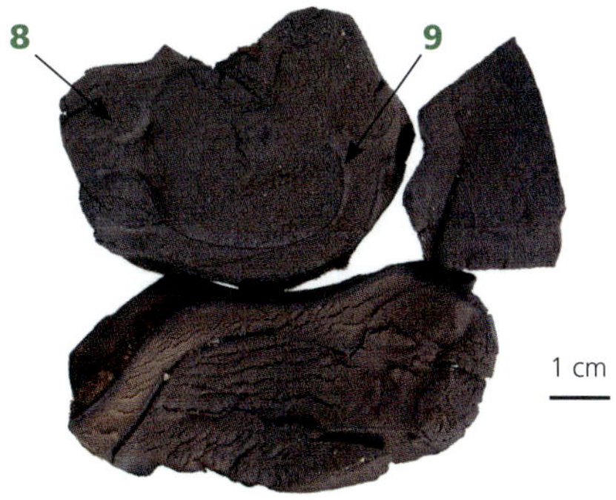

Decocting pieces-3 'Tang Gou Ji'

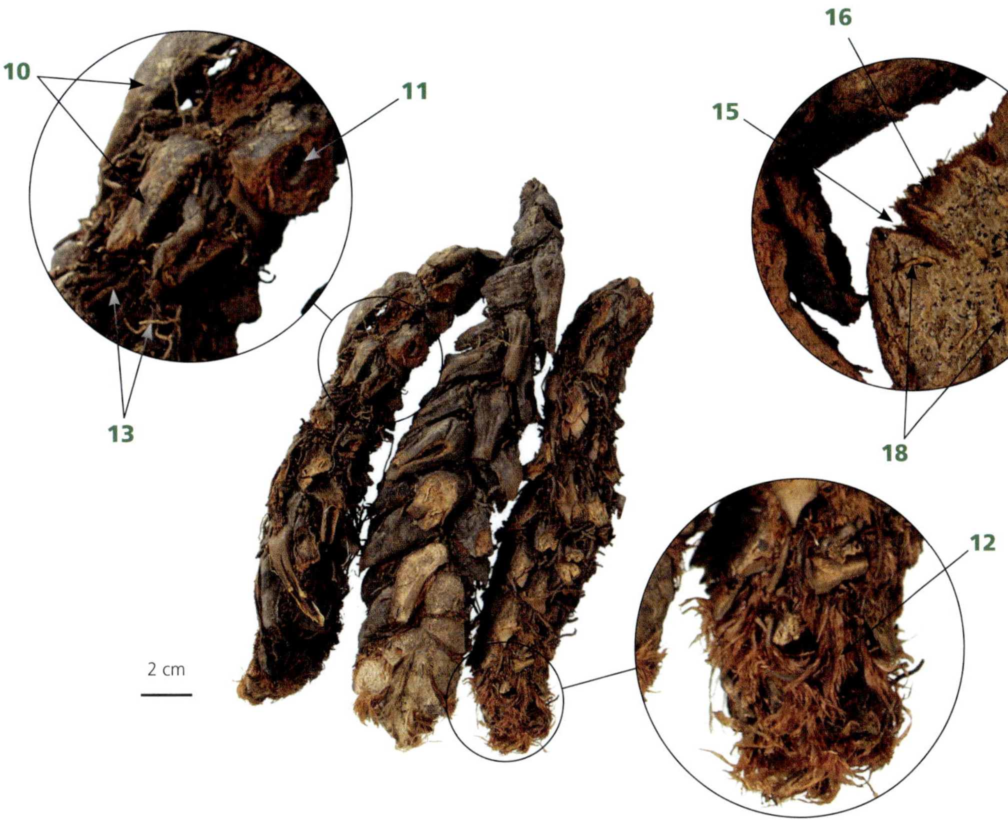

Dried rhizomes of *Woodwardia unigemmata*

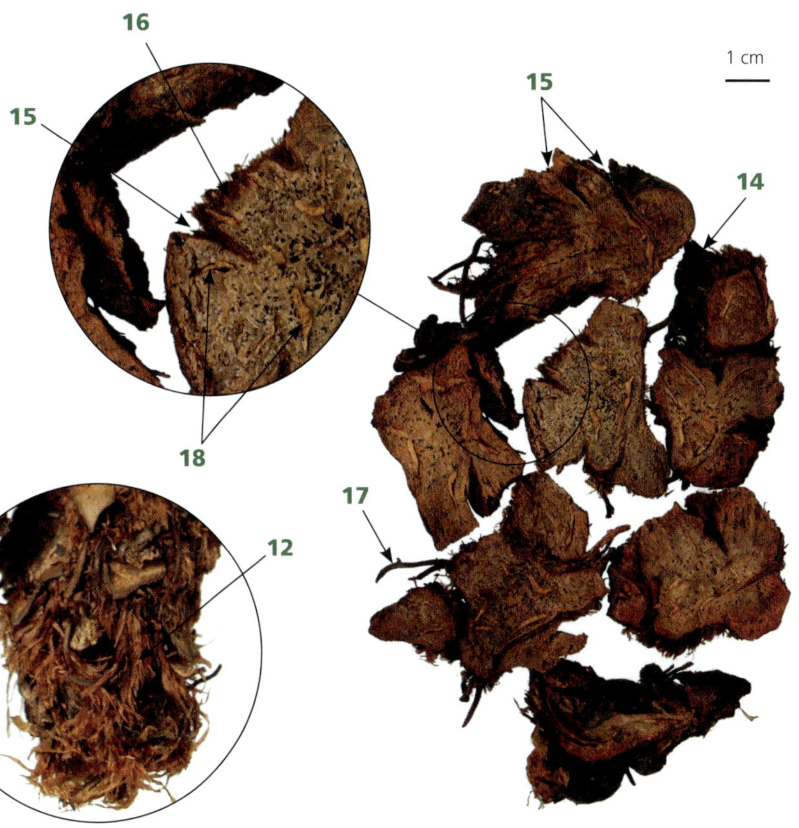

Dried, sliced and steamed rhizomes of *W. unigemmata*

He Shou Wu 何首乌 (or Sheng He Shou Wu 生何首乌)

Polygoni Multiflori Radix

Drug part: UNPROCESSED ROOT TUBER (rarely found commercially, see *Safety*)

(see also 'Zhi He Shou Wu' page 158)

PHARMACOPOEIA SPECIES

Fallopia multiflora (Thunb.) Haraldson

Polygonaceae

synonyms: *Fagopyrum multiflorum* (Thunb.) Grint.; *Polygonum multiflorum* Thunb. (name used in CP2015); *Reynoutria multiflora* (Thunb.) Moldenke (*Kew-MPNS* accepted name, 25/11/2016).

Species of *Fallopia* wild in China: 7–9

PLANT DESCRIPTION

A twining, herbaceous perennial. *Root tuber* variously-shaped, woody and blackish-brown; in transverse section vascular bundles arranged as 4–11 distinct and peripheral, pale yellowish or yellowish-pink tissues. These provide a diagnostic identification character to distinguish this species from its close relatives such as *F. aubertii* below (see also *Drug Morphology*). *Stems* slender and much branched. ***Leaves* alternate** and stalked; base of leaf stalks with an oblique stipule (ocrea) (3.5 mm long); leaf blades simple, ovate or narrowly ovate (3–7 cm long; 2–5 cm wide), bases heart-shaped or slightly notched, margins entire and tip long and tapering; virtually hairless. *Inflorescences* slender, branched and many-flowered terminal or axillary heads (panicles) (10–20 cm long). *Each flower* bisexual; white or greenish-white; sepals (3) resembling petals (2); sepals larger, and persisting in fruit to form wings; stamens 8; ovary superior; styles 3, very short with swollen heads. ***Fruit*** a dry, ovoid, 3-angled achene (2.5–3 mm long), shiny, blackish-brown and **smooth;** enclosed by orbicular wings (persistent sepals) **6–7 mm diam**. Fl. Jun–Oct, fr. Jul–Nov.

HARVESTING, SOURCING AND NATURAL RANGE

'He Shou Wu' is one of the most popular TCM drugs and widely revered for its powerful tonic properties. Old, large tubers fetch high prices and have attracted trophy status. The species is native to the mountains of central and southern China (200–3,000 m altitude) where it grows in thickets and rock crevices. It is cultivated as a medicinal crop throughout much of its natural range, especially in the Chinese provinces of Guangdong, Guangxi, Guizhou, Henan and Hubei with tubers harvested in autumn and winter after the leaves have withered; GAP-certified products have recently been produced in Guizhou. Overall, production volume is said to be around 50:50 (wild to cultivated). The species is also native in Japan.

CONSERVATION STATUS

China: Least Concern (MEPCAS, 2013, as 'var. *multiflora*')
Global: Not Evaluated (IUCN Red List, accessed 25/1/2016).

See also the processed root tuber 'Zhi He Shou Wu' page 158 sourced from this species. The separate TCM drug 'Shou Wu Tong' (Polygoni Multiflori Caulis) (*not included in this book*) is sourced from the stem of this species.

UNOFFICIAL SUBSTITUTE

F. aubertii (L. Henry) Holub

Polygonaceae

synonyms: *Polygonum aubertii* L. Henry; *Reynoutria aubertii* (L. Henry) Moldenke.
taxonomic/nomenclatural note: some taxonomic authorities (*FoC* and Stace, 2010) consider that this species may be the same as *F. baldschuanica* (Regel) Holub, a rampant European climbing plant also known as Russian vine.

PLANT DESCRIPTION

Very easily confused with PS1. In transverse section, **fresh root tuber with vascular tissues much smaller and scattered** with the larger and roundish **central xylem often woody. *Leaves* clustered at nodes (rarely solitary)** and slightly leathery. *Flowers* in clusters of 3–6, flower heads few branched. ***Fruit*** slightly larger (3.5–4 mm diam.) than that of PS1, with **a minutely granular surface**. Fl. Jul–Aug, fr. Aug–Sep.

SUBSTITUTION COMMENT. When the Pharmacopoeia species is wild-harvested, mis-identification with this common, very similar and closely related species is not unusual. Its tubers are remarkably similar too (see *'Drug Substitute' Morphology*).

CONSERVATION STATUS

Endemic to China: Least Concern (MEPCAS, 2013); Not Evaluated (IUCN Red List, accessed 16/8/2016).

TCM PROPERTIES AND USE

***properties*: 'sweet, bitter and neutral'**
***channels*: 'Liver, Heart and large Intestines'**
***actions*: 'tonifies jing, tonifies Blood, Liver and Kidney, eliminates toxins, moistens intestines'**

***indications*: constipation; sores, abscesses; chronic malaria**

Inflorescences of *Fallopia multiflora*

Medicinal crop of *F. multiflora*, Guangdong province

F. multiflora in full flower

Flowering plant of *F. aubertii*

Fresh root tubers of *F. aubertii* traded as official 'He Shou Wu' by a TCM wholesaler Guizhou province

Fallopia multiflora — Polygonaceae

DRUG MORPHOLOGY

CRUDE DRUG (DRIED UNPROCESSED ROOT TUBER, SOMETIMES HALVED): variably shaped root tuber **[1]**, often tapering at one or both ends and slightly or greatly contorted (6–15 cm long; 4–12 cm diam.). Externally dull reddish-brown or greyish-brown with rough, furrowed surface **[2]**, scattered raised transverse lenticels **[3]** and small rootlet scars. *Texture:* hard, dense and uneasily broken. ***Fracture:*** even, starchy, pale yellowish-brown or reddish-brown **[4]**; transverse section with **distinct pattern of 4–11 peripherally arranged** oval-shaped, pale yellowish or yellowish-pink **vascular tissues [5]** with a **larger central often ovate xylem [6]** (colour and shape of tissues often lost in dried material); *Odour:* slight. *Taste:* astringent and somewhat bitter and sweet. ***Good quality*:** firm, heavy and uneasily broken; externally reddish-brown, internally yellowish-brown.

DECOCTING PIECES (DRIED UNPROCESSED ROOT TUBER SLICES): irregularly shaped slices (c. 5 mm thick) **[7]**, transverse sections often with an undulating outline **[8]**. Uncut external surface externally dull reddish-brown or greyish with rough, furrowed surface **[9]**, scattered raised transverse lenticels, occasional. Cut surface pale, yellowish- or reddish-brown, dense and starchy **[10]**; transverse slices **often with distinct or more often indistinct pattern of 4–11 peripherally arranged roundish or ovate vascular tissues [5]** with a central woody xylem **[6]** (see also vascular tissue patterning in the fresh tubers). *Taste, odour* etc. as for crude drug.

PROCESSING

After removing the tuber ends, root tubers are washed, large pieces cut into slightly smaller ones, then dried (= crude drug); these are subsequently softened thoroughly in water before being thickly sliced (= unprocessed He Shou Wu or Sheng He Shou Wu: decocting pieces-1). Because of its toxicity, however, (see *Safety*) such unprocessed He Shou Wu is rarely found in commercial trade.

OTHER COMMON NAMES

干何首乌 Gan He Shou Wu; 干首乌 Gan Shou Wu; 生何首乌 Sheng He Shou Wu 'unprocessed black haired Mr He'; 生首乌 Sheng Shou Wu; 首乌 Shou Wu; fleeceflower; fleeceflower root (CP2010); fo-ti (SCN); Radix Polygoni Multiflori; unprocessed fo-ti.

SAFETY

Safety Class: 3; "not for use except under the supervision of an expert qualified in the appropriate use of this substance"; Interaction Class: A (Gardner & McGuffin, 2013). For reasons of safety this unprocessed drug is usually only used for short periods. Only the processed form ('Zhi He Shou Wu' (Polygoni Multiflori Radix Praeparata, page 158) is thought to be traded internationally.

F. aubertii — Polygonaceae

'DRUG SUBSTITUTE' MORPHOLOGY

DRIED ROOT TUBER, HALVED OR SLICED: the most diagnostic character to distinguish the root tuber of this species from the Pharmacopoeia species is best seen in the transverse section of the fresh tubers: **vascular tissues small, rounded, more numerous and scattered throughout the cut surface [11]**; these become much less distinct in dried tubers **[12]**; central or off-central xylem rounded or ovate and often woody **[13]**. Other characters are the same as those for the Pharmacopoeia species.

OTHER COMMON NAMES

木藤首乌 Mu Teng Shou Wu.

For **OTHER UNOFFICIAL SUBSTITUTES** see Zhi He Shou Wu page 159.

Crude drug of He Shou Wu (fresh root tubers of *F. multiflora*)

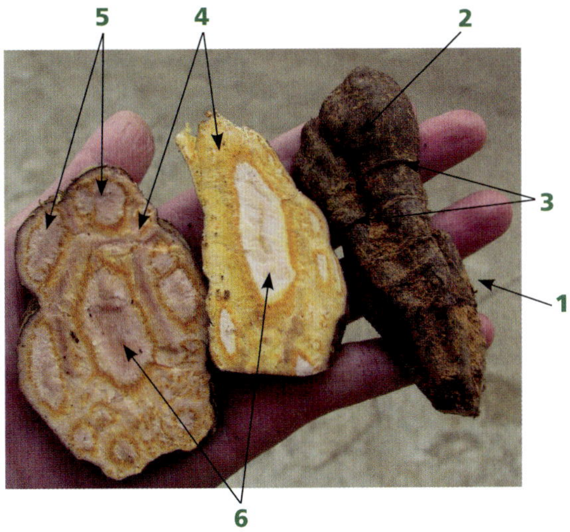

Crude drug of He Shou Wu
(transversely sliced fresh root tubers of *F. multiflora*)

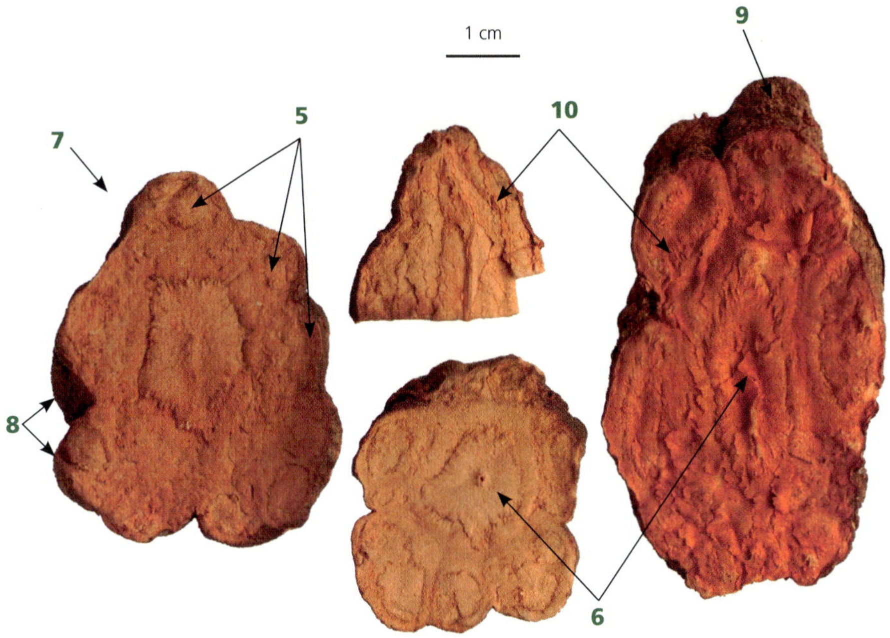

Unprocessed decocting slices of the He Shou Wu (*Fallopia multiflora*) showing indistinct pattern of vascular tissues (tissue pattern more obvious in fresh root tubers). Right-hand piece is a reddish-brown colour variant.

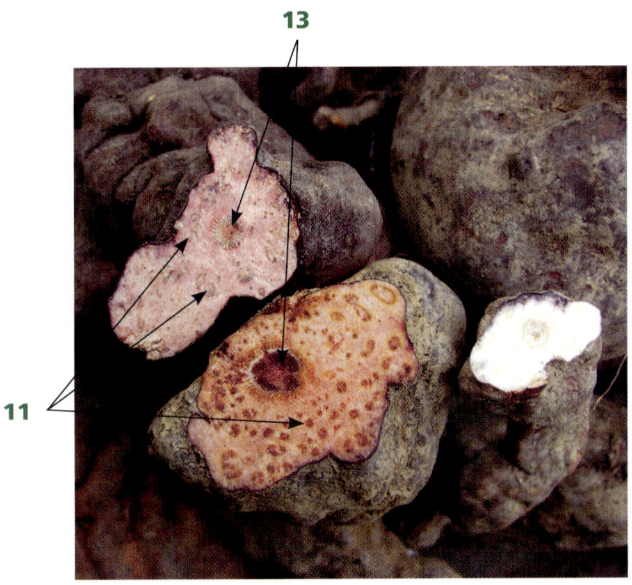

Transversely sliced fresh tubers of *F. aubertii*

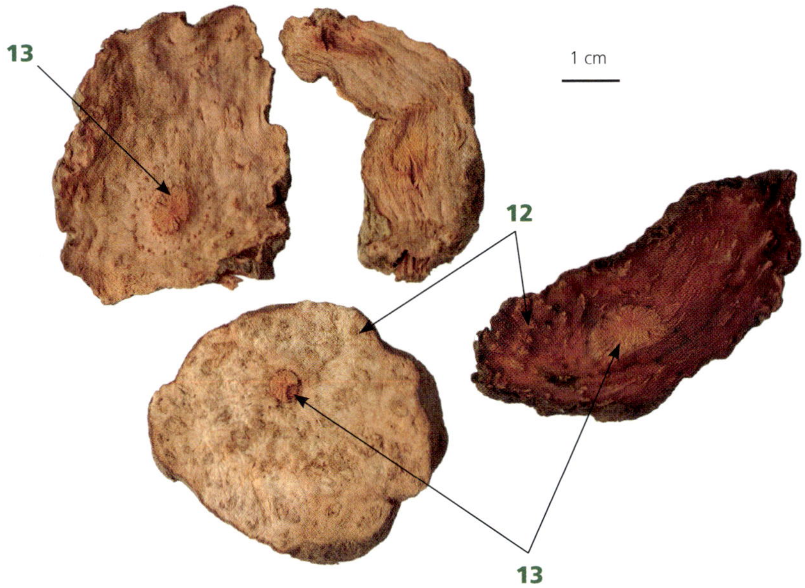

Dried transverse slices (left) and dried steamed transverse slices (right) of *F. aubertii* showing pattern of vascular tissues (less obvious than in fresh root tubers)

He Shou Wu – Zhi He Shou Wu 制何首乌

Polygoni Multiflori Radix Praeparata

Drug part: PROCESSED ROOT TUBER

(see also unprocessed root tuber page 154)

PHARMACOPOEIA SPECIES

Fallopia multiflora (Thunb.) Haraldson

Polygonaceae

PLANT DESCRIPTION

See unprocessed 'He Shou Wu' (Polygoni Multiflori Radix) page 154.

DRUG MORPHOLOGY

DECOCTING PIECES-1 (PROCESSED ROOT TUBER SLICES — STEAMED WITH BLACK SOYA BEAN JUICE): irregularly-shaped transverse or longitudinal slices **[1]** (c. 1 cm thick). Black or dark brown throughout **[2]**; processing usually causes the diagnostic arrangement of the vascular tissues to be lost (see unprocessed tubers *'He Shou Wu'* page 154). *Texture*: hard and difficult to break. *Fracture:* horny and **black.** *Odour*: indistinct. *Taste*: slightly sweet, bitter and astringent. ***Good quality*:** large, heavy, dense and brownish-black. Old, large tubers are highly sought after and sometimes traded as trophies.

DECOCTING PIECES-2 (PROCESSED ROOT TUBER SLICES — STEAMED): as for dp-1 above but orangish-brown or pinkish throughout **[3]** and, in transverse slices, the diagnostic arrangement of vascular tissues sometimes faintly discernible as 4–11 peripheral tissues **[4]** with a large central, somewhat ovate, non-woody xylem **[5]**.

HARVESTING, SOURCING AND NATURAL RANGE and CONSERVATION STATUS

See unprocessed 'He Shou Wu' page 154.

PROCESSING

Dried pieces of unprocessed 'He Shou Wu' (page 154) are mixed with black soya bean juice (made from *Glycine max*) then steamed until all the juice is absorbed and the tubers have turned brown throughout. These are then sun-dried, sliced and dried (= decocting pieces-1). The same procedure is sometimes followed without the use of soya bean juice (= decocting pieces-2).

OTHER COMMON NAMES

赤首乌 Chi Shou Wu; fo-ti (SCN); **He Shou Wu**; Radix Polygoni Multiflori Praeparata cum Succo Glycines Sotae; processed fleeceflower root.

SAFETY

Safety Class: 1; Interaction Class A (Gardner & McGuffin (2013). **Correct processing is essential to destroy toxins present in the unprocessed tubers** and which are described in the separate account 'He Shou Wu' page 154.

UNOFFICIAL SUBSTITUTE

F. aubertii (L. Henry) Holub

Polygonaceae

PLANT DESCRIPTION

see *Unofficial Substitute* in the unprocessed drug account 'He Shou Wu' page 154.

CONSERVATION STATUS

See Unofficial Substitute in unprocessed 'He Shou Wu' account page 154.

'DRUG SUBSTITUTE' MORPHOLOGY

PROCESSED ROOT TUBER SLICES — STEAMED WITH BLACK SOYA BEAN JUICE: macroscopically indistinguishable from those of the Pharmacopoeia species (whether transversely or longitudinally sliced); laboratory-based authentication methods are required for reliable identification.

PROCESSED ROOT TUBER SLICES — STEAMED: similar to the Pharmacopoeia species (dp-2 above) but distinguishable in transverse section if vascular tissues are visible. *Main differences*: **vascular tissues** (in transverse slices) much smaller and more numerous and **scattered [6]** throughout the cut surface **as round or oval rings [7]**; the **central or off-centre xylem much smaller,** sub-rounded and **woody [8]**. These characters are not distinguishable in longitudinally-sliced pieces.

> **SUBSTITUTION COMMENT.** The tubers of this easily confused and closely related species are frequently traded, often inadvertently, as genuine 'Zhi He Shou Wu' (see also 'He Shou Wu' substitutes page 159). Traditional processing with black soya juice will mask the pattern of vascular tissues making its macroscopic detection not possible.

OTHER COMMON NAMES

木藤首乌 Mu Teng Shou Wu.

TCM PROPERTIES AND USE

properties: **'sweet, bitter and slightly warm'**
channels: **'Liver and Kidney'**
actions: **'tonifies Liver and Kidney, tonifies Blood and jing, extinguishes wind'**

indications: **insomnia; dizziness; infertility; weakness in back and legs; rash, itchy skin**

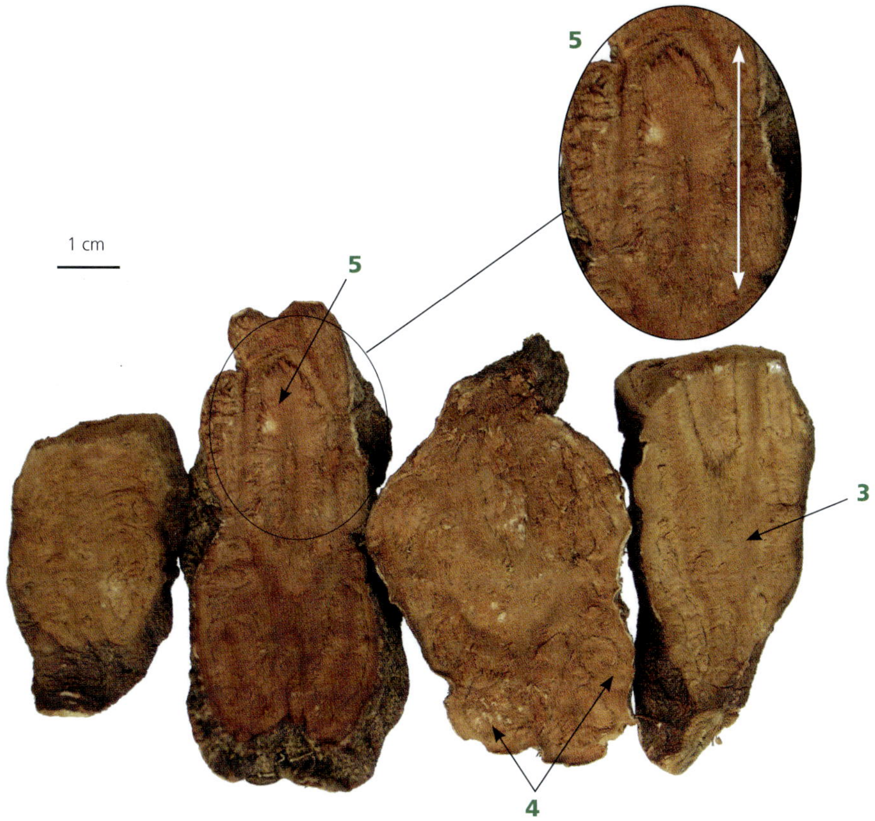

Decocting pieces-2 of Zhi He Shou Wu (longitudinally sliced and steamed without black bean juice) sourced from *F. multiflora*

Decocting pieces-1 of Zhi He Shou Wu (longitudinally sliced and processed with black bean juice) sourced from *Fallopia multiflora*

Decocting pieces-1 of Zhi He Shou Wu (cubed and processed with black bean juice) sourced from *F. multiflora*

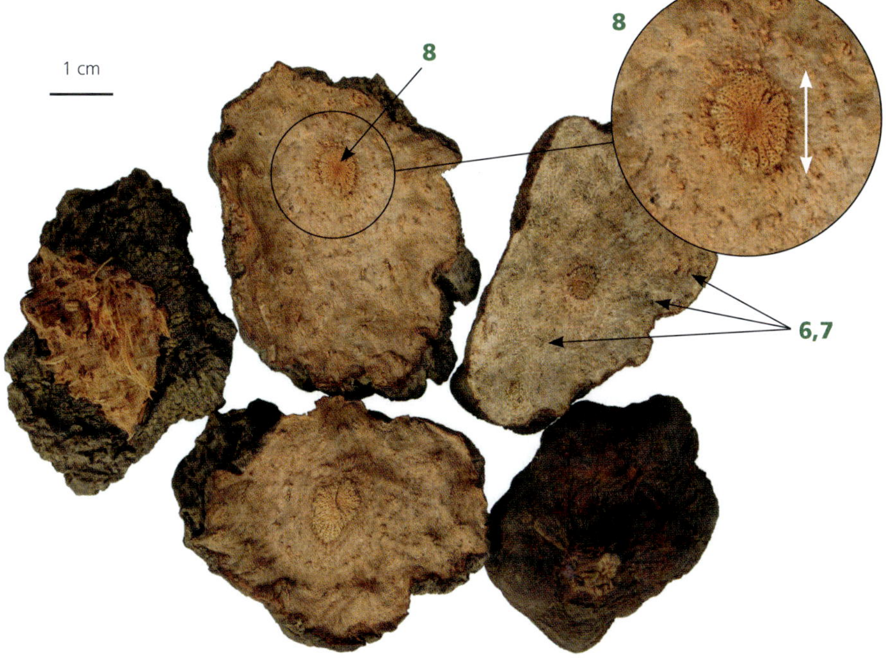

Processed (steamed) and transversely cut root slices of *F. aubertii* with vascular tissues still visible

OTHER UNOFFICIAL SUBSTITUTES/COUNTERFEITS

Large processed specimens of official He Shou Wu can fetch high prices making them susceptible to counterfeiting using tubers of other species. Materials used are likely to be processed with black soya juice (see *Processing* above) to mask the vascular tissues which enable identification of the official drug. These are unlikely to enter international trade due to improved policing of markets. Examples of counterfeit species reported (NICPBP, 1997–2002; Ho *et al.*, 2006; Xiao *et al.*, 2001–2002) include:

Polygonaceae:

1. *Pteroxygonum giraldii* Dammer & Diels (红药子 Hong Yao Zi). Leaves 2–4 clustered. Underground part a globose thickened root up to 15 cm diam.

Asclepiadaceae (plants with milky sap and opposite leaves):

2. *Cynanchum auriculatum* Royle ex Wight (牛皮消 Niu Pi Xiao). A high altitude species (2,800–3,600 m) from Sichuan, Tibet and Yunnan.
3. *C. bungei* Decne. (白首乌 Bai Shou Wu). Fresh tuber **much smaller** (3–7 cm long; 1.5–4 cm diam.); used in local (non-TCM) medicine.
4. *C. otophyllum* C. K. Schneid. (青羊参 Qing Yang Shen). Tuber used in local (non-TCM) medicine.
5. *C. wilfordii* (Maxim.) Hook. f. (隔山消 Ge Shan Xiao). Tuber much smaller than official drug (up to 10 cm long; c. 2 cm diam.) and used in local (non-TCM) medicine.

Dioscoreaceae:

6. *Dioscorea bulbifera* L. (黄独 Huang Du). Tuber black, ovoid or pear-shaped (4–10 cm diam.); used in local (non-TCM) medicine.

Hong Jing Tian 红景天 Rhodiolae Crenulatae Radix et Rhizoma

Drug parts: ROOT AND RHIZOME

Rhodiola crenulata (Hook. f. & Thomson) H. Ohba

Crassulaceae

synonym: *Sedum crenulatum* Hook. f. & Thomson (*Kew-MPNS* accepted name, 25/8/2016).

Species of *Rhodiola* wild in China: 55

PLANT DESCRIPTION

A perennial herb (5–20 cm tall) with a thick fleshy rhizome. Stems fleshy, rarely branched; non-flowering stems erect (4–17 cm); flowering stems many, yellowish or red (5–20 cm long); old rhizomes with persistent black remains of old flower stems. Lower stem leaves scale-like, widest towards tip (1–2 cm long) and pointed; all other leaves larger and clustered at top of stems alternately or in whorls, leaf blades elliptic-oblong to sub-orbicular (1.2–3 cm long; 1–2.2.cm wide), margins entire, wavy to scallop-edged, tips broad or shortly pointed. *Flowers* many, in terminal heads (c. 2 cm x 2–3 cm diam.), bracts present. Each flower single-sexed; sepals 5, narrowly triangular and persistent in fruit; petals 5, equal size and shape, broadly ovoid, widest towards tip, tip broadly pointed, red to purplish-red. Male flowers with 10 stamens; female flowers with a 5-lobed superior ovary. *Fruits* a group of dry erect pods (follicles), drying to red (8–10 mm long). *Seeds* ovoid and winged at both ends. Fl. Jun–Sep.

HARVESTING, SOURCING AND NATURAL RANGE

Roots and rhizomes are wild-harvested in autumn after flowering stalks have withered. This is a high-altitude species (2,800–5,600 m) growing in scrub, grassland and rocky places in the mountains of W and NW China (Qinghai, Sichuan, Tibet and Yunnan). Over-harvesting in recent years to meet global demand (see *Substitutes* below) threatens the survival of wild populations. [the species is also native in Bhutan, Nepal and Sikkim].

CONSERVATION STATUS

China: Endangered (MEPCAS, 2013); Endangered (Fu, 2011).
Global: Not Evaluated (IUCN Red List, accessed 7/11/2016).

Rhodiola crenulata with flowering stems in bud plus persistent black old flowering stems

R. crenulata: freshly uprooted plants

R. crenulata: wild plants in flower, Sichuan

TCM PROPERTIES AND USE

***properties*:** **'sweet, bitter and neutral'**
***channels*:** **'Lung and Heart'**
***actions*:** **'invigorates Blood, stops bleeding, clears Lung heat, alleviates cough'**

***indications*:** **fatigue; chest, heart pain; stroke**

DRUG MORPHOLOGY

CRUDE DRUG (DRIED ROOT AND RHIZOME): *Root and rhizome described together as often difficult to distinguish.* Stout, somewhat cylindrical, slightly curved and occasionally branching **[1]** (5–20 cm long; rhizome 2.9–4.5 cm diam.; root up to 1.5 cm diam.) Externally brown or greyish-brown **[2]**, often rough and coarsely wrinkled **[3]**, outermost epidermal layer yellowish, membranous and often flaking **[4]**; surface scattered with stem base scars ovate or irregular **[5]** sometimes with fibrous, linear and occasionally tufted stem remnants **[6]**. Lateral roots 10–30 cm long. *Rhizome fracture:* vivid pinkish or purplish-red **[7]** with scattered vascular bundles, often porous **[8]**. *Root fracture:* orangish-red or purplish red; often cracking. *Odour:* distinctly aromatic. *Taste:* somewhat biter and astringent. ***Good quality:*** rootstock large with a pinkish-red fracture.

DECOCTING PIECES (DRIED AND SLICED ROOT AND RHIZOME): irregularly ovate in outline **[9]** (2.9–4.5 cm wide; c. 3–10 mm thick). Uncut exterior rough and brown or greyish-brown **[2]** sometimes with membranous remnants of exfoliating yellowish epidermal layer **[4]**. Cut surface of rhizomes vivid, pinkish- or purplish-red with scattered vascular bundles **[10]**. Cut surface of roots **[11]** orangish-red or purplish-red, often with fissures **[12]**. *Odour* and *taste* as for crude drug.

PROCESSING

After removal of rough outer skin and fibrous rootlets, the roots and rhizomes are sun-dried (= crude drug) then sliced into decocting pieces.

OTHER COMMON NAMES

big flower rhodiola root (CP2010); plateau ginseng; Radix et Rhizoma Rhodiolae Crenulatae; Rhodiola crenulata (SCN).

SAFETY

An official healthfood in China (Anon., 2002a,b).

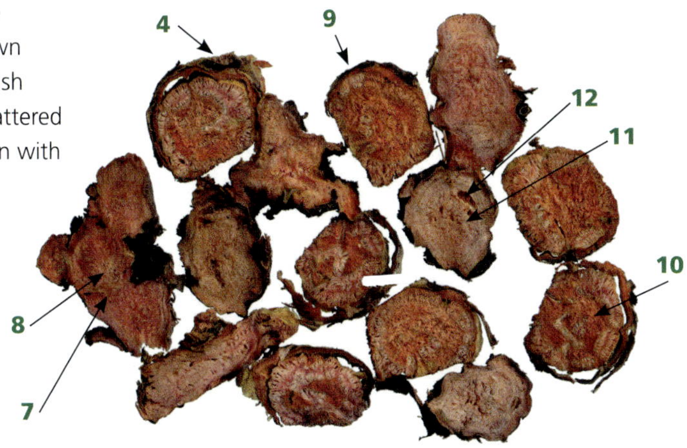

Decocting pieces (root and rhizome) of Hong Jing Tian, freshly sliced (*R. crenulata*)

Decocting pieces (mostly rhizome) of Hong Jing Tian (*R. crenulata*)

Crude drug of Hong Jing Tian (*Rhodiola crenulata*)

UNOFFICIAL SUBSTITUTES

The name Hong Jing Tian (红景天) is used for many Chinese plants, mostly for one or more of the other 54 species of *Rhodiola* that are native to China; this creates confusion in trade over the identity of items sold under this name. The Pharmacopoeia species (*R. crenulata*) is also known as Da Hua Hong Jing Tian (大花红景天) 'big flowered Hong Jing Tian' to distinguish it from other species but because rootstocks cannot be distinguished macroscopically this trade name may be mis-used in efforts to fetch higher prices. *R. crenulata* itself has been used to adulterate *R. rosea* L. (syn. *Sedum roseum* (L.) Scop.) for which there is a huge global demand; *R. rosea* is a more northern and westerly species used in food supplements and licensed herbal medicine products for treating stress-induced fatigue, exhaustion and anxiety (Edwards *et al.*, 2015). Because of the widespread reporting of *Rhodiola* adulteration in general (Booker *et al.*, 2016a; Li & Zhang, 2008) laboratory-based authentication methods are strongly recommended. Recent research to differentiate *Rhodiola* trade items to species level include the use of DNA sequencing (Xin *et al.*, 2015) and HPTLC coupled with NMR (Booker *et al.*, 2016b).

Huang Lian 黄连 Coptidis Rhizoma

Drug part: RHIZOME

Coptis chinensis Franch.

Ranunculaceae

Species of *Coptis* wild in China: 6

PLANT DESCRIPTION

A small, perennial herb (15–25 cm tall) with **3-6-branched** and nodular **rhizomes with short stolon internodes**; rhizomes bright golden-yellow internally. *Leaves* all basal, several, long stalked (3–5 cm) and ovate-triangular in outline; divided into 3 main leaflets, each further shallowly or deeply sub-divided into glossy segments; leaf blade ovate to triangular in outline (4–10 cm long; 4–10 cm wide); lateral segments obliquely ovate, not touching adjacent ones, margins acutely saw-toothed. *Flowering stems* 1 or 2, leafless and slightly taller than the leaves (12–25 cm tall). *Flowers* small and cryptic in terminal, 3–8-flowered heads; bracts usually 3, lance-shaped; each flower bisexual, with sepals 5, greenish-yellow and lance-shaped (6.5–12.5 mm long; 2–3 mm, wide); petals linear to lance-shaped (5–6.5 mm long) and gradually tapering; stamens many, 3–6 mm long, outer ones slightly shorter than petals. *Fruits* with 6–9, green, linear pods (6–8 mm long) on stalks radiating from a central point; outer end of each pod opening as a somewhat beaked mouth. Seeds brown and shiny. Fl. Feb–Mar, fr. Apr–Jun.

HARVESTING, SOURCING AND NATURAL RANGE

Rhizomes of this high-value drug are harvested in autumn almost exclusively from medicinal crop plants grown in the Chinese provinces of Chongqing, Gansu, Guizhou, Hubei, Shaanxi and Sichuan; traditional *Di Dao* production sites are located in Sichuan and Chongqing. A GAP standard has been developed for this species and certification awarded at selected production sites.

The wild species grows in shady, forested valleys (500–2,000 m altitude) and is endemic to C, S and W China (Anhui, Fujian, Guangdong, Guangxi, Guizhou, Hubei, Hunan, Shaanxi, Sichuan and Zhejiang). Over-harvesting of wild plants has dramatically reduced population sizes; the same is true for the other source species of this drug (PS2 and 3), and conservation action is urgently needed for all.

CONSERVATION STATUS

Endemic to China: Endangered (as 'var. *brevisepala*'); Vulnerable (as 'var. *chinensis*') (MEPCAS, 2013); Not Evaluated (IUCN Red List, accessed 6/7/2016).

Coptis chinensis cultivated on slopes under black shading mesh, Sichuan province

TCM PROPERTIES AND USE

properties: **'very bitter and very cold'**
channels: **'Heart, Spleen, Stomach, Liver, Gall Bladder, Large Intestine'**
actions: **'clears heat, dries damp, clears fire'**

indications: **diarrhoea; vomiting, nausea, epigastric pain, constipation; headaches, irritability, insomnia; (T)**

Medicinal crop of *C. chinensis*, Sichuan province

Feshly harvested plants of *C. chinensis* connected by short stoloniferous internodes, Sichuan province

Fruiting plants of *C. chinensis* in medicinal cultivation

Fruiting heads of *C. chinensis*

Traditional drying of *Coptis* rhizomes using heat from below

C. deltoidea C. Y. Cheng & P. K. Hsiao

Ranunculaceae

PLANT DESCRIPTION

Very similar to PS1 and PS3. **Rhizomes not branched or rarely branched** with **long** smooth **stolon internodes** (4–9 cm long; 0.5–1 cm diam.); lobes of leaf segments touching or slightly overlapping with adjacent ones; stamens about half as long as petals. Fl. Mar–Apr, fr. Apr–Jun.

HARVESTING, SOURCING AND NATURAL RANGE

Rhizomes are harvested in autumn almost exclusively from medicinal crop plants grown in western Sichuan (notably Mt Emei (Emei Shan) and Hongya district).

The species is endemic to China and grows wild in forests (1,600–2,000 m altitude) in the vicinity of its cultivated counterparts, but its wild poplulations have become so depleted they are no longer economic to harvest.

CONSERVATION STATUS

Endemic to China: Vulnerable (MEPCAS, 2013); Not Evaluated (IUCN Red List, accessed 6/7/2015).

C. teeta Wall.

Ranunculaceae

synonym: *C. teetoides* C. Y. Cheng

PLANT DESCRIPTION

Very similar to PS1 and PS2. *Main differences*: **rhizomes smaller than both PS1 and 2 and unbranched** 2–5 cm long; 2–4 mm diam.). Leaf segment lobes not touching adjacent ones. Petals spoon-shaped (spatulate) 5.4–5.9 mm long and stamens about half as long as petals.

HARVESTING, SOURCING AND NATURAL RANGE

Rhizomes sourced from this species rarely seen in trade today because wild populations are 'Critically Endangered' (resulting from medicinal over-harvesting). Cultivation of the spoecies as a medicinal crop has begun in recent years on a small scale in parts of its natural range (Yunnan province) to meet the high demand for this high-value medicinal herb.The major part of the species' wild distribution occurs over the border in Arunachal Pradesh (north-east India) where the species is equally threatened by over-harvesting as well as widespread deforestation (IUCN Red List, accessed, 6/7/2016).

CONSERVATION STATUS

China: Critically Endangered (MEPCAS, 2013).
Global: Endangered (IUCN Red List, accessed, 6/7/2016).

UNOFFICIAL SUBSTITUTES

Bensky *et al.* (2004) lists various local and historical substitutes and adulterants of 'Huang Lian'.These include rootstocks of:

1. Other ***Coptis*** species;
2. ***Beesia*** spp. (Ranunculaceae family);
3. ***Jeffersonia*** and ***Epimedium*** (Berberidaceae family);
4. ***Eomecon*** (Papaveraceae family);
5. ***Onychium*** (Pteridaceae family — a fern).

Their occurrence in international trade is uncertain. Due to high demand for this drug in Japan too, *Coptis japonica* (Thunb.) Makino (a native Japanese *Coptis*) has recently been brought into medicinal cultivation there as an additional local source and 4th official species of this drug in the Japanese Pharmacopoeia (16th ed., 2011).

Coptis japonica

Freshly uprooted plants of *Coptis deltoidea* showing stolon internodes up to 9 cm long

Medicinal crop of *C. deltoidea*, Sichuan province

Fruiting heads of *C. deltoidea*

Fruiting plants of *C. teeta*

C. chinensis

Ranunculaceae

DRUG MORPHOLOGY

CRUDE DRUG (DRIED RHIZOME): rhizomes 3–6-branched **[1]** and nodular **[2]** (3–6 cm long; 0.3–0.8 cm diam.), somewhat curved and often forming a tight cluster **[3]** (the latter is the origin of the colloquial terms Ying Zhua Huang Lian (鹰爪黄连) 'eagle-claw coptis' or Ji Zhua Huang Lian 鸡爪黄连 'chicken-claw coptis' traditionally used for this distinctive identification trait). External surface greyish-yellow or yellowish-brown, rough, with protuberances and rootlet remains **[4]**. Stolon internodes occasional, usually smooth and short (up to 2 cm long) and commonly called Guo Qiao 'bridging pieces' **[5]**. Apical ends of rhizomes **[6]** sometimes with brown scale-like remnants of leaf or flowering stems. *Texture*: hard. *Fracture*: uneven, bark orangish-red or dark brown, wood bright yellow or orangish-yellow **[7]** (radial in transverse fracture); central pith darker reddish-brown or hollow **[8]**. *Odour*: indistinct. *Taste*: very bitter.
Good quality: rhizomes with internodes short or apparently absent, devoid of rootlets, scaly leaf and stem bases. Uncut surface yellowish to yellowish-brown; interior yellow or orangish-yellow; extremely bitter.

DECOCTING PIECES-1 (SLICED DRIED RHIZOME): of irregular shapes and sizes. Longitudinally or transversely sliced, (mostly c. 3–5 cm long; 0.3–2 cm diam.; 1–2 mm thick), often branching or lobed **[9]**. Uncut external surface greyish-yellow or yellowish-brown, rough, with protuberances **[10]** and occasional rootlet remains **[11]**. Stolon internodes occasional and smooth. Apical part of rhizome (when present) often with brown remnants of leaf stems or flowering stems. Cut surfaces reveal a bright yellow, orangish-yellow or reddish-brown wood with many cavities **[7]**, often deeply indented at outer edge; central pith darker brown or mostly hollowed **[8]**. *Texture*: hard and somewhat fibrous. *Odour*: indistinct. *Taste*: very bitter.

DECOCTING PIECES-2 (WINE-PROCESSED RHIZOME): as for dried decocting pieces-1 but a darker colour throughout and slightly wine-scented.

DECOCTING PIECES-3 (GINGER-PROCESSED RHIZOME): as for dried decocting pieces-1 but brownish-yellow throughout and tasting of ginger.

DECOCTING PIECES-4 (TETRADIUM-PROCESSED RHIZOME): as for dried decocting pieces-1 but brownish-yellow throughout and with a pungent odour of *Tetradium* fruit.

PROCESSING

After removal of the rootlets, rhizomes are dried (= crude drug) then later cut into 1–2 mm thick slices (= decocting pieces-1). Optional further processing may include steeping the rhizomes in yellow rice wine (producing 'Jiu Huang Lian') (= decocting pieces-2) or in ginger juice (producing 'Jian Huang Lian') (= decocting pieces-3) or in a decoction made from the dried fruits of *Tetradium ruticarpum* ('Wu Zhu Yu') page 736 (producing 'Yu Huang Lian') then stir-frying to the point of dryness (= decoctong pieces-4).

OTHER COMMON NAMES

Commonly known in China as 味连 Wei Lian. Other names: 川黄连 Chuan Huang Lian 'Sichuan Huang Lian'; 鸡爪黄连 Ji Zhua Huang Lian 'chicken's claw coptis'; 鹰爪黄连 Ying Zhua Huang Lian 'eagle-claw coptis'; coptis (SCN); golden thread (CP2010); Rhizoma Coptidis.

SAFETY

Safety Class: 2b; Interaction Class: A (Gardner & McGuffin, 2013). Adverse effects reported after ingestion include transient diarrhoea, vomiting, nausea and bloating (Bensky *et al.,* 2004).

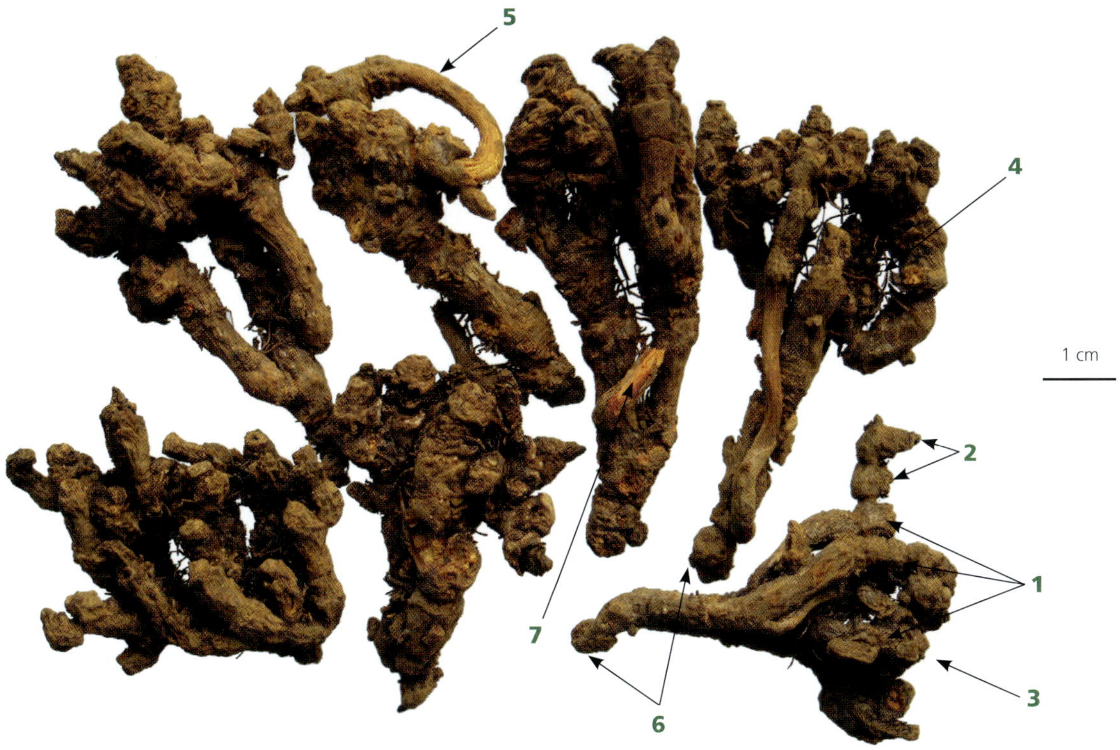

Crude drug of Huang Lian ('Wei Lian') sourced from *C. chinensis*

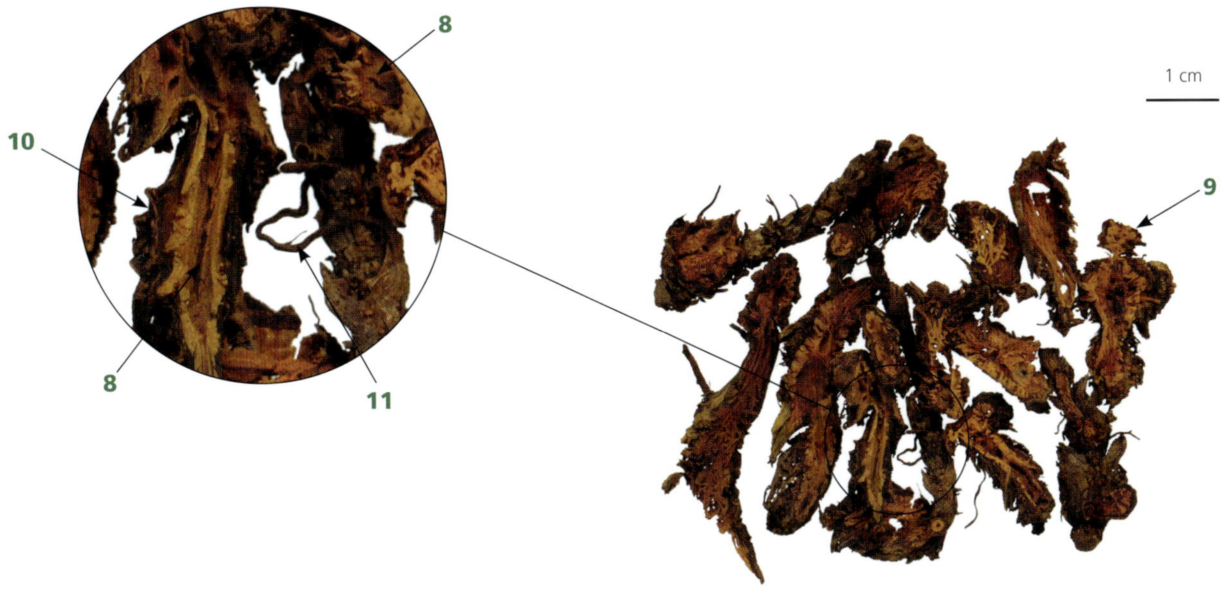

Longitudinally sliced decocting pieces of Huang Lian ('Wei Lian') sourced from *C. chinensis*

C. deltoidea

Ranunculaceae

DRUG MORPHOLOGY

CRUDE DRUG (DRIED RHIZOME): easily distinguished from PS1 and usually PS2; **unbranched or rarely branched [12]**, slightly curved **[13]** (4–8 cm long, 0.5–1 cm diam.) with **stolon internodes** smooth and **long** (up to 7 cm) **[14]**.

DECOCTING PIECES-1 (LONGITUDINALLY SLICED RHIZOME): as for the crude drug above but cut or broken surfaces bright yellow or reddish-yellow, and central pith darker brown or hollow. Radiating tissues occasional.

DECOCTING PIECES-2 (WINE-PROCESSED RHIZOME): as for decocting pieces-1 but a darker colour throughout and slightly wine-scented.

DECOCTING PIECES-3 (GINGER-PROCESSED RHIZOME): as for decocting pieces-1 but brownish-yellow throughout and tasting of ginger.

DECOCTING PIECES-4 (TETRADIUM-PROCESSED RHIZOME): as for decocting pieces-1 but brownish-yellow throughout and with a pungent odour of *Tetradium* fruit.

PROCESSING

As for PS1.

OTHER COMMON NAMES

Commonly known in China as 雅连 Ya Lian. Other names: 峨眉连 E Mei Lian; 三角叶黄连 San Jiao Ye Huang Lian (plant name); coptis (SCN); golden thread (CP2010); Rhizoma Coptidis.

SAFETY

Adverse effects as for PS1.

C. teeta

Ranunculaceae

DRUG MORPHOLOGY

CRUDE DRUG (DRIED RHIZOME): similar to PS2. Rhizomes **unbranched,** smaller than PS2 (shorter and narrower) (2–4 mm diam.) **[15]**, stolon internodes long, often ridged **[16]** and curvature more pronounced **[17]** than those of PS2.

Decocting pieces may be difficult to differentiate macroscopically from those of PS1 and PS2 since internode length, size and curvature may no longer be visible.

DECOCTING PIECES-1 (SLICED RHIZOME): as for the crude drug above but cut or broken surfaces bright yellow or reddish-yellow and central pith darker brown or hollow. Radiating tissues occasional.

DECOCTING PIECES-2 (WINE-PROCESSED RHIZOME): as for decocting pieces above but a darker colour throughout and slightly wine-scented.

DECOCTING PIECES-3 (GINGER-PROCESSED RHIZOME): as for dried decocting pieces above but brownish-yellow throughout and tasting of ginger.

DECOCTING PIECES-4 (TETRADIUM-PROCESSED RHIZOME): as for dried decocting pieces above but brownish-yellow throughout and with a pungent odour of *Tetradium* fruit.

PROCESSING

As for PS1.

OTHER COMMON NAMES

Most commonly known in China as 云连 Yun Lian. Other names: 云南黄连 Yun Nan Huang Lian (plant name); Indian goldthread (SCN); golden thread (CP2010); Rhizoma Coptidis.

SAFETY

Adverse effects as for PS1.

Crude drug of Huang Lian ('Ya Lian') sourced from *C. deltoidea*

Longitudinally sliced decocting pieces-1 of Huang Lian ('Ya Lian') sourced from *C. deltoidea*

Crude drug of Huang Lian ('Yun Lian') sourced from *C. teeta*

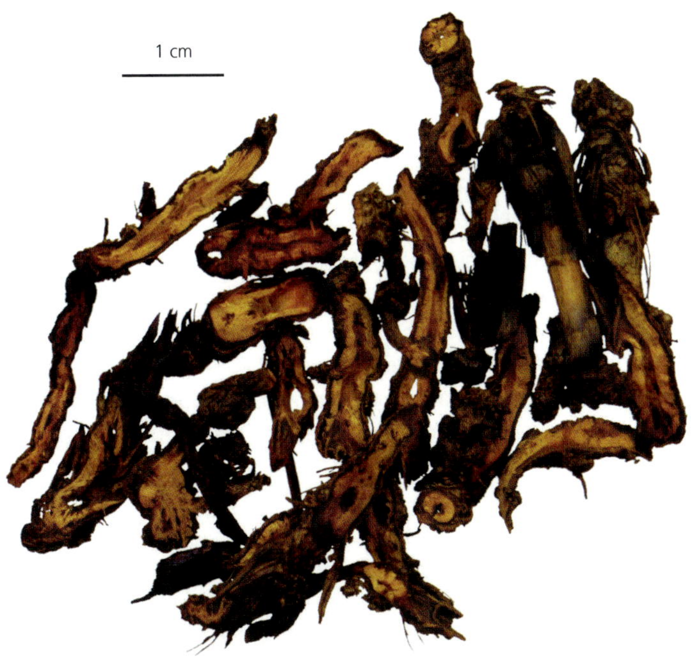

Longitudinally sliced decocting pieces-1 of Huang Lian ('Yun Lian') sourced from *C. teeta*

Huang Qi 黄芪 Astragali Radix

Drug part: ROOT

Astragalus mongholicus Bunge

Fabaceae

synonyms: *A. membranaceus* Fisch. ex Bunge; *A. membranaceus var. mongholicus* (Bunge) P. K. Hsiao; *Phaca macrostachys* Turcz.

taxonomic/nomenclatural note: the two taxa described as the source plant of Huang Qi in the CP2015 (as *A. membranaceus* (Fisch.) Bge. var. *mongholicus* (Bge.) Hsiao and *A. membranaceus* (Fisch.) Bge.) are treated here as synonyms (*FoC* and *Kew-MPNS*, 6/9/2016). The name *A. membranaceus* Fisch. ex Bunge is regarded as an illegitimate name ('nom. illeg.') because it is a later homonym of *A. membranaceus* Moench, and as such should not be used (*Kew-MPNS*).

Species of *Astragalus* wild in China: **401**

PLANT DESCRIPTION

Long established as a medicinal crop in China with many cultivars grown; their descriptions may vary slightly from the description below which is for the wild species.

A sprawling, herbaceous perennial **(25–60 cm tall;** taller when cultivated) with long, yellow, cylindrical tap roots. *Stems* hairless or loosely covered with spreading and appressed hairs. *Stipules* (at base of each leaf) **narrowly triangular** (4–10 mm long). *Leaves* alternate (6–15 cm long), almost stalkless, divided into 8–12 pairs of leaflets plus a terminal leaflet; **leaflet blade narrowly ovate to narrowly elliptic (5–22 mm long; 3–11 mm wide)**, tips broadly rounded and often with a shallow notch; margins entire. *Inflorescence* an unbranched, axillary, **loose, many-flowered head** (4–5 cm long), heads elongating to 9–14 cm when fruiting. Each flower distinctly stalked, laterally symmetrical, often pendant; bisexual; **calyx tubular to bell-shaped** with 5 apical teeth; petals 5, **yellowish-white (rarely pink),** unequal, with the standard (uppermost) petal widely elliptic (13–20 mm long; 7–9 mm wide), wing (lateral) petals 12–17 mm long; keel (lowermost) petals (11–16 mm long). ***Fruit* with a slender stalk 6–9 (–15) mm long**; each fruit a curved, inflated, papery pod (legume) (2–3 cm long; 8–12 mm high; 3–4 mm diam.), sharply keeled along one margin and rounded along the other; **in transverse outline semi-circular;** hairless or loosely covered with short hairs, both black or white. *Seeds* many per fruit, dark brown and kidney-shaped.

HARVESTING, SOURCING AND NATURAL RANGE

Roots are traditionally harvested in spring or autumn from 3–5 year old plants widely cultivated in parts of NE and E China: Heilongjiang, Hebei, Shandong and Jiangsu provinces. *Di Dao* quality material is traditionally sourced from Inner Mongolia and Shanxi. The species is also cultivated as a medicinal crop in parts of the USA.

Despite its large native distribution in N, W and NE China (Gansu, Hebei, Heilongjiang, Inner Mongolia, Jilin, Shaanxi, Shandong, Shanxi, Sichuan, Tibet and Xinjiang provinces) the species has declined in many areas due to medicinal over-harvesting. Its natural habitat is steppe, meadows and coniferous forests (800–2,000 m altitude) [also native to Kazakhstan, Mongolia, Russia (Far East, Siberia)].

CONSERVATION STATUS

China: Vulnerable (MEPCAS, 2013).
Global: Least Concern (IUCN Red List, accessed 3/2/2016).

Hedysarum polybotrys Hand.-Mazz.

Fabaceae

PLANT DESCRIPTION

Similar to the Pharmacopoeia species. *Main differences*: plants **clump-forming at base** (i.e. with many densely packed basal stems); height varies from 7–12 cm in wild specimens and up to c. 75 cm in medicinally cultivated ones (illustrated here); **rootstocks reddish-grey or reddish-brown;** stipules lance-shaped; **stems often dark red;** leaves with 6–10 pairs of leaflets, **leaflet blade usually larger** (15–30 mm long; 4–15 mm wide) than the Pharmacopoeia species; flowers yellow; **fruit distinctly constricted between seeds.** Fl. Jul–Aug, fr. Aug–Sep.

CONSERVATION STATUS

China: Least Concern (MEPCAS, 2013, as the varieties '*alaschanicum*' and '*polybotrys*').
Global: Least Concern (IUCN Red List, accessed 3/2/2016).

SUBSTITUTION COMMENT. *H. polybotrys* is a close relative of *A. mongholicus* (PS1) and the living plants as well as rootstocks are easily confused morphologically. Those of the former are the official source of a separate TCM drug 'Hong Qi' (Hedysari Radix) (CP2015) (not included in this book) and which, according to Bensky *et al.* (2004) has similar clinical properties to 'Huang Qi'; 'Hong Qi' is "somewhat weaker at tonifying the qi". This would explain the interchangeable dispensing of the 2 drugs in some TCM dispensaries (e.g. in Taiwan). It is unclear whether the interchanging of these two drugs is mostly due to clinical acceptance of their similar properties or mistaken identification. *H. polybotrys* is a common and widespread Chinese species endemic to N China (E and S Gansu, W Hebei, Inner Mongolia, Ningxia, Shanxi and NW Sichuan) and grows in stony places often at forest margins (1,200-3,200 m). 'Hong Qi' is sourced from medicinal crops with Di Dao-quality 'Hong Qi' grown in Gansu province (districts of Minxian, Wudu, Lintan and Dangchang).

TCM PROPERTIES AND USE

***properties*: 'sweet and slightly warm'**
***channels*: 'Spleen and Lung'**
***actions*: 'tonifies qi, raises yang, promotes healing, dispels pus, regulates water'**

***indications*: fatigue, poor appetite, diarrhoea, organ prolapse; respiratory disorders; wound regeneration**

Medicinal crop plant of *Astragalus mongholicus* in fruit, cultivated plant, Gansu province

A. mongholicus in fruit, cultivated plant, Gansu province

Freshly uprooted plants of *Hedysarum polybotrys* (left) and *Astragalus mongholicus* (right) showing different leaflet sizes and red stem of former

Astragalus mongholicus

Fabaceae

DRUG MORPHOLOGY

CRUDE DRUG (DRIED ROOT): cylindrical **[1]**, occasionally branched, and tapering slightly towards base (30–90 cm long; 1–3.5 cm diam.); root-head with **one or more** hard **narrow stem remnants** or stem scars **[2]**. Bark **pale brownish-yellow or pale brown [3]**, irregularly furrowed longitudinally, with scattered slightly raised transverse lenticels **[4]** and occasional sunken lateral root scars **[5]**. *Texture:* hard, pliable and difficult to break. *Fracture:* highly fibrous and starchy **[6]**; inner bark (cortex) yellowish-white **[7]**; cambium layer thin but distinct **[8]**; **xylem pale or deep yellow [9]** and, in older inferior roots, blackish-brown centrally or hollow. *Odour:* mild. *Taste:* starchy, somewhat sweet and slightly 'bean-like' when chewed. ***Good quality*:** opinions on morphological and chemical characters vary (see below).

DECOCTING PIECES-1 (TRANSVERSELY or OBLIQUE ROOT SLICES): sub-rounded or elliptical **[10]** (**1–3.5 cm diam**.; 2–4 mm thick). Outer bark pale brownish-yellow or pale brown, irregularly furrowed longitudinally **[11]**. Cut surfaces: **inner bark** (cortex) **yellowish-white [7]**; cambium layer thin but distinct **[8]**; central xylem pale yellow with distinct or indistinct radial fissures throughout **[9]**; poor quality roots blackish-brown in the middle, often with occasional clefts or hollow.

DECOCTING PIECES-2 (LONGITUDINALLY SLICED/PRESSED ROOT): long, flat and tongue-shaped **[12]** (8–18 cm long; 1.5–3 cm wide; 1–2 mm thick); cambium layer and radial fissures indistinct **[13]**. *Fracture, odour* and *taste* as for crude drug.

DECOCTING PIECES-3 (PROCESSED — HONEY-FRIED): as for dp-1 but darker throughout, with cut surfaces deep yellow **[14]**, slightly lustrous and occasionally charred **[15]**. *Odour:* honey-like. *Taste:* sweet, slightly sticky and bean-like when chewed.

***Good quality*:** Over the years, variations in cultivation methods of Huang Qi plus local environmental factors have resulted in a wide variability in the physical appearance including size of the commercially traded roots. A well-established root-grading system has been in place for many years in China designed to reflect this variability and provide indicators for quality and hence price. Recent research, however, challenges these traditional criteria such that these are now undecided and, in some cases, controversial (Upton, 1999a); further research is needed.

PROCESSING

After removal of the rootlets, roots are sun-dried (= crude drug) then quality-graded (see *Drug Morphology*) before softening then slicing into transverse or oblique decocting pieces-1 (2–4 mm thick). Another traditional and highly popular processing method involves soaking and pressing the roots into longitudinal slices to produce long, flat, tongue-shaped pieces (dp-2). Decocting pieces may also be steeped in honey then dry-fried (= Zhi Huang Qi or Radix Astragali Praeparata cum Melle).

OTHER COMMON NAMES

蒙古黄耆 Meng Gu Huang Qi (refers to the cultivar 'Meng Gu Huang Qi'); 膜荚黄耆 Mo Jia Huang Qi (refers to the cultivar 'Mo Jia Huang Qi'); milkvetch root (CP2010); Mongolian milk-vetch root; astragalus (SCN); Radix Astragali.

SAFETY

An official healthfood in China (Anon., 2002a,b). Safety Class: 1; Interaction Class: B (Gardner & McGuffin, 2013).

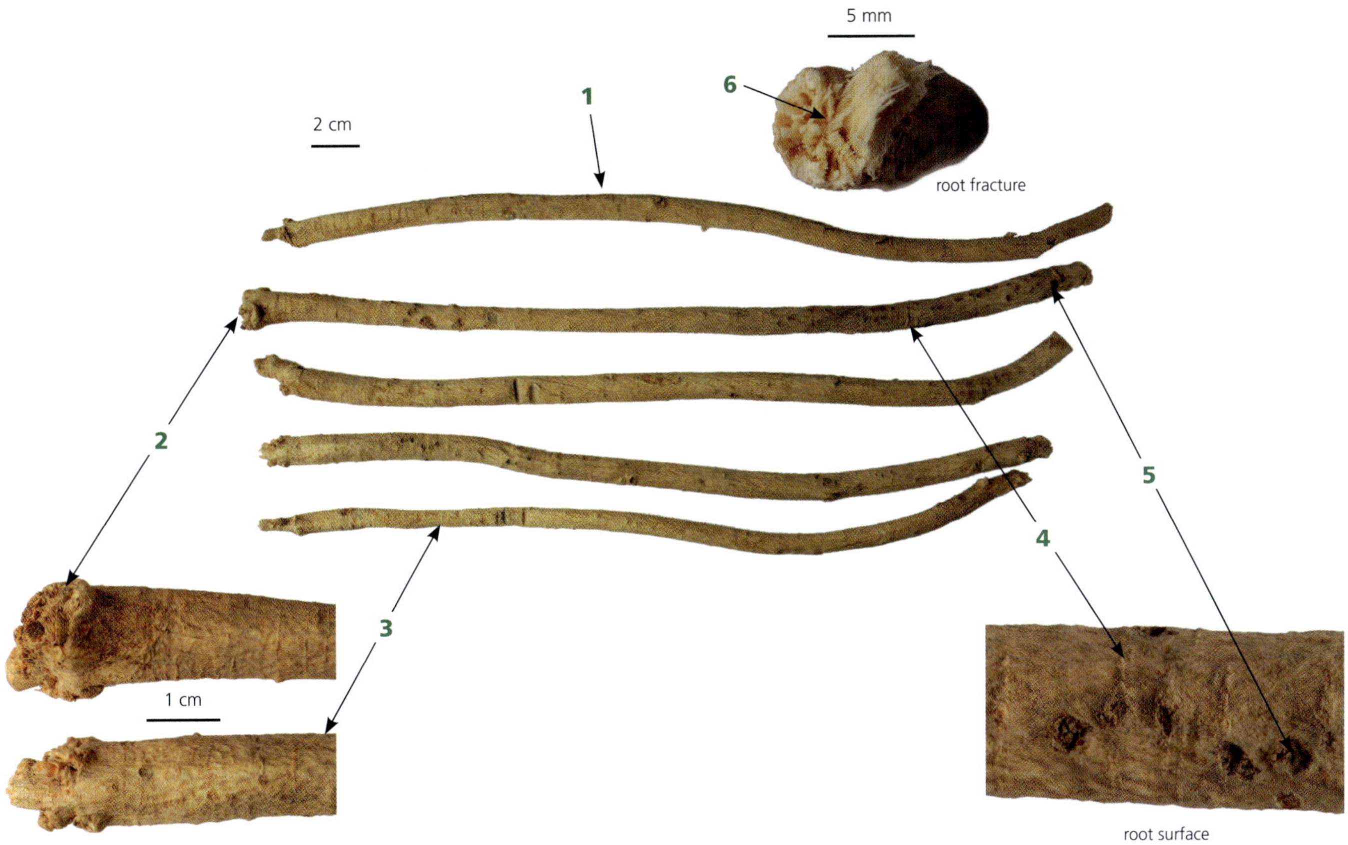

Crude drug of Huang Qi (sourced from *Astragalus mongholicus*)

Decocting slices of Huang Qi transversely/obliquely sliced (A. *mongholicus*)

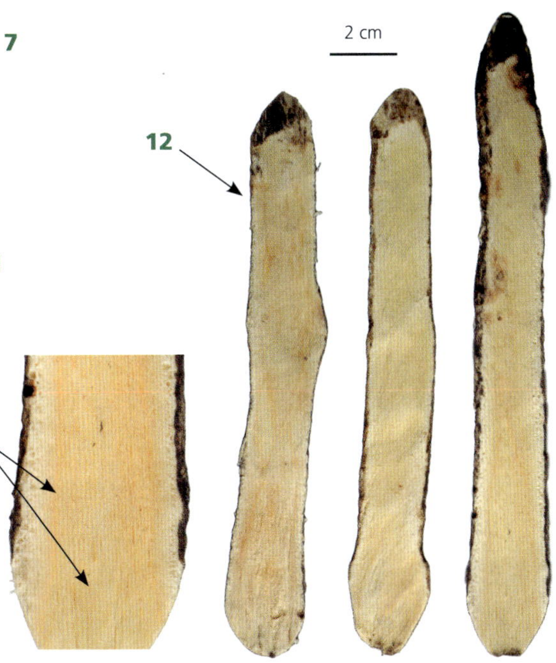

Decocting slices of Huang Qi longitudinally sliced and pressed (A. *mongholicus*)

Decocting pieces-3 of Huang Qi (honey-fried) sourced from *A. mongholicus*

Hedysarum polybotrys

Fabaceae

'DRUG SUBSTITUTE' MORPHOLOGY

Easily confused macroscopically with the official drug 'Huang Qi' especially as decocting pieces when the distinctive rootstock head of this species in no longer visible. For a more reliable identification of the latter, laboratory-based methods are advised.

DRIED ROOT (HONG QI — see *Substitution Comment*): outer bark **reddish-brown, tinged with grey [16]**; **root-head with many persistent stem bases [17] forming a head significantly broader** than that of the official drug; upper part of root with **outer bark often exfoliating to reveal brown reticulate fibres beneath [18]**; wood (xylem) pale yellowish-brown.

SLICED DRIED ROOT (HONG QI DECOCTING PIECES): transverse slices elliptical/sub-rounded and **often narrower [19]** (**0.4–1.5 cm diam.**; 2–4 mm thick); longitudinal slices oblong. Outer bark **reddish-brown [20] or, where outer bark exfoliated,** cortex yellowish-brown **[21]**; transverse or diagonally cut surface: inner bark (cortex) creamish-white.

OTHER COMMON NAMES

多序岩黄耆 Duo Xu Yan Huang Qi (plant name); 红芪 Hong Qi; 西芪 Xi Qi; Hedysari Radix (2015); hedysarum (SCN); Radix Hedysari.

Size-grading of official species (2 centre bundles) and substitute species of *H. polybotrys* (2 outermost bundles)

LESSER KNOWN UNOFFICIAL SUBSTITUTES AND ADULTERANTS

Now that the bulk of Huang Qi production from China and the USA is derived from cultivated plants, the use of other substitutes or adulterants (accidental or otherwise) sourced from other species is likely to be minimal. Local markets in China and SE Asia, however, are more prone to substitution; Guo Haiyun *et al.* (2010a) describe "twenty-three species from different genera including *Astragalus*, *Oxytropis*, *Hedysarum*, and *Glycyrrhiza*" to have been locally traded under the name 'Huang Qi'. Many laboratory-based methods for their differentiation (including DNA) have been published.

refs. NICPBP 1997–2002; Li *et al.*, 1994; Upton, 1999a; Xiao *et al.* 2001–2002.

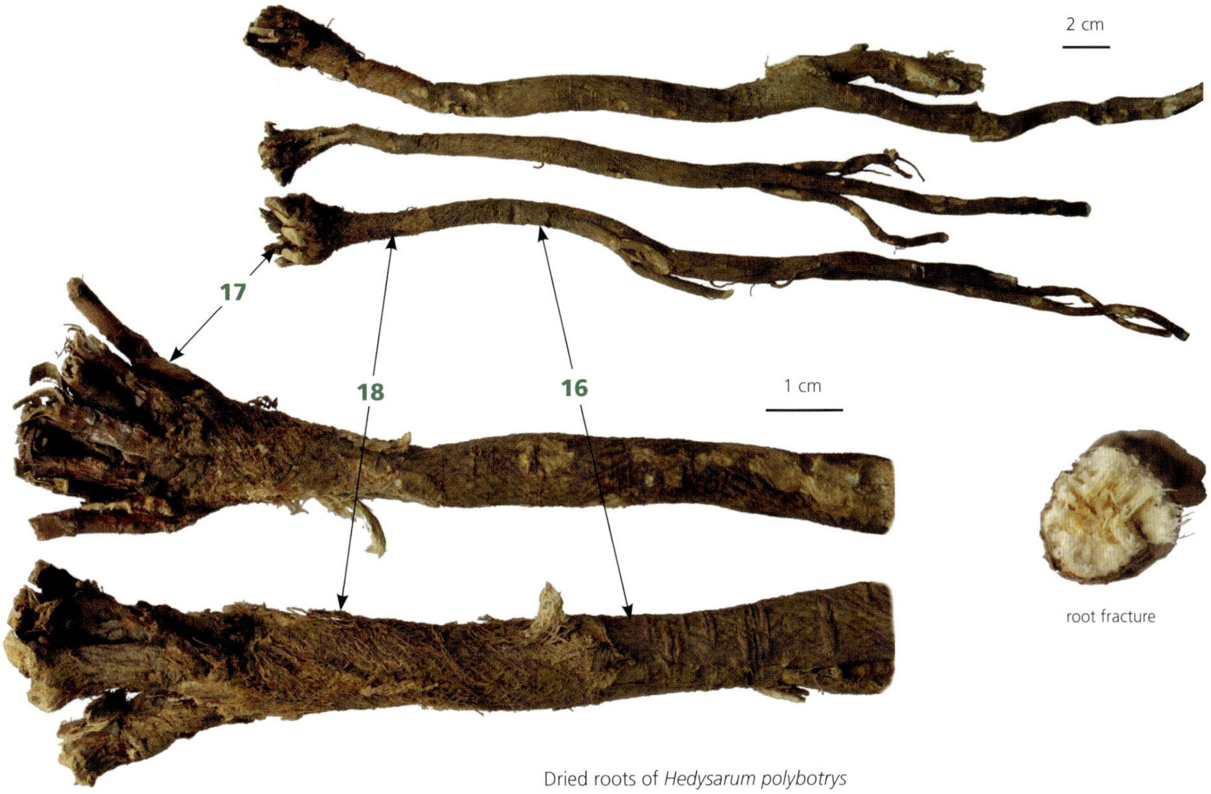

root fracture

Dried roots of *Hedysarum polybotrys*

Transversely sliced dried root of *H. polybotrys*

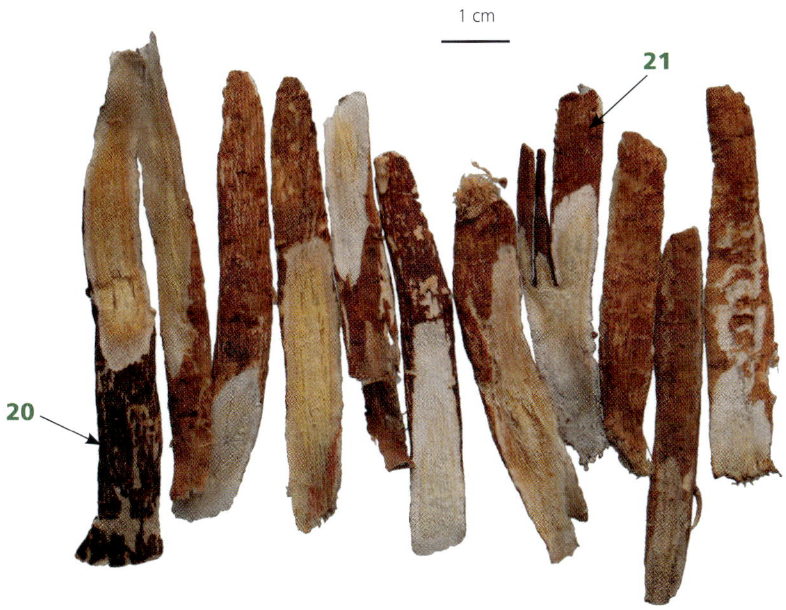

Dried and longitudinally sliced root of *H. polybotrys*

Huang Qin 黄芩 Scutellariae Radix

Drug part: ROOT

Scutellaria baicalensis Georgi

Lamiaceae

synonym: *S. macrantha* Fisch. ex Rchb.

Species of *Scutellaria* wild in China: 98

PLANT DESCRIPTION

An erect, herbaceous perennial (15–30 cm tall, sometimes taller) with a branched and somewhat twisted rootstock (1–3 cm diam.). *Stems* 4-sided, many, branching and shortly downy. *Leaves* opposite, with each pair at right angles to the next; stalkless; leaf blades simple, linear to lance-shaped (1.5–4.5 cm long; 0.5–1.2 cm wide); margins entire; lower leaf surfaces densely glandular. *Flowers* in many, dense, 1-sided terminal or axillary spikes (7–15 cm long). Each flower bisexual, laterally symmetrical and upright (2.3–3 cm long); calyx a short green tube (4 mm long) with 2 green, hood-like spurs (scutella) expanding in fruiting plants to 4 mm long; corolla conspicuous, purplish-blue or red, distinctly bent at base, densely hairy and glandular outside, gradually widening towards lobed mouth (6 mm wide); upper lobe distinctly hooded; lower lobe 3-lipped with a broad, triangular-ovate middle lobe (c. 7.5 mm wide); stamens 4 and unequal. *Fruit* with 4 dry, blackish-brown, ovoid nutlets (c. 1.5 mm long; 1 mm diam.). Fl. Jul–Aug, fr. Aug–Sep.

HARVESTING, SOURCING AND NATURAL RANGE

Mostly sourced from medicinal crop plants grown in the Chinese provinces of Shanxi, Hebei, Inner Mongolia and Liaoning. Harvesting takes place in spring or autumn (see *Drug Morphology*).

In the wild, the species is native to C and N China (Gansu, Hebei, Heilongjiang, Henan, Hubei, Jiangsu, Liaoning, Inner Mongolia, Shaanxi, Shandong and Shanxi provinces) where it grows in grassland and disturbed places (100–2,000 m altitude); [also native in Japan, Korean Peninsula, Mongolia and Russia].

CONSERVATION STATUS

China: Least Concern (MEPCAS, 2013).

Global: Not Evaluated (IUCN Red List, accessed 10/7/2016).

S. rehderiana Diels

Lamiaceae

PLANT DESCRIPTION

Easily confused and closely related to the Pharmacopoeia species. **Roots narrower (0.2–1 cm diam.)**; **aerial parts hairier (especially on stems)**; leaves broader (0.6–1.7 cm), stalked (2.8–9 mm), **leaf margins usually shallowly toothed**; **lower leaf surfaces with sunken glands few or, more often, absent**; flowers reddish to purplish-blue. Fl. May–Aug.

SUBSTITUTION COMMENT. *S rehderiana* is probably the most common species of *Scutellaria* to be used as a local substitute of *S. baicalensis* in China. The species are morphologically similar both as living plants and as dried roots. The Chinese name 甘肃黄芩 Gansu Huang Qin for *S. rehderiana* reflects the provenance of this species, predominantly from Gansu province, but the prefix 'Gansu' is not routinely used in trade leading to potential identification confusion.

The species is a common endemic of N and W China (Gansu, Shaanxi and Shanxi provinces) where it grows in open grassland between 1,300–3,200 m altitude.

CONSERVATION STATUS

Endemic to China: Least Concern (MEPCAS, 2013); Not Evaluated (IUCN Red List, acessed 10/7/2016).

EHTPA 4/14
GAP

TCM PROPERTIES AND USE

properties: 'bitter and cold'
channels: 'Lung, Gall Bladder, Stomach and Large Intestine
actions: 'clears heat, dries damp'

indications: fever, abdominal distension; diarrhoea; jaundice, ulcers, sore throat, cough, haematemesis; dysuria, threatened miscarriage

Medicinal crop of *Scutellaria baicalensis*, Hebei province

Flowering spikes of *S. baicalensis*

S. rehderiana: leaf margin shallowly toothed

Flowering spikes of *Scutellaria rehderiana*

Scutellaria baicalensis

Lamiaceae

DRUG MORPHOLOGY

CRUDE DRUG (DRIED ROOT):

Cultivated roots (bulk of trade): cylindrical, tapering, somewhat twisted **[1]** (8–25 cm long; 1–3 cm diam.), occasionally branched. **External surface yellowish-brown** with scattered, wart-like rootlet scars of varying size and shape **[2]** and occasional transverse oval lenticels **[3]**; upper (broader) parts quite rough and twisted, often with contorted stem bases **[4]**; finely wrinkled longitudinally; some parts reticulated. *Texture*: hard but easily broken. *Fracture characters of the wood (xylem) reflect 'quality'*: best quality roots 'Zi Qin', 'Zi Huang Qin' or 'Tiao Qin' typically with broad yellow wood **[5]** and often with a reddish-brown core **[6]** are sourced from young roots up to 3 years old. Lower quality roots (considered clinically inferior) are from older roots with a darker brown or brownish-black core **[7]**, sometimes withered or hollowed **[8]** (such roots are called 'Ku Qin' or 'Ku Huang Qin' (Ku (枯) = rotten). *Taste*: somewhat bitter. *Odour*: indistinct.

Wild roots: usually stouter and rarely branched; external surface dark yellowish. *Fracture*: more obviously horn-like. *Taste*: distinctly bitter. ***Good quality***: large, firm, golden-yellow roots without a withered or hollow core; bluish-green tissues **[11]** indicate poor processing and are considered 'clinically inferior'.

DECOCTING PIECES-1 (BOILED OR STEAMED ROOT SLICES): irregularly shaped and woody (1–2 mm thick). **Externally yellowish-brown or dark yellow [9],** rough and irregular; lenticels or root scars rare. Cut surface with narrow, yellowish-brown inner bark **[10]**, wood broad, yellow **[5]** (occasionally bluish-green **[11]**) with radial striations (in transversely cut slices) **[12]**, some clefts and either a reddish-brown core **[6]** (Zi Qin) or with a darker brown or brownish-black core **[7]**, sometimes withered or hollowed **[8]** (Ku Qin) (see also *crude drug*).

DECOCTING PIECES-2 (WINE-PROCESSED ROOT SLICES): as DP-1 above but darker **[13]** and smelling of yellow rice wine.

PROCESSING

After removal of aerial parts and rootlets, roots are sun-dried and beaten to remove the outer cork-like root-bark; the remaining inner root is sun-dried again (= crude drug). Subsequent boiling or steaming, then slicing thinly (1–2 mm) produces Huang Qin Pian 黄芩片 (decocting pieces-1); further dry-frying with yellow rice wine produces Jiu Huang Qin 酒黄芩 (= decocting pieces-2).

OTHER COMMON NAMES

酒炒黄芩 Jiu Chao Huang Qin 'wine processed Scutellaria'; 枯黄芩 Ku Huang Qin or 枯芩 Ku Qin 'rotten Scutellaria'; 条黄芩 Tiao Huang Qin or 条芩 Tiao Qin 'slender Scutellaria'; 子黄芩 Zi Huang Qin or 子芩 Zi Qin 'young Scutellaria'; Baical skullcap; Baical skullcap root (CP2010); Chinese skullcap (SCN); Radix Scutellariae; scute.

SAFETY

Safety Class: 1; Interaction Class: B (Gardner & McGuffin, 2013).

S. rehderiana

Lamiaceae

'DRUG SUBSTITUTE' MORPHOLOGY

DRIED ROOT: narrower (0.2–1.3 cm diam.) than the Pharmacopoeia species. **Externally greyish-brown [14]**; **rough and exfoliating in places [15]**; with wart-like rootlet scars more obvious and numerous **[16]**. *Fracture:* bright yellow **[17]**.

SLICED ROOT: cannot be reliably distinguished macroscopically from sliced root of the Pharmacopoeia species; laboratory-based authentication methods are recommended. In general, transversely sliced roots of this substitute tend to be narrower (not more than 1 cm diam.) and the woody xylem a brighter yellow **[17]**.

OTHER COMMON NAMES

甘肃黄芩 Gansu Huang Qin.

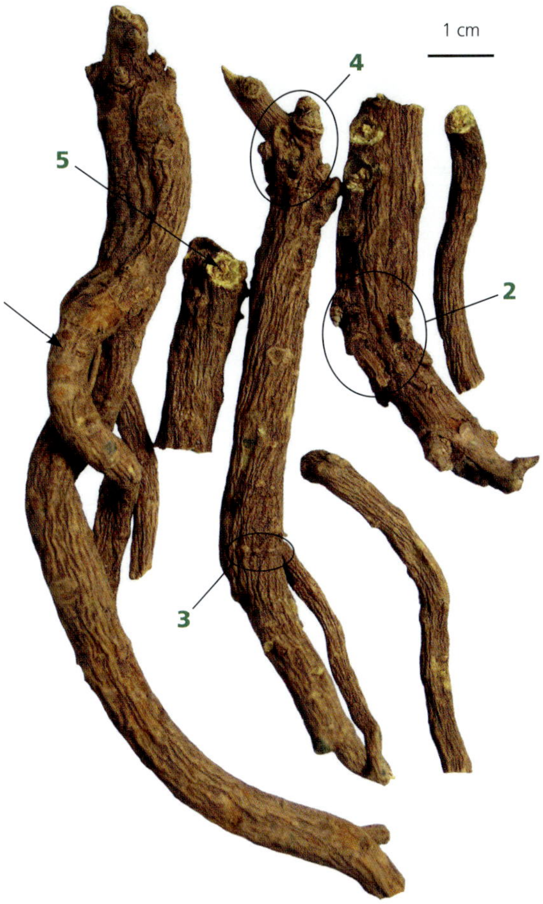

Crude drug of Huang Qin (cultivated *S. baicalensis*)

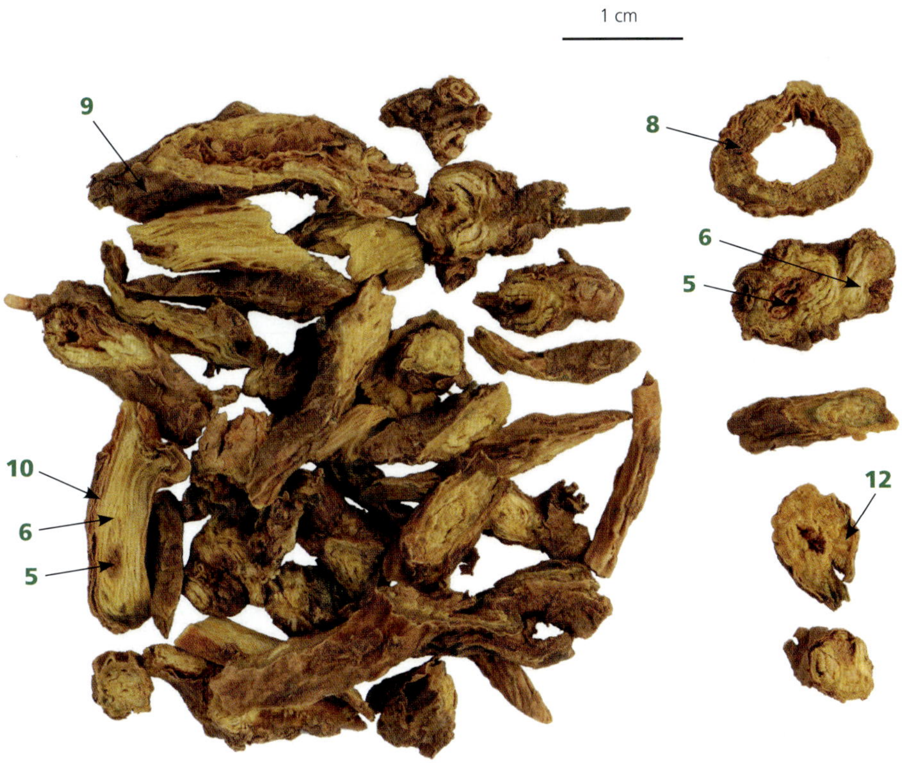

Steamed Huang Qin (*S. baicalensis*) = decocting pieces-1

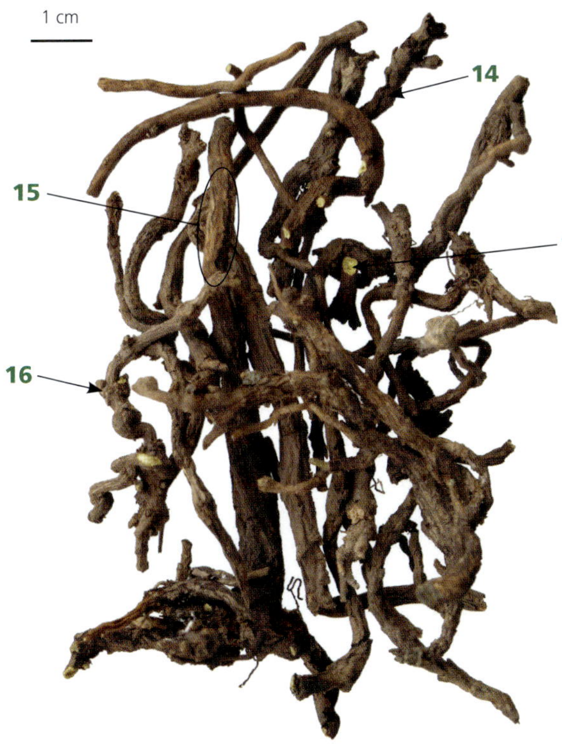

Dried roots of *S. rehderiana*

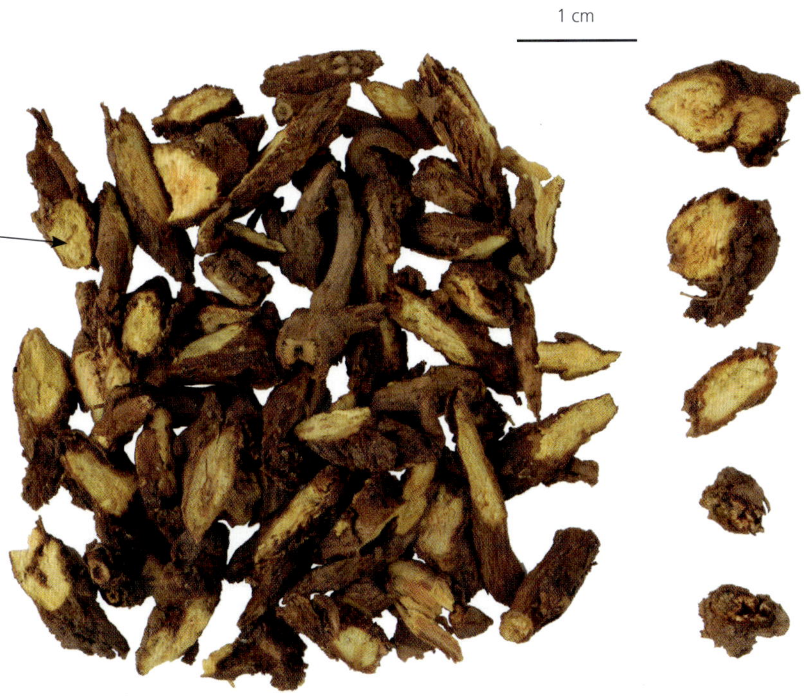

Dried and sliced roots of *S. rehderiana*

Grading dried roots of *Scutellaria baicalensis*, Shanxi province

LESSER KNOWN UNOFFICIAL SUBSTITUTES

The roots of many other species in the genus *Scutellaria* (of which 98 grow wild in China) are used as local Chinese substitutes of 'Huang Qin' (see below); their occurrence as substitutes in international trade is uncertain. Their roots tend to be narrower than those of the Pharmacopoeia species but for reliable root identification laboratory-based authentication methods are recommended. Many have been the focus of ethnopharmacological investigation in recent years because of their potential antitumour, hepatoprotective and *inter alia* antiviral activity (Shang *et al*., 2010).

1. ***Scutellaria amoena*** C. H. Wright (滇黄芩 Dian Huang Qin); roots forked; stems purplish, leaves stalked (1–2 mm), leaf blades oblong to oblong-ovate and tips rounded or broad; leaf margins with minutely rounded teeth; calyx purplish; endemic to Guizhou, Sichuan and Yunnan provinces.
2. ***S. hypericifolia*** H. Lév. (连翘叶黄芩 Lian Qiao Ye Huang Qin; 川黄芩 Chuan Huang Qin). Rootstock up to 2 cm diam.; stems purplish, white hairy at angles and bearded at nodes; leaf stalks 1–2 mm long; flowers white, greenish-white to purple or purplish-blue; endemic to Sichuan province.
3. ***S. likiangensis*** Diels (丽江黄芩 Li Jiang Huang Qin). Rootstocks 2–12 mm diam.; leaf margins often minutely round-toothed; leaf tips rounded or indented; flowers yellowish-white, yellow to greenish-yellow with purplish spots; endemic to Yunnan province.
4. ***S. tenax*** var. ***patentipilosa*** (Hand.-Mazz.) C. Y. Wu (展毛 Zhan Mao). Stems, leaves obviously stalked; leaf stalks and flowers densely white hairy; leaves with fine white bristles; leaf margins shallowly toothed; endemic to Sichuan and Yunnan provinces.
5. ***S. viscidula*** Bunge (粘毛黄芩 Zhan Mao Huang Qin). Rootstocks 2.5–4 (–18) mm diam.; leaf surfaces densely yellow glandular; flowers yellowish-white to white; endemic to N China: Hebei, Inner Mongolia, Shandong and Shanxi.

Huang Qin (*Scutellaria baicalensis*) showing examples of different processing and quality

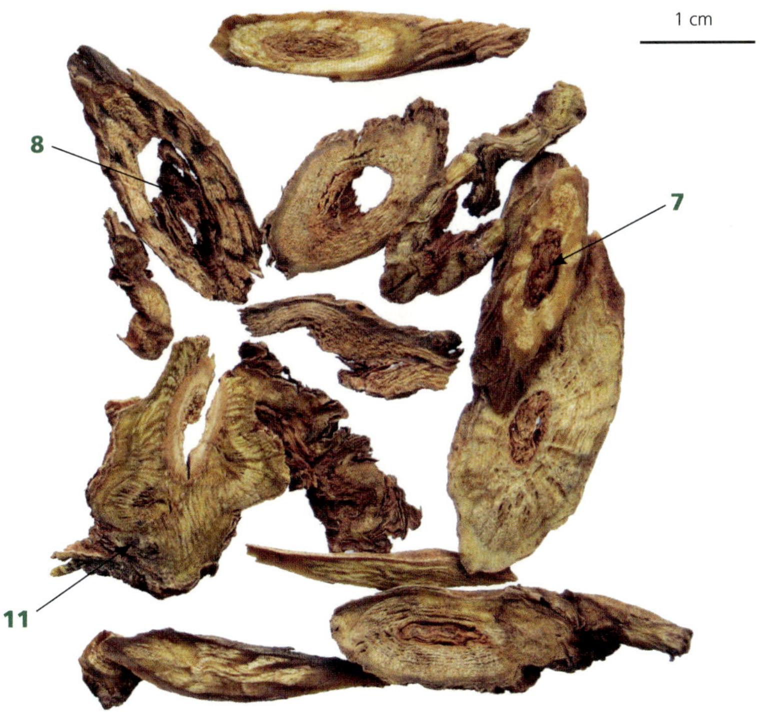

Steamed Huang Qin — poor quality (*S. baicalensis*) (= decocting pieces-1)

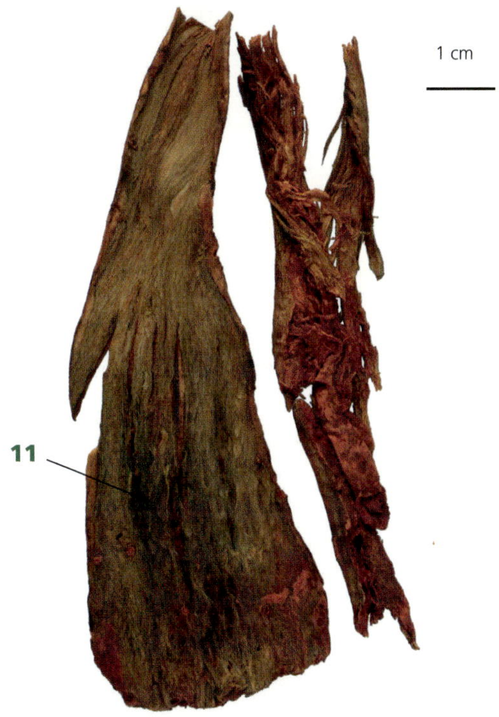

Poor quality (green) Huang Qi (longitudinal slices)

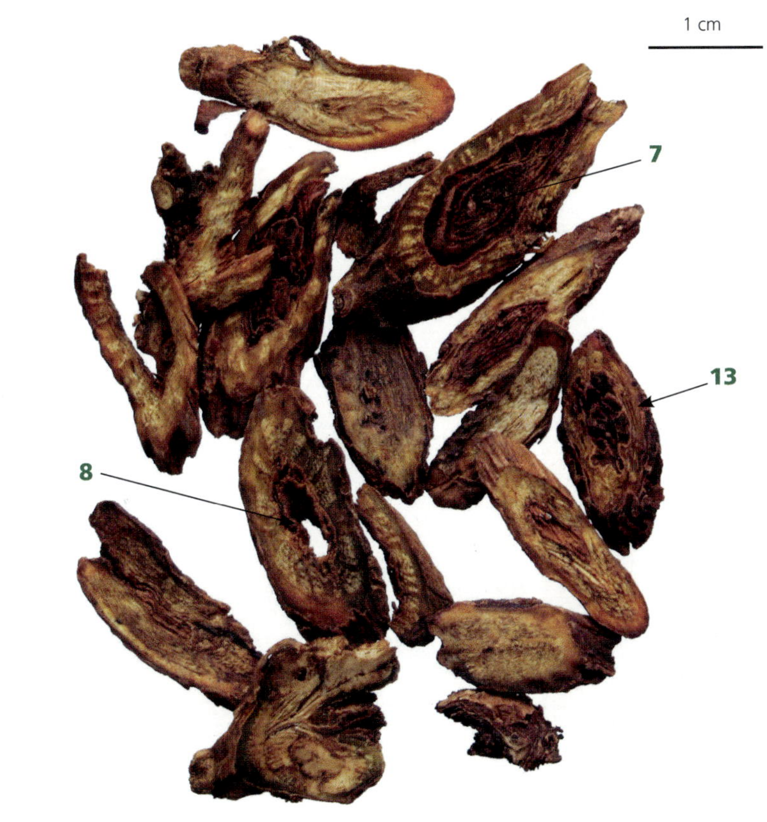

Wine-processed Huang Qin (Ji Huang Qin) sourced from *S. baicalensis* (= decocting pieces-2)

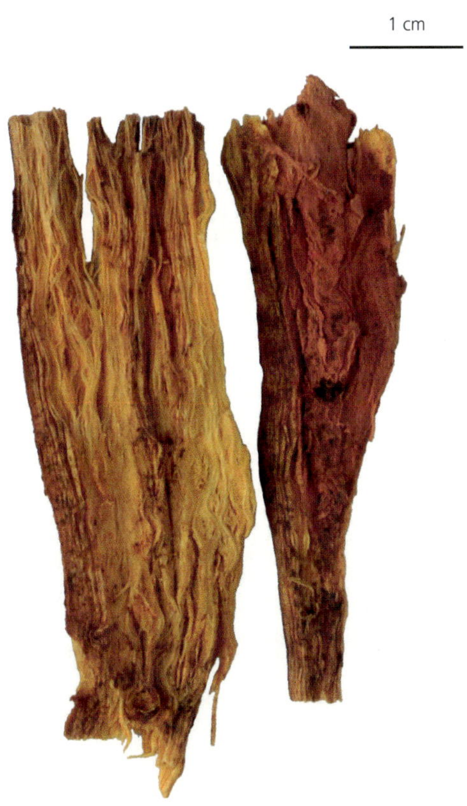

Good quality Huang Qin (longitudinal slices)

Hubei Bei Mu 湖北贝母 Fritillariae Hupehensis Bulbus

Drug part: BULB

Fritillaria monantha Migo

Liliaceae

synonym: *F. hupehensis* P. K. Hsiao & K. C. Hsia (name used in CP2015).

Species of *Fritillaria* wild in China: 24

PLANT DESCRIPTION

An erect, delicate-looking, herbaceous perennial (20–60 (sometimes up to 100) cm tall) with a slightly squat bulb (0.8–3.5 cm diam.; including dimensions of medicinally cultivated bulbs) with 2 or 3 fleshy scales. *Stem* unbranched. *Leaves* simple, opposite, whorled or alternate; leaf blades oblong-lance-shaped to lance-shaped (5–12 cm long; 1.5–3 cm wide) with tips slightly or strongly curled and margins entire. *Inflorescence* 1-flowered (rarely up to 4); bracts 3 (sometimes fewer), leaf-like and slightly or strongly curled. *Flower* bisexual, corolla bell-shaped and nodding (3.5–5 cm long; 1–2 cm wide); 'petals' 6, equal size, greenish-yellow to pale purple, spotted or densely chequered yellowish-brown or dark purple, with small 'shoulders' (raised nectaries) visible outside base of 'petals'; stamens 6; ovary superior; style lobes 3–8 mm long. *Fruit* a dry, ovoid, 6-angled, 3-valved capsule with broad (6–8 mm) membranous wings. *Seeds* many, flat, in 2 rows in each valve. Fl. Apr–Jun, fr. Jun–Jul.

HARVESTING, SOURCING AND NATURAL RANGE

Bulbs are harvested mostly from medicinal crop plants in early summer after aerial parts have withered. Endemic to China and confined to the provinces of Anhui, Henan, Hubei, Jiangxi, Sichuan and Zhejiang. *F. monantha* grows on limestone soils in forests and damp places such as flood plains (100–1,600 m altitude).

CONSERVATION STATUS

Endemic to China: Endangered (MEPCAS, 2013); Not Evaluated (IUCN Red List; accessed 10/8/2016).

Fritillaria monantha

F. monantha

TCM PROPERTIES AND USE

properties: **'slightly bitter and cool'**
channels: **'Lung and Heart'**
actions: **'clears heat, resolves phlegm'**

indications: **cough; sores and abscesses**

DRUG MORPHOLOGY

CRUDE DRUG AND DECOCTING PIECES (INTACT BULB): sub-globose and slightly squat, (0.8–2.2 cm tall; 0.8–3.5 cm diam.), with 2 or 3 fleshy, slightly kidney-shaped scale leaves almost equal in size or, more usually, of different sizes and clasping each other **[1]**. External surfaces pale yellowish-white to pale brown and relatively smooth **[2]**. Apex closed or slightly open to reveal a shrunken stem base **[3]** and small buds **[4]**. Basal end rounded and, centrally, slightly concave with remants of root bases **[5]**. *Texture*: fragile. *Fracture*: whitish and starchy. *Odour*: indistinct. Taste: bitter.

PROCESSING

Bulbs are washed, soaked in water or limewater then dried (= crude drug and decocting pieces).

OTHER COMMON NAMES

板贝 Ban Bei; 天目贝母 Tian Mu Bei Mu; 窑贝 Yao Bei; Bulbus Fritillariae Hupehensis; Hubei fritillary bulb (CP2010).

SAFETY

An official healthfood in China (Anon., 2002a,b); Safety Class: 1; Interaction Class: A (Gardner & McGuffin, 2013). .

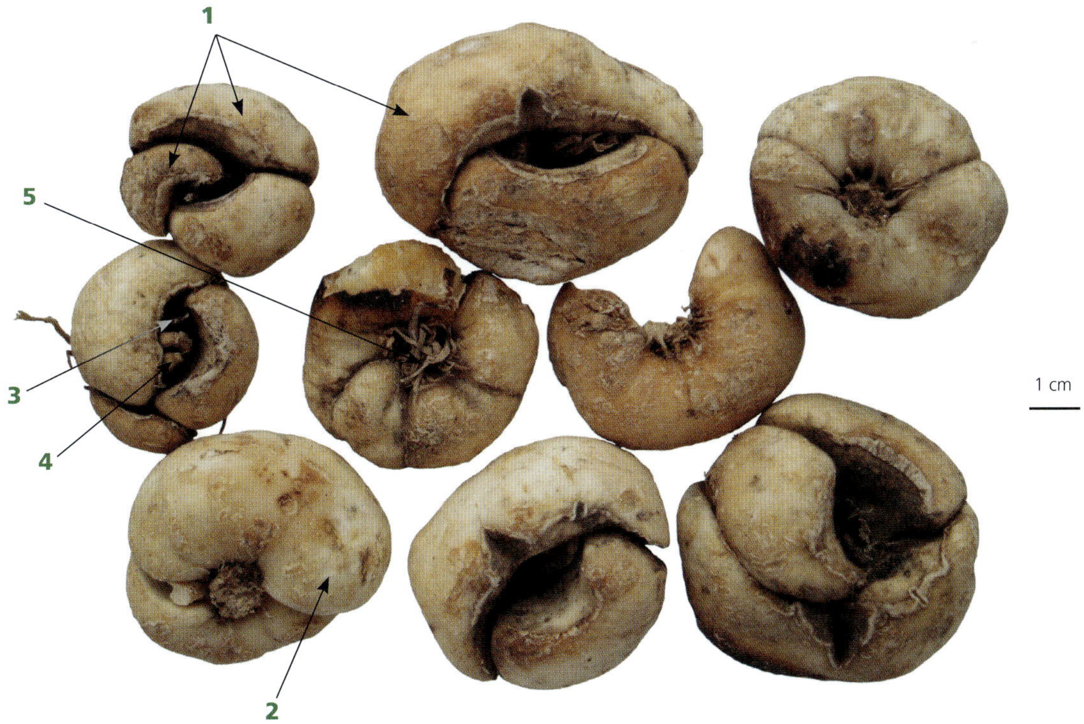

Crude drug/decocting pieces of Hubei Bei Mu (*F. monantha*)

F. monantha, freshly harvested bulbs of different ages

Jie Geng 桔梗

Platycodonis Radix

Drug part: ROOT

Platycodon grandiflorus (Jacq.) A. DC.

Campanulaceae

taxonomic/nomenclatural note: in CP2015 as '*Platycodon grandiflorum* (Jacq.) A. DC.'

synonym: *Campanula grandiflora* Jacq.

Species of *Platycodon* wild in China: 1

PLANT DESCRIPTION

An upright, perennial herb (20–120 cm tall) with a narrowly conical tuberous tap root. Virtually hairless and bluish-green almost throughout; stems usually unbranched. *Leaves* simple and stalkless, in whorls of 3–4 along the stems; leaf blades ovate, elliptical or lance-shaped (2–7 cm long; 0.5–3.5 cm wide), upper surface green, lower surface bluish-green; base broadly wedge-shaped or rounded, margin sharply toothed, tip short- or long-pointed. *Flowers* large and showy, terminal and solitary; calyx with 5 pointed lobes; petals fused into a blue or purplish-blue saucer-shaped corolla (1.5–4 cm diam.), balloon-shaped just prior to opening; stamens 5, free; ovary semi-inferior, style 1 with pollen-collecting hairs; stigma 5-parted. *Fruit* a dry, globose or egg-shaped capsule (0.7–2.5 cm long; 0.7–1.5 cm diam.); seeds many and black. Fl. Jul–Sep, fr. Aug–Oct.

HARVESTING, SOURCING AND NATURAL RANGE

Roots are harvested in spring and autumn mostly from plants widely cultivated for both food and medicine; main production areas are in northern and north-eastern China while traditional *Di Dao* quality material is sourced from Anhui, Jiangsu and Shandong provinces.

The wild species grows in sunny places in scrub and thickets up to 2,000 m altitude and has a wide distribution in China (Chongqing, Fujian, Guangdong, Guangxi, Guizhou, Hebei, Heilongjiang, Henan, Hubei, Hunan, Inner Mongolia, Jiangsu, Jiangxi, Jilin, Liaoning, Shaanxi, Shandong, Shanxi, Sichuan, Yunnan and Zhejiang [also native in Japan, Korean Peninsula and Russia (Far East)].

CONSERVATION STATUS

China: Least Concern (MEPCAS, 2013).

Global: Not Evaluated (IUCN Red List, accessed 21/7/2015).

Flowering medicinal crop of *Platycodon grandiflorus*, Shaanxi province

P. grandiflorus in flower and flowerbud

Leaves and stem of *P. grandiflorus*

TCM PROPERTIES AND USE

***properties*: 'bitter, pungent and neutral'**
***channel*: 'Lung'**
***actions*: 'resolves phlegm, dispels pus, raises qi'**

***indications*: cough; dyspnoea, phlegm; sores, abscess; diarrhoea; prolapse**

DRUG MORPHOLOGY

CRUDE DRUG (DRIED ROOT): roots long-cylindrical, tapering at base **[1]**, occasionally branched and slightly curved (7–20 cm long; 0.7–2 cm diam.). Outer peeled surface white or pale yellowish-white (or if unpeeled grey- or yellowish-brown), irregularly furrowed longitudinally with irregular and raised clusters of rootlet bases **[2]**; scattered throughout with raised oval transverse lenticels **[3]** and occasional blackish resin-like patches **[4]** and remnants of outer cork bark **[5]**. Apex often with several semi-circular sunken stem scars **[6]** and, immediately below the apex, a short zone of transverse annulations **[7]** often present. *Texture*: easily broken. *Fracture*: uneven; central wood (xylem) yellowish-white **[8]**, surrounded by a distinct slightly darker cambium ring **[9]** which in turn is surrounded by a narrow band of whitish inner bark with occasional clefts **[10]**. *Odour*: slight. *Taste*: slightly sweet, becoming bitter. ***Good quality*:** thick, white, heavy and bitter.

DECOCTING PIECES (SLICED ROOT): transverse or oblique slices elliptical or ovate in outline (2–4 mm thick) with an irregularly undulating margin. Cut surface with a broad central yellowish-white wood surrounded by a distinct and slightly darker cambium ring which in turn is surrounded by a narrow band of whitish inner bark with occasional clefts. Remnants of outer cork bark and blackish resin-like patches **[4]** sometimes present. *Texture:* easily broken. *Odour*: slight. *Taste*: slightly sweet, becoming bitter. Longitudinally cut slices usually tongue-shaped (c. 1 cm wide).

PROCESSING

After removal of rootlets, roots are dried immediately or peeled then dried (= crude drug); decocting pieces are made by slicing the roots (transversely, obliquely or longitudinally) into 2–4 mm thick pieces. In Hong Kong longitudinal slicing is more typical; these are usually pressed.

OTHER COMMON NAMES

白花桔梗 Bai Hua Jie Geng (white flowered Jie Geng); 北桔梗 Bei Jie Geng (northern Jie Geng); 粉桔梗 Fen Jie Geng (powdery Jie Geng); 甘桔梗 Gan Jie Geng; 苦桔梗 Ku Jie Geng (bitter Jie Geng); balloon flower; platycodon (SCN); Radix Platycodonis.

SAFETY

Safety Class: 2d; Interaction Class: A (Gardner & McGuffin, 2013; misspelt as '*P. grandiflorum*'). An official Chinese food and healthfood (Anon., 2002a,b; misspelt as '*P. grandiflorum*').

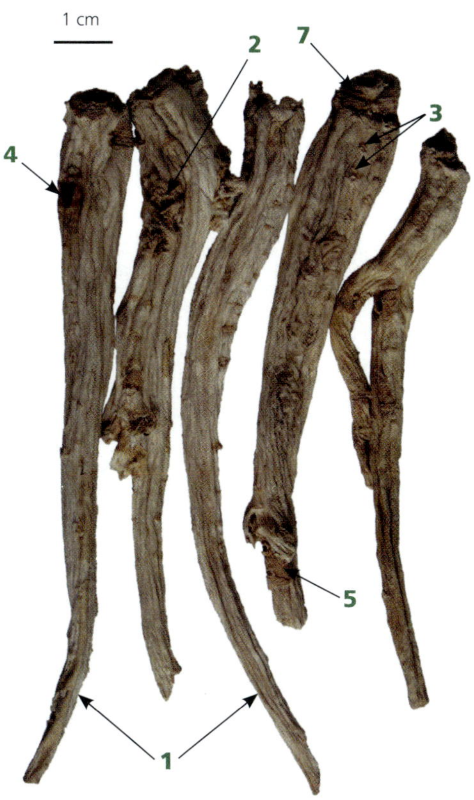

Crude drug of Jie Geng
(dried, peeled roots of *P. grandiflorus*)

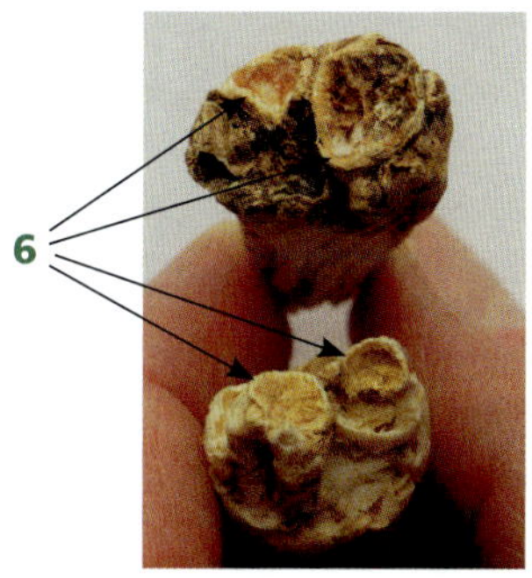

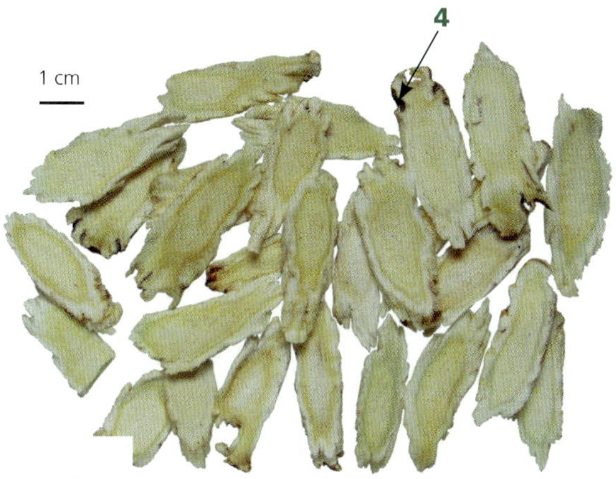

Decocting pieces of Jie Geng
(obliquely sliced roots of *P. grandiflorus*)

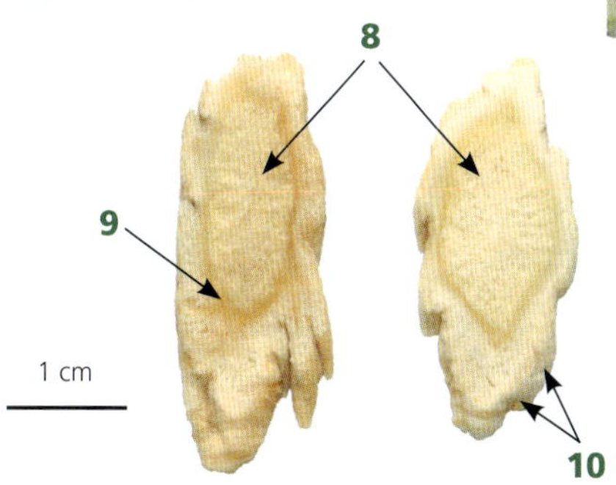

UNOFFICIAL SUBSTITUTES

Roots of other species have been reported as local and/or historical unofficial substitutes but substitution in international trade is unlikely given the wide availability of the official species as a cultivated medicinal crop:

1. *Adenophora* spp. — roots (family Campanulaceae)
2. *Gypsophila oldhamiana* Miq. — roots (family Caryophyllaceae)

Jin Guo Lan 金果榄 Tinosporae Radix

Drug part: ROOT TUBER

Tinospora sagittata (Oliv.) Gagnep. **Menispermaceae**

synonyms: *Limacia sagittata* Oliv.; *T. capillipes* Gagnep.

taxonomic/nomenclatural note: the CP2015 recognises the synonym *T. capillipes* Gagnep. as a separate species and alternative source of this drug. This account, however, follows the *Flora of China* and *Kew-MPNS* (28/6/2016) and treats it as a synonym.

Species of *Tinospora* wild in China: 6

PLANT DESCRIPTION

A creeping, evergreen, herbaceous vine. *Roots* with small, yellow, tuberous swellings (the medicinal part — 'Jin Guo Lan'); aerial roots absent. *Stems* slender and twining; lenticels small and orbicular. *Leaves* alternate and simple; leaf stalks long and hairy; leaf blades large (usually 7–15 cm long; 2.4–5 cm wide) longer than wide and arrowhead-shaped (sagittate) or sometimes with basal lobes pointing outwards (hastate); margins entire and veining palmate. *Flowers* tiny and inconspicuous; in stalked clusters (c. 2–10 cm long) arising from leaf axils and developing at same time as leaves; each flower creamy-white and single-sexed (with male and female flowers on separate plants); male flowers usually with 6 sepals (1–2 mm long) in 2 whorls (inner sepals larger, up to 5 mm long); petals 6, sub-rounded (1.4–2 mm. long) and often with a claw; female flowers with sepals like those of male; petals wedge-shaped (c. 0.4 mm long); staminodes 6; carpels 3; style short and fat; stigma reflexed with short pointed lobes. *Fruit* semi-spherical (6–8 mm diam.), fleshy and one-stoned. Fl. Apr; ft. autumn.

HARVESTING, SOURCING AND NATURAL RANGE

Root tubers are harvested in autumn and winter from wild plants which grow in forest edges and grassland in S, C and W China: Fujian, Guangdong, Guangxi, Guizhou, Hainan, Hubei, Hunan, Jiangxi, Shaanxi, Shanxi, Sichuan, Tibet and Yunnan [also native in Vietnam].

CONSERVATION STATUS

China: Least Concern (MEPCAS, 2013; except var. *sagittata* classified as Endangered).

Global: Not Evaluated (IUCN Red List, accessed 2/9/2015).

Twining habit of *Tinospora sagittata*

T. sagittata with young flower-heads

TCM PROPERTIES AND USE

properties: 'bitter and cold'
channels: 'Lung and Large Intestine'
actions: 'clears heat, eliminates toxins'

indications: **sore throat; abdominal distension; sores and abscesses; (T)**

DRUG MORPHOLOGY

CRUDE DRUG (DRIED ROOT TUBER): sub-spherical, ovoid or irregularly shaped (5–10 cm long; 3–6 cm diam.). Externally dull, brownish-yellow or brown **[1]**, often deeply and irregularly furrowed **[2]**; oval, protuberant lenticels often present **[3]**; raised rootlet scars occasional; root remnants often persistent at one or both ends **[4]**. Cut surface yellowish-white, starchy, with occasional cavities **[5]** and darker coloured vascular tissues in groups or arranged radially. *Texture:* hard and difficult to break. *Odour:* faint. *Taste:* bitter. ***Good quality*:** solid and heavy.

DECOCTING PIECES (DRIED SLICED ROOT TUBER): sub-rounded slices (3–6 cm diam.; 2–4 mm thick). Skin brownish-yellow or brown, usually rough and uneven **[6]**. Cut surface pale yellowish-white, sometimes with greyish-brown radial striations; clefts frequent **[5]**. *Odour*: indistinct. *Taste*: bitter.

PROCESSING

Rootlets are removed then the root tuber sun-dried (= crude drug); dried tubers are later softened thoroughly before slicing into 2–4 mm thick decocting pieces.

OTHER COMMON NAMES

地苦胆 Di Ku Dan; 金牛胆 Jin Niu Dan; 青牛胆 Qing Niu Dan (plant name); Radix Tinosporae; tinospora (SCN); tinospora root (CP2010).

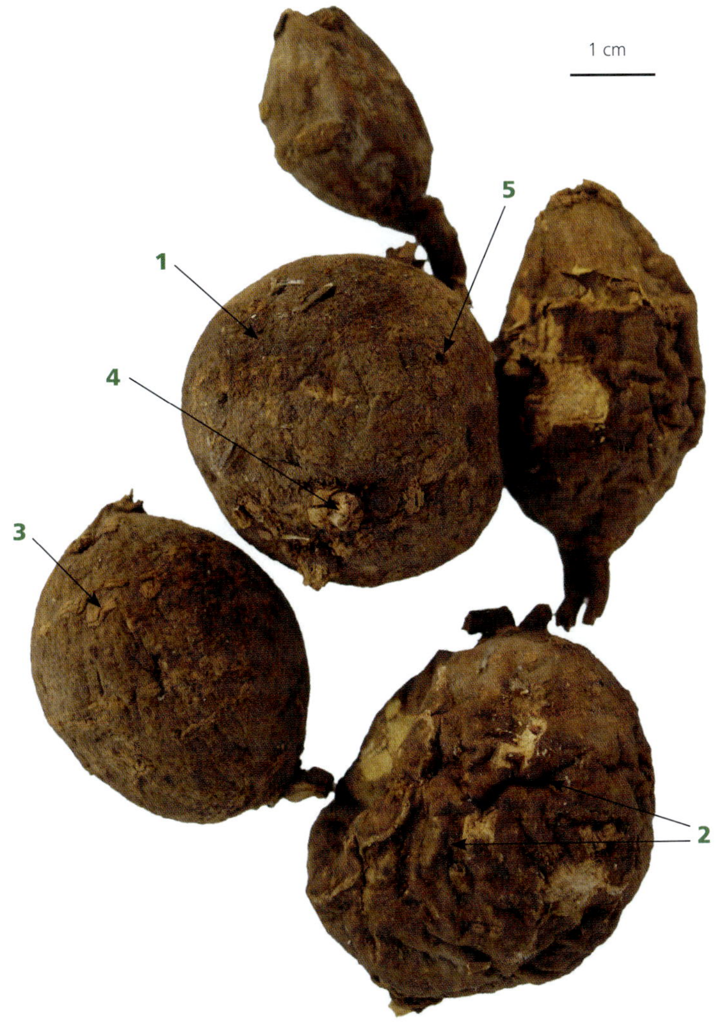

Crude drug of Jin Guo Lan
(dried root tubers of *Tinospora sagittata*)

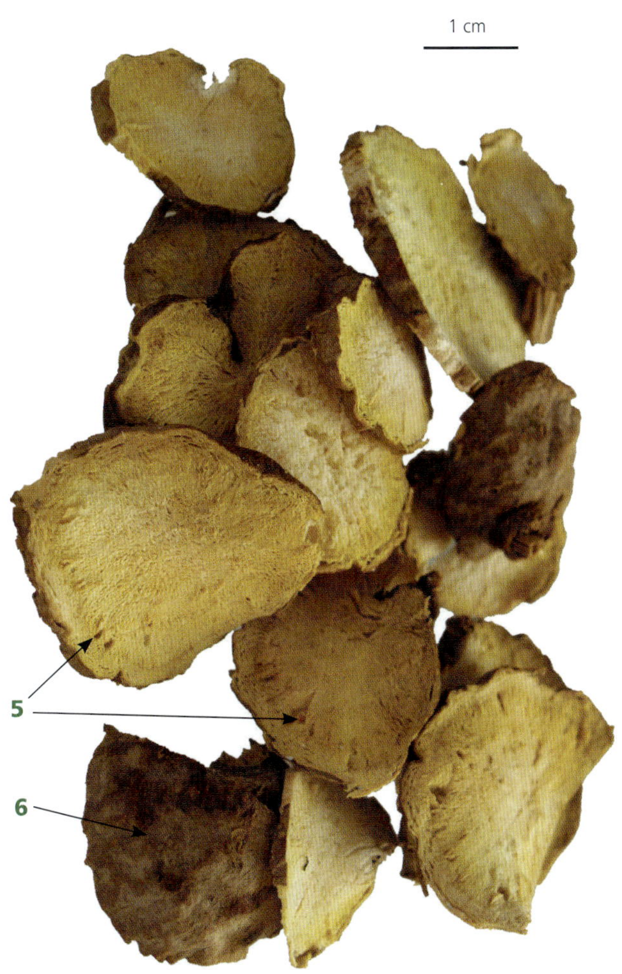

Decocting pieces of Jin Guo Lan
(dried and sliced root tubers of *T. sagittata*)

Jin Qiao Mai 金荞麦

Fagopyri Dibotryis Rhizoma

Drug part: RHIZOME

Fagopyrum dibotrys (D. Don) H. Hara

Polygonaceae

synonyms: *Fagopyrum acutatum* (Lehm.) Mansf. ex K. Hammer (*Kew-MPNS* accepted name, 25/5/2015); *Polygonum acutatum* Lehm.; *Polygonum dibotrys* D. Don.

Species of *Fagopyrum* wild in China: 10

PLANT DESCRIPTION

An upright, much-branched hairless, perennial herb (40–100 cm tall) with stout blackish-brown, woody rhizomes. *Stems* green or brownish. *Leaves* alternate, simple and stalked (stalk articulated at middle); sheath (ocrea) encircling stem at base of leaf stalk membranous and truncate (5–10 mm long); leaf blade triangular (4–12 cm long; 3–11 cm wide) with basal lobes pointing outwards (hastate); tip pointed; margin entire. *Flowers* in many-branched, many-flowered terminal or axillary heads; bract c. 3 mm long; each flower tiny, 'petals' 5, white, ovate and spreading (c. 3 mm long); stamens 8; styles 3. *Fruits* (achenes) dry, blackish-brown, broadly ovoid and 3-sided (c. 6–8 mm long). Fl. Apr–Oct, fr. May–Nov.

HARVESTING, SOURCING AND NATURAL RANGE

Rhizomes are harvested in winter from wild or cultivated plants. The main production areas are in the Chinese provinces of Jiangsu and Zhejiang.

The wild species grows in grassland in damp valleys (300–3,200 m altitude) in C, S and W China: Anhui, Fujian, Gansu, Guangdong, Guangxi, Guizhou, Henan, Hubei, Hunan, Jiangsu, Jiangxi, Shaanxi, Sichuan, Tibet, Yunnan and Zhejiang [also wild in Bhutan, India, Kashmir, Myanmar, Nepal, Sikkim and Vietnam].

CONSERVATION STATUS

China: Least Concern (MEPCAS, 2013).
Global: Not Evaluated (IUCN Red List, accessed 18/8/2016).

Flowering medicinal crop of *Fagopyrum dibotrys*

Foliage

Flowering plants

Inflorescence

Sheath (ocrea) at base of leaf stalk

Fresh rhizome

TCM PROPERTIES AND USE

properties: **'slightly pungent, sour and cool'**
channel: **'Lung'**
actions: **'clears heat, eliminates toxins, invigorates Blood, dispels wind damp'**

indications: **cough, sore throat; lung abscess**

DRUG MORPHOLOGY

CRUDE DRUG (DRIED RHIZOME): irregularly-shaped and woody, with squat and somewhat rounded nodules **[1]** (3–15 cm long, 1–4 cm diam.), often with stem remnants or scar present at apex **[2]**. External surface dull, greyish-brown or brown, with annular rings **[3]**, root scar cavities **[4]** and occasional rootlet remains **[5]**; bark thin **[6]**. *Texture*: hard and difficult to break. *Fracture*: pale pinkish-brown or yellowish-white, with radiating rays (in transverse fracture) **[7]** and a central, slightly darker or hollowed pith **[8]**. *Odour*: indistinct. *Taste*: slightly astringent.

DECOCTING PIECES (DRIED SLICED RHIZOME): sub-rounded or irregularly shaped slices (c. 1–4 cm diam; c. 4 mm thick). Bark thin **[6]**, greyish-brown or brown; annular rings **[3]**, root scar cavities and rootlet remains occasionally visible. Cut surface pale pinkish-brown or yellowish-white, with (in transverse section) radiating rays **[7]** and a central, slightly darker or hollowed pith **[8]**; cavities scattered **[9]**. Longitudinally cut slices with central pith sometimes visible as a long, straight slightly darker line or furrow **[10]**. *Odour*: indistinct. *Taste*: slightly astringent.

PROCESSING

After removal of stems and roots, rhizomes are sun-dried (= crude drug) then later softened and cut (in any direction) into thick (>4 mm) decocting slices.

OTHER COMMON NAMES

金荞 Jin Qiao (plant name); 苦荞头 Ku Qiao Tou; 荞麦三七 Qiao Mai San Qi; golden buckwheat rhizome (CP2010); Rhizoma Fagopyri Dibotryis.

SAFETY

An official healthfood in China (Anon., 2002a,b).

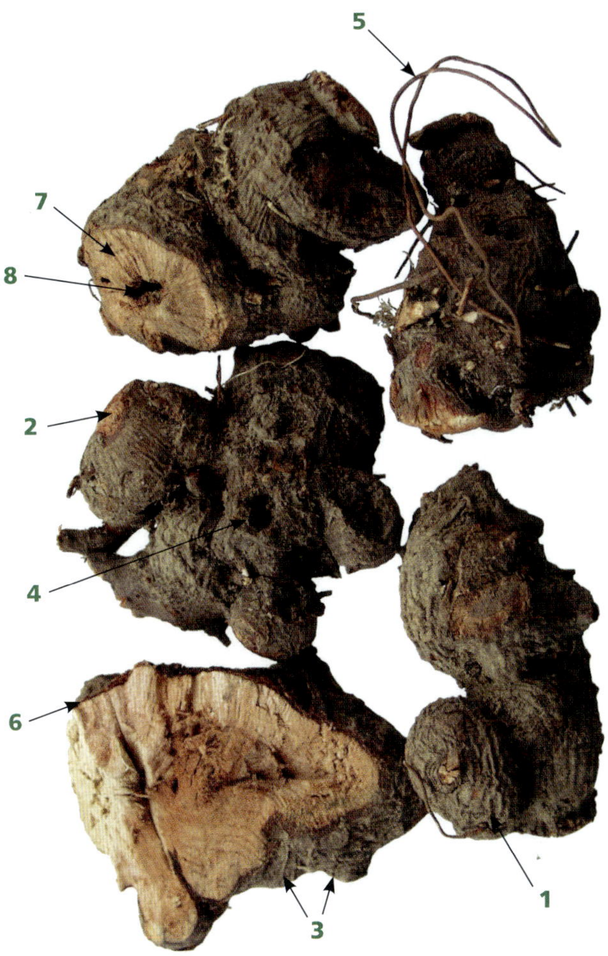

Crude drug of Jin Qiao Mai

1 cm

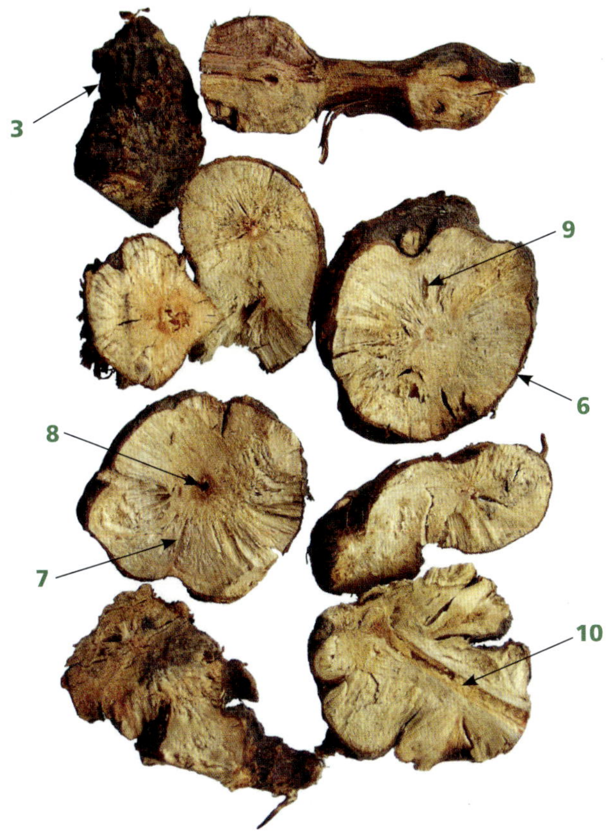

Decocting pieces of Jin Qiao Mai

PHARMACOPOEIA SPECIES

Ku Shen 苦参

Sophorae Flavescentis Radix

Drug part: ROOT

Sophora flavescens Aiton

Fabaceae

taxonomic/nomenclatural note: *FoC* and *Kew-MPNS* (16/8/2016) recognise 3 varieties of this species; this account assumes that 'Ku Shen' can be sourced from all.

Species of *Sophora* in China: 20

PLANT DESCRIPTION

A subshrub or perennial herb (1–2 m tall). *Stems* branching, green, dark green or greyish-brown, clothed in fine, dense yellow hairs when young. *Stipules* deciduous, tiny and lance-shaped (6–8 mm long). *Leaves* alternate (usually 20–25 cm long) with a slightly swollen base; leaf blade divided into 6–12 pairs of opposite to slightly alternate leaflets, and ending in a single terminal leflet; **leaflet blades elliptic, ovate, or lance-shaped (c. 3–4 cm long; 1.2–2 cm wide), papery** and margins entire. *Flowers* in long conspicuous terminal heads (15–25 cm long). Each flower widely spaced; flower-stalk slender (c. 7 mm long); bisexual and laterally symmetrical; sepals fused to form a pale yellowish-green, slightly oblique 'cup'; petals 5, white, pale yellow, purple-red, or red and forming a short, curved and round-tipped somewhat tube-shaped corolla (14–15 mm long); lower petals (keels) similar to wing petals; stamens 10, mostly separate (not fused); ovary superior; style slightly curved. *Fruit* a **long, narrow pod** (legume) (5–10 cm long), **tipped with a long beak,** 4-valved and **slightly constricted between each seed**. *Seeds* 1–5 per fruit, reddish-brown or purplish-brown. Fl. Jun–Aug, fr. Jul–Oct.

HARVESTING, SOURCING AND NATURAL RANGE

Roots are harvested from wild or cultivated medicinal crop plants in spring or autumn. The species is native throughout China where it grows in scrub and fields up to 1,500 m altitude. The main cultivated production areas are Shanxi (GAP), Hebei and Henan provinces [also wild in India, Japan, Korean Peninsula and Russia].

CONSERVATION STATUS

China: Least Concern (MEPCAS, 2013 as the varieties '*flavescens*', '*galegoides*' and '*kronei*').

Global: Least Concern (IUCN Red List, accessed online 20/6/2014).

PROTECTED STATUS

China: Cat. II (SEPA, 1999).

Medicinal crop of *Sophora flavescens* in flower

Part of a *S. flavescens* flower spike (2 flowers with petals removed to reveal stamens)

Fruiting plants of *S. flavescens*

Freshly harvested roots, Dangfeng, Shaanxi province

TCM PROPERTIES AND USE

properties: **'bitter and cold'**
channels: **'Heart, Liver and Stomach, Large Intestine and Bladder'**
actions: **'clears heat, dries dampness, dispels wind, promotes urination'**

indications: **diarrhoea; leucorrhoea; jaundice; stranguria; sores and abscesses; (T)**

DRUG MORPHOLOGY

CRUDE DRUG (DRIED ROOT): cylindrical (10–30 cm long; 1–6.5 cm diam.), slightly curved, often branching in lower parts, numerous stem bases or scars often present apically **[1]**. Outer bark greyish-brown or brownish-yellow **[2]**, distinctly wrinkled longitudinally; transverse raised lenticels occasional **[3]**; **outer root bark thin and exfoliating easily (sometimes recurved) [4] to reveal smooth and skin-like, yellow inner bark [5]**. *Texture*: hard and difficult to break. *Fracture*: fibrous and yellowish-white **[6]**, with radial wood and vascular tissues irregularly interspersed **[7]** or arranged concentrically **[9]**. *Odour:* indistinct. *Taste*: **strongly bitter**. ***Good quality:*** roots acutely bitter and yellowish-white in cross-section.

DECOCTING PIECES (DRIED TRANSVERSE ROOT SLICES): sub-rounded to ovate (1–6.5 cm diam.; 3–6 mm thick), **edges often indented or broken [8]**. Uncut edges usually with **smooth, thin yellow inner bark [5]**, occasionally with remnants of thicker greyish-brown or brownish-yellow outer bark **[2]**. Cut surface yellowish-white throughout with dense radial wood (xylem) and vascular tissues irregularly interspersed **[7]** or arranged concentrically **[9]**.

PROCESSING

After rootheads and rootlets are removed, roots are washed and dried (= crude drug) and subsequently cut into 3–6 mm thick slices or sliced while fresh, then dried (= decocting pieces).

OTHER COMMON NAMES

山槐根 Shan Huai Gen; 野槐根 Ye Huai Gen; Ku Shen Gen; light yellow sophora; Radix Sophorae Flavescentis; shrubby sophora (SCN).

SAFETY

Safety Class 1; Interaction Class: A (Gardner & McGuffin, 2013).

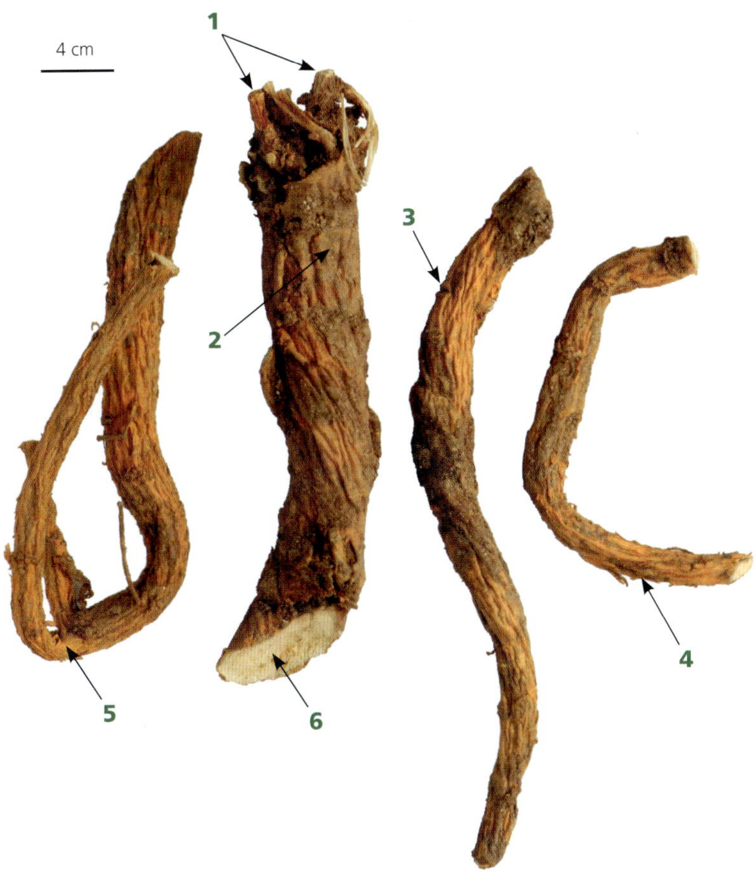

Crude drug

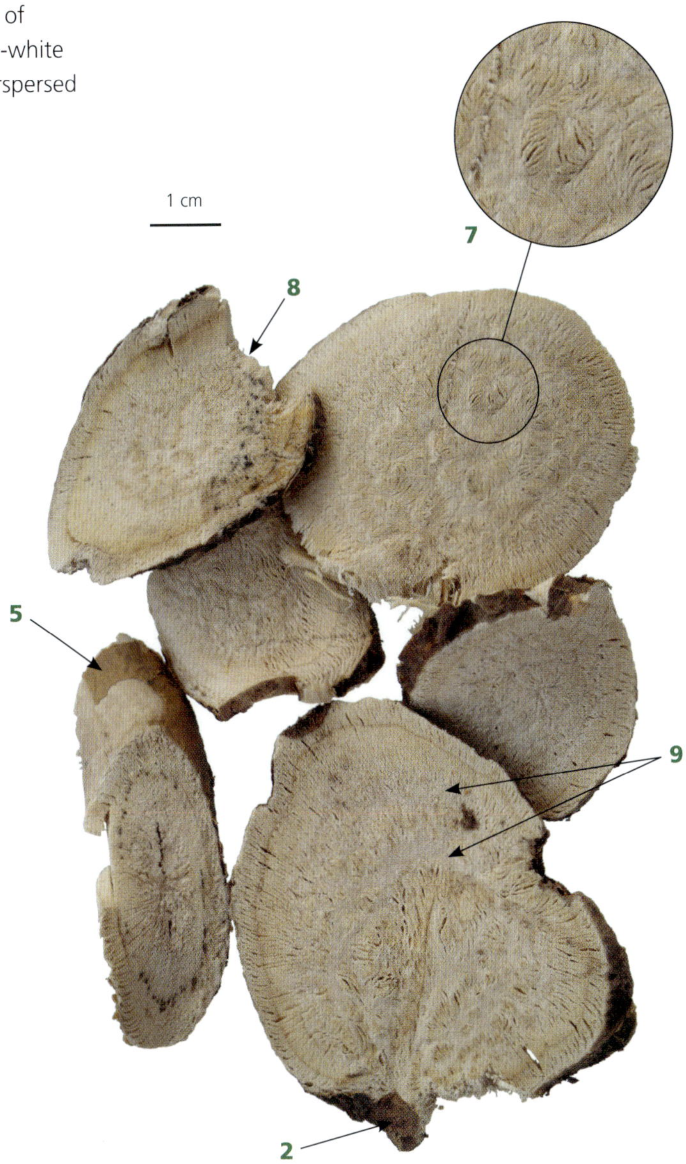

Decocting pieces

Liang Tou Jian 两头尖

Anemones Raddeanae Rhizoma

Drug part: RHIZOME

Anemone raddeana Regel

Ranunculaceae

Species of *Anemone* wild in China: 53

PLANT DESCRIPTION

A perennial herb (up to 30 cm tall) often forming ground cover, with tuberous, slightly nodular rhizomes (0.2–0.7 cm diam.). *True leaves* (not to be confused with the leaf-like bracts below each flower) solitary and basal, developing close to the base of the flowering stem; leaf stalk 5–20 cm long and hairless; leaf blade divided into 3 irregularly lobed and shortly stalked leaflets. *Flowering stem* 10–30 cm tall, leafless, but with 3 leaf-like bracts, deeply tri-lobed, sparsely hairy and 1–3 cm below each flower. *Flowers* solitary and large (c. 5 cm diam.) emerging above the foliage; petal-like sepals (true petals absent) 9–15, spreading, linear (12–20 mm long; 4–5 mm wide), white or white-tinged purple; stamens many; style short. *Fruits* many, inconspicuous, dry, narrowly ovoid achenes (c. 3 mm long; 1.5 mm diam.). Fl. Apr–May.

HARVESTING, SOURCING AND NATURAL RANGE

Rhizomes are harvested in summer from wild plants which grow on forest floors and in shady grassland (c. 800 m altitude). The species is native to NE China (Heilongjiang, Jilin, Liaoning and Shandong provinces); variety *lacerata* is confined to Zhejiang (500–1,000 m altitude). Also native to Japan, the Korean Peninsula and Russia (Far East).

CONSERVATION STATUS

China: Least Concern (MEPCAS, 2013).
Global: Not Evaluated (IUCN Red List, accessed 18/8/2016).

Anemone raddeana in the wild, Jilin province

UNOFFICIAL SUBSTITUTE

Zhao & Xiao (2009–2010) mention that the similar and closely related ***Anemone amurensis*** (Korsh.) Kom. (黑水银莲花 Hei Shui Yin Lian Hua) often grows wild in association with the Pharmacopoeia species in NE China. Because the pharmacological properties of *A. amurensis* rhizomes are quite different, care is needed to ensure that harvested batches of the official drug are not contaminated. Main differences: **rhizomes often branched, longer and narrower** (3–10 cm long; 1–3 mm diam.); basal leaf undivided or 2-lobed; **'petals' fewer (5–8), usually shorter and wider** (10–15 mm long; 5–7 mm wide); fruits slightly longer (4–5 mm). Flowering time is later (June). Batch contamination by this species is only likely to be an issue for the official drug when it is harvested from NE China as opposed to Zhejiang province in E China.

TCM PROPERTIES AND USE

***properties*: 'pungent and hot'**
***channel*: 'Spleen'**
***action*: 'dispels wind damp'**

***indications*: musculoskeletal pain, spasms, rheumatoid arthritis; carbuncles and abscesses; (T)**

DRUG MORPHOLOGY

CRUDE DRUG/DECOCTING PIECES (DRIED RHIZOMES): tuberous, usually slender (1–3 cm long; 0.2–0.7 cm diam.), narrowing at both ends (sub-fusiform) **[1]** and typically thicker towards (not at) one end **[2]**; often slightly curved. Externally greyish-brown or reddish-brown with a few distinctly prominent root remains **[3]**; annular rings 3-5 but very indistinct. *Fracture*: homogeneous, white or greyish-brown **[4]**. *Texture*: hard and brittle. *Odour*: indistinct. *Taste*: initially slight, then somewhat bitter and pungent **(see Safety)**.

PROCESSING

After removal of rootlets, the rhizomes are sun-dried (= crude drug/decocting pieces).

OTHER COMMON NAMES

多被银莲花 Duo Bei Yin Lian Hua; 竹节香附 Zhu Jie Xiang Fu (bamboo-shaped Xiang Fu); radde anemone rhizome (CP2010); Rhizoma Anemones Raddeanae.

SAFETY

Toxic (CP2015). Fresh rhizomes contain protoanemonin, a skin irritant, which is destroyed on drying (Barceloux, 2008).

Crude drug/decocting pieces

Lu Gen 芦根

Phragmitis Rhizoma

Drug part: RHIZOME

Phragmites australis (Cav.) Trin. ex Steud.

Poaceae

synonyms: *Phragmites communis* Trin. (name used in CP2015); *Arundo australis* Cav.; *A. phragmites* L.

Species of *Phragmites* wild in China: 3

PLANT DESCRIPTION

A highly variable species. A medium-tall perennial reed (up to 2 m) with rhizomes; often forming extensive stands. *Stems* c. 6 mm diam. and hollow; with mealy powder below each node. *Leaves* alternate and simple; leaf base sheathing the stem; ligule (structure at base of leaf blade) topped by fringe of tiny hairs (up to 0.6 mm long); leaf blade light green to bluish-green, large and lance-shaped (50 cm long; 1–3 cm wide) with a thread-like tip; margin smooth or rough. *Inflorescences* at top of plant, large and distinctive (20–50 cm long; 10 cm wide), soft, plume-like and pale purplish-brown with many spikelets (10–18 mm long); each spikelet with 2–5 florets surrounded by silky hairs, with 3 stamens. Fl. and fr. Jul–Nov.

HARVESTING, SOURCING AND NATURAL RANGE

A ubiquitous species of damp places notably marshes, lake-, river- and stream-sides and widespread throughout China and temperate and tropical regions worldwide. Rhizomes are harvested year round usually from wild plants. When used in the fresh state plants are cultivated locally in China for ease of supply.

CONSERVATION STATUS

China: Least Concern (MEPCAS, 2013).

Global: Least Concern (IUCN Red List, accessed 25/3/15).

Phragmites australis in the wild, Inner Mongolia

P. australis: inflorescence

TCM PROPERTIES AND USE

***properties*: 'sweet and cold'**
***channels*: 'Lung and Stomach'**
***action*: 'clears heat and fire, generates fluids'**

***indications*: fevers, rash; nausea, vomiting; stranguria; cough, phlegm**

DRUG MORPHOLOGY

CRUDE DRUG-1 (FRESH RHIZOME): cylindrical and hollow (except at nodes) and of varying length (usually c. 1–2 cm diam. with internodes c. 11 cm long). Externally pale yellow, internodes mostly smooth, slightly lustrous and faintly striated longitudinally; nodes dark brown and raised, often with buds, bud scars or fibrous roots. Transverse section with a large hollow centre, more than half the stem diameter, surrounded by a narrow ring of yellowish-white vascular tissue and an outer ring of almost circular vessels. *Texture*: light and strong, hard but flexible; very difficult to break. *Odour*: indistinct. *Taste*: sweetish.

CRUDE DRUG-2 (DRIED RHIZOME): as for the fresh but slightly compressed laterally, less lustrous and internodes more obviously wrinkled longitudinally. ***Good quality*:** broad, yellow and lustrous rhizomes.

DECOCTING PIECES (TRANSVERSE RHIZOME LENGTHS). *Fresh*: as for fresh crude drug but cut into short lengths; hollow or partially solid where node septa intact. *Dried*: as for dried crude drug but cut into short lengths; hollow or partially solid where node septa intact.

PROCESSING

After removal of buds, fibrous roots and membranous leaf bases, rhizomes are sun-dried or used fresh (= crude drug). Dried material is cut into transverse sections for decocting.

OTHER COMMON NAMES

苇根 Wei Gen; 鲜芦根 Xian Lu Gen; giant reed; phragmites (SCN); reed rhizome (CP2010); Rhizoma Phragmitis.

SAFETY

An official food and healthfood in China (Anon., 2002a,b).

Dried drug

Crude drug

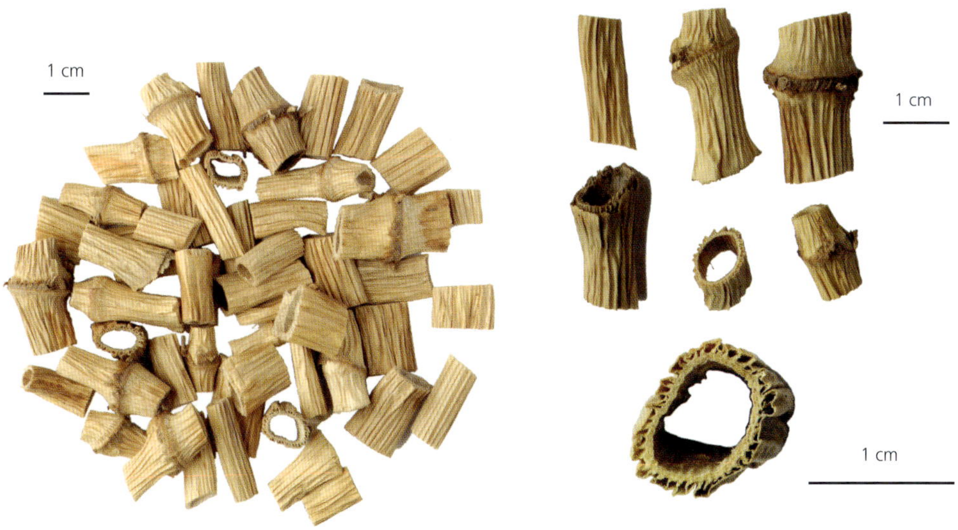

Decocting pieces

Fresh drug

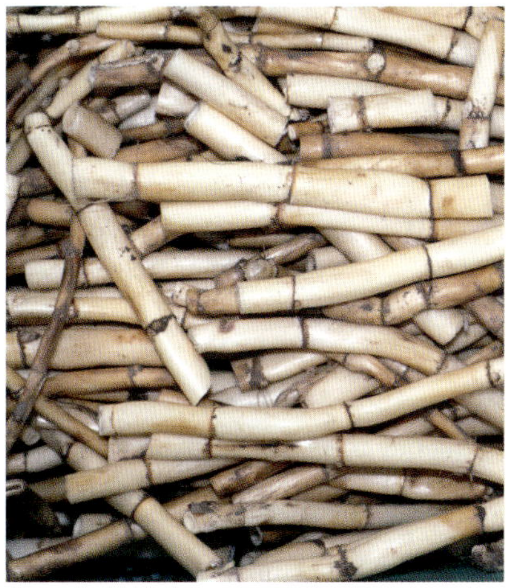

Lu Gen crude drug — fresh
(*P. australis* rhizomes)

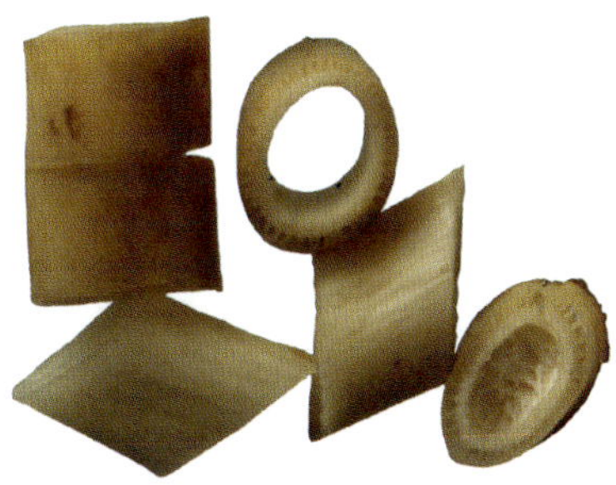

Lu Gen decocting pieces — fresh
(*P. australis* rhizomes)

Ma Huang Gen 麻黄根

Ephedrae Radix et Rhizoma

Drug parts: ROOT AND RHIZOME

PHARMACOPOEIA SPECIES 1

Ephedra sinica Stapf

Ephedraceae

synonym: *E. ma-huang* Tang S. Liu

Species of *Ephedra* wild in China: 14

PLANT DESCRIPTION

A sparsely branched subshrub (up to 40 cm tall). Lower *stems* woody, sometimes prostrate; upper stems green throughout (not woody) and jointed; stem internodes usually c. 3–4 cm long; c. 2 mm diam. *Leaves* reduced to 2 tiny, opposite membranous scales (3–4 mm long) at each node and fused along 1/3–2/3 their length with the free part (c. 5 mm long) tapering to a sharp point. *Male and female cones (reproductive parts)* on separate plants. Male (pollen) cones solitary or clustered at nodes, with ovoid heads of membranous bracts in 4 pairs; anthers 7 or 8, stalked or shortly stalked. Female (seed) cones solitary, oblong to ovoid, bracts in 4 pairs and, when mature, latter becoming red, fleshy and conspicuous (c. 8 mm long; 6–7 mm diam.). *Seeds* usually 2, blackish-red or greyish-brown. Pollination May–Jun, seed maturity Aug–Sep.

HARVESTING, SOURCING AND NATURAL RANGE

Although this species has been cultivated in recent years in Gansu and Ningxia provinces as a commercial source of 'Ma Huang' (page 454), underground parts continue to be wild-harvested and increasingly detrimental to the species' survival. Harvesting occurs in late autumn. The species grows in arid areas in sandy places, dry river beds, fields and mountain sides (700–1,600 m altitude) in northern China: Gansu, Hebei, Heilongjiang, Inner Mongolia, Jilin, Liaoning, Ningxia, Shaanxi and Shanxi [also native in Mongolia].

CONSERVATION STATUS

China: Near Threatened (MEPCAS, 2013); Vulnerable (Fu, 2011); 'rapidly decreasing' (Leung & Cheng, 2008).

Global: Least Concern (IUCN Red List, accessed 20/8/2016).

PHARMACOPOEIA SPECIES 2

E. intermedia Schrenk & C. A. Mey.

Ephedraceae

synonym: *E. intermedia* var. *tibetica* Stapf

PLANT DESCRIPTION

Easily confused with PS1. *Main differences:* **subshrub or shrub (up to 1 m tall)**; stems frequently branched; stem internodes 2–6 cm long, 1.5–3.5 mm diam.; **membranous scale leaves 3 (rarely 2),** fused at least 2/3 along their length; **bracts of seed cones in 2–5 pairs.**

For other TCM drugs sourced from *Ephedra* species see 'Ma Huang' (Ephedrae Herba) page 454.

HARVESTING, SOURCING AND NATURAL RANGE

Rootstocks harvested exclusively from wild plants in late autumn. The species grows between 800–4,600 m altitude in the Chinese provinces of Gansu, Hebei, Inner Mongolia, Liaoning, Ningxia, Qinghai, Shaanxi, Shandong, Shanxi, Tibet and Xinjiang [also native in Afghanistan, Kazakhstan, Kyrgyzstan, Mongolia, Pakistan, Russia, Tajikistan, Turkmenistan, Uzbekistan and SW Asia].

CONSERVATION STATUS

China: Near Threatened (MEPCAS, 2013). Declining locally due to over-harvesting for medicinal use (authors' pers. comm. 1999).

Global: Least Concern (IUCN Red List, accessed 20/8/2016).

TCM PROPERTIES AND USE

properties: 'sweet and neutral'
channel: 'Lung'
actions: 'astringes'

indications: spontaneous and night sweats; (T)

Female plant of *Ephedra sinica*; inset: pollen cone with yellow anthers

E. sinica with (inset) scale leaves in 2s and tips sharply pointed

E. sinica with mature fruit

E. intermedia in the wild

E. intermedia with scale leaves in 3s (rarely 2)

Fruiting plant of *E. intermedia*

Ephedra sinica

Ephedraceae

DRUG MORPHOLOGY

CRUDE DRUG (DRIED ROOT AND RHIZOME): predominantly of roots **[1]** with short lengths of rhizome **[2]** attached apically; roots long, cylindrical, infrequently branched and somewhat curved (8–25 cm long; 0.5–1.5 cm diam.). External surface is reddish-brown or greyish-brown **[3]**, with root branch remains **[4]** and rootlet scars **[5]**; root bark often splitting, flaking and fibrous **[6]** in places exposing similarly coloured tissues below. Rhizomes with internodes 0.7–2 cm long and elliptical transverse lenticels **[7]**. *Texture (root)*: light, hard and brittle. *Fracture*: pale yellow or yellow, with radiating rays **[8]**; rhizomatous parts with a white central pith. *Odour*: indistinct. *Taste*: slightly bitter. ***Good quality***: roots externally reddish-brown and devoid of cavities.

DECOCTING PIECES (SLICED ROOT AND RHIZOME): as for crude drug but sliced diagonally into 2–4 mm long slices.

PROCESSING

After removal of stems and rootlets, the rootstocks are sun-dried (= crude drug) then later softened and cut diagonally into 2–4 mm long pieces slices before drying again.

OTHER COMMON NAMES

草麻黄 Cao Ma Huang (plant name); ephedra (SCN); ephedra root (CP2010); Radix et Rhizoma Ephedrae.

SAFETY

The use of *Ephedra* species is legally restricted in many countries due to their 'ephedra alkaloids'. The latter have been associated with life-threatening adverse reactions, some fatal, following their use in dietary supplements (notably in weight loss products), and because of their widespread abuse as 'legal highs' (Edwards *et al.*, 2015; Federal Register, 2004; Martindale, 2011; MHRA, 2014; Barnes *et al.*, 2013. In TCM, strict adherence to dosing has been strongly recommended (Bensky *et al.*, 2004).

E. intermedia

Ephedraceae

DRUG MORPHOLOGY

Crude drug and decocting pieces are indistinguishable from those of PS1 using macroscopic characters; Laboratory-based methods are advised.

PROCESSING

As for PS1.

OTHER COMMON NAMES

草麻黄 Cao Ma Huang (plant name); 中麻黄 Zhong Ma Huang; ephedra (SCN); ephedra root (CP2010); Radix et Rhizoma Ephedrae.

SAFETY

As for PS1.

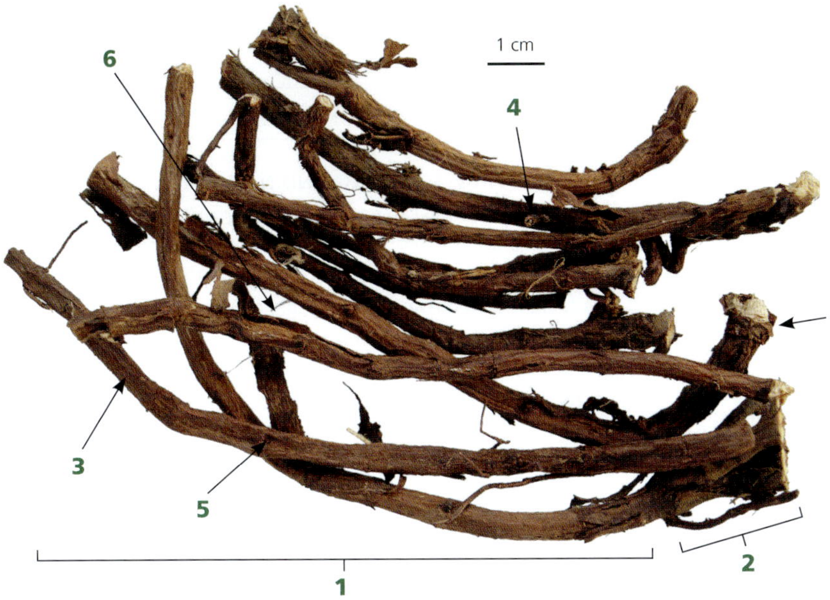

Crude drug of Ma Huang Gen (sourced from *Ephedra sinica*)

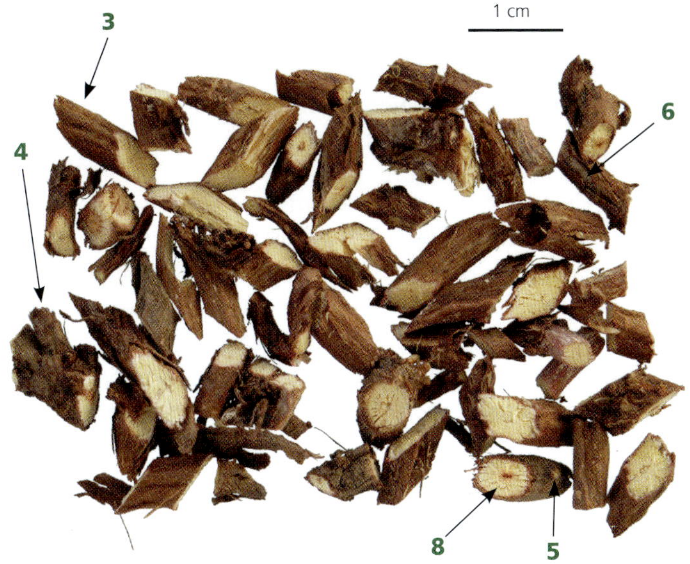

Decocting pieces of Ma Huang Gen (sourced from *E. sinica*)

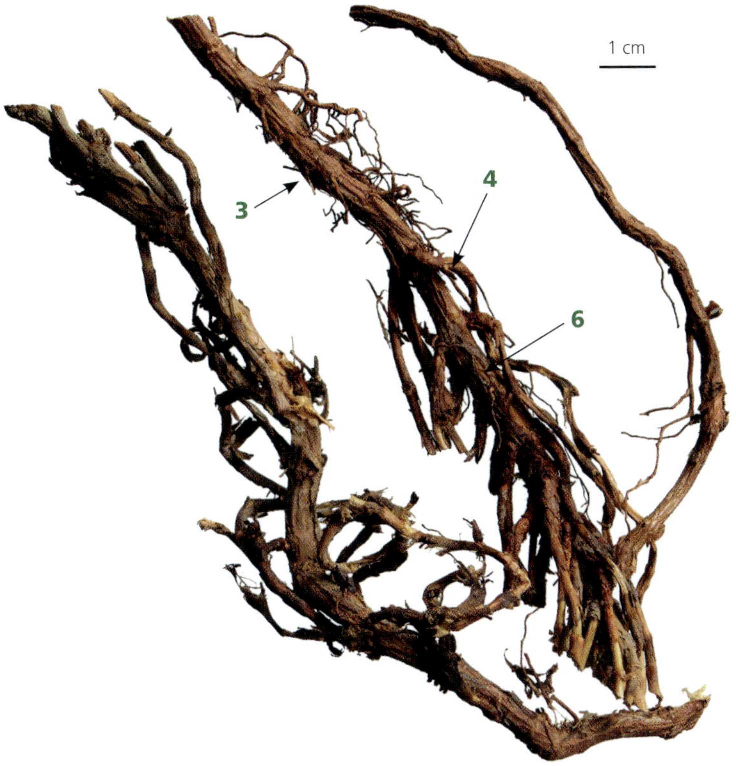

Crude drug of Ma Huang Gen (sourced from *E. intermedia*)

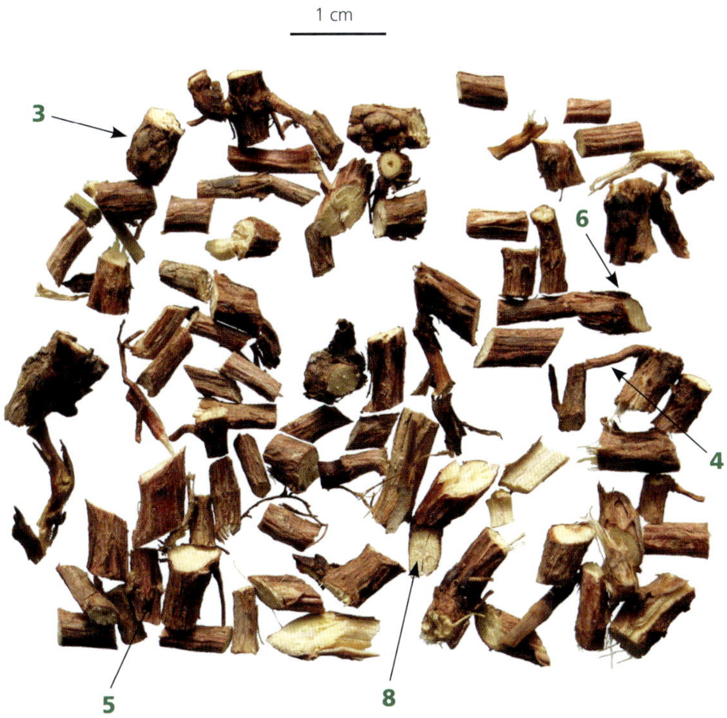

Decocting pieces of Ma Huang Gen (sourced from *E. intermedia*)

Mai Dong 麦冬 (or Mai Men Dong 麥門冬) Ophiopogonis Radix

Drug part: ROOT TUBER

PHARMACOPOEIA SPECIES

Ophiopogon japonicus (Thunb.) Ker Gawl.

Asparagaceae

synonym: *Convallaria japonica* Thunb.

taxonomic/nomenclatural note: in CP2015 as '*O. japonicus* (L.f.) Ker-Gawl.'

Species of *Ophiopogon* wild in China: 47

PLANT DESCRIPTION

A short, clump-forming perennial herb (10–50 cm tall) with **stolons** and tuberous roots. *Leaves* basal, stalkless and strap-like (10–50 cm long; 2–4 mm wide), 3- to 7-veined, margins minutely saw-toothed (plants with purple or variegated leaves are ornamental cultivars). *Flowers* in several- to many-flowered spikes (2–5 cm long) on leafless stems (6–15(–27) cm long) shorter than leaves; **flowering stems somewhat flattened in cross-section**; bracts lance-shaped. **Flowers usually solitary or in evenly spaced pairs**, short-stalked and **nodding; 'petals'** 6, **white or purplish**; stamens 6, filaments much shorter than anthers; style 1. *Fruits* 3-parted, bursting early to reveal fleshy, spherical **blue seeds** (7–8 mm diam.) often mistaken for fruits. Fl. May–Aug, fr. Aug–Sep.

HARVESTING, SOURCING AND NATURAL RANGE

The tuberous roots are harvested in summer mostly from plants cultivated on small-holdings or medicinal plant farms. Traditional *Di Dao* sources in Sichuan (producing 川麦冬 Chuan Mai Dong) and Zhejiang (producing 浙麦冬 Zhe Mai Dong or 杭麦冬 Hang Mai Dong).

The wild species grows in forests and dense scrub, in damp, shady places (200–2,800 m altitude) in C, S and W China (Anhui, Fujian, Guangdong, Guangxi, Guizhou, Hebei, Henan, Hubei, Hunan, Jiangsu, Jiangxi, Shaanxi, Shandong, Sichuan, Taiwan, Yunnan and Zhejiang provinces [also wild in Japan and the Korean Peninsula].

CONSERVATION STATUS

China: Not Evaluated.

Global: Not Evaluated (IUCN Red List, accessed 20/8/2016).

UNOFFICIAL SUBSTITUTE 1

Liriope spicata Lour.

Asparagaceae

synonym: *L. spicata* var. *prolifera* Y. T. Ma (name used in CP2015).

PLANT DESCRIPTION

Very similar to the Pharmacopoeia species. *Main differences*: **flowering stems round in cross-section**; flowers in groups of **3–5 (rarely 2)** and positioned **at many angles** on the stem (i.e. not all nodding); 'petals' pale purplish or bluish, sometimes white; filaments of stamens longer than or as long as anther. Fruits with **blackish seeds** (often mistaken for fruits). Fl. May–Jul, fr. Aug–Oct. (Differs from *L. muscari* below by having creeping stolons; see also 'Shan Mai Dong' page 264)

HARVESTING, SOURCING AND NATURAL RANGE

Root tubers (the source of 'Shan Mai Dong' (Liriopes Radix), page 264) are harvested mostly from medicinal crops grown in areas overlapping with those of *Ophiopogon* above.

UNOFFICIAL SUBSTITUTE 2

L. muscari (Decne.) L. H. Bailey

Asparagaceae

synonym: *L. platyphylla* F. T. Wang & Tang; *Ophiopogon muscari* Decne.

PLANT DESCRIPTION

Very similar to 'unofficial substitute 1'. *Main differences*: plants **without stolons**; flowers in groups of **4–8** (rarely 3).

Very similar to the Pharmacopoeia species. *Main differences*: flowering stems **round in cross-section**; flowers in groups of 4–8 (rarely 3); 'petals' pale purplish or bluish. Fruits with **blackish seeds** (often mistaken for fruits). See also 'Shan Mai Dong' page 264.

HARVESTING, SOURCING AND NATURAL RANGE

Similar to *L. spicata* above but with a smaller wild distribution and the main production area is in Fujian province (SE China). See also 'Shan Mai Dong' (Liriopes Radix) page 264.

TCM PROPERTIES AND USE

properties: 'sweet, slightly bitter and cool'
channels: 'Heart, Lung and Stomach'
action: 'tonifies yin, moistens Lungs, generates fluids, calms the shen, moistens the intestines'

indications: cough; epistaxis; irritability, memory loss, insomnia; constipation

Flowering plant of *Ophiopogon japonicus*

O. japonicus with immature and nodding fleshy seeds

Flowering plant of *Liriope spicata*

Flower head of *L. spicata*
(flower colour not reliable for identification)

L. spicata: ripe seeds not nodding; stem round

Flowering plant of *Liriope muscari*

Uprooted *Liriope muscari* showing fresh tubers

Ophiopogon japonicus

Asparagaceae

DRUG MORPHOLOGY

CRUDE DRUG/DECOCTING PIECES (DRIED TUBER): long-ovoid, narrowing towards both ends and often abruptly truncate at one or both ends **[1]** (1.5–3 cm long; 3–6 mm diam.); a central thread-like stele sometimes slightly emergent at one end **[2]**. Externally yellowish-cream and finely wrinkled longitudinally **[3]**. *Texture:* tough and slightly pliable. *Fracture:* even; translucent or horn-like, yellowish-white with paler central stele **[4]**. *Odour:* faintly aromatic. *Taste:* somewhat sweet and bitter; viscous when chewed. ***Good quality*:** yellowish-white, plump, translucent, aromatic and pliable.

PROCESSING

After washing, roots are partially (70–80%) sun- or oven-dried, rootlets removed then dried fully (= crude/decocting pieces).

OTHER COMMON NAMES

川麦冬 Chuan Mai Dong (Sichuan Mai Dong); 寸冬 Cun Dong; 麦冬 Mai Dong (plant name); 麥門冬 Mai Men Dong; 浙麦冬 Zhe Mai Dong or 杭麦冬 Hang Mai Dong (Zhejiang Mai Dong); dwarf lilyturf tuber (CP2010); Liriopis Tuber (KP, 2007); Ophiopogon (SCN); Ophiopogonis Tuber (JP, 2012); Radix Ophiopogonis.

SAFETY

Safety Class: 1; Interaction Class: A (Gardner & McGuffin, 2013). An official healthfood in China (Anon., 2002a,b).

Liriope spicata

Asparagaceae

'DRUG SUBSTITUTE' MORPHOLOGY

DRIED TUBER: cannot be reliably distinguished from 'Mai Dong' using macroscopic characters characters alone — laboratory based authentication methods are recommended. Some authorities consider the brittle nature of dried tubers of *Liriope* as opposed to the pliable ones of *Ophiopogon* to be diagnostic but such a character is not very robust. Furthermore, characters such as colour and plumpness are not helpful distinguishing charcaters; both 'Mai Dong' and 'Shan Mai Dong' are prone to insect damage and may be sulphured to deter insects and this may cause colour variation. Similarly variations in plumpness are most likely due to differing storage conditions as opposed to species' differences.

OTHER COMMON NAMES

寸冬 Cun Dong; 湖北麦冬 Hu Bei Mai Dong; 麦冬 Mai Dong; liriope root tuber (CP2010); Liriopes Radix (CP2010); Radix Liriopes.

SAFETY

An official healthfood in China (in Anon., 2002a,b, as the synonym *L. spicata* var. *prolifera* Y. T. Ma).

L. muscari

Asparagaceae

'DRUG SUBSTITUTE' MORPHOLOGY

As for *L. spicata* above.

OTHER COMMON NAMES

寸冬 Cun Dong; 短葶山麦冬 Duan Ting Shan Ma Dong; 麦冬 Mai Dong; liriope root tuber (CP2010); liriope tuber (KP, 2007); liriopis tuber (KP, 2007); Radix Liriopes.

Processing *Di-Dao*-quality Mai Dong in Sichuan province

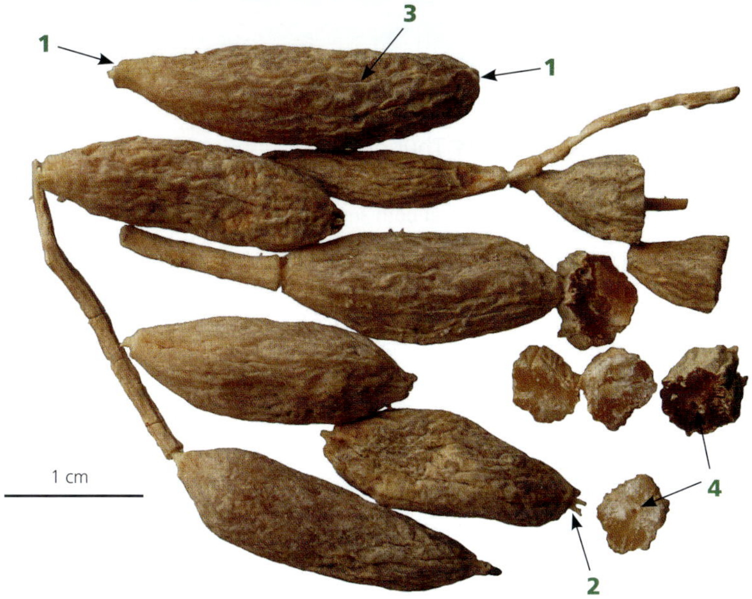

Crude drug/decocting pieces of Mai Dong (*Ophiopogon japonicus* root tubers)

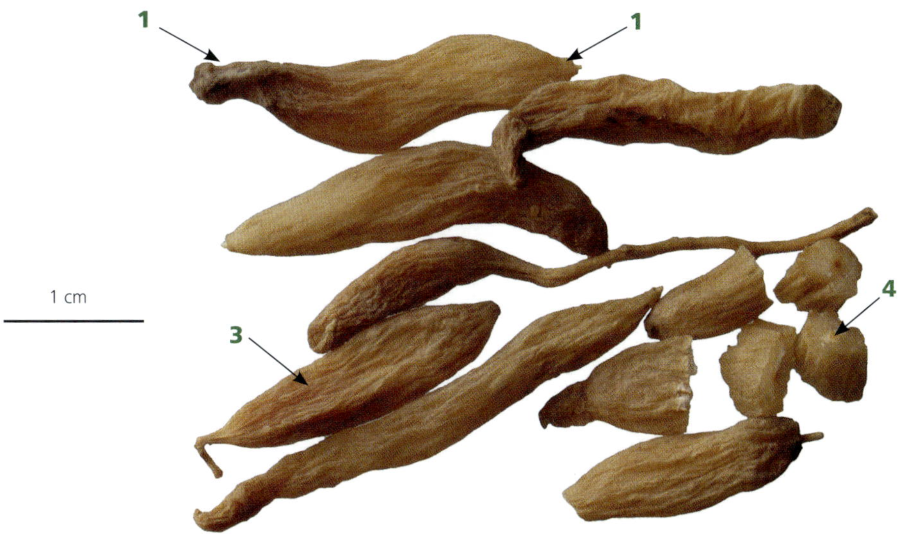

Crude drug/decocting pieces of Shan Mai Dong (*Lirope spicata* root tubers)

SUBSTITUTION COMMENT. Early editions of the CP included the tubers of these two species as alternative sources of Mai Dong. They are considered clinically inferior but recent editions of the Chinese Pharmacopoeia treat them as the source of a separate drug 'Shan Mai Dong' (Liriopes Radix) (page 264). Today they continue to be used interchangeably with Mai Dong in clinical settings. The name 'Shan Mai Dong' is often abbreviated to 'Mai Dong' creating confusion in the trade. Macroscopic characters are insufficient to distinguish 'Shan Mai Dong' from 'Mai Dong'.

Crude drug/decocting pieces of Shan Mai Dong (*L. muscari* root tubers)

Mao Zhao Cao 猫爪草

Ranunculi Ternati Radix

Drug part: ROOT TUBER

Ranunculus ternatus Thunb.

Ranunculaceae

taxonomic note: The *Flora of China* recognises two varieties (*ternatus* and *dissectissimus* (Migo) Hand.-Mazz.). This account assumes that both are used as the source of this drug.

Species of *Ranunculus* wild in China: 125

PLANT DESCRIPTION

A branching, perennial herb (5–18 cm tall) with a clustered *rootstock* of ovoid to spindle-shaped root tubers (3–5 mm. diam.). *Basal leaves* 5–10, stalked (2–6 cm long), blade usually divided into 3; pentagonal or broadly ovate in outline, sometimes simple, (0.6–1.5 cm long; 1–2.4 cm wide), leaflets rhombic and sometimes divided into 2 or 3 lobes. *Stem leaves* unstalked and smaller, divided into 3 lobes. *Flowers* solitary and terminal (1–1.6 cm diam.); sepals 5, ovate; petals 5, yellow, wedge-shaped with rounded tips, with a nectar pit at base of each petal; stamens many. *Fruit* an ovoid aggregate (2–4 mm long; 2–3 mm diam.) of hairless achenes each with a persistent apical style. Fl. Mar–May.

HARVESTING, SOURCING AND NATURAL RANGE

Rootstocks are harvested in spring and autumn usually from wild plants whose habitat is grassland, forests and field margins up to 500 m altitude. The species is na tive to C and S China: Anhui, Fujian, Guangxi, Henan, Hubei, Hunan, Jiangsu, Jiangxi, Taiwan and Zhejiang [also wild in Japan]. Cultivation as a medicinal crop takes place in central China in the provinces of Henan and Anhui.

CONSERVATION STATUS

China: Least Concern (MEPCAS, 2013); a common and weedy species (authors' pers. obs. 2013).
Global: Not Evaluated (IUCN Red List; accessed 20/3/2015).

Harvesting Mao Zhao Cao in the wild, Anhui province, China

R. ternatus, freshly uprooted plants showing tuberous roots

R. ternatus, flowering plants (inset: showing basal leaves divided into three, and small stem leaves)

TCM PROPERTIES AND USE

properties: **'sweet, pungent and warm'**
channels: **'Liver and Lung'**
actions: **'resolves phlegm, eliminates toxin, disperses nodules'**

indications: **hepatitis, jaundice and liver cancer; asthma; malaria; and rheumatism**

DRUG MORPHOLOGY

CRUDE DRUG/DECOCTING PIECES (DRIED ROOT TUBER): each rootstock (c. 4–20 mm long; c. 5–10 mm diam.) with many (c. 3–15) densely clustered root tubers on the basal side forming a nodular or claw-shaped mass **[1]**, colloquially described as a "cat's foot", sometimes with claws extended **[2]**; each tuber c. 3–12 mm long; 2–3 mm diam. Externally tubers yellowish-brown or greyish-yellow (darkening with time); apex sometimes with stem scars **[3]** or persistent stem bases **[4]**, surrounded by dot-like rootlet scars or rootlet remains **[5]**. *Texture*: compact. *Fracture*: creamish or yellowish-white, solid and starchy **[6]** or hollow. *Odour*: indistinct. *Taste*: slightly sweet. ***Good quality*:** rootstock large and plump.

PROCESSING

After removal of the rootlets, rootstocks are sun-dried (= crude drug/decocting pieces).

OTHER COMMON NAMES

猫爪草 Mao Zhua Cao; 小毛茛 Xiao Mao Gen (small Ranunculus); catclaw buttercup root (CP2010); Radix Ranunculi Ternati.

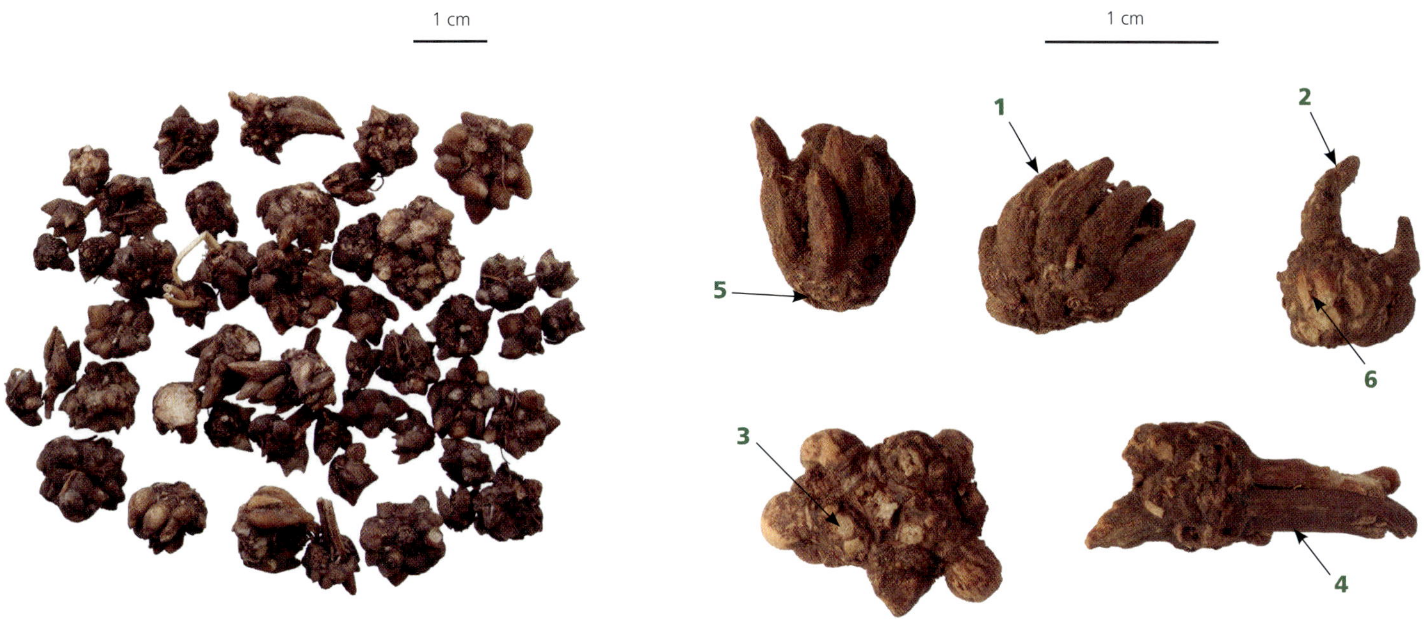

Crude drug/decocting pieces of Mao Zhao Cao (*Ranunculus ternatus* dried root tubers)

Mian Bi Xie 绵萆薢 Dioscoreae Spongiosae Rhizoma

Drug part: RHIZOME

PHARMACOPOEIA SPECIES 1

Dioscorea spongiosa J. Q. Xi, M. Mizuno & W. L. Zhao

Dioscoreaceae

identification note: prior to 2010 the CP used the name *D. septemloba* Thunb. but this was amended in line with the *FoC* which states that the name "*D. septemloba* should be restricted to Japanese material and the Chinese plants be treated as a distinct species, *D. spongiosa*."

Species of *Dioscorea* wild in China: 50

PLANT DESCRIPTION

A twining herbaceous perennial with a fleshy, horizontal, ginger-like rhizome **(c. 3–6 cm diam.)**; **soft and spongy with yellow flesh.** *Stems* twine to the left. *Leaves* alternate and stalked (stalks up to 12 cm long); leaf blades simple, triangular or ovate to heart-shaped with a rounded sinus (inner part of lobe), basal leaves sometimes 5–9-lobed; margins entire or distinctly toothed; blade **upper surface hairless**, lower surface **minutely hairy** [hl] and turning **greenish or greyish-white when dry.** *Flowers* tiny in linear spikes; male and female flowers on separate plants. Male spike loosely branched (up to 15 cm long) with male flowers solitary or in pairs, 'petals' orange; stamens 3. Female flower-spikes sometimes branched (12–30 cm long), with sterile stamens (staminodes) thread-like. *Fruits* in long, linear, pendant clusters; each fruit a dark brown, dry, 3-winged, sometimes shiny, capsule (13–16 mm diam.); wings 15–16 mm wide. *Seeds* winged. Fl. May–Aug, fr. Jun–Oct.

HARVESTING, SOURCING AND NATURAL RANGE

Rhizomes are wild-harvested in autumn and winter from plants growing in open forest and scrub (400–800 m altitude). The species is endemic to C and S China (Fujian, N Guangdong, E Guangxi, SW Hubei, Hunan, Jiangxi and Zhejiang) with the main source provinces being Zhejiang, Jiangxi and Fujian.

CONSERVATION STATUS

Endemic to China: Least Concern (MEPCAS, 2013). Not Evaluated by IUCN Red list, accessed 20/8/2016).

PHARMACOPOEIA SPECIES 2

D. futschauensis Uline ex R. Knuth

Dioscoreaceae

taxonomic/nomenclatural note: in CP2015 as '*Dioscorea futschauensis* Uline ex R. Kunth'.

PLANT DESCRIPTION

A closely related and very similar species to PS1. *Main differences*: **rhizome narrower** (1–3.5 cm diam.), **hard and rich in starch**; upper surface of leaf blades **sparsely covered in short, hard hairs [hl];** lower surface **densely covered in a mat of minute hairs** [hl]; leaves drying to **dark greyish.**

HARVESTING, SOURCING AND NATURAL RANGE

Rhizomes are harvested in autumn and winter from wild populations which grow in scrubby forest, mountain slopes and along roadsides (0–700 m altitude). The species is endemic to the Chinese provinces of: Fujian, N Guangdong, NE Guangxi, Hunan and S Zhejiang.

CONSERVATION STATUS

Endemic to China: Near Threatened (MEPCAS, 2013); Not Evaluated (IUCN Red list, accessed 20/8/2016).

UNOFFICIAL SUBSTITUTES

See also the TCM drug 'Fen Bi Xe' (Dioscoreae Hypoglaucae Rhizoma) (see page 134) used interchangeably with this drug in a clinical setting (Bensky *et al.*, 2004).

TCM PROPERTIES AND USE

***properties*: 'bitter and neutral'**
***channels*: 'Kidney and Stomach'**
***actions*: 'drains damp, dispels wind'**

***main indications*: urinary difficulty; leucorrhoea; musculoskeletal pain**

Dioscorea spongiosa (centre) fruiting in the wild, Zhejiang province

D. spongiosa in fruit

Soft and spongy wild-sourced rhizomes of *D. spongiosa*

D. spongiosa with lower leaf surface minutely hairy

D. futschauensis with inflorescences of tiny, orange male flowers

ABOVE
Palmately 7-lobed basal leaves of *D. futschauensis*

Narrower and harder fresh rhizomes of *D. futschauensis*

RIGHT
Lower leaf surface *D. futschauensis* densely hairy

PHARMACOPOEIA SPECIES 1

Dioscorea spongiosa

Dioscoreaceae

DRUG MORPHOLOGY

CRUDE DRUG/DECOCTING PIECES (DRIED RHIZOME SLICES): thin and irregularly-shaped (**3–6 cm diam. [1];** c. 2–5 mm thick). Outer skin, when present, thin, brownish-black or greyish-brown **[2]**, rough, with scattered and raised root-scars **[3]** or root remnants **[4]**; cut surfaces greyish-white to pale greyish-brown **[5]** with scattered yellowish or yellowish-brown patches of vascular bundles **[6]**. *Odour:* indistinct. *Texture:* spongy and slightly elastic. *Taste:* slightly bitter. ***Good quality:*** slices large, thin and slightly elastic.

PROCESSING

After removal of fibrous roots, rhizomes are washed, sliced and sun-dried (= crude drug/ decocting pieces).

OTHER COMMON NAMES

萆薢 Bi Xie; 绵萆薢 Mian Bi Xie (also the plant name); 山薯 Shan Shu; sevenlobed yam rhizome (CP2010).

PHARMACOPOEIA SPECIES 2

D. futschauensis

Dioscoreaceae

DRUG MORPHOLOGY

CRUDE DRUG/DECOCTING PIECES (DRIED AND SLICED RHIZOME): similar to PS1 above. *Main differences*: diameter of rhizome slices **narrower (<3 cm) [7]** and outer skin a brighter yellowish- brown **[8]**; cut surfaces **starchy [9]** and not elastic, **tissues starchy**.

PROCESSING

As for PS1.

OTHER COMMON NAMES

萆薢 Bi Xie; 福州薯蓣 Fu Zhou Shu Yu (plant name); Rhizoma Dioscoreae Septemlobae (CP2005); sevenlobed yam rhizome.

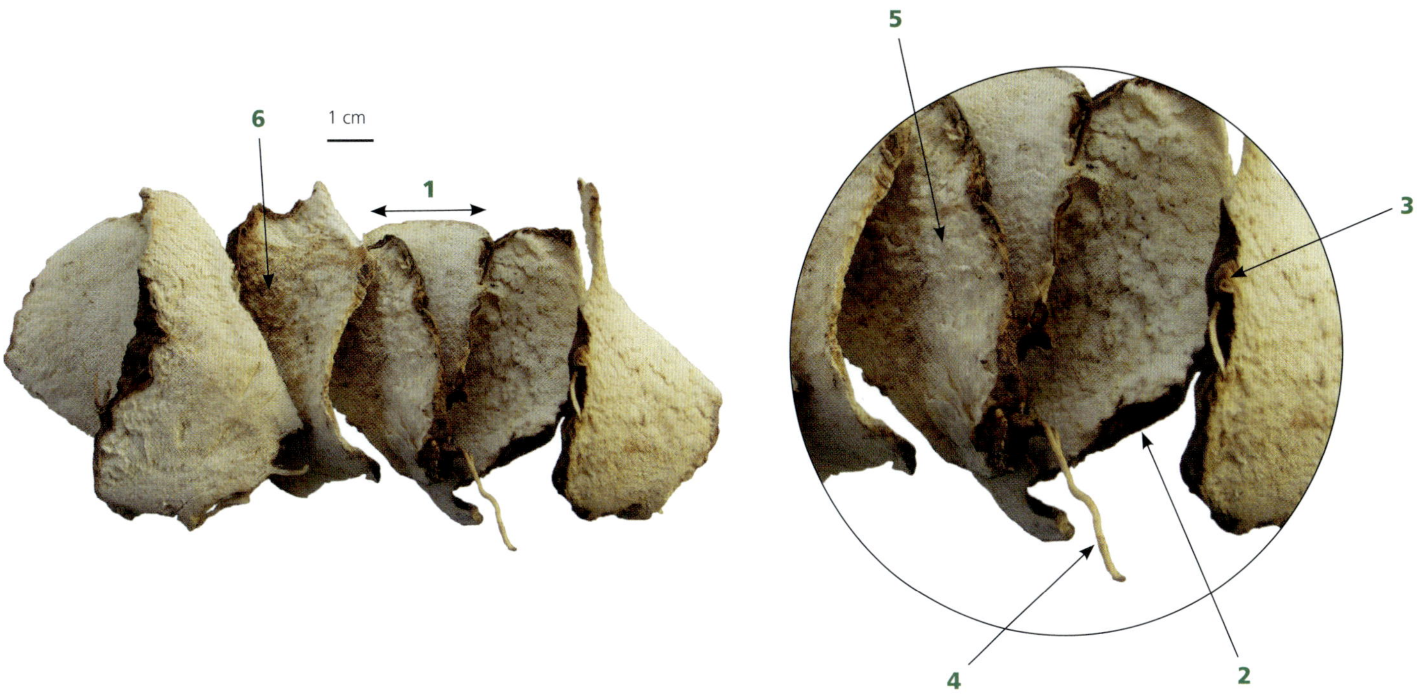

Crude drug/decocting slices of Mian Bi Xie sourced from *D. spongiosa*

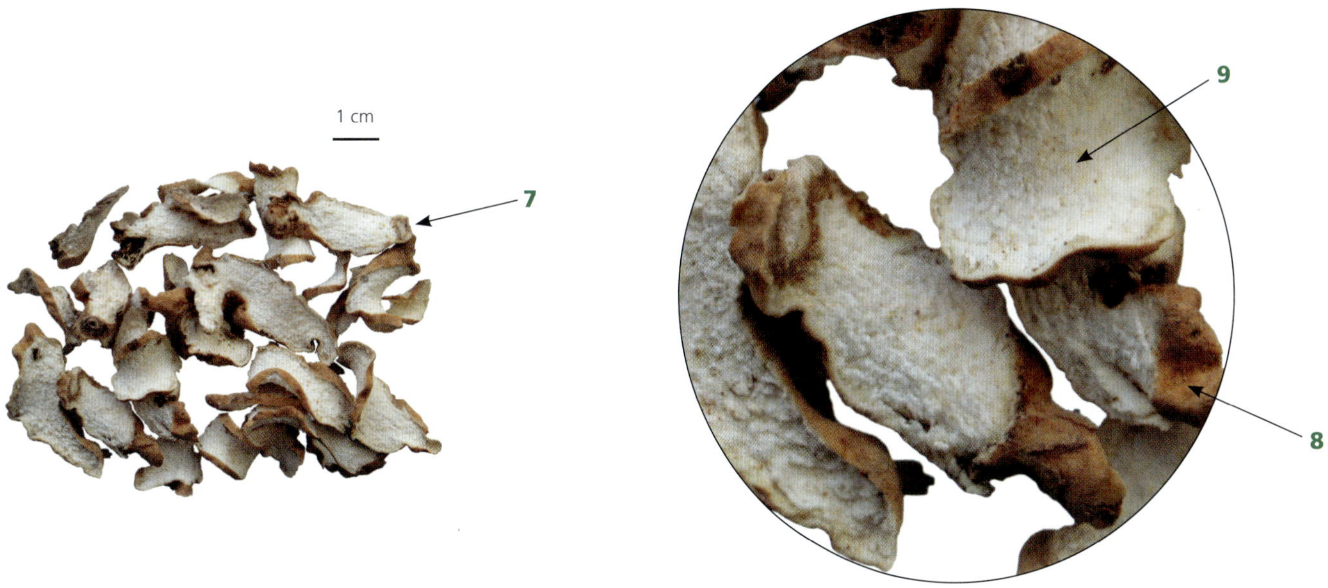

Crude drug/decocting slices of Mian Bi Xie sourced from *D. futschauensis*

Mu Xiang 木香 — Aucklandiae Radix

Drug part: ROOT

Aucklandia costus Falc.

Asteraceae

synonyms: *A. lappa* Decne. (name used in CP2015); *Saussurea costus* (Falc.) Lipsch.; *S. lappa* (Decne.) Sch. Bip.

Species of *Aucklandia* wild in China: 1 (excluding *A. costus*)

PLANT DESCRIPTION

A medium-tall perennial herb (0.4–)1–1.5(–2)m tall with a large and intensely aromatic rootstock (1–5 cm diam.). *Stems* stout, erect, hollow and often branching (1.5–2 cm diam.). *Leaves* alternate, simple and **mostly basal**, becoming smaller towards top of plant. Basal and lower stem leaves very large; **long-stalked (stalks 10-40 cm long and broadly winged)**; leaf blade lyre-shaped (20–50 cm long; 10–30 cm wide), heart-shaped or with 2 or more triangular outward-pointing basal lobes; **margin undulating and toothed**. The smaller middle and upper stem leaves stalked or stalkless, leaf blades lyre-shaped, ovate or triangular-ovate (usually 10–45 cm long; 5–25 cm wide); margin undulating and spiny-toothed. *Flower-heads* (*capitula*) at tip of aerial stems which rise above the leaves, solitary or in tight clusters of 2–5. Flower-heads thistle-like (2–4 cm diam.), outer part of each with 7–0 overlapping rows of spiny, cobwebby and often recurved bracts (phyllaries); outermost phyllaries narrowly triangular, middle ones ovate, inner ones linear. Inner part of each flowerhead with **flowers (florets) all similar**, tiny, **dark blackish-purple** and tubular (1.5–2 cm long) with a 2-lobed white style extending beyond the mouth of each floret. *Fruits* (achenes) tiny (6–8 mm long) and tightly packed into spherical heads; each fruit brown with black spots, somewhat conical to cylindrical, apically wrinkled, with a toothed crown and many straw-coloured **bristles** (pappus) (1.2–1.7 cm long). Fl. and fr. May–Aug.

HARVESTING, SOURCING AND NATURAL RANGE

Originally, this medicinal drug was imported into China from the Indian Himalayas (India, Kashmir and Pakistan) where the species is native and, since the 1920s, has been grown as a medicinal crop (Kuniyal *et al.*, 2015). In China, this imported 'Mu Xiang' coined the name 'Guang Mu Xiang' owing to its trade route via Guangzhou but since then the species has become established as a high altitude medicinal crop in western China (Yunnan province) at a height of 2,500–4,000 m. with its roots harvested during winter and spring. Its successful cultivation in China now extends to the provinces of Anhui, Fujian, Gansu, Guangxi, Guizhou, Shaanxi, Sichuan and Zhejiang and large volumes (some with GAP certification) are also exported to meet rising international demand for its root, root extract and its oil; yields are so successful that export volumes exceed those of its original source in NW India, Kashmir and northeast Pakistan (Schippmann, 2001).

Yunnan-grown 'Mu Xiang' is traded under the name 'Yun Mu Xiang' (云木香) to denote its provenance and *Di Dao*-quality (which has a high oil content).

CONSERVATION STATUS

China: not native (cultivated only).
India: Endangered (Nayar & Sastry, 1988); Critically Endangered (Molur & Walker, 1998).
Global: Critically Endangered (IUCN Red List, accessed 17/2/2016).

PROTECTED STATUS

CITES: Appendix I; EU Annex A (see Table 1 page 10).

TCM PROPERTIES AND USE

properties: **'pungent, bitter and warm'**
channels: **'Gall Bladder, Large Intestine, Spleen and Stomach**
actions: **'invigorates qi, tonifies Spleen, dispels damp heat'**

indications: **indigestion, poor appetite, abdominal distension and pain, diarrhoea**

Flowering crop of *Aucklandia costus*, Gansu province.

A. costus in flower

Immature flower-head of *A. costus*

Flower-heads of *A. costus*

UNOFFICIAL SUBSTITUTE 1

Dolomiaea souliei (Franch.) C. Shih **var. *souliei***

Asteraceae

synonyms: *Jurinea souliei* Franch.; *Vladimiria souliei* (C. Shih) Y. Ling

PLANT DESCRIPTION

Easily distinguished from the official species by its **height (not more than 30 cm), absence of aerial stems** and **leaves all forming a basal rosette**. Leaf margins are often lobed with spiny-toothed margins. **Flower/fruit-heads total 6–8, are stalkless and emerge directly from the basal rosette.** The tiny tubular flowers (florets) are **deep pinkish-red.** Fruits (achenes) with a **feathery** and longer (2.5–4 cm) pappus.

Fresh root of *D. souliei* var. *souliei*

> SUBSTITUTION COMMENT. Roots of this species are the source of the separate TCM drug 'Chuan Mu Xiang' (Vladimiriae Radix, in CP2015 as '*Vladimiria souliei* (Franch.) Ling'). They are sourced from wild populations endemic and widespread in Sichuan and Tibet and, like those of *Aucklandia costus* (official 'Mu Xiang') they grow at high altitude (3,700–4,800 m). Bensky *et al.* (2004) describe this drug as a cheaper and clinically 'weaker' one compared to official 'Mu Xiang'; they are used interchangeably in clinical settings knowingly or not. Both the fresh plant and the dried trade items of *D. souliei* bear some similarity with those of the Pharmacopoeia species. An added confusion is that *D. souliei* is often traded under the name 'Mu Xiang'.

CONSERVATION STATUS

Endemic to China: Least Concern (MEPCAS, 2013); Not Evaluated (IUCN Red List, accessed 11/1/2016).

UNOFFICIAL SUBSTITUTE 2

D. souliei **var. *cinerea*** (Y. Ling) Q. Yuan

Asteraceae

synonym: *Vladimiria souliei var. cinerea* Y. Ling (name used in CP2015)

PLANT DESCRIPTION

As for Unofficial Substitute 1 above but underside **of leaves greyish-white** with cobweb-like hairs and **dotted with glands.**

> SUBSTITUTION COMMENT. Like Substitute 1 above, roots of this species are the source species of the separate TCM drug 'Chuan Mu Xiang' (Vladimiriae Radix) (CP2015) sourced from wild populations endemic and widespread at high altitude (c. 3,800 m) in Sichuan, Tibet and Yunnan provinces. Roots are used interchangeably with official 'Mu Xiang' as described under Substitute 1 above.

CONSERVATION STATUS

Endemic to China: Least Concern (MEPCAS, 2013); Not Evaluated (IUCN Red List, accessed 11/1/2016).

Dolomiaea souliei var. *souliei* in the wild, Sichuan province

D. souliei var. *souliei* in fruit

Opened flowering heads of *D. souliei* var. *souliei*

D. souliei var. *cinerea* in the wild, Sichuan province (inset: basal rosette)

Fruiting head of *D. souliei* var. *cinerea*

Inula helenium L.

Asteraceae

PLANT DESCRIPTION

When not in flower, may be confused with PS1. A perennial herb (60–150(–250) cm tall, **softly hairy in most parts**; rootstock slightly (not intensely) aromatic. Basal leaves somewhat elliptic (usually 15–40 cm long; 10–20 cm wide), **undersides covered in velvety-woolly hairs and leaf bases extending down into the short but strongly ribbed leaf-stalks;** margins hard and toothed or entire. Stem leaf blades ovate or elliptic to lance-shaped (10–30 cm long; 4.5–12 cm wide), base of leaf blades heart-shaped, clasping the stem and with saw-toothed margins. **Flower-heads bright yellow** in lax, branched groups towards the top of the plant, each on a **6–12 cm stalk** emerging from a leaf axil. Involucre (below the yellow florets) a series of velvety green bracts becoming progressively narrower, drier and more scale-like towards the florets. **Florets of 2 types: outer yellow (ray) florets,** each with 1 long linear limb, together forming a radiating outer whorl, and **inner yellow (disk) florets,** each c. 9–11 mm long [hl] and **together creating a dense low-domed flower-head centre.** *Fruits (achenes)* tiny (3–4 mm long) and hairless [hl], crowned with a head of **c.50–60 barbed bristles (pappus) (6–10 mm long)** [hl]. Fl. Jun–Sep.

HARVESTING, SOURCING AND NATURAL RANGE

Although wild populations are confined to Xinjiang province of NW China the species is extensively cultivated in China, both as an ornamental and as a crop for its medicinal roots which are the source of the separate TCM drug 'Tu Mu Xiang' (Inulae Radix). Roots are harvested in summer with traditional *Di Dao*-quality sourced from Hebei province.

In the wild, the species grows along streamsides, roadsides and on disturbed ground up to 2,000 m. altitude [also wild in Russia, Tajikistan, Uzbekistan; SW Asia, Europe and North America].

CONSERVATION STATUS

China: Least Concern (MEPCAS, 2013).
Global: Not Evaluated (IUCN Red List, accessed 19/2/2016)

SUBSTITUTION COMMENT. The root of this species was the original source of 'Mu Xiang' before it was replaced by that of *Aucklandia costus* (today's Pharmacopoeia species) and with which it shares many clinical properties in addition to macroscopic root characters. It may be used as an adulterant of 'Mu Xiang', mostly perhaps when the latter is marketed as an extract or essential oil. Its roots have a long history of medicinal use in many other countries and where they continue to be used today for a broad range of medical conditions from bronchial and digestive conditions to hyper glycaemia and pain relief (Seca *et al.*, 2014). Macroscopic characters are usually sufficient to distinguish it from the Pharmacopoeia species but laboratory-based authentication may be needed for more reliable identification. It remains in the CP but as the separate drug 'Tu Mu Xiang' (Inulae Radix) (see also *Harvesting etc*......).

I. racemosa Hook. f.

Asteraceae

synonym: *Inula royleana* C. B. Clarke

PLANT DESCRIPTION

Very similar to *I. helenium* (Substitute 3 above) but flower-heads arranged on much shorter stalks **(0–4 cm long) along a central axis** (thus raceme-like)**.**

SUBSTITUTION COMMENT. A very closely related species to *I. helenium* above with early editions of the CP including it as an official source species of 'Tu Mu Xiang' (*I. helenium* above). Because of this and the similarity of its roots it may continue to be traded as 'Mu Xiang' or 'Tu Mu Xiang', especially locally in Xinjiang province (NW China) where it grows wild. A more specific trade name for this drug is Zong Zhuang Tu Mu Xiang (总状土木香). Macroscopic characters are usually sufficient to distinguish it from official 'Mu Xiang' (*A. costus*) but laboratory-based methods are needed to distinguish it from 'Tu Mu Xiang' above.

The species is also native to the NW Himalayan region where its roots are widely used in Ayurvedic medicine and more widely in the natural products industries.

CONSERVATION STATUS

China: Least Concern (MEPCAS, 2013).
Global: Not Evaluated (IUCN Red List, accessed 19/2/2016). Outside China, Himalayan populations are widely reported as endangered due to over-harvesting for medicinal use (Kala, 2005; Kaur *et al.*, 2010).

RIGHT
Inula helenium cultivated as a medicinal crop

RIGHT
Flowering plant of *Inula racemosa*

Aucklandia costus

Asteraceae

DRUG MORPHOLOGY

CRUDE DRUG (DRIED, SECTIONED AND DE-BARKED ROOT): almost cylindrical, **widest at upper end** and slightly twisted (5–10 cm long; 0.5–5 cm diam.). External surface dull, greyish- or reddish-brown, with longitudinal furrows **[1]** and root scars; **often branched [2];** many TCM texts describe the root as resembling 'dessicated bone' **[3]**. **Reticulated fibres below outer bark absent [4].** *Texture*: hard, but somewhat pliable. *Fracture*: mostly greyish-brown to dark brown. *Odour*: distinctively aromatic. *Taste*: somewhat bitter. ***Good quality*:** uniform, intensely aromatic oil glands.

DECOCTING PIECES (DRIED TRANSVERSE SLICES): ovate or elliptical (0.5–5 cm diam., c. 2–4 mm thick) with a thin skin, dark brown cambium zone **[5]** and a central yellowish-cream xylem with radiating vessels **[6]**. Dotted throughout **[7]** with dark oil glands.

PROCESSING

Fresh roots are cut into sections, dried, then the outer bark removed and discarded (= crude drug). Dried roots are then softened thoroughly before being cut into 2–4 mm thick transverse or oblique slices then shade-dried (= decocting slices); alternatively the roots are 'roasted' between paper to reduce the volatile oils, then sliced (= roasted decocting pieces called Wei Mu Xiang 煨木香). Longitudinal slicing is common in S China.

OTHER COMMON NAMES

广木香 Guang Mu Xiang; 煨木香 Wei Mu Xiang 'roasted Mu Xiang'; 云木香 Yun Mu Xiang 'Yunnan Mu Xiang'; 云木香属 Yun Mu Xiang Shu (plant name); costus (SCN); costus root; Kuth; puchuk; putchok; Radix Aucklandiae; Saussurea root.

SAFETY

An official healthfood in China (Anon., 2002a,b). Safety Class: 1; Interaction Class: A (Gardner & McGuffin, 2013).

Dolomiaea souliei var. *souliei*

Asteraceae

'DRUG SUBSTITUTE' MORPHOLOGY

DRIED ROOT: more uniformly cylindrical **(usually 1–3 cm diam. throughout its length)** and occasionally blotched with a black viscous substance called You Tou 油头 present in places. The outer bark often sheds to expose **reticulated (netted) fibres below [8].** *Texture*: not pliable but breaks more easily than the Pharmacopoeia species. *Odour*: slightly aromatic. *Taste*: bitter; viscous on chewing.

TRANSVERSELY SLICED ROOT: less dense than the Pharmacopoeia species with a more open, fibrous structure **[9]**; the central xylem frequently with **cavities [10] or with a 'rotted' appearance [11] and often splitting along one radial axis through to the epidermis [12].** Dark oil glands **less obvious** than those in the official species [hl].

Macroscopic characters above considered sufficient to reliably distinguish this substitute from the official drug.

OTHER COMMON NAMES

川木香 Chuan Mu Xiang 'Sichuan Mu Xiang'; 木香 Mu Xiang; 油头木香 You Tou Mu Xiang; Vladimiriae Radix.

See *Substitution Comment* on page 212.

Crude drug of Mu Xiang (*Aucklandia costus*)

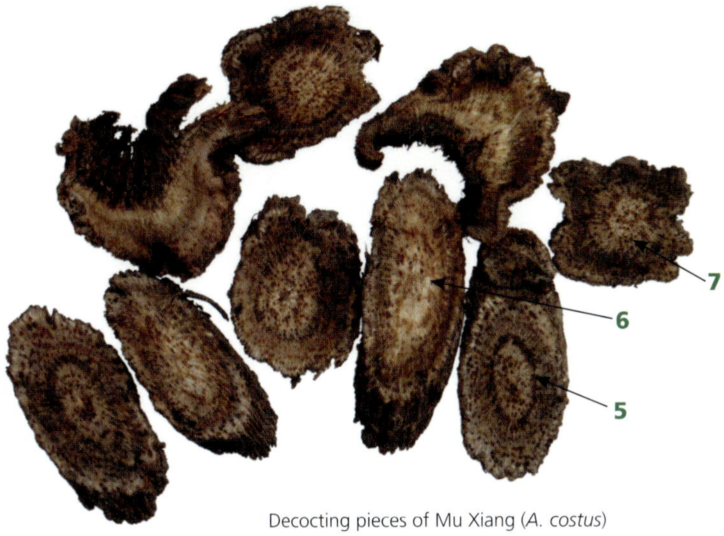

Decocting pieces of Mu Xiang (*A. costus*)

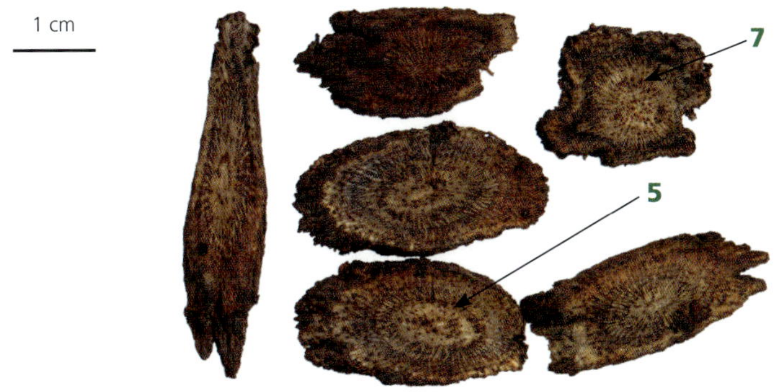

Decocting pieces of Wei Mu Xiang
(dried, sliced and 'roasted' roots of *A. costus*)

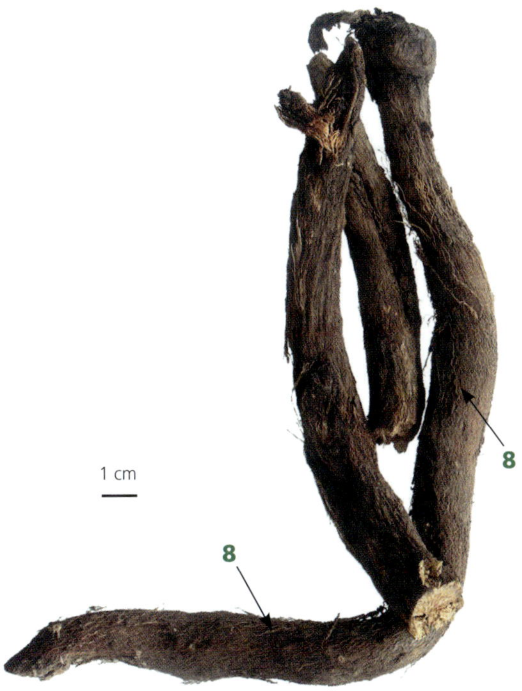

Dried root of *Dolomiaea souliei* var. *souliei* (Chuan Mu Xiang)

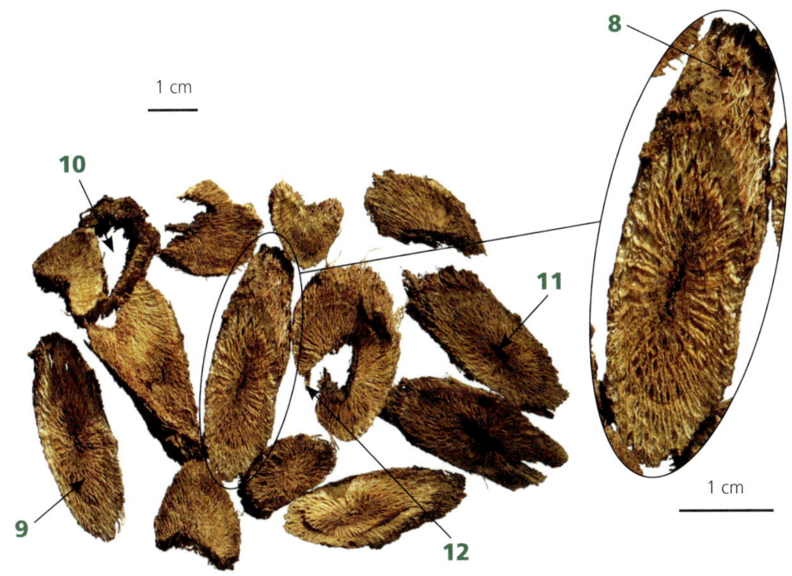

Sliced root of *D. souliei* var. *souliei* (Chuan Mu Xiang)

Dolomiaea souliei var. *cinerea*

Asteraceae

'DRUG SUBSTITUTE' MORPHOLOGY

Indistinguishable macroscopically from *D. soulliei* var. *souliei* (Substitute 1).

OTHER COMMON NAMES

川木香 Chuan Mu Xiang 'Sichuan Mu Xiang'; 灰毛川木香 Hui Mao Chuan Mu Xiang 'grey-haired Mu Xiang'; 木香Mu Xiang; 油头木香 You Tou Mu Xiang 'oily-headed Mu Xiang'; Vladimiriae Radix.

See *Substitution Comment* on page 212.

Inula helenium

Asteraceae

'DRUG SUBSTITUTE' MORPHOLOGY

DRIED ROOT: conical rather than cylindrical in outline (5–20 cm long) and **more tortuous [13]** than the official species. External surface yellowish-brown or dark brown. Apex **stout, with sunken stem scars [14] and leaf sheath remains [15].** *Texture*: hard, **not pliable.** *Fracture*: **yellow or yellowish-white [16]**; sunken oil glands visible [hl] but not as dark and obvious as the official species. *Odour*: slightly aromatic. *Taste*: bitter and pungent.

SLICED ROOT: slices ovate or irregular in outline (c. 2 mm thick). Skin yellowish-brown to dark brown and furrowed. Cut surfaces greyish-brown to dark brown with radial tissues and **central zone darker brown**; brown oil glands scattered throughout **but less dense than that of the official species**. *Odour*: slightly aromatic. *Taste*: bitter and pungent.

Laboratory-based authentication methods advisable for a more reliable identification.

OTHER COMMON NAMES

木香 Mu Xiang; 祁木香 Qi Mu Xiang; 土木香 Tu Mu Xiang; 藏木香 Zang Mu Xiang; elecampane (SCN); Helenii Rhizoma; Inula root (CP2010); Inulae Radix; Radix Inulae.

SAFETY

Safety Class: 1; Interaction Class: A (Gardner & McGuffin, 2013).

See *Substitution Comment* on page 214.

Inula racemosa

Asteraceae

'DRUG SUBSTITUTE' MORPHOLOGY

Whole and sliced root cannot be distinguished macroscopically from those of the substitute above (*Inula helenium*).

OTHER COMMON NAMES

木香 Mu Xiang; 总状土木香 Zong Zhuang Tu Mu Xiang (plant name); Indian elecampane (SCN); manu; pushkara; pushkarmool.

See *Substitution Comment* on page 214.

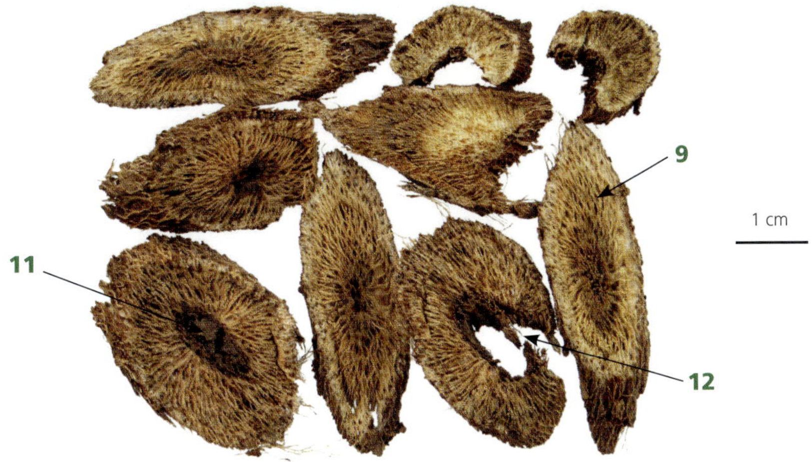

Sliced root of *D. souliei* var. *cinerea* (Chuan Mu Xiang)

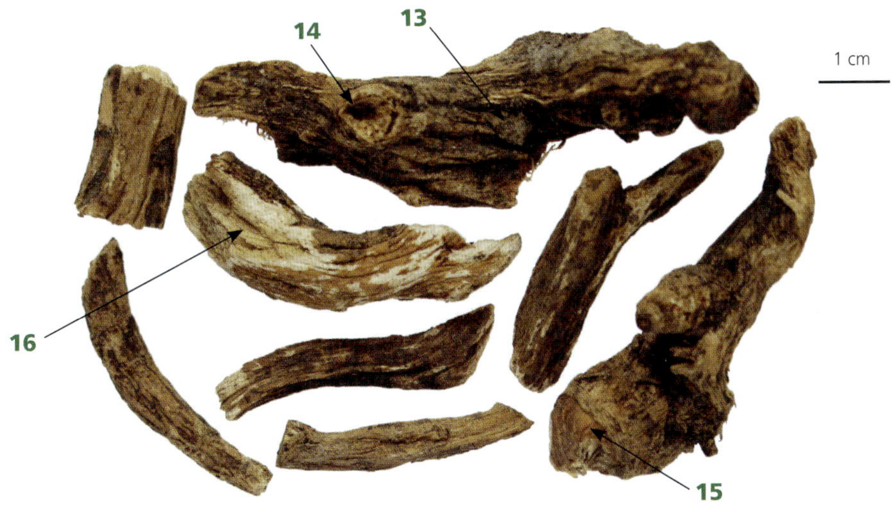

Sliced root of *Inula helenium* (Tu Mu Xiang)

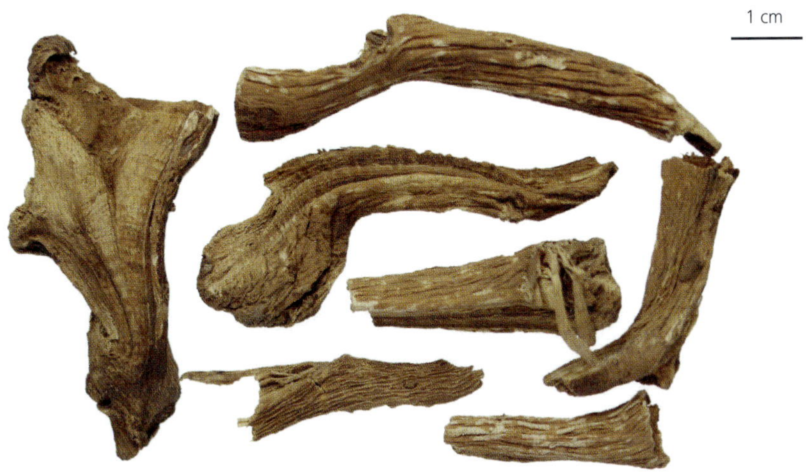

Sliced root of *I. racemosa* (Zong Zhuang Tu Mu Xiang)

Nan Ban Lan Gen 南板蓝根

Baphicacanthis Cusiae Rhizoma et Radix

Drug part: RHIZOME AND ROOT

Strobilanthes cusia (Nees) Kuntze

Acanthaceae

synonym: *Baphicacanthus cusia* (Nees) Bremek. (name used in CP2015); *Goldfussia cusia* Nees.

Species of *Strobilanthes* wild in China: 128

PLANT DESCRIPTION

An upright perennial herb (0.5–1.5 m tall) often woody at base. *Stems* virtually hairless. *Leaves* simple and opposite, with leaf stalks 0.5–7 cm long; leaf blades elliptic to ovate (4–20 cm long; 2–9 cm wide), hairless, paler green on lower surface, drying blackish, secondary veins 7–9 each side of mid-vein, base tapering, margin shortly saw-toothed and tip pointed. *Flowers* in terminal or axillary spikes, intermingled with leaf-like bracts, often aggregated (1–6 cm long). Each flower large, mauve or blue and conspicuous (3.5–5 cm long), corolla a curved funnel with 5 notched equal-sized petal lobes at mouth; stamens 4, unequal; style c. 3.2 cm long. *Fruit* a dry, elongate, 4-seeded capsule (1.5–2.2 cm long). Fl. Jul–Feb, fr. Dec–Feb.

HARVESTING, SOURCING AND NATURAL RANGE

Rhizomes and roots are harvested in summer and autumn, mostly from cultivated plants grown in Fujian, Guangdong, Guangxi and Sichuan. The species grows wild in damp woods (100–2,000 m altitude) in S and W China (Fujian, Guangdong, Guangxi, Guizhou, Hainan, Hunan, Sichuan, Taiwan, SE Tibet, Yunnan, Zhejiang) [also wild in Bangladesh, Bhutan, India, Laos, Myanmar, Thailand and Vietnam].

CONSERVATION STATUS

China: Least Concern (MEPCAS, 2013).
Global: Not Evaluated (IUCN Red List, accessed 19/6/2015).

Medicinal crop of *Strobilanthes cusia*, Fujian province

S. cusia in flower

TCM PROPERTIES AND USE

properties: **'bitter and cold'**
channels: **'Heart and Stomach'**
actions: **'clears heat, eliminates toxins, cools Blood'**
indications: **fever; sore throat; abscess, infections**

DRUG MORPHOLOGY

CRUDE DRUG (DRIED RHIZOME AND ROOT): mostly rhizomes (typically with small white central pith) with fewer roots (often still attached to rhizomes) of variable thickness but without white central pith. *Rhizomes*: irregularly shaped, often tortuous **[1]** and branched (10–30 cm long; 0.1–1.3 cm diam.); external surface finely wrinkled longitudinally with yellowish-brown bark often exfoliated to reveal characteristic bluish-grey tissues below **[2]**; nodes often swollen **[3]** (but shrinking on drying), rootlet scars occasional. *Texture*: hard and easily broken. *Fracture*: fibrous and uneven, wood bluish-grey to pale yellowish-brown surrounding central white pith. *Odour and taste*: indistinct. *Roots*: often curved and branched, exfoliating easily like rhizomes above; rootlets narrow and pliable. *Odour and taste*: indistinct. ***Good quality***: roots and rhizomes slightly bluish-brown.

DECOCTING PIECES (SLICED RHIZOME AND ROOT): sub-rounded transverse or obliquely cut slices of rhizome and root of varying width (0.1–1.3 cm). Thin, brown outer cork bark intact or exfoliating **[4]**; easily removed completely to reveal slightly thicker bluish-grey or pale yellowish-brown inner bark **[5]**; latter also easily removed. Cut surface predominantly with pale yellowish-brown or greyish-blue wood **[6]**; rhizome slices with a central white pith **[7]**. *Texture*: hard but brittle. *Odour* and *taste*: indistinct.

PROCESSING

After aerial parts are discarded, rootstocks are sun-dried (= crude drug) and subsequently softened, sliced thickly (2–4 mm) and dried again (= decocting pieces).

OTHER COMMON NAMES

板蓝 Ban Lan (plant name); 板蓝根 Ban Lan Gen; 广东大青叶 Guangdong Da Qing Ye; 马蓝 Ma Lan; 马蓝根 Ma Lan Gen; Rhizoma et Radix Baphicacanthis Cusiae; Strobilanthes cusia (SCN).

See also 'Ban Lan Gen' (Isatidis Radix) page 46.

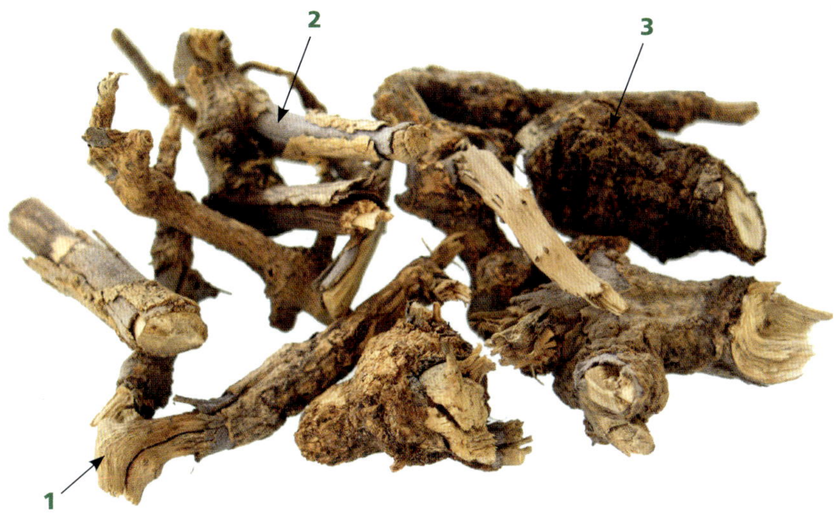

Crude drug of Nan Ban Lan Gen
(dried roots and rhizomes of *Strobilanthes cusia*)

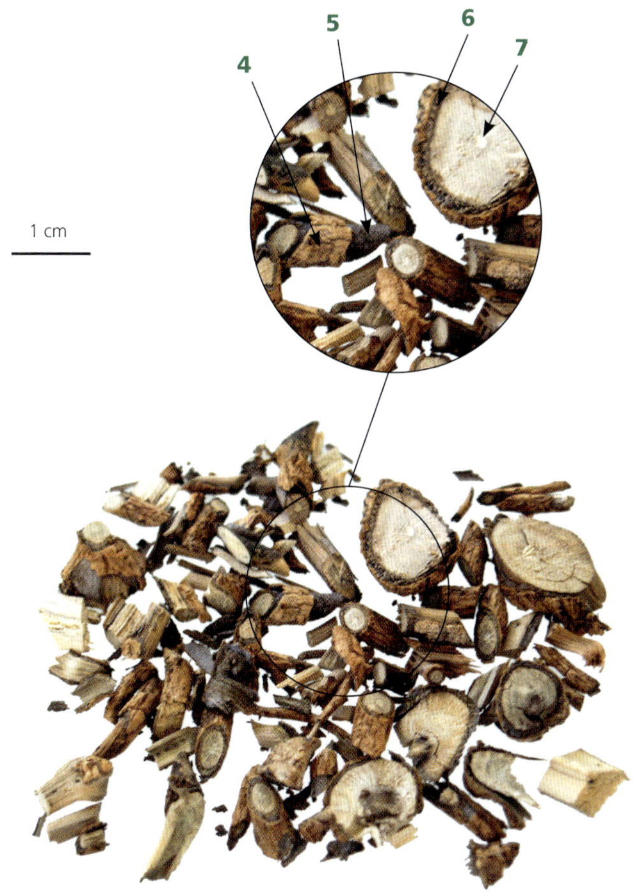

Decocting pieces of Nan Ban Lan Gen
(dried and sliced roots and rhizomes of *S. cusia*)

Nan Sha Shen 南沙参 Adenophorae Radix

Drug part: ROOT

Adenophora tetraphylla (Thunb.) Fisch.

Campanulaceae

synonyms: *Adenophora triphylla* (Thunb.) A. DC. (*Kew-MPNS* accepted name, 20/8/2016); *Adenophora verticillata* Fisch.; *Campanula tetraphylla* Thunb.; *C. triphylla* Thunb.

Species of *Adenophora* wild in China: 38

PLANT DESCRIPTION

An erect, virtually hairless, herbaceous perennial (up to 1.5 m tall) with a narrowly conical tuber (cylindrical, conical or bulbous). ***Leaves* in whorls of 3–6**, simple and virtually stalkless; basal leaves broadly oval to heart-shaped, becoming narrower in upper parts; margins sharply toothed. *Inflorescence* sparse, with flowers on branching lateral branches with central flower of each cluster opening first (a cyme) arranged almost in whorls up the stem. *Flowers* bisexual, with sepals (green) extending into 5 narrow linear lobes surrounding the base of each flower; corolla narrowly funnel-shaped (7–10 mm long with 5 short pointed corolla lobes at mouth, blue or purplish-blue (<11 mm wide); stamens 5; ovary inferior; style long (14–20 mm) and extending well beyond the corolla tube. *Fruit* a dry, ovoid-conical capsule (5–7 mm long; 4–5 mm diam.) with 5 persistent sepals. *Seeds* many, ellipsoid.

HARVESTING, SOURCING AND NATURAL RANGE

Roots are harvested in spring and autumn from wild plants found in scrub and grassy places (0–2,000 m altitude) in many Chinese provinces: Anhui, Fujian, Guangdong, Guangxi, Guizhou, Hebei, Heilongjiang, Henan, Hunan, Inner Mongolia, Jiangsu, Jiangxi, Jilin, Liaoning, Shandong, Shanxi, Sichuan, Taiwan, Yunnan and Zhejiang [also native in Japan, the Korean Peninsula, Laos, Russia (Far East, E Siberia) and Vietnam].

CONSERVATION STATUS

China: Least Concern (MEPCAS, 2013).
Global: Not Evaluated (IUCN Red List, accessed 19/8/2016).

A. stricta Miq.

Campanulaceae

taxonomic/nomenclatural note: *FoC* and *Kew-MPNS* recognise 4 subspecies.

PLANT DESCRIPTION

A shorter plant (40–80 cm tall); **most parts,** especially calyx, usually **covered with short stiff hairs**; stem leaves alternate (not in whorls); **flowers much larger** (1.5–2.5 cm long) and broadly bell-shaped with lobes almost half as long as corolla 'bell', blue or purple; fruit ellipsoid-globose. Fl. Aug–Oct.

HARVESTING, SOURCING AND NATURAL RANGE

Roots are harvested in spring and autumn from wild plants found in scrub, grassland and rocky crevices up to 3,300 m altitude in C, S and W China: Anhui, Chongqing, Fujian, Gansu, Guangxi, Guizhou, Henan, Hubei, Hunan, Jiangsu, Jiangxi, Shaanxi, Sichuan, Yunnan and Zhejiang [also native in the Korean Peninsula; naturalised in Japan].

CONSERVATION STATUS

China: Least Concern (MEPCAS, 2013).
Global: Not Evaluated (IUCN Red List, accessed 19/8/2016).

TCM PROPERTIES AND USE

properties: 'sweet, slightly bitter and cool'
channels: 'Lung and Stomach'
actions: 'tonifies yin, clears Lungs, resolves phlegm, generates fluids'

indications: dry coughs; chronic respiratory disorders; poor appetite, constipation and vomiting

Adenophora tetraphylla in the wild, China

Flowering spike of *A. tetraphylla*

Fresh root of *A. tetraphylla*

A. stricta in the wild, China

A. stricta with short fine hairs covering stem, calyx and flower 'ribs'

Adenophora tetraphylla

Campanulaceae

DRUG MORPHOLOGY

CRUDE DRUG (ROOT): size and shape highly variable **[1]** because all are wild harvested. Roots usually cylindrical or conical (7–27 cm long; 0.8–3 cm diam.) **[2]**, slightly curved and often bulbous towards the apex. Latter with 1 or 2 rhizome remnants **[3]** and, depending on age of the root, extending to form a narrow neck **[4]**. Externally yellowish-white or pale brownish-yellow with occasional dark brown root-bark remnants **[5]**. Lower (narrower) part of root longitudinally wrinkled **[6]**. *Texture:* lax and porous; easily broken. *Fracture:* easily broken, uneven, porous-looking with clefts of varying size **[7]**. *Odour:* faint. *Taste:* somewhat sweet. ***Good quality*:** thick, plump, yellowish-white roots, devoid of dark outer rootbark.

DECOCTING PIECES (TRANSVERSELY SLICED ROOT): irregularly ovate root slices of varying diameter **[8]** (0.8–3 cm); usually >2 mm thick; yellowish-white throughout **[9]**; externally with occasional dark brown root-bark remnants **[10]**; cut surfaces lax and porous with scattered clefts of varying size **[7]**. *Texture* etc. as for crude drug.

DECOCTING PIECES (LONGITUDINALLY SLICED ROOT): as for transverse slices but irregularly rectangular (c. 5–8 cm long).

PROCESSING

After removal of rootlets and rough outer skin the root is dried (= crude drug) then sliced transversely or for Hong Kong markets, longitudinally, into decocting pieces c. 2mm thick.

OTHER COMMON NAMES

轮叶沙参 Lun Ye Sha Shen (plant name); 沙参 Sha Shen; adenophora (SCN); fourleaf ladybell root (CP2010); Radix Adenophorae.

SAFETY

Occupational contact dermatitis has been reported by those handling the roots routinely (Bensky *et al.,* 2004). Safety Class: 1; Interaction Class: A (Gardner & McGuffin, 2013).

A. stricta

Campanulaceae

DRUG MORPHOLOGY

Crude drug and decocting pieces are indistinguishable macroscopically from those of PS1. Any apparent distinction is due to natural variation within this wild-harvested species.

PROCESSING

As for PS1.

OTHER COMMON NAMES

沙参 Sha Shen (plant name); Xing Ye Sha Shen; adenophora (SCN); fourleaf ladybell root (CP2010); Radix Adenophorae.

SAFETY

As for PS1.

UNOFFICIAL SUBSTITUTES

The identification of species in the genus *Adenophora* is challenging, not least because the group in is in need of taxonomic revision. Only some of its species are currently well-defined, notably *A. tetraphylla* and *A. stricta* described here. With 36 species native to China, of which around 30 are used in local Chinese folk medicine and all wild-sourced, there is a high likelihood, that materials traded as official 'Nan Sha Shen' may be accidentally substituted or contaminated by roots of other widely used *Adenophora* species. Roots of official 'Nan Sha Shen' traded on the international market may be similarly vulnerable and laboratory-based authentication methods are advised for their reliable identification and purity-testing.

See also 'Bei Sha Shen' (Glehniae Radix) page 62.

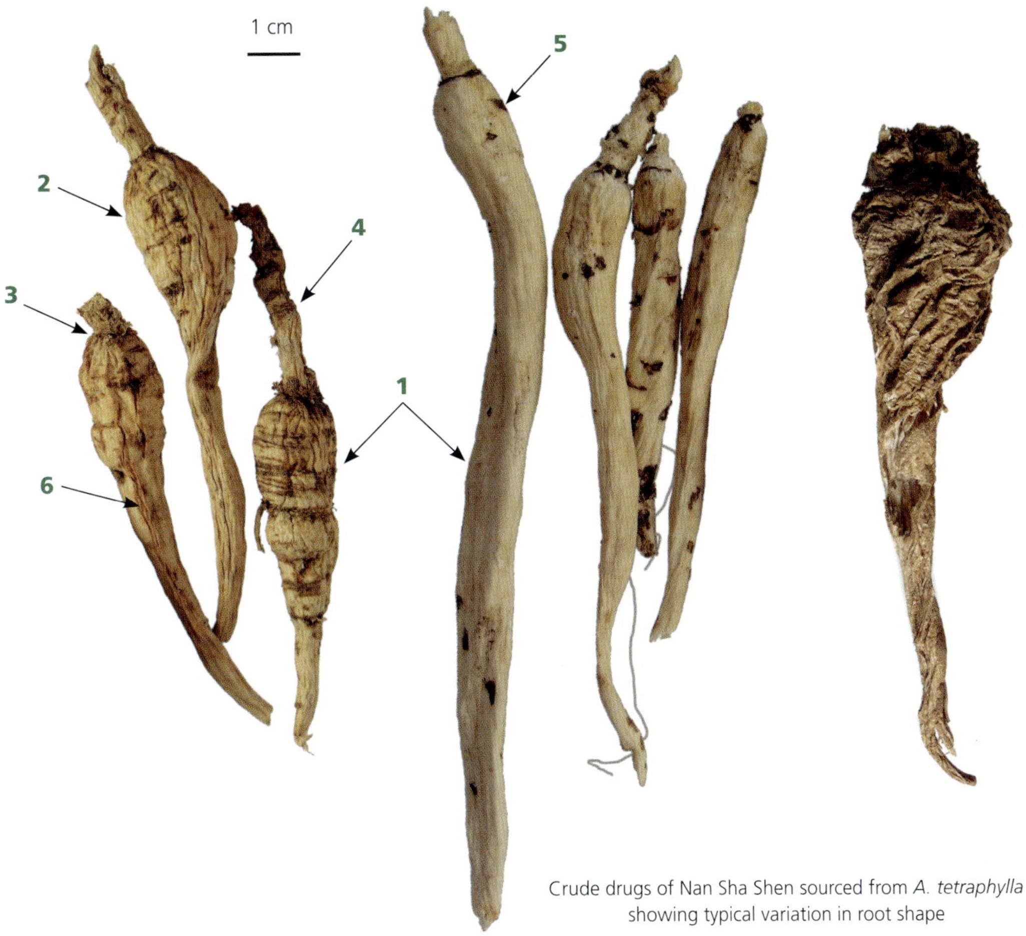

Crude drugs of Nan Sha Shen sourced from *A. tetraphylla* showing typical variation in root shape

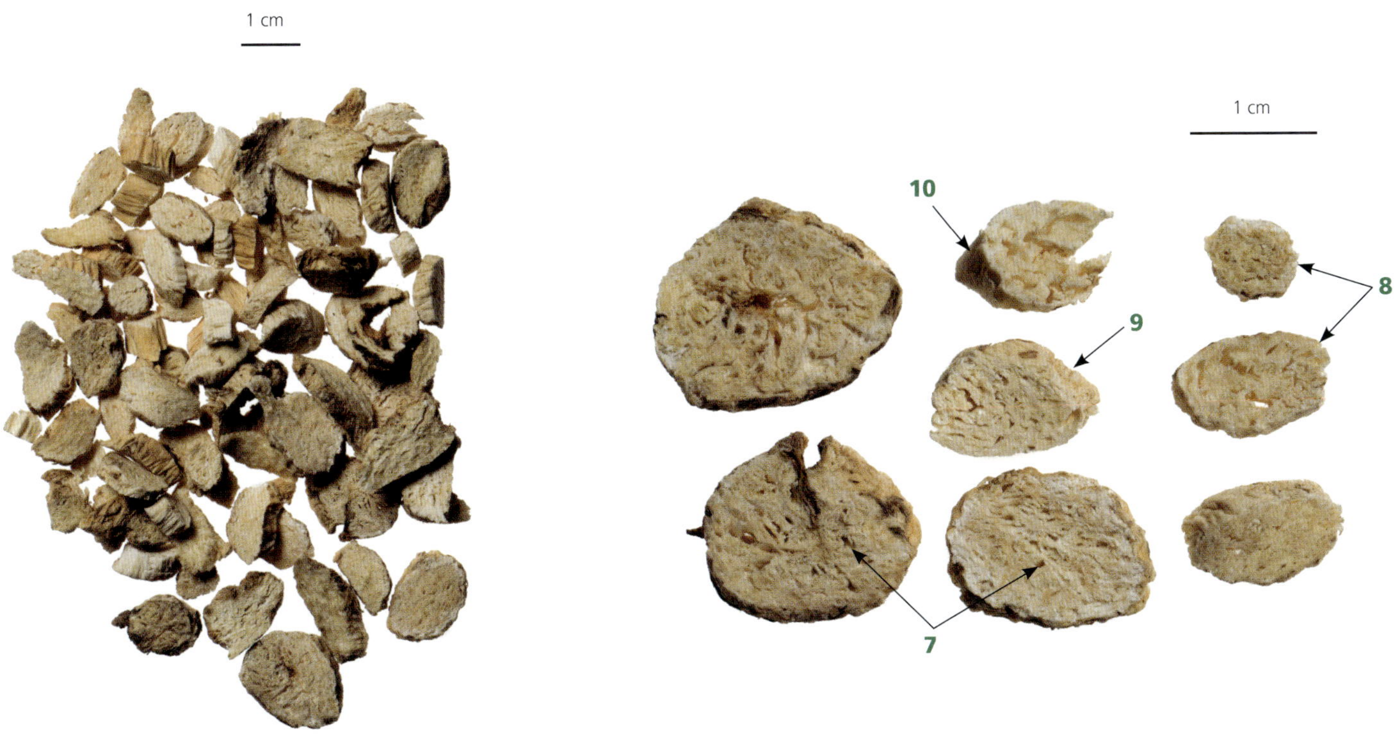

Decocting pieces sourced from *A. tetraphylla* (transversely sliced)

Niu Xi 牛膝 Achyranthis Bidentatae Radix

Drug part: ROOT

PHARMACOPOEIA SPECIES

Achyranthes bidentata Blume

Amaranthaceae

Species of *Achyranthes* wild in China: 3

PLANT DESCRIPTION

A highly variable herbaceous perennial (70–120 cm tall). *Stems* green or tinged purple, angular and oppositely branched. *Leaves* opposite, elliptic or elliptic/lance-shaped (4.5–12 cm long; 2–7.5 cm wide); narrowing at the tips; both surfaces **finely (but not softly) hairy.** *Flowers* mainly in terminal spikes (3–5 cm long or more), forming dense, **linear heads (3–5 cm long)**; individual flowers tiny (c. 5 mm diam.) and each with 1 outer membranous bract, 2 narrow, spiny bracteoles and 5 long, pointed, greenish-white, shiny membranous 'petals'; stamens 5, separated by tiny rectangular slightly saw-toothed lobes [hl]. ***Fruits* downward-pointing, pressed close to the stem**, dry, yellowish-brown, oblong and shiny (2–2.5 m long). *Seeds* many (c.1 mm long). Fl. Jul–Sep, fr. Sep–Oct.

HARVESTING, SOURCING AND NATURAL RANGE

Roots are harvested in winter after the aerial parts have withered. Sourcing is mostly from medicinal crops grown in Hebei and Henan provinces with traditional *Di Dao* quality drug sourced from the latter (districts of Bo'ai, Huixian, Qinyang, Wenxian, Wuzhi and Yuxian).

The species grows wild on hillsides between 200–1,800 m altitude and is widely distributed in W, C and S China (Anhui, Fujian, Hebei, Hubei, Hunan, Guangxi, Guizhou, Jiangsu, Shaanxi, Shanxi, Sichuan, Taiwan, Tibet and Zhejiang); [also wild in Bhutan, India, Indonesia, Japan, Korean Peninsula, Laos, Malaysia, Myanmar, Nepal, New Guinea, Philippines, Russia, Sikkim, Thailand and Vietnam].

CONSERVATION STATUS

China: Least Concern (*A. bidentata* var. *japonica* Miq.) (MEPCAS, 2013.)
Global: Not Evaluated (IUCN Red List, accessed 12/2/2015).

UNOFFICIAL SUBSTITUTE

Cyathula officinalis K. C. Kuan

Amaranthaceae

PLANT DESCRIPTION

***Leaves* softly hairy.** *Flowers* in dense **globose heads (1–1.5 cm diam.)**, pale green to whitish. *Fruits* dry, pale yellow and flask-shaped **(not pressed close to the stem).** *Seeds* many and glossy. Fl. Jun–Jul, fr. Aug–Sep.

HARVESTING, SOURCING AND NATURAL RANGE

See 'Chuan Niu Xi' (Cyathulae Radix) page 86.

LESSER KNOWN UNOFFICIAL SUBSTITUTES

The roots/rhizomes of other species have been reported as unofficial substitutes but their occurrence in international trade is uncertain (Ho *et al*., 2004, 2006; Xiao *et al*., 2001–2002):

1. ***Achyranthes aspera*** L. (土牛膝 Tu Niu Xi; prickly chaff flower (SCN)). An invasive pan-tropical weed and also used in Ayurvedia and Unani medicine. Very similar to the Pharmacopoeia species but leaves broadest above middle or elliptic-oblong and tips broad with a short, sharp terminal point. Staminodes (sterile stamens) bordered with long cilia [hl].
2. ***A. longifolia*** (Makino) Makino (柳叶牛膝 Liu Ye Niu Xi). Very similar to the Pharmacopoeia species but leaves lanceolate or broadly so (10–20 cm long; 2–5 cm wide), tips heart-shaped; bracteoles needle-shaped (c. 3.5 mm long) with margins fringed with hairs [hl]. *Kew-MPNS* treats this is as a synonym of the Pharmoacopoeia species (accessed 27/2/2015).
3. ***Silene baccifera*** (L.) Roth (syn. *Cucubalus baccifer* L.); (狗筋蔓 Gou Jin Man; 白牛膝 Bai Niu Xi) (Caryophyllaceae). A widespread hairy perennial herb with white, 5-petalled flowers and spherical black berries; dried roots thinner, greyish-brown and yellow internally.
4. Roots and rhizomes provisionally identified as ***Strobilanthes* sp.** (Acanthaceae): commonly traded in mainland China, Taiwan and the UK under the same Chinese name as the official drug 牛膝 Niu Xi (as well as 味牛膝 Wei Niu Xi) but they are macroscopically quite distinct. The TCM literature (Bensky *et al*., 2004; Zhao & Li, 2007) states the botanical source to be *S. forrestii* or *S. nemorosus*, however both names are under taxonomic review and the identity of this widespread substitute remains uncertain (for photo see separate account for 'Chuan Niu Xi' (Cyathulae Radix) page 86).

TCM PROPERTIES AND USE

properties: 'bitter, sour and neutral'
channels: 'Liver and Kidney'
actions: 'tonifies Liver and Kidney, invigorates Blood, dispels stasis'

indications: chronic or acute musculoskeletal pain; gynaecological disorders; dysuria; toothache, mouth sores

Achyranthes bidentata medicinal crop, Hebei province

A. bidentata in flower

Close-up of an *A. bidentata* flower

Linear *A. bidentata* flower-head

A. bidentata fruit-head

Cyathula officinalis medicinal crop, Sichuan province

C. officinalis in flower

C. officinalis flower-heads

Achyranthes bidentata

Amaranthaceae

DRUG MORPHOLOGY

CRUDE DRUG (DRIED ROOT): roots long and cylindrical, usually slightly curved (15–70 cm long; **0.4–1 cm diam. [1]**), **apex often with remains of persistent stem bases**. Externally **greyish-yellow or pale brown [2]**, with longitudinally often **slightly twisted striations [3]**, faint and scattered transverse lenticels and occasional raised or sunken rootlet scars. *Texture*: **hard, brittle and easily broken**. *Fracture*: even, slightly lustrous, with wood (xylem) containing a pale and central vascular bundle **[4]** surrounded by 2–4 loosely concentric rings of pale and punctiform vascular bundles **[5]**. *Odour*: indistinct. *Taste*: slightly sweet, slightly bitter and astringent. ***Good quality:*** long, fleshy roots, externally yellowish-grey; sweet-tasting.

DECOCTING PIECES-1 (SLICED ROOT): as for crude drug but cut into short, cylindrical pieces (5–12 mm long; **0.4–1 cm diam.**)

DECOCTING PIECES-2 (WINE-PROCESSED ROOT): as for root sections above but darker and with occasional charred areas; wine-scented.

PROCESSING

After removal of rootlets, roots are tied in small bundles to partially sun-dry then root-heads often removed too before drying is completed (= crude drug). Subsequent washing, thorough softening and cutting into short transverse sections produces dried decocting pieces. Optional further processing by moistening with yellow-rice wine then dry-frying produces 酒牛膝 Jiu Niu Xi (wine-processed decocting pieces).

OTHER COMMON NAMES

怀牛膝 Huai Niu Xi; achyranthes (SCN) root; Radix Achyranthis Bidentatae (CP2005).

SAFETY

Safety Class: 2b, 2d; Interaction Class: A (Gardner & McGuffin, 2013). An official Chinese food and healthfood (Anon., 2002a,b).

Cyathula officinalis

Amaranthaceae

'DRUG SUBSTITUTE' MORPHOLOGY

DRIED ROOT: as for *A. bidentata* but externally greyish-brown **[6]**, **strongly twisted, sometimes branched [7]** and **usually wider** (30–60 cm long; **0.5–3 cm diam**. **[8]**). *Texture*: strong, **pliable and difficult to break**. *Fracture*: **several concentric rings** of pale and punctiform, yellow vascular bundles; central vascular bundle often less obvious. *Odour*: indistinct. *Taste*: sweet.

SLICED ROOT: easily distinguished from the Pharmacopoeia species. Subrounded or elliptical and **often larger in diameter (0.5–3 cm diam.) [9]**. Cut surfaces with **several concentric rings** of punctiform vascular bundles **[10]**. *Taste*: sweet.

WINE-PROCESSED ROOT: as for root slices but externally brownish-black **[11]**. *Odour*: faintly of wine. *Taste:* sweet.

PROCESSING

See Chuan Niu Xi (Cyathulae Radix) page 86.

OTHER COMMON NAMES

川牛膝 Chuan Niu Xi (Sichuan Niu Xi); 天全牛膝 Tian Quan Niu Xi; 甜牛膝 Tian Niu Xi; Cyathulae Radix (CP2010); Radix Cyathulae (CP2005) see page 86.

SAFETY

Safety Class: 2b, **2d (other specific use restrictions exist)**; Interaction Class: A (Gardner & McGuffin, 2013). An official Chinese food and healthfood (Anon., 2002a,b).

SUBSTITUTION COMMENT. Roots of this species are the source of the separate TCM drug 'Chuan Niu Xi' (Cyathulae Radix) see page 86. It is reported to have similar clinical properties (Bensky *et al.,* 2004). When not in flower or fruit, aerial parts are easily confused with those of the Pharmacopoeia species.

Crude drug (dried root of *Achyranthes bidentata)*

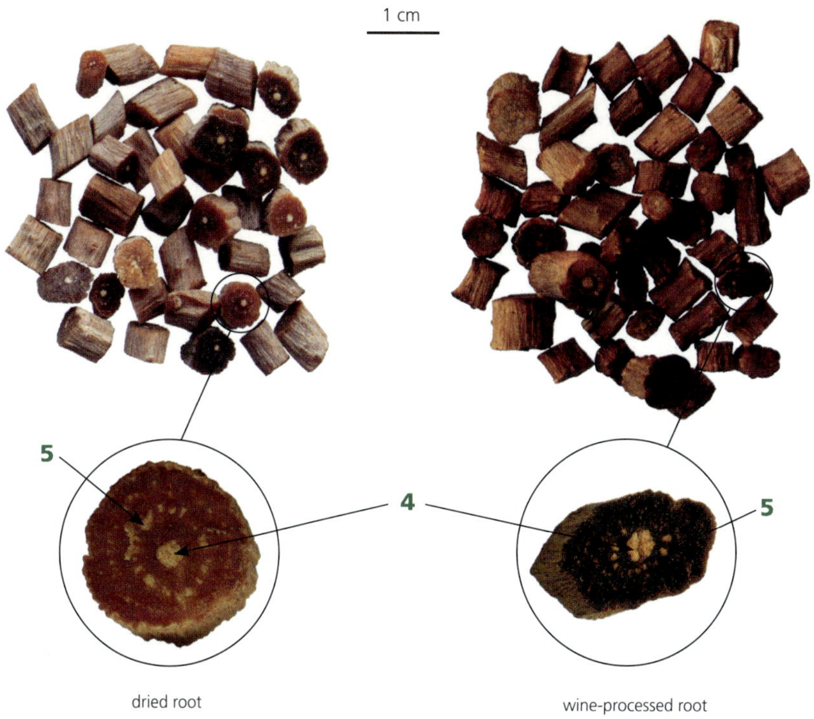

Decocting pieces

'Drug substitute' (crude) of Chuan Niu Xi sourced from dried roots of *Cyathula officinalis*

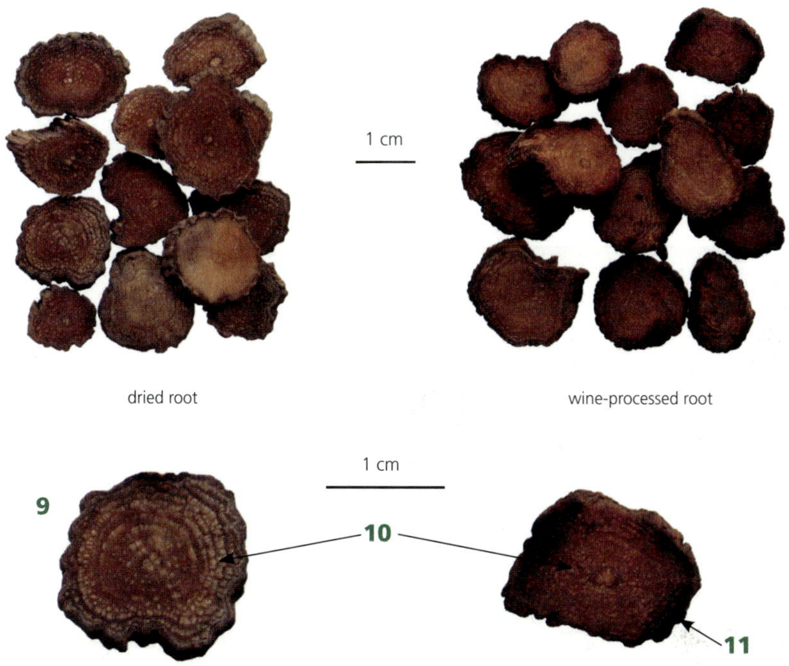

Decocting pieces of Chuan Niu Xi drug substitute sourced from dried roots of *C. officinalis*

Ou Jie 藕节 Nelumbinis Rhizomatis Nodus

Drug part: RHIZOME NODE

Nelumbo nucifera Gaertn.

Nelumbonaceae

synonyms: *Nelumbium speciosum* Willd.; *Nymphaea nelumbo* L.

Species of *Nelumbo* wild in China: 1

PLANT DESCRIPTION

An aquatic perennial herb with submerged, fleshy rhizomes; latter with spongy nodes containing large air sacs. *Leaves* large (25–90 cm in diam.) orbicular, emergent, with radiating veins and central leaf stalk; leaf blade bluish-green and water repellant; margin entire. *Flowers* solitary, large (10–23 cm diam.), emergent on long leafless stalks, with whorls of pink or white 'petals'; latter oblong-elliptic or broadest above the middle. *Fruit-head* thick and fibrous and resembling an up-turned shower-head (1.0–2.0 cm long; 7–15 cm wide), the flat top perforated to reveal tips of ovoid fruits within. Each fruit with a thick and hard pericarp; 1-seeded. Fl. Jun–Aug.

HARVESTING, SOURCING AND NATURAL RANGE

Rhizome nodes are a by-product of the widely cultivated vegetable "Ou" sourced from the fleshy internodes.

In the wild, the species grows in lakes, ponds and slow-flowing water in N and W China: Inner Mongolia, Qinghai and Tibet [also wild in Bhutan, India, Indonesia (Java), Japan, the Korean Peninsula, Malaysia, Myanmar, Nepal, New Guinea, Pakistan, Philippines, Russia (Far East), Sri Lanka, Thailand, Vietnam; SW Asia and Australia].

CONSERVATION STATUS

China: Not Evaluated.
Global: Not Evaluated (IUCN Red List, accessed 12/6/2015).

PROTECTED STATUS

China: Cat. II (SEPA,1999).

Nelumbo nucifera: flowering and fruiting crop, Guangxi province

Pink flowering form and developing fruit-heads

Flowering and fruiting plant

Flat-topped fruit-head with tips of fruits prominent

Freshly harvested nodes

TCM PROPERTIES AND USE

properties: **'sweet and neutral'**
channels: **'Lung, Stomach and Liver'**
actions: **'astringes, stops bleeding'**

indications: **haematemesis, epistaxis, haemoptysis**

DRUG MORPHOLOGY

CRUDE DRUG/DECOCTING PIECES-1 (RHIZOME NODE): short somewhat cylindrical pieces (2–4 cm long; 2 cm diam.), often with a swollen appearance **[1]**. Outer skin greyish-yellow to greyish-brown, finely striated longitudinally **[2]**, with distinct zone of prominent and crowded rootlet scars **[3]**) adjacent to a less distinct leaf scar ring **[4]** (occasionally fibrous) **[5]** encircling the node. Short sections of smooth-surfaced rhizome internode often still attached (**[6]**. Cut surface typically with up to 9 large, often distorted, air chambers (cavities) **[7]** surrounding a central one; each chamber up to 9 mm diam. *Texture*: very hard. *Fracture*: many sub-rounded cavities. *Odour*: indistinct. *Taste*: somewhat sweet and astringent. ***Good quality***: blackish-yellow nodes, creamish-white on transverse surfaces.

DECOCTING PIECES-2 (CARBONISED RHIZOME NODE — OU JIE TAN): as for above but external surface charred black or blackish-brown and inner surface yellowish-brown or brown. Some identification characters obscured.

PROCESSING

After removal of the fleshy internodes, the nodes are sun-dried and rootlets removed (= crude drug/decocting pieces-1). Subsequent carbonising may be performed by dry-frying until the flesh becomes blackish-brown or charred externally and yellowish-brown internally (= 藕节炭 Ou Jie Tan; decocting pieces-2).

OTHER COMMON NAMES

老藕节 Lao Ou Jie (old Ou Jie) and refers to nodes harvested in winter; 莲 Lian (lotus) (plant name); 南藕节 Nan Ou Jie (southern Ou Jie); lotus rhizome node (CP2010); Nodus Nelumbinis Rhizomatis; sacred lotus (SCN).

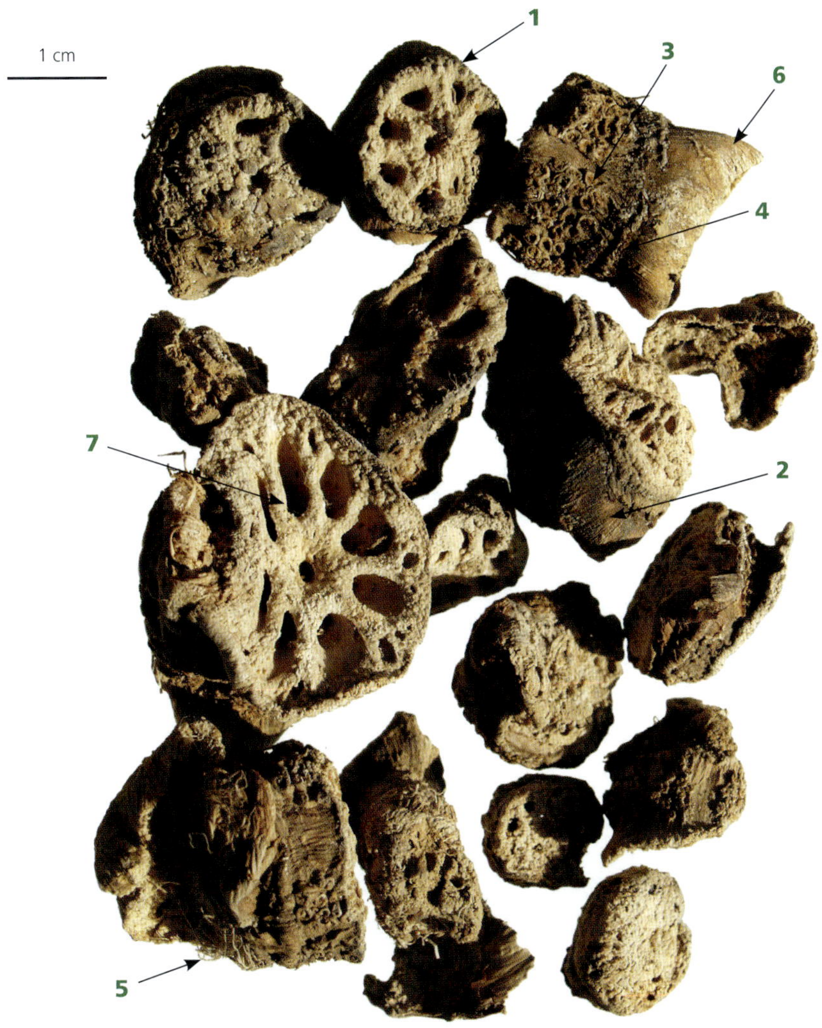

Crude drug/decocting pieces of Ou Jie

Carbonised Ou Jie (Ou Jie Tan)

Ping Bei Mu 平贝母

Fritillariae Ussuriensis Bulbus

Drug part: BULB

Fritillaria usuriensis Maxim.

Liliaceae

taxonomic/nomenclatural note: in CP2015, *FoC* and elsewhere as *'Fritillaria ussuriensis* Maxim.'.

Species of *Fritillaria* wild in China: 24

PLANT DESCRIPTION

An erect, delicate-looking herbaceous perennial (50–60 (rarely up to 100) cm tall) with a slightly squat bulb (1–1.5 cm diam.) with 2 fleshy scales, often with smaller bulbils at base. *Stem* unbranched. *Leaves* 14–17; basal ones in a whorl of 3, upper ones whorled or opposite; leaf blades linear to lance-shaped (7–14 cm long; 3–6.5 cm wide), bluish-green, tips slightly curled and margins entire. *Bracts* 2 and leaf-like. *Flowers* solitary or, rarely, in groups of up to 3; bisexual; corolla bell-shaped and nodding (c. 3.5 cm long; 1.5 cm diam.); 'petals' 6, equal-size, chequered brownish-violet and yellow (purple inside) with 'shoulders' of nectaries projecting at right angles at base of 'petals'; stamens 6; ovary superior; style lobes c. 5 mm long and covered in tiny papillae. *Fruit* a dry and wingless capsule, ovoid, 6-angled and 3-valved. Fl. May–Jun, fr. Jul.

HARVESTING, SOURCING AND NATURAL RANGE

Bulbs are harvested in spring, largely from medicinal crop plants grown in NE China in the provinces of Heilongjiang and Jilin, where traditional *Di Dao*-quality drug is produced and from where GAP-certified Ping Bei Mu has been grown. The species grows wild in the adjacent province of Liaoning in forests, scrub and grassland in wet and shady places (0–500 m altitude) but has declined dramatically in recent years. It is also native in the Korean Peninsula and Russia (Far East).

CONSERVATION STATUS

China: Vulnerable (MEPCAS, 2013).

Global: Not Evaluated (IUCN Red List; accessed 10/8/2016).

Fritillaria usuriensis, cultivated, Jilin province

F. usuriensis, medicinal crop in flower, Jilin province

Immature fruits of *F. usuriensis,*. Note absence of wings.

F. usuriensis, fresh bulbs with bulbils

TCM PROPERTIES AND USE

***properties*: 'bitter, sweet and cool'**
***channels*: 'Lung and Heart'**
***primary action*: 'clears heat, tonifies Lung, resolves phlegm, alleviates cough'**

***indications*: cough**

DRUG MORPHOLOGY

CRUDE DRUG/DECOCTING PIECES (INTACT BULB): broader than tall (0.5–1 cm height; 0.6–2 cm diam.) **[1]** with 2 fleshy scale leaves almost equal in size or one slightly larger and clasping the other **[2]**. Externally creamish-white or pale yellow; surface often mottled **[3]**. Apex slightly concave and open to reveal small buds **[4]** and stem remnant **[5]** within. Basal and slightly concave with shrivelled root remnants **[6]**. *Texture*: hard and brittle. *Fracture*: starchy. *Odour*: indistinct. *Taste*: bitter. ***Good quality*:** solid, dry, white with a powdery (not friable) texture.

PROCESSING

After removal of the fibrous roots, bulbs are dried at low temperature (= crude drug/decocting pieces).

OTHER COMMON NAMES

平贝 Ping Bei; Bulbus Fritillariae Ussuriensis; Ussuri fritillary bulb (CP2010).

SAFETY

An official healthfood in China (Anon., 2002a,b).

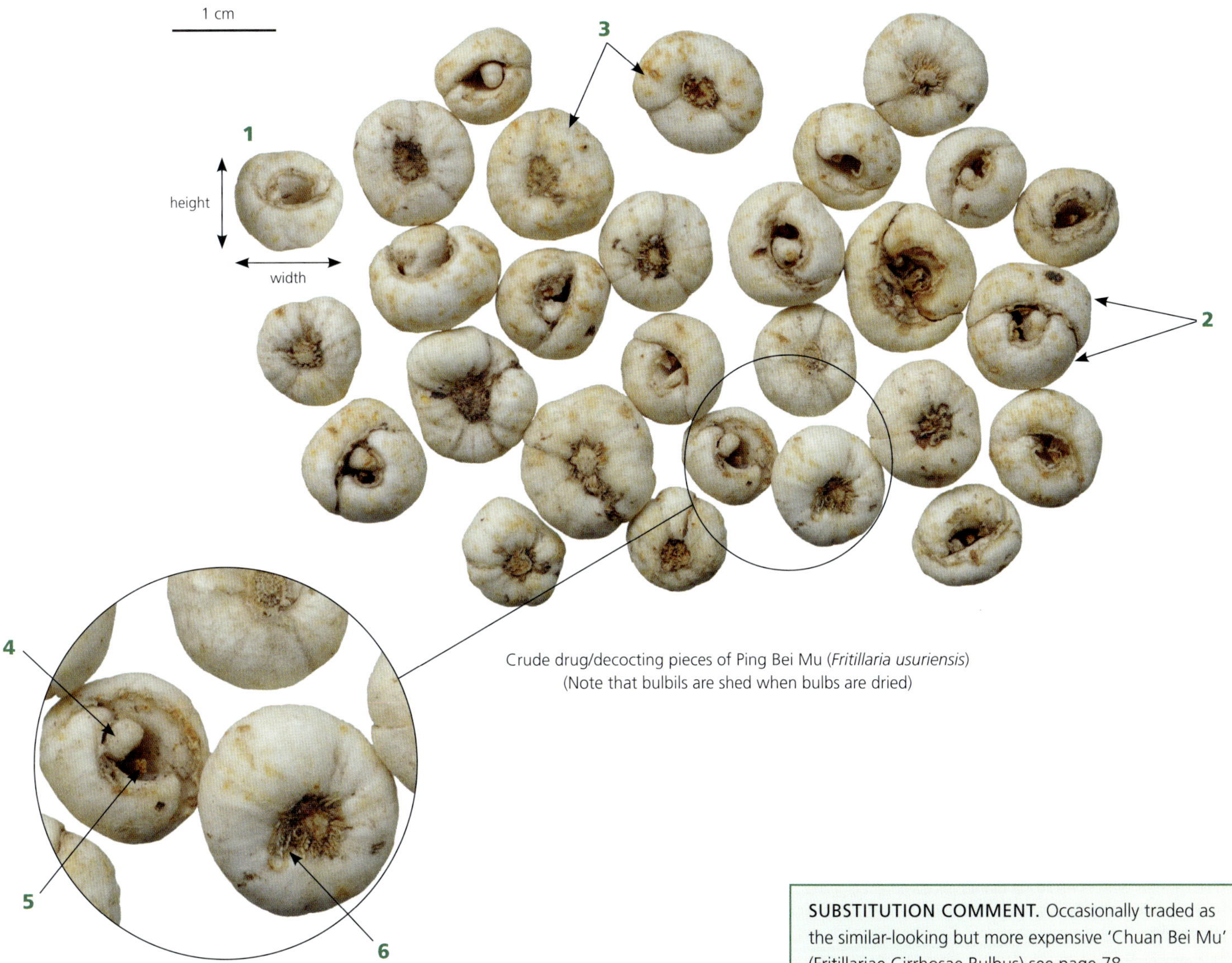

Crude drug/decocting pieces of Ping Bei Mu (*Fritillaria usuriensis*)
(Note that bulbils are shed when bulbs are dried)

SUBSTITUTION COMMENT. Occasionally traded as the similar-looking but more expensive 'Chuan Bei Mu' (Fritillariae Cirrhosae Bulbus) see page 78.

Qiang Huo 羌活 Notopterygii Rhizoma et Radix

Drug parts: RHIZOME AND ROOT

PHARMACOPOEIA SPECIES 1

Notopterygium incisum K. C. Ting ex H. T. Chang

Apiaceae

Species of *Notopterygium* wild in China: **6**

PLANT DESCRIPTION

A tall and erect, branching perennial herb (0.6–1.2 m tall) with a stout, woody and highly aromatic deep brown rootstock (= medicinal part 'Qiang Huo'). *Stems* hollow. *Leaves* alternate, rarely opposite (2–5 cm long; 0.5–2 cm wide; reduced in size up stem), **leaf stalks 5–12 cm long,** becoming wholly sheathing; **leaf blades finely divided** into lobed leaflets (3-pinnate) with outermost leaf segment oblong (2–5 cm long; 0.5–2 cm wide), margin **deeply lobed or narrowly toothed.** *Inflorescences* many, in loose, terminal and lateral compound umbels (3–13 cm diam.); main (terminal) umbel with bracts 3–6, small and undivided (1.5–3 cm long) and rays 7–20 (2–10 cm long); umbellets many-flowered and crowded with bracteoles 6–10, small and undivided. *Flowers* tiny, petals 5, **white or greenish-white**. *Fruit* in umbellate clusters; each fruit dry, oblong or ellipsoid (5–6 mm long; 2.5–3.5 mm diam.) splitting into 2 one-seeded carpels; all ribs winged and with 3 oil ducts in each furrow and **6 on the flat internal (commissural) surface of each carpel.** Fl. Jul–Aug, fr. Aug–Sep.

HARVESTING, SOURCING AND NATURAL RANGE

Rootstocks are harvested in spring or winter from wild plants which grow in forest margins and scrub (1,600–5,000 m altitude) in NC, W and NW China (Gansu, Qinghai, Shaanxi, Sichuan and Tibet). Traditional *Di Dao*-quality is sourced from Sichuan and traded as 'Chuan Qiang Huo' (Sichuan Qiang Huo).

CONSERVATION STATUS

Endemic to China: Near Threatened (MEPCAS, 2013); Vulnerable (VU A2c) (Wang & Xie, 2004); Not Evaluated (IUCN Red List, accessed 24/3/2015). Clinically, considered 'superior' to PS2 and has become threatened due to over-harvesting.

PHARMACOPOEIA SPECIES 2

N. franchetii H. Boissieu

Apiaceae

synonym: *N. forbesii* H. Boissieu

PLANT DESCRIPTION

Similar to *N. incisum* (PS1) but slightly taller (0.8–1.8 m) and rootstocks with many lateral roots. *Leaves:* **leaf stalks much shorter** (3–8 cm long); **leaf blades much less finely divided** with outermost leaf segment ovate to oblong-ovate (3–8 cm long; 1–3 cm wide); **margins entire or coarsely toothed;** bracts 1–3, small and undivided; bracteoles 4 or 5, undivided and very short (3–4 mm long); **petals pale yellow or yellowish-green**. Fruit with 3 or 4 oil ducts in each rib furrow with 4 on the commissural surfaces of each carpel. Fl. Jul–Aug, fr. Aug–Sep.

HARVESTING, SOURCING AND NATURAL RANGE

As for PS1 but a much more common species in the wild and more widely distributed (Gansu, Hubei, Inner Mongolia, Qinghai, Shaanxi, Shanxi, Sichuan and Yunnan), 1700–4800 m altitude.

CONSERVATION STATUS

Endemic to China: Least Concern (MEPCAS, 2013); Least Concern (authors' pers. comm., 2010); Not Evaluated (IUCN Red List, accessed 24/3/2015).

UNOFFICIAL SUBSTITUTES

Some local or historical substitution using the rootstocks of various Apiaceae and Rosaceae species has been reported (Bensky *et al.*, 2004; Wright, 2004). Such substitution in the international TCM trade is uncertain. Species reported:

1. *Angelica sylvestris* L. 新疆羌活 Xinjiang Qiang Huo (family Apiaceae). A common species of NW China (Xinjiang province), 900–1,100 m altitude. Roots conical, **probably thicker (3–8 cm)** than those of official 'Qiang Ho'; slightly aromatic. Plants 0.8–2 m. tall. **Umbels wider (10–20 cm diam.);** bracts absent or 1 or 2, linear; rays 15–30, softly hairy. **Petals white**. Fruit broad-ovoid (5–6 mm long; 3.5–5 mm diam.); dorsal ribs **thread-like**, 1 oil duct in each furrow, 2 on the joining faces of each carpel.
2. *Pleurospermum rivulorum* (Diels) M. Hiroe 蛇头羌活 She Tou Qiang Huo (family Apiaceae). Reported as a local substitute in NW Yunnan where it grows wild in damp grasslands at high altitude (3,100–4,100 m). Plants with **inflated sheaths of upper leaves** and when in flower, easily distinguished by their well-developed bracts and bracteoles. Flowers with conspicuous stalks; petals greenish-white. Fruit dark brown, ovoid-oblong (8–10 mm long; 4–6 mm diam.); ribs all narrowly winged; 1 oil duct in each furrow, 2 on commissure.
3. *Sanguisorba officinalis* L. (family Rosaceae). The root of this common species is the source of the separate TCM drug 地榆 'Di Yu' (Sanguisorbae Radix), see page114. Rootstock stout, brown or purplish-brown, usually fusiform, rarely cylindrical, **annulations absent, very hard**; transverse section yellowish-white or purple, not obviously fissured, concentric tissue zones indistinct, pith absent. While whole rootstocks are superficially similar to those of *N. incisum* the living plants are easily distinguished: basal leaves divided into **4–6 pairs** of stalked, ovate, oblong-ovate hairless leaflets with cordate to sub-cordate leaf bases, margins coarsely or sharply toothed. Flowers purple, red or tinged purple; stamens 4. Fruits hard, dry and with 4 longitudinal ribs (**not splitting into 2 mericarps**).

TCM PROPERTIES AND USE

properties: **'pungent, bitter and warm'**
channels: **'Bladder and Kidney'**
actions: **'releases wind cold, dispels wind damp'**

indications: **respiratory tract infections; fever; musculoskeletal pain**

Notopterygium incisum with immature fruit-heads; inset: immature fruit-head (umbellet)

N. incisum, freshly harvested rootstock Sichuan province

N. incisum in the wild, Sichuan province

N. franchetii, showing one mature leaf

Yellowish-green flower-heads of *N. franchetii* cultivated on a small-holding, Gansu province

Notopterygium incisum

Apiaceae

DRUG MORPHOLOGY

CRUDE DRUG (DRIED RHIZOME): cylindrical (4–13 cm long; 0.6–2.5 cm diam.) and slightly curved; externally brown to blackish-brown, somewhat yellowish where outer bark shed **[1]**; nodes with dotted or tuberculate root scars **[2]**; apex with aerial stem scars **[3]**. Two typical trade forms exist: **i)** rhizome internodes very short creating densely spaced annulations resembling a silkworm **[4]**, hence the name 蚕羌 Can Qiang (silkworm notopterygium); **ii)** rhizome internodes elongated **[5]** and with raised nodes **[6]**, resembling bamboo stems, hence the name 竹节羌 Zhu Jie Qiang (bamboo notopterygium). *Texture*: light and easily broken. *Fracture*: uneven **[7]**, radial fissures, cortex brown, xylem yellowish-white, pith yellowish-brown with numerous dark resin-filled cells. *Odour*: aromatic. *Taste*: slightly bitter and pungent. ***Good quality*:** Can Qiang 'silkworm notopterygium').

DECOCTING PIECES (DRIED, SLICED RHIZOME):

Transverse slices: subrounded or irregular (0.6–2.5 cm diam.), sometimes cut obliquely **[8]**. Externally brown to blackish-brown, rough and sometimes with dotted or tubercular root scars **[9]**. Cut surface usually with clearly demarcated concentric tissue rings: outer ring medium brown with radially interrupted vascular tissues **[10]** encircling a paler yellowish-white xylem ring **[11]** (both rings many-fissured); central pith yellowish-brown, dotted throughout with dark yellowish-orange resin-filled cells **[12]**. *Texture etc*: as for crude drug. ***Longitudinal slices:*** oblong rhizome slices (2–4 cm long; 0.6–2 cm diam.); uncut external surface brown to blackish-brown with densely spaced annulations **[4]** ('Can Qiang') or elongated longitudinally with furrowed internodes and raised bamboo-like nodes ('Zhu Jie Qiang') or without nodes ('Tiao Qiang'). Cut surface with two longitudinal yellowish-brown xylem bands **[11]** either side of central pith dotted throughout with yellowish-orange resin-filled cells **[12]**. *Texture etc*: as for crude drug.

PROCESSING

After removal of aerial parts and roots, rhizomes are washed and sun-dried (= crude drug) then transversely sliced into decocting pieces 2–4 mm thick. Longitudinal slices are often prepared for the Hong Kong market.

OTHER COMMON NAMES

Incised notopteryigum rhizome (CP2010); notopterygium (SCN); Rhizoma et Radix Notopterygii.

SAFETY

Safety Class: 1; Interaction Class: A (Gardner & McGuffin, 2013).

N. franchetii

Apiaceae

DRUG MORPHOLOGY

CRUDE DRUG (DRIED ROOT AND RHIZOME): externally greyish-brown. Two trade forms exist: **i)** 条羌 *Tiao Qiang (tail Qiang)* **[13]***:* rhizomes sub-cylindrical with persistent stem and leaf sheath bases; roots (8–15 cm long; 1–3 cm diam.) somewhat conical, longitudinally wrinkled with annulations especially dense near rhizome; ii) 大头羌 *Da Tou Qiang (large-headed Qiang)* **[14]***:* rhizomes large and stocky, irregularly nodular, bearing many stem bases; roots somewhat thin. *Texture*: lax and easily broken. *Fracture*: somewhat even. *Odour and taste*: slight.

DECOCTING PIECES (DRIED, SLICED ROOT AND RHIZOME):

Transverse slices: a **mixture of sliced roots and rhizomes and therefore of widely varying diameter**. Externally greyish-brown and rough **[15]**. Wide slices with central pith are rhizomes **[16]**; narrow slices without central pith are roots **[17]**. Cut surfaces usually with brown tissues of many shades irregularly demarcated (often with central xylem zone pale yellowish-cream), many-fissured and dotted throughout with yellowish-orange resin-filled cells. *Texture*: lax and easily broken. *Odour and taste*: slight.

PROCESSING

As for PS1 but roots are not removed.

OTHER COMMON NAMES

宽叶羌活 Kuan Ye Qiang Huo (broad-leaved Qiang Huo); notopteryium (SCN); Rhizoma et Radix Notopterygii.

SAFETY

Not known.

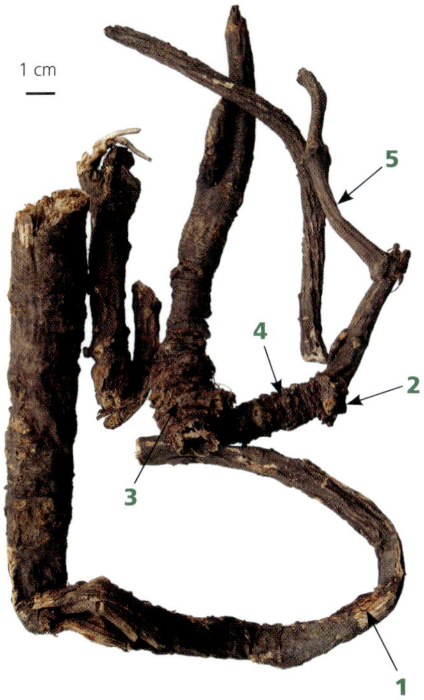

Crude drug (sourced from *N. incisum*)

Can Qiang

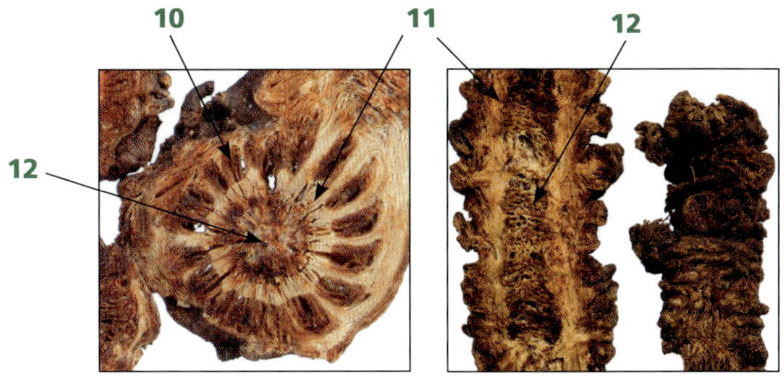

transverse slices

longitudinal slices

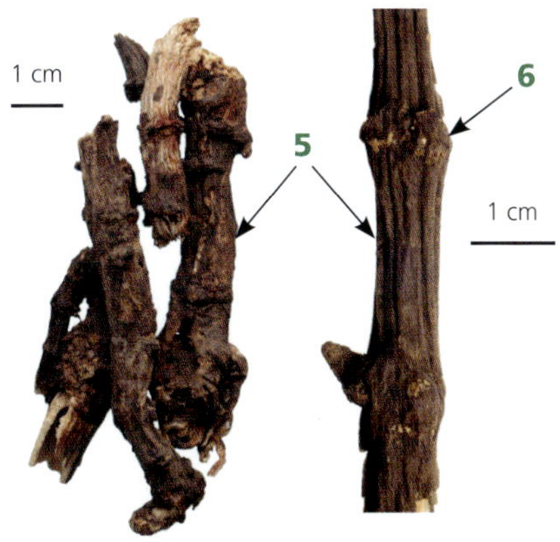

Zhu Jie Qiang

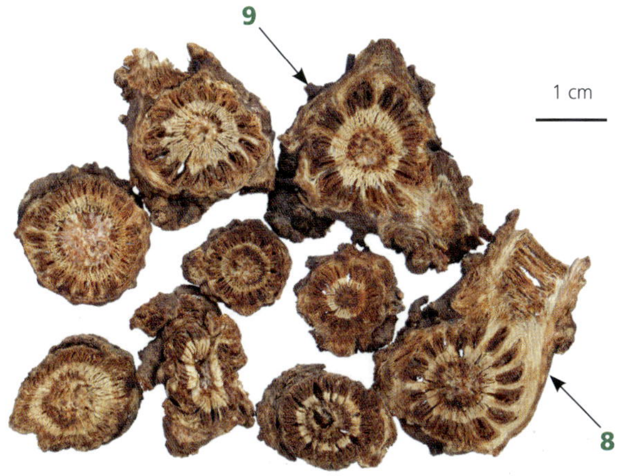

Decocting pieces (sourced from *N. incisum*)

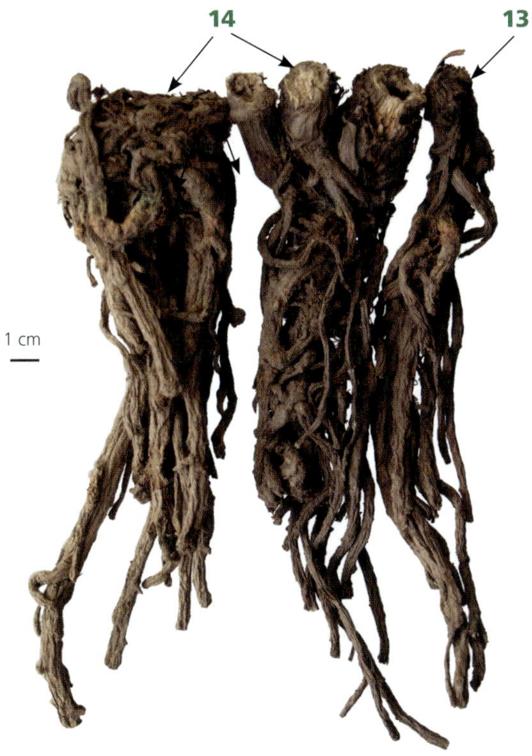

Crude drug (sourced from *N. franchetii*)

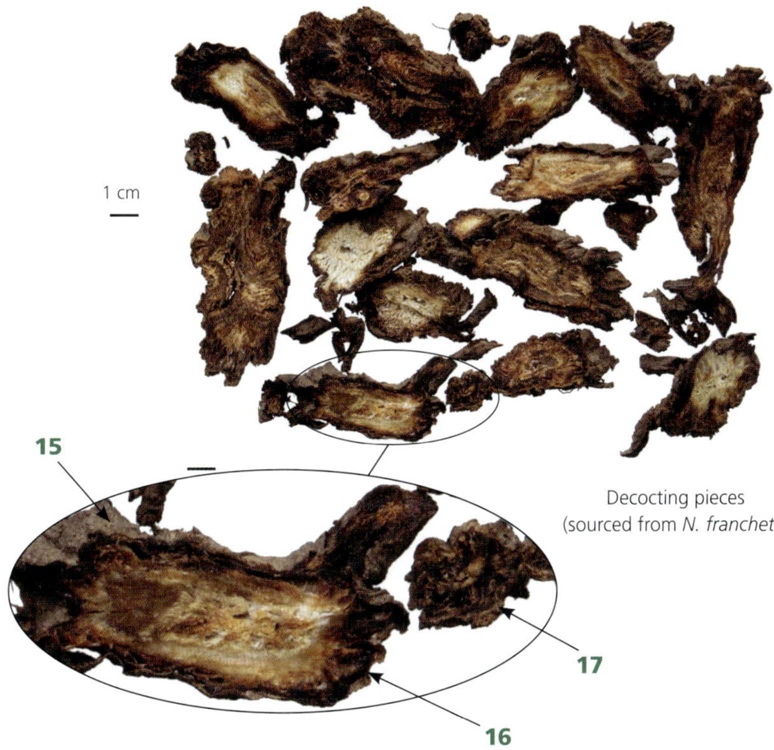

Decocting pieces (sourced from *N. franchetii*)

Qian Hu 前胡

Peucedani Radix

Drug part: ROOT

Peucedanum praeruptorum Dunn

Apiaceae

Species of *Peucedanum* wild in China: 40

PLANT DESCRIPTION

A perennial herb (10–100 cm tall) with an abruptly tapering taproot (c. 1 cm diam.) (= medicinal part 'Qian Hu'); crown of taproot densely clothed in fibrous leaf base remains. *Stem* single, slender and thinly ribbed; branching in upper parts. *Leaves* alternate and stalked; basal leaves with sheaths; leaf blades triangular-ovate in outline (1.5–6 cm long; 1.2–4 cm wide), 2–3-pinnate (usually in 3s); leaflets narrowly ovate (1.5–6 cm long; 1.2–4 cm wide) and distinctly stalked; margins irregularly saw-toothed and tips narrowly pointed. *Inflorescences* in terminal or lateral compound umbels; main umbels 3.5–9 cm diam. with bracts absent or few and undivided and rays 6–15, unequal (0.5–4.5 cm); umbellets 15–20-flowered with **bracteoles 8–12**, ovate to lance-shaped, rough and shorter than flower stalks. *Flowers* tiny, calyx absent, petals 5, **white,** curving inwards; styles 2 and short. *Fruits* in stalked umbellate clusters. Each fruit, dry, dorsally compressed and ovoid (4 mm long; 3 mm diam.); sparsely hairy, splitting into 2 one-seeded carpels (mericarps) with slender dorsal ribs and thickened and narrowly winged lateral ribs; 3–5 oil ducts in each furrow, 6–10 on the flat internal (commissural) surface. Fl. Aug–Sep; fr. Oct–Nov.

HARVESTING, SOURCING AND NATURAL RANGE

Predominantly wild harvested in China but medicinal crop production in Anhui province has become established over the last 5 years. Rhizomes are harvested during winter and spring after aerial parts of the plant have withered and before new growth has begun.

The species grows wild in woodland edges and grassland (200–2,000 m altitude) and is endemic to C, SE, SW and W China (Anhui, Fujian, Gansu, Guangxi, Guizhou, Henan, Hubei, Hunan, Jiangsu, Jiangxi, Sichuan and Zhejiang provinces).

CONSERVATION STATUS

Endemic to China: Least Concern (MEPCAS, 2013); locally common (LYL, pers. comm.) Not Evaluated (IUCN Red List, accessed 22/4/2015).

Flowering plant of *Peucedanum praeruptorum*, Anhui province

P. praeruptorum

P. praeruptorum with immature fruits

UNOFFICIAL SUBSTITUTES

Many incorrectly identified root materials are likely to be traded under the name 'Qiang Hu' since the genus *Peucedanum* has long been problematic to identify even by botanists (40 species grow wild in China); their rootstocks present an even greater challenge. With the bulk of 'Qian Hu' wild harvested, mis-identification of the source plant gives rise to many substitutes in trade. Reliable identification of commercial traded materials is best performed using laboratory based methods. Species most frequently reported or substitutes (Bensky *et al.,* 2004; Xiao *et al.*, 2001–2; Wright, 2004) include:

1. ***Angelica decursiva*** (Miq.) Franch. & Sav. (syn. *Peucedanum decursivum* (Miq.) Maxim., name used in CP2015, and whose roots are the source of the separate TCM drug 紫花前胡 'Zi Hua Qian Hu' (Peucedani Decursivi Radix) (purple-flowered peucedanum)). A common species with a root very similar to the official drug but **annular rings fewer and more angular [11]**; strongly aromatic. Stem taller (1–2 m) often purplish-green; **leaf sheaths purple; bracts 1–3, purplish**, ovate, sheath-like, reflexed; **rays 10–22; bracteoles 3–8**; petals and anthers dark purple or white; **fruit with 1–3 oil ducts in each furrow** and **4–6** on internal (commissural) surface. *Note:* a white-flowered form exists in north-east China, often called *A. decursiva* f. *albiflora* (Maxim.) Nakai.
2. ***Peucedanum medicum*** Dunn 华中前胡 Hua Zhong Qian Hu. A common species used as a regional substitute in China (Hebei Province) (Wright, 2004). A stout species 50–200 cm tall. Taproot rough; rootstock stout, greyish-brown, often purplish-tinged, **annular leaf scars many and conspicuous. Leaf blade 2–3-ternate,** rarely 2-pinnate; pinnae 3 pairs, upper surface shiny; umbels **mostly wider** (7–15 (–20) cm across); bracts absent or 1, deciduous; **rays 15–30; umbellets 10–30-flowered**; petals white; fruit ovoid-ellipsoid, **slightly larger** (6–7 mm long; 3–4 mm diam.), **3 oil ducts in each furrow, 8–10** on internal (commissure) surface.

TCM PROPERTIES AND USE

***properties*: 'pungent, bitter and cool'**
***channels*: 'Lung'**
***actions*: 'clears heat, resolves phlegm'**

***indications*: cough, phlegm; headache**

DRUG MORPHOLOGY

CRUDE DRUG (DRIED ROOT): irregularly cylindrical, conical or spindle-shaped (3–15 cm long; 1–2 cm diam.), often strongly curved **[1]** and branched and somewhat twisted. Outer surface greyish-yellow or greyish-brown usually with a characteristic zone of fibrous leaf bases **[2]**, stem scars and a zone of narrowly spaced annular rings **[3]** at the crown; the latter is colloquially known as 蚯蚓頭 Qiu Yin Tou (earthworm head); the whole surface of the central and lower root body is longitudinally wrinkled **[4]** with indistinct, raised transverse lenticels **[5]**. *Texture*: somewhat flexible, easily broken but becoming hard and brittle with age. *Fracture*: **markedly uneven**, predominantly pale yellowish-white **[6]** with a dark cambium ring and rays sometimes visible. *Taste*: slightly bitter with an acrid aftertaste. *Odour*: aromatic. ***Good quality:*** strong, thick and pliable roots; internally yellowish-white with abundant oil cells; intensely aromatic.

DECOCTING PIECES-1 (SLICED ROOT): ovate or irregularly-shaped slices (1–3 cm diam.; 1–2 mm thick). Bark greyish-yellow or greyish-brown **[7]** sometimes with indistinct, raised transverse lenticels; fibrous leaf bases **[8]** and annular rings occasionally visible. Cut surface pale yellow to yellowish-white with a darker cambium ring **[9]**, rays sometimes visible; dark, dot-like oil cells mostly scattered from the cambium ring outwards **[10]**. *Texture*: somewhat flexible, easily broken but becoming hard and brittle with age. *Taste*: slightly bitter with an acrid aftertaste. *Odour*: aromatic.

DECOCTING PIECES-2 (HONEYED ROOT — MI QIAN HU): as for sliced root but externally yellowish-brown, slightly moist and lustrous.

PROCESSING

After removal of rootlets, roots are sun-dried at low temperature (= crude drug); decocting pieces are made by subsequent softening, slicing thinly (1–2 mm) and sun-drying again. Further processing by steeping the slices in honeyed water then dry-frying until no longer sticky to the touch produces 蜜前胡 Mi Qian Hu (honeyed Qian Hu).

OTHER COMMON NAMES

白花前胡 Bai Hua Qian Hu 'white-flowered Peucedanum'); 前胡属 Qian Hu Shu (plant name); hogfennel root (CP2010); peucedanum (SCN); Radix Peucedani.

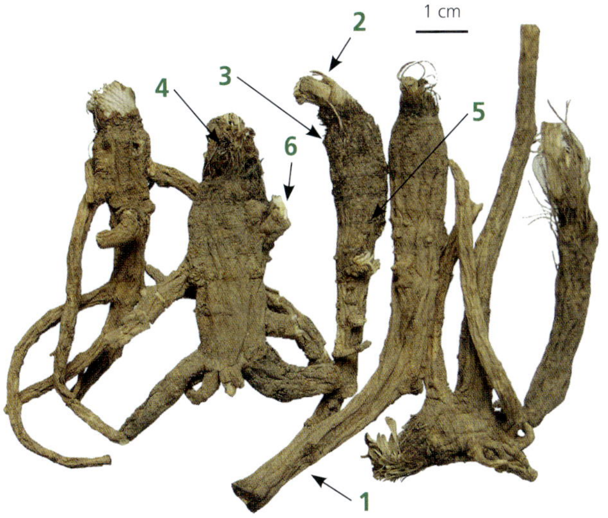

Crude drug of Qian Hu (whole dried roots of *Peucedanum praeruptorum)*

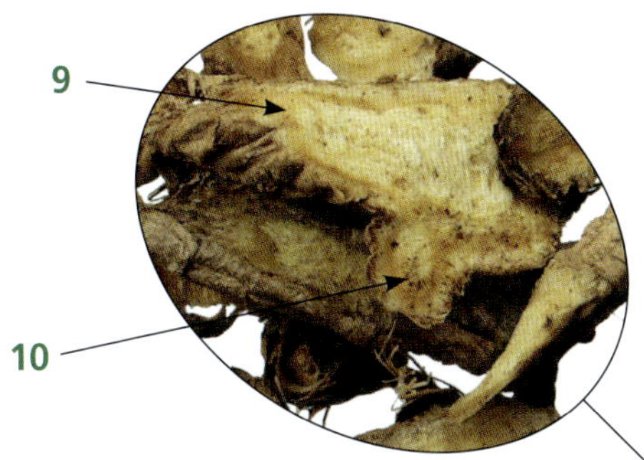

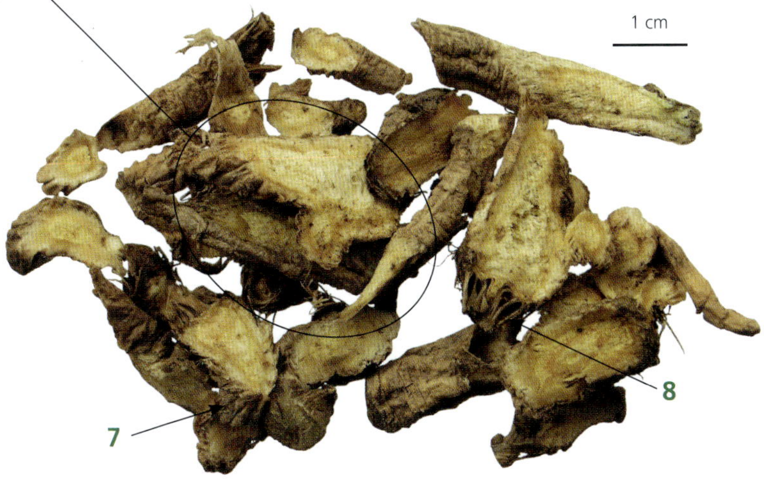

Decocting slices of Qian Hu (dried and sliced roots of *P. praeruptorum*)

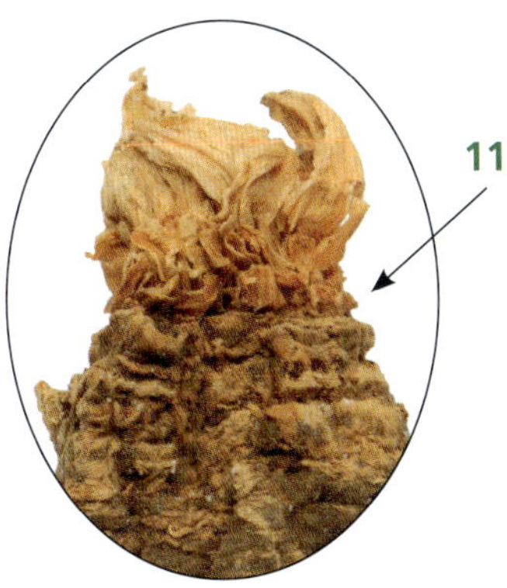

Root crown

Dried roots of the substitute *Angelica decursiva*

Ren Shen 人参

Ginseng Radix et Rhizoma

Asian ginseng

Drug parts: ROOT AND RHIZOME

Panax ginseng C. A. Mey. **Araliaceae**

Species of *Panax* wild in China: 6 (plus 1 introduced)

PLANT DESCRIPTION

An herbaceous perennial (30–60 cm tall). *Root* stout, fleshy, somewhat cylindrical or tapering and growing horizontally. *Stems* erect and rarely branched. *Leaves* at top of stem in one whorl of 3–6 (increasing in number in older plants); each leaf stalked, with 5 (rarely 3 or 7) stalked leaflets radiating from a single point (palmately-compound); leaflet blades oblong to widest above the middle (central leaflet 8–12 cm long; 3–5 cm wide; lateral leaflets 2–4 cm long; 1.5–3 cm wide), membranous, leaflet bases broadly wedge-shaped, **margins minutely saw-toothed,** tips narrowly pointed, **veins minutely bristly** on underside. ***Inflorescence*** a solitary, hemispherical **30–50-flowered** terminal head on an erect, 15–30 cm long stalk (peduncle), **peduncle longer than the leaf stalks**. *Each flower* tiny (2–3 mm diam.), bisexual and radially symmetrical; calyx green and cup-shaped, 5-lobed at mouth; petals 5, spreading, greenish or white with pointed tips; stamens 5, inserted onto a disk; ovary inferior; styles usually 2. *Fruit* (*drupe*) sub-globose (4–5 mm long; 6–7 mm wide), fleshy and, when mature, bright scarlet. *Seeds* white, kidney-shaped and laterally compressed.

UNOFFICIAL SUBSTITUTES AND ADULTERATION

This is an expensive drug and the roots of many cheaper species have been reported as unofficial substitutes (see 'Xi Yang Shen' (Panacis Quinquefolii Radix) page 310 and the examples below) (NICBP 1997–2002; Bensky *et al*., 2004; Court, 2000; Xiao *et al*., 2001–2002). Nevertheless, the substitute species listed below are unlikely to be found in international trade and problems are more likely to arise locally in China. The authenticity of highly prized wild ginseng is more likely to be compromised by mechanical adulteration; for example, extra stem scars can be inserted into the rhizome 'neck' to suggest that the material is of a greater age and therefore more valuable. Considerable expertise is needed to spot such ingenious methods of adulteration. Powdered ginseng is especially prone to adulteration, requiring chemical methods for authentication (Kite *et al*., 2003.

1. ***Adenophora*** spp. (family Campanulaceae) — the roots of some species are the source of the separate drug 'Nan Sha Shen' (Adenophorae Radix), see page 222.
2. ***Lactuca indica*** L. (family Asteraceae).
3. ***Hyoscyamus*** (family Solanaceae) — some species are **toxic**.
4. ***Mirabilis jalapa*** L. (family Nyctaginaceae).
5. ***Physochlaina infundibularis*** Kuang (family Solanaceae) — the roots are the source of the separate TCM drug 'Hua Shan Shen' (Physochlainae Radix); **toxic** (CP2015.)
6. ***Phytolacca acinosa*** Roxb. and ***P. americana*** L. (family Phytolaccaceae) — the roots are the source of the separate drug 'Shang Lu' (Phytolaccae Radix); **toxic** (CP2015).
7. ***Platycodon grandiflorus*** (Jacq.) A. DC. (family Campanulaceae) — the roots are the source of the separate drug 'Jie Geng' (Platycodonis Radix), see page 184.
8. ***Talinum paniculatum*** (Jacq.) Gaertn. (family Talinaceae).
9. ***Vigna vexillata*** (L.) A. Rich. (family Fabaceae).

HARVESTING, SOURCING AND NATURAL RANGE

The bulk of production is from cultivated plants that are grown extensively in NE China (Heilongjiang, Jilin and Liaoning provinces) and on the Korean peninsula. Rootstocks, usually of 5–7-year-old plants, called Yuan Shen (园参) 'field-cultivated ginseng', are harvested in the autumn. Traditional *Di Dao* material is sourced from the provinces of Jilin (Changbai, Fusong, Ji'an and Jingyu districts), Liaoning (Huanren, Kuandian, Qingyua and Xinbin districts) and Heilongjiang (Ningan and Yilan districts). To meet increasing demand in recent years, ginseng cultivation in China has extended south-west into Hebei and Shanxi provinces.

The species is native in China but over-harvesting of wild populations during the past century has caused their acute decline such that only relict, and now strictly protected, populations survive in mixed and deciduous broad-leaved forests in remote parts of NE China (Heilongjiang, Jilin and Liaoning), the Korean Peninsula and E Russia. Continued high demand for the reputedly 'superior' and highly prized wild-sourced material has, in part, been satisfied by 'semi-wild' ginseng (林下山参 Lin Xia Shan Shen or 子海 Zi Hai). The latter comes from cultivated plants that are transplanted into their natural habitat and tended under a strict management regime so as to reproduce some of the quality attributes of totally wild plants. The root-growth of semi-wild ginseng is much slower than that of field-cultivated plants, but its medicinal quality is reputedly much higher, as are the prices it fetches (Zhuravlev *et al*., 2008).

CONSERVATION STATUS

China: Critically Endangered (MEPCAS, 2013).
Korean Peninsula: Extinct (Zhuravlev *et al*., 2008).
Russia: 'threatened in Russia since 1975' (Zhuravlev *et al*., 2008).
Global: Not Evaluated (IUCN Red List, accessed 10/11/2016).

PROTECTED STATUS

CITES: Appendix II; EU Annex B (see Table 1 page 10).

TCM PROPERTIES AND USE

properties: **'sweet, slightly bitter and slightly warm'**
channels: **'Spleen, Lung, Heart and Kidney'**
actions: **'tonifies yuan qi, tonifies Spleen and Lung, generates fluids, calms the shen'**

indications: **fatigue after chronic illness; cough and dyspnoea; poor cognitive function; palpitations, insomnia; impotence and infertility**

Shaded crop of *Panax ginseng*

P. ginseng crop with developing flower-heads. Leaf spots are residues of recent pesticide spraying

Leaflet margins of *P. ginseng* minutely saw-toothed

P. ginseng with fruit-head

Freshly uprooted 5-year old plant of *P. ginseng*

Panax ginseng

Araliaceae

DRUG MORPHOLOGY

CRUDE DRUG (DRIED ROOT AND RHIZOME): *mostly of root* **[1]** *with rhizome* **[2]** *attached apically.*

Field-cultivated ginseng (Yuan Shen): *roots* — cylindrical or tapering, often slightly curved (3–15 (–25) cm long; 1–2 cm diam.) with basal part dividing into 2–3 long, often interwoven lateral roots **[3]** with many very slender adventitious roots ('Ding') **[4]**; lateral roots often forked and, when only 2, create the classic 'man-shaped' root. Externally greyish-yellow with main root longitudinally wrinkled **[5]** and with scattered shallow transverse, oval and slightly raised lenticels **[6]**. *Texture*: relatively hard. *Fracture* (see *Decocting Pieces* below). *Odour*: slight but distinctive. *Taste*: slightly bitter, becoming sweet. ***rhizomes*** **[2]** (芦头 Lu Tou) — still attached apically to root, short and narrow (1–4 cm long; 0.3–1.5 cm diam.), with nodular or sunken circular stem bases or scars ('Lu Wan') **[7]**. The number of stem scars indicate the plant's age (4 stem scars indicating a 5-year old root, 5 a 6-year old root etc.), roots less than 5 years old are considered chemically inferior, a view supported by analytical evidence (Choi *et al.*, 2007). ***Good quality*:** thick, long and unbroken with a thin cortex; quality varies enormously according to cultivation and processing methods.

Semi-wild ginseng (Lin Xia Shan Shen, Lin Xia Shen or Zi Hai) as for 'field-cultivated ginseng' but length of main root usually much shorter (1–6 cm long) and about the same length or shorter than the rhizome; cylindrical, rhomboid or V-shaped.

Wild ginseng (Ye Shan Shen): rarely seen in trade owing to its high price and protected status. As for field-and semi-wild ginseng but length of root plus rhizome usually much longer **[8]**, main root body often contorted and rootlets with tiny, wart-like tubercles **[9]**. Rhizome **[10]** more obviously linear with stem scars more distinctive **[11]** and varying in number. Because of the rootstock's slower growth compared to field-cultivated ginseng, a distinct zone of densely packed annular rings **[12]** clearly visibly immediately below the rhizome.

DECOCTING PIECES (FIELD- AND SEMI-WILD GINSENG): *transversely sliced roots:* sub-rounded or elliptical (1–2 cm diam.; 1–2 mm thick) sometimes with an undulating outline **[13]**. Outer bark thin, greyish-yellow **[14]**, furrows and lenticels occasionally visible. Cut surface starchy, cortex (outermost ring) white **[15]** with 1 or more concentric and interrupted rings of yellowish-brown or orange resin ducts clearly visible **[16]** and occasional radial clefts; cambium ring brownish-yellow **[17]** with occasional faint rays; central wood **[18]** (xylem) occupying c. $^2/_3$ of root diameter, yellowish-white, rays indistinct. *Texture:* starchy. *Odour*: slight but distinctive. *Taste:* slightly bitter, becoming sweet.

PROCESSING

Rootstocks are sun- or oven-dried and usually traded as whole roots (= crude drug). A wide range of traditional processing methods exist and there exists a complex Chinese nomenclature to reflect the resultant trade forms; some of the most commonly traded forms are illustrated opposite. Of these only steamed ginseng (Hong Shen) is described in the CP2015 (see separate entry for 'Hong Shen', page 244). Decocting pieces (thinly sliced, pulverised or broken) are usually prepared immediately prior to use.

COMMON NAMES

A huge complex of trade names exist for *P. ginseng* roots reflecting its many sources and processing types; those in widespread use include:

白人参 Bai Ren Shen (white ginseng); 朝鲜人参 Chao Xian Ren Shen 'Korean ginseng'; 吉林参 Ji Lin Shen (ginseng from Jilin province — reputedly top quality); 林下参 Lin Xia Shan Shen (semi-wild ginseng); 山参 Shan Shen (semi-wild ginseng); 鲜参 Xian Shen 'fresh ginseng'; 中国人参 Zhong Guo Ren Shen 'Chinese ginseng'; 子海 Zi Hai (semi-wild ginseng); 糖人参 Tang Ren Shen 'sugared ginseng'; 园参 Yuan Shen (cultivated ginseng); Asian ginseng (SCN); Chinese ginseng; Korean ginseng; Radix et Rhizoma Ginseng.

SAFETY

An official food and healthfood in China (Anon., 2002a,b). Safety Class: 1; Interaction Class: A (Gardner & McGuffin, 2013).

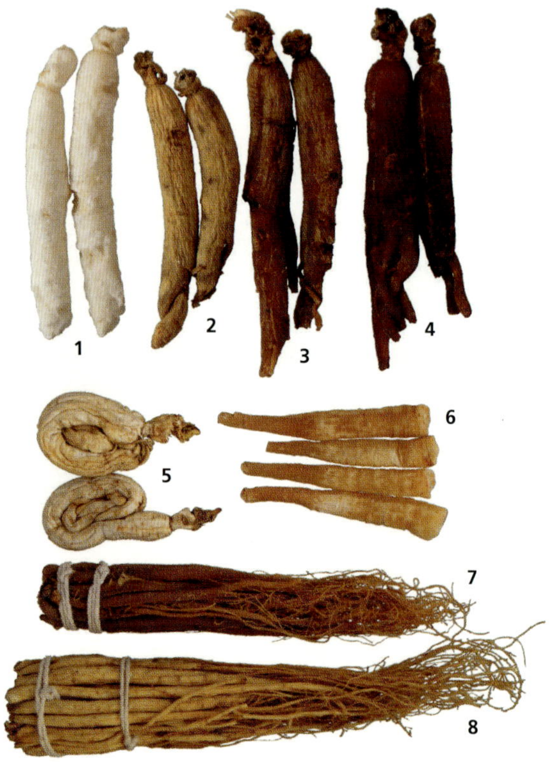

RIGHT: Typical range of traditional forms of processed *P. ginseng* roots: **1** Tang Shen (sugared); **2** Da Li Shen; **3** Hong Shen 'red' (steamed); **4** Gao Li Shen (Korean); **5** Qu Wei; **6** Chong Ding; **7** Hong Zhi Shen 'red straight lateral roots'; **8** Bai Zhi Shen 'white straight lateral roots'.

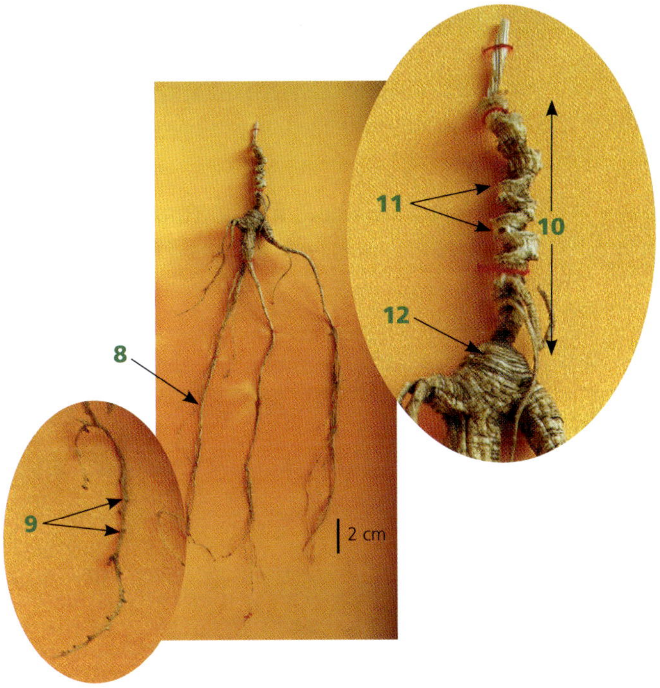

Crude drug (wild ginseng *P. ginseng* Ye Shan Shen)

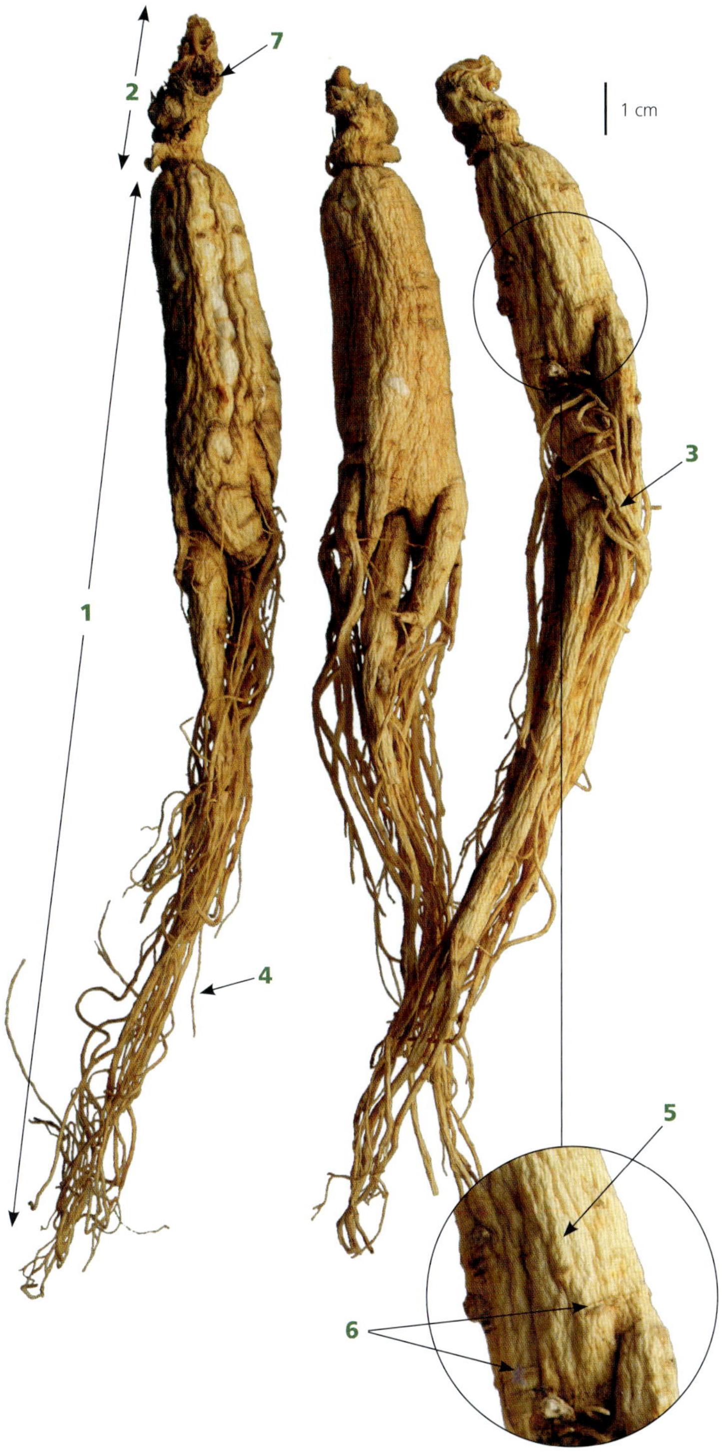

Crude drug (field-cultivated *P. ginseng* Yuan Shen)

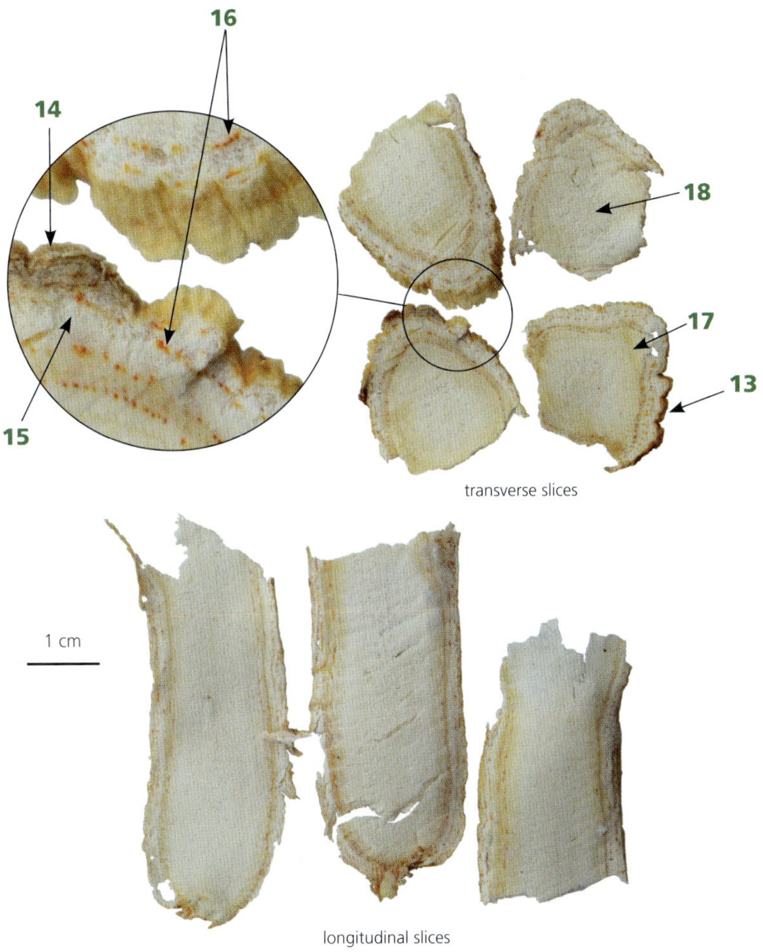

Decocting pieces (field-cultivated *P. ginseng* Yuan Shen)

Ren Shen – Hong Shen 红参 Ginseng Radix et Rhizoma Rubra

steamed Asian ginseng

Drug parts: ROOT AND RHIZOME

Panax ginseng C. A. Mey. **Araliaceae**

PLANT DESCRIPTION

see 'Ren Shen' (Ginseng Radix et Rhizoma) page 240.

HARVESTING, SOURCING AND NATURAL RANGE

see 'Ren Shen' (Ginseng Radix et Rhizoma) page 240.

CONSERVATION STATUS

see 'Ren Shen' (Ginseng Radix et Rhizoma) page 240.

Panax ginseng crop with developing flower-heads

P. ginseng with ripe fruit-head

Typical cultivation of *P. ginseng* on north-facing slopes (under black plastic shading), Jilin province, NE China

TCM PROPERTIES AND USE

***properties*: 'sweet, slightly bitter and warm'**
***channels*: 'Spleen, Lung, Heart and Kidney'**
***actions*: 'tonifies qi, Blood, yin, calms the shen'**
***indications*: cold limbs, faint pulse, metrorrhagia**

DRUG MORPHOLOGY

PROCESSED DRUG (STEAMED ROOT): trade forms vary in different regions of China but the following describes forms most commonly sold. Roots cylindrical, fusiform or columnar (3–10 cm long; 1–2 cm diam.). *Root:* externally reddish-brown and translucent **[1]** with fine longitudinal wrinkles, occasional dark yellowish-brown irregularly shaped spots **[2]**, scattered rootlet scars or rootlet remnants **[3]**. Upper parts sometimes with interrupted, rather indistinct annulations **[4]**; lower parts often with 2–3 intertwined lateral roots **[5]**. *Rhizome*: attached to root apically (1–2 cm long) with several sunken stem scars ('Lu Wan') **[6]** and occasional adventitious roots ('Ding'). *Texture*: hard and brittle. *Fracture*: even and horny. *Odour*: delicate but distinctive. *Taste*: sweet and slightly bitter. ***Good quality***: *thick, long, reddish-brown and translucent.*

DECOCTING PIECES: ovoid or rhomboidal slices of various sizes (1–2 mm thick) **[7]**; reddish-brown or brown and translucent. *Texture*: hard and brittle. *Odour*: fragrant. *Taste*: sweet and slightly bitter."

PROCESSING

The source material is 'Ren Shen' (Ginseng Radix et Rhizoma), page 240. This is steamed then dried. Decocting pieces are made by thorough softening then slicing into 1–2 mm thick slices, drying and breaking into pieces immediately prior to use.

OTHER COMMON NAMES

边条红参 Bian Tiao Hong Shen; 红直参 Hong Zhi Shen; Radix et Rhizoma Ginseng Rubra.

SAFETY

As for 'Ren Shen' page 240.

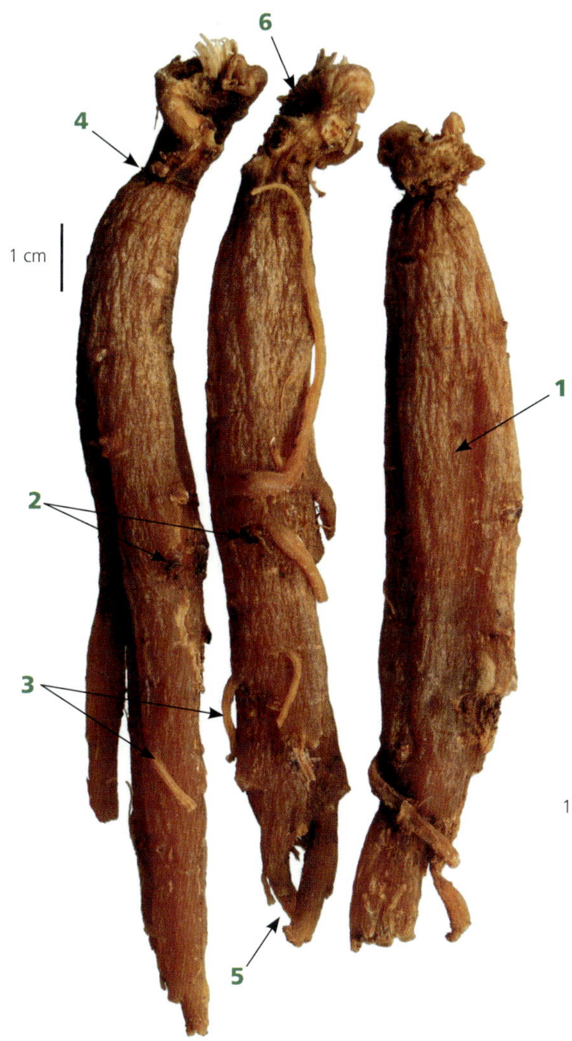

Hong Shen with small lateral roots removed

Hong Shen — an alternative trade form of lateral roots only

Hong Shen decocting pieces — flattened rhomboidal trade form

San Leng 三棱

Sparganii Rhizoma

Drug part: RHIZOME

Sparganium stoloniferum (Buch.-Ham. ex Graebn.) Buch.-Ham. ex Juz.

Typhaceae

synonyms: *S. ramosum* Huds. subsp. *stoloniferum* Buch.-Ham. ex Graebn.
taxonomic/nomenclatural note: in CP2015 as *'Sparganium stoloniferum* Buch.-Ham.'.

Species of *Sparganium* wild in China: 11

PLANT DESCRIPTION

A robust, perennial herb (70–120 cm tall or more) with creeping underwater rhizomes. *Leaves* all basal, arising directly from rhizome; leaf blades simple, long and strap-like (40–90 cm long; 0.7–1.6 cm wide) with lower surface strongly keeled; upper length of leaf flattened and margin entire. *Inflorescences* towards top of stems along 3–7 lateral branches with the minute single-sexed flowers densely crowded into globose heads; each lateral branch with 1 or 2 female flower-heads below a series of 7–11 (rarely just 1) male flower-heads. Sepals and petals modified into 3–6 paper-like scales. Male flowers white with 3 or 4 stamens. Female flowers with 3–4 mm long stigmas. *Fruit-heads* globose with many densely clustered, ribbed and conical nutlets (6–9 mm long). Fl. and fr. May–Oct.

HARVESTING, SOURCING AND NATURAL RANGE

Rhizomes are harvested from winter to spring mostly from cultivated crop plants sourced predominantly from Anhui, Jiangxi and Zhejiang provinces. The wild species grows in lakes, streams and bogs (up to 3,600 m altitude) in many provinces of China: ?Anhui, Gansu, Hebei, Heilongjiang, ?Henan, Hubei, Inner Mongolia, Jiangsu, Jiangxi, Jilin, Liaoning, Shaanxi, ?Shandong, Tibet, Xinjiang, Yunnan and ?Zhejiang [also native to Afghanistan, C Asia, Japan, Korean Peninsula and Russia].

CONSERVATION STATUS

China: Least Concern (MEPCAS, 2013).
Global: Not Evaluated (IUCN Red List, 20/8/2016).

Medicinal crop of *S. stoloniferum* ready for harvesting, Zhejiang province

Fruiting crop of *S. stoloniferum*

S. stoloniferum with male flowers and fruit

S. stoloniferum with developing fruit-heads

UNOFFICIAL SUBSTITUTES

Underground parts of other species have occasionally been traded as official San Leng 三棱. Their occurrence in international trade is uncertain.

1. Root tuber of ***Bolboschoenus yagara*** (Ohwi) Y. C. Yang & M. Zhan (synonyms: *Scirpus yagara* Ohwi; *S. fluviatilis* (Torrey) A. Gray var. *yagara* (Ohwi) T. Koyama) — family Cyperaceae). A similar-looking species to the source of 'San Leng' and growing in a similar habitat (lakes and streams). Its rhizomes are the source of a separate medicinal herb often traded on the Hong Kong market. Its similar names (Jing San Leng (荆三棱), Pao San Leng (炮三棱) or San Leng (三棱)) create occasional confusion in trade. Rhizomes slightly smaller (2–3 cm long; 3–4 cm wide) than those of *S. stoloniferum*; globose or depressed ovoid, externally blackish to purplish-brown with sparse, dented root scars; intact tubers very light and float in water (unlike those of *S. stoloniferum* which are much heavier and usually sink).
2. Rhizome of ***Sparganium subglosum*** Morong (Xi Ye Hei San Leng 细叶黑三棱; Xia Ye Hei San Leng 狭叶黑三棱); syn.: *S. stenophyllum* Maxim. ex Meinshausen. Wild in NE China.
3. Rhizome of ***S. emersum*** Rehmann (as the illegitimate name *'S. simplex'* Hudson in Bensky *et al.*, 2004); Xiao Hei San Leng 小黑三棱. Wild in NC, NE and NW China.

TCM PROPERTIES AND USE

***properties*: 'pungent, bitter and neutral'**
***channels*: 'Liver and Spleen'**
***actions*: 'breaks Blood, dispels stasis, invigorates qi, relieves pain'**

***indications*: abdominal pain, distension, masses; dysmenorrhoea, amenorrhoea**

DRUG MORPHOLOGY

CRUDE DRUG (DRIED RHIZOME): somewhat conical to polyhedral **[1]** (2–6 cm long; 2–4 cm diam.) with flat surfaces **[2]**, the result of removal by knife of the rhizome skin. Externally predominantly whitish- or greyish-yellow on cut faces; remnants of greyish-brown rhizome skin persistent in places **[3]** along with dense bands of fibrous roots **[4]** or protuberant root scars **[5]** and, more rarely, leaf bases. Internally yellowish-white and solid. *Texture*: hard, heavy and compact. *Odour*: indistinct. *Taste*: mild; slightly numbing when chewed. ***Good quality***: heavy, solid, and yellowish-white.

DECOCTING PIECES-1 (SLICED RHIZOME): ovate, circular or irregularly shaped slices (1–2 mm thick). Cut surfaces greyish-white or yellowish-white (**[6]**; often with a granular appearance and minutely dotted throughout with orange glands [hl] or rough with tissues somewhat lamellar **[7]**. Fibrous roots or root scars often visible on the unpeeled outer surface **[8]**. *Texture*: hard, heavy and compact. *Odour*: indistinct. *Taste*: mild; slightly numbing when chewed.

DECOCTING PIECES-2 (SLICED AND VINEGARED RHIZOME): as above but cut surfaces yellow or yellowish-brown **[9]** with occasional burnt areas **[10]**. *Odour*: slightly of vinegar.

PROCESSING

Rhizomes are washed, peeled and sun-dried (= crude drug); they are subsequently cut into 1–2 mm thick slices (decocting pieces-1) and may be further processed with vinegar then fried to produce Cu San Leng (醋三棱) (decocting pieces-2).

OTHER COMMON NAMES

京三棱 Jing San Leng; common bur-reed tuber (CP2010); Rhizoma Sparganii (CP2010).

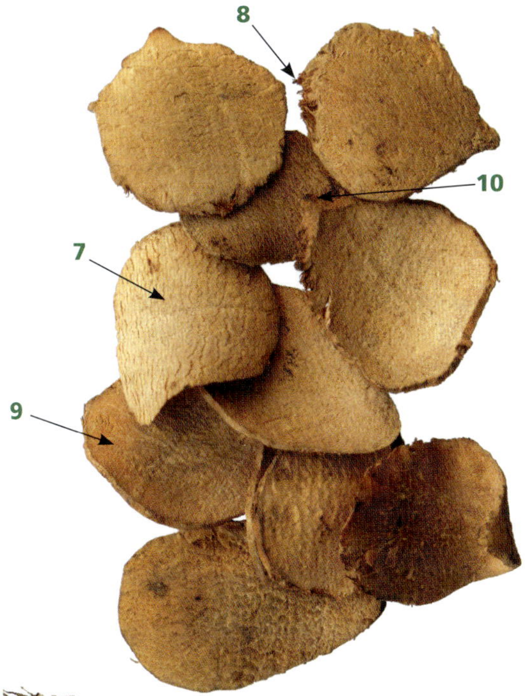

Decocting piece-2 (vinegar-processed rhizome of *S. stoloniferum*)

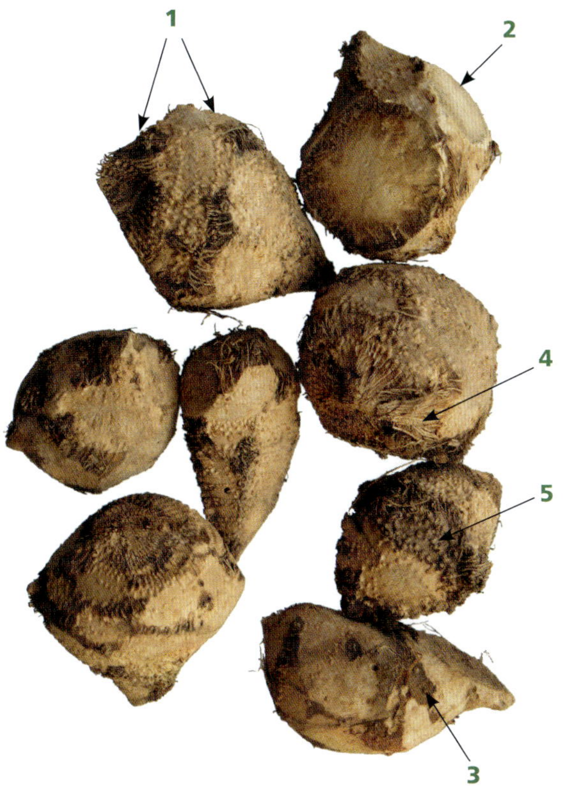

Crude drug of San Leng

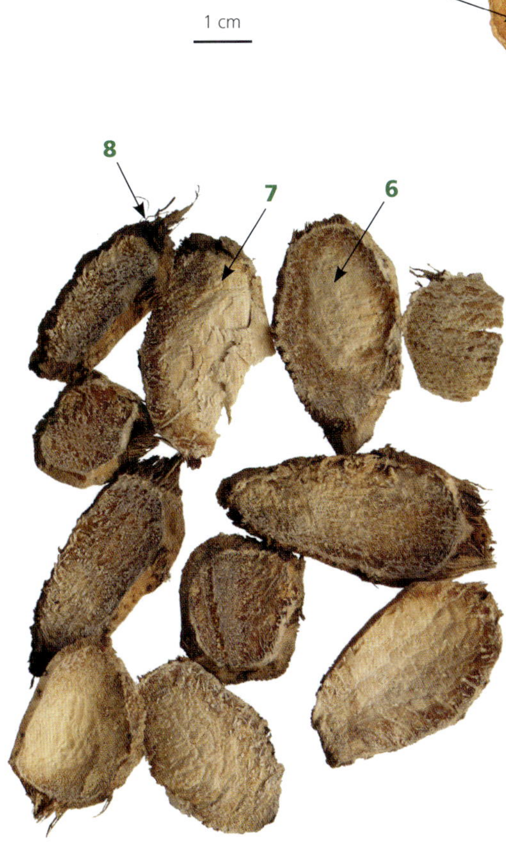

Decocting pieces-1 (peeled, dried and sliced rhizome of *S. stoloniferum*)

Cut surface of San Leng showing scattered orange glands

San Qi 三七 Notoginseng Radix et Rhizoma

Drug parts: ROOT AND RHIZOME

Panax notoginseng (Burkill) F. H. Chen

Araliaceae

synonyms: *Aralia quinquefolia* (L.) Decne. & Planch. var. *notoginseng* Burkill; *Panax pseudoginseng* Wall. var. *notoginseng* (Burkill) G. Hoo & C. J. Tseng

Species of *Panax* wild in China: 6

PLANT DESCRIPTION

Characters are based on FoC and those recently recorded from medicinally farmed plants long-cultivated in SW China; the latter display much morphological and DNA variation (Guo *et al.*, 2010a).

An herbaceous perennial **(20–60 cm tall).** *Rootstocks* (*rhizomes and roots*) fleshy yellow or purple, **1 to many, shortly cylindrical or club-shaped.** *Stems* erect, green, greenish-purple or purple. *Leaves* stalked, at top of stem in one whorl of **2–8 (usually 4 but increasing in number in older plants)**; each leaf with 5-7 stalked leaflets radiating from a single point (palmately-compound); leaflet blades oblong to widest above the middle (3.5–13 cm long; 1.5–7 cm wide), membranous, **with veins on both surfaces bristly**, base oblique, margin with bristles and alternating teeth of 2 sizes, tip tapering to a long point. ***Inflorescence*** a solitary, erect, hemispherical **80–100 (or more)-flowered** terminal head on an erect 7–25 cm long stalk (peduncle). *Each flower* tiny (2–3 mm diam.), stalked, bisexual and radially symmetrical; calyx green and cup-shaped, 5-lobed at mouth; petals 5, greenish or white; stamens 5 and inserted onto a disk; ovary inferior, styles 2. *Fruit (drupe)* compressed spherical, kidney-shaped or 3-sided (c. 1 cm diam.), when mature red, yellow or purple (with intermediates). ***Seeds*** 2 per fruit, **triangular-ovoid, slightly 3-ridged** (5–6 mm thick), not laterally compressed. Fl. Jul–Aug, fr. Aug–Oct. *Emboldened characters distinguish this species from other species of Panax.*

HARVESTING, SOURCING AND NATURAL RANGE

This popular and expensive herb is harvested totally from cultivated crops in autumn before plants flower. The species was first established in commercial cultivation back in the 1760s in south-west China (Tianyang district of Guangxi province) following the commercial exhaustion of wild populations due to over-harvesting. The Tianyang district continues to be a source of 'San Qi' today (producing a top quality *Di Dao* trade item called Tian Qi 田七) but the bulk of 'San Qi' production today (98%) is grown in the Wenshan mountains of south-east Yunnan province and is also designated as *Di Dao*-quality (Guo *et al.*, 2010a). Smaller cultivation sites are located in other parts of Yunnan as well as the provinces of Fujian, Jiangxi, Zhejiang and south-west Guangxi (the Jingxi district of the latter was a former *Di Dao* production area, Hong. *et al.*, 2010).

The species is native to the evergreen forests of south-east Yunnan (1,200–1,800 m altitude) but where today it is no w extinct. Also native in N Vietnam.

CONSERVATION STATUS

China: Extinct in the Wild (MEPCAS, 2013).
Global: Not Evaluated (IUCN Red List, accessed 10/10/2016).

Inflorescence of *P. notoginseng*

Fresh rootstock of *P. notoginseng*

TCM PROPERTIES AND USE

***properties*:** **'sweet, slightly bitter and warm'**
***channels*:** **'Liver and Kidney'**
***actions*:** **'stops bleeding, invigorates Blood, relieves pain'**
***indications*:** **epistaxis; metrorrhagia; (T)**

Young medicinal crop of *Panax notoginseng* typically shaded early in the season

P. notoginseng with immature fruit- heads

Roots of *P. notoginseng* sun-drying, Yunnan province

Semi-dried roots and rhizomes of *P. notoginseng*

Panax notoginseng Araliaceae

DRUG MORPHOLOGY

CRUDE DRUG-1 (DRIED MAIN ROOT — SAN QI): shortly cylindrical **[1]** or club-shaped **[2]**(1–6 cm long; 1–4 cm diam.). Externally greyish-brown or greyish-yellow **[3]**, with interrupted longitudinal wrinkles **[4]**; apex (widest end) with rhizome scar **[5]** surrounded by large, conspicuous nodular, often rounded lateral root scars **[6]** with smaller ones scattered elsewhere **[7]**. *Texture*: heavy and dense. *Fracture*: greyish-green, yellowish-green or greyish-white **[8]**; central xylem with indistinct rays. *Odour*: indistinct. *Taste*: bitter with a sweet aftertaste. ***Good quality***: large and heavy.

CRUDE DRUG-2 (DRIED RHIZOME — JIAN KOU): irregularly-shaped dense masses with several large and conspicuous, nodular, often rounded lateral root scars **[9]**; external surface with indistinct annulations **[10]** and occasional stem base scars **[11]** or stem remnants **[12]**. *Fracture*: centrally greyish-green or greyish-white, graduating into deep green or grey towards the margin.

CRUDE DRUG-3 (DRIED LATERAL ROOTS — JIN TIAO): cylindrical or conical (2–6 cm long, upper end 0.8 cm diam.; lower end 0.3 cm diam.), often curved **[13]**. Externally yellowish-grey, uniformly rough and wrinkled **[14]**, with occasional fibrous rootlet remains **[15]** and regularly spaced faint 'annular-like' rings **[16]**.

DECOCTING SLICES (MAIN ROOT — Hong Kong style): outline of longitudinal slices tapering or club-shaped and apically nodular **[17]** (up to 6 cm long); outline of transverse slices irregularly ovate and nodular **[18]** (transverse slices); both sections 1.5–3 cm wide; c. 3 mm thick. Cut surfaces often greenish-yellow or brownish-yellow **[19]**, sometimes greyish in outer parts **[20]** and more strongly yellow centrally **[21]**. *Fracture*: brittle. *Odour* and *taste*: as for the crude drug forms above.

PROCESSING

Rootstocks are washed then their parts are graded according to size of roots then divided into main roots ('San Qi' = crude drug-1), rhizomes ('Jian Kou' = crude drug-2) and lateral roots ('Jin Tiao' = crude drug-3). All are then dried and prior to decocting, powdered (San Qi Feng = decocting material). In Hong Kong and other parts of southern China, roots and rhizomes are usually sliced into decocting pieces. Rootstocks destined for Hong Kong markets are traditionally blackened using charcoal then brought to a lustrous shine using beeswax.

OTHER COMMON NAMES

参三七 Shen San Qi; 田七 Tian Qi; Notoginseng Radix; Notoginseng Radix et Rhizoma; Radix et Rhizoma Notoginseng; sanchi; sanqi; tienchi ginseng (SCN).

SAFETY

Safety Class: 1; Interaction Class: A (Gardner & McGuffin, 2013). The flowers of *P. notoginseng* (San Qi Hua 三七花) are a common Chinese folk medicine.

UNOFFICIAL SUBSTITUTES/ADULTERANTS AND COUNTERFEITS

San Qi is a high-value TCM herbal drug and, as such, is vulnerable to substitution by cheap look-alikes and counterfeiting.

Examples of source plants used for counterfeiting of San Qi:

1. ***Anredera cordifolia*** (Ten.) Steenis — 落葵薯 Luo Kui Shu (family Basellaceae). The thick rhizomes are used in Chinese folk medicine; not native to China but cultivated in Beijing, Fujian, Guangdong, Hainan, Jiangsu, Sichuan, Yunnan and Zhejiang provinces.
2. ***Bletilla striata*** (Thunb.) Rchb. f. (family Orchidaceae). Tuberous rhizomes are the source of the TCM drug 'Bai Ji' (Bletillae Rhizoma, see page 26).
3. ***Curcuma phaeocaulis*** Valeton (family Zingiberaceae). Rhizomes are the source of the separate TCM drug 'E Zhu' (Curcumae Rhizoma) see page 120.
4. ***C. longa*** L. (family Zingiberaceae). Rhizomes are the source of the separate TCM drug 'Jiang Huang' (Curcumae Longae Rhizoma) — turmeric.
5. ***Gynura japonica*** (Thunb.) Juel — 菊三七 Ju San Qi (family Asteraceae). Rootstocks thick and tuberous (3–4 cm diam.) with many fibrous roots.
6. ***Manihot esculenta*** Crantz (family Euphorbiaceae). Widely cultivated in southern China for its fleshy tuber used as edible starch (cassava, tapioca). Native to Brazil and cultivated throughout the tropics.
7. ***Panax stipuleanatus*** C. T. Tsai & K. M. Feng — Ye San Qi (family Araliaceae). A close relative of the Pharmacopoeia species and, at one time, used as a common substitute until its wild populations became Endangered (Zhou *et al.*, 2005; MEPCAS, 2013).
8. ***Schizocapsa plantaginea*** Hance — 裂果薯 Lie Guo Shu (syn. *Tacca plantaginea* (Hance) Drenth) family Taccaceae. Rhizome sub-cylindrical, short and thick and used as a Chinese folk medicine as an analgesic, antipyretic, anti-inflammatory and to treat wounds; shares some therapeutic properties with official 'San Qi'.

Confused species due to similar or shared Chinese names:

9. ***Phedimus aizoon*** (L.) 't Hart var. ***aizoon*** (syn. *Sedum aizoon* L.). Common name is 'Jing Tian San Qi'.
10. ?***Tupistra*** rhizome (family Asparagaceae): source of a medicinal herb from south-west China with a similar name 'Chuan San Qi 川三七'. Occasionally confused with official 'San Qi' in Hong Kong and US markets (Zhao & Li, 2007; Eric Brand and Zhao Zhong Zhen, pers. comm. 4/2016). Given the toxicity of many *Tupistra* species (Pan *et al.*, 2012; Shaw, 2010; Wong & Chan, 2005) care is needed to ensure these 2 drugs are not confused.

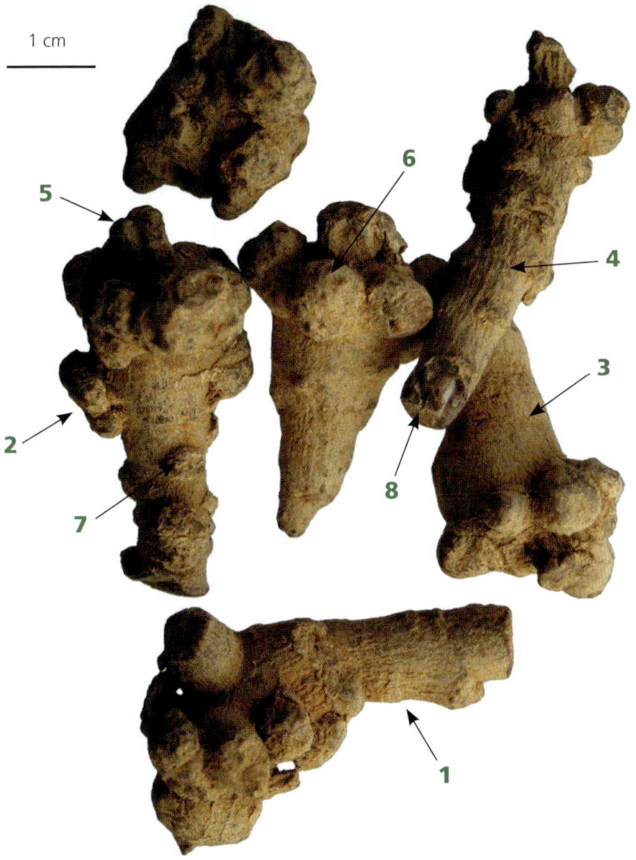

Crude drug-1 of San Qi sourced from main root of *Panax notoginseng*

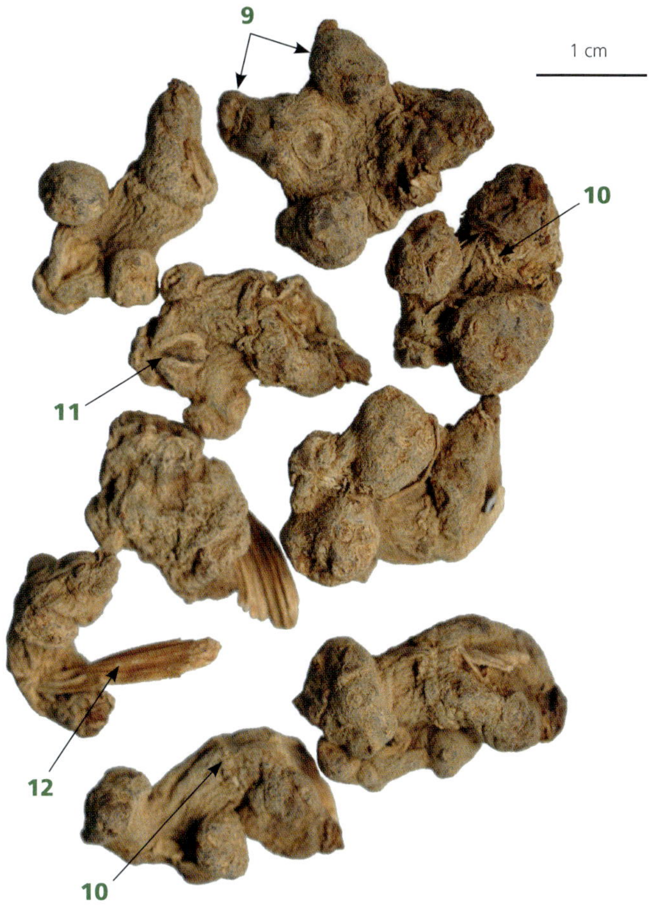

Crude drug-2 (Jan Kou type of San Qi; dried rhizomes of *P. notoginseng*)

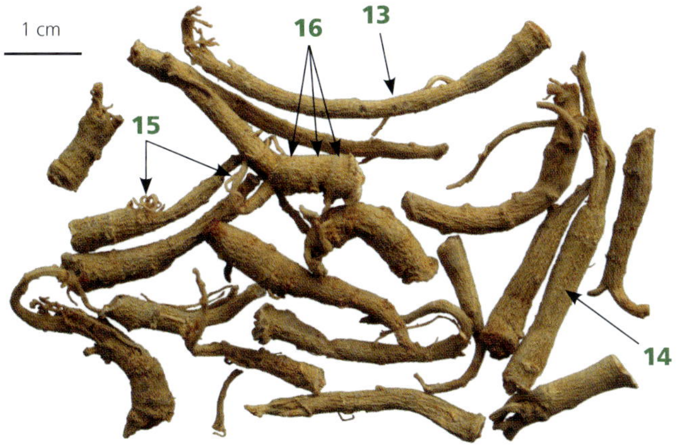

Crude drug-3 (Jin Tiao type of San Qi; dried lateral roots of *P. notoginseng*)

Hong Kong processing style of San Qi (root of *P. notoginseng*)

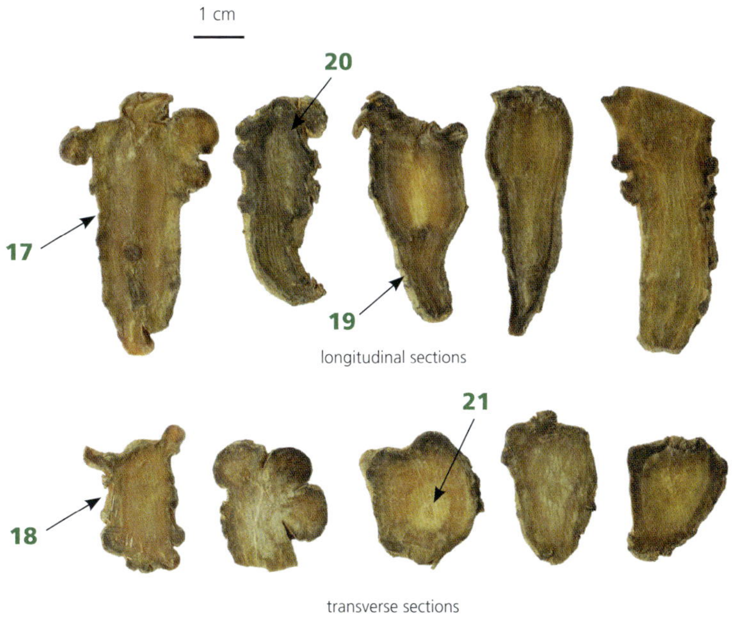

Hong Kong style of decocting pieces of San Qi
(sliced root of *P. notoginseng*)

Shan Ci Gu 山慈菇

Cremastrae Pseudobulbus; Pleiones Pseudobulbus

Drug part: PSEUDOBULB

PHARMACOPOEIA SPECIES 1

Cremastra appendiculata (D. Don) Makino

Orchidaceae

synonym: *Cymbidium appendiculatum* D. Don
taxonomic/nomenclatural note: *FoC* and *Kew-MPNS* recognise 2 varieties: *C. appendiculata* var. *variabilis* (Blume) I. D. Lund and the type variety.

Species of *Cremastra* wild in China: 3
Species of *Pleione* wild in China: 23

PLANT DESCRIPTION

A terrestrial orchid **(20–70 cm tall)** with an **underground, creamish-white**, ovoid or sub-globose, **pseudobulb** (1.5–3 cm tall; 1–3 cm diam). *Leaf* basal 1 (rarely 2), produced before the flowers; leaf blade large (18–34 cm long; 4–8 cm wide), narrowly elliptic, 3-veined, with a pointed tip and entire margin; hairless. *Flowering stem* arising from apex of pseudobulb, erect, solitary and leafless with several fragrant, **pendant, narrowly bell-shaped flowers** (up to 3.4 cm long). Each flower with sepals and petals similar, pale purple, pale yellow to orange-brown, narrowly linear (18–32 mm long), not spreading; lip (the 3rd and lower-most petal) 3-lobed, the middle lobe with a fleshy protuberance (callus) at base, callus somewhat wrinkled or warty. *Fruit* dry, ellipsoid and pendant (25–30 mm long; 8–13 mm diam.). Fl. May–Jun, fr. Sep–Dec.

HARVESTING, SOURCING AND NATURAL RANGE

Pseudobulbs are wild-harvested in summer and autumn from plants which grow in damp places in forests (400–2,900 m altitude).

The species has a wide distribution in China (Anhui, Gansu, Guizhou, Guangdong, Henan, Hubei, Hunan, Jiangsu, Jiangxi, Shaanxi, Shanxi, Sichuan, Taiwan, Tibet, Yunnan and Zhejiang); also wild in Bhutan, India, Japan, Nepal, Sikkim, Thailand and Vietnam.

CONSERVATION STATUS

China: Near Threatened (var. *appendiculata*); Least Concern (var. *variabilis*) (MEPCAS, 2013).
Global: Not Evaluated (IUCN Red List, accessed 19/2/2016).

PHARMACOPOEIA SPECIES 2

Pleione bulbocodioides (Franch.) Rolfe

Orchidaceae

synonym: *Coelogyne bulbocodioides* Franch.

PLANT DESCRIPTION

Very easily distinguished from *C. appendiculata* (PS1) above. *Main differences:* **much shorter** (7–20 cm tall), **pseudobulb above-ground, green or purple**, ovoid to almost conical **with a conspicuous flask-shaped neck** (1–2.5 cm long; 1–2 cm wide); the solitary **leaf often much shorter** (10–25 cm long) and appearing after the inflorescence; flower stalk produced from base of pseudobulb, covered by 3 tubular sheaths; the solitary **flower** (2 or more in cultivation) **much larger** (up to c. 8 cm wide) and **conspicuous;** sepals and petals 6, pointed and **spreading, pink to pale purple, lip** (lower-most petal) **with dark purple blotches;** callus of lip **with 4–5 irregular (nibbled looking) ridges.** Fl. Apr–Jun.

HARVESTING, SOURCING AND NATURAL RANGE

Pseudobulbs mostly wild-harvested in summer and autumn from plants which grow on cool, damp, mossy, rock faces at the edge of thickets in broad-leaved evergreen forest (900–3,600 m altitude).

The species is endemic to parts of S, C and W China (Anhui, Fujian, Gansu, Guangdong, Guangxi, Guizhou, Hubei, Hunan, Shaanxi, Sichuan, Tibet and Yunnan). The 'bulbs' are cultivated on medicinal plant farms in Guizhou province (S. China).

CONSERVATION STATUS

Endemic to China: Least Concern (MEPCAS, 2013); Not Evaluated (IUCN Red List, accessed 19/2/2016).

PROTECTED STATUS

CITES: Appendix II; EU Annex B (see Table 1 page 10).

TCM PROPERTIES AND USE

properties: **'sweet, pungent and cold'**
channels: **'Liver and Spleen'**
actions: **'clears heat, eliminates toxins, disperses nodules'**

indications: **swellings due to abscesses, stubborn boils, insect and snake bites; subcutaneous nodules; (T)**

Flowering and fruiting plants of *Cremastra appendiculata* growing wild on a forest floor, Sichuan province

C. appendiculata in full flower

Freshly harvested pseudobulbs *of C. appendiculata*

Typical habitat of *Pleione bulbocodioides*, Sichuan province

P. bulbocodioides, Sichuan province

Freshly uprooted plants of *P. bulbocodioides* with purple pseudobulbs

Pleione yunnanensis (Rolfe) Rolfe

Orchidaceae

synonym: *Coelogyne yunnanensis* Rolfe
taxonomic/nomenclatural note: in CP2015 as '*P. yunnanensis* Rolfe'

PLANT DESCRIPTION

Very similar to *P. bulbocodioides* (PS2) but **flowers smaller and paler** (usually pale purplish, pink or sometimes nearly white), petals more rounded and lip (lower-most petal) with broad, rounded side lobes, mid-lobe **deep red or with deep purple spots or streaks and a callus of 3–5 wavy parallel ridges.** Fl. Apr–May, fr. Sep–Oct.

HARVESTING, SOURCING AND NATURAL RANGE

Pseudobulbs are harvested in summer and autumn from wild and cultivated plants. Wild populations grow on shady, moss-covered rocks or in leaf litter in forest and forest margins (1,100–3,500 m altitude) in W and SW China (Sichuan, Guizhou, Tibet and Yunnan [also native in Myanmar]. Given the rapid depletion of wild populations in recent years due mostly to over-harvesting, the species has been established as a medicinal crop in parts of Guizhou province.

CONSERVATION STATUS

China: Vulnerable (MEPCAS, 2013).
Global: Not Evaluated (IUCN Red List, accessed 19/2/2016).

PROTECTED STATUS

CITES: Appendix II; EU Annex B (see Table 1 page 10).

Tulipa edulis (Miq.) Baker

Liliaceae

synonyms: *Amana edulis* (Miq.) Honda (*Kew-MPNS* accepted name, 19/2/2016); *Orithyia edulis* Miq.

PLANT DESCRIPTION

Easily distinguished from the 3 Pharmacopoeia species. *Main differences:* underground part is a true bulb (1.5–4 cm diam.) covered with a brownish papery outer skin; stem 10–25 cm tall; **leaves usually 2 and linear** (15–25 cm long; 5–9 mm wide) overtopping flower; flower terminal, solitary, usually pointing upwards; **sepals and petals all very similar in size and shape** (no lip), **white, flushed pink, outermost ones purplish-red externally; stamens free** (not fused into a column); **fruit** dry, **sub-globose** (5–7 mm diam.) with a long beaked tip. Fl. Mar–Apr, fr. Apr–May.

CONSERVATION STATUS

China: Least Concern (MEPCAS, 2013; rapidly declining (Ma *et al.*, 2014).
Global: Not Evaluated (IUCN Red List, accessed 19/2/2016).

SUBSTITUTION COMMENT. *T. edulis* bulbs have long been used as a local substitute of 'Shan Ci Gu' due to their similar clinical properties but their detection is essential given their potential cytotoxicity (Yin & Guo, 2009). Also called 'Guang Ci Gu' or 'Guang Shan Ci Gu' their bulbs are the source of a separate TCM (not in CP2015) widely used in breast cancer therapy (Lin *et al.*, 2015; Ma *et al.*, 2014; Miao, *et al.*, 2014). Over-harvesting of the species' wild populations (for cancer use) has prompted the commercial cultivation of this species (Bing *et al.*, 2008) but its occurrence as a 'Shan Ci Gu' substitute in international trade is uncertain.
T. edulis grows wild in China on grassy hillsides (0–1,700 m altitude) in Anhui, Hubei, Hunan, Jiangsu, Jiangxi, Liaoning, Shaanxi, Shandong and Zhejiang [also native to Japan and Korean Peninsula].

Iphigenia indica (L.) A. Gray ex Kunth

Colchicaceae

synonym: *Melanthium indicum* L.
taxonomic/nomenclatural note: *WCS* and *Kew-MPNS* accepted name, 16/10/2016; in *FoC* as '*I. indica* Kunth'.

PLANT DESCRIPTION

Easily distinguished from the 3 Pharmacopoeia species. *Main differences:* underground part a globose, solid corm (0.5–1.5 cm diam.); **stems leafy** (10–25 cm tall); **leaves several, linear** and with a distinct mid-vein, bases sheathing and clasping; flowers **2–10 at top of stem; sepals and petals linear, spreading and dark purplish-brown** (7–10 mm long; 0.7–1 mm wide); fruit dry, ovoid to oblong (6–8 mm long; 3–3.5 mm. diam.). Fl. and fr. Jun–Jul.

CONSERVATION STATUS

China: Least Concern (MEPCAS, 2013).
Global: Not Evaluated (IUCN Red List, accessed 19/2/2016).

SUBSTITUTION COMMENT. The corms of this unrelated medicinal species, native to tropical parts of southern China, are usually traded under the name Li Jiang Shan Ci Gu 丽江山慈菇, however they are also known as Shan Ci Gu 山慈菇, i.e. identical to the Pharmacopoeia drug name. This is a potential source of confusion and, although the corms of *Iphigenia indica* are easily distinguished macroscopically, their accurate identification is essential given their well-known toxicity (see *Safety*).
In China, *I. indica* grows wild in pine forests, damp grasslands and open fields of Yunnan and Hainan provinces (0–2,100 m altitude).

Pleione yunnanensis, wild in Yunnan PHOTO: PHILLIP CRIBB

Fruiting plant of *P. yunnanensis*

P. yunnanensis at London Orchid Show PHOTO: ANDRE SCHUITEMAN

Tulipa edulis

Cultivated plant of *Iphigenia indica* flowering and fruiting

Cremastra appendiculata

Orchidaceae

DRUG MORPHOLOGY

CRUDE DRUG (DRIED PSEUDOBULB): often squat-oblique **[1]** to shortly conical **[2]** or sometimes irregularly-shaped (1.8–3 cm tall; 1–2 cm diam.). Externally dark brownish-black, yellowish-brown or cream, indistinctly and shallowly wrinkled, typically with **2–3 distinct, narrow, light or dark brown, slightly raised annular rings [3],** with occasional filiform fibres **[4]**. Apical (narrowest) end with scar of flowering stem **[5]**, sometimes with a protuberant flower stem bud **[6]**; basal end wider and more obtuse with a few root remnants **[7]**. *Texture:* hard, uneasily broken, ***Fracture:*** greyish-white or yellowish-white, **slightly horny.** *Odour:* indistinct. *Taste:* mild and sticky. ***'Good quality':*** large, plump and solid; internally yellowish-white.

DECOCTING PIECES (TRANSVERSE PSEUDOBULB SLICES): outline irregularly oblong, elliptic or ovate (1.8–3 cm long; 1–2 cm diam.) **[8]**. Uncut external surface with occasional root remnants **[9]**; annular rings sometimes visible. Cut surface blackish-brown around perimeter **[10]**, become yellowish centrally **[11]**. *Texture:* horny and slightly translucent.

PROCESSING

After removal of the aerial parts, pseudobulbs are graded by size, then steamed thoroughly before drying (= crude drug). These are later softened thoroughly to facilitate cutting into thin transverse slices (1–2 mm thick) (=decocting pieces); or pulverised prior to decocting.

OTHER COMMON NAMES

杜鹃兰 Du Juan Lan (plant name); Mao Ci Gu 毛慈菇 'hairy Ci Gu'; Pseudobulbus Cremastrae seu Pleiones.

Pleione bulbocodioides and *P. yunnanensis*

Orchidaceae

DRUG MORPHOLOGY

The pseudobulbs (drug part) of these two species (PS2 and 3) are morphologically indistinguishable so the below description applies to both. The crude drug of both species are easierly distinguishable macroscopically from that of PS1; decocting pieces often distinguishable.

CRUDE DRUG (DRIED PSEUDOBULB): *easily distinguished from those of PS1.* Long- **[12]** or squat-conical **[13]** or irregularly-shaped **with a distinctive flask-shaped 'neck' [14]**, often laterally flattened (1.5–3 cm long; 1–2 cm diam.). When unpeeled external pseudobulb **surface rough**, brown, cream or pinkish with fine, longitudinal wrinkles **[15]**; **annular rings not visible;** yellowish-white when peeled. Apex (narrowest end) with basal remnants of flower stem **[16]** (c. 1–3 mm long) with a pale and distinct circumferential ring **[17]**. Basal end somewhat obliquely rounded **[18]**, often dented **[19]** and usually with persistent rootlet remains. *Texture:* hard and difficult to break. ***Fracture:*** yellowish, **horny and translucent.**

DECOCTING PIECES (TRANSVERSE PSEUDOBULB SLICES: irregularly shaped, part broken or sliced pieces of pseudobulb (up to 3 cm diam.) **with occasional features discernible, such as the distinctive flask-shaped 'neck' [14]**, apical circumferential ring **[17]** and basal dent with rootlet remains **[18]**. Cut or broken surfaces yellow, cream or pinkish, granular **[20]**; externally cream, yellow, brown or pinkish, finely wrinkled **[15]**.

PROCESSING

As for PS1.

OTHER COMMON NAMES

PS2: 冰球子 Bing Qiu Zi; 独蒜兰 Du Suan Lan (plant name); Pseudobulbus Cremastrae seu Pleiones.

PS3: 冰球子 Bing Qiu Zi; 云南独蒜兰 Yunnan Du Suan Lan (plant name); Pseudobulbus Cremastrae seu Pleiones.

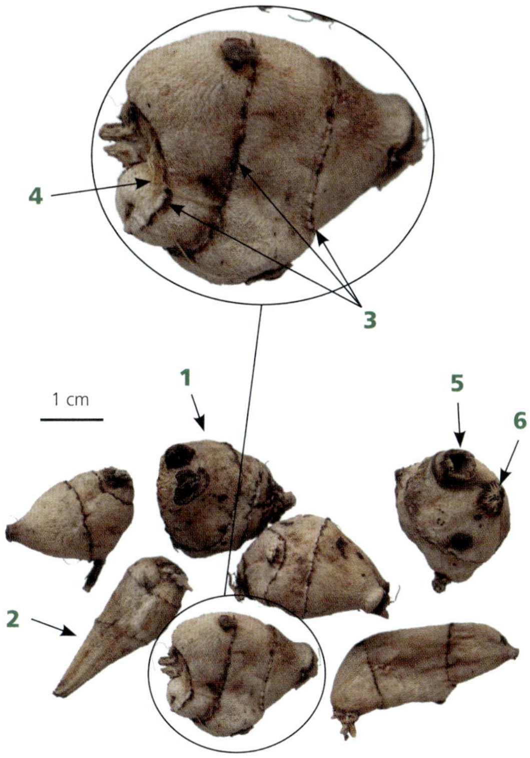

Crude drug of Shan Ci Gu (Mao Ci Gu) sourced from *C. appendiculata* — creamish-white form

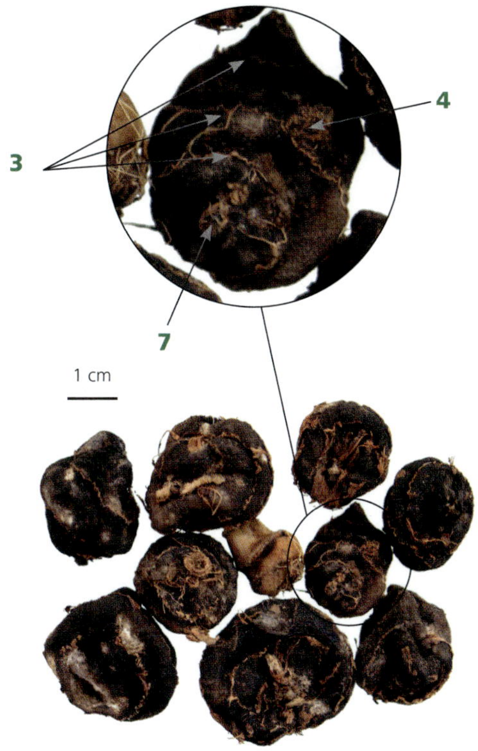

Crude drug of Shan Ci Gu (Mao Ci Gu) sourced from *C. appendiculata* — dark brown form

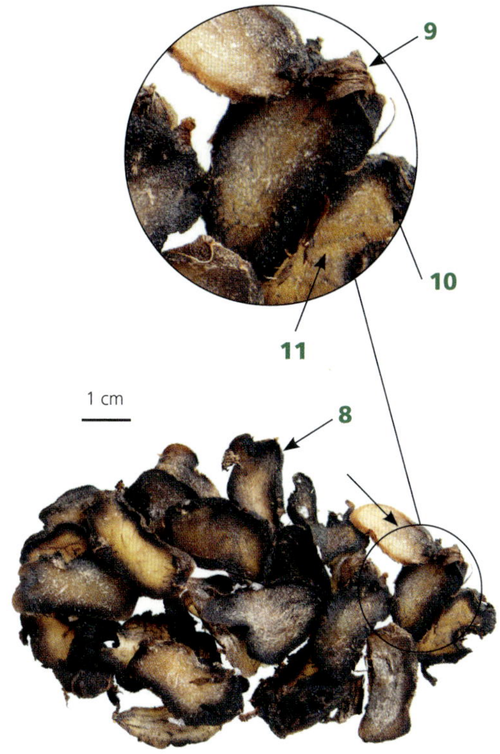

Decocting pieces of Shan Ci Gu (Mao Ci Gu) sourced from *C. appendiculata*

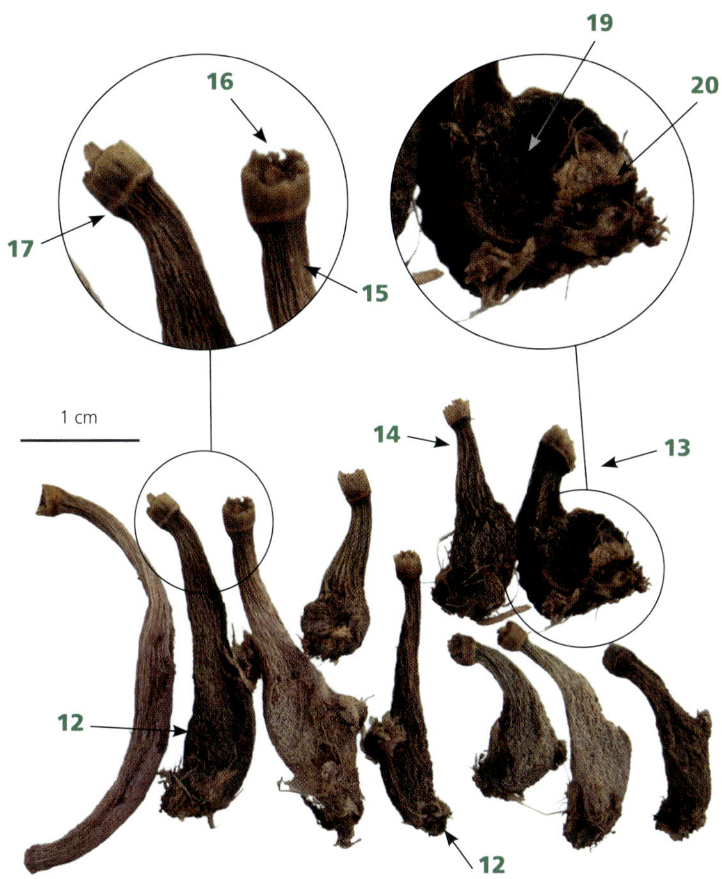

Crude drug Shan Ci Gu sourced from *P. bulbocodioides*

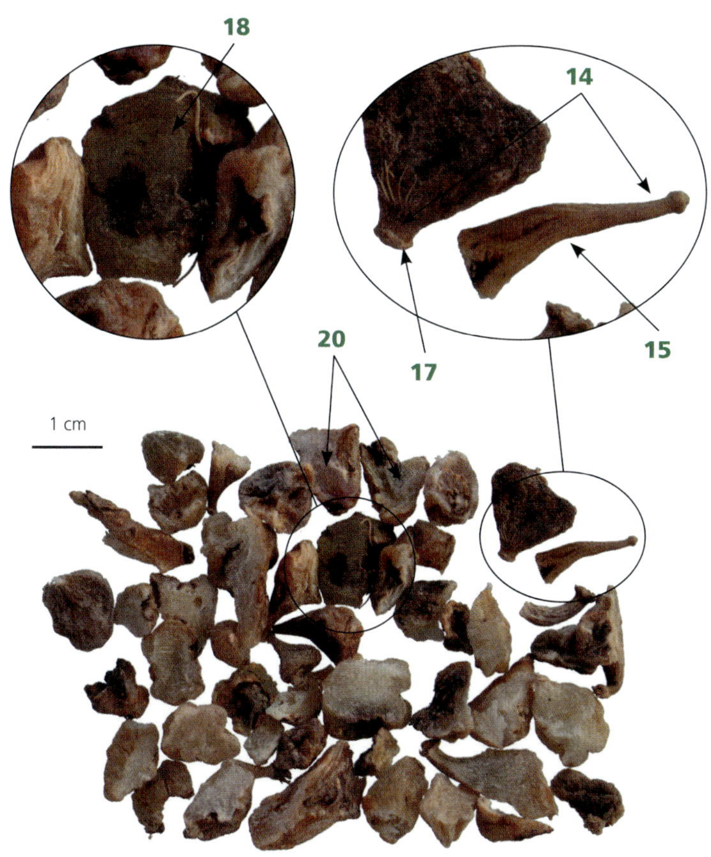

Decocting pieces of Shan Ci Gu sourced from *P. bulbocodioides*

UNOFFICIAL SUBSTITUTE 1

Tulipa edulis

Liliaceae

'DRUG SUBSTITUTE' MORPHOLOGY

Relatively easy to distinguish from all three official species. Bulbs ovoid to shortly conical (c. 0.8–2 cm tall; 0.6–1.6 cm diam.). **Apex acute and beak-like [21]**, base broadly truncate or broadly oblique with ovate slightly sunken scar and fibrous roots **[22]**. Externally (once outer tunic removed prior to drying) brown, creamish- or yellowish-white, **surface smooth [23]**, usually with a (sometimes indistinct) **shallow longitudinal furrow enlarging to base to form 2 small rounded 'shoulders' [24]; annular rings absent.** *Texture*: hard, uneasily broken. ***Fracture*: powdery [25]**; **centrally with small longitudinal hollow [26]**. *Odour:* indistinct. *Taste*: indistinct.

OTHER COMMON NAMES

慈菇 Ci Gu; 光慈菇 Guang Ci Gu; 光山慈菇 Guang Shan Ci Gu; 老鸦瓣 Lao Ya Ban (plant name); 山慈菇 Shan Ci Gu.

SAFETY

Potentially toxic (Yin & Guo, 2009).

UNOFFICIAL SUBSTITUTE 2

Iphigenia indica

Colchicaceae

'DRUG SUBSTITUTE' MORPHOLOGY

The corm of this species is easily distinguished from the pseudobulbs of the 3 official species. When brown outer tunic **[27]** removed, **surface smooth [28]. Usually smaller** (1–1.5 cm long; 0.5–1.5 cm diam.) and **irregularly globose to ovoid, apex rounded** with central stem remnants, **base slightly concave [29]**, with a **shallow longitudinal furrow [30]**, externally yellowish-white. Circumferential annular rings and 'bottle-nack' **absent**. *Odour:* indistinct. *Taste:* bitter and numbing.

OTHER COMMON NAMES

山慈菇 Shan Ci Gu (plant name); 丽江山慈菇 Li Jiang Shan Ci Gu; Indian grass lily.

SAFETY

Toxic. Contains very high levels high of colchicine (Mukhopadhyay *et al.*, 2002); colchicine can be fatal in very small doses (c. 7 mg) (Milne & Meek, 1998).

COMMENT. In the UK, at the time of writing, it is not unusual to encounter TCM clinics dispensing 'Shan Ci Gu' as a mixture of *Cremastra* and *Pleione* and it is likely that such mixed batches are commonplace elsewhere in trade (C. Leon, pers. obs, 2015).

LESSER KNOWN UNOFFICIAL SUBSTITUTES

The following species have been reported as intentional or accidental substitutes of 'Shan Ci Gu' locally in China; their occurrence as such in international trade is unlikely.

1. ***Asarum sagittarioides*** C. F. Liang. Family Aristolochiaceae. An unrelated and easily distinguished species, however, confusion may arise because one of the Chinese names of this local medicinal species has the same pronunciation as Shan Ci Gu (although its Chinese characters differ: 山茨菇). **Potentially toxic** due to the presence of aristolochic acids (Hashimoto *et al.*, 1999). The species is endemic to Guangxi province (S China).
2. ***Bletilla*** species. Family Orchidaceae. Tuberous rhizomes of this genus bear a superficial resemblance to the pseudobulbs of 'Shan Ci Gu'. When the latter is in short supply and prices are high, *Bletilla* rhizomes may be traded as a cheaper and readily available look-alike and widely cultivated as the source of the separate TCM drug 'Bai Ji' (Bletillae Rhizoma) see page 26.
3. ***Tinospora sagittata*** Gagnep. (syn. *T. capillipes*), Qing Niu Dan 青牛胆; family Menispermaceae. Roots of this species have small yellow tuberous swellings that superficially resemble the pseudobulbs of 'Shan Ci Gu' and may be traded as cheap look-alikes.

Dried bulbs of *Tulipa edulis* with outer tunics removed

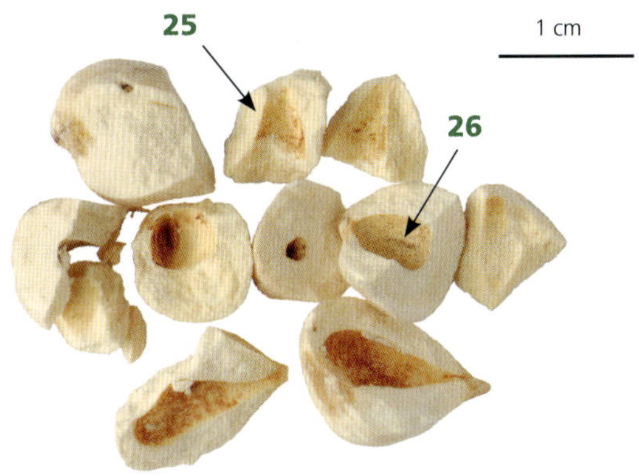

Dried and broken bulbs of *T. edulis*

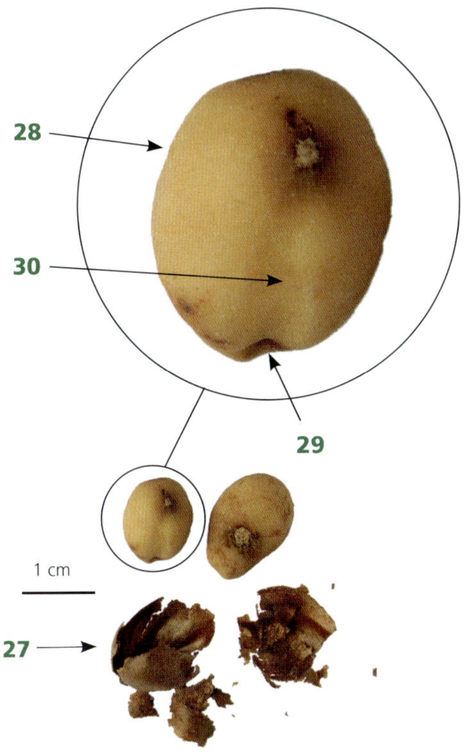

Dried corms of *Iphigenia indica*
with outer tunics removed

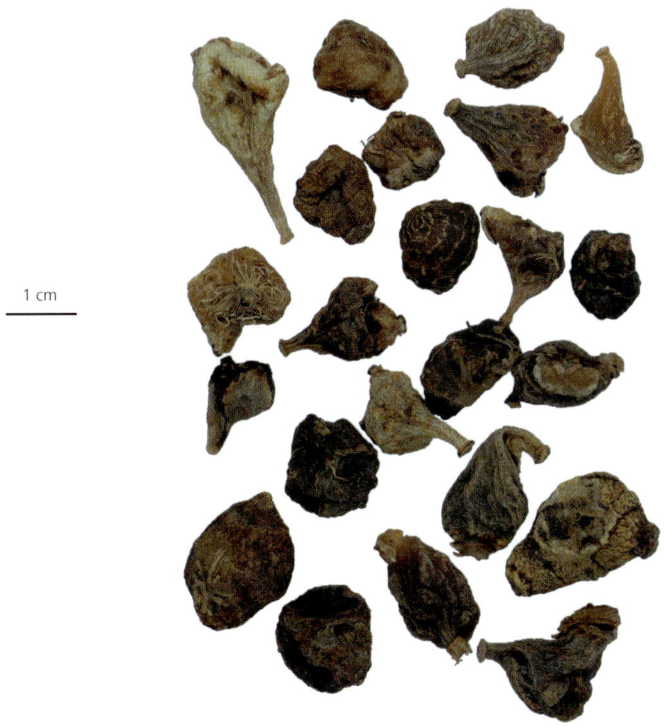

Trade sample of 'Shan Ci Gu' sold in the UK market containing a mixture of pseudobulbs resembling those of *Pleione* and *Cremastra*.

Shan Dou Gen 山豆根

Sophorae Tonkinensis Radix et Rhizoma

Drug part: ROOT AND RHIZOME

PHARMACOPOEIA SPECIES

Sophora tonkinensis Gagnep.

Fabaceae

taxonomic/nomenclatural note: *FoC* and *Kew-MPNS* (16/8/2016) recognise 3 varieties of this species; this account assumes that 'Shan Dou Gen' can be sourced from all.

Species of *Sophora* wild in China: 20

PLANT DESCRIPTION

A much branched **woody shrub** (up to 1.3 m tall). *Stipules* present. *Leaves* alternate (10–15 cm long) with **5–16 (rarely 19) pairs of opposite to alternate leaflets plus a single larger terminal leaflet**; **leaflet blades ovate, oblong or lance-shaped (15–25 mm long; 10–15 mm wide)**, **leathery** and margins entire. *Flowers* in loosely cylindrical many-flowered heads (10–30 cm long) arising from the leaf axils. Each flower laterally symmetrical and bisexual; sepals fused to form a 'cup' with 5 small apical teeth; petals 5, yellow or dark purple (10–12 mm long) of unequal size and shape; wing petals slightly longer than the upper (standard) petal; stamens 10; ovary superior. *Fruit* a **long pod** (legume) (3–5 cm long)**, strongly constricted between seeds.** *Seeds* 1–3, black. Fl. May–Jul, fr. Aug–Dec.

HARVESTING, SOURCING AND NATURAL RANGE

Rootstocks are harvested in autumn, traditionally from wild populations which grow in open scrub and forest on stony, calcareous or sandstone mountain slopes (1,000–2,000 m altitude) in SW China (Guangxi, Guizhou and Yunnan provinces). Since the 1950s, rising medicinal demand has largely exhausted the wild supplies of two of the three varieties in China and, more recently, populations of the same species native to neighbouring Vietnam (see *Conservation Status*). Establishing sustainable supplies from commercial cultivation has proved challenging due to the species' exacting ecological needs; tissue culture research is underway in an attempt to establish a sustainable supply from propagated plants (Ngyen Tien Ban, 2007; Wei *et al.*, 2013).

CONSERVATION STATUS

China: Vulnerable (var. *tonkinensis*); Near Threatened (var. *polyphylla*); Least Concern (var. *purpurascens*) (MEPCAS, 2013).
Vietnam: Vulnerable (Ngyen Tien Ban, 2007).
Global: Not Evaluated (IUCN Red List, accessed 20/8/2016).

UNOFFICIAL SUBSTITUTE

Menispermum dauricum DC.

Menispermaceae

PLANT DESCRIPTION

Easily distinguished from the Pharmacopoeia species. *Main differences*: **trailing and non-woody.** Leaf stalks often reddish and attached off-centre to the underside of each leaf blade. **Leaves simple** (not divided into leaflets), **shallowly lobed (3–12 cm long; 3–12 cm wide); flowers tiny and inconspicuous, in small heads**; **pale creamish-yellow** with male and female flowers on separate plants. Fruit **spherical and purplish-black (c. 1 cm diam**.).

SUBSTITUTION COMMENT. Supply problems of official 'Shan Dou Gen' has seen the rhizomes of *M. dauricum* become a popular substitute owing to their similar clinical properties (Bensky *et al.*, 2004). *M. dauricum* rhizomes are the source of the separate TCM drug 'Bei Dou Gen' (Menispermi Rhizoma) page 60; its similar Chinese name 'Bei Dou Gen' (meaning northern Dou Gen) reflects its more northerly distribution compared to 'Shan Dou Gen' however the latter name is widely used in trade to refer to either drug and is a potential source of confusion. Note that 'Bei Dou Gen' is wild harvested and it too may be vulnerable to substitution due perhaps to mis-identification of source plants (see 'Bei Dou Gen' page 60).

CONSERVATION STATUS

China: Least Concern (MEPCAS, 2013).
Global: Not Evaluated (IUCN Red List, accessed 20/8/2016).

TCM PROPERTIES AND USE

properties: 'bitter and cold'
channels: 'Lung and Stomach'
actions: 'clears heat, eliminates toxins'

indications: sore throat and gums; sores and abscesses

Sophora tonkinensis in flower

Fruiting plant of *S. tonkinensis*

Fresh root of *S. tonkinensis*

Typical trailing growth form of *Menispermum dauricum* with flower-heads in bud (centre)

Fully open flower-head of *M. dauricum*

Sophora tonkinensis

Fabaceae

DRUG MORPHOLOGY

CRUDE DRUG (DRIED ROOT AND RHIZOME): roots represent the main part of the crude drug; these are long, cylindrical, often branched and of variable length (10–15 cm long; 0.7–1.5 cm diam.); outer bark brown to greyish-brown with distinct longitudinal wrinkles **[1]** and raised transverse lenticels **[2]** (c. 3–4 mm wide). *Texture*: **hard**, woody and **difficult to break**. Rhizomes occasional, irregularly nodular **[3]**, with stem remnants at apex and several roots still attached below. *Fracture* (root) (*see photographs of Decocting Pieces*): inner bark (cortex) pinkish-brown **[4]** (c. 3 mm wide), central wood (xylem) pale creamish-yellow **[5]** (4–8 mm diam.). *Odour*: strongly pungent. *Taste*: very bitter. ***Good quality***: thick, solid and bitter.

DECOCTING PIECES (DRIED ROOT AND RHIZOME SLICES): irregularly shaped or sub-rounded transversely or obliquely cut slices (0.7–1.5 cm diam.; c. 2–4 mm thick). Outer bark very thin and brown or greyish-brown **[6]**; longitudinally wrinkled. Cut surface with inner bark pinkish-brown **(c. 3–4 mm wide) [4]**, central wood pale creamish-yellow and **radially striated [hl] [5]** (4–8 mm diam.); rigid silver-white hairs randomly arranged and occasionally visible throughout cut surface [hl] **[7]**.

PROCESSING

Rootstocks are dried (= crude drug) then later softened thoroughly and cut into 2–4 mm thick slices for decocting.

OTHER COMMON NAMES

广豆根 Guang Dou Gen 'bushy Sophora' from Guangxi province; 越南槐 Yue Nan Huai; Hoe Bac Bo (Vietnamese name); Radix et Rhizoma Sophorae Tonkinensis; Vietnamese sophora root (CP2010).

SAFETY

Toxic (CP2015; Hong Kong, 1999a).

Menispermum dauricum

Menispermaceae

'DRUG SUBSTITUTE' MORPHOLOGY

This substitute is easily distinguished from official 'Shan Dou Gen' (above) primarily because it is sourced from rhizomes in contrast to Shan Dou Gen which is sourced primarily from roots, with small amounts of rhizome. Rhizomes are easily recognised by the presence of a **white central pith [8]** (clearly visible in the transverse decocting slices of *M. dauricum* opposite).

Other distinguishing characters of M. dauricum rhizomes: rhizome diameter usually smaller (3– 8 mm) **[9]**, **outer bark yellowish-brown to brown [10]**, fibrous roots or root scars many **[11]**, texture of rhizome **pliable** (as crude drug); wood of transverse decocting pieces more fibrous, **distinctly radial [12], pale yellow, evenly dotted with round vessels [13] alternating with vascular tissues or cavities [14].** *Odour*: indistinct. *Taste*: bitter.

OTHER COMMON NAMES

北豆根 Bei Dou Gen; 蝙蝠葛 Bian Fu Ge; 山豆根 Shan Dou Gen; Asiatic moonseed (CP2010); Menispermae Rhizoma.

SAFETY

Slightly toxic (CP2015).

LESSER KNOWN UNOFFICIAL SUBSTITUTES

Acute shortages of Shan Dou Gen in recent years are likely to result in an increasing range of substitutes and adulterants on the market including those that are clinically related and those that are simply look-alikes and clinically inappropriate. Examples of either type are listed below (Fang & Guo,1992). Their occurrence as Shan Dou Gen substitutes in international TCM trade is uncertain.

Fabaceae family:

1. ***Indigofera amblyantha*** Craib (多花木蓝 Duo Hua Mu Lan).
2. ***I. carlesii*** Craib (苏木蓝 Su Mu Lan).
3. ***I. fortunei*** Craib (华东木蓝 Hua Dong Mu Lan).
4. ***I. decora*** var. ***ichangensis*** Y. Y. Fang & C. Z. Zheng (庭藤 Ting Teng).
5. ***I. kirilowii*** Maxim. ex Palib. (花木蓝 Hua Mu Lan).
6. ***I. szechuensis*** Craib (四川木蓝 Si Chuan Mu Lan), as '*Indigofera potaninii* Craib.' in Fang & Guo, 1992.

Ranunculaceae family:

7. ***Beesia calthifolia*** (Maxim. ex Oliv.) Ulbr. (铁破锣 Tie Po Luo), as '*Beesia calthaefolia* (Maxim.) Ulbr.' in Fang & Guo, 1992.

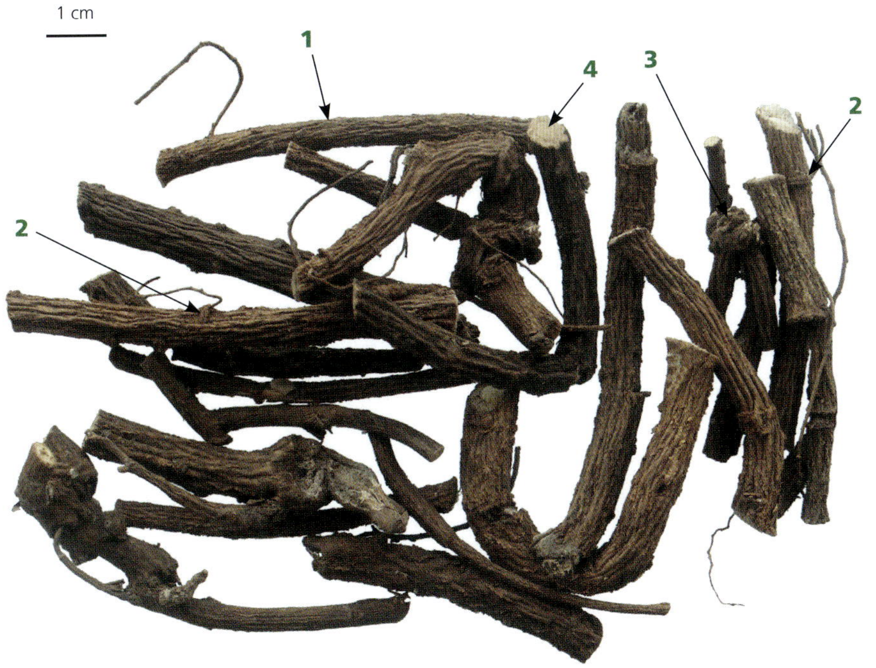

Crude drug of Shan Dou Gen (*S. tonkinensis*)

Decocting pieces (transverse) of Shan Dou Gen (*S. tonkinensis*)

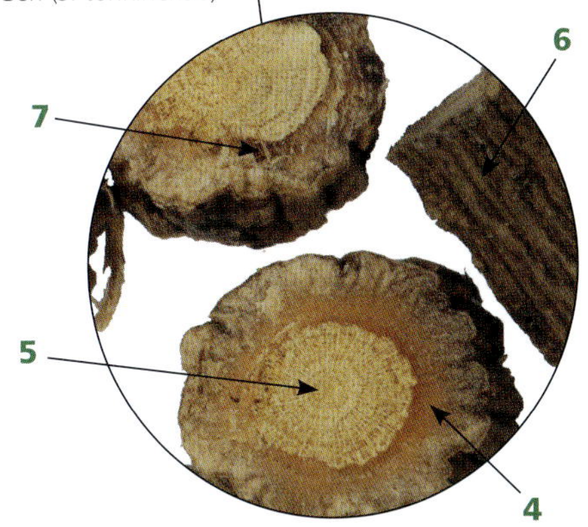

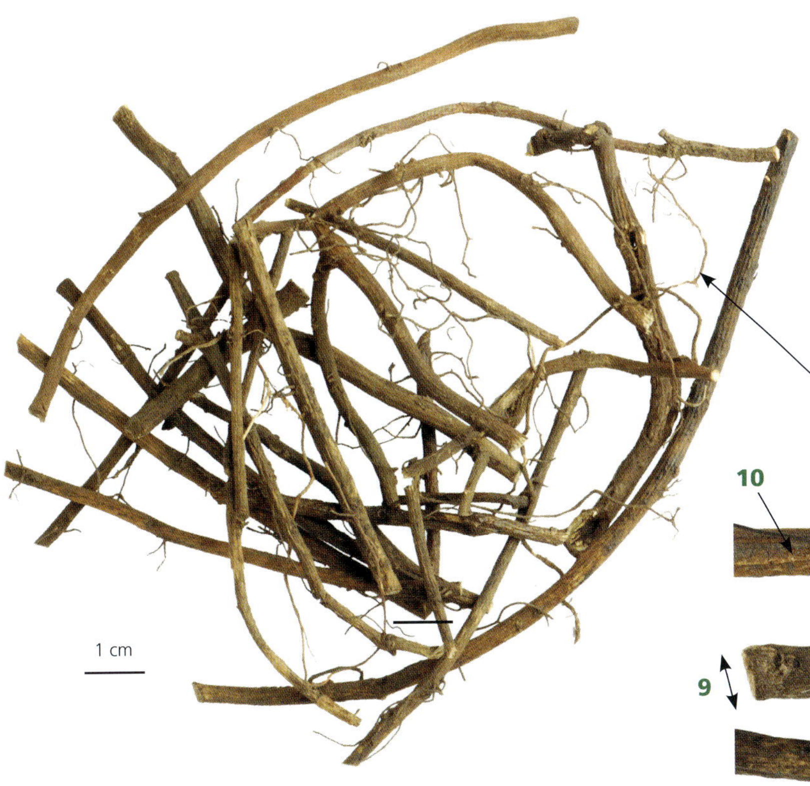

Dried rhizomes of *Menispermum dauricum* (Bei Dou Gen)

Decocting pieces of Bei Dou Gen (*M. dauricum*)

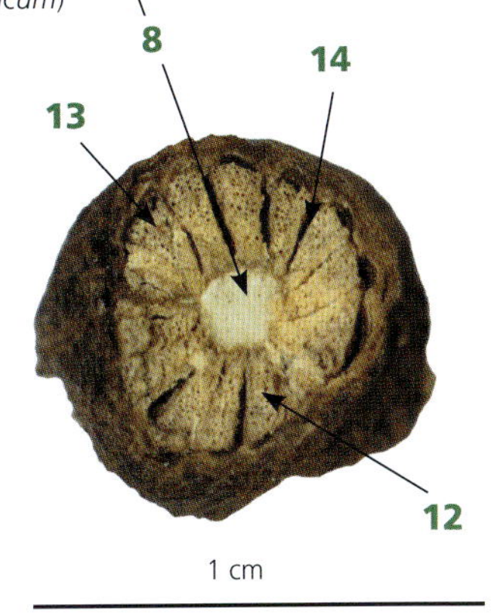

Shan Mai Dong 山麦冬

Liriopes Radix

Drug part: ROOT TUBER

PHARMACOPOEIA SPECIES 1

Liriope spicata Lour.

Asparagaceae

synonym: *L. spicata* var. *prolifera* Y. T. Ma (name used in CP2015 as '*L. spicata* (Thunb.) Lour. var. *prolifera* Y. T. Ma'

Species of *Liriope* wild in China: 6

PLANT DESCRIPTION

A short, tufted, herbaceous perennial **with creeping stolons;** roots with fleshy tubers narrowing at both ends. *Leaves* basal, stalkless and strap-like (25–60 cm long; 4–8 mm diam.), lower surface bluish-green and **distinctly 5-veined,** margins minutely saw-toothed (plants with purple or variegated leaves are ornamental cultivars). ***Flowers* in groups of 3–5 (rarely 2)** towards top of leafless stems (25–65 cm tall) and positioned at many angles on stem; stems round in cross-section; bracts lance-shaped. Each flower tiny and bisexual; 'petals' 6, spreading, pale purplish or bluish (sometimes white); stamens 6, filaments of stamens about the same length as 'petals'; ovary superior; style 1, column-like. *Fruits* burst early to reveal round blackish seeds (often mistaken for fruits) (c. 5 mm diam.). Fl. May–Jul, fr. Aug–Oct.

HARVESTING, SOURCING AND NATURAL RANGE

Tuberous roots are harvested in early summer, mostly from plants cultivated on small-holdings and medicinal plant farms with traditional *Di Dao* sources located in Hebei province (districts of Xiangyang, Gucheng and Laohekou).

The species grows wild in forests and grasslands on hillsides and in damp places (up to 1,800 m altitude); its natural range extends across C, S and W China (Anhui, Fujian, Gansu, Guangdong, Guangxi, Guizhou, Hainan, Hebei, Henan, Hubei, Hunan, Jiangsu, Jiangxi, Shaanxi, Shandong, Shanxi, Sichuan, Taiwan, Yunnan and Zhejiang provinces) [also wild in Japan, the Korean Peninsula, Vietnam].

CONSERVATION STATUS

China: Least Concern (MEPCAS, 2013).
Global: Not Evaluated (IUCN Red List, accessed 22/8/2016).

PHARMACOPOEIA SPECIES 2

L. muscari (Decne.) L. H. Bailey

Asparagaceae

synonyms: *L. platyphylla* F. T. Wang & Tang; *Ophiopogon muscari* Decne.

PLANT DESCRIPTION

Easily confused with PS1. Plants **without stolons** (creeping horizontal stems which produce new plants); leaves sometimes wider (up to 2 cm, rarely to 3.5 cm); **usually 9–11 veined; flowers in groups of 4–8 (rarely 3)**.

HARVESTING, SOURCING AND NATURAL RANGE

As for PS1 but the main production area is Fujian province (districts of Hui'an and Xianyou). In the wild in China, the species has a slightly narrower distribution: Anhui, Fujian, Guangdong, Guangxi, Guizhou, Henan, Hubei, Hunan, Jiangsu, Jiangxi, Shandong, Sichuan, Taiwan and Zhejiang [also wild in Japan].

CONSERVATION STATUS

China: Least Concern (MEPCAS, 2013).
Global: Not Evaluated (IUCN Red List, accessed 22/8/2016).

TCM PROPERTIES AND USE

properties: 'sweet, slightly bitter and cool'
channels: 'Heart, Lung and Stomach'
action: 'tonifies yin'

indications: dry cough; constipation; anxiety, insomnia

Large-scale cultivation of *Liriope spicata,* Hubei province

L. spicata in the wild, Anhui province

L. spicata (pale purple flowering form; sometimes also white)

L. spicata (pale purple flowering form)

Typical stolons of *L. spicata*

Freshly uprooted plant of *L. spicata* with root tubers

Seeding head of *L. spicata*

Flowering plant of *Liriope muscari*

Fresh rootstock of *L. muscari* (just washed)

Liriope spicata

Asparagaceae

DRUG MORPHOLOGY

CRUDE DRUG/DECOCTING PIECES (DRIED ROOT TUBER): long-ovoid, narrowing towards both ends **[1]** (1.2–3 cm long; 0.4–0.7 cm diam.) and often abruptly truncate at one or both ends **[2]**), slightly curved; remains of slender unswollen part of root sometimes persistent at one or both ends **[3]**. Externally yellowish-cream to brownish-yellow, with irregular longitudinal wrinkles **[4]**. *Texture*: pliable when fresh, becoming hard and brittle when dry. *Fracture*: pale yellow or brownish-yellow, horn-like **[5]**, with a small central stele **[6]**). *Odour*: indistinct. *Taste*: sweet; sticky when chewed.
***Good quality*:** plump, yellowish-white and somewhat translucent.

PROCESSING

Tubers are dried in strong sun (usually in heaps) until partially dried, then fibrous roots are removed berfore drying thoroughly (= crude/decocting pieces).

OTHER COMMON NAMES

寸冬 Cun Dong; 湖北麦冬 Hu Bei Mai Dong; 麦冬 Mai Dong; liriope root tuber (CP2010); Radix Liriopes.

SAFETY

An official healthfood in China (Anon., 2002a,b, as the synonym '*L. spicata* var. *prolifera* Y. T. Ma).

L. muscari

Asparagaceae

DRUG MORPHOLOGY

CRUDE DRUG/DECOCTING PIECES (DRIED ROOT TUBER): cannot be reliably differentiated from those of PS1 using macroscopic characters alone. Laboratory-based methods are required.

PROCESSING

As for PS1.

OTHER COMMON NAMES

寸冬 Cun Dong; 短葶山麦冬 Duan Ting Shan Ma Dong; 麦冬 Mai Dong; liriope root tuber (CP2010); liriope tuber (KP, 2007); liriopis tuber (KP, 2007); Radix Liriopes.

See also Mai Dong (Ophiogonis Radix) page 200.

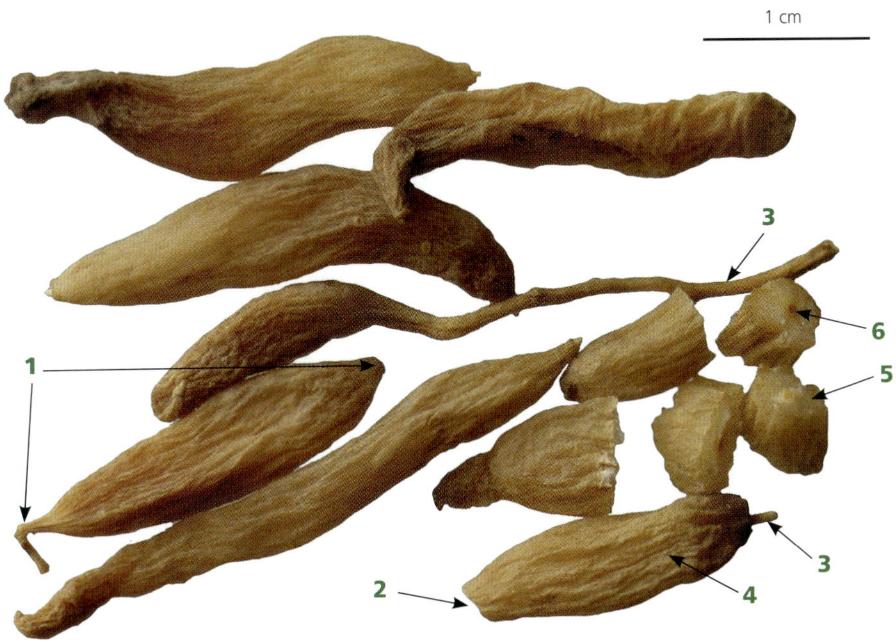

Crude drug/decocting pieces of Shan Mai Dong (sourced from *L. spicata*)

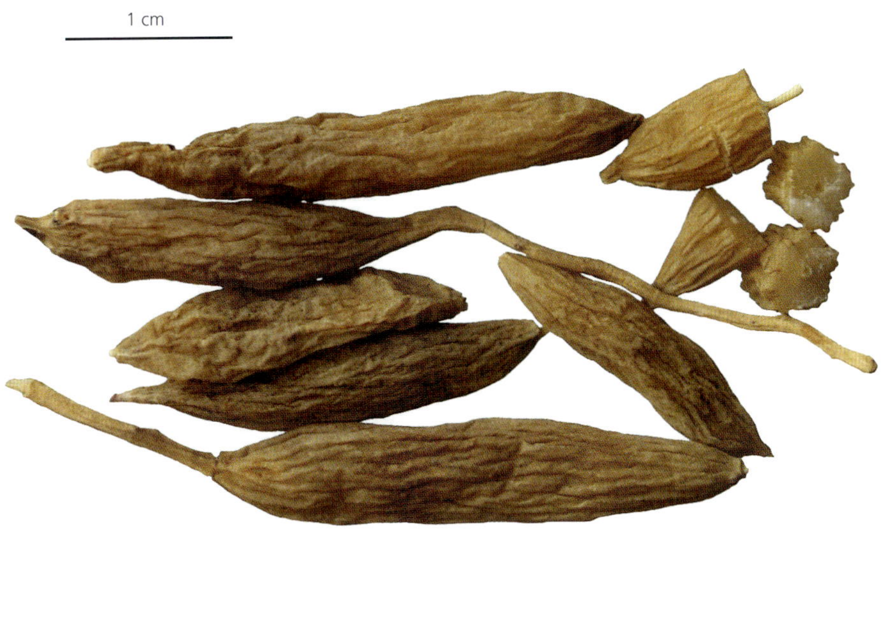

Crude drug/decocting pieces of Shan Mai Dong (sourced from *L. muscari*)

UNOFFICIAL SUBSTITUTES

1. ***Lophatherum gracile*** Brongn. (family Poaceae) — rhizomes of this grass are an occasional adulterant in local trade in China. Rhizomes narrower, often longer and slightly curved. *Taste*: mild. The species is common in the wild throughout W, C and S China and aerial parts are the source of the separate TCM drug Dan Zhu Ye (Lophatheri Herba) page 346.
2. ***Ophiopogon bodinieri*** H. Lév. (family Liliaceae) — root tubers. Endemic to C, W and S China. Stoloniferous. Leaves with **longitudinal whitish stripes**; flowers **nodding,** solitary or paired; fleshy seeds **bluish-purple.**

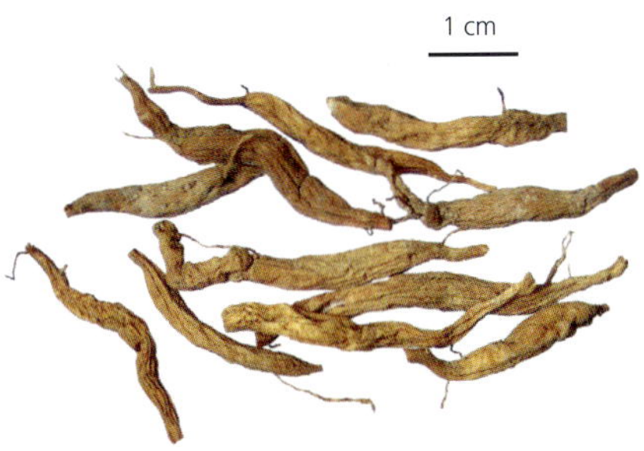

Rhizomes of Lophatherum gracile

Shan Yao 山药

Dioscoreae Rhizoma

Drug part: RHIZOME

Dioscorea polystachya Turcz.

Dioscoreaceae

Species of *Dioscorea* wild in China: 50

synonyms: *D. opposita* sensu auct., non Thunb. (name used in CP2015 as '*D. opposita* Thunb.'); *D. oppositifolia* sensu auct., non L.; *D. batatas* Decne.; *D. potaninii* Prain & Burkill; *D. rosthornii* Diels

Identification comment: the names '*D. opposita* Thunb.' and '*D. oppositifolia* L.' have been consistently mis-applied to the source species of 'Shan Yao'. *D. oppositifolia* L. (syn. *D. opposita* Thunb.) is, however, a separate species native to India and the Indian subcontinent (see photo of illustrative specimen below). The source of 'Shan Yao' is actually *D. polystachya* Turcz. (first pointed out in *FoC* followed by *FNA*) was confirmed in 2009 by M. Gilbert (*FoC* taxonomist, Missouri BG) and identified by P. Wilkin (*Dioscorea* taxonomist, RBG Kew) using herbarium voucher material (TCMK93 Lin & Leon) collected by the authors in China.

PLANT DESCRIPTION

A twining herbaceous perennial with a large, cylindrical vertically-growing, rhizome with white flesh (up to 1 m long). *Stems* often purplish red, hairless and twining to the right. Conspicuous fleshy bulbils (aerial tubers) sometimes present in leaf axils. *Leaves*, alternate or opposite; stalked (2–3 cm long); **leaf blades simple, ovate to triangular or arrowhead-like with 2 basal lobes**; margins entire; leaf blades often drying greyish or greenish. *Flowers* c. 1 mm long, creamish-white in linear spikes (**simple inflorescences**) with male and female flowers on separate plants. Male flower-spike in groups of 2–8, with a zigzag stem (when mature) (2–8 cm long); male flowers stalkless; stamens 6. Female flower-spike solitary or in groups of 2–3. *Fruit* a dry 3-winged capsule, waxy when immature, turning light brown when mature. *Seeds* winged. Fl. May–Aug, fr. Jun–Oct.

HARVESTING, SOURCING AND NATURAL RANGE

Rhizomes are harvested in winter mostly from extensively farmed medicinal crop plants after stems and leaves have withered. Material of *Di Dao*-quality is traditionally sourced from Henan province.

In the wild the species grows in forests and scrub on mountain slopes and along river- and roadsides (100–2,500 m altitude) and has a wide distribution in China (Anhui, Fujian, E Gansu, N Guangdong, Guangxi, Guizhou, Hebei, Henan, Hubei, Hunan, Jiangsu, Jiangxi, Jilin, Liaoning, S Shaanxi, Shandong, Sichuan, Taiwan, N Yunnan and Zhejiang. It is also native in Japan and the Korean Peninsula.

CONSERVATION STATUS

China: Least Concern (MEPCAS, 2013).

Global: Not Evaluated (IUCN Red List, accessed 11/6/2016).

RIGHT Herbarium specimen of *D. oppositifolia* L. with lanceolate leaves and compound (branched) inflorescences. In contrast, *D. polystachya* has leaves ovate or arrow-head-like with 2 basal lobes and simple (unbranched) inflorescences. See *Identification comment* at top of page.

TCM PROPERTIES AND USE

properties: 'sweet and neutral'
channels: 'Kidney, Lung and Spleen'
actions: 'tonifies qi, tonifies yin, tonifies Spleen, Lung, Kidney'
indications: fatigue, poor appetite, diarrhoea; cough, dyspnoea; leucorrhoea; spermatorrhoea, nocturnal emissions

Medicinal crop of. *D. polystachya*, Henan province

D. polystachya with stalkless male flowers

D. polystachya with warty bulbil

D. polystachya with developing capsules (fruits)

Dioscorea polystachya Dioscoreaceae

DRUG MORPHOLOGY

CRUDE DRUG-1 (DRIED RHIZOME — MAO SHAN YAO): irregularly cylindrical **[1]**, somewhat curved or flattened **[2]** (15–30 cm long; 1.5–6 cm diam.). Externally yellowish-white **[3]** with remnants of orangish-yellow outer skin **[4]**, occasional longitudinal grooves **[5]**, and scattered fibrous root scars. *Texture:* heavy, dense, tough and difficult to break. *Fracture:* white and starchy **[6]**. *Odour:* indistinct. *Taste:* slightly sour; viscous when chewed. ***Good quality:*** heavy, white and powdery. 'Guang Shan Yao' (below) is considered superior to 'Mao Shan Yao' (see *Processing*).

CRUDE DRUG-2 (DRIED PLANED RHIZOME — GUANG SHAN YAO): as for Mao Shan Yao but neatly cylindrical with sharply truncate ends **[7]** (9–18 cm long; 1.5–3 cm diam.); externally white or yellowish-white and smooth slices **[8]**; unsulphured rhizomes with darker markings of outer skin still visible **[9]**.

DECOCTING PIECES-1 (FRESH-SLICED RHIZOME PIECES): ovate, rounded or slightly undulant in outline **[10]** (1.5–6 cm diam., 2–4 mm thick) with cut surfaces white or pale yellowish-white, starchy and somewhat rough with vascular tissues shallowly domed **[11]**. *Texture* etc. as for crude drug.

DECOCTING PIECES-2 (RHIZOME SLICES): ovate **[12]** (1.5–6 cm diam.; 2–4 mm thick) with cut surfaces white or pale yellowish-white and starchy **[13]**. *Texture* etc. as for crude drug; easily broken.

DECOCTING PIECES-3 (BRAN-PROCESSED CUT RHIZOME SLICES — FU CHAO SHAN YAO): as for decocting slices above but cut surfaces darker **[14]** with occasional areas charred. *Odour*: somewhat charred.

Baskets of graded *D. polystachya* rhizome prior to polishing to become Guang Shan Yao, Henan province

PROCESSING

Rhizomes are washed and peeled then roots and rootlets removed before drying; alternatively rhizomes can be sliced while fresh, then dried (both types are called Mao Shan Yao 毛山药 'hairy yam' = crude drug-1). An alternative and more uniform trade item is made from large straight rhizomes which are planed into long, neatly cylindrical straight-ended pieces; these are sun-dried and sometimes whitened using sulphur before being polished (= Guang Shan Yao 光山药 'polished yam', crude drug-2). Both forms of crude drug are cut into 2–4 mm thick slices (= decocting pieces-1 and -2). Optional processing by dry-frying with wheat bran until yellow produces Fu Chao Shan Yao 麸炒山药 (= decocting pieces-3).

OTHER COMMON NAMES

麸炒山药 Fu Chao Shan Yao 'dry-fried mountain herb'; 光山药 Guang Shan Yao 'polished mountain herb'; 怀山药 Huai Shan Yao — from Henan province; 淮山药 Huai Shan Yao — from Jiangxi province; 毛山药 Mao Shan Yao 'hairy mountain herb'; Chinese yam (SCN, as '*D. oppositifolia*'); cinnamon vine; common yam rhizome (CP2010); Rhizoma Dioscoreae; Shu Yu.

SAFETY

An official food and healthfood in China (Anon., 2002a,b, as '*D. opposita*'). Safety Class: 1; Interaction Class: A (Gardner & McGuffin, (2013 as '*D. oppositifolia*').

UNOFFICIAL SUBSTITUTES

The genus *Dioscorea* is a large and difficult one requiring botanical expertise to reliably identify plants to species.The rhizomes of many other species of *Dioscorea* have been reported in historical texts as accidental or intentional substitutes in China (NICBP, 1997–2002; Bensky *et al*., 2004) but their occurrence in domestic or international trade today is uncertain. Of the 52 species of *Dioscorea* that grow wild in China, *D. persimilis* Prain & Burkill, *D. fordii* Prain & Burkill and *D. alata* L. are reported to be the most commonly encountered since they are used for both food and medicine.

COUNTERFEITS

In Hong Kong and Taiwan there have been reports of cassava root tubers (*Manihot esculenta* Crantz, family Euphorbiaceae) traded as counterfeit 'Shan Yao' (Ho & Chang, 2006) and in mainland China the same is true of sweet potato (*Ipomoea batatas* (L.) Lam., family Convolvulaceae) (Bensky *et al*., 2004). Laboratory-based authentication methods may be needed where counterfeiting is suspected.

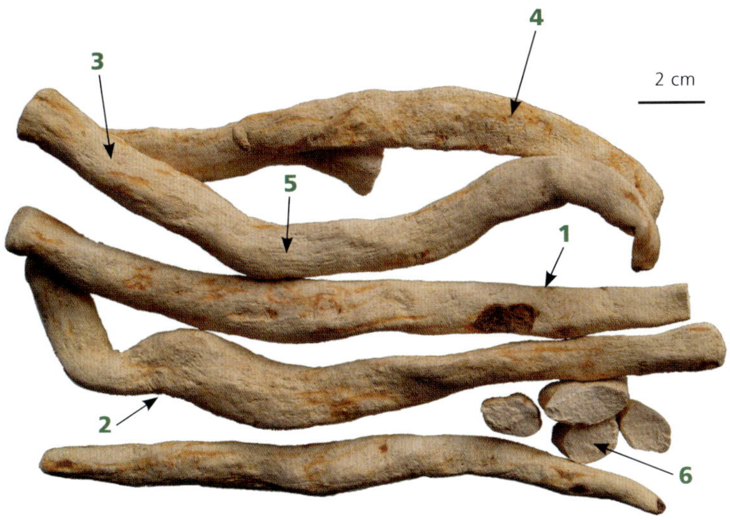

Mao Shan Yao (*D. polystachya*) = crude drug-1

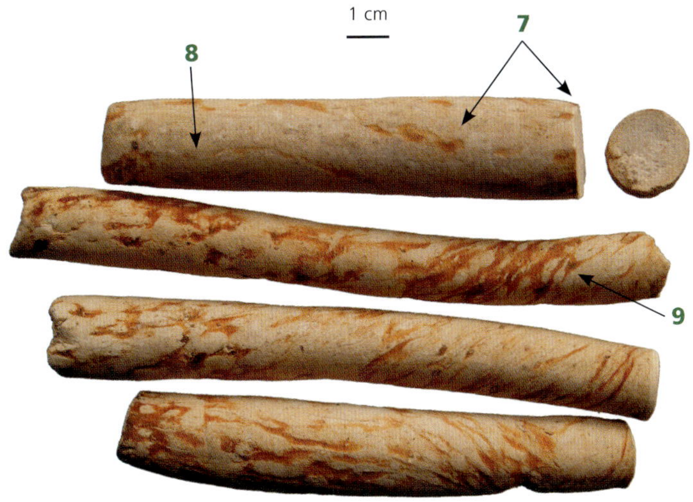

Guang Shan Yao (*D. polystachya*) = crude drug-2 (unsulphured)

Mao Shan Yao decocting pieces-1 (*D. polystachya*)

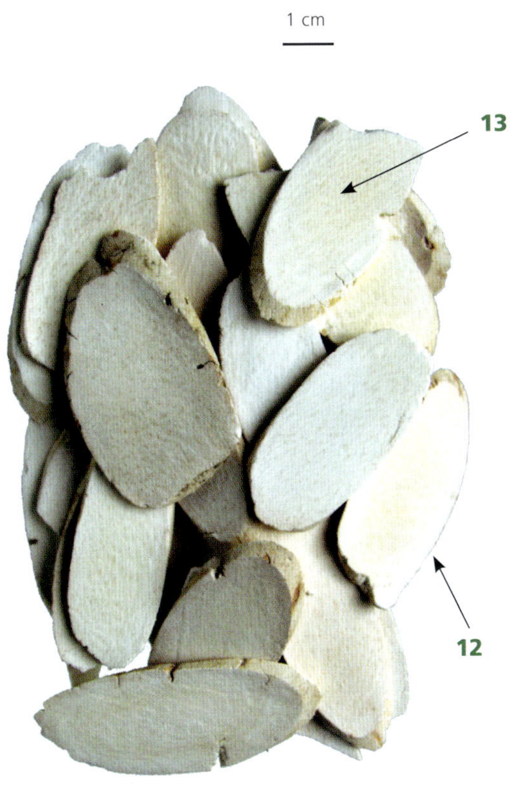

Decocting pieces-2 (*D. polystachya*) — sulphured

Bran-processed Shan Yao (Fu Chao Shan Yao) = decocting pieces-3

She Gan 射干

Belamcandae Rhizoma

Drug part: RHIZOME

PHARMACOPOEIA SPECIES

Belamcanda chinensis (L.) Redouté

Iridaceae

synonyms: *Iris domestica* (L.) Goldblatt & Mabb. (*Kew-MPNS* accepted name, 27/8/2016); *Ixia chinensis* L.

taxonomic/nomenclatural note: in CP2015 as '*Belamcanda chinensis* (L.) DC.'.

Species of *Belamcanda* wild in China: 1

PLANT DESCRIPTION

A herbaceous perennial with pale brown, nodular rhizomes. *Leaves* mostly basal, sword-shaped (20–60 cm long; 2–4 cm wide), parallel-veined with mid-vein obscure and **leaf blades oriented edgewise to flowering stems**; margins entire. ***Flowering stems* (1–1.5 m tall)** leafy and taller than basal leaves. *Flowers* large (3–4 cm diam.), conspicuous and few, in loose, branching heads; each flower shallowly bowl-shaped; **'petals'** 6, **orange**, obovate or elliptic and with dark red blotches; stamens usually 3; ovary 3-loculed; style 1. *Fruit* a dry many-seeded capsule **(2.5–3 cm long; 1.5–2.5 cm diam.)**. *Seeds* black and glossy. Fl. Jun–Aug, fr. Jul–Sep.

HARVESTING, SOURCING AND NATURAL RANGE

Sourced mainly as a medicinal crop which is grown throughout much of its natural range (Anhui, Fujian, Gansu, Guangdong, Guangxi, Guizhou, Hainan, Hebei, Heilongjiang, **Henan**, **Hubei**, Hunan, **Jiangsu**, Jiangxi, Jilin, Liaoning, Ningxia, Shaanxi, Shandong, Shanxi, Sichuan, Taiwan, Xizang, Yunnan and Zhejiang).

Dry mountain grassland near riverbeds, 0–2,200 m altitude.

CONSERVATION STATUS

China: Least Concern (MEPCAS, 2013).

Global: Not Evaluated (IUCN Red List, accessed 2/7/16).

UNOFFICIAL SUBSTITUTE

Iris tectorum Maxim.

Iridaceae

PLANT DESCRIPTION

Base of plant with many fibres (**old leaf bases**), flowering stems **shorter (20–40 cm); flowers large (c. 10 cm in diam.), 'petals' bluish-violet**; the 3 large-lobed, outer 'petals' with white, crested centres and the inner 'petals' more upright; fruits **larger** (**4.5–6 cm long; 2–2.5 cm diam.**); an ellipsoid or obovoid capsule Fl. Apr–May, fr. Jun–Aug.

HARVESTING, SOURCING AND NATURAL RANGE

Rhizomes are harvested throughout the year from wild and naturalised populations, which grow along forest margins, on sunny banks in mountain meadows and in damp places (500–3,500 m altitude) in the provinces of Anhui, Fujian, Gansu, Guangdong, Guangxi, Guizhou, Hainan, Hubei, Hunan, Jiangsu, Jiangxi, Qinghai, Shaanxi, Shanxi, Sichuan, Xizang, Yunnan and, Zhejiang [also wild in Japan, the Korean Peninsula and ?Myanmar]. Popularly grown as an ornamental in China and also available horticulturally in the West (Waddick & Zhao, 1992).

CONSERVATION STATUS

China: Least Concern (author's pers. obs. 2006).

Global: Not Evaluated (IUCN Red List accessed 8/7/2016).

SUBSTITUTION COMMENT. Long used as a substitute for 'She Gan' since the rhizomes and aerial parts closely resemble those of *B. chinensis*. Since 2005, its rhizomes have been officially recognised as the source of a separate TCM drug 'Chuan She Gan' (Iridis Tectori Rhizoma) (CP2005).

TCM PROPERTIES AND USE

properties: 'bitter and cold'
channel: 'Lung'
actions: 'clears heat, eliminates toxins, resolves phlegm'
indications: sore throat, coughs, dyspnoea

Belamcanda chinensis fruiting and flowering in cultivation

Flower of *B. chinensis*

Iris tectorum in the wild, Sichuan province

Flower of *I. tectorum*

Belamcanda chinensis Iridaceae

DRUG MORPHOLOGY

CRUDE DRUG (DRIED RHIZOME): nodular, tortuous and irregularly shaped (3–10 cm long; 1–2 cm diam.). Externally dull **yellowish-brown, dark brown or blackish-brown**, rough, with occasional large stem scar depressions (0.5–1 cm diam.) **[1]**, stem bases sometimes persistent **[2]**, zones of smaller crater-like root scars (c. 2–3 mm diam.) **[3]**; sometimes fibrous remnants of roots persistent **[4]** and zones of annular growth rings **[5]**. *Texture*: hard. *Fracture*: **yellow or yellowish-orange**, lax or granular. *Odour*: faint. *Taste*: bitter and slightly acrid. ***Good quality***: well-dried rhizomes devoid of fibrous roots and yellow/orange in cross-section.

DECOCTING PIECES (SLICED RHIZOME): thin slices (1–2 mm thick) with an irregular outline (1–2 cm diam.). Skin of rhizome rough, **dull yellowish-brown, dark brown or blackish-brown** with occasional crater-like root scars **[3]**, fibrous root bases **[4]** and annular rings **[5]**. Cut surface **pale yellow, yellowish-orange [6]** or (if poorly processed) yellowish-brown or grey) lax and fibrous or dense and granular **[7]**. *Odour*: indistinct. *Taste*: bitter and slightly acrid.

PROCESSING

After removal of fibrous roots, rhizomes are sun-dried (= crude drug) and subsequently softened before slicing thinly (1–2 mm). Sliced fresh rhizomes quickly lose the yellowish-orange colour of their flesh and are considered inferior.

OTHER COMMON NAMES

扁竹 Bian Zhu; 蝴蝶花 Hu Die Hua.

Iris tectorum Iridaceae

'DRUG SUBSTITUTE' MORPHOLOGY

DRIED RHIZOME: very similar to that of the Pharmacopoeia species; **often more conical** and externally yellowish- or **greyish-brown [8]**. *Taste*: sweet and bitter. The most reliable character is the fracture colour: **grey, yellowish-white or yellowish-brown [9]. Note:** poorly processed rhizomes of the Pharmacopoeia species (see note under *B. chinensis*) are virtually indistinguishable from this substitute and laboratory-based authentication methods may be needed.

DECOCTING PIECES (SLICED RHIZOME): as for Pharmacopoeia species but external (uncut) surface greyish- or yellowish-brown or brown. Cut surface **grey, yellowish-white or yellowish-brown [9]**. *Taste*: sweet and bitter.

OTHER COMMON NAMES

Chuan She Gan (Sichuan She Gan); 鸢尾 Yuan Wei (plant name).

LESSER KNOWN SUBSTITUTES

1. ***Iris dichotoma*** Pall. (野鸢尾 Ye Yuan Wei). *Rhizomes* greyish-brown, very short and stout. *Flowers* violet, pale blue, or cream with purplish-brown markings. *Fruit* yellowish-green, cylindrical (3.5–5 cm long; 1–1.2 cm diam.) with winged seeds.
2. ***I. japonica*** Thunb. (蝴蝶花 Hu Die Hua). *Rhizomes* suberect and thick or long, slender and creeping. *Leaves* yellowish-green and, basally, reddish-purple; glossy on one surface. *Flowers* pale bluish with prominent yellow crest. Widely cultivated; possibly naturalised.
3. ***Polygonatum cirrhifolium*** (Wall.) Royle (卷叶黄精 Juan Ye Huang Jing). *Rhizomes* cylindrical and tuberous or bead-like (1–2 cm diam.). *Leaves* in many-leaved whorls along stems. *Flowers* tiny, petals white, greenish or pale purple.

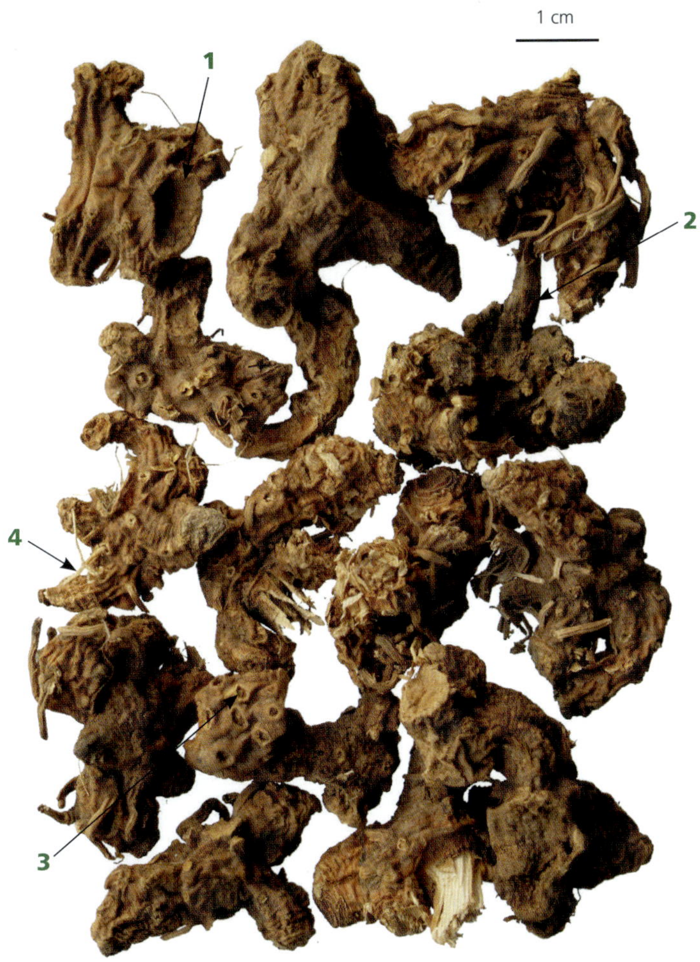

Crude drug She Gan (dried rhizomes of *Belamcanda chinensis*)

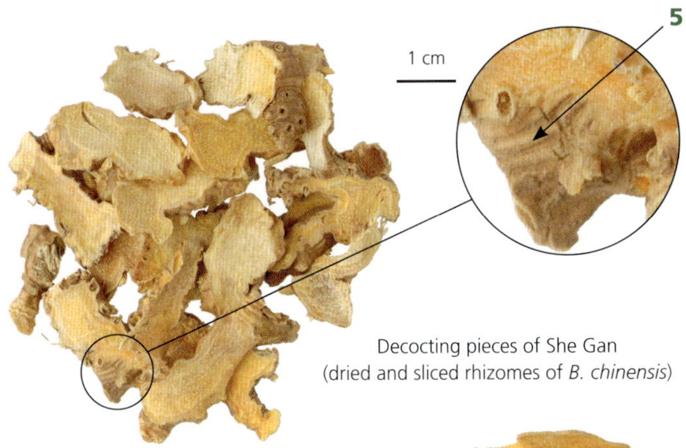

Decocting pieces of She Gan
(dried and sliced rhizomes of *B. chinensis*)

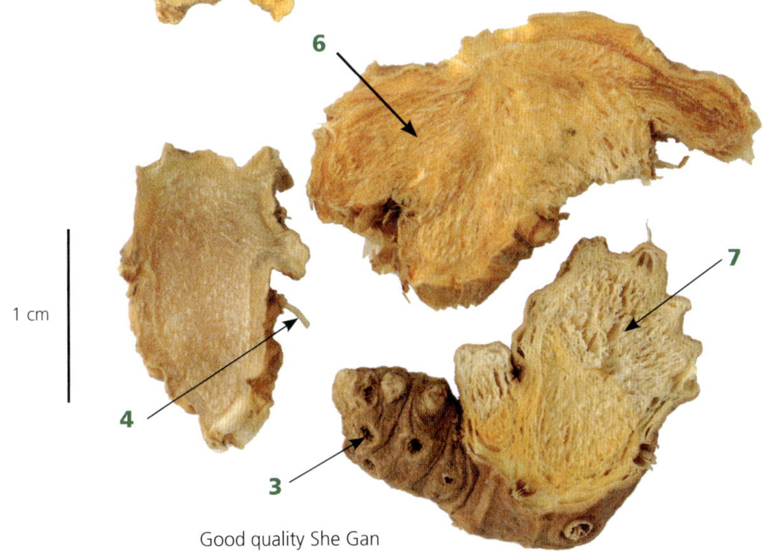

Good quality She Gan

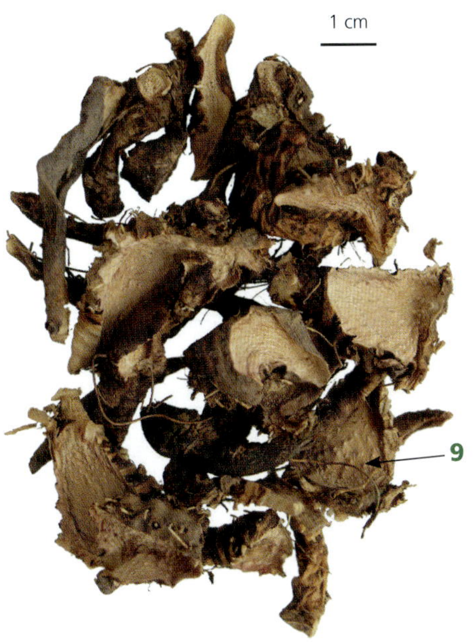

Crude drug Chuan She Gan
(dried rhizomes of *Iris tectorum*)

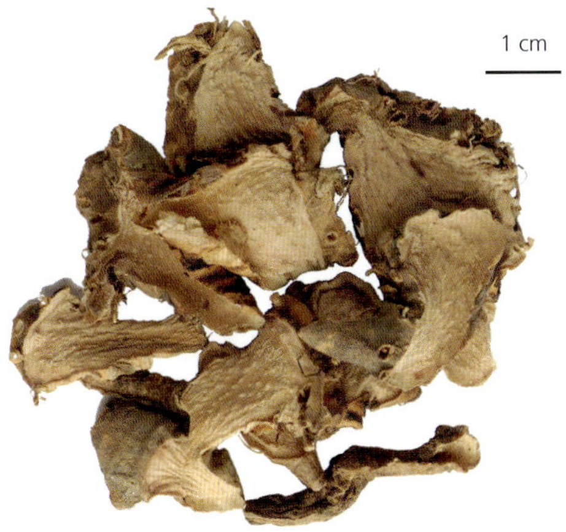

Decocting pieces of Chuan She Gan
(dried and sliced rhizomes of *I. tectorum*)

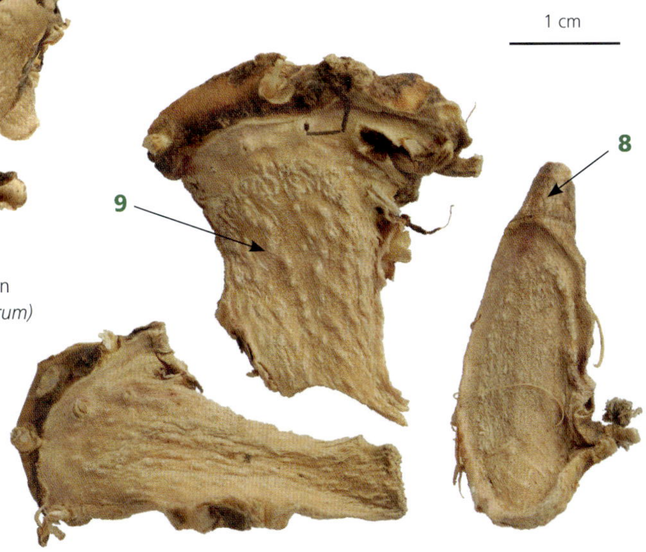

Iris tectorum (typical quality)

Tian Dong 天冬 (or Tian Men Dong 天门冬) Asparagi Radix

Drug part: ROOT TUBER

Asparagus cochinchinensis (Lour.) Merr.

Asparagaceae

synonym: *Melanthium cochinchinense* Lour.

Species of *Asparagus* wild in China: 29

PLANT DESCRIPTION

A medium-sized, scrambling perennial herb (1–2 m tall), with dense clusters of swollen and elongated root tubers (3–5 cm long; 1–2 cm diam.). *Stems* arching, angled or slightly winged, with occasional spines (up to 3.5 mm). *'Leaves'* (modified stems called cladodes) needle-like (0.5–8 cm long; 1–2 mm wide), flat or slightly 3-angled in widely-spaced **clusters of 3**. *Flowers* tiny, usually in pairs in 'leaf' axils, petals pale green (2.5–3 mm long); male and female flowers on separate plants. *Fruit* fleshy, spherical (6–7 mm diam.), **pale green when ripe**. *Seeds* 1 or 2. Fl. May–Jun, fr. Sep.

HARVESTING, SOURCING AND NATURAL RANGE

Root tubers are harvested in autumn and winter usually from medicinal crops grown predominantly in Guangxi, Guizhou and Sichuan provinces.

The wild species has a large natural range in China (Anhui, Fujian, Gansu, Guangdong, Guangxi, Guizhou, Hainan, Hebei, Henan, Hubei, Hunan, Jiangsu, Jiangxi, Shaanxi, Shandong, Shanxi, Sichuan, Taiwan, Tibet, Yunnan and Zhejiang) where it grows in open forest, roadsides and waste places up to 1,700 m altitude [also wild in Japan, Korean Peninsula, Laos and Vietnam].

CONSERVATION STATUS

China: Least Concern (MEPCAS, 2013); Least Concern.
Global: Not Evaluated (IUCN Red List accessed, 7/4/2015).

A medicinal crop of *Asparagus cochinchinensis*, Guangxi province

Flowering stems of *A. cochinchinensis* (male plant)

Spiny stem of *A. cochinchinensis*

Green, ripe fruit of *A. cochinchinensis*

Fresh root tubers of *A. cochinchinensis*

Peeled root tubers of *A. cochinchinensis* sun-drying, Guangxi province

TCM PROPERTIES AND USE

properties: **'sweet, bitter and very cold'**
channels: **'Lung and Kidney'**
actions: **'clears Lung heat, clears fire, tonifies yin, moistens dryness'**

indications: **coughs, sore throat, fever, constipation**

DRUG MORPHOLOGY

CRUDE DRUG (BOILED, STEAMED AND DRIED TUBER): tapering at both ends (fusiform) **[1]**, often curved (5–18 cm long; 0.5–2 cm diam.). Externally translucent yellowish-white or yellowish-brown; shallowly or deeply furrowed longitudinally **[2]** with occasional remnants of greyish-brown outer skin **[3]**. *Texture*: hard or mucilaginous. *Fracture:* horny, yellowish-white throughout. *Odour*: indistinct. *Taste*: sweet and somewhat bitter. ***Good quality***: tubers yellowish-white, translucent and pliable.

DECOCTING PIECES (SLICED TUBER): subrounded, semi-opaque and somewhat translucent yellowish-white or yellowish-brown throughout (0.5–2 cm diam.; 1–2 mm thick) **[4]** with a central pale or dark stele **[5]**.

PROCESSING

After washing, then removal of stem bases and rootlets, root tubers are boiled or steamed until the central vascular tissues are no longer white. They are then peeled while hot, then washed and dried (= crude drug). Decocting pieces are thin transverse or longitudinal slices (1–2 mm thick).

OTHER COMMON NAMES

天门冬 Tian Men Dong; Chinese asparagus (SCN); cochinchinese asparagus root (CP2010); Radix Asparagi.

SAFETY

Safety Class: 1; Interaction Class: A (Gardner & McGuffin, 2013). An official healthfood in China (Anon., 2002a,b).

Crude drug of Tian Dong (steamed root tubers of *Asparagus cochinchinensis)*

Transversely sliced decocting slices of Tian Dong

LOCAL UNOFFICIAL SUBSTITUTE

Asparagus filicinus D. Don (羊齿天门冬 Yang Chi Tian Men Dong) — root tubers (Asparagaceae family). A wild species of C, W and E China. Stems without spines, leaves in groups of 5–8 and flowerstalks longer. Root tubers macroscopically indistinguishable from those of the Pharmacopoeia species such that laboratory-based authentication methods are recommended. Given the widespread medicinal cultivation of the Pharmacopoeia species, however, international trade in this local substitute is considered unlikely.

Tian Ma 天麻 — Gastrodiae Rhizoma

Drug part: RHIZOME (stem tuber)

Gastrodia elata Blume

Orchidaceae

Species of *Gastrodia* wild in China: 15

PLANT DESCRIPTION

An erect and leafless orchid (30–100 cm tall, rarely up to 2 m) devoid of chlorophyll and dependent on its fungal partner (the honey fungus *Armillaria mellea*) and its host of decidous trees for its nutrients. *Rhizomes* large, fleshy and ellipsoid (8–12 cm long; 3–5(–7 cm diam.) with many transverse annulations dotted with tiny, broadly ovate or triangular bud scales. *Leaves* absent. *Aerial parts* a solitary, unbranched stem bearing 20–50 inconspicuous flowers towards its tip. *Stem* (peduncle) orange, yellow or sometimes greyish-brown or greenish throughout with membranous sheaths at the base. *Flowers* becoming denser towards the stem tip; bract (at base of each flower) short, membranous and narrowly pointed (10–16 mm long); flower stalk fleshy (7–12 mm long), same colour as stem; sepals and petals fused to form a smooth, pouch (8–10 mm long; 5–7 mm wide), orange, pale yellow, bluish-green or yellowish-white with a bulbous base and mouth constricted where a small but deep sinus separates the lateral 'petals'; lip (enclosed in the upper side of the flower) with irregularly-fringed margins, somewhat swollen; anther 1, deep inside each flower and modified into 2 pollen masses (pollinia). *Fruit* dry and ovoid (14–18 mm long; 9 mm diam.). Fl. and fr. May–Jul.

Mature inflorescence of *G. elata* (cultivated), Sichuan province

Fresh tubers of *G. elata* with stem remnant at apex, growth rings and (on middle tuber) remnants of honey fungus

Close-up of *G. elata* flowers

HARVESTING, SOURCING AND NATURAL RANGE

Today, most 'Tian Ma' is sourced from artificially propagated rhizomes with best quality ('Dong Ma' i.e. winter Tian Ma) harvested in November. Harvesting continues throughout winter but rhizomes sprouting in early spring are considered inferior and traded as 'Chun Ma' (spring Tian Ma). Wild-sourced rhizomes command a higher price than both their cultivated counterparts since traditional knowledge considers them clinically superior.

Despite the species' large natural distribution in China where it grows between 400–3,200 m altitude in open forests and thickets, the 1970s and 80s saw its wild populations plummet due to over-harvesting and extensive deforestation; *G. elata* is totally reliant on its host plants, deciduous trees, along with the honey fungus, for its survival. Successful artificial propagation was not established on a commercial scale until the late 1980s satisfying the huge demand for this drug and undermining the previously vast industry that had develped in 'Tian Ma' counterfeits (see *Unofficial substitutes* below). It was China's national logging ban of 1998 that provided the turning point for the recovery of the species' wild populations and today, wild harvesting, following sustainable harvesting practices, is permitted in some localities.

Di Dao-quality 'Tian Ma' is produced in Guizhou province 'Gui Tian Ma' (in the districts of Anshun, Bijie, and Zunyi) and a GAP medicinal cultivation standard has been developed with GAP-certified bases established in Shaanxi province. The species is also native to Bhutan, NE India, Japan, the Korean Peninsula, Nepal and Russia (Far East).

CONSERVATION STATUS

China: Data Deficient (MEPCAS, 2013).

Global: Vulnerable (IUCN Red List, accessed 26/8/2016). See also *Harvesting* ... above.

PROTECTION

CITES: Appendix II; EU Annex B (see Table 1 page 10).

TIAN MA COUNTERFEITS

Tubers and rhizomes of readily available edible or ornamental species have long been used to make counterfeits of 'Tian Ma' before commercial cultivation of *G. elata* made such activites uneconomic. They included:

1. ***Adenostyles*** spp. (formerly *Cacalia*) — Asteraceae family
2. ***Arnoglossum*** spp. (formerly *Cacalia*) — Asteraceae family
3. ***Canna indica*** L. (syn. *C. edulis*) (canna lily) – Cannaceae family
4. ***Dahlia pinnata*** Cav. — Asteraceae family
5. ***Ipomoea batatas*** (L.) Lam. (sweet potato) — Solanaceae family
6. ***Mirabilis jalapa*** L. (marvel of Peru) — Nyctaginaceae family
7. ***Solanum tuberosum*** L. (potato) — Solanaceae family

Other forms of adulteration to increase the weight (hence price) of authentic rhizomes have included the insertion of metal weights.

TCM PROPERTIES AND USE

properties: **'sweet and neutral'**
channel: **'Liver'**
action: **'extinguishes wind, pacifies yang'**

indications: **spasms, tremors, convulsions, stroke; paralysis, numbness; dizziness, headache, hypertension**

DRUG MORPHOLOGY

CRUDE DRUG (BOILED OR STEAMED RHIZOME): irregularly cylindrical, slightly compressed laterally, often slightly curved (3–15 cm long; 1.5–6 cm diam.; 0.5–2 cm thick). Externally pale yellowish-brown or yellowish-white, longitudinally wrinkled and often furrowed **[1]**; widely spaced annular rings throughout the rhizomes' length **[2]** usually dotted with pale broadly ovate or triangular bud scales **[3]**. Remnants of honey fungus occasionally persist as thin black mycelial threads. Apex with brown, reddish-brown or pale red stem remnant, sometimes angled and indistinct **[4]**. Basal end somewhat truncate with roundish, sunken or crater-like scar **[5]**. *Texture*: hard and difficult to break. *Fracture*: yellowish-cream to brown, often with a horn-like lustre; relatively even. *Odour*: indistinct. *Taste*: somewhat sweet. ***Good quality*:** rhizomes large and dense, externally yellowish-white, internally glossy and translucent.

DECOCTING PIECES (SLICED RHIZOME): irregularly shaped, often oblong slices (1–2 mm thick). Cut surface pale yellowish-white to pale yellowish-brown, translucent and often with a horn-like lustre **[6]**; vascular tissues as indistinct striations or dots **[7]**. Brown, reddish-brown or pale red stem remnant occasionally visible **[8]**. *Odour*: indistinct. *Taste*: somewhat sweet.

PROCESSING

Rhizomes are boiled or steamed until soft thoroughout then flattened and dried in stages (= crude drug). Dried rhizomes are thinly sliced (1–2 mm) then dried agan prior to decocting.

OTHER COMMON NAMES

川天麻 Chuan Tian Ma (Sichuan-sourced Tian Ma); 春天麻 Chun Tian Ma; 冬天麻 Dong Tian Ma (winter Tian Ma); 贵天麻 Gui Tian Ma (Guizhou-sourced Tian Ma); 明天麻 Ming Tian Ma; gastrodia (SCN); Rhizoma Gastrodiae; tall gastrodia tuber (CP2010).

SAFETY

Safety Class: 1; Interaction Class: A (Gardner & McGuffin, 2013). An official healthfood in China (Anon., 2002a,b).

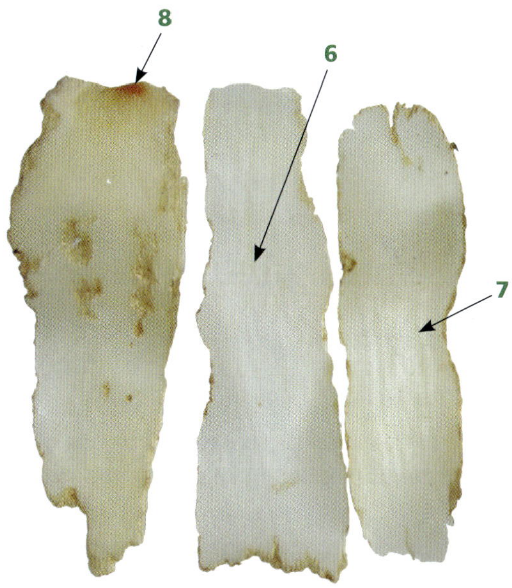

Longitudinal decocting pieces of Tian Ma (*G. elata*)

Crude drug of Tian Ma (*G. elata*)

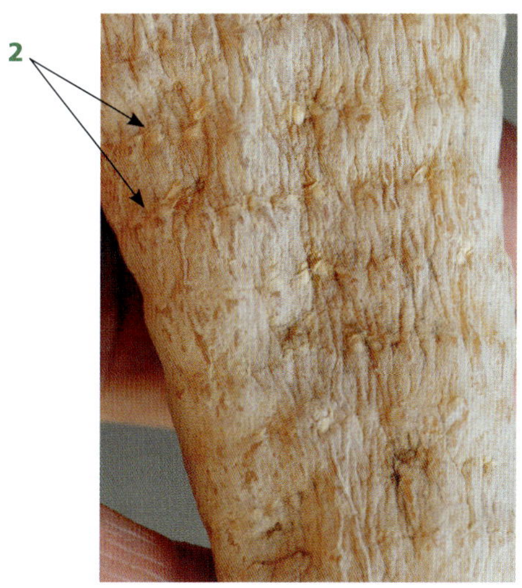

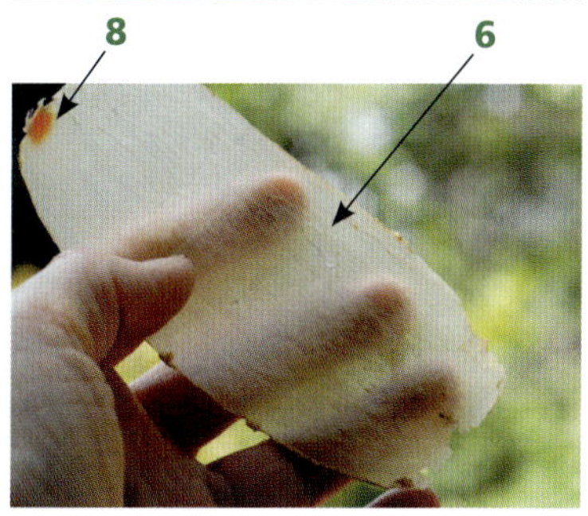

Tian Nan Xing 天南星

incl. Zhi Tian Nan Xing 制天南星

Arisaematis Rhizoma

Drug part: UNPROCESSED RHIZOME

PHARMACOPOEIA SPECIES 1

Arisaema erubescens (Wall.) Schott

Araceae

synonym: *Arum erubescens* Wall.

Species of *Arisaema* wild in China: 78

PLANT DESCRIPTION

A perennial herb (up to 120 cm tall) with a squat, subglobose rhizome (2–7 cm diam.). ***Leaves* 1 (rarely 2)** with long, bare brownish-purple spotted leaf 'stems' (up to 100 cm long; c. 2 cm diam.); leaf blades large and divided into **18–23 radiating leaflets**, oblong to elliptic, (16–28 cm long; 0.2–2 cm wide), **leaflet tips slender and thread-like (7–10 cm long).** *Flowers* in a single flower-spike (appearing below the leaf blades) enclosed by a partly tubular spathe (green with white or purple stripes or green throughout) with a long thread-like purple pendant tip up to 35 cm long. Flower spike (spadix) a straight fleshy column bearing the densely packed upper zone of male flowers (dark violet) above the lower zone of female flowers (green); spadix tip upright, cylindrical, pale green and devoid of flowers. *Fruit-head* nodding; berries red. Fl. Jun–Jul.

HARVESTING, SOURCING AND NATURAL RANGE

Rhizomes are wild-harvested in autumn and winter after stems and leaves have withered. The species is widely distributed in China (Anhui, Fujian, Gansu, Guangdong, Guangxi, Guizhou, Hebei, Henan, Hubei, Hunan, Jiangxi, Shaanxi, Shandong, Shanxi, Sichuan, Taiwan, Yunnan and Zhejiang) and grows in temperate pine and mixed forest, thickets and grassy slopes, often between rocks (up to 3,200 m altitude). Also native in Bhutan, NE India, Laos, Myanmar, Nepal, N Thailand and Vietnam.

CONSERVATION STATUS

China: Least Concern (MEPCAS, 2013).
Global: Not Evaluated (IUCN Red List, accessed 20/2/2016).

PROTECTED STATUS

EU Annex D (EU Wildlife Trade Legislation).

PHARMACOPOEIA SPECIES 2

A. heterophyllum Blume

Araceae

PLANT DESCRIPTION

Rhizomes 2–6 cm diam. Leaflets **spaced at intervals of 0.5–5 cm** (**not radiating** from a central point); **central leaflet much smaller than adjacent lateral leaflets**. Spathe pale green to pale yellow; spadix either with male and female flower zones or male zone only; tip of spadix S-shaped and gradually tapering. Berries yellowish-red or red.

HARVESTING, SOURCING AND NATURAL RANGE

Harvesting as for PS1. The species is widely distributed in the wild throughout much of China, except Tibet, and grows in grassland and thickets in temperate mixed forest prone to frequent flooding (up to 2,700 m altitude).

Also native in Japan and the Korean Peninsula.

CONSERVATION STATUS

China: Least Concern (MEPCAS, 2013) but populations declining due its fragile habitat (Crook, 2013).
Global: Not Evaluated (IUCN Red List, accessed 20/2/2016).

TCM PROPERTIES AND USE

properties: **'bitter, pungent'and warm'**
channels: **'Liver, Lung and Spleen'**
actions: **'clears heat, resolves phlegm', 'extinguishes wind' (Zhi Tian Nan Xing)**

indications: **Tian Nan Xing — abscesses (topical use only); Zhi Tian Nan Xing — coughs with phlegm; convulsions, stroke; (T)**

Arisaema erubescens flowering in the wild

Inflorescence of *A. erubescens*

A. erubescens wild in Yunnan province

Arisaema heterophyllum in the wild, Anhui province, China

A. heterophyllum with developing flower-head, Anhui province, China

Fresh unprocessed rhizomes of *A. heterophyllum*

Arisaema amurense Maxim.

Araceae

PLANT DESCRIPTION

Similar to PS1 and PS2 but **rhizomes larger** c. 6–7 cm diam.; leaves divided into **3–5 or more leaflets**; **leaflets ovate** and often widest towards the tips **(7–11 cm long; 4–7 cm)**, sometimes with fine whitish markings, margins entire or toothed and tips shortly pointed. Spathe greenish or dark purple often with pale purple stripes **with tip shortly pointed**. **Spadix tip almost extending beyond the encircling spathe**. Berries red, narrowly conical; seeds red and ovoid. Fl. May, mature fr. Sep.

HARVESTING, SOURCING AND NATURAL RANGE

Harvesting as for PS1. Wild populations are confined to N and NE China (provinces of Hebei, Heilongjiang, Henan, Inner Mongolia, Jilin, Liaoning, Ningxia, Shandong and Shanxi where they grow along streamsides in temperate deciduous and mixed forests (up to 200 m altitude). [Also in Korean Peninsula, Russia (Siberia)].

CONSERVATION STATUS

China: Least Concern (MEPCAS, 2013).
Global: Not Evaluated (IUCN Red List, accessed 20/2/2016).

Pinellia pedatisecta Schott

Araceae

PLANT DESCRIPTION

Easily mistaken for any of the official species. *Main differences*: **rhizomes** (up to 4 cm diam.) **often with tubercles** ('daughter' rhizomes) growing on the outermost edge, especially when mature. **Leaves** 1–5, **divided into 6–11 lobed or successively divided leaflets. Spathe narrowly ellipsoid** (20–50 cm long), gaping at base. **Spadix much longer than spathe and thread-like at tip;** female zone of flowers fused to the spathe. **Berries green yellowish-green or whitish.**

SUBSTITUTION COMMENT. Rhizomes of this species are often traded as a cheap and readily available closely related look-alike of official Tan Nan Xing. It is a common species in the wild in C and E China and is easy to cultivate, but although purported to have similar clinical properties, pharmalogically they are considered inferior.

CONSERVATION STATUS

China: Not Evaluated.
Global: Not Evaluated (IUCN Red List, accessed 20/2/2016).

LESSER KNOWN UNOFFICIAL SUBSTITUTES

Many other wild species of *Arisaema* (78 are wild in China) are easily mistaken for the 3 Pharmacopoeia species at point of harvest. The frequency of such substitution (accidental or intentional) in local and international markets is unclear.Laboratory-based identification methods are needed to differentiate the rhizomes of this complex group. It is likely the toxicity of such substitutes will be similar to that of the official species.

A. amurense flowering in the wild, Liaoning province, China

A. amurense in full flower

Pinellia pedatisecta (in full flower) grown on a small-holding in Shaanxi province, central China

Flowering plant of *P. pedatisecta*

Arisaema erubescens

Araceae

DRUG MORPHOLOGY

UNPROCESSED DRIED RHIZOME 'SHEN TIAN NAN XING': *Rarely seen in trade owing to its toxicity.* Rhizomes squat, sub-globose (1–2 cm high; 1.5–6.5 cm diam.). Externally yellowish-brown and somewhat smooth in places **[1]**; apical end with a sunken dorsal depression (leaf stem scar) **[2]** surrounded by a small to extensive rough zone **[3]** with cream, pitted root scars **[4]**; basal end rounded and relatively smooth **[5]**. Small, squat, lateral buds (tubercles) occasional **[6]**. *Texture*: hard and difficult to break. *Fracture*: uneven, white and powdery **[7]** (see *A. heterophyllum* photo). *Odour*: acrid. *Taste*: numbing and pungent. ***Good quality***: large and white; powdery in cross section.

UNPROCESSED DRIED RHIZOME SLICES OR POWDER 'SHEN TIAN NAN XING': not seen.

PROCESSED AND SLICED RHIZOME 'ZHI TIAN NAN XING': sub-rounded or irregularly shaped, thin slices of rhizome (1.5–6.5 cm diam.; 1–2 mm thick). Yellow or pale brown. *Texture:* brittle and easily broken. *Fracture:* horny. *Taste:* slightly numbing and astrigent. ***Good quality***: cross-section **horny, not powdery.**

PROCESSING

After removal of the fibrous outer tissues, rhizomes are dried (= unprocessed crude drug) and later sliced then powdered (= Shen Tian Nan Xing 'unprocessed Tian Nan Xing'; for topical use only, see *Safety*). To reduce the toxicity of unprocessed rhizomes they typically undergo careful processing involving soaking in several changes of water before subsequent boiling with fresh alum and ginger; rhizomes are then sliced when semi-dried, then dried thoroughly and powdered before application (= Zhi Tian Nan Xing i.e. 'processed Tian Nan Xing').

OTHER COMMON NAMES

Unprocessed rhizome: 生天南星 Sheng Tian Nan Xing; 一把伞南星 Yi Ba San Nan Xing.
Processed rhizome: 制南星 Zhi Nan Xing; 制天南星 Zhi Tian Nan Xing; Shou Tian Nan Xing.
Either processed or unprocessed: Chinese arisaema (SCN); Jack in the pulpit tuber (CP2010); Rhizoma Arisaematis.

A. heterophyllum and *A. amurense*

Araceae

DRUG MORPHOLOGY

Macroscopic characters are insufficiently robust to reliably differentiate the unprocessed and processed drug from those of PS1. Laboratory-based methods are recommended.

PROCESSING

As for PS1.

OTHER COMMON NAMES

Unprocessed rhizome: 生天南星 Sheng Tian Nan Xing.
Processed rhizome: 制南星 Zhi Nan Xing; 制天南星 Zhi Tian Nan Xing.
Either processed or unprocessed: Chinese arisaema (SCN); Jack in the pulpit tuber (CP2010); Rhizoma Arisaematis.
A. amurense is also called: 东北南星 Dong Bei Nan Xing 'north-east Nan Xing'.

SAFETY

As for PS1.

Pinellia pedatisecta

Araceae

'DRUG SUBSTITUTE' MORPHOLOGY

Rhizomes of this species are easily mistaken for the processed or unprocessed drugs of the 3 Pharmacopoeia species (PS1–3) especially when immature. ***P. pedatisecta* rhizomes seldom grow larger than 4 cm diameter.** Mature rhizomes typically develop **1–4 lateral buds (tubercles) [8]** and the name Zhang Nan Xing 虎掌 南星 'tiger's paw Nan Xing' is used traditionally to describe their shape; tubercles are only occasionally seen in the official drug. Laboratory-based methods are strongly recommended to differentiate these rhizomes from those of official Tian Nan Xing.

OTHER COMMON NAMES

掌叶半夏 Zhang Ye Ban Xia (plant name); 虎掌 Hu Zhang 'tiger's paw'; Hu Zhang Nan Xing 虎掌 南星 'tiger's paw Nan Xing'.

SAFETY

Probably a similar toxicity to the medicinal rhizomes of its close relative *Pinellia ternata* (see Ban Xia (Pinelliae Rhizoma) (page 50) the **unprocessed rhizomes** of which are **toxic** (CP2015; Bensky *et al.*, 2004).

SAFETY

Processed and unprocessed rhizomes are **toxic** (CP2015). Fresh unprocessed rhizomes should be handled with gloves to avoid contact irritation. *Dried unprocessed rhizomes* are **for topical application only** (Bensky *et al.*, 2004; Chen & Chen, 2004; Gardner & McGuffin, 2013). *Processed rhizomes* are to be used **'only under the supervision of an expert qualified in the appropriate use of this substance' (Safety Class 3);** Interaction class: A (Gardner & McGuffin, 2013). Processed and unprocessed rhizomes are officially forbidden in China both as food and healthfood (Anon., 2002a,b).

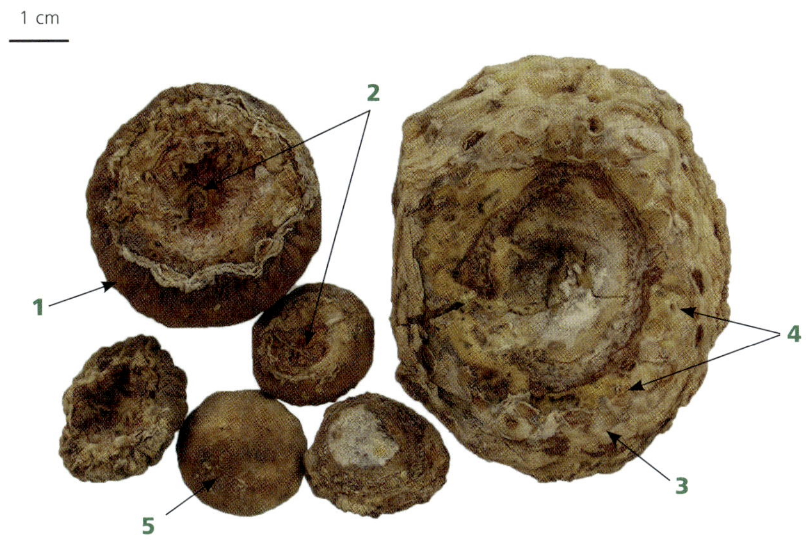

Unprocessed dried rhizomes of Tian Nan Xing (Shen Tian Nan Xing) of *Arisaema erubescens*

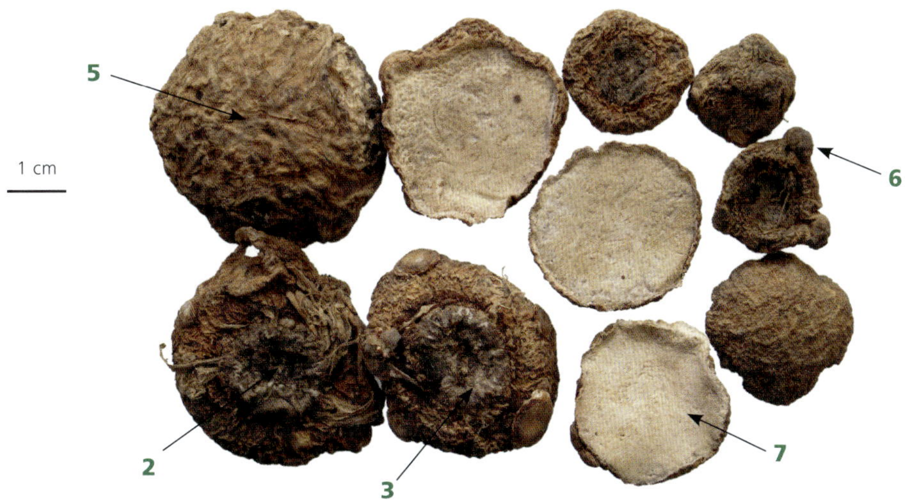

Unprocessed rhizomes of *Arisaema heterophyllum*

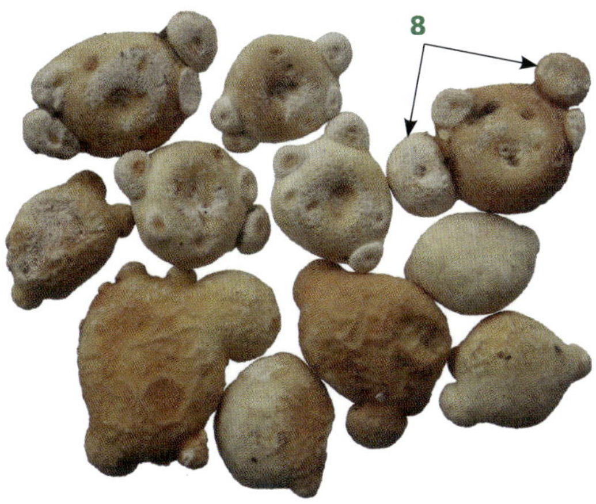

Dried and sulphured mature rhizomes 'drug' of *Pinellia pedatisecta*

1 cm

Dried immature rhizomes of *P. pedatisecta*; some rhizomes typically with 'daughter' rhizomes developing

Tu Fu Ling 土茯苓 Smilacis Glabrae Rhizoma

Drug part: RHIZOME

PHARMACOPOEIA SPECIES

Smilax glabra Roxb.

Smilacaceae

Species of *Smilax* wild in China: 79

PLANT DESCRIPTION

A thornless, woody perennial vine (up to 4 m) with well-developed tendrils and a stout, fleshy and nodular rhizome. *Leaves* simple and alternate; leaf stalk 5–15 (rarely 30) cm long and narrowly winged along part of length; leaf blades somewhat leathery, narrowly or broadly oval (6–15 cm long; 1–7 cm wide), main veins 3–7, margins entire. *Flowers* in a solitary, 10–30-flowered cluster (umbel), on a short stalk (1–5 mm long) in leaf axil; bracteoles many. Male and female flowers on separate plants. Male flowers tiny (c. 2 mm long; 3 mm diam.), slightly 6-angled, greenish-white, 'petals' with a ragged-toothed margin, hooded and hardly opening; stamens 6. Female flowers tiny, 'petals' with an entire margin, very short style and 3 stigmas. *Fruits* in spherical heads; each fruit a spherical bluish-black berry (6–10 mm diam.) with a white powdery surface. **Fl. Jul–Nov, fr. Nov–Apr.**

HARVESTING, SOURCING AND NATURAL RANGE

Rhizomes are harvested year-round, mostly from wild plants. The species grows on lightly forested slopes in scrub, in valleys and along riversides (300–1,800 m altitude) in S, C and W China: Anhui, Fujian, S. Gansu, Guangdong, Guangxi, Guizhou, Hainan, Hubei, Hunan, Jiangsu, Jiangxi, Sichuan, Taiwan, Tibet, Yunnan and Zhejiang [also wild in India, Myanmar, Thailand, Vietnam].

CONSERVATION STATUS

China: Least Concern (MEPCAS, 2013); plants not easily destroyed due to their deep rhizomes, parts of which often survive *in situ* after harvesting (authors' pers. obs.).
Global: Not Evaluated (IUCN Red List, accessed 25/5/2015).

UNOFFICIAL SUBSTITUTE 1

S. china L.

Smilacaceae

PLANT DESCRIPTION

Similar to the official species but **stems sparsely prickly**; leaves **elliptic to orbicular** (3–10 cm long; 1.5–6(–10) cm. wide); flowers yellowish-green, emerging before leaves mature; **fruits red** (0.6–1.5 cm diam.); **fl. Feb-May; fr. Sep–Nov.**

SUBSTITUTION COMMENT. Rhizomes of this closely related species are the official source of 'Ba Qia' (Smilacis Chinae Rhizoma) page 58 and often used as a cheap and easy source of 'Tu Fu Ling' owing to their shallower-growing and similar-looking rhizomes which are more easily harvested.

UNOFFICIAL SUBSTITUTE 2

Fagopyrum dibotrys (D. Don) H. Hara

Polygonaceae

synonyms: *Fagopyrum acutatum* (Lehm.) Mansf.ex K. Hammer (*Kew-MPNS* accepted name, 28/8/2016); *Polygonum dibotrys* D. Don.

PLANT DESCRIPTION

Aerial parts easily distinguished from those of the Pharmacopoeia species: an upright, non-woody, much branched herbaceous perennial (40–100 cm tall); **tendrils absent; leaf blades triangular (4–12 cm long; 3–11 cm wide);** margins entire; flowers 5-petalled, **white** (c. 2.5 mm long), in **irregularly shaped heads** (not spherical) from spring to late summer; fruits **inconspicuous,** dry, blackish-brown and **3-sided** (c. 6–8 mm long). Rhizomes similar to official species but blackish-brown and distinctly woody.

SUBSTITUTION COMMENT. Rhizomes similar to those of the official drug; they are the source of the separate TCM drug 'Jin Qiao Mai' (Fagopyri Dibotryis Rhizoma) see page 188.

TCM PROPERTIES AND USE

properties: **'sweet, bland and neutral'**
channels: **'Liver and Stomach'**
actions: **'clears heat, eliminates toxins, cools Blood'**

indications: **syphilis and gonorrhea; diarrhoea and dysentery; leucorrhoea; arthralgia; and skin conditions (e.g. psoriasis, sores and ringworm)**

Smilax glabra with ripe fruit

Freshly harvested rhizomes of *S. glabra*, Yunnan province

Female flowering plant of *Smilax china* in the wild, Anhui province

Fruiting plant of *S. china* with freshly harvested rhizome, Anhui province

Fruiting plant of *S. china*

Flowering plants of *Fagopyrum dibotrys*

Flower-head of *F. dibotrys*

F. dibotrys showing straight-topped stem sheath (ochrea)

Smilax glabra

Smilacaceae

DRUG MORPHOLOGY

CRUDE DRUG (DRIED RHIZOME): rhizomes stout, irregularly-shaped solid masses, often somewhat shortly cylindrical and slightly flattened (5–22 cm long; 2–5 cm diam.); short branches occasional; often nodular. External surface yellowish- or greyish-brown, rough and uneven, irregular fissures occasional, outer layers often flaking **[1]**; surface with hard, scattered and fibrous root remains **[2]**, large oval stem scars, stem remnants and stem buds. *Texture*: somewhat tough, producing powdery dust when broken; slippery when moistened. *Fracture*: whitish to pale reddish-brown. *Odour*: none. *Taste*: somewhat sweet and astringent. ***Good quality***: pale brown externally and powdery internally.

DECOCTING PIECES (SLICED RHIZOME): oval, oblong or irregularly-shaped whitish or pale pinkish-brown slices (1–5 mm thick); outline uneven. Bark thin, yellowish- or greyish-brown, rough and uneven. Cut surface whitish to pale reddish-brown, vascular tissues starchy, powdery and, in transverse slices, undifferentiated **[3]**; in longitudinal slices vascular tissues are visible as fine lines or dots. *Texture*: **slightly pliable [4]**; slippery when moistened. *Fracture*: **producing powdery dust** (粉尘 Fen Chen) **when broken [4]**. *Odour*: none. *Taste*: somewhat sweet and astringent.

PROCESSING

After removal of aerial parts and fibrous roots, rhizomes are sun-dried (= crude drug) then later softened thoroughly before thinly slicing transversely or longitudinally into decocting pieces; alternatively rhizomes are sliced when fresh then dried.

OTHER COMMON NAMES

The name 茯苓 Fu Ling is incorporated into many common names for this drug owing to its similar appearance to true 茯苓 Fu Ling (Poria) (not included in this book).

红土茯苓 Hong Tu Fu Ling (red local poria); 红土苓 Hong Tu Ling (red poria); 土苓 Tu Ling; Chinese smilax (SCN); glabrous greenbrier rhizome (CP2010); Rhizoma Smilacis Glabrae.

SAFETY

An official healthfood in China (Anon., 2002a,b).

S. china

Smilacaceae

'DRUG SUBSTITUTE' MORPHOLOGY

WHOLE DRIED RHIZOME: similar to that of the Pharmacopoeia species but overall shape more cylindrical (2–4 cm diam.), **root bases often distinctly conical [5]**, **external surface smooth and slightly lustrous [6]**. *Texture*: **hard [7]**. *Fracture*: starchy (**not** producing powdery dust when broken **[7]**). Margins of cut slices convoluted **[8]**.

OTHER COMMON NAMES

菝葜 Ba Qia; 金刚头 Jin Gang Tou; Chinaroot greenbrier rhizome (CP2010); Rhizoma Smilacis Chinae; Smilacis Chinae Rhizoma.

Fagopyrum dibotrys

Polygonaceae

'DRUG SUBSTITUTE' MORPHOLOGY

DRIED RHIZOME: **woody**, ringed externally with dotted lines of lenticels; sliced decocting pieces **woody throughout [9]**, **not powdery**, with radiating fissures **[10]** (transverse slices) and a central darker pith **[11]**.

OTHER COMMON NAMES

金荞 Jin Qiao (plant name); 金荞麦 Jin Qiao Mai; 苦荞头 Ku Qiao Tou; 荞麦三七 Qiao Mai San Qi; golden buckwheat rhizome (CP2010).

OTHER UNOFFICIAL SUBSTITUTE

Heterosmilax japonica Kunth (syn.: *H. gaudichaudiana* (Kunth) Maxim.) (rhizomes); 肖菝葜 Xiao Ba Qia (plant name); family Smilacaceae. A common and widespread species native to 11 Chinese provinces. Plants climbing and thornless; tendrils well developed; flowers a tiny bottle-shaped tube; **leaf stalks shorter** (1–3 cm long); **leaf blades usually much broader** (2.5–12 cm wide); **fruits black**; fr. Jul–Nov. Since the rhizomes, especially when sliced, are virtually indistinguishable from those of the Pharmacopoeia species, **laboratory based authentication methods are recommended**.

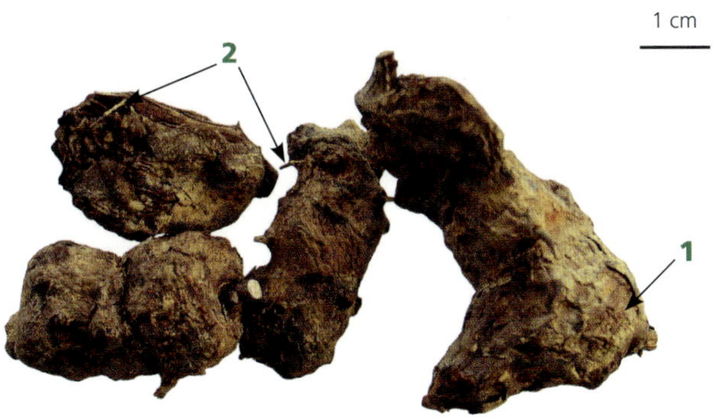

Tu Fu Ling crude drug
(dried rhizomes of *Smilax glabra*)

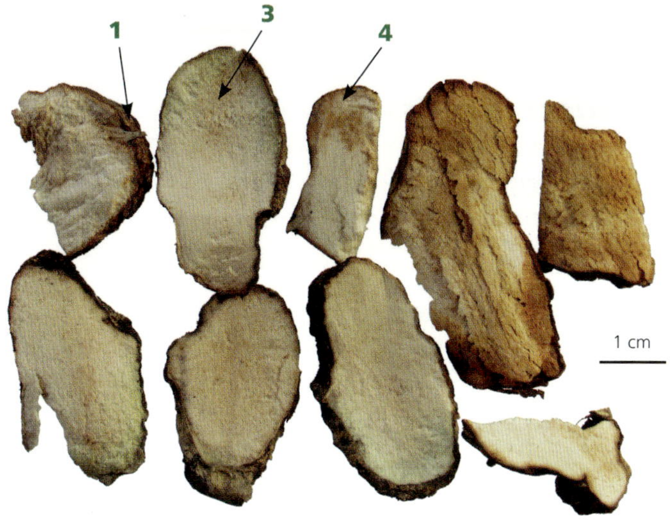

Tu Fu Ling decocting slices
(dried and sliced rhizomes of *S. glabra*)

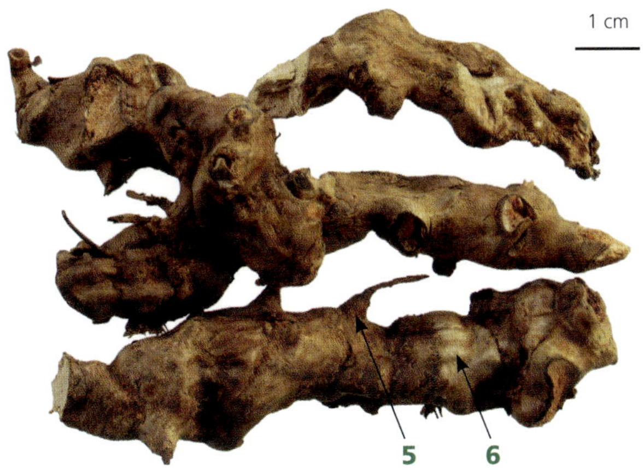

Dried rhizome of *S. china*

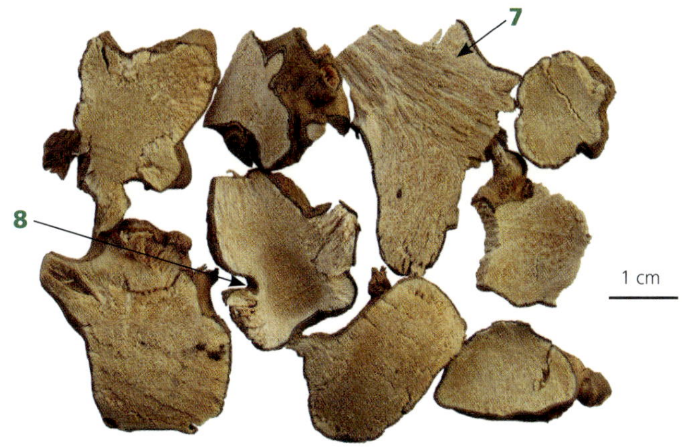

Dried, sliced rhizome of *S. china*

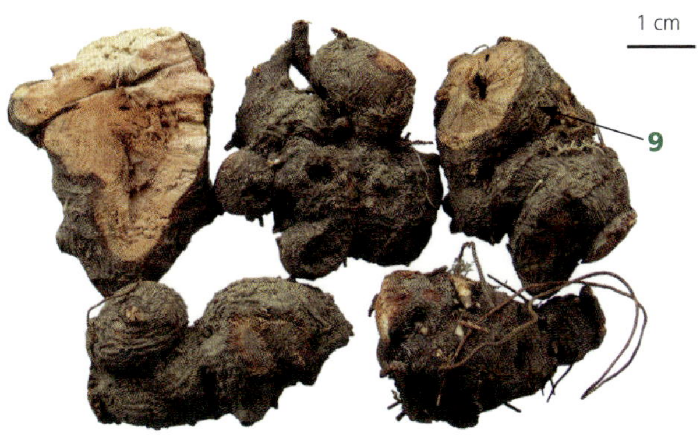

Dried and cut rhizome of *Fagopyrum dibotrys*

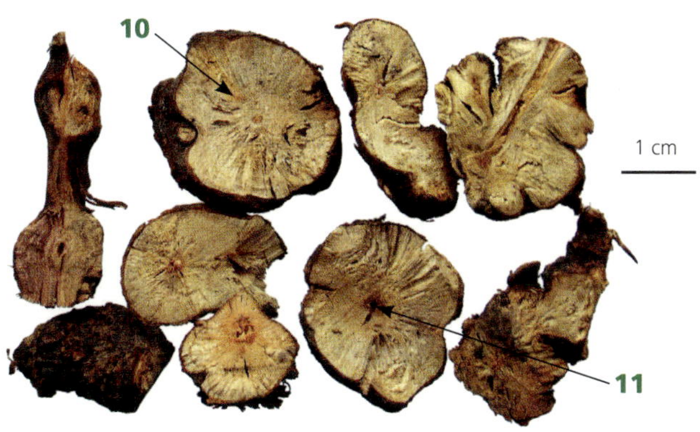

Dried, cut and sliced rhizome of *F. dibotrys*

Wei Ling Xian 威灵仙 Clematidis Radix et Rhizoma

Drug parts: ROOT AND RHIZOME

Clematis chinensis Osbeck

Ranunculaceae

synonym: *C. sinensis* Lour.

taxonomic/nomenclatural note: 3 varieties are recognised in *FoC*, whilst 6 varieties are recognised in *Kew-MPNS*, 28/8/2016.

Species of *Clematis* wild in China: 147

PLANT DESCRIPTION

A woody climber, turning black on drying. *Stems* branching, with 8–14 longitudinal grooves. *Leaves* opposite and pinnate with **5 stalked leaflets**; leaflet blades ovate to linear lance-shaped (1.5–9.5 cm long; **0.7–6.4 cm wide**), papery, virtually hairless or densely covered in fine hairs on lower surface, bases rounded, broadly wedge-shaped or somewhat heart-shaped, margins entire, tips pointed or tapering to a narrow point, veins on lower surfaces somewhat raised. *Flowers* conspicuous, in terminal or axillary many-flowered heads; bracts present, small and linear. Each flower bisexual and radially symmetrical (1.2–2.2 cm diam.), petal-like sepals 4–6, white, obovate to oblong, spreading; true petals absent; stamens many; style 1, densely hairy. *Fruit-head* a tight cluster of achenes, latter *not* strongly flattened laterally *nor* with a clearly defined margin; style persistent, long and curved. Fl. Jun–Sep, fr. Aug–Nov.

HARVESTING, SOURCING AND NATURAL RANGE

Rootstocks are harvested in autumn from wild populations. The species grows in scrub, on slopes and along streams (100–1,500 m altitude) in S, C and W China in the provinces of Anhui, Fujian, Guangdong, Guangxi, Guizhou, Hainan, Henan, Hubei, Hunan, Jiangsu, Jiangxi, Shaanxi, Sichuan, Taiwan, Yunnan and Zhejiang; it is also native species in Japan and Vietnam.

CONSERVATION STATUS

China: Least Concern (MEPCAS, 2013, as the varieties *anhweiensis, chinensis* and *vestita*).
Global: Not Evaluated (IUCN Red List, accessed 7/8/2016).

C. hexapetala Pall.

Ranunculaceae

taxonomic/nomenclatural note: 2 varieties are recognised in *FoC* and *Kew-MPNS*, 28/8/2016.

PLANT DESCRIPTION

Similar to PS1. *Main differences*: **erect (not climbing)**; **leaves leathery** and deeply divided (pinnatisect) into **unstalked leaf-like lobes (pinnae)** of greatly varying width (0.1–2 (rarely up to 2.6) cm) and **always much narrower** than the leaflets of PS1. *Flower-heads* 1–3-flowered. *Fruit* obovate with a distinctly feathered persistent style. Aerial parts not becoming black when dried.

HARVESTING, SOURCING AND NATURAL RANGE

Rootstocks are harvested in autumn from wild populations which grow in grasslands and dunes (100–1,300 m altitude). The species' natural distribution is C, N and E China (Gansu, Hebei, Heilongjiang, Henan, Hubei, Inner Mongolia, Jiangsu, Jilin, Liaoning, Ningxia, Shaanxi, Shandong and Shanxi); also a native species of the Korean Peninsula, Mongolia and Russia (E Siberia).

CONSERVATION STATUS

China: Least Concern (MEPCAS, 2013, as the two varieties *hexapetala* and *tchefouensis*).
Global: Not Evaluated (IUCN Red List, accessed 7/8/2016).

TCM PROPERTIES AND USE

properties: **'pungent, salty and warm'**
channel: **'Bladder'**
actions: **'dispels wind damp, opens collaterals, relieves pain'**

indications: **musculoskeletal pain**

Clematis chinensis in flower

C. chinensis in fruit

C. chinensis: fruit-heads with fruits (achenes) slightly flattened laterally, margins indistinct

Clematis hexapetala in cultivation

C. hexapetala fruiting in the wild, Inner Mongolia

C. hexapetala: styles of fruits distinctly feathered

Clematis terniflora DC. var. *mandshurica* (Rupr.) Ohwi

Ranunculaceae

synonym: *C. mandshurica* Rupr. (name used in CP2015 as '*Clematis manshurica* Rupr.')

PLANT DESCRIPTION

Difficult to distinguish from PS1: fruits (achenes) strongly flattened laterally and with a distinct margin. Plants not becoming black when dried.

Easily distinguished from PS2: **woody and climbing**; **leaves divided into 5–7 short-stalked leaflets** (not stalkless lobes); leaflets papery to slightly leathery.

HARVESTING, SOURCING AND NATURAL RANGE

Rootstocks are harvested in autumn from wild populations which grow in scrub, often on slopes (200–800 m altitude) in N and NE China (Heilongjiang, Inner Mongolia, Jilin, Liaoning); also native in Korean Peninsula, Mongolia and Russia (Siberia).

CONSERVATION STATUS

China: Least Concern (MEPCAS, 2013).
Global: Not Evaluated (IUCN Red List, accessed 7/8/2016).

Sinopodophyllum hexandrum (Royle) T. S. Ying

Berberidaceae

synonyms: *Podophyllum hexandrum* Royle (*Kew-MPNS* accepted name, 28/8/2016) (in CP2010 as '*Podophyllum hexandrum* Royle Ying'); *Sinopodophyllum emodi* (Wall. ex Honigberger) T. S. Ying.

PLANT DESCRIPTION

The living plant is easily distinguished from the official species. Stem solitary. **Leaves 2, basal**, long-stalked (10–25 cm); **leaf blades much larger** (11–20 cm long; 18–30 cm wide), **orbicular**, 3–5-dissected to about midway or more, with each dissection 2- or 3-lobed; margin entire or coarsely toothed. **Flowers** conspicuous, **appearing before the leaves, solitary**; **petals 6, pink** and equal. **Fruit conspicuous (bright red), solitary and large** (4–7 cm long; 2.5–4 cm diam.), ovoid to globose; succulent; seeds ovoid to triangular. Fl. May–Jun, fr. Jul–Sep.

CONSERVATION STATUS

China: Least Concern (MEPCAS, 2013).
Himalayas: 'threatened' (Nadeem *et al.*, 2000).
Global: Not Evaluated (IUCN Red List, accessed 10/11/2016).

PROTECTED STATUS

China: Cat III (SEPA, 1999).
CITES: Appendix II; EU Annex B (see Table 1 page 10).

SUBSTITUTION COMMENT. Rootstocks of this highly toxic species have been widely reported in Hong Kong as an adulterant of Wei Ling Xian resulting in serious adverse reactions (see *Safety*). Rootstocks have long been used in NW China as a traditional non-TCM medicine called 'Tao Er Qi' used to treat a variety of conditions but in small doses owing to its strong purgative action and associated toxicity. In recent years, Western markets have created a new and much larger demand for the rootstocks of the species which provide a major source of podophyllotoxin whose semi-synthetic derivatives (e.g. etoposide) are used in cancer therapy (Gordaliza *et al.*, 2004). The context of the above adverse reactions is not clear but, given the large volumes of *P. hexandrum* rootstocks exported via Hong Kong and the similarity of these to the popularly traded Wei Ling Xian, identification confusion is the most likely explanation.

The fruits of this species are the source of a separate but seldom-used TCM drug called 'Xiao Ye Lian' (Sinopodophylli Fructus) [not included in this book]; fruits are described as 'slightly toxic' (CP2015).

S. hexandrum is harvested from the temperate forests and meadows of N and W China (Gansu, Qinghai, Shaanxi, Sichuan, Tibet and Yunnan) between 2,200–4,300 m altitude); over-harvesting has depleted wild populations such that this species is now CITES-listed (see *Protected Status)* but despite high demand its medicinal cultivation has yet to become economically viable due to its slow rate of establishment (Kharkwal *et al.*, 2008). The species is also native in Afghanistan, Bhutan, N India, Kashmir, Nepal and Pakistan.

Clematis terniflora var. *mandshurica* in fruit (fruits strongly flattened laterally; margin distinct)

C. terniflora var. *mandshurica*, cultivated (yet to start climbing; becoming woody at base)

Sinopodophyllum hexandrum in flower; leaves still unfurling (cultivated)

S. hexandrum fruiting plants

Clematis chinensis

Ranunculaceae

DRUG MORPHOLOGY

CRUDE DRUG (DRIED ROOT AND RHIZOME): a mass of slender cylindrical roots **[1]** (usually still attached to rhizome) **[2]**, somewhat curved (7–15 cm long, **1–2 mm diam.**) **[3]**, externally yellowish-brown or blackish-brown, finely wrinkled longitudinally and **root-bark easily shed [4]** to reveal yellowish-white cortex below. *Root texture*: hard and brittle; *root fracture*: cortex whitish or cream, often clearly demarcated from the central yellowish-white porous wood (xylem) **[5]** with the ratio of cortex to wood variable. ***Rhizomes*** somewhat cylindrical (5–10 cm long; 0.3–1.5 cm diam.), shorter and broader than roots **[6]**; **outer skin easily shed [7] revealing fibrous tissues below [8]**; externally pale brownish-yellow; apex with cluster of stem bases **[9]**; lower parts with roots usually still attached. Roots and rhizomes: *odour* and *taste* indistinct. ***Good quality*:** roots externally blackish-brown; fracture powdery.

DECOCTING PIECES (CUT LENGTHS OF ROOT AND RHIZOME): as for the crude drug above with roots cut into short, slender lengths (c.1–3 cm long, **1–2 mm diam.**) **[3]** and rhizome pieces occasional, irregularly-shaped **with internal tissues fibrous [8]**.

PROCESSING

After removal of soil, rootstocks are sun-dried (= crude drug) then later softened thoroughly and cut into short lengths for decocting.

OTHER COMMON NAMES

铁脚威灵仙 Tie Jiao Wei Ling Xian; 老虎须 Lao Hu Xu; 威灵仙 Wei Ling Xian (plant and drug name); Radix et Rhizoma Clematidis; Chinese clematis (SCN); Chinese clematis root (CP2010).

SAFETY

Safety Class: 1; Interaction Class: A (Gardner & McGuffin, 2013).

C. hexapetala

Ranunculaceae

DRUG MORPHOLOGY

CRUDE DRUG (DRIED ROOT AND RHIZOME): very similar to the crude drug of PS1. Roots 4–20 cm long, externally brown to brownish-black. Main differences from the rootstocks of *Sinopodophyllum hexandrum* (*Unofficial Substitute*): **root diameter 1–2 mm [3] root-bark easily shed [4]**; **outer skin of rhizomes easily shed [7] revealing fibrous tissues below [8]**.

DECOCTING PIECES (CUT LENGTHS OF ROOT AND RHIZOME): as for the crude drug of this species (above) but cut into short sections (usually 1–5 cm long).

Laboratory-based authentication methods are needed for reliable discrimination from the other two Pharmacopoeia species and the Unofficial Substitute *S. hexandrum*.

PROCESSING

As for PS1.

OTHER COMMON NAMES

棉花铁线莲 Mian Hua Tie Xian Lian (plant name); 山棉花 Shan Mian Hua; Chinese clematis root (CP2010); Radix et Rhizoma Clematidis.

SAFETY

As for PS1.

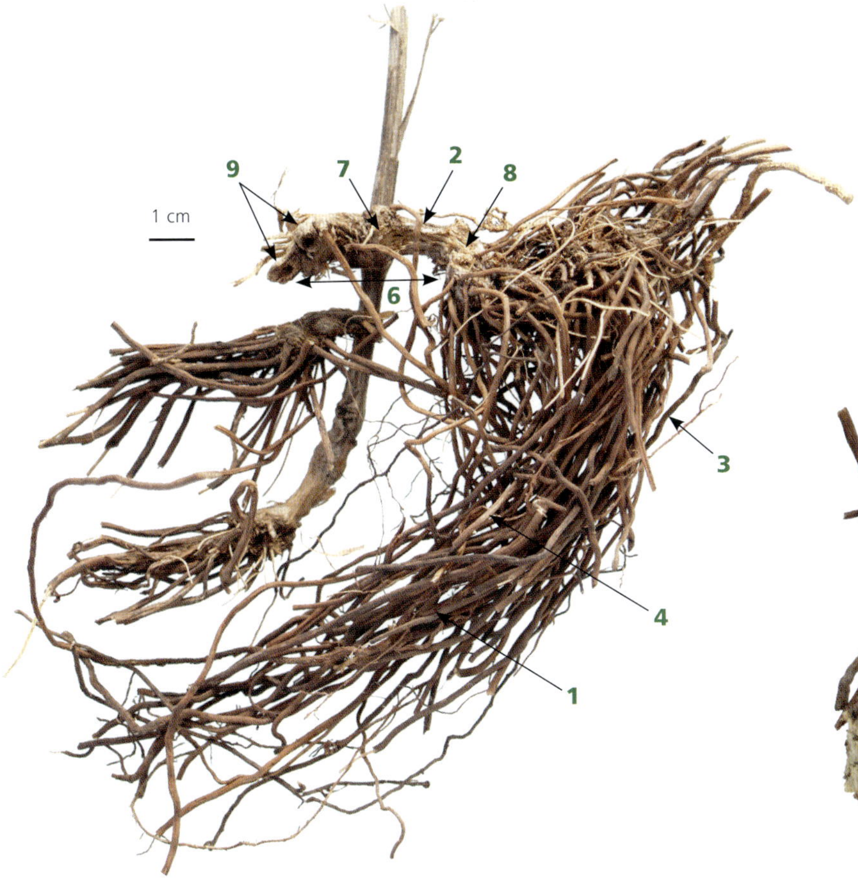

Crude drug of Wei Ling Xian sourced from *Clematis chinensis*

Transverse root sections of *C. chinensis*

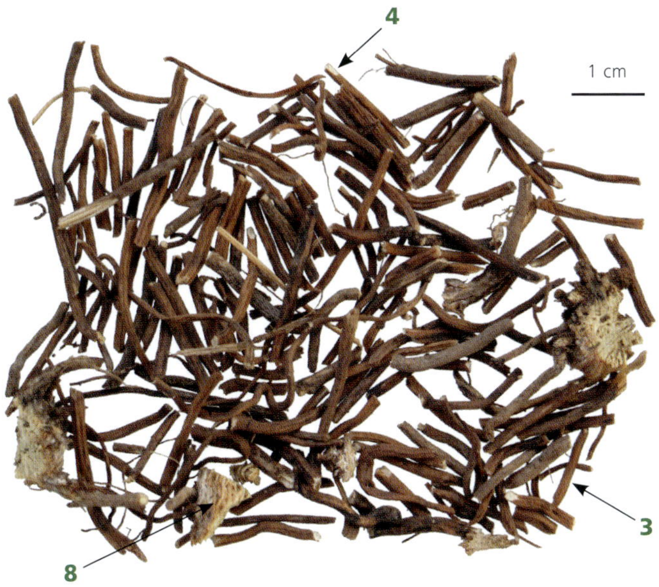

Decocting pieces of Wei Ling Xian sourced from *C. chinensis*

Crude drug of Wei Ling Xian sourced from *C. hexapetala*

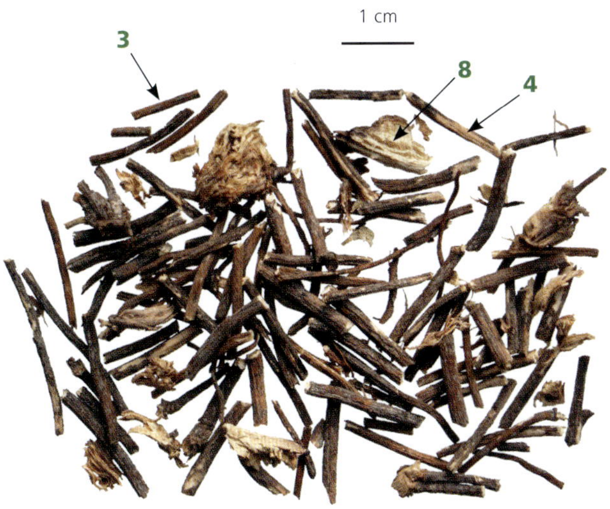

Decocting pieces of Wei Ling Xian sourced from *C. hexapetala*

Clematis terniflora var. *mandshurica*

Ranunculaceae

DRUG MORPHOLOGY

CRUDE DRUG (DRIED ROOT AND RHIZOME): very similar to the crude drug of PS1. Roots sometimes longer (5–23 cm), externally brownish-black. Rhizomes cylindrical (1–11 cm long). *Taste*: pungent. *Main differences* from the rootstocks of *S. hexandrum* (*Unofficial Substitute*): **root diameter 1–2 mm [3] root-bark easily shed [4]**; **outer skin of rhizomes easily shed revealing fibrous tissues below [8]**.

DECOCTING PIECES (CUT LENGTHS OF ROOT AND RHIZOME): as for the crude drug of this species (above) but cut into short sections (usually 1–5 cm long).

Laboratory-based authentication methods are needed for reliable discrimination from the other two Pharmacopoeia species and the Unofficial Substitute *S. hexandrum*.

PROCESSING

As for PS1.

OTHER COMMON NAMES

辣蓼铁线莲 La Liao Tie Xian Lian (plant name); 山棉花 Shan Mian Hua; Chinese clematis root (CP2010); Radix et Rhizoma Clematidis.

SAFETY

As for PS1.

Sinopodophyllum hexandrum

Berberidaceae

'DRUG SUBSTITUTE' MORPHOLOGY

DRIED ROOT AND RHIZOME: easily confused with those of the Pharmacopoeia species. *Main differences:* **roots slightly broader (2–3 mm diam.) [10]** and **root-bark not easily shed [11]**; ***rhizome*** **skin not easily shed and internal tissues not fibrous [12]**. *Taste*: bitter. (Lin Yu-Lin, pers. obs. 2016; Xiao *et al.*, 2001–2).

CUT LENGTHS OF ROOT AND RHIZOME: as for the above of this species but cut into short lengths. *Taste*: **bitter**.

WARNING: Because of the acute toxicity of this adulterant, macroscopic characters alone are not considered sufficient for reliable differentiation from the Pharmacopoeia species. **Laboratory based authentication methods are strongly recommended.**

OTHER COMMON NAMES

桃儿七 Tao 'Er Qi; Himalayan mayapple (SCN as Podophyllum hexandrum); Radix et Rhizoma Sinopodophylli.

SAFETY

All parts **highly toxic** except the ripe fruit (Barceloux, 2008; Bensky *et al.*, 2004; Jellin *et al.*, 2012). Serious adverse effects and fatalities have been reported in Asian countries (1995–96) following the adulteration of Wei Ling Xian with rootstocks of this species (But, 1994). Listed as **toxic** on Schedule 1 of the Hong Kong SAR Chinese Medicine Ordinance, 6/8/1999 (Hong Kong, 1999a). **Safety Class: 3**; Interaction Class: A (Gardner & McGuffin, 2013).

OTHER UNOFFICIAL SUBSTITUTES

The rootstocks of various other species have been recorded in the TCM literature as local or historical substitutes and contaminants of Wei Ling Xian. Many of these materials are from other species of *Clematis* that grow wild in China (147 species in total); their identification to species requires considerable expertise. Given that the 3 Pharmacopoeia species of Wei Ling Xian are all wild sourced, mis-identification of the living plants at the point of harvest is the most likely explanation. The occurrence of such substitutes and adulterants in internationally traded Wei Ling Xian remains unclear.

Crude drug of Wei Ling Xian sourced from *C. terniflora* var. *mandshurica*

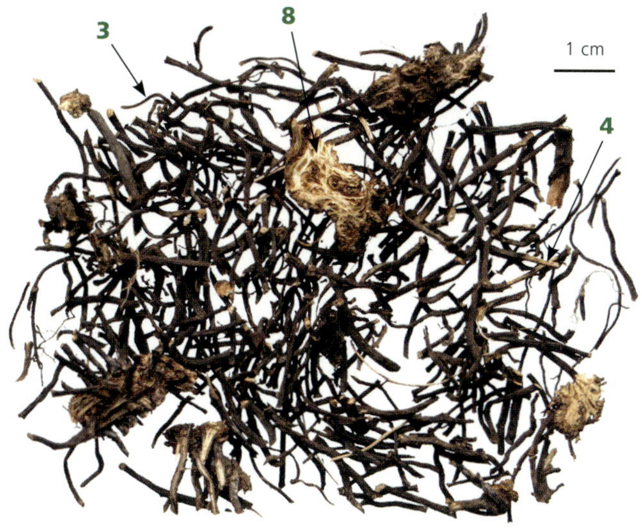

Decocting pieces of Wei Ling Xian sourced from *C. terniflora* var. *mandshurica*

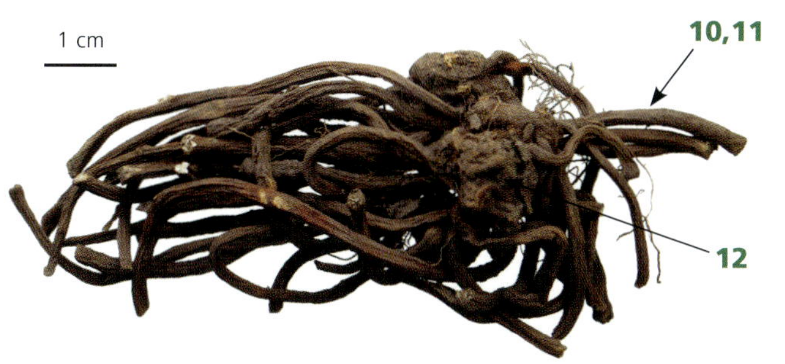

Dried rootstock of *Sinopodophyllum hexandrum*

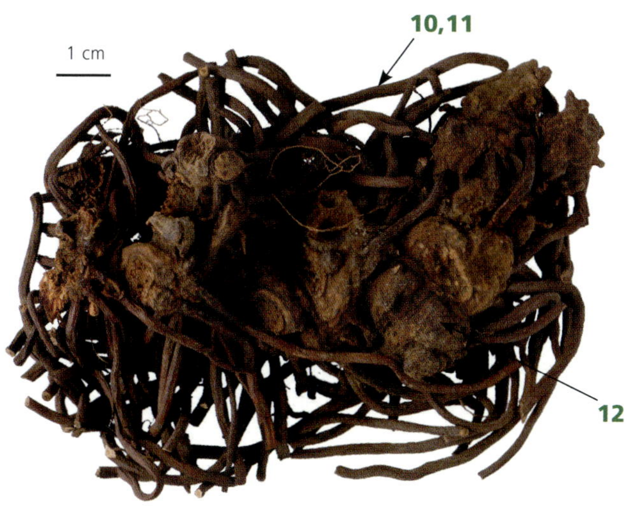

Old and mature rootstock of *S. hexandrum*

Dried rootstock of young plant of *S. hexandrum*

Transverse root sections of *S. hexandrum*

Xiang Fu 香附 Cyperi Rhizoma

Drug part: RHIZOME

PHARMACOPOEIA SPECIES

Cyperus rotundus L.

Cyperaceae

synonym: *C. rotundus* var. *quimoyensis* L. K. Dai

Species of *Cyperus* wild in China: 62 (incl. 4 introduced)

PLANT DESCRIPTION

An upright, perennial herb (a sedge) (15–90 cm tall) with ellipsoidal tuberous rhizomes (2.5–5 cm long; 0.5–1 cm wide) and slender stolons. *Leaves* all basal, in 3 ranks, linear, bluish-green and grass-like (2–5 mm wide, mostly flat), as long as or shorter than flowering stem; leaf margin entire; leaf base with a brown sheath often becoming fibrous. *Flowering stem* solitary (15–90 cm tall), distinctly triangular in cross-section. Involucral bracts leaf-like, 2 or 3, at base of petal-less inflorescence. *Flowers* tiny, purplish-brown and petal-less, clustered in 3–10 linear heads (spikes) (1–3 cm long; 1.5–2 mm wide) forming a loose star-shaped inflorescence at top of the flowering stem. Each flower petal-less, with **deep reddish to purplish-brown husks,** 3 stamens, a long style with 3 slender stigmas. *Fruit* a dry, 3-sided brown, single-seeded 'nutlet'. Fl. and fr. May–Nov.

See also *Lesser Known Substitutes*.

HARVESTING, SOURCING AND NATURAL RANGE

Rhizomes are harvested in autumn from wild plants which are a common weed of damp places such as grasslands, stream-sides and ditches; it grows up to 2,000 m altitude throughout much of China in the provinces of Anhui, Fujian, Gansu, Guangdong, Guangxi, Guizhou, Hainan, Hebei, Henan, Hubei, Jiangsu, Jiangxi, Liaoning, Shaanxi, Shandong, Shanxi, Sichuan, Taiwan, Yunnan and Zhejiang. Traditional *Di Dao*-quality 'Xiang Fu' is traditionally sourced from Henan and Shandong provinces.

CONSERVATION STATUS

China: Least Concern (MEPCAS, 2013); a common and invasive weed.

Global: Least Concern (IUCN Red List, accessed 18/8/2016); subject to statutory control in some countries.

UNOFFICIAL SUBSTITUTE

Anemone raddeana Regel

Ranunculaceae

PLANT DESCRIPTION

See also *Substitution Comment* (page 301).

A smaller perennial herb (up to 30 cm tall when in flower) **often forming ground cover in deciduous woodland**, with ellipsoidal tuberous and **slightly nodular rhizomes** (2–7 mm diam.) (see also *Drug Substitute Morphology* (page 300). ***Leaves*** all basal **(not grass-like),** somewhat pentagonal in outline (2–5 cm × 2–4 cm), and **divided into 3 irregularly lobed, short-stalked leaflets.** Flowering stems with conspicuous leaf-like bracts just below the flower. ***Flowers*** large, **white** and **conspicuous (c. 5 cm diam.)** and opening above the foliage; each flower with 9–15 linear, spreading, white or white-tinged purple 'petals' (true petals absent); stamens many; style short. *Fruits* inconspicuous, dry, narrowly ovoid achenes (c. 3 mm long; 1.5 mm diam.). Fl. Apr–May.

CONSERVATION STATUS

China: Least Concern (MEPCAS, 2013).

Global: Not Evaluated (IUCN Red List, accessed 18/8/2016).

LESSER KNOWN SUBSTITUTES

There are 62 species of *Cyperus* growing wild in China and their reliable identification usually requires botanical expertise. Fortunately *C. rotundus* is one of the most ubiquitous Chinese species in this genus and its distinctive reproductive parts (notably its reddish-purple flower spikelets) together with its tuberous rhizomes enable its relatively easy identification compared to many others.

Cyperus stoloniferus Retz. — rhizomes (粗根茎莎草 Cu Gen Jing Suo Cao; 大香附 Da Xiang Fu) (family Cyperaceae). Closely related to the official source of 'Xiang Fu'; common in swamps, open grassland and moist places near the sea in S and SE China (Fujian, Guangdong, Hainan, Taiwan and Paracel Islands). Rhizomes thought to be slightly larger; when in flower or fruit, living plants often shorter (8–22m cm tall); leaves **usually folded**; linear spikes of flowers fewer (3 or 4) and **flower spikelets multi-coloured yellow, brownish-red and green.**

TCM PROPERTIES AND USE

properties: 'pungent, slightly bitter, slightly sweet and neutral'
channels: 'Liver, Spleen and san jiao'
actions: 'regulates qi, courses the Liver, relieves pain'
indications: abdominal distension and pain, poor appetite; hernia pain; menstrual disorders, breast pain

Cyperus rotundus, growing wild in a ditch, Shaanxi province

Flower-heads of *Cyperus rotundus*

Flowering plants of *Anemone raddeana* in the wild, Jilin province

Freshly uprooted plants of *A. raddeana*

Flowering plants of *A. raddeana*

Cyperus rotundus

Cyperaceae

DRUG MORPHOLOGY

CRUDE DRUG-1 (MAO XIANG FU) — see *Processing*: ellipsoidal (2.3–5 cm long; 0.5–1 cm diam.) often tapering at one or both ends **[1]**. Externally dark brown or blackish-brown, indistinctly longitudinally striated, **6–10 slightly raised and widely spaced annular rings [2]** and covered in fibrous roots **[3]**. *Texture*: hard but brittle. *Fracture*: white and starchy **[4]** with an **obvious endodermis ring [5] surrounding a slightly darker centre dotted with vascular tissues.** *Odour*: aromatic. *Taste*: somewhat bitter.

CRUDE DRUG-2 (GUANG XIANG FU) — see *Processing*: as for Crude Drug-1 but external surface smooth in places and devoid of fibrous roots **[6]**. *Fracture*: yellowish- or reddish-brown and horn-like.

DECOCTING PIECES-1:

Mao Xiang Fu: oval, or irregularly shaped slices often tapering at one or both ends **[1]** (2.3–5 cm long; 0.5–1 cm wide). External uncut surface dark brown or blackish-brown with **slightly raised and widely spaced annular rings sometimes visible [2]**; covered in fibrous roots **[3]**. Cut surfaces white or cream and starchy **[4]**, with distinct endodermis ring **[5]**; central area dotted with vascular tissues. *Texture*: hard. *Odour*: aromatic. *Taste*: somewhat bitter.

Guang Xiang Fu: as for 'Mao Xiang Fu' above but uncut surface smooth in places and devoid of fibrous roots. Cut surfaces yellowish- or reddish-brown and horn-like.

DECOCTING PIECES-2: as for decocting pieces of 'Mao Xiang Fu' and 'Guang Xiang Fu' but all surfaces darker **[7]**. *Odour*: slightly of vinegar.

PROCESSING

Two types of product are traded: 1. Mao Xiang Fu 'hairy Xiang Fu' — rhizomes sun-dried without removal of fibrous roots (= crude drug-1); 2. Guang Xiang Fu 'smooth Xiang Fu' — fibrous roots removed (usually burnt off) before rhizomes briefly boiled or thoroughly steamed then sun-dried (= crude drug-2). The different processing methods affect the rhizome tissues such that sun-dried 'Mao Xiang Fu' has a starchy fracture while steamed or boiled 'Guang Xiang Fu' has a horn-like fracture. The latter is considered superior.

Decocting pieces are made by slicing or pulverising the above-processed rhizomes (decocting pieces-1) or by further processing with rice vinegar then dry-frying (= 醋香附 Cu Xiang Fu; decocting pieces-2).

OTHER COMMON NAMES

莎草根 Sha Cao Gen; 香附子 Xiang Fu Zi; coco grass; nutgrass; nutgrass galingale rhizome (CP2010); Rhizoma Cyperi. Many other trade names exist for this drug reflecting its medicinal use in other countries, notably India and Japan (see *Kew-MPNS*).

SAFETY

An official healthfood in China (Anon., 2002a,b). Safety Class: 1; Interaction Class: A (Gardner & McGuffin, 2013).

Anemone raddeana

Ranunculaceae

'DRUG SUBSTITUTE' MORPHOLOGY

CRUDE DRUG/DECOCTING PIECES (DRIED RHIZOME): superficially similar to crude drugs-1 and -2 of 'Xiang Fu' (above) but macroscopically easy to distinguish.

Main differences: overall shape slightly different (often slightly narrower (0.2–0.7 cm diam.) and **often thickening towards (not at) one end) [8]**; externally greyish-brown or reddish-brown with **a few distinctly prominent root remains [9]**; annular rings 3–5 but very indistinct; ***fracture:* homogeneous [10]** white or greyish-brown, **endodermis ring absent [11]**; slightly horn-like; *texture:* hard but brittle; ***odour:* indistinct**; *taste:* initially slight, then somewhat bitter and pungent.

A further contrast to 'Xiang Fu' is that when *A. raddeana* rhizomes are traded as 'Liang Tou Jian' (Anemones Raddeana Rhizoma), they are neither sliced nor vinegar-processed.

OTHER COMMON NAMES

多被银莲花 Duo Bei Yin Lian Hua; 两头尖 Liang Tou Jian 'both ends tapering'; 竹节香附 Zhu Jie Xiang Fu 'bamboo-shaped Xiang Fu'; radde anemone rhizome (CP2010); Rhizoma Anemones Raddeanae.

SAFETY

Toxic (CP2015). Fresh rhizomes contain protoanemonin, a skin irritant, which is destroyed on drying (Barceloux, 2008).

Crude drug-1 (Mao Xiang Fu)

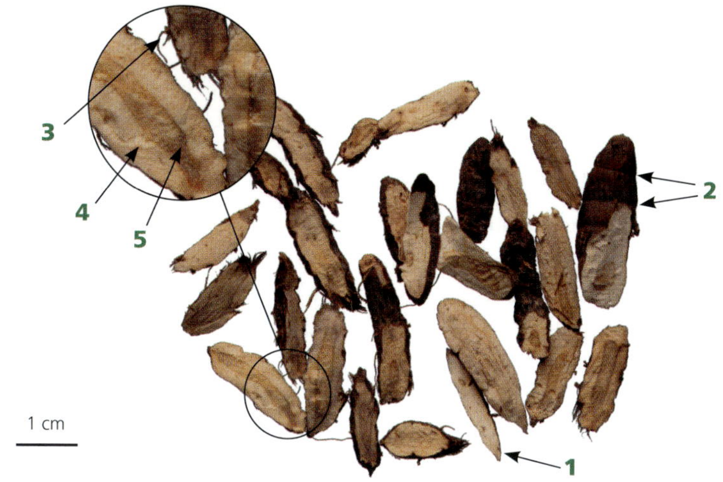

Decocting pieces-1 (Mao Xiang Fu)

Crude drug-2 (Guang Xiang Fu)
(colour difference with crude drug-1 not a reliable character)

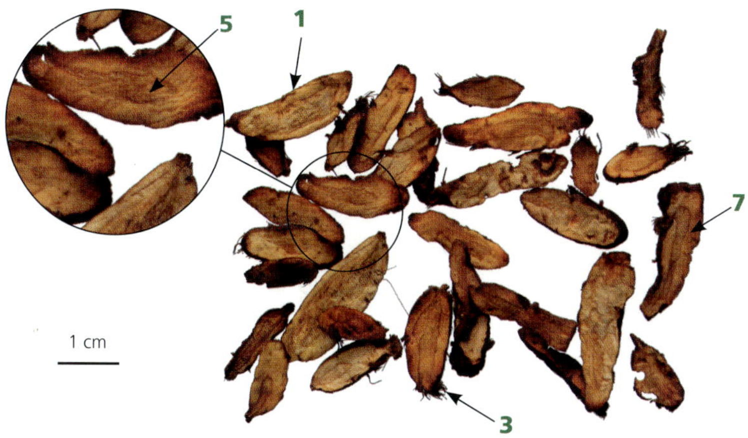

Decocting pieces-2 (Cu Xiang Fu) — vinegar processed

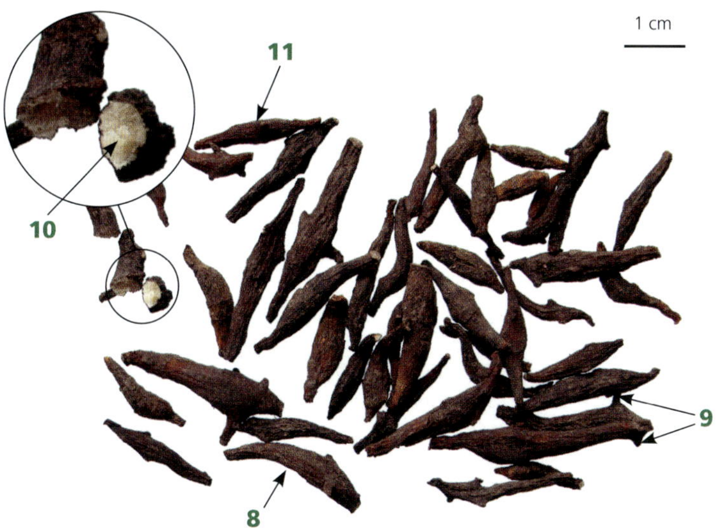

Dried rhizomes of *Anemone raddeana* (crude)

SUBSTITUTION COMMENT. Living plants of *A. raddeana* are not confused with those of the Pharmacopoeia species at point of harvest since they are morphologically completely different; they belong to a different family and have conspicuous brilliant white flowers and 3-lobed leaves (see *Plant Description* and photos). Their rhizomes, however, the source of the separate TCM drug 'Liang Tou Jian' (Anemones Raddeanae Rhizoma) page 192, may inadvertently be traded as the crude drug (whole rhizome) of 'Xiang Fu' owing to their superficial similarity or confusion arising from a similar trade name: 竹节香附 Zhu Jie Xiang Fu 'bamboo-shaped Xiang Fu'. Trade in this unofficial substitute is only likely to occur in local markets in NE China where the source plant is abundant in the wild.

Importantly, the CP2015 describes this drug as '**toxic**' and gives a recommended dosage of 1–3 g (this is significantly smaller than the dosage for 'Xiang Fu' (6–10 g). **Those performing quality assurance procedures need to be cognisant of these differences.**

Xian Mao 仙茅

Curculiginis Rhizoma

Drug part: RHIZOME

Curculigo orchioides Gaertn.

synonym: *C. orchioides* var. *minor* Benth.

Hypoxidaceae

Species of *Curculigo* wild in China: 7

PLANT DESCRIPTION

A small unbranched, upright perennial herb (up to 45 cm; rarely 90 cm) with cylindrical, tuberous rhizomes (c. 10 cm long; 1 cm diam.). *Leaves* several, all basal, simple, unstalked or short-stalked, often folded, lance-shaped (10–45 cm long; 0.5–2.5 cm wide) and bright-green; margin entire and tip narrowly pointed. *Flowering stems* (6–7 cm long) sheathed by leaf-bases. Bracts lance-shaped, fringed with hairs. *Flowers* in groups of 4–6; each flower with 6 'petal-lobes', spreading, bright yellow, oblong to lance-shaped (8–12 mm long; 2.5–3 mm wide) and distinctively fringed with soft white hairs; stamens 6; ovary inferior; stigma 3-lobed, longer than style. *Fruits* somewhat spindle-shaped (1.2–1.5 cm long; c. 0.6 cm diam.); with a (2.5 mm long) beak, divided internally into 3 compartments. *Seeds* small. Fl. and fr. Apr–Sep.

HARVESTING, SOURCING AND NATURAL RANGE

Rhizomes are harvested in autumn and winter mostly from wild plants which grow in forests and open grassy slopes (up to 1,600 m altitude) in southern China (Fujian, Guangdong, Guangxi, Guizhou, Hunan, Jiangxi, Sichuan, Taiwan and Zhejiang). Also grown on small-holdings. The species' natural range extends to Cambodia, India, Indonesia, Japan, Laos, Myanmar, Pakistan, Papua New Guinea, Philippines, Thailand and Vietnam.

CONSERVATION STATUS

China: Least Concern (MEPCAS, 2013).
Global: Not Evaluated (IUCN Red List, accessed 16/4/2015).

Curculigo orchioides in the wild, Guizhou province

C. orchioides in flower

Freshly uprooted plant of *C. orchioides*

UNOFFICIAL SUBSTITUTES

Roots of **toxic** species in the genus *Aconitum* have been reported as local and/or historical unofficial substitutes but their occurrence in international TCM trade is unlikely:

1. ***Aconitum pendulum*** N. Busch (plant name: 铁棒锤 Tie Bang Chui; family Ranunculaceae).
2. ***A. flavum*** Hand.-Mazz. (plant name: 伏毛铁棒锤 Fu Mao Tie Bang Chui; family Ranunculaceae).

TCM PROPERTIES AND USE

properties: **'pungent and hot'**
channel: **'Kidney'**
actions: **'tonifies Kidney, warms yang, dispels cold damp'**

indications: **incontinence; impotence, premature ejaculation; infertility, menstrual disorders; diarrhoea; musculoskeletal pain**

DRUG MORPHOLOGY

CRUDE DRUG (DRIED RHIZOME): somewhat cylindrical (3–10 cm long; 4–12 mm diam.) and slightly curved. Externally brown or blackish-brown, rough and irregularly furrowed **[1]**, scattered remnants of fibrous rootlets **[2]** and pitted rootlet scars **[3]**. *Texture*: hard and brittle. *Fracture*: uneven, cortex brown to pale brownish-white with a darker brown core (xylem) **[4]**. *Odour*: faintly aromatic. *Taste*: somewhat bitter and pungent. ***Good quality***: long and thick; externally greyish-black.

DECOCTING PIECES (CUT RHIZOME LENGTHS): irregularly shaped, shortly cylindrical or sub-rounded slices. Outer bark thin, dull brown or blackish-brown, rough and irregularly furrowed; fibrous rootlets and pitted rootlet scars occasional. Cut surface with brown to pale brownish-white cortex densely scattered with oil cells and surrounding a darker central xylem **[4]**.

PROCESSING

Main stems, rhizome heads and fibrous roots removed and discarded, then rhizomes sun-dried (= crude drug) and cut into short decocting lengths or decocting slices.

OTHER COMMON NAMES

地棕根 Di Zong Gen; common Curculigo rhizome (CP2010); Curculigo (SCN); Rhizoma Curculiginis.

SAFETY

Toxic (CP2015). Safety Class: 2b and 2d 'not for long-term use'; Interaction Class: A (Gardner & McGuffin, 2013). Serious side-effects reported in over-dose (Bensky *et al.*, 2004).

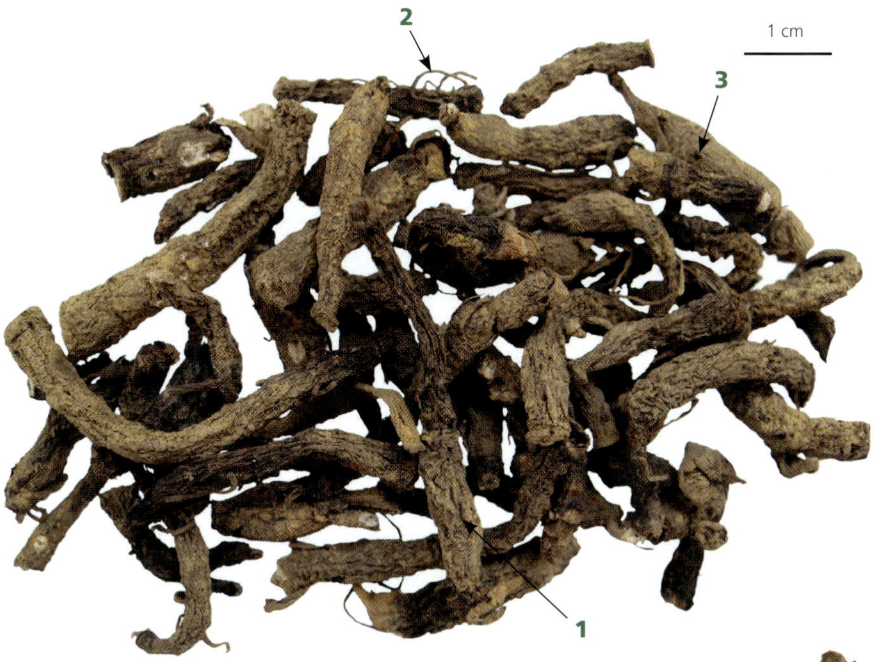

Crude drug of Xian Mao (dried rhizomes of *Curculigo orchioides*)

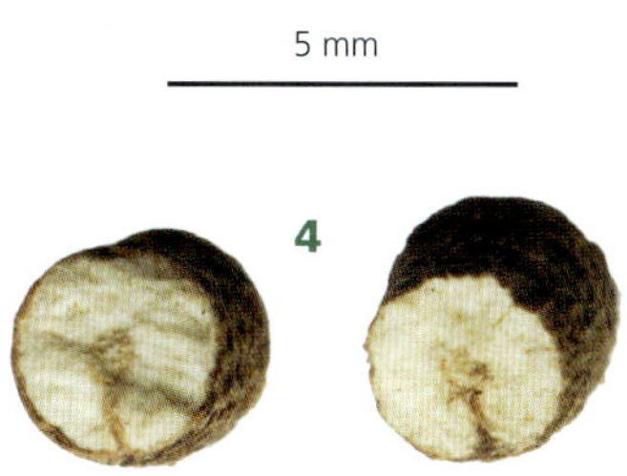

Cortex of *C. orchioides*

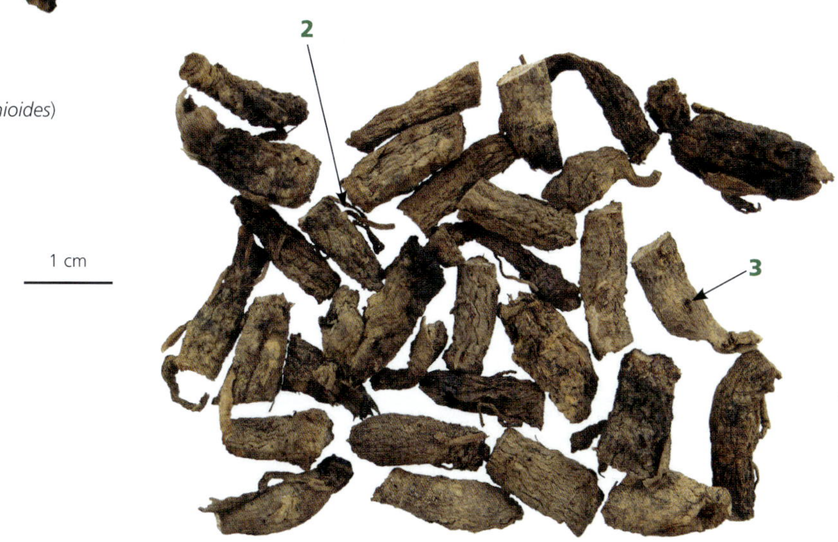

Decocting pieces of Xian Mao (dried and cut rhizome lengths of *C. orchioides*)

Xia Tian Wu 夏天无 — Corydalis Decumbentis Rhizoma

Drug part: RHIZOME (stem tuber)

Corydalis decumbens (Thunb.) Pers.

Papaveraceae

synonym: *Fumaria decumbens* Thunb.

Species of *Corydalis* wild in China: 357

PLANT DESCRIPTION

A small, delicate-looking perennial herb (10–25 cm tall), with juicy sap and a rounded tuberous rhizome (0.5–3 cm long; 0.5–2.5 cm diam.), latter growing directly from old rhizome and mainly in axils of basal leaves. *Stems* several, emerging from each tuber, slender with 2 or 3 scale leaves. *Leaves* alternate, twice- or three-times tri-lobed; leaflets entire or deeply lobed. *Flowers* fragrant, in lax, 3–10-flowered heads with flowers arranged along a central axis. Each flower bisexual, petals 4 (unequal size and shape), upper petal with a narrow crest and a straight or slightly curved spur (14–18 mm long); lower petal broadly spatula-shaped; smaller inner petals with rounded crests are fused at their apex; stamens 4; ovary superior. *Fruit* a many-seeded linear capsule (13–18 mm long). *Seeds* black, keeled and covered with papillae. Fl. and fr. Feb–May.

HARVESTING, SOURCING AND NATURAL RANGE

Rhizomes are harvested from wild plants in spring and early summer when plants are in bud. The species grows in disturbed places on mountain slopes, roadsides and on coastal wastelands (100–300 m altitude) in parts of E and S China (Anhui, Fujian, Hubei, Hunan, Jiangsu, Jiangxi, Shaanxi, Taiwan and Zhejiang provinces) [it is also native in Japan].

CONSERVATION STATUS

China: Least Concern (authors' pers. obs., 2011).
Global: Not Evaluated (IUCN Red List, accessed 9/7/2015).

Corydalis decumbens in the wild, Anhui province

TCM PROPERTIES AND USE

properties: **'slightly pungent, bitter and warm'**
channel: **'Liver'**
actions: **'invigorates Blood and qi, dispels stasis, relieves pain'**

indications: **headache, traumatic injury; musculoskeletal pain; paralysis**

DRUG MORPHOLOGY

CRUDE DRUG (DRIED RHIZOME): somewhat rounded to angular; often nodular **[1]** (0.5–3 cm long; 0.5–2.5 cm diam.). Externally yellowish-grey to dark green or brown, often smooth **[2]** but distinctly and finely wrinkled in places **[3]**; dorsal side often with shrivelled stem remnants **[4]** and pale leaf scars; larger root scars irregularly scattered **[5]**. *Texture*: hard. *Fracture*: yellowish-white or yellow, granular or horny, sometimes starchy. *Odour*: indistinct. *Taste*: bitter.

Rhizomes used as a substitute Yan Hu Suo (Corydalis Rhizoma) see page 318.

PROCESSING

After removal of stems, leaves and roots, rhizomes are washed and dried (= crude drug) and powdered prior to oral adminstration.

OTHER COMMON NAMES

土元胡 Tu Yuan Hu; decumbent corydalis rhizome (CP2010); Rhizoma Corydalis Decumbentis.

SAFETY

Chen & Chen (2004) describe the potential for interaction with anti-coagulent or anti-platelet medication.

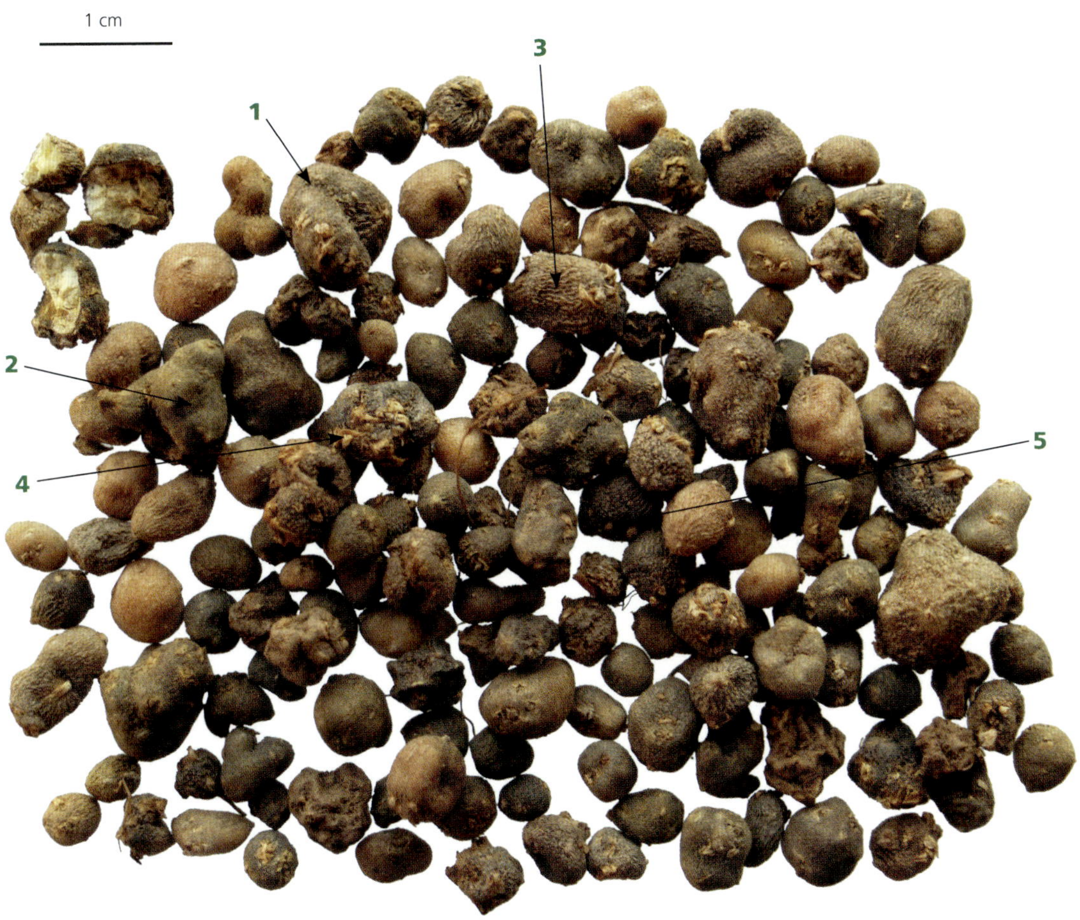

Crude drug of Xia Tian Wu (rhizome of *C. decumbens*)

Xi Xin 细辛

Asari Radix et Rhizoma

Drug part: ROOT AND RHIZOME

PHARMACOPOEIA SPECIES 1

Asarum heterotropoides F. Schmidt.

Aristolochiaceae

synonyms: *A. heterotropoides* var. *mandshuricum* (Maxim.) Kitag. (name used in CP2015)

Species of *Asarum* wild in China: 39

PLANT DESCRIPTION

A perennial herb (10–30 cm tall) with horizontal rhizomes (2–3 mm diam.; **rhizome internodes 0.1–0.5 cm long)** and roots, stems and leaves with oil cells. *Leaves* all basal, simple and paired; stalked (1–15 cm long); leaf blade uniformly coloured, ovate to heart-shaped or somewhat kidney-shaped (4–9 cm long; 5–13 cm wide), lower surface with dense mat of short, soft hairs, base heart-shaped, **lateral lobes 3–4 cm long; 4–5 cm wide, tip broad or shortly pointed; margins entire,** scale-leaves (cataphylls) subrounded (c. 0.8 cm long; 0.8 cm wide). ***Flower*** solitary on a leafless, often purplish (rarely purplish-green), stem (3–5 cm long) and usually emerging below the leaves; sepals and petals deep purple, fused to form a squat urn-shape flower (1–1.5 cm long; 1 cm diam.) externally longitudinally ribbed, **flower lobes** 3, broad, **recurved** and purple (c. 0.7 cm long; 0.9 cm wide); stamens 12; ovary superior; styles 6, free and short. Seeds many. Fl. May.

HARVESTING, SOURCING AND NATURAL RANGE

The huge market for 'Xi Xin' globally has led to the over-harvesting of wild populations and their rapid decline in recent years. Today the drug is mostly harvested from medicinal crop plants grown extensively in parts of the species' natural range in the provinces of Jilin and Liaoning in NE China. Roots and rhizomes are harvested when plants are fruiting (in summer) or in early autumn with Xi Xin sourced from Liaoning considered the best quality and called 'Liao Xi Xin' (Liaoning Asarum) or 'Bei Xi Xin' (northern Asarum) (Zhou, 2006).

The natural distribution of the species in NE China is the neighbouring provinces of Heilongjiang, Jilin and Liaoning provinces where the plant's remaining populations grow in mountain forests and valleys in damp shady places. According to the *FoC* a form called *A. heterotropoides* f. *heterotropoides* is restricted to Japan.

CONSERVATION STATUS

China: Vulnerable (MEPCAS, 2013).
Global: Not Evaluated (IUCN Red List, accessed 9/11/2016).

PHARMACOPOEIA SPECIES 2

A. sieboldii Miq.

Aristolochiaceae

synonyms: *A. sieboldii* var. *seoulense* Nakai (name used in CP2015); *Asiasarum sieboldii* (Miq.) F. Maek.

taxonomic/nomenclatural note: although the CP2015 treats *A. sieboldii* Miq. var. *seoulense* Nakai as a separate taxon, *Kew-MPNS* and *FoC* treat this name as a synonym of *A. sieboldii* Miq. hence the present work recognises only two sources of Xi Xin.

PLANT DESCRIPTION

Flowers are needed in order to reliably distinguish this species from PS1. *Main differences:* **rhizome internodes longer** (1–2 cm long) (but this character seems to vary); **lateral lobes of leaves smaller (1.5–4 cm long; 2–2.5 cm wide)**; blade tip long- or shortly-pointed. **Flower lobes erect or spreading.** Fl. Apr–May.

HARVESTING, SOURCING AND NATURAL RANGE

Harvesting as for PS1. Mostly sourced from medicinal crop plants grown on a large scale in Liaoning province (NE China). Wild populations, heavily over-harvested for medicinal use, grow in forests and damp valleys in Liaoning province. The species is also wild in the Korean Peninsula.

CONSERVATION STATUS

China: Vulnerable (MEPCAS, 2013); 'rapidly decreasing' Leung and Cheng (2008).
Global: Not Evaluated (IUCN Red List, accessed 9/11/2016).

UNOFFICIAL SUBSTITUTES

Many unofficial substitutes of Xi Xin have been reported in the past (Bensky *et al*., 2004) but with the bulk of Xi Xin sourced today from medicinal cultivation, substitutes are less likely, certainly in international trade. Past examples include many of the 37 other *Asarum* species that grow wild in China (whose identification requires considerable botanical expertise). Unofficial substitutes from other families have included *Chloranthus fortunei* (A. Gray) Solms, *C. henryi* Hemsl. (Chloranthaceae), *Ligularia hodgsonii* Hook. f. (Asteraceae) and *Erigeron breviscapus* (Vaniot) Hand.-Mazz. (Asteraceae).

TCM PROPERTIES AND USE

***properties*:** **'pungent and warm'**
***channels*:** **'Lung, Heart and Kidney'**
***actions*:** **'dispels wind, disperses cold, relieves pain, resolves phlegm'**

***indications*:** **headache, joint pain, abdominal pain; respiratory tract infection, coughs, clears nose, dyspnoea; (T)**

Medicinal crop of *Asarum heterotropoides*, Liaoning province

Freshly uprooted cultivated plants of *A. heterotropoides*; note reflexed flower lobes

Medicinal crop of *Asarum sieboldii*

Flowering plant of *A. sieboldii*

Freshly uprooted plant of *A. sieboldii*; note more pointed leaves and flower lobes *not* reflexed

Asarum heterotropoides

Aristolochiaceae

DRUG MORPHOLOGY

CRUDE DRUG (DRIED ROOT AND RHIZOME): mostly roots **[1]** with rhizomes apically **[2]**. *Roots*: a dense and partially contorted mass **[3]**, slender and unbranched (10–20 cm long; 1 mm diam.), slightly curved in upper parts and normally still attached apically to rhizomes. Externally, roots yellowish or greyish-yellow, smooth or longitudinally wrinkled; some fibrous rootlets **[4]** or rootlet scars present. *Rhizomes*: irregularly cylindrical **[5]** (1–10 cm long; 2–4 mm diam.), sometimes shortly branched, externally greyish-brown, with a few raised annulations **[6]**, **internodes 2–3 mm long [7];** apical end with ovate stem scar **[8]**. *Texture*: brittle. *Fracture*: even, yellowish-white or white. *Odour*: aromatic. *Taste*: pungent and tongue-numbing. ***Good quality*:** narrow greyish-yellow roots and rhizomes with a hot, acrid taste.

DECOCTING PIECES (ROOT AND RHIZOME CUT LENGTHS): mostly of roots **[9]** with occasional slightly larger masses of cut rhizome **[10]** with root lengths attached. *Roots:* somewhat cylindrical (variable length; 1 mm diam.), unbranched, externally yellowish or greyish-yellow, smooth or longitudinally wrinkled, some fibrous rootlets or rootlet scars present. Cut surface yellowish-white or white **[11]**. *Rhizome*: 2–4 mm diam. but rarely intact thus difficult to see the key character, the internode length; **if intact then internodes 2–3 mm long.** See *crude drug* for other less important rhizome characters. *Odour:* aromatic. *Taste:* pungent and tongue-numbing.

PROCESSING

Roots and rhizomes are shade-dried (= crude drug). Decocting pieces are made by softening the rootstocks with sprayed water before cutting them into sections and shade-drying.

OTHER COMMON NAMES

北细辛 Bei Xi Xin 'northern Asarum'; 了细辛 Liao Xi Xin 'Liao [ning] Asarum'; Chinese wild ginger (SCN); Radix et Rhizoma Asari; Manchurian wildginger root (CP2010).

SAFETY

Slightly toxic (Chang & But, 1986). Aristolochic acids (AAs), which are nephrotoxic and carcinogenic, have been detected in *Asarum* species but in "negligible amounts"; aerial parts containing higher quantities than underground parts (Zhao *et al.*, 2008). Many countries have restricted or banned the medicinal supply and use of plants containing AAs (owing to a spate of serious adverse reactions to *Aristolochia* species (in the same family) in the 1990s but which contained much higher levels of AAs. The need for regulations on the medicinal use of *Asarum* species have also been considered in recent years but owing to the negligible levels detected, regulations and pharmacopoeia guidance vary from country to country; e.g. editions of the CP from 2005 onwards specify that **only roots and rhizomes of Xi Xin are permitted** (as opposed to all parts previously) whereas Hong Kong SAR, Switzerland and the UK introduced a voluntary ban on the use of any parts (Anon., 2003a).

A. sieboldii

Aristolochiaceae

DRUG MORPHOLOGY

CRUDE DRUG (DRIED ROOT AND RHIZOME): as for PS1 but with rhizome **internodes longer (1–2 cm long) [12]**. For reliable discrimination from the drug of PS1 laboratory-based authentication methods are required.

Xi Xin with leaves included. Sourced from *A. sieboldii* (flowers also present)

PROCESSING

As for PS1.

OTHER COMMON NAMES

汉城细辛 Han Cheng Xi Xin; 华细辛 Hua Xi Xin; Chinese wild ginger (SCN); Radix et Rhizoma Asari; Manchurian wildginger root (CP2010).

SAFETY

As for PS1.

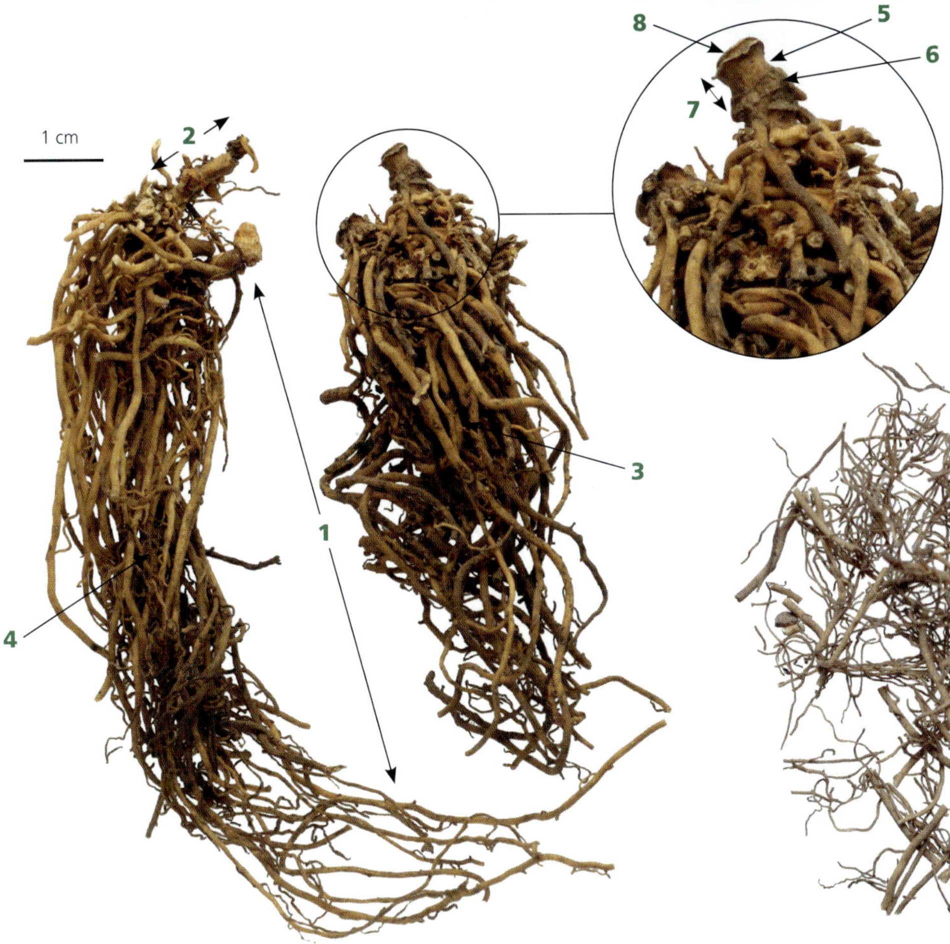

Crude drug of Xi Xin sourced from *Asarum heterotropoides*

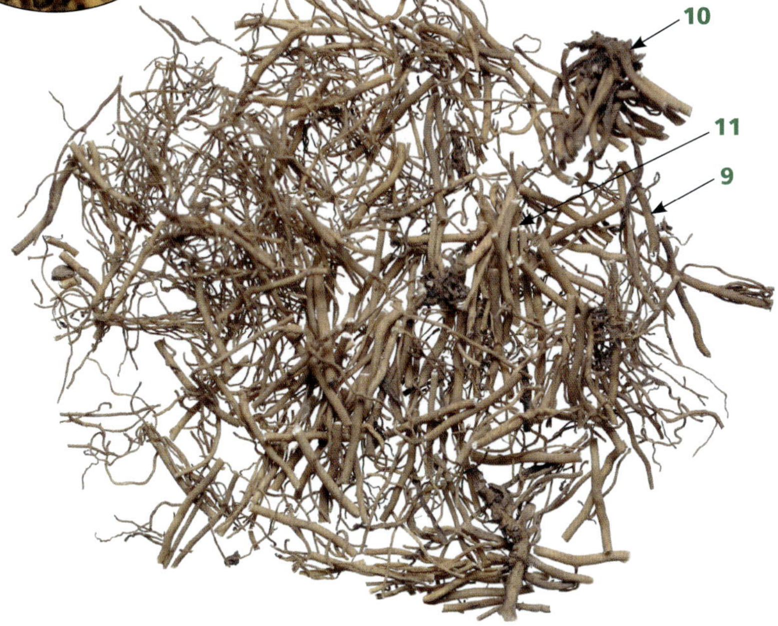

Decocting pieces of Xi Xin sourced from *A. heterotropoides*

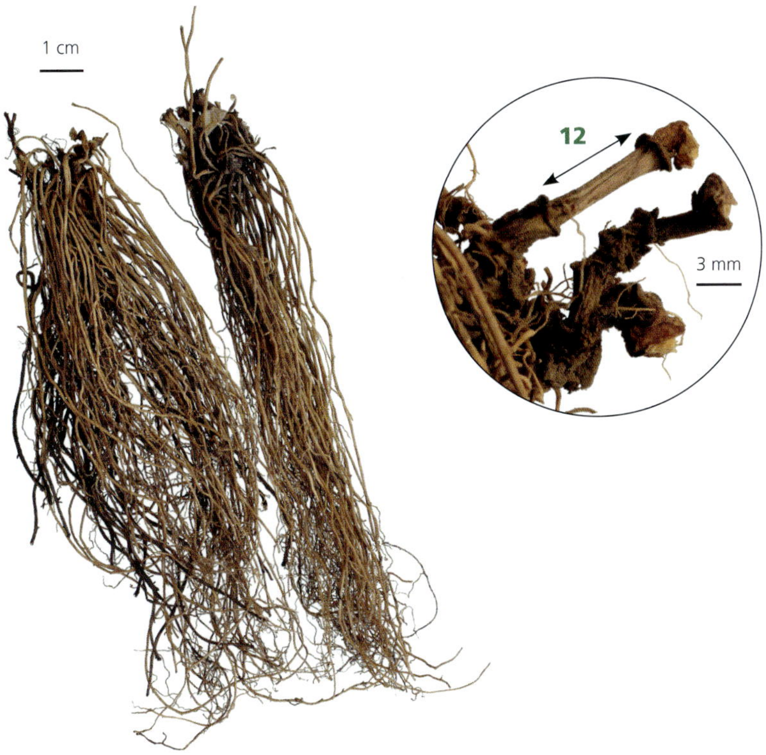

Crude drug of Xi Xin sourced from *A.sieboldii*

Decocting pieces of Xi Xin sourced from *A. sieboldii*

Xi Yang Shen 西洋参

Panacis Quinquefolii Radix

American ginseng

Drug parts: ROOT AND RHIZOME

Panax quinquefolius L..

Araliaceae

synonym: *Aralia quinquefolia* (L.) Decne & Planch.
taxonomic/nomenclatural note: in CP2015 as '*P. quinquefolium* L.'

Species of *Panax* wild in China: 6

PLANT DESCRIPTION

An herbaceous perennial (20–50 cm tall). *Root* stout, fleshy, branched and aromatic; slightly tapering and growing horizontally. *Stem* solitary, erect and rarely branched. *Leaves* stalked, at top of stem in one whorl of 3–6 (increasing in number in older plants); each leaf with 5 (rarely 3 or 7) stalked leaflets radiating from a single point (palmately-compound); leaflet blades oblong to widest above the middle (6–15 cm long; (2–) 2.5–8 cm wide), lateral leaflets much smaller, veins sparsely bristly, sometimes hairless on upper surface, leaflet bases rounded, **margins coarsely saw-toothed** and tips abruptly or obviously narrow-pointed. ***Inflorescence*** a solitary, hemispherical **6–20-flowered** terminal head on an erect, 1–25 cm-long stalk (peduncle), **peduncle shorter than the leaf stalks**. *Each flower* tiny (2–3 mm diam.), stalked, bisexual, radially symmetrical; calyx green and cup-shaped, 5-lobed at mouth; petals 5, spreading, greenish or white with pointed tips; stamens 5 and inserted onto a disk; ovary inferior, styles usually 2. *Fruit (drupe)* sub-globose (c. 10 mm diam.), fleshy and, when mature, bright red. *Seeds* creamish-white, kidney-shaped and laterally compressed.

HARVESTING, SOURCING AND NATURAL RANGE

Panax quinquefolius (American ginseng) is not native in China. It originates from eastern North America (USA and Canada) where it grows wild in temperate woodland on well drained soils. Owing to the depletion by the 1760s of China's own wild ginseng (*P. ginseng*, source of 'Ren Shen', page 240), and which at that time was thought to be the same species, dried roots of American ginseng were imported into China and incorporated in 1765 into the Chinese *materia medica* (Upton, 2012). Once its separate identity was established along with its slightly different, and more widely sought after, therapeutic properties, the cultivation of American ginseng China took off. The bulk of production today for the TCM industry is sourced from China's provinces of Heilongjiang, Jiangxi, Jilin, Liaoning, Shaanxi and Shandong, and, in recent years, Guizhou (SW China) has also begun production. Roots of the highly prized 'wild' American ginseng (in N America called 'wild-simulated' ginseng because it is grown under semi-natural conditions) are still imported into China in small quantities to meet demand for its allegedly superior therapeutic properties. The latter, however, have yet to be validated analytically (Upton, 2012).

CONSERVATION STATUS

China: not native.
N. America: uncommon to rare and threatened in many states of the USA (McGraw *et al.*, 2013).
Canada: Endangered (Government of Canada).
Global: Not Evaluated (IUCN Red List, accessed 8/2/2016).

PROTECTED STATUS

CITES: Appendix II; EU Annex B (see Table 1 page 10).

P. ginseng C. A. Mey.

Araliaceae

PLANT DESCRIPTION

Difficult to distinguish from *P. quinquefolius*. Considerable expertise is required to do so reliably. *Main differences:* **leaflet margins minutely** (not coarsely) **saw-toothed** and hairless; ***inflorescence*** **a 30–50-flowered** terminal head; stalk of flower- and fruit-head **(peduncle) longer than leaf stalks.**

SUBSTITUTION COMMENT. Roots of *P. ginseng* are the source of the separate TCM drug Asian ginseng ('Ren Shen' (Ginseng Radix et Rhizoma), see page 240) and are sometimes difficult to distinguish macroscopically from American ginseng (above). Given that both Asian and American ginseng are high-value herbal products and that their prices fluctuate independently from each other from year to year, unscrupulous traders may substitute one for the other in order to capitalise on the highest market prices.

CONSERVATION STATUS

China: Critically Endangered (MEPCAS, 2013).
Korean Peninsula: ?Extinct (Zhuravlev *et al.*, 2008).
Russia: 'threatened in Russia since 1975' (Zhuravlev *et al.*, 2008).
Global: Not Evaluated (IUCN Red List, accessed 10/11/2016).

PROTECTED STATUS

CITES: Appendix II; EU Annex B (see Table 1 page 10).

TCM PROPERTIES AND USE

***properties*:** **'sweet, slightly bitter and cool'**
***channels*:** **'Heart, Lung and Kidney'**
***actions*:** **'tonifies qi and yin, clears fire, generates fluids'**

***indications*:** **fatigue; cough, haemoptysis; thirst, dry mouth; haematochezia**

Ripening crop of *Panax quinquefolius*, Jilin province

Coarsely toothed leaf margins of *P. quinquefolius*

P. quinquefolius with ripening fruit-head

Young inflorescence of *P. quinquefolius*

Leaflet margins of *P. ginseng* minutely (not coarsely) toothed

P. ginseng crop with developing flower-heads. Leaf spots are residues of recent pesticide spraying

P. ginseng with fruit-head

Panax quinquefolius Araliaceae

DRUG MORPHOLOGY

CRUDE DRUG (DRIED ROOT AND RHIZOME): *Root* **[1]** (main body of rootstock) cylindrical or conical (3–12 cm long; 0.8–2 cm diam.) with one or several lateral roots in basal parts but usually broken off **[2]**. Externally pale yellowish-brown or yellowish-white, subtly clothed throughout in minute, soft hairs; with many long and distinct transverse-striations **[3]** and much shorter, slightly protuberant lenticels **[4]** and occasional rootlet scars; longitudinal wrinkles less obvious. *Rhizome* (Lu Tou) **[5]** often short, compact, densely annulated **[6]** and with rounded or semi-rounded stem scars **[7]**. Upper parts of root **densely annulated** **[8]** and with rounded or semi-rounded stem scars (Lu Wan) **[7]**; small adventitious roots (Ding) sometimes still attached. *Texture:* hard and difficult to break. *Fracture:* even, yellowish-white, slightly starchy **[8]** (see *Decocting Pieces* description). *Odour*: slight but distinctive. *Taste:* slightly bitter and sweet. ***Good quality***: solid, heavy, yellowish-white and densely annulated; aromatic, with a sweet and bitter taste.

DECOCTING PIECES: *transversely sliced roots:* sub-rounded or elliptical slices (1–2 cm diam.; 1–2 mm thick) with undulating margin **[9]**; outer bark thin, pale yellowish-brown **[10]**, cut surface starchy **[8]**, cortex (outermost ring) white or pale yellowish-white **[11]**, typically with 1 or more concentric rings of yellowish-brown or orange dot-like resin ducts **[12]** and occasional radial clefts; cambium ring brownish-yellow **[13]** with occasional faint rays; central wood **[14]** (xylem) occupying c. 2/3rds of the root diameter, yellowish-white, rays faint. *Texture:* starchy. *Odour and taste*: as for crude drug.

PROCESSING

Rootstocks are dried in the sun or at low temperature (= crude drug); the roots are then softened thoroughly before being sliced thinly (1–2 mm thick) or pounded into small pieces (= decocting pieces).

OTHER COMMON NAMES

广东人参 Guang Dong Ren Shen; 花旗参 Hua Qi Shen; 西参 Xi Shen; 西洋人参 Xi Yang Ren Shen; 洋参 Yang Shen; American ginseng (SCN); North American ginseng; Radix Panacis Quinquefolii.

SAFETY

Heavy metal and pesticide levels require careful monitoring since cultivated plants may contain high levels. An official food and healthfood in China (Anon., 2002a,b, as '*P. quinquefolium*'). Safety Class: 1; **Interaction Class: B** (Gardner & McGuffin, 2013).

P. ginseng Araliaceae

'DRUG SUBSTITUTE' MORPHOLOGY

Differentiating rootstocks of *P. ginseng* from those of *P. quinquefolius*: *Ginseng experts claim to be able to distinguish rootstocks of* P. ginseng *from those of* P. quinquefolius *using organoleptic characters alone:* rhizome of *P. ginseng* tends to be more pronounced; roots less hard and less dense, **not clothed in minute, soft hairs**; and upper parts are **not densely annulated.** Subtle differences in taste, colour and especially odour are also used to separate the two species: in general, *P. ginseng* has a more distinctive initial sweetness and *P. quinquefolius* has a stronger bitter aftertaste. Non-specialists are advised against using such characters, however, especially since trade items are highly variable. For robust differentiation of the rootstocks laboratory-based authentication methods are strongly advised; many techniques have been developed (e.g. Kite *et al.*, 2003).

Other main characters of *P. ginseng* (for detailed description see 'Ren Shen' page 240)**:**

CRUDE DRUG (DRIED ROOT AND RHIZOME) 'YUAN SHEN': *Root* **[15]** cylindrical or tapering (3–15 cm long; 1–2 cm diam.) dividing basally into 2–3 long, often interwoven lateral roots **[16]** with slender rootlets **[17]**; lateral roots often forked. Externally greyish-yellow with scattered transverse lenticels **[18]**. *Rhizome* **[19]** attached at root apex, short and narrow (1–4 cm long; 0.3–1.5 cm diam.), contorted with stem bases or scars **[20]**.

DECOCTING PIECES 'YUAN SHEN': *transversely sliced roots:* sub-rounded or elliptical slices (1–2 cm diam.; 1–2 mm thick) sometimes with an undulating margin **[21]**. Outer bark thin, greyish-yellow **[22]**, cut surface starchy, cortex (outermost ring) white **[23]** with 1 or more interrupted rings of yellowish-brown or orange resin ducts **[24]**; cambium ring brownish-yellow **[25]**, (xylem) occupying c. 2/3 of root diameter, yellowish-white **[26]**.

OTHER COMMON NAMES

see 'Ren Shen' (Ginseng Radix et Rhizoma) page 240.

SAFETY

see 'Ren Shen' (Ginseng Radix et Rhizoma) page 240.

LESSER KNOWN UNOFFICIAL SUBSTITUTES

Owing to the high prices that American ginseng drug can fetch, unofficial substitution, adulteration and counterfeiting is a concern on the Chinese and international markets; and the general name 'ginseng' has been widely mis-applied in retail outlets. In the US, legislation passed in 2004 requires that only species in the genus *Panax* can be marketed as 'ginseng' (Upton, 2012). Chemical adulteration may include the spraying of look-alike rootstocks of other species with ginsenosides specific to *P. quinquefolius* in attempt to dupe quality-control laboratories; thus the presence of American ginsenosides alone is no guarantee of species identity. Powdered ginseng is especially prone to adulteration. Examples of species used as counterfeits include:

1. ***Adenophora*** species (Campanulaceae) — some roots are the source of the TCM drug 'Nan Sha Shen' (page 222).
2. ***Codonopsis pilosula*** (Franch.) Nannf. and *C. tangshen* Oliv. (Campanulaceae) — source of TCM drug 'Dang Shen' (page 102).
3. ***Lactuca indica*** L. (Asteraceae) — 'Chi Guo'.
4. ***Physochlaina infundibularis*** Kuang (Solanaceae) — 'Lou Dou Pao Nang Cao'.
5. ***Phytolacca acinosa*** Roxb. and ***P. americana*** L. (Phytolaccaceae) — source of the separate TCM drug 'Shang Lu' (Phytolaccae Radix).
6. ***Talinum paniculatum*** (Jacq.) Gaertn. (Portulacaceae) — 'Tu Ren Shen'.
7. ***Vigna vexillata*** (L.) A. Rich. (Fabaceae) — 'Ye Jiang Dou'.

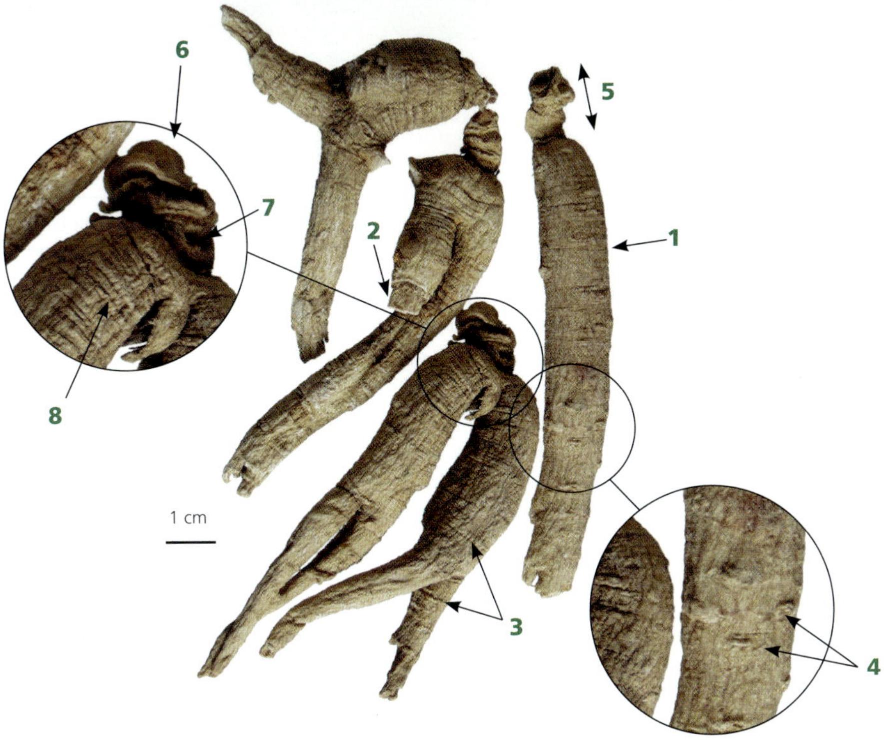

Crude drug of Xi Yang Shen from cultivated *Panax quinquefolius*

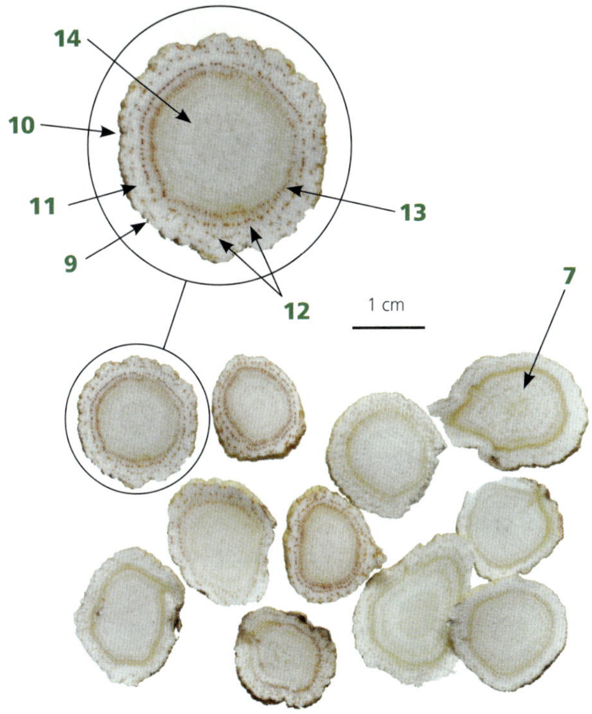

Decocting pieces (transversely sliced) of Xi Yang Shen

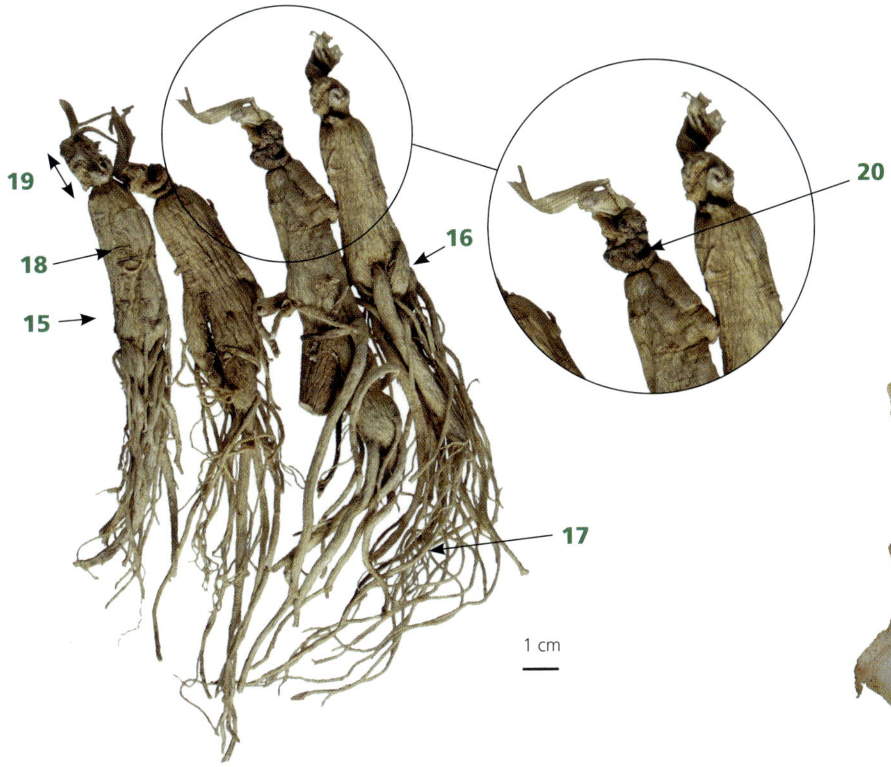

Crude druge of cultivated *P. ginseng* (Ren Shen/Yuan Shen)

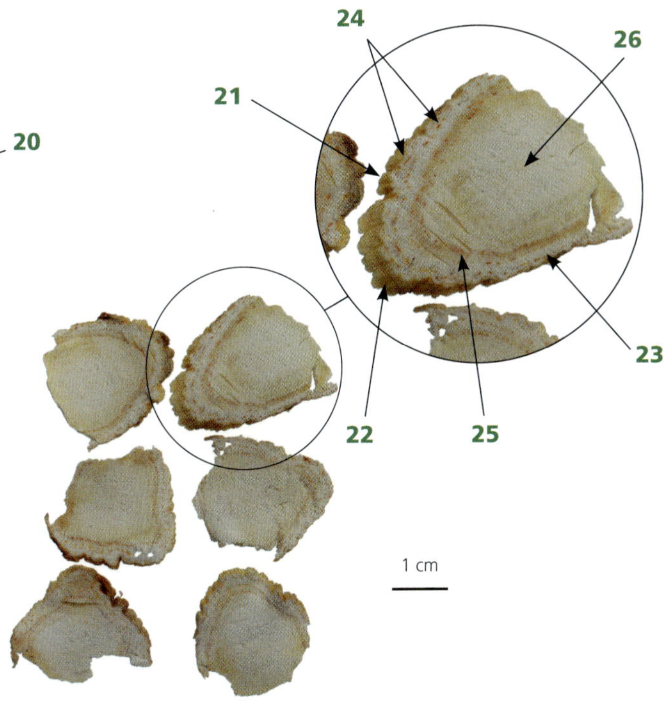

Decocting pieces of Ren Shen/Yuan Shen sourced from *P. ginseng*

The vernacular name 'ginseng' is used worldwide to refer to a number of other species (in unrelated families) with tonic/adaptogenic properties. Although most are not used as substitutes for American or Asian ginseng, the use of shared similar common names may lead to confusion, e.g. 'Brazilian ginseng' (*Hebanthe eriantha* (Poir.) Pedersen (family Amaranthaceae), 'blue ginseng' *Gynostemma pentaphyllum* (Thunb.) Makino (Cucurbitaceae); 'Siberian ginseng' (*Eleurotherococcus senticosus* (Rupr. & Maxim.) Maxim. (family Araliaceae) with parts used as the source of the separate TCM drug 'Ci Wu Jia'; Indian ginseng (*Withania somnifera* (L.) Dunal (family Solanaceae), also commonly called ashwagandha.

Xuan Shen 玄参

Scrophulariae Radix

Drug part: ROOT

Scrophularia ningpoensis Hemsl.

Scrophulariaceae

Species of *Scrophularia* wild in China: 36

PLANT DESCRIPTION

An upright perennial herb (up to 1.5 m tall) with a branching, tuberous rootstock of spindle-shaped to conical roots. *Stems* 4-sided and shallowly grooved longitudinally, hairless or with white hairs. *Leaves* opposite (sometimes alternate towards top of the plant), simple and stalked; leaf blades large: lower ones usually ovate, upper ones ovate to lance-shaped (up to 30 cm long; 19 cm wide), base wedge-shaped, rounded or somewhat heart-shaped; margin minutely saw-toothed. *Flowers* in stalked groups of 2–4 within a much larger, loose, many-flowered erect spike (thyrse) (up to 50 cm long). Each flower small, inconspicuous and bisexual; on a long glandular hairy stalk; calyx 5-lobed (2–3 mm long) with rounded, membranous margins; petals fused into a squat, brownish-purple cup-like corolla tube (8–9 mm long) opening into 5 sub-orbicular lobes of different size and shape; middle lobe (lip) shorter than the four lateral and upper lobes; ovary superior; style with a pinhead-like stigma (c. 3 mm long); stamens 4 and slightly shorter than lower lip. *Fruit* a dry, ovoid capsule (8–9 mm long); seeds many and minute. Fl. Jun–Oct, fr. Sep–Nov.

HARVESTING, SOURCING AND NATURAL RANGE

Roots are harvested in winter, after aerial parts have withered. Sourcing is mostly from medicinal crops and small-holdings in the Chinese provinces of Zhejiang, Sichuan, Shaanxi, Hubei and Jiangxi. *Di Dao*-quality 'Xuan Shen' is sourced from sites in Zhejiang province. Some trade materials are produced to GAP standard. In the wild the species grows in bamboo forests, tall grassland, thickets and streamsides (up to 1,700 m altitude) and is endemic to C and S China (Anhui, Fujian, Guangdong, Guizhou, Hebei, Henan, Jiangsu, Jiangxi, Shaanxi, Shanxi, Sichuan and Zhejiang).

CONSERVATION STATUS

China: Not Evaluated.

Global: Not Evaluated (IUCN Red List, accessed 19/8/2016).

Crop of *Scrophularia ningpoensis*, Zhejiang province (leaves turn purple before they wither)

Inflorescence with immature (green) fruits

Flower of *S. ningpoensis*

Fresh rootstock of *S. ningpoensis* (cultivated as a medicinal crop, Zhejiang province)

TCM PROPERTIES AND USE

properties: **'bitter, sweet, salty and cold'**
channels: **'Lung, Stomach and Kidney'**
actions: **'clears heat, cool blood, tonifies yin, eliminates toxins'**

indications: **fevers, thirst, cough; constipation, painful eyes; sore throat, nodules; purpura; sores and abscesses**

DRUG MORPHOLOGY

CRUDE DRUG (DRIED ROOT): sub-cylindrical (6–20 cm long; 1–3 cm diam.), somewhat curved and widest either towards the middle or apex (end with stem scars) or narrowing at both ends. Externally dull, greyish-brown or greyish-yellow **[1]**, with a small number of deep, irregular and rounded longitudinal furrows **[2]** and scattered throughout with transverse and slightly raised lenticels **[3]** and rootlet remnants or oval rootlet scars **[4]**. *Texture*: dense and difficult to break. *Fracture*: black or brownish-black **[5]**, often lustrous. *Odour*: distinctive; has been likened to burnt sugar (caramel). *Taste*: slightly bitter and somewhat sweet. ***Good quality*:** large, solid roots; internally black and devoid of fissures.

DECOCTING PIECES (SLICED ROOT): sub-ovate slices (c. 2.5–6 cm long; 1–3 cm wide; 1–3 mm thick) often with an irregular and deeply convoluted outline **[6]**. Outer bark thin, dull, greyish-brown or greyish-yellow **[7]**; raised lenticels and oval rootlet scars occasional. Cut surface black or brownish-black **[5]**, sometimes lustrous; vascular tissues sometimes visible as fine, yellow and wavy rays **[8]**. *Texture:* dense, tough and brittle. *Fracture:* densely black and even. *Odour* and *taste* as for the crude drug.

PROCESSING

After removal of rhizomes and rootlets, roots undergo a lengthy process of drying over 3–6 days (= crude drug) and are subsequently thinly sliced (often obliquely) into decocting pieces.

OTHER COMMON NAMES

黑玄参 Hei Xuan Shen; 润玄参 Run Xuan Shen; 乌玄参 Wu Xuan Shen; 元参 Yuan Shen; 浙玄参 Zhe Xuan Shen; figwort root (CP2010); Radix Scrophulariae; scrophularia (SCN).

SAFETY

An official healthfood in China (Anon., 2002a,b).

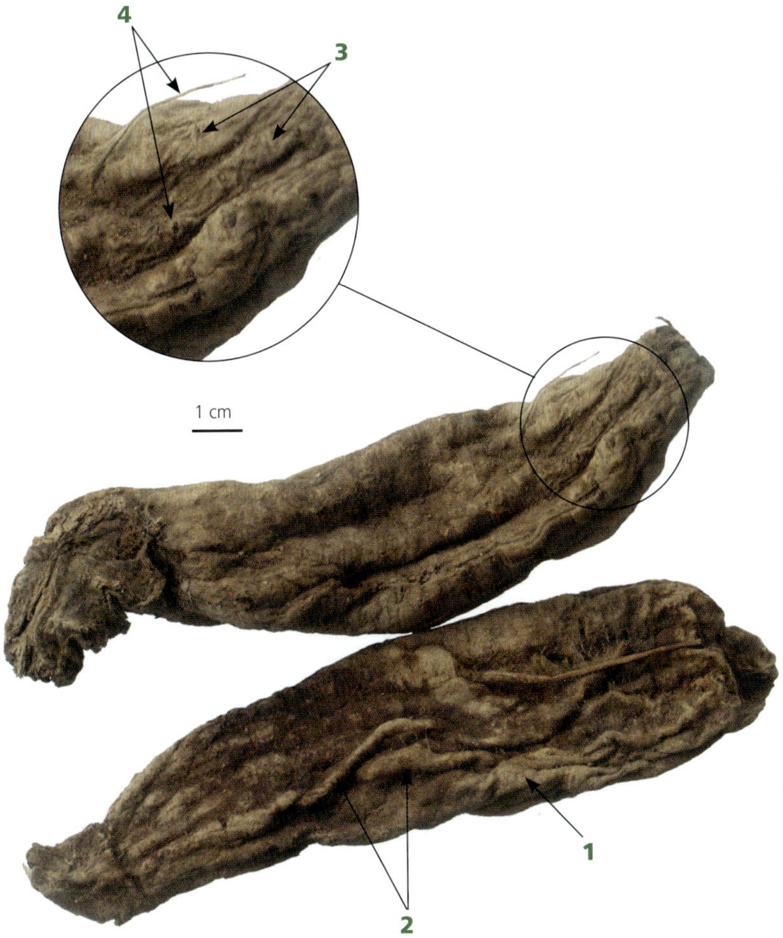

Crude drug of Xuan Shen (dried roots of *S. ningpoensis*)

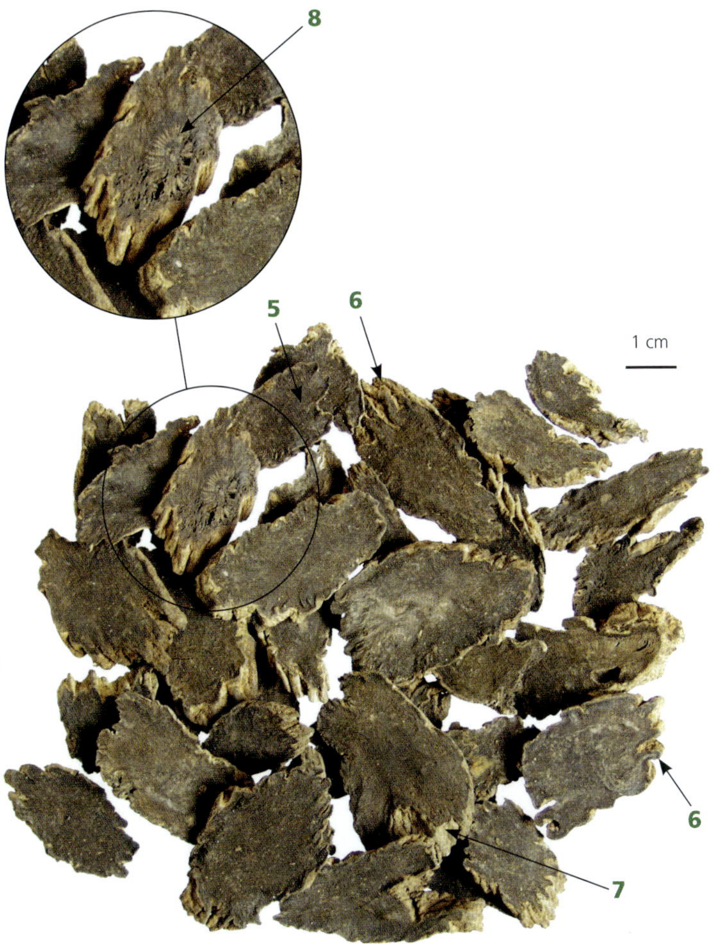

Decocoting slices of Xuan Shen (sliced, dried roots of *S. ningpoensis*)

Xu Chang Qing 徐长卿 Cynanchi Paniculati Radix et Rhizoma

Drug parts: ROOT AND RHIZOME

Cynanchum paniculatum (Bunge) Kitag. ex H. Hara

synonyms: *Asclepias paniculata* Bunge; *Vincetoxicum mukdenense* Kitag. (*Kew-MPNS* accepted name, 28/8/2016).

taxonomic/nomenclatural note: in *FoC* and CP2015 as *Cynanchum paniculatum* (Bunge) Kitag.

Apocynaceae

Species of *Cynanchum* wild in China: 57

PLANT DESCRIPTION

An erect perennial herb (up to 1 m tall) with a densely fibrous rootstock (medicinal part). *Stems* slender, rigid, hairless (except in lower parts) and unbranched or, in upper parts, sparsely branched. *Leaves* opposite, simple and short-stalked (stalks c. 3 mm long); leaf blades lance-shaped to linear (5–13 cm long; 0.5–1.5 cm wide); texture like stiff paper; hairless or minutely hairy on upper surface, margins hairy; base and tip of blade shortly pointed; lateral veins indistinct. *Flowers* in few-flowered loosely branched heads (up to 7 cm long) at top of plant or in upper leaf axils on the upper stem; flower-head stalks (peduncles) 2.5–4 cm long. Each flower bisexual and radially symmetrical; sepals 5, free and tiny; petals 5 with the widely spreading lobes yellowish-green and ovate (4–5.5 mm long) and fused basally to form a short tube; stamens 5 and fused to the style to form a complex structure (gynostegium) which is surrounded by a series of small, fleshy lobes (the corona); ovary superior. *Fruit* a long, dry follicle (4–8 cm long; 3–8 mm diam.); seeds many with a conspicuous tuft of hairs (1.5–3 cm long) at one end. Fl. May–Jul, fr. Aug–Dec.

HARVESTING, SOURCING AND NATURAL RANGE

Rootstocks are harvested in autumn, mostly from wild populations which grow on sunny mountain slopes in the Chinese provinces of Anhui, Fujian, Gansu, Guangdong, Guangxi, Guizhou, Hebei, Henan, Hubei, Hunan, Inner Mongolia, Jiangsu, Jiangxi, Liaoning, Shaanxi, Shandong, Shanxi, Sichuan, Taiwan, Yunnan and Zhejiang [also wild in Japan, the Korean Peninsula and Mongolia].

CONSERVATION STATUS

Wild populations in China are suspected to be declining; the species is seldom cultivated (authors' pers. obs. 2001).

China: Least Concern (MEPCAS, 2013); Near Threatened (authors' pers. obs 2001).

Global: Not Evaluated (IUCN Red List, accessed 26/8/2016).

Medicinal crop of Xu Chang Qing (*Cynanchum paniculatum*), Jiangsu province

A developing fruit of *C. paniculatum*

Flowering plant of *C. paniculatumx* in the wild

Fresh rootstock

TCM PROPERTIES AND USE

***properties*: 'pungent and warm'**
***channels*: 'Stomach and Liver'**
***actions*: 'dispels wind damp, invigorates Blood, relieves pain, eliminates toxins'**

***indications*: musculoskeletal, traumatic pain; oedema, itchy skin**

DRUG MORPHOLOGY

CRUDE DRUG (DRIED ROOT AND RHIZOME): a tangled rootstock, mostly of slender roots **[1]** attached to the slightly broader and similarly-coloured rhizomes **[2]**. Roots slender and curved (10–16 cm long; 1–1.5 mm diam.); externally brown, brownish-yellow or pale yellowish-white, finely wrinkled longitudinally and interspersed with groups of tiny fibrous rootlets **[3]** Rhizomes mostly cylindrical and somewhat knotty (0.5–3.5 cm long; 2–4 mm diam.) with slightly broader hollow stem remnants sometimes attached apically **[4]**. *Root texture:* fragile and easily broken. *Root fracture:* starchy, bark white or yellowish-white, cambium ring pale brown and central xylem small. *Odour:* aromatic. *Taste:* pungent and cool.

DECOCTING PIECES (DRIED LENGTHS OF ROOT AND RHIZOME): as for the crude drug but cut into short lengths **[5]** (c. 1–3 cm long); separation of rhizomes and roots more difficult to decipher but former distinguished by a central pith and with nodes giving rise to a mass of roots **[6]**.

PROCESSING

Roots and rhizomes are shade-dried (= crude drug), then briefly washed and cut into decocting lengths.

OTHER COMMON NAMES

寥刁竹 Liao Diao Zhu; 寥习竹 Liao Xi Zhu; 徐长卿 Xu Chang Qing (plant name); 枝香 Zhi Xiang; paniculate cynanchum; paniculate swallowwort root (CP2010); Radix et Rhizoma Cynanchi Paniculati.

Crude drug of Xu Chang Qing (*C. paniculatum*)

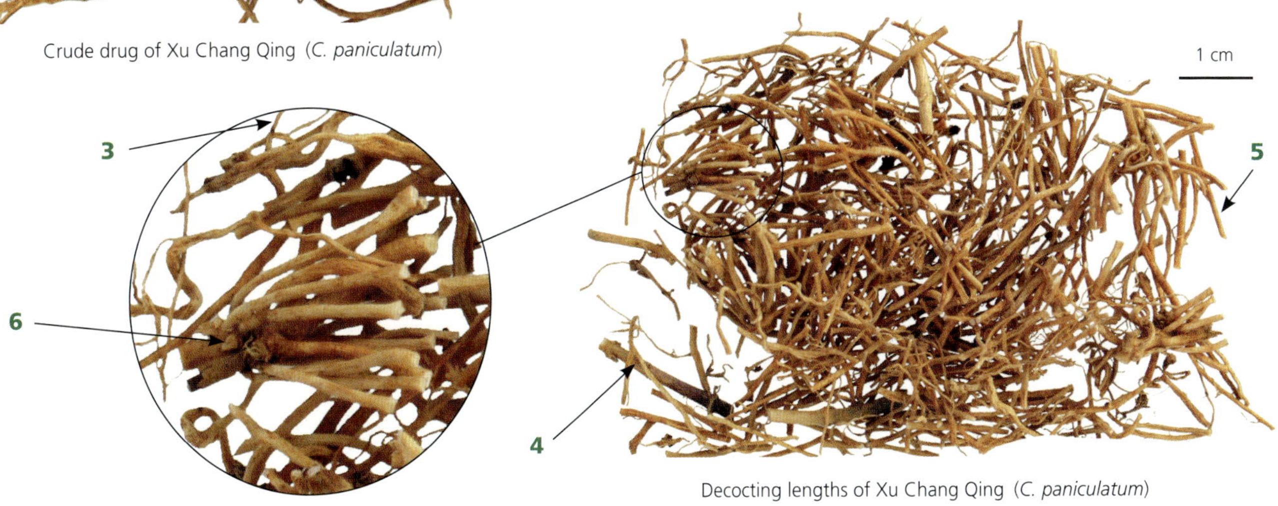

Decocting lengths of Xu Chang Qing (*C. paniculatum*)

Yan Hu Suo 延胡索

Corydalis Rhizoma

Drug part: RHIZOME (stem tuber)

Corydalis yanhusuo (Y. H. Chou & Chun C. Hsu) W. T. Wang ex Z. Y. Su & C. Y. Wu

Papaveraceae

synonyms: *C. turtschaninovii* Besser f. *yanhusuo* Y. H. Chou & Chun C. Hsu.
taxonomic/nomenclatural note: in CP2015 as *Corydalis yanhusuo* W. T. Wang.

Species of *Corydalis* wild in China: 357

PLANT DESCRIPTION

A small, delicate-looking perennial herb (10–30 cm tall) with juicy sap and a rounded, yellowish and tuberous rhizome (1–2.5 cm (rarely 0.5 cm) diam.). *Stems* with 1 (rarely 2) scale leaves and 2, 3 or 4 stem leaves. *Leaves* alternate, twice- or three-times tri-lobed; leaflets (2.25 cm long; 5–8 mm wide) entire or deeply divided into 3 lobes. *Flowers* in lax, 5–15-flowered heads arranged along a central axis. Each flower with a 10 mm long stalk (up to 20 mm when in fruit) bilaterally symmetrical, bisexual; sepals small and falling; petals 4 (unequal size and shape) magenta (deep rose-red); spur of upper petal upwardly curved and cylindric (11–13 mm long); nectary present along c. half spur length; lower petal shortly clawed; smaller inner petals 8–9 mm long; stamens 2; ovary superior; stigma orb-shaped. *Fruit* a many-seeded linear capsule (20–28 mm long) with a persistent style. *Seeds* black and glossy; in 1 row. Fl. and fr.: Apr–Jun.

HARVESTING, SOURCING AND NATURAL RANGE

Rhizomes are harvested entirely from cultivated plants in early summer after aerial parts have withered. Main production areas are Beijing, Gansu, Shaanxi, Sichuan, Yunnan and Zhejiang provinces with the traditional *Di Dao* sources located in Zhejiang (Dongyang and Pan'an districts).

The wild species grows in hill grasslands and is endemic to C and E China (Anhui, Henan, Hubei, Hunan, Jiangsu and Zhejiang provinces).

CONSERVATION STATUS

Endemic to China: Vulnerable (MEPCAS, 2013); Not Evaluated (IUCN Red List, accessed 13/5/2015).

Medicinal crop of *Corydalis yanhusuo*, Anhui province

Planting rhizomes of *C. yanhusuo*, Zhejiang province

C. yanhusuo in flower

UNOFFICIAL SUBSTITUTES/ADULTERANTS

The extensive cultivation of 'Yan Hu Suo' as a medicinal crop in China usually makes it uneconomic for cheaper look-alike, inferior substitutes or adulterants to be traded under this name. When market conditions are favourable, however, trade in rhizomes of many other wild-harvested *Corydalis species* may become popular, especially on local Chinese markets. A notable example is the use of *C. turtschaninovii* Besser (齿瓣延胡索 Chi Ban Yan Hu Suo) native to NE China (near the N Korean border); its rhizomes are the official source of 'Corydalis Tuber' in the Korean Pharmacopoeia (9th ed., 2007), which uses the synonym *C. ternata* (Nakai) Nakai.

Although less common, substitution and adulteration has also been reported on the international market:

1. ***Corydalis decumbens*** (Thunb.) Pers. A common species in China whose wild-harvested rhizomes are the source of the separate TCM drug 'Xia Tian Wu' (Corydalis Decumbentis Rhizoma) see page 304. **Rhizomes larger (0.5–3 cm long; 0.5–2.5 cm diam.),** sometimes quite angular; rhizome skin **yellowish-grey to dark green or brown,** skin often much smoother but finely wrinkled in places; apex with stem remnants surrounded by pale leaf scars and rootlet scars. *Fracture*: yellowish-white or yellow**, granular or horny; sometimes starchy.** *Taste*: bitter.
2. ***Curcuma longa*** L. Rhizomes are the source of the separate TCM drug 'Jiang Huang' (turmeric — Rhizoma Curcumae Longae) (not included in this book). **Rhizomes larger (2–5 cm long, 1–3 cm diam.);** skin rough, with distinct annulations; cut surface with distinct endodermis ring and many scattered vascular bundles. *Odour*: characteristic of turmeric unless rhizomes have already been extracted. *Taste*: bitter and pungent.

TCM PROPERTIES AND USE

***properties*: 'pungent, bitter and warm'**
***channels*: 'Liver and Spleen'**
***actions*: 'invigorates Blood and qi, relieves pain'**
***indications*: pain in the chest and abdomen; dysmenorrhoea; traumatic injury**

DRUG MORPHOLOGY

CRUDE DRUG (BOILED AND DRIED RHIZOME): sub-globose or rounded to nodular (0.5–1.5 cm diam.). Externally brown or yellowish-brown, rough and wrinkled **[1]**, sometimes with leaf scars. Apex slightly depressed with a stem scar **[2]**; base often tuberculate **[3]**. *Texture*: hard but brittle. *Fracture*: yellow or dark brown, lustrous and wax-like **[4]**. *Odour*: indistinct. *Taste*: bitter. ***Good quality*:** hard and brittle; in section pale yellow and lustrous.

DECOCTING PIECES-1 (SLICED RHIZOME): halved rhizomes, rounded or long-ovate in outline **[5]** (0.5–1.5 cm diam.; c. 4–6 mm thick). Rhizome skin brown or yellowish-brown, rough and wrinkled **[6]**, sometimes with leaf scars. Apex, if present, slightly depressed with stem scar **[7]**; base, if present, often tuberculate **[3]**. Cut surface yellow or dark brown, translucent and sometimes lustrous **[4]**. *Texture*: hard and brittle. *Odour*: indistinct. *Taste*: bitter.

DECOCTING PIECES-2 (VINEGAR-PROCESSED SLICED RHIZOME): as for sliced rhizome above but cut surfaces usually a more vibrantly yellowish-brown **[8]**. *Texture*: hard. *Odour*: vinegar-scented.

PROCESSING

After removal of fibrous roots, the rhizomes are boiled and sun-dried (= crude drug), then cut into decocting slices (c. 4–6 mm thick) or halved. Alternatively the crude drug is steeped in rice vinegar then dry-fried before slicing or breaking into pieces to produce Cu Yan Hu Suo (醋延胡索) 'vinegared Yan Hu Suo'.

OTHER COMMON NAMES

延胡 Yan Hu; 元胡 Yuan Hu; Corydalis yanhusuo (SCN); Rhizoma Corydalis; yanhusuo (CP2010 and plant name).

SAFETY

Safety Class: 2b; Interaction Class: A (Gardner & McGuffin, 2013).

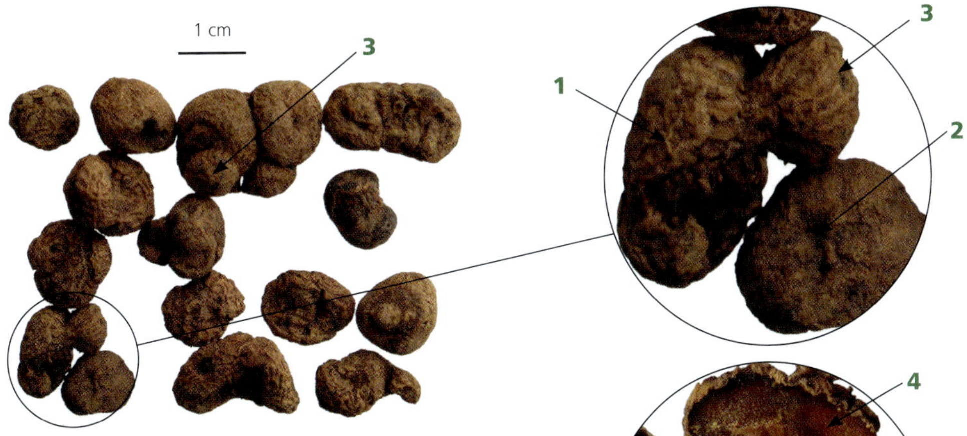

Crude drug of Yan Hu Suo (dried stem tubers of *C. yanhusuo*)

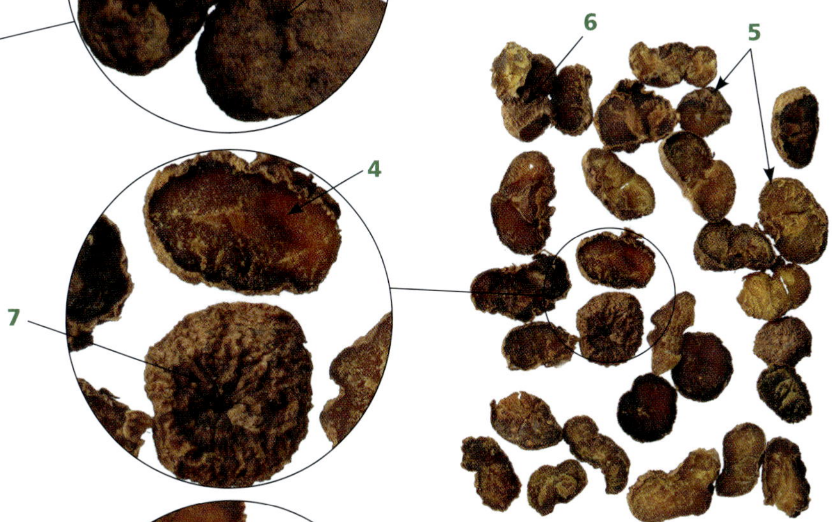

Decocting pieces of Yan Hu Suo (halved tubers of *C. yanhusuo*)

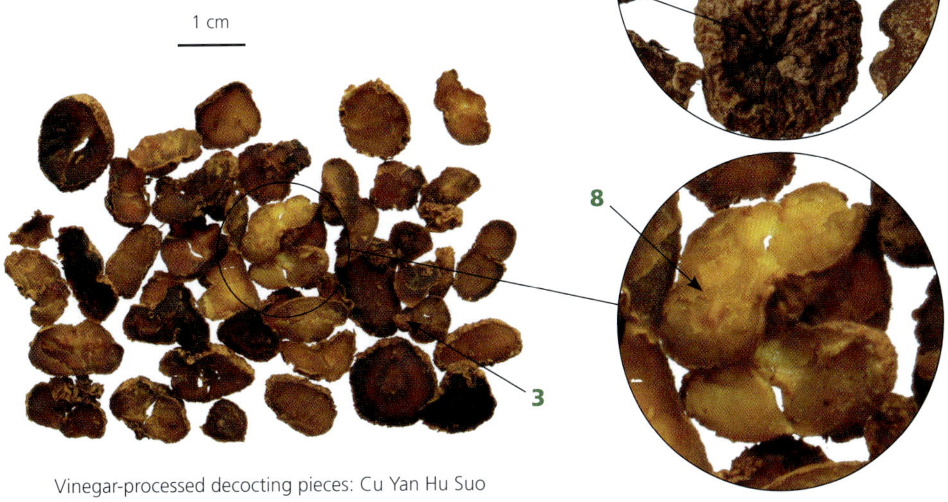

Vinegar-processed decocting pieces: Cu Yan Hu Suo

Yi Bei Mu 伊贝母 Fritillariae Pallidiflorae Bulbus

Drug part: BULB

PHARMACOPOEIA SPECIES 1

Fritillaria walujewii Regel

Liliaceae

Species of *Fritillaria* wild in China: 24

PLANT DESCRIPTION

An erect, herbaceous perennial (20–50 cm tall), with a sub-globose, slightly squat bulb (1–2.5 cm diam.) with 2 fleshy scales. *Stem* unbranched. *Leaves* 7–13, opposite, alternate or in whorls; leaf blades simple, linear to lanceolate (5.5–10 cm long; 0.2–0.9 mm wide), tips narrow and slightly curled; margins entire. ***Inflorescence* 1- (rarely 2 or more) flowered; bracts 3**, leaf-like, and strongly curled. *Flower* bisexual, corolla bell-shaped and nodding (3–5 cm long; 1–1.5 cm diam.); 'petals' 6, equal size, **usually purplish, sometimes whitish-green**, spotted and slightly chequered purple with small 'shoulders' (raised nectaries) towards base of 'petals'; stamens 6; ovary superior; **style lobes 2–3 mm long**. *Fruit* a dry, ovoid, 6-angled, 3-valved capsule with broad (4–5 mm) membranous wings. Fl. May–Jun, fr. Jul–Aug.

HARVESTING, SOURCING AND NATURAL RANGE

Bulbs are harvested during May–July but are rarely seen traded as a source of 'Yi Bei Mu' today, owing to controls on their wild harvesting and inadequate volume yet to be sourced from their medicinal cultivation. The remaining and increasingly vulnerable wild populations native to Xinjiang province (NW China) and neighbouring Kazakhstan grow in glades in *Picea* forests, scrub, meadow and steppe (1,300–2,000 m altitude).

CONSERVATION STATUS

China: Vulnerable (MEPCAS, 2013); Vulnerable (Wang & Xie, 2004); Vulnerable (Kaisa Sulaiman, pers. comm. 2012).
Global: Not Evaluated (IUCN Red List, accessed 18/8/2016).

PHARMACOPOEIA SPECIES 2

F. pallidiflora Schrenk

Liliaceae

PLANT DESCRIPTION

Easily distinguished from *F. walujewii* (PS1) by bulbs **often larger (1–4 cm diam.)**, a more upright and robust habit, **much broader leaves** (2–4 cm wide), **flowers** usually in groups of **2–5 (rarely solitary)**, bracts **solitary** and 'petals' **pale yellow**, sometimes with darker veins and spots. Fl. May–Jun, fr. Sep.

HARVESTING, SOURCING AND NATURAL RANGE

Bulbs are harvested in May–July from medicinal crop plants recently cultivated extensively in the Yili district of north-west Xinjiang province.They provide the main supply of traded 'Yi Bei Mu'. The wild species grows in mountain forests, scrub, grassland and steppe (1,300–2,500 m altitude) in the same district of Xinjiang province but its populations are seriously declining due to habitat degradation and previous over-harvesting. [Also native in Kazakhstan].

CONSERVATION STATUS

China: Vulnerable (MEPCAS, 2013); Near Threatened (Wang & Xie, 2004).
Global: Not Evaluated (IUCN Red List; accessed 20/3/2015); Vulnerable (Walter & Gillett, 1998).

TCM PROPERTIES AND USE

properties: **'bitter, sweet and cool'**
channels: **'Lung and Heart'**
actions: **'clears heat, resolves phlegm, relieves cough'**
indications: **dry cough; bloody phlegm**

Fritillaria walujewii in the wild, Xinjiang province

F. walujewii, inside of flower of whitish-green form

F. walujewii PHOTO: MARTYN RIX

Fritillaria pallidiflora fruiting medicinal crop, Xinjiang province

F. pallidiflora medicinal crop, Yili district, Xinjiang province, NW China PHOTO: KAISA SULAIMAN

Fritillaria walujewii

Liliaceae

DRUG MORPHOLOGY

CRUDE DRUG/DECOCTING PIECES (INTACT BULB): bulbs sub-globose and **slightly squat** (0.5–1.5 cm tall), with 2 fleshy slightly kidney-shaped scale leaves almost equal in size or of different sizes and clasping **[1]**. External surfaces smooth and white **[2]** unless membranous sheaths still present **[3]**, latter usually thin and torn. Apex closed or slightly open **[4]** to reveal a shrunken stem base and small buds **[5]**. Basal end **[6]** rounded and, centrally, slightly concave with root remains. *Texture:* hard but brittle. *Fracture:* white and starchy. *Odour:* faint. *Taste:* slightly bitter.

PROCESSING

Bulbs are sun-dried, then fibrous roots and outer membranous sheaths removed (= crude drug/decocting pieces).

OTHER COMMON NAMES

新疆贝母 Xinjiang Bei Mu; Sinkiang fritillary (SCN); Sinkiang fritillary bulb (CP20120); Bulbus Fritillariae Pallidiflorae.

Fritillaria pallidiflora

Liliaceae

DRUG MORPHOLOGY

CRUDE DRUG/DECOCTING PIECES (INTACT BULB). *Cultivated bulbs*: similar to those of PS1 but **slightly or significantly larger (up to 2.5 cm tall; 2.7 cm diam.),** fleshy scales 2, **pale yellowish-white [7]**, almost equal in size or of different sizes and clasping. Apex pointed **[8]**, closed or slightly open and concave with remnants of stem base or leaves **[9]** and small buds. Basal end **[10]** somewhat flat and oblique with many fibrous root remains. Externally **somewhat rough.** *Wild bulbs*: as for cultivated but **smaller and conical** rather than squat.

PROCESSING

As for PS1.

OTHER COMMON NAMES

伊贝母 Yi Bei Mu (named after Yili town, Xinjiang); Sinkiang fritillary (SCN); Sinkiang fritillary bulb; Bulbus Fritillariae Pallidiflorae.

> SUBSTITUTION COMMENT. Small bulbs of *Fritillaria pallidiflora* are occasionally traded unofficially as the more expensive but very similar looking bulbs of 'Chuan Bei Mu' (Fritillariae Cirrhosae Bulbus) see page 78.

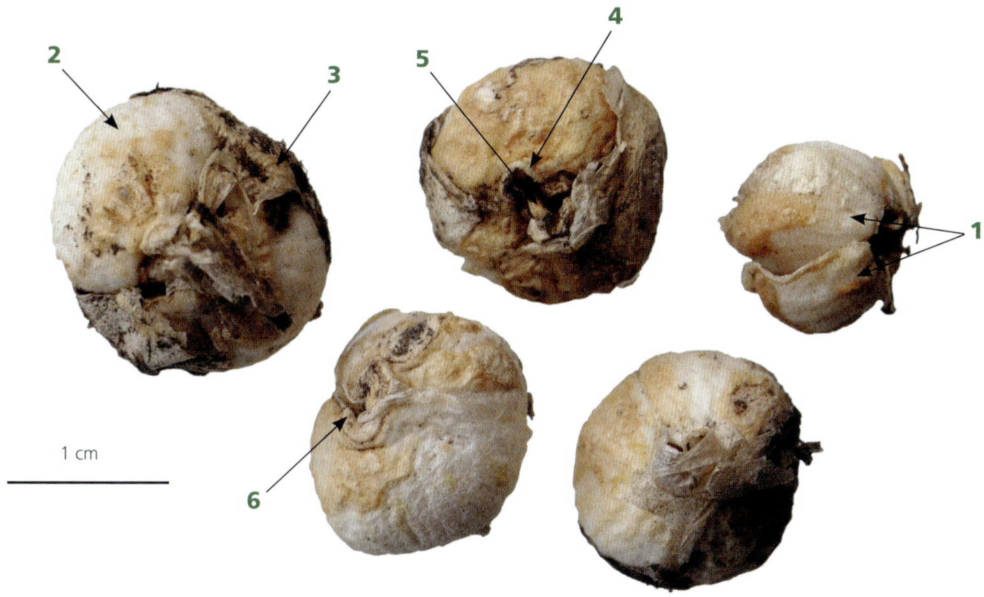

Yi Bei Mu with membranous sheath intact (sourced from *F. walujewii*)

Wild bulbs of Yi Bei Mu (sourced from *F. pallidiflora*)

Yin Chai Hu 银柴胡

Stellariae Radix

Drug part: ROOT

Stellaria dichotoma L. var. ***lanceolata*** Bunge

Caryophyllaceae

Species of *Stellaria* wild in China: 64

PLANT DESCRIPTION

A perennial herb (15–30 (rarely up to 60) cm tall) covered in glandular hairs, with stout, cylindrical roots; often forming large cushion-like clumps. *Stems* tufted and oppositely branched. *Leaves* simple, opposite (decussate) and stalkless; leaf blades linear to lance-shaped or oblong to lance-shaped (0.5–2.5 cm long; 1.5–5 mm wide), both surfaces glandular hairy, base rounded, somewhat clasping, margin entire, apex tapering to a long tip. *Flowers* many, in lax, branching, terminal heads. Each flower bisexual, stalked, sepals 5, petals 5, white and conspicuous (c. 4 mm long), spreading, equal-sized and deeply notched; stamens 10; styles 3; ovary superior. *Fruit* a dry, ovoid capsule (c. 3 mm long), shorter than persistent sepals and splitting into 6 valves when ripe; seeds 1–5. Fl. Jun–Jul, Fr. Jul–Aug.

HARVESTING, SOURCING AND NATURAL RANGE

Harvesting of the roots is mostly from wild plants (considered superior to cultivated ones) during spring and summer as the aerial parts emerge or in autumn after aerial parts have withered. The plant grows in dry, stony steppes (1,300–3,100 m altitude) and is distributed in N China (Gansu, Inner Mongolia, Liaoning, Ningxia and Shaanxi); [also wild in Mongolia and Russia]. Cultivated plants are harvested in the middle of September in their 3rd year of growth or in mid-April in their 4th year. The traditional *Di Dao* source of this drug is the Yinchuan district of Ningxia province; hence the name '*Yin* Chai Hu'.

CONSERVATION STATUS

China: "wild resources … almost exhausted" (Zhao & Xiao, 2009–2010).

Global: Not Evaluated (IUCN Red List, accessed 8/7/2015).

Stellaria dichotoma var. *lanceolata* flowering in the wild

Freshly exposed root of *S. dichotoma* var. *lanceolata*

UNOFFICIAL SUBSTITUTES

Owing to the over-harvesting and decline of the Pharmacopoeia species in the wild, the use of similar-looking roots of other species in the same family (Caryophyllaceae) has been reported as a growing problem in recent years (Zhao & Xiao, 2009–2010; Xiao *et al*., 2001–2002). These include:

1. ***Arenaria juncea*** M. Bieb. (Lao Niu Jin 老牛筋)
2. ***Gypsophila oldhamiana*** Miq. (Chang Rui Shi Tou Hua 长蕊石头花)
3. ***Silene fortunei*** Vis. (He Cao 鹤草)
4. ***S. jeniseensis*** Willd. (Shan Ma Zha Cao 山蚂蚱草)

The *Flora of China* recognises two additional varieties of *Stellaria dichotoma*: var. *linearis* Fenzl (wild in Inner Mongolia and Shaanxi) and var. *dichotoma* (with a much wider distribution compared to the Pharmacopoeia variety (Gansu, Hebei, Heilongjiang, Inner Mongolia, Liaoning, Qinghai and Xinjiang). The use of these varieties as an alternative sources of 'Yin Chai Hu' is unclear. Laboratory-based authentication research is needed to determine the frequency of the above as substitutes or adulterants in trade.

TCM PROPERTIES AND USE

properties: **'sweet and cool'**
channels: **'Liver and Stomach'**
actions: **'clears heat and deficient heat'**
indications: **night sweats; fever; infantile malnutrition due to heat**

DRUG MORPHOLOGY

CRUDE DRUG (DRIED ROOT): roots of wild and cultivated plants differ in appearance. ***Wild roots***: cylindrical, rarely branched; sometimes twisted and slightly curved (15–40 cm long; 0.5–2.5 cm diam.). Externally pale yellow, greyish-yellow or yellowish-white, shallow to deeply wrinkled longitudinally **[1]** with scars of lateral roots forming characteristic depressions (colloquially called 'Sha Yan') **[2]**; root-head short and slightly swollen with the colloquial term 'Zhen Zhu Pan' (tray of pearls) used to describe the densely packed stem-buds and nodular stem remnants which are dotted white **[3]** where the outer bark has been removed. 'Sha Yan'and 'Zhen Zhu Pan' are features of wild harvested roots which are typically older than cultivated ones. *Texture*: hard; easily broken. *Fracture*: uneven and splitting, mostly lax; wood with alternating yellow and white rays. *Odour*: indistinct. *Taste*: sweet. ***Good quality*** wild roots: long, thick, greyish-yellow and uniform with a solid yellowish-white (never black) interior.
Cultivated roots: branching; twisted mostly in lower parts (0.6–1.2 cm diam.). Externally pale yellowish-brown; longitudinal wrinkles fine and distinct; 'Sha Yan' depressions (see above) virtually absent. Root-head as for wild roots. *Fracture*: dense, hardly splitting, somewhat starchy; wood with indistinct ray tissues. *Taste*: somewhat sweet.

DECOCTING PIECES (SLICED DRIED ROOT):
Transverse slices — wild source: sub-rounded or ovate (5–2.5 cm diam.; 2–4 mm thick). Bark thin and pale yellow, greyish-yellow or yellowish-white, shallow to deeply wrinkled longitudinally **[4]** with 'Sha Yan' (see above) visible only in apical slices. Cut surface mostly lax with alternating yellow and white xylem rays. *Texture*: hard; but easily broken and splitting. *Odour*: faint. *Taste*: sweet. ***Transverse slices — cultivated source***: as for wild source but cut surface dense, somewhat starchy, with indistinct radiating wood rays **[5]**. ***Longitudinal slices***: sub-oblong, often tongue-shaped slices (3–12 cm long; 2–4 cm wide; 2–4 mm thick). Outer bark pale yellow, greyish-yellow or yellowish-white, shallow to deeply wrinkled longitudinally **[6]** with occasional 'Sha Yan' (see above) and 'Zhen Zhu Pan' visible when root apex present **[7]**. Cut surface mostly lax with some cavities **[8]**. *Texture*: hard; easily broken. *Odour*: indistinct. *Taste*: sweet.

PROCESSING

After removal and discarding of aerial parts and rootlets, roots are sun-dried (= crude drug) and later softened with water before slicing transversely into decocting pieces. In Hong Kong and parts of southern China longitudinal slicing is more typical

OTHER COMMON NAMES

Radix Stellariae; starwort root (CP2010).

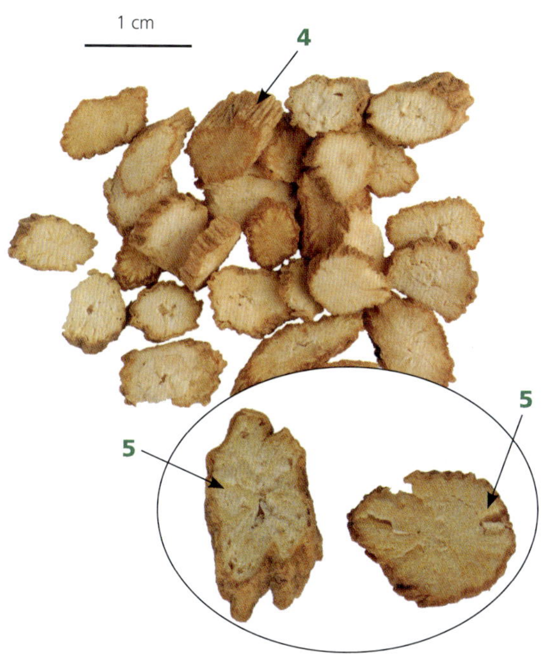

Transverse decocting slices of cultivated Yin Chai Hu from roots of *S. dichotoma* var. *lanceolata*

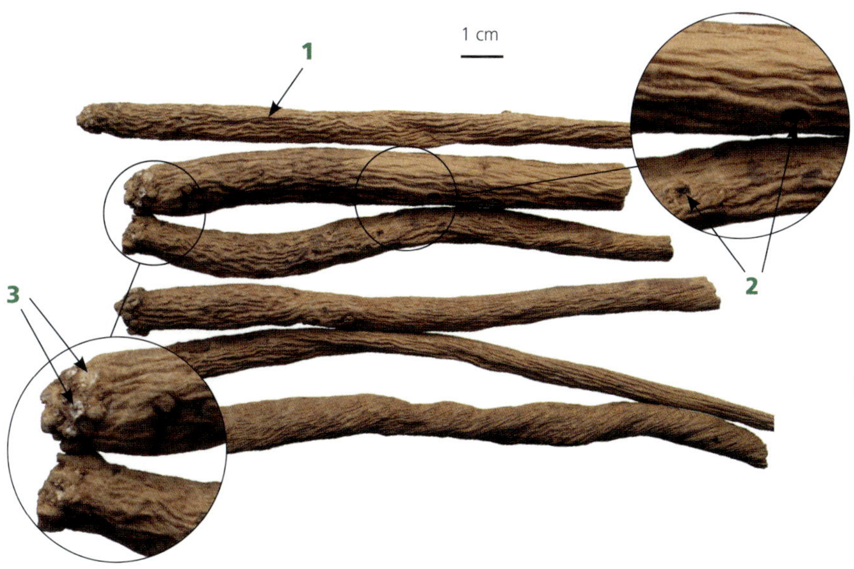

Crude drug (wild roots) of *S. dichotoma* var. *lanceolata*

Dried, longitudinally sliced roots of *S. dichotoma* var. *lanceolata*

Yu Jin 郁金

Curcumae Radix

Drug part: ROOT TUBER

Curcuma wenyujin Y. H. Chen & C. Ling

Zingiberaceae

taxonomic/nomenclatural note: *Kew-MPNS* accepted name, 10/10/2016; *C. aromatica* Salisb. [which *FoC* considers to be a separate species].

Species of *Curcuma* wild in China: 11

PLANT DESCRIPTION

An erect, hairless perennial herb (0.8–1.6 m tall). *Rhizomes* fleshy and aromatic; **when fresh externally whitish**; **internally yellow or whitish**. *Root tubers* (source of 'Yu Jin') fleshy and spindle-shaped. *Leaves* basal; leaf stalks c. 30 cm long; blades large or oblong to lance-shaped (35–75 cm long;14–22 cm wide) with prominent mid-vein, lateral veins **hairless**; margins entire. *Flowers* in tight conical heads (20–30 cm long) on separate shoots arising from rhizomes. **Lower bracts green** and fertile (with flowers); **upper petal-like bracts large and conspicuous (5–8 cm long), tips rose-tinged** and sterile (flowerless). Each flower (inside lower bracts) inconspicuous with flower tube white; lip (labellum) reflexed, yellow, with a darker centre. *Fruit* ellipsoid and 3-valved. Fl. May–Jun.

HARVESTING, SOURCING AND NATURAL RANGE

Root tubers harvested in winter, after stems and leaves have withered, entirely from medicinal crops grown in S China (Guangdong, Guangxi and Zhejiang) and traded as 'Wen Yu Jin' (after Wenzhou town in Zhejiang province, the source of *Di Dao*-quality 'Yu Jin'). The wild species is endemic to these three provinces. ***Rhizomes*** harvested at same time are the source of 'E Zhu' (Curcumae Rhizoma) see page 120.

CONSERVATION STATUS

Endemic to China: Not Evaluated (IUCN Red List, accessed 20/8/2016).

C. longa L.

synonym: *C. domestica* Valeton

PLANT DESCRIPTION

Similar to the other 3 Pharmacopoeia species. *Main differences*: **flesh of fresh rhizome orange or bright yellow. Leaf blades 15–18 cm long; hairless on both surfaces.** Lower bracts pale green (3–5 cm long) with broad tip; upper petal-like bracts white and green, sometimes tinged reddish-purple with pointed tip. Flower lip (labellum) pale yellow with darker central band.

HARVESTING, SOURCING AND NATURAL RANGE

Harvesting as for PS1; sourced entirely from medicinal crops grown in S and W China (Fujian, Guangdong, Guangxi, Sichuan, Taiwan, Tibet and Yunnan) and traded as 'Huang Si Yu Jin'. The species is only known from cultivation. ***Rhizomes*** are the source of turmeric.

CONSERVATION STATUS

China: not native.
Global: Not Evaluated (IUCN Red List, accessed 20/8/2016).

C. kwangsiensis S. G. Lee & C. F. Liang

synonym: *C. chuanyujin* C. K. Hsieh & Hao Zhang

PLANT DESCRIPTION

Similar to the other 3 Pharmacopoeia species. *Main differences*: **usually smaller. Flesh of fresh rhizome milky-white or cream.** Leaf blades narrower (4.5–9.5 cm wide); **hairy on both surfaces.** Lower bracts greenish (c. 4 cm long); **upper (sterile) petal-like bracts pale red.** Flowers lobes red; lip (labellum) pale yellow or yellowish-white. Fl. May–Jul.

HARVESTING, SOURCING AND NATURAL RANGE

Harvesting as for PS1; sourced entirely from medicinal crops grown in S and C China (Guangdong, Guangxi, Sichuan and Yunnan) and traded 'Gui Yu Jin'; *Di Dao* production areas occur in Guangxi province. The wild species is endemic to China and grows in grassland and scrub in these four provinces. ***Rhizomes*** harvested at the same time are the source of 'E Zhu' (Curcumae Rhizoma) see page 120.

CONSERVATION STATUS

Endemic to China: Near Threatened (authors' pers. obs., 2006); Not Evaluated (IUCN Red List, accessed 20/8/2016).

C. phaeocaulis Valeton

PLANT DESCRIPTION

Similar to the other 3 Pharmacopoeia species. *Main differences*: **Flesh of freshly cut rhizomes pale blue, pale green, yellowish-green or yellow.** Leaves 25–60 cm long, 10–15 cm wide; upper surface hairless **with obvious purple band** along mid-vein; **lower leaf surface with dense mat of short, soft hairs.** Lower bracts whitish-green, tips pink and pointed; upper petal-like bracts white with deep red tips. Flower tube white, flower lobes red; lip (labellum) pale yellow with darker central band.

HARVESTING, SOURCING AND NATURAL RANGE

Harvesting as for PS1; sourced from medicinal crops grown in S China (Fujian, Guangdong, Guangxi and Sichuan) and traded as 'Lu Si Yu Jin'. The wild species is native to Yunnan [also Indonesia and Vietnam]. ***Rhizomes*** harvested at the same time are the source of 'E Zhu' (Curcumae Rhizoma) see page 120.

CONSERVATION STATUS

China: Not Evaluated.
Global: Not Evaluated (IUCN Red List, accessed 20/8/2016).

TCM PROPERTIES AND USE

properties: **'pungent, bitter and cold'**
channels: **'Heart, Liver and Gall Bladder'**
actions: **'invigorates Blood and qi, relieves pain, clears heat, cools Blood'**

indications: **abdominal, chest pain; dysmenorrhoea, metrorrhagia; haematemesis, epistaxis, haematuria; anxiety, mania, convulsions**

Curcuma wenyujin ready for harvesting

Hairless leaf of *C. wenyujin*

Fresh rootstocks of *C. wenyujin*

C. longa crop in flower

Hairless leaf of *C. longa*

Fresh rootstocks of *C. longa*

C. kwangsiensis in flower

Leaves of *C. kwangsiensis* hairy on both surfaces

Fresh rootstocks of *C. kwangsiensis*

C. phaeocaulis in flower

Leaves of *C. phaeocaulis*

Fresh rootstocks of *C. phaeocaulis*

PHARMACOPOEIA SPECIES 1

Curcuma wenyujin

Zingiberaceae

DRUG MORPHOLOGY

CRUDE DRUG (STEAMED, BOILED ROOT TUBER): ovoid or oblong (3.5–7 cm long; 1.2–2.5 cm diam.), slightly curved; tapering at both ends **[1]** with a root scar at each **[2]**. Externally pale greyish-brown or brown and irregularly wrinkled **[3]**. *Texture*: hard and compact. *Fracture*: horn-like, yellowish, greyish-brown or brown **[4]** with a distinct and paler endodermis ring **[5]**. *Odour*: indistinct. *Taste*: somewhat bitter. ***Good quality*:** solid, heavy greyish-brown or greyish-yellow horn-like fracture.

DECOCTING PIECES (SLICED ROOT TUBER): long elliptical slices (3.5–7 cm long; 1.2–2.5 cm diam.**[6]**; 1–2 mm thick). External surface as for crude drug. Cut surface as for crude drug fracture.

PROCESSING

After removal of rootlets, tubers are steamed or boiled, then dried (= crude drug). Tubers are later softened, sliced thinly (1–2 mm) then dried again (= decocting pieces).

Note: the drug 'E Zhu' (Curcumae Rhizoma) is derived from the rhizomes of the same species and is harvested at the same time (see page 120).

OTHER COMMON NAMES

黑郁金 Hei Yu Jin 'black Yu Jin'; 温郁金 Wen Yu Jin; Curcuma wenyujin (SCN); Radix Curcumae; turmeric root tuber (CP2010).

PHARMACOPOEIA SPECIES 2

C. longa

Zingiberaceae

DRUG MORPHOLOGY

CRUDE DRUG (STEAMED, BOILED ROOT TUBER): similar to that of the other species (PS1, 3 and 4). **Ovoid (2.5–4.5 cm long; 1–1.5 cm in diam.) [7]**, slightly curved, tapering at one or both ends; externally brownish-grey or greyish-yellow **[8]**, with fine wrinkles **[9]**. ***Fracture*: orange [10]**, somewhat brownish-yellow to brownish-red outside the endodermis ring **[11]**. *Texture*: hard. *Odour*: aromatic. *Taste*: pungent.

DECOCTING PIECES (SLICED ROOT TUBER): ovoid to tapering slices (2.5–4.5 cm long; 1–1.5 cm diam.; 1–2 mm thick) **[12]**. External surface as for crude drug. Cut surface as for crude drug fracture **[8, 9, 10]**.

PROCESSING

As for PS1 (except the rhizome of this species (true culinary turmeric) is not used as an alternative source of 'E Zhu').

OTHER COMMON NAMES

川郁金 Chuan Yu Jin; 黄丝郁金 Huang Si Yu Jin; curcuma root (KP, 2007); Curcumae Longae Radix (KP, 2007); Radix Curcumae; turmeric root tuber (CP2010).

PHARMACOPOEIA SPECIES 3

C. kwangsiensis

Zingiberaceae

DRUG MORPHOLOGY

CRUDE DRUG (STEAMED, BOILED ROOT TUBER): similar to that of the other species (PS1, 2 and 4). **Long ovoid or oblong (2–6.5 cm long; 1–1.8 cm diam.)**. Externally **less wrinkled [13]** than all the other Pharmacopoeia species, wrinkles occasional, shallow or coarse and reticulated. *Fracture*: brown. *Texture*: horn-like and hard. *Odour*: indistinct. *Taste*: bitter and somewhat pungent.

DECOCTING PIECES (SLICED ROOT TUBER): ovoid or oblong slices (2–6.5 cm long; 1–1.8 cm diam.; 1–2 mm thick). External surface as for crude drug of PS1. Cut surface as for crude drug fracture of PS1.

PROCESSING

As for PS1.

OTHER COMMON NAMES

广郁金 Guang Yu Jin; 桂郁金 Gui Yu Jin; curcuma root (KP, 2007); Curcumae Longae Radix (KP, 2007); Radix Curcumae; turmeric root tuber (CP2010).

PHARMACOPOEIA SPECIES 4

C. phaeocaulis

Zingiberaceae

DRUG MORPHOLOGY

CRUDE DRUG (STEAMED, BOILED ROOT TUBER): root tubers similar to that of the other species (PS1, 2 and 3) but **usually shorter (1.5–3.5 cm long; 1–1.2 cm diam.) [14]** and **not tapering [15]**. *Texture*: hard. *Odour*: indistinct. ***Taste*: indistinct**.

DECOCTING PIECES (SLICED ROOT TUBER): elliptical slices (1.5–3.5 cm long **[14]**; 1–1.2 cm diam.; 1–2 mm thick). External surface as for crude drug of PS1. Cut surface as for crude drug fracture of PS1.

Laboratory-based authentication methods are needed for reliable differentiation between the drugs (crude and decocting) of these 4 species.

PROCESSING

As for PS1.

OTHER COMMON NAMES

川郁金 Chuan Yu Jin; 绿丝郁金 Lu Si Yu Jin; curcumae root (KP, 2007); Curcumae Longae Radix (KP, 2007); Radix Curcumae; turmeric root tuber (CP2010).

See also E Zhu (Curcumae Rhizoma) page 120.

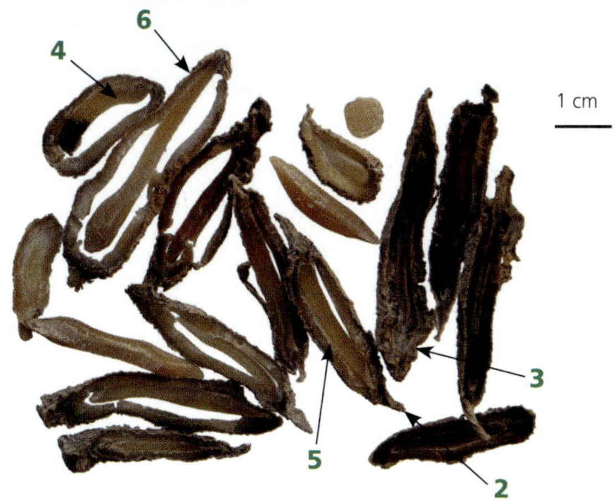

Crude drug and decocting pieces of Yu Jin sourced from *Curcuma wenyujin* (= Wen Yu Jin)

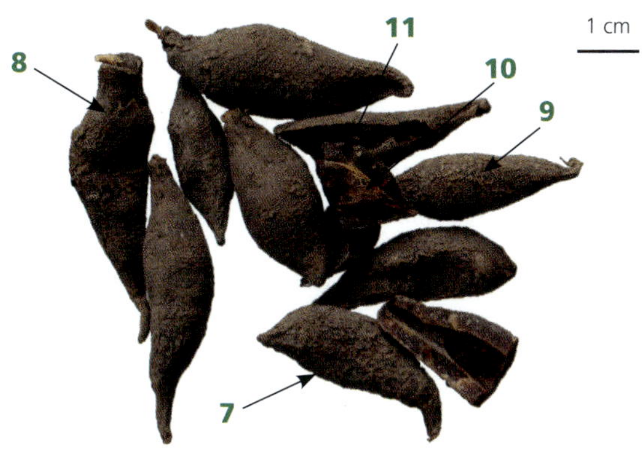

Crude drug and decocting pieces of Yu Jin sourced from *C. longa* (= Huang Si Yu Jin)

Crude drug of Yu Jin sourced from *C. kwangsiensis* (= Gui Yu Jin)

Crude drug and decocting pieces of Yu Jin sourced from *C. phaeocaulis* (= Lu Si Yu Jin)

Ze Xie 泽泻

Alismatis Rhizoma

Drug part: RHIZOME (stem tuber)

Alisma orientale (Sam.) Juz.

Alismataceae

synonyms: *A. plantago-aquatica* L. var. *orientale* Sam.; *A. plantago-aquatica* subsp. *orientale* (Sam.) Sam. (*Kew-MPNS* accepted name, 28/8/2016).

taxonomic/nomenclatural note: often misspelt as '*Alisma orientalis* (Sam.) Juz.'

Species of *Alisma* wild in China: 6

PLANT DESCRIPTION

A marginal aquatic herbaceous perennial, with a swollen sub-globose rhizome (1–2 cm diam.; medicinally cultivated ones larger). *Leaves* all basal, simple, long-stalked, large and lance-shaped to elliptical (3.5–11 cm long; 1.3–7 cm wide or larger); blades distinctly 5–7-veined; base heart-shaped or rounded, margins entire. *Flowering stems* leafless, usually taller (20–70 cm) than rest of plant with 3–9 widely spaced whorls of flower-bearing branches with delicate flowers maturing from the bottom upwards. Sepals 3, green; petals 3, much larger than sepals, white or pinkish, sub-orbicular with **wavy margins,** stamens 6; carpels in a single **irregularly arranged** whorl; styles **c. 0.5 mm long.** *Fruit* a cluster of dry, laterally compressed 1- or 2-grooved achenes; style persistent. Fl. and fr. May–Sep.

HARVESTING, SOURCING AND NATURAL RANGE

Rhizomes are harvested during winter after the stems have withered. Sourcing is mostly from medicinal crop plants grown predominantly in Fujian and Sichuan provinces. Traditional *Di Dao* material is sourced from Fujian (Jian'ou, Pucheng and Jiangyang districts) and Sichuan provinces. In recent years the species has also been cultivated in Guangxi province for its flower heads, eaten as a popular vegetable.

The species grows wild throughout much of China in damp places along margins of marshes, ponds, lakes and ditches, up to c. 2,500 m altitude. Natural range: Anhui, Fujian, Gansu, Guangdong, Guangxi, Guizhou, Hebei, Heilongjiang, Henan, Hubei, Hunan, Inner Mongolia, Jiangsu, Jiangxi, Jilin, Liaoning, Ningxia, Qinghai, Shaanxi, Shandong, Shanxi, Sichuan, Xinjiang, Yunnan and Zhejiang [also wild in India, Japan, Kashmir, Korean Peninsula, Mongolia, Myanmar, Nepal, Russia and Vietnam].

CONSERVATION STATUS

China: Least Concern (MEPCAS, 2013).
Global: Not Evaluated (IUCN Red List, accessed 12/2/2015).

Alisma orientale medicinal crop, Sichuan province

A. orientale young cultivated plants

A. orientale flowering head

TCM PROPERTIES AND USE

***properties*: 'sweet, bland and cold'**
***channels*: 'Kidney and Bladder'**
***actions*: 'drains damp, regulates water, clears deficient heat'**

***indications*: urinary disorders; oedema; diarrhoea**

DRUG MORPHOLOGY

CRUDE DRUG (DRIED RHIZOME): rhizome creamish-yellow or yellowish-brown ovate or sub-spheroidal (2–7 cm long; 2–6 cm diam.). Externally often with transverse pitted annular furrows **[1]**, dotted elsewhere with small oval root scars or short curved root remains **[2]**; basal parts often with large nodular buds **[3]**; apex with one or more raised or sunken stem scars **[4]**. *Texture*: hard and dense; difficult to break. *Fracture*: yellowish-white, starchy, with many small pores **[5]**. *Odour*: slight. Taste: slightly bitter. ***Good quality*:** large, heavy, yellowish-white rhizomes with a powdery, pale yellow interior.

DECOCTING PIECES-1 (SLICED RHIZOME): sub-circular or elliptical slices (2–7 cm diam.; 2–4 mm thick). Outer skin surface yellowish-white or pale yellowish-brown with occasional raised scars of roots **[2]** or bulbous buds **[3]**. Cut surfaces pale yellowish-white, starchy and scattered with tiny pores **[5]**. *Odour*: indistinct. *Taste*: slightly bitter.

DECOCTING PIECES-2 (SLICED & SALTED RHIZOME): as for dried slices above but all surfaces darker and sometimes scorched **[6]**. *Taste*: salty.

PROCESSING

Rhizomes are washed, dried and fibrous roots removed (= crude drug) then cut transversely into thick decocting slices; subsequent processing with salt produces 'Yan Ze Xie' (= salted Ze Xie).

OTHER COMMON NAMES

川泽泻 Chuan Ze Xie (Sichuan Ze Xie); 建泽泻 Jian Ze Xie (Fujian Ze Xie — top quality); Asian water plantain (SCN); Rhizoma Alismatis.

SAFETY

Safety Class: 1; Interaction Class: A (Gardner & McGuffin, 2013).

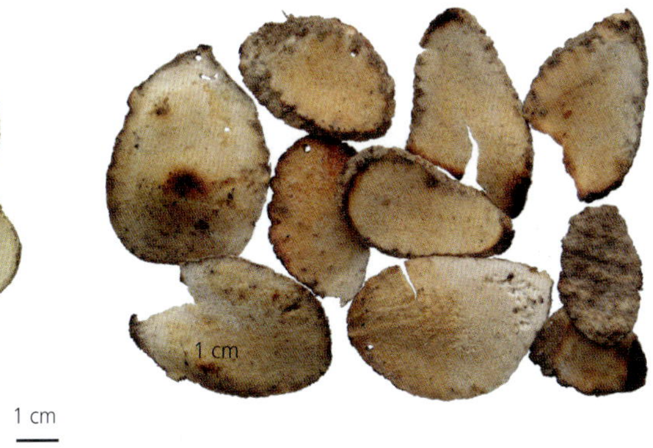

Crude drug

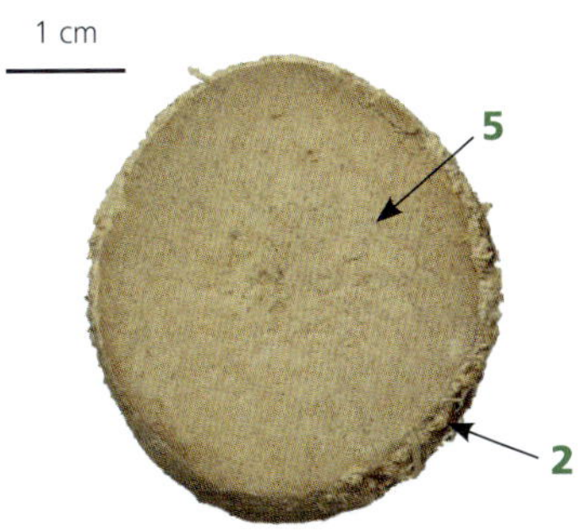

Decocting pieces-1 (transversely sliced)

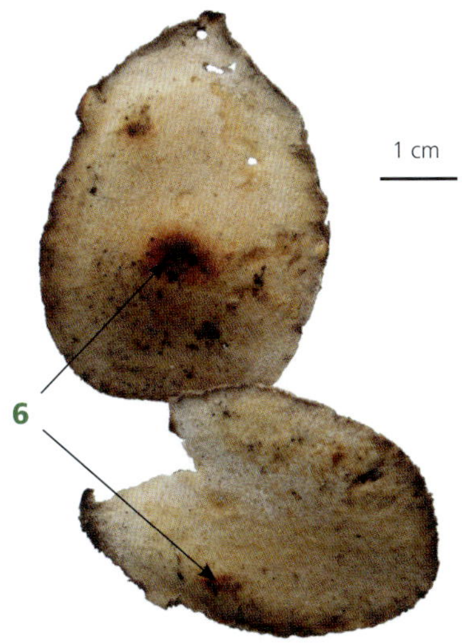

Decocting pieces-2 (longitudinally sliced and salted)

Zhe Bei Mu 浙贝母

Fritillariae Thunbergii Bulbus

Drug part: BULB

Fritillaria thunbergii Miq.

Liliaceae

taxonomic/nomenclatural note: includes the type variety (var. *thunbergii*) and var. *chekiangensis* P. K. Hsiao & K. C. Hsia which are recognised by *FoC* but not by *Kew-MPNS*, 28/8/2016.

Species of *Fritillaria* wild in China: 24

PLANT DESCRIPTION

F. thunbergii **Miq. var.** ***thunbergii***: An erect, herbaceous perennial (50–80 cm tall) with a slightly squat bulb (1.5–3 cm diam.) of 2 (or 3) fleshy scales. *Stem* unbranched. *Leaves* 12–20, mostly alternate and simple; stem leaves unstalked; leaf blades linear to lance-shaped (7–11 cm long; 1–2.5 cm wide) with tips often slightly curled and margins entire. *Inflorescence* 1–6-flowered; bracts 2–4, leaf-like, linear and typically curled. *Flower* bisexual, corolla bell-shaped and nodding (2.5–3 cm long); 'petals' 6, equal size, yellowish-white veined with green, sometimes tinged pale purple or chequered purplish-brown; nectaries small; stamens 6; ovary superior; style lobes 1.5–2 mm long. *Fruit* a dry, ovoid, 6-angled, 3-valved capsule with broad (6–8 mm) membranous wings. *Seeds* many and flat; in 2 rows in each valve. Fl. Mar–Apr, fr. May–Jun.

F. thunbergii **var.** ***chekiangensis*** P. K. Hsiao & K.C. Hsia: A shorter plant (15–30 cm.) with a slightly smaller bulb (c. 1 cm diam.) with 3 fleshy scales. Leaves mostly opposite.

HARVESTING, SOURCING AND NATURAL RANGE

Bulbs are harvested entirely from medicinal crop plants in early summer, after the aerial parts have withered. The species has a long history of medicinal cultivation in China dating back c. 700 years with the main production areas for the type variety (var. *thunbergii*) being the eastern Chinese provinces of Anhui, Hunan, Jiangsu and Zhejiang. The latter is the source of traditional *Di Dao*-quality (hence the name 'Zhe Bei Mu') and where variety *chekiangensis* is also grown.

Uncertainty exists over the native range of the species with the *FoC* describing the type variety and variety *chekiangensis* as endemic to eastern China while other authorities state that the species is only native much further west in the Tarbagatai Mountains of Kazakhstan (adjacent to Xinjiang in NW China) (Leon *et al.*, 2009).

CONSERVATION STATUS

China: doubtfully native. Variety *chekiangensis* is listed as Vulnerable (MEPCAS, 2013).

Global: Not Evaluated (IUCN Red List; accessed 10/8/2016).

Medicinal crop of Zhejiang province *Fritillaria thunbergii*

F. thunbergii: flowering plant

Bulbs (white) of *F. thunbergii* sun-drying, Zhejiang province

F. thunbergii in flower, Zhejiang

SUBSTITUTION COMMENT. Owing to their similar size and shape, bulbs of *Fritillaria thunbergii* var. *chekiangensis* are sometimes traded as an unofficial substitute of the more expensive Chuan Bei Mu (Fritillariae Cirrhosae Bulbus) see page 78.

TCM PROPERTIES AND USE

***properties*: 'bitter and cold'**
***channels*: 'Lung and Heart'**
***actions*: 'resolves phlegm, relieves cough, clears heat, disperses nodules'**

***indications*: productive cough with thick phlegm; sores and abscesses**

DRUG MORPHOLOGY

LIMED AND DRIED BULBS SOURCED FROM:

F. thunbergii var. ***thunbergii***:

i) 'Zhu Bei': intact bulbs (1–1.5 cm tall; 1–2.5 cm diam.), slightly squat, with 2 fleshy, slightly kidney-shaped scale leaves almost equal in size or one slightly larger and clasping the other **[1]**. Externally smooth, white or pale yellow **[2]**. Apex slightly concave and open to reveal a shrunken stem base **[3]** and a tiny bud **[4]**. Basal end rounded and, centrally, slightly concave with remnants of root bases **[5]**. *Texture*: hard and powdery. *Fracture*: starchy. *Odour*: indistinct. *Taste*: slightly bitter. (= crude drug and decocting pieces).

ii) 'Da Bei': separated fleshy scales of larger bulbs (1–2 cm tall; 2–3.5 cm diam.), somewhat crescent shaped **[6]**. Outer surface smooth, white or pale yellow **[7]**; inner surface white or pale brown and powdery **[8]**. *Texture, Odour and Taste* as for Zhu Bei. (= decocting pieces).

iii) 'Zhe Bei Pian': fleshy scales (1–2 cm diam.) elliptical, sub-rounded or kidney-shaped in outline **[9]**. Uncut external surface pale yellow **[10]**; cut surface white, smooth and powdery **[11]**. Central buds and stem bases mostly removed **[12]**. *Texture*: hard and brittle. *Fracture*: starchy. (= decocting pieces).

F. thunbergii var. ***chekiangensis***:

iv) 'Dong Bei': intact bulbs like those of 'Zhu Bei' above but smaller (1–1.5 cm tall; 1.25 cm diam.) **[13]**. (= crude drug and decocting pieces).

***Good quality*:** solid, dry and white with a powdery (not friable) texture.

PROCESSING

Bulbs are graded by size then limed using powdered shell to remove moisture before being sun- or oven-dried. Three kinds of Zhe Bei Mu are recognised in trade:

1. 'Zhu Bei' (珠贝): intact bulbs, sourced from both varieties, but excluding these larger bulbs (> 2.5 cm diam.) of var. *thunbergii*;
2. Da Bei (大贝): separated outer fleshy scales of var. *thunbergii* only and sourced from bulbs > 2.5 cm diam.;
3. 'Zhe Bai Pian': bulbs of variety *thunbergii* sliced longitudinally when fresh, with the central buds removed.

OTHER COMMON NAMES

大贝母 Da Bei Mu 'big Bei Mu'; 苏贝 Su Bei (sourced from Jiangsu province); 香贝母 Xiang Bei Mu; 元宝贝 Yuan Bao Bei; Bulbus Fritillariae Thunbergii; Thunberg fritillary bulb (CP2010); Zhejiang fritillary (SCN).

SAFETY

An official healthfood in China (Anon., 2002a,b). Safety Class: 1; Interaction Class: A (Gardner & McGuffin, 2013).

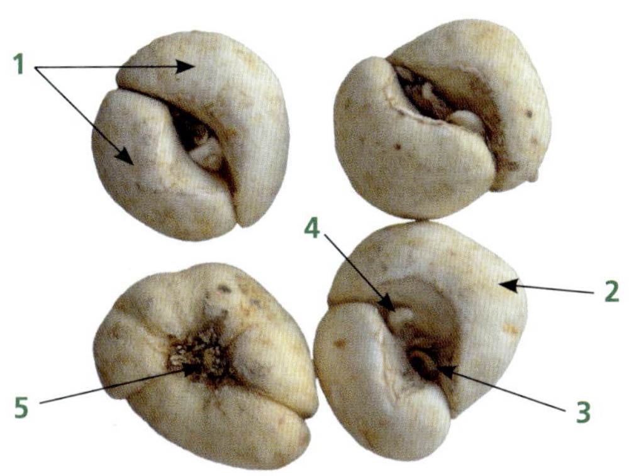

Zhu Bei (sourced from *Fritillaria thunbergii* var. *thunbergii*)

Da Bei (sourced from *F. thunbergii* var. *thunbergii*)

1 cm

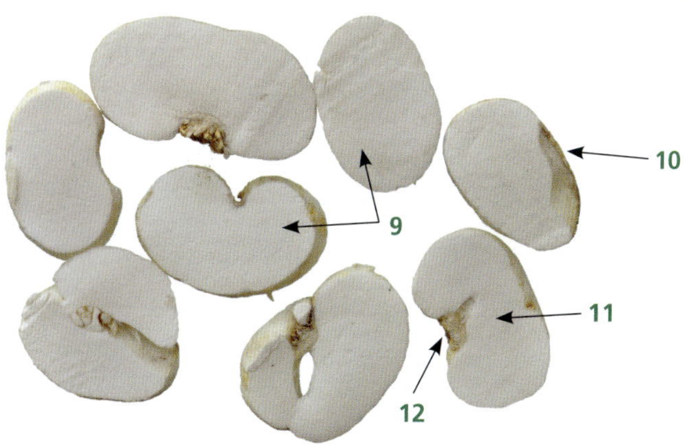

Zhe Bei Pian (sourced from *F. thunbergii* var. *thunbergii*)

Dong Bei (sourced from *F. thunbergii* var. *chekiangensis*)

Zhi Mu 知母

Anemarrhenae Rhizoma

Drug part: RHIZOME

Anemarrhena asphodeloides Bunge

Asparagaceae

PLANT DESCRIPTION

A clump-forming perennial herb (20–100 cm tall) with stout, horizontal, creeping rhizomes (c. 10 cm long; 0.5–1.7 cm diam.). *Leaves* all basal, simple, long and narrow (grass-like) (10–60 cm long; 0.15–1 cm wide); leaf blades hairless and margins rough. Flowering stems erect and, when mature, c. twice the height of the leaves (20–100 cm), leafless and unbranched with many, tiny, inconspicuous flowers clustered (in a raceme) along the upper part. Bracts small and membranous with a pointed or thread-like tip. *Flowers* solitary or in groups of 2 or 3; short-stalked or virtually stalkless; bisexual; 6 'petals' fused to form a pink, pale purple or whitish, narrow, linear funnel (5–10 mm long; c. 1.5 mm wide); stamens 3; ovary superior and 3-loculed; style short (c. 1 mm long). *Fruit* a 6-angled, narrowly ovoid-ellipsoid, many-seeded capsule (8–15 mm long including beak; 3–6 mm diam.), maturing from green to pale brown. *Seeds* black (7–12 mm long; 2.5–3 mm diam.). Fl. and fr. Jun–Sep.

Species of *Anemarrhena* wild in China: 1

HARVESTING, SOURCING AND NATURAL RANGE

Rhizomes are harvested in spring or autumn from wild or medicinally farmed crop plants. In the wild in China, the species grows in scrub, steppe and open sandy grasslands from sea level to 1,500 m altitude from NE to SC China (Gansu, Guizhou, Hebei, Heilongjiang, Inner Mongolia, Jiangsu, Jilin, Liaoning, Shanxi, Shandong, Shanxi and Sichuan provinces) [the species is also native to the Korean Peninsula and Mongolia]. The main production areas are Hebei, Shaanxi and Inner Mongolia with '*Di Dao*-quality' material traditionally sourced from Yixian county in Hebei province.

The rhizomes are also grown to supply China's healthfood industry.

CONSERVATION STATUS

China: locally common (Lin Yu-Lin pers. obs. 2008); but rapidly decreasing (Leung & Cheng, 2008).

Global: Not Evaluated (IUCN Red List, accessed 12/2/2015).

Extensive stand of wild plants showing rotational harvesting (bare areas) to allow regrowth and sustainability, Inner Mongolia.

Medicinal crop, Shaanxi province

Flowering spikes

BELOW
Fruiting spike

TCM PROPERTIES AND USE

***properties*: 'bitter, sweet and cold'**
***channels*: 'Lung, Stomach and Kidney'**
***actions*: 'clears heat and fire, tonifies yin and moistens dryness'**

***indications*: fever, thirst, dry coughs; menopause; night sweats**

DRUG MORPHOLOGY

CRUDE DRUG-1 (UNPEELED RHIZOME — MAO ZHI MU): rhizomes cylindrical, slightly curved and sometimes branched (3–15 cm long; 0.8–1.5 cm diam.). Dorsal side with an indistinct longitudinal groove **[1]** and dense annular rings **[2]**, both often obscured by a dense covering of fibrous yellowish-brown leaf bases **[3]**. Ventral surface more rounded and with evenly spaced, oval root scars **[4]**. *Texture*: hard, easily broken. *Fracture*: yellowish-white with a hint of green **[5]**. *Odour*: indistinct. *Taste*: somewhat sweet and bitter, becoming viscous on chewing. ***Good quality***: large, soft and compact, yellowish-white in section with a hint of green.

CRUDE DRUG-2 (PEELED RHIZOME — GUANG ZHI MU): as for crude drug-1 but narrower and externally yellowish-white, twisted and furrowed **[6]**. Stem **[7]**, leaf and root scars **[8]** occasional. Devoid of fibrous leaf bases.

DECOCTING PIECES-1 (SLICED RHIZOME): irregularly ovate (1–4 cm long; 2–3 cm wide; 2–4 mm thick). Externally yellowish-brown or brown with occasional fibrous stem bases **[9]** and raised or sunken evenly spaced root scars **[4]** and annular rings **[2]**. Cut surface yellowish-white or yellow **[10]**, finely dotted throughout with vascular tissues. *Odour* and *taste* as for the crude drug.

DECOCTING PIECES-2 (SALT-PROCESSED RHIZOME — YAN ZHI MU): as decocting pieces above but yellow with occasional charred spots.

PROCESSING

Two forms of crude drug are traded: 1. sun-dried rhizomes with leaf bases intact (i.e. unpeeled) called 毛知母 Mao Zhi Mu (hairy Anemarrhena) (= crude drug-1); normally for China's domestic market; 2. peeled (fibrous roots and leaf bases removed) and sun-dried rhizomes called 光知母 Guang Zhi Mu (naked Anemarrhena) (= crude drug-2); normally for export. Decocting slices are prepared by soaking the crude drug in water until thoroughly softened then cutting into 2–4 mm thick slices; further optional processing with salt then dry-frying produces 盐知母 Yan Zhi Mu.

OTHER COMMON NAMES

盐炒知母 Yan Chao Zhi Mu 'salted and dry-fried anemarrhena'; 知母肉 Zhi Mu Rou 'anemarrhena flesh' (this is an alternative name for Guang Zhi Mu above); anemarrhena (SCN) Rhizoma Anemarrhenae.

SAFETY

Safety Class: 2d; Interaction Class: A (Gardner & McGuffin, 2013). An official Chinese healthfood (Anon., 2002a,b).

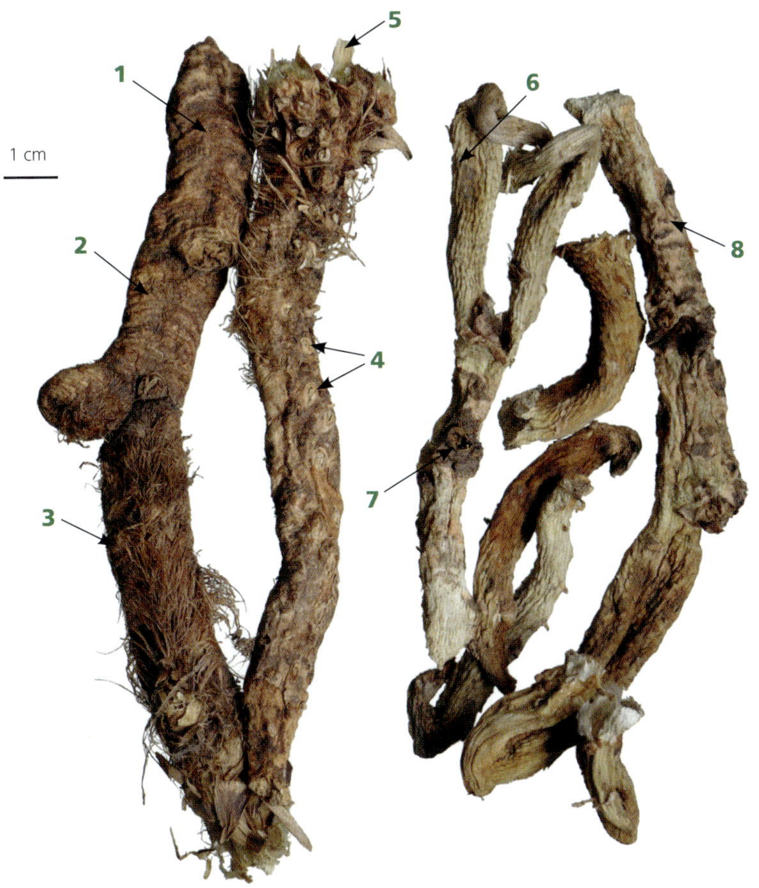

Crude drug (left: Mao Zhi Mu; right: Guang Zhi Mu)

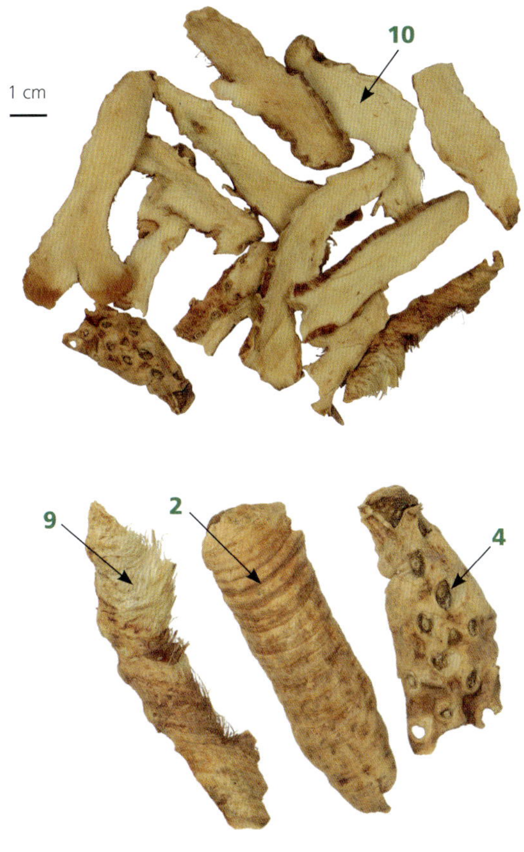

Decocting pieces (Guang Zhi Mu — export quality)

Zi Cao 紫草

(literal: purple herb)

Drug part: ROOT

Arnebiae Radix

PHARMACOPOEIA SPECIES 1

Arnebia euchroma (Royle ex Benth.) I. M. Johnst.

Boraginaceae

synonym: *Lithospermum euchromon* Royle ex Benth.
taxonomic/nomenclatural note: in CP2015 as '*A. euchroma* (Royle) Johnst.'

Species of *Arnebia* wild in China: 6

PLANT DESCRIPTION

A perennial herb (15–40 cm tall) with a stout rootstock (up to 2 cm diam.) with abundant purple dye. *Stems* branching above, with spreading white or pale yellowish hairs and sheathing leaf remnants at base. *Leaves* alternate, simple, basal leaves longest (7–20 cm long; 0.5–1.5 cm wide), **linear to lance-shaped**; leaf bases virtually clasping; margins entire with long hard, white hairs; tip of leaf tapering to a sharp point. *Flowers* in many-flowered one-sided terminal heads (2–6 cm long when in full flower); bracts lance-shaped. Flowers bisexual, sepals 5, petals fused to form a **dark purple** (sometimes pale yellow tinged purplish-red), tube (1–1.4 cm long) opening into 5 much shorter flower lobes; lobes spreading and slightly overlapping; stamens 5; ovary superior. *Fruit* a dry, 1–4-seeded nutlet. Fl. and fr. Jun–Aug.

HARVESTING, SOURCING AND NATURAL RANGE

Roots are traditionally harvested in spring or autumn from wild populations growing in gravelly marshes and meadows and amongst scree on rocky slopes. Xinjiang province has been the main wild harvesting area for many years but the species has become critically threatened due to over-harvesting of its roots for medicine and as a purple dye used in the food industry (see Conservation status). The species is also native to western Tibet and, outside China, to Afghanistan, NW India, Kazakhstan, Kyrgyzstan, Nepal, Pakistan, Russia, Tajikistan, Turkmenistan, Uzbekistan and parts of SW Asia.

CONSERVATION STATUS

China: Endangered (MEPCAS, 2013). Efforts to cultivate the species commercially, notably in Xinjiang province, have so far proved unsuccessful such that, today, wild harvesting levels are strictly controlled (C. Leon pers. obs. 2012).
Global: Not Evaluated (IUCN Red List, accessed 16/7/2015).

PHARMACOPOEIA SPECIES 2

A. guttata Bunge

Boraginaceae

synonym: *Lithospermum guttatum* (Bunge) I. M. Johnst.

PLANT DESCRIPTION

Similar to PS1, **usually shorter (10–25 cm)** and often much branched. Leaves **spatula-shaped to linear** (1.5–5.5 cm long; 3–11 mm wide), densely hairy with broad tips. Flower-heads 3–10 cm long; **flower tubes and lobes yellow;** lobes frequently purple-spotted. Fl. and fr. Jun–Oct.

HARVESTING, SOURCING AND NATURAL RANGE

A threatened and overharvested species much like PS1, growing in a similar habitat and with a similar distribution: W Gansu, N Hebei, Inner Mongolia, Ningxia, Tibet and Xinjiang. [Also wild in Afghanistan, NW India, Kazakhstan, Kyrgyzstan, Mongolia, Pakistan, Russia, Tajikistan, Turkmenistan and Uzbekistan].

CONSERVATION STATUS

China: Vulnerable (MEPCAS, 2013).
Global: Not Evaluated (IUCN Red List, accessed 16/7/2015).

UNOFFICIAL SUBSTITUTE

Lithospermum erythrorhizon Siebold & Zucc.

Boraginaceae

synonym: *L. officinale* L. subsp. *erythrorhizon* (Siebold & Zucc.) Hand.-Mazz.

PLANT DESCRIPTION

A taller plant (40–90 cm), leaves broader (3–8 cm long; **0.7–1.7 cm wide, ovate-lance-shaped to broadly lance-shaped); flowers white** (7–9 mm diam.); flower tube shorter (c. 4 mm long) and **almost equalling length of flower lobes**; inner part of petal lobes (at top of throat) with semi-globose appendages such that the central zone of lobes **resemble wheel spokes.** Fl. and fr. Jun–Sep.

CONSERVATION STATUS

China: Least Concern (MEPCAS, 2013); "rapidly decreasing" (Leung & Cheng, 2008).
Global: Not Evaluated (IUCN Red List, accessed 17/7/2015).

TCM PROPERTIES AND USE

properties: **'sweet, salty and cold'**
channels: **'Heart and Liver'**
actions: **'cools Blood, clears heat, eliminates toxins'**

indications: **rashes, abscesses, sores, burns; epistaxis, haematuria; (T)**

Arnebia euchroma, flowering in the wild, Xinjiang province (NW China) PHOTO: KAISA SULAIMAN

Inflorescences of *A. euchroma* PHOTOS: KAISA SULAIMAN

Whole plant of *A. euchroma* freshly harvested

Desert habitat of *A. guttata,* Karakorum mountains, SW Xinjiang

Flowering plants of *A. guttata* in the wild (right photo: with deep purple rootstock)

Lithospermum erythrorhizon in the wild, Liaoning province

Inflorescence of *L. erythrorhizon*, Liaoning province

PHARMACOPOEIA SPECIES 1

Arnebia euchroma

Boraginaceae

DRUG MORPHOLOGY

CRUDE DRUG (DRIED ROOT): an irregularly twisted, somewhat cylindrical or conical rootstock (7–20 cm long; 1–2.5 cm diam.), sometimes curved. Externally deep pink, purplish-red or purplish-brown with **cortex easily exfoliating into 10 or more paper-like layers [1]**. Apex often with branching remnants of bright pinkish stem **[2]**. *Fracture*: uneven; **wood relatively narrow, yellowish-white or yellow [3]**. *Texture*: light, lax, soft and easily broken **[4]**. *Odour*: distinctive. *Taste*: astringent and slightly bitter. ***Good quality*:** large, long, soft roots, with a small creamish-white centre.

DECOCTING PIECES (CUT ROOT LENGTHS OR SLICES): mostly of layers of deep purple, papery cortex [1], easily fragmented or still intact and surrounding a relatively narrow central yellowish-white or yellow wood [2]. *Texture* etc. as for crude drug above.

PROCESSING

Roots are dried (= crude drug) then cut into sections or sliced thickly as decocting pieces.

OTHER COMMON NAMES

软紫草 Ruan Zi Cao (soft purple herb); 新紫草 Xin Zi Cao; 新疆紫草 Xinjiang Zi Cao; 紫草根 Zi Cao Gen; arnebia (SCN); arnebia root (CP2010); Radix Arnebiae; Xinjiang arnebia.

SAFETY

Legal restrictions and recommendations exist in many countries about the use of pyrrolizidine alkaloid-containing plants, such as *A. euchroma*, because long term internal use has proven to be hepatotoxic and carcinogenic (Roeder, 2000). Such restrictions and advice are currently in a state of flux as new research is awaited (EMA, 2014; HMPC, 2016).

PHARMACOPOEIA SPECIES 2

A. guttata

Boraginaceae

DRUG MORPHOLOGY

CRUDE DRUG (DRIED ROOT): this drug tends to be **harder and brittle [5]** compared to that of PS1 which is relatively soft and flaking; the fracture is **relatively even and the wood often broader [6]**. Laboratory-based methods are advised for more reliable differentiation.

DECOCTING PIECES (CUT ROOT LENGTHS OR SLICES): cannot be reliably differentiated from those of PS1 using macroscopic characters alone; laboratory-based methods are advised.

PROCESSING

As for PS1.

OTHER COMMON NAMES

黄花软紫草 Huang Hua Ruan Zi Cao (yellow flowered soft purple herb); 紫草根 Zi Cao Gen; arnebia (SCN); arnebia root; Radix Arnebiae; red root gromwell.

SAFETY

As for PS1 above.

UNOFFICIAL SUBSTITUTE

Lithospermum erythrorhizon

Boraginaceae

'DRUG SUBSTITUTE' MORPHOLOGY

Much more woody [7] compared to that of PS1 and PS2. Outer bark thin and rough with distinct longitudinal furrows **[8]** (peeling easily but not peeling off in lamellae); wood pinkish-white creamish-white and broader than outer cortex **[9]**.

OTHER COMMON NAMES

山紫草 Shan Zi Cao; 硬紫草 Ying Zi Cao; 紫草 Zi Cao; red-root lithospermum (SCN).

SAFETY

As for PS1 above.

SUBSTITUTION COMMENT. A widespread meadow species native to northern and central China. Traditionally roots of this species were considered the botanical drug of choice for the drug 'Zi Cao' but clinical preference for the two *Arnebia* species above saw this species replaced by them in the 2005 edition of the CP onwards. Subsequent over-harvesting in the wild of both PS1 and PS2 for dye and medicine is likely to see *L. erythrorhizon* increasingly traded as a 'Zi Cao' substitute.

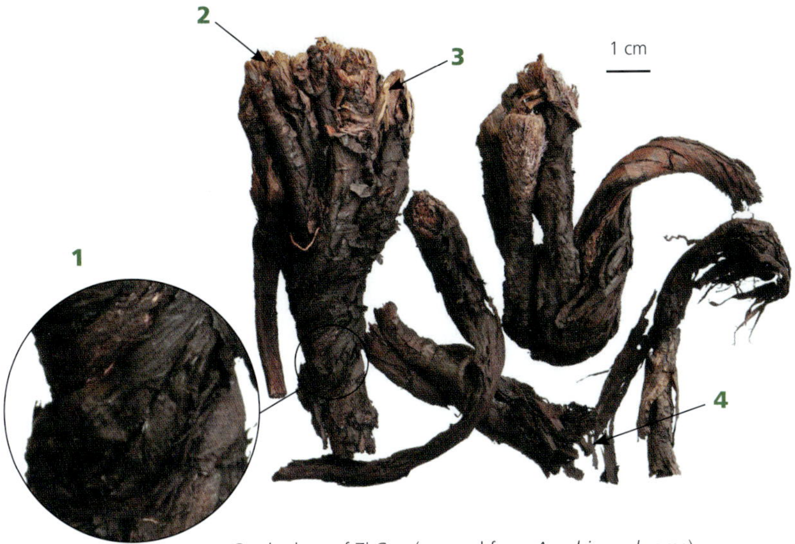

Crude drug of Zi Cao (sourced from *Arnebia euchroma*)

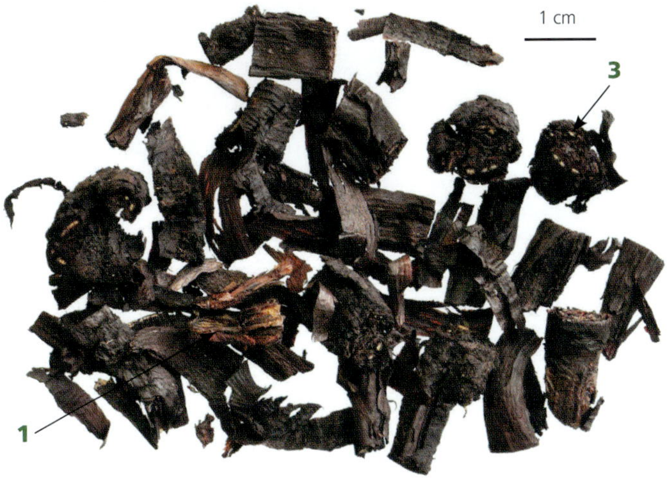

Decocting pieces of Zi Cao (sourced from *A. euchroma*)

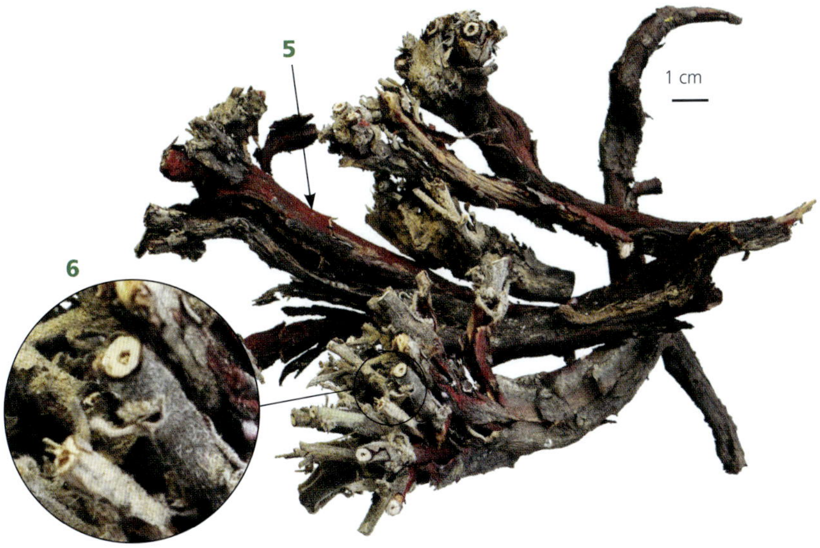

Crude drug of Zi Cao (sourced from *Arnebia guttata*)

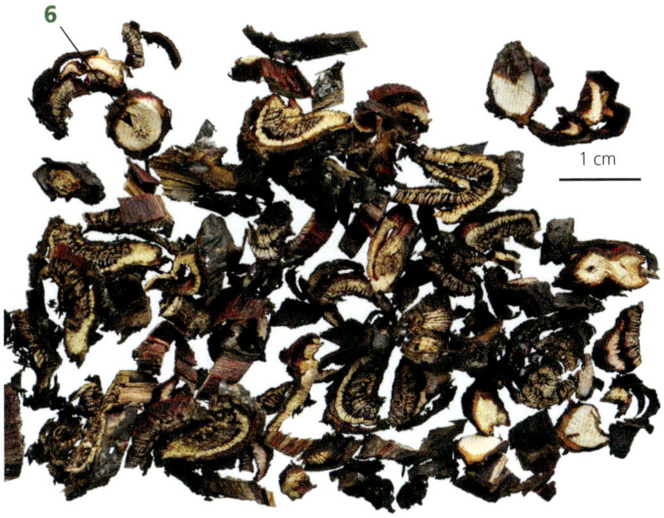

Decocting pieces of Zi Cao (sourced from *A. guttata*)

Dried roots of *Lithospermum erythrorhizon*

LESSER KNOWN UNOFFICIAL SUBSTITUTES

1. ***Potentilla chinensis*** Ser. (family Rosaceae). Aerial parts are the source of the separate TCM drug 'Wei Ling Cai' (Potentillae Chinensis Herba) not included in this book. A likely cause of confusion is its similar Chinese name Bei Zi Cao 北紫草 used in Hong Kong markets.
2. ***Onosma*** species (family Boraginaceae): root-bark of some species traded as 'Zi Cao' as a cheaper and readily available substitute. The broad yellow woody xylem is often discarded prior to sale to avoid detection as an unofficial substitute.

Wei Ling Cai (*Potentilla chinensis*)

Eupatorium fortunei

aerial parts and whole plants

Da Ji 大蓟

Cirsii Japonici Herba

Drug part: AERIAL

PHARMACOPOEIA SPECIES

Cirsium japonicum DC.

Asteraceae

synonym: *Carduus japonicus* (DC.) Franch.; *Cnicus japonicus* (DC.) Maxim.
taxonomic/nomenclatural note: in CP2015 as '*Cirsium japonicum* Fisch. ex DC.'

Species of *Cirsium* wild in China: 46

PLANT DESCRIPTION

An erect, spiny perennial herb (30–80 (rarely to 150) cm tall). *Roots* clustered, tuberous and spindle-shaped. *Stems* greyish-white, branched or unbranched above, ribbed and densely clothed in long filamentous hairs; not winged. *Basal leaf rosette* densely spiny; blade pinnately divided into 6–12 deeply lobed segments on each side (8–20 cm long; 2.5–8 cm wide); veins with long sparse hairs; margins toothed, teeth with marginal and apical spines (2–6 mm long). Lower and middle stem leaves alternate and similar to those in basal rosette, becoming smaller upwards with bases clasping the stem (amplexicaul). *Flower-heads* (capitula) 'thistle-like', solitary or few, clustered at the top of the single flowering stem. Each flower head surrounded by a bell-shaped involucre (c. 3 cm diam.) of c. 6 rows of overlapping scales (phyllaries), hairless or sparsely cobwebby and scattered with dark resinous glands [hl]; outer and middle phyllaries triangular to narrowly triangular, tip of each with a 1–2 mm spine; inner phyllaries lance-shaped to linear-lance-shaped and softly spiny; flowers (florets) bisexual, red to purple and very densely packed (c. 2.1 cm long). *Fruit-heads* a dense mass of achenes (single-seeded fruits) c. 4 mm long, each with 3 or 4 rows of pale brown feathery bristles forming a pappus at one end (c. 2 cm long). Fl. and fr. Apr–Nov.

HARVESTING, SOURCING AND NATURAL RANGE

Aerial parts are harvested from wild plants when in flower or fruit during summer and autumn. The species grows in forest margins, grassland, scrub, roadsides and waste places (400–2,100 m altitude) and has a large distribution in China: Chongqing, Fujian, Guangdong, Guangxi, Guizhou, Hebei, Hubei, Hunan, Inner Mongolia, S Jiangsu, Jiangxi, Qinghai, Shaanxi, Shandong, Sichuan, Taiwan, Yunnan and Zhejiang provinces [also wild in Japan, the Korean Peninsula, Russian Far East and Vietnam].

CONSERVATION STATUS

China: Least Concern (MEPCAS, 2013).
Global: Not Evaluated (IUCN Red List, accessed 25/8/2016).

SAFETY

An official healthfood in China (Anon., 2002a,b).

UNOFFICIAL SUBSTITUTE

The **root** of the official species (*C. japonicum*)

PLANT DESCRIPTION

See above.

SUBSTITUTION COMMENT. Some trade outlets continue to sell Da Ji as the **roots** of this species due to a carry-over from earlier editions of the CP when roots alone were recognised as the official drug.

LESSER KNOWN UNOFFICIAL SUBSTITUTES AND ADULTERANTS

Othe species in the Asteraceae family have been reported as local or historical substitutes or adulterants in China; their occurrence in international trade is uncertain:

1. ***Cirsium arvense*** var. ***integrifolium*** Wimm. & Grab. Aerial parts are the source of the separate TCM drug 'Xiao Ji' (Cirsii Herba) see page 392.
2. ***Carduus crispus*** L. (飞廉 Fei Lian). An adulterant. Fresh plants are easily distinguished from the Pharmacopoeia species but the commercial trade items are a challenge to distinguish since many key characters are lost when fragmented.Key characters: **stems winged and toothed**; teeth with spiny margins and tips. Involucre subglobose, narrower (1.5–2(–2.5) cm diam.). Inner phyllaries linear. **Pappus bristles shorter** (up to 1.3 cm).
3. ***Cirsium chlorolepis*** Petr. (两面蓟 Liang Mian Ji). An adulterant. Lower surface of leaf blades sparsely covered with tiny spines (<1 mm long), upper surfaces rough and densely covered with tiny spines (1–2 mm long). Involucre broadly campanulate (3.5–4 cm diam.). Phyllaries **all of similar length,** apical portions **needle-like** (6–10 mm long) and **spreading to recurved.** Pappus bristles brownish.

TCM PROPERTIES AND USE

properties: 'sweet, bitter and cool'
channels: 'Liver and Heart'
actions: 'cools Blood, stops bleeding, dispels stasis'

indications: metrorrhagia, haematuria, epistaxis; reduces swelling in abscesses; oedema; jaundice; hypertension

Densely spiny basal rosette of leaves

C. japonicum with emerging flowers

Flower-heads of *C. japonicum*

Fresh rootstocks of *C. japonicum*

Cirsium japonicum

Asteraceae

DRUG MORPHOLOGY

CRUDE DRUG (DRIED AERIAL PARTS): *stems* cylindrical, greyish-white or greenish-brown (up to 1.2 cm diam.), distinctly ribbed longitudinally and densely clothed in long filamentous hairs **[1]**; fracture greyish-white, pith lax or hollowed. *Leaves* crumpled, often broken, densely spiny and pinnately divided **[2]** (when expanded each blade with 6–12 deeply lobed segments on each side), upper surface greyish-green or yellowish-brown **[3]** with lower surface paler **[4]**; veins with long, sparse, greyish-white filamentous hairs; margins toothed, teeth with marginal and apical spines (2–6 mm long) **[5]**. Flower-heads (capitula) thistle-like, globose or ellipsoidal (c. 3 cm diam.). Involucre yellowish-brown, bell-shaped with c. 6 rows of overlapping scales (phyllaries); outer and middle phyllaries triangular to narrowly triangular each with a 1–2 mm apical spine; inner phyllaries lance-shaped to linear-lance-shaped and softly spiny. *Flowers* (*florets*) densely packed. Fruit-heads a dense mass of achenes (single-seeded fruits) **[6]**, each c. 4 mm long and with a feathery, pale brown pappus of bristles at one end (c. 2 cm long) **[7]**. *Odour:* slight. *Taste:* indistinct. ***Good quality*:** abundant greyish-green leaves; roots absent.

DECOCTING PIECES (SHORT LENGTHS OF DRIED AERIAL PARTS): as for crude drug but cut into lengths (c. 1–2 cm long). Flower- and fruit-heads may be absent.

DECOCTING PIECES (CARBONISED — Da Ji Tan): as for decocting pieces above but externally brownish or brownish-black and internally brown.

PROCESSING

Aerial parts are sun-dried (= crude drug) then later cut into short decocting sections. These may be further processed by dry-frying until carbonised (= 大戟炭 Da Ji Tan).

OTHER COMMON NAMES

大戟炭 Da Ji Tan (carbonised Da Ji); 蓟 Ji (plant name); Herba Cirsii Japonici; Japanese thistle (SCN); Japanese thistle herb (CP2010).

SAFETY

An official healthfood in China (Anon., 2002a,b).

C. japonicum — *root*

'DRUG SUBSTITUTE' MORPHOLOGY

Roots long and spindle-shaped (tapering at both ends) **[8]** (5–10 cm long; c.1 cm diam.), often slightly curved **[9]**; externally dark brown and longitudinally wrinkled. *Texture*: hard and brittle. *Fracture:* uneven, bark thin and wood (xylem) yellowish-whitish and granular. *Odour:* distinctive. *Taste:* slightly bitter and pungent.

OTHER COMMON NAMES

大蓟草 Da Ji Cao; 牛口刺 Niu Kou Ci.

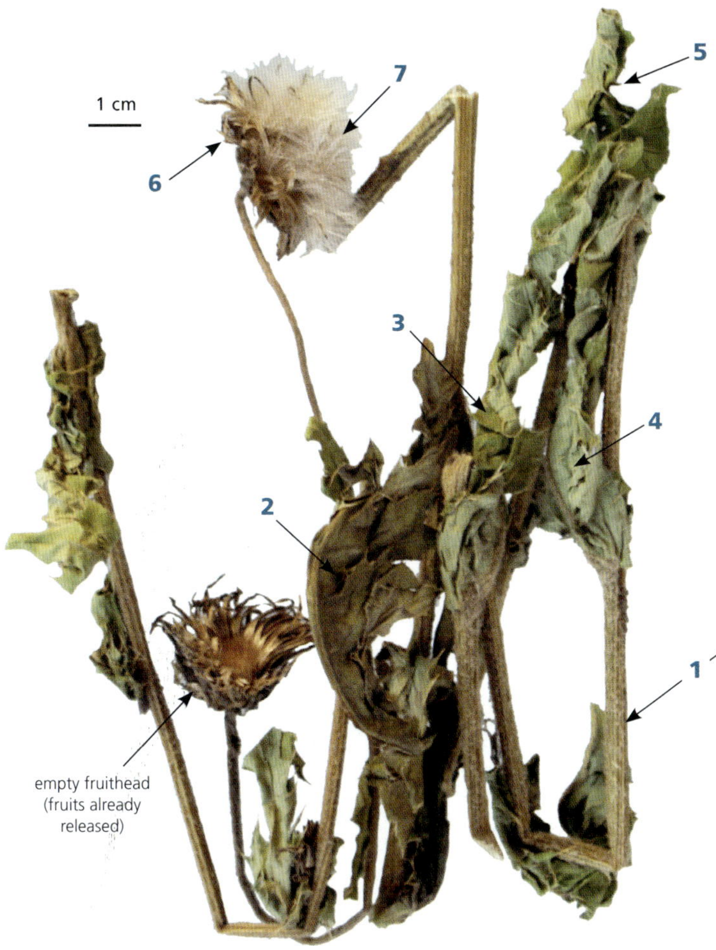

Crude drug with fruit-heads

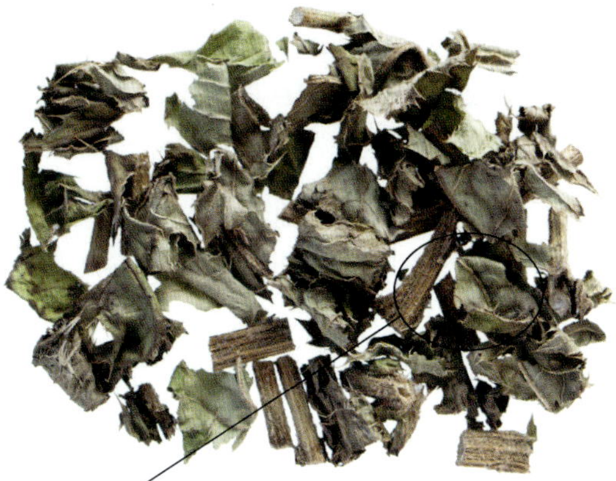

Decocting pieces

Carbonised drug (Da Ji Tan)

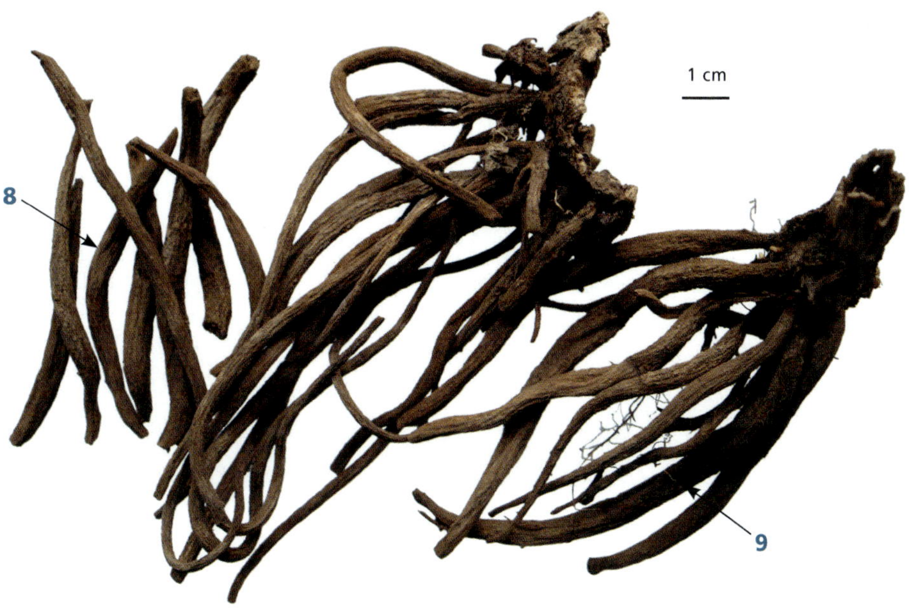

Roots of *C. japonicum*

Dan Zhu Ye 淡竹叶

Lophatheri Herba

Drug parts: LEAF AND STEM

Lophatherum gracile Brongn.

Poaceae

PLANT DESCRIPTION

A perennial grass (60–150 cm tall) with a knotty rootstock and tuberous spindle-shaped roots. *Stems* stiffly erect, slender and unbranched. *Leaves* alternate and simple; leaf base sheathing the stem with ligule (structure at base of leaf blade) brown and hairy on the outer side; leaf blade narrowly lance-shaped (5–30 cm long; 1–5 cm wide), venation distinctively rectangular; margin often in-rolled; hairless or with long hairs on both surfaces. *Inflorescence* inconspicuous (10–25 cm long) with loose or dense narrowly lance-shaped spikelets (flowerheads) (7–12 m long; 5–2.5 mm wide) each with 1 floret; lowermost lemma with straight back. Fl. and fr. Jun–Sep.

Species of *Lophatherum* wild in China: 2

HARVESTING, SOURCING AND NATURAL RANGE

Aerial parts are harvested from wild populations in summer before plants flower. The species grows in shady places along forest margins and on disturbed ground in C, SE, S and W China: Anhui, Fujian, Guangdong, Guangxi, Guizhou, Hainan, Hubei, Hunan, Jiangsu, Jiangxi, Sichuan, Taiwan, Yunnan and Zhejiang provinces [also wild in Cambodia, India, Indonesia, Japan, the Korean Peninsula, Malaysia, Myanmar, Nepal, New Guinea, Pacific Islands (Polynesia), Philippines, Sri Lanka, Thailand, Vietnam and Australia].

CONSERVATION STATUS

China: Least Concern (MEPCAS, 2013).
Global: Not Evaluated (IUCN Red List, accessed 25/3/2015).

Lophatherum gracile in the wild, Anhui province

L. gracile in cultivation

TCM PROPERTIES AND USE

properties: **'sweet, bland and cold'**
channels: **'Heart, Stomach and Small Intestine'**
actions: **'clears heat and fire'**
indications: **irritability, dysuria**

DRUG MORPHOLOGY

CRUDE DRUG (LEAF AND STEM): *stems* unbranched, slender and cylindrical **[1]** (25–75 cm long) yellowish-green or pale yellow with distinct nodes **[2]**. *Leaves* alternate (if still attached) and simple, greyish-green or yellowish-green; leaf base sheathing; ligule brown and hairy on the outer side. The stem leaf blades crumpled or inrolled or (when intact and expanded) narrowly lance-shaped **[3]** (5–30 cm long; 1–5 cm wide), sometimes with stalk-like base, venation distinctively rectangular **[4]**; margin often in-rolled **[5]**; hairless or with long hairs on both surfaces; *Texture*: light and pliable. *Fracture:* hollow. *Odour:* indistinct. *Taste:* indistinct. ***Good quality*:** leaves many, large and green and devoid of roots and flowering/fruiting spikes.

DECOCTING PIECES (CUT LENGTHS OF STEM AND LEAF): as for the crude drug but cut into short and often fragmented pieces with loss of many identification characters.

PROCESSING

Aerial parts are sun-dried (= crude drug) then cut into sections for decocting.

OTHER COMMON NAMES

竹叶麦冬 Zhu Ye Mai Dong (bamboo-leaved Mai Dong); Herba Lophatheri; lophatherum herb (CP2010); lophatherum (SCN).

SAFETY

An official food and healthfood in China (Anon., 2002a,b).

Crude drug

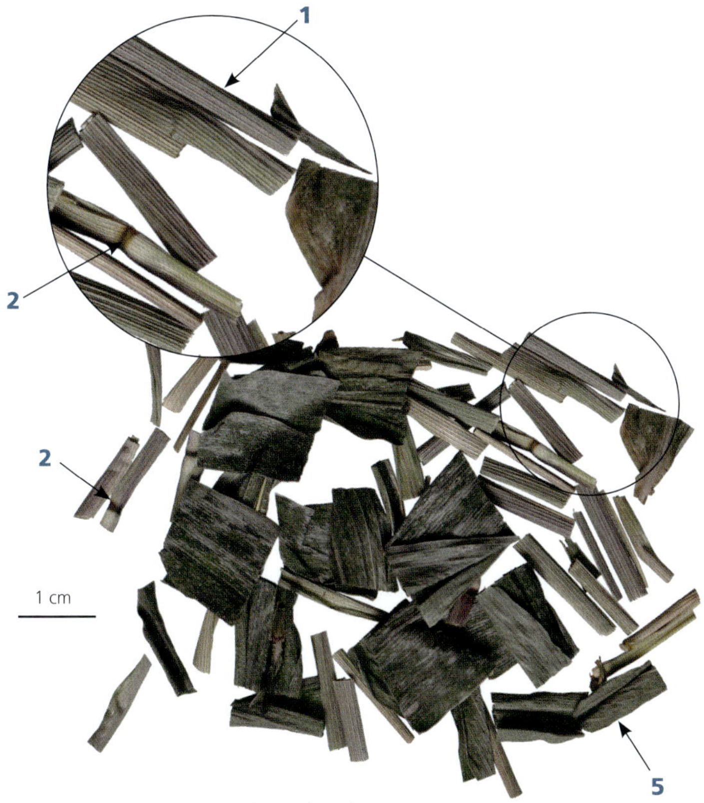

Decocting pieces

Guang Huo Xiang 广藿香

Pogostemonis Herba

patchouli

Drug part: AERIAL

Pogostemon cablin (Blanco) Benth.

Lamiaceae

synonyms: *Mentha cablin* Blanco; *P. javanicus* Backer ex Adelb.; *P. patchouly* Pellet.

Species of *Pogostemon* wild in China: 16

PLANT DESCRIPTION

An erect, many branched, herbaceous perennial or small shrub becoming woody with age (30–100 cm tall), **softly hairy throughout** and intensely aromatic. *Stems* somewhat 4-sided and solid. *Leaves* opposite and simple; **leaf stalks densely hairy** (1–6 cm long); leaf blades circular to broadly-ovate (2–10.5 cm long; 1–8.5 cm wide), **leaf undersides covered in a dense mat of short, soft hairs,** leaf bases usually wedge-shaped, **margins irregular.** *Flowers* in dense, hairy, 10 to many-flowered whorls (verticillasters) arranged along **terminal and axillary flower spikes** (4–6.5 cm long; 1.5–1.8 cm diam.). Flowers subtended by densely hairy bracts and bracteoles not longer than the calyces. Each flower bisexual and laterally symmetrical; calyx tubular (7–9 mm long) with crystals and 5 short, pointed teeth; corolla pinkish-purple (c. 1 cm long), shaggily hairy outside, opening into a 2-lipped mouth (upper lip 3-lobed; lower lip undivided); stamens, 4, bearded and extending beyond the corolla. *Fruits* smooth, dry, somewhat flattened nutlets. Fl. April. *Plants rarely flower in cultivation.*

HARVESTING, SOURCING AND NATURAL RANGE

Not native to China but widely cultivated as a medicinal crop and harvested when foliage is lush. The main production areas are in S China: Fujian, Guangdong, Guangxi, Hainan and Taiwan with traditional *Di Dao* sites in Guangdong and Hainan. Crops of this species are also grown for their aromatic leaves, which are rich in essential oils and are also the main ingredient in patchouly oil used in the world's fragrance industry. *P. cablin* is native in China; native in India, Indonesia, Malaysia, Philippines and Sri Lanka.

CONSERVATION STATUS

China: not native

Global: Not Evaluated (IUCN Red List, accessed 18/8/2016).

Agastache rugosa (Fisch. & C. A. Mey.) Kuntze

Lamiaceae

synonym: *Lophanthus rugosus* Fisch. & C. A. Mey.

PLANT DESCRIPTION

Easily distinguished from the Pharmacopoeia species. *Main differences:* often taller (up to 1.5 m); **virtually hairless throughout** (leaf hairs mostly confined to veins); older stems *not* becoming woody; leaf stalks often shorter (1.5–3.5 cm long), leaf blades heart-shaped/ovate to lance-shaped (4.5–11 cm long; 3–6.5 cm wide); **leaf undersides virtually hairless;** leaf bases heart-shaped or rounded-truncate; **margins saw-toothed; flower spikes terminal only**, more obviously cylindrical and compact in outline (2.5–12 cm long; 1.8–2.5 cm diam.); corollas purplish-blue with a spreading 3-lobed lower lip.

SUBSTITUTION COMMENT. The similar clinical properties of this and the official drug have seen these two species used interchangeably in TCM practice for many years. The name 'Huo Xiang' 藿香 has, confusingly, been applied to both species while their similar appearance as living plants and TCM items, especially when flowers are absent, adds a further confusion.

A. rugosa continues to be widely grown in China as a medicinal and is also popularly grown in home gardens for use as a culinary herb.

The species is native in Japan, Korea, Russia and the USA.

CONSERVATION STATUS

China: not native.

Global: Not Evaluated (IUCN Red List, accessed 18/8/2016).

TCM PROPERTIES AND USE

***properties*: 'pungent and slightly warm'**
***channels*: 'Liver, Spleen and Stomach'**
***actions*: 'transforms damp, releases the exterior'**

***indications*: abdominal distension, poor appetite; nausea, vomiting; fever; sinus congestion**

Non-flowering medicinal crop of *Pogostemon cablin* (basal stems woody), Guangdong province

Young leaves of *P. cablin*

P. cablin with young flowering branch

Agastache rugosa in full flower (medicinal crop)

Young leaves of *A. rugosa*

A. rugosa flower spikes

Pogostemon cablin

Lamiaceae

DRUG MORPHOLOGY

CRUDE DRUG (DRIED AERIAL PARTS): a mass of branching, slightly curved, stems (30–60 cm long) and crumpled leaves; **softly greyish-white hairy (tomentose) throughout** (flowering and fruiting spikes usually absent). Stems of two types with one or both types present: *young stems (from top of plant)* non-woody (6–7 mm diam.), externally greyish-brown or greyish-yellow **[1]**, covered in very short, soft, greyish-white hairs, somewhat 4-sided with rounded corners **[2]** with a broad white pith **[3]**, *texture:* light and fragile; *older stems* (*basal*) broader (1–1.2 cm diam.) with greyish-brown cork bark **[4]**, sometimes exfoliating and white pith narrower, *texture*: light and difficult to break. Leaves opposite; leaf stalks 1–6 cm long **[5]**, densely covered in short, soft hairs; leaf blades usually tightly curled **[6]**, **covered in a dense mat of short, soft hairs (especially on leaf undersides) [7]**; leaf blades (when intact and expanded) circular to broadly-ovate (2–10.5 long; 1–8.5 cm wide), bases wedge-shaped and **leaf margins irregular [8]** (not saw-toothed). *Odour*: aromatic. *Taste*: slightly bitter.

DECOCTING PIECES (DRIED AERIAL LENGTHS): as for the crude drug but stems cut into short lengths **[9]**; externally, young stems either greyish-brown, greyish-yellow or reddish-brown **[10]**. *Odour*: aromatic. *Taste*: slightly bitter.

PROCESSING

Aerial parts are sun-dried and protected overnight; this is repeated until thoroughly dry (= crude drug); subsequently stems are cut into decocting lengths and mixed with the leaves.

OTHER COMMON NAMES

藿香 Huo Xiang; 南藿香 Nan Huo Xiang 'southern Huo Xiang'; Herba Pogostemonis; cablin patchouli herb (CP2010); patchouli (SCN).

SAFETY

Safety Class: 1; Interaction Class: A (Gardner & McGuffin, 2013).

Agastache rugosa

Lamiaceae

'DRUG SUBSTITUTE' MORPHOLOGY:

Both the crude drug and decocting pieces are easily distinguished macroscopically from the Pharmacopoeia species. *Main differences:* **virtually hairless in all parts. Stems distinctly 4-sided** and **longitudinally furrowed [11], never woody, hollow [12]**. Leaf stalks **[13]** and leaf undersides **[14]** virtually hairless and **margins distinctly saw-toothed [15].**

OTHER COMMON NAMES

藿香 Huo Xiang; 土藿香 Tu Huo Xiang; Agastachis Herba (KP, 2002); Chinese giant hyssop (SCN); giant hyssop.

SAFETY

An official food and healthfood in China (Anon., 2002a,b). Safety Class: 1; Interaction Class: A (Gardner & McGuffin, 2013).

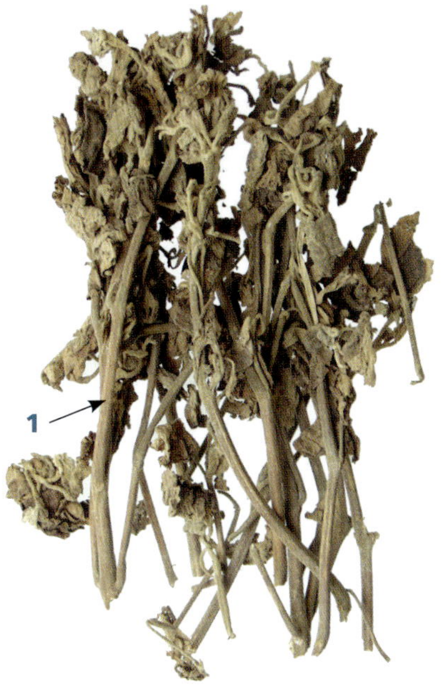

Crude drug of Guang Huo Xiang (*P. cablin*)
(sample here includes only the
young non-woody stems)

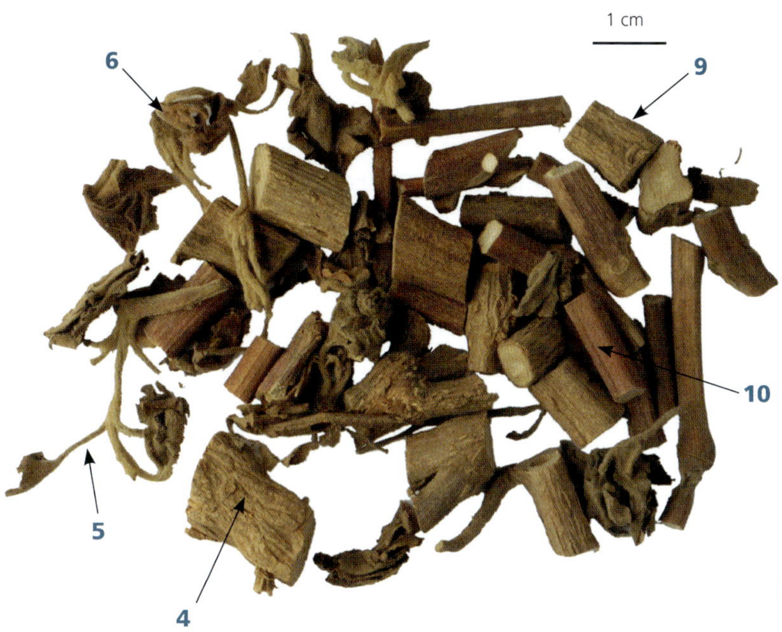

Decocting pieces of Guang Huo Xiang (*P. cablin*)
(sample with young and old stems)

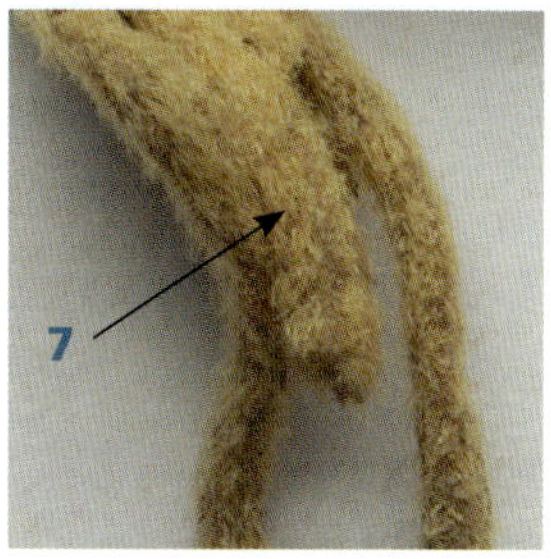

close-up of leaf stalk

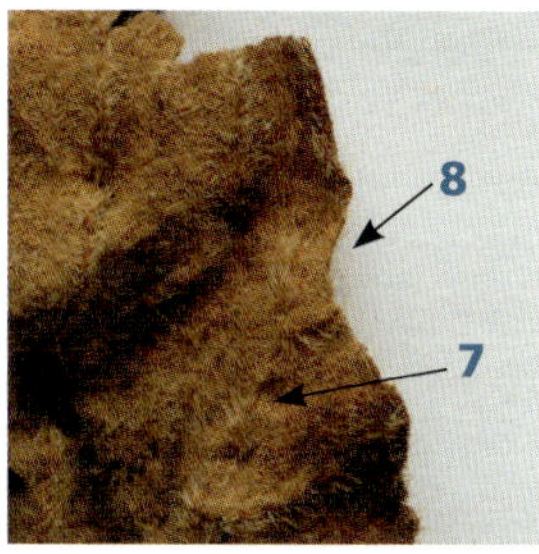

close-up of leaf underside

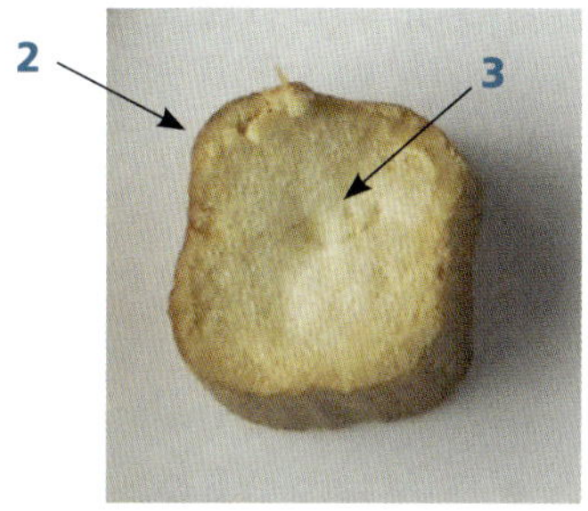

young stem
(transverse section)

Dried stems and leaves of Huo Xiang
(*Agastache rugosa*)

Dried stem lengths and leaves of Huo Xiang (*A. rugosa*)

Close-up of leaf
underside

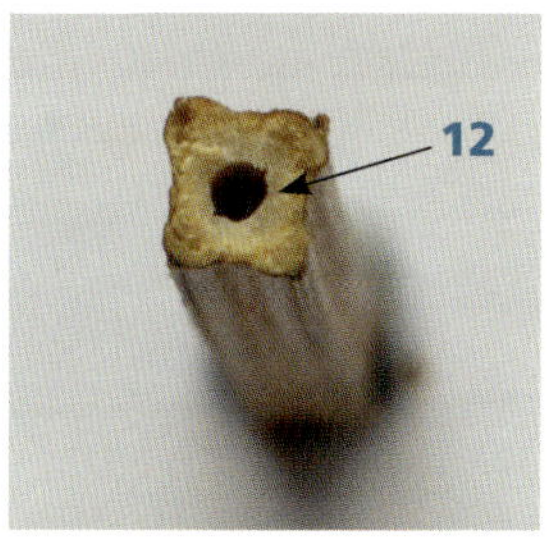

Stem transverse
section

Guang Jin Qian Cao 广金钱草 Desmodii Styracifolii Herba

Drug part: AERIAL

Desmodium styracifolium (Osbeck) Merr.

Fabaceae

synonym: *Hedysarum styracifolium* Osbeck

Species of *Desmodium* wild in China: 29

PLANT DESCRIPTION

A much-branched, creeping sub-shrub (30–100 cm tall) with young branches densely covered in soft white or yellowish hairs. *Leaves* alternate, divided into 3 leaflets (trifoliolate); leaf stalk 1–2 cm long and densely hairy; stipules 2, brown, lance-shaped and membranous (c. 8 mm long); leaflet blades almost circular (2–4.5 cm long; 2–4.5 cm wide), base heart-shaped or rounded, tips rounded or indented; blade undersides densely covered in a soft, white matt of hairs with 8–10 slightly raised, curved and parallel lateral veins; margins densely soft, white hairy. *Flowers* tiny (c. 4 mm long), bisexual and clustered in loosely conical inflorescences (1–3 cm long) at tips of branches or in leaf axils; axis of inflorescence (rachis) densely covered in soft, silky hairs; each flower laterally symmetrical with a 4-lobed calyx and 5 pink to purplish-red petals of different sizes; the two lower petals (keels) longer than lateral petals (wings), strongly curved with a long claw; stamens 10. *Fruit* a 3–6-segmented pod (1–2 cm long; c. 2.5 mm wide), covered in soft hairs. Fl. and fr. Jun–Sep.

HARVESTING, SOURCING AND NATURAL RANGE

Aerial parts are harvested in summer (May–July) and autumn (Aug–Oct) mostly from medicinal crop plants grown in southern China, notably Guangdong and Guangxi provinces, and where traditional *Di Dao* sources are located. The wild species grows in grassland and thickets on hillsides below 1,000 m altitude in the provinces of Guangdong, Guangxi, Hainan, Hubei and Yunnan. Also native in Cambodia, India, Malaysia, Myanmar, Sri Lanka, Thailand and Vietnam.

CONSERVATION STATUS

China: Least Concern (MEPCAS, 2013)
Global: Not Evaluated (IUCN Red List, accessed 25/8/2016).

Medicinal crop of *Desmodium styracifolium*

D. styracifolium with emerging inflorescence

Flowers and young fruits

TCM PROPERTIES AND USE

***properties*:** **'sweet, bland and cool'**
***channels*:** **'Liver, Kidney and Bladder'**
***actions*:** **'clears heat, dispels damp'**

***indications*:** **stranguria, urinary stones; jaundice**

DRUG MORPHOLOGY

DRIED AERIAL PARTS (CRUDE DRUG): a loosely crumpled mass of mainly leaves and stems (flowers and fruits occasional). Stems pale brown **[1]**, cylindrical (up 1 m long; c. 2 mm diam.), ridged and densely clothed in soft, spreading, pale yellow hairs **[2]**; internodes 0.8–3 cm long and easily broken to reveal a small, central, white pith. *Leaves* (when still attached) alternate, divided into 3 leaflets (trifoliolate); leaflet stalks 1–2 cm long and densely hairy **[3]**; stipules **[4]** at base of leaf stalks, brown, lance-shaped and membranous (c. 8 mm long); leaflet blades (when intact and expanded) almost circular **[5]** (2–4.5 cm long; 2–4.5 cm diam.), base shallowly heart-shaped or rounded, tips rounded or indented **[6]**; leaflet blade upper surface greyish-green or yellowish-green and hairless **[7]**, lower surface densely covered in a soft, greyish-white white matt of hairs **[8]** with 8–10 slightly raised, curved and parallel lateral veins **[9]**; margins densely edged soft, greyish-white hairy **[10]**. *Flowers* tiny **[11]** (c. 4 mm diam.) and (when still attached) clustered in loosely conical inflorescences (1–3 cm long) at tips of branches or in leaf axils; each flower with 5 dark purplish petals of different sizes. *Fruit* a 3–6-segmented pod **[12]** (1–2 cm long; c. 2.5 mm wide), covered in soft hairs. *Odour*: slightly aromatic. *Taste*: slightly sweet. ***Good quality*:** leaves many and pale green; flowers, fruits and roots absent.

PROCESSING

Aerial parts are sun-dried (= crude drug) then cut into short lengths for decocting.

OTHER COMMON NAMES

铜钱草 Tong Qian Cao; coin-leaf desmodium (SCN); Herba Desmodii Styracifolii; snowball-leaf tickclover herb (CP2010).

COMMENT

See also 'Jin Qian Cao' (Lysimachiae Herba) page 358, and for which this drug is a common substitute.

Crude drug of Guang Jin Qian Cao (dried aerial parts of *D. styracifolium*)

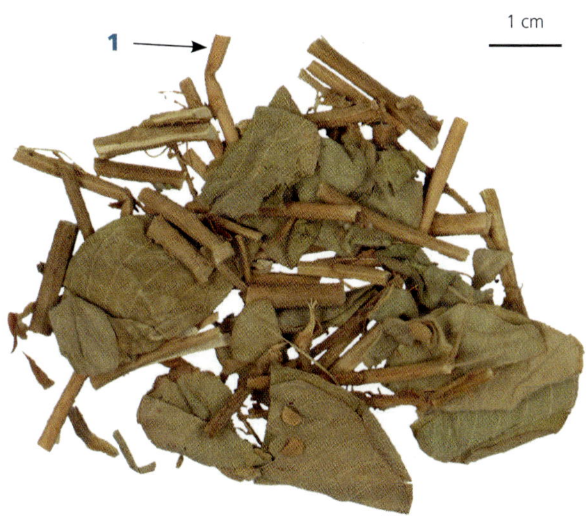

Decocting pieces of Guang Jin Qian Cao

Hu Ji Sheng 槲寄生

Visci Herba

coloured mistletoe

Drug parts: WHOLE PLANT

Viscum coloratum (Kom.) Nakai

Santalaceae

synonym: *V. album* L. subsp. *coloratum* Kom.
taxonomic/nomenclatural note: in CP2010 as '*V. colouratum* (Komar.) Nakai'

Species of *Viscum* wild in China: 12

PLANT DESCRIPTION

Green or yellowish-green, parasitic evergreen perennial, forming loose and somewhat tangled clusters (30–80 cm diam.) on deciduous trees; most parts opposite. Stems cylindrical, green and non-woody, pliable and branching at right angles; stem nodes articulated and slightly swollen. *Leaves* opposite, elliptic or oblong-lanceolate (3–7 cm long; 0.7–2 cm wide), leathery; margin entire. *Flowers* small and inconspicuous with male and female on separate plants. *Fruits* spherical (c.1 cm diam.); yellow, red or orange; flesh sticky and mucilaginous. Fl. Apr–May, fr. Sep–Nov.

HARVESTING, SOURCING AND NATURAL RANGE

The plant is harvested during winter to spring, usually from wild plants which parasitise a wide variety of wild or cultivated deciduous trees (*Alnus, Malus, Populus, Prunus, Pterocarya, Pyrus, Quercus, Salix* and *Tilia*) between 500–2,200 m altitude. The species grows wild throughout China except in Guangdong, Tibet, Yunnan and Xinjiang [also native to Japan, the Korean Peninsula and E Russia].

CONSERVATION STATUS

China: Near Threatened (*authors' pers. obs.* 2009)
Global: Not Evaluated (IUCN Red List, accessed 25/8/2016).

Viscum coloratum heavily infesting a host tree of *Liquidambar formosana*, Anhui province, China

V. coloratum mature, fruiting plant

V. coloratum with mature and developing leaves and young fruits

UNOFFICIAL SUBSTITUTES/ADULTERATION

The separate TCM drug 'Sang Ji Sheng' (Taxillii Herba) page 382 is used interchangeably with this drug in a clinical setting as are their Chinese names; this gives rise to much confusion. The 'Other Unofficial Substitutes' listed under 'Sang Ji Sheng' also apply to this drug, creating yet further confusion.

TCM PROPERTIES AND USE

***properties*:** **'bitter and neutral'**
***channels*:** **'Liver and Kidney'**
***actions*:** **'dispels wind damp, tonifies Liver and Kidney'**

***indications*:** **musculoskeletal pain; metrorrhagia, threatened miscarriage; dizziness, vertigo**

DRUG MORPHOLOGY

CRUDE DRUG (DRIED WHOLE PLANT): *stems* yellowish-green or yellowish-brown, cylindrical **[1]** (c. 30 cm long; 0.3–1 cm wide), 2–5 branched, longitudinally wrinkled, with nodes articulated and slightly swollen **[2]**; and occasional stem scars at nodes **[3]**. Leaves yellowish-green or yellowish-brown **[4]**, opposite (easily shed), without a true stalk, leaf blade elliptical-lanceolate (2–7 cm long; 0.5–1.5 cm wide), base wedge-shaped, apex broadly rounded **[5]** and margin entire; finely wrinkled throughout and 5-veined (middle 3 distinct) **[6]**. *Texture:* leaves leathery; stems light and brittle. *Odour* indistinct. *Taste:* somewhat bitter; sticky when chewed. *Fracture:* stems uneven, wood pale yellow with radiating rays; pith obscured or off-centre. ***Good quality*:** young growth with many leaves and yellowish-brown stems.

DECOCTING PIECES (SLICED PLANT): as for crude drug but cut into short pieces.

PROCESSING

After removal of old thick stems, plants are dried or steamed (= crude drug), then later softened with water before cutting into 2–4 mm thick slices (= decocting pieces).

OTHER COMMON NAMES

北寄生 Bei Ji Sheng: northern parasite; 北桑寄生 Bei Sang Ji Sheng northern mulberry parasite; 寄生 Ji Sheng; coloured mistletoe (CP2010); Herba Visci.

SAFETY

Dependent on toxicity of host tree since chemical compounds can be sequestered from host to parasite.

All parts of the very closely related *Viscum album* L. (European mistletoe) are classified as Safety Class: 2b and 2d Interaction Class: A (Gardner & McGuffin, 2013).

Crude drug of Hu Ji Sheng (dried *Viscum coloratum*)

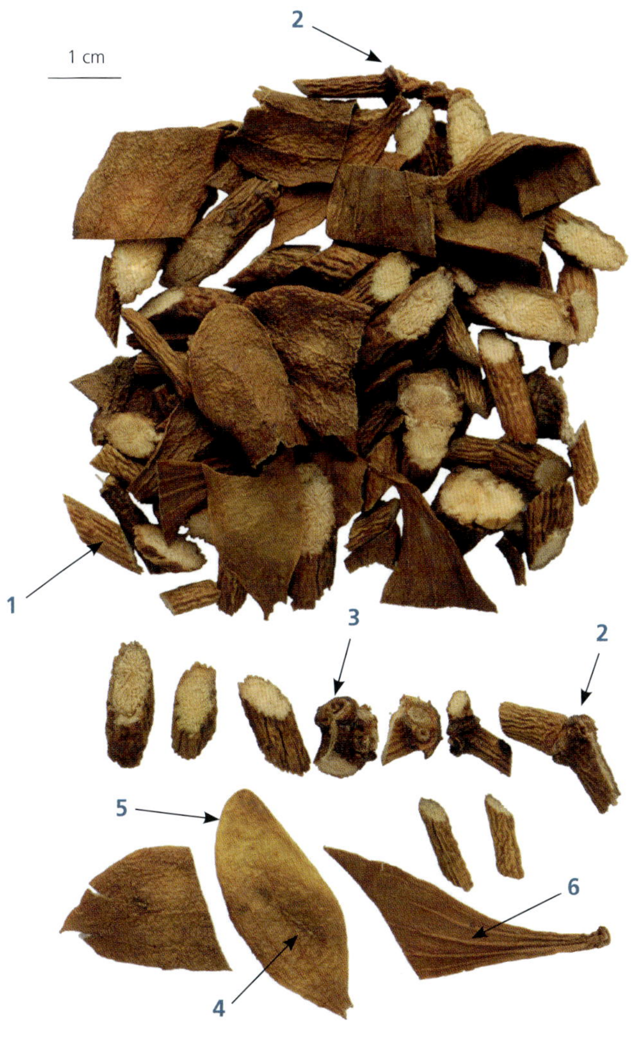

Decocting pieces of Hu Ji Sheng (dried and sliced *V. coloratum*)

Jing Jie 荆芥
incl. Jing Jie Tan 荆芥炭

Schizonepetae Herba
incl. Schizonepetae Herba Carbonisata

Drug part: AERIAL

Nepeta tenuifolia Benth.

Lamiaceae

synonym: *Schizonepeta tenuifolia* (Benth.) Briq. (name used in CP2015, as '*S. tenuifolia* Briq.')

Species of *Nepeta* wild in China: 42

PLANT DESCRIPTION

An aromatic, herbaceous annual (30–100 cm tall). *Stems* 4-sided, with many opposite branches; softly, greyish-white hairy. *Leaves* opposite, stalked and deeply 3- or 5-lobed with middle lobe longest; lobes lance-shaped, margins entire, tips pointed (entire leaf in outline c. 1–3.5 cm long; 1.5–2.5 cm wide); leaf blade glandular, lower surface greyish-green. *Flowers* many and tiny in dense whorls (verticillasters) along terminal spikes (2–15 cm long), each whorl with 2 virtually opposite flower clusters, each cluster with a leaf-like bract and minute linear bracteoles; clusters becoming densely packed towards top of flower spike. Each flower bisexual and laterally symmetrical; calyx tubular or bell-shaped (c. 3 mm long), with 5 narrow triangular teeth; corolla pale violet (c. 4.5 mm long), softly hairy and 2-lipped (upper lip 2-lobed, flat or concave; lower lip 3-lobed with middle lobe larger); stamens 4; ovary superior; style with 2-cleft stigma. *Fruit* (enclosed by a persistent calyx) with 4 nutlets, dry, brown, 3-angled and oblong (c. 1.5 mm long; 0.7 mm diam.). Fl. Jul–Sep, fr. Aug–Oct.

NOTE — not to be confused with *Nepeta tenuiflora* Diels, a **perennial** herb up to 160 cm high, with opposite, oblong-lanceolate, serrated leaves (**not divided into lobes**), and **larger (1.3–1.5 cm) purplish-blue** flowers.

HARVESTING, SOURCING AND NATURAL RANGE

Aerial parts are harvested in summer and autumn when flowering/fruiting spikes are mostly turning green (i.e. beginning to fruit). Mostly sourced from medicinal crops grown in the Chinese provinces of Hebei, Henan, Shandong and Zhejiang. GAP-certified crops have recently been grown in the Yutian region of Hubei.

The wild species grows along forest margins and valleys (500–2,700 m altitude) in N, C and SW China in the provinces of Gansu, Guizhou, Hebei, Heilongjiang, Henan, Liaoning, Qinghai, Shaanxi, Shanxi and Sichuan; [also native in the Korean Peninsula].

CONSERVATION STATUS

China: Least Concern (MEPCAS, 2013); Least Concern (authors' pers. comm. 2011).

Global: Not Evaluated (IUCN Red List, accessed 19/8/2016).

For other TCM drugs sourced from this species see 'Jing Jie Sui' (Schizonepetae Spica) page 662.

Medicinal crop of *Nepeta tenuifolia*

N. tenuifolia in flower bud

Young flower-heads of *N. tenuifolia*

Deeply lobed leaves of *N. tenuifolia*

TCM PROPERTIES AND USE

properties**: 'pungent and slightly warm'**
[Jing Jie Tan: 'pungent, sour and warm']
channels**: 'Lung and Liver'**
action**: 'releases the exterior, dispels wind, stops bleeding'**
[Jing Jie Tan: 'stops bleeding']

indications**: respiratory tract infection; rashes, sores; haemoptysis, epistaxis, haematemesis and other skin eruptions or sores**
[Jing Jie Tan: blood in the stools, menstrual flooding, postpartum faintness]

DRUG MORPHOLOGY

CRUDE DRUG (INTACT AERIAL PARTS): stems 4-sided **[1]** (50–80 cm long; 2–4 mm diam.) with leaves and flowering or fruiting heads attached **[2]** or separate. Stems with opposite branches **[3]**; externally greyish-, yellowish- or pinkish-green **[4]** with soft, fine greyish-white hairs. *Texture*: light and brittle. *Fracture*: almost white. Leaves opposite, mostly shed and fragmented **[5]**; when intact, stalked and deeply 3- or 5-lobed **[6]** (1–3.5 cm long; 1.5–2.5 cm wide); leaf blade glandular, lower surface greyish-green, margin entire. Occasional flower remnants may persist as tiny 2-lipped and softly hairy corollas (4.5 mm long) **[7]**. Fruiting spikes cylindrical (2–15 cm long; c. 7 mm diam.) with fruits in dense whorls (verticillasters) **[8]**; short sections of stem often visible between whorls in lower part of spike, each whorl formed from 2 indistinctly opposite fruit clusters with a green leaf-like bract often still attached to the base of each cluster **[9]**; bracts becoming smaller towards apex of spike. Each fruit a group of 4 nutlets totally enclosed in a persistent, papery and softly hairy calyx **[10]** (c. 3 mm long; 1.2 mm diam.), greenish-grey or yellowish-green, fragile in texture and with 5 sepal teeth [hl]; nutlets dry, 3-angled, brownish-black and oblong or long-ovoid **[11]** (c. 1.5 mm long; 0.7 mm diam.). *Odour*: aromatic. *Taste*: faintly astringent, pungent and cool. ***Good quality***: pale green and strongly aromatic, with densely packed flower- and fruit-heads.

DECOCTING PIECES-1 (CUT AERIAL LENGTHS): as for crude drug but cut into short lengths and more fragmented.

DECOCTING PIECES-2 (CUT AND CHARRED AERIAL LENGTHS): as for DP-1 above but charred black externally **[12]** and yellow internally. *Odour*: slightly aromatic. *Taste*: bitter and pungent.

PROCESSING

Aerial parts are sun-dried (= crude drug) and subsequently cut into short decocting lengths (= decocting piece-1); the latter may subsequently be carbonised by dry-frying until charred black externally and charred yellow internally 'jing Jie Tan' (= decocting pieces-2).

OTHER COMMON NAMES

荆芥草 Jing Jie Cao; 裂叶荆芥 Lie Ye Jing Jie; 香荆芥 Xiang Jing Jie; Jing Jie Tan (carbonised Jing Jie); Herba Schizonepetae; fineleaf Schizonepeta herb (CP2010); schizonepeta (SCN).

SAFETY

Contraindicated in fully erupted measles or open sores; some reports of allergic reactions of the skin and digestive system reported (Bensky *et al.*, 2004).

Jing Jie (crude drug) sourced from *N. tenuifolia*

Jing Jie Tan (decocting pieces-2) sourced from *N. tenuifolia*

Jin Qian Cao 金钱草 Lysimachiae Herba

Drug parts: ALL

PHARMACOPOEIA SPECIES

Lysimachia christinae Hance

Primulaceae

taxonomic/nomenclatural note: in *FoC* as '*L. christiniae* Hance'.

Species of *Lysimachia* wild in China: 138

PLANT DESCRIPTION

A variable species as regards its hairiness and the size and shape of its leaves and calyx lobes. A perennial and prostrate herb (20–60 cm tall). *Leaves* opposite, simple and stalked; bracts present; leaf blades ovate, sub-orbicular or kidney-shaped (1.5–)6(–8) cm long; 1–4(–6) cm wide with **transparent glandular lines visible against the light with lines becoming black and distinct when dry; hairless to** densely hairy; base truncate to heart-shaped, tip shortly pointed, broad or rounded; margins entire and **hairless**. *Flowers* in axils of leaves, solitary, bisexual and radially symmetrical; bright yellow and conspicuous; flower stalks 1–5 cm long; calyx 5-parted with lobes lance-shaped to linear, hairless or hairy; corolla yellow, basally fused to form a short tube (2–4 mm long) and opening into **5 narrowly ovate to lance-shaped petal lobes** (5–11 mm long) with long black glandular lines; stamens 5; ovary superior; style 6–8 mm long. *Fruit* a dry sub-globose capsule (4–5 mm diam.) with black glandular lines and a persistent calyx; when ripe, capsule divides into 5 valves. Fl. May–Jul.

HARVESTING, SOURCING AND NATURAL RANGE

Plants are harvested in summer and autumn from wild populations which grow in damp areas along streamsides, open forests and forest margins (500–2,300 m altitude). The species is endemic to C, S and W China: Anhui, Fujian, Guangdong, Guangxi, Guizhou, Henan, Hubei, Hunan, Jiangsu, Jiangxi, Shaanxi, Sichuan, Yunnan and Zhejiang.

CONSERVATION STATUS

Endemic to China: Least Concern (authors' pers. obs, 2005); Not Evaluated (IUCN Red List, accessed 25/8/2016).

UNOFFICIAL SUBSTITUTE 1

L. congestiflora Hemsl.

Primulaceae

PLANT DESCRIPTION

Easily confused with the Pharmacopoeia species. *Main differences:* **flowers in groups of 2-4; petal lobes ovate-elliptic to oblong.**

CONSERVATION STATUS

China: Least Concern (authors' pers. obs. 2001).

Global: Not Evaluated (IUCN Red List, accessed 25/8/2016).

UNOFFICIAL SUBSTITUTE 2

Desmodium styracifolium (Osbeck) Merr.

Fabaceae

synonym: *Hedysarum styracifolium* Osbeck

PLANT DESCRIPTION

Easily distinguished from the official species. *Main differences:* resembles a small creeping sub-shrub (30–100 cm tall); **softly hairy throughout.** Leaves 1- (usually) or 3- (rarely) parted; 'leaf' blades **not glandular, lower surfaces and margins densely covered in silky white hairs;** lateral veins on 'leaf' undersides white and raised. Flowers densely clustered into conical heads at each node with flower stems densely covered in soft, silky hairs. Each flower laterally symmetrical (c. 4 mm diam.) with 5 **purplish-pink petals** of different sizes; lower petal (keel) longer than lateral petals, strongly curved with a long claw. Fruit a 3–6 jointed pod (1–2 cm long; c. 2.5 mm wide), covered in soft hairs. Fl. and fr. Jun–Sep.

CONSERVATION STATUS

China: Least Concern (MEPCAS, 2013); authors' pers. obs. 2010).

Global: Not Evaluated (IUCN Red List, accessed 25/6/2016).

TCM PROPERTIES AND USE

properties: **'sweet, salty and cool'**
channels: **'Bladder, Gall Bladder, Kidney and Liver'**
actions: **'regulates water, dispels damp heat, promotes urination'**

indications: **stranguria, urinary stones; jaundice; toxic swellings and sores**

Lysimachia christinae flowering in the wild

L. congestiflora flowering in the wild

Desmodium styracifolium in flower and fruit (cultivated as a medicinal crop — source of Guan Jin Qian Cao)

Lysimachia christinae

Primulaceae

DRUG MORPHOLOGY

CRUDE DRUG (DRIED AERIAL PARTS): a twisted mass, predominantly of fine stems and crumpled leaves **[1]**; all structures hairless, sparsely hairy or, occasionally, densely hairy. Stems twisted (1–3 mm diam.), externally brown or dark brownish-red, longitudinally striated **[2]**, fine rootlets occasional on nodes of lower parts of stems; fracture solid. Leaves usually crumpled and broken **[3]**; opposite **[4]**, simple and stalked **[5]** (stalks 1–4 cm long); leaf blades, when whole and expanded, ovate, sub-orbicular or kidney-shaped **[6]** (1.5–)6(–8) cm long; 1–4(–6) cm wide, upper surface greyish-green or dark brown, lower surface paler with **midrib distinctly prominent [7]; thin black or brown gland lines present on either surface [8];** hairless to densely hairy; base truncate to heart-shaped, tip shortly pointed, broad or rounded; margins entire and **hairless [9]**. Flowers sometimes present; solitary, on long slender stalks **[10]** (1–5 cm long) arising from a leaf axil **[11]** (usually opposite another flower that arises in the opposite leaf axil); sepals 5 **[12]**, petals 5, orangish **[13]** narrowly ovate to lance-shaped [hl] (5–11 mm long) **with long black glandular lines [14]**; stamens 5 [hl]. Fruits occasional, dry, long-stalked and sub-globose **[15]** (4–5 mm diam.) with black glandular lines and a persistent calyx **[16]**; when ripe, capsule divides into 5 valves **[17]** [hl]. *Odour:* indistinct. *Taste*: indistinct.

DECOCTING PIECES (DRIED SHORT LENGTHS OF AERIAL PARTS): as for crude drug but cut into short lengths; some leaf blade characters no longer visible due to fragmentation.

OTHER COMMON NAMES

过路黄 Guo Lu Huang (plant name); 大金钱草 Da Jin Qian Cao; 铜钱草 Tong Qian Cao; Christina loosestrife (CP2010); Herba Lysimachiae; lysimachia (SCN).

Crude drug of Jin Qian Cao (*Lysimachia christinae*)

Decocting pieces of Jin Qian Cao (*L. christinae*)

Flowering stem of *L. christinae*

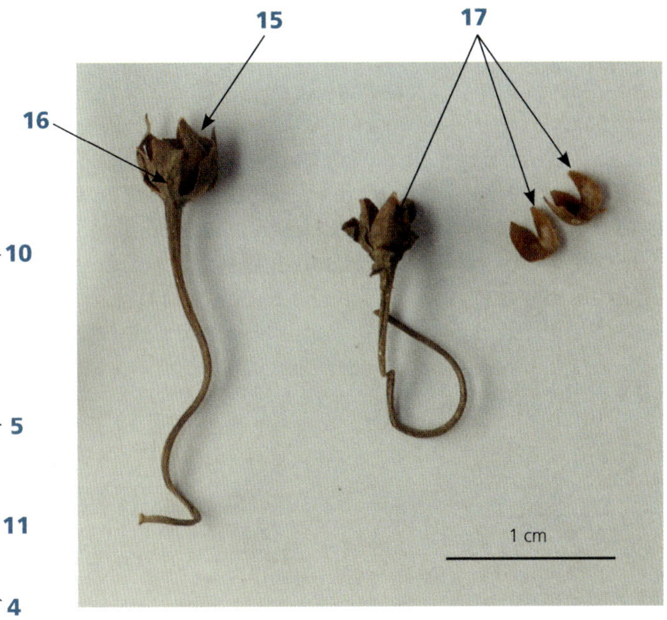

2 fruits (left and centre) and shed fruit valves (right)

Lysimachia congestiflora

Primulaceae

'DRUG SUBSTITUTE' MORPHOLOGY

Very similar and easily confused with the crude drug and decocting pieces of the Pharmacopoeia species. *Main differences:* leaves **without linear glands [hl] [18]** dark dot-like glands occasionally visible near leaf margins [hl]; **flowers** (when still attached to stems) **in groups of 2–4 [19]**; when expanded, petal lobes ovate-elliptic to oblong.

OTHER COMMON NAMES

临时救 Lin Shi Jiu (plant name); 聚花过路黄 Ju Hua Guo Lu Huang.

> SUBSTITUTION COMMENT. A widespread, common and very similar species to the Pharmacopoeia species and with which it is easily confused both at point of harvest and in TCM herbal trade. It is used as a local folk medicine in China.
>
> The species grows in damp places such as ditches, roadsides, margins of damp forests and rice paddies (200–1,200 m altitude) in C, S and W China: Anhui, Fujian, S Gansu, Guangdong, Guangxi, Guizhou, Hainan, Hubei, Hunan, Jiangsu, Jiangxi, Qinghai, S Shaanxi, Sichuan, Taiwan, Tibet, Yunnan and Zhejiang [also native to Bhutan, NE India, Myanmar, Nepal, Sikkim, Thailand and Vietnam].

Desmodium styracifolium

Fabaceae

'DRUG SUBSTITUTE' MORPHOLOGY

Easily distinguished from the Pharmacopoeia drug (both as crude drug and decocting pieces). *Main differences:* much hairier throughout; **'leaves' larger** (2–4.5 cm long; 2–4.5 cm diam.) and more obviously sub-orbicular **[20]** with upper surfaces hairless and **lower surfaces [21] and margins [22] densely covered in soft greyish-white hairs;** 'leaf' blades without glands; **lateral veins on blade undersides conspicuous (curving, white, raised and parallel) [23]**; flowers occasional and (when still attached) in clusters along a central flower axis **[24]**; each flower c. 4 mm diam., petals 5 and dark purplish **[25]**; fruits 3–6-jointed pods **[26]** (rarely present) (1–2 cm long; c. 2.5 mm wide), covered in soft hairs.

OTHER COMMON NAMES

广金钱草 Guang Jin Qian Cao; 铜钱草 Tong Qian Cao; coin-leaf desmodium (SCN); Herba Desmodii Styracifolii; Desmodii Styracifolii Herba.

> SUBSTITUTION COMMENT. This species is the source of the separate TCM drug Guang Jin Qian Cao 广金钱草 (Desmodii Styracifolii Herba), see page 352, and is a frequent substitute of 'Jin Qian Cao'. This is probably due to the similar clinical attributes of the two drugs, however, the common abbreviation of the Chinese names of both drugs to 'Qian Cao' may be a source of confusion in trade for those unfamiliar with the 2 species despite the fact that their trade forms are easily distinguished macroscopically.
>
> 'Guang Jin Qian Cao' is readily available from medicinal cultivation (unlike 'Jin Qian Cao' which is wild-harvested) with 'Guang' indicating its main sourcing from the south China provinces of Guangdong and Guangxi. See also *Harvesting, Sourcing and Natural Range* in the account for 'Guang Jin Qian Cao'.

Crude drug of *L. congestiflora*

Decocting pieces of *L. congestiflora*

Dried aerial parts of *D. styracifolium* (Guang Jin Qian Cao)

Ku Di Ding 苦地丁 Corydalis Bungeanae Herba

Drug part: WHOLE PLANT

Corydalis bungeana Turcz.

Papaveraceae

PLANT DESCRIPTION

A soft, clump-forming annual or biennial herb (10–20 (rarely up to 50) cm tall) with a slender pale yellowish-brown tap root. *Stems* branching, spreading and 5-ridged. *Leaves* basal and along stem, greyish-green and stalked; 2-pinnate with many leaflets deeply divided into 3-10 short, narrowly obtuse segments. *Flowers* on upper parts of stems in loose, 6–10-flowered heads (racemes). Each flower bisexual and laterally symmetrical (11–15 mm long) with leaf-like bracts at base (c.3–7 mm long); sepals 2, petal-like, deltoid and toothed, usually falling; petals 4, multi-coloured (pink, mauve, white and purple) with upper and lower petals shortly crested and enclosing 2 inner, dark-tipped ones fused at their apex; upper petal with a curved spur at base (4–5 mm long); lower petal with a lip; stamens 2; ovary superior. *Fruit* a dry, oblong or elliptical capsule (12–20 mm long; c. 4 mm wide), compressed and pendant, tip pointed. *Seeds* 8–15 in 2 rows, sharply keeled, black and glossy when mature. Fl. Apr-May.

Species of *Corydalis* wild in China: 357

HARVESTING, SOURCING AND NATURAL RANGE

Sourced from medicinal crop plants grown in Hebei province (Anguo district). Whole plants are harvested when in flower or fruit in summer and autumn. The wild species grows in open, stony and sandy places, often on slopes (up to 1,500 m altitude) and has a wide distribution in northern and eastern China (Gansu, Hebei, Henan, Hunan, Inner Mongolia, Jiangsu, Jilin, Liaoning, Ningxia, Shaanxi, Shandong and Shanxi; also native in the Korean peninsula, SE Mongolia and Russia (Far East).

CONSERVATION STATUS

China: Least Concern (MEPCAS, 2013).
Global: Not Evaluated (IUCN Red List, accessed 16/8/2016).

C. bungeana with leaf-like bracts, flowers (lower-most opened to show papillate stigma) and a ripening fruit (with persistent stigma)

Typical clump-forming habit of *Corydalis bungeana* in full flower

Freshly harvested *C. bungeana* with pale yellowish-brown tap roots

EHTPA 4/14
GAP

TCM PROPERTIES AND USE

***properties*: 'bitter and cold'**
***channels*: 'Hear, Liver and Large Intestine'**
***actions*: 'clears heat, eliminates toxins'**

***indications*: respiratory tract infection; sore throat; sores, abscesses, swelling and pain; (T)**

DRUG MORPHOLOGY

Crumpled greyish-green or yellowish-green masses (10–30 cm long) mostly of leaves and stems with some flowers, fruits and roots. *Tap root* conical and yellowish-brown **[1]**. *Stems* slender, with 5 longitudinal ridges **[2]**, frequently branched and hollow. *Leaves* mostly crumpled, broken and dark green or greyish-green **[3]**, when whole and expanded, stalked **[4]** blades 2-pinnate **[5]**, with many leaflets deeply divided into 3–10 short, narrowly obtuse segments **[6]**. *Flowers* (when still attached) in loose, 6–10-flowered heads (racemes) **[7]**; each flower **[8]** (bisexual and laterally symmetrical (11–15 mm long); sepals petal-like, usually fallen off; petals 4 (when fresh pink, mauve, white and purple) with upper and lower petals **[9]** shortly crested and enclosing 2 inner, dark-tipped ones **[10]** fused at their apex; upper petal with a curved spur at base **[11]** (4–5 mm long) but seldom discernible in trade material; lower petal with a lip **[12]**; stamens 2; ovary superior. *Fruit* a dry, oblong or elliptical capsule **[13]** (12–20 mm long; c. 4 mm wide), compressed and pendant, tip pointed. *Seeds* 8–15 in 2 rows **[14]**, sharply keeled, black and glossy when mature. Fl. Apr–May. *Texture (stems):* soft and hollow. *Odour:* indistinct. *Taste:* strongly bitter.

DECOCTING PIECES (CUT LENGTHS OF WHOLE PLANT): as for the crude drug but cut into short lengths; many leaf and flower characters lost or obscured due to fragmentation. The distinctive fruits and seeds (*see above*) provide the most useful identification characters.

PROCESSING

Whole herbs are sun-dried (= crude drug), then later cut into decocting pieces.

OTHER COMMON NAMES

紫花地丁 Zi Hua Di Ding; Herba Corydalis Bungeanae.

> **COMMENT.** 'Ku Di Ding' is often reported as a substitute or adulterant of the separate TCM herb 'Zi Hua Di Ding' (Violae Herba) see page 412.

Expanded 2-pinnate leaf of Ku Di Ding (*C. bungeana*)

Crude drug of Ku Di Ding (*C. bungeana*)

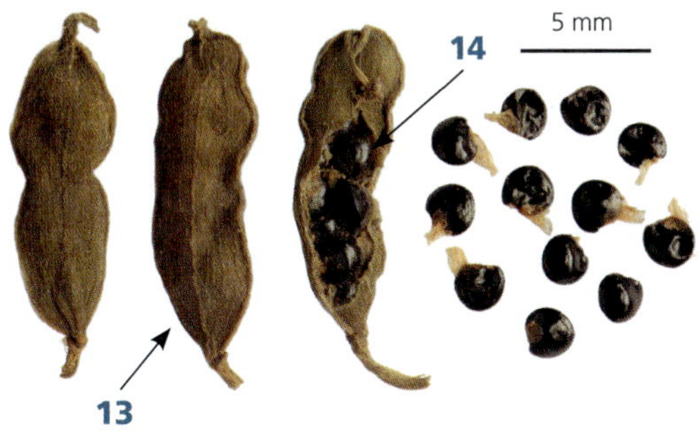

Fruits and seed

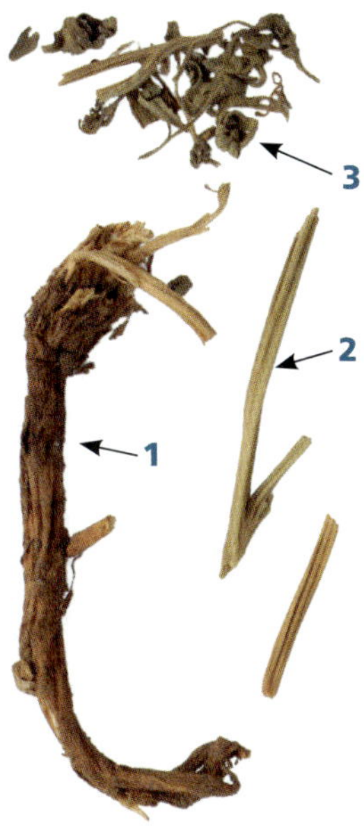

Stems

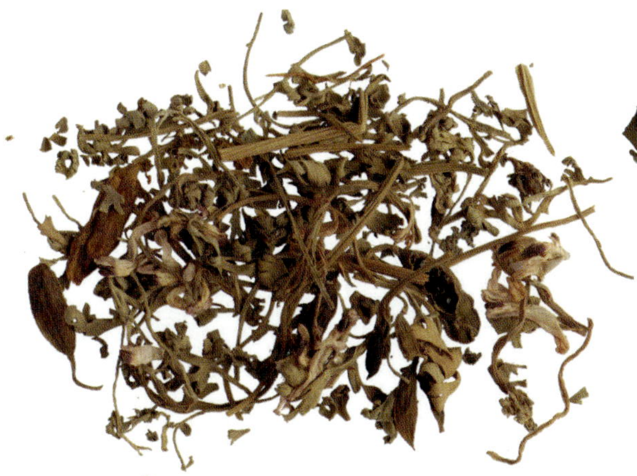
Decocting pieces of Ku Di Ding (*C. bungeana*)

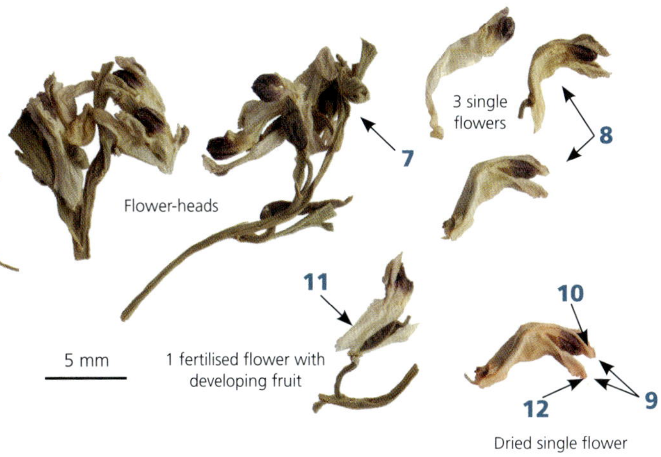

Flowers

Ma Bian Cao 马鞭草

Verbenae Herba

Drug part: AERIAL

Verbena officinalis L.

Verbenaceae

Species of *Verbena* wild in China: 1

PLANT DESCRIPTION

An erect, annual or short-lived perennial herb (30–140 cm tall) with a lax, somewhat untidy and branching growth; softly hairy throughout. *Stem* 4-sided. *Leaves* opposite and simple; most leaves with blades ovate or oblong in outline with margins coarsely toothed or often deeply divided or 3-lobed; upper stems leaves ovate, unlobed and with margins entire. *Flowers* in long and slender terminal or lateral spikes, basal flowers opening first. Each flower tiny, subtended by a green, scale-like bract (sometimes fallen off) and largely enclosed by a tiny green and glandular 5-parted calyx; petals pink or blue, fused basally into a tube which opens into 5 spreading unequal-sized lobes (c. 4–8 mm diam. overall); stamens 4. *Fruit* a dry, oblong capsule (c. 2 mm long) with four 1-seeded nutlets, partly enclosed by a persistent calyx. Fl. and fr. Jul–Oct.

HARVESTING, SOURCING AND NATURAL RANGE

Aerial parts are harvested June to August from wild populations found in open grassy places in mountains (100–1800 m altitude) in many provinces of China (except in the north): Anhui, Fujian, Gansu, Guangdong, Guangxi, Guizhou, Hainan, Hubei, Hunan, Jiangsu, Jiangxi, Shaanxi, Shanxi, Sichuan, Taiwan, Tibet, Xinjiang, Yunnan and Zhejiang.

CONSERVATION STATUS

A widespread weed of temperate and tropical zones world-wide.

China: Least Concern (authors' pers. obs. 2014).

Global: Not Evaluated (IUCN Red List; accessed 23/8/2016).

Verbena officinalis growing wild in Gansu province, China

V. officinalis in flower

Close-up of flowers of *V. officinalis*

TCM PROPERTIES AND USE

properties: **'bitter and cool'**
channels: **'Liver and Stomach'**
actions: **'invigorates Blood, dispels stasis, clears heat, eliminates toxins'**

indications: **amenorrhoea and dysmenorrhoea; malaria; diphtheria; boils and sores**

DRUG DESCRIPTION

CRUDE DRUG (DRIED AERIAL PARTS): mostly of stems with crumpled leaves and some flowering or fruiting spikes; covered throughout in very fine hairs. Stems with or without pith, 0.5–1 m long, with opposite branches **[1]**, greyish-green, slightly rough, 4-sided **[2]** and with many longitudinal furrows **[3]**. *Stem texture*: hard and brittle. Leaves opposite **[4]**; greenish-brown, coarsely hairy especially along veins on underside of leaf blades **[5]**; most leaves, when unbroken and expanded, ovate to oblong in outline (2–8 cm long; 1–5 cm wide), often 3-lobed **[6]** with margins strongly toothed **[7]** except upper leaves which are smaller, un-lobed and with margins entire **[8]**. Flower spikes long and slender **[9]**, with many, small, stalkless flowers arranged alternately along the stem **[10]**; corolla usually shrivelled but calyx persistent (1–4 mm long) and often with green scale-like bract at base **[11]**. Fruits dry, oblong capsules (c. 2 mm long) with four 1-seeded nutlets partly enclosed by the persistent calyx. *Odour*: indistinct. *Taste*: bitter. ***Good quality***: green and devoid of roots.

DECOCTING PIECES (CUT LENGTHS OF AERIAL PARTS): as for crude drug but cut into 5–15 mm lengths.

PROCESSING

Aerial parts are sun-dried (= crude drug), then washed and briefly softened before cutting into short lengths and sun-drying again (= decocting pieces).

OTHER COMMON NAMES

铁马鞭 Tie Ma Bian; common vervain; European vervain (SCN), as '*V. officinalis* ssp. *officinalis*'); Herba Verbenae; verbena herb. Many other trade names exist for this drug reflecting its medicinal use in other countries (see *Kew-MPNS*).

SAFETY

Safety Class 2b; Interaction Class A (Gardner & McGuffin, 2013).

Crude drug of Ma Bian Cao

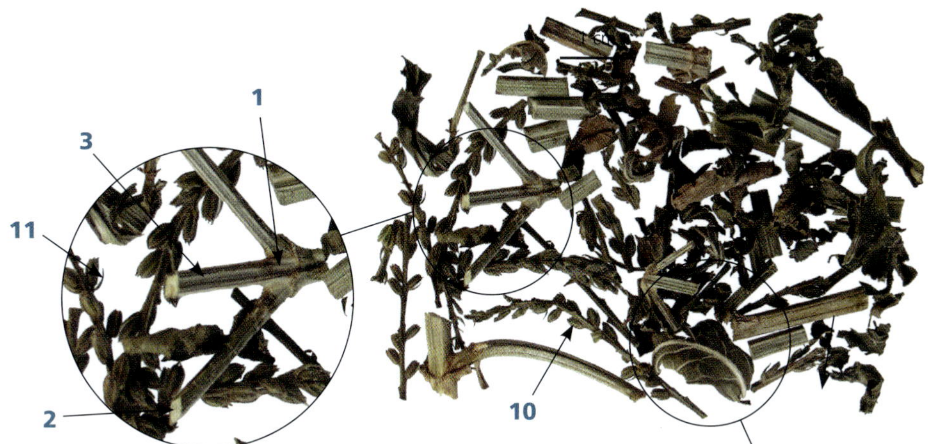

Decocting pieces of Ma Bian Cao

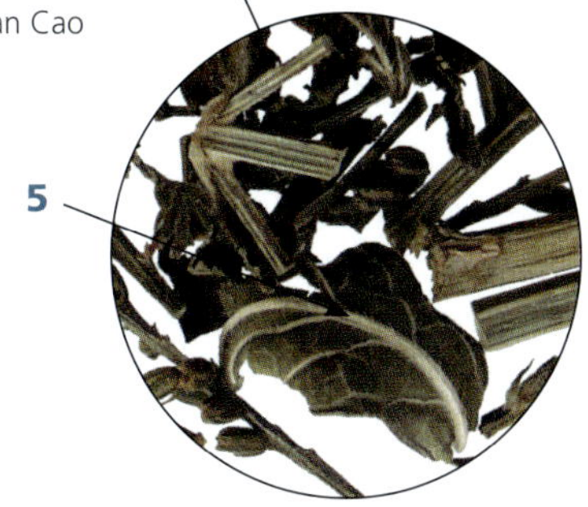

Ma Chi Xian 马齿苋

Portulacae Herba

Drug parts: AERIAL

Portulaca oleracea L.

Portulacaceae

Species of *Portulaca* wild in China: **5**

PLANT DESCRIPTION

A prostrate, branching annual herb (10–30 cm tall). *Stems* fleshy and often reddish. *Leaves* alternate (except below flowers where they are whorled), simple, fleshy and virtually stalk-less; a few inconspicuous stiff bristles often present in leaf axils; leaf blades ovate (10–30 mm long; 5–15 mm wide) and broadest at the tips which are rounded or truncate; margins entire. *Flowers* small but conspicuous **(bright yellow)** in terminal **clusters of 2-5** (often opening one at a time per cluster). Each flower radially symmetrical (0.4–0.5 cm diam.), stalkless and bisexual; sepals 2, green and fused; petals 5 (short-lived), free, equal, yellow and spreading, sometimes slightly notched apically; stamens 7–12; ovary half-inferior; style 1 with a 2–9-lobed stigma. *Fruit* a dry, ovoid and thin-walled capsule, **with an obvious ridge around the circumference.** *Seeds* many, black, orbicular or kidney-shaped. Fl. May–Aug, fr. Jun–Sep.

HARVESTING, SOURCING AND NATURAL RANGE

Aerial parts are mostly wild-harvested in summer and autumn. The species grows throughout China where it is a common weed of cultivation and dsiturbed sites, especially in urban areas. It is also native to tropical and temperate regions worldwide.

CONSERVATION STATUS

China: Least Concern (authors' pers. obs., 2001).
Global: Not Evaluated (IUCN Red List, accessed 20/8/2106)

Bacopa monnieri (L.) Wettst.

Plantaginaceae

synonym: *Lysimachia monnieri* L.
taxonomic/nomenclatural note: in *FoC* and elsewhere as '*B. monnieri* (L.) Pennell'.

PLANT DESCRIPTION

Superficially similar but botanically unrelated to the Pharmacopoeia species. Dried trade material is more challenging to identify compared to the living plants; the presence of fruits in the former is likely to prove the most useful identification aid.

Main differences from the Pharmacopoeia species: **leaves opposite**, tips rounded; **flowers solitary** on a long slender stalk (0.5–3.5 cm. long) as long as or longer than the leaves; fruits solitary, on a slender stalk and **enveloped in a persistent 5-parted papery calyx; narrowly ovoid, with a pointed tip and without an obvious ridge around the circumference.**

SUBSTITUTION COMMENT. A common and native species of southern China confined to the tropical and subtropical parts of Fujian, Guangdong, Guangxi, Hainan, Taiwan and Yunnan and grows by water or in wet and often sandy places. Easily harvested in error from the wild in mistake for the Pharmacopoeia species especially when not in flower (see *Plant Description*). One of its Chinese names (假马齿苋 Jia Ma Chi Xian 'false Ma Chi Xian') suggests its mistaken identification is well-known. See also *Safety*.

CONSERVATION STATUS

A pan-tropical and sub-tropical weed.
China: Least Concern (authors' pers. obs.).
Global: Least Concern (IUCN Red List, accessed 20/8/2016).

TCM PROPERTIES AND USE

properties: 'sour and cold'
channels: 'Liver and Large Intestine'
actions: 'clears heat, eliminates toxins, cools Blood, stops bleeding'
indications: diarrhoea, haematochezia; haemorrhoids; dysuria; sores and abscesses; (T)

Portulaca oleracea, in clusters of 2–5 with reddish stems and yellow flowers

Non-flowering plants of *Bacopa monnieri*

Solitary flower (white, blue or purple) of *B. monnieri* PHOTO: R. FANG

Portulaca oleracea

Portulacaceae

DRUG MORPHOLOGY

CRUDE DRUG (STEAMED/BLANCHED AERIAL PARTS): a crumpled mass of greyish-green leaves **[1]**, paler, slender stems **[2]**, fruits **[3]** and, very occasionally, flowers. *Stems* cylindrical (up to 30 cm long; 1–2 mm diam.) and longitudinally wrinkled**.** *Leaves* hairless, brittle, slightly fleshy, virtually stalkless **[4]** and often turning blackish when dry; when intact and expanded leaf blades simple and ovate (10–30 mm long; 5–15 mm wide), broadest towards the tips with tips rounded or truncate **[5]** and margins entire. Flowers usually shrivelled and difficult to decipher; stalkless; sepals 2. ***Fruits* often in groups of 2–5 [6]** and (when intact) brown, membranous and **broadly ovoid [7] (c. 5 mm. long) with a distinct circumscissile ridge [8] halfway along its length which splits at maturity; the distinctive goblet-shaped 'lids' [9] of dehisced fruits often scattered throughout a sample.** *Seeds* many, orbicular or kidney-shaped seed (0.6–1.2 mm long) glossy and black when mature. *Odour:* indistinct. *Taste:* slightly sour.

DECOCTING PIECES (AERIAL LENGTHS): as for the crude drug but cut into short sections 5–15 mm long; some leaf characters no longer be discernible; fruit characters provide those most useful for identification *(see numbered annotations above for the photographs opposite).*

PROCESSING

Aerial parts are blanched or briefly steamed, then sun-dried (= crude drug); they are subsequently softened with water and cut into 5–10 mm or 10–15 mm sections before sun-drying again (= decocting pieces).

OTHER COMMON NAMES

长寿菜 Chang Shou Cai 'long-life plant'; 耐旱菜 Nai Han Cai 'drought-surviving plant'; Herba Portulacae; purslane. Many other trade names exist for this drug reflecting its medicinal use in other countries, notably India in Ayurvedic and Unani medicine.

SAFETY

Safety Class: 2b; Interaction Class: A (Gardner & McGuffin, 2013). An official healthfood in China (Anon., 2002a,b).

Bacopa monnieri

Plantaginaceae

'DRUG SUBSTITUTE' MORPHOLOGY

Crude drug and decocting pieces superficially very similar to those of the Pharmacopoeia species but can be reliably distinguished using macroscopic characters. ***Leaves* (if still attached) obviously opposite [10]**, tips of leaf blades rounded. *Flowers* often discernible **[11],** solitary, on long, slender stalks (0.5-3.5 cm. long), sepals 5 and narrowly ovate. ***Fruits* [12]** on long slender stalks **always solitary, narrowly ovoid with a pointed tip and almost completely enveloped in a persistent papery calyx [13]**; an obvious ridge around the circumference **absent [hl].**

OTHER COMMON NAMES

假马齿苋 Jia Ma Chi Xian 'false ma chi xian'; brahmi; mandukaparni.

SAFETY

B. monnieri is a widespread and popular herb in India where it is used for many medical conditions e.g. skin complaints, fevers and as a nervine (Aguilar, 2001). Gardner & McGuffin (2013) classify this species as 'Safety Class: 1' and 'Interaction Class: A'; however its pharmacological profile may be quite different from that of the Pharmacopoeia species described here such that its inadvertent use may present a safety concern when used in a TCM context.

A separate concern is the potential of this species to absorb heavy metals from contaminated soils and groundwater (PROSEA, 1999).

See also *Substitution Comment.*

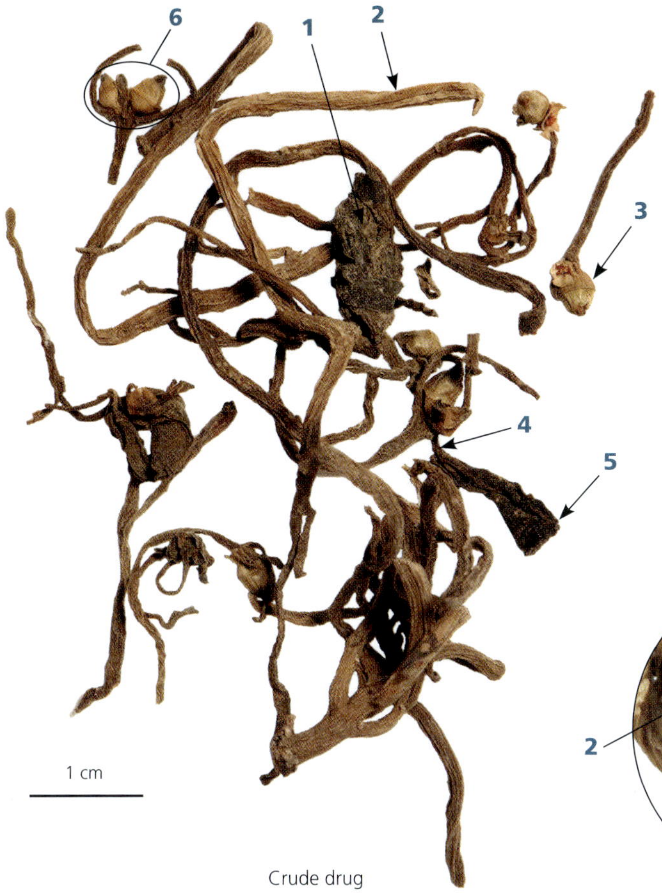

Crude drug

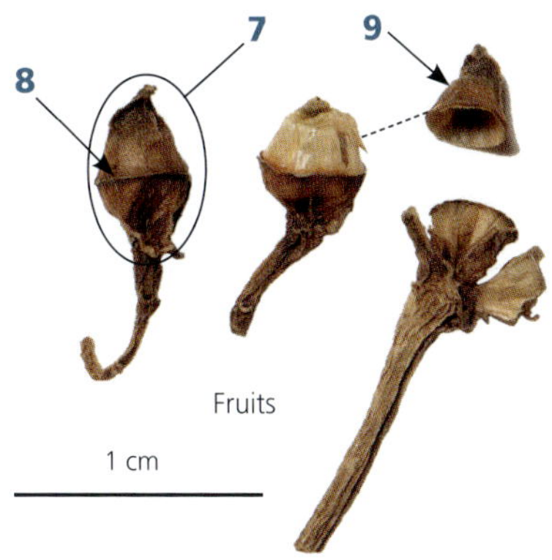

Fruits

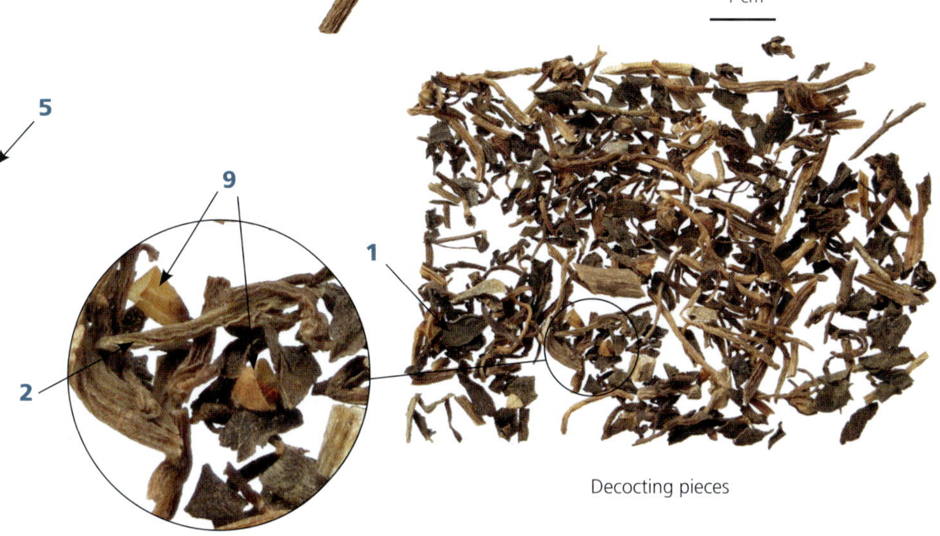

Decocting pieces

Dried aerial parts of *Bacopa monnieri* (crude)

Fruits

Decocting pieces

Mo Han Lian 墨旱莲 (or Han Lian Cao 旱莲草) Ecliptae Herba

Drug parts: AERIAL

Eclipta prostrata (L.) L.

Asteraceae

synonyms: *E. alba* (L.) Hassk.; *Verbesina prostrata* L.

taxonomic/nomenclatural note: in CP2015 as '*E. prostrata* L.'

Species of *Eclipta* wild in China: 1

PLANT DESCRIPTION

An annual herb up to 60 (rarely 100) cm tall with a straggly habit and mostly covered in short, stiff white hairs. *Stems* cylindrical, 4-sided or flattened; reddish-pink and branching; releasing white sap when broken and which turns black on contact with the air. *Leaves* opposite and lance-shaped (3–10 cm long; 0.5–2.5 cm wide) with **both surfaces covered in stiff white whites**; margins entire or minutely toothed. Daisy-like, white hemispherical 'flowers' (flower-heads/capitula), c. 6 mm diam., comprise 5 or 6 leaf-like bracts (often tipped reddish-pink) beneath a ring of white, strap-like 'petals' (ray florets), 2.5–3 mm long, which in turn surround a hub of very short, tubular and densely packed flowers (disk florets) [hl]. *Fruits* inconspicuous, tiny and 1-seeded.

HARVESTING, SOURCING AND NATURAL RANGE

Aerial parts are wild-harvested when plants are in flower. The species grows in disturbed, damp places and abandoned ponds often near habitation and roadsides (up to 1,600 m altitude). A common and widespread species in throughout much of China; [also native to Central, North, and South America; introduced in Africa, Asia, Australia, Europe, and Pacific islands].

CONSERVATION STATUS

China: Least Concern (authors' pers. obs., 2011).
Global: Data Deficient (IUCN Red List, accessed 9/11/2016).

Typical habitat of *Eclipta prostrata* growing along the edge of a rice paddy

E. prostrata in the wild

Flowering plants of *E. prostrata*

UNOFFICIAL SUBSTITUTES

Aerial parts of species in other families have been reported as local unofficial substitutes or adulterants of 'Mo Han Lian' in China (NICBP, 1997–2002; Bensky *et al.*, 2004; Ho *et al.*, 2006). Those below are very easily distinguished from the Pharmacopeia species and their use as substitutes is likely to have arisen due to some shared clinical properties (Perry, 1980). Their occurrence as 'Mo Han Lian' substitutes in international trade is unlikely.

1. *Hypericum ascyron* L. (紅旱蓮 Hong Han Lian; giant St John's wort), family Hypericaceae. Used locally as a medicine and for many other uses in China. **Much taller** (0.5–1.3(–2) m); **flowers vivid yellow** and much larger 3–7(–8) cm diam.
2. *Potentilla supina* L. (朝天委陵菜 Chao Tian Wei Ling Cai), family Rosaceae. A locally common native throughout much of China, with many medicinal and other uses. **Leaves 3-foliolate or pinnate** with 2–5 pairs of leaflets and not covered in long white hairs; flowers yellow, 5-petalled.

TCM PROPERTIES AND USE

***properties*:** **'sweet, sour and cold'**
***channels*:** **'Kidney and Liver'**
***actions*:** **'tonifies yin, stops bleeding, clears heat'**

***indications*:** **dizziness, tinnitus, vertigo; musculoskeletal pain; traumatic bleeding; epistaxis, haematuria, haematemesis**

DRUG MORPHOLOGY

CRUDE DRUG (AERIAL PARTS): a crumpled mass of stems, leaves, flower- and fruit-heads. Most parts, **covered in short, stiff white hairs [1]**. *Stems* cylindrical (c. 2–5 mm diam.) **[2]**, greenish-brown or dark green, branching, sometimes longitudinally wrinkled. *Leaves* mostly fragmented, greyish-green, with short, stiff white hairs on both surfaces **[3]**, margins entire or shallowly toothed **[4]**; when still attached and intact, leaves almost stalkless, blades lance-shaped. *Flower-heads* (capitula) (2–6 mm. diam.) with 5 or 6 green phyllaries surrounding a dense head of tiny flowers (florets); *fruiting-heads* larger, with, phyllaries persistent and membranous **[5]** surrounding a dense low-domed head **[6]** of 20 or more fruits (achenes); each achene **[7]** brown or golden-brown, sub-oblong and variously-sized (2–3 mm long; 1–1.5 mm diam.), sometimes [hl] 3-sided **[8]**, smooth or warty, with a minute bristle-like tuft (pappus) at one end **[9]**. *Stem fracture:* brittle; pale yellowish-white, sometimes with a white pith or hollow. *Odour:* indistinct. *Taste:* indistinct, slightly salty. ***Good quality:*** dry, green and pungent.

DECOCTING PIECES (SHORT LENGHTS OF DRIED AERIAL PARTS): as for the crude drug; some characters lost or obscured due to fragmentation.

PROCESSING

Aerial parts are sun-dried (= crude drug) then later cut into 5–10 or 10–15 mm long decocting sections.

OTHER COMMON NAMES

黑旱莲 Hei Han Lian 'black Han Lian' (commonly used to differentiate the official drug from a substitute with a similar name: 紅旱莲 Hong Han Lian (see *Unofficial Substitutes*); Herba Ecliptae. Many other trade names exist for this drug reflecting its medicinal use in other countries (notably in Ayurveda) as well as for other purposes (e.g. as a dye) (see *Kew-MPNS*).

SAFETY

An official healthfood in China (Anon., 2002a,b). Safety Class: 1; Interaction Class: A (Gardner & McGuffin, 2013). See also *Other Common Names*.

Crude drug of Mo Han Lian (*Eclipta prostrata*)

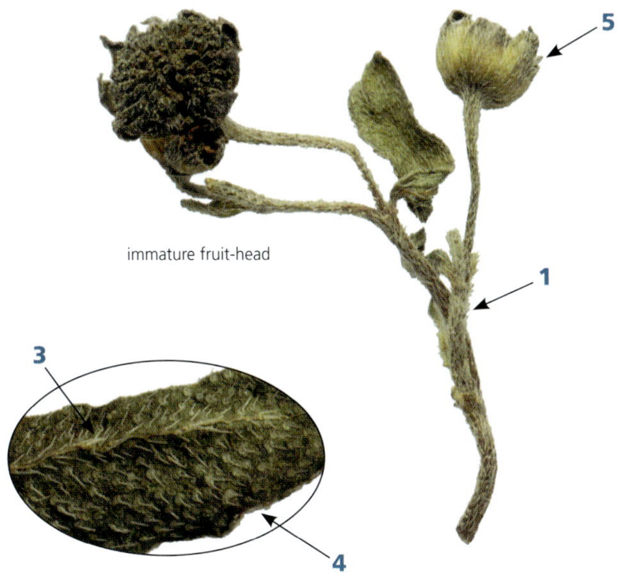

immature fruit-head

underside of broken leaf

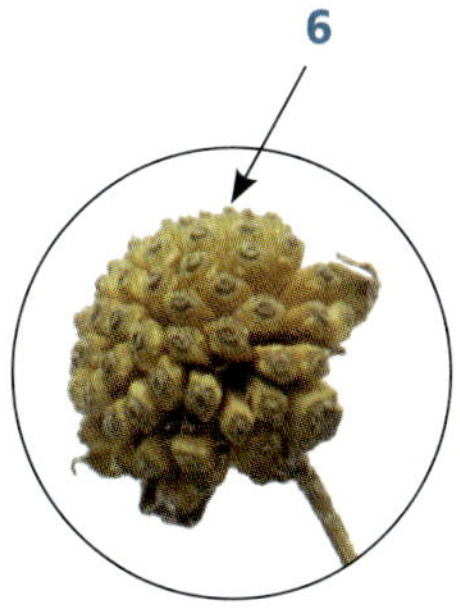

mature fruit-head

dehisced fruit-head

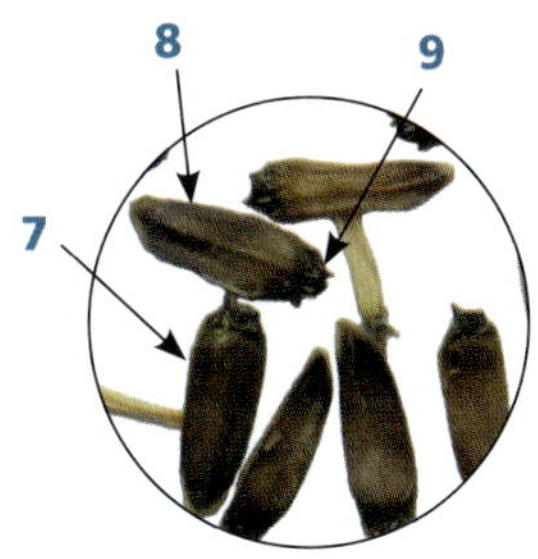

achenes

Pei Lan 佩兰

Eupatorii Herba

Drug parts: AERIAL

Eupatorium fortunei Turcz.

Asteraceae

Species of *Eupatorium* wild in China: 14

PLANT DESCRIPTION

An erect, herbaceous perennial (40–100 cm tall), highly aromatic when crushed; rhizomes reddish-brown and almost creeping. *Stems* cylindrical, few-branched, green or purplish. *Stem leaves* opposite, short-stalked and hairless; upper and lower stem leaves simple, blades linear or ovate; leaves distinctly 3-lobed with middle lobe much larger (5–10 cm long; 1.5–2.5 cm wide) than lateral lobes; margins irregularly or indistinctly toothed, either coarsely or shallowly. Flowerheads many-branched, lax (3–6 (rarely up to 10) cm diam.) at top of plant. Each flowerhead (capitulum) with a few tiny white, pink or reddish flowers (florets); outer part of each capitulum with overlapping greenish or purplish-red, scale-like phyllaries (c. 7 mm long); outer phyllaries shorter than middle ones; inner phyllaries narrowly elliptic; hairs and glands absent; each floret bisexual, narrowly funnel-shaped and 5-lobed; stamens 5 (hidden inside floret); style 1 with 2 conspicuous long, curving, thread-like lobes emerging to about same length as floret. *Fruits* (achenes) dry, blackish-brown and 5-angled (3–4 mm long); apex with white hairs (pappus) (c. 5 mm long). Fl. and fr. Jul–Nov.

HARVESTING, SOURCING AND NATURAL RANGE

Aerial parts are harvested in summer and autumn mostly from medicinal crop plants grown in the Chinese provinces of Jiangsu, Hebei, Anhui and Shandong. The species grows wild in thickets and roadside ditches (c. 2,000 m altitude) in C, S and W China: ?Anhui, ?Fujian, Guangdong, Guangxi, Guizhou, Hainan, Henan, Hubei, Hunan, Jiangsu, Jiangxi, Shaanxi, Shandong, Sichuan, Yunnan and Zhejiang provinces [possibly introduced in Japan, the Korean Peninsula, Thailand and Vietnam].

CONSERVATION STATUS

China: Least Concern (MEPCAS, 2013).
Global: Not Evaluated (IUCN Red List, accessed 26/8/2016).

Medicinal crop of *Eupatorium fortunei* beginning to flower

Mature inflorescence of *E. fortunei*

UNOFFICIAL SUBSTITUTES

This drug 'Pei Lan' is frequently confused with another TCM drug 'Ze Lan' (Lycopi Herba); see *Substitution Comment* about *Eupatorium fortunei on* page 408.

Lesser Known Unofficial Substitutes:

Aerial parts of other species in the same family (Lamiaceae) have been reported as local or historical substitutes in China; their occurrence as such in international trade is uncertain:

1. ***Eupatorium japonicum*** Thunb. (大叶佩兰 Dan Ye Pei Lan; 红泽兰 Hong Ze Lan). Native to China (18 provinces). Leaves rough.
2. ***E. lindleyanum*** DC; (轮叶泽兰 Lun Ye Ze Lan; 野马追 Ye Ma Zhui). Native to China (throughout, except Xinjiang). Leaves rough on both surfaces.
3. ***E. heterophyllum*** DC. Native to central and western China (9 provinces). Leaves softly hairy on both surfaces.
4. ***Ocimum basilicum*** L. var. ***pilosum*** (Willd.) Benth. Usually cultivated (13 provinces). Stems square; flowers in whorls at intervals along upper stem.

Underside of 3-lobed middle-stem leaf

TCM PROPERTIES AND USE

properties: **'pungent and neutral'**
channels: **'Spleen, Stomach and Lung'**
action: **'dispels damp heat, clears summerheat'**

indications: **abdominal distension; poor appetite; chest oppression; nausea, vomiting**

DRUG MORPHOLOGY

CRUDE DRUG (AERIAL PARTS): *stems* few-branched, cylindrical (30–100 cm long; 2–5 mm diam.), green or purplish, distinctly ridged longitudinally **[1]**; nodes slightly swollen **[2]**, sometimes purplish, virtually hairless. *Leaves* opposite, stalked (0.7–1 cm long) **[3]**, greenish-brown and hairless; stem leaves crumpled and often broken; when whole, either 3-lobed or undivided (blades and lobes linear or ovate); 3-lobed leaves, when intact, with middle lobe largest (5–10 cm long; 1.5–2.5 cm wide), narrowly elliptic, elliptic to lance-shaped or broader above the middle (5–10 cm long; 1.5–2.5 cm wide), tip pointed; lateral lobes as for middle lobe but smaller; margins irregularly or indistinctly toothed, either coarsely or shallowly **[4]**. *Flowers and fruits* usually absent. *Texture*: light and fragile; stem fracture white or pith hollowed **[5]**. *Odour*: aromatic. *Taste*: slightly bitter. ***Good quality***: leaves many, intensely aromatic; flowers (if present) still in bud.

DECOCTING PIECES (DRIED AERIAL LENGTHS): as for crude drug but cut into short sections with many key characters (notably those of leaves) often lost due to fragmentation.

PROCESSING

Aerial parts are sun-dried (= crude drug) then later, softened briefly, before cutting into decocting lengths.

OTHER COMMON NAMES

大泽兰 Da Ze Lan; 山泽兰 Shan Ze Lan; Chinese eupatorium (SCN); Fortune eupatorium herb (CP2010); Herba Eupatorii.

SAFETY

An official healthfood in China (Anon., 2002a. b). Legal restrictions and recommendations exist in many countries about the internal use of pyrrolizidine alkaloid-containing plant species (such as *Eupatorium* species), because long-term internal use has proven to be hepatotoxic and carcinogenic. (Stewart & Steenkamp, 2001). Such restrictions and advice are currently in a state of flux worldwide as new research is awaited (EMA, 2014; HMPC, 2016).

Crude drug of Pei Lan (*Eupatorium fortunei*)

Decocting pieces of Pei Lan (*E. fortunei*)

Pu Gong Ying 蒲公英

Taraxaci Herba

Drug parts: WHOLE PLANT

The definition of 'Pu Gong Ying' in the CP2015 is ambiguous and states the source species of this drug to be "*Taraxacum mongolicum* Hand.-Mazz. or *T. borealisinense* Kitam. or several other species of the same genus". Here, the two CP2015 named species are described. The genus *Taraxacum* requires considerable expertise to identify accurately to species level.

Taraxacum mongolicum Hand.-Mazz.

Asteraceae

Species of *Taraxacum* wild in China: 113

PLANT DESCRIPTION

A perennial herb (8–25 cm tall). *Leaves* in a basal rosette; leaf blades broadening towards rounded tips, pinnately lobed (6–15 cm long; **2–3.5 cm wide**); lobes broadly triangular and cut almost to the mid-vein, often slightly backward-pointing; margins entire to sparsely toothed. *Flowering stems* 1 to many, arising from base and slightly taller than leaves; unbranched, hollow, leafless and upper part (directly below flower-head) with **dense greyish-white woolly hairs**. *Flower-heads* (*capitula*) solitary (3–4 cm diam.). Phyllaries (enclosing flower-head) in two whorls; linear, greenish and leaf-like; outer phyllaries 9–15, flushed purplish-pink and c. half the length of the inner phyllaries; **tips with a greenish-purplish horn**, fringed with hairs; inner phyllaries (1.1–1.5 cm long) with flat or horned tips. *Flowers (florets)* bright yellow in a radiating whorl; each floret strap-shaped and 5-toothed; teeth of central florets often blackish-purple; stigmas pale green. *Fruit-head* with many long-beaked and somewhat brown to greyish-brown, achenes, body **tuberculate throughout** (5.2–5.8 mm long; beak 7–10 mm); tip of beak with yellowish bristles (pappus) (c. 6 mm long). Fl. spring, sometimes late summer.

HARVESTING, SOURCING AND NATURAL RANGE

Whole plants of this Chinese endemic are harvested almost exclusively from the wild from spring to autumn as plants come into flower, although fruiting plants may also be harvested. The main sourcing areas are Hebei, Shanxi, Sichuan and Tibet where the species grows in grassland and disturbed open places (800–2,000 m altitude (sometimes higher). Recent years have seen the species brought into medicinal cultivation, notably in the Anguo district of Hebei province (NE China).

CONSERVATION STATUS

Endemic to China: Least Concern (authors' pers. obs. 2010); Not Evaluated (IUCN Red List, accessed 22/5/2015).

T. sinicum Kitag.

Asteraceae

synonym: *T. borealisinense* Kitam. (name used in CP2015)

PLANT DESCRIPTION

Easily confused with PS1. *Main differences*: **leaves narrower (0.6–1 cm wide)**; phyllaries below flower-heads in 3 whorls (not 2); **tips of outer phyllaries always flat (not horned)**; outer teeth of inner florets yellow to greyish; stigmas greenish-grey. Fruits (achenes) light greyish (3.5–4.4 mm long; 0.9–1 mm diam.), body **sometimes spiny with upper ones curved backwards**; beak slightly shorter then PS1 (5–6.5 mm long); tip of beak with yellowish-white hairs (pappus) 6.5–7 mm long.

HARVESTING, SOURCING AND NATURAL RANGE

T. sinicum is harvested exclusively from wild populations during spring to autumn. The species grows in slightly saline pastures and seasonally wet grasslands (600–2,000 m altitude) in northern and western China (provinces of Gansu, Hebei, Heilongjiang, Inner Mongolia, Jilin, Liaoning, Qinghai, Shaanxi and Shanxi); also native to Kyrgyzstan, Mongolia and Russia (Altai).

CONSERVATION STATUS:

China: Not Evaluated (probably Least Concern).
Global: Not Evaluated (IUCN Red List, accessed 22/5/2015).

TCM PROPERTIES AND USE

properties: **'bitter, sweet and cold'**
channels: **'Liver and Stomach'**
actions: **'clears heat, eliminates toxins, dispels damp heat, promotes urination'**

indications: **sores and abscesses; tonsillitis; jaundice; dysuria; painful eyes**

Fruiting medicinal crop of *Taraxacum mongolicum*

Wild flowering plant of *T. mongolicum*

T. mongolicum showing underside of flower-head and top of flower-stem

Medicinal crop of *T. sinicum* growing in Anguo, Hubei province

T. sinicum in the wild

T. sinicum

Taraxacum mongolicum

Asteraceae

DRUG MORPHOLOGY

CRUDE DRUG (DRIED WHOLE PLANT): crumpled masses predominantly of leaves and roots; stems, flower-heads and fruits occasional. *Rootstocks* often with brown or yellowish-white hairs; roots conical **[1]**, often curved (3–7 cm long), externally brown. ***Leaves*** often crumpled and broken; greenish-brown or dark greyish-green and, if still attached, all basal **[2]**; when intact and expanded blades pinnately lobed broadening towards rounded tips (6–15 cm long; **2–3.5 cm wide**); lobes broadly triangular and cut almost to the raised mid-vein, often slightly backward-pointing, margins entire to sparsely toothed **[3]**. Flower stems hollow and upper part with dense white, woolly hairs **[4]**. *Flower-heads (capitula)* **[5]** mostly young and, when intact up to 3–4 cm diam.; underside of flower-heads with 2 whorls of linear, greenish, leaf-like phyllaries **[6] usually tipped with a greenish-purplish horn [7]** fringed with hairs. *Flowers (florets)* in a radiating whorl, yellowish-brown or pale yellowish-white, each comprising a strap-shaped, 5-toothed ligule with teeth of central florets often blackish-purple. *Fruit-heads* **[8]** occasional; intact or broken, with many brown to greyish-brown achenes **[9]** (5.2–5.8 mm long) each with a long-beak (7–10 mm long) and a whitish-yellow feathery pappus **[10]** (c. 6 mm long). *Odour*: indistinct. *Taste*: slightly bitter.

DRIED SECTIONED HERB (DECOCTING PIECES): as for crude drug (above) but cut into short lengths (some features lost or obscured due to fragmentation). ***Good quality*:** abundant greyish-green leaves and stout rootstocks.

PROCESSING

Entire plants are sun-dried (= crude drug) then cut into short lengths for decocting.

OTHER COMMON NAMES

蒙古蒲公英 Meng Gu Pu Gong Ying (plant name); dandelion (SCN; CP2010); Herba Taraxaci; Mongolian dandelion; Radix cum Herba Taraxaci.

SAFETY

An official food and healthfood in China (Anon., 2002a,b).

T. sinicum

Asteraceae

DRUG MORPHOLOGY

CRUDE DRUG (DRIED PARTS): easily confused with that of *T. mongolicum* (PS1) above. *Main differences:* **leaves narrower (0.6–1 cm. wide) [11]**; phyllaries below flower-heads in 3 whorls (not 2); **tips of outer phyllaries always flat (not horned) [12];** outer teeth of inner ligules yellow to greyish; stigmas greenish-grey. Fruits (achenes) slightly shorter (3.5–4.4 mm long), sometimes spiny with upper spines backward-pointing. *Odour:* indistinct. *Taste:* slightly bitter.

DRIED PARTS (DECOCTING PIECES): As for crude drug but cut into short lengths.

PROCESSING

As for PS1.

OTHER COMMON NAMES

华蒲公英 Hua Pu Gong Ying (plant name); 碱地蒲公英 Jian Di Pu Gong Ying; Chinese dandelion (SCN); Herba Taraxaci; Radix cum Herba Taraxaci.

SAFETY

As for PS1.

UNOFFICIAL SUBSTITUTES

Many species of *Taraxacum* have been reported as substitutes or adulterants of 'Pu Gong Ying' in China but the CP2015 is unclear as to which may be legitimately used (see above). The taxonomic complexity of the genus creates a further challenge and to reliably identify species in commercial trade, laboratory-based methods are recommended. Key macroscopic identification characters of substitutes/adulterants reported from other genera in the Asteraceae family are outlined below:

1. *Emilia sonchifolia* (L.) DC. (点红 Yi Dian Hong or 土公英 Tu Gong Ying). Described by Bensky *et al.* (2004) as an adulterant of 'Pu Gong Ying'. Living plants: taller than PS1 and PS2 (25–40 cm), leaves at intervals along flowering stems (not in a basal rosette), often becoming purple and with a large rounded terminal lobe; phyllaries 8–9 and sharply pointed; flowers pink or purplish; bristles of fruits snow-white.
2. *Elephantopus scaber* L. (地胆草 Di Dan Cao). Leaves not lobed; flower-heads on branching stems, surrounded by leaf-like bracts; florets 4, 5-lobed, purplish or pink; fruits 10-ribbed.
3. *Picris divaricata* Vaniot (云南毛连菜 Yun Nan Mao Lian Cai). 10–40 cm tall; leaves basal and along stem; leaf margins wavy or toothed (not lobed) (3–10 cm long; 0.5–2 cm), with short, rigid hooked hairs on leaf mid-vein; fruits with stiffly hairy bristles. Endemic to Tibet and Yunnan.
4. *Ixeris chinensis* (Thunb.) Kitag. Stem leaves sometimes present; flower stems wiry (not fleshy); phyllaries in several whorls; florets bright yellow, pale yellow, white, or purplish; fruits with a slender 2.5–3 mm long beak.

Crude drug of Pu Gong Ying sourced from *T. mongolicum*

Decocting pieces (*T. mongolicum*)

Crude drug of Pu Gong Ying sourced from *T. sinicum*

Decocting pieces (*T. sinicum*)

Qing Hao 青蒿

Artemisiae Annuae Herba

Drug part: AERIAL

Artemisia annua L.

Asteraceae

PLANT DESCRIPTION

An erect many-branched annual (0.7–1.6 m tall; cultivated plants up to 2 m), virtually hairless and strongly aromatic. *Leaves* alternate, gland-dotted with middle stem leaves 2–3-times pinnately compound (2–3-pinnatipartite) imparting a soft, feathery appearance to the whole plant. *Inflorescences* arranged along outer parts of the branching stems with many, small, inconspicuous flower-heads (capitula). Each flower-head (capitulum) globose (1.5–2.5 mm diam.), short-stalked, nodding and yellowish-green with many minute florets: outer female florets 10–18; inner bisexual (disk) florets 10–30. *Fruiting heads*: clusters of tiny, dry, ellipsoid to ovoid achenes. Fl. and fr. Aug–Oct.

Artemisia annua in the wild, China

A. annua leaves

Dried fruiting branch

Species of *Artemisia* wild in China: 186

HARVESTING, SOURCING AND NATURAL RANGE

Aerial parts are harvested from wild and cultivated plants when in flower, mostly in autumn. Medicinal crops of selected chemotypes are extensively grown in China and other countries to meet the demand for plant material with a high artemisinin content (WHO, 2006; Wright, 2002).

The wild species grows in forest steppes, forest margins, dry flood plains and rocky, saline and disturbed places (2,000–3,700 m altitude). It is native throughout much of China (Anhui, Fujian, Gansu, Guangdong, Guangxi, Guizhou, Hebei, Heilongjiang, Henan, Hubei, Hunan, Inner Mongolia, Jiangsu, Jiangxi, Jilin, Liaoning, Ningxia, Qinghai, Shaanxi, Shandong, Shanxi, Sichuan, Taiwan, Xinjiang, Tibet, Yunnan and Zhejiang provinces) [also native in N Africa, Asia, Europe, and North America].

CONSERVATION

China: Least Concern (MEPCAS, 2013).
Global: Not Evaluated (IUCN Red List, accessed 22/7/2015).

Flowering head

UNOFFICIAL SUBSTITUTES

Artemisia is a taxonomically complex genus with 186 species wild in China. Expert botanical skills are required for identification to species level.

When wild-harvested, easily confused species of *Artemisia* include (Xiao *et al.*, 2001–2002):

1. ***A. japonica*** Thunb. (牡蒿 Mu Hao). Middle stem leaves shaped like a small spatula (spatulate) or wedge-shaped; obliquely divided or cleft, with margins slightly saw-toothed at tip.
2. ***A. eriopoda*** Bunge (南牡蒿 Nan Mu Hao). Middle stem leaves once- or twice-divided often to the midrib.
3. ***A. gmelinii*** Weber ex Stechm. (细裂叶莲蒿 Xi Lie Ye Lian Hao). A tufted perennial with woody rhizomes.

EHTPA 4/14
GAP

TCM PROPERTIES AND USE

properties**: 'bitter, pungent and cold'**
channels**: 'Liver, Gall Bladder and Stomach'**
actions**: 'clears heat, summer heat and deficient heat'**

indications**: fevers, malaria (*****also as a pharmaceutical derivative*****: artemisinin), jaundice, epistaxis, red eyes**

DRUG MORPHOLOGY

CRUDE DRUG (DRIED AERIAL PARTS): stems thin and cylindrical, longitudinally ridged and branched in upper parts (30–80 cm long; 2–6 mm diam.) **[1]**; externally yellowish-green or yellowish-brown, hairless (*texture*: light and brittle). *Fracture*: even, with white pith **[2]**. *Leaves* alternate, dark green or brownish-green and strongly curled and crumpled; when whole and expanded pinnately compound, 2–3 pinnatisect (divided to the midrib) or pectinatisect (divided into narrow, oblong, parallel segments) **[3]**; leaf surfaces hairless when mature and minutely gland-dotted; when young, softly grey tomentose on both surfaces. *Flower-heads* (capitula) yellowish-green, globose (1.5–2.5 mm diam.) **[4]** sometimes still attached by a short stalk **[5]** to the branching stems; each flower-head (capitulum) with many minute florets and which, when dissected (under low magnification) reveal 10–18 outer female florets and 10–30 inner bisexual (disk) florets (for an annotated illustration see 'Yin Chen' (Artemisiae Scoparia) page 404. *Odour*: distinctively aromatic. *Taste*: slightly bitter.

DECOCTING PIECES (DRIED SECTIONED AERIAL PARTS): as for crude drug but cut into short sections.

PROCESSING

After removal and discarding of older stems, aerial parts are shade-dried (= crude drug) then later softened briefly with water before cutting into decocting sections and dried in the sun.

OTHER COMMON NAMES

臭蒿 Chou Hao; 黄蒿 Huang Hao; 黄花蒿 Huang Hua Hao (plant name); 青蒿叶 Qing Hao Ye (sweet wormwood leaves); 香青蒿 Xiang Qing Hao (aromatic sweet wormwood); annual wormwood; Chinese wormwood; sweet wormwood (SCN).

SAFETY

Safety Class: 2b; Interaction Class: A (Gardner & McGuffin, 2013).

Crude drug of Qing Hao (dried aerial parts of *Artemisia annua*)

Decocting pieces of Qing Hao (sectioned *A. annua*)

Capitula attached (right) and separated (left)

Stem fragments

Leaflet fragments

Sang Ji Sheng 桑寄生 Taxilli Herba

Drug parts: TWIG AND LEAF

PHARMACOPOEIA SPECIES

Taxillus chinensis (DC.) Danser — Loranthaceae

synonyms: *Loranthus chinensis* DC.; *Scurrula chinensis* (DC.) G. Don

Species of *Taxillus* wild in China: 18

PLANT DESCRIPTION

A woody hemi-parasite (0.5–1 m tall) growing directly from thewoody branches of a variety of host trees. *Stems* woody and branching; young stems and leaves with soft, reddish-brown hairs; nodes not articulated. *Leaves* virtually opposite, short-stalked, papery or leathery, ovate to oblong (2.5–6 cm long; 1.5–4 cm wide) with margin entire. *Flowers* quite large (2.5–2.7cm long), stalked; usually in pairs; flower bud with a swollen and ovoid tip; petals fused to form a long tube (2.5–2.1cm long), brownish or yellowish-green; inflated at the base and split along one side; apex with 4–6 short, reflexed petal lobes; ovary inferior. *Fruit* ellipsoid or subglobose (8–10 mm long; 5–6 mm diam.), yellowish and leathery with wart-like projections, becoming smooth when mature; mucilaginous internally. Fl. and fr. Feb–Dec.

HARVESTING, SOURCING AND NATURAL RANGE

Plants are harvested during winter and spring from wild plants which grow on a variety of host trees (wild or cultivated) notably mulberry '*Morus alba*', between 100–400 m altitude and in the Southern China provinces of Fujian, Guangdong, Guangxi and Hainan [the species is also native in Cambodia, Indonesia, Laos, Malaysia, Philippines, Thailand and Vietnam].

CONSERVATION STATUS

China: Not Evaluated.
Global: Not Evaluated (IUCN Red List, accessed 28/8/2016).

UNOFFICIAL SUBSTITUTE

Viscum coloratum (Kom.) Nakai — Santalaceae

synonym: *V. album* L. subsp. *coloratum* Kom.

PLANT DESCRIPTION

A parasitic/hemi-parasitic species (like the official species) but easily distinguished: *stems* **green or yellowish-green non-woody**, *stem nodes* **articulated and slightly swollen**; *leaves* elliptic or oblong-lanceolate **(3–7 cm long; 0.7–2 cm wide)** and leathery; *flowers* **tiny and inconspicuous**; fruits sticky, spherical (c.1 cm diam.), yellow, red or orange.

CONSERVATION

China: Near Threatened (authors' pers. obs. 2009).
Global: Not Evaluate (IUCN Red List, accessed 12/3/2015).

OTHER UNOFFICIAL SUBSTITUTES

The official drug is frequently substituted or contaminated with a wide range of other parasitic species in the families Loranthaceae and Santalaceae. Whilst *Viscum coloratum* (above) is one of the commonest substitutes and easily differentiated macroscopically, to reliably eliminate other substitutes and contaminants microscopic examination of the leaf cuticles and/or DNA bar-coding is advised. Chemical authentication is unlikely to be reliable given the parasitic nature of these species; distinguishing the parasites' compounds from those of its hosts may be too complex. Substitute/contaminant species reported in the literature include:

1. ***Helixanthera*** species: **distinctly flattened internodes** and greenish-white berries; epi parasitic on many species including *Scurrula parasitica* and *Taxillus chinensis*.
2. ***Scurrula parasitica*** L. A troublesome parasite of fruit trees with many hosts in many families. Calyx pear-shaped (pyriform) or turbinate, base shortly pointed; fruit base narrow or long attenuate. **Toxic when parasitising *Coriaria nepalensis* Wall. or *C. intermedia* Matsumura**.
3. ***Taxillus nigrans*** (Hance) Danser. (毛叶钝果寄生 Mao Ye Dun Guo Ji Sheng). **Leaves larger (6–11 cm long; 3–5 cm wide)** and **lower surfaces densely covered (often somewhat matted) in short, soft hairs (tomentose)**. Hosts: *Camellia oleifera*, *Cinnamomum camphora*, *Morus alba*, *Quercus* and *Salix* spp.
4. ***T. sutchuenensis*** (Lecomte) Danser (四川寄生 Sichuan Ji Sheng; 桑寄生 Sang Ji Sheng). **Lower surface of leaves persistently and densely covered** (often somewhat matted) **in short, soft hairs (tomentose)**. **Toxic when parasitising *Coriaria sinica*** Maxim.
5. ***Viscum liquidambaracola*** Hayata. Hosts: *Diospyros kaki*, *Liquidambar formosana*, *Vernicia fordii*, *V. montana*, and species of Fagaceae; occasionally epi parasitic *on Viscum* species.

TCM PROPERTIES AND USE

properties: **'bitter, sweet and neutral'**
channels: **'Liver and Kidney'**
actions: **'dispels wind damp, tonfies Liver and Kidney'**

indications: **musculoskeletal pain; weakness of back and knees; threatened miscarriage**

Taxillus chinensis on host tree *Camellia oleifera* C. Abel ('tea-oil plant'), China (*Taxillus* encircled)

T. chinensis in flower

Viscum coloratum mature and fruiting plant

V. coloratum

Taxillus chinensis

Loranthaceae

DRUG MORPHOLOGY

CRUDE DRUG (DRIED OR STEAMED LEAF AND STEM LENGTHS): *twigs:* **woody [1]** cylindrical, slightly curved (3–4 cm long; 0.2–1 cm diam.) externally greyish-brown or reddish-brown (young twigs with a reddish-brown pubescence) **[2]**, finely wrinkled longitudinally with many and scattered brown dot-like lenticels **[3]**; leaf scars slightly raised and opposite **[4]**; *texture* hard; *fracture* uneven and wood pale reddish-brown. *Leaves:* frequently broken and curled **[5]**, opposite, short-stalked **[6]**, yellowish-brown or yellowish-green (young leaves with a reddish-brown pubescence) **[7]**, blades (when whole and expanded) ovate to oblong (2.5–6 cm long; 1.5–4 cm wide), base rounded or wedge-shaped **[8]**, apex obtusely rounded; margin entire **[9]**; lateral veins in 3 or 4 pairs **[10]**; flowers or flower-buds occasional **[11]** (see *Plant Description*). *Texture:* leathery. *Fracture:* uneven. *Odour:* indistinct. *Taste:* astringent. ***Good quality*:** twigs young and tender and leaves many and reddish-brown.

DECOCTING PIECES (SLICED LEAF AND STEM): as for the crude drug but shorter lengths and more fragmented.

PROCESSING

Thick stems are discarded then remaining leaves and stems cut into lengths and dried, or steamed then dried (= crude drug); decocting pieces are made by further cutting and slicing.

OTHER COMMON NAMES

广寄生 Guang Ji Sheng; Herba Taxilli; Chinese Taxillus herb (CP2010).

SAFETY

May cause serious or life-threatening adverse reactions because of the ability of this hemi-parasitic species to sequester toxins (notably cardiac glycosides) from host trees (e.g. *Coriaria* species; *Nerium oleander* L.). For this reason some regulatory authorities require analytical evidence to demonstrate the absence of toxins such as cardiac glycosides in this drug (Okuda *et al.* 1987).

Viscum coloratum

Santalaceae

'DRUG SUBSTITUTE' MORPHOLOGY

Stems **not woody** but **yellowish-green or yellowish-brown [12]** (c. 0.3–1 cm diam.), stem nodes swollen into a raised ring **(articulated) [13]; leaves unstalked [14]**, yellowish-green or yellowish-brown, **blades narrower** (**elliptic-lanceolate** (2–7 cm long; **0.5–1.5 cm wide) [15]**, bases wedge-shaped, apices broadly rounded; finely wrinkled throughout, 5-veined (middle 3 distinct **[16]**) and easily shed. *Texture:* leaves leathery **[17]**; stems light and brittle. *Odour* indistinct. *Taste:* somewhat bitter; sticky when chewed. *Fracture:* stems uneven, **wood pale yellow [18]** with radiating rays; pith obscured or off-centre **[19]**.

OTHER COMMON NAMES

北寄生 Bei Ji Sheng; 北桑寄生 Bei Sang Ji Sheng (northern mulberry parasite); 槲寄生 Hu Ji Sheng; 寄生 Ji Sheng; coloured mistletoe; Herba Visci; Visci Herba.

SAFETY

Dependent on toxicity of host tree. All parts of the closely related *Viscum album* L. (European mistletoe) are classified as Safety Class: 2b and 2d Interaction Class: A (Gardner & McGuffin, 2013).

SUBSTITUTION COMMENT. *V. coloratum* is the source of the separate TCM drug 'Hu Ji Sheng' (Visci Herba), page 354, and despite being recognised by the CP2015 as a separate drug it is often used interchangeably with 'Sang Ji Sheng' in a clinical setting; their activities are reported to be similar but not identical (Bensky *et al.*, 2004). Given their different range of hosts and that some may give rise to safety concerns (see above), for this reason clear discrimination of the two drugs and detection of mixed trade batches is vital.

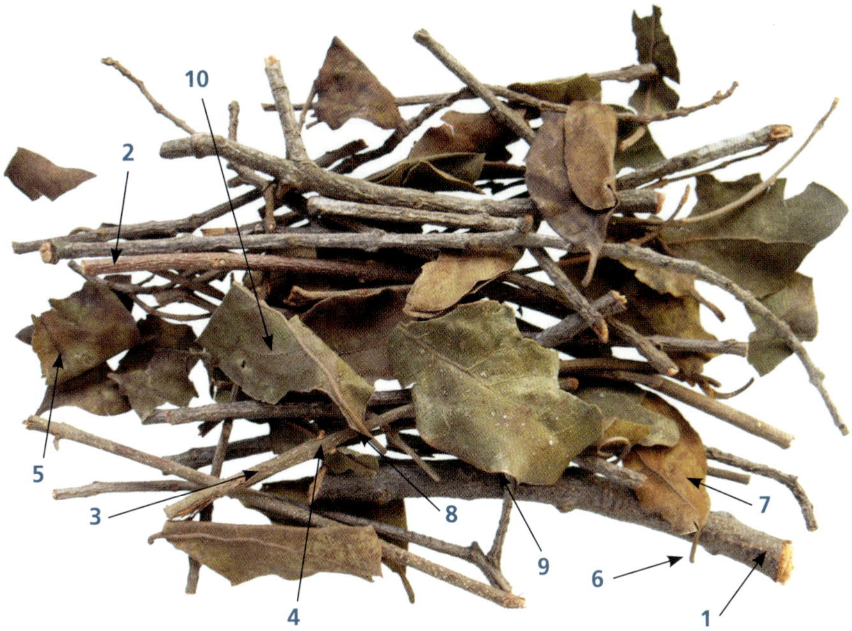

Crude drug of Song Ji Sheng

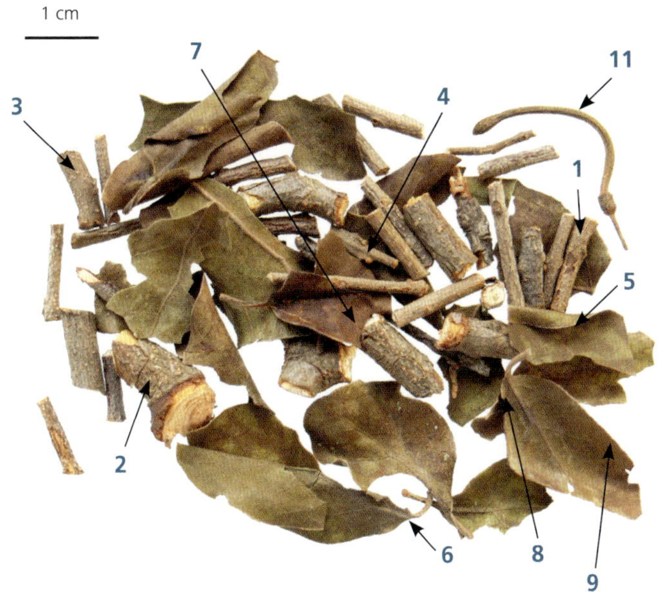

Decocting pieces of Song Ji Sheng

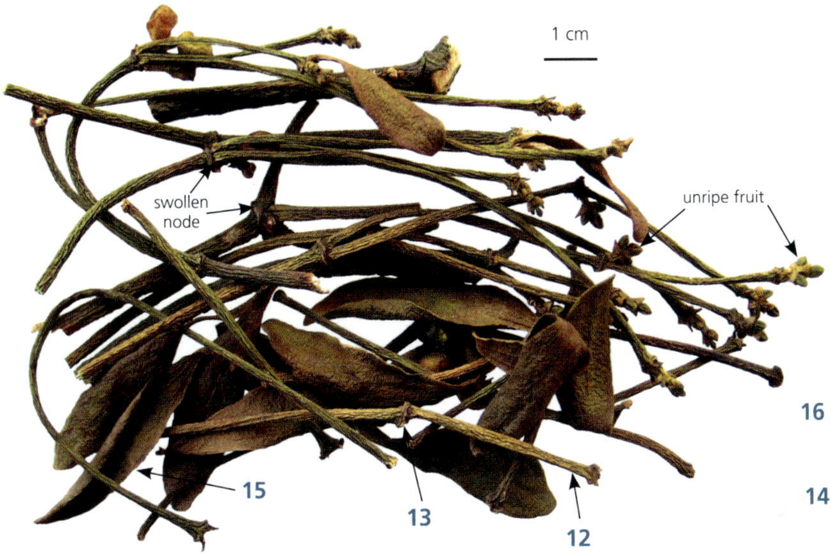

Crude drug of substitute Hu Ji Sheng (*Viscum coloratum*)

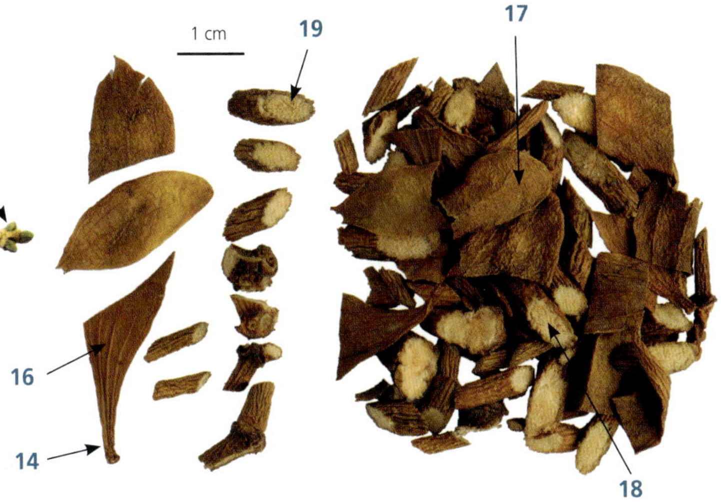

Decocting pieces of substitute Hu Ji Sheng

CONTAMINANT

Viscum articulatum Burm. f.; 扁枝槲寄生 Bian Zhi Hu Ji Sheng (family Santalaceae). This species is epi-parasitic on the official species hence care is needed when harvesting to avoid contamination. Easily identified by its distinctly flattened internodes **[20]** (stems) and leaves reduced to pairs of tiny, thin and dry scales at nodes **[21]**.

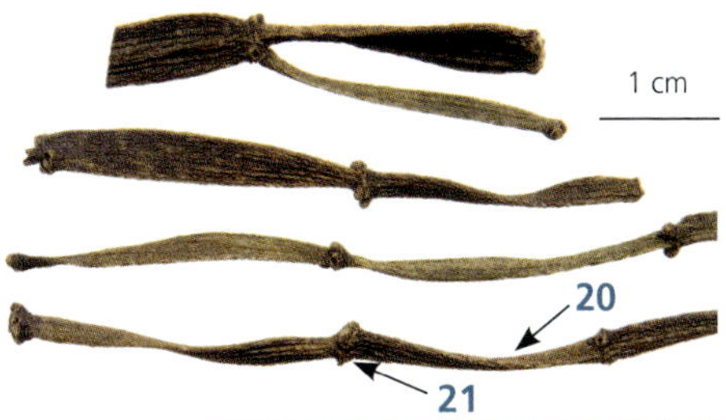

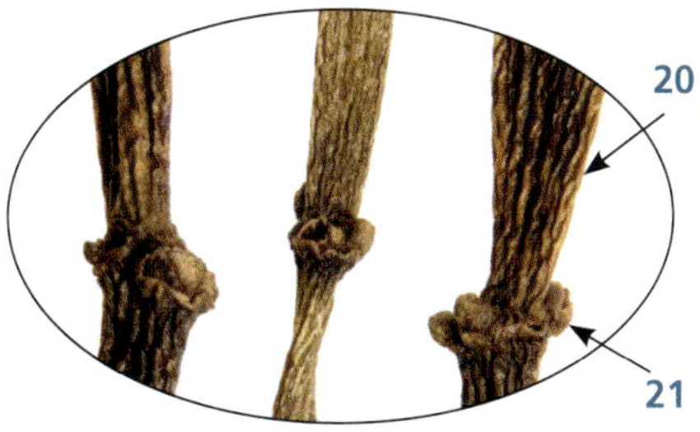

Xiang Ru 香薷

Moslae Herba

Drug part: AERIAL

PHARMACOPOEIA SPECIES

Mosla chinensis Maxim.

Lamiaceae

Species of *Mosla* wild in China: 12

PLANT DESCRIPTION

The names Qing Xiang Ru (青香茹) and Shi Xiang Ru (石香薷) are commonly used for 'Xiang Ru' that is wild harvested. The higher quality (*Di Dao*) material is called 'Jiang Xiang Ru' and is cultivated in Jiangxi province.

Qing Xiang Ru: an erect herbaceous aromatic annual (9–50 cm tall) covered throughout in fine, soft, white hairs and most parts becoming purplish-red with maturity. *Stems* square, much branched from the base or unbranched. Stipules absent. *Leaves* simple, opposite, very short-stalked; **leaf blade linear, narrowly ovate to lance-shaped (1.3–2.8 cm long; 2–4 (rarely up to 7) mm wide)**, sparsely hairy, base narrowly to broadly wedge-shaped, tips pointed; surfaces densely pitted with glands; margins virtually entire. **Flowers** in terminal and axillary head-like clusters **(1–3 cm long)**. Bracts overlapping, round-ovate or reversed round-ovate, margins fringed with hairs. *Flowers* tiny, bisexual and laterally symmetrical, with a 10-veined pale purplish-red or greyish-green, calyx with a woolly throat surrounding a purplish-red, reddish to white flower tube (c. 5 mm long); **stamens not extending beyond the flower tube**, 4 (only 2 obvious; others small and sterile); style 2-cleft. *Fruits* with 4 globose or subspherical, greyish-brown nutlets (0.7–1.2 mm diam.) [hl]. Fl. Jun–Sep, fr. Jul–Nov.

Jiang Xiang Ru: as for *Qing Xiang Ru* but slightly taller (55–66 cm). Most parts (stems, leaves, flower- and fruiting-heads) yellowish-green, leaf blade margins with shallow 5–9 widely spaced lobes.

See also *Drug Morphology*.

HARVESTING, SOURCING AND NATURAL RANGE

Mostly sourced from medicinal crops widely cultivated in the Chinese provinces of Guangdong, Guangxi, Fujian and Hunan. The *Di Dao* cultivar 'Jiang Xiang Ru' 江香茹 has been grown for hundreds of years and is sourced exclusively from Jiangxi province. Aerial parts are harvested on fine days in summer when foliage is lush and plants are flowering and fruiting.

In the wild the species grows in grassland and in forests (up to 1,400 m altitude) in C, E and S China (Anhui, Fujian, Guangdong, Guangxi, Guizhou, Hubei, Hunan, Jiangsu, Jiangxi, Shandong, Sichuan, Taiwan and Zhejiang) [also native in Vietnam].

CONSERVATION STATUS

China: Least Concern (MEPCAS, 2013).
Global: Not Evaluated (IUCN Red List, accessed 10/8/2016).

UNOFFICIAL SUBSTITUTE

Elsholtzia splendens Nakai ex F. Maek.

Lamiaceae

PLANT DESCRIPTION

Superficially similar and sometimes confused with *M. chinensis* (Pharmacopoeia species). *Main differences:* leaves **ovate to lance-shaped, usually much larger** (3–6 cm long; 0.8–2.5 cm wide) and **saw-toothed; flower-/fruit-heads longer (3.5–4.5 cm long)** and **arranged on one side of the stem**; bracts (between the flowers) distinctive (broadly ovate, c. 5 × 6– 7 mm, with tip abruptly narrowed into a short, hard spine; edges densely fringed with white hairs); corolla rose-purple; stamens **extending well beyond the corolla tube;** nutlets tuberculate. Fl. and fr. Sep–Nov.

SUBSTITUTION COMMENT. According to Bensky *et al.* (2004) much confusion surrounding this species has occurred historically within the TCM community owing to *E. splendens* being "mistakenly identified as the main herb for Xiang Ru". Although closely related, the two species can easily be distinguished as living plants but their trade items require more careful examination possibly with the help of a handlens (x10). The occurrence of *E. splendens* in international trade today is uncertain but its local use in China is thought to continue probably under the name Hai Zhou Xiang Ru (海州香薷).

In the wild the species grows in grassy places on hills (200–300 m altitude) in NE and SE China: Guangdong, Hebei, Henan, Hubei, Jiangsu, Jiangxi, Liaoning, Shandong and Zhejiang [also native in the Korean Peninsula].

See also *Lesser Known Unofficial Substitutes*.

CONSERVATION STATUS

China: Not Evaluated.
Global: Not Evaluated) IUCN Red List, accessed 10/8/2016).

TCM PROPERTIES AND USE

properties: **'pungent and slightly warm'**
channels: **'Lung and Stomach'**
actions: **'dispels wind cold damp, promotes urination'**
indications: **respiratory tract infections; vomiting, diarrhoea; dysuria; oedema**

Freshly harvested crop of *Mosla chinensis*, Anhui province (purple form)

Flowering heads of *M. chinensis*, Anhui province (purple form)

Flower- and fruit-heads of *M. chinensis*

Wild plant of *Elsholtzia splendens*, Jiangxi province

Herbarium specimen of *E. splendens* showing leaves and flower-heads

E. splendens showing elongated, 1-sided flower-head

Mosla chinensis

Lamiaceae

DRUG MORPHOLOGY

CRUDE DRUG OF QING XIANG RU or SHI XIANG RU (wild or cultivated): aerial parts mostly stems **[1]** with some flower-/fruit-heads **[2]** and occasional leaves **[3]**; entire plant covered with fine white hairs. Stems 30–50 cm long, purplish-red, yellowish-green or pale yellow; 4-sided **[4]** (1–2 mm diam.); nodes distinct; internodes 4–7 cm long; opposite-branched **[5]**; *texture:* brittle. Leaves mostly crumpled or fallen; when intact leaf blades simple, dark green, yellowish-green or purple, **linear [6] to narrowly ovate [7]** (1.3–2.8 cm long; 2-4 mm wide); **margins virtually entire [8]**. **Flower-/fruit-heads** terminal or axillary **(1–3 cm long) [9]**; bracts round-ovate, overlapping, usually fallen; **calyx,** pale purplish-red or greyish-green **[10]**, **apex with 5 narrowly pointed teeth [11]**, densely hairy; individual flowers tiny and shriveled and difficult to see with **stamens hidden within the flower tube [12].** Fruits with persistent 5-pointed calyx enclosing 4 one-seeded, globose or sub-spherical, greyish-brown nutlets (0.7–1.1 mm diam.). *Odour*: intensely aromatic. *Taste*: cool and slightly pungent.

CRUDE DRUG OF JIANG XIANG RU (cultivated): as for Qing Xiang Ru above but slightly longer (55–66 cm), most parts (stems, leaves, flower- and fruiting heads) yellowish-green, leaf blade margins with 5–9 indistinct saw-like lobes.

DECOCTING PIECES (AERIAL PARTS CUT INTO SHORT LENGTHS): as for both types of crude drug but cut into short 1–5 cm lengths forming a tangled mass of stems with flower-/fruit-heads often still attached.

PROCESSING

Aerial parts are shade dried (= crude drug) then cut into short lengths (= decocting pieces).

OTHER COMMON NAMES

江香茹 Jiang Xiang Ru; 青香茹 Qing Xiang Ru; 细香薷 Xi Xiang Ru; Chinese mosla (CP2010); Herba Moslae.

SAFETY

Decoctions drunk hot (unless in combination with certain other herbs) may cause vomiting (Bensky *et al.*, 2004; Chen & Chen, 2004).

Elsholtzia splendens

Lamiaceae

'DRUG SUBSTITUTE' MORPHOLOGY

Easily distinguished from that of the *M. chinensis* (Pharmacopoeia species) if leaves or flower-/fruit-heads present and intact. **Leaves usually much larger (3–6 cm long; 0.8–2.5 cm wide) [13]**, ovate to lance-shaped with margins **saw-toothed [14], flower-/fruit-heads longer (3.5–4.5 cm long) [15]; bracts** (between the flowers) **distinctive [16]** (broadly ovate c. 5 mm long; 6–7 mm wide, **with tip abruptly narrowed into a short, hard spine [17]**; edges densely fringed with white hairs; **stamens extending well beyond the corolla tube [18]**.

OTHER COMMON NAMES

海州香薷 Hai Zhou Xiang Ru; 嫩香薷 Nen Xiang Ru 'tender Xiang Ru'; Haizhou elsholtzia (SCN).

LESSER KNOWN UNOFFICIAL SUBSTITUTES

Elsholtzia ciliata (Thunb.) Hyl. (Lamiaceae family). Like *E. splendens* this species has been used mistakenly in the past as the official source of 'Xiang Ru'. Its other names are 土香薷 Tu Xiang Ru and 嫩香薷 Nen Xiang Ru 'tender Xiang Ru'. It is a common plant and occurs throughout China except in Qinghai and Xinjiang provinces. **Leaves much larger (3–9 cm long; 1–4 cm wide) ovate to elliptic/lance-shaped**; leaf stalks narrowly winged; flower-/fruit-spikes often longer (2–7 cm long; 1.3 cm diam.) in one-sided heads; flowers purplish; nutlets yellowish-brown, oblong (c. 1 mm diam.). Fl. Jul–Oct, fr. Oct–Jan.

Crude drug of Xiang Ru (Jiang Xiang Ru)
(*Mosla chinensis*)

Decocting pieces of Xiang Ru (Jiang Xiang Ru) (*M. chinensis*)

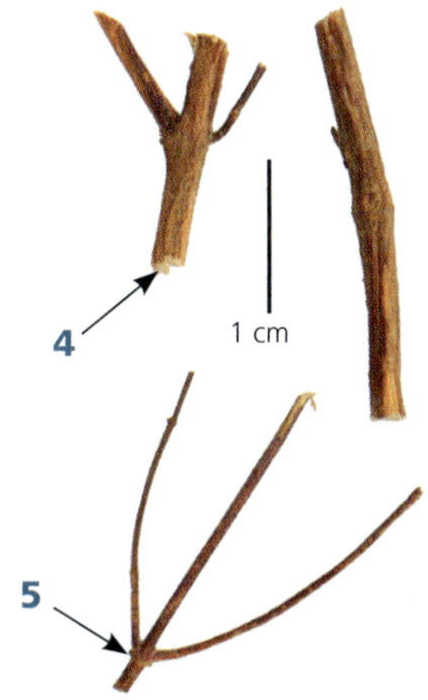

Stems of Xiang Ru (*M. chinensis*)

Flowers of Xiang Ru (*M. chinensis*)

Leaves and flower/fruit-head of Xiang Ru

Aerial parts of *E. splendens*

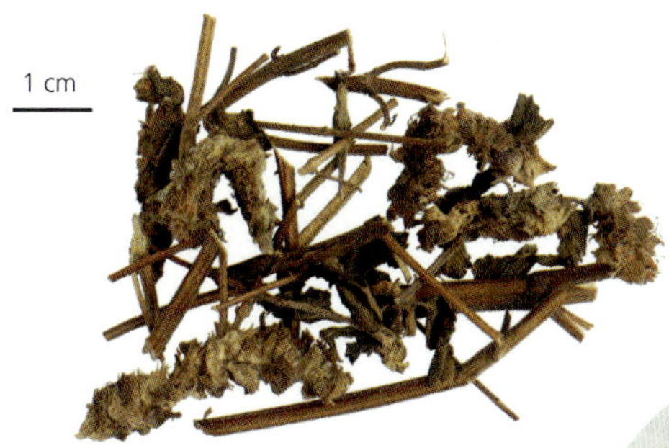

Aerial parts (dried and sectioned)
of *E. splendens*

Stems of *E. splendens*

Flowers and bracts of *E. splendens*

Leaves and flower/fruit-head of *E. splendens*

Xian He Cao 仙鶴草

Agrimoniae Herba

Drug part: AERIAL

Agrimonia pilosa Ledeb.

Rosaceae

Species of *Agrimonia* wild in China: 4

PLANT DESCRIPTION

A herbaceous perennial (30–120 cm tall) with a short tuberous, creeping rhizome and slender, flowering stems extending well above the basal leaves. *Stems*, especially in upper parts, softly hairy. *Stipules*, at base of each leaf, green, leaf-like, usually toothed and clasping stem nodes. *Basal leaves* alternate, divided into 3 or 4 (rarely 2) opposite pairs of leaflets plus a terminal one, often with very reduced leaflets between main leaflets; upper leaves with only 3 leaflets. Leaflets virtually unstalked, ovate, elliptic or lance-shaped (1.5–5 cm long; 1–2.5 cm wide), veins on lower leaf surface covered in short, thin appressed hairs; margins saw-toothed. *Flowers* in terminal and lateral, long, slender spikes; flowering stems sparsely or densely hairy. Each flower small (6–9 mm diam.) and very short-stalked; sepals 5, green and triangular; petals 5, larger than sepals, yellow and oblong; stamens 8–15 (rarely 5); style thread-like; stigma club-shaped. *Fruits* dry and somewhat conical with a crown of hooked prickles above a 10-ribbed receptacle (7–8 mm long; 3–4 mm diam. including prickles); single seeded. Fl. and fr. May–Dec.

HARVESTING, SOURCING AND NATURAL RANGE

Aerial parts are harvested in summer or autumn from wild plants which grow wild throughout China in forest margins, thickets, meadows and along stream- and road-sides (100–3,800 m altitude) [also wild in Bhutan, India, Japan, Laos, Korean Peninsula, Mongolia, Myanmar, Nepal, Russia, Sikkim, Thailand, Vietnam and Europe].

CONSERVATION STATUS

China: Least Concern (MEPCAS, 2013).
Global: Not Evaluated (IUCN Red List, accessed 17/8/2016).

Agrimonia pilosa — cultivated plants flowering and fruiting

Flowering spike

Fruiting spike

Leaf stalk

TCM PROPERTIES AND USE

***properties*: 'bitter, sour and neutral'**
***channels*: 'Heart and Liver'**
***actions*: 'stops bleeding, astringes, binds the intestines, tonifies qi and Blood'**

***indications*: haemoptysis; metrorrhagia; haematochezia; malaria; (T)**

DRUG MORPHOLOGY

CRUDE DRUG (DRIED AERIAL PARTS): crumpled aerial parts mostly of leaves and stems; flowers and fruits sometimes present. Stems (50–100 cm long) covered in fine white hairs **[1]** and often with membranous remains of small toothed and leaf-like stipules at nodes **[2]**; stem upper parts (if distinguishable) greenish-brown, square and slightly concave **[3]**; lower stem parts reddish-brown **[4]**; *texture* light and brittle; *fracture* hollow. *Leaves* green, crumpled and often fragmented; when whole and expanded leaves alternate with 3 or 4 (rarely 2) opposite pairs of leaflets plus a terminal one; upper leaves with only 3 leaflets. Leaflets virtually unstalked; if whole and expanded then, ovate, elliptic or lance-shaped (1.5–5 cm long; 1–2.5 cm wide); larger leaflets with wedge-shaped bases and pointed tips; veins on lower leaf surface covered in short, thin, appressed hairs **[5]**; margins saw-toothed **[6]**. Very small extra leaflets present along leaf stems between the main leaflet pairs. Flowers (rarely present) small (6–9 mm diam.) and short-stalked (1–5 mm long). *Fruits* dry, and somewhat conical and with a distinctive crown of hooked prickles **[7]** above a 10-ribbed receptacle **[8]** (7–8 mm long; 3–4 mm diam. including prickles); single seeded. *Odour:* indistinct. *Taste:* slightly bitter. ***Good quality*:** leaves many and stems reddish-brown.

DECOCTING PIECES (DRIED AERIAL PARTS CUT INTO SHORT LENGTHS): as for the crude drug but cut into short lengths; often fragmented.

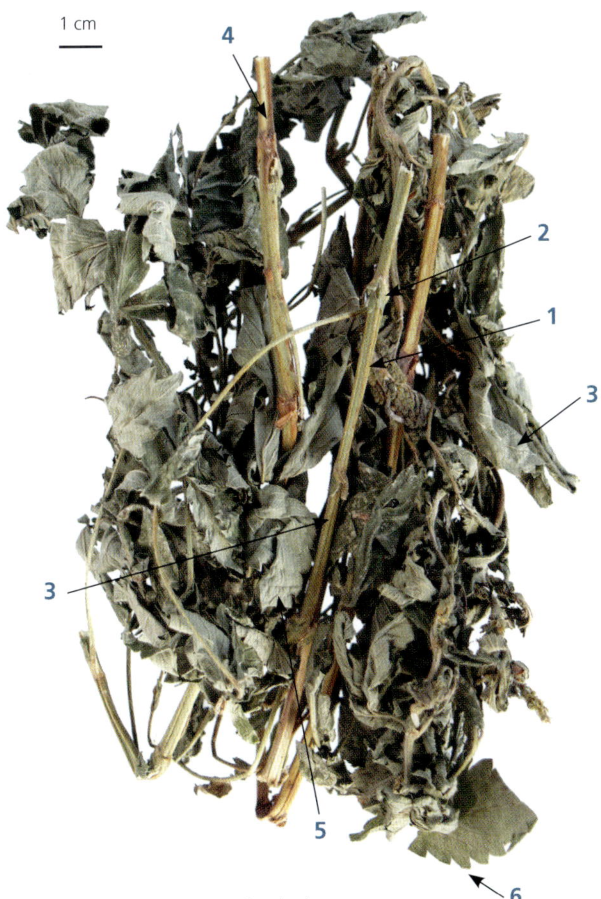

Crude drug

PROCESSING

Aerial parts are dried (= crude drug) and subsequently softened before cutting into sections (= decocting pieces).

OTHER COMMON NAMES

瓜香草 Gua Xiang Cao; 黄龙尾 Huang Long Wei; 龙牙草 Long Ya Cao; agrimony (SCN); hairy vein agrimonia herb (CP2010); Herba Agrimoniae.

SAFETY

Safety Class: 1; Interaction Class: A (Gardner & McGuffin, 2013).

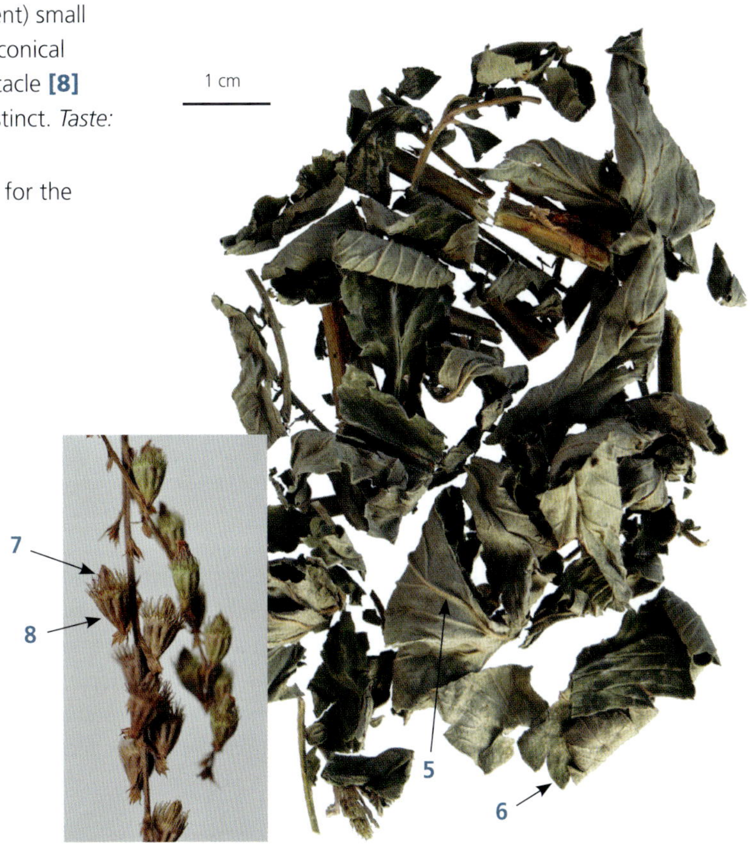

Decocting pieces

UNOFFICIAL SUBSTITUTES

Easily confured and closely related species have occasionally been traded as 'Xian He Cao' in local Chinese markets:

1. *Agrimonia eupatoria* subsp. *asiatica* (Juzep.) Skalicky. **Flowers larger (12–13 mm diam.)** Wild populations confined to Xinjiang province (NW China); also native to N and W Africa.
2. *A. coreana* Nakai. **Stamens 17–24**. Confined to NE and E China (Jilin, Liaoning, Shandong and Zhejiang provinces) [also native to Japan, Korean Peninsula and Russia].

Xiao Ji 小蓟

Cirsii Herba

Drug part: AERIAL

PHARMACOPOEIA SPECIES

Cirsium arvense var. ***integrifolium*** Wimm. & Grab.

Asteraceae

synonyms: *Cephalonoplos setosus* (Willd.) Kitam.; *Cirsium segetum* Bunge; *C. setosum* (Willd.) M. Bieb. (name used in CP2015).

taxonomic/nomenclatural note: *Kew-MPNS* accepted name 28/8/2016, *C. arvense* (L.) Scop.

Species of *Cirsium* wild in China: 46

PLANT DESCRIPTION

A vigorously creeping perennial herb (30–160 cm tall). *Rootstock* far-creeping, forming extensive populations. *Stem* erect, branched, spineless, furrowed, not winged, **hairless or sparsely felted**. **Basal leaves not forming a rosette.** *Leaves* alternate, becoming smaller upwards; lower and middle stem leaves usually stalkless; **leaf blades** narrowly elliptic to elliptic-lance-shaped (7–15 cm long; 1.5–10 cm wide), **undivided or sometimes pinnately lobed**, base usually wedge-shaped, [tip broad to rounded; margin entire or irregularly toothed or densely fringed with small, soft spines (spinules)]; upper surface hairless and smooth; lower surface green to greyish-green and hairless or sparsely cobwebby. Male and female flowers on separate plants. *Flower-heads* (capitula) usually many, in terminal loose flat-topped groups (corymbs); spineless; outer part of each capitulum (the involucre) hairless with 5–7 rows of over-lapping pointed green scales (phyllaries); outer and middle phyllaries triangular to ovate with an apical spinule (<0.5 mm. long); inner phyllaries elliptic to lance-shaped to almost linear; tips narrowly pointed, thin and dry (not green). *Flowers* densely packed; corolla pale pinkish-purple (rarely white); female florets 1.6–2.4 cm long (petal tube 1.3–1.8 cm); male florets 1.5–1.8 cm long (petal tube 0.9–1.2 cm). *Fruit-heads* (capitulum) a dense mass of yellow achenes (3–4.5 mm long) **each with a very long pappus (2.5–3.5 cm)** of dirty-white feathery hairs. Fl. and fr. May–Sep.

HARVESTING, SOURCING AND NATURAL RANGE

Aerial parts are harvested entirely from wild populations in summer and autumn when in flower. The species grows on mountain slopes, by rivers and waterways and as a weed on agricultural land (100–2,700 m altitude). It is native throughout China [also wild in Japan, the Korean Peninsula, Mongolia, Russia; SW Asia and Europe]. It is regarded as a noxious weed in some parts of the world.

CONSERVATION STATUS

China: Least Concern (MEPCAS, 2013; as the species not the variety).

Global: Not Evaluated (IUCN Red List, accessed 25/8/2016).

UNOFFICIAL SUBSTITUTE

C. japonicum DC.

Asteraceae

synonym: *Carduus japonicus* (DC.) Franch.; *Cnicus japonicus* (DC.) Maxim.

PLANT DESCRIPTION

Easily distinguished from the Pharmacopoeia species. *Main differences*: *roots* clustered and spindle-shaped; **stems** densely clothed in **a mat of short hairs interspersed with long filamentous ones**; **basal leaves forming a rosette** of densely spiny leaves; **leaf blades pinnately divided** into 6–12 segments each side (8–20 cm long; 2.5–8 cm wide); lower and middle stem leaves similar to those in basal rosette. *Flowers* (florets) pinkish-red to purple; fruits with long pappus of pale brown bristles (c. 2 cm long). Fl. and fr. Apr–Nov.

CONSERVATION STATUS

China: Least Copncern (MEPCAS, 2013)

Global: Not Evaluated (IUCN Red List, accessed 19/5/2015).

LESSER KNOWN UNOFFICIAL SUBSTITUTES

The aerial parts of some closely related and very similar-looking species have been reported to be traded as 'Xiao Ji' in local Chinese markets probably due to mis-identification. Their occurrence as such in international trade requires further study.

1. *Cirsium botryodes* Petr. (as '*Cirsium botryoides*' in Bensky *et al.*, 2004). Stem greyish-white with long, cobwebby hairs. **Leaves rigid, densely white-felted on lower surface; margins densely spiny**; many leaves pinnately divided and with a 3–7 mm long spine at tip. Involucre c. 3.5 cm diam. Outer phyllaries ending in 8–13 mm spine. Florets bisexual and white. Achene with brownish pappus bristles c. 2 cm long.
2. *Cirsium lineare* (Thunb.) Sch. Bip. (线叶蓟 Xian Ye Ji — plant name). Leaf blade entire or obscurely lobed. Involucre somewhat narrowly ovoid (1–2.5 cm diam.). Phyllaries each with a **dark resinous gland**; inner phyllaries each with a narrowly pointed pale or purplish pink, irregularly toothed appendage at tip. Florets purplish-red to pink; bisexual. Achene c. 2.5 mm. Pappus brownish to dirty white, **shorter (c. 1.5 cm long).**
3. *Sonchus brachyotus* DC. (长裂苦苣菜 Chang Lie Ku Ju Cai — plant name). Stems and leaves exuding white latex when broken. With a taproot and lateral roots (**not creeping**). Stem with stalked glandular hairs. Leaf margins **not** densely fringed with small, soft spines. Florets **yellow.** Achenes longitudinally ribbed; pappus white, a mix of soft outer hairs and thicker inner ones.

TCM PROPERTIES AND USE

properties: **'sweet, and cool'**
channels: **'Liver and Heart'**
action: **'cools Blood, stops bleeding, invigorates Blood'**

indications: **bleeding disorders including haematuria and haemorrhoids; sores and abscesses**

LEFT
Dehiscing fruit-head of *C. arvense* var. *integrifolium* with long feathery pappus of achenes

Flowering plants of *Cirsium arvense* var. *integrifolium*

Flowering plants of *Cirsium japonicum*

Cirsium arvense var. *integrifolium*

Asteraceae

DRUG MORPHOLOGY

CRUDE DRUG (DRIED AERIAL PARTS): *stems* cylindrical, greyish-green or purplish, often branched in upper parts (5–30 cm long, 2–5 mm in diam.), longitudinally furrowed **[1]**, **hairless or covered in a short sparse mat of hairs**. *Texture* brittle. *Fracture* hollow. *Leaves* crumpled, often broken, alternate, unstalked or with a short stalk; leaf blade, when intact and expanded, simple, **narrowly elliptic to elliptic-lance-shaped (3–12 cm long; 0.5–8 cm wide) [2], undivided or sometimes pinnately lobed**; **leaf margin entire or irregularly toothed or densely fringed with small, soft spines (spinules) [3]**; upper surface hairless and smooth; lower surface green to greyish-green and hairless or sparsely cobwebby. *Flower-heads* (capitula) usually many, in terminal loose flat-topped groups (corymbs); spineless; outer part of each capitulum (the involucre) hairless with 5–7 rows of unequal-sized, tightly over-lapping pointed green scales (phyllaries); outer and middle phyllaries triangular to ovate with an apical spinule (<0.5 mm. long); inner phyllaries elliptic to lance-shaped to almost linear; tips narrowly pointed and thin. *Flowers* densely packed; female florets 1.6–2.4 cm long (petal tube 1.3–1.8 cm); male florets 1.5–1.8 cm long (petal tube 0.9–1.2 cm). *Fruit-heads* a dense mass of achenes (3–4.5 mm long) each with a **very long (2.5–3.5 cm) pappus [4]** of dirty-white feathery hairs. *Odour*: indistinct. *Taste*: slightly bitter. ***Good quality***: abundant green leaves.

DECOCTING PIECES-1 (SHORT LENGTHS OF DRIED AERIAL PARTS): as for crude drug but cut into lengths (c. 1–2 cm long). Characters such as stem branching and leaf shape are no longer visible.

DECOCTING PIECES-2 (CARBONISED AERIAL PARTS — XIAO JI TAN): as for decocting pieces above but externally brownish or brownish-black and internally brown.

PROCESSING

Aerial parts are sun-dried (= crude drug) then later softened and cut into sections for decocting (dp-1). Sections may be subsequently carbonised (= 小蓟炭 Xiao Ji Tan; decocting pieces-2).

OTHER COMMON NAMES

刺儿菜 Ci'er Cai; 刺菜 Ci Cai; 小蓟草 Xiao Ji Cao; 小蓟炭 Xiao Ji Tan; Canadian thistle (SCN); creeping thistle; field thistle herb (CP2010); Herba Cephalanoplos; Herba Cirsii; setose thistle.

SAFETY

An official food and healthfood in China (Anon., 2002a,b).

C. japonicum

Asteraceae

'DRUG SUBSTITUTE' MORPHOLOGY

Stems greyish-white, often thicker (up to 1.2 cm diam.), distinctly ribbed longitudinally **[5]** and **densely clothed in a mat of short hairs interspersed with long filamentous ones [6]**; fracture greyish-white, pith lax or hollowed. ***Leaves* densely spiny [7] and pinnate** (when intact, each leaf blade with 6–12 deeply lobed segments each side), upper surface greyish-green or yellowish-brown and the lower surface lighter in colour; veins with sparse, long, greyish-white slender hairs; **leaf margin toothed,** and the **teeth with marginal and apical spines (2–6 mm long) [8]. *Fruits* with shorter pappus of pale brown bristles (c. 2 cm long).**

OTHER COMMON NAMES

蓟 Ji (plant name); Herba Cirsii Japonici; Japanese thistle (SCN); Japanese thistle herb (CP2010).

SAFETY

An official healthfood in China (Anon., 2002a,b).

> **SUBSTITUTION COMMENT.** Aerial parts are the source of the separate TCM drug 'Da Ji' (Cirsii Japonici Herba) see page 342. The 2 drugs are often used interchangeably owing to their similar clinical properties and they appear together in many TCM texts. While fresh plants of the two species are easily distinguished their dried trade items are not.

Crude drug of Xiao Ji
(*Cirsium arvense* var. *integrifolium*

Carbonised drug (Xiao Ji Tan)

Decocting pieces of Xiao Ji

Crude drug of Da Ji (*C. japonicum*)

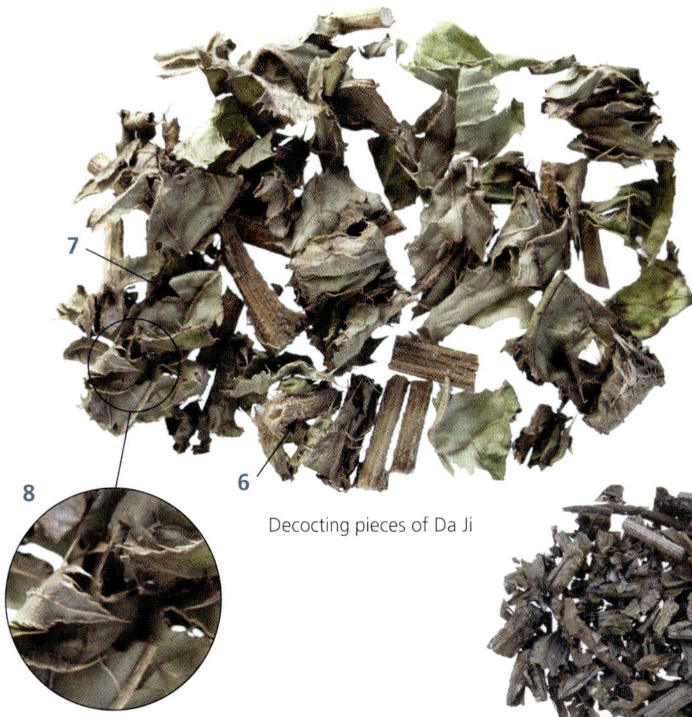

Decocting pieces of Da Ji

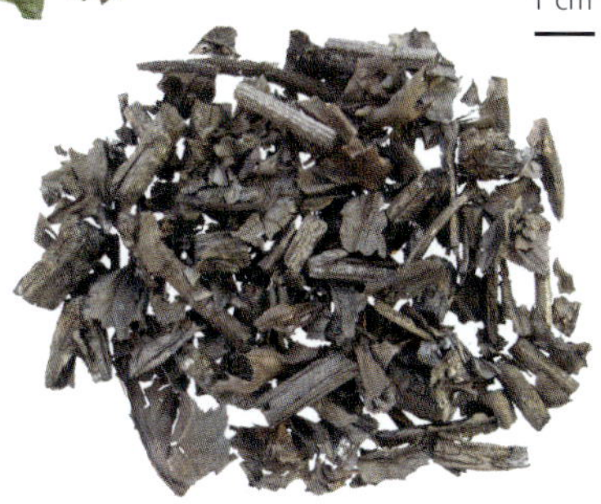

Carbonised drug (Da Ji)

Xi Xian Cao 豨莶草 Siegesbeckiae Herba

Drug part: AERIAL

Sigesbeckia orientalis L.

Asteraceae

accepted name in *Kew-MPNS* (7/9/2016) and *FoJ*.

taxonomic/nomenclatural note: owing to the variable and intermediate characters of the three CP2015 source species (as *Siegesbeckia orientalis* L., *S. glabrescens* Makino and *S. pubescens* Makino) this account follows *Kew-MPNS* and *FoJ* which treat these species as subspecies of *Sigesbeckia orientalis* L. The genus is frequently spelt '*Siegesbeckia*'.

Species of *Sigesbeckia* wild in China: 1 (*Flora of Japan* taxonomy)

PLANT DESCRIPTION

An annual herb (20–120 cm tall) with a tap root. *Stems* obtusely angular, often covered (sometimes densely) with short, soft hairs. *Leaves* opposite and simple; leaf blades ovate to triangular (5–14 cm long; 3–12 cm wide), upper leaves becoming smaller and narrower; leaf bases base truncate or wedge-shaped; one or both surfaces (usually lower) densely covered in fine hairs and **usually scattered with gland exudates (clear or yellow);** margins toothed or shallowly lobed. *Flowers* minute, in many-flowered heads (capitula) (c. 5 mm diam.), inflorescence stalks (peduncles) sometimes distinctly glandular. Each flower-head with 5 radiating, narrow, often spoon-shaped outer **bracts (phyllaries), densely covered in gland-tipped hairs**; bracts in two whorls, the inner whorl much shorter than the outer and partially enclosing the minute yellow florets within. *Fruits (achenes)* dry, dark grey or black, 4-angled (2–3.5 mm long) each with a white apical groove and lacking apical bristles (i.e. a pappus). Fl. Apr–Oct, fr. Jun–Nov.
Based on FoJ; emboldened characters denote key genus characters.

HARVESTING, SOURCING AND NATURAL RANGE

Aerial parts are harvested from wild populations in summer or autumn before or during flowering. The species grows in grassland, scrub, forests and forest margins (100–3,400 m altitude) and is native throughout much of China: Anhui, Fujian, Gansu, Guangdong, Guangxi, Guizhou, Hainan, Hebei, Henan, Hubei, Hunan, Inner Mongolia, Jiangsu, Jiangxi, Jilin, Liaoning, Shaanxi, Sichuan, Taiwan, Tibet, Yunnan and Zhejiang [also native in Bhutan, India, Japan, Korean Peninsula, Laos, Malesia, Nepal, Russia, Thailand, Vietnam; Africa, tropical America, Australia and Oceania].

CONSERVATION STATUS

China: Not Evaluated.
Global: Not Evaluated (IUCN Red List, accessed 25/8/2016).

UNOFFICIAL SUBSTITUTES

Many other Chinese plants are called Xi Xian Cao 豨莶草 so careful identification is needed of this wild-harvested herbal drug so as to avoid the use of accidental substitutes and mixed species batches. Genuine mis-identification may also be a cause for concern. Examples reported in the TCM literature (Bensky *et.al.*, 2004; Xiao *et al.*, 2001–2; Ho *et al.*, 2006; Zhao & Li, 2007) include:

1. *Bidens bipinnata* L. (婆婆针 Po Po Zhen; 鬼针草 Gui Zhen Cao); family Asteraceae. **Leaves** usually hairless and **deeply 2- or 3-lobed; phyllaries** (involucral bracts surrounding flower-heads) **not glandular,** 8–12; **fruits with 2 barbed awns.**
2. *B. tripartita* L. (狼杷草 Lang Pa Cao); family Asteraceae. **Leaves deeply 3-lobed; phyllaries** (involucral bracts surrounding flower-heads) **not glandular;** fruits barbed.
3. *B. parviflora* Willd. (小花鬼针草 Xiao Hua Gui Zhen Cao); family Asteraceae. Stems hairless or with indistinct hairs; leaves **bi- or tri-pinnatisect (divided almost to the midrib)**; phyllaries (involucral bracts surrounding flower-heads) **not glandular**; fruits barbed.
4. *Anisomeles indica* (L.) Kuntze (syn. *Epimeredi indicus* (L.) Rothm.); (豨莶草 Xi Xian Cao); family Lamiaceae. **Strong smelling;** stems distinctly 4-sided; leaves broadly ovate, margins irregularly dentate, hairy; **flowers in tight whorled clusters up the stem**; flowers purplish and 2-lipped.
5. *Phlomis umbrosa* (Turcz.) Kamelin & Makm. (*Kew-MPNS* accepted name, 21/8/2016, *Phlomoides umbrosa* (Turcz.) Kamelin & Makhm.); (糙苏 Cao Su; 豨莶草 Xi Xian Cao); family Lamiaceae. Stems distinctly 4-sided, purplish-red with backward-pointing hairs; leaves ovate or round; **leaf margins with rounded symmetrical teeth; flowers in tight whorled clusters up the stem**; flowers pink to purplish-red, 2-lipped and densely silky.

bi-pinnate leaves of *Bidens bipinnata*

TCM PROPERTIES AND USE

properties: **'pungent, bitter and cold'**
channels: **'Kidney' and 'Liver'**
actions: **'dispels wind damp, open collaterals, clears heat, eliminates toxins'**

indications: **musculoskeletal pain, spasms, rashes and itchy skin; hypertension**

S. orientalis (hairy example)

Inflorescences of S. *orientalis*

Leaf of S. *orientalis*

Bidens tripartita

Anisomeles indica

fruit (achenes) of *B. tripartita* with 2 barbed awns

Phlomis umbrosa

Sigesbeckia orientalis Asteraceae

DRUG MORPHOLOGY

CRUDE DRUG (DRIED AERIAL PARTS): stems **[1]** obtusely angular, greyish-green, yellowish-brown or purplish-brown, (30–110 cm long; 0.3–1 cm diam.) with many opposite branches **[2]** and longitudinal striations; sparsely or densely covered in short, soft hairs **[3]**; nodes distinct and slightly enlarged **[4]**; *texture* brittle; *fracture* uneven, yellowish-white or greenish; pith whitish, broad and hollowed **[5]**. Leaves greyish-green and (if still attached) opposite; blades usually crumpled and fragmented **[6]**; when whole and expanded, ovate to triangular in outline (5–14 cm long; 3–12 cm wide); one or both surfaces (usually lower) densely covered in fine hairs **[7]** and often **with scattered, clear or yellow gland exudates [hl]** **[8]**; margins toothed or shallowly lobed. Flower-/fruit-heads **[9]** occasional, hemispherical or subglobose (c. 5 mm diam.); each flower-/fruit -head with 5 leaf-like, **densely glandular** and narrowly oblong (often spoon-shaped) phyllaries **[10]** in two whorls; the outer whorl much longer than the inner and surrounding several tiny and shrivelled, yellow florets **[11]**. Young fruits (achenes) 4-angled (2–3.5 mm long) **[12]** *Odour*: indistinct. *Taste*: slightly bitter. ***Good quality*:** many dark green leaves; stems fragile.

DECOCTING PIECES-1 (AERIAL LENGTHS): as for crude drug but cut into short lengths (c. 2–4 cm long). Stem sections vary greatly in diameter from narrow fragile ones **[13]** (1 mm diam.) to uneasily broken, almost woody ones **[14]** (c. 5 mm diam.); larger stems more angular and hollow.

DECOCTING PIECES-2 (WINE-PROCESSED AERIAL LENGTHS): as for decocting pieces above but brownish-green or blackish-green and with a wine-like aroma.

PROCESSING

Aerial pars are sun-dried (= crude drug); then softened before cutting into short lengths 2–4 cm long (= decocting pieces-1). Further optional processing involves steaming with yellow rice wine (= decocting pieces-2).

OTHER COMMON NAMES

毛梗豨莶 Mao Geng Xi Xian (*FoC* plant name for *S. orientalis* subsp. *glabrescens*); Xi Xian 豨莶 (*FoC* plant name for *S. orientalis* subsp. *orientalis*); 腺梗豨莶 Xian Geng Xi Xian (plant name for *P. orientalis* subsp. *pubescens*); Herba Siegesbeckiae; Siegesbeckia herb (CP2010).

Leaf underside of Xi Xian Cao (*S. orientalis*)

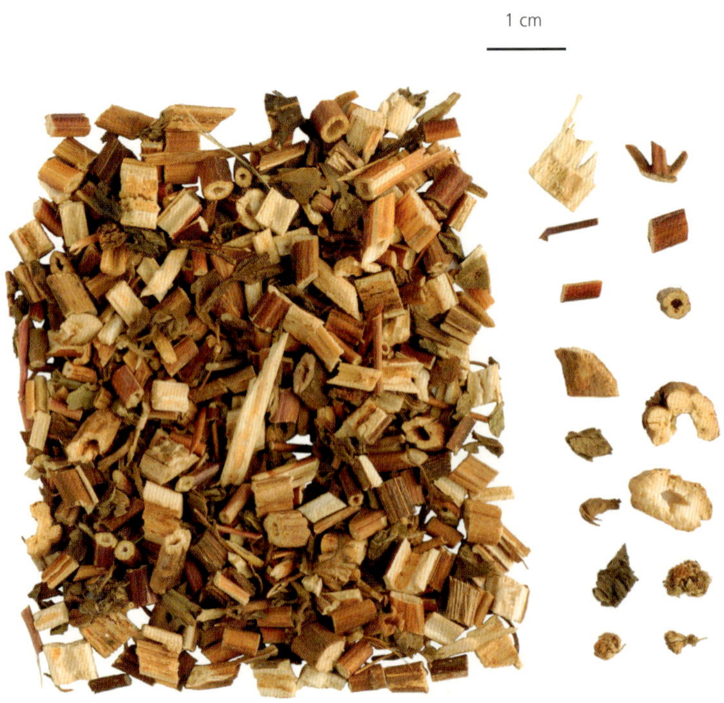

Poor quality Xi Xian Cao containing little leaf material

Crude drug of Xi Xian Cao (*S. orientalis*)

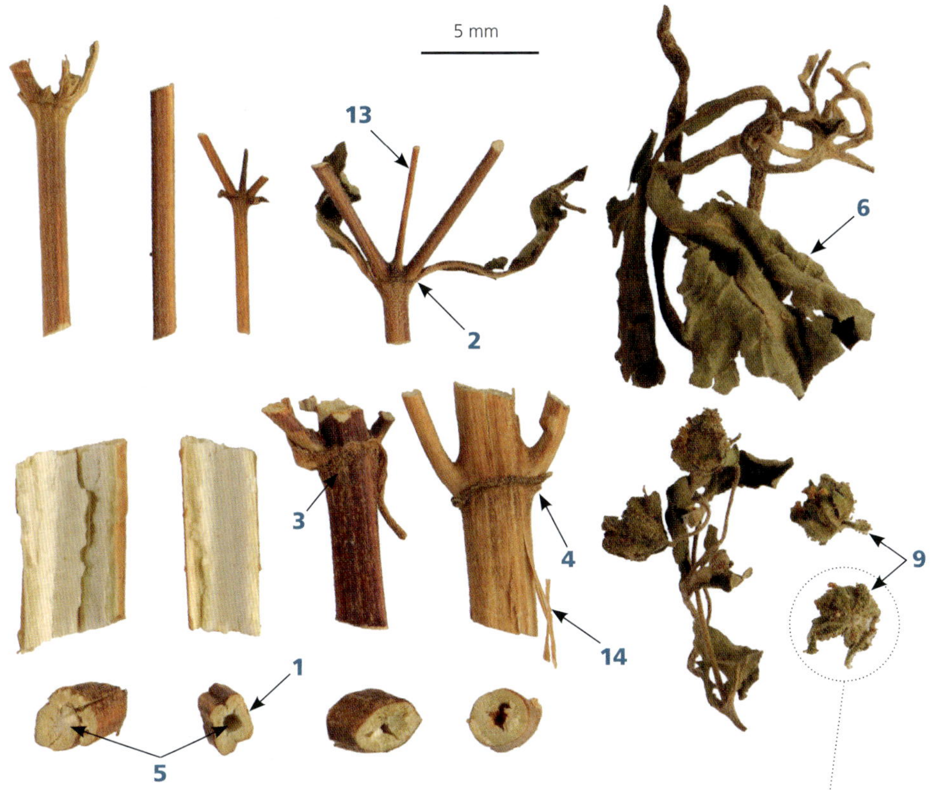

Constituent parts of Xi Xian Cao (*S. orientalis*); left: stems with opposite branching; top right: leaves; bottom right: glandular flower-/fruit-heads

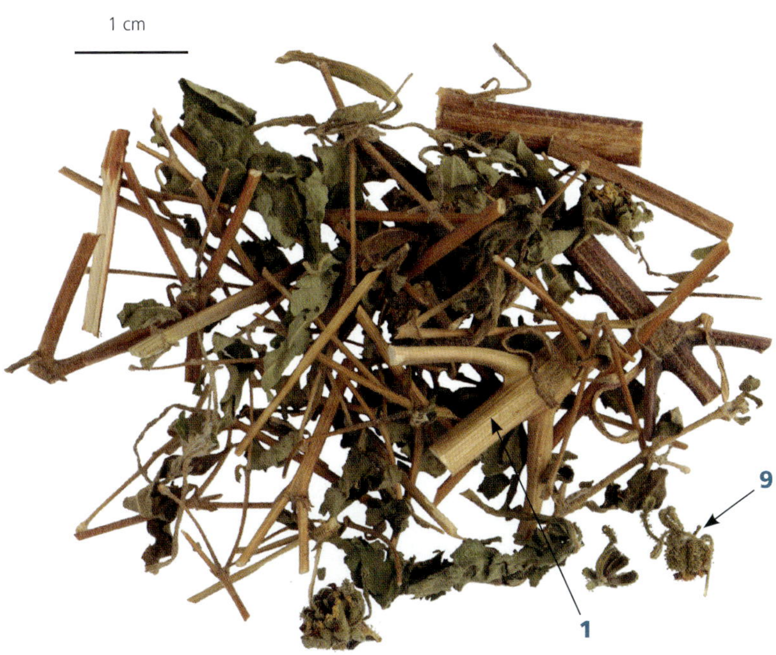

Decocting pieces of Xi Xian Cao (*S. orientalis*)

Flower-/fruit-heads (capitula) of *S. orientalis* with dissected achenes with remains of the flower

Ya Zhi Cao 鸭跖草 — Commelinae Herba

Drug parts: AERIAL

Commelina communis L.

Commelinaceae

Species of *Commelina* wild in China: 8

PLANT DESCRIPTION

A prostrate annual herb, with many branched stems up to 1 m long; internodes 3–9 cm long; rooting at nodes. *Leaves* alternate and simple; leaf blade lance-shaped to ovate-lance-shaped (3–9 cm long; 1.5–2.5 cm wide) with base encircling the stem by a membranous sheath; margin of leaf blade entire. A leafy, curved, longitudinally folded and shortly pointed spathe (1.2–2.5 cm long) is borne opposite each leaf and encloses an *inflorescence* of stalked male and bisexual flowers. Each *flower*, when open, extends beyond the spathe; flower small, bright and conspicuous, laterally symmetrical and lasting 1 day only; sepals 3, boat-shaped and membranous (c. 5 mm long); petals 3, with 2 much larger (9–10 mm long), orbicular and brilliant blue; the 3rd petal is much reduced (c. 5 mm), often white; fertile stamens 3; infertile stamens shorter and with distinctive 4-lobed, butterfly-shaped tips. *Fruit* on a recurved stalk, (and largely concealed within the spathe) a dry, oblong-ellipsoid, 2-valved capsule (5–7 mm diam.). *Seeds* 2 per valve; brownish-yellow, semi-ellipsoid (2–3 mm long), flat on 1 surface, irregularly pitted and one end truncate.

HARVESTING, SOURCING AND NATURAL RANGE

Aerial parts are harvested in summer from wild plants which grow in damp and disturbed places throughout China except Qinghai, Xinjiang, and Tibet.

The species is also wild in Cambodia, Japan, the Korean Peninsula, Laos, ?Malaysia, Russia (Far East), Thailand and Vietnam and has become naturalised in Europe and Eastern USA where many variegated and other flower-colour forms exist.

CONSERVATION STATUS

China: Least Concern (authors' pers. obs., 2007).
Global: Not Evaluated (IUCN Red List, accessed 26/7/2016).

Creeping habit of *Commelina communis* in the wild, Fujian province

Close-up of flower of *C. communis*

Flowering plants of *C. communis*

TCM PROPERTIES AND USE

properties: 'sweet, bland and cool'
channels: 'Bladder, Liver and Stomach'
actions: 'clears heat, cool Blood, eliminate toxins'

indications: toxic swellings; dysuria; (T)

DRUG MORPHOLOGY

CRUDE DRUG (DRIED AERIAL PARTS): yellowish-brown or yellowish-green tangled mass of leaves, stems and flowers **[1]**, up to 60 cm long. *Stems* c. 2 mm diam., longitudinally ridged **[2]**; internodes 3–9 cm long; nodes enlarged; frequently branched and often with fibrous roots persisting **[3]**. *Leaves* alternate, often shrunken and torn; when expanded, intact leaf blades lance-shaped or ovate to lance-shaped **[4]** (3–9 cm long; 1–2.5 cm wide); the leaf base encircles the stem with a membranous sheath; leaf tip pointed (acute) and leaf margin entire. *Sheath* large and distinctive (1.2–2.5 cm long) **[5]**, opposite each leaf, longitudinally folded, curved and shortly pointed. *Flowers* (male or bisexual) usually fallen **[1]**; petals usually crumpled, sometimes with a trace of blue persisting. *Seeds* 2 per valve; brownish-yellow, semi-ellipsoid (2–3 mm long), flat on 1 surface, irregularly pitted and one end truncate. *Stem fracture:* medullated. *Odour:* indistinct. *Taste:* bland. ***Good quality*:** aerial parts mostly yellowish-green.

DECOCTING PIECES: as for the crude drug but cut into short sections.

PROCESSING

Aerial parts are harvested from wild plants in summer and autumn, then sun-dried (= crude drug), these are then cut into sections before decocting.

OTHER COMMON NAMES

蓝花草 Lan Hua Cao; common dayflower herb (2010); Herba Commelina.

Crude drug

Decocting pieces

Yi Mu Cao 益母草

Leonuri Herba

Drug part: FRESH OR DRIED AERIAL

Leonurus japonicus Houtt.

Lamiaceae

synonym: *L. heterophyllus* Sweet

Species of *Leonurus* wild in China: 12

PLANT DESCRIPTION

An erect, herbaceous annual or biennial (30–120 cm tall) with a taproot and dense fibrous rootlets. *Stems* 4-sided, deeply furrowed longitudinally, branching in lower parts and minutely hairy on nodes and angles. *Leaves* opposite with each pair at right angles to the next; ***leaves of 2 main shapes:*** **i. stem leaves deeply divided into 3 lobes with each lobe more than 3 mm wide,** oblong to linear with a narrowly wedge-shaped base; **ii.** leaves in flowering spikes long and thin (3–12 mm long; 2–8 mm wide), *not divided into lobes,* with margins entire or weakly toothed. *Flowers* in dense 8–15-flowered whorls; each whorl 2–2.5 cm diam. arranged at intervals along a long flowering spike with stiff, sharp spines (bracteoles) between the flowers. Each flower bisexual and laterally symmetrical; calyx funnel-shaped (longer than bracteoles), **5-veined**, *minutely hairy* [hl], with 5 triangular, spine-tipped lobes (2 lobes slightly larger than others); corolla white, reddish or purplish-red (1–1.2 cm long), 2-lipped (upper lip straight, *oblong* and concave (c. 7 mm long; 4 mm wide), fringed with hairs; lower lip slightly shorter, divided into 3 lobes, spotted or striped, with middle lobe much larger than lateral lobes); stamens 4; ovary superior and style long-pointed. *Fruit* with 4 dry nutlets, strongly 3-angled, truncate and enclosed in a persistent calyx. Fl. Jun–Sep, fr. Sep–Oct.

HARVESTING, SOURCING AND NATURAL RANGE

Sourced from wild and medicinal crop plants grown widely in China. Aerial parts are harvested from spring to early summer before flowering for trading or fresh 'Yi Mu Cao'. The dried drug is harvested in summer when there is abundant leaf growth and before the flowers are fully open.

The species grows wild in sunny, mostly disturbed places (0–3,400 m altitude) throughout much of China: Anhui, Fujian, Gansu, Guangdong, Guangxi, Guizhou, Hainan, Hebei, Heilongjiang, Henan, Hubei, Inner Mongolia, Hunan, Jiangsu, Jiangxi, Jilin, Liaoning, Ningxia, Qinghai, Shaanxi, Shandong, Shanxi, Sichuan, Taiwan, Xinjiang, Tibet, Yunnan and Zhejiang [also native in Cambodia, Japan, Korean Peninsula, Laos, Malaysia, Myanmar, Thailand, Vietnam; Africa, North and South America].

CONSERVATION STATUS

China: Not Evaluated.

Global: Not Evaluated (IUCN Red List, accessed 21/8/2016).

For other TCM drugs sourced from this species see 'Chong Wei Zi' (Leonuri Fructus) page 618.

Leonurus japonicus showing unlobed linear leaves between the fruit whorls

Flowering spike of *L. japonicus*

UNOFFICIAL SUBSTITUTES

The TCM literature reports various local substitutes of 'Yi Mu Cao' in China sourced from closely related genera (Lamiaceae family). Reasons behind their use may be mis-identification or local clinical preference/availability; their occurrence as substitutes of 'Yi Mu Cao' in international trade is unclear. Most frequently cited species:

Leonurus sibiricus

1. ***Lagopsis supina*** (Steph. ex Willd.) Ikonn.-Gal. (夏至草 Xia Zhi Cao). Common weed of waste places in N and C China (0–2,600 m altitude); used as a regional substitute in Inner Mongolia and traded as 'Yi Mu Cao' (same as official). Easily confused with the official species: **much shorter** (15–35 cm tall); **leaves circular** (1.5–2 cm long; 1.5–2 cm wide); whorls few-flowered; **calyces 5–10-veined;** flowers white (rarely soft pink) and **covered in woolly hairs.**
2. ***Leonurus sibiricus*** L. (细叶益母草 Xi Ye Yi Mu Cao; Siberian motherwort). Easily confused with the Pharmacopoeia species as living and dried plant material; the **deeply lobed floral leaves of *L. sibiricus*** are one of its key distinguishing characters *(see right-hand photo).* The name '*L. sibiricus* L.', furthermore, has been widely mis-applied to *L. japonicus* Houtt. (Harley & Paton, 2001) creating much confusion about its alleged medicinal use and pharmacology.
3. ***Panzerina lanata*** (L.) Soják (绒毛脓疮草 Rong Mao Nong Chuang Cao). Common species of desert steppes in N China. Many-branched perennial with a woody base; leaf undersides conspicuously covered in white hairs; stem leaves 5-lobed; flower spikes short and dense; floral leaves 3-lobed; calyx 10-veined; flowers yellowish-white with dense silky hairs.

TCM PROPERTIES AND USE

***properties*: 'bitter, pungent and cool'**
***channels*: 'Heart, Liver and Bladder'**
***actions*: 'invigorates blood, dispels stasis, promotes urination, clears heat, eliminates toxins'**

***indications*: dysmenorrhoea, infertility; abdominal pain; dysuria; sores and abscesses**

DRUG MORPHOLOGY

CRUDE DRUG-1 (FRESH AERIAL PART): as for the dried crude drug below but stems dark green and tender and flower-buds tightly closed. ***Good quality:*** young plants with abundant greyish-green foliage.

CRUDE DRUG-2 (DRIED AERIAL PART): predominantly leaves, with some stems and flowering spikes. *Leaves and stems* greyish-green or yellowish-green **[1]**. *Stems* light and pliable, 4-sided and deeply furrowed longitudinally **[2]**, minutely hairy [hl], internally with a white pith **[3]**. *Leaves* mostly crumpled or fragmente **[4]**; usually fallen off; if still attached to stem, leaf bases opposite **[5]**. When intact and expanded, **leaf blades of 2 types: i) stem leaves [6]** (below flowering spikes) deeply divided into 3 lobes **each more than 3 mm wide;** lobes oblong or ovate; **ii) floral leaves** (between flower whorls) **long, thin and undivided [7]** (3–12 mm long; 2–8 mm wide), margins entire or weakly toothed. *Flowers* (when attached to stems) in **8–15-flowered** whorls **[8]** (each whorl 2–2.5 cm diam.) at intervals up the long flowering spikes; whorls with short and often curved spines (bracteoles) **[9]** between flowers; calyx **10]** membranous, funnel-shaped, **5-veined, minutely hairy** [hl] with 5 spiny teeth **[11]**. Flower **[12]** usually faded to yellowish-brown but if freshly dried white, reddish or purplish-red (1–1.2 cm long when fresh), laterally symmetrical and 2-lipped; stamens 4; style long-pointed and apically 2-cleft. *Odour:* indistinct. *Taste*: slightly bitter.

DECOCTING PIECES (DRIED AERIAL LENGTHS: as for the dried crude drug above but cut into short lengths with many leaf characters destroyed due to fragmentation making differentiation from closely related species in the genus (e.g. *L. sibiricus* illustrated opposite) no longer possible. *Odour:* indistinct. *Taste*: slightly bitter.

PROCESSING

Aerial parts are sun-dried (= crude drug) then cut into short decocting lengths or cut first then dried.

OTHER COMMON NAMES

红花艾 Hong Hua Ai; Chinese motherwort (SCN); Herba Leonuri; motherwort herb (CP2010).

SAFETY

An official healthfood in China (Anon., 2002a,b). Safety Class: 2b; Interaction Class: A (Gardner & McGuffin, 2013).

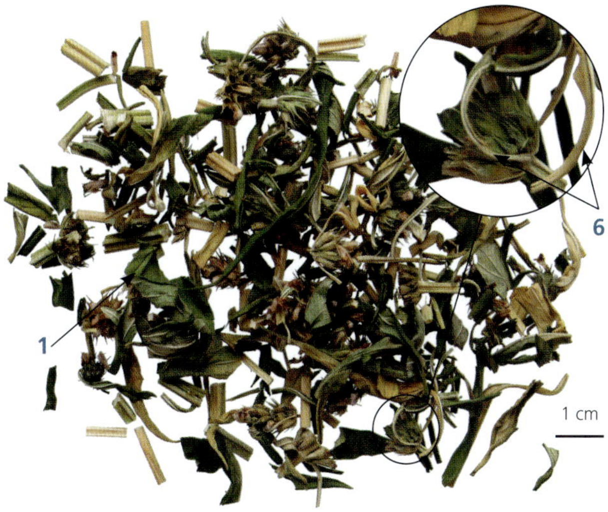

Decocting pieces of Yi Mu Cao (*L. japonicus)*

Dried crude drug of Yi Mu Cao (*L. japonicus)*

Stem fragments

Flowering spike of *L. japonicus* showing linear, unlobed floral leaves

Verticillaster (flower whorl) of *L. japonicus*

Yin Chen 茵陈 (or Yin Chen Hua 茵陈蒿) Artemisiae Scopariae Herba

Drug part: AERIAL

PHARMACOPOEIA SPECIES 1

Artemisia scoparia Waldst. & Kit.

Asteraceae

PLANT DESCRIPTION

An annual, biennial or perennial herb (40–90 (rarely 130) cm tall); strongly aromatic. *Stems* erect and branching throughout. *Leaves* alternate, with lower and middle stem leaves usually covered in soft, silky hairs. In mature plants, 3 leaf-shapes present up the stem. *Lower stem leaves* with 2–4 cm long stalks, blades ovate-oblong or elliptic (1.5–3.5 cm long; 1–3 cm wide) or deeply divided into paired and branching lobes. *Middle stem leaves* stalkless, blades oblong or ovate-oblong (1–2 cm long; 0.5–1.5 cm wide) and deeply divided into slender, thread-like and often curved lobes. *Upper stem leaves* and leaf-like bracts tiny and divided into 3–5 thread-like lobes. *Inflorescences* loose and branching with many, small, inconspicuous flower-heads (capitula). Each flower-head sub-globose (1.2–1.5 mm long; 1–1.5 mm diam.), short-stalked, with 3–4 rows of ovate scale-like phyllaries below the tiny tubular florets; outer florets female, 5–7; inner (disk) florets bisexual, 4–10. *Fruiting heads*: clusters of tiny, dry, ovoid achenes. Fl. and fr. Jul–Oct.

This description excludes the morphology of the leaves of spring seedlings, the source of Mian Yin Chen, and whose morphology is described under 'Drug Morphology' below.

Species of *Artemisia* wild in China: 186

HARVESTING, SOURCING AND NATURAL RANGE

Aerial parts are harvested from wild and medicinally cultivated plants predominantly sourced from Shaanxi, Shanxi and Anhui provinces. Two trade products can be distinguished according to the season in which they are harvested. Spring or early summer (May and June) harvesting of seedlings (c. 6–10 cm tall) yields 'Mian (seedling) Yin Chen'; late summer to autumn harvesting of the larger and somewhat woody flowering plants yields 'Hao (winter) Yin Chen'. Different clinical properties have traditionally been associated with the two products (Cha *et al.*, 2005); the CP2015 includes them in one monograph suggesting they are interchangeable. Wild plants are typically found in dry places (up to 3,200 m altitude) throughout much of China [also native in Afghanistan, India, Japan, the Korean Peninsula, Pakistan, Russia, Thailand; C & SW Asia and Europe].

CONSERVATION STATUS

China: Least Concern (authors' pers. obs. 2010).
Global: Not Evaluated (IUCN Red List, accessed 3/9/2015).

PHARMACOPOEIA SPECIES 2

A. capillaris Thunb.

Asteraceae

PLANT DESCRIPTION

Extremely difficult to distinguish from PS1 and requires dissection of the flower-heads under magnification. *Main differences*: always perennial; several roots; middle stem leaves with lobes narrowly linear and not more than 0.5–1 mm wide; flower-head (capitulum) slightly larger (1.75–2.25 mm long; 1–1.5 mm wide) with **3–5 outer female florets** and 5–7 inner bisexual (disk) florets [hl or other magnification].

HARVESTING, SOURCING AND NATURAL RANGE

As for PS1 with two product types produced. Aerial parts are harvested from wild populations which grow in moist places along roadsides, terraces and river banks (100–2,700 m altitude). The species is native throughout China except the most northerly provinces; [also native in Cambodia, Indonesia, Japan, the Korean Peninsula, Malaysia, Philippines, E Russia and Vietnam].

CONSERVATION STATUS

China: Least Concern (authors' pers. obs. 2010).
Global: Not Evaluated (IUCN Red List, accessed 3/9/2015).

UNOFFICIAL SUBSTITUTE

Origanum vulgare L.

Lamiaceae

PLANT DESCRIPTION

Easily distinguished from the Pharmacopoeia species. *Main differences*: An herbaceous perennial (25–60 cm tall). ***Stems* 4-angled, covered in soft white hairs**, purplish. ***Leaves* opposite, entire (not lobed), ovate to oblong-ovate** (1–4 cm long; 0.4–1.5 diam.) **with dense dot-like glands. *Flowers*** larger, **shortly trumpet-shaped** and 2-lipped, **purplish-red to white**. Fl. Jul–Sep, fr. Oct–Dec.

HARVESTING, SOURCING AND NATURAL RANGE

Aerial parts are harvested from wild populations which grow in grassland and forest margins; 500–3,600 m altitude. The species is native throughout much of China: Anhui, Fujian, Gansu, Guangdong, Guizhou, Henan, Hubei, Hunan, Jiangsu, Jiangxi, Shaanxi, Sichuan, Taiwan, Tibet, Xinjiang, Yunnan and Zhejiang [also native in Kazakhstan, Kyrgyzstan, Russia; Africa, Europe and introduced in North America]

CONSERVATION STATUS

China: Least Concern (authors' pers. obs. 2010).
Global: Least Concern (authors' pers. obs. 2010); Not Evaluated (IUCN Red List, accessed 28/8/2015).

TCM PROPERTIES AND USE

properties: **'bitter and cool'**
channels: **'Liver, Spleen and Gall Bladder'**
actions: **'dispels wind damp, releases the exterior'**

indications: **jaundice; dysuria, infectious hepatitis, liver cirrhosis; fever; skin conditions; (T)**

Typical untidy habit of *Artemisia scoparia*

Young seedling of *A. scoparia* used as the source of Mian Yin Chen

Branch of *A. scoparia* in full flower

A. capillaris in the wild

Branch of *A. capillaris* in flower-bud

Inflorescence of *Origanum vulgare*

LESSER KNOWN SUBSTITUTES

Artemisia is a notoriously difficult genus to identify to species level. The most commonly confused species reported in the TCM literature are listed below; their occurrence as substitutes in international trade is uncertain.

1. *Artemisia frigida* Willd. (Leng Hao 冷蒿). **Lower leaves with stalks shorter** (0.5–2 cm long) and **blades smaller** (0.8–1.5 cm long; 0.8–1.5 cm wide); **middle leaves smaller** (5–7 mm long; 5–7 mm wide); **capitula broader** (usually 2.5–4.5 mm diam.); **female florets 8–13** and **bisexual florets 20–30.**
2. *A. gmelinii* Weber ex Stechmann (Xi Lie Ye Lian Hao 细裂叶莲蒿). A **tufted subshrub with a woody rhizome;** stems branching in upper parts only; leaves gland-dotted; **middle stem leaves stalked** (1–5 cm) and often much larger (2–10 cm long; 2–8 cm wide) with a serrated central axis and lobes with comb-like teeth; **capitula larger** 2–3.5(–5) mm diam.; **marginal female florets 10–12; central bisexual florets 20–40.**
3. *A. anethoides* Mattf. (Shi Luo Hao 莳萝蒿). Used as fodder for livestock. Scale-like phyllaries of capitula densely covered in soft hairs; **marginal female florets 3–6; central bisexual florets 8–16.**
4. *A. fauriei* Nakai (Hai Zhou Hao 海州蒿). Coastal beaches and sandbanks of estuaries. A **much shorter** herb (20–40 cm tall) **covered in a mat of short dense hairs**. Basal leaves much larger (11–18 cm long; 8–16 cm wide). **Marginal female florets 2–5; central bisexual florets 8–15.**

Artemisia scoparia

Asteraceae

DRUG MORPHOLOGY

CRUDE DRUG (DRIED AERIAL PARTS): the two trade products 'Mian Yin Chen' and 'Hua Yin Chen' are macroscopically distinct but identification to species level is only possible for the latter since this requires the dissection (under magnification) of the flower-heads (flower-heads are not present in 'Mian Yin Chen' since this is seedling material): *Mian Yin Chen* (*seedling stage*): a mass of tightly curled greyish-or silvery-white immature leaves **[1]** densely covered in soft, silky hairs **[2]**. Short woody basal stems sometimes present, light and brittle. When whole and expanded leaves stalked and blades 1–3 cm long, c. 1 cm wide and divided 1–3-pinnately (almost to midrib) into ovate lobes **[3]** with pointed tips **[4]**. *Odour*: aromatic. *Taste*: slightly bitter.

Hua Yin Chen (*mature stage with flowers or fruits*): stems cylindrical, slender and many-branched (30–100 cm long; 2–8 mm diam.) **[5]**, externally purple or yellowish-brown **[6]**, longitudinally ridged and softly hairy; *texture*: light and brittle; *fracture*: cream. Usually only upper stem leaves present: tiny and divided into 3–5 thread-like lobes (0.2–0.5(–1) mm wide) **[7]**. Inflorescence loose and branching with many, small and inconspicuous flower-heads (capitula). Each capitulum sub-globose (1.2–1.5 mm long; 1–1.2 mm diam.) **[8]**, short-stalked, with 3–4 rows of ovate scale-like phyllaries **[9]** surrounding the tiny tubular florets [hl]; **outer female florets 5–7 [10]**; **inner bisexual florets 4–10 [11] [hl or other magnification required]**. *Fruiting heads*: clusters of tiny, dry, ovoid, yellowish-brown achenes. *Odour*: aromatic; *taste*: slightly bitter. ***Good quality***: young, soft and greyish-white; intensely aromatic.

DECOCTING PIECES (DRIED LENGTHS OF AERIAL PARTS): *Mian Yin Chen* and *Hua Yin Chen*: as for the crude drug; latter broken or cut into short sections.

PROCESSING

Aerial parts of both types are sun-dried (mature stems of 'Hua Yin Chen' are removed beforehand) (= crude drug). 'Hua Yin Chen' is then cut or broken into short decocting pieces while 'Mian Yin Chen' is simply sifted to remove foreign matter.

OTHER COMMON NAMES

茵陈蒿 Yin Chen Hua; 猪毛蒿 Zhu Mao Hua (plant name); Herba Artemisiae Scopariae; virgate wormwood herb (CP2010); Yin-Chen wormwood (SCN).

SAFETY

Safety Class: 1; Interaction Class: A (Gardner & McGuffin, 2013).

A. capillaris

Asteraceae

DRUG MORPHOLOGY

CRUDE DRUG AND DECOCTING PIECES (DRIED AERIAL PARTS): extremely difficult to distinguish macroscopically from PS1 and requires dissection of the flower-heads (capitula) under magnification. *Main differences from PS1*: flower-head (capitulum) slightly larger (1.75–2.25 mm long; 1–1.5 mm wide) with **outer female florets 3–5** and inner bisexual ones 5–7 [hl or other magnification required].

PROCESSING

As for PS1.

OTHER COMMON NAMES

茵陈蒿 Yin Chen Hua (plant name); Bin Hao; Herba Artemisiae Scopariae; virgate wormwood herb (CP2010); Yin-Chen wormwood (SCN).

SAFETY

As for PS1.

Origanum vulgare

Lamiaceae

'DRUG SUBSTITUTE' MORPHOLOGY

DRIED AERIAL PARTS: easily distinguished from the drug of both Pharmacopoeia species: **stems 4-angled [12], covered in soft white hairs [13]** and purplish. **Leaves much larger and undivided [14]**: opposite and, when expanded and whole, ovate to oblong-ovate (1–3.5 cm long; 0.4–1 diam.) **densely dotted with glands [15]**. Flowers in terminal and axillary clusters, much larger, shortly trumpet-shaped and 2-lipped. See also *Plant Description*.

PROCESSING

Aerial parts are sun-dried.

OTHER COMMON NAMES

广茵陈 Guang Yin Chen; 牛至 Niu Zhi (plant name); 土茵陈 Tu Yin Chen; marjoram; oregano.

SUBSTITUTION COMMENT. A widespread species of temperate China used in folk medicine and, worldwide, as a culinary herb ('pot marjoram'). Accidental substitution occasionally arises due to confusion of its similar trade name Tu Yin Chen 土茵陈.

Crude drug/decocting pieces of Mian Yin Chen
(seedlings of *A. scoparia*)

Expanded leaf of Mian Yin Chen

Crude drug of Hua Yin Chen (*A. scoparia*)

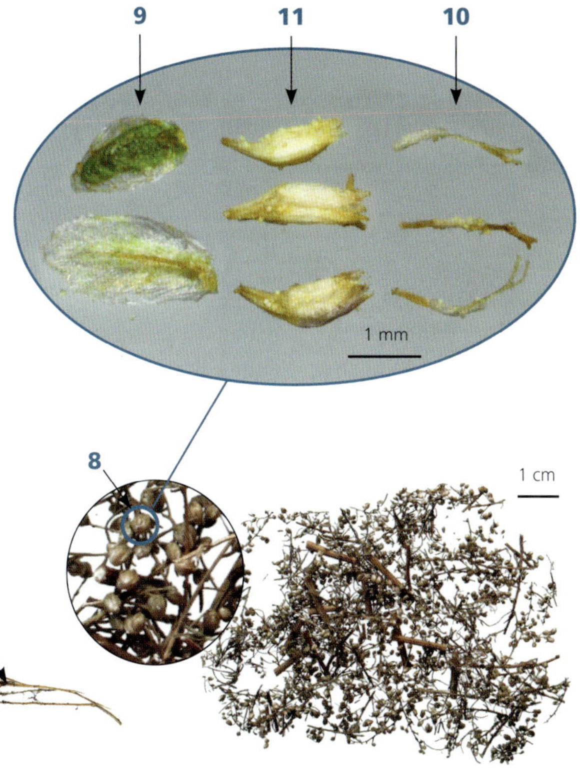

Decocting pieces of Hua Yin Chen (*A. scoparia*)

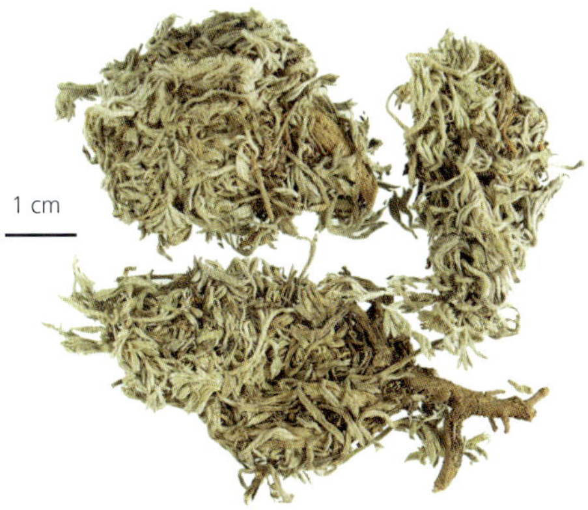

Crude drug/decocting pieces (*A. capillaris*)

Crude drug (*A. capillaris*)

Decocting pieces (*A. capillaris*)

Crude drug of *O. vulgare*

Decocting pieces of *O. vulgare*

Ze Lan 泽兰

Lycopi Herba

Drug part: AERIAL

PHARMACOPOEIA SPECIES

Lycopus lucidus Turcz. ex Benth. **var. *hirtus*** (Regel) Makino & Nemoto

Lamiaceae

synonym: *L. lucidus* f. *hirtus* Regel

taxonomic/nomenclatural note: in CP2015 as '*Lycopus lucidus* Turcz. var. *hirtus* Regel'

Species of *Lycopus* wild in China: 4

PLANT DESCRIPTION

An erect, usually unbranched herbaceous perennial (20–70 cm tall) with an apically swollen rhizome. ***Stem* 4-sided**, distinctly furrowed longitudinally, **often purplish-red, nodes densely hairy** and angles minutely hairy. Leaves opposite with each pair at right angles to adjacent ones; floral leaves similar to stem leaves; **leaf blades simple**, dark green, **lance-shaped** (4–10 cm long; 1.2–2.5 cm diam.), much longer than stem internodes; tips narrowly pointed; margins entire or shallowly lobed; lower surfaces densely gland-dotted; margins of main stem leaves coarsely saw-toothed while those of upper stem are entire. Flowers in dense, globose, many-flowered whorls (verticillasters) (1.2–1.5 cm diam.) along the flowering stem with short, spine-like bracteoles between the flowers. Each flower bisexual and laterally symmetrical; calyx bell-shaped with 5 spine-tipped teeth; corolla white, bell-shaped and opening into a 2-lipped mouth (the slightly larger upper lip undivided; the lower lip 3-lobed with middle lobe of lower lip larger than lateral ones); ovary superior; style 1, extending beyond the flower and apically 2-cleft. Fruits squat, brown, 4-sided nutlets with a flat base and truncate apex; nutlets shorter than persistent calyx. Fl. Jun–Sep, fr. Aug–Nov.

HARVESTING, SOURCING AND NATURAL RANGE

Aerial parts are harvested from wild plants, often when in flower in summer and autumn and when the foliage is abundant. The species grows in marshy places (300–2,400 m altitude) throughout much of China: Anhui, Fujian, Gansu, Guangdong, Guangxi, Guizhou, Hebei, Heilongjiang, Hubei, Hunan, Inner Mongolia, Jiangsu, Jiangxi, Jilin, Liaoning, Shaanxi, Shandong, Shanxi, Sichuan, Taiwan, Yunnan and Zhejiang [also wild in Japan and Russia].

CONSERVATION STATUS

China: Least Concern (MEPCAS, 2013).

Global: Not Evaluated (IUCN Red List, accessed 26/8/2016).

UNOFFICIAL SUBSTITUTE

Eupatorium fortunei Turcz.

Asteraceae

PLANT DESCRIPTION

Easily differentiated from the official source of Ze Lan (above). *Main differences*: highly aromatic when crushed; ***stems* cylindrical, without dense white hairs at nodes**; ***flowers*** white, pink or reddish **in loose, branching terminal heads** (not in whorls up the stems); *leaves* similar **but middle stem leaves distinctly 3-lobed with middle lobe much larger (5–10 cm long; 1.5–2.5 cm wide) than lateral lobes.**

HARVESTING, SOURCING AND NATURAL RANGE

Aerial parts are harvested in summer and autumn from plants grown on medicinal plant farms. The species grows wild in thickets and roadside ditches (c. 2,000 m altitude) in C, S and W China: ?Anhui, ?Fujian, Guangdong, Guangxi, Guizhou, Hainan, Henan, Hubei, Hunan, Jiangsu, Jiangxi, Shaanxi, Shandong, Sichuan, Yunnan and Zhejiang provinces [possibly introduced in Japan, the Korean Peninsula, Thailand and Vietnam].

CONSERVATION STATUS

China: Least Concern (MEPCAS, 2013).

Global: Not Evaluated (IUCN Red List, accessed 26/8/2016).

SUBSTITUTION COMMENT. Aerial parts are the source of the separate TCM drug 'Pei Lan' (Eupatorii Herba) see page 374. The widespread confusion between these two drugs is most likely due to: i) an earlier shared Chinese name Ze Lan 泽兰; and ii) possible identification confusion of the fresh plants (when not in flower) and the drugs (especially as decocting pieces). Given that the TCM properties and indications of the two drugs are quite different, intentional substitution for clinical reasons seems very unlikely.

TCM PROPERTIES AND USE

properties: **'bitter, pungent and neutral'**
channels: **'Liver and Spleen'**
actions: **'invigorates Blood, regulates water'**

indications: **dysmenorrhoea, amenorrhoea; traumatic injury; oedema; sores and abscesses**

Typical waterside habitat of *Lycopus lucidus* var. *hirtus*, Gansu province

Flowering plants of *L. lucidus* var. *hirtus*

Flowering spike of *L. lucidus* var. *hirtus*

Eupatorium fortunei in full flower

Young inflorescence of *E. fortunei*

Leaves on middle part of stem 3-lobed

LESSER KNOWN UNOFFICIAL SUBSTITUTES

The local use in China of closely related and easily confused varieties and species in the same genus (*Lycopus*) has been reported. This is probably due to mis-identification when plants are wild-harvested.

1. *L. lucidus* var. ***lucidus*** (Di Sun; 地瓜儿苗 Di Gua Er Miao). Stems hairless or **very sparsely hairy on nodes; leaves hairless.** These tiny differences and ease of availability, especially in N China and Yunnan where its rhizomes are eaten as a boiled or salted vegetable, may have led to its as yet undetected occurrence in international trade as 'Ze Lan'. Further research is needed.
2. *L. lucidus* var. ***maackianus*** Maxim. ex Herder (异叶变种 Yi Ye Bian Zhong; 狭叶地瓜儿苗 Xia Ye Di Gua Er Miao). Stems slender and **lower leaves shallowly lobed.** Locally uncommon in the wild and endemic to Heilongjiang in NE China.
3. ***L. europaeus*** L. (欧地瓜儿苗 Ou Di Gua Er Miao). Native to Hebei, Shaanxi and Xinjiang. Lower and middle stems leaves with **basal part of margins shallowly lobed**; upper leaves coarsely saw-toothed.
4. ***L. parviflorus*** (Benth.) Maxim. (*Kew-MPNS* accepted name, 24/8/2015, *Lycopus uniflorus* Michx.); 地瓜儿苗 Di Gua Er Miao; 小花地笋 Xiao Hua Di Sun). Calyx lobes **shortly pointed** (not spine-tipped); nutlets **longer than persistent calyx.** A locally common species but confined to Heilongjiang and Jilin (endemic to China).

Lycopus lucidus var. *hirtus*

Lamiaceae

DRUG MORPHOLOGY

CRUDE DRUG (AERIAL PART): *stems* usually yellowish-green, yellowish-brown or purple, unbranched, **4-sided**, markedly 4-furrowed longitudinally (5–10 cm long; 2–6 mm diam.) **[1]** and minutely hairy [hl] along stem angles; **nodes often purple and densely covered in fine hairs [2]**; *stem texture:* brittle and *fracture* yellowish-white, with hollowed pith. *Leaves* folded or crumpled; opposite; simple and short- or unstalked; if still attached to stem then often emerging from centre of a flower cluster (verticillaster) which surrounds a stem node; both surfaces of leaves greyish-green (sometimes blackish on upper); upper surface minutely hairy [hl]; **lower surface densely gland-dotted [hl] [3]** and minutely hairy [hl] on veins; when whole and expanded blades lance-shaped or oblong (5–10 cm long; 1.2–2.5 cm wide), tips narrowly pointed and **margins distinctly saw-toothed** (main stem leaves **[4]**), or entire (apical leaves). *Flowers* tiny (2–3 mm diam.), shrivelled or fallen off; when present, brown or cream and 2-lipped [hl] **in tight, many-flowered clusters (verticillasters) along stems [5]**. If flowers have fallen then cup-shaped calyx tube persists; greyish-green or yellowish-brown with 5 spine-tipped lobes. A single (floral) leaf may also persist emerging from the middle of each flower cluster; these are similar in size and shape to the lower stem leaves. *Odour*: indistinct. *Taste*: indistinct. ***Good quality***: unfragmented plants, with many leaves (yellowish-green or greyish-green) and flowers.

DECOCTING PIECES (AERIAL PARTS CUT INTO SHORT LENGTHS): as for the crude drug but leaves and stems cut into short (1–2 cm) lengths. Characters such as stem branching and leaf size, shape and position along stem will be lost.

PROCESSING

Aerial parts are sun-dried (= crude drug), then later, after softening with water, are cut into short decocting lengths.

OTHER COMMON NAMES

地瓜儿苗 Di Gua Er Miao; 地笋 Di Sun; Herba Lycopi; hirsute shiny bugleweed herb (CP2010); shiny bugleweed (SCN).

SAFETY

An official healthfood in China (Anon., 2002a,b).

Eupatorium fortunei

Asteraceae

'DRUG SUBSTITUTE' MORPHOLOGY

AERIAL PART: stems cylindrical [6] and without whorls of flowers, calyces or bracteoles; **nodes slightly swollen and virtually hairless [7]**; **leaves without glands [hl] [8]** or hairs, blades simple or 3-lobed with middle lobe much larger (5–10 cm long; 1.5–2.5 cm wide); **leaf margins less distinctly toothed [9]**. *Odour:* aromatic. *Taste:* slightly bitter.

SHORT LENGTHS OF AERIAL PARTS: stems cylindrical [6] and without whorls of flowers, calyces or bracteoles; **nodes (if present) slightly swollen and virtually hairless [7]**; **leaves without glands [8]** or hairs; leaf margins irregularly or indistinctly toothed **[9]**. *Odour*: aromatic. *Taste*: slightly bitter.

PROCESSING

Aerial parts are sun-dried (= crude drug) then later, softened briefly before cutting into decocting lengths.

OTHER COMMON NAMES

大泽兰 Da Ze Lan; 佩兰 Pei Lan; 山泽兰 Shan Ze Lan; Chinese eupatorium (SCN); Fortune eupatorium herb (CP2010); Herba Eupatorii.

SAFETY

An official healthfood in China (Anon., 2002a,b). Legal restrictions and recommendations exist in many countries about the internal use of pyrrolizidine alkaloid-containing plant species (such as *Eupatorium* species) due to their assocation with liver damage and cancer (Stewart & Steenkamp, 2001). Such restrictions and advice are currently in a state of flux worldwide reflecting new lines of research (EMA, 2014; HMPC, 2016).

Crude drug of Ze Lan (*Lycopus lucidus* var. *hirtus*)

Decocting lengths of Ze Lan (*L. lucidus* var. *hirtus*)

Crude drug of Pei Lan (*Eupatorium fortunei*)

Decocting pieces of Pei Lan (*E. fortunei*)

Zi Hua Di Ding 紫花地丁

Violae Herba

Drug part: WHOLE PLANT

PHARMACOPOEIA SPECIES

Viola philippica Cav. var. ***philippica***

Violaceae

synonym: *V. yedoensis* Makino (name used in CP2015).

taxonomic/nomenclatural note: *FoC* also recognises a second variety: *V. philippica* var. *pseudojaponica* (Nakai) Y. S. Chen which is restricted to Taiwan and Japan.

Species of *Viola* wild in China: 96

PLANT DESCRIPTION

A short perennial herb (4–14 cm tall up to 20 cm when fruiting) with a short, stout, white to brownish rhizome. Creeping horizontal shoots absent. *Leaves* in a basal rosette, simple; leaf stalks narrowly winged in upper part, becoming broadly winged at fruiting; *stipules* membranous (1.5–2.5 cm long) and fused almost throughout their length to base of leaf stalks; upper leaves oblong to lance-shaped or triangular-ovate (1.5–10 cm long; 0.5–4 cm wide) and usually smaller than lower ones, lower ones oblong, narrowly oblong-lance-shaped or oblong-ovate (1.5–4 cm long; 0.5–1 cm wide); leaf bases ending abruptly or wedge-shaped (rarely heart-shaped), surfaces hairless or slightly hairy, tip blunt, margin shallowly scalloped. *Flowers* solitary on leafless stems emerging above the leaves; laterally symmetrical and bisexual; sepals 5 with tips tapering and base with a single green ear-like lobe; petals 5, purple-violet or purplish, rarely white, ovate, dorsal petal largest and with a straight tubular spur (3–8 mm long; 2–5 times as long as sepals; c.1.5 mm diam.); style club-shaped. *Fruit* a dry, ellipsoid, 3-valved capsule (5–12 mm long). Seeds many, yellowish, ovoid-globose (c. 1.8 mm diam.). Fl. Apr–May, fr. May–Sep.

HARVESTING, SOURCING AND NATURAL RANGE

Plants are harvested from wild populations in spring or autumn where they grow in open grassy places, fields, forest margins and thickets (up to 1,700 m.). The species is widely distributed in China (Anhui, Chongqing, Fujian, Gansu, Guangdong, Guangxi, Guizhou, Hainan, Hebei, Heilongjiang, Henan, Hubei, Inner Mongolia, Jiangsu, Jiangxi, Jilin, Liaoning, Ningxia, Shaanxi, Shandong, Shanxi, Sichuan, Taiwan, Yunnan and Zhejiang) and extends to Cambodia, India, Indonesia, Japan, Korean Peninsula, Laos, Mongolia, Philippines and Vietnam.

CONSERVATION STATUS

China: Least Concern (MEPCAS, 2013) as *V. philippica* variety *pseudojaponica*.

Global: Not Evaluated (IUCN Red List, accessed 10/10/2016).

UNOFFICIAL SUBSTITUTE 1

V. prionantha Bunge

Violaceae

PLANT DESCRIPTION

Easily confused with PS1. Leaf stalks **not winged**, leaf blades at flowering, ovate to lance-shaped or narrowly ovate (1–4.5 cm long; 0.6–0.2 cm wide) and becoming much larger at fruiting (10 cm long; 4 cm wide), triangular to ovate, broadest near middle; **leaf bases slightly or broadly heart-shaped, truncate or broadly wedge-shaped; sepal tips shortly pointed** (not tapering), flower spur slightly broader (1.5–2.5 mm diam.) and often slightly pointing upwards at base.

SUBSTITUTION COMMENT. One of the more common *Viola* species with which the Pharmacopoeia species is regularly confused at point of harvest especially since it grows in a similar habitat. See also *Other Unofficial Substitutes* page 416**.**

It grows wild in N and W China (up to 2,800 m altitude) in the provinces of Gansu, Hebei, Heilongjiang, Henan, Hubei, Inner Mongolia, Jilin, Liaoning, Ningxia,Qinghai, Shaanxi, Shandong, Shanxi, Sichuan. The species is also native in the Korean Peninsula and Russia.

CONSERVATION STATUS

China: Not Evaluated.

Global: Not Evaluated (IUCN Red List, accessed 10/10/2016).

TCM PROPERTIES AND USE

properties: **'bitter and cold'**
channels: **'Liver and Heart**
actions: **'clears heat, eliminates toxins'**

indications: **deep and toxic abscesses, sores, insect and snake bites**

Viola philippica var. *philippica* in the wild, Hebei province

Flowers of *V. philippica* var. *philippica* (left): sepal tips tapering; spur straight; and *V. prionantha* (right): sepal tips shortly pointed; spur broader and pointing upwards

Viola prionantha in the wild, Hebei province

Corydalis bungeana Turcz.

Papaveraceae

PLANT DESCRIPTION

A soft, clump-forming herb (10–20 (–50) cm tall) with **branching, 5-ridged stems.** Leaves basal and along stems **(not in a basal rosette),** 2-pinnate **with a feathery appearance and with many leaflets deeply divided into 3-10 short narrow obtuse segments.** Flowers **slender and tube-like** (11–15 mm long) on upper stems and arising in loose 6–10-flowered heads; pinnately divided **leaf-like bracts** below each flower; petals multi-coloured (pink, mauve, white and purple). *Fruit* a dry, **oblong or elliptical capsule (12–20 mm long; c. 4 mm diam.),** compressed and pendant; tip pointed; **seed black, glossy** and sharply keeled when mature.

SUBSTITUTION COMMENT. The name Zi Hua Di Ding 紫花地丁 has been used down the ages in China to refer to a variety of species which may be quite unrelated taxonomically but which may share similar clinical properties. *Corydalis bungeana* is one such example and is a member of the Papaveraceae (poppy) family; its aerial parts are the source of the separate TCM drug 'Ku Di Ding' (Corydalis Bungeanae Herba) page 364, and some TCM practitioners prefer it clinically to *Viola philippica* (Lin Yu-Lin, pers. obs., 2011). The medicinal cultivation of this species is on the rise and it is not difficult to find it traded under the name Zi Hua Di Ding 紫花地丁in local Chinese and overseas markets (for example in the UK, Hongwen Yu, pers. comm., 2000; C. Leon, pers. obs., 2013, 2014). The casual use (mis-use) of this name is likely to cause considerable confusion and clarity over the trade names of these 2 species is needed.

CONSERVATION STATUS

China: Least Concern (MEPCAS, 2013).
Global: Not Evaluated (IUCN Red List, accessed 10/10/2016).

Gueldenstaedtia verna (Georgi) Boriss.

Fabaceae

synonyms: *Astragalus vernus* Georgi; *Gueldenstaedtia multiflora* Bunge

PLANT DESCRIPTION

Easily distinguished from the Pharmacopoeia species: leaves **pinnate** (with 7–9 pairs of opposite, ovate to linear, virtually stalkless leaflets plus a terminal one; **covered throughout in short, soft hairs)**; greyish-green. Flowering stems **densely clothed in soft white hairs**; flowers **in few-flowered heads** (not solitary); sepals forming a calyx tube and densely covered in long hairs; petals 5, the 2 lowermost forming a boat-shaped keel, half as long as wing petals; petal colour purplish-red, purple, pink, rose, or white; **without a spur.** Fruit a swollen, **elongated**, many-seeded pod without septa. Fl. (Mar–)May, fr. (May–) Jun–Jul (–Nov).

SUBSTITUTION COMMENT. This unrelated species is easily distinguished in fresh and dried state from the Pharmacopoeia species. The likely reasons for its occurrence as an unofficial substitute may be due to its similar clinical properties to those of the official species (Qin *et al.*, 1995).

It is a common and widespread native species of C, W and N China, growing in open sandy places in grassland, field margins, roadsides and sea bluffs (up to 2,600 m altitude) an dhas been recorded from Gansu, Hebei, Heilongjiang, Henan, Hubei, Inner Mongolia, Jiangsu, Jiangxi, Jilin, Liaoning, Ningxia, Qinghai, Shaanxi, Shandong, Shanxi, Sichuan, Tianjin, Yunnan and Zhejiang; it is also native in India, Laos, Mongolia, Myanmar, Pakistan and Russia.

CONSERVATION STATUS

China: Least Concern (MEPCAS, 2013).
Global: Not Evaluated (IUCN Red List, accessed 10/10/2016).

Corydalis bungeana in the wild, China

C. bungeana with flowers and developing fruit

RIGHT
C. bungeana with leaf-like bracts, flowers (lowermost opened to show papillate stigma)

Gueldenstaedtia verna flowering in the wild, Jiangxi province

Flowers of *G. verna*

Viola philippica var. *philippica*

Violaceae

DRUG DESCRIPTION

CRUDE DRUG (WHOLE HERB): a crumpled mass of leaves and roots, sometimes with flowers and/or fruits. *Main roots* stout (1–3 mm diam.), pale yellowish-brown, finely wrinkled longitudinally **[1]**. *Leaves* all basal, greyish-green **[2]**; **leaf stalks narrowly winged in upper part [3], becoming broadly winged at fruiting**; when intact, leaf blades either i) (upper leaves) oblong **[4]** to lance-shaped or ovate **[5]**(1.5–10 cm long; 0.5–4 cm wide) or ii) (lower leaves) narrowly oblong **[6]** to lance-shaped or ovate (1.5–4 cm long; 0.5–1 cm wide); **leaf bases ending abruptly [7] or wedge-shaped** (rarely heart-shaped), surfaces hairless or slightly hairy, tips blunt **[8]**; both surfaces hairless or slightly hairy; margins broadly and shallowly scalloped **[9]**. *Flowers* on slender, leafless stalks **[10]**, **sepals** 5 with **tips tapering** and base with a single green ear-like lobe; petals 5, ovate, purple **[11]** or pale brown **[12]**, unequal; spur tubular (3–8 mm long **[13]**; 2–5 times as long as sepals; c.1.5 mm diam.); ovary superior. *Fruit* a dry and ellipsoid capsule (5–12 mm long), opening when ripe into **3 spreading, boat-shaped valves [14]**; sepals persistent; seeds many, pale yellow, ovoid-globose **[15]** (c. 1.8 mm diam.). *Odour:* indistinct. *Taste:* somewhat bitter and sticky. ***Good quality*:** leaves and stems greyish-green; roots yellowish; unfragmented.

DECOCTING PIECES (HERB CUT INOT SHORT LENGTHS): as for the crude drug but cut into smaller pieces; leaves seldom intact.

PROCESSING

Whole plants are sun-dried (= crude drug) then cut into small decocting pieces.

OTHER COMMON NAMES

琉球堇菜 Liu Qiu Jin Cai (*V. philippica* var. *pseudojaponica*); Herba Violae; Tokyo violet (SCN); Tokyo violet herb (CP2010).

SAFETY

Probably similar to other *Viola* species: Safety Class: 1; Interaction Class: A (Gardner & McGuffin, 2013).

V. prionantha

Violaceae

'DRUG SUBSTITUTE' MORPHOLOGY

Easily confused with the official drug and much more difficult to discriminate than in the fresh state. When intact, **leaf stalks not winged**; **leaf bases slightly or broadly heart-shaped, truncate or broadly wedge-shaped; sepal tips shortly pointed** (not tapering).

OTHER COMMON NAMES

早开堇菜 Zao Kai Jin Cai (plant name).

OTHER UNOFFICIAL SUBSTITUTES/CONFUSED SPECIES

The genus *Viola* contains over 90 species native in China and they are a notoriously difficult group to reliably identify to species level (Xu *et al.*, 1997). For this reason, given that Zi Hua Di Ding is entirely wild-harvested it is likely that material traded under this name contains more than one species of *Viola* (Qin *et al.* 1995; Xiao et al., 2001-02); Zeng *et al.*, 1996; Zhao & Xiao (2009-10) aside from the other substitutes described above and below.

Other substitutes and adulterants reported in the literature include:

1. ***Gentiana loureiroi*** (G. Don) Griseb.(华南龙胆 Hua Nan Long Dan, 龙胆地丁 Long Dan Di Ding; 广地丁 Guang Di Ding). Family Gentianaceae. Until its wild populations became depleted, this species was used widely as the source of Zi Hua Di Ding in south China. Roots fleshy; basal leaves *stalkless*, elliptic (1.5–3 cm long; 3–5 mm wide); stems *present* with opposite, lance-shaped to elliptic widely spaced leaves (4–8 mm long; 1–2.5 mm wide); leaf margins *entire*; leaf tips *shortly pointed*; sepals *fused* to form a calyx tube; flowers arising *in leaf axils*, bell-shaped, solitary or in few-flowered clusters; petals 4–5, *overlapping, equal-sized* and *fused in lower parts to form a flower tube*; fruit *2-valved* (1.5–1.8 cm long).
2. ***G. yokusai*** Burkill (灰绿龙胆 Hui Lu Long Dan, 龙胆地丁 Long Dan Di Ding). Family Gentianaceae. As for *G. loureiroi* above but stem leaves ovate or cordate (4–12 mm long; 3–6 mm wide); leaf margins *thick and cartilaginous*, leaf tips *sharply pointed*; fruit narrowly obovoid (3–6.5 mm long).
3. ***Hypecoum leptocarpum*** Hook. (family Papaveraceae).
4. ***Osbeckia chinensis*** L. (family Melastomataceae).
5. ***Polygala*** spp. (family Polygalaceae). Flowering stems with leaves. Sepals without 'ears'. Fruits 2-valved; seeds 2.
6. ***Striga asiatica*** (L.) Kuntze (family Orobanchaceae).
7. ***Vicia amoena*** Fisch. ex Ser. (family Fabaceae).
8. ***Zornia diphylla*** (L.) Pers. (family Fabaceae).

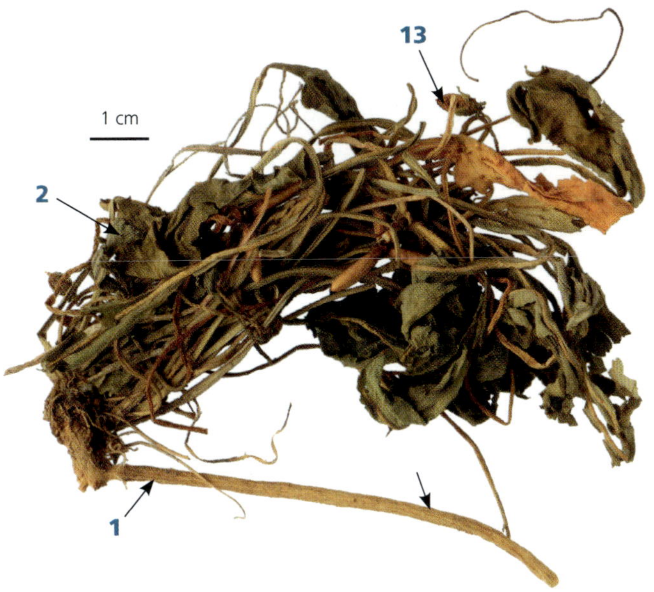

Crude drug of Zi Hua Di Ding (sourced from *Viola philippica* var. *philippica*)

Decocting pieces of Zi Hua Di Ding (sourced from *V. philippica* var. *philippica*)

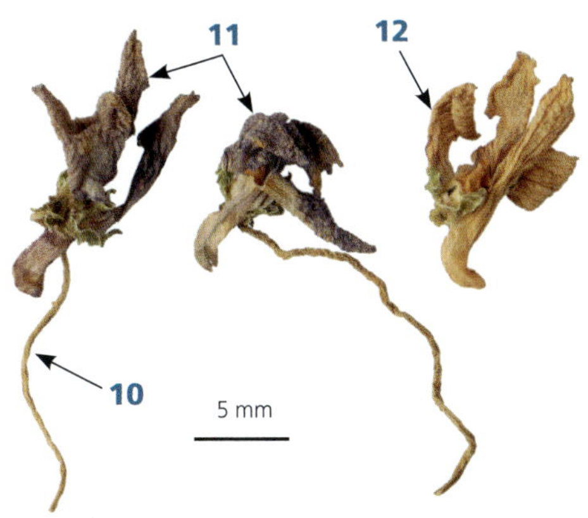

3 dried flowers of Zi Hua Di Ding (sourced from *V. philippica* var. *philippica*) with tubular spur at base of each and green 'eared' sepals

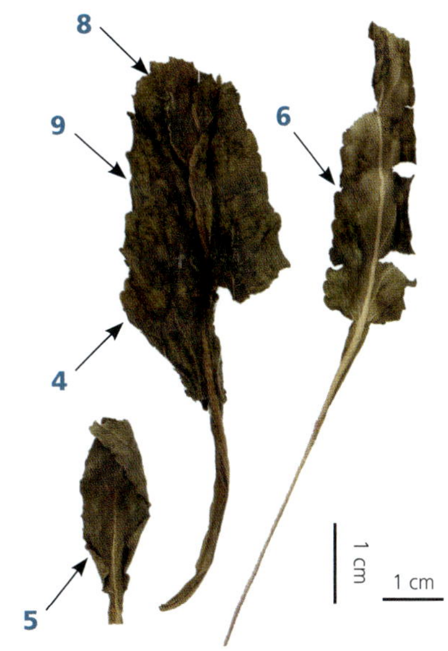

Expanded dried leaves of *V. philippica* var. *philippica*

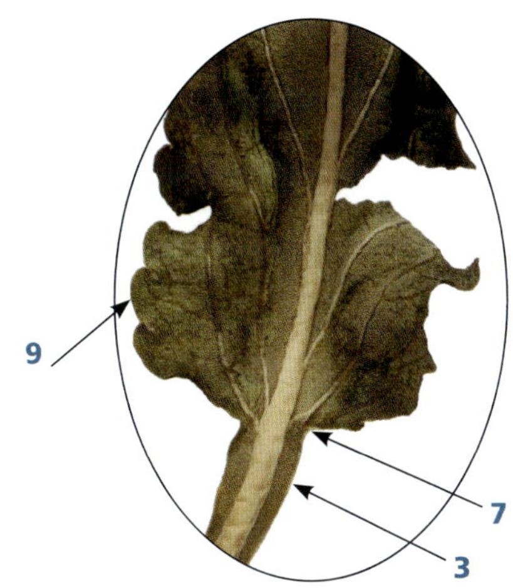

Narrowly winged upper part of leaf stalk of *V. philippica* var. *philippica*

Dried ripe fruits of *V. philippica* var. *philippica*; top right: dehiscing 3-valved fruit exposing ripe seeds within and long thin fruit-stalk still attached; top left: ripe fruit with 1 valve broken off; bottom: intact 3-valved fruit with seeds removed

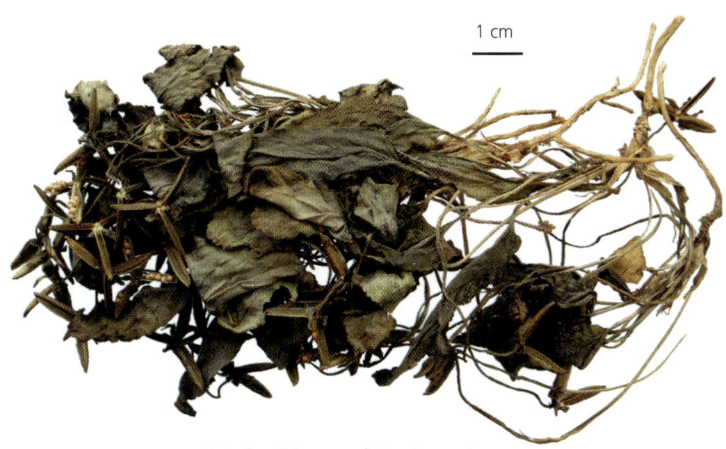

Dried aerial parts of *V. prionantha*

Corydalis bungeana

Papaveraceae

'DRUG SUBSTITUTE' MORPHOLOGY

This is the source of the separate drug 'Ku Di Ding' (see also page 364). In dried crude drug form it is superficially similar to the official drug of 'Zi Hua Di Ding' (Viola philippica *var.* philippica) *described here, but easily distinguished when separated into its constituent parts.*

Main differences: stems with **5 longitudinal ridges [16]**, frequently branched and hollow. **Leaves 2-pinnate with leaflets deeply divided into 3–10 short, narrow obtuse segments [17]**. *Flowers* (when still attached) in loose, 6–10-flowered heads **[18]** with divided **leaf-like bracts [19]** below each flower; flowers slender and tubular **[20]** (11–15 mm long); petals 4, with the 2 upper and lower petals **[21]** cream, shortly crested and enclosing 2 inner, dark, purple-tipped petals **[22]** fused at their apex; upper petal with a curved spur at base **[23]** (4–5 mm long) but seldom discernible in trade material. Fruit a dry, **oblong or elliptical capsule [24]** (12–20 mm long; c. 4 mm wide), tip pointed; seeds in 2 rows, **black and glossy [25]**. *Taste:* strongly bitter. *Decocting pieces:* as for the crude drug above but more fragmented.

OTHER COMMON NAMES

苦地丁 Ku Di Ding; Corydalis Bungeanae Herba; Herba Corydalis Bungeanae.

Comparison of flowers of *Viola philippica* var. *phillippica* with flowers of Unofficial Substitutes 2 and 3

Gueldenstaedtia verna

Fabaceae

'DRUG SUBSTITUTE' MORPHOLOGY

As dried crumpled aerial parts (i.e. crude drug trade item) this drug substitute appears superficially similar to that of the official drug (Viola philippica *var.* philippica)*; it is easily distinguished, however, when separated into its constituent parts (below). In decocting form (i.e. short cut lengths) many key characters are lost through fragmentation but intact flowers and fruits are likely to persist and provide useful identification characters.*

Main differences: leaves greyish-green and **pinnate [26])** (with 7–9 pairs of opposite, ovate to linear, almost stalkless leaflets **[27]** plus a terminal one; **leaflets covered throughout in short, soft hairs [28]**. Flowering stems **densely clothed in soft white hairs [29]**; flowers **in few-flowered heads** (not solitary) **[30]**; sepals fused into a densely hairy calyx tube **[31]** opening apically into 5 long narrow teeth **[32]**; petals 5, traces of petal colour sometimes visible: purplish red, purple, pink, rose, or white **[33]**; **spur absent.** Fruit a swollen, **elongated**, many-seeded pod **[34]** without septa.

OTHER COMMON NAMES

少花米口袋 Shao Hua Mi Kou Dai (plant name); 甜地丁 Tian Di Ding.

Unofficial substitute (Ku Di Ding) crude drug sourced from *Corydalis buneana*)

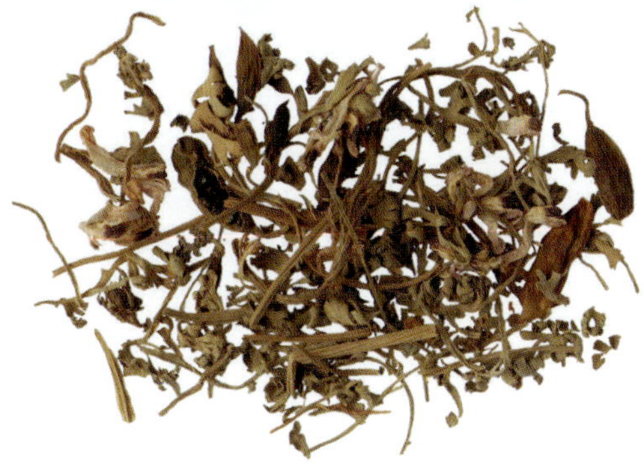

Unofficial substitute (Ku Di Ding) decocting pieces (*C. bungeana*)

fruits

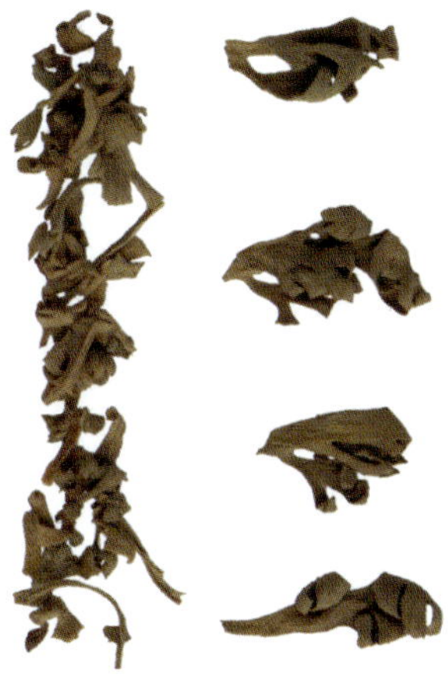

crumpled leaves

stem

expanded leaf

inflorescences

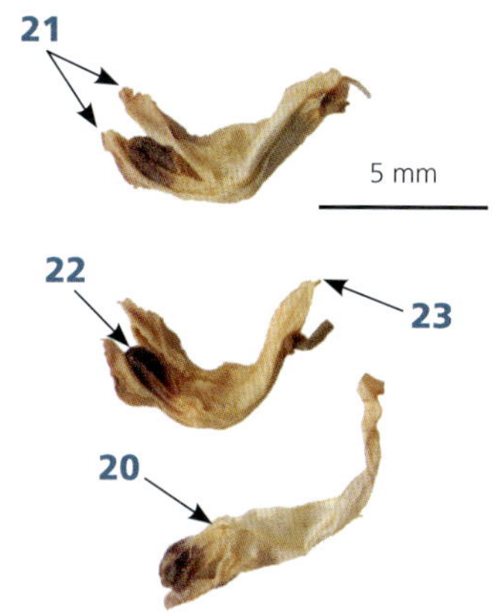

individual flowers

Separated aerial parts of Unofficial Substitute 2 'Ku Di Ding' (*C. bungeana*)

Dried whole plant of *Gueldenstaedtia verna*

Dried whole plant of *G. verna* cut into small pieces

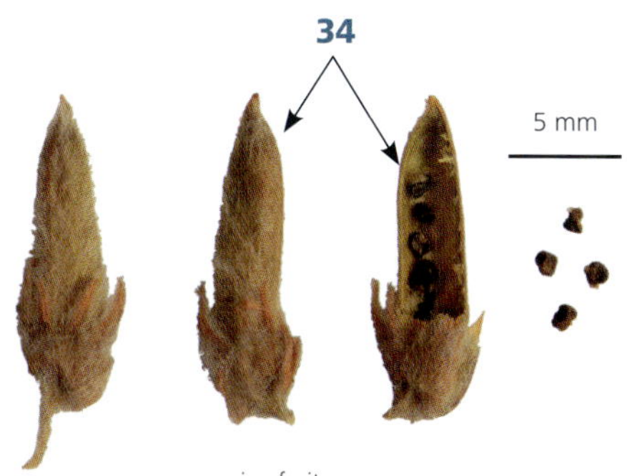

ripe fruits

pinnate leaf (some leaflets fallen)

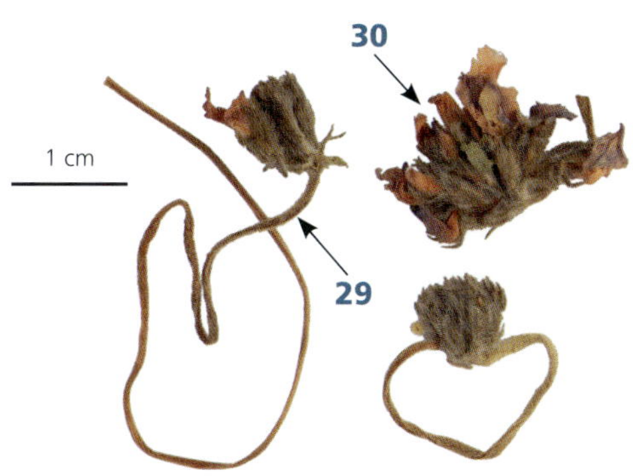

Inflorescences

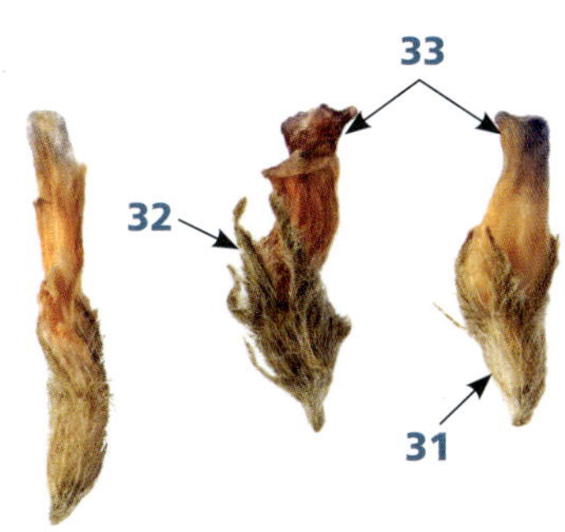

flowers (unopened)

Separated aerial parts of Unofficial Substitute 3 'Tian Di Ding' (*G. verna*)

Uncaria macrophylla

stems and woods

A Wei 阿魏

Ferulae Resina

Drug part: RESIN

Ferula sinkiangensis K. M. Shen

Apiaceae

Species of *Ferula* wild in China: 26

PLANT DESCRIPTION

An erect, perennial herb (0.5–1.5 m tall) with a stout, conical taproot; highly aromatic (onion-smelling). *Stem* branching, rough, **with fine, dense hairs;** basal parts often purplish-red and with membranous sheaths. *Leaves* alternate and distinctly greyish-green; leaf-blades triangular in outline and deeply segmented (3-pinnatisect); margins shallowly lobed or toothed; lower surfaces densely hairy. *Inflorescences* large and conspicuous with simple and compound yellow or greenish-yellow umbels (each 8–12 cm diam.) contrasting sharply with the greyish-green foliage; lateral umbels long-stalked and exceeding the length of terminal umbels. Flowers male, female or bisexual, with all types present on the same plant; bracts absent; rays 5–25; bracteoles lance-shaped and deciduous; petals 5, yellow, pale yellow or greenish-yellow. ***Fruit*** dry and ellipsoid (**10–12 mm long**; 5–6 mm diam.), **equalling or shorter than fruit stalk**, splitting into 2 one-seeded carpels (mericarps) at maturity; each carpel strongly compressed dorsally, dorsal ribs thread-like, lateral ribs winged with 3–4 dark oil ducts in each furrow, 10–14 on the commissural (inner) surface. Fl. and fr. Apr–Jun.

HARVESTING, SOURCING AND NATURAL RANGE

Resin is harvested from wild or cultivated plants during late spring or early summer (when plants are in flower or beginning to fruit). The stems are incised and, over a 2–3 month period, the oleogum resin is tapped. The species is endemic to NW China (Yining district of Xinjiang province) where it grows in semi-desert gravels between 800–900 m altitude.

Wild populations have become acutely endangered (see *Conservation Status)* mainly due to the indiscriminate harvesting of their rootstocks which are used as a substrate for growing a highly prized local Xinjiang medicinal delicacy 'A Wei mushroom' (*Pleurotus* spp.) (pers. comm. K. Sulaiman, Aug. 2011). In addition, the medicinal properties of the resin are said to be similar to those of the more well-known but closely-related Iranian species *F. assa-foetida* L. (asafetida) widely used in Ayurvedic medicine (Iranshahy & Iranshahi, 2011) and for which the resin of *F. sinkiangensis* is used as a local substitute in Xinjiang.

CONSERVATION STATUS

Endemic to China: Critically Endangered (MEPCAS, 2013); Not Evaluated (IUCN Red List, accessed 20/2/2016).

Ferula fukanensis K. M. Shen

Apiaceae

PLANT DESCRIPTION

Easily confused with *F. sinkiangensis* (PS1). *Main differences:* **stems virtually hairless;** leaves ovate in outline; **mature fruits longer than the fruit stalks** and compared with those of PS1 **slightly longer** (12–16 mm long).

HARVESTING, SOURCING AND NATURAL RANGE

Similar to that for *F. sinkiangensis* (PS1). The species is endemic to NW China (Xinjiang province) where it grows on valley slopes in marginal desert zones up to c. 700 m altitude.

CONSERVATION STATUS

Endemic to China: Endangered (MEPCAS, 2013); Not Evaluated (IUCN Red List, accessed 20/2/2016).

TCM PROPERTIES AND USE

***properties*: 'bitter, pungent and warm'**
***channels*: 'Lung and Stomach'**
***action*: 'dispels stasis'**

***indications*: food stagnation; intestinal worms; (T)**

Ferula sinkiangensis in the wild, Xinjiang province PHOTO: KAISA SULAIMAN

F. sinkiangensis in flower PHOTO: KAISA SULAIMAN

Ripening fruits of *F. sinkiangensis*, Xinjiang province PHOTO: KAISA SULAIMAN

Medicinal resin exuding from damaged stems of *F. sinkiangensis* PHOTO: KAISA SULAIMAN

Trial cultivation of *F. fukanensis*, Xinjiang province PHOTO: KAISA SULAIMAN

F. fukanensis in flower PHOTO: KAISA SULAIMAN

Fruiting plant of *F. fukanensis* PHOTO: KAISA SULAIMAN

Ferula sinkiangensis

Apiaceae

DRUG MORPHOLOGY

CRUDE DRUG/DECOCTING PIECES (DRIED RESIN): traded either as viscid or hard resin. ***Hard resin pieces***: irregularly-shaped, hard and angular **[1]**; externally translucent with colour varying in intensity from yellowish- to dark brown. Cut surfaces pale, darkening with age. *Texture*: light and waxy. *Fracture*: slightly porous. ***Viscid resin***: greyish-white, glutinous or sticky. *Odour*: strong and distinctly onion-like. *Taste*: pungent, with a burning sensation when chewed.

PROCESSING

The resin is shade-dried (= crude drug/decocting material).

OTHER COMMON NAMES

新疆阿魏 Xinjiang A Wei; Chinese asafetida (CP2010); Ferula sinkiangensis (SCN); Resina Ferula.

SAFETY

Probably similar to the closely related *F. assa-foetida* classified as Safety Class: 2b and 2d; Interaction Class: A (Gardner & McGuffin, 2013).

F. fukanensis

Apiaceae

DRUG MORPHOLOGY

CRUDE DRUG AND DECOCTING PIECES: indistinguishable macroscopically from those of PS1. Laboratory-based authentication methods are required.

PROCESSING

As for PS1.

OTHER COMMON NAMES

阜康阿魏 Fu Kang A Wei; Chinese asafetida (CP2010); Resina Ferula.

SAFETY

As for PS1.

Crude drug/decocting material of A Wei (hard resin) sourced from *F. sinkiangensis*

UNOFFICIAL SUBSTITUTES

Resin from many other species of Chinese *Ferula* (26 native in China) have been reported to be traded as 'A Wei' owing to confusion among non-experts about their field identification (Xiao *et al.*, 2001–2). Research is underway on the medicinal potential of these unofficial species (Sahebkar & Iranshahi, 2011). Common substitutes include:

1. ***Ferula conocaula*** Korovin (圆锥茎阿魏 Yuan Zhui Jing A Wei). SW Xinjiang [also Kyrgyzstan]. Stems very stout and conical; leaves oblong-ovate, larger, and lobes longer than 5 cm.
2. ***F. ferulioides*** (Steud.) Korovin (syn. *F. resinosa* Siev.).
3. ***F. krylovii*** Korovin (托里阿魏 Tuo Li A Wei), NW Xinjiang [also Russia; C Asia]. Segments of leaves lanceolate; umbellets 10–13-flowered; fruit c. 14 mm long, 2 oil glands in each furrow, 6 on commissure.

Chen Xiang 沉香

Aquilariae Lignum Resinatum

Chinese agarwood

Drug part: RESIN-CONTAINING HEARTWOOD

Aquilaria sinensis (Lour.) Spreng.

Thymelaeaceae

synonym: *Ophispermum sinense* Lour.

taxonomic/nomenclatural note: in CP2015 as '*A. sinensis* (Lour.) Gilg'

Species of *Aquilaria* wild in China: 2

PLANT DESCRIPTION

An evergreen tree (5–15 m tall) with smooth, dark grey bark and white inner bark. *Leaves* alternate and short-stalked (5–7 mm); leaf blades simple, elliptic to oblong, sometimes obovate (5–9 cm long; 2.8–6 cm wide), base broadly wedge-shaped, tip tapering to a long point or ending in an abrupt point; leathery, hairless, dark or purplish-green, upper surface glossy, lower surface light green, pinnately veined with 15–20 pairs of lateral veins; margins entire, sometimes with very fine hairs. *Flowers* in a terminal, loosely branching inflorescence. Each flower bisexual and fragrant after dark; short-stalked (5–10 mm) and densely covered in yellowish-grey hairs; calyx tube (5–6 mm long) yellowish-green, narrowly bell-shaped, with 5 reflexed or erect lobes; petals 10, equal and yellowish-green; stamens 10; ovary superior; style virtually absent; stigma resembling a pin-head. *Fruit* a dry, egg-shaped, 2-valved capsule (2–3 cm long; c. 2 cm diam.), densely clothed in yellow hairs and base enclosed in a persistent calyx, tip with short sharp point; fruit-wall thin; seeds ovoid and dark brown. Fl. Spring–summer; ft. summer–autumn.

HARVESTING, SOURCING AND NATURAL RANGE

Wild populations of this endemic Chinese tree have declined dramatically in recent decades due to over-harvesting by the highly lucrative market in its resinous heartwood (agarwood) which is highly sought after both as a TCM and for its aromatic properties by the incense and perfumery industries (Barden, 2000). Habitat loss has also contributed to its decline. Once common in the semi-evergreen tropical forests in S China (Fujian, Guangdong, Guangxi, Hainan and Yunnan) the species is now extinct in many of its former locations and heavily protected in its very few remaining localities. Many other non-Chinese species of *Aquilaria* have suffered a similar fate; *A. malaccensis* Lam., for example, was imported from India into China in recent years as a substitute for Chinese-sourced 'Chen Xiang' but this too is now Vulnerable (IUCN Red List, 18/2/2016). In an effort to protect the wild populations of all *Aquilaria* species the entire genus is listed on CITES (*see below*).

Extensive efforts in recent years to artificially trigger resin-formation in plantation-grown trees has been successful in producing low-quality resin; the bulk of Chen Xiang used in TCM today is sourced from such plantations grown primarily in Thailand, Laos and other parts of SE Asia (Mei *et al.*, 2013). The production of top-end agarwood, however, for the perfume industry has proven technically challenging (Liu *et al.*, 2013) such that demand continues to threaten the wild species.

CONSERVATION STATUS

Endemic to China: Vulnerable (MEPCAS, 2013); Vulnerable (IUCN Red List, accessed 11/10/2016).

PROTECTED STATUS

China: Cat. II (SEPA, 1999).

CITES: Appendix II; EU Annex B (see Table 1 page 10).

Plantation of young trees of *Aquilaria sinensis*, Guangxi province, China

Flowering branch of *A. sinensis*

A. sinensis with immature fruit

TCM PROPERTIES AND USE

properties: **'pungent, bitter and slightly warm'**
channels: **'Spleen, Stomach and Kidney'**
actions: **'invigorates qi, relieves pain, warms Kidney, Spleen and Stomach'**

indications: **distension, oppression and pain in the chest and abdomen; vomiting, hiccup**

DRUG MORPHOLOGY

CRUDE DRUG (RESIN-FILLED HEARTWOOD): heartwood pieces of irregular shape and size, often large lumps **[1]** or smaller shards **[2]**. External surface uneven, rounded or sharply angular, sometimes with knife marks **[3]** and holes **[4]** where non-resinous wood has been removed; fissured areas sometimes resemble rotten wood **[5]**. Wood pale yellowish-grey to greyish-brown **[6]** and typically impregnated with blackish-brown resin **[7]** creating a characteristic mottled or streaked appearance. An important diagnostic character of the wood (xylem) of *Aquilaria** species is the presence of islands of '*included phloem*' **[8]** visible in transverse section [x 10 hl] as short dark, interrupted fusiform lines. This character requires validation under higher magnification as illustrated in the anatomical section below (Gasson, 2011; Gasson *et al.*, 2011; *Inside Wood* 2004 onwards: http://insidewood.lib.ncsu.edu/search). *Texture*: relatively compact and hard. *Fracture*: splintering. *Odour*: aromatic. *Taste*: bitter. ***Good quality*:** heavy, brownish-black, oily and lustrous; releases oil and intense aroma on burning.

DECOCTING PIECES (WOOD SHARDS OR PULVERISED POWDER): dependent on the size of the pieces, islands of '*included phloem*' may or may not be discernible.

*Islands of 'included phloem' also occur in the wood of the closely related genus *Gyrinops* in the same family (Gasson, 2011); this genus is native to Indonesia and Indian Subcontinent and both *Aquilaria* and *Gyrinops* are regulated by CITES.

PROCESSING

After removal of the non-resinous outer wood, the resin-filled heartwood is shade-dried (= crude drug) and subsequently split into small pieces for decocting or pulverised into powder.

OTHER COMMON NAMES

海南沉香 Hai Nan Chen Xiang; 女儿香 Nu Er Xiang; 土沉香 Tu Chen Xiang (plant name); A ga ru (Tibetan name); agarwood; aloewood; Chinese agarwood (SCN); Chinese eaglewood wood (CP2010); gaharu; Lignum Aquilariae Resinatum. Many other trade names exist for this expensive product (see *Kew-MPNS*).

IDENTIFICATION COMMENT: the identification of *Aquilaria* to species level has yet to be achieved using anatomical or chemical methods but DNA bar-coding approaches are actively being researched and the arrival of validated methods looks promising.

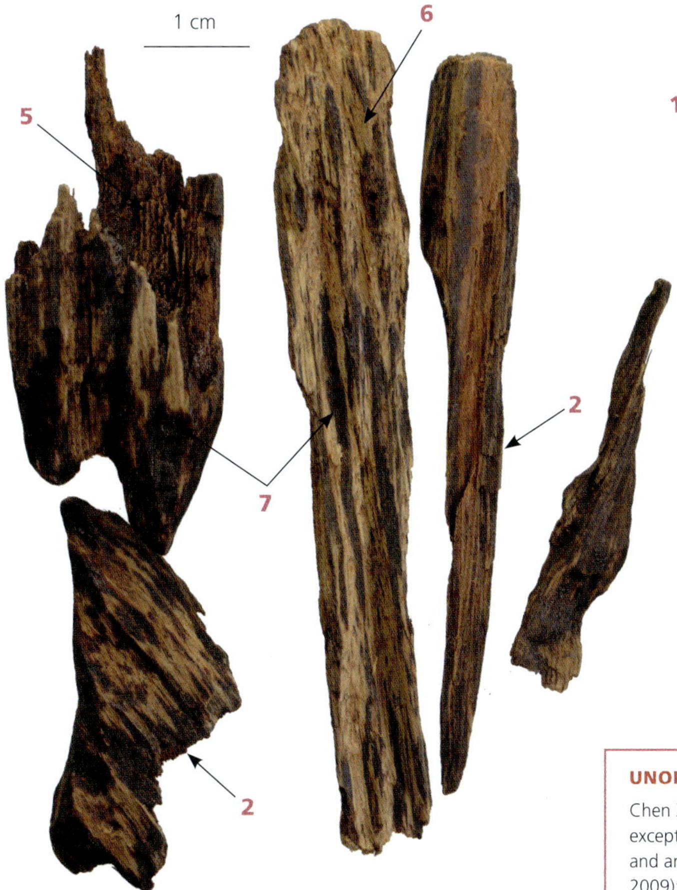

Crude drug of Chen Xiang sourced from *Aquilaria sinensis*

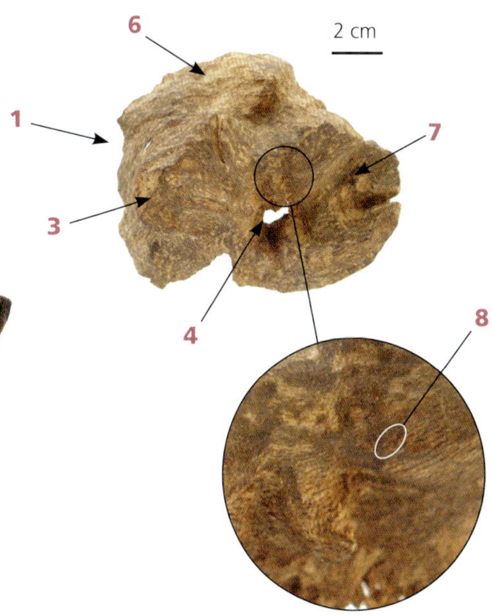

Crude drug of Chen Xiang sourced from *Aquilaria sinensis*

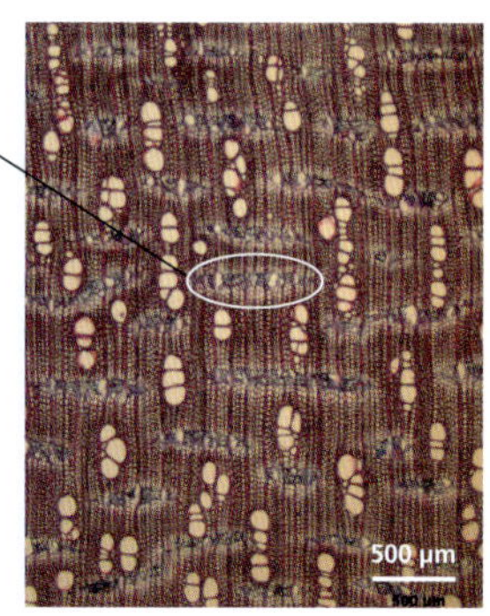

Transverse wood section of *Aquilaria* showing islands of 'included phloem'

UNOFFICIAL SUBSTITUTES AND COUNTERFEITS

Chen Xiang is a high-value TCM herbal drug and can fetch exceptional prices; counterfeiting using a wide variety of dyes and aromas is commonplace (Leon and Lin, pers. obs., 1999, 2009); see also *Harvesting Sourcing and Natural Range*.

Right: example of a counterfeit sold as 'Chen Xiang' in a TCM clinic in Chinatown, London — unidentified wood painted with a black substance supposedly mimicking the resin of 'Chen Xiang' (C. Leon pers. obs., 2004).

Chuan Mu Tong 川木通 Clematidis Armandii Caulis

Drug part: STEM

Clematis armandii Franch.

Ranunculaceae

taxonomic/nomenclatural note: *FoC* and *Kew-MPNS* recognise three varieties.

Species of *Clematis* wild in China: 147

PLANT DESCRIPTION

A woody climber. *Stems* branching; with 10–16 distinct longitudinal grooves. *Leaves* alternate and stalked; leaf blades divided into 3 stalked leaflets; leaflets narrowly ovate, lance-shaped or ovate (5–16 cm long; 1.5–7 cm wide), both surfaces hairless, base rounded to sub-heart-shaped or broadly wedge-shaped, margins entire, tips gradually narrowing. *Inflorescences* 1- to many-flowered, axillary; bracts 2, narrowly oblong to linear (0.7–2 cm long), undivided or 3-lobed. *Each flower* conspicuous (2–8 cm diam.), radially symmetrical and bisexual; true petals absent; petal-like sepals usually 4 (sometimes up to 6), equal, narrowly oblong (1.2–4.7 cm long; 0.2–2 cm wide), tip broads or rounded, white (sometimes tinged pink) and spreading; stamens many; style 1, densely covered in soft hairs. *Fruit-head* a tight cluster of dry achenes; each achene (4–5 mm long; 2–2.6 mm diam.) with a persistent and softly feathered style (1.6–4.8 cm long). Fl. Mar–Apr, fr. Apr–Jul.

HARVESTING, SOURCING AND NATURAL RANGE

Stems are harvested in spring and autumn exclusively from wild plants which grow in forests, forest margins, scrub and along streams (100–2,400 m altitude) in S, C and W China (Fujian, S Gansu, N Guangdong, Guangxi, Guizhou, Hubei, Hunan, Jiangxi, S Shaanxi, Sichuan, E Tibet, Yunnan and S Zhejiang) [also native in Myanmar] *Di Dao*-quality drug is traditionally sourced from Sichuan provinces.

CONSERVATION STATUS

China: Least Concern (MEPCAS, 2013).
Global: Not Evaluated (IUCN Red List, accessed 25/1/2016).

See also the closely related drug 'Mu Tong' (Akebiae Caulis) page 458.

C. montana Buch.-Ham. ex DC.

Ranunculaceae

taxonomic/nomenclatural note: in CP2015 as '*C. montana* Buch.-Ham.' *FoC* recognises 6 varieties.

PLANT DESCRIPTION

Easily distinguished from PS1. *Main differences*: **stems with 4–10 longitudinal grooves. *Leaves*** divided into 3 **much smaller leaflets**, rhombic to elliptic (1.8–7 (rarely up to 14) cm long; 1–5 cm wide), **not leathery**, each leaflet undivided or 3-lobed with margins sparsely toothed or entire. **Petal-like sepals usually much broader (ovate),** 1.3–6.5 cm long; 1–2.5 cm wide, white (sometimes tinged pink). Fl. Apr–Sep, fr. Jul–Sep.

HARVESTING, SOURCING AND NATURAL RANGE

Very similar to PS1 but this species grows at higher altitude (1,000–4,000 m) and has a larger natural distribution in China (Anhui, Fujian, Gansu, Guangxi, Guizhou, Henan, Hubei, Hunan, Jiangxi, Ningxia, Qinghai, Shaanxi, Sichuan, Taiwan, Xinjiang, Yunnan and Zhejiang) and is also native in Afghanistan, Bhutan, India, Kashmir, Myanmar, Nepal, Pakistan and Sikkim. *Di Dao*-quality drug is traditionally sourced from Sichuan province.

CONSERVATION STATUS

China: Least Concern (MEPCAS, 2013).
Global: Not Evaluated (IUCN Red List, accessed 25/1/2016).

EHTPA 4/14

TCM PROPERTIES AND USE

***properties*:** **'bitter and cold'**
***channels*:** **'Heart, Small Intestine and Bladder'**
***actions*:** **'drains damp, clears heat'**

***indications*:** **urinary conditions; mouth ulcers; amenorrhoea; poor milk production**

Leaf *C. armandii* narrowly with 3 ovate leaflets

C. armandii flowering in the wild, Sichuan province

Fruiting heads of *C. armandii*

Freshly harvested stem of *Clematis armandii* with leaves and fruits for a herbarium voucher, Guizhou province

Rhombic to elliptic leaflets of *C. montana*

Flowering branch of *C. montana* in the wild, China

Clematis montana in the wild (rambling white-flowered plants in centre of photo); 3700 m, Sichuan province

Cut stem of living plant of *C. montana*

Aristolochia manshuriensis Kom.

Aristolochiaceae

synonyms: *Hocquartia manshuriensis* (Kom.) Nakai; *Isotrema manshuriense* (Kom.) H. Huber

PLANT DESCRIPTION

Easily distinguished from PS1 and PS2 above. *Main differences:* ***stem*** with shallow cork-like bark, **strongly aromatic when crushed**. ***Leaves*** **much larger** (15–29 cm long; 13–28 cm wide), **leaf blades not divided into leaflets;** usually heart-shaped; margins entire. ***Flower*** solitary or in pairs; a **yellow to greenish-yellow pipe-shaped utricle** (4.5–5.5 cm long) **with 3 yellow deltoid lobes at mouth**. ***Fruit*** a **much larger**, solitary and **cylindrical pod** (9–11 cm long; 3–4 cm diam.), ripening from green to brown; many-seeded. *Fl.* Jun –Jul; *fr.* Aug–Sep.

SUBSTITUTION COMMENT. In the 1950s, the stem of this species (Guan Mu Tong) became legitimised as an alternative to other species in the 'Mu Tong' herbal group (including *C. armandii* and *C. montana*, PS1 and 2). This was for reasons of similar clinical function coupled with lower prices and higher yields (Zhang *et al.,* 2012b) and *A. manshuriensis* soon became the main source plant of 'Mu Tong/Chuan Mu Tong' in trade. Like all *Aristolochia* species, however, *A. manshuriensis* contains aristolochic acids which are both nephrotoxic and carcinogenic but it was not until the 1990s, following a spate of serious adverse reactions (some fatal) in Europe that *Aristolochia* species (*A. manshuriensis* and *A. fangchi*) were proven to be the cause (Shaw *et al.,* 2012; Zhu *et al.*, 2010; Debelle *et al.*, 2008). A worldwide ban quickly followed (e.g. UK Government, 2001) along with their removal from the Chinese Pharmacopoeia (SFDA, 2003) and, although robust aristolochic acid detection techniques have been developed, their rigorous enforcement is essential if the trade in these species is to be regulated.

CONSERVATION STATUS

Endemic to China: Near Threatened (MEPCAS, 2013); Not Evaluated (IUCN Red List, accessed 24/1/2016).

LESSER KNOWN UNOFFICIAL SUBSTITUTES

Given that the two Pharmacopoeia sources of 'Chuan Mu Tong' are species entirely wild harvested and that there are another 145 species of *Clematis* growing wild in China, there is a high likelihood of mis-identification, substitution or contamination. To date, over 12 species have been reported as local or historical substitutions in China but their occurrence as such in international trade remains uncertain as does their pharmacological suitability. Recent research points to promising developments in the authentication of this entire group using DNA markers as well as their pharmacological activity (Hao *et al.*, 2013; Guo *et al.*, 2010b).

Aristolochia manshuriensis — showing vigorous, climbing habit of wild plant

A. manshuriensis showing stems, leaves and flowers

A. manshuriensis in flower

Clematis armandii and *C. montana*

Ranunculaceae

The crude drug and decocting pieces of these two species cannot be distinguished macroscopically. They are described together below.

DRUG MORPHOLOGY

CRUDE DRUG (DRIED, DE-BARKED STEM): hard, cylindrical (50–100 cm long; 0.5–3.5 cm diam.) **[1]**, slightly twisted and sometimes branching. **Bark usually removed.** External (de-corticated) surface pale yellow, yellowish-brown or greyish-brown **(not spongy) [2]** with distinct longitudinal furrows, **readily cracking longitudinally [3]**; oval and pitted leaf and branch scars occasional **[4]** along with **remnants of thin bark [5].** Nodes swollen **[6]**. *Texture*: light, hard, **fibrous [7]** and difficult to break transversely. *Fracture*: uneven; thin bark remnants attached or fallen; wood (xylem) pale yellow or pale yellowish-brown **[8]** (*see also cut surfaces of Decocting Pieces*). ***Odour*: indistinct.** *Taste*: bland.

DECOCTING PIECES (DE-BARKED TRANSVERSE STEM SLICES): sub-rounded (2–3.5 cm diam.; 2–4 mm thick). **Outer and inner bark usually absent**; perimeter of **slices shallowly scalloped [9] or irregularly notched edge [10]** with yellowish-brown remnants of bark occasional or absent **[11]**. Cut surface (wood) pale yellow or pale yellowish-brown **[12]**, densely packed throughout with round vessels arranged in concentric rings and creating a regularly porous appearance **[13]**; wood interrupted radially at regular intervals with much narrower medullary rays **[14]** often extending to the stem perimeter; central pith small, pale yellowish-white **[15]** or sometimes absent (i.e. hollow) **[16]**. A dark, blackish-brown centre usually denotes poor quality **[17]**. *Fracture*: brittle; breaking along the rays. ***Odour*: indistinct.** *Taste*: bland.

PROCESSING

After removal of the coarse outer (cork) bark, stems are sun-dried (= crude drug); and later softened thoroughly before cutting into 2–4 mm thick transverse slices sometimes stems are sliced while fresh, then dried (= decocting pieces).

OTHER COMMON NAMES

Names for PS1 and PS2: 花木通 Hua Mu Tong; 小木通 Xiao Mu Tong (plant name); Armand clematis stem (CP2010); Armand's clematis (SCN); Armand virgin bower; Caulis Clematidis Armandii. PS2 is also called 绣球藤 Xiu Qiu Teng (plant name).

SAFETY

Generally regarded as safe herbs (Bensky *et al.*, 2004). Through association, however, with the toxic and at one time clinically interchangeable herb *Aristolochia manshuriensis* 'Guan Mu Tong' (see below), *Clematis armandii* and *C. montana* have become subject to legal restrictions in many countries in the interests of public safety (UK Government, 2001). See *Substitution Comment* page 430.

Aristolochia manshuriensis

Aristolochiaceae

'DRUG SUBSTITUTE' MORPHOLOGY

Crude drug and decocting slices usually easily distinguished from those of PS1 and PS2 by the presence of ***spongy and aromatic cork bark*** *and a slightly darker, greyish-yellow wood.*

CRUDE DRUG (DRIED STEM): hard, cylindrical and sometimes broader (1–2 m long; 0.8–6 cm diam.); slightly twisted. **Outer and inner bark usually still attached [18]. Outer bark** pale greyish-brown or greyish-yellow, **spongy [19]**, with irregular, oblong or rhomboidal longitudinal furrows; **inner bark greyish-yellow (up to 4 mm thick) [20]**, easily removed; branch scars occasional. Nodes slightly swollen **[21]**. *Texture*: light, hard **(not fibrous) [22]** and difficult to break. *Fracture*: (*see cut surfaces of Decocting Pieces*). ***Odour*: cork bark aromatic (camphor-like) when crushed.** *Taste*: mildly bitter and pungent.

DECOCTING PIECES (TRANSVERSE STEM SLICES): sub-rounded (0.8–6 cm diam.; 2–4 mm thick); sometimes becoming convex when drying. **Outer bark slightly spongy,** pale greyish-brown or greyish-yellow **[19]**; **inner bark greyish-yellow (up to 4 mm thick) [20]**, easily removed (bark : wood ratio c. 1: 5–7). Cut surface with wood pale greyish-yellow **[23]** densely packed throughout with round vessels arranged in concentric rings and creating a strongly porous appearance; wood interrupted by regularly spaced and much narrower medullary rays **[24]** of a similar colour; pith rather small and inconspicuous **[25]**, same colour as wood. *Fracture*: brittle. ***Odour*: cork bark aromatic (camphor-like) when crushed.** *Taste*: mildly bitter and pungent.

SAFETY

Toxic. Species in this genus (*Aristolochia*) contain aristolochic acids which are nephrotoxic and carcinogenic (Debelle *et al.*, 2008; Upton & Graff, 2006). Their medicinal use is prohibited or restricted in many countries. See *Substitution Comment* page 430.

OTHER COMMON NAMES

关木通 Guan Mu Tong; 木通 Mu Tong; Manchurian birthwort (SCN); Caulis Aristolochiae Manshuriensis; Aristolochiae Manshuriensis Caulis; Manchurian Dutchman's pipe.

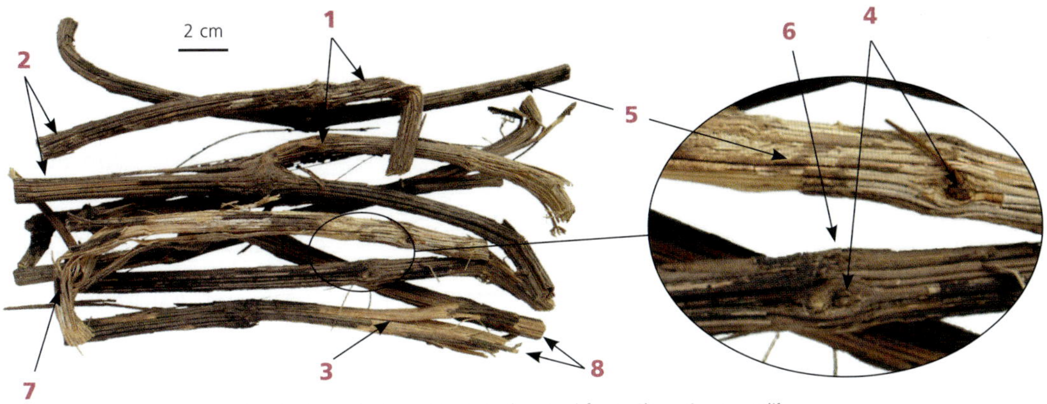

Crude drug of Chuan Mu Tong (sourced from *Clematis armandii*)

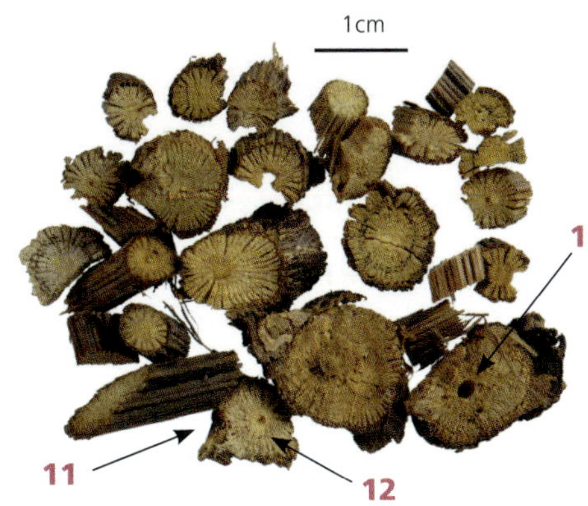

Decocting pieces (roughly sliced after drying) sourced from *C. armandii*

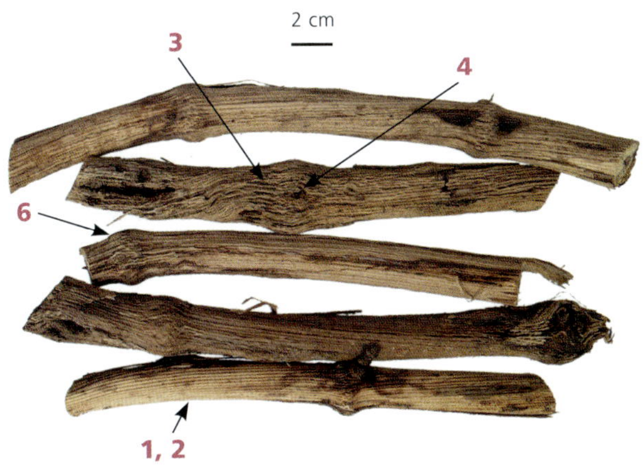

Crude drug of Chuan Mu Tong (sourced from *C. montana*)

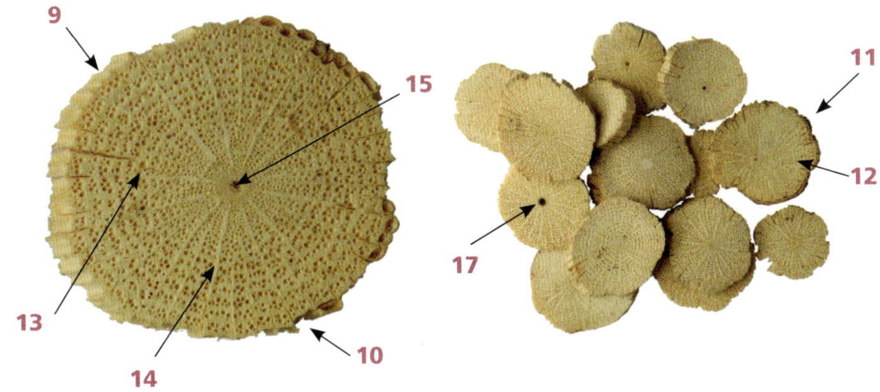

Decocting pieces (thinly sliced when fresh) sourced from *C. armandii*

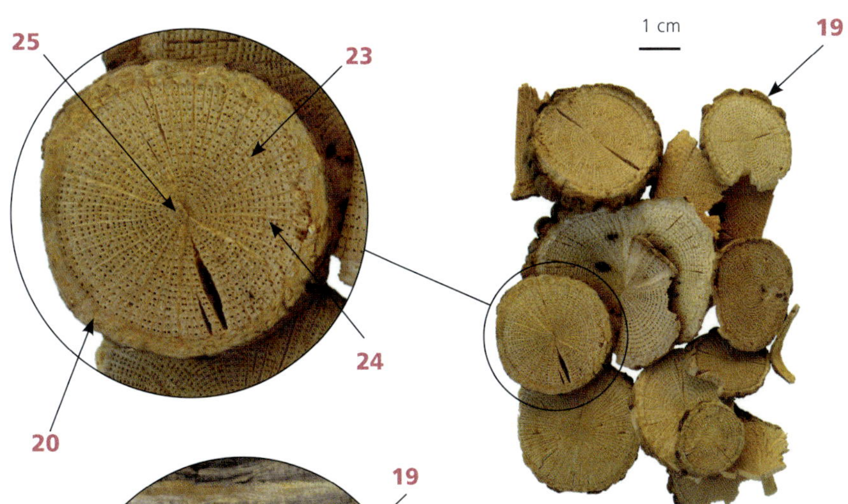

Decocting pieces (transverse stem) of *A. manshuriensis*

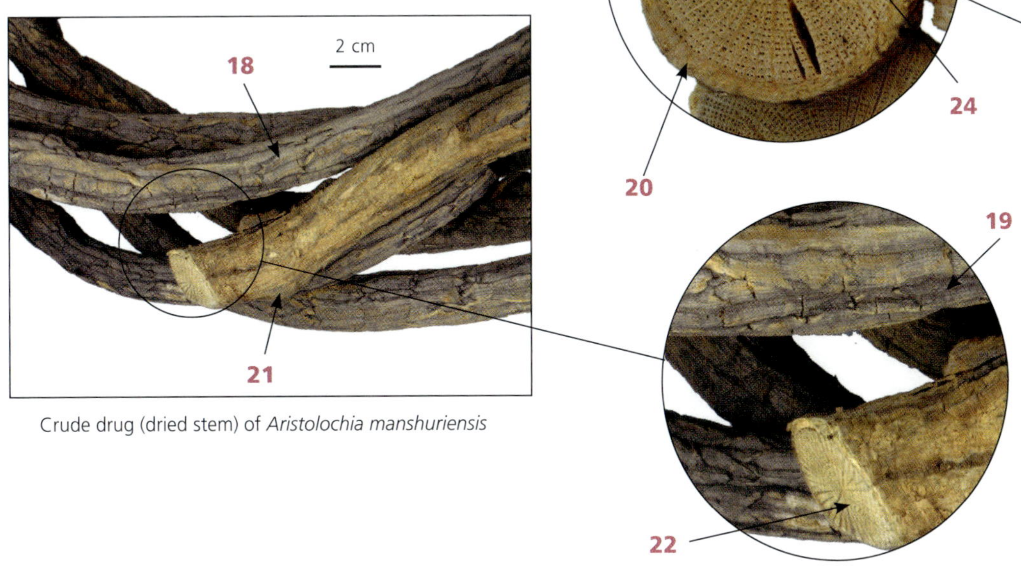

Crude drug (dried stem) of *Aristolochia manshuriensis*

Deng Xin Cao 灯心草

Junci Medulla

Drug part: STEM PITH

Juncus effusus L.

Juncaceae

Species of *Juncus* wild in China: 76

PLANT DESCRIPTION

A densely tufted perennial rush (25–90 cm tall, or taller), unbranched and grass-like. *Stems* green, cylindrical, smooth and pliable (1–5 mm diam.) with a creamish-white continuous pith. *Leaves* (cataphylls) all basal, sheathing and scale-like, reddish-brown to chestnut-brown (2–22 cm long). *Flowers* in a dense or lax lateral head, towards the tip of the stem. Each flower pale brown, tiny and inconspicuous; stamens 3 (rarely 6), 'petals' usually 6 and pale brown; style 1 and very short; wind-pollinated. *Fruit* a dry, brown, ovoid to oblong 3-valved capsule. *Seeds* many. Fl. Apr–Jul, fr. May–Sep.

HARVESTING, SOURCING AND NATURAL RANGE

Stems are harvested in late summer or autumn mainly from plants grown as medicinal crops. The main production areas are in the Chinese provinces of Fujian, Guizhou, Jiangsu and Sichuan with *Di Dao*-quality traditionally sourced from Suzhou in Sichuan. Cultivation also occurs throughout much of China and where it grows wild: Anhui, Gansu, Guangdong, Guangxi, Guizhou, Hebei, Heilongjiang, Henan, Hubei, Hunan, Jiangxi, Jilin, Liaoning, Shandong, Taiwan, Tibet, Yunnan and Zhejiang. The species grows in forest margins, damp grassland, lakes, river-sides and rice fields (200–3,400 m altitude) and is a widespread species worldwide in temperate and tropical montane regions.

CONSERVATION STATUS

China: Least Concern (authors' pers. obs., 2011).
Global: Least Concern (IUCN Red List, accessed 25/8/2016).

Typical tufted growth of *Juncus effusus*

J. effusus PHOTO: D. SIMPSON

Fruiting plants of *J. effusus*

TCM PROPERTIES AND USE

***properties*: 'sweet, bland and cool'**
***channels*: 'Heart, Lung and Small Intestine'**
***actions*: 'promotes urination, clears heat and fire, cools Blood**

***indications*: restlessness, insomnia; dysuria, mouth ulcers, epistaxis**

DRUG MORPHOLOGY

CRUDE DRUG (PITH): long, cylindrical and slender with truncated ends (up to 90 cm long; 0.1–0.3 cm diam.); white or pale yellowish-white throughout; externally with fine longitudinal striations **[1]**. *Texture*: light, soft and pliable. *Odour*: indistinct. *Taste*: none. ***Good quality*:** long, white and elastic with outer green cortex **[2]** totally removed (see *Safety*).

DECOCTING PIECES-1 (CUT LENGTHS OF PITH): as for the crude drug but cut into 10–15 mm lengths.

DECOCTING PIECES-2 (CARBONISED PITH): as for decocting pieces-1 above but externally black. *Texture*: light and easily broken. *Odour*: indistinct. *Taste*: somewhat astringent.

PROCESSING

Stems are sun-dried then the green outer epidermis removed and the pith straightened and bundled (= crude drug); it is then cut into 10–15 mm lengths for decocting. Further processing may be performed by carbonising the pith to produce Deng Xin Tan 灯心炭. In some parts of China the green epidermis is only partially removed and reproductive parts are also included as part of the drug too (see *Safety*).

OTHER COMMON NAMES

灯草 Deng Cao; 灯心 Deng Xin; 水灯心 Shui Deng Xin; common rush (CP2010); Medulla Junci; soft rush (SCN).

SAFETY

Because of the ability of *Juncus effusus* to adsorb heavy metals from contaminated groundwater (Deng *et al*., 2004), this species is prone to heavy metal accumulation. Some authorities (Zhao & Xiao, 2009–2010) suggest avoiding the use of plants known to be harvested from contaminated sites while others recommend the routine removal of the green epidermis of the stem (as described in the CP2015) where such harmful contaminants are likely to accumulate.

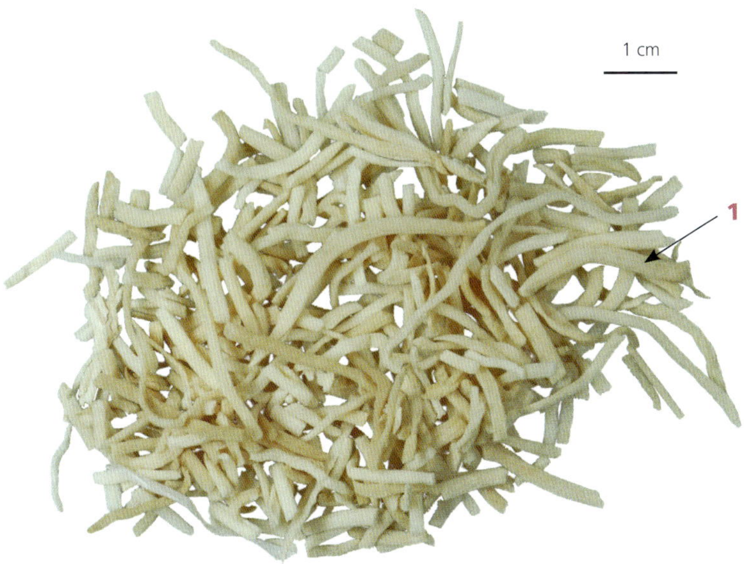

Decocting pieces of Deng Xin Cao (*J. effusus* pith)

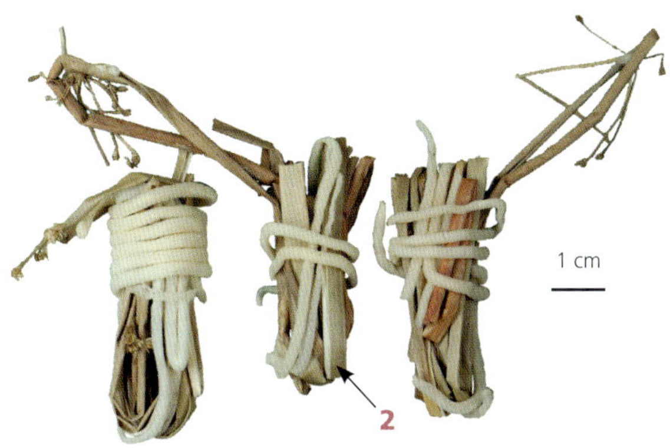

Poor quality Deng Xin Cao with green cortex still present

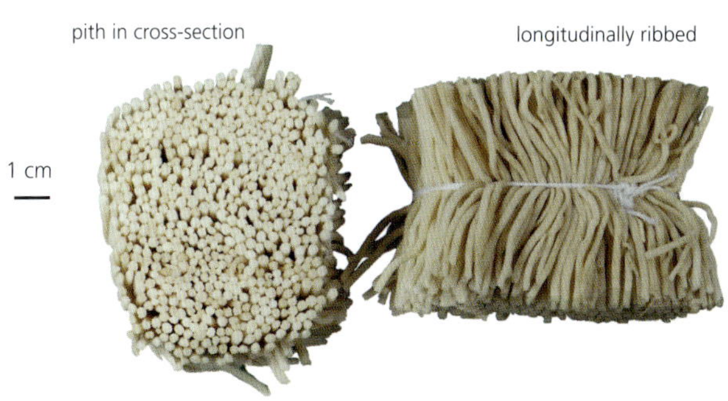

Decocting pieces in typical trade bundles (*J. effusus* pith)

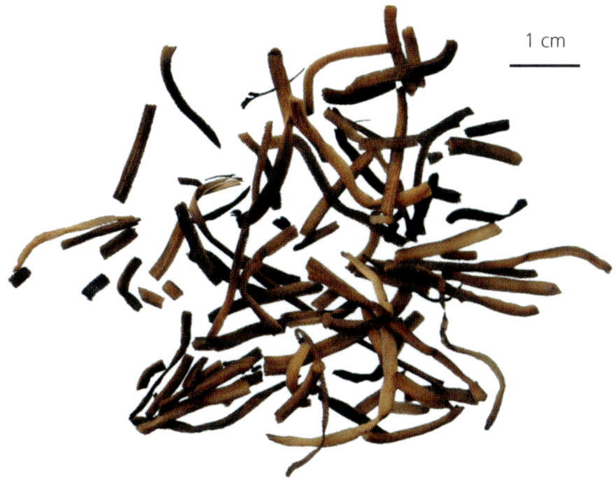

Processed Deng Xin Tan (carbonised pith of *J. effusus*)

Gong Lao Mu 功劳木

Mahoniae Caulis

Drug part: STEM

Mahonia bealei (Fortune) Carrière

Berberidaceae

synonym: *Berberis bealei* Fortune (*Kew-MPNS* accepted name, 25/8/2016).

Species of *Mahonia* wild in China: 31

PLANT DESCRIPTION

A spineless evergreen shrub or small tree (0.5–4 m tall, occasionally to 8 m). *Leaves* alternate and pinnate (27–51 cm long, 10–20 cm wide) with 4–10 pairs of leaflets and a single larger terminal one; leaf stalks 0.5–2.5 cm long; leaflets oblong to ovate, stalkless, hard and leathery, with 2–7 spiny teeth along each margin; lower leaflets shorter and wider than upper ones; upper surface dark greyish-green and shiny. *Flower-heads* are many-flowered cylindrical spikes (7–26 cm long) arising in clusters of 3–9 from the stem tips. Each flower with 2 whorls of 3 yellow, hooded 'petals' with distinct basal glands; stamens 6; ovary superior. *Fruits* ovoid berries (c. 1.5 cm long; 1–1.2 cm diam.) dark purple or dark blue with a dusty coating and persistent styles; 1–7-seeded. Fl. Sep–Jun of following year; ft: Mar–May.

HARVESTING, SOURCING AND NATURAL RANGE

Stems are harvested from wild plants year round, where they grow in forests, forest margins, scrub, stream- and roadsides; 500–2,000 m altitude. Endemic to NC, SC and SE China: Anhui, Fujian, Guangdong, Guangxi, Henan, Hubei, Hunan, Jiangsu, Jiangxi, Shaanxi, Sichuan and Zhejiang [cultivated and naturalised elsewhere].

CONSERVATION STATUS

Endemic to China: Vulnerable (authors' pers. obs.; over-collected for medicinal use); Not Evaluated (IUCN Red List, accessed 6/7/2016.

M. fortunei (Lindl.) Fedde

Berberidaceae

synonym: *Berberis fortunei* Lindl. (*Kew-MPNS* accepted name, 25/8/2016).

PLANT DESCRIPTION

As for *B. bealei* but shorter (usually only 2 m); **leaves 10–28 cm long; 8–18 cm wide**; **leaf stalks longer (2–9 cm); leaflets in 2–5 pairs, longer and narrower**, with **5–10 pairs of spiny teeth**; flower spikes much shorter **(3–7 cm).**

HARVESTING, SOURCING AND NATURAL RANGE

Stems are harvested from wild plants, year-round where they grow in forests, scrub and marginal areas; 300–2,000 m altitude. Endemic to NC, SC and SE China: Guangxi, Hubei, Jiangxi, Sichuan, Taiwan, Zhejiang; also widely cultivated as an ornamental in China [cultivated in Indonesia, Japan, and USA].

CONSERVATION STATUS

Endemic to China: Vulnerable (authors' pers. obs.; over-collected for medicinal use); Not Evaluated (IUCN Red List. accessed 6/7/2016).

TCM PROPERTIES AND USE

properties: 'bitter and cold'
channels: 'Liver, Stomach and Large Intestine'
actions: 'clears heat, dispels damp, clears fire, eliminates toxins'
indications: diarrhoea; dysentery; jaundice; toothache, sores, boils, abscesses; (T)

Wild plant of *Mahonia bealei* with immature fruits

M. bealei with immature fruits

Young fruiting plant of *Mahonia fortunei*

Flowering plant of *M. fortunei* in the wild

Mahonia bealei

Berberidaceae

DRUG MORPHOLOGY

CRUDE DRUG (DRIED STEM): long, hard, cylindrical woody stems (occasionally curved or angled) and of varying length; 1–2 cm diam. Outer bark (1–2 mm thick) greyish-yellow to dark brown, rugged and deeply fissured into rectangular blocks **[1]** or, when young, smooth and lustrous; remains of obliquely attached, hard leaf bases present at occasional and alternate intervals **[2]**. Outer bark easily removed to reveal a thin, yellowish inner bark (phloem region) below **[3]**. *Fracture*: hard, difficult to break; cut surface reveals dense yellow wood (xylem) with several concentric rings and radial tissues **[4]** and a larger central pith **[5]**. *Odour*: indistinct. *Taste*: bitter. ***Good quality***: yellow in cross-section with a strong bitter taste.

DECOCTING PIECES (SLICED STEM): circular or oblique woody stem slices (usually 1–2 cm diam.; c. 4–6 mm thick). Cork bark greyish-yellow to dark brown and deeply fissured (c. 2–4 mm wide) **[1]**. Cut surface as for crude drug. *Odour:* indistinct. *Taste* bitter.

PROCESSING

Stems are cut into short lengths and dried (= crude drug), then sliced transversely or obliquely into short pieces (= decocting pieces).

OTHER COMMON NAMES

阔叶十大功劳 Kuo Ye Shi Da Gong Lao; 十大公劳 Shi Da Gong Lao; Caulis Mahoniae; chinese mahonia stem; Mahonia bealei (SCN).

M. fortunei

Berberidaceae

DRUG MORPHOLOGY

CRUDE DRUG (DRIED STEM): stem diameter usually smaller than that for PS1 but to reliably differentiate between the drugs sourced from these two species laboratory-based methods are recommended.

DECOCTING PIECES (SLICED STEM): as for PS1.

PROCESSING

As for PS1.

OTHER COMMON NAMES

Caulis Mahoniae; Mahonia fortunei (SCN); Xi Ye Shi Da Gong Lao.

UNOFFICIAL SUBSTITUTES

Given that this drug is wild-sourced, and macroscopic characters are too few to differentiate stems (the drug part) of the two official species from the many other wild species of *Berberis* and *Mahonia* in China, labortatory-based methods are recommended for authenticating trade items of Gong Lao Mu 功劳木. Further research is needed to determine which species are entering the trade.

Gong Lao Mu crude drug and decocting pieces
sourced from dried and sliced stem of *Mahonia bealei*

Gong Lao Mu crude drug sourced
from dried sliced stem of *Mahonia fortunei*

Gong Lao Mu decocting pieces sourced
from dried, sliced stem of *M. fortunei*

Gou Teng 钩藤 Uncariae Ramulus cum Uncis

Drug parts: HOOK-BEARING TWIGS

PHARMACOPOEIA SPECIES 1

Uncaria rhynchophylla (Miq.) Miq. ex Havil.

Rubiaceae

synonym: *Nauclea rhynchophylla* Miq.

Species of *Uncaria* wild in China: 12

PLANT DESCRIPTION

A liana with slender, weakly or markedly 4-sided stems, opposite branches, and armed with a pair of downward-curving hook-like spines. *Leaves* opposite; stipules (between the leaf stalks of each pair of opposite leaves) 4–15 mm long, divided for up to 4/5 of length into two linear or triangular lobes; often falling off; leaf blades simple, ovate or lance-shaped (5–12 cm long; 3–7 cm wide), hairless and, when dried, papery and often dark red or reddish-brown; margins entire. *Inflorescences* in leaf axils, distinctly spherical (1.2–1.5 cm diam.) with many densely packed, tiny flowers; each flower stalkless or nearly so; bisexual; calyx lobes 5, c. 1 mm long; flower tube 5.5–6 mm long, broadening at mouth into 5 spreading lobes (probably pale green); stamens 5. *Fruiting heads* globose (1–2 cm diam.) with base of each fruit attached to a central point; each fruit stalkless or shortly stalked, ovoid or tapering at both ends. Fl. and fr. May–Dec.

HARVESTING, SOURCING AND NATURAL RANGE

Aerial parts are wild-harvested in autumn and winter from plants which grow in open forests, thickets and riversides in deep valleys (0–1,000 m altitude). The species is native in SC and SE China in the provinces of Fujian, Guangdong, Guangxi, Guizhou, Hubei, Hunan, Jiangxi, Yunnan and Zhejiang [also native in Japan].

CONSERVATION STATUS

China: Least Concern (MEPCAS, 2013); 'rapidly decreasing' (Leung & Cheng, 2008).
Global: Not Evaluated (IUCN Red List, accessed 16/8/2016).

PHARMACOPOEIA SPECIES 2

U. macrophylla Wall.

Rubiaceae

PLANT DESCRIPTION

Distinguished from PS1 by **stems, leaf stalks and flower/fruit stalks softly hairy**; **leaves broader** (10–16 cm long; **6–12 cm wide**) and, **when dry, thin and leathery with upper surfaces yellowish-brown**; stipules **divided 1/2 to 2/3 of length** into 2 triangular or ovate lobes; latter often falling off; **flowers and fruits stalked**.

HARVESTING, SOURCING AND NATURAL RANGE

Harvesting as for PS1 from wild populations which grow in secondary forests (300–900 m altitude) in SC and SE China in the provinces of Guangdong, Guangxi, Hainan and Yunnan [also native in Bangladesh, Bhutan, India, Laos, Myanmar, N Thailand and Vietnam].

CONSERVATION STATUS

China: Least Concern (MEPCAS, 2013).
Global: Not Evaluated (IUCN Red List, accessed 16/8/2016).

TCM PROPERTIES AND USE

properties: 'sweet and cool'
channels: 'Liver and Pericardium'
actions: 'extinguishes wind, clears heat, pacifies yang'

indications: convulsions; high fever, irritability, restlessness; headache; bloating; dizziness; vertigo; hypertension; tinnitus, insomnia

Typical growth form of *Uncaria rhynchophylla* (centre foreground) in the wild, Guizhou province

U. rhynchophylla with opposite pairs of ovate leaves with two deeply lobed stipules and immature hooks at base of each

Stipule (circled) of *U. rhynchophylla* divided into 2 linear or triangular lobes

U. macrophylla with immature, spherical flower-heads

U. macrophylla with flowering and fruiting heads

Flower-head of *U. macrophylla* broken open (single flower circled)

Uncaria hirsuta Havil.

Rubiaceae

PLANT DESCRIPTION

Distinguished from PS1 by **stiff hairs** on stems and lower leaf surfaces and hook-like spines; **stipules** (at base of leaf pairs) **deeply divided (2/3 of length) into 2 broadly ovate lobes** and the **tiny flowers almost stalkless**.

HARVESTING, SOURCING AND NATURAL RANGE

Harvesting as for PS1 from wild populations which grow in scrub or forest along valley streamsides (up to 500 m altitude) and are endemic to SC and SE China: Fujian, Guangdong, Guangxi, Guizhou and Taiwan.

CONSERVATION STATUS

Endemic to China: Least Concern (MEPCAS, 2013; Not Evaluated (IUCN Red List, accessed 16/82016).

U. sinensis (Oliv.) Havil.

Rubiaceae

synonym: *Nauclea sinensis* Oliv.

PLANT DESCRIPTION

Distinguished from PS1, 2 and 3 by **stipules persistent** (not falling) and **not clearly divided into 2 lobes** but broadly triangular, ovate, orbicular or with a distinct notch; distinguished from PS1 by larger inflorescences (c. 3 cm diam.) and fruit-heads (2–3 cm diam.).

HARVESTING, SOURCING AND NATURAL RANGE

Harvesting as for PS1 from wild populations which grow in open or wet, secondary forests (900–1,100 m altitude) and are endemic to NC, SC and SE China: Gansu, Guangxi, Guizhou, Hubei, Hunan, Shaanxi, Sichuan and Yunnan.

CONSERVATION STATUS

Endemic to China: Least Concern (MEPCAS, 2013); Not Evaluated (IUCN Red List, accessed 16/8/2016).

U. sessilifructus Roxb.

Rubiaceae

PLANT DESCRIPTION

Distinguished from PS1 by **stipules typically deciduous** (falling early), when present **divided 2/3 or more of length** into 2 narrowly triangular or sharply pointed, hairless or densely hairy lobes (7–10 mm long); the **tiny flowers and fruits stalkless**. Almost hairless throughout.

HARVESTING, SOURCING AND NATUAL RANGE

Harvesting as for PS1 from wild populations which grow in dense forests or thickets in valleys (300–1,500 m altitude) in SC China: Guangxi and Yunnan [also native to Bangladesh, Bhutan, India, Laos, Myanmar, Nepal and Vietnam].

CONSERVATION STATUS

China: Least Concern (MEPCAS, 2013).
Global: Not Evaluated (IUCN Red List, accessed 16/8/2016).

Stipule (circled) of *Uncaria hirsuta* divided up to 2/3 of length into two broadly ovate lobes

Stem and leaf undersides of *U. hirsuta* covered in stiff hairs

U. sinensis (centre) growing wild in dense secondary forest

Stipule (circled) of *U. sinensis* **not** divided into 2 lobes, with small apical notch

Typical scrambling growth form of *U. sessilifructus* in the wild

Branch of *U. sessilifructus* with immature fruit-head (circled) bearing **stalkless**, spindle-shaped fruits and **stipule fallen (typically deciduous)**

Uncaria rhynchophylla — Rubiaceae

DRUG MORPHOLOGY

CRUDE DRUG/DECOCTING PIECES (hook-bearing twigs): short brown, cylindrical or subsquare twigs (2–3 cm long; 2–5 mm diam.) with hooked spines. Externally variable in colour: reddish-brown, purplish-red, yellowish-green or greyish-brown; longitudinally wrinkled **[1]**, patches of dotted lenticels occasional **[2];** yellowish-brown hairs occasional. Two (usually) opposite and strongly curved hooks above each node **[3]**; hooks slightly flattened laterally or rounded, narrowing from base to a sharply pointed tip. Dotted scars of leaf stalks and ring-shaped scars of stipules **[4]** present immediately below bases of the hooks. *Texture*: hard and tough. *Fracture*: bark fibrous, pith yellowish-white **[5]** or hollow. *Odour*: indistinct. *Taste*: weak. ***Good quality*:** hooks and twigs purplish- or reddish-brown and glossy throughout; hooks solid, anchor-shaped and in pairs.

PROCESSING

After removal of the leaves the hook-bearing parts of the twigs are cut into short sections then sun-dried (= crude drug/decocting pieces).

OTHER COMMON NAMES

嫩钩藤 Nen Gou Teng; 双钩 Shuang Gou; Gambir plant (CP2015); Ramulus Uncariae cum Uncis; *Uncaria rhynchophylla* (SCN).

U. macrophylla — Rubiaceae

DRUG MORPHOLOGY

Macroscopically indistinguishable from those of the other four Pharmacopoeia species. Laboratory-based techniques are required for their reliable differentiation.

PROCESSING

As for PS1.

OTHER COMMON NAMES

大叶钩藤 Da Ye Gou Teng 'big leaved hook vine'; Gambir plant (CP2015); Ramulus Uncariae cum Uncis.

U. hirsuta — Rubiaceae

DRUG MORPHOLOGY

As for PS2.

PROCESSING

As for PS1.

OTHER COMMON NAMES

毛钩藤 Mao Gou Teng 'hairy hook vine'; Gambir plant (CP2015); Ramulus Uncariae cum Uncis.

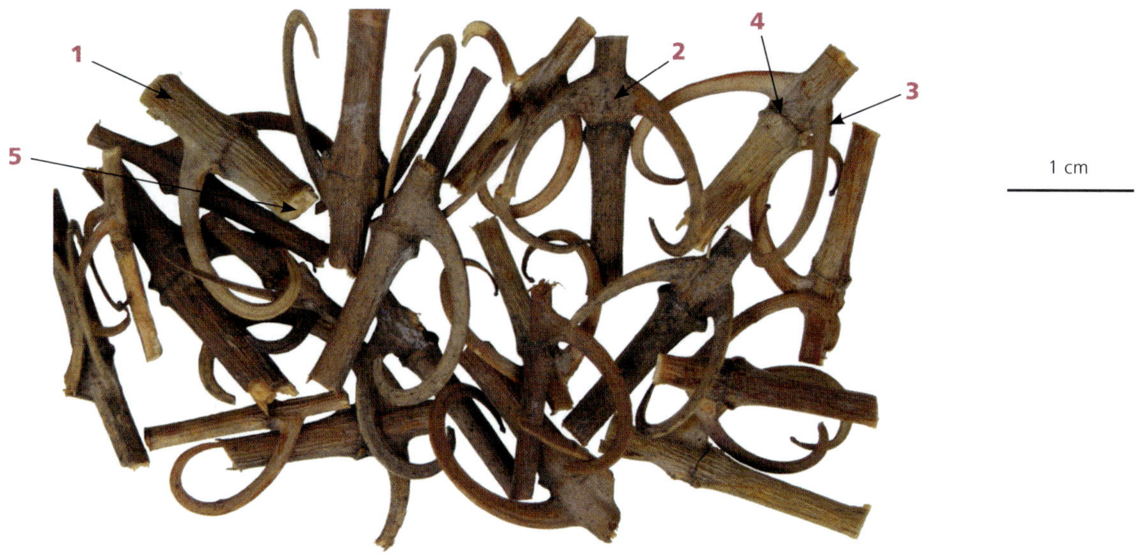

Crude drug/decocting pieces of Gou Teng (sourced from *U. rhynchophylla*)

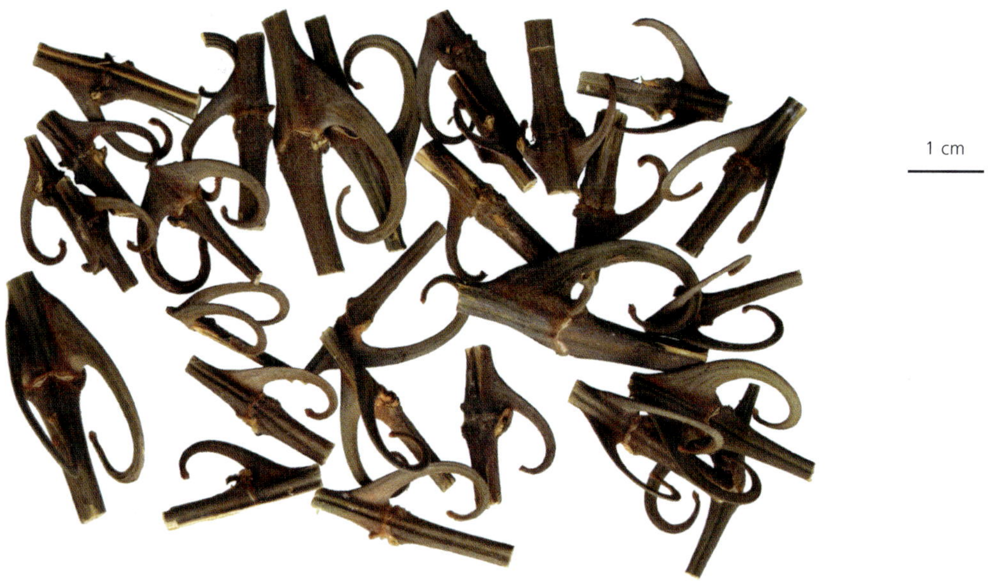

Crude drug/decocting pieces of Gou Teng (sourced from *U. macrophylla*)

Crude drug/decocting pieces of Gou Teng (sourced from *U. hirsuta*)

Uncaria sinensis

Rubiaceae

DRUG MORPHOLOGY

As for PS2.

PROCESSING

As for PS1.

OTHER COMMON NAMES

华钩藤 Hua Gou Teng (plant name); Gambir plant (2015); Ramulus Uncariae cum Uncis.

U. sessilifructus

Rubiaceae

DRUG MORPHOLOGY

As for PS2.

PROCESSING

As for PS1.

OTHER COMMON NAMES

白钩藤 Bai Gou Teng 'white hook vine'; 无柄钩藤 Wu Bing Gou Teng; Gambir plant (CP2015); Ramulus Uncariae cum Uncis.

UNOFFICIAL SUBSTITUTES

The most likely substitutes to be traded as 'Gou Teng' are the hook-bearing twigs of the seven other *Uncaria* species that grow wild in China because, at the time of writing, 'Gou Teng' is entirely wild-sourced and these closely related species are easily confused with the five Pharmacopoeia species. Differentiating the 'hooks' of official from unofficial species requires laboratory-based methods and Zhao & Xiao (2009–10) suggest research is needed to determine whether their pharmacology differs. Examples of some of these unofficial substitute species are listed below. Whether or not such unofficial substitutes of 'Gou Teng' occur in international trade is unclear.

1. ***Uncaria scandens*** (Sm.) Hutch. (攀茎钩藤 Pan Jing Gou Teng). Endemic to south China: Guangdong, Guangxi, Hainan, Sichuan, Tibet and Yunnan.
2. ***U. lancifolia*** Hutch. (倒挂金钩 Dao Gua Jin Gou). Native in Yunnan province but where populations have declined; the species is now listed as 'Vulnerable' in China (MEPCAS, 2013). Also native in N Vietnam.
3. ***U. laevigata*** Wall. ex G. Don (平滑钩藤 Ping Hua Gou Teng). Native in S China (Guangxi, Taiwan and Yunnan). Also native in Bangladesh, India, Laos, Myanmar, Thailand, Vietnam.
4. ***U. yunnanensis*** K. C. Hsia (possibly synonymous with *U. gambir* (W. Hunter) Roxb. (云南钩藤 Yun Nan Gou Teng). Endemic to Yunnan province (Xishuangbanna district) of SW China and where populations have declined due to over-harvesting; the species is now listed as 'Vulnerable' in China (Wang & Xie, eds 2004). Widely cultivated in Sarawak for the production of gambir for tanning and for chewing with betel nut.

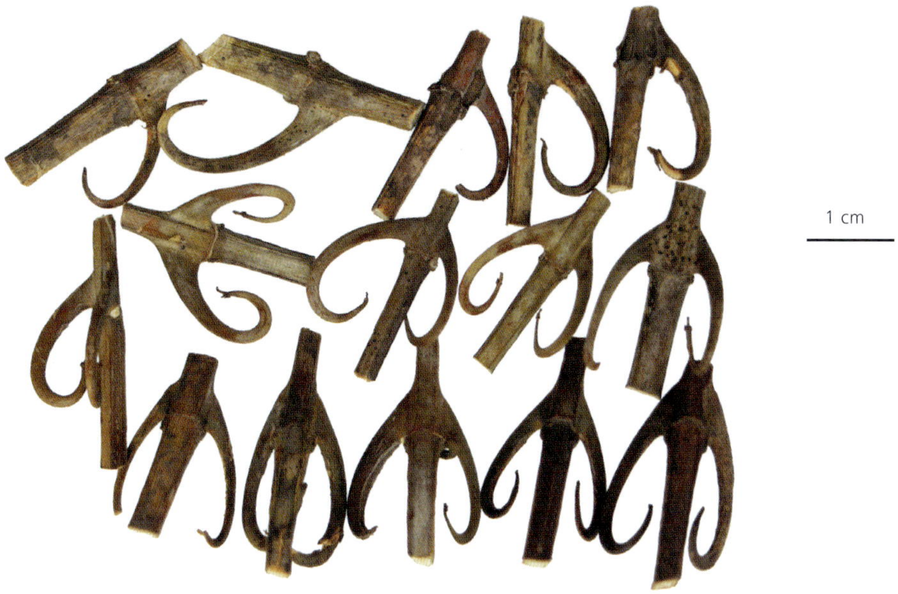

Crude drug/decocting pieces of Gou Teng (sourced from *U. sinensis*)

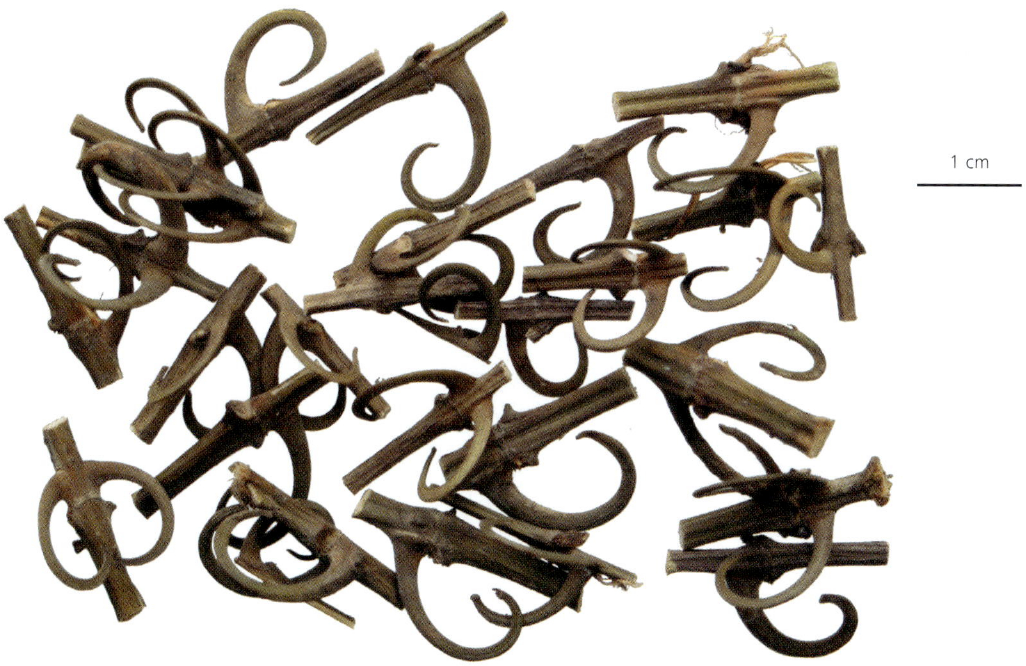

Crude drug/decocting pieces of Gou Teng (sourced from *U. sessilifructus*)

Gui Zhi 桂枝

Cinnamomi Ramulus

Chinese cinnamon/cassia twig

Drug part: TWIG

Cinnamomum cassia (L.) J. Presl

Lauraceae

synonyms: *Cinnamomum aromaticum* Nees; *Laurus cassia* L.; *Persea cassia* (L.) Spreng.
taxonomic/nomenclatural note: in CP2015 as '*C. cassia* Presl.'

Species of *Cinnamomum* wild in China: 48

PLANT DESCRIPTION

An evergreen, medium-sized tree with bark greyish-brown up to 13 mm thick at maturity. Current year's branchlets yellowish-brown, becoming dark brown, with longitudinal grooves and slightly downy. *Leaves* alternate or almost opposite, simple and stalked (1.2–2 cm); blade narrowly elliptic to lance-shaped (8–34 cm long; 4–9.5 cm wide), base sharply narrowing, shiny dark green above, minutely hairy below, conspicuously 3-veined and with tufts of hairs in axils of lateral veins, aromatic and leathery; margins entire, tough and rolled inwards. *Flowers* in branched axillary or terminal clusters; flower tube about c. 2 mm long; 'petals' 6, tiny, inconspicuous, white and covered in short yellowish-brown hairs; 'petals' often falling off before stamens developed; stamens 12 (9 obvious); ovary superior; style 1, stigma 1. *Fruit* fleshy and ellipsoid (10 mm long; 7–9 mm diam.), blackish-purple when mature, hairless; with persistent 'cup' at base. Fl. Jun–Aug, fr. Oct–Dec.

HARVESTING, SOURCING AND NATURAL RANGE

Twigs are harvested in spring and summer from cultivated trees widely grown as a plantation crop in tropical and subtropical parts of south and south-eastern China (Fujian, Guangdong, Guangxi, Guizhou, Hainan, Taiwan and Yunnan) as well as India, Indonesia, Laos, Malaysia, Thailand and Vietnam. The main production areas in China are the provinces of Guangdong, Guangxi and Fujian.

Although, today, the species is only known from cultivation, the wild species is thought to originate from south China (*FoC*); its parentage may include *C. austrosinense* Hung T. Chang (M. Gilbert, pers. comm. 17/4/2015).

CONSERVATION STATUS

China/Global: only known from cultivation.

Cinnamomum cassia plantation, China

Branch of *C. cassia* with immature fruit

SUBSTITUTION COMMENT. Market demand for 'Gui Zhi' is much lower than that of 'Rou Gui' (the bark of the same species, see page 520) but it is nevertheless prone to similar substitution and adulteration since, as a cheap, by-product of 'Rou Gui', it too may be sourced from closely related unofficial substitutes. See *Unofficial Substitutes and Adulterants* of 'Rou Gui' (Cinnamomi Cortex), page 520.

TCM PROPERTIES AND USE

***properties*: 'pungent, sweet and warm'**
***channels*: 'Heart, Lung and Bladder'**
***actions*: 'releases the exterior, tonifies yang'**

***indications*: chills and fevers; musculoskeletal pain; urinary and menstrual disorders**

DRUG MORPHOLOGY

CRUDE DRUG (DRIED TWIG): cylindrical twigs (30–75 cm long; 0.3–1 cm diam.), with remains of alternate side branches **[1]**. Externally dull, greyish-brown or purplish-brown with fine longitudinal striations **[2]**, lenticels oval and scattered **[3]**, and with leaf-scars **[4]**, branch-scars and bud-scars **[5]**. *Texture*: hard and brittle. *Fracture*: dense and solid throughout **[6]** with a thin dark outer bark (c. 1 mm) surrounding a thin mid-brown or reddish cambium, a broad yellowish-white or cream xylem and a darker central sub-square pith **[7]**. *Odour*: intensely and characteristically aromatic. *Taste*: sweet and slightly pungent, strongest in the bark. ***Good quality*:** thin, tender twigs, brownish-yellow and strongly aromatic.

DECOCTING PIECES (SLICED TWIG): slices elliptical to circular (2–4 mm thick); bark thin purplish-brown or greyish-brown **[2]** (sometimes with oval lenticels), easily removed and highly aromatic when broken. Cambium zone narrow, mid-brown or reddish **[8]**, xylem wide and dense **[9]**, yellowish-white or cream, surrounding an often indistinct semi-circular or sub-square pith **[10]**. *Taste*: as for crude drug.

PROCESSING

After removal of the leaves, twigs are sun-dried (= crude drug) then transversely or diagonally sliced into decocting pieces (2–4 mm thick) or sliced when fresh.

OTHER COMMON NAMES

Liu Gui; 嫩桂枝 Nen Gui Zhi 'tender cinnamon twigs'; cassia (SCN); cassia twig (CP2010); Chinese cinnamon; Ramulus Cinnamomi.

SAFETY

Safety Class: 2b; Interaction Class: A (Gardner & McGuffin, 2013). Allergic reactions relatively common (Gardner & McGuffin, 2013).

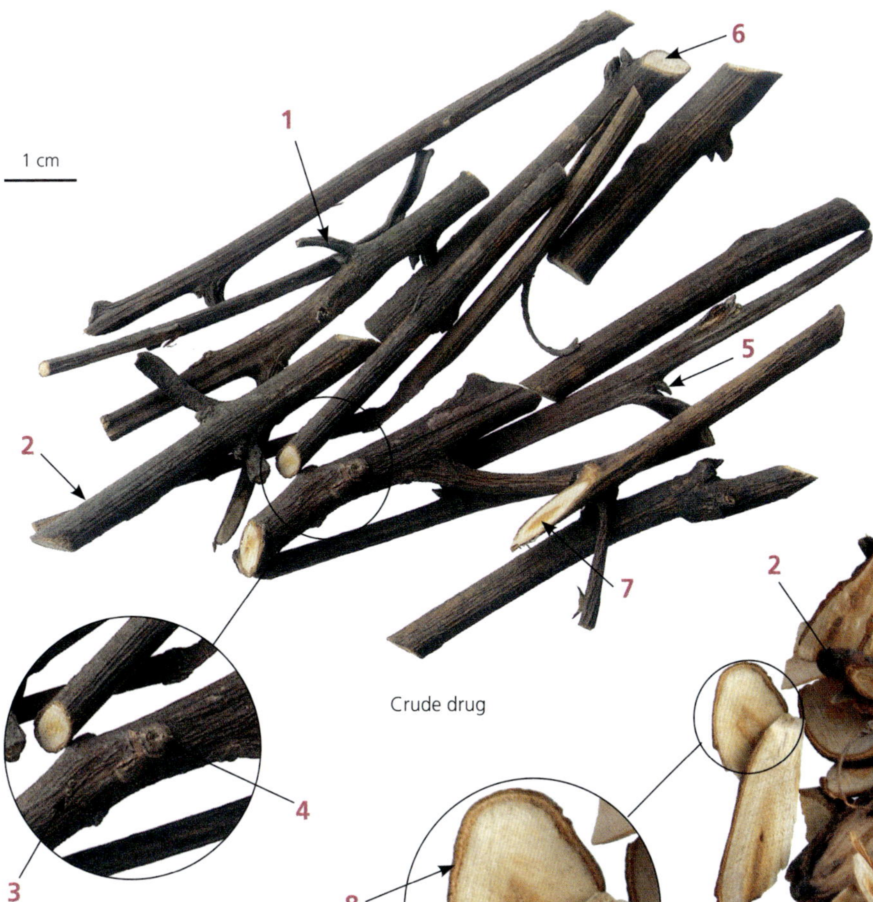

Crude drug

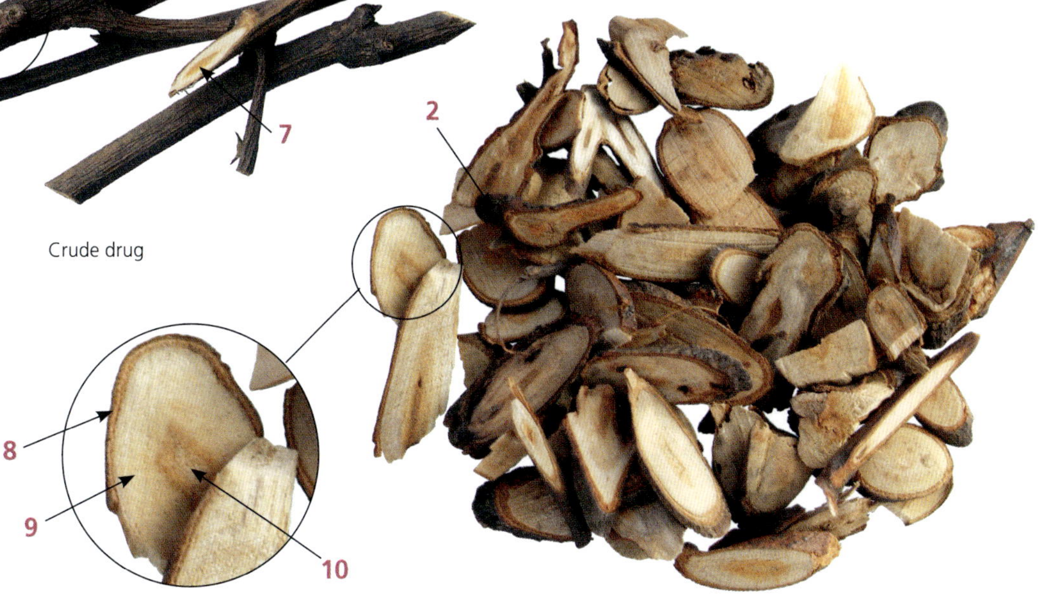

Decocting pieces

Ji Xue Teng 鸡血藤 Spatholobi Caulis

Drug part: STEM

PHARMACOPOEIA SPECIES

Spatholobus suberectus Dunn

Fabaceae

Species of *Spatholobus* wild in China: 10

PLANT DESCRIPTION

A vigorous, woody climber to 15 m; transverse cut stems display **resin-filled tissues laid down in an obliquely concentric pattern with pith off-centre**. *Stipules* present. *Leaves* alternate, swollen at base of stalk base (pulvinate), divided into 3 leaflets with **leaflet veins pinnate** (with 6–8 pairs of veins radiating off a central mid-vein); *terminal leaflet:* symmetrical and broadly elliptic (9–19 cm long; 5–14 cm wide); *lateral leaflets:* with stalks much shorter than that of terminal leaflet, blades somewhat asymmetrical, slightly narrower than terminal one with a broadly wedge-shaped base and undersides with tufts of hairs in vein axils. *Flowers* 0.5–1 cm long, in many-flowered axillary or terminal heads along central axis. Bracts small and often shedding. Each flower **bisexual and laterally symmetrical**, calyx with 5 rounded or slightly obtuse teeth, petals 5 white or pale yellow, flushed pink at base (**unequal size and shape**), (one large posterior petal, 2 lateral and parallel and 2 lower petals); ovary superior. ***Fruit* a long, somewhat sickle-shaped flat, pod** (8–11 cm long; 2–3.3 cm diam.) covered in dense velvety hairs and with a single flattened seed (17 mm long; 8 mm diam.). Fl. Jun., fr. Nov–Dec.

HARVESTING, SOURCING AND NATURAL RANGE

Stems are harvested in autumn (Aug–Oct) and winter (Nov–Jan) from wild plants which grow in open scrub and sunny forest glades on mountain slopes and valleys (800–1,700 m altitude). The species is endemic to south China (Fujian, Guangdong, Guangxi and Yunnan) but wild populations have suffered a rapid decline in recent years due to over-harvesting for medicinal use with Guangxi province witnessing the collapse of its former multi-million dollar industry. Today markets are increasingly supplied by imported material from neighbouring countries (Miao Jian-hua pers. comm., 10/2008). Some re-introduction projects have been proposed (Zeng, 2008).

CONSERVATION STATUS

Endemic to China: Vulnerable (MEPCAS, 2013); Not Evaluated (IUCN Red List, accessed 11/11/2016).

UNOFFICIAL SUBSTITUTE

Sargentodoxa cuneata (Oliv.) Rehder & E. H. Wilson

Lardizabalaceae

synonyms: *S. simplicifolia* S. Z. Qu & C. L. Min; *Holboellia cuneata* Oliv.

PLANT DESCRIPTION

Easily confused with the Pharmacopoeia species. **Main differences: leaflet veins palmate** (radiating from base of leaflet) and transverse stem sections with **resin-filled phloem tissues in broad radial striations** (not in oblique concentric rings). *Flowers* **unisexual, radially symmetrical**, petals 6 (**equal size and shape**), yellow to yellowish-green, not flushed pink at base with ***fruit* a sub-spherical head** of 20–40 carpels turning dull blue when ripe. Seed ovoid and blackish (5 mm long).

CONSERVATION STATUS

China: Least Concern (MEPCAS, 2013).
Global: Not Evaluated (IUCN Red List, accessed 11/11/2016).

SUBSTITUTION COMMENT. A common and widespread vine whose stems are the source of the separate TCM drug 'Da Xue Teng' (Caulis Sargentodoxae) and which are often substituted or mixed with 'Ji Xue Teng' due to their superficial similarity or confusion about their similar Chinese names or the mistaken belief that they have very similar clinical functions.

Chen & Chen (2004) and Zhao & Xiao (2009–10) state that while the 2 drugs share some clinical properties their differences are such that they should not be used interchangeably in a clinical setting. In early Chinese herbals, such as the *Tu Jing Ben Cao* (1061) the less specific name 'Xue Teng' was used for both species; it was not until the mid-1980s that the Chinese Pharmacopoeia formally coined the prefixes 'Ji' and 'Da' to discriminate these two drugs. Enforcing the routine use of s uch vernacular names, however, across all markets is not straightforward and this name confusion persists even today. Finally, there may also be confusion at the point of harvest of these 2 species owing to the similarity of their growth form, leaves, habitat and distribution. The distinctive arrangement of the resin-filled tissues in their stems (whether fresh or dried) (see '*Drug substitute' morphology*) is the most reliable character to distinguish them but the easy availability of this substitute compared to the official species (see *Harvesting* .... and *Conservation status*) is likely to see it persist in local and international markets.

TCM PROPERTIES AND USE

properties**: 'bitter, slightly sweet and warm'**
channels**: 'Liver'**
actions**: 'invigorates Blood, tonifies Blood, dispels stasis, opens the channels and collaterals'**

indications**: dysmenorrhoea, amenorrhoea; musculoskeletal pain, numbness; weak extremities; T**

Spatholobus suberectus in the wild with typical climbing habit (centre of photo), Guangxi province, S. China

Leaves of *S. suberectus*

Cut stems of *S. suberectus* with obliquely-concentric resin-filled phloem tissues

Leaves of *Sargentodoxa cuneata*

Cut stems of *S. cuneata* with **radiating** resin-filled phloem tissues

Spatholobus suberectus

Fabaceae

DRUG MORPHOLOGY

CRUDE DRUG DECOCTING PIECES (SLICED STEM): elliptical, oblong or oblique slices (0.3–1 cm thick); remnants of greyish-brown or greyish-white outer bark with distinctive longitudinal furrows occasional and easily removed to reveal darker brownish-red layer beneath **[1]**. Cut surface with distinctive reddish-brown to blackish-brown resin-filled phloem arranged as **several elliptical or obliquely concentric rings [2]** alternating with wider and paler bands of **reddish-brown or brown xylem [3]**. Pith small and **off-centre [4]**. *Texture*: hard and dense. *Taste*: astringent. *Odour*: indistinct. ***Good quality***: transverse slices with 3–8 oblique, reddish-brown resin-rich tissues.

PROCESSING

After removal of leaves and young stems, mature stems are obliquely sliced then sun-dried (= crude drug and decocting pieces).

OTHER COMMON NAMES

密花豆 Mi Hua Dou (plant name); Caulis Spatholobi; spatholobus (SCN); suberect spatholobus stem (CP2010).

Ji Xue Teng sun drying (*Spatholobus suberectus*)

Sargentodoxa cuneata

Lardizabalaceae

'DRUG SUBSTITUTE' MORPHOLOGY

DRIED, SLICED STEM: outer bark easily shed to expose the dark reddish-brown inner bark below **[5]**. **Resin-filled tissues (phloem) arranged radially throughout the cut surface [6]** between the woody **yellowish-white xylem [7]** producing a distinctive wheel-like effect.

OTHER COMMON NAMES

大血藤 Da Xue Teng; 大血藤属 Da Xue Teng Shu (plant name); 红藤 Hong Teng 'red vine'; Sargentodoxae Caulis; Caulis Sargentodoxae; sargent glory-vine stem (CP2010).

LESSER KNOWN UNOFFICIAL SUBSTITUTES

The resinous stems of many other climbing shrubs in the same (Fabaceae) and other families have been reported as local and/or historical unofficial substitutes probably due to identification confusion at the point of harvest (NICBP, 1997–2002; Xiao *et al.* 2001–2002); XPG-III; Bensky *et al.*, 2004; Zhao & Li, 2007). Their occurrence as substitutes in international trade is uncertain and further research is warranted; in the meantime laboratory-based authentication methods may prove useful for their detection in trade.

1. ***Callerya dielsiana*** (Harms) P. K. Lôc ex Z. Wei & Pedley var. ***dielsiana*** (as the syn. *Millettia dielsiana* in Bensky *et al.*, 2004) 山鸡血藤 Shan Ji Xue Teng — family Fabaceae. Endemic to S and C China.
2. ***C. nitida*** (Benth.) R. Geesink var. ***nitida*** (as '*Millettia nitida*' in Bensky *et al.*, 2004) 亮叶鸡血藤 Liang Ye Ji Xue Teng — family Fabaceae. Endemic to S China.
3. ***Mucuna sempervirens*** Hemsl. 常春油麻藤 Chang Chun You Ma Teng — family Fabaceae. S and C China.
4. ***Mucuna birdwoodiana*** Tutcher 白花油麻藤 Bai Hua You Ma Teng — family Fabaceae. Endemic to S and C China (6 provinces).
5. ***Rourea microphylla*** (Hook. & Arn.) Planch. 红叶藤 Hong Ye Teng — family Connaraceae. S China.

Crude drug/decocting pieces of Ji Xue Teng (*Spatholobus suberectus*)

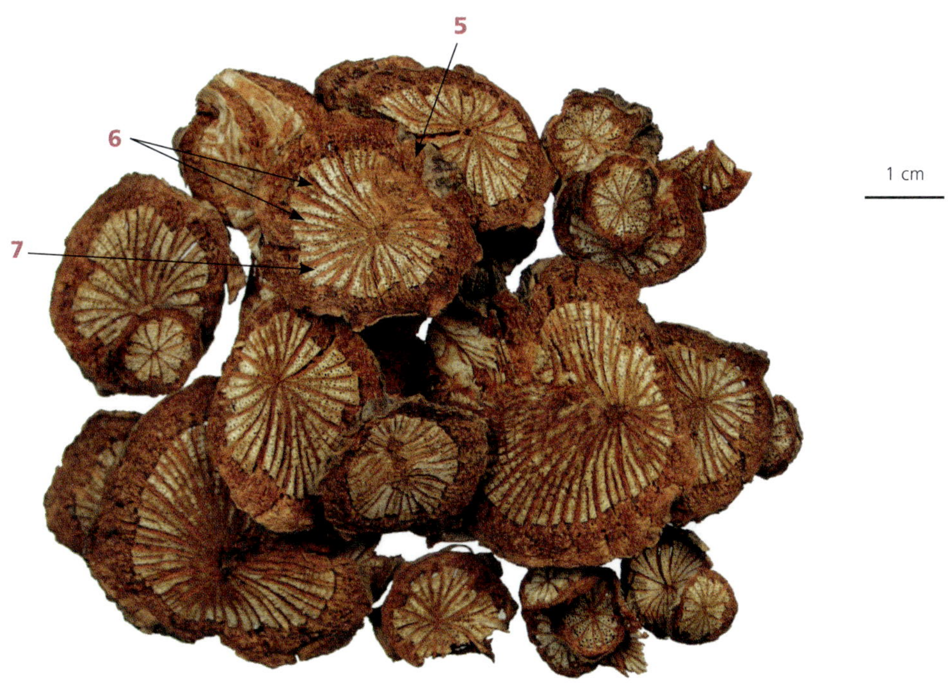

Crude drug/decocting of unofficial substitute Da Xue Teng (*Sargentodoxa cuneata*)

Ma Huang 麻黄 Ephedrae Herba

Drug part: STEM

PHARMACOPOEIA SPECIES 1

Ephedra sinica Stapf

Ephedraceae

synonym: *E. ma-huang* Tang S. Liu

Species of *Ephedra* wild in China: 14

PLANT DESCRIPTION

A sparsely branched subshrub (up to 40 cm tall). Lower *stems* woody, sometimes prostrate; upper stems green throughout (not woody) and jointed; stem internodes usually c. 3–4 cm long; c. 2 mm diam. *Leaves* reduced to 2 tiny, opposite membranous scales (3–4 mm long) at each node and fused along 1/3–2/3 their length with the free part (c. 5 mm long) tapering to a sharp point. *Male and female cones (reproductive parts)* on separate plants. Male (pollen) cones solitary or clustered at nodes, with ovoid heads of membranous bracts in 4 pairs; anthers 7 or 8, stalked or shortly stalked. Female (seed) cones solitary, oblong to ovoid, bracts in 4 pairs and, when mature, latter becoming red, fleshy and conspicuous (c. 8 mm long; 6–7 mm diam.). *Seeds* usually 2, blackish-red or greyish-brown. Pollination May–Jun, seed maturity Aug–Sep.

HARVESTING, SOURCING AND NATURAL RANGE

Non-woody stems are harvested in autumn and, until recently, entirely from wild populations which grow in arid, sandy places and dry river beds (700–1,600 m altitude), in China's northern provinces of Gansu, Hebei, Heilongjiang, Inner Mongolia, Jilin, Liaoning, Ningxia, Shaanxi and Shanxi. Habitat loss and over-harvesting has seen wild populations decline rapidly in recent years leading to extensive medicinal cultivation becoming established in Gansu and Ningxia provinces.

CONSERVATION STATUS

China: Near Threatened (MEPCAS, 2013); Vulnerable (Fu, 2011); 'rapidly decreasing' (Leung & Cheng, 2008).
Global: Least Concern (IUCN Red List, accessed 20/8/2016).

PHARMACOPOEIA SPECIES 2

E. intermedia Schrenk & C. A. Mey.

Ephedraceae

synonym: *E. intermedia* var. *tibetica* Stapf

PLANT DESCRIPTION

Easily confused with PS1 and PS3. *Main differences:* **subshrub or shrub (up to 1 m tall)**; stems frequently branched; stem internodes 2–6 cm long,1.5–3.5 mm diam.; **membranous scale leaves 3 (rarely 2),** fused at least 2/3 along their length; **bracts of seed cones in 2–5 pairs.**

HARVESTING, SOURCING AND NATURAL RANGE

Stems harvested exclusively from wild plants which grow in dry, sandy or rocky places (usually 800–4,600 m altitude) in northern China: Gansu, Hebei, Liaoning, Inner Mongolia, Ningxia, Qinghai, Shaanxi, Shandong, Shanxi, Tibet and Xinjiang [also native in Afghanistan, Kazakhstan, Kyrgyzstan, Outer Mongolia, Pakistan, Russia, Tajikistan, Turkmenistan, Uzbekistan and SW Asia].

CONSERVATION STATUS

China: Near Threatened (MEPCAS, 2013). Declining locally due to over-harvesting for medicinal use (authors' pers. comm. 1999).
Global: Least Concern (IUCN Red List, accessed 20/8/2016).

PHARMACOPOEIA SPECIES 3

E. equisetina Bunge

Ephedraceae

PLANT DESCRIPTION

Easily confused with PS1 and PS2. *Main differences:* **shrubs to 1(–1.5) m tall**; **woody stems well developed** and frequently branched; **stem internodes shorter and narrower (1–3 cm long; 1–1.5 mm diam.)** than those of PS1 and PS2; **membranous scale leaves 2 (opposite),** brownish (**1.5–3 mm long),** fused for c. 3/4 their length, free part bluntly triangular; **bracts of seed cones in 3 pairs.**

For other TCM drugs sourced from *Ephedra* species see 'Ma Huang Gen' (Ephedrae Radix et Rhizoma) page 196.

HARVESTING, SOURCING AND NATURAL RANGE

Stems harvested from wild plants which grow in dry, rocky places (800–3,000 m altitude) in northern China: Gansu, Hebei, Inner Mongolia, Ningxia, Qinghai, Shanxi and Xinjiang [also wild in Afghanistan, Kazakhstan, Kyrgyzstan, Mongolia, Russia, Tajikistan, Turkmenistan and Uzbekistan].

CONSERVATION STATUS

China: Least Concern (MEPCAS, 2013, as the two varieties '*equisetina*' and '*monoica*').
Global: Not Evaluated (IUCN Red List, accessed 20/8/2016).

TCM PROPERTIES AND USE

properties: **'pungent, bitter and warm'**
channels: **'Lung and Bladder'**
action: **'releases wind cold'**

indications: **cough, respiratory tract infection, chest distension; oedema**

Ephedra sinica, wild in Inner Mongolia

E. sinica with (inset) scale leaves in 2s and tips sharply pointed

Female plant of *E. sinica*; inset: pollen cone with yellow anthers

E. sinica with mature fruit

E. intermedia in the wild

E. intermedia with scale leaves in 3s (rarely 2)

Fruiting plant of *E. intermedia*

Mature plants of *E. equisetina* with well-developed woody stems

E. equisetina with scale leaves in 2s and tips bluntly triangular

Fruiting plant of *E. equisetina*

Ephedra sinica

Ephedraceae

DRUG MORPHOLOGY

CRUDE DRUG (NON-WOODY STEM): long, slender, **rarely branching** (1–2 mm diam.) with **internodes usually 3–4 cm long [1]**; woody stems occasionally present. Non-woody stems matt, greyish-green or yellowish-green, **slightly rough**, with fine, longitudinal furrows **[2]**. **Nodes with 2 opposite [3]** pale brown or reddish-brown leaves reduced to membranous **scale leaves (3–4 mm long)** and fused along 1/3–2/3 their length, free part (c. 5 mm long) tapering to a sharp point (leaf-scales may fall off in trade items). *Texture*: light and brittle. *Fracture*: slightly fibrous, pith sub-rounded and reddish-brown. *Odour*: indistinct. *Taste*: astringent and slightly bitter. ***Good quality*:** difficult to break, pale green or yellowish-green stems, reddish-brown internally; tasting bitter and astringent.

DECOCTING PIECES-1 (SHORT STEM LENGTHS): as for crude drug but cut into short lengths (up to c. 2 cm long).

DECOCTING PIECES-2 (HONEY-PROCESSED STEM LENGTHS): as for dp-1 but dark yellow and slightly lustrous **[4]**. *Odour:* honey-scented. *Taste:* sweet.

PROCESSING

Non-woody stems are sun-dried (= crude drug) then cut into short decocting sections up to c. 2 cm long. Stems are sometimes treated with honey, then dry-fried to produce 'Mi Ma Huang' or 'Mi Zhi Ma Huang'.

OTHER COMMON NAMES

草麻黄 Cao Ma Huang (plant name); 蜜麻黄 Mi Ma Huang or 蜜炙麻黄 Mi Zhi Ma Huang (honey-processed Ma Huang); Chinese ephedra; ephedra (SCN and CP2010); Herba Ephedrae.

SAFETY

The internal use of *Ephedra* species is legally restricted in many countries due to their 'ephedra alkaloids'. The latter have been associated with life-threatening adverse reactions, some fatal, following their use in dietary supplements (notably weight loss products) and widespread abuse as 'legal highs' (Edwards *et al*., 2015; Federal Register, 2004; Martindale, 2011; MHRA, 2014). In TCM, strict adherence to dosing is strongly recommended (Bensky *et al*., 2004).

E. intermedia

Ephedraceae

DRUG MORPHOLOGY

CRUDE DRUG (NON-WOODY STEM): easily confused with PS1 and PS3. *Main differences*: **stems frequently branched [5]**, **rough** (1.5–3 mm diam.); membranous **scale leaves usually in 3s (rarely 2s) [6]**, 2–3 mm long, fused for at least 2/3 their length with pointed tips, often twisted. *Fracture*: pith triangular to rounded.

DECOCTING PIECES-1 (SHORT STEM LENGTHS): as for the crude drug of this species but cut into short lengths (up to c. 2 cm).

DECOCTING PIECES-2 (HONEY-PROCESSED STEM LENGTHS): as for dp-1 but dark yellow and slightly lustrous **[7]**. *Odour*: honey-scented. *Taste*: sweet.

PROCESSING

As for PS1.

OTHER COMMON NAMES

As for PS1 and 中麻黄 Zhong Ma Huang (plant name).

E. equisetina

Ephedraceae

DRUG MORPHOLOGY

CRUDE DRUG (NON-WOODY STEM): easily confused with PS1 and PS2. *Main differences*: **stems frequently branched [8]**, **not rough** (1–1.5 mm diam.), **internodes 1.5–3 cm long [9]; scale leaves usually in 2s [10] (rarely 3s)**, 1.5–3 mm long, fused along c. 3/4 their length but often splitting when dry, free parts bluntly triangular, sometimes recurved, with bases brownish-red to brownish-black.

DECOCTING PIECES-1 (SHORT STEM LENGTHS): as for the crude drug of this species but cut into short lengths (up to c. 2 cm).

DECOCTING PIECES-2 (HONEY-PROCESSED STEM LENGTHS): as for dp-1 but dark yellow and slightly lustrous **[11]**. *Odour*: honey-scented. *Taste*: sweet.

PROCESSING

As for PS1.

OTHER COMMON NAMES

As for PS1 and 木贼麻黄 Mu Zei Ma Huang (plant name).

Crude drug of Ma Huang (sourced from *Ephedra sinica*)

Decocting pieces-1 of Ma Huang (*E. sinica*)

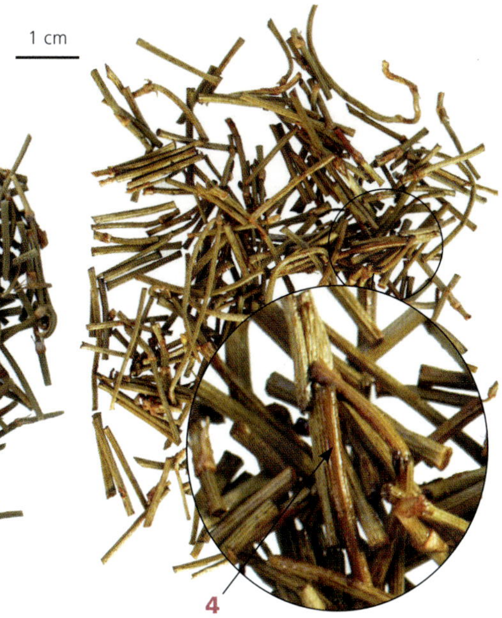

Honeyed decocting pieces-2 of Ma Huang (Mi Ma Huang) (*E. sinica*)

Crude drug of Ma Huang (sourced from *E. intermedia*)

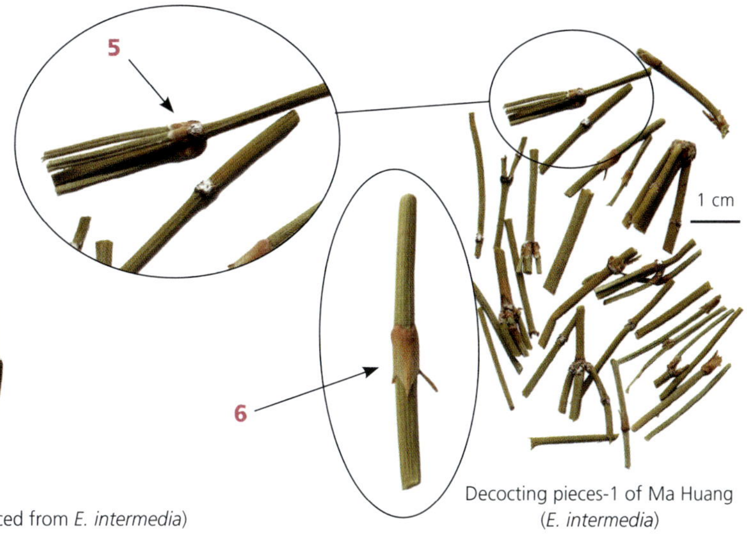

Decocting pieces-1 of Ma Huang (*E. intermedia*)

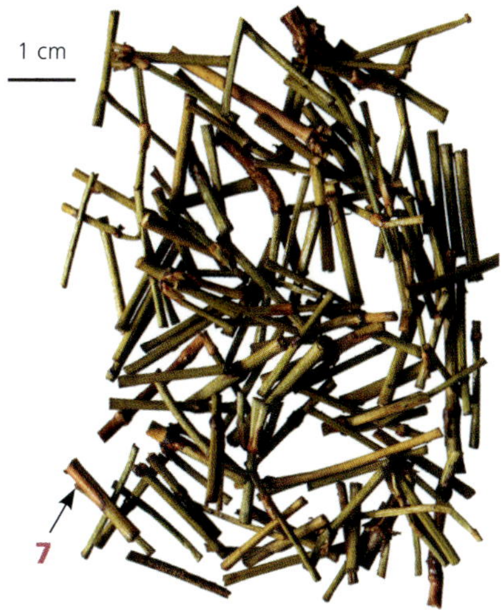

Honeyed decocting pieces-2 of Ma Huang (Mi Ma Huang) (*E. intermedia*)

Crude drug of Ma Huang (sourced from *E. equisetina*)

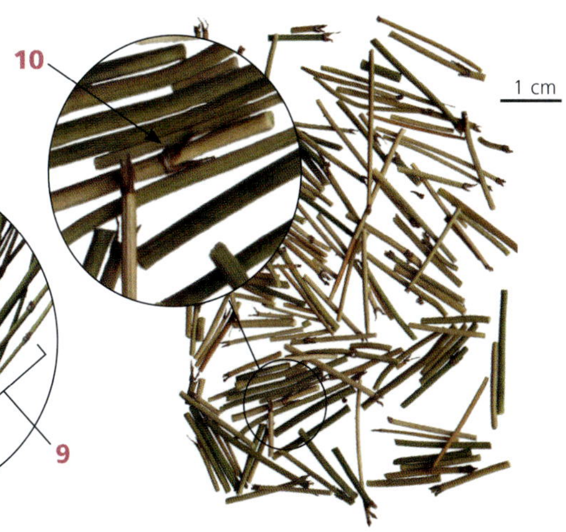

Decocting pieces-1 of Ma Huang (*E. equisetina*)

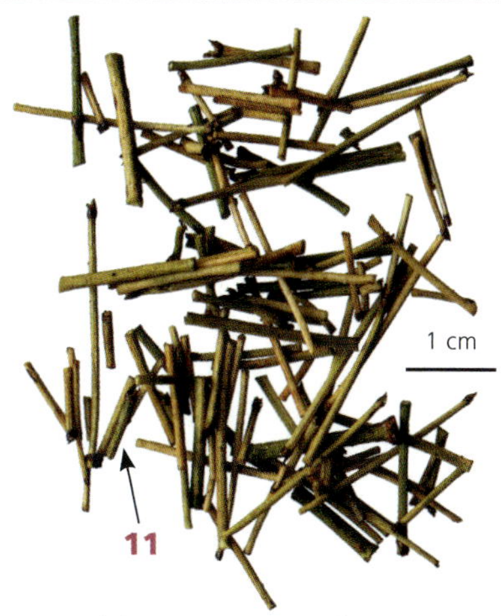

Honeyed decocting pieces-2 of Ma Huang (Mi Ma Huang) *E. equisetina*

Mu Tong 木通

Akebiae Caulis

Drug part: STEM

Akebia quinata (Houtt.) Decne.

Lardizabalaceae

synonym: *Rajania quinata* Houtt.

taxonomic/nomenclatural note: in CP2015 as '*Akebia quinata* (Thunb.) Decne.'

Species of *Akebia* wild in China: 4

PLANT DESCRIPTION

A deciduous, woody climber. *Stem* slender, greyish-brown (pinkish-grey when young) with prominent lenticels. Stipules absent. *Leaves* long-stalked and divided into 5 (rarely 3 or 7) short-stalked leaflets; leaflet blades paper-like, broadest above the middle or elliptic, upper surface dark green, lower surface bluish-green, base rounded to broadly wedged-shaped, tip rounded and usually minutely notched, margins entire; lateral leaflets 2–5 cm long; 1.5–2.5 cm wide; terminal leaflet 2.5–5 (–7) cm long; leaf and leaflet stalks swollen at both ends. *Flowers* single-sexed, slightly fragrant, in clusters of 4–8 (rarely 11) forming 6–12 cm long inflorescences [racemes] arising from leaf axils; petals absent; sepals resemble petals. Male flowers towards tip of inflorescence with 3 (rarely 5) petal-like pale purple sepals, sometimes pale green or white (6–8 mm long; 4–6 mm wide), distinctly hooded and with a rounded tip; stamens 6 (rarely 7) and free. Female flowers solitary or 2, at base of inflorescence and larger than male flowers; sepals 3 dark purple (rarely green or white) and broadly elliptical (1–2 cm long; 0.8–1.5 cm wide); ovary superior and carpels 3–9, purple and cylindrical; with each stigma resembling a pin-head. *Fruit* fleshy, oblong to ellipsoid (5–8 cm long; 3–4 cm diam.) and purplish at maturity. *Seeds* many in several rows within pulp. Fl. Apr–May, fr. Jun–Aug.

The fruit of *A. quinata* are the source of the separate TCM drug 'Yu Zi Zhi' (not included in this book).

HARVESTING, SOURCING AND NATURAL RANGE

According to classical Chinese herbal texts, *Akebia quinata* (along with PS2/3 below) was the original source of 'Mu Tong' (Zhang *et al.* 2012b; Zhu, 2002). From the 1950s these species were officially complemented by the addition of 'Guan Mu Tong' *Aristolochia manshuriensis* (see *Unofficial Substitute* 1) but the demise of the latter in recent years (owing to its toxicity) has seen demand return to *Akebia* and which today is the sole official source of 'Mu Tong' in the CP2015.

Akebia quinata is sourced from wild and medicinal crop plants. The wild species grows in broad-leaved deciduous forests in C, W and SE China (300–1,500 m altitude) with stems harvested in autumn. The species is also native in Japan and the Korean Peninsula.

CONSERVATION STATUS:

China: Least Concern (MEPCAS, 2013).

Global: Not Evaluated (IUCN Red List, accessed 3/9/2016).

A. trifoliata (Thunb.) Koidz.

Lardizabalaceae

synonyms: *A. lobata* Decne. var. *australis* Diels; *A. trifoliata* subsp. *australis* (Diels) T. Shimizu; *A. trifoliata* var. *australis* (Diels) Rehder; *Clematis trifoliata* Thunb.

taxonomic/nomenclatural note: *Kew-MPNS* accepted name, 21/10/2016; in CP2015 as two taxa: '*A. trifoliata* (Thunb.) Koidz.' and '*A. trifoliata* (Thunb.) Koidz. var. *australis* (Diels) Rehd.' This account, however, follows *Kew-MPNS* (21/10/2016) and does not recognise the variety.

PLANT DESCRIPTION

As for PS1 but **leaves divided into 3 leaflets** (trifoliolate), ovate to broadly ovate (3–8 cm long; 1.5–6 cm wide), margins entire or sinuous; petal-like sepals of both male and female flowers slightly smaller than those of PS1 (2–3 mm long and 0.9–1.2 cm long respectively; cylindrical carpels in female flowers 5–7. ***Fruit* yellowish-brown at maturity**, often slightly larger (6–8 cm long; 3–5 cm diam.) than those of PS1. Fl. Apr–May, fr. Jun–Sep.

The fruit of *A. trifoliata* are the source of the separate TCM drug 'Yu Zi Zhi' (not included in this book).

HARVESTING, SOURCING AND NATURAL RANGE

Similar to *A. quinata* (PS1) above but this species is endemic to China growing wild in a similar habitat in the provinces of Anhui, Fujian, Guangdong, Guizhou, Henan, Hubei, Hunan, Jiangsu, Jiangxi, S Shaanxi, S Shanxi, Sichuan, Taiwan, Yunnan and Zhejiang.

CONSERVATION STATUS

Endemic to China: Least Concern (MEPCAS, 2013); Not Evaluated (IUCN Red List, accessed 3/9/2016).

TCM PROPERTIES AND USE

properties: 'bitter and cold'
channels: 'Heart, Small Intestine and Bladder'
actions: 'drains damp, clears heat, promotes urination'
indications: promote urination, menstruation, lactation; mouth sores

Inflorescence of *Akebia quinata*

Male and female flowers of *A. quinata*

Ripening fruit

Ripening fruit (part of fruit wall removed to reveal seeds)

Flowering branch of *Akebia trifoliata*

Flowering plant of *A. trifoliata*

UNOFFICIAL SUBSTITUTE 1

Aristolochia manshuriensis Kom.

Aristolochiaceae

synonyms: *Hocquartia manshuriensis* (Kom.) Nakai; *Isotrema manshuriense* (Kom.) H. Huber

PLANT DESCRIPTION

Although a woody climber like PS1, 2 and 3 this species is easily distinguished. **Stem with outer (cork) bark strongly fissured longitudinally, often spongy and strongly aromatic** (camphor-like) when crushed. **Leaves much larger** (15–29 cm long; 13–28 cm wide), **usually heart-shaped** (not divided into leaflets) and margins entire. Flowers solitary or in pairs, **petals fused into a distinctive yellow to greenish-yellow pipe-shaped utricle** (4.5–5.5 cm long), **with 3 yellow deltoid lobes at mouth**; bisexual. Fruit pendant, dry, narrowly cylindrical (9–11 cm long; 3–4 cm diam.) and 6-valved. Fl. Jun–Jul; fr. Aug–Sep.

CONSERVATION STATUS

Endemic to China: Near Threatened (MEPCAS, 2013); Not Evaluated (IUCN Red List, accessed 24/1/2016).

SUBSTITUTION COMMENT. Since the 1950s, this species (Guan Mu Tong) was formally recognised as an alternative source of 'Mu Tong'. The toxicity of this genus (*Aristolochia*), however, was not recognised until the late 1990s, following cases of renal failure in Europe (Debelle *et al.*, 2008; Shaw *et al.*, 2012; Zhu *et al.*, 2010). A worldwide ban on its supply and use soon followed along with its removal from the Chinese Pharmacopoeia (SFDA, 2003). Aristolochic acid (AA) testing (present in *Aristolochia* and proven to be nephrotoxic and carcinogenic) has become a standard quality control method for identifying this genus in medicinal trade items. *Aristolochia manshuriensis* is endemic to NE China and from where it has been wild harvested as 'Mu Tong' until its sudden fall from grace. See *Safety* page 464.

UNOFFICIAL SUBSTITUTE 2

Clematis armandii Franch.

Ranunculaceae

PLANT DESCRIPTION

Easily distinguished from PS1, 2 and 3. **Stems distinctly fibrous and grooved longitudinally**. **Leaves divided into 3 leaflets** (trifoliolate) like those of PS2/3 but much larger **(5–16 cm long; 1.5–7 cm wide),** ovate, **leathery,** with tips gradually narrowing (not notched). ***Flowers* white, much larger** and bisexual, petal-like sepals 4 (occasionally up to 6), narrowly oblong (1.2–4.7 cm long; 0.2–2 cm wide). Fruit a cluster of dry achenes, each with a persistent and softly feathery style.

CONSERVATION STATUS

China: Least Concern (MEPCAS, 2013).

Global: Not Evaluated (IUCN Red List, accessed 25/1/2016).

SUBSTITUTION COMMENT. Due to the similar clinical properties of the stems of this species to those of *Akebia* (PS1 and 2), this species has long been regarded as an alternative 'Mu Tong' and formally recognised as such in earlier editions of the CP. Today the species is recognised as the source of a separate CP drug (Chuan Mu Tong — Clematidis Armandii Caulis, page 428), along with *C. montana* below, but, given the supply shortage of *Akebia* (PS1 and 2) it is also widely used in clinical circles as a 'Mu Tong' substitute. See also 'Chuan Mu Tong' page 428.

UNOFFICIAL SUBSTITUTE 3

C. montana Buch.-Ham. ex DC.

Ranunculaceae

PLANT DESCRIPTION

Easily distinguished from PS1 and PS2. **Stems distinctly fibrous and grooved longitudinally.** Leaves divided into 3 **leaflets** (trifoliolate) like those of PS2 (1.8–7 (rarely up to 14) cm long; 1–5 cm wide), **tips shortly or long-pointed** and margins often lobed or sparsely dentate. ***Flowers* white, much larger** and bisexual, petal-like sepals 4 (occasionally up to 6), broadly oblong (1.3–6.5 cm long; 1–2.5 cm wide). *Fruit* similar to *C. armandii* above.

CONSERVATION STATUS

China: Least Concern (MEPCAS, 2013).

Global: Not Evaluated (IUCN Red List, accessed 25/1/2016).

SUBSTITUTION COMMENT. As for *C. armandii* above.

Aristolochia manshuriensis — showing climbing habit

Flowering stem of *A. manshuriensis*

Flower of *A. manshuriensis*

Clematis armandii — leaf with 3 leaflets

C. armandii flowering in the wild, Sichuan province

Freshly harvested stem and fruiting branch of *C. armandii*

C. montana leaf

C. montana flowering branch

Cut stem of living plant of *C. montana*

Akebia quinata

Lardizabalaceae

DRUG MORPHOLOGY

CRUDE DRUG (DRIED STEM): hard and cylindrical **[1]** (0.3–0.7 m long; 0.5–2 cm diam.), slightly twisted and with nodes often distinctly swollen and nodular **[2]** on one side (scars of lateral branches). Outer bark surface greyish-brown (mature stems) or pinkish-grey (young stems) **[3]**, finely striated longitudinally with many prominent pale brown small wart-like bumps **[4]** and occasional orbicular lenticels (c. 1 mm diam.) **[5]**; bark of older stems tends to exfoliate to reveal darker stem tissues beneath **[6]** see photograph of *A. trifoliata*). **Inner bark distinctive [7]** (see *Decocting Pieces*). *Texture*: light and difficult to break. *Fracture*: uneven; bark thin, wood broad **[8]**, pale creamish-yellow or pinkish-yellow with narrow medullary rays **[9]**; central pith small, white or yellowish-white and spongy **[10]** (occasionally hollow). ***Odour*: indistinct.** *Taste*: slightly bitter and astringent.

DECOCTING PIECES (TRANSVERSE STEM LENGTHS OR SLICES): in outline, round to oblique slices (or short lengths) occasionally with remains of woody and nodular lateral branches at nodes **[11]**. Outer cork bark usually present (not spongy) **[3]** and not separating easily from inner bark unless in older stems (see *Crude Drug*). Cut surfaces of young stems with bark to wood ratio c. 1: 4–5, becoming 1:3 in mature stems. **Inner bark** narrow (up to 2 mm) brown or pale yellow, **often separating from wood to form a distinctive ring of scallop-shaped lacunae around the stem's circumference [7]**; **wood broad [8], pale creamish-yellow or pinkish-yellow** with many scattered vessels creating a porous appearance **[12]**, interrupted by narrow and indistinct medullary rays **[13]**; central pith small white or yellowish-white and spongy **[10]** (occasionally hollow) **[14]**. ***Odour*: indistinct**. *Taste*: slightly bitter and astringent.

PROCESSING

After removal of very young stems, mature stems are shade-dried (= crude drug) then softened thoroughly before transversely slicing into short pieces and drying again (= decocting pieces).

OTHER COMMON NAMES

五叶木通 Wu Ye Mu Tong '5-leaved Mu Tong'; akebia (SCN); akebia stem (CP2010); Caulis Akebiae; five-leaf akebia.

SAFETY

Generally regarded as a safe herb (Bensky *et al*., 2004), however, through association with the toxic and at one time clinically interchangeable herb *Aristolochia manshuriensis* 'Guan Mu Tong' (see below), *Akebia quinata* is subject to legal restrictions in many countries in the interests of public safety (Zhang *et al*., 2012b). See *Substitution Comment* page 460.

A. trifoliata

Lardizabalaceae

DRUG MORPHOLOGY

CRUDE DRUG (DRIED STEM) AND DECOCTING PIECES (TRANSVERSE STEM LENGTHS OR SLICES): cannot be reliably differentiated macroscopically from PS1. Differences between photographs opposite are most likely due to the maturity of the stems rather than species differences.

PROCESSING

As for PS1.

OTHER COMMON NAMES

白木通 Bai Mu Tong 'white Mu Tong'; 三叶木通 San Ye Mu Tong '3-leaved Mu Tong'; Caulis Akebiae; akebia (SCN).

SAFETY

As for PS1.

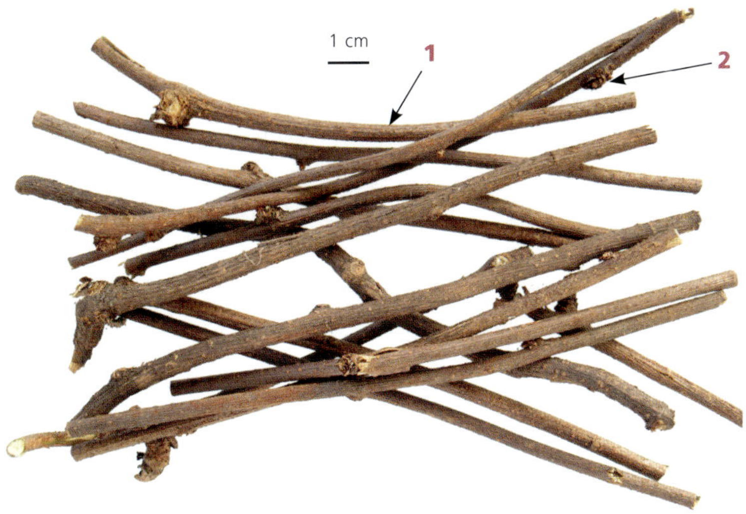

Crude drug of Mu Tong (young stems) sourced from *Akebia quinata*

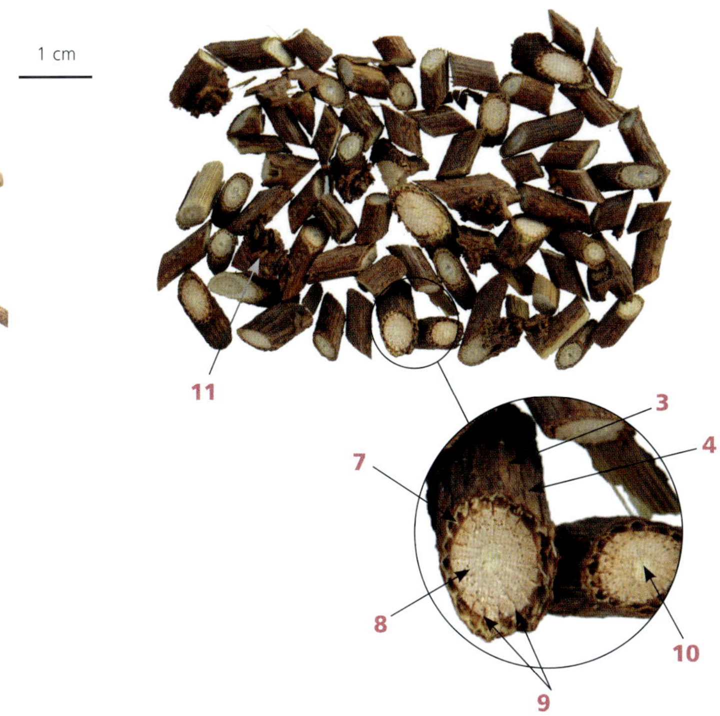

Decocting pieces of Mu Tong (young stems) sourced from *A. quinata*

Surface of young stem of Mu Tong (*A. quinata*)

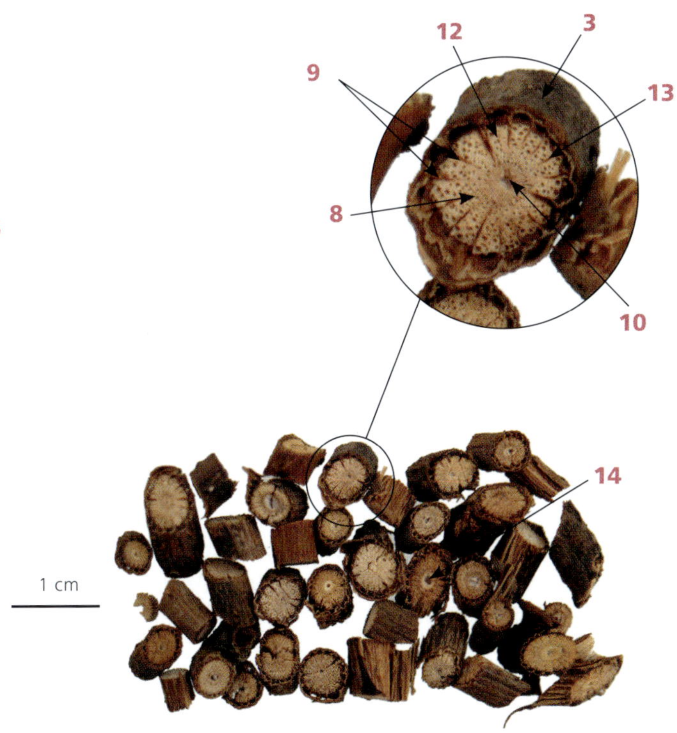

Crude drug of Mu Tong (mature stems) sourced from *Akebia trifoliata*

Decocting pieces — *A. trifoliata*

Aristolochia manshuriensis

Aristolochiaceae

'DRUG SUBSTITUTE' MORPHOLOGY

Stem lengths and transverse slices usually easily distinguished macroscopically from those of PS1 and PS2.

DRIED STEM LENGTHS: hard and cylindrical **[15]** (1–2 m long; 0.8–6 cm diam.) (usually broader than PS1 and 2), slightly twisted and **nodes indistinctly swollen [16]** (nodular lateral branch scars absent). **Remnants of outer (cork)** sometimes present, **pale greyish-brown or greyish-yellow, slightly spongy**, with **large longitudinal furrows [17]**. **Inner bark (up to 4 mm thick)** greyish-yellow, **dense** and easily removed **[18]**. *Fracture*: *see cut surfaces of sliced stem below*. *Texture*: light, hard (not fibrous) and difficult to break. ***Odour*: cork bark aromatic (camphor-like) when crushed.** *Taste*: mildly bitter and pungent.

TRANSVERSE STEM SLICES: in outline sub-rounded (0.8–6 cm diam.; 2–4 mm thick) (often sliced much more thinly than those of PS1 and 2). **Remnants of outer** pale **greyish-brown or greyish-yellow, spongy bark** may persist **[19]**, **inner bark dense** and greyish-yellow (up to 4 mm thick), easily removed (18). Remainder of cut surface mostly of wood (xylem) **[20] (inner bark to wood ratio c. 1: 5–7); wood pale greyish-yellow**, densely packed throughout with round vessels arranged in concentric rings, creating a strongly porous appearance and interrupted by regularly spaced and much narrower medullary rays of a similar colour **[21]**; pith rather small and inconspicuous **[22]**, same colour as wood. *Fracture*: brittle. ***Odour*: cork bark aromatic (camphor-like) when crushed.** *Taste*: mildly bitter and pungent.

OTHER COMMON NAMES

关木通 **Guan Mu Tong**; 木通 Mu Tong; Manchurian birthwort (SCN); Aristolochiae Manshuriensis Caulis; Caulis Aristolochiae Manshuriensis; Manchurian Dutchman's pipe.

SAFETY

Toxic. Over the last 20 years, *Aristolochia* species (including *A. manshuriensis*) have been proven to cause over 100 cases in the West of serious adverse reaction (some fatal) mostly due to renal failure (Nortier *et al*., 2003; Vanherweghem *et al*., 1993). The genus contains aristolochic acids which are nephrotoxic and carcinogenic (Heinrich *et al*., 2009; Debelle *et al*., 2008; Upton & Graff, 2006) and permanent banning orders covering the sale and medicinal supply of *Aristolochia* species exist in many countries including China (SFDA, 2003; MHRA, 2012; Upton *et al*., 2013; USFDA, 2000, 2001a, 2001b). In 2003, *A. manshuriensis* was removed from the Chinese Pharmacopoeia on safety grounds (SFDA, 2003) See also *Substitution Comment* page 460.

Clematis armandii and *C. montana*

Ranunculaceae

'DRUG SUBSTITUTE' MORPHOLOGY

Crude drug and decocting slices of these two unofficial substitutes are not distinguishable macroscopically so they are described together below. Both, however, are easily distinguished from those of the two Pharmacopoeia species.

CRUDE DRUG 'CHUAN MU TONG' (DECORTICATED STEM): often broader (up to 3.5 cm), external surface (i.e. inner bark) **pale yellow, yellowish-brown [23]** or greyish-brown **with distinct longitudinal furrows and cracks [24]**; remnants of dark grey cork-bark occasional **[25]**. *Texture*: distinctly fibrous **[26]** and difficult to break. ***Fracture***: uneven, **wood yellowish-cream or yellowish-brown [27] with vessels more evenly distributed throughout fractured surface**; pale yellow medullary rays more distinctive **[28]** (than in PS1 and 2) and extending clearly to the central pith. Pith small pale yellowish-white **[29]**, sometimes absent or hollow. *Taste*: bland.

DECORTICATED TRANSVERSE STEM SLICES: sub-rounded (2–3.5 cm diam.; 2–4 mm thick) **[30]**. Outer perimeter of cut surface (where cork-bark has been removed) **shallowly scalloped or irregularly notched [31]**; **wood yellowish-cream or yellowish-brown [27]**; pale yellow medullary rays more numerous and distinctive **[28]** and extending clearly to the central pith; **wood vessels** mostly of equal size and **regularly distributed throughout fractured surface** in concentrically aligned rings **[32]** and creating an evenly porous appearance. Pith small pale yellowish-white **[29]**, sometimes absent or hollow. *Fracture:* brittle and breaking along the medullary rays. *Taste*: bland.

OTHER COMMON NAMES

For both *C. armandii* and *C. montana*: 川木通 **Chuan Mu Tong**; 花木通 Hua Mu Tong; Caulis Clematidis Armandii.

C. armandii: 小木通 Xiao Mu Tong (plant name); Armand clematis stem (CP2010); Armand's clematis (SCN); Armand virgin bower.

C. montana: 绣球藤 Xiu Qiu Teng (plant name).

SAFETY

Generally regarded as a safe herb (Bensky *et al*., 2004), however, through association with the toxic and at one time clinically interchangeable herb *Aristolochia manshuriensis* 'Guan Mu Tong' (see above), *Clematis armandii* and *C. montana* are subject to legal restrictions in many countries in the interests of public safety (Zhang *et al*., 2012b). See *Substitution Comment* page 460.

LESSER KNOWN UNOFFICIAL SUBSTITUTE

Aristolochia moupinensis Franch. — stem (family Aristolochiaceae). Endemic to China (Fujian, Guizhou, Hunan, Jiangxi, Sichuan, Yunnan and Zhejiang). **Toxic**. Stem similar to that of *A. manshuriensis* but bark usually densely covered with greyish-yellow hairs.

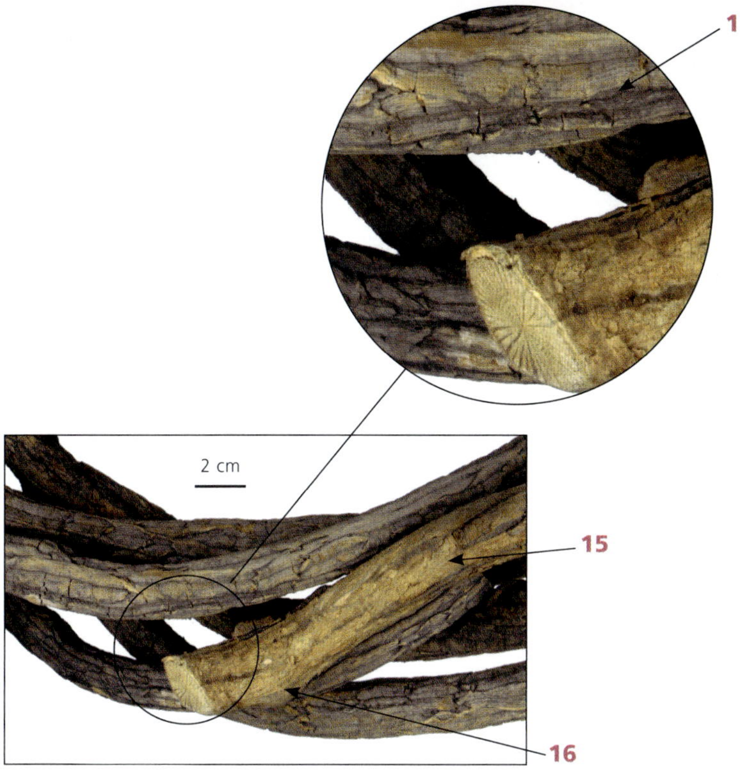

Dried stem of *Aristolochia manshuriensis*

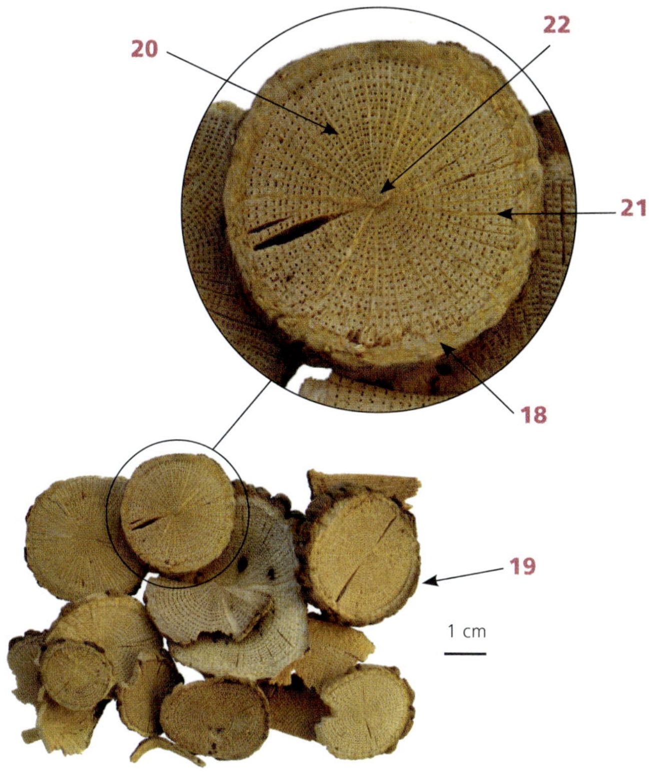

Sliced stem of *A. manshuriensis*

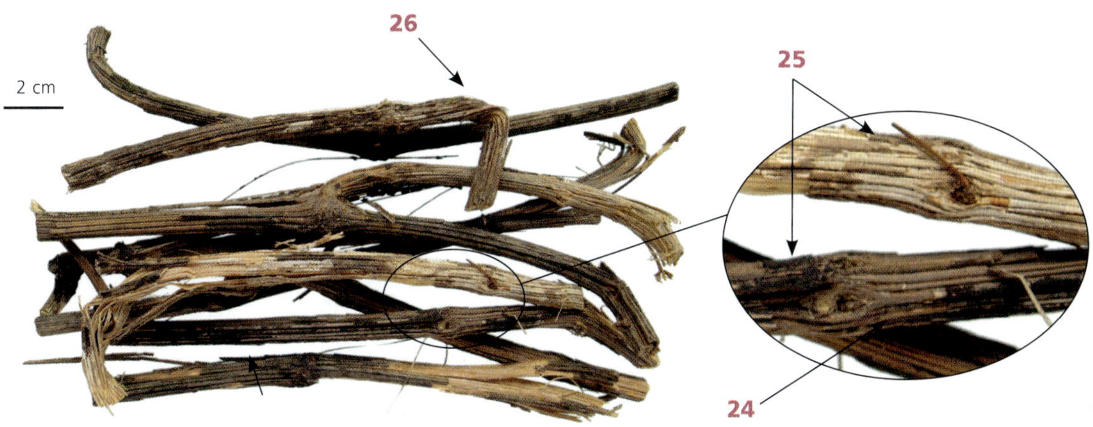

Crude drug sourced from *Clematis armandii*

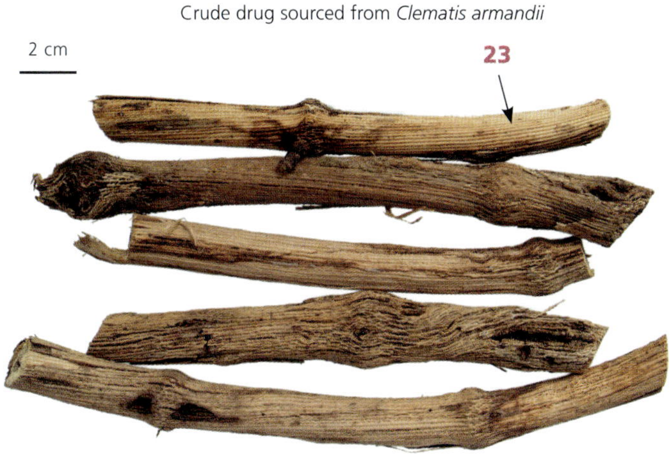

Dried and de-barked stem of *C. montana*

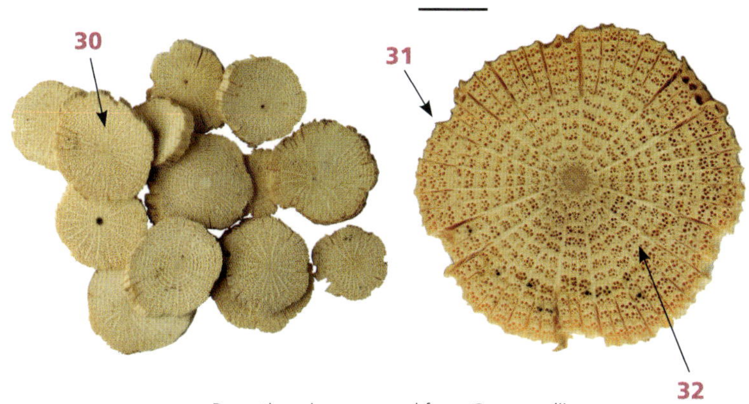

Decocting pieces sourced from *C. armandii*

Ren Dong Teng 忍冬藤

Lonicerae Japonicae Caulis

Japanese honeysuckle stem

Drug part: STEMS

Lonicera japonica Thunb.

Caprifoliaceae

taxonomic note: two varieties are recognised by *FoC:* the 'type' variety *L. japonica* var. *japonica* and the less common *L. japonica* var. *chinensis* (P. Watson) Baker

Species of *Lonicera* wild in China: 57

PLANT DESCRIPTION

A vigorous, semi-evergreen climber. *Stems* with yellowish-brown, stiff, spreading hairs. *Leaves* opposite, ovate to lance-shaped (3–8 cm long); **undersides sparsely to densely hairy but not totally covering leaf surface** [hl] and **margins fringed with hairs (ciliate)** [hl]. A pair of **leaf-like bracts, ovate to elliptic (1–3 cm long)** present at base of flowerheads. *Flowers* in pairs within branching flowerheads. Sepals fused into a **hairless** calyx tube ending in 5 small **calyx lobes with densely hairy margins** [hl]. Petals 5, fused narrow into a tube (**3–5 cm long)**, with spreading lobes and 2 lips; externally white, becoming yellow, orange or purple; those changing to purple are *L. japonica* var. *chinensis*. *Fruits* spherical and bluish-black.

For other TCM drugs sourced from this species see 'Jin Yin Hua' (Lonicera Japonica Flos) page 572.

HARVESTING, SOURCING AND NATURAL RANGE

Young and older (over-wintered) stems are harvested from wild and cultivated plants in autumn and winter. The wild the species grows in open forests and scrub, in stony places, mountain slopes and roadsides (up to 1,500 m altitude) and is distributed in Anhui, Fujian, Gansu, Guangdong, Guangxi, Guizhou, Hebei, Henan, Hubei, Hunan, Jiangsu, Jiangxi, Jilin, Liaoning, Shaanxi, Shandong, Shanxi, Sichuan, Taiwan, Yunnan and Zhejiang. The species is also wild in Japan and the Korean Peninsula and introduced and invasive in North America.

CONSERVATION STATUS

China: Least Concern (MEPCAS, 2013).
Global: Not Evaluated (IUCN Red List, accessed 8/10/2015).

A medicinal crop of *Lonicera japonica*

Flowering branch of *L. japonica*

TCM PROPERTIES AND USE

***properties*: 'sweet and cold'**
***channels*: 'Lung, Large Intestine and Stomach'**
***actions*: 'clears heat, releases exterior, eliminates toxins'**

***indications*: abscesses, sores; musculoskeletal pain; abdominal pain; metrorrhagia**

DRUG MORPHOLOGY

CRUDE DRUG (DRIED STEMS): thin (1.5–6 mm diam.), woody **[1]**, somewhat cylindrical (often cut into 10–20 cm lengths), frequently branching, slightly twisted and often tied into bundles. Internodes long **[2]** (c. 6–9 cm long), reddish-brown **3]**, dark brown or greyish-green, finely striated **[4]**, hairless or (if stems young) covered in dense soft, fine hairs (pubescent). Nodes hard and swollen with branch and leaf scars **[5]**; nodes of young stems densely hairy **[6]**; shrivelled leaf remains sometimes present. *Fracture:* outer bark thin and easily removed **[7]**; yellowish-white and hollow **[8]**. *Texture:* hard and brittle. *Odour:* indistinct. *Taste:* young branches indistinct; older branches slightly bitter. ***Good quality:*** uniformly sized reddish-brown stems. Young (usually hairier) stems that have yet to over-winter are considered superior.

DECOCTING PIECES (DRIED STEM LENGTHS): as for the crude drug above but cut into short lengths (c.10-15 mm long, 1.5-6 mm diam.).

PROCESSING

Stems are sun-dried (= crude drug) then cut into decocting lengths 10–15 mm long.

OTHER COMMON NAMES

二花藤 Er Hua Teng; 金银藤 Jin Yin Teng; Lao Teng (old vine; refers to over-wintered stems); 银花藤 Yin Hua Teng 'silver flowered vine'; Nen Teng (tender vine; refers to young stems not over-wintered); Caulis Lonicerae Japonicae; Japanese honeysuckle (SCN).

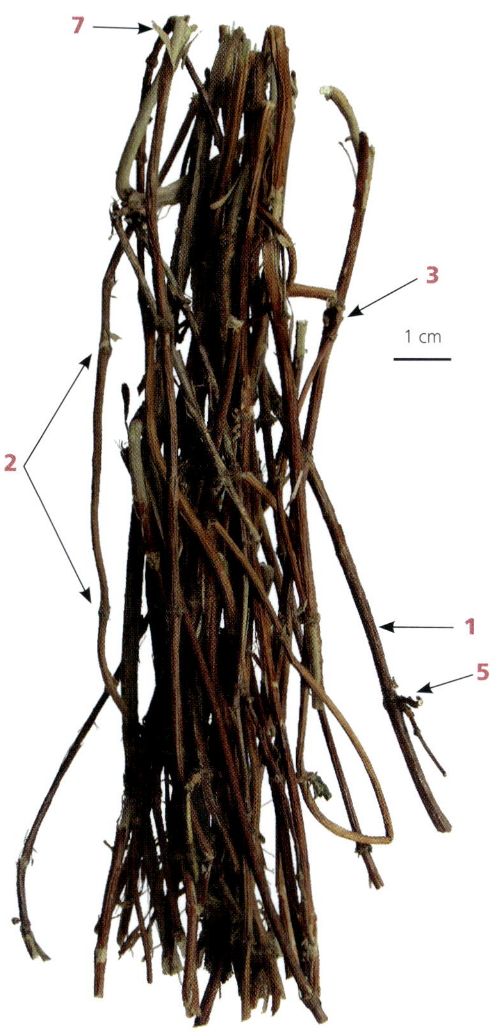

Crude drug of Ren Dong Teng (dried stems of *Lonicera japonica*)

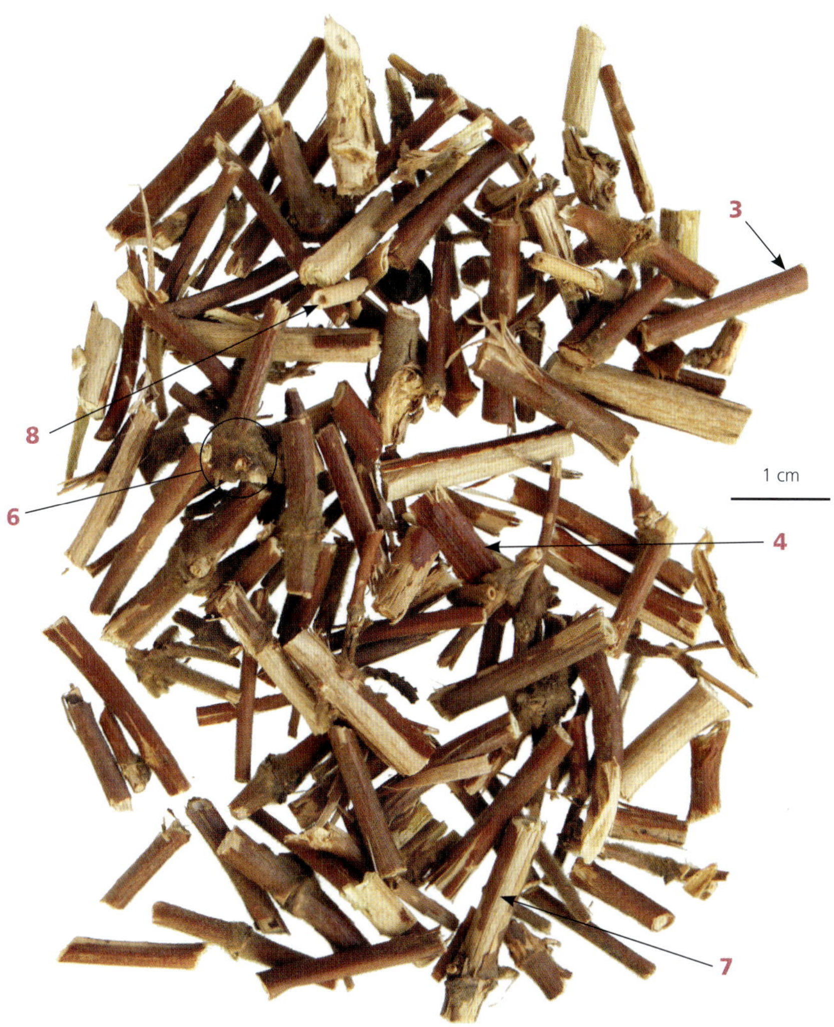

Decocting pieces of Ren Dong Teng (dried, lengths of *L. japonica*)

Rou Cong Rong 肉苁蓉 Cistanches Herba

Drug part: FLESHY STEM WITH SCALES

Identification comment. The genus *Cistanche* is in need of taxonomic revision (see PS2). Key morphological characters are found in their inflorescences but, since the latter are usually absent in commercial trade items of 'Rou Cong Rong', laboratory-based authentication methods are advised for their reliable identification. Given the parasitic nature of this genus a DNA-based approach is likely to be more reliable than a chemical one. Furthermore, with many species of *Cistanche* host-specific, the creation of herbarium vouchers and photographs of host plants in addition to those for the drug plant will aid the taxonomic reaearch needed.

Species of *Cistanche* wild in China: 5

Cistanche deserticola Ma

Orobanchaceae

taxonomic/nomenclatural note: in CP2015 as '*C. deserticola* Y.C. Ma'

PLANT DESCRIPTION

An upright, fleshy parasitic perennial without chlorophyll; only the inflorescences and upper parts of stem emerge above ground (0.4–1.6 m tall). *Stems* fleshy (2–10 cm diam.), usually unbranched (rarely 2–4-branched). *Leaves* modified into hairless, overlapping scales; those on lower part of stem ovate or triangular-ovate (0.5–1.5 cm long; 1–2 cm wide), those on upper part longer and narrower (2–4 cm long; 5–10 mm wide). *Inflorescence* many, densely packed into a large, cylindrical head (15–50 cm long); drying brown. **Bracts linear to lance-shaped**, about the **same length as the mature corolla tube** (3–4 cm long). Flowers stalkless; corolla funnel-shaped and white with 5 corolla lobes, short, broad and **tipped pale purple or whitish-yellow;** stamens 4 with **anthers** densely villous and lowest parts **sharply pointed (mucronate) [hl]**; style slender and stigma 2-lobed. *Fruit* a dry egg-shaped or globose capsule with 2 or 3 valves. Fl. early summer.

This species is host specific to woody trees and shrubs in the Chenopodiaceae family (notably *Haloxylon ammodendron* (C. A. Mey.) Bunge ex Fenzl. and *H. persicum* Bunge ex Boiss. & Buhse): many-branched, with whitish-grey bark, jointed branchlets, opposite and distinctive scale-like leaves which are cottony in their axils. Flowers small and solitary, in leaf axils, with 2 bractlets, 5 membranous 'petals', 5 stamens and 2–5 stigmas. *Fruit* hemispherical and concave with fleshy, winged arms.

Freshly unearthed flowering plant of *C. deserticola* (note: this stem not suitable for medicinal use because good quality is harvested before flowering)

Haloxylon sp. (family Chenopodiaceae), host of *C. deserticola*

HARVESTING, SOURCING AND NATURAL RANGE

The fleshy stems of *C. deserticola*, a parasitic species native to the sandy deserts of Xinjiang province (NW China) have been used medicinally in China for almost 2,000 years. They are traded as a high-value medicinal product with a rapidly growing international market. In recent decades wild populations have become seriously depleted (see *Conservation Status*) due to over-harvesting for medicinal use while their host plants (mainly *Haloxylon* species) have also suffered from over-use by local herds-people who use them as a source of fuel-wood. This threatens the already fragile habitat of *C. deserticola* and accelerates desertification.

Since the 1980s, however, innovative research has led to a large-scale conservation recovery programme with the re-introduction of host plants to suitable habitats and their subsequent infection with *C. deserticola*. This has proved highly successful in, for example, Xinjiang province and where sustainable harvesting practices of *C. deserticola* are widely encouraged as part of a long-term management regime (Ma *et al*., 2006; Xu *et al*., 2009) and, more recently, some production areas have been awarded GAP certification (Zhang *et al*., 2010).

The stems of wild, farmed and re-introduced plants are harvested in spring (occasionally autumn), before or just as they emerge above ground (i.e. before they flower). The species grows in close proximity to its host plants (see '*Plant description*') in open and sandy desert locations (at 200–1,200 m altitude) in Gansu, Inner Mongolia, Ningxia and Xinjiang with *Di Dao-quality* 'Rou Cong Rong' traditionally sourced from sites in Inner Mongolia.

CONSERVATION STATUS

China: Endangered (MEPCAS, 2013); Critically Endangered (Wang & Xie, 2004); Endangered (Fu, 1992).
Global: Not Evaluated (IUCN Red List, accessed 25/1/2016).

PROTECTED STATUS

China: Cat. II (SEPA, 1999).
CITES: Appendix II; EU Annex B (see Table 1 page 10).

TCM PROPERTIES AND USE

EHTPA 4/14
GAP

properties: **'sweet, salty and warm'**
channels: **'Kidney and Large Intestine'**
actions: **'tonifies Kidney yang, tonifies Blood, moistens intestines'**

indications: **impotence, infertility, lack of libido; chronic constipation**

Inflorescence of *Cistanche deserticola* with leafless *Haloxylon* host (behind)

C. deserticola in full flower in the wild, Xinjiang province, NW China
PHOTO: KAISA SULAIMAN

Inflorescence of *C. deserticola*

Expanded flower of *C. deserticola* with (inset) pointed anthers (visible with a x10 hand lens)

Cistanche 'Tamarisk'

Orobanchaceae

Identification comment: the name '*Cistanche tubulosa* (Schenk) Wight' is used in the CP2015 (like earlier editions) to describe the official alternative source of 'Rou Cong Rong'. This name has long been used in the wider TCM literature and, to discriminate it from *C. deserticola* (PS1), sometimes traded as Guan Hua Rou Cong Rong 管花肉苁蓉. The botanical community, however, is uncertain about the botanical name of this material; e.g. *FoC* states that *C. tubulosa* is not native to China and that wild plants deposited by botanists in herbaria under that name 'have been mis-identified' and in fact are '*C. mongolica* Beck.' Botanical descriptions of the latter, however, are not a good match* with the source plant of 'Guan Hua Rou Cong Rong' originating from Xinjiang (its native Chinese province) but they are a good match with that for *C. tubulosa* in *Fl. USSR*. Despite this, the *Fl USSR* description (Schischkin, 2000) is not a good match with African specimens of *C. tubulosa* (consulted by M.Gilbert and C. Leon in the Herbarium of RBG Kew, 6/2016). These African specimens are a very good match with the original botanical ('type') description of *C. tubulosa* which was published using whole plant specimens from Egypt. As such, the source species of 'Guan Hua Rou Cong Rong' appears to be an undescribed species. Further work, including DNA profiling is needed to name this entity. The provisional name "*Cistanche* 'Tamarisk' " is coined here because of its specificity to host plants in the genus *Tamarix*. These observations highlight the need for an up-to-date taxonomic revision of this genus.

PLANT DESCRIPTION

Closely related to *C. deserticola* (PS1) and, easily distinguished. *Main differences:* aerial parts (inflorescences) 0.3–1 m tall; over-lapping stem scales (leaves) slightly larger (up to 2–3 cm long) and ovate to lance-shaped; **bracts ovate or oblong to lance-shaped, much shorter than mature corolla tube** (latter 2.8–4 cm long); **corolla** funnel-shaped with **tube creamish-white to very pale yellow** with **corolla lobes, 5, broadly ovate to deltoid and white.** Anthers densely villous with **both ends rounded (obtuse)**.

In contrast to *C. deserticola* (PS1), *C.* 'Tamarisk' appears **host-specific to *Tamarix* species** (family Tamaricaceae). The genus *Tamarix* is made up of shrubs or small trees typically with minute, alternate scale-like leaves and many-flowered inflorescences in long drooping panicles.

HARVESTING, SOURCING AND NATURAL RANGE

Wild populations of what is called in this account "*Cistanche* 'Tamarisk'" (see '*Identification comment*') are considered native to the Dzhungaria-Kashgaria area of southern Xinjiang and Tibet where they grow with their host plants in sandy desert situations (c. 1,200 m altitude). Widespread over-harvesting of this species for the medicinal plant trade in recent years has threatened the survival of its wild populations (see *Conservation Status*) but an extensive re-introduction programme in the plant's native habitat, managed on a plantation basis, has proved highly successful. Today, the bulk of domestic and international trade in 'Rou Cong Rong' is reported to be sourced from these re-introduced plants (Wang *et al.*, 2012).

CONSERVATION STATUS

China: Vulnerable (MEPCAS, 2013).
Global: Not Evaluated (IUCN Red List, accessed 25/1/2016).

PROTECTED STATUS

China: Cat. II (SEPA, 1999).

**C. mongolica* has rose- or violet-white corollas with corolla lobes sub-rounded whereas source plants of Guan Hua Rou Cong Rong (traditionally called '*C. tubulosa*') have whitish-yellow flowers with corolla lobes broadly ovate to deltoid (and appear host-specific to *Tamarix* species).

C. salsa (C. A. Mey.) Beck

Orobanchaceae

synonyms: *C. salsa* var. *albiflora* P. F. Tu & Z. C. Lou; *Phelypaea salsa* C. A. Mey.

PLANT DESCRIPTION

Closely related to PS1 and PS2 and especially difficult to distinguish from the former. ***Main differences from both are its height and its host species*** (see below).

Usually **much shorter (10–45 cm tall)** than PS1; unbranched (rarely 2–3-branched); **bracts about half the length of the mature corolla tube** (length of latter 2.5–4 cm); corolla tube white or yellow-white with **corolla lobes tipped pale purple or purple,** sub-rounded; flowers **drying yellowish-white** (not brown as in PS1); lowermost part of anthers **sharply pointed.**

In contrast to PS and PS2, ***C. salsa* has a broad range of hosts,** notably in the genera *Kalidium*, *Salsola* and *Suaeda* (Chenopodiaceae family) and *Nitraria* (Nitrariaceae family) (Xu *et al.*, 2009); they are easily distinguished from the main host of *C. deserticola* (PS1) because none of these genera have jointed branches nor scale-like leaves. Identification of the host plant (at least to genus level) often provides valuable information in support of an identification of *Cistanche* to species level.

SUBSTITUTION COMMENT. Closely related and very similar to both the plants and the drugs of PS1 and PS2. *C. salsa* is considerably more widespread than the Pharmacopoeia species and with a wider range of host plants, this species is widely reported as a common and probably cheaper substitute for both official species. *Cistanche deserticola* and the 'clinically inferior' *C. salsa* are difficult to distinguish macroscopically. They also require differentiation because *C. deserticola* is nationally protected and requires CITES permits when traded internationally (see Protected Status). Although not recognised by the Pharmacopoeia as an official source of 'Rou Cong Rong', *C. salsa* is farmed because it is easier to grow; it is often planted alongside the official species and may be traded as mixed batches (Leon and Lin, pers. obs., Xinjiang, 2011). The wild species is native to N and NW China: Gansu, Inner Mongolia, Qinghai and Xinjiang (700–2,700 m) and adjacent countries of SC Asia.

CONSERVATION STATUS

China: Least Concern (MEPCAS, 2013).
Global: Not Evaluated (IUCN Red List, accessed 25/1/2016).

Plantation-grown *C.* 'Tamarisk' emerging close to host (*Tamarix* sp.), Hetian district, Xinjiang province PHOTO: KAISA SULAIMAN

Inflorescence of *C.* 'Tamarisk' with leaves of *Tamarix* host behind
PHOTO: KAISA SULAIMAN

Flowers and flower-buds of *C.* 'Tamarisk'
PHOTO: KAISA SULAIMAN

Inflorescence of *C. salsa* growing as a farmed medicinal crop on one of its host plants (*Suaeda* sp.), Xinjiang province (NW China)

Flowers and flower-buds of *C. salsa*

Suaeda sp. (family Chenopodiaceae)— one of the host plants of *C. salsa*

Cynomorium songaricum Rupr.

Cynomoriaceae

synonyms: *C. coccineum* L. subsp. *songaricum* (Rupr.) J. Léonard (*Kew-MPNS* accepted name, 21/9/2016); *C. purpureum* Rupr.

PLANT DESCRIPTION

Like *Cistanche, Cynomorium* is also parasitic, grows in a similar habitat, is devoid of chlorophyll and only the inflorescence emerges above ground. It belongs to a different family and is easily distinguished as a living plant. *Main differences from PS1 and PS2*: underground stem basally thickened; **stem scales usually shed with age and restricted towards the growing tip of the pre-flowering stem; inflorescences much shorter** (5–16 cm long; 2–6 cm diam.), **deep reddish-brown, dense and somewhat club-shaped. Flowers very much smaller** with male and female flowers on separate plants: **male flowers purplish-red and tipped white (3–6 mm long)** with a single thick dark-red stamen; **female flowers (1–2 mm long)** with a purplish-red club-shaped style and a flat stigma. Fruit a dry nutlet (c. 1 mm long), whitish with persistent yellow styles. Fl. May–Jul.; fr. Jun–Jul.

In China the host families are Nitrariaceae (genus *Nitraria*), Tamaricaceae (genera *Reaumuria* and *Tamarix)* and Chenopodiaceae (genus *Salsola*).

SUBSTITUTION COMMENT. The stems of *Cynomorium songaricum* are officially recognised in the Chinese Pharmacopoeia as the source of a separate TCM drug 'Suo Yang' (Cynomorii Herba). Although belonging to a different plant family, this species reputedly shares some clinical properties with the Pharmacopoeia species but is considered 'less potent'. Fresh plants are easily discernible from the official ones, but when dried, the flowering heads bear a superficial similarity to the Pharmacopoeia trade item and have been traded internationally as 'Rou Cong Rong' (C. Leon, pers. obs.) partly as a cheap, readily available look-alike. *C. songaricum* populations have declined in recent years, however, making such practice no longer economic. The frequency of its trade as an unofficial substitute in today's international market remains unclear.

Wild populations grow in sandy desert close to standing or running water (at 500–700 m altitude) and are distributed in N and NW China (Gansu, Inner Mongolia, Ningxia, Qinghai, Shaanxi and Xinjiang); the species' natural range extends to Afghanistan, Mongolia, C Asia and SW Asia (Iran).

CONSERVATION STATUS

China: Vulnerable (MEPCAS, 2013).
Global: Not Evaluated (IUCN Red List, accessed 25/1/2016).

Cistanche deserticola

Orobanchaceae

DRUG MORPHOLOGY

CRUDE DRUG (DRIED STEM): **cylindrical [1]** (not tapering), slightly flattened and curved (3–15 cm long; 2–8 cm diam.). **Externally dark brown or greyish-brown [2]**, densely covered with over-lapping somewhat fleshy, scale-like leaves **[3]** with tips often broken. *Texture*: dense and heavy, slightly pliable but difficult to break. *Transverse and oblique fractures* (see *Decocting Pieces* below and photographs). *Odour*: indistinct. *Taste:* somewhat sweet and also slightly bitter. *Note*: flowers occasionally present (difficult to detect) but, when dissected, **anthers** densely villous **[4]** with lowermost ends **sharply pointed [5]** [hl or light microscope]. ***Good quality***: large, fat, greyish-brown or brownish-black; oily.

DECOCTING PIECES-1 (SLICED STEM): somewhat oblong slices **[6]** (2–8 cm long; c. 2 cm diam.; c. 3 mm thick) with an irregular, often deeply undulating outline denoting the remnants of dark brown or greyish-brown scale-like leaves **[7]**. Transversely or obliquely cut slices **typically with a narrow [8] or broad [9] sinuous ring or rings (sometimes interrupted) of paler brown or brownish-yellow vascular tissues distinctly alternating with adjacent darker tissues [10]**. *Odour* etc. as for the crude drug. See *Crude Drug* above.

DECOCTING PIECES-2 (SLICED AND WINE-PROCESSED STEM): as for decocting pieces-1 but darker brown or blackish-brown throughout **[11]**. *Texture*: soft and pliable. *Odour*: pungent. *Taste*: sweet and slightly bitter.

PROCESSING

After removal of immature inflorescences (if present), stems are cut into long sections and sun- or oven-dried (= crude drug). These are subsequently softened thoroughly with water and cut into slices c. 3 mm thick (decocting pieces-1). Further processing is sometimes performed by steaming or boiling in yellow-rice wine until fully absorbed (= decocting pieces-2).

OTHER COMMON NAMES

Cong Rong 苁蓉; Da Yun 大芸; Herba Cistanches; desert ginseng; desert-living cistanche.

SAFETY

Safety Class: 1; Interaction Class: A (Gardner & McGuffin, 2013). In recent years stems have also become popular as a Chinese healthfood (Wang *et al*., 2012).

Cynomorium songaricum in the wild with host plants nearby, Inner Mongolia

Inflorescences of *C. songaricum*

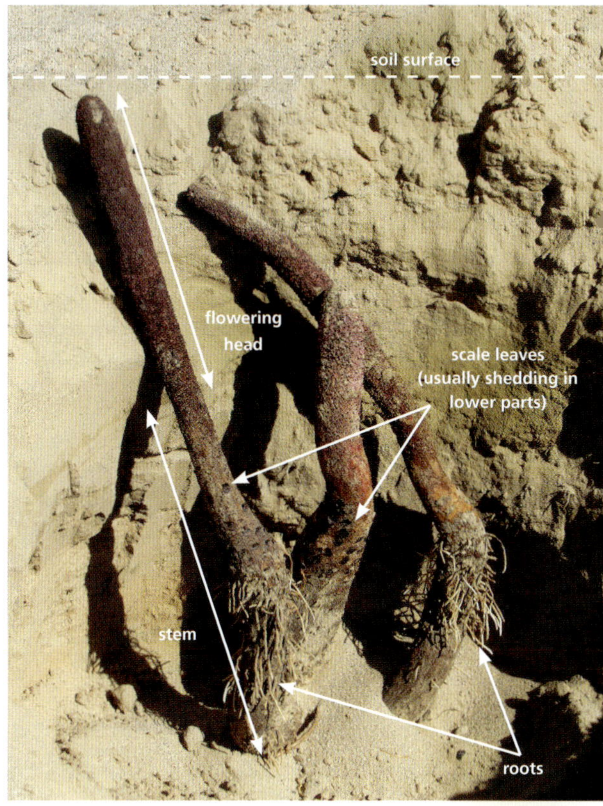

Unearthed plants of *C. songaricum* showing subterranean growth

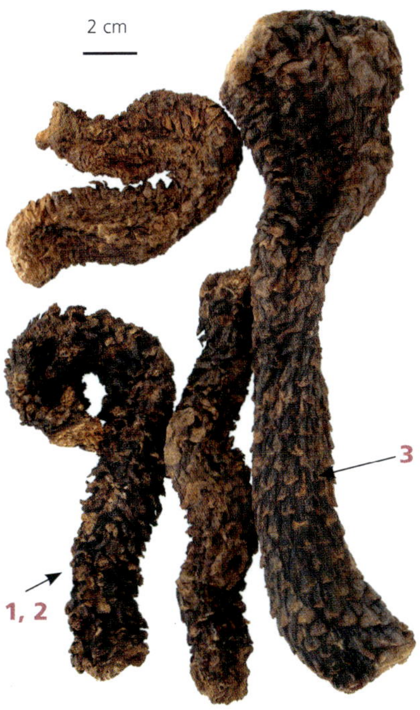

Crude drug of Rou Cong Rong (sourced from *Cistanche deserticola*)

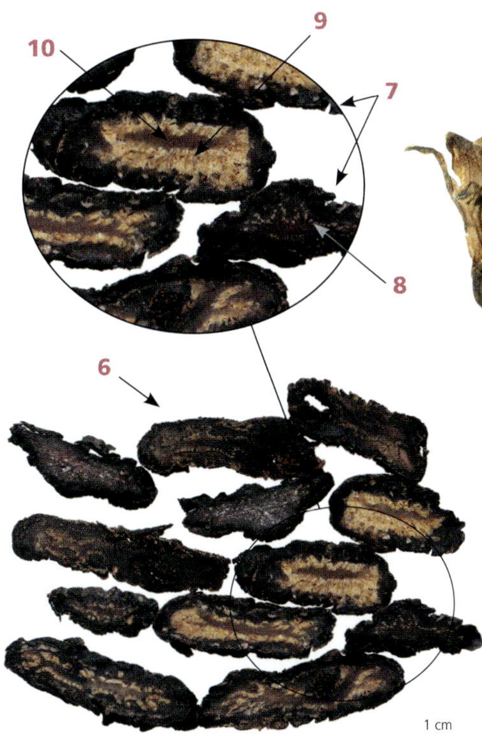

Oblique decocting slices-1 of Rou Cong Rong (sourced from *C. deserticola*)

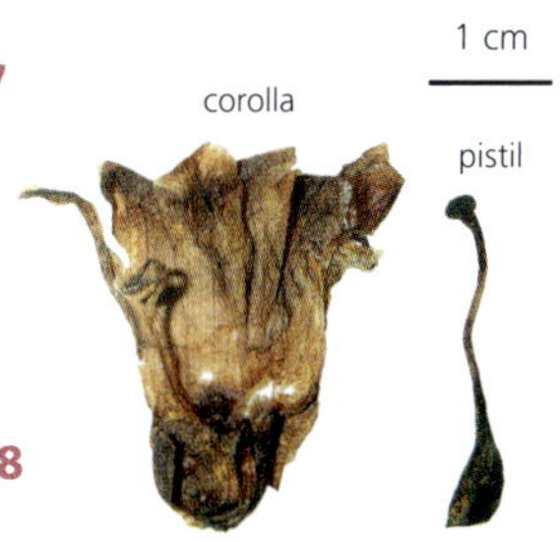

Dissected flower

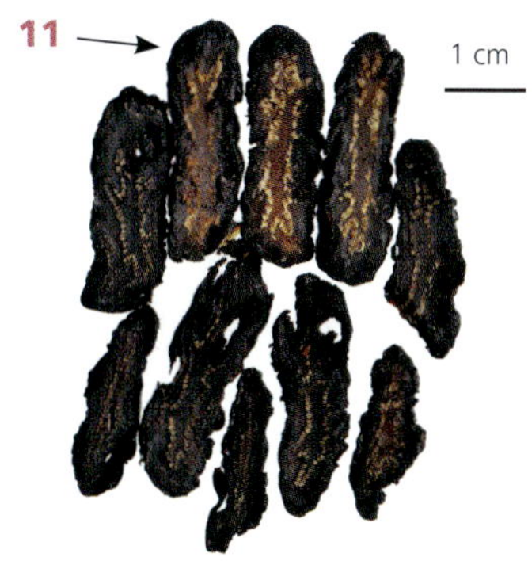

Oblique decocting slices-2 (wine-processed, sourced from *C. deserticola*)

Anther

PHOTO: LIU BO, INSTITUTE OF BOTANY (PE), BEIJING, SPECIMEN: PE01542427

Cistanche 'Tamarisk'

Orobanchaceae

DRUG MORPHOLOGY

CRUDE DRUG (DRIED STEM): Easily confused with that of PS1. *Main differences*: Stem somewhat tapering at both ends **[12]** (5–25 cm long; 2.5–9 cm diam.). **Transverse fracture with pale cream vascular tissues evenly scattered throughout entire surface creating a speckled appearance [13]**. If flowers present, **anthers** densely villous **with upper and lower ends rounded (obtuse)**.

DECOCTING PIECES-1 (SLICED STEM): as for PS2 but **cut surfaces with pale-cream vascular tissues scattered throughout entire surface creating a speckled appearance [13]** (not forming an undulating ring).

DECOCTING PIECES-2 (SLICED AND WINE-PROCESSED STEM): as for decocting pieces-1 but darker brown or blackish-brown **[14]** with **evenly scattered, dot-like vascular tissues [13]** (not forming an undulating ring). *Texture*: soft and pliable. *Odour*: pungent. *Taste*: sweet and slightly bitter.

PROCESSING

As for PS1.

OTHER COMMON NAMES

苁蓉 Cong Rong; 大芸 Da Yun; 管花肉苁蓉 Guan Hua Rou Cong Rong; 新疆肉苁蓉 Xinjiang Rou Cong Rong; Herba Cistanches; desert-living cistanche.

C. salsa

Orobanchaceae

'DRUG SUBSTITUTE' MORPHOLOGY

Macroscopic characters of the stem of *Cistanche salsa* (whole or sliced) are generally considered indistinguishable from those of *C. deserticola* (PS1). Reliable identification of the trade items of these 3 species however is most reliably performed using laboratory-based methods, notably DNA profiling.

Laboratory-based authentication methods are required.

OTHER COMMON NAMES

盐生肉苁蓉 Yan Sheng Rou Cong Rong; 肉苁蓉 Rou Cong Rong; broomrape (SCN).

SAFETY

Safety Class: 1; Interaction Class: A (Gardner & McGuffin, 2013).

Cynomorium songaricum

Cynomoriaceae

'DRUG SUBSTITUTE' MORPHOLOGY

DRIED STEM: superficially similar to the crude drugs of *C. deserticola* (PS1) and C. 'Tamarisk' (PS2). Somewhat cylindrical and slightly compressed laterally (5–15 cm long; 1.5–5 cm diam.). **Externally brown rather than greyish-brown [15], distinctly pitted and furrowed longitudinally [16]**. **Scale leaves [17]** (when intact, ovate to triangular (0.5–1.5 cm long; 0.5–1.2 cm wide), sharply pointed) **occasionally scattered over the stem surface or, more usually, densely overlapping near the rounded stem tips [18]**. *Texture*: heavy, hard and difficult to break. *Fracture*: pale brown or brown with **scattered yellowish-brown, triangular or angular vascular bundles [19]**. *Odour*: indistinct. *Taste*: sweet and astringent.

STEM SLICES: irregularly shaped and sized slices (c. 3–6 mm thick); uncut external edges brown, rough and irregular **[20]**; **transverse cut surfaces** brown, **scattered throughout with paler, yellowish-brown, triangular or angular vascular bundles [19]**; diagonal or longitudinally cut surfaces with brown cortex flecked with interrupted yellowish-brown lines of lines of vascular tissues **[21]**. *Texture*: *brittle*. *Odour*: indistinct. *Taste*: sweet and astringent.

OTHER COMMON NAMES

锁阳 Suo Yang; Cynomorii Herba; Herba Cynomorii; Songaria Cynomorium herb; Wu Lang Gao Yao (Mongolia).

SAFETY

Safety Class: 1; Interaction Class: A (Gardner & McGuffin, 2013). Considered 'highly nutritious' with potential as a healthfood (Zhao & Xiao, 2009–10).

Criude drug of Rou Cong Rong (sourced from *Cistanche* 'Tamarisk')

Decocting slices-1 of Rou Cong Rong (sourced from *C.* 'Tamarisk')

Oblique decocting slices-2 (wine-processed sourced from *C.* 'Tamarisk')

LESSER KNOWN UNOFFICIAL SUBSTITUTES

Other closely related parasitic species in the same or other genera have also been traded as 'Rou Cong Rong', these include:

1. *Cistanche sinensis* Beck — 沙苁蓉 Sha Cong Rong. Endemic to Gansu, Inner Mongolia, Ningxia and NE Xinjiang with wild populations declining throughout. **Plants** 15–70 cm tall with flowerheads **smaller than those of PS1 and PS2** (5–15 cm long, 4–6 cm diam.), **flowers yellow, drying to dark blue; anthers** long-ovoid, densely villous **with lowermost parts sharply pointed.**
2. *Boschniakia rossica* (Cham. & Schltdl.) B. Fedtsch. — 草苁蓉属 Cao Cong Rong Shu (family Orobanchaceae). A much smaller plant than PS1 and PS2 and native to Inner Mongolia and NE China in forests and riversides where it is parasitic on *Alnus* spp. (alder) and the source of a local medicine with similar properties to the official 'Rou Cong Rong'. Plant fleshy (15–35 cm tall) with a cylindrical horizontal rootstock and 2–3 stems; **inflorescences** 7–22 cm long,1.5–2.5 cm diam. with **dark purplish-red,** bell-shaped flowers with a short, 3-lobed lower lip.

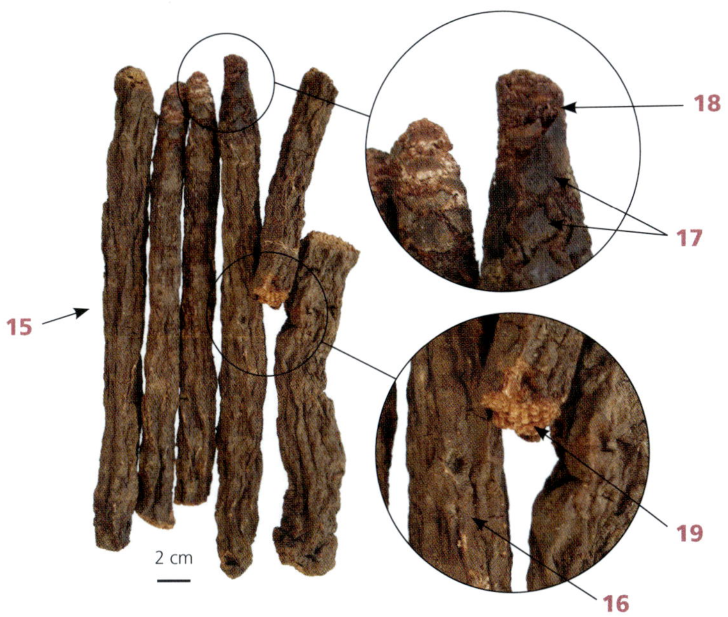

Dried stem pieces of *C. songaricum* showing tips with fewer scale leaves

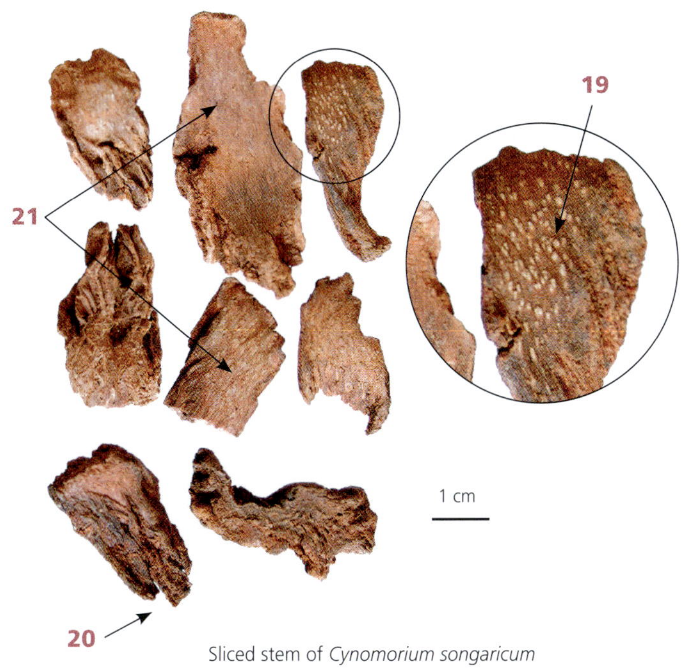

Sliced stem of *Cynomorium songaricum*

Sang Zhi 桑枝

Mori Ramulus

white mulberry twig

Drug part: YOUNG TWIG

Morus alba L.

Moraceae

taxonomic note: two varieties and many cultivars exist and all are thought to be used as the source of this drug.

Species of *Morus* wild in China: 10

PLANT DESCRIPTION

A deciduous shrub or small tree (3–10 m tall). Bark grey and shallowly furrowed. *Stipules* (at base of leaf stalks) lance-shaped (2–3.5 cm long, usually falling). *Leaves* alternate, simple, stalked and large (5–30 cm long; 5–12 cm wide), ovate or irregularly lobed; base rounded or heart-shaped, margin coarsely toothed or shallowly scalloped, tip sharply pointed or broad; **leaf undersides sparsely hairy along mid-veins or in tufts in vein axils**. *Male* and *female flowers* in short cylindrical catkins arising from the leaf axils and usually on separate plants. Male catkins pendent and densely white hairy (2–3.5 cm long) with tiny male flowers each with 4 pale green, overlapping calyx lobes and 4 stamens; petals are absent. Female catkins 1–2 cm long (shorter than male), softly hairy and packed with tiny female flowers; each flower with 4 green ovoid and overlapping calyx lobes (becoming fleshy in fruit), style absent and stigma with 2 distinctively white-edged and diverging branches [hl]; petals absent. *Fruit* a cylindrical or ovoid multiple fruit (syncarp), resembling a blackberry and becoming very succulent; depending on the cultivar, ripening from red to either blackish-purple, purple or greenish-white. Fl. Apr–May, fr. May–Aug.

For other drugs sourced from this species see 'Sang Bai Pi' (page 522), 'Sang Shen' (page 700) and 'Sang Ye' (page 554).

HARVESTING, SOURCING AND NATURAL RANGE

Young branches are harvested in late spring to early summer from cultivated plants. The latter are grown throughout China primarily for their leaves for the rearing of silkworms for the silk industry. The main source of medicinal twigs is from plants grown in Anhui, Hunan, Jiangsu, Sichuan and Zhejiang provinces (Zhao & Chen, 2014).

The wild species was originally considered endemic to central and northern China.

CONSERVATION STATUS

China: Not Evaluated.

Global: Not Evaluated (IUCN Red List, accessed 25/8/2016).

M. alba branch with young fruit

Crop of *Morus alba*, Anhui province

Leaf underside of *M. alba* showing white hair tufts in axils of midvein and primary lateral veins

TCM PROPERTIES AND USE

***properties*: 'slightly bitter and neutral'**
***channel*: 'Liver'**
***actions*: 'dispels wind damp, opens the collaterals'**
***indications*: musculoskeletal pain; post stroke paralysis; dysuria**

DRUG MORPHOLOGY

CRUDE DRUG (DRIED TWIG): long, cylindrical twigs of varying length (0.5–1.5 cm diam.), occasionally branching. *Bark* greyish-yellow or greyish-brown, with fine longitudinal striations; scattered ovate lenticels **[1]** usually pale yellow and raised; leaf buds alternate occasional, pale yellow **[2]** and with a sunken scar of last year's leaf immediately below **[3]**. *Fracture*: fibrous; bark thin **[4]**; wood broad, yellowish-white and dense with indistinct rays **[5]**; pith white or yellowish-white. *Texture*: hard and difficult to break. *Odour*: indistinct. *Taste*: indistinct. ***Good quality*:** easily broken twigs, devoid of leaves.

DECOCTING PIECES-1 (TRANSVERSELY OR OBLIQUELY SLICED TWIG): ovate or elliptical woody slices (0.4–2 cm diam.; 0.2–0.5 mm thick). Bark thin **[4]**, greyish-yellow or yellowish-brown with fine longitudinal striations; lenticels scattered, ovate pale yellow and slightly raised **[1]**; leaf buds occasionally present **[2]**, pale yellow with sunken scar of last year's leaf immediately below. Cut surface with pale yellowish-white inner bark **[6]**, broad zone of dense yellowish-white wood with indistinct rays **[5]**; and a central white or yellowish-white pith **[7]**. *Texture*: hard and difficult to break. *Odour*: indistinct. *Taste*: indistinct.

DECOCTING PIECES-2 (DRY-FRIED TWIG): as for decocting pieces above but all surfaces darker with occasional charring. *Odour*: aromatic.

PROCESSING

After removal of the leaves, young twigs are sun-dried (= crude drug) then sliced into 0.2–0.5 mm thick decocting pieces or sliced when fresh then dried. Subsequent dry-frying produces 'Zhi Sang Zhi' (processed mulberry twig).

OTHER COMMON NAMES

老桑枝 Lao Sang Zhi; 嫩桑枝 Nen Sang Zhi; 桑 Sang (plant name); 童桑枝 Tong Sang Zhi; Ramulus Mori; white mulberry (SCN).

SAFETY

Safety Class: 1; Interaction Class: A (Gardner & McGuffin, 2013). An official healthfood in China (Anon., 2002a,b).

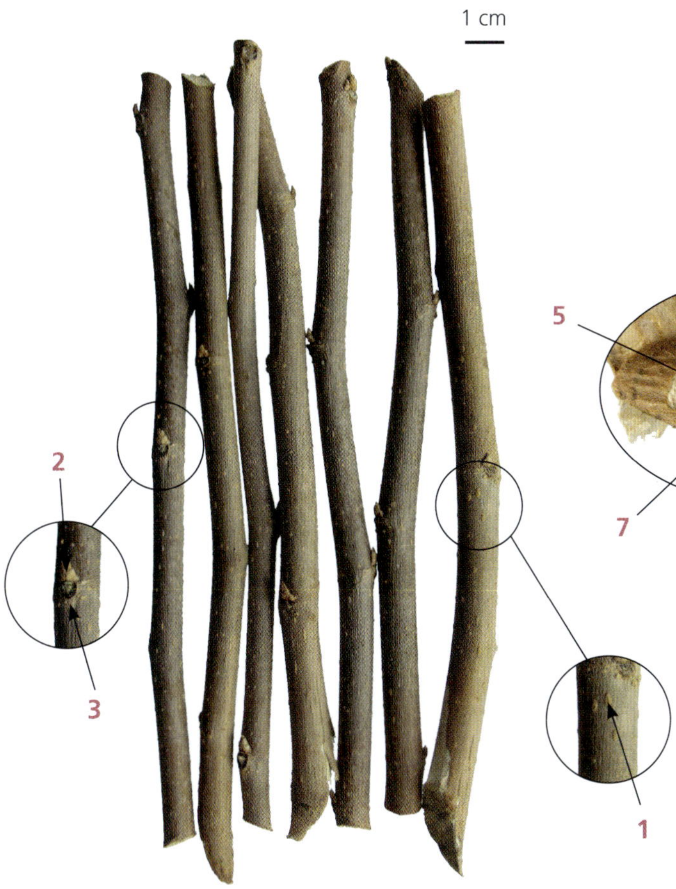

Crude drug

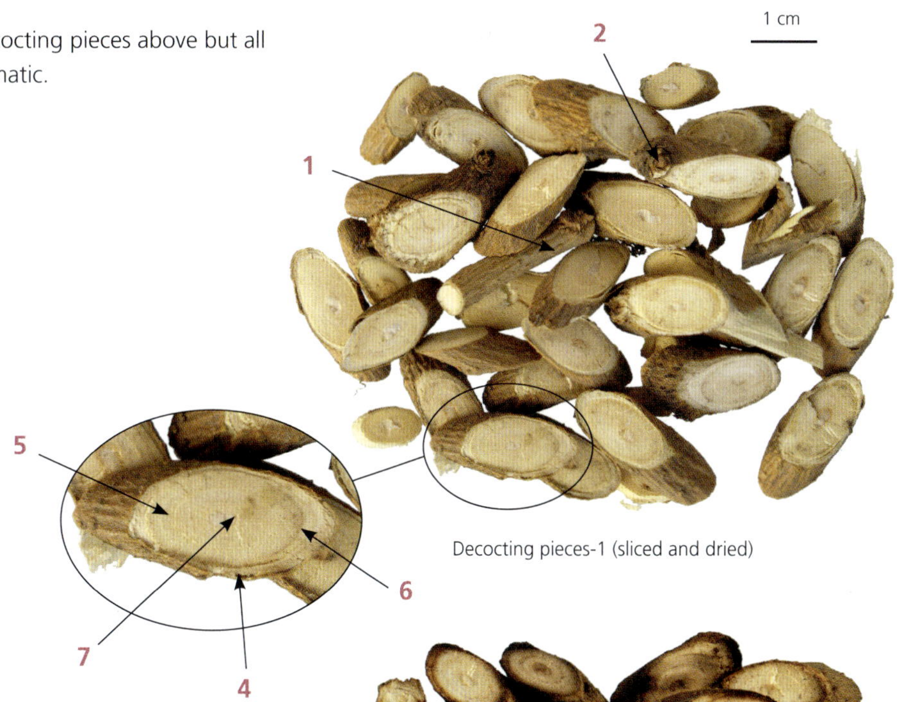

Decocting pieces-1 (sliced and dried)

Decocting pieces-2 (sliced, dried then dry-fried)

Shi Hu 石斛 Dendrobii Caulis

Drug part: STEM (FRESH OR DRIED)

Pharmacopoeia definition and identification issues. This high-value herbal drug is officially sourced from the stems of *Dendrobium nobile* Lindl., *D. chrysotoxum* Lindl., *D. fimbriatum* Hook. and "similar species from the same genus" (CP2015). This imprecise definition derives from the difficulty in accurately distinguishing the stems of the many species of *Dendrobium* (78) native to China. Stems of some of the more widely traded species can, to the experienced eye, be distinguished using organoleptic characters alone but in recent years a range of DNA-barcoding methods have been developed to address this broader identification need (e.g. Lau *et al.*, 2001). Two of the three CP-named species are included below as examples

PHARMACOPOEIA SPECIES 1

Dendrobium nobile Lindl. — Orchidaceae

PLANT DESCRIPTION

A stout, many-stemmed perennial orchid (10–60 cm tall), epiphytic on rocks and tree trunks. *Stems* erect, fleshy, unbranched and cylindrical (up to 1.3 cm diam.) with internodes 2.5–4 cm long; drying to golden yellow. *Leaves* alternate and unstalked; leaf blades simple, oblong (6–11 cm long; 1–3 cm wide), leathery, bases often sheathing, tip broad and unequally bi-lobed and margins entire. *Flowers* in loose 1–4-flowered heads lateral along stems; each flower large and conspicuous, 6-petalled, 5 petals spreading and broadly ovate; usually white, pale purple and pale purplish-red, often darker coloured at tips; dorsal and lateral 'petals' 5-veined; lip (the 'lower petal') broadly ovate (2.5–3.5 cm long; 2.2–3.2 cm wide), with a large dark purple or white central spot and densely covered in short, soft hairs; column short, green and stout (c. 5 mm long). *Fruit* a dry capsule with numerous dust-like seeds.

See also the closely related TCM drug 'Tie Pi Shi Hu' (Dendrobii Officinalis Caulis) page 486.

Species of *Dendrobium* wild in China: 78

HARVESTING, SOURCING AND NATURAL RANGE

Stems are harvested throughout the year from cultivated or wild 'fostered' plants; 'fostering' involves the sustainable management of young plants re-introduced to their native habitat. The main production areas are in Guizhou and Yunnan provinces.

The species is native in S and SW China (Guangxi, Guizhou, Hainan, Hong Kong, Hubei, Sichuan, Taiwan, SE Tibet and Yunnan); where it grows as an epiphyte on tree trunks or rocks in mountain forests and valleys (500–1,700 m altitude). Also native in Bhutan, India, Laos, Myanmar, Nepal, N Thailand and Vietnam.

CONSERVATION STATUS

China: Vulnerable (MEPCAS, 2013).
Global: Not Evaluated (IUCN Red List, accessed 28/6/2016).

PROTECTED STATUS

CITES: Appendix II; EU Annex B (see Table 1, page 10)

PHARMACOPOEIA SPECIES 2

D. fimbriatum Hook. — Orchidaceae

synonym: *D. fimbriatum* var. *oculatum* Hook.

PLANT DESCRIPTION

Easily distinguished from PS1. *Main differences:* **stems pendant, longitudinally grooved** (50–100 cm long); internodes slightly longer (3.5–4.8 cm long); leaves often slightly larger (8–15.5 cm long; 2–3.6 cm wide), drying to pale yellow or pale yellowish-brown; flower slightly fragrant, **'petals' all golden yellow, lip petal with a deep reddish-brown central spot** and **margin of lip petal distinctly fringed**.

HARVESTING, SOURCING AND NATURAL RANGE

Stems are harvested throughout the year from wild and cultivated plants. The species is native to SW China (provinces of Guangxi, Guizhou and Yunnan) where it grows on tree trunks and damp rocks in thick forest in mountain valleys; 600–1,700 m altitude [also native in Bhutan, India, Laos, Myanmar, Nepal, Thailand and Vietnam].

CONSERVATION STATUS

China: Vulnerable (MEPCAS, 2013).
Global: Not Evaluated (IUCN Red List, accessed 28/6/2016).

PROTECTED STATUS

CITES: Appendix II; EU Annex B (see Table 1, page 10).

TCM PROPERTIES AND USE

***properties*: 'sweet and slightly cold'**
***channels*: 'Stomach and Kidney'**
***actions*: 'tonifies stomach, generates fluids, tonifies yin, clears heat'**

***indications*: fevers after illness, chronic dry cough, poor appetite; constipation and thirst; blurred vision**

Stems and leaves of *Dendrobium nobile*; inset: leaf tip unequally bi-lobed

D. nobile, Guizhou province

Medicinally 'fostered' plants of *D. nobile* Guizhou province

D. nobile

Freshly harvested stems of *D. fimbriatum* cultivated for Chinese medicine, Laos PHOTO: ANDRE SCHUITEMAN

D. fimbriatum PHOTO: ANDRE SCHUITEMAN

D. fimbriatum PHOTO: ANDRE SCHUITEMAN

Flickingeria spp.

Orchidaceae

synonym: *Ephemerantha* spp. (name widely used in the TCM literature)

taxonomic/nomenclatural note: treated by *WCS* and *Kew-MPNS* as part of *Dendrobium*.

PLANT DESCRIPTION

Easily distinguished from the Pharmacopoeia species (*Dendrobium* spp.). *Main differences:* stems climbing or pendulous, **terminating in a conspicuous 1-leaved, pseudobulb (swollen stem**), laterally compressed; leaf bases not sheathing; **flowers solitary,** fragile and seldom lasting more than one day; arising on short tufted stalks from base of a leaf; 'petals' of **various colours** white, pink, yellow, brownish or purple; spotted or unspotted.

SUBSTITUTION COMMENT. Owing to high prices of official 'Shi Hu', stems of many *Flickingeria* species are used as cheap substitutes and can be seen regularly traded as 'Shi Hu' in local and international markets. It is unclear how their clinical actions compare with the official drug. The genus is widely grown as an ornamental and it is likely that the sale of their pseudobulbs as a cheap 'Shi Hu' substitute serves as a lucrative by-product. Like the Pharmacopoeia species, *Flickingeria* are epiphytes and in the wild they frequent the sub-tropical and tropical parts of S China, notably Guangxi, Hainan, Yunnan and Taiwan.

CONSERVATION STATUS

China: most species 'Least Concern but some 'Vulnerable' or 'Critically Endangered' (MEPCAS, 2013).

Global: Not Evaluated (IUCN Red List, accessed 28/6/2016).

PROTECTED STATUS

CITES: Appendix II; EU Annex B (see Table 1, page 10).

Phragmites australis (Cav.) Trin. ex Steud.

Poaceae

synonyms: *Arundo australis* Cav.; *P. communis* Trin.

PLANT DESCRIPTION

Easily distinguished from the official species. **A tall (up to 2 m), slender grass (reed) with a stout and vigorous underground rhizome** often forming extensive stands along lakes and river margins. *Stems* c. 6 mm diam., with mealy powder below nodes. *Leaves* alternate and stalkless; leaf bases sheathing with a ligule topped by a fringe of tiny hairs (up to 0.6 mm long); leaf blades simple and lance-shaped (50 cm long; 1–3 cm wide), usually drooping, margins smooth or rough, light green to bluish-green, tips thread-like. ***Flowerheads*** large and conspicuous (20–50 cm long; 10 cm wide), **plume-like, soft and pale purplish-brown with many silky spikelets** (10–18 mm long), each with 2–5 florets. Fl. and fr. Jul–Nov.

SUBSTITUTION COMMENT. The high price of 'Shi Hu' sometimes creates incentives for look-alike materials to be traded in their place. One infrequent example is the rhizomes of *P. australis*, a ubiquitous reed in the grass family, and the source of a separate TCM drug 'Lu Gen' (Phragmitis Rhizoma) page 194. This substitute is easily distinguished macroscopically from the official 'Shi Hu' but it is only likely to be encountered at a local level in China.

CONSERVATION STATUS

China: Least Concern (MEPCAS, 2013).

Global: Least Concern (IUCN Red List, accessed 28/6/2016).

Flickingeria albopurpurea Seidenfaden, artificially propagated
PHOTO: STEPHAN GALE, KADOORIE FARM AND BOTANIC GARDEN

Flickingeria sp. PHOTO: LEONID AVERYANOV

Typical stands of plants of *Phragmites australis*, Inner Mongolia

P. australis

Dendrobium nobile

Orchidaceae

DRUG MORPHOLOGY

CRUDE DRUG-1 'XIAN SHI HU' (FRESH STEM): cylindrical or slightly flattened-cylindrical (ca 30 cm long; 0.4–1.2 cm diam.), often curving. Externally yellowish-green, smooth or with longitudinal furrows **[1]**. Nodes darkly coloured **[2]** and with membranous remnants of leaf sheaths **[3]**; internodes 2.5–4 cm long. Cut surface **[4]** even, green and fleshy; dotted with cream vascular bundles. *Texture:* fleshy and easily broken. *Odour:* indistinct. *Taste:* somewhat bitter, becoming sweet and viscous when chewed.

DECOCTING PIECES-1 'XIAN SHI HU' (FRESH STEM LENGTHS): as for the fresh crude drug but cut into short lengths (0.4–1.2 cm diam.).

CRUDE DRUG-2 (DRIED STEM): flattened-cylindrical **[5]** (20–40 cm long; 4–6 mm diam.) with **internodes 2.5–4 cm long [6].** Externally **[7]** golden or greenish-yellow, often shiny and distinctly furrowed longitudinally. *Texture:* hard and brittle. *Fracture:* quite even. *Taste:* bitter.

DECOCTING PIECES-2 (DRIED STEM LENGTHS): tightly spiralled stems (c. 1 cm long) **[8]**. Externally **[9]** golden, greenish-yellow or brownish-yellow, lustrous, deeply furrowed or angled longitudinally. Leaf sheath remnants sometimes visible **[10]**. Cut surface yellowish-white to yellowish-brown, dotted with vascular bundles. *Texture:* hard and brittle. *Fracture:* quite even. *Taste:* indistinct or slightly bitter and viscous when chewed.

DECOCTING PIECES-3 (DRIED STEM LENGTHS): as for DP-2 above but stems in short lengths (c.0.4–0.8 mm long) with outline of cut surfaces distinctly undulating **[11]**.

PROCESSING

Stems are used fresh or dried. Fresh stems ('Xian Shi Hu' = crude drug-1) are prepared by removing the roots then cutting into short lengths (= decocting pieces-1). Dried stems are either blanched or gently baked leaf sheaths and roots removed, then either dry-fried until the stems spiral (= crude drug-2) or cut into lengths (decocting pieces-2).

OTHER COMMON NAMES

扁黄草 Bian Huang Cao; 大黄草 Da Huang Cao; 金钗石斛 Jin Chai Shi Hu; Caulis Dendrobii. Chinese orchid; dendrobium (SCN; CP 2010).

SAFETY

An official healthfood in China (Anon., 2002a,b). Safety Class: 1; Interaction Class: A (Gardner & McGuffin, 2013).

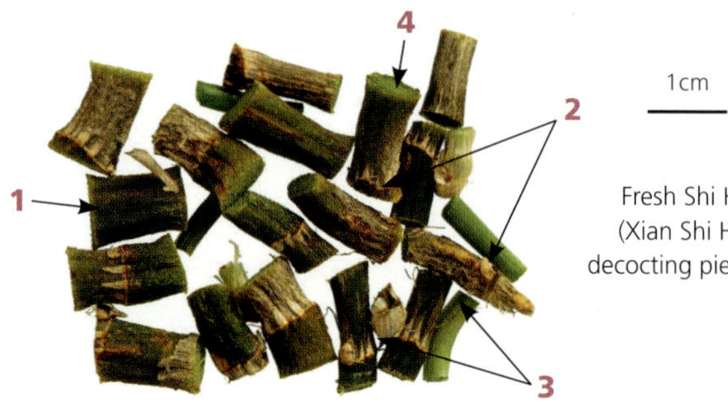

Fresh Shi Hu (Xian Shi Hu) decocting pieces-1

Dendrobium fimbriatum

Orchidaceae

DRUG MORPHOLOGY

Crude drug differences from that of PS1: **stems not flattened, longer** (40–100 cm long) and **wider** (5–8 mm diam.); internodes 3–4.5 cm long. Externally yellow to dark yellow, distinctly furrowed longitudinally. *Texture*: lax. ***Fracture:* fibrous.** *Taste*: slightly bitter. Decocting pieces differ in their narrower diameter, with cut surfaces more lax and fibrous. *It is advised that macroscopic identification be supported by laboratory-based analytical testing.*

PROCESSING

As for PS1 but because the stems of this species are too broad the spiralled processing form is not made.

OTHER COMMON NAMES

流苏石斛 Liu Su Shi Hu (plant name); 马鞭杆 Ma Bian Gan; Caulis Dendrobii; Chinese orchid; dendrobium (SCN; CP2010).

SAFETY

An official healthfood in China (Anon., 2002a,b).

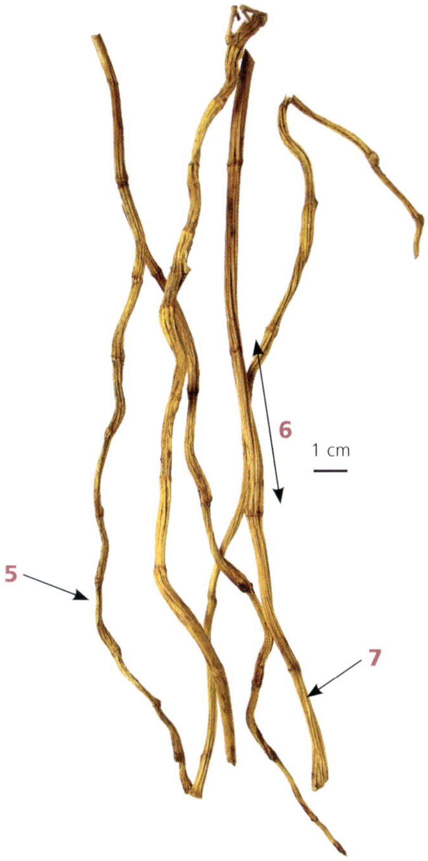

Crude drug-2 of Shi Hu sourced from *D. nobile*

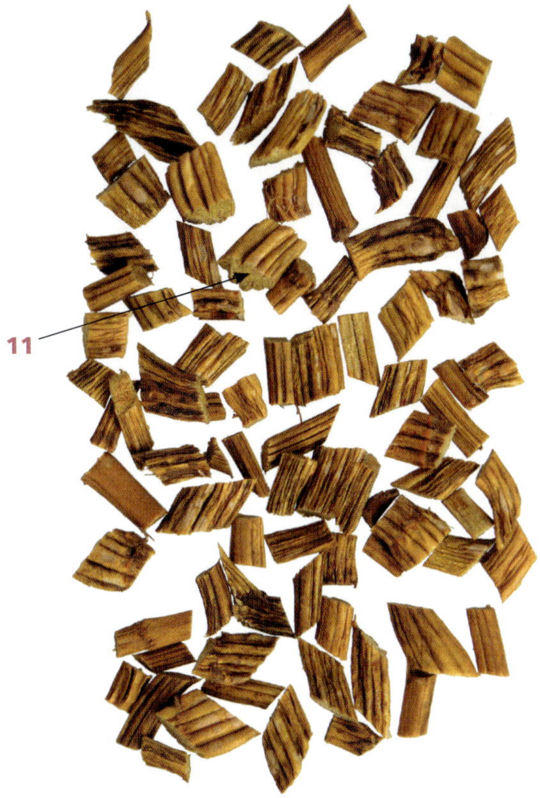

Decocting pieces-3 (dried and cut *D. nobile*)

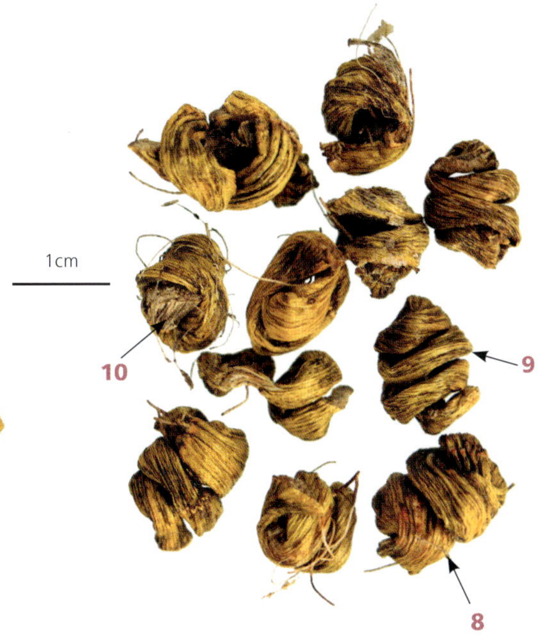

Decocting pieces-2 (dry-fried *D. nobile*)

Crude drug-2 of Shi Hu sourced from *D. fimbriatum*

Bulk wild-harvesting of *D. fimbriatum* stems in Laos (2007), destined for traditional medicine markets in China. Shading is used to keep plants alive so as to encourage further stem production prior to export.
PHOTO: ANDRE SCHUITEMAN

Flickingeria spp.

Orchidaceae

'DRUG SUBSTITUTE' MORPHOLOGY

Intact stems are easily distinguished from stems of official Shi Hu by the distinctly swollen pseudobulbs **[12]**; sometimes branched and often laterally compressed and club-shaped. When stems are cut into short lengths this character may be less obvious making differentiation dependent on laboratory-based authentication.

OTHER COMMON NAMES

金石斛属 Jin Shi Hu Shu; 有瓜石斛 You Gua Shi Hu; Herba Ephemeranthae Fimbriatae.

Phragmites australis

Poaceae

'DRUG SUBSTITUTE' MORPHOLOGY

Intact and sliced **rhizomes** easily distinguished from the official drug. *Main differences:* internodes sometimes much longer (up to 11 cm long) and **diameter much larger [13] (c. 1–2 cm);** transverse section reveals a **large hollow centre [14]**, more than half the rhizome diameter, with a ring of yellowish-white vascular tissue surrounded by **an outer ring of interrupted tissue with regular oval cavities [15]**. *Texture:* light and strong, hard but flexible; **very difficult to break.** *Odour:* indistinct. *Taste:* sweetish.

OTHER COMMON NAMES

芦根 Lu Gen; phragmites; common reed.

SAFETY

A food and healthfood in China (Anon., 2002a,b).

OTHER UNOFFICIAL SUBSTITUTES

As a high-value medicinal product 'Shi Hu' is prone to widespread substitution and adulteration by many other orchid genera notably the stems and pseudobulbs of species in the genus *Pholidota;* these are easily identified by their inflated, fleshy pseudobulbs similar to those of *Flickingeria* above.

Pholidota chinensis

PHOTO: STEPHAN GALE, KADOORIE FARM AND BOTANIC GARDEN

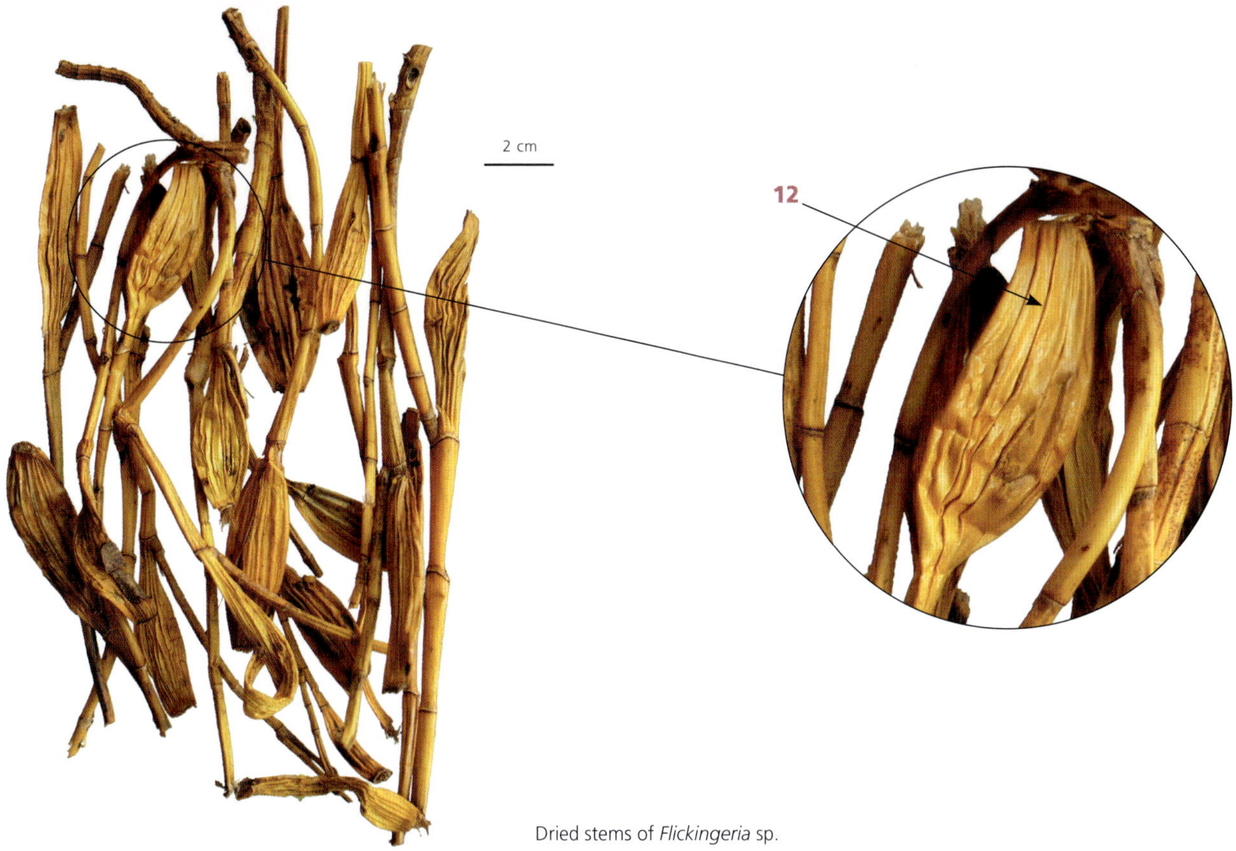

Dried stems of *Flickingeria* sp.

Dried rhizomes of *Phragmites australis* (crude drug)

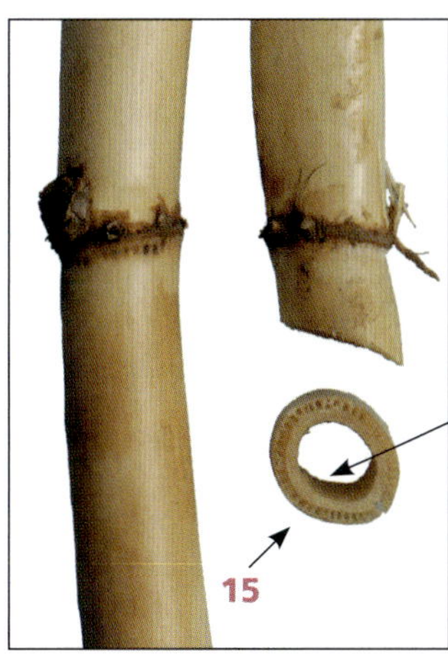

Fresh rhizome of *P. australis*; lower right in cross-section

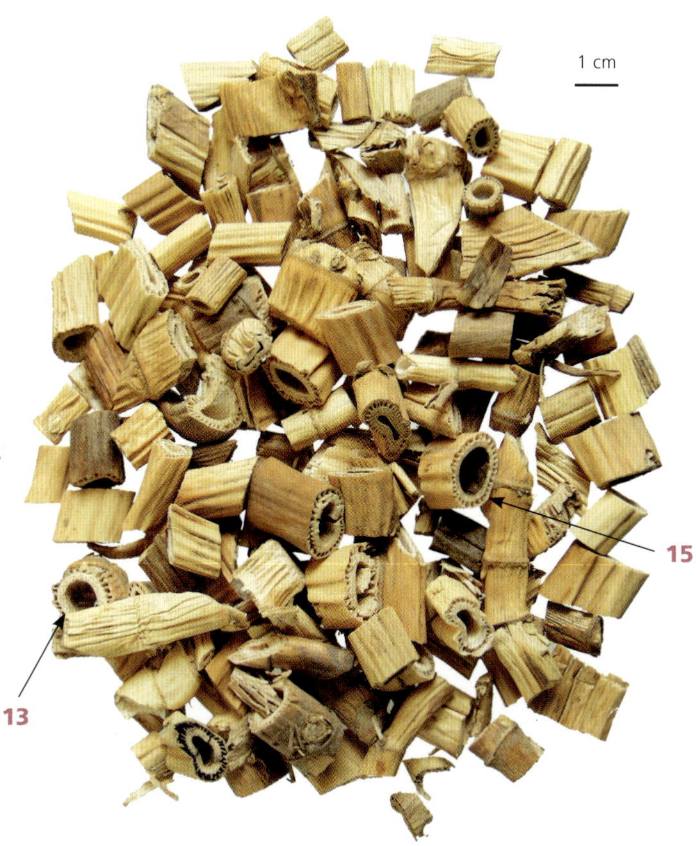

Dried cut rhizomes of *P. australis* (yin pian)

Tie Pi Shi Hu 铁皮石斛 Dendrobii Officinalis Caulis

Drug parts: STEM

Dendrobium catenatum Lindl.

synonym: *D. officinale* Kimura & Migo (name used in CP2015).

Orchidaceae

Species of *Dendrobium* wild in China: 78

PLANT DESCRIPTION

A perennial herb growing over rocks, with a short rhizome. *Multi-stemmed* with stems cylindrical, unbranched (3–35(–60) cm long; 2–18 mm diam.), erect or pendulous; many-noded with internodes (0.3–)1.3–4 cm long, pale yellowish-green, sometimes pale purple dotted; becoming pale yellow when dry. *Leaves* alternate, 2–5, mostly along upper part of stem, strap- or lance-shaped (3–21 cm long; 0.5–1.5 cm wide), stalkless, tips broad or slightly unequal (minutely bi-lobed); margins entire; leathery. *Inflorescences* 1–3 per stem, lateral, (1–)3–8-flowered arising from leafy or old leafless stems. *Each flower* large, 'petals' 6, conspicuous and spreading, yellowish-green, pale yellowish-white or white, sometimes becoming creamish-yellow and (except for lip) tips pointed and 3-5-veined (12–17 mm long; 4–7 mm wide); lip (lower-most petal) much shorter than sepals and petals and slightly wedge-shaped at base, with 2 large purplish-red markings.

See also the closely related TCM drug 'Shi Hu' (Dendrobii Caulis) page 478.

HARVESTING, SOURCING AND NATURAL RANGE

Stems are harvested from November to March from commercially propagated plants grown exclusively for the TCM market. The final product 'Tie Pi Shi Hu' or, when spiralled, 'Tie Pi Feng Dou', is regarded as one of China's superior medicinal herbs and is highly prized for its therapeutic properties. Until recently, 'Tie Pi Shi Hu' was considered more challenging to propagate than *D. nobile*, the main source species of the closely related 'Shi Hu' (Dendrobii Caulis, see page 478) and which has similar clinical properties. Accordingly, 'Tie Pi Shi Hu' has traditionally fetched much higher prices (upwards of US$ 3,000/kg) but, in recent years, following much investment in more efficient propagation methods, yields have increased and prices have begun to fall (E. Brand, pers. comm, 11/ 2016; Ding *et al.*, 2002, 2008).

In the wild in China, the species grows epiphytically on rocks and trees in damp places in the mountains of Anhui, Fujian, Guangxi, Sichuan, Taiwan, Yunnan and Zhejiang (c. 1,600 m altitude) but its indiscriminate harvesting for medicinal use coupled with habitat loss over several years has led to its populations becoming critically endangered (Fu,1992). The species is also native to Japan.

CONSERVATION STATUS

China: Critically Endangered (MEPCAS, 2013; Fu, 1992).
Global: Critically Endangered (IUCN Red List, accessed 20/2/2016; as the synonym *D. officinale*)

PROTECTED STATUS

CITES: Appendix II; EU Annex B.

Flowering branch of *D. catenatum*

Young propagated plants of *Dendrobium catenatum*, Guizhou province

TCM PROPERTIES AND USE

properties: **'sweet and slightly cold'**
channels: **'Stomach' and 'Kidney'**
actions: **'to boost the stomach and engender fluid, nourish yin and clear heat'**

indications: **fevers after illness, chronic dry cough, poor appetite; constipation and thirst; blurred vision**

DRUG MORPHOLOGY

CRUDE DRUG/DECOCTING PIECES:

Tie Pi Shi Hu: cylindrical or flattened-cylindrical (up to 30 cm long; 0.4–1.2 cm diam.). Externally yellowish-green, smooth or wrinkled longitudinally **[1]**, nodes distinct, slightly swollen, often darker than internodes **[2]** and sometimes with remains of membranous leaf sheaths **[3]**. *Texture*: fleshy, succulent and easily broken. *Odour*: indistinct. *Taste*: slightly bitter; sweet and viscous when chewed.

Tie Pi Feng Dou: tightly spiraled **[4]**, 3.5–8 cm long when straightened; 2–3 mm diam. Externally yellowish-green, narrowly striated longitudinally **[5]**. *Texture*: dense, easily broken. *Fracture:* even. *Taste*: viscous on chewing.

PROCESSING

After removal of fibrous roots, stems are dry-fried until they spiral, then baked dry (= 'Tie Pi Feng Dou' or 'Er Huan Shi Hu' crude drug-1/decocting pieces-1). Alternatively stems are cut into varying lengths, then dried or baked at low temperature (= 'Tie Pi Shi Hu' crude drug-2/decocting pieces-2). 'Tie Pi Feng Dou' normally fetches higher prices than 'Tie Pi Shi Hu'.

OTHER COMMON NAMES

黑节草 Hei Jie Cao; 黄石斛 Huang Shi Hu 'yellow Shi Hu'; 铁皮 Tie Pi; Caulis Dendrobii Officinalis; dendrobium (SCN).

Crude drug/decocting pieces of Tie Pi Shi Hu sourced from *Dendrobium catenatum*

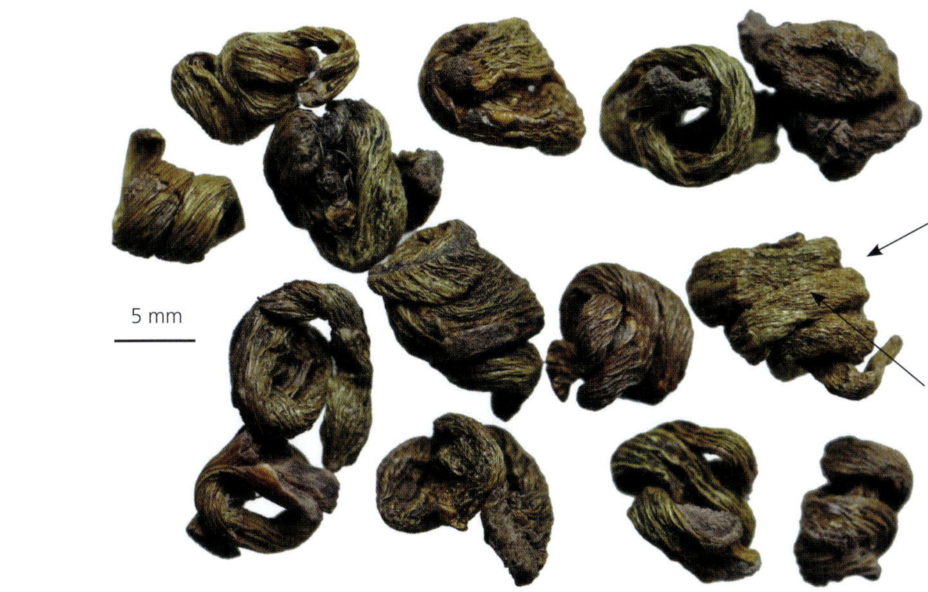

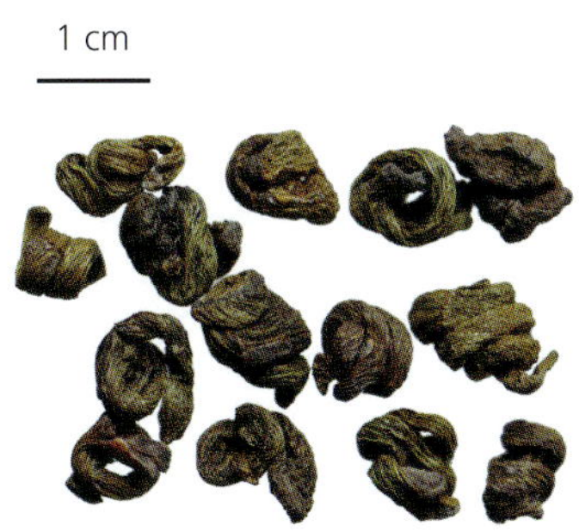

Crude drug/decocting pieces of Tie Pi Shi Hu (Tie Pi Feng Dou) sourced from *D. catenatum*

UNOFFICIAL SUBSTITUTES/ADULTERANTS

Many substitutes and adulterants have been reported in the TCM literature for this high-value drug (Bensky et al., 2004; Ding *et al.*, 2002; Xiao *et al.*, 2001-2) perhaps the most common being other species of *Dendrobium* readily available from commercial-grown sources. Examples include:

1. ***Dendrobium devonianum*** Paxton
2. ***D. loddigesii*** Rolfe
3. ***D. moniliforme*** (L.) Sw.
4. ***D. nobile*** Lindl.

Zi Su Geng 紫苏梗

Perillae Caulis

Drug part: STEM

Perilla frutescens (L.) Britton

Lamiaceae

synonym: *Ocimum frutescens* L.

taxonomic/nomenclatural note: 3 varieties are recognised by *FoC* and *Kew-MPNS*, 20/8/2016.

Species of *Perilla wild* in China: 1

PLANT DESCRIPTION

An erect herbaceous annual (0.3–2 m tall). *Stems* 4-sided, green or purple, sparsely or densely covered in fine hairs. *Leaves* opposite with each pair at right angles to the next; stalks 3–5 cm long; **leaf blades** simple, **broadly ovate to circular** (4.5–13 cm long; 2.8–10 cm wide), green, purplish, or purplish-black, bases rounded to broadly wedge-shaped, tips short or narrowly pointed and margins narrowly to coarsely saw-toothed. *Flowers* tiny, in densely hairy, sub-cylindrical many-flowered spikes (1.5–15 cm long) arising from the leaf axils; flowers subtended by an ovate shield-like bract (c. 4 mm long; 4 mm wide), green or purple, membranous, pointed and of similar length to the flower. Each flower bisexual, laterally symmetrical and short-stalked; calyx cup-shaped (c. 3 mm long), green to yellowish, glandular, densely hairy at its base and with 5 pointed teeth at its apex; corolla white to purplish-red, bell-shaped and opening into a 2-lipped mouth (lower lip 3-lobed and slightly longer than upper lip); **stamens 4, shorter than corolla;** ovary superior; style 1. *Fruit* with 4 dry, sub-globose nutlets; each nutlet **(c.1.5 mm diam.)**, greyish-brown, greyish-yellow or purplish and **with a netted surface**. Fl. Aug–Nov, fr. Aug–Dec.

For other drugs sourced from this species see 'Zi Su Ye' (Perillae Folium) page 566, 'Zi Su Zi' (Perillae Fructus) page 556.

HARVESTING, SOURCING AND NATURAL RANGE

Stems are usually harvested in summer after plants have flowered or in autum after fruits have ripened. Sourcing is mainly from cultivated crop plants well-established in China with many cultivars bred for medicinal or other purposes; most are thought to be used in TCM. The main production areas for 'Zi Su Geng' are the central and eastern Chinese provinces of Jiangsu, Zhejiang, Hebei, Hubei and Henan. The purple-leaved cultivars are widely used as a condiment and salad vegetable.

In the wild, the species is common on disturbed ground and has a wide distribution in China (Fujian, Guangdong, Guangxi, Guizhou, Hebei, Hubei, Jiangsu, Jiangxi, Shanxi, Sichuan, Taiwan, Tibet, Yunnan and Zhejiang [also native in Bhutan, Cambodia, India, Indonesia (Java), Japan, Korean Peninsula, Laos and Vietnam].

CONSERVATION STATUS

China: Least Concern (MEPCAS, 2013, as '*P. frutescens* var. *purpurascens*').

Global: Not Evaluated (IUCN Red List, accessed 18/8/2016).

Medicinal crop of *Perilla frutescens*

Inflorescence of *P. frutescens*

P. frutescens (form with purple leaf undersides)

TCM PROPERTIES AND USE

***properties*:** **'pungent, sweet and slightly warm'**
***channels*:** **'Lung and Spleen'**
***actions*:** **'regulates qi, relieves pain'**
***indications*:** **abdominal distension; morning sickness**

DRUG MORPHOLOGY

CRUDE DRUG (DRIED STEM): long, hard, 4-sided stems **[1]** of variable length (up to c. 1 m long; 0.5-1.5 cm diam.) and slightly curved. Externally purplish-brown, dark purple or greenish-brown, longitudinally furrowed, sparsely or densely covered in fine hairs **[2]**, with branch scars and remains of 2 opposite leaf stalks **[3]** at intervals up the stem; nodes slightly swollen **[4]**. *Texture:* light, smooth and hard. *Transverse fracture:* uneven; rectangular in outline with rounded corners **[5]**; outer bark thin **[6]**, wood (xylem) yellowish-white **[7]** with fine, densely packed rays [hl] surrounding a large central white pith **[8]**, the latter loose. *Odour*: slightly aromatic. *Taste*: indistinct. ***Good quality***: aromatic, purplish-brown or purplish-black.

DECOCTING PIECES (SLICED STEM): 4-sided **[1]**, corners rounded **[5]**, diagonally or transversely sliced (c. 4 mm long; 0.5–1.5 cm diam.); externally purplish-brown, dark purple or greenish-brown **[9]**, longitudinally furrowed and finely striated **[10]**; opposite branch scars or branch bases occasional **[11]**. Cut surface: see *transverse fracture* of crude drug.

PROCESSING

Leaves and, if present, fruits are removed, then stems are stripped of their leaves, fruits etc. then either sun-dried (= crude drug) and sliced diagonally or transversely into c. 4 mm long decocting pieces, or sliced when fresh, then dried.

OTHER COMMON NAMES

苏梗 Su Geng; Caulis Perillae (CP2010).

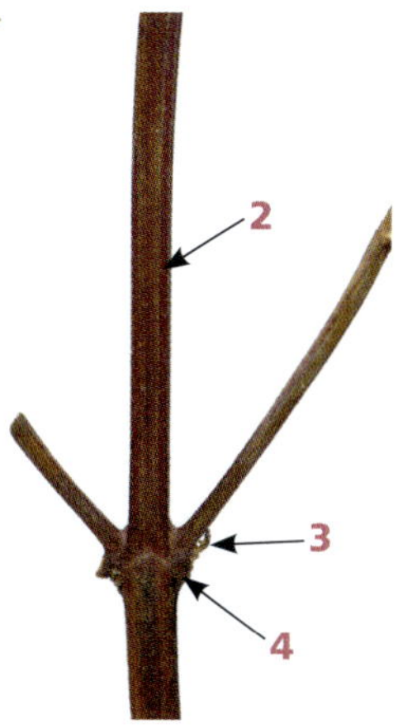

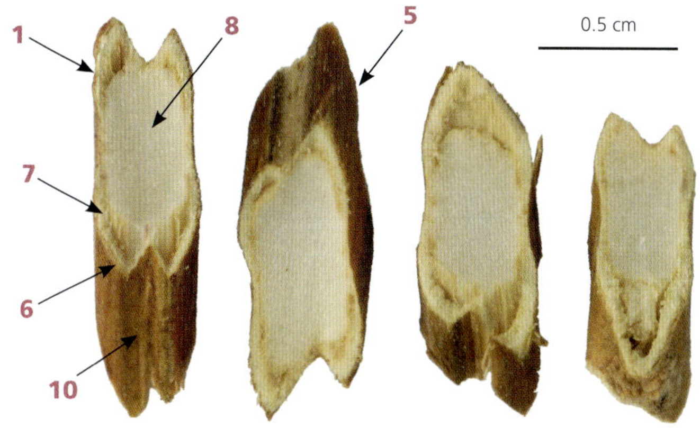

Crude drug of Zi Su Geng (*Perilla frutescens*)

Decocting pieces (transverse) of Zi Su Geng (*P. frutescens* — purplish-brown colour form)

Decocting pieces (oblique) of Zi Su Geng (*P. frutescens* — greenish-brown colour form)

Ailanthus altissima

stem-barks and
root-barks

Bai Xian Pi 白鲜皮

Dictamni Cortex

Drug part: ROOT-BARK

Dictamnus dasycarpus Turcz.

Rutaceae

synonyms: *D. albus* L. subsp. *dasycarpus* (Turcz.) N. A. Winter; *D. albus* var. *dasycarpus* (Turcz.) T. N. Liou & Y. H. Chang

taxonomic/nomenclatural note: *Kew-MPNS* accepted name 10/10/2016, *D. albus* L.

Species of *Dictamnus* wild in China: 1

PLANT DESCRIPTION

An erect, perennial herb (40–100 cm tall), most parts rich in aromatic oils. *Leaves* alternate and pinnate (divided into 3–6 pairs of opposite leaflets, plus terminal leaflet); central axis of pinnate leaf narrowly winged; each leaflet with scattered, almost transparent oil glands; leaflet blades ovate, elliptic or oblong (3–12 cm long; 1–5 cm wide), lateral ones stalkless, terminal one stalked; margins with rounded, symmetrical teeth. Flowers and fruits scattered with dark, stalked or unstalked glands. *Flowers* bisexual, many, in few-flowered, stalked clusters along erect terminal spikes (up to 30 cm long) above the leaves; each flower laterally symmetrical; sepals 5; petals 5, large, ovate (2–2.5 cm long; 0.5–0.8 cm wide), pale pink with dark pink veining; stamens 10, at right angles to petals, with broad pink filaments curving upwards; ovary superior; style 1. *Fruit* a group of 5 dry pods (follicles) (1–2 cm long). *Seeds* black, almost spherical and shiny. Fl. May, fr. Aug–Sep. Many cultivars are grown as ornamentals in the West.

HARVESTING, SOURCING AND NATURAL RANGE

Roots are harvested in spring and autumn from wild plants mainly sourced from the northeastern provinces of Liaoning, Hebei and Shandong, although the species grows throughout China except the tropical south (Anhui, Gansu, Hebei, Heilongjiang, Henan, Hubei, Inner Mongolia, Jiangsu, Jiangxi, Jilin, Liaoning, Ningxia, Shaanxi, Shandong, Shanxi, Sichuan and Xinjiang); also wild in the Korean Peninsula, Mongolia and Russia (Far East). The species' habitat is thickets, grassland and open temperate forests.

CONSERVATION STATUS

China: Least Concern (MEPCAS, 2013).
Global: Not Evaluated (IUCN Red List, accessed 10/4/2015).

Dictamnus dasycarpus with emerging flower spikes

D. dasycarpus in flower

Flower with dark oil glands clearly visible on sepals, petal undersides and stamens

D. dasycarpus with developing fruit head

Densely glandular fruiting head

TCM PROPERTIES AND USE

***properties*:** **'bitter and cold'**
***channels*:** **'Spleen, Stomach and Bladder'**
***actions*:** **'clears heat, eliminates toxins'**
***indications*:** **skin sores, abscess; jaundice, haematuria; vaginal itching; (T)**

DRUG MORPHOLOGY

CRUDE DRUG/DECOCTING PIECES-1 (DRIED ROOT-BARK): root in cylindrical quills (5–15 cm long, 1–2 cm diam., 2–5 mm thick) **[1]**; cut or broken ends frequently oblique **[2]**. Externally somewhat smooth or slightly rough, greyish-white or pale greyish-yellow **[3]**, with indistinct longitudinal wrinkles, occasional rootlet scars **[4]**, and zones of small raised granular dots **[5]**. Internal surface almost white, finely striated longitudinally. *Texture*: brittle; producing fine dust when broken. *Fracture*: almost white, uneven and sometimes with concentric fractures; minutely crystalline. *Odour*: 'goat-like'. *Taste*: slightly bitter. ***Good quality*:** thick root-bark with a greyish-white cortex.

DECOCTING PIECES-2 (DRIED, TRANSVERSELY SLICED ROOT-BARK): ring-shaped or strongly curled (quilled or double-quilled) (1–2 cm diam., 2–5 mm thick). Outer uncut surface smooth or slightly rough, greyish-white or pale greyish-yellow **[7]**; cut surface even or sometimes lamellar (with concentric fractures) **[6]** inner surface almost white, finely striated longitudinally. *Texture* etc. as for the crude drug.

PROCESSING

After removal of coarse outer bark and central xylem the inner root-bark is dried (= crude drug decocting pieces-1); sometimes the root-bark is then sliced transversely into 3–10 mm long pieces (decocting pieces-2).

OTHER COMMON NAMES

北鲜皮 Bei Xian Pi; 八股牛 Ba Gu Niu; Cortex Dictamni; dense-fruit dittany (SCN).

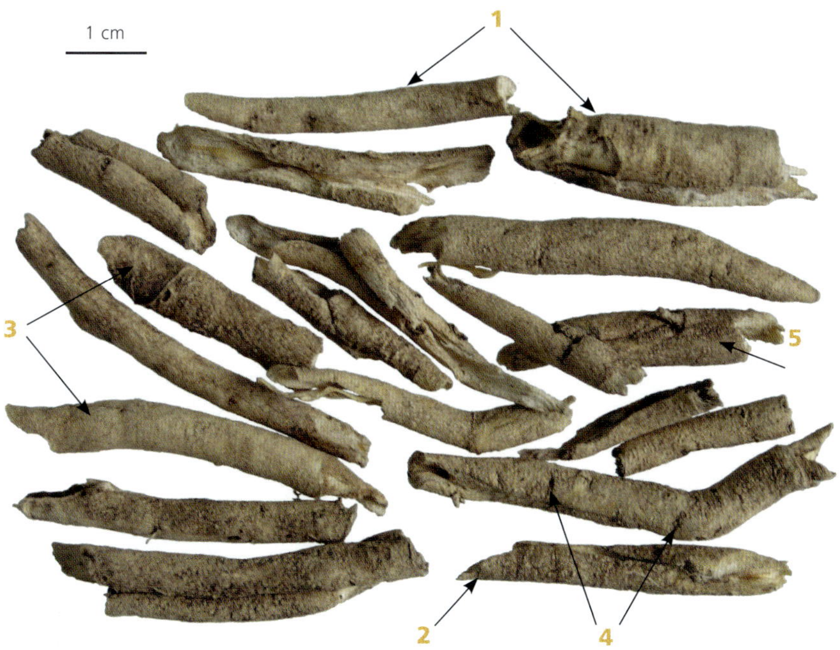

Crude drug/decocting pieces-1

Decocting pieces-2: transverse slices of root-bark

UNOFFICIAL SUBSTITUTES

These two are easily distinguished (as whole plants) from the official species but the occurrence of their root-barks as substitutes or adulterants of Bai Xian Pi in international trade is uncertain:

1. *Cynanchum chinense* R. Br. (family Apocynaceae). A common climbing plant to 4 m with white latex; leaves heart-shaped (not divided into lobes); flowers white. **Possibly toxic**.
2. *Caragana sinica* (Buc'hoz) Rehder (family Fabaceae). A common shrub to 2 m tall. Leaves with 4 leaflets; flowers solitary, yellow.

Chun Pi 椿皮

Ailanthi Cortex

Drug part: STEM-BARK or ROOT-BARK

Ailanthus altissima (Mill.) Swingle

Simaroubaceae

synonym: *Toxicodendron altissimum* Mill.

taxonomic/nomenclatural note: 3 varieties are recognised in the *FoC* and *Kew-MPNS* 25/8/2016.

Species of *Ailanthus* wild in China: **6**

PLANT DESCRIPTION

A deciduous tree (up to 20 m tall) with dark grey bark. *Leaves* large, alternate and compound (40–60 cm long) with 6–13 pairs of virtually opposite leaflets ending in a terminal leaflet; leaflet blades ovate to lance-shaped, often with 2 small glands towards base of each leaflet forming a slightly angular basal shoulder; aromatic when rubbed. Tiny *flowers* are packed into large and conspicuous, branched clusters (10–30 cm long) arising from leaf axils. *Flowers* single-sexed or bisexual, radially symmetrical and short-stalked; sepals 5 and tiny (0.5–1 mm long); petals 5, radiating, slightly larger petals (2–2.5 mm long), pale green and valve-shaped; stamens 10 and longer than petals; styles 2–5; stigmas fused and 5-lobed. *Fruit* a dry, oblong pod (samara) (3–4.5 cm long; 1–1.2 cm diam.) with a membranous wing and a single flattened-globose *seed* in middle of the wing. Fl. Apr–May, fr. Aug–Oct.

HARVESTING, SOURCING AND NATURAL RANGE

Stem-bark or root-bark is harvested year round from wild or cultivated trees with the best quality harvested in spring. The species is endemic to China and wild throughout most of the country except the provinces of Hainan, Heilongjiang, Jilin, Ningxia and Qinghai. It grows in many habitats between 100–2,500 m altitude and is widely naturalised and cultivated as an ornamental worldwide.

CONSERVATION STATUS

Endemic to China: Least Concern (MEPCAS, 2013); Not Evaluated (IUCN Red List, accessed 20/8/2016).

Ailanthus altissima fruiting in the wild in China

Fruiting branch of *A. altissima*

Close-up of branching inflorescence

Branch with dry, winged, one-seeded fruits

Leaflet bases

TCM PROPERTIES AND USE

properties: **'bitter, sour and cold'**
channels: **'Large Intestine, Stomach and Liver'**
actions: **'clears heat, dries damp, binds the intestines, stops bleeding'**

indications: **leucorrhoea; diarrhoea, blood in stools**

DRUG MORPHOLOGY

CRUDE DRUG (DRIED BARK)

Root-bark: pieces of variable size, sometimes quilled (0.3–1 cm thick). External surface greyish **[1]** (if outer coarse bark still intact) or pale yellow to yellowish-brown (if outer coarse bark removed); the former is rough with many irregular longitudinal and transverse fissures **[2]** and occasional raised lenticels **[3]**. Internal surface pale yellow **[4]**, smoother and densely pitted throughout with small rounded pores or spindle-shaped fissures [hl] **[5]**. *Stem-bark*: similar to root-bark but usually flattish not quilled (0.5–2 cm thick). *Texture*: hard but fragile. *Odour*: indistinct. *Taste*: bitter. ***Good quality***: harvested in spring; inner bark yellowish-white and outer cork layer removed.

DECOCTING PIECES-1 (DRIED SLICED BARK): as for the crude drug but cut into smaller slivers of variable shape and size (0.3–1 cm thick) and outer surface pale yellowish **[6]** with traces of lenticels occasional **[7]**; depending on the style of slicing, remnants of dark, coarse outer bark may be visible.

DECOCTING PIECES-2 (BRAN-FRIED SLICED BARK): as for decocting pieces-1 but cut surfaces darker yellow or yellowish-brown **[8]** and slightly aromatic. Depending on style of slicing, remnants of dark, coarse outer bark **[9]** sometimes visible.

PROCESSING

The coarse outer bark is sometimes removed before it is dried in the sun (= crude drug) then later softened thoroughly to slice or cut into decocting pieces before drying again. The latter may subsequently be dry-fried with bran (= 麸炒椿皮 Fu Chao Chun Pi).

OTHER COMMON NAMES

椿杆皮 Chun Gan Pi (stem-bark); 椿根皮 Chun Gen Pi (root-bark); 樗皮 Chu Pi ; 樗白皮 Chu Bai Pi; 椿皮根 Chun Pi Gen; Cortex Ailanthi; tree-of-heaven bark (CP2010).

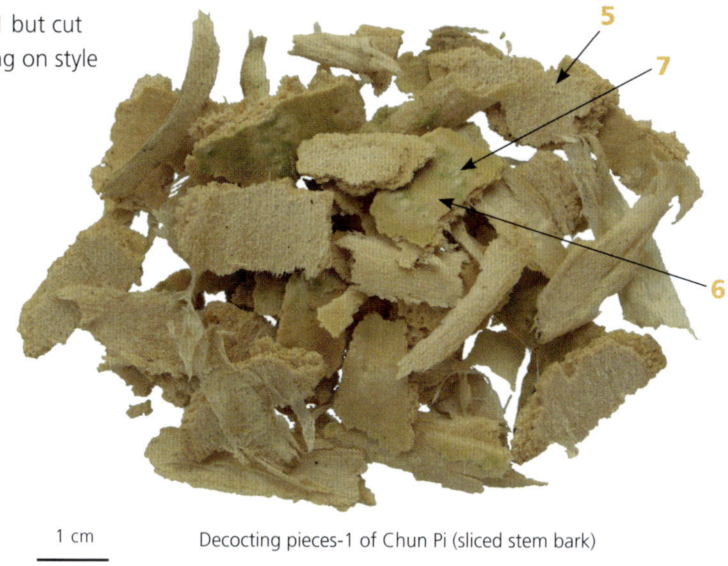

Decocting pieces-1 of Chun Pi (sliced stem bark)

Crude drug of Chun Pi (stem bark)

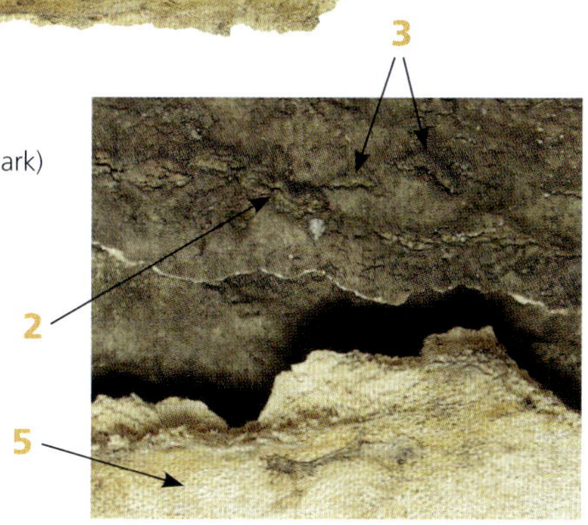

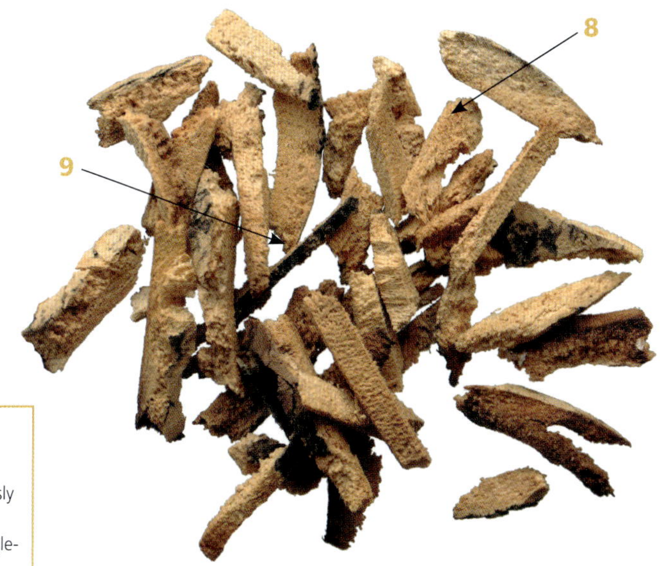

Decocting pieces-2 of Fu Chao Chun Pi (bran-fried stem bark)

UNOFFICIAL SUBSTITUTE

Toona sinensis (Juss.) M. Roem. (香椿 Xiang Chun); family Meliaceae. A common and widely cultivated tree easily confused with the Pharmacopoeia species owing to the similarity of its leaves. *Bark* 'obnoxiously pungent' (*FoC*); **inner bark pinkish;** sapwood cream (highly aromatic). ***Leaves* with 8–20 pairs of leaflets**; basal glands absent (emerging leaves aromatic and widely used locally in cooking). *Flowers* single-sexed; petals white or flushed pink; stamens 5 or 6. *Fruits* not winged, many-seeded.

Di Gu Pi 地骨皮 Lycii Cortex

Drug part: ROOT-BARK

PHARMACOPOEIA SPECIES 1

Lycium chinense Mill.

Solanaceae

taxonomic/nomenclatural note: *FoC* and *Kew-MPNS* recognise two varieties

Species of *Lycium* wild in China: 7

PLANT DESCRIPTION

A deciduous, many-branched and thorny perennial shrub (0.5–1 m tall). *Leaves* simple, solitary or clustered, ovate, rhombic, lance-shaped or linear lance-shaped (1.5–5 cm long; 0.5–2.5 wide cm [larger in medicinal cultivars: 10 cm long; 4 cm wide]); margins entire. *Flowers* solitary or clustered, arising from leaf axils and stalked; basal part of each flower with a green **3-5-lobed calyx** (3–4 mm long); flower pale purple (0.9–1.2 cm diam.) with a corolla tube and 5 spreading corolla lobes **(tube shorter or almost shorter (8–10 mm long) than the corolla lobes;** stamens 5; style 1. *Fruit* a fleshy, often glossy, red berry, oblong or ovoid (0.7–1.5 cm long; 5–8 mm diam. [larger in medicinal cultivars: c. 2.2 cm long; 1 cm diam.]; calyx persistent. *Seeds* many, yellowish, 2.5–3 mm long.

HARVESTING, SOURCING AND NATURAL RANGE

Root-bark is harvested from wild plants which grow on waste ground and in disturbed places, often near habitation. The species is native to China: Anhui, Fujian, Gansu, Guangdong, Guangxi, Guizhou, Hainan, Hebei, Heilongjiang, Henan, Hubei, Hunan, Inner Mongolia, Jiangsu, Jiangxi, Jilin, Liaoning, Ningxia, Qinghai, Shaanxi, Shanxi, Sichuan, Taiwan, Yunnan and Zhejiang [also native in Japan, Korean Peninsula, Mongolia, Nepal, Pakistan, Thailand; SW Asia and Europe].

CONSERVATION STATUS

China: Least Concern (MEPCAS, 2013).

Global: Not Evaluated (IUCN Red List, accessed 10/9/2016).

PHARMACOPOEIA SPECIES 2

L. barbarum L.

Solanaceae

taxonomic/nomenclatural note: *FoC* and *Kew-MPNS* recognise the varieties 'var. *auranticarpum* K. F. Ching' and the type variety 'var. *barbarum*'; *Kew-MPNS* also recognises a third variety (published after the *FoC* account) 'var. *implicatum* T. Y. Chen & Xu L. Jiang'

PLANT DESCRIPTION

A very similar species to PS1 and often difficult to distinguish owing to its widespread cultivation in China which has produced many cultivars.

Main differences: **calyx usually 2-lobed; corolla tube 8–10 mm long** and **obviously longer than lobes.**

HARVESTING, SOURCING AND NATURAL RANGE

Root-bark is thought to be mostly harvested from medicinal crop plants (primarily grown for their fruit, the source of the separate TCM drug 'Gou Qi Zi' (Lycii Fructus), page 632). The three varieties of the wild species are endemic to China and grow in disturbed places often near habitation (for further details see 'Gou Qi Zi' page 632).

CONSERVATION STATUS

China: Least Concern (var. *barbarum*); Data Deficient (var. *auranticarpum*) (MEPCAS, 2013).

Global: Not Evaluated (IUCN Red List, accessed 10/9/2016).

UNOFFICIAL SUBSTITUTE

Periploca sepium Bunge

Apocynaceae

PLANT DESCRIPTION

Easily distinguished from PS1 and PS2. A thornless shrub with milky sap. **Leaves opposite, much larger** (5–9 cm long; 1.5–2.5 cm wide) with 20–25 pairs of distinct lateral veins; **flowers distinctly globe-shaped** (c. 1.5 cm diam.) with petals fused and reflexed; **fruit long, dry and cylindrical (7–12 cm long; c. 5 mm diam.) with 2 long slender follicles (fused at tip), dry and brown**; seeds many, each with a distinctive long, silk-like coma (c. 3 cm long).

CONSERVATION STATUS

Endemic to China: Least Concern (MEPCAS, 2013); Not Evaluated (IUCN Red List, accessed 10/9/2016).

SUBSTITUTION COMMENT. The root-bark of *P. sepium* is the source of the separate TCM drug 'Xiang Jia Pi' (Periplocae Cortex) page 530, and likely reasons for its use as a substitute include its ease of availability from the wild (it is common throughout most of the country) and its macroscopic similarity to the official drug (Di Gu Pi). Its clinical properties, however, are quite different as is its safety profile (see *Safety*) while its frequency of occurrence in international trade as a potentially harmful substitute is unclear. The species is endemic to China and occurs in all province except Guangdong, Guangxi, Hainan, and Taiwan.

TCM PROPERTIES AND USE

properties: **'sweet, bland and cold'**
channels: **'Lung and Kidney'**
actions: **'clears heat, deficient heat, cools Blood'**

indications: **fevers, night sweats; cough, haemoptysis, epistaxis**

Lycium chinense in flower

L. chinense in flower

L. chinense in fruit

Flowering plant of *L. barbarum*

Fruiting plant of *L. barbarum*

Periploca sepium in the wild, Shaanxi province

P. sepium in flower

P. sepium in fruit

Lycium chinense

Solanaceae

DRUG MORPHOLOGY

CRUDE DRUG/DECOCTING PIECES (DRIED ROOT-BARK): quilled or channelled **[1]** (3–10 cm long; 0.5–1.5 cm diam.; **1–3 mm thick [2]**). Externally yellowish-brown or greyish-yellow **[3]**, rough with numerous irregular fissures and scale-like layers easily removed **[4]**. Inner surface paler and **greyish-white to greyish-yellow [5]** with fine longitudinal striations. *Fracture*: uneven, easily broken and light. *Odour*: indistinct. *Taste*: sweet then bitter. ***Good quality***: thick, fleshy pieces devoid of the woody inner xylem.

PROCESSING

Roots are harvested in early spring or late autumn, the inner xylem removed and discarded and the root-bark sun-dried (= crude drug/decocting pieces).

OTHER COMMON NAMES

枸杞根皮 Gou Qi Gen Pi; 骨皮 Gu Pi; 枸杞 Gou Qi (plant name); 枸杞属 Gou Qi Shu; 南骨皮 Nan Gu Pi; Chinese boxthorn; Chinese wolfberry; Cortex Lycii; lycium (SCN).

SAFETY

An official healthfood in China (Anon., 2002a,b). Safety Class: 1); Interaction Class: A (Gardner & McGuffin, 2013).

L. barbarum

Solanaceae

DRUG MORPHOLOGY

CRUDE DRUG/DECOCTING PIECES (DRIED ROOT-BARK): as for *L. chinense* and cannot be distinguished using macroscopic characters alone.

PROCESSING

As for *L. chinense*.

OTHER COMMON NAMES

枸杞根皮 Gou Qi Gen Pi; 骨皮Gu Pi; 南骨皮 Nan Gu Pi; 宁夏枸杞 Ning Xia Gou Qi (plant name); Chinese wolfberry; Chinese wolfberry root-bark (CP2010); Cortex Lycii; lycium (SCN); matrimony vine.

SAFETY

An official healthfood in China (Anon., 2002a,b). Safety Class: 1; Interaction Class: A (Gardner & McGuffin, 2013). .

Periploca sepium

Apocynaceae

'DRUG SUBSTITUTE' MORPHOLOGY

DRIED ROOT-BARK: closely resembles root-bark of the two official species. *Main differences*: slightly thicker (2–4 mm **[6]** rather than 1–3 mm thick), externally yellowish brown or greyish brown **[7]**sometimes double-quilled **[8]**, the inner surface more **yellow [9]**, **odour strong and distinctive (vanilla-like)** and **taste bitter.** The last two characters are the easiest to discern but these are too subjective to be reliable. **Laboratory-based authentication methods are recommended.**

SUBSTITUTION COMMENT. The root-bark of this species is the source of the separate TCM drug 'Xiang Jia Pi' (Periplocae Cortex) see page 530. Likely reasons for its use as a substitute include its competitive price, ease of availability from cultivation, and morphological similarity to PS1 and PS2. Its clinical properties are quite different.

OTHER COMMON NAMES

杠柳 Gang Liu (plant name); **香加皮 Xiang Jia Pi**; 北五加皮 Bei Wu Jia Pi (northern Wu Jia Pi); Chinese silk vine; Cortex Periplocae; Periploca sepium (SCN).

SAFETY

Toxic (CP2015). Compared to the official drug (Di Gu Pi) the root-bark of this separate medicinal species (Xiang Jia Pi, see *Substitution Comment*) has a narrow therapeutic safety window (it contains cardiac glycosides and is not for long-term use). For this reason **its detection as an unofficial substitute is essential.** Xiang Jia Pi is officially forbidden as a food or healthfood in China (Anon., 2002).

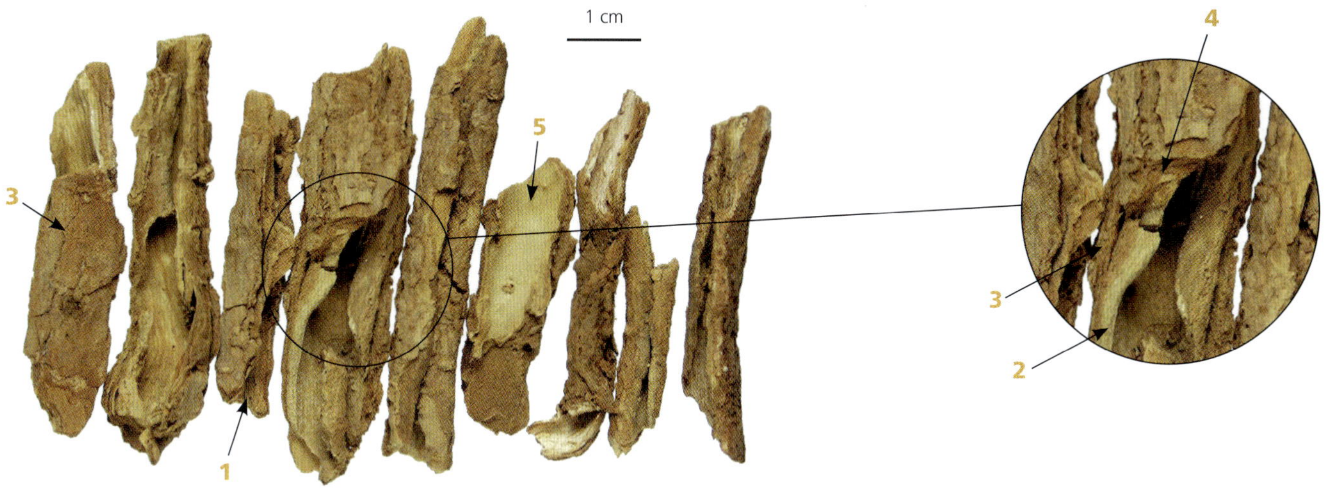

Crude drug/decocting pieces of Di Gu Pi (sourced from dried rootbark of *Lycium chinense*)

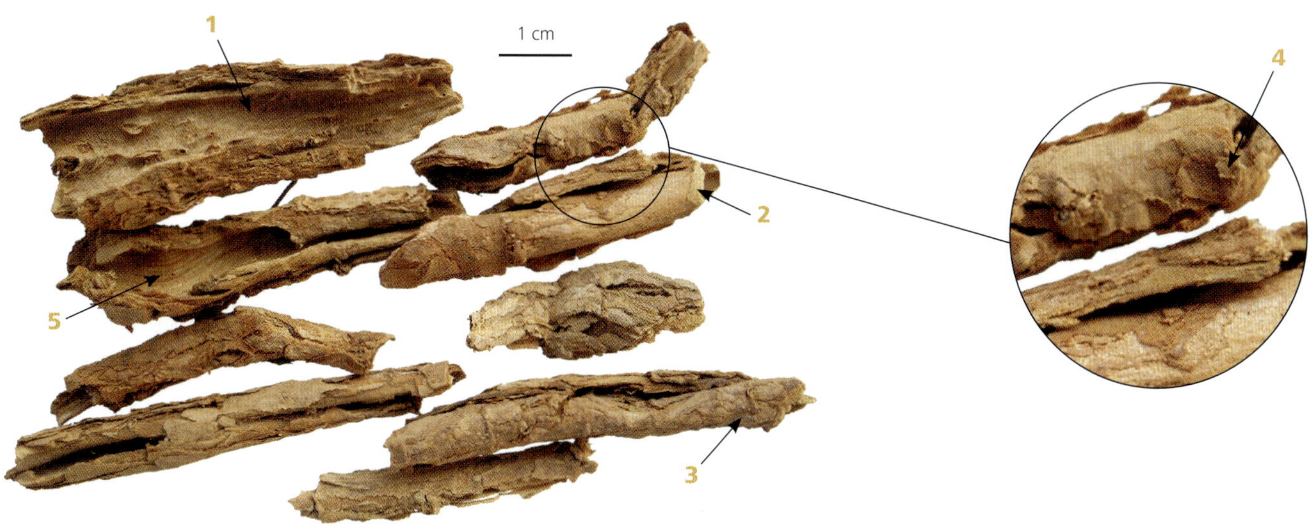

Crude drug/decocting pieces of Di Gu Pi (sourced from dried rootbark of *L. barbarum*)

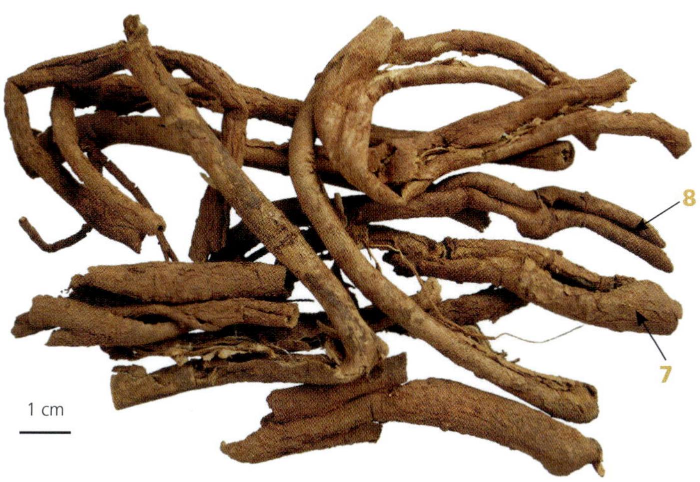

Dried root-bark of *Periploca sepium* (crude drug of 'Xiang Jai Pi')

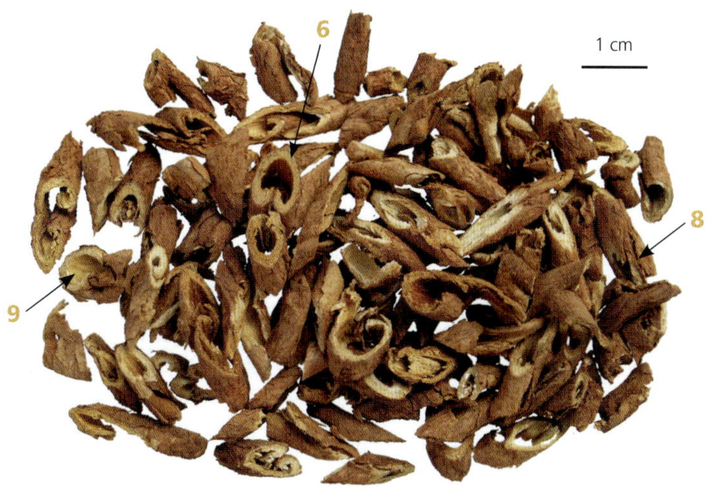

Sliced root-bark of *P. sepium* (decocting pieces of 'Xiang Jia Pi')

Du Zhong 杜仲

Eucommiae Cortex

Drug part: STEM-BARK

Eucommia ulmoides Oliv.

Eucommiaceae

PLANT DESCRIPTION

A deciduous tree (up to 20 m tall; up to 50 cm d.b.h.). *Bark* greyish-brown and rough. Distinctive silver-grey latex in most parts, easily visible as elastic parallel strands in cut bark and severed leaves. *Leaves* alternate, simple, stalked; blades usually elliptic, sometimes ovate (5–15 cm long; 2.5–7 cm wide), base rounded or wedge-shaped, tip abruptly pointed, lateral veins in 6-9 pairs, margins densely saw-toothed and teeth tipped with tiny glands. *Flowers* without petals (wind-pollinated), inconspicuous and superficially leaf-like; appearing before leaves; male and female flowers on separate trees. Male flowers in axillary clusters, stamens 5–12, linear (c. 1 cm long). Female flowers solitary, with a 2-lobed apex and 2 backward-pointing stigmas; ovary c. 1 cm long. *Fruits* in pendant clusters; each fruit (samara) dry, narrowly oblong or elliptical (2.5–3.5 cm long; 1–1.3 cm wide), flattened and winged. Single *seed*, linear and flattened. Fl. Mar–May, fr. Jun–Nov.

Species of *Eucommia* wild in China: 1

HARVESTING, SOURCING AND NATURAL RANGE

Bark is harvested from cultivated trees mostly grown extensively on plantations throughout the natural range of the species (also naturalised). Traditional *Di Dao* sources: Guizhou, Shaanxi and Sichuan.

The wild species is endemic to China (Anhui, Gansu, Guizhou, Henan, Hubei, Hunan, Jiangxi, Shaanxi, Shanxi, Sichuan, Yunnan and Zhejiang) and grows in mixed, open forests and thickets in dry ravines (100–2,000 m altitude).

CONSERVATION STATUS

Wild populations have declined markedly due to habitat loss, unsustainable harvesting of the medicinal bark, felling for fuel and timber and the many uses of the latex in dentistry and the electrical and construction industries.

Endemic to China: Lower Risk/Near Threatened "v.2.3" (IUCN Red List accessed 8/4/2015); Vulnerable (MEPCAS, 2013).

Mature trees of *Eucommia ulmoides*

Torn leaf of *E. ulmoides* revealing silver-grey latex

Close-up of bark of *E. ulmoides*

Fruiting branch of *E. ulmoides*

TCM PROPERTIES AND USE

properties: 'sweet and warm'
channels: 'Kidney and Liver'
actions: 'tonifies Kidney, warms yang, invigorates Blood'

indications: lower back pain, fractures; threatened miscarriage; impotence, incontinence; hypertension

DRUG MORPHOLOGY

CRUDE DRUG (BARK): dried bark pieces of variable sizes (usually c. 3–7 mm thick); flattened or margins inrolled (quilled). Externally pale greyish-brown with longitudinal striations or fissures [1]. Lenticels distinct when outer bark still present [2]. Inner surface smooth and purplish-brown. *Texture*: fragile. *Fracture*: reveals silver-grey elastic thread-like latex [3] 'Jiao Si' (glue threads). *Odour*: indistinct. *Taste*: slightly bitter. **Good quality:** thick bark with purplish-brown cambium [4] (inner surface) and abundant silvery-grey thread-like latex.

DECOCTING SLICES-1 (SLICED BARK): as for crude drug (with or without outer coarse bark removed) and cut into flat, narrow or broadly oblong slices; latex visible.

DECOCTING SLICES-2 (SALTED BARK — YAN DU ZHONG): as for decocting slices above but latex less obvious [5], shape often slightly curled, surface sometimes charred and taste slightly salty [6].

PROCESSING

Stem-bark is harvested in rectangular strips from April to June, the coarse outer bark (rhytidome) sometimes removed and discarded, then sun-dried in heaps until the inner bark becomes purplish-brown (= crude drug). Decocting pieces are prepared by scoring or cutting the bark in various styles. Further processing by frying with salt water until the latex is no longer visible is called 盐杜仲 Yan Du Zhong (salted Du Zhong). It is common practice today not to remove the coarse outer bark (see photo below left).

OTHER COMMON NAMES

Chuan Du Zhong (Eucommia sourced from Sichuan or Guizhou); Du Zhong Tan (charred Eucommia); Chinese rubber tree; Cortex Eucommiae; eucommia (SCN); Eucommia bark (CP2010).

SAFETY

Safety Class: 1; Interaction Class: A (Gardner & McGuffin, 2013).

Crude drug of Du Zhong in quill form with outer coarse bark intact (dried bark of *Eucommia ulmoides*)

Decocting squares of Du Zhong (sliced dried bark of *E. ulmoides*)

Decocting strips of Du Zhong (sliced dried bark of *E. ulmoides*)

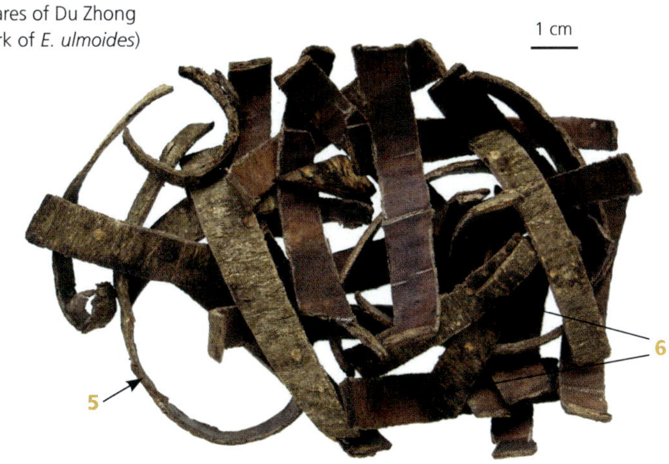

Salted Du Zhong (Yan Du Zhong)

Guan Huang Bo 关黄柏 Phellodendri Amurensis Cortex

Drug part: DE-CORKED STEM-BARK

Phellodendron amurense Rupr.

Rutaceae

PLANT DESCRIPTION

A deciduous tree (10–30 m tall) with a spreading canopy. *Outer bark* greyish-brown and with a **thick, fissured cork layer**, revealing a bright yellow phloem layer when removed. *Leaves* opposite and pinnate, with 3–6 opposite pairs of ovate-elliptic or lance-shaped leaflets arranged along a **slender** central axis (rachis); terminal leaflet present; leaflet blades papery or thinly papery with oil glands along margins; margins entire or with minute rounded teeth; leaflet tip narrowly pointed. Male and female flowers on separate plants. *Flowers* in **lax**, terminal heads (8.5–13.5 cm long; 6.5–9 cm wide), main stalk **slender** (5–8.5 cm long), individual flower stalks (pedicels) **slender and at least 1 cm long**; each flower yellowish-green, small and inconspicuous; sepals 5; petals 5. Fruit-heads **lax**, with a slender stalk (peduncle); *fruit* spherical (1 cm diam.), fleshy, black or purplish-black. Fl. May–Jun, fr. Sep–Oct.

Species of *Phellodendron* wild in China: 2

HARVESTING, SOURCING AND NATURAL RANGE

Bark is harvested mainly from plantation-grown trees primarily in NE China (Heilongjiang, Jilin and Liaoning provinces).

The wild species grows in forests and thickets in mountains and river valleys in C, N and NE China (Anhui, Hebei, Heilongjiang, Henan, Inner Mongolia, Jilin, Liaoning, Shandong and Shanxi); [also native to Japan, Korean Peninsula and Russia (Far East)]. In China, its primarily north-eastern distribution distinguishes it from the more southerly *P. chinense* and with whose bark this drug is regularly confused (see Huang Bo — Phellodendri Chinensis Cortex page 512).

CONSERVATION STATUS

China: Vulnerable (MEPCAS, 2013); 'rapidly declining' (Leung & Cheng, 2008); Vulnerable (Fu, 1992).
Global: Not Evaluated (IUCN Red List, accessed 25/4/2015).

PROTECTED STATUS

China: Cat. II (SEPA, 1999).

Female tree of *P. amurense* with young leaves and persistent fruits of last season

Thick cork bark with bright yellow phloem below

Pinnate leaves and immature fruits

UNOFFICIAL SUBSTITUTES

The most common substitution arises from confusion with the bark of the closely related tree ***Phellodendron chinense*** C. K. Schneid. from which the separate (but very similar) TCM drug Huang Bo is sourced (for a more detailed explanation see 'Huang Bo' (Phellodendri Chinensis Cortex) page 512.

Lesser known substitutes include:
Berberis spp. — family Berberidaceae (the bright yellow wood of the stems has occasionally been used locally in China).
Ipomoea batatas (L.) Lam. — family Convolvulaceae (dyed sweet potato tubers have been reportedly traded as counterfeit Guan Huang Bo)

TCM PROPERTIES AND USE

properties: **'bitter and cold'**
channels: **'Kidney, Bladder and Large Intestine'**
actions: **'clears heat, drains damp, clear fire, eliminates toxins'**

indications: **diarrhoea; jaundice; dysuria; leucorrhoea; night sweats; sores and abscesses; (T)**

DRUG MORPHOLOGY

CRUDE DRUG (DRIED BARK): curved or rectangular bark pieces of variable length and width [1] (c. 2–6 mm thick). Outer surface (phloem) relatively smooth, yellow, yellowish-green or yellowish-brown [2] with occasional remnants of brown or greyish-brown outer cork bark typically raised, fissured and slightly spongy [3]. Lenticels occasionally visible in the outer bark [4]. Inner surface smooth and yellow or yellowish-brown throughout with fine longitudinal striations [5]. *Texture*: light and hard. *Fracture*: fibrous; bright yellow or yellowish-green. *Odour*: indistinct. *Taste*: very bitter; becoming viscous when chewed.

DECOCTING PIECES-1 (SLICED BARK): transversely cut slices (c.1–6 mm thick) either slightly curved or semi-quilled [6], longitudinally cut slices flatter and often rectangular. All other characters as for crude drug. Greyish-brown remnants of corky outer bark [7] a key character to distinguish this drug from the very similar Huang Bo (Phellodendri Chinensis Cortex) page 512.

DECOCTING PIECES-2 (SALT-PROCESSED BARK — YAN GUAN HUANG BO): as for sliced bark but with occasional charred areas [8], inner surface darker yellow and taste slightly salty and very bitter.

DECOCTING PIECES-3 (CARBONISED BARK — GUAN HUANG BO TAN): as for sliced bark but charred black externally (obscuring most identification characters [9]) but corky outer remnants still often discernible [10]; internally dark brown or brownish-black. *Fracture*: charred brown and brittle. *Taste*: bitter and astringent.

PROCESSING

The coarse outer cork layer is removed, leaving the inner bark to be sun-dried (= crude drug) and which is then later cut into transverse or longitudinal decocting pieces; further optional processing by steeping in salt-water then dry-frying produces 盐关黄柏 Yan Guan Huang Bo or carbonising produces 关黄柏炭 Guan Huang Bo Tan.

OTHER COMMON NAMES

黄柏 Huang Bo; Amur cork-tree (CP2010); Guang Bai; Guang Huang Bai; Huang Bai; Cortex Phellodendri Amurensis; phellodendron (SCN).

SAFETY

Safety Class: 1; Interaction Class: A (Gardner & McGuffin, 2013).

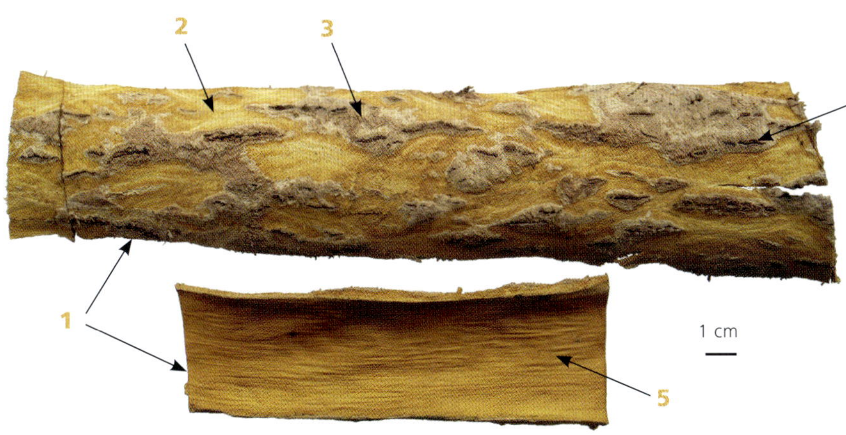

Crude drug

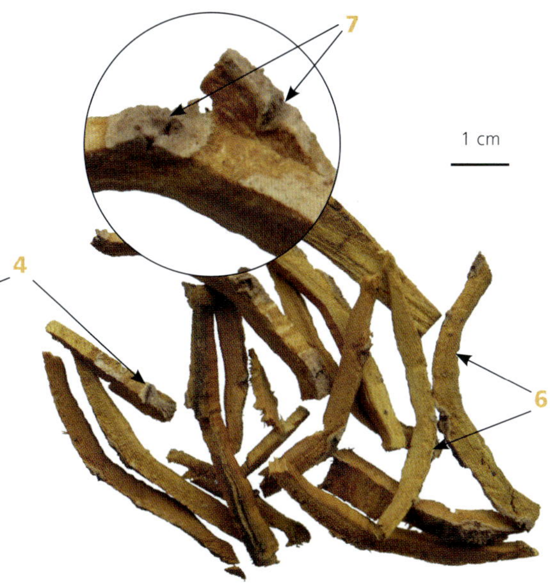

Decocting pieces-1 (transversely sliced)

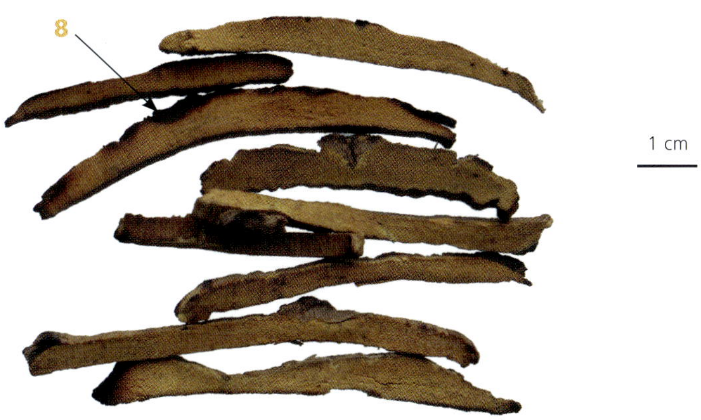

Decocting pieces-2 — Yan Guan Huang Bo (salt-processed)

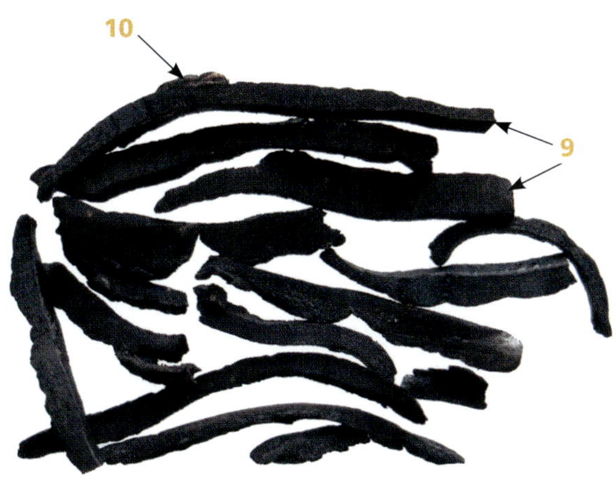

Decocting pieces-3 — Guan Huang Bo Tan (carbonised)

He Huan Pi 合欢皮

Albiziae Cortex

Drug part: STEM-BARK

PHARMACOPOEIA SPECIES

Albizia julibrissin Durazz.

Fabaceae

PLANT DESCRIPTION

A deciduous tree with spreading crown (up to 16 m tall). *Stipules* falling, smaller than leaflets. *Leaves* alternate, divided into **4–12 pairs of pinnae** (sometimes 20 in cultivated plants), **each pinna with 10–30 pairs of linear to oblong leaflets** (0.6–1.2 cm long; 0.1–0.4 cm wide); leaflets sharply tipped, margins fringed with hairs. *Inflorescences* conspicuous, globose, **predominantly pink** (top part of filaments of stamens) and long-stalked. *Each flower* with calyx funnel-shaped and 5-toothed (c. 3 mm long); corolla funnel-shaped and 5-lobed (c. 8 mm long), pale yellowish-green; stamens many with long **filaments** (c. 2.5 cm) **basally white, becoming bright pink in outer parts.** *Fruit* a long flat pod (9–15 cm long; 1.5–2.5 cm wide). Fl. May–Jul, fr. Aug–Oct.

Species of *Albizia* wild in China: 14

HARVESTING, SOURCING AND NATURAL RANGE

Stem bark is harvested in summer and autumn usually from naturalised (not native) trees whose habitat is river banks and disturbed places at low altitude. Native throughout much of China [C, E and SW Asia] and widely cultivated in Anhui, Fujian, Gansu, Guizhou, Henan, Hubei, Hunan, Jiangsu, Jiangxi, ?Liaoning, ?Shaanxi, Shandong, Shanxi, Taiwan, Yunnan and Zhejiang.

CONSERVATION STATUS

China: Introduced.

Global: Not Evaluated (IUCN Red List; accessed 10/8/2016). An invasive species in the USA and Japan.

UNOFFICIAL SUBSTITUTE

A. kalkora (Roxb.) Prain

Fabaceae

synonym: *Mimosa kalkora* Roxb.

PLANT DESCRIPTION

Very similar to the Pharmacopoeia species. *Main differences*: a shrub or small tree (3–8 m tall) with **inflorescences predominantly white, turning yellow** (filaments of the stamens). *Leaves* divided into **fewer pinnae (2–4 pairs)** and **fewer (5–14) but larger (0.8–4.5 mm long; 0.7–2 mm wide)**. *Calyx* and *petal tubes* **densely clothed in long soft brown hairs**. Fl. May–Jun, fr. Aug–Oct.

HARVESTING, SOURCING AND NATURAL RANGE

A common and widely distributed species of thickets and open forests (0–2,600 m altitude); N, SC and SE China (Anhui, Fujian, Gansu, Guangdong, Guangxi, Guizhou, Hainan, Henan, Hubei, Hunan, Jiangsu, Jiangxi, Shaanxi, Shandong, Shanxi, Sichuan, Taiwan and Zhejiang) [also native in India, Japan, Myanmar and Vietnam].

CONSERVATION STATUS

China: Least Concern (MEPCAS, 2013).

Global: Not Evaluated (IUCN Red List; accessed 10/8/2016).

TCM PROPERTIES AND USE

properties: **'sweet and neutral'**
channels: **'Heart and Liver'**
actions: **'calms the shen, regulates qi, invigorates Blood'**

indications: **irritability, depression, insomnia; pain, oedema due to physical injury; (T)**

Cultivated *Albizia julibrissin* tree in full flower

A. julibrissin branches

A. julibrissin

Albizia kalkora in full flower, Sichuan province

A. kalkora flowering branches

Albizia julibrissin

Fabaceae

DRUG MORPHOLOGY

CRUDE DRUG (DRIED STEM-BARK): long rectangular, channelled [1] and slightly curved. Outer surface of bark greyish-brown and densely covered in orangish-brown or brown, transverse, **elliptical or oval lenticels** [2]. Inner bark pale yellowish-white or yellowish-brown, smooth and finely striated longitudinally [3]. *Texture*: brittle. *Fracture*: fibrous, pale yellowish-white or yellowish-brown. *Odour*: slightly aromatic. *Taste:* slightly irritant and somewhat astringent, with unpleasant sensation in throat. ***Good quality***: lenticels obvious and outer cork bark removed.

DECOCTING PIECES (SLICED STEM-BARK): as for the crude drug but cut into 4–5 mm wide transverse slices; curved, single- or doloble-quilled [4]. Lenticels may or may not be present in individual pieces

PROCESSING

Bark is sun-dried (= crude drug) then washed, soaked in water, sliced and dried again (= decocting pieces).

OTHER COMMON NAMES

合欢树 He Huan Shu; 夜合树 Ye He Shu; Cortex Albiziae; pink silk tree; silk tree (SCN).

SAFETY

Safety Class: 2b; Interaction Class: A (Gardner & McGuffin, 2013).

A. kalkora

Fabaceae

'DRUG SUBSTITUTE' MORPHOLOGY

Transverse lenticels larger and more oblong [5]. This character is very subjective, however, and therefore unreliable. Laboratory-based authentication methods are recommended both for large stem bark pieces and smaller decocting-size transverse slices.

PROCESSING

As for the Pharmacopoeia species.

OTHER COMMON NAMES

山合欢 Shan He Huan.

SUBSTITUTION COMMENT. While the stem-bark of this and the official drug are easily confused, the living trees of these two species are easily distinguished (when leaves [and flowers] are present) suggesting that substitution may be intentional unless harvesting occurs after leaves have fallen when the two species are easily confused.

Crude drug of He Huan Pi sourced from *Albizia julibrissin*

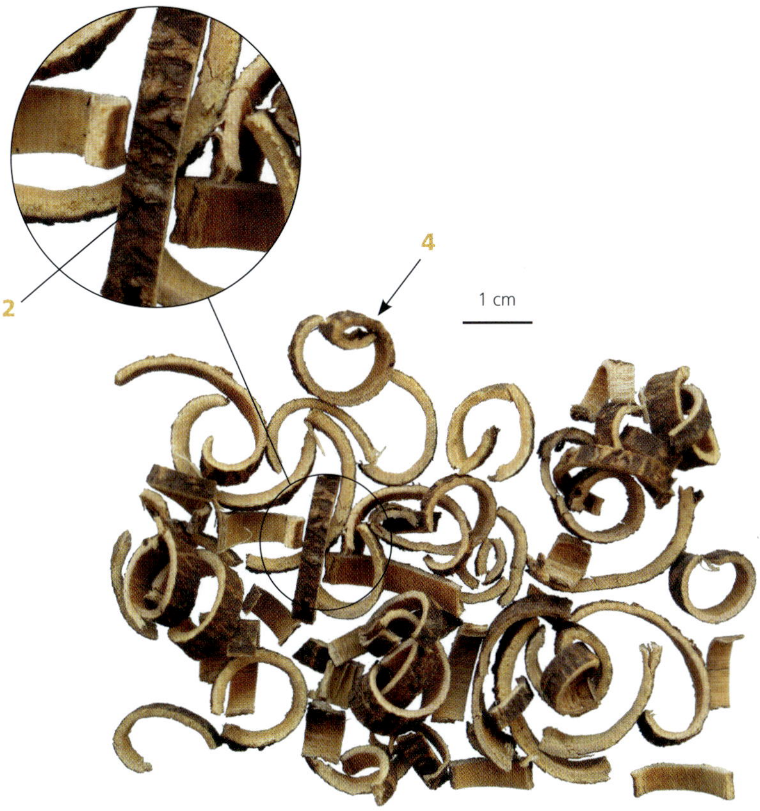

Decocting pieces of He Huan Pi sourced from *A. julibrissin*

Bark of *Albizia kalkora*

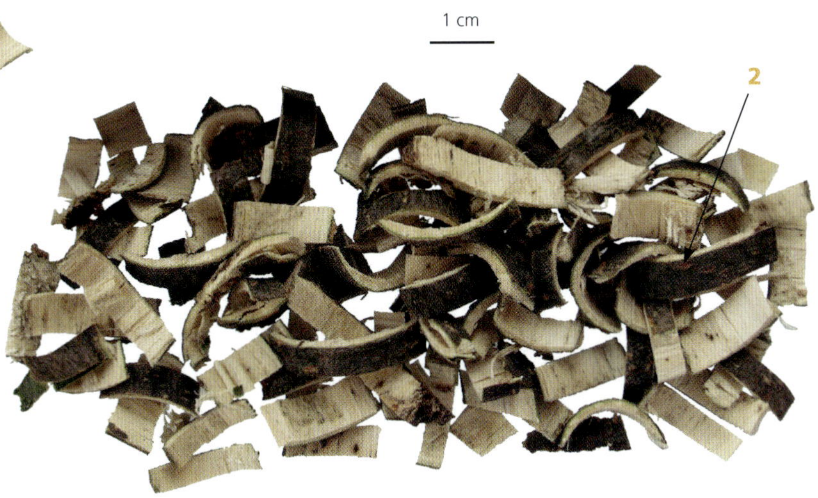

Sliced bark of *A. kalkora*

PHARMACOPOEIA SPECIES 1

Magnolia officinalis Rehder & E. H. Wilson var. *officinalis*

Magnoliaceae

synonym: *Houpoëa officinalis* (Rehder & E. H. Wilson) N. H. Xia & C. Y. Wu (*FoC* accepted name)
taxonomic/nomenclatural note: *Kew-MPNS* accepted name; 23/8/2016; in CP2015 as '*M. officinalis*' Rehd. et Wils.'

Species of *Magnolia* wild in China: 3

PLANT DESCRIPTION

A deciduous tree (up to 20 m tall); bark brown, thick and relatively smooth. *Leaf stalks* thick and strong (2.5–4 cm long) with stipular scar along c. 2/3rds of its length. *Leaves* spirally arranged, usually in clusters of 7–9 at ends of twigs, folded when in bud; leaf blades large (22–45 cm long; 10–24 cm wide) oblong-obovate, slightly leathery, lower surface greyish-green or bluish-green, softly downy; upper surface green and hairless; base wedge-shaped, margin entire or slightly wavy; tip broadly rounded with a tiny short central point. *Flowers* solitary, bisexual, large (10–15 cm diam.), showy and fragrant; 'petals' 9–12 (rarely up to 17), mostly white, thickly fleshy, outer 3 'petals' sometimes with flushes of pale green and usually spreading before inner ones; inner 'petals' obovate to spoon-shaped (7–8.5 cm long; 3–4.5 cm wide); stamens many and densely packed (2–3 cm long) below the pistils with filaments thick, flat and red; stamens often falling off; pistils many, forming an ovoid mass above the stamens (2.5–3 cm long). *Fruit* when mature, a cylindrical head (9–15 cm long) with many distinct, leathery or somewhat woody follicles (carpels), each mature follicle with a 3–4 mm beak. *Seeds* exposed, red and fleshy (c. 1 cm long), 1–12 per follicle. Fl. May–Jun, fr. Aug–Oct.

HARVESTING, SOURCING AND NATURAL RANGE

Bark is harvested from April to June from trees grown extensively on medicinal plantations in Chongqing, Fujian, Guizhou, Henan, Hubei and southern Sichuan; traditional *Di Dao* sites occur in Chongqing (districts of Chengkou, Fengdu, Guangyuan and Yingjing). Also cultivated as an ornamental and for its timber.

The species is endemic to China and grows in broad-leaved deciduous forests (300–1500 m altitude) in Gansu, Guizhou, Henan, Hubei, Hunan, Shaanxi and Sichuan.

CONSERVATION STATUS

Endemic to China: Lower Risk/Near Threatened (IUCN Red List, accessed 2/9/2015); Least Concern (MEPCAS, 2013; as *Houpoëa officinalis*). Declining due to bark-stripping for medicinal use.

PROTECTED STATUS

China: Cat. II (SEPA, 1999).

PHARMACOPOEIA SPECIES 2

Magnolia officinalis var. *biloba* Rehder & E. H. Wilson

Magnoliaceae

synonyms: *M. biloba* (Rehder & E. H. Wilson) W. C. Cheng & Y. W. Law; *M. officinalis* subsp. *biloba* (Rehder & E. H. Wilson) W. C. Cheng & Y. W. Law
taxonomic/nomenclatural note: *Kew-MPNS* accepted name, 23/8/2016

PLANT DESCRIPTION

Similar to PS1 but leaves notched at apex to form two shallowly rounded lobes. Fl. Apr.–May; Fr. Oct.

HARVESTING, SOURCING AND NATURAL RANGE

As for PS1, with the main cultivated production provinces being Jiangxi and Zhejiang.

The distribution of this wild endemic variety is Anhui, Fujian, Guangdong, Guangxi, Hunan, Jiangxi and Zhejiang provinces.

CONSERVATION STATUS:

Endemic to China: as for PS1.

PROTECTED STATUS

China: Cat. II (SEPA, 1999).

Flowers of both varieties are the source of the separate TCM drug 'Hou Po Hua' (Magnolia Officinalis Flos), not included in this book.

EHTPA 4/14

TCM PROPERTIES AND USE

properties: **'bitter, pungent, aromatic and warm'**
channels: **'Spleen, Stomach, Lung and Large Intestine'**
actions: **'dries damp, regulates qi, dispels stasis, resolves phlegm'**

indications: **abdominal distension, vomiting, diarrhoea; cough, phlegm**

Magnolia officinalis var. *officinalis*, flowering branch

Flower of *M. officinalis* var. *officinalis*

Bark and bark-stripping

M. officinalis var. *biloba* about to flower, Sichuan province

Magnolia officinalis var. *officinalis*

Magnoliaceae

DRUG MORPHOLOGY

CRUDE DRUG (DRIED BARK):

Stem bark: when in single or double quills (30–35 cm long, 2–7 mm thick), it is commonly known as 'Tong Po'. Outer surface greyish-brown, slightly rough, sometimes scaly and easily exfoliated, with distinct elliptical lenticels [1] and longitudinal wrinkles [2], appearing yellowish-brown when the coarse bark peeled; inner surface purplish-brown or dark purplish-brown [3], relatively smooth [3], with fine and dense longitudinal striations [4] and exhibiting oily traces when scored. *Texture*: hard, uneasily broken. *Fracture*: granular, greyish-brown in the outer layer and purplish-brown or brown in the inner layer, oily, sometimes with minute scattered crystals [hl] [5]. *Odour:* aromatic. *Taste*: pungent and slightly bitter. ***Good quality*:** bark thick and oily with inner surface dark purplish; intensely aromatic; spicy-tasting.

Root bark (Gen Po): quilled singly or in irregularly shaped pieces. *Texture:* hard, easily broken. *Fracture*: fibrous.

Branch bark (Zhi Po): quilled singly, 10–20 cm long, 1–2 mm thick. *Texture:* fragile and easily broken. *Fracture:* fibrous.

DECOCTING PIECES (DRIED SLICED BARK): transverse slices curved or in single or double quills. Otherwise as for Crude Drug.

PROCESSING

The root bark and the branch bark are shade-dried; the stem bark is blanched in boiling water then stacked in a dark and humid place to 'sweat' until its inner surface becomes purplish-brown or dark brown, then softened by steaming before being rolled into a tube and dried (= crude drug). Decocting pieces are made by removing the coarse bark then slicing transversely. Further optional processing involves stir-frying with ginger to produce Jiang Hou Po 姜厚朴.

OTHER COMMON NAMES

川朴 Chuan Po; 油朴 You Po; Cortex Magnoliae Officinalis; magnolia (SCN); officinal magnolia bark (CP2010).

SAFETY

Safety Class: 2b; Interaction Class: A (Gardner & McGuffin, 2013). An official healthfood in China (Anon., 2002a,b).

Magnolia officinalis var. *biloba*

Magnoliaceae

DRUG MORPHOLOGY

CRUDE DRUG (DRIED BARK): as for PS1.

DECOCTING PIECES (DRIED SLICED BARK): as for PS1.

PROCESSING

As for PS1.

OTHER COMMON NAMES

凹叶厚朴 Ao Ye Hou Po (plant name); 川朴 Chuan Po; 油朴 You Pu; magnolia (SCN).

SAFETY

Safety Class: 2b; Interaction Class: A (Gardner & McGuffin, 2013). An official healthfood in China (Anon., 2002a,b).

Crude drug (source plant *Magnolia officinalis* var. *officinalis*)

Crude drug of Hou Po (stem bark);
source species *Magnolia officinalis* var. biloba

Decocting pieces of Hou Po

External surface of outer bark of Hou Po

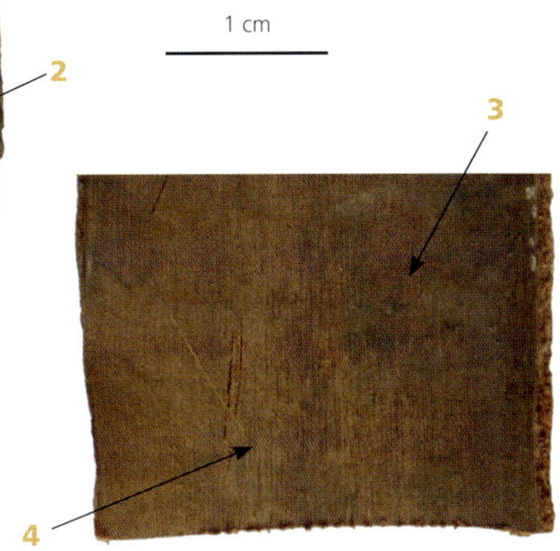

Internal surface of outer bark of Hou Po

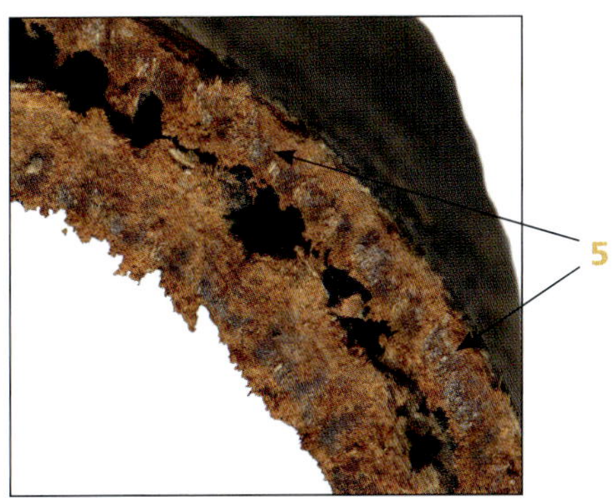

Cut surface of outer bark with minute crystals

Huang Bo 黄柏

Phellodendri Chinensis Cortex

Drug part: STEM-BARK

PHARMACOPOEIA SPECIES

Phellodendron chinense C. K. Schneid.

Rutaceae

taxonomic/nomenclatural note: in CP2010 as '*Phellodendron chinensis* Schneid.'

Species of *Phellodendron* wild in China: 2

PLANT DESCRIPTION

A deciduous tree (5–20 (rarely 25 m) tall) with a spreading canopy. *Bark* greyish-brown, **with many transverse lenticels but without thick cork; inner bark (phloem) bright yellow.** *Leaves* opposite and pinnate with 7–15 leaflets in opposite pairs, with a terminal leaflet, and arranged along a robust, densely brown-hairy to hairless central axis (rachis) (var. *glabriusculum* has few or no hairs on the rachis); leaflet blades ovate-elliptic or oblong to lance-shaped, papery; margins entire or with minute rounded teeth and oil glands, leaf tips narrowly pointed. *Male and female flowers* on separate plants. **Flowers in compact, terminal heads** (6.5–9.6 cm long; 4–6.5 cm wide), main stalk (peduncle) robust (4–6 cm long) with branches short to almost absent. *Flowers* yellowish-green, small and inconspicuous; sepals 5; petals 5. ***Fruit-heads* compact** with short, robust stalks; fruit sub-spherical (1–1.5 cm diam.), fleshy, black or purplish-black. Fl. May–Jun, fr. Sep–Nov.

HARVESTING, SOURCING AND NATURAL RANGE

Bark is harvested mainly from cultivated plantation trees grown in the central Chinese provinces of Sichuan, Chongqing and Guizhou.

The wild species grows in mixed broad-leaved forest (800–1,500 (rarely up to 3,000) m altitude) and is endemic to S, C and E China (provinces of Anhui, Fujian, Gansu, Guangdong, Guangxi, Guizhou, Henan, Hubei, Hunan, Jiangsu, Shaanxi, Sichuan, Yunnan and Zhejiang). Its wild distribution does not overlap with the more northerly species *P. amurense* (see *Unofficial Substitute* below).

CONSERVATION STATUS

Endemic to China: Least Concern (MEPCAS, 2013); Critical (Fu De-Zhi, 2011); Not Evaluated (IUCN Red List, accessed 25/4/2015).

PROTECTED STATUS

China: Cat. II (SEPA, 1999).

UNOFFICIAL SUBSTITUTE

P. amurense Rupr.

Rutaceae

PLANT DESCRIPTION

Similar to *P. chinense* but **outer bark with a thick layer of cork**; stalks of leaves, flowers and fruits slender (not robust); **flower- and fruit-heads lax** (not compact) (8.5–13.5 cm long; 6.5–9 cm wide); flower-head branches at least 1 cm long. Fl. May–Jun, fr. Sep–Nov (based on *FoC* and Ma *et al.,* 2006).

SUBSTITUTION COMMENT. *P. amurense* is a close relative of *P. chinense* and both species have long been used in TCM with their barks used interchangeably in clinical settings (often unknowingly due to their similar appearance). Editions of the CP prior to 2005 monographed the 2 species as alternative sources of Huang Bo (sometimes called Huang Bai) but since then they have been monographed separately (with different trade and Pharmacopoeia names). Their TCM 'Channels', 'Actions' and 'Indications' remain identical (CP2010) as do their 'Safety and Interactions' classification in Gardner & McGuffin (2013).

HARVESTING, SOURCING AND NATURAL RANGE

Bark is harvested mainly from plantation-grown trees in NE China (provinces of Heilongjiang, Jilin and Liaoning).

The wild species grows in forests and thickets in mountains and river valleys in C, N and NE China (provinces of Anhui, Hebei, Heilongjiang, Henan, Inner Mongolia, Jilin, Liaoning, Shandong and Shanxi). Its primarily north-eastern distribution distinguishes it from the more southerly *P. chinense* (above) and with whose bark this drug is regularly confused; their respective trade names (Guan Huang Bai 'north-eastern' and Chuan 'Sichuan') reflect their distributional differences. *P. amurense* is also native to Japan, the Korean Peninsula and Russia (Far East).

CONSERVATION STATUS

China: Vulnerable (MEPCAS, 2013; Vulnerable (Fu, 1992).
Global: Not Evaluated (IUCN Red List, accessed 25/4/2015).

PROTECTED STATUS

China: Cat. II (SEPA, 1999).

TCM PROPERTIES AND USE

properties: 'bitter and cold'
channels: 'Kidney, Bladder and Large Intestine'
actions: 'clears heat, dries damp, clears fire, eliminates toxins, clear deficient heat'

indications: diarrhoea; jaundice; dysuria; leucorrhoea; night sweats; sores and abscesses; (T)

Mature tree of *Phellodendron chinense* (cultivated), China

Flowering branch of male *P. chinense*

Trunk of *P. chinensis*

Cultivated female tree of *P. amurense* with young leaves and previous season's clusters of persistent fruits

Branch of *P. amurense* showing pinnate leaves and infructescence with immature fruits

Trunk of *P. amurense*

Phellodendron chinense

Rutaceae

DRUG MORPHOLOGY

CRUDE DRUG (DRIED BARK, sometimes de-corked): shallowly quilled or rectangular bark pieces **[1]** of variable length and width (c. 1–6 mm thick); outer surface greyish-brown **[2]** or yellowish (latter dependent on whether any thin outer cork bark is removed) with irregularly scattered raised transverse lenticels **[3].** Inner surface smooth and yellow or yellowish-brown throughout with fine longitudinal striations **[4].** *Texture*: light and hard. *Fracture*: fibrous, yellow or yellowish-green **[5]**; older barks thicker and somewhat layered **[6]**. *Odour*: indistinct. *Taste*: very bitter; becoming viscous when chewed.

DECOCTING PIECES-1 (SLICED BARK): bark slices (c.1–6 mm thick); transversely cut slices slightly curved or semi-quilled **[7]**, longitudinally cut slices flat and often rectangular **[8]**. All other characters as for the crude drug.

DECOCTING PIECES-2 (SALT-PROCESSED BARK): as for sliced bark (DP-1 above) with occasional charred areas **[9]** and inner surface darker yellow. *Taste*: slightly salty and very bitter.

DECOCTING PIECES-3 (CARBONISED BARK): as for sliced bark (DP-1 above) but externally charred black **[10]** (obscuring identification characters); internally dark brown or brownish-black. *Fracture*: charred brown and brittle. *Taste*: bitter and astringent.

PROCESSING

Bark is sun-dried (sometimes occasional coarse outer bark may need to be removed) (= crude drug); the bark is later sliced transversely or longitudinally prior to decocting (= decocting pieces-1. Subsequent (optional) processing by steeping in salt-water then dry-frying produces 盐黄柏 Yan Huang Bo; alternatively carbonising produces 黄柏炭 Huang Bo Tan.

OTHER COMMON NAMES

川黄柏 Chuan Huang Bo (Sichuan Huang Bo); 黄柏 Huang Bo; Bai Pi; Chinese corktree (CP2010); Cortex Phellodendri Chinensis; Phellodendron (SCN).

Note: the character 川 (Chuan) denotes the main production area from Sichuan province.

SAFETY

Safety Class: 1; Interaction Class: A (Gardner & McGuffin, 2013).

P. amurense

Rutaceae

'DRUG SUBSTITUTE' MORPHOLOGY

CRUDE DRUG (DE-CORKED DRIED BARK): shallowly quilled or rectangular bark pieces of variable length and width; thickness c. 1–6 mm (see *Note* below). The typical removal of the thick outer cork bark of this species (see *Processing*) exposes the bright yellow inner bark (phloem layer) below **[11].** This distinguishes it from the crude drug of *P. chinense* (which is not thick and corky and therefore seldom removed.) External (i.e. de-corked) surface of *P. amurense* **usually yellow, yellowish-green or yellowish-brown, with occasional remnants of the brown or greyish-brown outer cork bark [12], the latter typically raised and fissured [13], sometimes slightly spongy and with occasional lenticels.** Internal surface of inner bark smooth and yellow or yellowish-brown throughout with fine longitudinal striations. *Texture*: light and hard. *Fracture*: fibrous, yellow or yellowish-green. *Odour*: indistinct. *Taste*: very bitter; becoming viscous when chewed.

Note: thickness of the inner bark (yellow phloem layer) is highly variable in both species and is not a useful identification character because it reflects the age of the tree not its identity.

DECOCTING PIECES: differentiating the various styles of decocting pieces from those of *P. chinense* is often difficult owing to loss of key characters due to slicing, dry-frying and/or carbonising (see *Processing*). The most reliable macroscopic characters of *P. amurense* decocting pieces are the presence of **remnants of thick outer cork bark which is typically raised, fissured and often slightly spongy [14]** and yellow uncut exterior surface **[15]**. Laboratory-based authentication methods combined with macroscopy are recommended for robust identification of decocting pieces. *For a more detailed description see 'Guan Huang Bo' (Phellodendri Amurensis Cortex)* page 502.

PROCESSING

The coarse outer cork layer is removed, leaving the inner bark to be sun-dried (= crude drug) then later cut into transverse or longitudinal decocting pieces (-1); further optional processing by steeping in salt-water then dry-frying produces Yan Guan Huang Bo (decocting pieces-2) or carbonising produces Guan Huang Bo Tan (= decocting pieces-3).

OTHER COMMON NAMES

黄柏 Huang Bo; Amur corktree (CP2010); Guan Bai; Guang Huang Bai; Huang Bai; Cortex Phellodendri Amurensis; Phellodendron (SCN). *Note*: the character 关 (Guang) denotes the main production area from NE China.

SAFETY

Safety Class: 1; Interaction Class: A (Gardner & McGuffin, 2013).

Crude drug: Huang Bo

Phellodendron chinense

Decocting pieces: Huang Bo

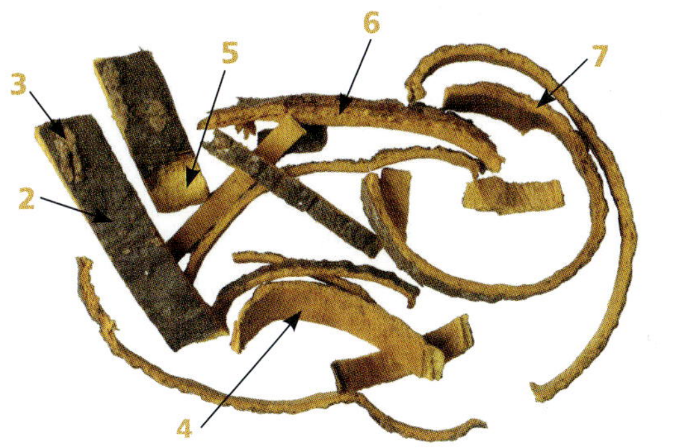

Dried (*P. chinense*)

1 cm

Longitudinal (*P. chinense*)

Salt-processed (*P. chinense*)

Carbonised (*P. chinense*)

Crude drug: Guang Huang Bo

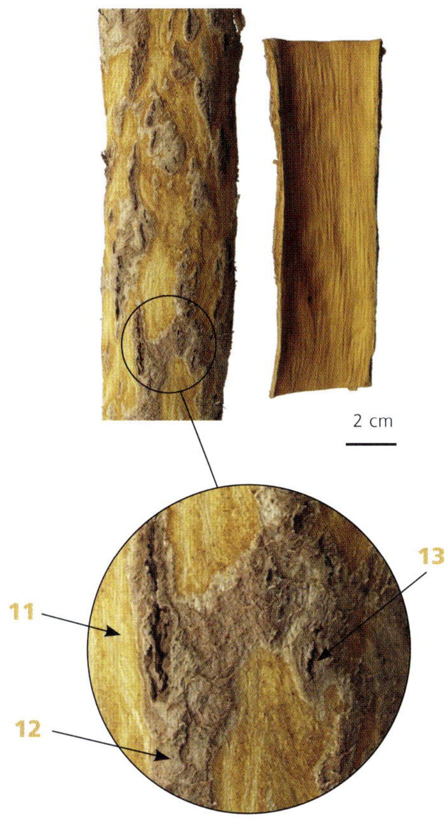

Phellodendron amurense

Decocting pieces: Guang Huang Bo

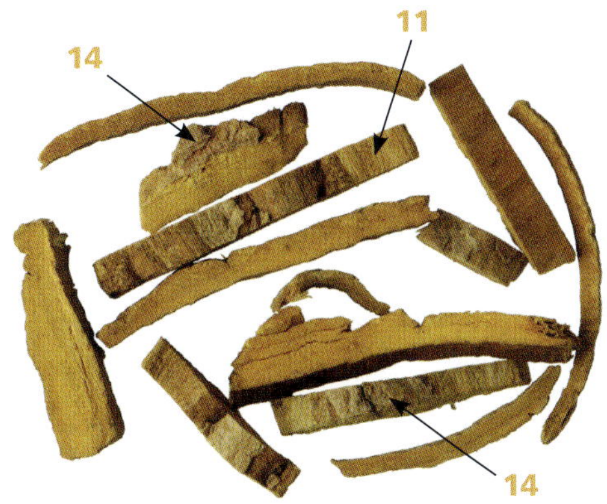

Dried (*P. amurense*)

1 cm

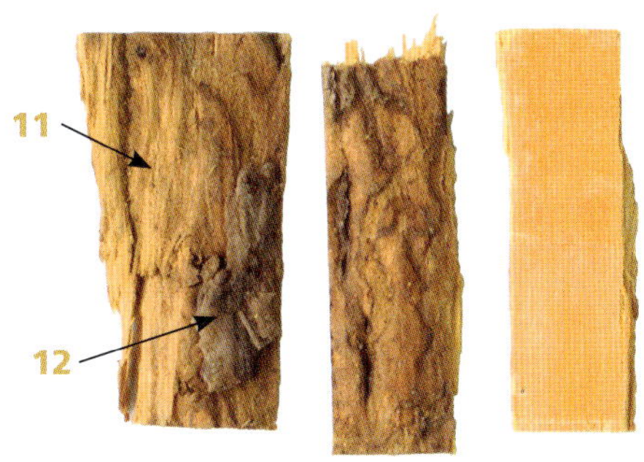

Longitudinal (*P. amurense*)

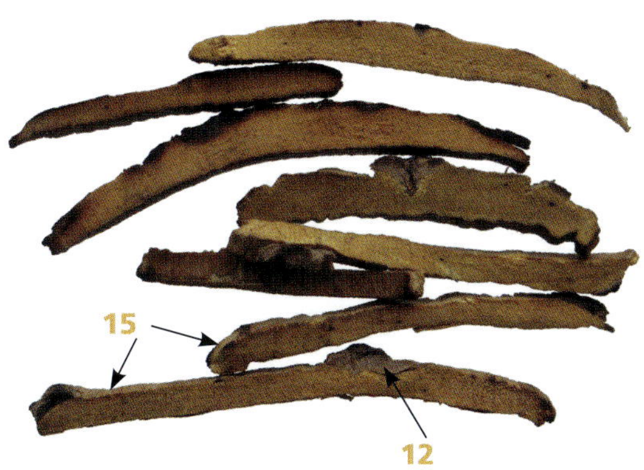

Salt-processed (*P. amurense*)

Carbonised (*P. amurense*)

Mu Dan Pi 牡丹皮

Moutan Cortex

Drug part: ROOT-BARK

Paeonia suffruticosa Andrews

Paeoniaceae

PLANT DESCRIPTION

The exact botanical identity of the plant referred to as ***Paeonia suffruticosa*** Andrews in the CP2015 has been confused — see 'taxonomic note' (right). Entities included in this vast taxonomic group are easily distinguished from *P. ostii*: leaves with **9 leaflets (arranged in 3 groups of 3),** flowers with 5 bracts, and petals (in single as opposed to double flowers) 5–11, white, red or pinkish-purple. The wild species is endemic to China, on cliff faces in Anhui and Henan provinces.

Species of *Paeonia* wild in China: 15 (*FoC*; Hong 2010)

taxonomic/nomenclatural note: CP2015 states the source of 'Mu Dan Pi 'as *Paeonia suffruticosa* Andrews. However, recent taxonomic research (Hong & Pan, 1999), now adopted by the *Flora of China*, indicates that the source plant of most (*but not all*) 'Mu Dan Pi' in domestic and international trade has been mis-identified and the correct identification is *P. ostii* T. Hong & J. X. Zhang. The name '*P. suffruticosa*' refers to a huge group of plants, largely made up of hundreds of ornamental cultivars long revered in China for their stunning blooms but of no medicinal value (Hong, 2010). One of the authors here (Lin Yu-Lin), however, has found that a small proportion of 'Mu Dan Pi' in trade is still sourced from this '*P. suffruticosa*' group, but it is unclear what scientific name to ascribe to this material; further research is needed. In the meantime the inexact name '*P. suffruticosa*' is used here.

P. ostii T. Hong & J. X. Zhang

Paeoniaceae

synonym: *P. suffruticosa* Andrews subsp. *ostii* (T. Hong & J. X. Zhang) Halda
taxonomic/nomenclatural note: see *P. suffruticosa* above .

PLANT DESCRIPTION

A small **shrub** or 'tree peony' (up 1.5 m tall) with thick fleshy roots. *Leaves* alternate, simple, **divided into 11–15 ovate, occasionally lobed, leaflets**; margins entire. Bracts 1–4, green and leaf-like. *Flowers* solitary, large (c. 11 cm diam.); sepals 3 or 4, greenish-yellow; petals c. 11, **white** and showy with outermost part broadest, tips notched; stamens many, with purplish-red filaments and yellow anthers; carpels 5, bulbous, surrounded by a distinctive purple rim (disk); stigmas red. *Fruits* with 5 oblong pods (follicles), densely covered in soft brownish-yellow hairs. Fl. Apr–May, fr. Aug.

HARVESTING, SOURCING AND NATURAL RANGE

Roots are harvested in autumn from medicinal plant farms reputedly established up to 1,000 years ago and where four well established cultivars (supposedly all derived from *Paeonia ostii*) are recognised as the traditional *Di Dao* sources: 粉丹皮 Fen Dan Pi (white phoenix) and 药丹皮 Yao Dan Pi (both from Anhui province); 湘丹皮 Xiang Dan Pi (from Hunan province) and 川丹皮 Chuan Dan Pi (from Sichuan province) (Bensky *et al.*, 2004; Haw, 2001).

The wild species is endemic to China, growing in broad-leaved deciduous forest and thickets (800–1,600 m altitude) and confined to western Henan (FoC). Haw (2006) describes its natural range to include Anhui, Hunan and Sichuan too, but it may now be extinct in these provinces.

CONSERVATION STATUS

China: Critically Endangered (Fu, 2011); 'Endangered' (Hong, 2010); Lin Yu-Lin pers. obs.).
Global: Not Evaluated (IUCN Red List, accessed 10/4/2015).

TCM PROPERTIES AND USE

***properties*:** **'bitter, pungent and cool'**
***channels*:** **'Heart, Liver and Kidney'**
***actions*:** **'clears heat, cools blood, invigorates blood'**

***indications*:** **night sweats, fever; spermatorrhoea; menstrual disorders; haematemesis; epistaxis; irritability; hypertension; traumatic injury**

Flowering plant of *Paeonia suffruticosa*, cultivated in Sichuan province (note 9 leaflets)

Bi-pinnate leaf of *P. suffruticosa* with 9 leaflets

Medicinal crop of *Paeonia ostii*, Shaanxi province

Bi-pinnate leaf of *P. ostii* with 11–15 leaflets

Flower of *P. ostii*

Underside of flower of *P. ostii*

Paeonia ostii

Paeoniaceae

DRUG MORPHOLOGY

CRUDE DRUG (DRIED ROOT-BARK)

Gua Dan Pi or *Fen Dan Pi* (export grade): root-bark with **outer root-bark removed**, quilled **[1]** or double-quilled **[2]**, always hollow **[3]**, mostly cylindrical (5–20 cm long; 5–12 mm diam.; 1–4 mm thick), slightly curved; inner edges of cut cortex inrolled **[4]**. Externally pinkish- or yellowish-brown often with areas of exposed inner bark showing creamish-white or greyish-yellow **[5]**; oval scars of root branches and smaller rootlets occasional **[6]**; large transverse raised lenticels occasional **[7]**. Inner root-bark surface creamish-white or greyish-yellow with fine longitudinal striations and occasionally with bright crystals [hl]. *Texture*: hard but easily broken. *Fracture*: even, pale pink and friable. *Odour*: aromatic. *Taste*: slightly bitter and astringent. ***Good quality*:** root-bark thick, white and starchy; intensely aromatic.

Lian Dan Pi (domestic grade in China): as for Gua Dan Pi but **outer root-bark present [8]**, externally slightly rough, greyish-brown or yellowish-brown revealing paler pinkish-brown inner bark **[9]** in places.

DECOCTING PIECES (SLICED ROOT-BARK)

Gua Dan Pi or *Fen Dan Pi* (export grade): near circular (quilled), curled or double-quilled slices (5–10.5 mm diam.; slices 1–6 mm thick). **Outer root-bark removed**, external uncut surface pinkish, yellowish-brown or creamish-white **[10]**; oval scars of root branches and smaller rootlets occasional **[11]**; large transverse raised lenticels occasional. Cut surface creamish-white and even **[12]**. Inner surface of root-bark creamish-white or greyish-yellow with fine longitudinal striations and occasionally with bright crystals [hl]. *Texture:* hard but easily broken. *Odour:* aromatic. *Taste:* slightly bitter and astringent.

Lian Din Pi (domestic grade in China): as for decocting pieces of 'Gua Dan Pi' but external uncut surface with bark intact, slightly rough, greyish-brown or yellowish-brown **[13]**; cut surfaces pinkish-cream. Diameter of root very slightly larger (5–12 mm diam.).

Freshly processed *Paeonia ostii* roots: *lower left,* discarded xylem core; *top centre*, Gua Dan Pi/Fen Dan Pi; *lower right*, Lian Dan Pi

PROCESSING

Rootlets are removed and discarded and the woody core of xylem tissue stripped out. The root-bark that remains is the drug; for domestic trade within China this is simply sun-dried to produce Lian Dan Pi. For export purposes, the outer root-bark (greyish-brown) is removed and discarded, to leave the inner root-bark 刮丹皮 Gua Dan Pi. Both types are transversely cut into thin slices (1–2 mm thick) and dried (= decocting pieces).

OTHER COMMON NAMES

凤丹 Feng Dan; 凤丹皮 Feng Dan Pi; Cortex Moutan; tree peony (SCN); tree peony bark (CP2010).

SAFETY

An official healthfood in China (Anon., 2002a,b). Safety Class: 2b and 2d; Interaction Class: A (Gardner & McGuffin, 2013). Occasional reports of transient side effects (nausea and dizziness) (Bensky *et al*., 2004).

UNOFFICIAL SUBSTITUTES

The use of root-bark of other *Paeonia* species has been reported but these are unlikely to occur as substitutes in international trade; species 1 and 2 are scarce in the wild.

1. ***Paeonia delavayi*** Franch. (滇牡丹 Dian Mu Dan). Leaves 17–25-lobed. Flowers 1–3 per shoot; yellow, red, purplish-red, occasionally white or greenish-yellow;
2. ***P. decomposita*** Hand.-Mazz. subsp. ***decomposita*** (syn. *P. szechuanica* W. P. Fang). Flowers pink.
3. ***P. anomala*** L. subsp. ***veitchii*** (Lynch) D. Y. Hong & K. Y. Pan. Leaves pinnate, flowers 2–4 per shoot, petals pink rarely white. Source of the separate TCM drug 'Chi Shao' (Paeoniae Radix Rubra) *see* page 74.

Crude drugs (sourced from *P. ostii*)

1 cm

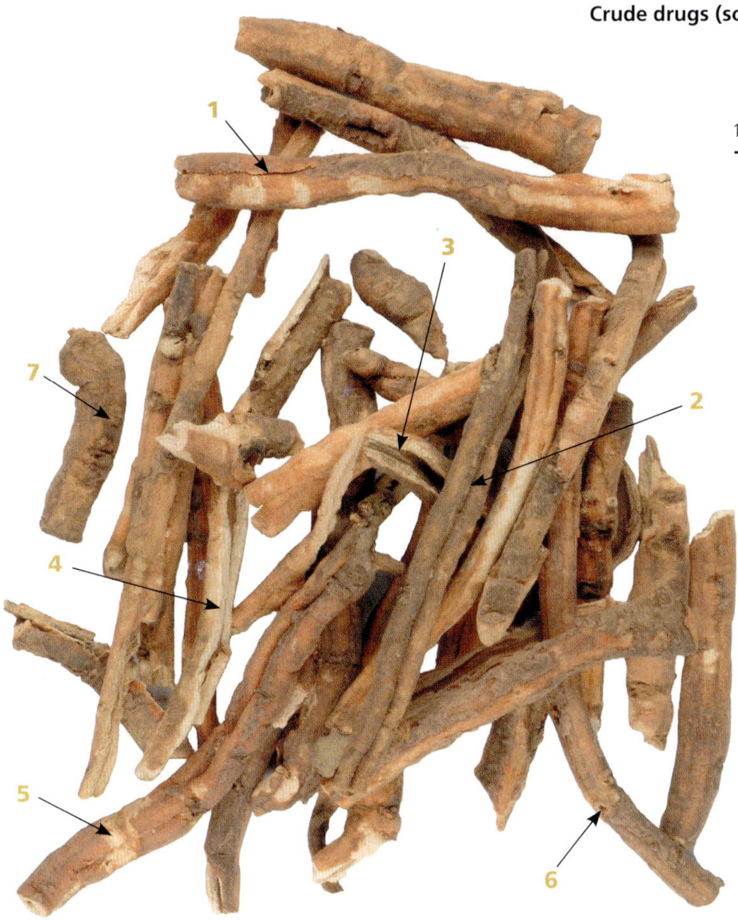

Export grade

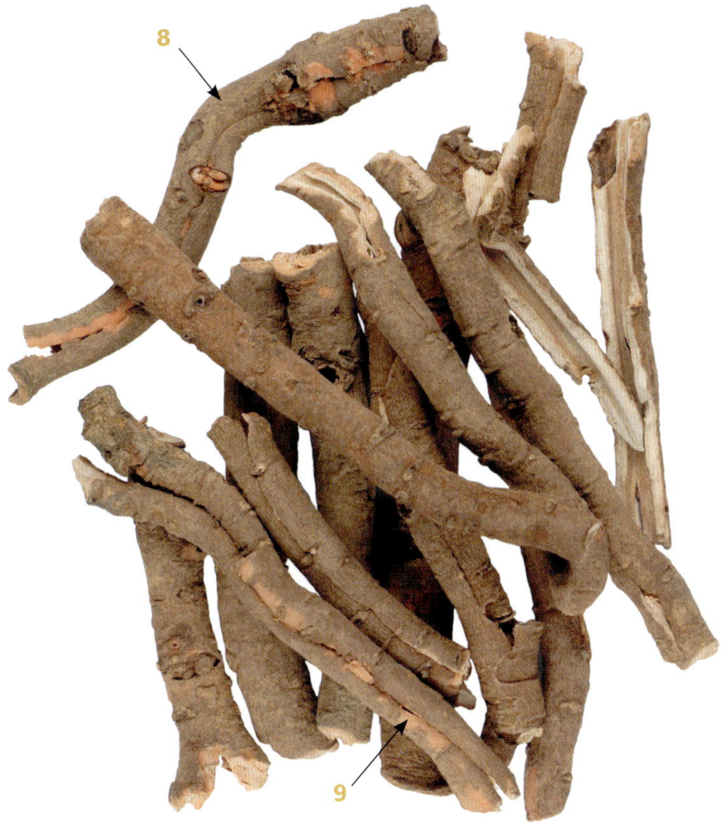

Domestic (China) grade

Decocting pieces (sourced from *P. ostii*)

1 cm

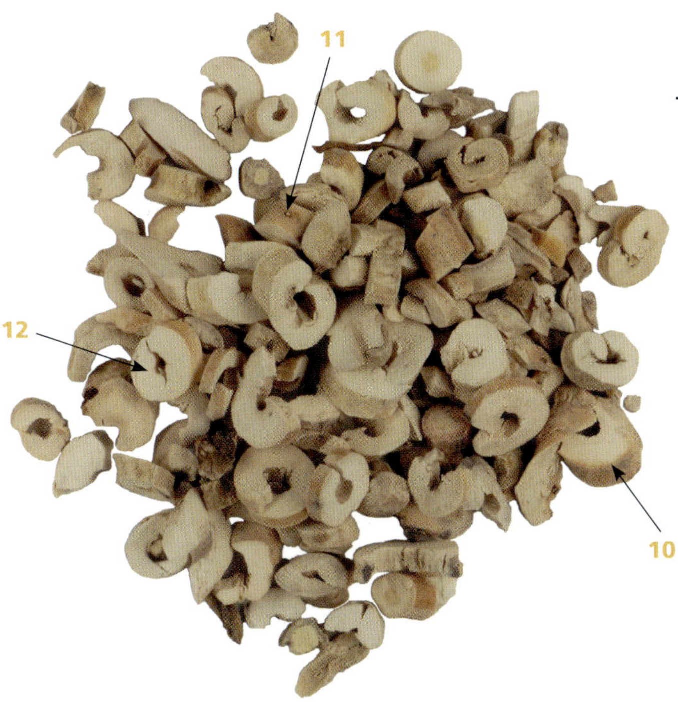

Export grade

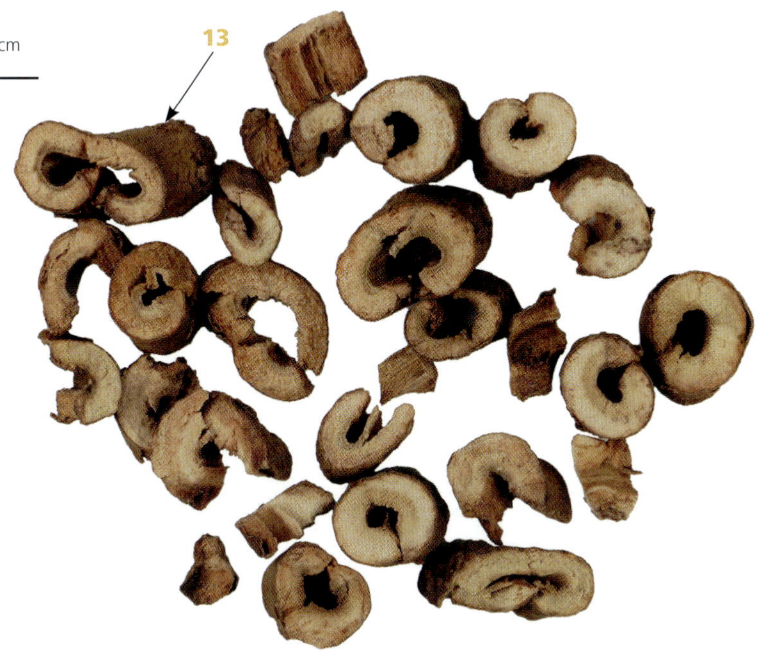

Domestic (China) grade

Rou Gui 肉桂

Cinnamomi Cortex

Chinese cinnamon/cassia bark

Drug part: STEM-BARK

Cinnamomum cassia (L.) J. Presl

Lauraceae

synonyms: *Laurus cassia* L.; *Cinnamomum aromaticum* Nees; *Persea cassia* (L.) Spreng.

Species of *Cinnamomum* wild in China: 48

PLANT DESCRIPTION

An evergreen, medium-sized tree with bark greyish-brown up to 13 mm thick at maturity. Current year's branchlets yellowish-brown, becoming dark brown, with longitudinal grooves and slightly downy. *Leaves* alternate or almost opposite, simple and stalked (1.2–2 cm); blade narrowly elliptic to lance-shaped (8–34 cm long; 4–9.5 cm wide), base sharply narrowing, shiny dark green above, minutely hairy below, conspicuously 3-veined and with tufts of hairs in axils of lateral veins, aromatic and leathery; margins entire, tough and rolled inwards. *Flowers* in branched axillary or terminal clusters; flower tube about c. 2 mm long; 'petals' 6, tiny, inconspicuous, white and covered in short yellowish-brown hairs; 'petals' often falling off before stamens developed; stamens 12 (9 obvious); ovary superior; style 1, stigma 1. *Fruit* fleshy and ellipsoid (10 mm long; 7–9 mm diam.), blackish-purple when mature, hairless; with persistent 'cup' at base. Fl. Jun–Aug, fr. Oct–Dec.

Young branches are the source of the separate TCM drug 'Gui Zhi' (Cinnamom Ramulus) page 448.

HARVESTING, SOURCING AND NATURAL RANGE

Bark is harvested in autumn and spring from cultivated trees widely grown as a plantation crop in tropical and subtropical parts of south and south-eastern China (Fujian, Guangdong, Guangxi, Guizhou, Hainan, Taiwan, and Yunnan) as well as India, Indonesia, Laos, Malaysia, Thailand, and Vietnam. The main production areas in China are in Guangdong, Guangxi, Hainan and Fujian, however 'Rou Gui' sourced from Vietnamese-grown trees is considered to be the highest quality.

Although, today, the species is only known from cultivation, the wild species is thought to originate from south China (*FoC*); its parentage may include *C. austrosinense* Hung T. Chang (M. Gilbert, pers. comm. 17/4/2015).

CONSERVATION STATUS

China/Global: only known from cultivation.

Cinnamomum cassia plantation, China

Branch of *C. cassia* with immature fruit

UNOFFICIAL SUBSTITUTES AND ADULTERANTS

Due to the large demand for 'Rou Gui' (both as a culinary spice and as medicine), the bark of many other species (examples below) in the family Lauraceae (mostly *Cinnamomum* species) have been widely reported as unofficial substitutes on the Chinese market (NICPBP, 1997–2002; Bensky *et al.*, 2004) with pulverised bark being especially prone to adulteration. Chinese *Cinnamomum* species are notoriously difficult to identify to species level even as whole trees since flower and fruit characters are usually required but are seldom seen on the tree at the same time. The characters below aim to help distinguish some of the more widely reported adulterant species. Discriminating bark to species level however, requires laboratory-based authentication methods.

1. *Cinnamomum heyneanum* Nees (as *C. burmannii* in Bensky *et al.*, 2004). Leaves **papery**; leaf blade **gradually narrowing at base;** flowers **greenish-white**.
2. *C. tamala* (Buch.-Ham.) T. Nees & Eberm. (柴桂 Chai Gui; Indian cassia). Leaf stalks **0.5–1.3 cm, hairless;** flowers **whitish-green.**
3. *C. bejolghota* (Buch.-Ham.) Sweet (钝叶桂 Dun Ye Gui). Bark is used as a spice in Hainan. Bark **green**; flowers **yellow**; fruit **green** when fresh, with a flower cup persistent, purplish-red and tinged yellow
4. *C. austrosinense* Hung T. Chang (华南桂 Hua Nan Gui). Used for similar purposes to the Pharmacopoeia species and may represent its wild form (*FoC*). Leaf stalks and leaf undersides **densely covered in short grey-brown hairs; flowers yellowish-green** with greyish-brown hairs. Fruit 10 mm long; 9 mm diam.
5. *C. japonicum* Siebold (天竺桂 Tian Zhu Gui). Leaves greyish-green beneath, **hairless on both surfaces**; leaf base **tapering gradually.** Flower stalks **hairless and smooth**. Protected species in China: Cat. II (SEPA, 1999).
6. *C. wilsonii* Gamble (川桂 Chuan Gui). Leaf blade greyish-green, ovate, with silky white hairs when young; veins **yellowish** when dry; leaf base gradually narrowing, **extending down leaf stalk**.
7. *C. subavenium* Miq. (香桂 Xiang Gui). Tree to 20 m tall, 50 cm dbh. Bark **grey**, smooth; leaf stalks **5–15 mm;** leaf blade yellowish-green beneath; leaf base **gradually tapering**. Flowers **yellowish** (3–4 mm).
8. *Lindera obtusiloba* Blume (三桠乌药 San Ya Wu Yao). **Deciduous** tree or shrub (3–10 m tall). Bark **blackish-brown**; leaves **rounded, mostly 3-lobed**; **male and female flowers on separate plants**.

TCM PROPERTIES AND USE

***properties*: 'pungent, sweet and hot'**
***channels*: 'Heart, Liver, Kidney, Spleen'**
***actions*: 'tonifies yang, warms Spleen, disperses cold, relieves pain'**

***indications*: poor libido, impotence; cough, dyspnoea; abdominal, musculoskeletal pain; diarrhoea**

DRUG MORPHOLOGY

CRUDE DRUG (DRIED BARK): bark pieces either curved, channelled or quilled (single, double and compound); 30–40 cm long, 3–10 cm wide, 0.2–0.8 cm thick. Outer surface dull, greyish-brown or pinkish-grey **[1]**, slightly rough with fine irregular wrinkles **[2]**, scattered and raised lenticels (either oval and crater-like or larger and transverse **[3]**); occasional larger leaf, bud or branch scars present as solitary round or oval holes **[4]**. *Irregularly shaped mottled zones on the outer bark surface may sometimes be due to lichen encrustation.* Internal surface brown or purplish-brown **[5]**, mostly smooth and finely striated longitudinally **[6]**; oily when scratched. *Texture*: hard and brittle. *Fracture*: uneven; outer layer brown; inner layer reddish-brown and oily; a yellowish-brown layer separates the two. *Odour*: strongly and characteristically aromatic. *Taste*: sweet and pungent. ***Good quality*:** thick, heavy and oily, purplish-red internally and intensely aromatic.

DECOCTING PIECES: *transverse slices* strongly curled (3–7 cm long, c. 2–8 mm wide); most other characters as for the crude drug.

PROCESSING

Stem-bark is shade dried, then any rough outer bark removed and discarded (= crude drug). Pulverised, transversely sliced or cut into irregular-shaped pieces for decocting.

OTHER COMMON NAMES

板桂 Ban Gui; 官桂 Guan Gui; 桂皮 Gui Pi; 企边桂 Qi Bian Gui; cassia (SCN); cassia bark (CP2010); Cortex Cinnamomi.

SAFETY

An official food and healthfood in China (Anon., 2002a,b). Safety Class: 2b; Interaction Class: A (Gardner & McGuffin, 2013).

Crude drug of Rou Gui (*Cinnamomum cassia*)

Decocting material of *C. cassia*

Lin Yu-Lin examining Guangdong-sourced Rou Gui in a wholesale store

Sang Bai Pi 桑白皮

Mori Cortex

white mulberry root-bark

Drug part: ROOT-BARK

PHARMACOPOEIA SPECIES

Morus alba L.

Moraceae

taxonomic note: two varieties and many cultivars exist and all are thought to be used as the source of this drug.

Species of *Morus* wild in China:10

PLANT DESCRIPTION

A deciduous shrub or small tree (3–10 m tall). *Bark* grey and shallowly furrowed. *Stipules* (at base of leaf stalks) lance-shaped (2–3.5 cm long, usually falling). *Leaves* alternate, simple, stalked and large (5–30 cm long; 5–12 cm wide), ovate or irregularly lobed; base rounded or heart-shaped, margin coarsely toothed or shallowly scalloped, tip sharply pointed or broad; **leaf undersides sparsely hairy along mid-veins or in tufts in vein axils**. *Male and female flowers* in short cylindrical catkins arising from the leaf axils and usually on separate plants. Male catkins pendant and densely white hairy (2–3.5 cm long) with tiny male flowers each with 4 pale green, overlapping calyx lobes and 4 stamens; petals are absent. Female catkins 1–2 cm long (shorter than male), softly hairy and packed with tiny female flowers; each flower with 4 green ovoid and overlapping calyx lobes (becoming fleshy in fruit), style absent and stigma with 2 distinctively white-edged and diverging branches [hl]; petals absent. *Fruit* a cylindrical or ovoid multiple fruit (syncarp), resembling a blackberry and becoming very succulent; depending on the cultivar, ripening from red to either blackish-purple, purple or greenish-white. Fl. Apr–May, fr. May–Aug.

For other drugs sourced from this species see 'Sang Shen' (page 700), 'Sang Zhi' (page 476) and 'Sang Ye' (page 554).

HARVESTING, SOURCING AND NATURAL RANGE

Today, white mulberry is only known from cultivation and is grown throughout China mostly for its leaves which are used to rear silkworms for the silk industry. The root-bark is harvested from crop plants in late autumn after leaves have fallen, or in early spring.

The wild species was originally considered endemic to central and northern China.

CONSERVATION STATUS

China: Not Evaluated.
Global: Not Evaluated (IUCN Red List, accessed 25/8/2016).

UNOFFICIAL SUBSTITUTE

Broussonetia papyrifera (L.) L'Hér. ex Vent.

Moraceae

synonym: *Morus papyrifera* L.

PLANT DESCRIPTION

Easily distinguished from the Pharmacopoeia species: **a deciduous tree 10–20 m tall**, leaves variable in shape (young leaves typically deeply lobed); lower leaf surfaces densely covered in very short, soft, fine hairs; male flowers in pendant catkins, female flowers in spherical heads (1–2 cm diam.). The ***fruit* is a globose, brilliant orangish-red aggregate (1.5–3 cm diam.) of fleshy 1-seeded drupelets**. Fl. Apr–May, fr. Jun–Jul.

SUBSTITUTION COMMENT. The inner **stem-bark** of this highly invasive and closely related tree closely resembles official 'Sang Bai Pi' and provides a cheap and readily available unofficial look-alike. The inner bark has long been used in paper-making (hence its common name 'paper mulberry tree'). Its bright orange fruits are the source of a separate TCM drug 'Chu Shi Zi' (Broussonetiae Fructus) not included in this book.

The species is native throughout much of China, except the far north; it grows in secondary vegetation, forest margins and mixed deciduous and evergreen woodland (up to 2,300 m altitude) [also native to Cambodia, Japan, the Korean Peninsula, Laos, Malaysia, Myanmar, Sikkim, Thailand, Vietnam and the Pacific Islands].

CONSERVATION STATUS

China: Least Concern (MEPCAS, 2013); an invasive species.
Global: Not Evaluated (IUCN Red List, accessed 25/8/2016).

TCM PROPERTIES AND USE

properties: 'sweet and cold'
channel: 'Lung'
actions: 'clears Lung heat, relieves cough, resolves phlegm, promotes urination'

indications: cough, dyspnoea; water retention, oedema; hypertension

ABOVE *Morus alba* crop, Anhui province
TOP RIGHT Branch of *M. alba* with female flowers
RIGHT Leaves and young fruit of *M. alba*
FAR RIGHT Branch of *M. alba* with ripening fruit

Fruiting tree of *Broussonetia papyrifera*

Varying leaf shapes of *B. papyrifera* with ripe fruit-head

Fruiting branch of *B. papyrifera*

Morus alba Moraceae

DRUG MORPHOLOGY

CRUDE DRUG (DRIED ROOT-BARK): longitudinal, channelled **[1]**, quilled **[2]**, flat or twisted pieces of inner bark of varying length and width; 1–4 mm thick. Externally white or creamish-white, relatively smooth **[3]** or verrucose with occasional darker rootlet scars **[4]**; orangish-yellow or brownish-yellow remnants of outer bark sometimes persistent **[5]**. Internal surface greyish-yellow or yellowish-white with fine longitudinal striations **[6]**. *Texture*: light, pliable and fibrous; autumn-harvested bark is more powdery; easily torn longitudinally. *Fracture*: fibrous. *Odour*: indistinct. *Taste*: slightly sweet. ***Good quality*:** thick, white inner root-bark devoid of orangish- or brownish-yellow, coarse outer bark

DECOCTING PIECES-1 (SLICED INNER ROOT-BARK): usually strongly curved **[7]** (c. 3–5 cm long; 0.5 cm wide; 1–4 mm thick); yellowish-cream or white fibrous throughout; outer surface with occasional fragments of orangish- or brownish-yellow coarse outer bark **[8]**.

DECOCTING PIECES-2 (HONEY-PROCESSED ROOT-BARK): as for sliced root-bark above but darker throughout, charred in places **[9]** and, when freshly processed, lustrous **[10]**. *Taste*: sweet.

PROCESSING

After removal of the coarse, yellowish-brown or orangeish-yellow outer bark, the inner bark is cut longitudinally and sun-dried (= crude drug); the bark is subsequently sliced into smaller longitudinal or transverse decocting pieces. Subsequent honey-processing involves steeping in warm diluted honey until thoroughly moistened, then stir-frying (= 'Zhi Sang Bai Pi' (processed Sang Bai Pi).

OTHER COMMON NAMES

桑 Sang (plant name); 桑根白皮 Sang Gen Bai Pi; 桑皮 Sang Pi; Cortex Mori; white mulberry (SCN); white mulberry root-bark.

SAFETY

An official healthfood in China (Anon., 2002a,b). Safety Class: 1; Interaction Class: A (Gardner & McGuffin, 2013).

For other TCM drugs sourced from *Morus alba* see 'Sang Shen' (Mori Fructus) page 700, 'Sang Ye' (Mori Folium) page 554 and 'Sang Zhi' (Mori Ramulus) page 476.

Broussonetia papyrifera Moraceae

'DRUG SUBSTITUTE' MORPHOLOGY

It is the stem-bark of this species that is used as an Unofficial Substitute. Compared to the official drug in general it is **more fibrous [11]**, and the **remains of the orange outer bark is less obvious [12]**, laboratory-based methods are recommended to reliably differentiate this bark from that of the Pharmacopoeia species.

OTHER COMMON NAMES

构树 Gou Shu (plant name); broussonetia (SCN); paperbark mulberry.

OTHER UNOFFICIAL SUBSTITUTES

The root-bark of other species of *Morus* (of which 10 are wild in China) have been traded as 'Sang Bai Pi' due to mis-identification of source plants; their occurrence in trade, however, is only likely to be seen at a local level in China. Reliable differentiation of these root-barks from official 'Sang Bai Pi' is best performed using laboratory-based authentication methods. Living plants can be identified as follows:

1. ***M. mongolica*** (Bureau) C. K. Schneid. Leaves hairless; teeth of leaf margins with short, needle-like points.
2. ***M. australis*** Poir. Lower leaf surface sparsely clothed with thick hairs; leaf margins serrate or entire (smooth).
3. ***M. cathayana*** Hemsl. Female flowers with a very short style and leaf undersides densely clothed in a covering of white or yellowish-grey hairs.

Crude drug of Sang Bai Pi
(fibrous texture indicates harvesting in spring)

Honey-processed Sang Bai Pi

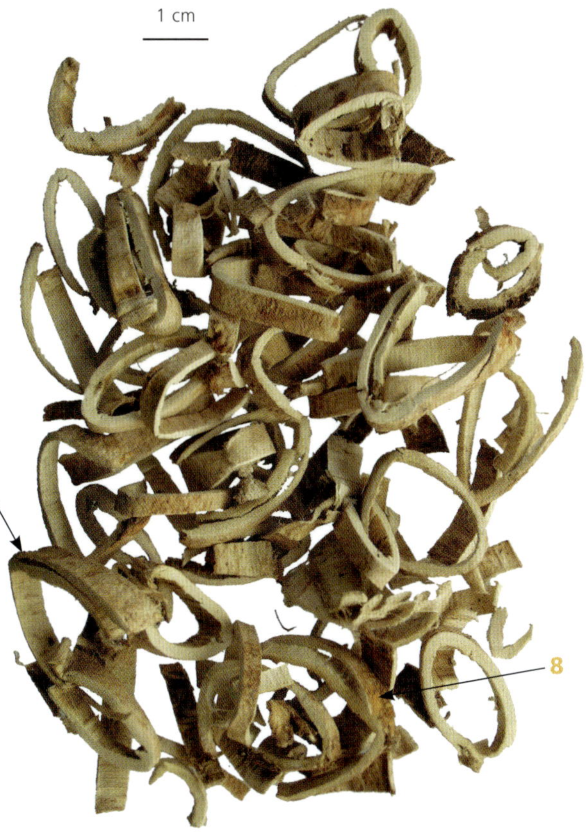

Decocting slices of Sang Bai Pi

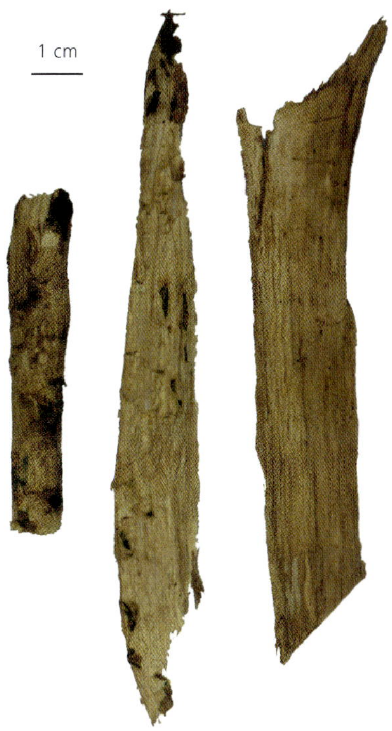

Dried, inner root-bark of *Broussonetia papyrifera*

B. papyrifera: transverse fracture
of dried sliced rootbark

B. papyrifera: dried, sliced inner root-bark

Wu Jia Pi 五加皮 Acanthopanacis Cortex

Drug part: ROOT-BARK

PHARMACOPOEIA SPECIES

Eleutherococcus nodiflorus (Dunn) S. Y. Hu

Araliaceae

synonyms: *Acanthopanax gracilistylus* W. W. Sm. (name used in CP2015); *A. nodiflorus* Dunn

Species of *Eleutherococcus* wild in China: 18

PLANT DESCRIPTION

A deciduous multi-stemmed shrub up to 3 m tall, sometimes climbing. *Stems* slightly pendulous, hairless, with short, scattered, reflexed prickles at nodes, rarely unarmed. *Leaves* alternate, palmately compound with 5 ovate leaflets broadest towards their tips (3–8 cm long; 1–3.5 cm wide), leaflet bases wedge-shaped, margins finely serrate or shallowly rounded. *Inflorescences* numerous, spherical, many-flowered heads in leaf axils; each head solitary or 2 or 3 together. *Flowers* tiny, stalked, radially symmetrical, bisexual or male; calyx a low 5-parted rim; petals 5, yellowish-green, touching (not over-lapping); ovary inferior; stamens 5; styles 2–5. *Fruit* black (at maturity), fleshy and globose (c. 6 mm diam.); each fruit usually 5-seeded. Fl. Apr–Jul, fr. Jun–Oct.

HARVESTING, SOURCING AND NATURAL RANGE

Roots are harvested in summer and autumn from wild plants which grow along forest margins and in scrub, often by streams and roadsides (up to 3,000 m altitude). Endemic to N, C and E China: Anhui, Fujian, Gansu, Guangdong, Guangxi, Guizhou, Henan, Hubei, Hunan, Jiangsu, Jiangxi, Shaanxi, Shanxi, Sichuan, Taiwan, Yunnan and Zhejiang.

CONSERVATION STATUS

Endemic to China: Least Concern (MEPCAS, 2013); declining due to over-harvesting for medicinal use (Lin and Leon, pers. obs., 2006); Not Evaluated (IUCN Red List, accessed 28/9/2016).

UNOFFICIAL SUBSTITUTE 1

Periploca sepium Bunge

Apocynaceae

PLANT DESCRIPTION

Easily distinguished from the Pharmacopoeia species. Main differences: **rootstock highly aromatic**. Most aerial parts with **milky latex. *Leaves* opposite and simple** (not divided into radiating leaflets) with **leaf blades narrowly ovate** (5–9 cm long; 1.5–2.5 cm wide) and tips pointed; leaf blades with conspicuous veins in 20–25 lateral pairs. ***Flowers* much larger** (c. 15 mm diam.), **pinkish-purple, somewhat bulbous-shaped** with 5 claw-like appendages and covered in white, woolly hairs. ***Fruit* a dry, cylindrical pod** (7–12 cm long; c. 5 mm diam.) with 2 curved follicles often connected apically.

CONSERVATION STATUS

Endemic to China: Least Concern (MEPCAS, 2013); Not Evaluated (IUCN Red List, accessed 10/9/2016).

SUBSTITUTION COMMENT. The root-bark of this botanically unrelated species is the source of a separate TCM drug 'Xiang Jia Pi' (see page 530) whose safety profile is quite different (see *Safety*) from 'Wu Jia Pi' described here. It is a frequent substitute of Wu Jia Pi (knowingly or not) both in China and in world markets (it is commonly traded as 'Wu Jia Pi' in the UK, C. Leon, pers. obs.). Possible reasons for its use as a substitute include a legacy from earlier CP editions when this root-bark species was a legitimate clinical alternative to 'Wu Jia Pi' and its easy availability (and hence cheaper price) from the wild (It is common throughout China) See 'Drug Substitute' Morphology for distinguishing macroscopic characters.

UNOFFICIAL SUBSTITUTE 2

Eleutherococcus giraldii (Harms) Nakai

Araliaceae

synonym: *Acanthopanax giraldii* Harms

PLANT DESCRIPTION

Closely resembling the Pharmacopoeia species but stems **densely covered with bristle-like prickles** and leaflet margins **irregularly bi-serrate.** Fl. Jun–Jul, fr. Sep–Oct.

SUBSTITUTION COMMENT. The stem-bark (not root-bark) of this closely related species is often traded on world markets as 'Wu Jia Pi'. The reasons are unclear but confusion between their similar Chinese names may be one, with 'Hong Mao Wu Jia' 红毛五加 (for *E. giraldii*) sometimes abbreviated to 'Wu Jia' 五加. The species is a widespread endemic of central and western China where it grows in mountain scrub (1,300–3,500 m altitude).

CONSERVATION STATUS

Endemic to China: Least Concern (Lin and Leon, pers. obs., 2002); Not Evaluated (IUCN Red List, accessed 28/9/2016).

TCM PROPERTIES AND USE

properties: 'pungent, slightly bitter and warm'
channels: 'Kidney and Liver'
primary action: 'dispels wind damp, tonifies Liver and Kidney, regulates water'
indications: musculoskeletal pain, oedema

Eleutherococcus nodiflorus — flowering branch

Fruiting branches of *E. nodiflorus*

Periploca sepium in flower (inset: single flower)

Typical habitat, Shaanxi province

P. sepium fruiting branch

Leaves of *P. sepium*

Eleutherococcus giraldii with young fruiting head

E. giraldii

Eleutherococcus nodiflorus

Araliaceae

DRUG MORPHOLOGY

CRUDE DRUG (DRIED ROOT-BARK): sub-cylindrical, slightly curve quills (1) (5–15 cm long; 0.4–1.4 cm diam.; c.0.2 cm thick) [ ]. **External surface (periderm) greyish-brown**, covered with **scales but not easily removed** [2] and scattered, raised, transverse lenticels [3]. Internal surface of cortex with fine, longitudinal striations. *Texture:* light and easily broken. ***Fracture*: pale yellowish-white to greyish-white** [4]; uneven. *Odour*: **slightly aromatic.** *Taste*: slightly acrid and bitter. ***Good quality*:** quilled pieces with a thick cortex.

DECOCTING PIECES (SLICED ROOT-BARK): as for the crude drug but cut into short pieces (c.0.5–2 cm long) often rhomboid, hollow and cylindrical.

PROCESSING

Roots are harvested from wild plants in summer and autumn, root-bark removed and sun-dried (= crude drug) then cut into thick (>2 cm long) slices (= decocting pieces-1).

OTHER COMMON NAMES

南五加皮 Nan Wu Jia Pi 'southern Wu Jia Pi'; Cortex Acanthopanacis; Eleutherococcus gracilistylus (SCN).

SAFETY

Bensky *et al*. (2004) state that any safety issues reported probably relate to the inadvertent use of the unofficial substitute 'Xiang Jia Pi' (sourced from *Periploca sepium*), see below.

Periploca sepium

Asclepiadaceae

'DRUG SUBSTITUTE' MORPHOLOGY

The root-bark is often difficult to distinguish macroscopically from that of the official species; its strong aroma is a helpful but subjective character such that laboratory-based authentication methods are advised for robust identification especially where substitution or adulteration is suspected.

DRIED ROOT-BARK: sub-cylindrical, slightly curved quills (3–10 cm long; 1–2 cm diam.; 0.2–0.4 cm thick). **External surface yellowish-, pale orangish- or greyish-brown** [5], with periderm **soft, loose and readily flaking** [6]; internal surface smooth and finely striated longitudinally. *Texture:* more brittle than the official species. ***Fracture:* pale yellow or brownish-yellow** [7], uneven. *Odour:* **highly aromatic**. *Taste:* bitter.

DECOCTING PIECES: as above but sliced or cut into short pieces (0.2–2 cm long); larger pieces are rhomboidal, hollow and cylindrical.

OTHER COMMON NAMES

北五加皮 Bei Wu Jia Pi 'northern Wu Jia Pi'; **香加皮 Xiang Jia Pi**; Chinese silk vine; Cortex Periplocae; Periplocae Cortex; Periploca sepium (SCN).

SAFETY

Toxic (CP2015). Compared to the official drug ('Wu Jia Pi') the root-bark of this separate medicinal species ('Xiang Jia Pi', see *Substitution Comment*) has a narrow therapeutic safety window (it contains cardiac glycosides and is not for long-term use). For this reason **its detection as an unofficial substitute is essential**. 'Xiang Jia Pi' is officially forbidden as a food or healthfood in China (Anon., 2002a,b).

Eleutherococcus giraldii

Araliaceae

'DRUG SUBSTITUTE' MORPHOLOGY

It is the stem not root-bark that is traded as the substitute. Stem-bark thin [8] and brittle and **densely covered with spreading or backward-pointing bristle-like prickles** [9].

OTHER COMMON NAMES

红毛五加 Hong Mao Wu Jia.

OTHER UNOFFICIAL SUBSTITUTES

1. ***Eleutherococcus henryi*** Oliv. (糙叶五加 Cao Ye Wu Jia); family Araliaceae; root-bark. A common local substitute in China with densely hairy leaf stalks.
2. ***E. senticosus*** (Rupr. & Maxim.) Maxim. (syn. *Acanthopanax senticosus* (Rupr. & Maxim.) Harms); root-, rhizome- or stem-bark is the source of the separate drug Ci Wu Jia 刺五加 (Acanthopanacis Senticosi Radix et Rhizoma seu Caulis), not included in this book.
3. ***Oldenlandia hedyotidea*** (DC.) Hand.-Mazz. (syn. *Hedyotis hedyotidea* (DC.) Merr., family Rubiaceae. The woody stems of this unrelated species (provisionally named here since authors have not seen source plant) are widely sold as 'Wu Jia Pi' in North American markets for treating influenza, coughs and eczema (Zhao & Li, 2007; Eric Brand pers. comm. 6/2016). Its other Chinese name Tu Wu Jia 土五加 may, in part, be the cause of confusion with official 'Wu Jia Pi' because these materials are easily distinguished macroscopically. Also called Niu Bai Teng 牛白藤. Once a common substitute on the UK market but now rarely encountered (Yu Hongwen pers. comm. 2008; Zhao Kaicun pers. comm. 2010).

Oldenlandia hedyotidea

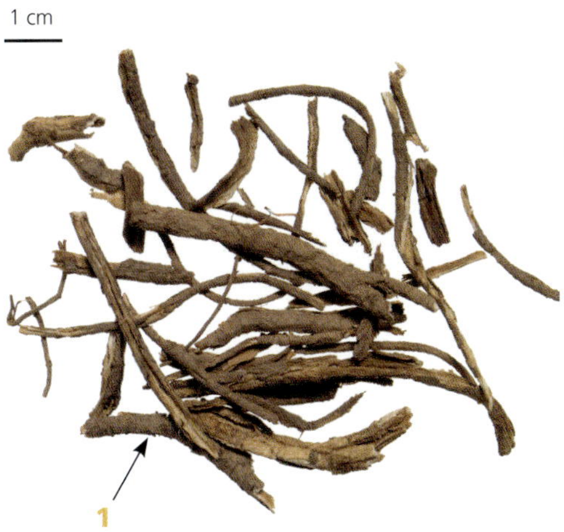

Crude drug of Wu Jia Pi (*Eleutherococcus nodiflorus*)

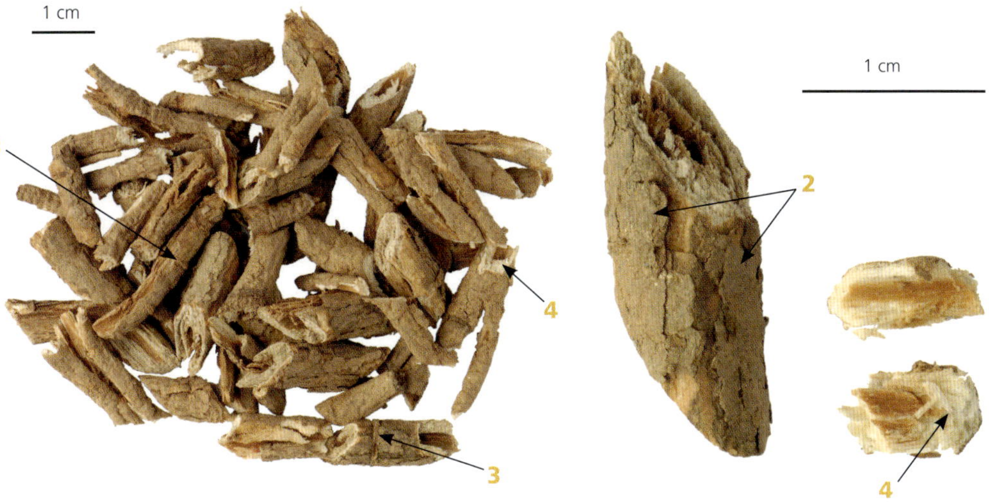

Decocting pieces of Wu Jia Pi (*E. nodiflorus*)

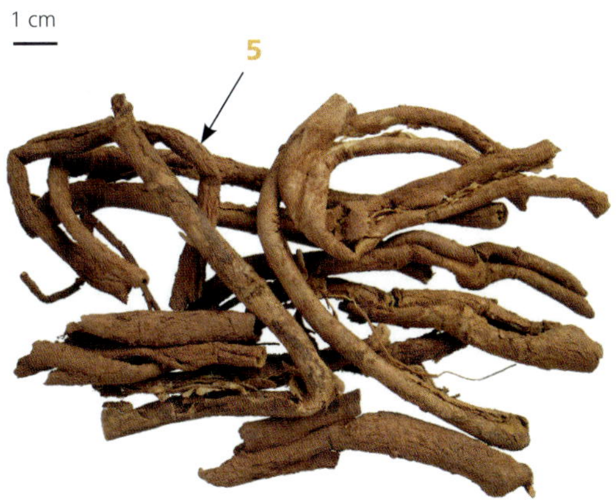

Crude drug of unofficial substitute Xiang Jia Pi (*Periploca sepium*)

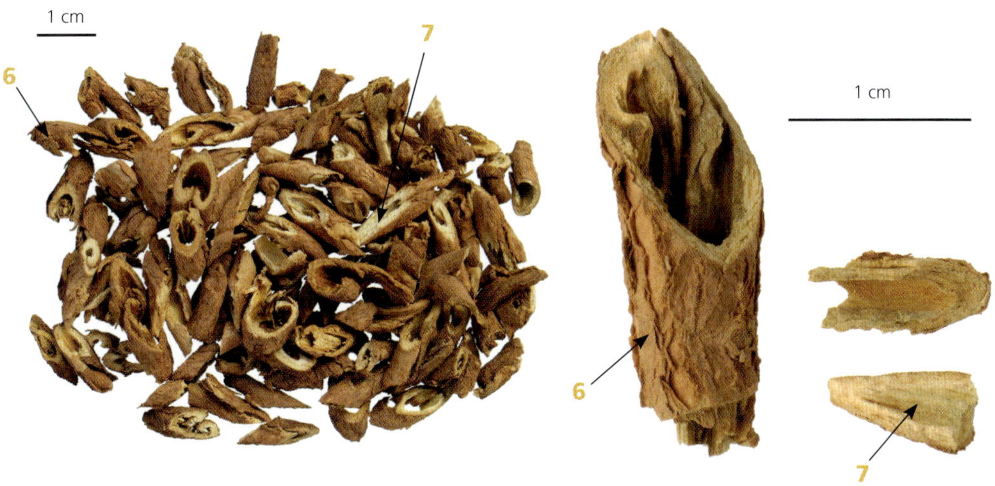

Decocting pieces of unofficial substitute Xiang Jia Pi (*P. sepium*)

Dried stem of unofficial substitute Hong Mao Wu Jia (*Eleutherococcus giraldii*)

Decocting pieces of unofficial substitute Hong Mao Wu Jia (*E. giraldii*)

Xiang Jia Pi 香加皮 Periplocae Cortex

Drug part: ROOT–BARK

Periploca sepium Bunge

Asclepiadaceae

Species of *Periploca* wild in China: 5

PLANT DESCRIPTION

A rambling, deciduous, shrub (up to 4 m) with a highly aromatic rootstock. Most parts with milky latex. *Leaves* opposite, short-stalked, blades long and pointed (5–9 cm long; 1.5–2.5 cm wide), margins entire, leaf veins distinct in 20–25 lateral pairs. *Flowers* in loosely branching groups of 2–3; flower bulbous-shaped (c. 15 mm diam.), 5-parted, covered in white, wooly hairs and with 5 pinkish claw-like appendages (corona). *Fruit* dry, comprising 2 curved, cylindrical pods, sometimes connected at tips; seeds many, each with a distinctive long, silk-like coma (c. 3 cm long). Fl. May–Jun, fr. Jun–Jul.

HARVESTING, SOURCING AND NATURAL RANGE

Root-bark is harvested in spring and autumn from wild plants which are common along forest margins and on roadside banks throughout China except the far south (Guangdong, Guangxi, Hainan and Taiwan). The bulk of harvesting is reported to occur in the more northerly provinces of Shanxi, Henan, Hebei, Shandong and Gansu (Zhao & Chen, 2014).

CONSERVATION STATUS

Endemic to China: Least Concern (MEPCAS, 2013); Not Evaluated (IUCN Red List, accessed 10/9/2016).

Typical habitat of *Periploca sepium*, Shaanxi province

P. sepium in flower

Flowers of *P. sepium* (lower purple one fully developed)

Fruiting branch of *P. sepium*

Fresh roots of *P. sepium*

TCM PROPERTIES AND USE

properties: **'pungent, bitter and warm'**
channels: **'Liver, Kidney and Heart'**
actions: **'dispels wind damp, promotes urination'**

indications: **musculoskeletal pain; oedema; stranguria**

DRUG MORPHOLOGY

CRUDE DRUG (DRIED ROOT-BARK): quilled, double-quilled, channelled or irregularly shaped **[1]** (3–10 cm long; 1–2 cm diam.; 0.2–0.4 cm thick). Externally yellowish-brown or greyish-brown **[2]**, with outermost bark often loose and flaking easily **[2]**; inner surface smooth, pale yellow or brownish-yellow, finely striated longitudinally **[3]**. *Texture*: light, fragile and brittle. *Fracture:* uneven, pale yellow **[4]**. ***Odour*: highly aromatic**. *Taste*: bitter. ***Good quality*:** light, brittle and intensely aromatic.

DECOCTING PIECES (DRIED SLICED ROOT-BARK): as for the crude drug but cut into short pieces (0.5–2 cm long); often rhomboid **[5]**, hollow and cylindrical.

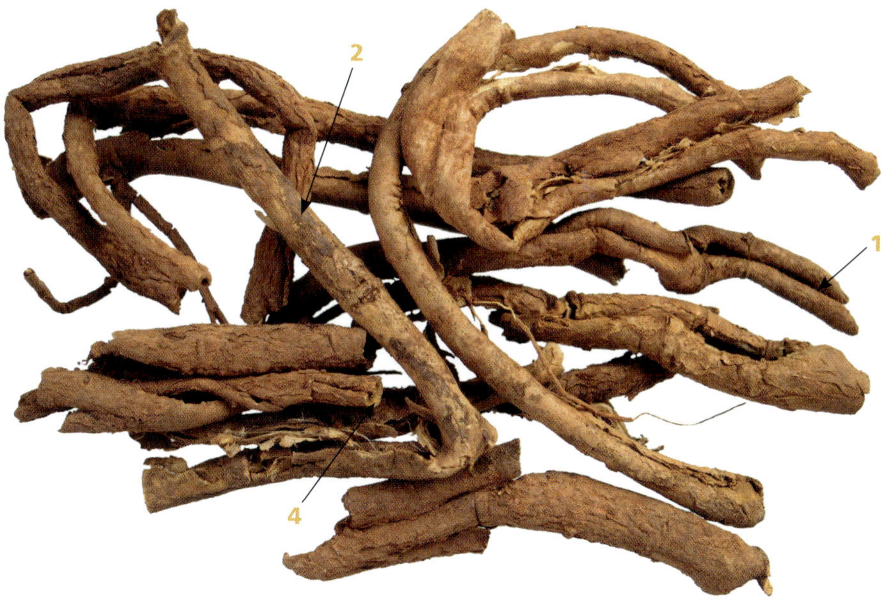

Crude drug of Xiang Jia Pi (sourced from *Periploca sepium*)

PROCESSING

After removal and discarding of the inner xylem, the remaining root-bark is sun-dried (= crude drug) then latex cut into decocting pieces.

OTHER COMMON NAMES

杠柳 Gang Liu (plant name); 北五加皮 Bei Wu Jia Pi 'northern Wu Jia Pi'; Chinese silk vine; Cortex Periplocae; Periploca sepium (SCN).

SAFETY

Toxic (CP2015); contains cardiac glycosides and is not for long-term use. Adverse events are most likely to occur either when taken in overdose or as an inadvertent adulterant in the popular drink 'Wu Jia Pi Liquor' (Bensky *et al.*, 2004; Chen & Chen 2004; Chang & But, 1986; Martindale & Reynolds, 1996; Wichtl, 2004). Officially forbidden as a food or health-food in China (Anon., 2002).

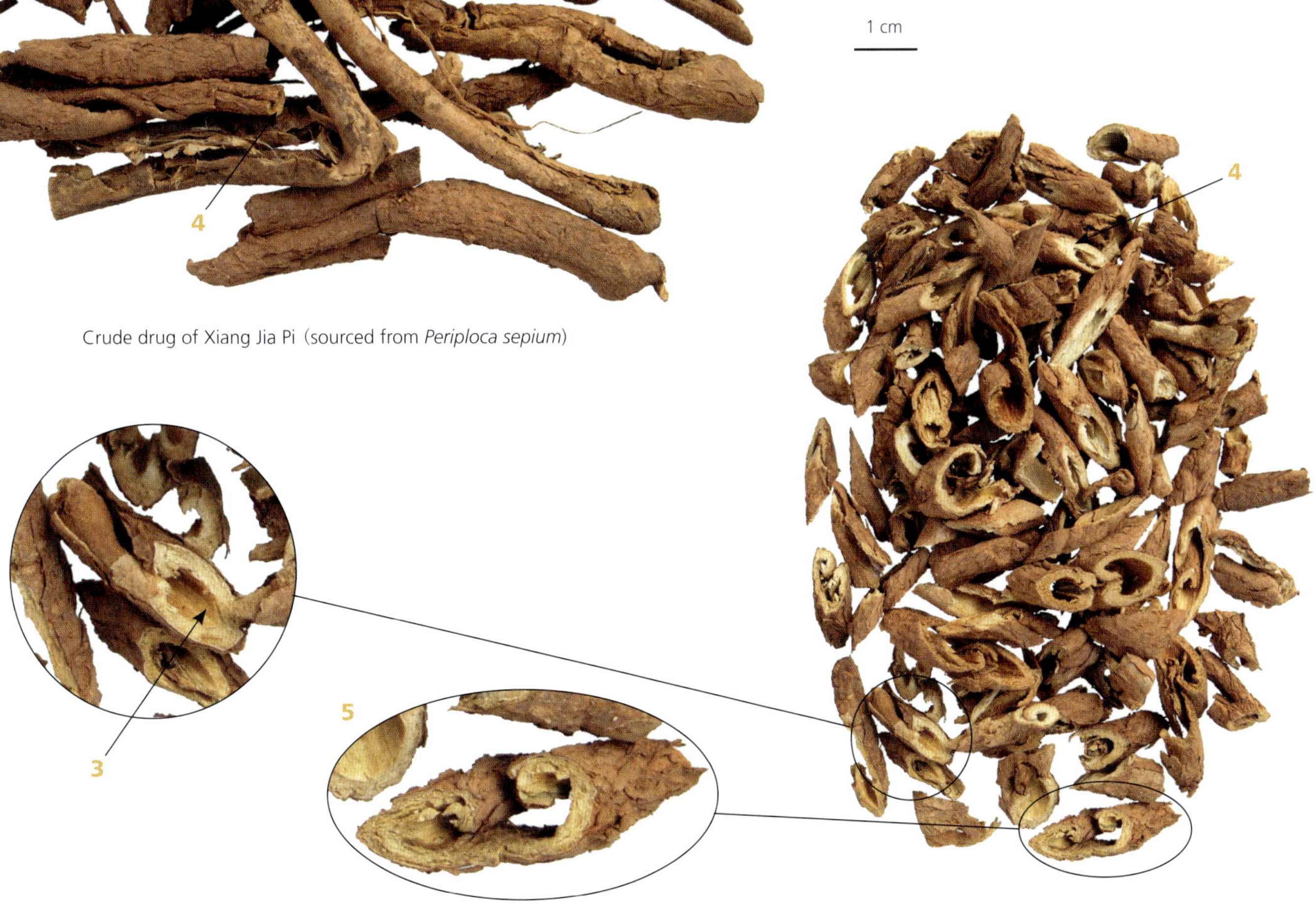

Decocting slices of Xiang Jia Pi (sourced from *P. sepium*)

Yuan Zhi 远志

Polygalae Radix

Drug part: ROOT-BARK

Polygala tenuifolia Willd.

Polygalaceae

synonyms: *P. sibirica* L. var. *angustifolia* Ledeb.; *P. sibirica* f. *tenuifolia* (Willd.) Backer & Moore

Species of *Polygala* wild in China: 44

PLANT DESCRIPTION

A delicate-looking perennial herb (15–50 cm tall), forming clumps; main root stout and fleshy. *Stems* many branched, softly hairy. *Leaves* alternate, almost stalkless; blade narrowly linear to linear-lance shaped (10–30 mm long; 0.5–1 (rarely up to 3) mm wide), base wedge-shaped, tip shortly pointed; mid-vein raised on lower surface; margin entire and down-rolled. *Flowers* widely spaced in 1-sided terminal spikes (5–7 cm long). Bracts and bracteoles present, very small. Each flower, solitary, sepals 5 (inner 2 'wings' large (c. 5 mm long), petal-like and dull purplish-pink); petals 3 (2 much smaller than 'wings') purplish-pink, lower-most 'keel' forming a tube (enclosing the stamens) and open on one side; apex with a purplish-blue, very finely divided spreading, feather-like appendage; stamens 8; ovary superior. *Fruit* dry and disk-shaped, with 2 lobes (c. 4 mm across); seed 2 and black. Fl. and fr. May–Sep.

HARVESTING, SOURCING AND NATURAL RANGE

Roots are harvested in spring or autumn, mostly from medicinally cultivated plants grown in Shanxi and Shaanxi provinces. Traditional *Di Dao*-quality drug is sourced from the latter in the districts of Yangao, Wenxi, Yuci, Ruicheng and Wanrong.

The wild species is native to N, W and E China (Gansu, Hebei, Heilongjiang, Henan, Inner Mongolia, Jiangsu, Jiangxi, Liaoning, Ningxia, Qinghai, Shaanxi, Shanxi and Sichuan) where it grows in open grasslands, forests and scrub, usually 500–2,300 m altitude. It is also wild in the Korean Peninsula, Mongolia and Russia.

CONSERVATION STATUS

China: Least Concern (MEPCAS, 2013: as 'variety sibirica' only).

Global: Not Evaluated (IUCN Red List, accessed, 7/4/15).

P. sibirica L.

Polygalaceae

PLANT DESCRIPTION

Similar to PS1 but shorter (10–30 cm), more upright. *Stems*, *branchlets* and *inflorescences* **softly hairy.** Lower *leaves* ovate (c. 6 mm; 4 mm wide), upper leaves **wider** than those of PS1 and lance-shaped or elliptic/lance-shaped (10–20 mm long; 3–6 mm wide); tips broad. *Flower spikes* not terminal; fewer flowered. The two inner petal-like sepals **greenish** (c. 7.5 mm long); petals **bluish-purple**; stamens all similar. *Fruit* slightly notched at base. Fl. Apr–Jul, fr. May–Aug.

HARVESTING, SOURCING AND NATURAL RANGE

Unlike PS1, roots of this species are wild harvested. The species grows in limestone mountains of N, W and SW China (Gansu, Guizhou, Hebei, Heilongjiang, Henan, Inner Mongolia, Jilin, Liaoning, Ningxia, Qinghai, Shaanxi, Shanxi, Sichuan, Tibet, Yunnan) in open forests, forest margins and grassy slopes and roadsides (1,100–3,300 m altitude). Also wild in Bhutan, India, Japan, Kashmir, Korean Peninsula, Mongolia, Myanmar, Nepal, Russia; SW Asia, Australia and Europe.

CONSERVATION STATUS

China: Least Concern (MEPCAS, 2013: as 'variety *sibirica*' only); Least Concern (authors' pers. comm. 2004.)

Global: Not Evaluated (IUCN Red List, accessed 7/4/15).

TCM PROPERTIES AND USE

properties: **'pungent, bitter and slightly warm'**
channels: **'Heart and Lung'**
actions: **'calms the shen, tonifies Heart qi, resolves phlegm'**

indications: **insomnia; anxiety; palpitations; restlessness; confusion; cough, phlegm, dyspnoea; sores and abscesses**

Medicinal crop of *Polygala tenuifolia* growing in a *Di Dao* locality on the border of Shaanxi and Shanxi provinces

P. tenuifolia, close-up of leaves

P. tenuifolia with linear leaves

P. tenuifolia, in flower in the wild, Hebei province

P. tenuifolia, developing fruits

Polygala sibirica with oval leaves showing raised mid-vein and down-rolled margins

P. sibirica in flower bud; note broader leaves compared to *P. tenuifolia*

P. sibirica: freshly uprooted wild plant

Polygala tenuifolia

Polygalaceae

DRUG MORPHOLOGY

CRUDE DRUG (DRIED ROOT): somewhat cylindrical, solid, and slightly curved (3–15 cm long; 3–8 mm wide). Root-bark thin yellowish-grey to greyish-brown, smooth in places [1], older roots densely covered in transverse furrows, sometimes becoming nodular [2]; lenticels occasional, transverse, linear and raised [3]. *Texture*: hard and brittle. *Fracture:* cortex brownish-yellow; central xylem pale yellowish-white. *Odour*: faint. *Taste*: bitter, slightly pungent, irritates throat on chewing.

DECOCTING PIECES-1 (CUT LENGTHS OF ROOT-BARK): as for crude drug but shorter (c. 1.5–5 cm long) and hollow. *Texture:* hard and brittle. *Odour:* faint. *Taste:* bitter, slightly pungent, irritates throat on chewing. ***Good quality*:** roots hollow and pale; outer bark thin and cortex thick.

DECOCTING PIECES-2 (LIQUORICE-PROCESSED ROOT-BARK): as for unprocessed decocting lengths but darker (yellowish-brown) and slightly sweet.

PROCESSING

After the rootlets have been removed and discarded, roots are sun-dried (= crude drug) the central woody xylem removed and discarded and remaining parts cut into short lengths (= decocting pieces). The latter may be further processed by decocting with liquorice (Gan Cao — Glycyrrhizae Radix, not included in this book) then dried again (= 制远志 Zhi Yuan Zhi).

OTHER COMMON NAMES

远志 Yuan Zhi (plant name); 远志棍 Yuan Zhi Gun; 远志筒 Yuan Zhi Tong; 远志肉 Yuan Zhi Rou; Chinese senega; thin-leaf milkroot root (CP2010); polygala (SCN).

SAFETY

Official Chinese healthfood (Anon., 2002a,b). Safety Class: 2b and 2d; Interaction Class: A (Gardner & McGuffin, 2013).

P. sibirica

Polygalaceae

DRUG MORPHOLOGY

CRUDE DRUG (DRIED ROOT-BARK): as for PS1.

DECOCTING PIECES-1: (CUT LENGTHS OF ROOT-BARK): as for PS1.

DECOCTING PIECES-2: (LIQUORICE-PROCESSED ROOT-BARK): as for PS1.

The herbal drugs of PS1 and PS2 cannot be distinguished macroscopically other than by the diameter of their roots and root cortex. Roots of PS1 are usually wider as they are normally from cultivation.

PROCESSING

As for PS1.

OTHER COMMON NAMES

卵叶远志 Luan Ye Yuan Zhi (plant name); 远志棍 Yuan Zhi Gun; 远志肉 Yuan Zhi Rou; Chinese senega; Polygala (SCN) root; Siberian milkwort; Siberian polygala; Radix Polygalae; thin-leaf milkwort root (CP2010).

SAFETY

As for PS1.

UNOFFICIAL SUBSTITUTES

ALL PLANT PARTS of both Pharmacopoeia species above are used locally in China (Bensky *et al.*, 2004; NICBP 1997–2002).

All plant parts of *P. sibirica* (dried)

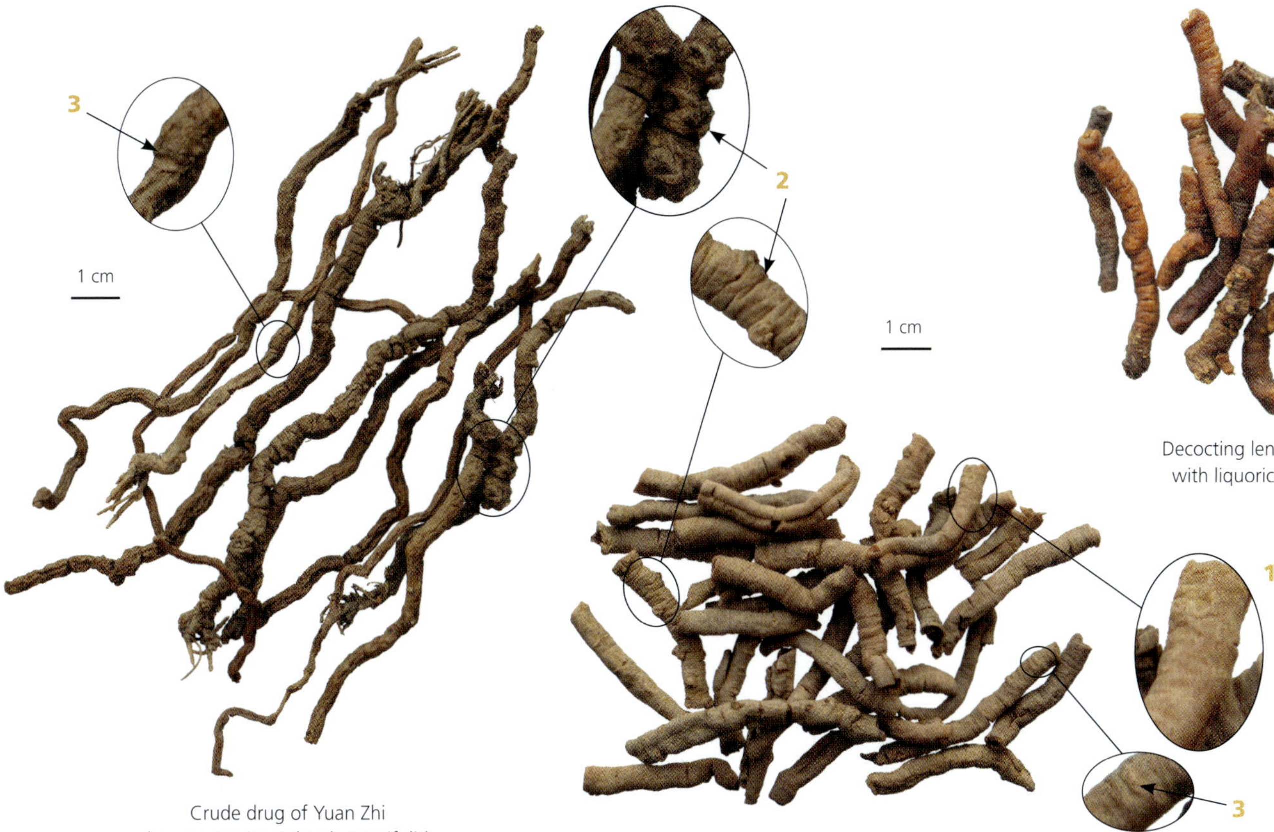

Crude drug of Yuan Zhi
(source species *Polygala tenuifolia*)

Decocting lengths of Yuan Zhi
(source species *P. tenuifolia*)

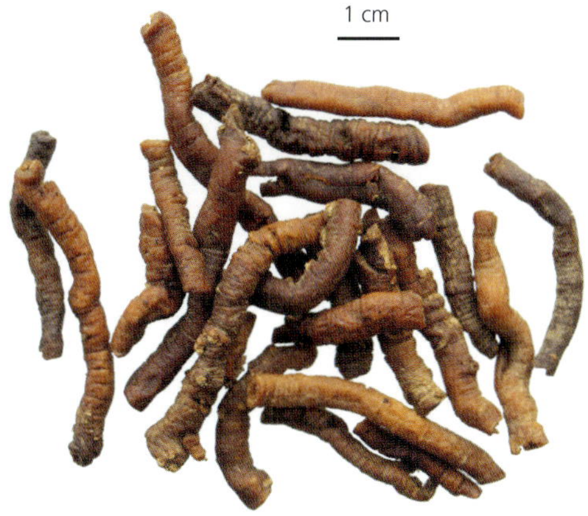

Decocting lengths of Zhi Yuan Zhi processed
with liquorice (source species *P. tenuifolia*)

Crude drug of Yuan Zhi
(source species *Polygala sibirica*)

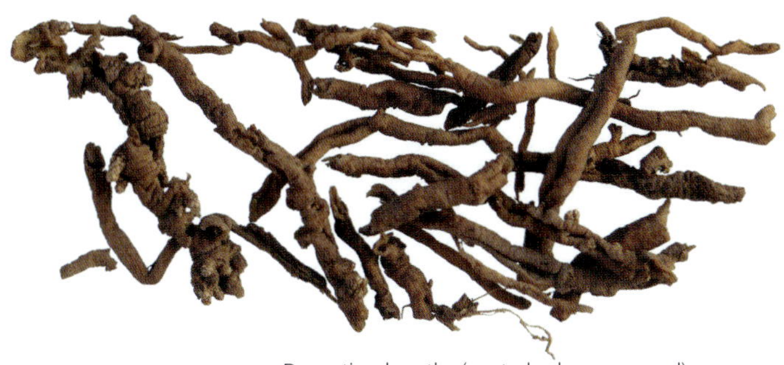

Decocting lengths (central xylem removed)
of Yuan Zhi processed with liquorice
(sourced from roots of *P. sibirica*)

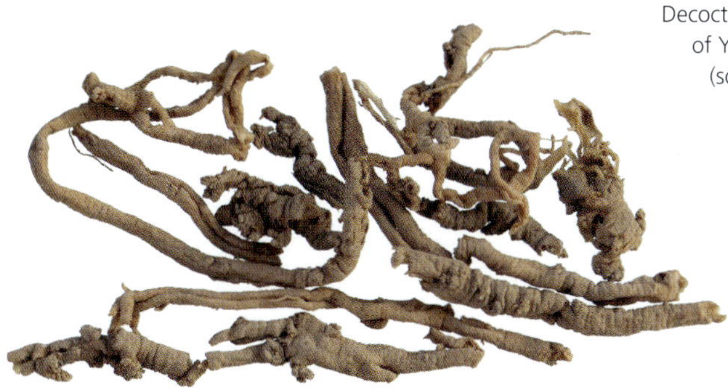

Decocting lengths (central xylem removed) of Yuan Zhi
(sourced from roots of *P. sibirica*)

Eriobotrya japonica plantation

leaves

Ai Ye 艾叶 Artemisiae Argyi Folium

Drug part: LEAF

Artemisia argyi H. Lév. & Vaniot

Asteraceae

Species of *Artemisia* wild in China: 186

PLANT DESCRIPTION

An erect, perennial herb or subshrub 0.8–1.5 (rarely 2.5) m tall; aromatic. *Stems* with greyish-white 'bloom', branching apically. *Leaves* alternate and variable in shape and degree of lobing; middle stem leaves with short stalks, usually broadly ovate or triangular (5–8 cm long; 4–7 cm wide), upper surface with minute, pale gland dots [hl] and sparse soft white hairs [hl], lower surface covered with a dense mat of soft, grey hairs; margins 1- or 2-pinnatiform with shallow or deep lobing; leaf base tapering along leaf stalk; upper stem leaves smaller, unlobed or 3-pinnatiform. The green leaf-like bracts are much smaller (1–2 cm long; 2–4 mm wide), virtually stalkless, usually elliptic or lance-shaped with entire margins. *Inflorescences* lax, leafy and with branches up to 10 cm long. *Flowers* tiny in dense, inconspicuous many-flowered, greyish heads (capitula) (2–4 mm. diam.); each capitulum with 6–10 outer female florets and 8–12 inner bisexual florets. *Fruits* are dry ovoid or oblong achenes. Fl. and fr. Jul-Oct.

HARVESTING, SOURCING AND NATURAL RANGE

Leaves are harvested mostly from wild plants before flowering in spring or early summer; the species is also frequently grown on small-holdings for many other uses too. In the wild the species grows in open forests and disturbed places up to 1,500 m altitude throughout much of China: Anhui, Fujian, Gansu, Guangdong, Guangxi, Guizhou, Hebei, Heilongjiang, Henan, Hubei, Hunan, Inner Mongolia, Jiangsu, Jiangxi, Jilin, Liaoning, Ningxia, Shaanxi, Shanxi, Shandong, Sichuan, Yunnan and Zhejiang provinces [also native in the Korean Peninsula, Mongolia and Russia].

CONSERVATION STATUS

China: Least Concern (authors' pers. obs., 2011).
Global: Not Evaluated (IUCN Red List, accessed 20/8/2016).

Artemisia argyi in the wild

Upper stem of *A. argyi* in fruit

Fresh Ai Ye for sale, Shaanxi province

UNOFFICIAL SUBSTITUTES

With over 180 species wild in China, the genus *Artemisia* is a large and difficult group to identify to species level. The leaves of 4 other *Artemisia* species have been reported as local and/or historical unofficial substitutes (Xiao *et al.*, 2001–2002; Bensky *et al.*, 2004); their occurrence as substitutes in international trade is uncertain:

1. ***Artemisia stolonifera*** (Maxim.) Kom. N, NE and E China (12 provinces). Not strongly aromatic; **middle stem leaves unstalked**.
2. ***A. mongolica*** (Fisch. ex Besser) Nakai. Throughout most of China. **Leaves without white gland dots**.
3. ***A. princeps*** Pamp. N, S, E and W China (19 provinces). **Leaves without white gland dots**; outermost lobe larger than lateral lobes; flower-heads nodding.
4. ***A. indica*** Willd. N, S, E and W China (24 provinces). **Leaves without white gland dots**; middle stem leaves with 3 (or 4) pairs of segments, outermost lobe larger than lateral lobes.

TCM PROPERTIES AND USE

properties: **'bitter, pungent and warm'**
channels: **'Spleen, Liver and Kidney'**
actions: **'stops bleeding, dispels cold and damp, relieves pain'**

indications: **dyspnoea; infertility; threatened miscarriage; abdominal pain; topically for itching; (T)**

DRUG MORPHOLOGY

CRUDE DRUG/DECOCTING PIECES-1 (DRIED LEAF): a mass of crumpled leaves, sometimes broken. When intact and expanded, leaves variable in shape and degree of lobing **[1]**; most leaves are those from the middle stem region of the plants and typically are short-stalked **[2]**, broadly ovate or triangular (5–8 cm long; 4–7 cm wide), upper surface with minute, pale gland dots [hl] **[3]** and sparse soft white hairs [hl] **[4]**, lower surface covered with a dense mat of soft, grey hairs **[5]**; margins 1- or 2-pinnatiform and shallowly or deeply lobed; leaf base tapering along leaf stalk. Occasional much smaller, often unlobed upper stem leaves present **[6]**. *Texture*: soft. *Odour*: aromatic. *Taste*: bitter. ***Good quality***: thick, soft leaves, white tomentose on lower surfaces and intensely aromatic.

DECOCTING PIECES-2 (CARBONISED & VINEGARED LEAVES — CU AI TAN): as for above but blackish-brown throughout. *Odour*: vinegar-scented.

PROCESSING

Leaf-stalks are removed and discarded then leaves sun-dried (= crude drug/decocting pieces-1). Often further processed by carbonising then spraying with vinegar before being dried again (Cu Ai Tan 醋艾炭 = decocting pieces-2).

OTHER COMMON NAMES

Ai Tan 'charred mugwort'; Ai Ye Tan 'charred mugwort leaves'; 大叶艾 Da Ye Ai; 灸草 Jiu Cao; Folium Artemisiae Argyi; Chinese mugwort (SCN); sweet wormwood herb (CP2010).

SAFETY

Slightly toxic (CP2015).

Crude drug/decocting pieces-1 of Ai Ye (*Artemisia argyi*)

Cu Ai Tan (decocting pieces-2) leaves of *A. argyi*, carbonised and vinegared

Close-up of lower leaf surface

Close-up of upper leaf surface

Expanded leaves of Ai Ye showing variation in size, shape and lobing

Da Qing Ye 大青叶 Isatidis Folium

Drug part: LEAF

PHARMACOPOEIA SPECIES

Isatis tinctoria L.

Brassicaceae

synonym: *I. indigotica* Fortune ex Lindl. (name used in CP2015 as '*Isatis indigotica* Fort.').
taxonomic note: many TCM texts treat *I. indigotica* as a separate species but *FoC* recognises a huge variability in *I. tinctoria* and includes *I. indigotica* (like *Kew-MPNS*, 17/6/2016) within it.

Species of *Isatis* wild in China: 4

PLANT DESCRIPTION

A highly variable species. A biennial herb (40–100 (rarely 30–150) cm tall). *Stem* branching above, often bluish-green, hairy or hairless; membranous sheath (ocrea) absent. *Leaves* simple and **alternate**. Basal leaves (if present) often in a rosette, stalked, **blades oblong or broadest above the middle** (usually 5–20 cm long; 1.5–6 cm wide), **bases tapering**, margins wavy or sometimes toothed; **tips broadly rounded**; occasionally hairy. Middle stem leaves unstalked, becoming smaller up the stem, blades lance-shaped (usually 3–7 cm long; 0.8–2.5 cm wide), sometimes with 2 auricles (backward-pointing 'ears') or clasping the stem, margins entire, tips pointed. ***Flowers*** (appearing in second year) in many-branched sub-rounded heads; latter axillary and terminal. Each flower with 4, ovate or oblong sepals; petals 4, bright yellow, equal, wedge-shaped and alternating with sepals; stamens 6 (in 2 whorls); style 1. *Fruit* a dry, 2-valved capsule, flattened, pendant and variously shaped (slightly broader towards the tip), black or purplish-brown; fruit-stalk (pedicel) swollen below the fruit. Fl. Apr–Jun, fr. May–Jul. *Since ancient times the leaves have also been used as an important source of (known as 'woad' in Europe).*

HARVESTING, SOURCING AND NATURAL RANGE

I. tinctoria is not native to China but has been long cultivated there for use as a medicine (roots and leaves) and dye (leaves); many forms or cultivars are thought to exist. 'Da Qing Ye' is sourced from basal leaves of the plant before it has flowered with medicinal crops mostly grown in Anhui, Hebei, Jiangsu and Shaanxi provinces (Zhao & Chen, 2014; Zhu, 1998). The species has become widely naturalised throughout much of temperate China in fields and other disturbed places between 600–2,800 m altitude [it is native in Europe].

CONSERVATION STATUS

China: Least Concern (MEPCAS, 2013).
Global: Not Evaluated (IUCN Red List, accessed 17/6/2015). An invasive plant in some localities.

For other TCM herbal drugs sourced from this species see 'Ban Lan Gen' (Isatidis Radix) page 46.

UNOFFICIAL SUBSTITUTE 1

Polygonum tinctorium Aiton

Polygonaceae

synonym: *Persicaria tinctoria* (Aiton) Spach (Kew-MPNS accepted name, 29/10/2016).

PLANT DESCRIPTION

A branching annual herb (50–80 cm tall). Stem nodes slightly swollen, **typically surrounded by a membranous, tubular and hairy-fringed sheath (ochrea) (1–1.5 cm long).** Leaves ovate or broadly elliptic (3–8 cm long; 2–4 cm wide), **bases broadly wedge-shaped**, apex broad or slightly pointed, margin entire and fringed with short hair. Flowers tiny (2.5–3 mm diam.) in **slender 3–5 flowered spikes (2–5 cm long)**; 'petals' 5, **pinkish-white**. Fruits dry, tiny and 3-angled in terminal heads (2–2.5 mm long). Fl. Jun–Sep, fr. Aug–Oct.

SUBSTITUTION COMMENT. Early editions of the CP recognised *P. tinctorium* as an official source of 'Da Qing Ye' and although it has similar TCM properties it is now treated as a separate TCM drug 'Liao Da Qing Ye' (Polygoni Tinctorii Folium) see page 546 (CP2015).

CONSERVATION STATUS

China: Least Concern (Lin Yu Lin pers. obs.)
Global: Not Evaluated (IUCN Red List, accessed 28/8/2016).

UNOFFICIAL SUBSTITUTE 2

Strobilanthes cusia (Nees) Kuntze

Acanthaceae

synonym: *Baphicacanthus cusia* (Nees) Bremek.; *Goldfussia cusia* Nees

PLANT DESCRIPTION

An upright perennial (0.5–1.5 m tall) **often woody** at base. **Leaves opposite, blades elliptic to ovate** (4–20 cm long; 2–9 cm wide) and **tips pointed**. Flowers **much larger (3.5–5 cm long)** than those of the Pharmacopoeia species, **blue, conspicuous** and **funnel-shaped** with 5 notched equal-sized lobes. Flowers intermingled with leaf-like bracts. Fruit a dry, elongate, 4-seeded capsule. Fl. Jul–Feb, fr. Dec–Feb.

CONSERVATION STATUS

China: Least Concern (MEPCAS, 2013).
Global: Not Evaluated (IUCN Red List, accessed 20/8/2016).

EHTPA 4/14
GAP

TCM PROPERTIES AND USE

properties: **'bitter and very cold'**
channels: **'Heart, Lung and Stomach'**
actions: **'clears heat, eliminate toxins, cools Blood'**

indications: **fevers; sore throat; rash; haematemesis**

Medicinal crop of *Isatis tinctoria*, Hebei province (ready for harvest as 'Da Qing Ye' (i.e. before flowering)

Flower-heads of *I. tinctoria*

Stem leaves of *I. tinctoria* clasping or 'eared'

Flowering crop of *I. tinctoria*

Freshly harvested crop of Da Qing Ye (*I. tinctoria*), sun-drying

Flowering crop of *Polygonum tinctorium*, Hebei province

Flowering plant of *P. tinctorium*

SUBSTITUTION COMMENT. As the Chinese name for *S. cusia* suggests, 'Nan Da Qing Ye' (southern Da Qing Ye) is a species from southern China and, owing to its similar TCM properties, was included in earlier editions of the CP as an official alternative of 'Da Qing Ye'. The CP2015 no longer recognises this species as 'Da Qing Ye' but the species continues to be used medicinally in southern China where its continued trade under both names 'Nan Da Qing Ye' and 'Da Qing Ye' is a source of confusion on domestic and international markets (Lin, 1990–1996; Zhao, 2007). *Underground parts of this species are the official CP2015 source of another TCM herbal drug 'Nan Ban Lan Gen'* (see page 220).

Isatis tinctoria

Brassicaceae

DRUG MORPHOLOGY

CRUDE DRUG (DRIED LEAF): Leaf blades crumpled, curled and often broken **[1]**. Apparent long, narrow leaf stalks are actually **long tapering leaf bases (4–10 cm long) yellowish-white or pale brownish [2]** (the leaves are unstalked). **Leaf blades with a broad yellowish-white mid-vein [3];** when intact and expanded, leaf blades oblong or oblanceolate (widest above the middle) (**5–20 cm long;** 2–6 cm wide) with upper surface bluish-green or dark greyish-green **[4]** and lower surface darker **[5]**; margins wavy or entire and **tips broadly rounded**. *Texture:* fragile and brittle. *Odour:* indistinct. *Taste:* bitter and astringent. ***Good quality:*** leaves thick and bluish-green; stems absent.

DECOCTING PIECES (DRIED CUT LEAVES): as for the crude drug but cut into small pieces; key characters such as leaf base shape may be obscured.

PROCESSING

Leaves are sun-dried (= crude drug) then washed, cut into small pieces and dried again (= decocting pieces).

OTHER COMMON NAMES

菘蓝 Song Lan (plant name); 菘蓝大青叶 Song Lan Da Qing Ye; dyer's woad (SCN for *Isatis tinctoria*) dyer's woad leaf (CP2010); Folium Isatidis; isatis (SCN for *Isatis indigotica*).

Polygonum tinctorium

Polygonaceae

'DRUG SUBSTITUTE' MORPHOLOGY

Care is need in order to distinguish dried leaves of this species from those of *I. tinctoria* (the Pharmacopoeia species above). A useful character is the presence of pale membranous and tubular stem sheaths (ocrea) (1–1.5 cm long) **[6]** with one end fringed with long hairs (cilia); sheaths may be attached or detached from stems. Leaf blades, when whole and expanded, ovate or broadly elliptic (3–8 cm long; 2–4 cm wide), **bases broadly wedge-shaped** (not tapering), apex broad or slightly pointed, margin entire and fringed with short hairs. Laboratory-based authentication methods may be needed for more reliable identification.

OTHER COMMON NAMES

蓼大青叶 Liao Da Qing Ye; 蓼蓝大青叶 Liao Lan Da Qing Ye; Polygoni Tinctorii Folium; Indigo plant leaf (CP2010); polygonum indigo (SCN).

Strobilanthes cusia

Acanthaceae

'DRUG SUBSTITUTE' MORPHOLOGY

Leaves stalked [7] and **usually much shorter** (0.5–7 cm long) **[8]**; blades, when whole and expanded, **elliptic to ovate** (4–20 cm long; 2–9 cm wide), hairless, lower surface **[9]** paler than upper, **drying blackish-brown**, secondary veins 7–9 each side of mid-vein, base tapering, **tips pointed, margins shortly saw-toothed**.

OTHER COMMON NAMES

板蓝 Ban Lan (plant name); 广东大青叶 Guangdong Da Qing Ye; 马蓝 Ma Lan; Nan Da Qing Ye 'southern Da Qing Ye'; strobilanthes cusia (SCN).

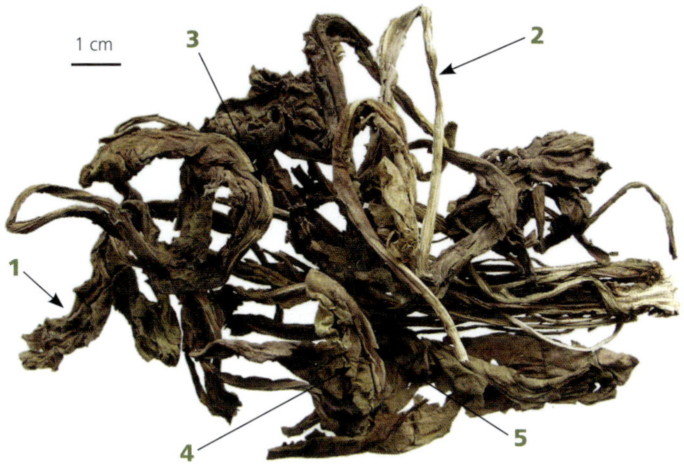

Crude drug of Da Qing Ye (*Isatis tinctoria*)

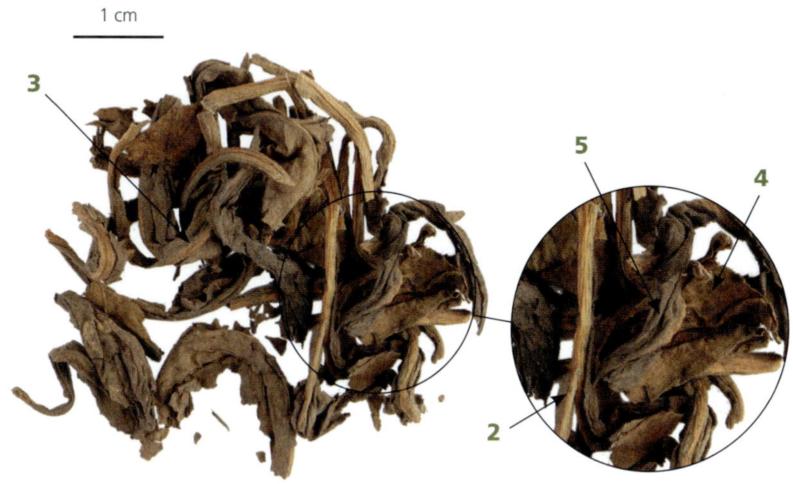

Decocting pieces of Da Qing Ye (*I. tinctoria*)

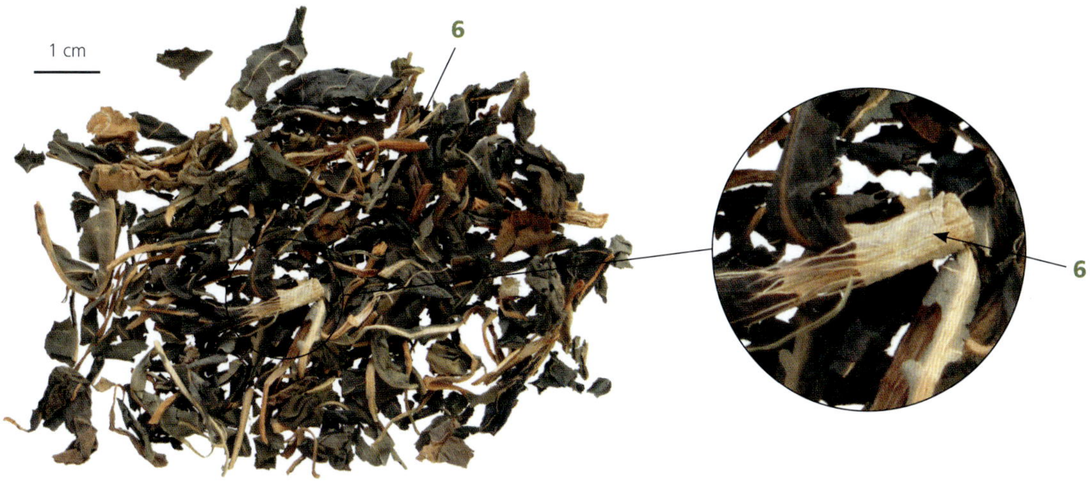

Dried leaves of *Polygonum tinctorium* (substitute)

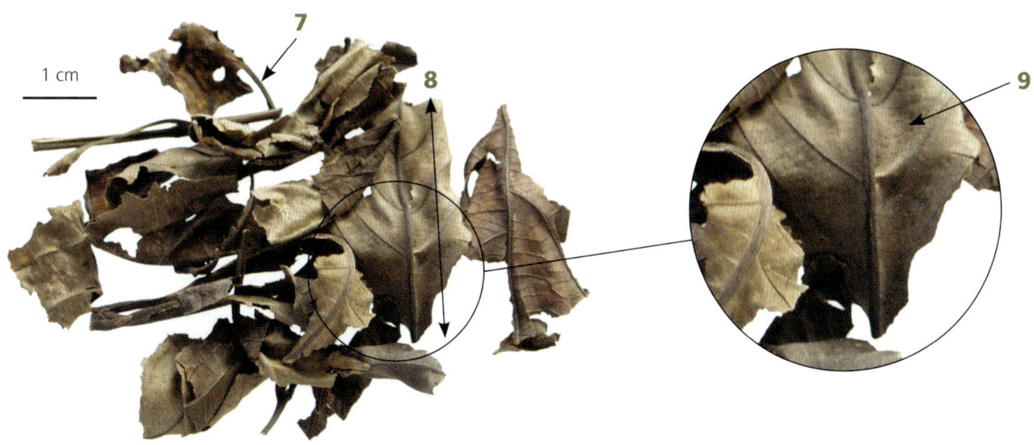

Dried leaves of *Strobilanthes cusia*

Du Zhong Ye 杜仲叶

Eucommiae Folium

Drug part: LEAF

Eucommia ulmoides Oliv.

Eucommiaceae

Species of *Eucommia* wild in China: **1**

PLANT DESCRIPTION

A deciduous tree (up to 20 m tall; up to 50 cm d.b.h.). *Bark* greyish-brown and rough. Distinctive silver-grey latex in most parts, easily visible as elastic parallel strands in cut bark and severed leaves. *Leaves* alternate, simple, stalked; blades usually elliptic, sometimes ovate (5–15 cm long; 2.5–7 cm wide), base rounded or wedge-shaped, tip abruptly pointed, lateral veins in 6– 9 pairs, margins densely saw-toothed and teeth tipped with tiny glands. *Flowers* without petals (wind-pollinated), inconspicuous and superficially leaf-like; appearing before leaves; male and female flowers on separate trees. Male flowers in axillary clusters, stamens 5–12, linear (c. 1 cm long). Female flowers solitary, with a 2-lobed apex and 2 backward-pointing stigmas; ovary c. 1 cm long. *Fruits* in pendant clusters; each fruit (samara) dry, narrowly oblong or elliptical (2.5–3.5 cm long; 1–1.3 cm wide), flattened and winged. Single seed, linear and flattened. Fl. Mar–May, fr. Jun–Nov.

HARVESTING, SOURCING AND NATURAL RANGE

Leaves are harvested when mature in summer and autumn from cultivated trees. See also 'Du Zhong' (Eucommiae Cortex) page 500.

CONSERVATION STATUS

see 'Du Zhong' (Eucommiae Cortex) page 500.

Mature, cultivated trees of *Eucommia ulmoides*, Beijing province

Torn leaf of *E. ulmoides* revealing silver-grey latex

TCM PROPERTIES AND USE

***properties*: 'pungent and warm'**
***channels*: 'Kidney and Liver'**
***actions*: 'tonifies Liver and Kidneys, tonifies yang'**

***indications*: dizziness, vertigo; bone fractures**

DRUG MORPHOLOGY

CRUDE DRUG/DECOCTING PIECES (DRIED LEAF): whole or fragmented. When whole and expanded, usually elliptic, sometimes ovate or broadest nearer the tip (7–15 cm long; 3.5–7cm wide); stalks often still attached (1–2.5 cm long); leaf bases rounded or wedge-shaped **[1]**; tip abruptly pointed **[2]**; blades yellowish-brown or bluish-green, upper surface obviously darker than the lower; lateral veins in 6–9 pairs **[3]**; margins densely saw-toothed and tipped with tiny glands **[4]**. *Texture*: thin and brittle; silvery elastic latex threads sometimes present. *Odour*: indistinct. *Taste*: slightly bitter.

PROCESSING

Leaves are dried in the sun or at low temperature (= crude drug/decocting pieces).

OTHER COMMON NAMES

Eucommia leaf (CP2010); Folium Eucommiae.

Crude drug/decocting pieces of Du Zhong Ye (dried leaves of *Eucommia ulmoides*)

Du Zhong Ye (*Eucommia ulmoides*)

Liao Da Qing Ye 蓼大青叶 Polygoni Tinctorii Folium

Drug part: LEAF

Polygonum tinctorium W. T. Aiton

Polygonaceae

synonym: *Persicaria tinctoria* (Aiton) Spach (*Kew-MPNS* accepted name, 29/10/2016).

Species of *Polygonum* wild in China: 113

PLANT DESCRIPTION

A branching annual herb (50–80 cm tall). *Stem* nodes slightly swollen and typically surrounded by a membranous, tubular and hairy-fringed sheath (ochrea) (1–1.5 cm long). *Leaves* simple, alternate, ovate or broadly elliptic (3–8 cm long; 2–4 cm wide), bases broadly wedge-shaped, margins entire, fringed with short hairs, tips broad or slightly pointed. *Flowers* in 3–5-flowered terminal spikes (2–5 cm long). Each flower tiny (2.5–3 mm diam.), bisexual, 'petals' 5 and pinkish-white; stamens usually 7 or 8; styles 2 or 3; bracts and bracteoles membranous. *Fruit* dry and 3-angled (2–2.5 mm long) with persistent membranous flower parts. Fl. Jun–Sep, fr. Aug–Oct.

HARVESTING, SOURCING AND NATURAL RANGE

Aerial parts are harvested twice yearly in summer and autumn, mostly from cultivated plants grown throughout much of China. The wild species grows in damp places often along stream-sides (200–1,000 m altitude) and is widespread throughout the country (also native in Indo-China and occasionally naturalised elsewhere).

This species is also used widely as a traditional dye plant.

CONSERVATION STATUS

China: Least Concern (authors' pers. obs 2012.)
Global: Not Evaluated (IUCN Red List, accessed 17/6/2016).

Flowering crop of *Polygonum tinctorium*, Hebei province

Flowering plants of *P. tinctorium*

TCM PROPERTIES AND USE

properties: **'bitter and cold'**
channels: **'Heart and Stomach'**
actions: **'clears heat, removes toxins, cools Blood'**

indications: **fevers; abscesses, sores; cough, sore throat**

DRUG MORPHOLOGY

CRUDE DRUG/DECOCTING PIECES (DRIED LEAF): a mass of crumpled and usually fragmented leaves with blades bluish-green or blackish-blue **[1]**. Whole and expanded leaves simple, broadly elliptical or ovate (3–8 cm long; 2–4 cm wide), bases broadly wedge-shaped, margins entire and fringed with short hairs, apex broad or slightly pointed. Leaf veins pale yellowish-brown and distinctly paler than leaf blades **[2]**; slightly prominent on lower surface. Leaf stalks flattened. Membranous tubular and long-fringed stem sheaths (ochreas) (1–1.5 cm long) occasionally present **[3]**, sometimes still attached to stem remnants. *Texture*: brittle. *Odour*: indistinct. *Taste*: slightly astringent and bitter.

PROCESSING

After removal of stems, leaves are dried (= crude drug/decocting pieces).

OTHER COMMON NAMES

蓼蓝大青叶 Liao Lan Da Qing Ye; Folium Polygoni Tinctorii; Indigo plant leaf (CP2010); polygonum indigo (SCN).

Crude drug/decocting pieces of Liao Da Qing Ye (*Polygonum tinctorium*)

Luo Bu Ma Ye 罗布麻叶

Apocyni Veneti Folium

Drug part: LEAF

Apocynum venetum L.

synonym: *Trachomitum venetum* (L.) Woodson

Apocynaceae

Species of *Apocynum* wild in China: 2

PLANT DESCRIPTION

A perennial herb (up to 4 m) with white latex. *Leaves* simple, usually opposite, shortly stalked, blade narrowly elliptic to narrowly ovate (1–8 cm long; 0.5–2.2 cm wide); margin finely toothed. *Flowers* in branched loose terminal heads; scented; sepals 5, narrowly elliptic or narrowly ovate (c. 1.5 mm long), purplish red or pink; petals pink, whitish-pink or purple, fused into a bell-shaped tube (6–8 mm long), deeper than wide, with a wide throat, lobes 5 (3–4 mm); stamens 5 and joined to pistil head. *Fruits* usually in pairs, dry and slender (8–20 cm long; 2–3 mm wide), opening along one side; seeds many, each with a conspicuous white tuft of hairs. Fl. Apr–Sep; fr. Jul–Dec.

HARVESTING, SOURCING AND NATURAL RANGE

Leaves are harvested in summer from wild or cultivated plants. Main crop growing areas are in NW, N and NE China. Wild plants grow in salt pans, desert margins, alluvial zones and riversides in N, C and W China (Gansu, Hebei, Henan, Inner Mongolia, Jiangsu, Liaoning, Qinghai, Shaanxi, Shandong, Shanxi, Tibet and Xinjiang). [The species is also wild in India, Japan, Mongolia, Pakistan, Russia; SW Asia and Europe].

CONSERVATION STATUS

China: Least Concern (MEPCAS, 2013).

Global: Not Evaluated (IUCN Red List, accessed 12/2/2015).

Apocynum venetum, medicinal crop coming into flower

Flowering plant

Broken stem of *A. venetum* exposing latex

A. venetum, close-up of flower-head

A. venetum with developing fruit

TCM PROPERTIES AND USE

***properties*: 'sweet, bitter and cool'**
***channel*: 'Liver'**
***actions*: 'extinguishes wind, clears heat'**
***indications*: dizziness, palpitations, insomnia; hypertension; urinary disorders**

DRUG MORPHOLOGY

CRUDE DRUG/DECOCTING PIECES (DRIED LEAF): shrivelled, with some broken; when intact and expanded leaves simple, narrowly elliptic to narrowly ovate **[1]** (2–5 cm long; 0.5–2 cm wide); obtuse or with a very short, sharp point (mucronate **[2]**, base obtuse or wedged -shaped **[3]**; margin finely toothed; hairless throughout; lower surface with pale, raised midrib **[4]** and lateral veins pinnate (feather-shaped) in 7–10 pairs **[5]**; leaf stalks thin and short (c. 4 mm long) and often still attached **[6]**. *Texture*: brittle. *Taste*: indistinct. ***Good quality***: leaves pale green and unbroken.

PROCESSING

Leaves are sun-dried (= crude drug/decocting pieces).

OTHER COMMON NAMES

红麻 Hong Ma; 罗布麻 Luo Bu Ma; 野茶 Ye Cha; Folium Apocyni Veneti; dogbane leaf (CP2010).

SAFETY

Although an official healthfood in China (Anon., 2002a,b), side effects have been reported (nausea, vomiting and diarrhoea and (less often) additional epigastric effects; hypokalaemia may arise from chronic use) (Bensky *et al.*, 2004).

Crude drug/decocting pieces

UNOFFICIAL SUBSTITUTES

Two other species have been reported as local and/or historical unofficial substitutes (NICPBP, 1997–2002; Zhao, 2004); their occurrence as substitutes in international trade is uncertain:

1. ***Urena lobata*** L. (地桃花 Di Tao Hua) — whole plant or root (family Malvaceae); wild in 15 provinces in China.
2. ***Eriobotrya japonica*** (Thunb.) Lindl. (枇杷叶 Pi Pa Ye) — leaves (family Rosaceae); tree or large shrub; native in Chongqing and Hubei provinces and widely cultivated for food, cosmetics and medicines. The leaves are the source of a separate TCM drug 'Pi Pa Ye' (Eriobotryae Folium) see page 552.

Man Shan Hong 满山红

Rhododendri Daurici Folium

Drug part: LEAF

Rhododendron dauricum L.

Ericaceae

Species of *Rhododendron* wild in China: 571

PLANT DESCRIPTION

A many-branched, semi-evergreen shrub (0.5–2 m tall). *Young branches* thin, curved and hairy. *Leaves* simple and alternate, sometimes clustered at tips of branches; leaf stalks (2–10 mm long), softly hairy; leaf blades elliptic or oblong (1–7.5 cm long; 1–3 cm wide), somewhat leathery, base wedge-shaped, tip broadly rounded, margin entire or minutely toothed; leaf blade underside pale green with dense scaly glands; leaf blade upperside darker green and sparsely scaly. *Flowers* large and showy, in tight unbranched groups of 4, **usually opening before the leaves appear**. Each flower short-stalked (0.2–0.8 cm); calyx with 5 overlapping densely scaly lobes; petals fused to form a broad, 5-lobed (rarely up to 8-lobed) funnel, pink or purplish-red (1.3–2.3 cm diam.), externally often softly hairy; stamens 5–10, **filaments pubescent in lower parts**, anthers purplish-red; style longer than corolla. Bracteoles paired and basal. *Fruit* a dry, brown, cylindrical cone-shaped or ovoid capsule (10–15 mm long; c. 5 mm diam.), apex 5-lobed, calyx persistent. *Seeds* many. Fl. May–Jun, fr. Jul.

HARVESTING, SOURCING AND NATURAL RANGE

Leaves are harvested from wild plants in summer and autumn. The species grows in birch and larch forests in northern and north-eastern China (Heilongjiang, Inner Mongolia and Jilin); [also native in Japan, the Korean Peninsula, Mongolia and Russia].

CONSERVATION STATUS

China: Least Concern (MEPCAS, 2013); Least Concern (Wang & Xie, 2004).
Global: Not Evaluated (IUCN Red List, accessed 13/6/2016).

TCM PROPERTIES AND USE

properties: 'pungent, bitter and cold'
channels: 'Lung and Spleen'
actions: 'relieves cough, resolves phlegm'

indications: cough, phlegm

DRUG MORPHOLOGY

CRUDE DRUG/DECOCTING PIECES (DRIED LEAF): on drying the leaves typically become recurved **[1]**, with older leaves often becoming cylindrical **[2]**; occasionally they are broken into small fragments. When whole and expanded, leaf blades elliptic or oblong (1–7.5 cm long; 1–3 cm wide), tip broadly rounded, base usually wedged-shaped **[3]**, margin entire or minutely toothed **[4]**; somewhat leathery; lower surface greyish-green and densely covered in scaly glands **[5]**; upper surface darker brownish-green or green with scattered scale-like pale yellow glands **[6]**; leaf stalks often still attached (2–10 mm long) **[7]**. *Texture:* somewhat leathery. *Odour:* aromatic. *Taste*: somewhat bitter. ***Good quality***: aromatic.

PROCESSING

Leaves are shade-dried (= crude drug/decocting pieces).

OTHER COMMON NAMES

兴安杜鹃 Xing'An Du Juan; Dahurian rhododendron leaf (CP2010); Folium Rhododendri Daurici

Decocting pieces of Man Shan Hong (dried leaves of *Rhododendron dauricum)*
(*left:* older leaves; *right*: fresher leaves)

Pi Pa Ye 枇杷叶

Eriobotryae Folium

Drug part: LEAF

Eriobotrya japonica (Thunb.) Lindl.

Rosaceae

synonyms: *Crataegus bibas* Lour.; *Mespilus japonica* Thunb.

Species of *Eriobotrya* wild in China: 14

PLANT DESCRIPTION

A small spreading, evergreen tree (up to 10 m high) with stout stems. *Branchlets* yellowish-brown, densely covered in orangish-brown hairs. *Stipule* (at base of leaf stalk) stout and needle-like (1–1.5 cm long). *Leaves* alternate and simple, with stalks short and covered in a dense mat of greyish-brown hairs. *Leaf blade* leathery and large (12–30 cm long; 3–9 cm wide), lance-shaped, ovate or elliptic-oblong and often widest towards the pointed tips, margin shortly saw-toothed; leaf upper surface rough, lower surface densely covered in soft greyish-brown or yellowish-brown hairs (tomentose) with veining pinnate and midrib distinctly raised. *Flowers* at tips of branches in somewhat compact many-flowered, branching clusters (10–19 cm long); flower-stalk densely covered in orangish-brown hairs. Flowers fragrant, 1.2–2 cm diam., basal cup (hypanthium) shallow and hairy; sepals 5, hairy and triangular; petals 5, equal size and shape, oblong or ovate, creamy-white; stamens 20; ovary superior; styles 5. *Fruit* (loquat) fleshy, short-stalked, globose or egg-shaped (1–1.5 cm diam.), yellow or orangish-yellow, becoming virtually hairless with age, sepals persistent; edible. *Seeds* 1 or 2, within a membranous core. Fl. Jun, fr. Jul–Aug.

HARVESTING, SOURCING AND NATURAL RANGE

Leaves are harvested throughout the year from cultivated plants (also grown for their edible fruit) in Anhui, Fujian, Gansu, Guangdong, Guangxi, Guizhou, Henan, Hubei, Hunan, Jiangsu, Jiangxi, Shaanxi, Taiwan, Yunnan and Zhejiang provinces and throughout SE Asia. In China, the species grows wild in Chongqing and Hubei provinces.

CONSERVATION STATUS

China: Not Evaluated.
Global: Not Evaluated (IUCN Red List, accessed 9/4/2015).

Eriobotrya japonica plantation, Fujian province

Flowering branch of *E. japonica*

Fruiting branch of *E. japonica*

Flowering bush of *E. japonica*

TCM PROPERTIES AND USE

***properties*:** **'bitter and cold'**
***channels*:** **'Lung and Stomach'**
***actions*:** **'resolves phlegm, relieves cough'**
***indications*:** **coughs; nausea, vomiting**

DRUG MORPHOLOGY

CRUDE DRUG (LEAF): leaf blades (when intact) lance-shaped, ovate or elliptic-oblong (12–30 cm long; 3–9 cm wide) and often widest towards the pointed tips **[1]**, base narrowly wedge-shaped, margin shortly saw-toothed. Upper leaf surface smooth **[2]**, yellowish-brown, greyish-green or reddish-brown; lower surface densely covered in soft greyish-brown or yellowish-brown hairs **[3]** and veining pinnate **[4]** midrib distinctly raised **[5]**. Leaf stalks very short and softly hairy. *Texture:* leathery but brittle. *Odour:* indistinct. *Taste:* slightly bitter. ***Good quality*:** unbroken green or reddish-brown leaves (sunlight causes browning) **[6]**.

DECOCTING PIECES-1 (SLICED LEAF): leaf fragments of varying shapes and sizes according to cutting style. As for the crude drug except for those characters lost due to slicing.

DECOCTING PIECES-2 (HONEY-PROCESSED LEAF): as for decocting pieces above but leaf upper surface always yellowish-brown or reddish-brown, somewhat lustrous and with a slightly viscous coating **[7]**. *Odour:* honey-like. *Taste:* slightly sweet.

PROCESSING

When partly dry, leaves are tied into small bundles then sun-dried (= crude drug). Hairs are subsequently removed from hairy-leaved cultivars (see Safety), then leaves sliced into decocting pieces of varying shape. Leaves are sometimes honey-processed by moistening them in a boiled honey solution then dry-frying them over a gentle heat until no longer sticky (= 蜜炙枇杷叶 Mi Zhi Pi Pa Ye [honey-processed Pi Pa Ye]).

OTHER COMMON NAMES

Folium Eriobotryae; loquat (SCN); loquat leaf (CP2010).

SAFETY

Cultivars with hairy leaves may irritate the mucosa and larynx; their removal and/or decoction in muslin bags is sometimes suggested to minimise such exposure. Safety Class: 1; Interaction Class: A (Gardner & McGuffin, 2013).

Crude drug (good quality)

Crude drug
(poor quality — brown)

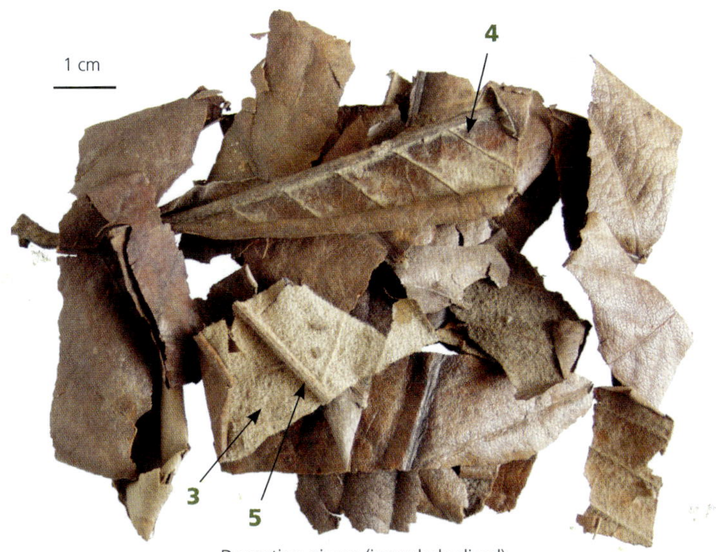

Decocting pieces (irregularly sliced)

Decocting pieces
(regularly sliced)

Decocting pieces (honey-processed)

Sang Ye 桑叶

Mori Folium

white mulberry leaf

Drug part: LEAF

Morus alba L.

Moraceae

taxonomic note: two varieties and many cultivars exist and most are thought to be used as the source of this drug.

Species of *Morus* wild in China: 10

PLANT DESCRIPTION

A deciduous shrub or small tree (3–10 m tall). *Bark* grey and shallowly furrowed. *Stipules* (at base of leaf stalks) lance-shaped (2–3.5 cm long, usually falling). *Leaves* alternate, simple, stalked and large (5–30 cm long; 5–12 cm wide), ovate or irregularly lobed; base rounded or heart-shaped, margin coarsely toothed or shallowly scalloped, tip sharply pointed or broad; **leaf undersides sparsely hairy along mid-veins or in tufts in vein axils.** *Male and female flowers* in short cylindrical catkins arising from the leaf axils and usually on separate plants. Male catkins pendent and densely white hairy (2–3.5 cm long) with tiny male flowers each with 4 pale green, overlapping calyx lobes and 4 stamens; petals are absent. Female catkins 1–2 cm long (shorter than male), softly hairy and packed with tiny female flowers; each flower with 4 green ovoid and overlapping calyx lobes (becoming fleshy in fruit), style absent and stigma with 2 distinctively white-edged and diverging branches [hl]; petals absent. *Fruit* a cylindrical or ovoid multiple fruit (syncarp), resembling a blackberry and becoming very succulent; depending on the cultivar, ripening from red to either blackish-purple, purple or greenish-white. Fl. Apr–May, fr. May–Aug.

For other drugs sourced from *M. alba* see 'Sang Bai Pi' (Mori Cortex) page 522, 'Sang Shen' (Mori Fructus) page 700 and 'Sang Zhi' (Mori Ramulus) page 476.

HARVESTING, SOURCING AND NATURAL RANGE

This species is only known from cultivation and where it is grown throughout China primarily for its leaves for the rearing of silkworms for the silk industry. The main source of medicinal leaves are from plants grown in Anhui, Hunan, Sichuan and Zhejiang (Zhao & Chen, 2014). The CP2010 advises medicinal leaf harvesting take place in autumn.

The wild species was originally considered endemic to central and northern China.

CONSERVATION STATUS

China: Not Evaluated.

Global: Not Evaluated (IUCN Red List, accessed 25/8/2016).

Morus alba, cultivated in Anhui province

Leaf underside of *M. alba* showing white hair tufts in axils of midvein and primary lateral veins

M. alba with ripening fruit

Leaves and young fruit of *M. alba*

TCM PROPERTIES AND USE

***properties*:** **'sweet, bitter and cold'**
***channels*:** **'Lung and Liver'**
***actions*:** **'dispels wind heat, clears Lung heat, moistens dryness, cools the Blood'**

***indications*:** **respiratory tract infections, cough; haematemesis; dizziness, vertigo; painful eyes**

DRUG MORPHOLOGY

CRUDE DRUG (DRIED LEAF): crumpled, broken or flattened leaf masses. When intact and unfolded, leaf blade ovate to broadly ovate in outline (see plant photos opposite) (8–15 cm long, 7–13 cm wide); tip either shortly pointed (acute), long pointed (acuminate) or blunt (obtuse); base truncate, rounded or heart-shaped, margin often ragged and broken, but when intact, coarsely saw-toothed (see plant photos opposite) or shallowly scalloped (crenate); upper leaf surface yellowish-green or pale yellowish-brown, hairless, occasionally with wart-like protuberances **[1]**; lower surfaces paler, with prominent, cream veins **[2]** with a sparsely hairy (pubescent) mid-vein **[3]** and alternating lateral and sub-lateral veins, the latter with a netted appearance; mid-vein often with tufts of white hairs in axils of mid- and lateral veins (plant photos opposite). Curved leaf stalk sometimes present **[4]** (1.5–5.5 cm long). *Texture*: fragile, quickly disintegrating when handled. *Odour*: indistinct. *Taste*: mild, slightly bitter and astringent.
***Good quality*:** unbroken, yellowish-green leaves.

DECOCTING PIECES (CRUSHED LEAF): fragmented leaves of varying size and shape. Upper leaf surface yellowish-green or pale yellowish-brown, hairless, occasionally with wart-like protuberances **[1]**; lower surface with prominent, cream veins **[2]** and often with tufts of white hairs in axils of main veins and lateral veins. Other occasional features: leaf blade tip either shortly pointed, long pointed or blunt (obtuse); leaf blade base truncate, rounded or heart-shaped.

PROCESSING

Leaves are sun-dried (= crude drug) then, after removal of the leaf stalks, are crushed by hand into fragmented decocting pieces.

OTHER COMMON NAMES

冬桑叶 Dong Sang Ye; 老桑叶 Lao Sang Ye; 桑 Sang (plant name); Shuang Sang Ye; Folium Mori; white mulberry (SCN).

SAFETY

An official food and healthfood in China (Anon., 2002a,b). Safety Class: 1; Interaction Class: A (Gardner & McGuffin, 2013).

Crude drug

Decocting pieces

Zi Su Ye 紫苏叶

Perillae Folium

Drug part: LEAF

Perilla frutescens (L.) Britton

Lamiaceae

synonym: *Ocimum frutescens* L.

taxonomic/nomenclatural note: 3 varieties are recognised by *FoC* and *Kew-MPNS*, 20/8/2016.

Species of *Perilla* wild in China: 1

PLANT DESCRIPTION

An erect, herbaceous annual (0.3–2 m tall). *Stems* 4-sided, green or purple, sparsely or densely covered in fine hairs. *Leaves* opposite with each pair at right angles to the next; stalks 3–5 cm long; **leaf blades** simple, **broadly ovate to circular** (4–13 cm long; 2.8–10 cm wide), green, purplish, or purplish-black, bases rounded to broadly wedge-shaped, tips pointed and margins narrowly to coarsely saw-toothed. *Flowers* tiny, in densely hairy, sub-cylindrical many-flowered, 1-sided spikes (1.5–15 cm long) arising from the leaf axils; each flower subtended by an ovate bract (c. 4 mm long; 4 mm wide), green or purple, membranous, pointed and of similar length to the flower. Each flower bisexual, laterally symmetrical and short-stalked; calyx cup-shaped (c. 3 mm long), green to yellowish, glandular, densely hairy at its base and 2-lipped at apex with 3 pointed teeth in the upper lip and 2 pointed teeth in the lower lip; corolla white to purplish-red, bell-shaped and opening into a 2-lipped mouth (lower lip 3-lobed and slightly longer than upper lip); **stamens 4, shorter than corolla;** ovary superior; style 1. *Fruit* with 4 dry, sub-globose nutlets; each nutlet **(c.1.5 mm diam.)**, greyish-brown, greyish-yellow or purplish and **with a netted surface**. Fl. Aug–Nov, fr. Aug–Dec.

For other drugs sourced from this species see 'Zi Su Geng' (Perillae Caulis) page 488 and 'Zi Su Zi' (Perillae Fructus) page 766.

HARVESTING, SOURCING AND NATURAL RANGE

Leaves are usually harvesed in summer, when the foliage is lush, from plants grown as medicinal crops or on small-holdings. The species has long been cultivated in China with many cultivars bred for medicinal or other purposes; most are thought to be used in TCM. The main production areas for Zi Su Ye are the central and eastern Chinese provinces of Jiangsu, Zhejiang, Hebei, Hubei and Henan. The purple-leaved cultivars are widely used as a condiment and salad vegetable. In the wild, the species is common on disturbed ground and has a wide distribution in China (Fujian, Guangdong, Guangxi, Guizhou, Hebei, Hubei, Jiangsu, Jiangxi, Shanxi, Sichuan, Taiwan, Tibet, Yunnan and Zhejiang [also native in Bhutan, Cambodia, India, Indonesia (Java), Japan, Korean Peninsula, Laos and Vietnam].

CONSERVATION STATUS

China: Least Concern (MEPCAS, 2013, as '*P. frutescens* var. *purpurascens*').

Global: Not Evaluated (IUCN Red List, accessed 18/8/2016).

Medicinal crop of *Perilla frutescens*

Inflorescence of *P. frutescens*

P. frutescens (form with purple leaf undersides)

TCM PROPERTIES AND USE

***properties*:** **'pungent and warm'**
***channels*:** **'Lung and Spleen'**
***action*:** **'releases the exterior'**

***indications*: respiratory tract infection; nausea, vomiting, morning sickness, seafood poisoning**

DRUG MORPHOLOGY

CRUDE DRUG (WHOLE LEAF): leaves flattened, crumpled or rolled; often neatly pressed together **[1]**, stacked and tied. When whole and expanded, leaf blade simple, broadly ovate to circular **[2]** (4–13 cm long; 2.8–10 cm wide); leaf tips abruptly pointed or tapering to a point; leaf bases rounded to broadly wedge-shaped **[3]**; margins narrowly to coarsely saw-toothed **[4]** with teeth pointing towards leaf tips; leaf surfaces both green or purple or lower surface purple **[5]**; sparse white hairs occasionally on both surfaces [hl] **[6]**; lower surface dotted with tiny glands [hl]; veins pinnate **[7]** and often raised on lower surface **[8]**. Leaf stalks usually still attached **[9]** 3–5 cm long, purple, purplish-green or brown. *Texture*: fragile. *Odour:* aromatic. *Taste*: faintly pungent. ***Good quality*:** intensely aromatic, unbroken and stalkless.

DECOCTING PIECES (TRANSVERSELY SLICED LEAF): irregularly shaped as rhomboidal leaf slices **[10]**, often crumpled or folded. Many identification characters lost due to fragmentation such that macroscopic identification not possible. Characters visible include colour, texture, hairiness, odour and taste; for details of these see the crude drug description.

PROCESSING

Leaves are sun-dried (= crude drug) then cut into decocting pieces.

OTHER COMMON NAMES

家苏叶 Jia Su Ye; 苏叶 Su Ye; beefsteak plant; Folium Perillae (CP2010).

SAFETY

Bensky *et al.* (2014) state that large doses can raise blood sugar levels and should be not be used by diabetics.

Crude drug of Zi Su Ye (leaves of *P. frutescens* neatly pressed together; form with purple leaf undersides illustrated here)

Decocting pieces of Zi Su Ye

Lonicera japonica

flowers and flower parts

Ding Xiang 丁香

Caryophylli Flos

clove

Drug part: FLOWER BUD

PHARMACOPOEIA SPECIES

Syzygium aromaticum (L.) Merr. & L. M. Perry

Myrtaceae

synonyms: *Caryophyllus aromaticus* L.; *Eugenia caryophyllata* Thunb. (name used in CP2015).

Species of *Syzygium* wild in China: 78

PLANT DESCRIPTION

A slender, evergreen tree with a dense canopy (conical when young) (up to 20 m tall); cultivated trees usually smaller and branched from the base. *Leaves* opposite, simple, oblong to elliptical (5–13 cm long; 2.5–6 cm wide); margins entire; hairless and leathery; upper surface shiny; lower surfaces gland-dotted. *Inflorescences* terminal, in branched, 2–20 (rarely 40)-flowered clusters (panicles) (c. 5 cm long); flowers usually in groups of 3. When in bud, flowers (cloves) exude oil when squeezed; mature flowers (1–2 cm long) with a long, waxy calyx tube, with 4 short apical teeth; petals 4, pinkish-white and triangular (shorter than calyx tube and falling soon after opening); stamens many, forming a showy yellowish-white spray; fragrant. *Fruit* a dark red, fleshy drupe (1–2.5cm long). *Seed* usually solitary (1.5 cm long). Fl. Sep–Mar (description based on Elzebroek & Wind, 2008 and Verheij & Snijders, 1999).

HARVESTING, SOURCING AND NATURAL RANGE

Flower-buds are harvested during the winter when they are changing colour from green to orangish-red. Harvesting is entirely from cultivated plants long established in southern China: Guangdong, Hainan and Yunnan. The species does not grow wild in China; it is native to Indonesia and widely cultivated as a popular spice worldwide.

CONSERVATION STATUS

China: introduced.
Global: Not Evaluated (IUCN Red List, accessed 18/5/2015).

UNOFFICIAL SUBSTITUTE

S. aromaticum — *fruit*

Myrtaceae

PLANT DESCRIPTION

As above.

HARVESTING

Fruits are collected when almost ripe.

CONSERVATION STATUS

As above.

TCM PROPERTIES AND USE

properties: **'pungent and warm'**
channels: **'Kidney, Spleen and Stomach'**
actions: **'warms interior, tonifies yang'**

indications: **vomiting, nausea; poor appetite, diarrhoea, stomach pain; impotence; leucorrhoea and fungal infections; (T)**

Syzygium aromaticum in flower bud ready for harvesting, Hainan province

Flowering branch of *S. aromaticum*

Flower-buds of *S. aromaticum* sun-drying

Fruiting branch of *S. aromaticum*

Fresh fruit broken open to reveal cotyledons

Syzygium aromaticum

Myrtaceae

DRUG MORPHOLOGY

CRUDE DRUG (DRIED FLOWER BUD — WHOLE CLOVE): club-shaped (1–2 cm long; up to 0.6 cm diam. at widest part). Calyx tube (main 'stem' of clove) somewhat cylindrical, slightly flattened and/or curved (0.7–1.4 cm long; 0.3–0.6 cm diam.), solid, reddish-brown or dark brown **[1]**; basal end slightly tapering **[2]** with a shallow flower-stalk scar at base **[3]**; apex with 4 fleshy, spreading calyx teeth **[4]**. Four tightly overlapping petals form a globose head **[5]** (0.3–0.6 cm diam.); petals membranous, dark brown to brownish-yellow; stamens many with yellow anthers, style 1; stigma 1. Occasionally the closed petals fall off to leave a shallow saucer-shaped surface with a sunken centre **[6]**. *Texture:* hard and oily. *Odour*: strongly aromatic. *Taste*: pungent and numbing.
Good quality: plump, intensely aromatic, reddish-brown flower-buds. When floated in water, flower buds hang vertically with the calyx tube below the closed petal head.

PROCESSING

Flower-buds are sun-dried (= crude drug) then pulverised immediately prior to decocting.

OTHER COMMON NAMES

大花丁香 Da Hua Ding Xiang 'big-flowered clove'; 丁子香 Ding Zi Xiang or 子丁香 Zi Ding Xiang 'small clove'; 公丁 Gong Ding or 公丁香 Gong Ding Xiang 'male clove'; clove (English, CP2010, SCN); Flos Caryophylli.

SAFETY

Safety Class: 2d; Interaction Class: A (Gardner & McGuffin, 2013). Do not exceed recommended dose (Bensky *et al.*, 2004). An official Chinese food and healthfood (Anon., 2002a and 2002b).

S. aromaticum — *fruit*

Myrtaceae

'DRUG SUBSTITUTE' MORPHOLOGY

Easily distinguished macroscopically from the flower bud (above). Fruits **ovoid-oblong** (1.5–3.0 cm long; **0.5–1.0 cm diam**.), externally yellowish-brown **to dark brown**, rough and finely wrinkled **[7]**. Apex with four persistent, bent and somewhat hook-shaped calyx teeth (often shrivelled) **[8]**; base with a fruit-stalk scar **[9]**. Internally with 1 or 2 oblong **seeds** each with 2 brown or dark brown cotyledons, densely packed with oil glands; the inner surface of each cotyledon bears a longitudinal furrow. *Texture*: hard, dry and **difficult to break.** *Odour*: aromatic. *Taste*: less pungent and numbing than Ding Xiang.

PROCESSING

Fruits are sun-dried (=crude drug) then pulverised immediately prior to decocting.

OTHER COMMON NAMES

雌丁香 Ci Ding Xiang; 母丁 Mu Ding; clove fruit (CP2010).

SUBSTITUTION COMMENT. The mature dried **fruits** of this species have been reported as occasional and inferior substitutes (Ho *et al.*, 2006). They are a separate TCM drug in the CP2015 under the name 母丁香 'Mu Ding Xiang' (female clove) (Caryophylli Fructus) and used for the same medical conditions.

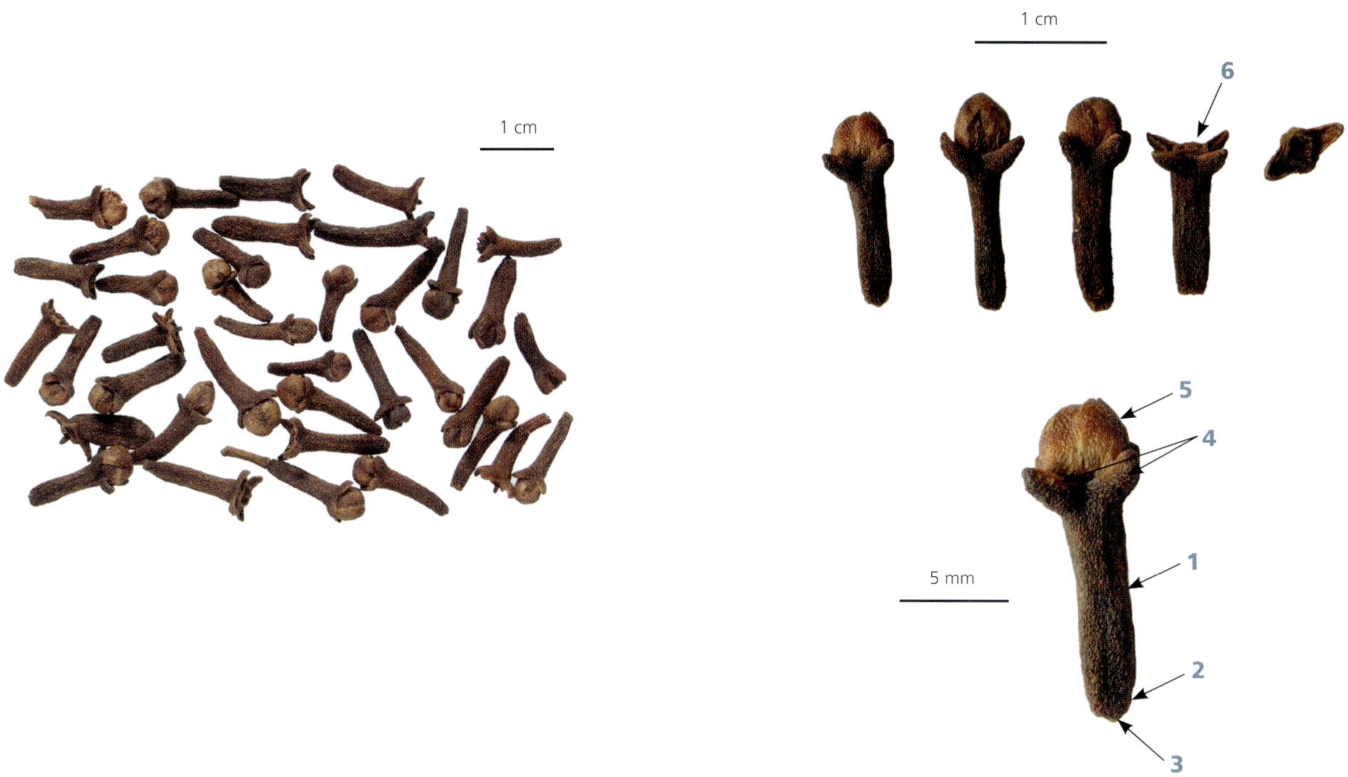

Crude drug (flower-buds of *Syzygium aromaticum*)

Crude drug (fruit of *S. aromaticum*)

He Huan Hua 合欢花

Albiziae Flos

Drug part: FLOWER-HEAD (INFLORESCENCE)

PHARMACOPOEIA SPECIES

Albizia julibrissin Durazz.

Fabaceae

PLANT DESCRIPTION

A deciduous tree with spreading crown (up to 16 m tall). *Stipules* falling, smaller than leaflets. *Leaves* alternate, divided into **4–12 pairs of pinnae** (sometimes 20 in cultivated plants), **each pinna with 10–30 pairs of linear to oblong leaflets** (0.6–1.2 cm long; 0.1–0.4 cm wide); leaflets sharply tipped, margins fringed with hairs. *Inflorescences* conspicuous, globose, **predominantly pink** (top part of filaments of stamens) and long-stalked. *Each flower* with calyx funnel-shaped and 5-toothed (c. 3 mm long); corolla funnel-shaped and 5-lobed (c. 8 mm long), pale yellowish-green; stamens many with long **filaments** (c. 2.5 cm) **basally white, becoming bright pink in outer parts.** *Fruit* a long flat pod (9–15 cm long; 1.5–2.5 cm wide). Fl. May–Jul, fr. Aug–Oct.

Species of *Albizia* wild in China: 14

HARVESTING, SOURCING AND NATURAL RANGE

Inflorescences are harvested on sunny days in summer from wild trees when in full flower ('He Huan Hua') or in flower-bud ('He Huan Mi'). For sourcing and natural range see 'He Huan Pi' (Albiziae Cortex) page 504.

CONSERVATION STATUS

China: Introduced.
Global: Not Evaluated (IUCN Red List; accessed 10/8/2016). An invasive species in the USA and Japan.

UNOFFICIAL SUBSTITUTE 1

A. kalkora (Roxb.) Prain

Fabaceae

synonym: *Mimosa kalkora* Roxb.

PLANT DESCRIPTION

Very similar to the Pharmacopoeia species. *Main differences*: a shrub or small tree (3–8 m tall) with **inflorescences predominantly white, turning yellow** (top part of filaments of stamens). *Leaves* divided into **fewer pinnae (2–4 pairs)** and **fewer (5–14)** but **larger (0.8–4.5 mm long; 0.7–2 mm wide) leaflets**. *Calyx* and *petal tubes* **densely clothed in long soft brown hairs**. Fl. May–Jun, fr. Aug–Oct.

HARVESTING, SOURCING AND NATURAL RANGE

For sourcing and natural range see 'He Huan Pi' (Albiziae Cortex) page 504.

CONSERVATION STATUS

China: Least Concern (MEPCAS, 2013).
Global: Not Evaluated (IUCN Red List accessed 10/8/2016).

UNOFFICIAL SUBSTITUTE 2

Celastrus orbiculatus Thunb.

Celastraceae

PLANT DESCRIPTION

Easily distinguished from the Pharmacopoeia species. *Main differences*: a deciduous **woody vine** with **simple, broadly ovate or orbicular leaves** (5–13 cm long; 3–9 cm wide), margins saw-toothed; inflorescences small and relatively inconspicuous; flowers many, **minute, green**, 5-petalled; fruits bright yellow, sub-spherical (8–13 mm diam.), 3-valved; seeds reddish-brown surrounded by remains of soft dark orange aril. Fl. May–Jun, fr. Jul–Oct.

HARVESTING, SOURCING AND NATURAL RANGE

A common vine growing wild in mixed forests, thickets and on grassy slopes between 400–2,200 m altitude in N and SE China [also native in Japan and Korean Peninsula]. Cultivated as an ornamental in many countries and in some (USA) it has become an invasive plant (Harrington *et al.,* 2003).

CONSERVATION STATUS

China: Least Concern (MEPCAS, 2013).
Global: Not Evaluated (IUCN Red List; accessed 10/8/2016).

UNOFFICIAL SUBSTITUTE 3

Magnolia coco (Lour.) DC.

Magnoliaceae

synonyms: *Lirianthe coco* (Lour.) N. H. Xia & C. Y. Wu (*FoC* accepted name); *Liriodendron coco* Lour.

PLANT DESCRIPTION

Easily distinguished from the Pharmacopoeia species. *Main differences*: an evergreen tree or shrub (2–4 m tall); **leaves simple** (not divided into leaflets), elliptic and **much larger** (7–14 cm long; 2–4.5 cm wide), texture leathery and upper surface glossy. **Flowers solitary**, bisexual, globose (3–4 cm diam.); **petals 9, white, fleshy, concave and overlapping**; stamens many; fruits somewhat woody (c. 3 cm diam.).

HARVESTING, SOURCING AND NATURAL RANGE

A wild but acutely endangered species of forests (600–900 m altitude) in SC and SE China: Fujian, Guangdong, Guangxi, Taiwan, Yunnan, Zhejiang [also native in Vietnam]. Widely cultivated in S China as an ornamental and, in Guangdong, for its medicinal flowers (see also *Substitution Comment*).

CONSERVATION STATUS

China: Endangered (MEPCAS, 2013); Vulnerable A2c (Wang & Xie, 2004).
Global: Data Deficient (IUCN Red List; accessed 12/2/2015).

TCM PROPERTIES AND USE

properties: **'sweet and neutral'**
channels: **'Heart and Liver'**
actions: **'calms the shen, regulates qi'**
indications: **depression, irritability, anxiety, insomnia**

Albizia julibrissin cultivated tree in full flower

A. julibrissin flowering branch

Albizia kalkora in full flower

A. kalkora flowering branch

Leaves of *Celastrus orbiculatus*

C. orbiculatus showing minute female flowers

C. orbiculatus fruiting branch

Magnolia coco with typical globose flower

Albizia julibrissin

Fabaceae

DRUG MORPHOLOGY

CRUDE DRUG/DECOCTING PIECES (FLOWER-HEADS — HE HUAN HUA): very light, soft and feathery crumpled masses of inflorescences, yellow or yellowish-brown. Long flower-head stalks usually still attached, longitudinally ridged, yellowish-green and sparsely hairy (3–4 cm long; c. 1 mm diam.). Inflorescences laterally compressed (globose when fresh); usually with c. 10 tightly packed flowers. **Flowers almost stalkless [1]** and slightly curved (7–13 mm long); **calyx** tubular (3 mm long), pale yellow or yellowish-brown, 5-toothed at apex and **sparsely hairy [2]**; **petal tube** (partially enclosed by the calyx tube) more than twice as long as calyx tube **sparsely hairy [3]**, with 5 lance-shaped lobes; stamens many, filaments very long (c. 2.5 cm) and thread-like (extending 1–2 cm beyond top of petal tube but often coiled into a mass), yellowish-brown or dark orangish-brown. *Odour*: indistinct. *Taste*: weak. ***Good quality*:** inflorescences yellowish-brown or dark orangish-brown; aromatic.

CRUDE DRUG /DECOCTING PIECES (FLOWER-BUDS — HE HUAN MI): light and soft masses of mostly unopened inflorescences. Stalks of inflorescences as for 'He Huan Hua' above. Inflorescences laterally compressed (globose when fresh), usually with c. 10 densely packed flower-buds. Flower-buds virtually stalkless and slightly curved; calyx tubular, pale yellow or yellowish-brown, 5-toothed at apex and **sparsely hairy;** petal tube extending beyond the similar coloured calyx tube to form a **sparsely hairy** unopened club-shaped flowerbud (2–6 mm long, c. 2 mm diam. at widest point). When dissected, flower-bud contains many stamens with very long, thread-like filaments (c. 2.5 cm long). *Odour*: indistinct. *Taste*: weak.

PROCESSING

Flower-heads are sun-dried (= crude drug/decocting pieces).

OTHER COMMON NAMES

夜合花 Ye He Hua (Hong Kong drug name); Flos Albiziae (CP2005); pink silk tree (SCN).

SAFETY

Safety Class: 1; Interaction Class: A (Gardner & McGuffin, 2013).

A. kalkora

Fabaceae

'DRUG SUBSTITUTE' MORPHOLOGY

Similar to the official drug but **calyx- and petal tubes densely covered in soft, mat-like brownish hairs [4]. Each flower shortly stalked (1.5–2 mm long) [5]**. Filaments of stamens yellowish-brown. The colour of the long filaments of the stamens is not a reliable character; this varies according to drying and storage conditions.

OTHER COMMON NAMES

山合欢花 Shan He Huan Hua.

SUBSTITUTION COMMENT. A very similar species, widespread in the wild in China, with white inflorescences (turning yellowish) suggesting substitution is intentional as this character is obvious on the living tree.

Crude drug/decocting pieces of He Huan Hua
(*Albizia julibrissin* inflorescences)

Close-up of dried inflorescences of *A. julibrissin*

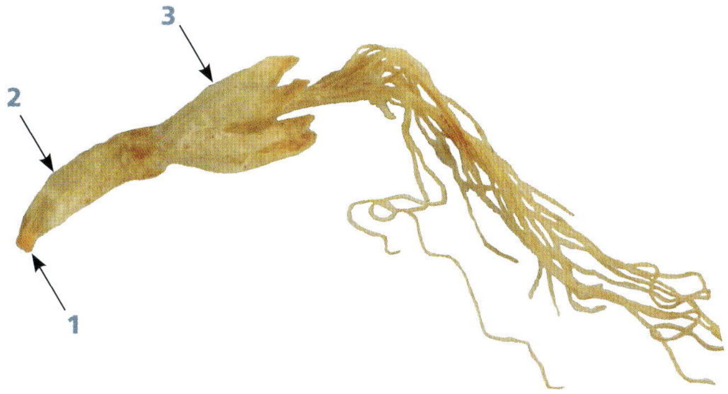

Single flower of *A. julibrissin*

Albizia kalkora dried inflorescences

A. kalkora inflorescence with mature thread-like flowers

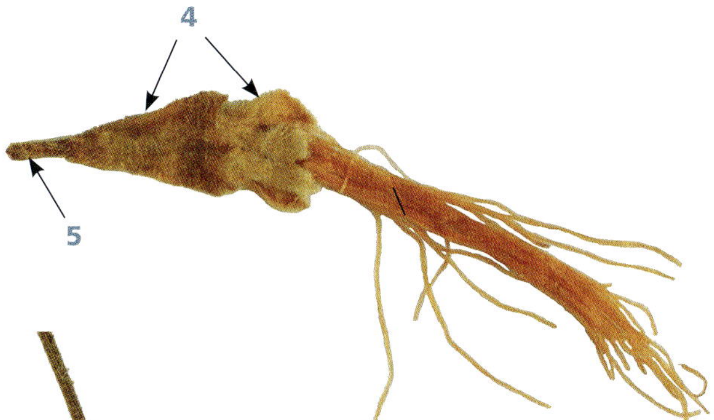

Single flower of *A. kalkora*

A. kalkora inflorescence in bud with 2 flowers open to reveal thread-like stamens within

Celastrus orbiculatus — Celastraceae

'DRUG SUBSTITUTE' MORPHOLOGY

*It is the **fruits** of this species that are often traded as He Huan Hua (the official drug).* **Fruits bright yellow**, sub-spherical (8–13 mm diam.), 3-valved **[6]** and often broken open to reveal the **soft dark orange aril [7] surrounding reddish-brown**, **elliptic and slightly flattened seeds [8]** (4–5 mm long; 2.5–3 mm wide).

OTHER COMMON NAMES

南蛇藤果 Nan She Teng Guo (oriental bittersweet).

SAFETY

Fruits are **slightly toxic** and known to contain gastro-intestinal irritants which may cause vomiting, diarrhoea, nausea and cramping (Burrows & Tyrl, 2013; Wink & van Wyk, 2008).

SUBSTITUTION COMMENT. Fruits of this species are considered a clinical alternative to the Pharmacopoeia drug in south China and where it is traded under the same name 'He Huan Hua'. They are described under this name in a classic TCM text of south China, the *Guang Dong Zhong Yao Zhi* (Lin, 1990-1996). Such substitution has also been seen in overseas trade (C. Leon pers. obs., in the UK) creating much confusion.

Magnolia coco — Magnoliaceae

'DRUG SUBSTITUTE' MORPHOLOGY

Flowers easily distinguished macroscopically from the official drug. *Main differences*: a **globose** (when whole) solitary flower **[9]**, **much larger (3–4 cm diam.)**; **petals 9, brown, fleshy and overlapping [10]**; stamens many, flattened and somewhat tongue-shaped **[11]**. When fresh, intensely aromatic.

OTHER COMMON NAMES

广东夜合花 Guang Dong Ye He Hua; 夜合花 Ye He Hua.

SUBSTITUTION COMMENT. Flowers of *M. coco* are the source of a totally unrelated medicinal drug in south China (Lin, 1990–1996). According to Zhao and Li (2007) this drug and He Huan Hua described here have long been confused because they share the name 夜合花 Ye He Hua in different TCM texts. This confusion dates back to the Song Dynasty (late 13th c.) yet persists today. This is classic example of the need for nomenclatural standardisation within TCM and in a wider context.

LESSER KNOWN UNOFFICIAL SUBSTITUTE

Euonymus maackii Rupr. (syn. *E. bungeanus* Maxim.) 白杜 Bai Du; 明夜合花 Ming Ye He Hua; family Celastraceae. Fruits are used as a local substitute in China. Fruits rhombic with 4 angles and deep grooves; brown or yellowish-brown to reddish-brown (c. 1 cm diam.); seeds reddish-brown (Bensky *et al.*, 2014).

Celastrus orbiculatus fruits, mostly broken open to reveal the dark orange aril and seed within

Magnolia coco flowers, dried and fragmented

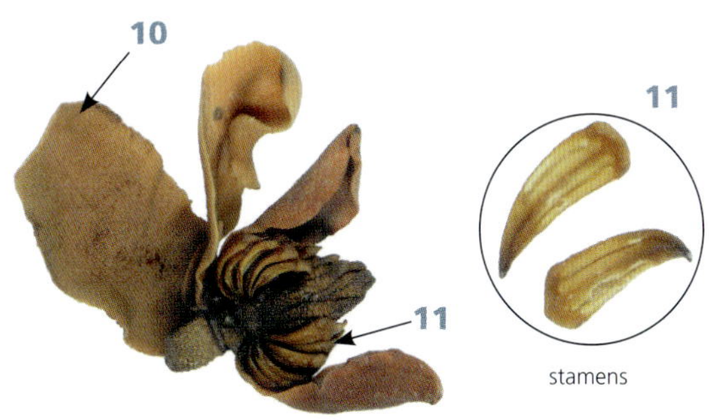

Magnolia coco close-up of laterally dissected flower

Hong Hua 红花

Carthami Flos

safflower

Drug part: FLOWER

Carthamus tinctorius L.

Asteraceae

Species of *Carthamus* wild in China: 0

PLANT DESCRIPTION

An annual herb (50–100 (rarely up to 150) cm tall. *Stem* erect, branching in upper parts; smooth, whitish and hairless. *Leaves* alternate, simple, rigid, leathery and hairless. Lowermost and middle stem leaves unstalked, lance-shaped, ovate or elliptic (7–15 cm long; 2.5–6 cm wide); margin entire or spine-toothed. Upper stem leaves lance-shaped, becoming smaller up the stem; margin minutely spine-toothed. *Flower-heads* (capitula) few to many in flat-topped groups, rarely solitary; ovoid to ellipsoid. Involucre surrounding each flower-head, with 5 rows of phyllaries. Outer phyllaries rigid and leaf-like (2–4 cm long; c. 1 cm wide) with margins spiny; middle phyllaries with leaf-like tips and green-netted veining (6–20 mm long; 4–7 mm wide); inner phyllaries stiff and dry (c. 2.2 cm long; 0.5 cm wide) with tips gradually narrowing. Corolla red, orange or yellow (to 2.8 cm long) with all florets bisexual and the same size and shape; corolla tube slender. *Fruit* a 4-angled achene; pappus absent. Fl. and fr. May–Aug.

HARVESTING, SOURCING AND NATURAL RANGE

An introduced and widely cultivated species in temperate parts of China (Gansu, Guizhou, Hebei, Heilongjiang, Inner Mongolia, Jiangsu, Jilin, Liaoning, Qinghai, Shaanxi, Shandong, Shanxi, Sichuan, Tibet, Xinjiang and Zhejiang) and occasionally naturalised on saline and alkaline soils, in cold, dry locations. Widely cultivated for culinary purposes too (seeds). Native origin unknown. Flower-heads are harvested exclusively from cultivated plants in summer when flower-heads change from yellow to red.

CONSERVATION STATUS

China: Introduced.

Global: Not Evaluated (IUCN Red List, accessed 20/5/2015).

Medicinal crop of *Carthamus tinctorius* with flower-heads changing with maturity from yellow, to orange to red, Xinjiang province PHOTO: KAISER SULAIMAN

Single flower-head (capitulum) of *C. tinctorius* with each floret spreading into 5 lobes

Maturing flowering heads of *C. tinctorius*, Xinjiang province

Fruiting-head with developing fruits (pale grey, 4-angled achenes)

TCM PROPERTIES AND USE

properties: **'pungent and warm'**
channels: **'Heart and Liver'**
actions: **'invigorates Blood, dispel stasis, opens channels'**

indications: **menstrual, postpartum bleeding; abdominal pain, traumatic injuries, musculoskeletal pain, skin conditions**

DRUG MORPHOLOGY

CRUDE DRUG/DECOCTING PIECES (DRIED FLOWERS/FLORETS): flowers/florets red, orange or yellow; flower tube slender **[1]** (1–2 cm long; c. 0.5–1 mm diam.), spreading apically into 5 strap-shaped lobes **[2]** (5–8 mm long). Stamens 5, yellowish-white; filaments attached to lower part of inner flower tube wall. Stigma divided into 2 short branches. *Texture*: delicate and pliable. *Odour*: slightly aromatic. *Taste*: slightly bitter. ***Good quality***: flowers/florets soft, bright red or orangish-red and slightly aromatic.

PROCESSING

Flowers/florets are shade- or sun-dried (= crude drug/decocting pieces).

OTHER COMMON NAMES

草红花 Cao Hong Hua; Flos Carthami; safflower (CP2010; SCN).

SAFETY

Flos Carthami is an official Chinese healthfood (Anon., 2002a,b). Safety Class: 2b, 2d; Interaction Class: B (Gardner & McGuffin, 2013).

COMMENT. Widely sold as a cheap unofficial substitute of Xi Hong Hua (Croci Stigma, saffron) see page 596.

Crude drug/decocting pieces of Hong Hua (红花) = florets of *Carthamus tinctorius*

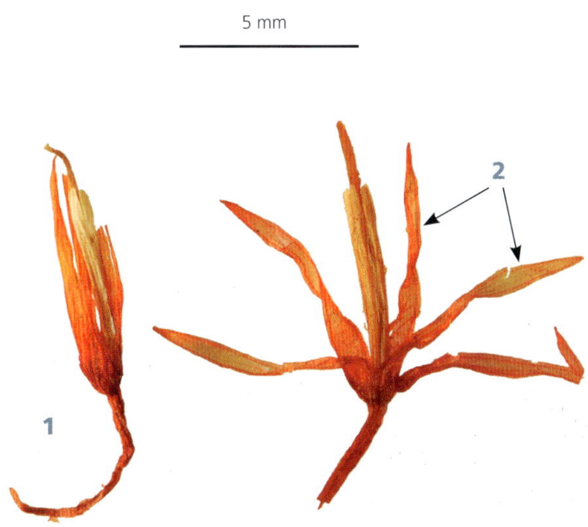

2 florets of *C. tinctorius* (left: closed; right: open)

Jin Yin Hua 金银花

Lonicerae Japonicae Flos

Japanese honeysuckle

Drug parts: FLOWER-BUD

Lonicera japonica Thunb.

Caprifoliaceae

taxonomic note: two varieties are recognised by *FoC*: *L. japonica* var. *japonica* and the less common *L. japonica* var. *chinensis* (P. Watson) Baker

Species of *Lonicera* wild in China: 57

PLANT DESCRIPTION

A vigorous, semi-evergreen climber. *Stems* with yellowish-brown, stiff, spreading hairs. *Leaves* opposite, ovate to lance-shaped (3–8 cm long); **undersides sparsely to densely hairy but not totally covering leaf surface** [hl] and **margins fringed with hairs (ciliate)** [hl]. A pair of **leaf-like bracts, ovate to elliptic (1–3 cm long)** present at base of flowerheads. *Flowers* in pairs within branching flowerheads. Sepals fused into a **hairless** calyx tube ending in 5 small **calyx lobes with densely hairy margins** [hl]. Petals 5, fused to form a narrow tube (**3–5 cm long)**, with spreading lobes and 2 lips; externally white, becoming yellow, orange or purple; those changing to purple are *L. japonica* var. *chinensis*. *Fruit* a spherical, bluish-black shiny berry.

For other TCM drugs sourced from this species see 'Ren Dong Teng' (Lonicera Japonicae Caulis) page 466.

HARVESTING, SOURCING AND NATURAL RANGE

Flower-buds are mostly harvested in early summer from plants widely grown on medicinal plant farms. Traditional *Di Dao*-quality drug is sourced from sites in Henan (Pingyi) and Shandong (Mixian) provinces in C China.

In the wild the species grows in open forests and scrub, in stony places, mountain slopes and roadsides (up to 1,500 m altitude) and is distributed in Anhui, Fujian, Gansu, Guangdong, Guangxi, Guizhou, Hebei, Henan, Hubei, Hunan, Jiangsu, Jiangxi, Jilin, Liaoning, Shaanxi, Shandong, Shanxi, Sichuan, Taiwan, Yunnan and Zhejiang. The species also grows wild in Japan and Korean Peninsula and has been introduced and become invasive in North America.

CONSERVATION STATUS

China: Least Concern (MEPCAS, 2013).

Global: Not Evaluated (IUCN Red List, accessed 1/9/2016).

A medicinal crop of *Lonicera japonica*.

Flowering branch of *L. japonica*

Dried aerial parts of *L. japonica*

TCM PROPERTIES AND USE

***properties*: 'sweet and cold'**
***channels*: 'Large Intestine, Lung, Heart and Stomach'**
***actions*: 'clears heat, eliminates toxins'**
***indications*: rash, sores, abscesses; diarrhoea; respiratory tract infections**

DRUG MORPHOLOGY

CRUDE DRUG/DECOCTING PIECES (FLOWER-BUDS and FLOWERS). *Flower-buds*: club-shaped, slightly curved, gradually tapering towards the base (2–3.4 cm long; 1.5–3 mm diam.). Bud surface yellowish-white or greenish-white, darkening with age or if poorly dried, **densely covered in soft, short hairs**, sometimes with stalked orange glands **[1]**. **Leaf-like bracts ovate to elliptic (0.8–3 cm long)** occasionally present. **Calyx tube hairless [2]**; margins of the 5 pointed **calyx lobes densely hairy** [hl]. *Opening flowers:* becoming white (3–5 cm long); apex 2-lipped with upper lip 4-lobed; stamens 5 with lower parts attached to walls of flower tube. *Odour*: sweetly aromatic. *Taste*: weak and slightly bitter. ***Good quality*:** yellow unbroken flowers (in bud or just opened).

PROCESSING

Flower-buds, some opened, are dried (= crude drug and decocting pieces).

OTHER COMMON NAMES

二花 Er Hua; 忍冬 Ren Dong (plant name); 忍冬花 Ren Dong Hua; flos lonicerae japonicae; Japanese honeysuckle (SCN); Japanese honeysuckle flower (CP2010).

SAFETY

An official Chinese food and healthfood (Anon., 2002a,b).

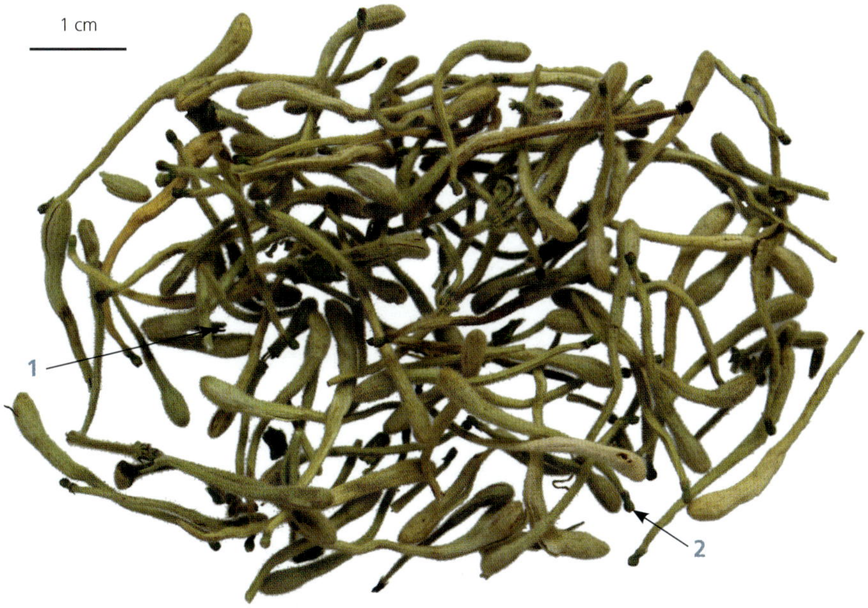

Crude drug/decocting pieces of Jin Yin Hua (flower-buds of *Lonicera japonica*); greenish colour indicates freshly dried sample

L. japonica flower-buds and opening flowers (some with hairless calyx tube still attached); fragments of leaf-like bracts (right)

Base of *L. japonica* flower-bud

UNOFFICIAL SUBSTITUTES

Given that this herbal drug is grown extensively in China as a medicinal crop, opportunities for its substitution or adulteration are minimised, especially if sourced from a GAP-certified production site. However, like all high value drugs, this drug is prone to substitution and adulteration with much cheaper and 'less potent' look-alikes and the remaining 56 wild species of Chinese ***Lonicera*** provide a ready supply. The flower-buds of other medicinally cultivated but cheaper ***Lonicera*** species (such as those of the separate drug 'Shan Yin Hua' (Lonicerae Flos), see page 588) are also known to be traded as genuine 'Jin Yin Hua'. Identification of flower-buds to species level however requires laboratory-based authentication methods (notably DNA) e.g. Sun *et al.*, 2011; macroscopy alone is unreliable with much variation in appearance according to the storage conditions used and the age of the drug.

Source plants of the 3 main substitutes (source of 'Shan Yin Hua') can be identified as follows:

1. ***Lonicera confusa*** DC. Leafy bracts much narrower (linear to lance-shaped, 5–14 mm long); leaf margins **not obviously fringed with hairs**; calyx tube **usually densely hairy,** occasionally sparsely hairy or hairless.
2. ***L. hypoglauca*** Miq. Stems **greyish, soft and downy.** Leaf undersides bluish-green. Leafy bracts much narrower (linear to lance-shaped, 5–14 mm long); leaf undersides with **stalkless orange glands [hl].**
3. ***L. macrantha*** (D. Don) Spreng. Leafy bracts much narrower (linear to lance-shaped, 5–14 mm long); stems densely clothed with **yellowish-brown hairs**; hairiness of leaf undersides highly variable; calyx tube **hairless or sparsely hairy**; fruits white to black.

Lesser known wild-sourced substitutes:

4. ***L. bournei*** Hemsl.
5. ***L. calcarata*** Hemsl.
6. ***L. reticulata*** Champ.
7. ***L. similis*** Hemsl.
8. ***L. tragophylla*** Hemsl.

Counterfeit/adulterant:

Radish rootlets (***Raphanus sativus***) have a superficial resemblance to Jin Yin Hua and have occasionally been reported as a cheap bulking agent.

Ju Hua 菊花

Chrysanthemi Flos

Drug part: FLOWERHEAD (capitulum)

Chrysanthemum × morifolium (Ramat.) Hemsl.

Asteraceae

synonym: *Matricaria × morifolia* Ramat.

taxonomic/nomenclatural note: in CP2015 as *'Chrysanthemum morifolium* Ramat.'. This plant is part of an enormous hybrid complex, more widely known in the global horticultural trade as '*C. grandiflorum*', with c. 500 predominantly ornamental but also medicinal cultivars.

PLANT DESCRIPTION

The following description includes characters shared by the four main medicinal cultivars of *C. × morifolium* cultivated in China as the source of Ju Hua: Bo Ju (亳菊), Chu Ju (滁菊), Gong Ju (贡菊) and Hang Ju (杭菊). These names reflect both the provenance of Ju Hua as well as their local growing conditions and processing styles and all are recognised as *Di Dao* drugs.

A perennial herb (30–90 cm tall). *Stems* erect, woody at base, much branched, densely white pubescent, slightly purplish-red. *Leaves*, simple and alternate, with furrowed leaf stalks; blades ovate to lance-shaped (5–15 cm long; 3–4 cm wide), tips broad or shortly pointed, bases heart- or wedge-shaped, deeply lobed often to the midrib, lobes oblong-ovate to sub-rounded; margins deeply cut or saw-toothed; upper surfaces dark green; lower surfaces paler; both surfaces with fine hairs. *Inflorescences* (2.5–15 cm diam.) solitary or aggregated at the ends of branches. Each capitulum with involucral bracts in 3–4 layers: outer bracts ovate or ovate-lance-shaped, green with membranous margins; inner bracts oblong, margins brown, broad and membranous; ray (ligulate) florets yellow or white, disk (tubular) florets yellow. Fl. Sep–Oct. (based on Chen, 2010a).

Species of *Chrysanthemum* wild in China: 22

HARVESTING, SOURCING AND NATURAL RANGE

The medicinal cultivars are only known from cultivation; *Di Dao* sites: 'Hang Ju' from Tongxiang, Haining, Jiaxing and Wuxing (Zhejiang prov.); 'Bo Ju' from Bozhou and Woyang (Anhui prov.); 'Gong Ju' from Shexian (Anhui prov.); 'Chu Ju' from Chuxian, Quanjiao and Huizhou (Anhui prov.) and 'Huai Ju' from Wuzhi, Boai, Xinyang (Henan province).

CONSERVATION STATUS

China: Cultivated.
Global: Not Evaluated (IUCN Red List, accessed 11/9/2015).

C. morifolium 'Gong Ju'

Medicinal crop of *Chrysanthemum × morifolium* 'Hang Ju'

C. × morifolium 'Gong Ju'

C. morifolium 'Hang Ju'

Grading freshly dried capitula of *C. × morifolium* 'Gong Ju'

TCM PROPERTIES AND USE

***properties*: 'pungent, sweet, bitter and cool'**
***channels*: 'Lung and Liver'**
***actions*: 'releases wind heat, calms Liver yang, clears heat, eliminates toxins'**

***indications*: respiratory tract infection, headache, fever; dizziness; red eyes; sores, swellings**

DRUG MORPHOLOGY

CRUDE DRUG/DECOCTING PIECES (DRIED FLOWER-HEAD)

Bo Ju: obconical or cylindrical, sometimes slightly compressed to fan-shade, 1.5–3 cm in diameter, separated. Involucre dish-shaped; bracts 3–4 layers, ovate or elliptical, herbaceous, yellowish-green or brownish-green, the outer surface covered with white hairs, margin membranous. Receptacle hemispherical, without paleas or palea haris. Ligulate (outer) florets in several rows, with female marginal florets, whitish, linear **[1]**, longitudinally folded and wrinkled. Tubular florets **abundant**, hermaphrodite, occurring in the centre, concealed by the ligulate florets, yellow, with 5 terminal teeth **[2]**. Achenes undeveloped, pappus absent. *Texture*: light and soft, lax and brittle when dried. *Odour*: pleasant and aromatic. *Taste*: sweet, slightly bitter.

Chu Ju: irregularly spherical or oblate, 1.5–2.5 cm in diameter; ligulate florets whitish, irregularly **twisted [3]**, curved inward, margin wrinkled, sometimes pale brown glandular spots visible; tubular florets mostly concealed **[4]**.

Gong Ju: oblate or irregularly spherical, 1.5–2.5 cm in diameter; ligulate florets white or whitish, obliquely upward, the upper part folded outward, margins slightly curved inward **[5]** and wrinkled, glandular spots usually absent; tubular florets few, exposed **[6]**.

Hang Ju: dish-shaped or oblate, 2.5–4 cm in diameter, often several flower-heads flattened and pressed together to form plates; ligulate florets whitish or yellow, spread flat or folded slightly; glandular spots usually absent; tubular florets abundant and often clearly visible **[7]**. ***Good quality***: unbroken, yellowish-white flower-heads with a fresh aroma.

PROCESSING

Flower-heads (capitula) with or without stems are harvested from September to November from medicinal crop plants with different processing methods employed according to each medicinal cultivar. Trade items are named according to their production area.

Bo Ju: entire flowering branches are harvested then shade-dried before flower-heads are removed.

Chu Ju: flower-heads sun-dried to semi-dryness, then usually sulphured, then processed until globe-shaped before drying is completed.

Gong Ju: flower-heads oven-baked.

Hang Ju: flower-heads steamed and sulphured then sun-dried.

OTHER COMMON NAMES

白菊花 Bai Ju Hua; 甘菊花 Gan Ju Hua; 怀菊 Huai Ju; chrysanthemum flower; Flos Chrysanthemi.

SAFETY

An official food and healthfood in China (Anon., 2002a,b). Safety Class: 1; Interaction Class: A (Gardner & McGuffin, 2013).

Types of Ju Hua from four medicinal cultivars of *Chrysanthemum* × *morifolium*

Ling Xiao Hua 凌霄花 — Campsis Flos

Drug part: FLOWER

PHARMACOPOEIA SPECIES 1

Campsis grandiflora (Thunb.) K. Schum.

Bignoniaceae

synonyms: *Bignonia grandiflora* Thunb.; *B. chinensis* Lam.*C. chinensis* (Lam.) Voss; *Tecoma chinensis* (Lam.) K. Koch

Species of *Campsis* wild in China: 1

PLANT DESCRIPTION

A vigorous woody, deciduous vine (up to 6 m tall). *Leaves* opposite (4–13 cm long) with 3–4 pairs of opposite, short-stalked oval or lance-shaped leaflets (plus a terminal leaflet) (usually 3–6 cm long; 1.5–3 cm wide), hairless, base broadly wedge-shaped, margin saw-toothed, tip tapering to a long point; lateral veins 6 or 7 on each side of midrib. *Flowers* in short, terminal, branching clusters. Each flower large, showy, red to orangish-red and trumpet-like (c. 3–5 cm long); calyx shallowly bell-shaped, green, divided almost half way into 5 calyx teeth; corolla c. 5 cm long and with 5 semi-rounded spreading corolla lobes; stamens 4–5 (but rarely all fertile), arising from corolla base, anthers yellow; style thread-like (c. 3 cm long); stigma 2-lobed. *Fruit* a long, ridged, 2-valved capsule with a broad tip. *Seeds* many, with membranous wings. Fl. May–Aug.

HARVESTING, SOURCING AND NATURAL RANGE

Flowers are harvested from cultivated plants grown as ornamentals in summer and autumn and widely cultivated as an ornamental. Top medicinal quality is sourced from Suzhou in Jiangsu province (E China) (Bensky *et al.,* 2004). The wild species is distributed in Fujian, Guangdong, Guangxi, Hebei, Shandong and Shanxi provinces (also wild in India, Japan, Pakistan and Vietnam).

CONSERVATION STATUS

China: Least Concern (MEPCAS, 2013).
Global: Not Evaluated (IUCN Red List, accessed 22/6/2015).

PHARMACOPOEIA SPECIES 2

C. radicans (L.) Bureau

Bignoniaceae

synonyms: *Bignonia radicans* L.; *Tecoma radicans* (L.) Juss.
taxonomic/nomenclatural note: in CP2015 as '*C. radicans* (L.) Seem.'

PLANT DESCRIPTION

Similar to PS1. Main differences: up to 10 m tall; leaflets in 3–6 pairs and slightly narrower, with midrib on lower surface pubescent. *Flowers* in more tightly packed heads (in groups of 4–12); **calyx orange**, wax-like, and **divided to less than half the length of the calyx tube,** midveins not obvious; flowers longer (up to 7 cm) and **narrower** at mouth, reddish-orange in the limb (trumpet) with a basal orange tube.

HARVESTING, SOURCING AND NATURAL RANGE

Introduced in China and widely cultivated as an ornamental in gardens chiefly in Beijing, Hunan, Guangdong and Jiangsu provinces. Native to North America and widely naturalised elsewhere (GRIN online database, accessed 22/6/2015).

CONSERVATION STATUS

China: Least Concern (MEPCAS, 2013).
Global: Not Evaluated (IUCN Red List, accessed 22/6/2015).

UNOFFICIAL SUBSTITUTE

Paulownia tomentosa (Thunb.) Steud.

Paulowniaceae

synonym: *Bignonia tomentosa* Thunb.

PLANT DESCRIPTION

Easily distinguished from PS1 and PS2. **A large deciduous tree (up to 20 m tall)**. *Leaves* opposite, **simple (not divided into leaflets)**, large (up to 40 cm long). *Flowers* in large erect pyramidal 3–5(–8)-flowered heads, **pale purple/lilac** and heavily scented. Each flower with a bell-shaped and **densely hairy calyx (tomentose)** with 5 teeth, corolla trumpet shaped 5–7.5 cm long, ridged ventrally; mouth of corolla 5-lobed; stamens 4; style 1. ***Fruit*** a dry woody 2-valved capsule. Fl. Apr.–May, fr. Aug–Sep.

HARVESTING, SOURCING AND NATURAL RANGE

The wild species is endemic to C and NE China (Anhui, Hebei, Henan, Hubei, Hunan, Jiangsu, Jiangxi, S Liaoning, Shaanxi, Shandong and Shanxi). (up to 1,800 m altitude) and widely cultivated in China for many reasons [also native in, North America, Europe, Japan, and the Korean Peninsula (Nagata *et al.*, 2013].

CONSERVATION STATUS

China: Not Evaluated.
Global: Not Evaluated (IUCN Red List, accessed 22/6/2015); considered an invasive species in parts of the USA.

TCM PROPERTIES AND USE

properties: 'sweet, sour and cold'
channels: 'Liver and Pericardium'
actions: 'invigorates Blood, dispels wind, clears fire'
indications: abdominal masses, menstrual irregularities; postpartum breast lumps; itchy rashes

Campsis grandiflora

C. grandiflora

C. radicans

C. radicans

Poaulownia *tomentosa* var. *tomentosa*

Leaves of *P. tomentosa* var. *tomentosa*

P. tomentosa var. *tomentosa* PHOTO: PETER GASSON

Campsis grandiflora

Bignoniaceae

DRUG MORPHOLOGY

DRIED FLOWERS (CRUDE DRUG/DECOCTING PIECES): flowers shrivelled, slightly curved, yellowish brown to orange-brown and, when intact **3–5 cm long**. Calyx bell-shaped (2–2.5 cm long) **[1]**, **hairless and membranous** with **5 teeth extending to half the length of the calyx [2]** and with **midveins conspicuous [3]**. Flower tube trumpet-shaped **[4]**, yellowish-brown to orange-brown **[5]**, **without ridges on the ventral side**; flower lobes 5, of **similar size and length [6]**. Inner surface of flower tube with **distinct longitudinal stripes [7]**. **Stamens 4 or 5, not obviously twisted at base**, bilobed **[8]**, length unequal **[9]** and only some fertile, style 1 and thread-like, stigma 2-lobed **[10]**. *Odour*: slightly aromatic. *Taste*: slightly bitter and sour.

PROCESSING

Flowers are dried (= crude drug/decocting pieces).

OTHER COMMON NAMES

凌霄 Ling Xiao (towering clouds) (plant name); 紫葳 Zi Wei (purple luxuriance); Flos Campsis; trumpet creeper flower.

C. radicans

Bignoniaceae

DRUG MORPHOLOGY

DRIED FLOWERS (CRUDE DRUG/DECOCTING PIECES): very similar to PS1. Main differences: calyx somewhat fleshy (not membranous) **[11]**, teeth <½ the length of the calyx **[12]**, midveins absent **[11]**.

PROCESSING

As for PS1

OTHER COMMON NAMES

凌霄 Ling Xiao (plant name); 美国凌霄 Mei Guo Ling Xiao (American trumpet creeper); 紫葳 Zi Wei; Flos Campsis; trumpet creeper flower.

Paulownia tomentosa

Paulowniaceae

'DRUG SUBSTITUTE' MORPHOLOGY

Easily confused with PS1 and PS2. **Calyx densely covered in soft, short brown hairs (tomentose) [12]**. **Flower tube and lobes sometimes pale purple [13]** and with longitudinal ridges on the ventral side of the tube. When dissected, **flower lobes of unequal length (2 shorter, upper lip [14]) , (3 longer, lower lip [15]) ; inner surface of flower tube distinctly covered in deep purple, almost black, glandular dots [16]**; stamens distinctly twisted near base **[17]**.

OTHER COMMON NAMES

毛泡桐 Mao Pao Tong (plant name); 泡桐花 Pao Tong Hua; princess tree.

> SUBSTITUTION COMMENT. The bark, flowers and flower-buds are used in local medicine and, when market prices are favourable, the flowers which resemble those of PS1 and PS2 when dried may be traded as such as a cheap but incorrect substitute.

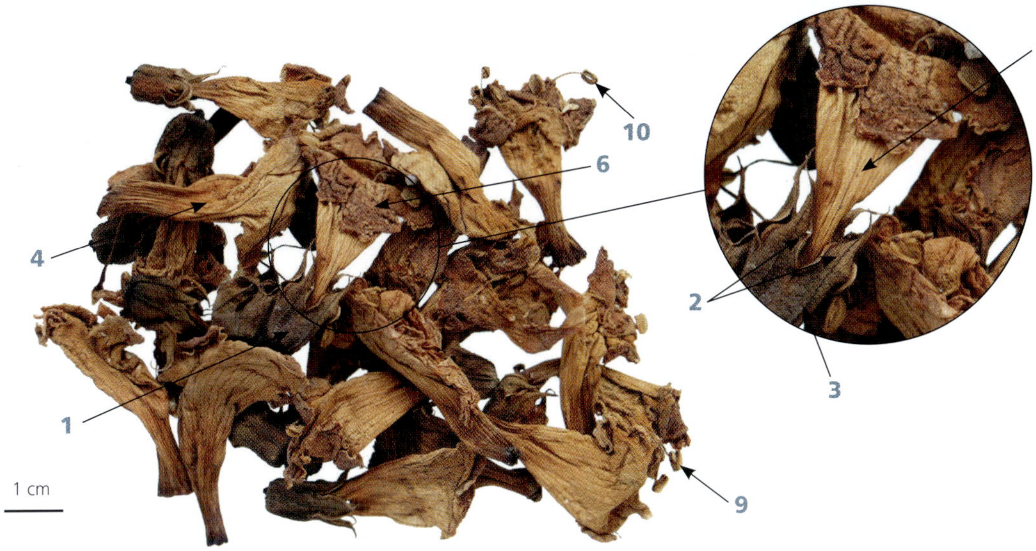

Crude drug/decocting pieces (*Campsis grandiflora*)

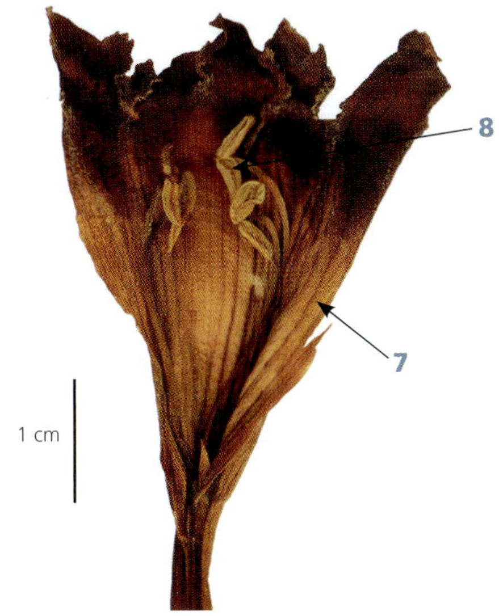

Dissected flower of *C. grandiflora*

Crude drug/decocting pieces (*C. radicans*)

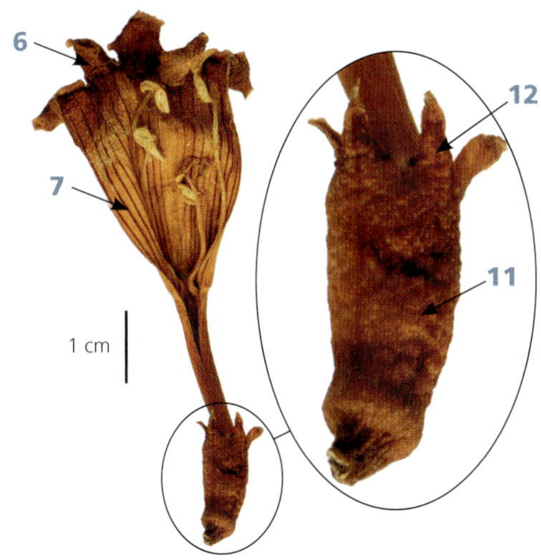

Dissected flower of *C. radicans*

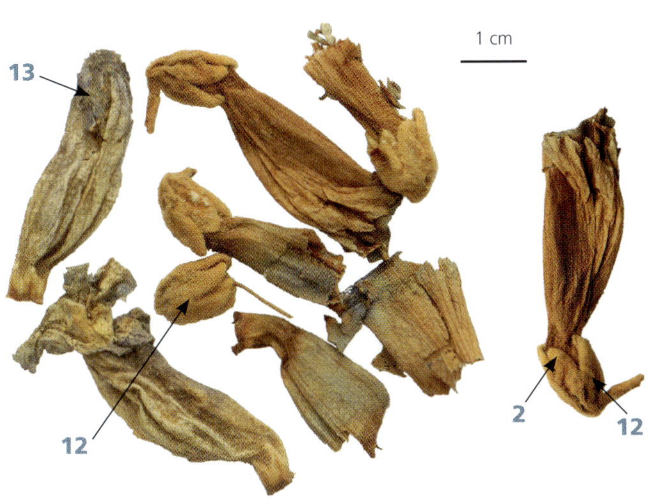

Dried flowers of *Paulownia tomentosa*

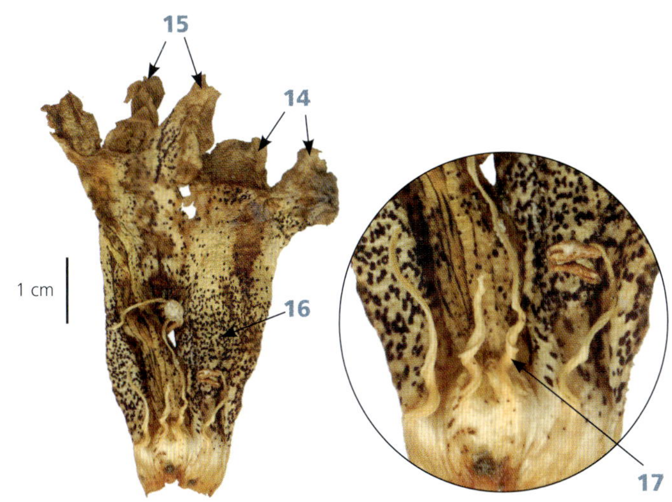

Dissected flower of *P. tomentosa*

Nao Yang Hua 闹羊花 Rhododendri Mollis Flos

Drug part: FLOWER

PHARMACOPOEIA SPECIES

Rhododendron molle (Blume) G. Don

Ericaceae

synonym: *Azalea mollis* Blume
taxonomic/nomenclatural note: in CP2015 as '*R. molle* G. Don'

Species of *Rhododendron* wild in China: 571

PLANT DESCRIPTION

A branching deciduous shrub (0.5–2 m tall). *Branches* covered in short, dense greyish-white hairs. *Leaves* alternate and short-stalked; leaf blades simple, oblong to oblong-lance-shaped (5–11 cm long; 1.5–3.5 cm wide), somewhat **papery,** base wedge-shaped, margin fringed with hairs, tip broad with a short sharp point; blade underside with dense greyish-white hairs, yellowish-brown along midrib. ***Flowers*** large and showy, **in terminal clusters**, opening with or before the leaves. Each flower with **calyx reduced to a hairy rim**; petals fused to form a broad, 5-lobed corolla, **yellow or golden yellow**, flecked darker; **corolla tube (up to 2.5 cm long)** tapering towards base and minutely hairy outside; stamens 5, unequal; style up to 6 cm long, hairless. *Fruit* a conical to cylindrical 5-ribbed capsule (25–35 mm long). Fl. Mar–May, fr. Jul–Aug.

HARVESTING, SOURCING AND NATURAL RANGE

Flowers are harvested in spring from wild populations which grow in mountain *Pinus* forests, scrub and in grassland (0–2,500 m altitude). The species is native to C, S and W China (Anhui, Fujian, Guangdong, Guangxi, Guizhou, Henan, Hubei, Hunan, Jiangsu, Jiangxi, Sichuan, Yunnan and Zhejiang provinces), with most harvesting occuring in Jiangsu, Zhejiang, Henan and Jiangxi (Zhao & Chen, 2014). Also cultivated widely as an ornamental; many cultivars exist.

CONSERVATION STATUS

China: Least Concern (MEPCAS, 2013).
Global: Not Evaluated (IUCN RED List, accessed 14/6/2015).

UNOFFICIAL SUBSTITUTE

Datura metel L.

Solanaceae

synonym: *D. alba* F. Muell.

PLANT DESCRIPTION

Easily distinguished from the Pharmacopoeia species. An upright **annual** (0.5–1.5 m tall). *Leaves* ovate or broadly ovate and **larger** (5–20 cm long; 4–15 cm wide), margins not fringed with hairs. **Flowers solitary, white**, sometimes yellowish or pale purple; **calyx long, green and tubular (4–9 cm);** petals fused into a trumpet-shaped **corolla, much longer and narrower** (14–20 cm long). Fruit a dry, sub-globose and densely tuberculate or spiny capsule. Fl. and fr. Mar–Dec.

CONSERVATION STATUS

China: Not Evaluated.
Global: Not Evaluated (IUCN Red List, accessed 10/9/2016).

TCM PROPERTIES AND USE

properties: **'pungent and warm'**
channel: **'Liver'**
actions: **'dispels wind damp'**
indications: **musculoskeletal pain; headache; traumatic injury; (T)**

Rhododendron molle, golden-yellow flowering form (inset yellow flowering form)

Fruit of *R. molle*

Datura metel in flower with young fruit

D. metel with ripening fruit

Rhododendron molle

Ericaceae

DRUG MORPHOLOGY

CRUDE DRUG/DECOCTING PIECES (DRIED FLOWERS): shrivelled greyish-yellow to brownish-yellow flowers either single (if fallen off) **[1]** or several together (if still attached) **[2]**. Flower stem hairy **[3]** (1–2.5 cm long). Calyx reduced to a hairy rim **[4]**. Petals fused to form a broad, 5-lobed corolla **[5]**, **yellow or golden yellow**, flecked darker; **corolla (up to 2.5 cm long)** tapering towards base; corolla lobes elliptic or ovate-oblong **[6]**; stamens 5, **unequal, filaments curved [7]** with lower parts softly hairy; anthers reddish-brown; style 1 (up to 6 cm long) with stigma ending in a head. *Taste*: **slightly numbing.**

PROCESSING

Flowers are sun- or shade-dried (= crude drug/decocting pieces).

OTHER COMMON NAMES

黄杜鹃花 Huang Du Juan Hua; 黄喇叭花 Huang La Ba Hua; 羊踯躅 Yang Zhi Zhu; Flos Rhododendri Mollis; yellow azalea flower (CP2010).

SAFETY

Highly toxic. (CP2015, Hong Kong 1999a). Short-term use only.

Datura metel

Solanaceae

'DRUG SUBSTITUTE' MORPHOLOGY

Flowers easily distinguished from those of the official drug. *Main differences*: **flower stalk shorter** (c. 1 cm) **[8]**; **calyx much longer and forming a conspicuous tube [9]** almost half the length of the entire corolla, with 5 pointed apical teeth **[10]** and 5 longitudinal veins prominent at the base **[11]**; **corolla much longer (10–20 cm long), forming a long narrow trumpet [12]**.

OTHER COMMON NAMES

广东洋金花 Guang Dong Yang Jin Hua; 南洋金花 Nan Yang Jin Hua 'southern datura flower'; 洋金花 Yang Jin Hua; Daturae Flos; datura flower (CP2010); Flos Daturae; horn-of-plenty (SCN).

SAFETY

Toxic (CP2015, Hong Kong 1999a). Officially forbidden in China as food or healthfood (Anon., 2002b).

See also 'Yang Jin Hua' (Daturae Flos) page 600.

SUBSTITUTION COMMENT. Dried flowers of this species are the source of the separate but also **toxic** TCM drug 'Yang Jin Hua' (Daturae Flos). Although the fresh plants and trade items are easily distinguished from those of *R. molle* (above) it shares one of its common names Nao Yang Hua 闹羊花, causing potential confusion in the trade.

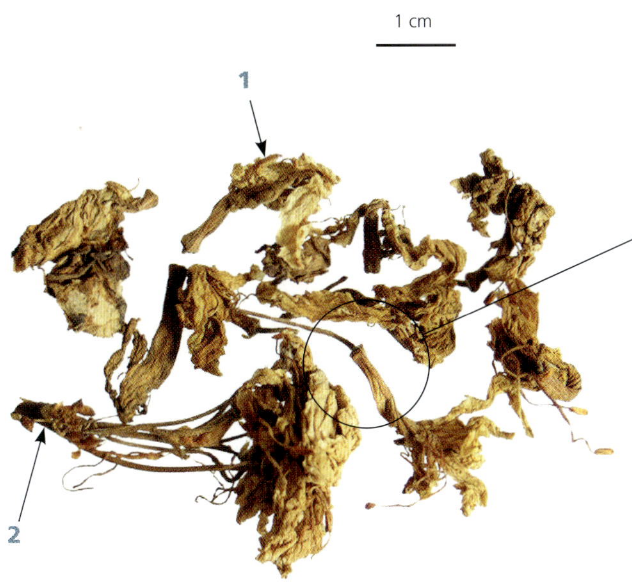

Crude drug/decocting pieces of Nao Yang Hua (*R. molle*)

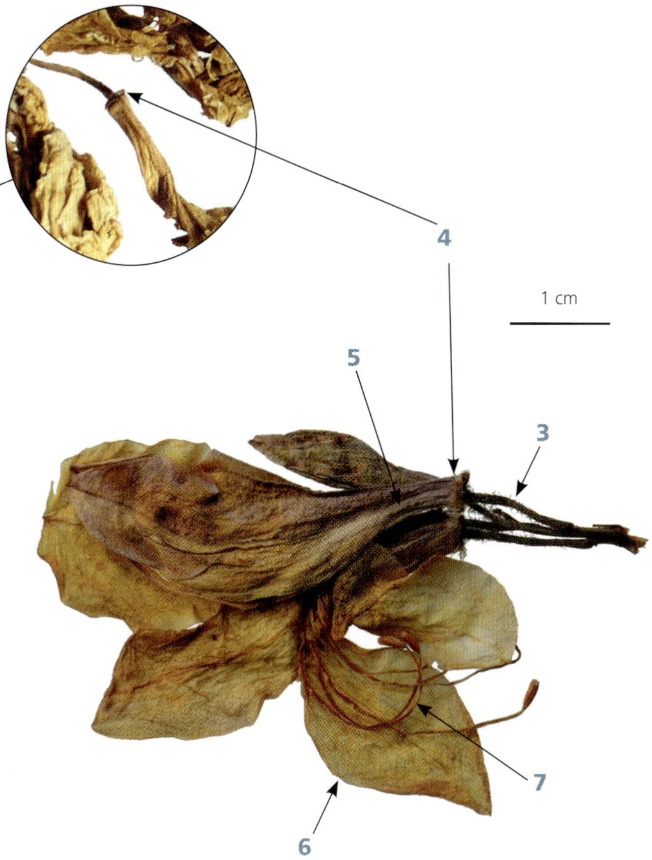

Expanded flowers of *R. molle*

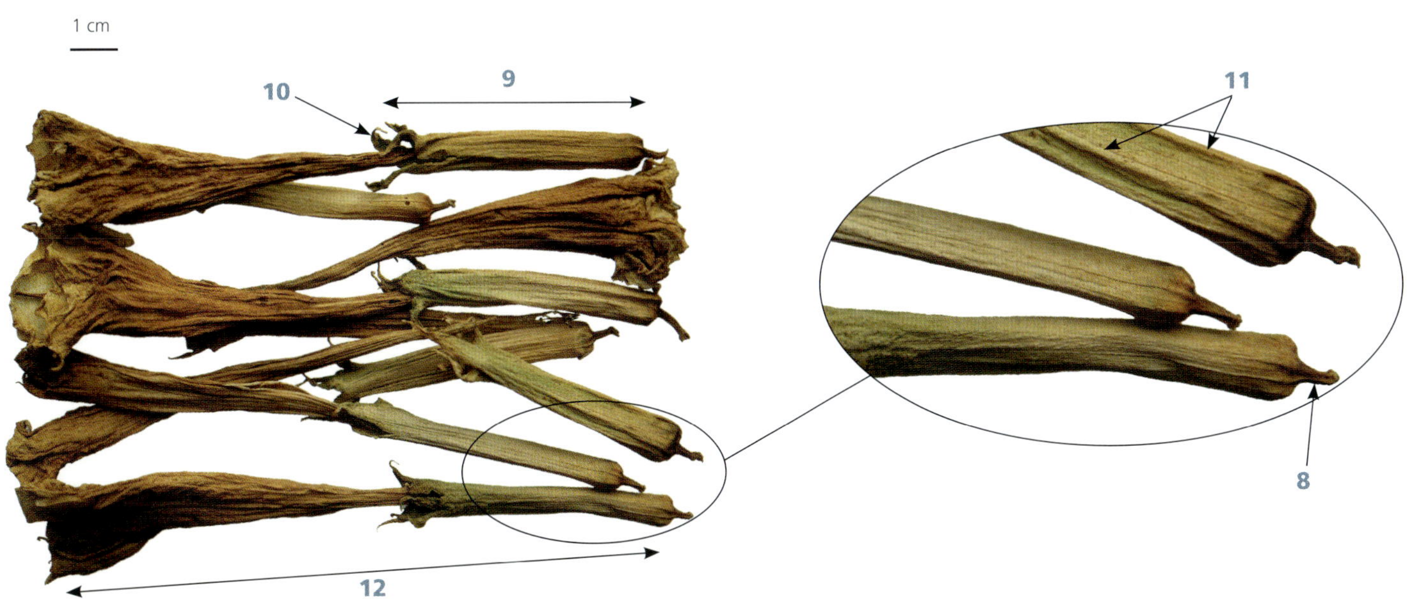

Crude drug/decocting pieces of substitute Yang Jin Hua (*D. metel*)

Pu Huang 蒲黄

Typhae Pollen

Drug part: POLLEN

The CP2015 states that any species in the genus *Typha* can be used as the source of Pu Huang. Four of the most common and widespread species are included here.

Species of *Typha* wild in China: 12

Typha angustifolia L.

Typhaceae

PLANT DESCRIPTION

A tall upright aquatic/waterside perennial (1.5–3 m tall) with creeping rhizomes. *Leaves* simple, long and linear (**52–120 cm long**; 0.4–0.9 cm wide) mostly basal on two opposite sides of the stem, with sheathing bases; outer surface convex and, in cross-section, semi-circular; margin entire. *Flowers* single-sexed, minute and densely crowded into a cylindrical spike with a lower stout, brown and sponge-like female zone (5–30 cm long) and an upper, narrower male flower zone (c. 8 cm long); these 2 zones separated by a short non-flowering gap (axis) 2.5–7 cm long. Sepals and petals absent. Each male flower with 1–3 stamens, encircled by hairs, short filaments, anthers c. 2 mm long and pollen grains solitary or in fours [visible with a light microscope]. Each female flower with slender, thread-like bracteoles and a broad stigma. *Fruits* minute, long-elliptical and nut-like. Fl. and fr. Jun–Sep.

HARVESTING, SOURCING AND NATURAL RANGE

Male inflorescences are harvested when yellow from wild populations during summer. The species grows in shallow water along margins of lakes, ponds and rivers throughout much of China except in the tropical south: Gansu, Hebei, Heilongjiang, Henan, Hubei, Inner Mongolia, Jiangsu, Jilin, Liaoning, Shaanxi, Shandong, Taiwan, Xinjiang and Yunnan [also native in India, Indonesia, Japan, Nepal, Pakistan, Philippines, Russia; America, Australia and Europe].

CONSERVATION

China: Least Concern (authors' pers. obs., 2009).
Global: Least Concern (IUCN Red List, acessed 20/8/2016).

PLANT DESCRIPTIONS

Typhaceae

T. orientalis C. Presl

Similar to PS1 but **leaves usually shorter (40–70 cm long**; 0.4–0.9 cm wide); gap between female flower zone and male flower zone **absent**; male flower zone 2.7–9 cm long, with 1–3 deciduous bracts at base or middle. Fl. and fr. May–Aug.

T. latifolia L.

Similar to PS1. **Leaves 45–95 cm long and 0.5–1.5 cm wide**; gap between female flower zone and male flower zone **absent**; female flower zone 5–23 cm long; female flowers without bracteoles. Fl. and fr. May–Aug.

T. domingensis Pers.

synonym: *T. angustata* Bory & Chaub. (name used in CP2010)

Similar to PS1. **Stems stout (0.7–2.5m tall)**; leaves 40–150 ccm long; 0.3–0.8 cm wide; **male flower zone 7–30 cm long** with 1 or 2 bracts up to 32 cm long (deciduous), **distinct gap between male and female flower zones**; fruits tapering at both ends. Fl. and fr.: Jun–Aug.

HARVESTING, SOURCING AND NATURAL RANGE

The pollen of these 3 species is harvested in the same way as that for PS1 and their wild distributions in China are also very similar; none are endemic.

CONSERVATION (for all 3 species)

China: Least Concern (authors' pers. obs., 2009) MEPCAS, 2013).
Global: Least Concern (IUCN Red List, accessed 20/8/2016).

TCM PROPERTIES AND USE

properties: 'sweet and neutral'
channels: 'Liver and Pericardium'
actions: 'stops bleeding, invigorates Blood, resolves stasis'

indications: metrorrhagia; haematuria, dysuria; abdominal pain; haematemesis, epistaxis; haemoptysis; (T)

Typha angustifolia in the wild, Inner Mongolia

T. angustifolia

T. angustifolia

T. orientalis

T. latifolia

T. domingensis

T. domingensis

Typha angustifolia

Typhaceae

DRUG MORPHOLOGY

CRUDE DRUG (UNSIEVED POLLEN): a yellow powder, sometimes clumped and sometimes with other flower parts such as stamens. *Texture*: satin-like when fingered. High (e.g. SEM) magnification is required to confirm identification to genus. ***Good quality***: dry, bright yellow and smooth.

DECOCTING PIECES-1 (SIEVED POLLEN): as for the crude drug but devoid of clumps and other contaminant flower parts.

PROCESSING

Male inflorescences are sun-dried before being ground (= crude drug) and sieved (decocting pieces-1); subsequent and optional carbonising produces 'Pu Huang Tan' (= decocting pieces-2).

OTHER COMMON NAMES

蒲棒 Pu Bang; 蒲花 Pu Hua; bulrush; cattail (SCN); narrow-leaf cattail; Pollen Typhae.

T. orientalis

Typhaceae

DRUG MORPHOLOGY

Not possible to distinguish macroscopically from other *Typha* species.

PROCESSING

As for PS1.

OTHER COMMON NAMES

蒲棒 Pu Bang; 蒲花 Pu Hua; 东方香蒲 Dong Fang Xiang Pu; cattail (SCN); oriental cattail; Pollen Typhae.

T. latifolia

DRUG MORPHOLOGY

Not possible to distinguish macroscopically from other *Typha* species.

PROCESSING

As for PS1.

OTHER COMMON NAMES

蒲棒 Pu Bang; 蒲花 Pu Hua; bulrush; cattail (SCN); common cattail; common reed mace; Pollen Typhae.

T. domingensis

DRUG MORPHOLOGY

Not possible to distinguish macroscopically from other *Typha* species.

PROCESSING

As for PS1.

OTHER COMMON NAMES

蒲棒 Pu Bang; 蒲花 Pu Hua; 长苞香蒲 Chang Bao Xiang Pu; cattail (SCN); southern cattail; Pollen Typhae.

5 mm

Crude drug of Pu Huang (sourced from *Typhe angustifolia*)

Decocting pieces-1 of Pu Huang (sourced from *T. angustifolia*)

Carbonised Pu Huang Tan (decocting pieces-2)

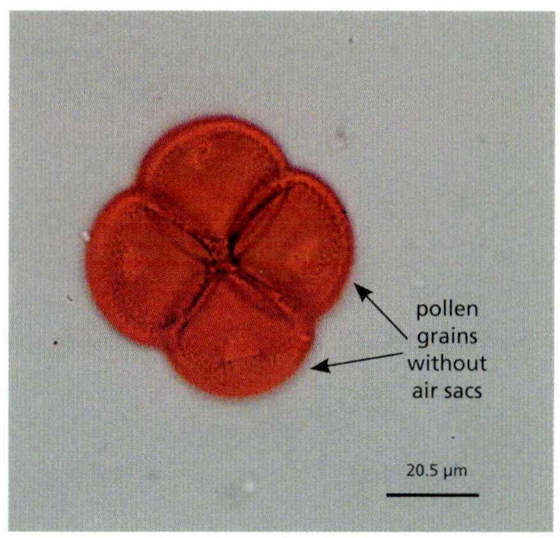

Pollen grain of *T. latifolia*

UNOFFICIAL SUBSTITUTES

Although pollen identification requires high magnification (and therefore arguably is prone to substitution) 'Pu Huang' pollen is unlikely to be substituted or adulterated with that from other families or genera since few other species produce pollen in sufficient quantity to be commercially viable. One exception, however, is pollen from *Pinus* species (see 'Song Hua Fen' (Pini Pollen page 592).

Using a light microscope pollen of *Typha* and *Pinus* are easily distinguished: *Pinus* pollen grains each have two air sacs whereas those of *Typha* have none.

Song Hua Fen — crude drug/decocting material (dried pollen of *Pinus tabuliformis*)

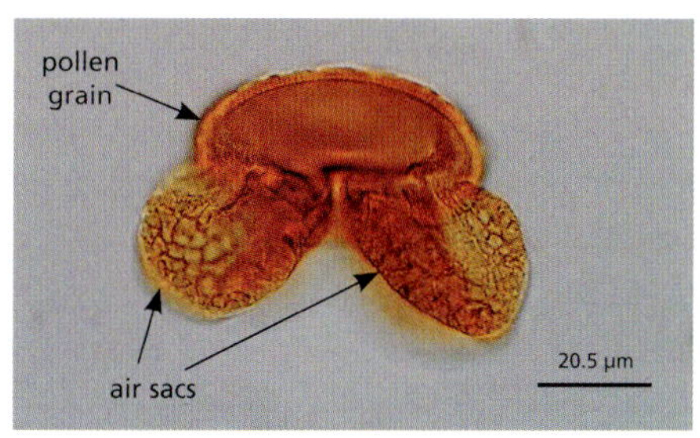

Pollen grain of a species of *Pinus* (*P. krempfii*) always with 2 air sacs

Shan Yin Hua 山银花

Lonicerae Flos

honeysuckle

Drug part: FLOWER-BUD

taxonomic/nomenclatural note: two of the official source species according to the CP2015, *Lonicera fulvotomentosa* P. S. Hsu & S. C.Cheng and *L. macranthoides* Hand.-Mazz., are considered by *FoC* and *Kew-MPNS* to be the same species with the accepted name *L. macrantha* (D. Don) Spreng. (PS3) see below.

PHARMACOPOEIA SPECIES 1

Lonicera hypoglauca Miq.

Caprifoliaceae

synonym: *L. hypoglauca* subsp. *nudiflora* P. S. Hsu & H. J. Wang

Species of *Lonicera* wild in China: 57

PLANT DESCRIPTION

A semi-evergreen climber. *Stem* covered in short, dense, greyish, downy hairs. *Leaves* opposite, simple, stalked (3–15 mm), ovate to lance-shaped (3–8 cm long); **undersides bluish-green** with many, scattered, **stalkless, orange glands [hl];** margin with minute fringe of hairs. *Flowers* fragrant, in pairs in leaf axils along a central axis towards tips of branchlets; *bracts* **leaf-like, ovate to elliptic (0.8–3 cm long)**. Each flower 2.5–4.5 cm long with sepals fused to form a **hairless or very slightly hairy calyx tube** with 5 lance-shaped lobes (calyx teeth) at outer end; petals fused to form a long, narrow tube (3–5 cm long) dividing into 2 spreading lobes more than ¼ length of tube and changing colour from white to yellow when flower fertilised; stamens 5; style slender; stigma head like a pin-head. *Fruit* black and spherical (7–12 mm diam.). Seeds few, ovoid (5–7 mm long). Fl. Apr–May (–Jun), fr. Oct–Nov.

HARVESTING, SOURCING AND NATURAL RANGE

Flower-buds, some just opening, are harvested in early summer mostly from cultivated plants grown as medicinal crops with the main production area in Guizhou province. This common species is native to central, southern and south-eastern China (S Anhui, Fujian, Guangdong, Guangxi, Guizhou, SW Hubei, Hunan, Jiangxi, Sichuan, Taiwan, Yunnan and Zhejiang) where it grows in scrub and open forest between 200–700 m (rarely to up 1,800 m) altitude. The species is also native to Japan and possibly Kashmir and Nepal.

CONSERVATION STATUS

China: Least Concern (MEPCAS, 2013)
Global: Not Evaluated (IUCN Red List, accessed 17/2/2016)

PHARMACOPOEIA SPECIES 2

L. confusa DC.

Caprifoliaceae

PLANT DESCRIPTION

As for *L. hypoglauca* (PS1) but stems either covered in short, dense, greyish, downy hairs or hairless. Leaf undersides hairless or soft and downy; **stalkless orange glands absent.** Calyx tube **usually densely hairy,** sometimes sparsely hairy or hairless. Berries black. Fl. Mar–May, fr. Aug–Oct.

HARVESTING, SOURCING AND NATURAL RANGE

The flower-buds (with some just opening) of this common and widespread species of southern China (Guangdong, Guangxi, Hainan and Yunnan provinces) are mainly wild-harvested. Plants grow in scrub and mixed forest, often on slopes and along road- and river-sides (300–800 m altitude). Some cultivation takes place in Guizhou. The species also grows wild in neighbouring Nepal and N Vietnam.

CONSERVATION STATUS

China: Least Concern (MEPCAS, 2013).
Global: Not Evaluated (IUCN Red List, accessed 17/2/2016)

PHARMACOPOEIA SPECIES 3

L. macrantha (D. Don) Spreng.

Caprifoliaceae

synonyms: *L. fulvotomentosa* P. S. Hsu & S. C. Cheng (name used in CP2015); *L. macranthoides* Hand.-Mazz. (name used in CP2015); *Caprifolium macranthum* D. Don

PLANT DESCRIPTION

As for *L. hypoglauca* (PS1) but stems with **long, stiff and spreading yellowish-brown hairs and minute glandular hairs [hl]** ***or*** **densely covered in short, soft yellowish-brown hairs (tomentose).** Leaf undersides with rather **coarse stiff hairs. Calyx tube hairless to sparsely covered in coarse stiff hairs.** Berries white to black. Fl. Apr–May, fr. Jul–Aug.

HARVESTING, SOURCING AND NATURAL RANGE

Another common wild species of central and southern China (S Anhui, Fujian, Guangdong, Guangxi, Guizhou, ?Hainan, Hubei, Hunan, Jiangxi, Sichuan, Taiwan, Tibet, Yunnan and Zhejiang provinces) from where flower-buds are harvested in early summer from plants growing in mixed forests, scrub and mountain valleys (300–1,800 m altitude). The distribution of the wild species also extends to Bhutan, India and Nepal. Some *L. macrantha* is also sourced from medicinally-grown crops in Guizhou province.

CONSERVATION STATUS

China: Least Concern (MEPCAS, 2013).
Global: Not Evaluated (IUCN Red List, accessed 17/2/2016)

TCM PROPERTIES AND USE

properties: 'sweet and cold'
channels: 'Lung, Heart and Stomach'
actions: 'releases wind heat, eliminates toxins'
indications: abscesses, sores; dysentery; respiratory tract infections

Lonicera hypoglauca in the wild, Fujian province

Stalkless orange glands on lower leaf surface of *L. hypoglauca*

Branch of *L. hypoglauca* with immature fruits, Guizhou province

L. confusa in the wild, Guizhou province

L. confusa with hairy calyces at base of each flower

Medicinal crop of *L. macrantha*, Guizhou province

L. macrantha with flower-buds

L. macrantha showing white velvety leaf underside with prominent veins and hairless calyx tubes.

Lonicera hypoglauca

Caprifoliaceae

DRUG MORPHOLOGY

CRUDE DRUG/DECOCTING PIECES (FLOWER-BUDS): club-shaped, slightly curved, gradually tapering towards the base (2–3.4 cm long; 1.5–3 mm diam. Bud surface yellowish-white or greenish-white, darkening with age, **densely covered in short, fine soft hairs**, sometimes with stalked orange glands. Leaf-like bracts occasionally present, ovate to elliptic (0.8–3 cm long). Calyx tube **hairless or only very slightly hairy [1];** margins of the 5 lance-shaped calyx lobes **densely hairy** [hl]. *Opening flowers:* becoming white; 3–5 cm long; apex 2-lipped with upper lip 4-lobed; stamens 5 with lower parts attached to walls of petal tube. *Odour*: sweetly aromatic. *Taste*: weak and slightly bitter. Laboratory-based methods are recommended for reliable authentication.

PROCESSING

Flower-buds are dried (= crude drug/decocting pieces).

OTHER COMMON NAMES

菰腺忍冬 Gu Xian Ren Dong (plant name); Tu Yin Hua; Flos Lonicerae; honeysuckle (SCN).

SAFETY

An official Chinese food and healthfood (Anon., 2002a,b).

L. confusa

Caprifoliaceae

DRUG MORPHOLOGY

CRUDE DRUG/DECOCTING PIECES (FLOWER-BUDS): cannot be reliably distinguished from that of the other official species. Laboratory-based methods are recommended for reliable authentication. Calyx tube, flower tube and petals **usually densely covered with greyish-white hairs.**

PROCESSING

As for PS1.

OTHER COMMON NAMES

华南忍冬 Hua Nan Ren Dong; Flos Lonicerae; honeysuckle (SCN).

SAFETY

An official Chinese food and healthfood (Anon., 2002a,b)

L. macrantha

Caprifoliaceae

DRUG MORPHOLOGY

CRUDE DRUG/DECOCTING PIECES (FLOWER-BUDS): cannot be reliably distinguished from that of the other official species. Laboratory-based methods are recommended for reliable authentication.

PROCESSING

As for PS1.

OTHER COMMON NAMES

灰毡毛忍冬 Hui Zhan Mao Ren Dong; Flos Lonicerae.

UNOFFICIAL SUBSTITUTES

Since *Lonicera* species require considerable botanical expertise to identify them reliably even as fresh whole plants and, given that harvesting is predominantly from wild populations, there is a high likelihood of confusion with the many other wild species in this genus. The making of herbarium voucher specimens which include all available aerial parts is recommended at the point of harvesting to aid subsequent identification by a competent botanist and to minimise mixed batches entering trade. Harvesting from cultivated stock will also help to reduce the use of unofficial species assuming good farming practices such as GAP are implemented.

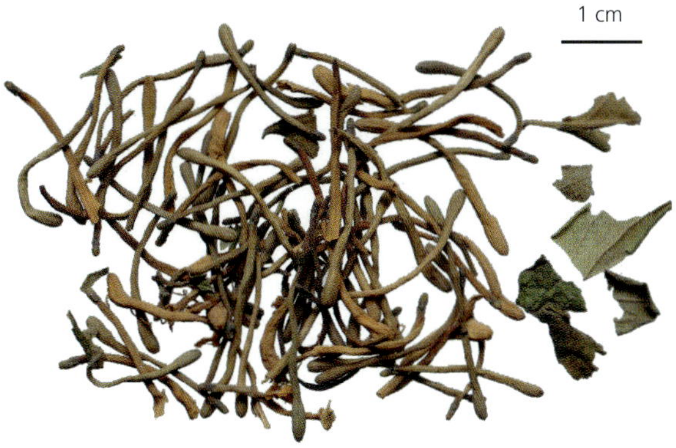

Shan Yin Hua sourced from *Lonicera hypoglauca* — leafy bracts remnants on right

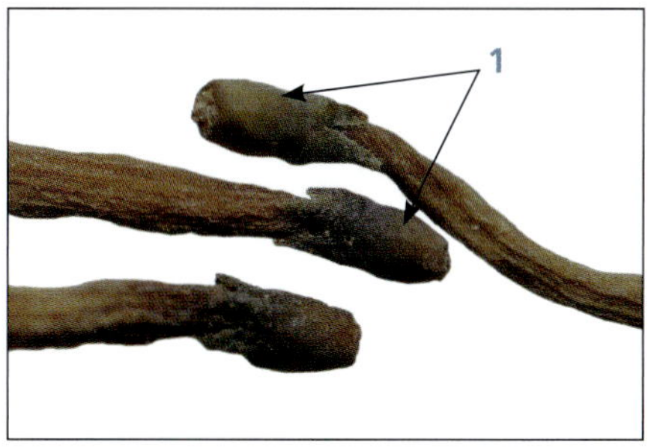

L. hypoglauca

L. hypoglauca, fragments of leaf-like bracts occasionally present

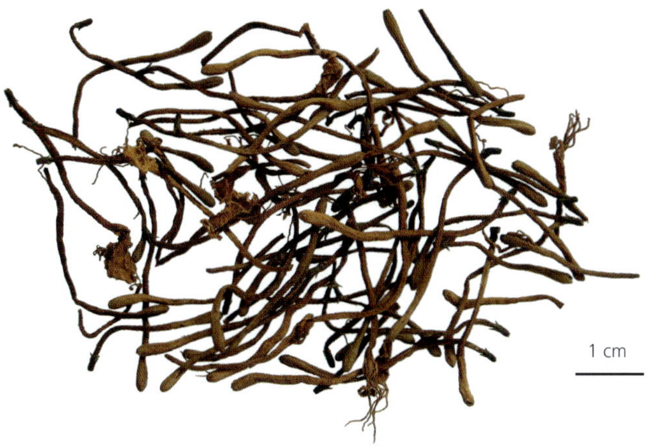

Shan Yin Hua sourced from *Lonicera confusa*

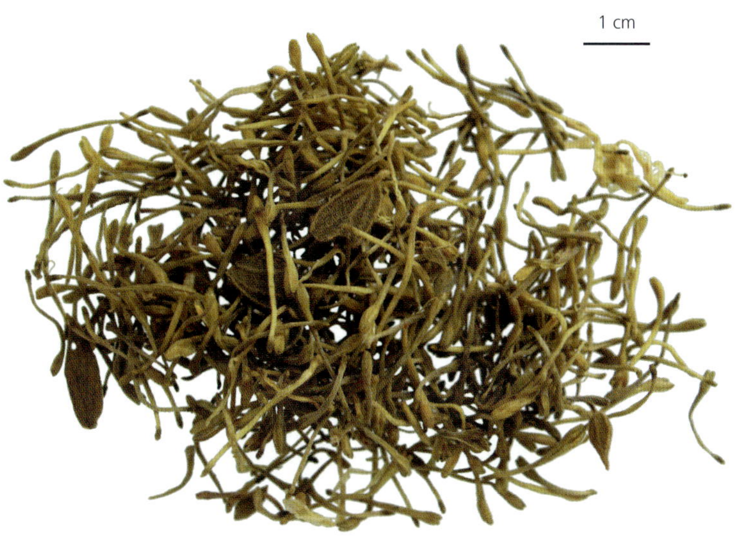

Shan Yin Hua sourced from *Lonicera macrantha*

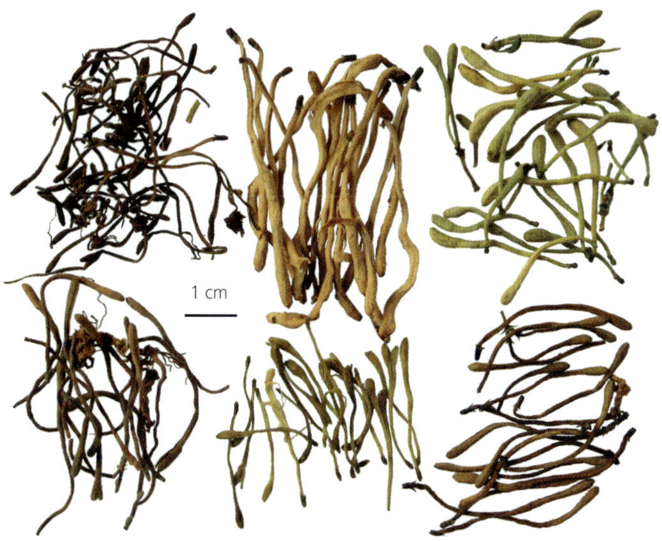

Flower-buds harvested from a range of official and unofficial species. Although they vary in size, shape, colour and hairiness these characters are not consistent between species for reliable identification; many of these characters reflect the maturity of the flower-buds and may therefore be misleading. Laboratory-based authentication is strongly recommended.

Song Hua Fen 松花粉 Pini Pollen

Drug parts: POLLEN

The CP2015 states that 'several other plants of the same genus (*Pinus*)' can be used as the source of 'Song Hua Fen'. The two species named in the CP2015 are described here.

Species of *Pinus* wild in China: 39

Pinus massoniana Lamb.

Pinaceae

taxonomic/nomenclatural note: three varieties are recognised by *FoC* and *Kew-MPNS*.

PLANT DESCRIPTION

A tall, evergreen tree (up to 45 m) with a broad pyramidal or umbrella-like crown. *Trunk* to 1.5 m diam. with irregular, flaking, reddish-brown bark. *Leaves* terminal on very short branchlets, needle-shaped **(12–20 cm long)**, soft to the touch, normally in twos (rarely threes), **slightly twisted** and base enclosed by a sheath; leaf cross-section reveals **4–8 resin canals** [hl]. *Reproductive parts* develop as separate male and female cones on the same plant. Male (pollen) cones in erect spike-like cylindrical-ovoid clusters. Female (seed) cones solitary and pendent, short-stalked and ovoid to conical (2.5–7 cm long; 2.5–5 cm wide); chestnut-brown when mature (in the second year); seed scales thick and woody with the apex of each scale bearing a **central blunt or shortly pointed protrusion** (umbo). *Seeds* 2 per scale, narrowly ovoid; terminal wing 1.6–2.1 cm long. Pollination Apr–May, seed maturity Oct–Dec of the second year.

HARVESTING, SOURCING AND NATURAL RANGE

Pollen is harvested from wild trees in spring. The species grows on hills and plains, 0–2,000 m altitide and is endemic to SC and SE China: Anhui, Fujian, Guangdong, Guangxi, Guizhou, Hainan, W Henan, Hubei, Hunan, S Jiangsu, Jiangxi, SE Shaanxi, Sichuan, Taiwan, E Yunnan and Zhejiang. Most 'Song Hua Fen' is likely to be sourced from the type variety (*massoniana*) since it is widespread and common compared to the other two varieties (var. *shaxianensis* is confined to Fujian province and var. *hainanensis* to the tropical island of Hainan. A common tree of afforestation projects in S China owing to its many uses (gum resin, timber, wood pulp, medicine etc.).

CONSERVATION STATUS

Endemic to China: Least Concern (IUCN Red List, accessed 20/8/2016); MEPCAS (2013) assigns the 3 varieties different IUCN conservation categories: var. *massoniana* (Least Concern); var. *hainanensis* W. C. Cheng & L. K. Cheng (Vulnerable) and var. *shaxianensis* D. X. Zhou (Data Deficient).

P. tabuliformis Carrière

Pinaceae

taxonomic/nomenclatural note: in CP2010 as *'P. tabulaeformis* Carr.' and in CP2015 as *'P. tabulieformis* Carr.'.

PLANT DESCRIPTION

Similar to PS1 but shorter (up to 25 m tall), crown flat topped; needles **straight and shorter (5–12 cm long), slightly thicker and more rigid** with **5–9 resin canals** in cross-section [hl]; female (seed) cones ovoid to ovoid-globose (2.5–9 cm long; 4–9 cm wide), centre of each woody scale (umbo) **extending into a spine.** Pollination Apr–May, seed maturity Oct of 2nd year.

HARVESTING, SOURCING AND NATURAL RANGE

Harvesting as for PS1. The species grows on hills and mountains (100–2,600 m altitiude) in NC, NE and SE China: Gansu, Hebei, Henan, Hubei, Hunan, Inner Mongolia, Jilin, Liaoning, Ningxia, Qinghai, Shaanxi, Shandong, Shanxi and Sichuan [also native in the Korean Peninsula].

CONSERVATION STATUS

China: MEPCAS (2013) assigns the 4 varieties individual IUCN conservation categories: var. *tabuliformis* (Least Concern); var. *mukdensis* (Uyeki ex Nakai) Uyeki (Data Deficient) and var. *umbraculifera* Liou & Q. L. Wang (Data Deficient).

Global: Least Concern (IUCN Red List, accessed 20/8/2016) with var. *umbraculifera* categorised as Data Deficient.

TCM PROPERTIES AND USE

properties: **'sweet and warm'**
channels: **'Liver and Spleen'**
actions: **'resolves damp, promotes healing, stops bleeding'**

indications: **topical use only: bleeding; traumatic injury; sores**

Trees of *Pinus massoniana* in the wild, China

Immature male (pollen) cones of *P. massoniana*

P. massoniana with current and previous year's seed cones

Pinus tabuliformis in cultivation

Mature male (pollen) cones of *P. tabuliformis*

P. tabuliformis with current and previous year's female (seed) cones

Seed cone of *P. tabuliformis*

Pinus massoniana

Pinaceae

DRUG MORPHOLOGY

CRUDE DRUG/DECOCTING MATERIAL: fine yellowish powder. Pollen grains of all *Pinus* species typically have two air sacs (visible under high magnification – see SEM photos). *Texture:* light, easily becoming air-borne and, when twisted in the hand, exerts little friction. *Odour*: indistinct. *Taste*: faint. ***Good quality*:** yellow pollen, well-sieved to remove foreign matter.

PROCESSING

Pollen is sun-dried then sieved to remove any foreign matter (= crude drug/decocting material).

OTHER COMMON NAMES

Chinese red pine; Masson pine (SCN); Pollen Pini.

P. tabuliformis

Pinaceae

DRUG MORPHOLOGY

As for PS1.

PROCESSING

As for PS1.

OTHER COMMON NAMES

Chinese pine (SCN); Pollen Pini.

1 cm

Song Hua Fen sourced from *P. massoniana*

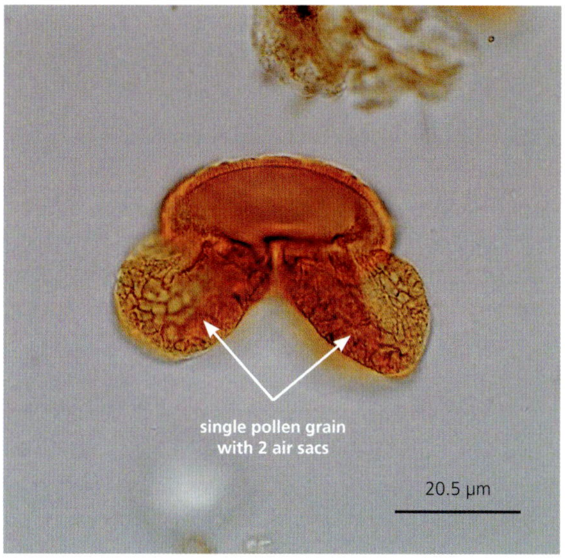

Pollen grain of *Pinus krempfii* PHOTO: H. BANKS

1 cm

Song Hua Fen sourced from *P. tabuliformis*

Xi Hong Hua 西红花

Croci Stigma

saffron

Drug part: STYLE AND STIGMA

PHARMACOPOEIA SPECIES

Crocus sativus L.

Iridaceae

PLANT DESCRIPTION

A small perennial herb with a squat corm (c. 3 cm diam.) covered in a light yellow or yellowish-brown, fibrous 'tunic'. ***Leaves* all basal**, 9–15, **narrowly linear (9–20 cm long; 2–3 mm wide)**, longer than the flowers, with the base of each leaf surrounded by a membranous sheath; upper leaf surface with pale, central, longitudinal stripe, margins entire. ***Flowers* 1 or 2, emerging from below ground, large, showy and fragrant**, flower tube long and slender, **'petals' 6, purple** and often veined, equal, broadest above the middle, tip broad or rounded (4–5 cm long); stamens usually 3 (c. 2.5 cm long), anthers yellow; style divided half-way along into 3 deep red and recurved branches (each 2.5 cm long); stigmatic surfaces tiny, yellow and confined to tip of each style branch. The cultivar grown in China is sterile. Fl. autumn.

Styles and stigmas have been used worldwide since ancient times for medicine, cooking and dye.

Species of *Crocus* wild in China: 1 (not *C. sativus*)

HARVESTING, SOURCING AND NATURAL RANGE

An introduced species in China, widely cultivated. Flowers are harvested when in full bloom in autumn. Crops with GAP certification have been grown in Shanghai province (Zhang *et al.*, 2010).

CONSERVATION STATUS

China: introduced.

Global: Not Evaluated (IUCN Red List, accessed 21/5/2015). Origin uncertain, probably a hybrid.

Medicinal crop coming into flower, Zhejiang province

UNOFFICIAL SUBSTITUTE

Carthamus tinctorius L.

Asteraceae

PLANT DESCRIPTION

Easily distinguished from the Pharmacopoeia species and not confused as whole plants. *C. tinctorius* is **much larger (usually up to 1 m tall)** with smooth, whitish stems, branching above. Underground storage organ (e.g. corm) absent. *Leaves* simple, unstalked, lance-shaped, ovate or elliptic, **much wider (mostly 7–15 cm long; 2.5–6 cm wide)**; margins entire or spiny toothed. ***Flowers* in dense many-flowered semi-spherical heads** (capitula) (c. 2.8 cm diam.), towards the top of the plant. Individual **florets red, orange or yellow** (changing with maturity) with a slender flower tube (up to 2.8 cm long) opening into 5 linear lobes; stamens 5. *Fruits* are tiny 4-angled achenes. Fl. and fr. May–Aug.

HARVESTING, SOURCING AND NATURAL RANGE

Introduced species. Widely cultivated as a commercial crop in China as a source of cooking oil (seeds), medicine and dye (flowers).

CONSERVATION STATUS

See 'Hong Hua' (Carthami Flos) page 570.

TCM PROPERTIES AND USE

properties: **'sweet and neutral'**
channels: **'Heart and Liver'**
actions: **'invigorates Blood, clears heat, tonifies Blood'**

***indications*: amenorrhoea, postpartum stasis and obstruction; rashes with fever; emotional distress**

Crocus sativus in full flower emerging through straw-protected soil

Medicinal crop of *Carthamus tinctorius*, Xinjiang province CREDIT: KAISA SULAIMAN

C. tinctorius in flower and flower bud, yellow stage inset

Crocus sativus — Iridaceae

DRUG MORPHOLOGY

DRIED STYLE AND STIGMA (CRUDE DRUG/DECOCTING PIECES): the drug is the style, typically deep red and 3-branched [1], together with the minute yellow, stigmatic surface [hl] at the branch tips [2]. Style branches (usually broken off) long, slender and tubular and opening longitudinally into a broad flattened tip (c. 3 cm long; 0.5–2 mm diam.), branch tips ragged with tiny, pale yellow stigmatic surface **[2]** (1–2 mm long). Lower, yellowish-cream unbranched part of the style occasionally present (but usually removed during processing), very slender (<1 mm wide). *Texture:* soft and light; pliable when fresh, brittle when dry. *Odour:* distinctively aromatic. *Taste:* slightly irritating and bitter; turns saliva yellow. ***Good quality***: dark red.

Poor quality trade sample (lower part of styles [yellow] not removed

PROCESSING

Styles and stigmas are removed from the flower and the 'petals' discarded. The lower yellowish-cream section of each style is also removed and discarded. The remaining parts of each style (including stigma) are then dried (= crude drug/ decocting pieces).

Note: this drug is the same as saffron, traded worldwide as a culinary spice and as a source of dye. Many texts erroneously state that it is from the stigma, but botanically it is the style; the tiny stigmatic surface is confined to the tip of each style branch.

OTHER COMMON NAMES

番红花 Fan Hong Hua (foreign red flower); 藏红花 Zang Hong Hua (Tibetan red flower); saffron (CP2010; SCN); Stigma Croci.

SAFETY

Safety Class: 2b; Interaction Class: A (Gardner & McGuffin, 2013).

Carthamus tinctorius — Asteraceae

'DRUG SUBSTITUTE' MORPHOLOGY

The substitute is a single floret (tiny flower) derived from the composite flower-head (capitulum). Its structure is therefore completely different from that of 'Xi Hong Hua'. Main differences: **shorter (c. 1 cm long), reddish- or orangish-yellow with a cylindrical flower tube [3] opening into 5 red strap-shaped floret lobes [4]** [hl], **enclosing 5 anthers fused to form a creamish-yellow tube which in turn surrounds a long orange, club-shaped pistil [5] (style and stigma).** *Odour:* indistinct. *Taste:* slightly bitter.

OTHER COMMON NAMES

草红花 Cao Hong Hua; 红花 Hong Hua (plant name); bastard saffron; Flos Carthami; safflower (CP2010; SCN).

SAFETY

Safety Class: 2b, 2d; Interaction Class: B (Gardner & McGuffin, 2013). Carthami Flos is an an official Chinese healthfood (Anon., 2002a,b).

SUBSTITUTION COMMENT. Florets of *C. tinctorius* are the source of the separate TCM drug 'Hong Hua' (Carthami Flos) (see page 570). Compared to 'Xi Hong Hua', this drug is much cheaper and clinically 'less potent', but it is frequently traded as genuine 'Xi Hong Hua'. A further confusion in trade arises when genuine 'Xi Hong Hua' is abbreviated to 'Hong Hua'.

LESSER KNOWN SUBSTITUTES/ADULTERANTS

Macroscopic characters (visible with a x10 hand lens) provide a quick, easy and reliable authentication method for most situations, except where mixed batches are suspected or for powdered 'Xi Hong Hua'or 'Hong Hua'. Many laboratory-based authentication methods (mostly using chemical or DNA markers) are available for powdered materials and as an adjunct to macroscopy. Powdered material is especially vulnerable to adulteration. Substitutes or bulking agents reported in the literature (Marieschi *et al.*, 2012; Wichtl, 2004) include:

1. *Calendula officinalis* L. (Asteraceae) — florets
2. *Chrysanthemum x morifolium* (Ramat.) Hemsl. (Asteraceae) — the tubular ray florets, part of the TCM drug 'Ju Hua' (Chrysanthemi Flos) see page 574.
3. *Curcuma longa* L. (Zingiberaceae) — powdered rhizome (turmeric)
4. *Nelumbo nucifera* Gaertn. (Nelumbonaceae) — stamens of the TCM drug 'Lian Xu' (Nelumbinis Stamen), not included in this book.
5. *Zea mays* L. (Poaceae) — long silky styles (fruit-heads are maize)

In markets outside China the following have been reported:

6. *Arnica montana* L. (Asteraceae) — flowers
7. *Crocus vernus* (L.) Hill (Iridaceae) — stigmas and styles
8. *Hemerocallis* species (Xanthorrhoeaceae) — 'petals'
9. *Tagetes* species (Asteraceae) — flowers

Xi Hong Hua (dried styles and stigmas of *Crocus sativus*)

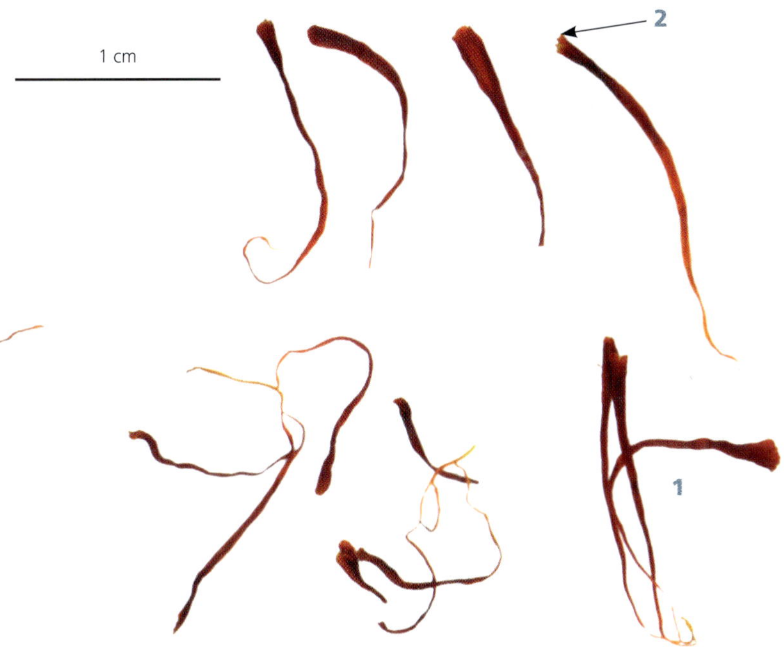

Xi Hong Hua (dried styles and stigmas of *C. sativus*)

Carthamus tinctorius florets (Hong Hua, Carthami Flos)

C. tinctorius: left closed floret; right: dissected floret opened

Counterfeit (unidentified material) sold as Xi Hong Hua by a herbal medicine wholesaler in Urumqi, Xinjiang province (C. Leon pers. obs., 2012)

Genuine Xi Hong Hua (*C. sativus*)

Substitute/adulterant *Calendula officinalis* florets before being dyed red

Substitute/adulterant *C. tinctorius* florets

Yang Jin Hua 洋金花 Daturae Flos

Drug part: FLOWER

Datura metel L.

Solanaceae

synonym: *Datura alba* F. Muell.

Species of *Datura* wild in China: 3

PLANT DESCRIPTION

An upright annual (0.5–1.5 m tall) with a spreading habit. **Stems often purple.** *Leaves* simple, alternate and stalked (stalk 2–6 cm long); leaf blade ovate or broadly ovate (5–20 cm long; 4–15 cm wide), membranous, base ending abruptly or wedge-shaped, asymmetrical, margin irregularly wavy-toothed, lobed or entire, tip narrowly pointed. *Flowers* solitary, in leaf axils, large and symmetrical, erect, white, yellowish or pale purple; calyx long, green and tubular (4–9 cm) **minutely hairy througout**; petals fused into a trumpet-shaped corolla (**14–20 cm long**), limb 6–10 cm diam.; stamens 5, style 1. *Fruit* a dry, sub-globose, solitary capsule (slightly at an angle) (c. 3 cm diam.) and **conspicuously covered in protuberances or broad-based prickles**; internally irregularly 4-valved. *Seeds* many, pale brown, shiny and flat (c. 3 mm diam.). Fl. and fr. Mar –Dec.

HARVESTING, SOURCING AND NATURAL RANGE

Young flowers (from white-flowering plants) are harvested from April to November, usually in the morning after the dew has dried, using cultivated plants often grown in cities. Not native to China but widely naturalised in open, grassy places, close to habitation (1,200–2,100 m altitude) in S China (Fujian, Guangdong, Guangxi, Guizhou, Hainan, Taiwan and Yunnan provinces). Native to the Americas.

CONSERVATION STATUS

China: Introduced.
Global: Not Evaluated (IUCN Red List, accessed 8/6/2015).

D. stramonium L.

Solanaceae

PLANT DESCRIPTION

Closely resembles the Pharmacopoeia species but **flowers shorter** (<11 cm long), calyx tube shorter (3–5 cm long) with **5 distinctly raised longitudinal ribs;** fruits **point upwards** and have **longer and sharper prickles.** Fl. Jun–Oct.; fr. Jul–Nov.

SUBSTITUTION COMMENT. An introduced species commonly cultivated as an ornamental and now naturalised; it is used as a local medicine for conditions similar to those of the Pharmacopoeia species. Whole plants and dried flowers of *D. stramonium* are similar to those of the official and are easily confused. The species is native to Mexico.

CONSERVATION STATUS

China: Introduced
Global: Not Evaluated (IUCN Red List, accessed 8/6/2015).

D. innoxia Mill.

Solanaceae

taxonomic/nomenclatural note: the spelling '*D. inoxia*' (with one 'n') is sometimes used in error.

PLANT DESCRIPTION

Closely resembles official species and *D. stramonium* (above) but **stems, leaf undersides and calyces densely covered in short, soft, greyish hairs.** Fruits densely armed with **long slender unequal-length spines clothed in white hairs.** Differs from *D. stramonium* in that fruits **hang downwards** and calyces do not have raised longitudinal ribs. Corolla trumpet 15–18 cm long; corolla limb 7–10 cm diam. Fl. and fr. Jun–Sep.

SUBSTITUTION COMMENT. Introduced to China as an ornamental and used as a local medicinal substitute for the Pharmacopoeia species in China's northern provinces of Hebei, Henan, Inner Mongolia, Liaoning, Shaanxi and Shandong (Zhao & Chen, 2014) and hence its Chinese name 'Bei Yang Jin Hua' (northern Datura flower). The species is native to the Americas.

CONSERVATION STATUS

China: Introduced
Global: Not Evaluated (IUCN Red List, accessed 7/6/2016).

TCM PROPERTIES AND USE

properties: 'pungent and warm'
channels: 'Lung and Liver'
action: 'relieves cough, relieves pain'

indications: cough, bronchitis, asthma; musculoskeletal pain; (T)

Medicinal crop of *Datura metel* in fruit

D. metel with hairless stems, leaves and calyx

D. metel

D. metel with ripening fruit

Flowering plants of *Datura stramonium*

Unripe fruit of *D. stramonium*

D. stramonium with fruit covered in numerous sharp spines

D. innoxia with softly hairy stems, leaf undersides and calyces

Fruiting plant of *D. innoxia*

Hairy underside of leaves of *D. innoxia*

Datura metel

Solanaceae

DRUG MORPHOLOGY

CRUDE DRUG/DECOCTING PIECES (DRIED FLOWERS): shrivelled long, narrow flowers, **9–15 cm long [1]** when intact. Calyx tube greyish-green or greyish-yellow (4–9 cm long and almost half the length of the entire corolla tube), with 5 pointed apical teeth and 5 longitudinal veins prominent at the base **[2]**; **minutely hairy throughout [3]**. Corolla pale yellow or yellowish-brown and trumpet-shaped, corolla tip with 5 short, pointed lobes with slightly furrowed longitudinal striations. Stamens 5, filaments attached to corolla walls. Style 1. *Texture:* fragile (when sun-dried); pliable (when shade-dried). *Odour:* indistinct (sun-dried); distinctive (shade-dried). *Taste:* slightly bitter. ***Good quality*:** yellowish.

PROCESSING

Young flowers (from white-flowering plants) are harvested from April to November, then sun-dried or dried at low temperature.

OTHER COMMON NAMES

广东洋金花 Guang Dong Yang Jin Hua; 南洋金花 Nan Yang Jin Hua (southern datura flower); Datura flower (CP2010); Flos Daturae; horn-of-plenty (SCN).

SAFETY

All parts **Toxic** (CP2015, Hong Kong 1999a), especially the seeds. Most cases of adverse effect arise from the inadvertent use of this drug as a substitute of 'Nao Yang Hua' (see page 580) (Zhao & Xiao, 2009–10). The recreational use of the seeds for their hallucinogenic effects has also been reported although this activity is more widely associated with the abuse of *D. stramonium* seed (see below). Officially forbidden in China as food or healthfood (Anon., 2002a,b).

D. stramonium

Solanaceae

'DRUG SUBSTITUTE' MORPHOLOGY

Flowers similar to those of *D. metel* but **shorter, <11 cm long [4]** and **calyx shorter, 3–5 cm long** with **5 raised longitudinal ribs [5]**.

OTHER COMMON NAMES

曼陀罗 Man Tuo Luo; Jamestown weed; Jimson weed (SCN); stinkweed; thorn-apple.

SAFETY

All parts **Toxic**, especially the seed (Barceloux, 2008). Worldwide, most cases of adverse effect arise from the recreational use of the seeds for their hallucinogenic effects (Boumba *et al.,* 2004; Klein-Schwartz & Oderda, 1984).

D. innoxia

Solanaceae

'DRUG SUBSTITUTE' MORPHOLOGY

Flower similar to the Pharmacopoeia species and *D. stramonium* above but calyces **densely covered in soft, greyish and glandular hairs [6]**, longitudinally striated but without raised ribs **[6]**. *The length of the flower is **not** a useful character since this varies considerably.*

OTHER COMMON NAMES

北洋金花 Bei Yang Jin Hua (northern Datura flower); 毛曼陀罗 Mao Man Tuo Luo; downy thorn-apple; sacred datura (SCN).

SAFETY

Toxic.

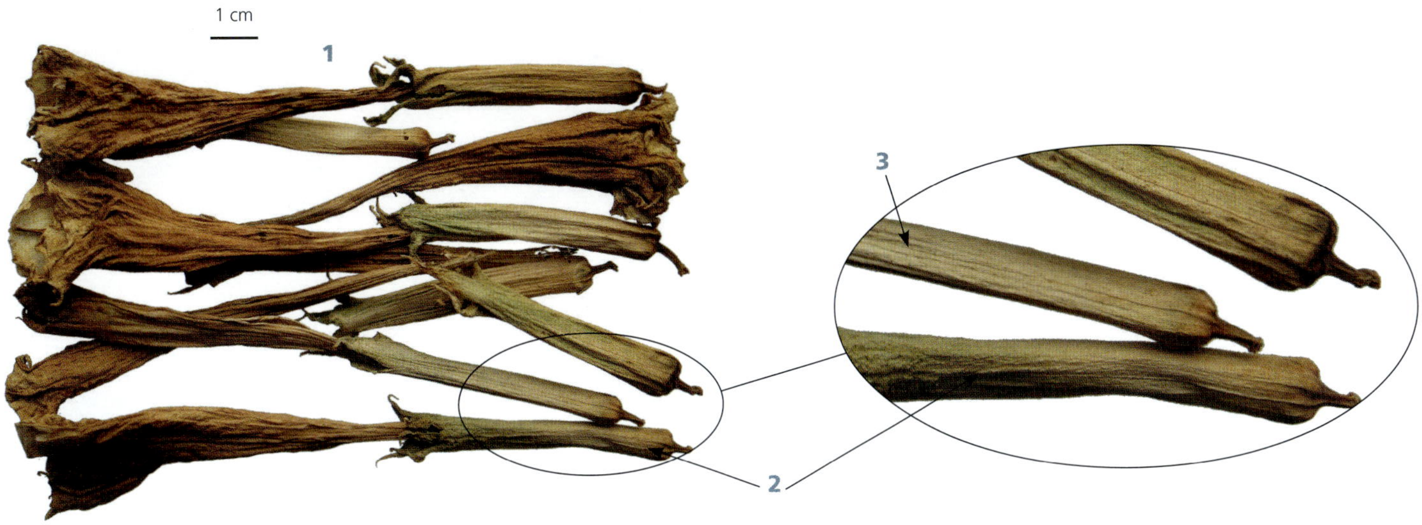

Yang Jin Hua (dried flowers of *D. metel*)

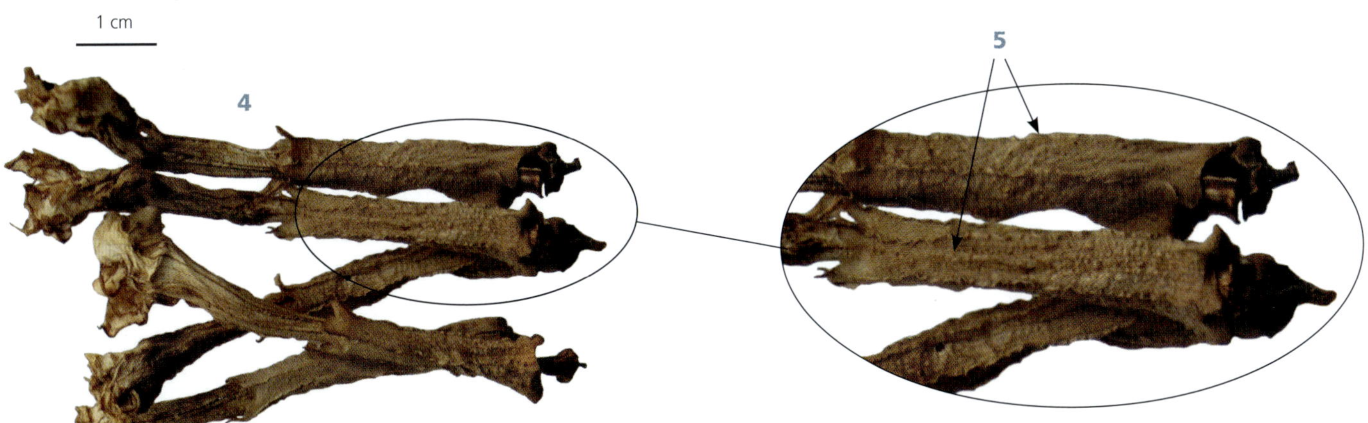

Calyx with 5 raised longitudinal ribs

Dried flowers of *D. stramonium*

Dried flowers of *D. innoxia*

Citrus x reticulata

fruits, seeds and other fruit parts

Cang Er Zi 苍耳子 Xanthii Fructus

Drug part: BURR (fruit with hard, spiny bract)

Xanthium strumarium L.

Asteraceae

synonyms: over 80 (Hind, 2011), including *X. sibiricum* Patrin ex Widder (name used in CP2015 as '*Xanthium sibiricum* Patr.').

Species of *Xanthium* wild in China: 0 (2 introduced)

PLANT DESCRIPTION

A highly variable species with different plants displaying a wide range of leaf shapes, leaf margins and size of burrs. A stout, annual herb (20–120 cm tall) with a taproot. *Stems* without spines. *Leaves* alternate and simple, blades broad, ovate, sub-orbicular or kidney-shaped (9–25 cm long), base **often shallowly heart-shaped or triangular,** papery, **both surfaces same colour** and distinctly rough with a dense covering of short stiff hairs; margin irregularly toothed and sometimes shallowly 3–5-lobed. *Inflorescences* single-sexed but male and female on the same plant. Male inflorescences terminal and globose; each male flower minute and petal-less with stamens fused into a tube. Female inflorescences in leaf axils with an enveloping shield-like bract enclosing 2 tiny tubular florets; style branches extending beyond the mouth of each floret. *Burrs* spindle-shaped, oblong or ovoid (0.8–1.8 cm long; 0.6–1.2 cm diam.), brownish or yellowish-brown when ripe and **densely covered in stout, hooked spines;** apical end of burr with **2 stout beaks** often incurved and sometimes fused; basal end of burr with a **truncate stalk scar**. Burr interior distinctively 2-chambered, with a single fruit (achene) in each chamber. Fl. Jul–Aug, fr. Sep–Oct. (based on Cronquist, 1994; *FoC* (online) and Hind, 2011).

HARVESTING, SOURCING AND NATURAL RANGE

Not native to China but introduced and widely naturalised as a weed of disturbed places, such as field margins and waste ground especially if damp or seasonally damp. The spiny burrs are harvested in autumn when ripe (brown) with best quality said to originate from Shandong and Jiangsu provinces (Zhao & Chen, 2014). The species probably originates from the New World; today it is a pantropical weed (Mitch, 1987).

CONSERVATION STATUS

China: introduced.

Global: Not Evaluated (IUCN Red List, accessed 12/8/2015).

Xanthium strumarium

Developing burrs of *X. strumarium*

Typical leaf variation on different plants of *X. strumarium*

LOCAL UNOFFICIAL SUBSTITUTE

Glycyrrhiza pallidiflora Maxim. (刺果甘草 Ci Guo Gan Cao 'spiny-fruited liquorice'). Family Fabaceae. A common species in N, NW and NE China quite distinct in appearance from the Pharmacopoeia species but whose fruits superficially resemble 'Cang Er Zi' and are reported to be occasionally traded as such in local markets (NICBP, 1997–2002). Fruits are similar in shape and size (ovoid, 10–17 mm long; 6–8 mm diam.) but **the spines are more rigid** and the **fruit tip is narrowly and sharply pointed**. Seeds 2, black and c. 2 mm long. Its occurrence as a substitute in international trade is unlikely.

TCM PROPERTIES AND USE

properties: 'bitter, pungent and warm'
channel: 'Lung and Liver'
action: 'dispels wind cold damp'
indications: sinus congestion; musculoskeletal pain; itchy skin

DRUG MORPHOLOGY

CRUDE DRUG (DRIED BURR): burr tapering at both ends, oblong or ovoid (variable in size: 0.8–1.8 cm long; 0.6–1.2 cm diam.) **[1]**. External surface brown, yellowish-brown or yellowish-green and densely covered throughout with long, acutely hooked spines; apical end with **2 stout beaks** **[2]** (occasionally fused) ; basal end of burr somewhat truncate with a small stalk scar **[3]**. *Texture*: hard but pliable. *Longitudinal fracture*: distinctively 2-chambered **[4]**, with a single fruit (achene) in each chamber **[5]**. Achenes tapering at both ends, somewhat flattened on one side, greyish-black, finely striated longitudinally **[6]**, with style remnants at apex **[7]**. The single seed within each achene with 2 yellowish, oily cotyledons **[8]**. *Odour*: none. *Taste*: slightly bitter. ***Good quality*:** fruit plump, brownish-yellow and containing ripe seeds.

DECOCTING PIECES (DRY-FRIED BURR): as for the crude drug but usually darker (yellowish-brown) **[9]** and sometimes charred; burr surface covered throughout with protuberant spine bases **[10]**. The 2 stout beaks usually still present **[2]**.

Crude drug

Dry-fried drug

PROCESSING

After removal of stalks, burrs are dried (= crude drug). Decocting pieces are prepared by dry-frying the burrs until they darken and the spines scorch and fall off; burrs are then sieved to remove any remaining spines (Chao Cang Er Zi 煸炒 'dry-fried' Cang Er Zi) (=decocting pieces).

OTHER COMMON NAMES

苍耳 Cang Er 'deep green ear' (refers to the leaves); 苍耳蒺藜 Cang Er Ji Li; cocklebur fruit; Fructus Xanthii; Siberian cocklebur fruit; xanthium (SCN); xanthium fruit.

SAFETY

Unprocessed burrs (i.e. fresh or dried) are **toxic** (CP2015). It is normal TCM practice to dry-fry the burrs before they are used internally since heating reduces their toxicity while removal of the spines is considered equally important by some authorities since these contain a different toxin not destroyed by heat (Kamboj & Saluja, 2010; Bensky *et al*., 2004). Overdose or long-term internal use even of the processed burrs (as described) is strongly discouraged to avoid serious adverse reactions (Chang & But, 1986).

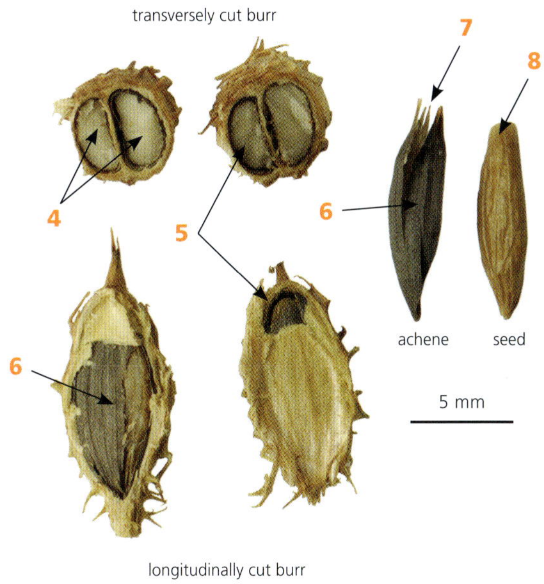

Dissected burrs

Cao Dou Kou 草豆蔻

Alpiniae Katsumadai Semen

Drug part: SEED

Alpinia hainanensis K. Schum.

synonym: *A. katsumadae* Hayata (name used in CP2015 as '*A. katsumadai* Hayata').

Zingiberaceae

Species of *Alpinia* wild in China: 51

PLANT DESCRIPTION

An erect perennial herb with thick creeping rhizomes. '*Stems*' (pseudostems — densely overlapping leaf bases) up to 3 m tall. *Leaves* alternate, in two rows on opposite sides of the 'stem'; leaf stalk short (up to 2 cm) or absent; ligule (at base of each leaf blade) 5–10 mm (rarely 3 mm) long with outermost surface hairy; leaf blades simple, linear to lance-shaped (20–65 cm long; 2–12 cm wide), hairless or sparsely hairy on lower surface, veins pinnate with many lateral ones, margins entire, tips sharply pointed and twisted. ***Flowers*** bisexual, in large, erect, conspicuous and somewhat cylindrical, **terminal spikes** (10–30 cm long) **at top of 'stems'; before opening, each flower spike enclosed by a large bract** and each flower by a bracteole; the latter is petal-like, concave and milky-white or pink (2–3.5 cm long); calyx bell-shaped (2–3 cm long) and split along one side, tip 2-toothed; petals fused to form a very short 3-lobed tube (0.8–1 cm long) fringed with hairs; **the broad and distinctive petal-like lip or labellum (3.5–4.5 cm long), yellowish with dark red stripes** radiating to the margins and its tip shallowly 2-cleft; ovary inferior; one style with prominent stigma. *Fruit* a dry, globose capsule (2–3 cm diam.), covered in fine, yellow hairs. *Seeds* many; densely packed within a whitish membrane. Fl. Apr–Jun, fr. May–Aug.

HARVESTING, SOURCING AND NATURAL RANGE

Fruits are harvested in summer and autumn mostly from wild plants which grow in dense or sparse sub- and tropical forests in the southern Chinese provinces of Guangdong, Guangxi and Hainan. *Di Dao*-quality 'Cao Dou Kou' is traditionally sourced from Hainan province. The species is also native to Vietnam.

CONSERVATION STATUS

China: Least Concern (MEPCAS, 2013).
Global: Not Evaluated (IUCN Red List, accessed 20/8/2016).

Alpinia hainanensis flowering plants in the wild in China

Flower spike with lowest flowers opening to reveal a yellow and red-striped lip

Fruit of *A. hainanensis*

TCM PROPERTIES AND USE

***properties*: 'pungent and warm'**
***channels*: 'Spleen and Stomach'**
***actions*: 'warms the interior, dries damp'**

***indications*: painful abdominal distension with belching, nausea, vomiting, poor appetite**

DRUG MORPHOLOGY

CRUDE DRUG (DRIED SEED MASSES): sub-spheroidal densely packed seed masses (1.5–2.7 cm diam.) **[1]**. Externally greyish-brown, somewhat smooth, with yellowish-white membranous septa (arising centrally) **[2]** dividing each seed mass into 3 groups **[3]**. Each seed irregular to many-sided (3–5 mm long; c. 3 mm diam.) and covered in a greyish membranous aril **[4]**; a distinctive raphe forms a longitudinal furrow on one seed surface extending along 2/3 the length of the seed **[5]**; a large, oval and sunken hilum is positioned at the outer end of the raphe **[6]**. *Seed texture and fracture*: hard; endosperm greyish-white and, when seed cut longitudinally, the inward-growing seed coat (testa) occupies c. half of the contents of the seed. *Odour*: aromatic. *Taste*: pungent and slightly bitter. ***Good quality*:** full, heavy and with an intense aroma.

DECOCTING PIECES (SEPARATED SEEDS): see seed description above.

PROCESSING

Pericarps (fruit-walls) are removed and discarded, then contents (seed mass plus aril) sun-dried (= crude drug); alternatively fruits are blanched then semi-dried before pericarps are removed and contents thoroughly dried. The crude drug is broken into pieces prior to decocting.

OTHER COMMON NAMES

草蔻仁 (草扣仁) Cao Kou Ren; 草蔻 (草扣) Cao Kou; katsumada galangal (SCN); Semen Alpiniae Katsumadai.

Crude drug of Cao Dou Kou
(dried seed masses of *Alpinia hainanensis*)

Decocting pieces of Cao Dou Kou
(separated seeds of *A. hainanensis*)

UNOFFICIAL SUBSTITUTES

The seeds of two other species in the same family have been reported as local or historical unofficial substitutes for reasons of clinical substitution or confusion (Anon., 1; Bensky *et al.*, 2004).

1. ***Alpinia blepharocalyx*** K. Schum. (云南草蔻 Yun Nan Cao Kou). Ligule c. 6 mm, tip with long shaggy hairs; flower spikes drooping; **flowers mainly flesh-coloured with lower lip (labellum) smaller** (3–3.5 cm long; c. 3 cm wide), **red with a yellow centre**; fruit ellipsoid, with short soft hairs; seeds in **smaller** globose masses (1.2–1.6 cm diam.); seeds greyish-yellow to dull brown. Wild in Guangdong, Guangxi and Yunnan.
2. ***A. platychilus*** K. Schum. (宽唇山姜 Kuan Chun Shan Jiang). A shorter species ('stems' up to c. 2 m); ligule c. 1 cm, with yellow shaggy hairs; leaf blades wider (c. 60 cm long; 16 cm wide), **lower surfaces velvety**; flower spikes **25 cm long or more**; bracteoles reddish, flower mainly **white, lower lip (labellum) larger** (4.5–7 cm long; 8–9 cm wide), **yellow, tinged with red.** Endemic to Yunnan (800–1,600 m altitude).

Cao Guo 草果

Tsaoko Fructus

Drug part: DE-HUSKED FRUIT

Amomum tsao-ko Crevost & Lemarié

Zingiberaceae

Species of *Amomum* wild in China: 38

PLANT DESCRIPTION

An erect perennial herb with creeping rhizomes resembling those of ginger. *'Stems'* (pseudostems — densely overlapping leaf bases) up to 3 m tall. *Leaves* alternate and many; ligule (at base of each leaf blade) entire and broad (0.8–1.2 cm long); leaf blades simple, large, elliptic or oblong (40–70 cm long; 10–20 cm wide), margins drying membranous and tips sharply pointed. *Flowers* bisexual, in dense 5–30-flowered spikes (13–18 cm long; c. 5 cm wide) **at base of plant and growing directly from the rhizomes;** fruit stalk at least 10 cm long; scale-like sheaths dense and brown when dry, narrowly elliptic (5.5–7 cm long); **flower spike *not* surrounded by a bract before opening** but each flower with a tubular bracteole (c. 3 cm long) shed soon after the flower opens; calyx tubular with a 3-toothed tip; petals orangish-red, fused into a cylindrical 3-lobed tube (c. 2.5 cm long), central lobe erect and wider than lateral lobes; **lip (labellum) elliptic, orangish-red and conspicuous (c. 2–7 cm long; 1.4 cm wide) with a slightly toothed apex.** *Fruit* a dry, sub-globose capsule (2.5–4.5 cm long; 1–2.5 cm diam.), red when ripe, drying to dull greyish- or reddish-brown with numerous longitudinal ridges; aromatic (see also *Drug Morphology*). *Seeds* many, multi-angled, 4–6 mm diam. and densely packed in a yellowish-white membrane. Fl. Apr–Jun, fr. Sep–Dec.

HARVESTING, SOURCING AND NATURAL RANGE

Fruits are harvested when ripe (red) in autumn, mostly from widely cultivated crop plants grown mainly in China's south-western provinces of Yunnan, Guangxi and Guizhou. The wild species is endemic to Yunnan where it grows in open forests (1,100–1,800 m altitude).

CONSERVATION STATUS

Endemic to China: Near Threatened (IUCN Red List, accessed 23/8/2016).

Medicinal crop of *Amomum tsao-ko*, Yunnan province

Mature crop of *A. tsao-ko*, Yunnan province

Mature fruit growing at base of plant

TCM PROPERTIES AND USE

properties: **'pungent and warm'**
channels: **'Spleen and Stomach'**
actions: **'dries damp, warms interior'**

indications: **abdominal pain, distension; vomiting, diarrhoea, fever**

DRUG MORPHOLOGY

CRUDE DRUG (DE-HUSKED AND DRY-FRIED SEED MASSES): when intact, oblong to egg-shaped (2–4 cm long; 1–2.5 cm wide) with many densely packed seeds **[1]** sometimes enclosed in a greyish- or yellowish-white membranous aril **[2]**. Seeds arranged in 3 compartments of 8–11 seeds per compartment **[3]** and separated by greyish-white membranous septa **[4]**; seeds hard, reddish-brown, polyhedral (4–6 mm diam.) **[5]** with a crater-like hilum at the apex [hl] **[6]** often obscured by membranous septa; endosperm white **[7]**; embryo cream **[8]**. *Odour*: distinctly aromatic. *Taste*: slightly bitter and pungent. ***Good quality*:** dense masses of brownish-red intensely aromatic seeds with husks removed.

GINGER PROCESSED SEED: loose seeds as for those for crude drug but deep yellowish brown **[9]** and with occasional charring; slightly ginger-flavoured and ginger-scented.

PROCESSING

Ripe fruit are dried in the sun or at low temperature then dry-fried until brown and slightly inflated; the husks (fruit-wall) are then removed and discarded (= crude drug). Subsequent processing with ginger juice before dry-frying until deep yellow produces Jiang Cao Guo Ren (姜草果仁) (= ginger processed Cao Guo). Seed masses are broken into pieces prior to decocting.

OTHER COMMON NAMES

草果仁 Cao Guo Ren 'tsao-ko kernels'; 草果子 Cao Guo Zi 'tsao-ko seeds'; 炒草果仁 Chao Cao Guo Ren 'dry-fried tsao-ko kernels'; Amomi Tsao-Ko Fructus (KP, 2007); 姜草果仁 Jiang Cao Guo Ren 'ginger-processed tsao-ko'; Fructus Tsaoko; tsao-ko amomum (SCN).

SAFETY

Safety Class 1; Interactions Class: A. (Gardner & McGuffin, 2013). In China this herb is also used as a culinary spice (McGuffin *et al.*, 2000).

Pre-crude drug (i.e. husks still intact)

Ginger-processed seeds (Jiang Cao Guo Ren)

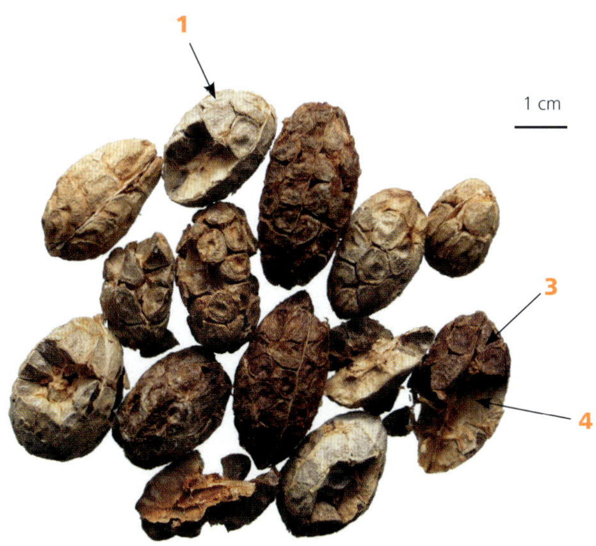

Crude drug (husks removed)

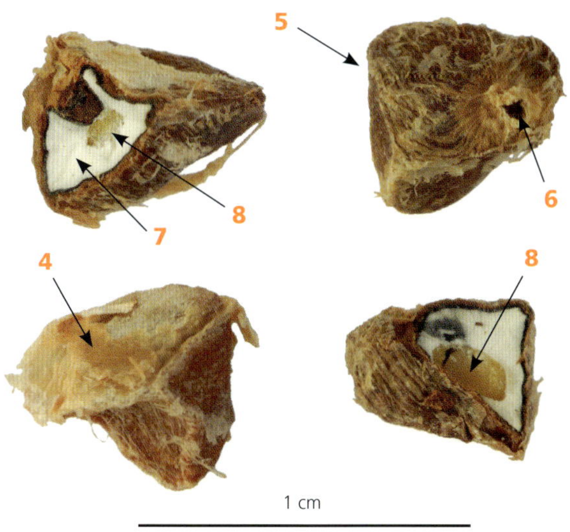

Individual seeds of Cao Guo removed from a seed mass (dry-fried)

Chen Pi 陈皮

Citri Reticulatae Pericarpium

tangerine peel

Drug part: PEEL OF RIPE FRUIT

Citrus × reticulata Blanco

Rutaceae

taxonomic/nomenclatural note: in CP2015 as '*C. reticulata* Blanco'; *Kew-MPNS* accepted name 7/9/2016, *C.* × *aurantium* L., but *FoC* considers it to be a separate species; four medicinal cultivars are specifically used as the source of Chen Pi (see *Harvesting, Sourcing* etc.).

Species and hybrids of *Citrus* wild in China: 11 (+ 5 hybrid cultivated species)

PLANT DESCRIPTION

A small evergreen tree with many branchlets and occasional spines. *Leaves* alternate and simple with **stalks usually narrowly winged or with wing remnants**; blade somewhat leathery, lance-shaped, elliptic or broadly ovate; basal articulated part narrow; margin shallowly scalloped, broadly so near the tip. *Flowers* solitary or in groups of up to 3; fragrant. Each flower with calyx irregularly 3–5-lobed, petals 4–5, white, spreading and equal (up to 1.5 cm long); stamens 20–25; style long and slender; stigma club-shaped. *Fruits* (tangerines/mandarin oranges); squat or subglobose, peel leathery, smooth or coarse, thin or thick, pale yellow, orange or red, **and easily removed**; pith spongy; flesh with 7–14 segments (rarely more), **sweet**. Fl. Apr–May, fr. Oct–Dec.

See also other drugs sourced from this hybrid: 'Ju He' (page 664), 'Ju Hong' (page 666) and 'Qing Pi '(page 696).

HARVESTING, SOURCING AND NATURAL RANGE

Fruits are harvested when ripe (pale orange, yellow or red) entirely from cultivated crop plants sourced from the cultivars: 'Da Hong Pao', 'Un Shui' and 'Tangerina'. All are traded as typically representative 'Chen Pi'. Trees of these cultivars are grown extensively south of the Qin Ling mountains (NC China) on forested hills at low elevations. A higher quality but much less common product called Guang Chen Pi is sourced from a fourth exclusively medicinal cultivar 'Cha Chi'. This is mainly cultivated in Guangdong province (S China), districts of Guangzhou, Jiangmen, Sihui and Xinhui with the latter locality the traditional *Di Dao* source of Chen Pi.

The original wild species may possibly be native to south-east China and Taiwan.

CONSERVATION STATUS

China: Least Concern (MEPCAS, 2013) but FoC claims its native status is uncertain.

Global: Not Evaluated (IUCN Red List, accessed 24/6/2015).

Di Dao crop of *Citrus* × *reticulata* 'Cha Chi'

Ripe fruit of *C.* × *reticulata* 'Cha Chi' (source of top quality Chen Pi)

Ripe fruit of *C.* × *reticulata* 'Un Shui' (source of Chen Pi of typical quality)

Freshly processed peel of *C.* × *reticulata* 'Cha Chi'

Leaves of *C.* × *reticulata* 'Un Shui'

TCM PROPERTIES AND USE

***properties*: 'pungent, bitter and warm'**
***channels*: 'Lung and Spleen'**
***actions*: 'regulates qi, tonifies Spleen, resolves damp and phlegm**

***indications*: abdominal distension; nausea, vomiting; abdominal pain; diarrhoea; indigestion; fatigue; cough**

DRUG MORPHOLOGY

CRUDE DRUG (DRIED PEEL):

Chen Pi grade: irregular-shaped, often lobed, pieces of peel (exocarp), slightly or tightly curled inwards (c. 1–4 mm thick) **[1]**. External surface pale or dark orange, orangish-red to reddish-brown, densely covered throughout with sunken oil ducts **[2]**; base with round and sunken stalk scar **[3]**. Inner surface of peel creamish-white or yellowish-white with rough pith remnants **[4]**. ***Texture*: somewhat hard and brittle**. *Odour*: aromatic. *Taste*: pungent and bitter.

Guang Chen Pi grade (*top quality*): easily distinguished from other medicinal varieties. Peel regularly tri-lobed **[5]**, joined basally, c. 1 mm thick throughout. **Sunken oil-ducts [6] larger than those of Chen Pi grade, and transparent against the light**. ***Texture*: relatively soft.** *Taste:* slightly bitter.

DECOCTING PIECES (DRIED, SLICED PEEL):

Chen Pi grade (*top quality*): as for crude drug but cut into slivers (c. 5–30 mm long; c. 3–5 mm wide; c. 2–3 mm thick), usually curved **[7]**. Sunken stalk scar unlikely to be visible.

Guang Chen Pi grade: as for crude drug of Guang Chen Pi above but cut into slivers (c. 5–30 mm long; c. 3–5 mm wide; 1 mm thick throughout).

PROCESSING

Peel (fruit wall) is removed from the ripe fruit and dried in the sun or at low temperature (= crude drug); decocting pieces are made by softening the peel thoroughly using sprayed water before slicing thinly and drying in the shade.

OTHER COMMON NAMES

川橘皮 Chuan Ju Pi; 大红袍 Da Hong Pao; 柑橘 Gan Ju (plant name); 广陈皮 Guang Chen Pi; 橘皮 Ju Pi; 新会皮 Xin Hui Pi; tangerine; dried Tangerine peel (CP2010); Pericarpium Citri Reticulatae; tangerine (SCN).

SAFETY

Safety Class: 1; Interaction Class: A (Gardner & McGuffin, 2013). An official food and healthfood in China (Anon., 2002a,b).

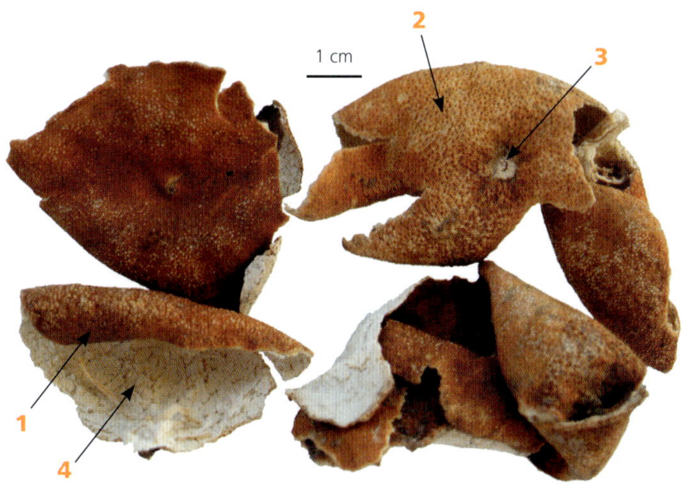

Crude drug of Chen Pi sourced from *C.* × *reticulata* 'Tangerina'

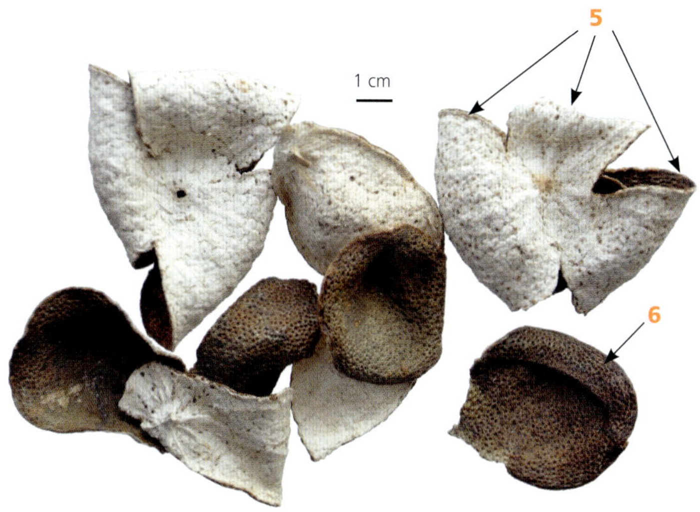

Crude drug of Chen Pi sourced from *Di Dao* quality cultivar *C.* × *reticulata* 'Cha Chi'

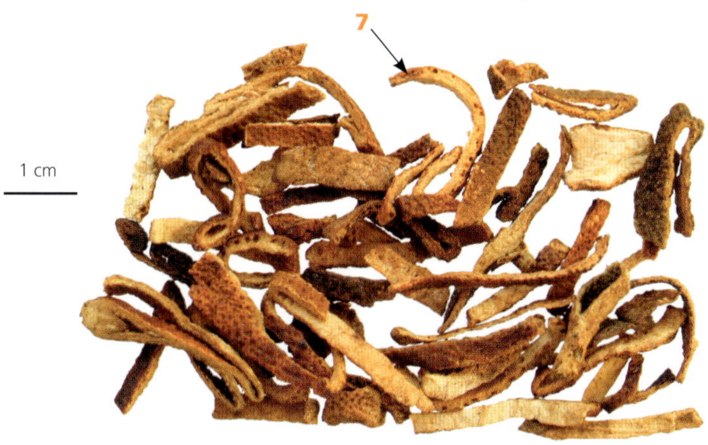

Decocting pieces of Chen Pi sourced from *C.* × *reticulata* 'Tangerina'

Che Qian Zi 车前子 Plantaginis Semen

Drug part: SEED

PHARMACOPOEIA SPECIES 1

Plantago asiatica L.

Plantaginaceae

Species of *Plantago* wild in China: 27

PLANT DESCRIPTION

A short perennial herb (up to 30 cm) **with many fibrous roots**. *Leaves* stalked and in a basal rosette; leaf stalks not winged at middle; leaf blades broadly ovate to elliptic (4–12 cm long; 2.5–6.5 cm wide), sparsely hairy, margin entire, veins 3–7 and curved like a bow. *Flowers* in upright, narrow cylindrical spikes (3–40 cm long) on long stems (5–30 cm); each flower tiny, with 1 bract, bisexual, with female parts developing before the male; sepals 4, tiny (2–4 mm long), overlapping and persistent [hl]; petals tiny (as long as or slightly longer than the sepals), white, thin and dry, fused and pointed; ovary superior; style 1; stamens white and conspicuous with very long slender filaments and **white ovoid anthers (1–1.2 mm long)**. *Fruit* a tiny, dry goblet-shaped capsule (3–6 mm long), with lid releasing at maturity. *Seeds* 5–15 per capsule; blackish-brown **(1.2–2 mm long; 1 mm wide)**. Fl. Apr–Aug, fr. Jun-Sep.

HARVESTING, SOURCING AND NATURAL RANGE

Fruiting spikes are harvested in summer and autumn when seeds are ripe from wild and medicinally cultivated crop plants.

The species grows in grassland, along riverbanks and in disturbed places (up to 3,800 m altitude) and is a widespread native throughout China: Anhui, Chongqing, Fujian, Gansu, Guangdong, Guangxi, Guizhou, Hainan, Hebei, Heilongjiang, Henan, Hubei, Hunan, Inner Mongolia, Jiangsu, Jiangxi, Jilin, Liaoning, Qinghai, Shandong, Shanxi, Sichuan, Taiwan, Xinjiang, Xizang, Yunnan and Zhejiang [also native to Bangladesh, Bhutan, India, Indonesia, Japan, the Korean Peninsula, Malaysia, Nepal and Sri Lanka].

CONSERVATION STATUS

China: Least Concern (authors' pers. obs. 2010).
Global: Not Evaluated (IUCN Red List, accessed 3/9/2015).

The 'herb' of both species is the source of the TCM drug 'Che Qian Cao' (Plantaginis Herba), not included in this book.

PHARMACOPOEIA SPECIES 2

P. depressa Willd.

Plantaginaceae

PLANT DESCRIPTION

Aerial parts are difficult to distinguish from PS1 above; the **long straight tap root** of *P. depressa*, however, provides an easy differentiating character. Other less obvious aerial characters include the leaf blade (sometimes longer and narrower (5–14 cm long; 2–3 cm wide), flowering spikes (often shorter, 6–12 cm); **seeds (only 4 per fruit capsule; seed almost half the size (0.9–1.7 mm long; 0.6–0.9 mm diam.)**.

HARVESTING, SOURCING AND NATURAL RANGE

Very similar in all respects to PS1 above but harvesting is entirely from the wild and the species is absent from the tropical and sub-tropical parts of S China.

The species is native to N, C and W China (Anhui, Gansu, Hebei, Heilongjiang, Henan, Hubei, Inner Mongolia, Jiangsu, Jiangxi, Jilin, Liaoning, Ningxia, Qinghai, Shandong, Shanxi, Sichuan, Xinjiang, Xizang and Yunnan) where it grows in damp and disturbed places (up to 4,500 m altitude); [also native to Afghanistan, Bhutan, India, Kashmir, Kazakhstan, the Korean Peninsula, Kyrgyzstan, Mongolia, Pakistan and E Russia].

CONSERVATION STATUS

China: Least Concern (authors' pers. obs. 2010).
Global: Not Evaluated (IUCN Red List, accessed 28/8/2016).

TCM PROPERTIES AND USE

properties: 'sweet and cold'
channels: 'Kidney, Bladder, Liver and Lung'
actions: 'regulates water, dispels damp heat'
indications: oedema; dysuria; dizziness; cough; painful eyes

Flowering and fruiting crop plants of *Plantago asiatica*

Medicinal crop of *P. asiatica*, Sichuan province

Flowering plant of *P. depressa* in the wild, Jiangxi province

Wild plants of *P. depressa* typically growing in the middle of a track

Plantago asiatica

Plantaginaceae

DRUG MORPHOLOGY

CRUDE DRUG/DECOCTING PIECES-1 (DRIED SEED): ellipsoidal (1.2–2 mm long; 1 mm wide) and irregularly angled **[1]**. Externally yellowish-brown to dark brown, finely wrinkled, with a whitish or grey sunken hilum on one side **[2]**. *Texture*: hard. *Odour*: slight. *Taste:* weak. ***Good quality*:** hard, plump and dark brown.

DECOCTING PIECES-2 (SALT-PROCESSED SEED): difficult to distinguish from above. All seeds dark brown and slightly aromatic. *Taste:* somewhat salty.

PROCESSING

Seeds are sun-dried (= crude drug/decocting pieces-1); subsequent dry-frying with salt produces (Yan Che Qian Zi 盐车前子 salt-processed Che Qian Zi).

OTHER COMMON NAMES

车前实 Che Qian Shi; 大车前子 Da Che Qian Zi; Asian plantain (SCN); psyllium; plantain seed (CP2010); Semen Plantaginis.

SAFETY

Class: 2d; Interaction Class: A (Gardner & McGuffin, 2013). An official healthfood in China (Anon., 2002a,b).

P. depressa

Plantaginaceae

DRUG MORPHOLOGY

CRUDE DRUG/DECOCTING PIECES-1 (DRIED SEED): difficult to distinguish macroscopically from PS1 above without magnification. Almost half the size (0.9–1.7 mm long; 0.6–0.9 mm diam.) **[3]** of PS1 and less angular [hl] **[4]**.

DECOCTING PIECES-2 (SALT-PROCESSED SEED): difficult to distinguish from the dried seed of this species. Dark brown throughout and slightly aromatic. *Taste:* somewhat salty.

PROCESSING

As for PS1 above.

OTHER COMMON NAMES

车前实 Che Qian Shi; 平车前子 Ping Che Qian Zi; 小车前子 Xiao Che Qian Zi; depressed plantain (SCN); plantain seed (CP2010); Semen Plantaginis.

SAFETY

An official healthfood in China (Anon., 2002a,b).

UNOFFICIAL SUBSTITUTES AND CONTAMINATION

With 25 other species of *Plantago* growing wild in China, many of which are easily confused with the Pharmacopoeia species, wild harvested 'Che Qian Zi' is prone to accidental contamination. Most frequently reported contaminant species on the Chinese market have included *P. major* L., *P. camtschatica* Link, *P. lanceolata* L. and *P. minuta* Pallas and the similar-looking seeds of many *Codonopis* species (Campanulaceae). Nutlets (1-seeded fruits) of *Nepeta* species (family Lamiaceae) have also been reported to be traded as 'Che Qian Zi' (NICPBP 1997–2002; Xiao *et al.*, 2001–2002).

A recent UK-based TCM market study (2012) detected high contamination levels (>50%) of some 'Che Qian Zi' batches traded in the UK (Van der Valk & Leon, in prep.). Commonest contaminants resembled fruits of yet further species (Apiaceae and Amaranthaceae families) together with many small stones; the size and colour similarity to 'Che Qian Zi' of these UK contaminants suggests intentional bulking up (see trade sample right).

Che Qian Zi traded on Chinese and international markets, therefore, warrants close inspection to detect possible contamination and substitution. Macroscopic examination (using at least x10 magnification) is recommended for an initial quality assurance assessment and, where batches are suspected to be wild-harvested, laboratory-based authentication methods are advised. Further research may be needed to develop sufficiently robust identity- and purity-test methods.

- green: authentic *Plantago asiatica* or *P. depressa* seed
- blue: contaminant — Apiaceae fruit
- purple: contaminant stones
- orange: unidentified contaminant

Crude drug/decocting pieces-1 of Che Qian Zi
(sourced from seed of *Plantago asiatica*)

Dried seed of *P. asiatica*

Decocting pieces-2
(salt-processed seeds of *P. asiatica*)

Crude drug/decocting pieces-1 of Che Qian Zi
(sourced from dried seed of *P. depressa*)

Dried seed of *P. depressa*

Decocting pieces-2
(salt-processed seeds of *P. depressa*)

Chong Wei Zi 茺蔚子 Leonuri Fructus

Drug part: FRUIT (nutlet)

Leonurus japonicus Houtt.

Lamiaceae

synonym: *L. heterophyllus* Sweet

Species of *Leonurus* wild in China: 12

PLANT DESCRIPTION

An erect, herbaceous annual or biennial (30–120 cm tall) with a taproot and dense fibrous rootlets. *Stems* 4-sided, deeply furrowed longitudinally, branching in lower parts and minutely hairy on nodes and angles. *Leaves* opposite with each pair at right angles to the next; ***leaves of 2 main shapes*****: i. stem leaves deeply divided into 3 lobes with each lobe more than 3 mm wide,** oblong to linear with a narrowly wedge-shaped base; **ii.** leaves in flowering spikes long and thin (3–12 mm long; 2–8 mm wide), *not divided into lobes,* with margins entire or weakly toothed. *Flowers* in dense 8–15-flowered whorls; each whorl 2–2.5 cm diam. arranged at intervals along a long flowering spike with stiff, sharp spines (bracteoles) between the flowers. Each flower bisexual and laterally symmetrical; calyx funnel-shaped (longer than bracteoles), **5-veined**, *minutely hairy* [hl], with 5 triangular, spine-tipped lobes (2 lobes slightly larger than others); corolla white, reddish or purplish-red (1–1.2 cm long), 2-lipped (upper lip straight, *oblong* and concave (c. 7 mm long; 4 mm wide), fringed with hairs; lower lip slightly shorter, divided into 3 lobes, spotted or striped, with middle lobe much larger than lateral lobes); stamens 4; ovary superior and style long-pointed. *Fruit* with 4 dry nutlets, strongly 3-angled, truncate and enclosed in a persistent calyx. Fl. Jun–Sep, fr. Sep–Oct.

For other TCM drugs sourced from this species see 'Yi Mu Cao' (Leonuri Herba) page 402.

HARVESTING, SOURCING AND NATURAL RANGE

Aerial parts are harvested in autumn from wild and medicinally cultivated plants. The species grows wild in sunny places (0–3,400 m altitude) throughout much of China: Anhui, Fujian, Gansu, Guangdong, Guangxi, Guizhou, Hainan, Hebei, Heilongjiang, Henan, Hubei, Hunan, Inner Mongolia, Jiangsu, Jiangxi, Jilin, Liaoning, Ningxia, Qinghai, Shaanxi, Shandong, Shanxi, Sichuan, Taiwan, Xinjiang, Tibet, Yunnan and Zhejiang [also wild in Cambodia, Japan, Korean Peninsula, Laos, Malaysia, Myanmar, Thailand, Vietnam; Africa, North and South America].

CONSERVATION STATUS

China: Not Evaluated.

Global: Not Evaluated (IUCN Red List, accessed 21/8/2016).

Leonurus japonicus flowering in the wild, Beijing province

L. japonicus showing unlobed linear leaves between the fruit whorls

Flowering plant of *L. japonicus*

TCM PROPERTIES AND USE

properties: **'bitter, pungent and cool'**
channels: **'Liver, Pericardium and Spleen'**
action: **'invigorates blood, dispels wind, clears heat'**

indications: **dysmenorrhoea; headaches; painful eyes**

DRUG MORPHOLOGY

CRUDE DRUG/DECOCTING PIECES (FRUIT): distinctly 3-angled longitudinally **[1]** with smooth, flat sides (2–3 mm long; c. 1.5 mm diam.), sharply truncate at broadest end **[2]** and bluntly wedge-shaped at other end **[3]**. External surface (pericarp) greyish-brown, thin and often irregularly spotted **[4]**. Internally 1-seeded, with cotyledons oily and whitish. *Odour*: indistinct. *Taste*: bitter. ***Good quality***: large, plump.

PROCESSED DECOCTING PIECES (DRY-FRIED FRUIT 'CHAO CHONG WEI ZI'): as for above but slightly darker **[5]** sometimes cracked.

PROCESSING

Aerial parts are dried then threshed to release the fruit (= crude drug/decocting pieces); fruits may be subsequently dry-fried (= Chao Chong Wei Zi 炒茺蔚子).

OTHER COMMON NAMES

益母草子 Yi Mu Cao Zi; Chinese motherwort (SCN); Fructus Leonuri; motherwort fruit (CP2010).

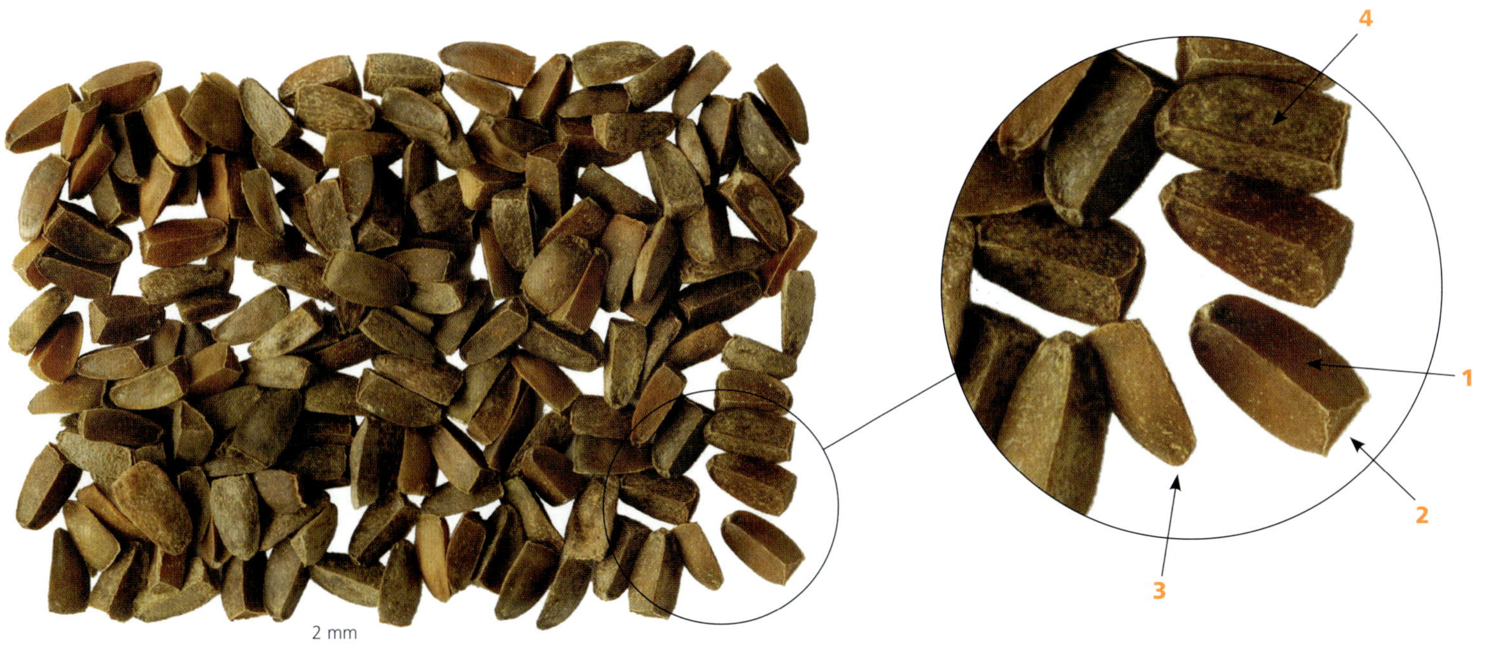

Crude drug/decocting pieces of Chong Wei Zi (*Leonurus japonicus*)

CONTAMINANTS

In recent years, relatively high levels of contamination have been detected in Chong Wei Zi traded on the UK TCM market with most frequent contaminants including fruit of species in the Apiaceae family (possibly the genus *Cnidium*) (Van der Valk & Leon, pers. obs. 2013). Note that *C. monnieri* fruit are wild harvested as the source of the separate TCM drug 'She Chuang Zi' (Cnidii Fructus) see page 716. Further research into the frequency and source of such contamination may be worthwhile.

fruit of *?Cnidium* sp.

other contaminants

2 mm

Chong Wei Zi (*L. japonicus*)

Chao Chong Wei Zi
(dry-fried fruit of *L. japonicus*)

Chuan Lian Zi 川楝子

Toosendan Fructus

Drug part: FRUIT

Melia azedarach L.

Meliaceae

synonym: *M. toosendan* Siebold & Zucc. (name used in CP2015).

taxonomic/nomenclatural notes: i) many herbal medicine texts erroneously treat the above synonym as a separate species; **ii)** the source species of Chuan Lian Zi (*Melia azedarach*) should not be confused with the closely related 'neem' tree which is a separate species and whose Latin scientific name is very similar: *Melia azadirachta* L. (accepted name *Azadirachta indica* A. Juss.). The latter is not native in China, nor is it used in TCM.

PLANT DESCRIPTION

A spreading, deciduous tree (up to 10 m tall). *Leaves* in spirals, 2 or 3 pinnate (20–40 cm long) with opposite pairs of short-stalked, ovate, elliptic or lance-shaped leaflets (3–7 cm long; 2–3 cm wide), terminal leaflet slightly larger; both surfaces with star-shaped hairs when young; leaflet bases somewhat oblique to wedge-shaped, margins entire or scalloped, tips shortly pointed. *Flowers* conspicuous in loose axillary clusters, bisexual and fragrant; sepals 5, small and pointed; petals 5, lilac or white, long-ovate and spreading (0.9–1.3 cm long); stamens fused into a purple 10-ribbed tube (7–8 mm long), longitudinally striped, with 10 narrow lobes at apex; anthers 10, alternating with the purple lobes of the staminal tube; style with a pin-like head. *Fruit* (drupe) globose to ellipsoid (1–3 cm long; 0.8–1.5 cm diam.), sub-fleshy, golden when ripe. *Seeds* pendant, ellipsoid, hard and chitinous. Fl. Mar–May, fr. Oct–Dec.

The bark of this species is the source of the separate TCM drug 'Ku Lian Pi' (Meliae Cortex), not included in this book.

Species of *Melia* wild in China: 1

HARVESTING, SOURCING AND NATURAL RANGE

Ripe fruits are harvested in winter from cultivated, wild and naturalised trees. The species is widely cultivated in China for medicine, timber and as an ornamental.

In the wild, the species grows in mixed evergreen and deciduous forests and disturbed places such as field margins (500–2,100 m altitude) and is distributed widely (except in N China) in the provinces of Anhui, Fujian, S Gansu, Guangdong, Guangxi, Guizhou, Hainan, S Hebei, Henan, Hubei, Hunan, Jiangsu, Jiangxi, S Shaanxi, Shandong, S Shanxi, Sichuan, Taiwan, SE Tibet, Yunnan and Zhejiang. Since the species is widely naturalised its exact distribution is uncertain (also native to Bhutan, India, Indonesia, Laos, Nepal, Papua New Guinea, Philippines, Sri Lanka, Thailand, Vietnam; tropical Australia and the Pacific islands).

CONSERVATION STATUS

China: Least Concern (MEPCAS, 2013 but see above)).
Global: Not Evaluated (IUCN Red List, accessed 19/8/2016).

Cultivated trees of *Melia azedarach*, Guangxi province

Flower-head of *M. azedarach*

Branch with immature fruit

TCM PROPERTIES AND USE

***properties*: 'bitter and cold'**
***channels*: 'Liver, Small Intestine and Bladder'**
***action*: 'invigorates qi, relieves pain'**

***indications*: abdominal pain and distension; hernia pain; irritability; parasites and fungal infections; (T)**

DRUG MORPHOLOGY

CRUDE DRUG (DRIED WHOLE FRUIT): almost spherical (2–3.2 cm diam.). Externally golden or brownish-yellow, slightly lustrous and often dented **[1]**; fruit wall hard, thin and chitinous **[2]**, and regularly scattered with dark brown dots **[3]**; basal end with fruit stalk or fruit scar **[4]**, often sunken; apex with style remains **[5]**. Outer wall often separating **[6]** from the granular or spongy, pale yellow inner wall 'flesh' (mesocarp) with the latter forming the bulk of the fruit **[7]**; flesh becoming viscous when wet. Transverse fracture reveals a slightly darker wavy-edged central husk **[8]** with 6–8 locules **[9]**), each locule with one brownish-black oblong seed (often removed in decocting pieces) **[10]**. Longitudinal fracture **[11]** shows roughly ovate outline of husk with 1 or 2 locules visible, with or without seed. Husk usually inseparable from the flesh. *Odour:* indistinct. *Taste:* sour and bitter. ***Good quality***: fruit plump, externally golden-yellow, internally yellowish-white.

DECOCTING PIECES-1 (HALVED OR BROKEN FRUIT): as for the crude drug but hemispherical or irregularly shaped with inner fruit structure exposed at cut surface.

DECOCTING PIECES-2 (PROCESSED DRY-FRIED FRUIT): as for decocting pieces-1 but externally dark reddish-brown **[12]** or charred; cut surface greyish- to dark yellow **[13]**. *Odour:* slightly aromatic.

PROCESSING

Whole fruits are dried (= crude drug) then broken into pieces or sliced in half (= decocting pieces-1); optional further processing involves cutting the fruits into thick slices before dry-frying (= Chao Chuan Lian Zi 炒川楝子; decocting pieces-2).

OTHER COMMON NAMES

川楝 Chuan Lian; 金铃子 Jing Ling Zi 'little golden bells'; 苦楝子 Ku Lian Zi; 楝 Lian (plant name); bead tree; Chinatree; Fructus Toosendan; melia (SCN); pagoda tree; Persian lilac; Szechwan chinaberry fruit (CP2010).

SAFETY

Fresh (unprocessed) fruits are **toxic** (Chen & Chen, 2004) and ranked as categories **'Safety Class: 3;** Interaction Class: A' (Gardner & McGuffin, 2013). Dried or dry-fried fruits are **slightly toxic** (CP2010; Chen & Chen, 2004). In severe cases, ingestion of 6–8 fresh (unprocessed) fruits may cause unconsciousness (Chen & Chen, 2004).

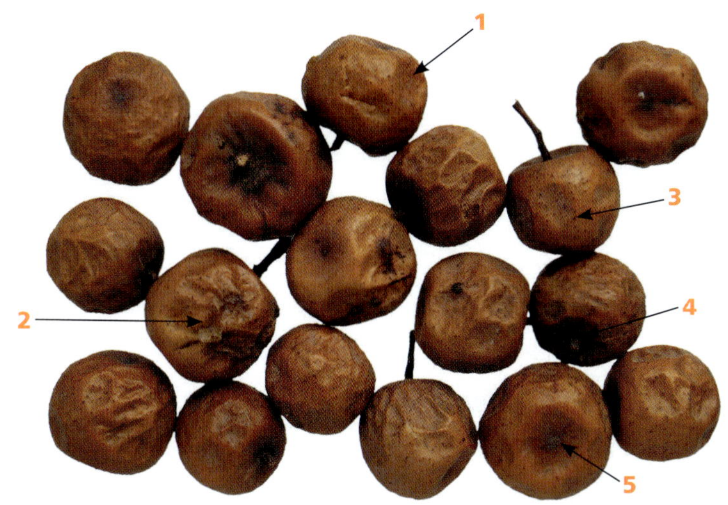

Crude drug (dried whole fruit)

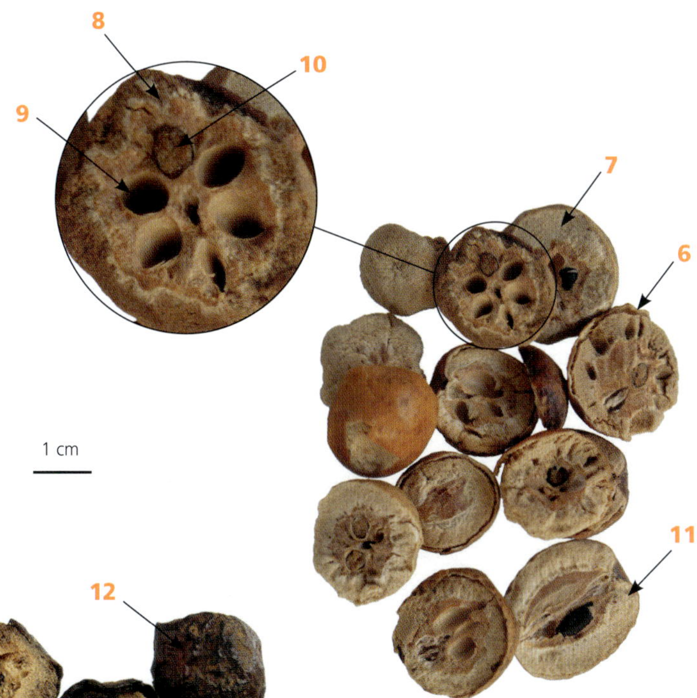

Decocting pieces-1 (halved fruit)

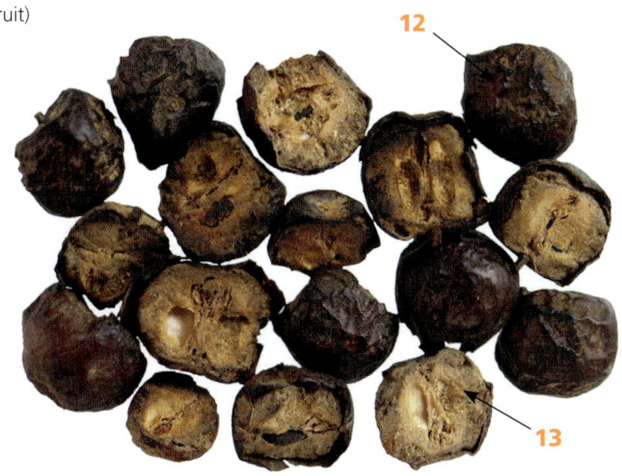

Decocting pieces-2: Chao Chuan Lian Zi (dry-fried fruit)

Di Fu Zi 地肤子

Kochiae Fructus

Drug part: FRUIT

Kochia scoparia (L.) Schrad.

Amaranthaceae

synonyms: *Bassia scoparia* (L.) A. J. Scott (*Kew-MPNS* accepted name 20/2/2015); *Chenopodium scoparium* L.

Species of *Kochia* wild in China: 7

PLANT DESCRIPTION

A highly variable species with many subspecies, varieties and ornamental forms. A much branched, often untidy-looking annual (50–100 cm tall). *Branch stems* ribbed, light green or reddish-purple. *Leaves* alternate, simple, virtually stalkless, linear to lance-shaped (2–5 cm long; 3–7 mm wide) and usually with 3 distinct veins (upper leaves smaller and 1-veined). *Flowers* solitary or up to 3 in sparse groups along a central stem, bisexual or female; each flower tiny, without obvious petals but with 5 triangular wing-like appendages [perianth segments], pale-green or pink with wavy margins; male flowers with 5 stamens; female flowers with ovary superior, a very short style and 2 brownish-purple stigmas. *Fruits* (utricles) small (1–3 mm diam.), dry and hard with 5 persistent membranous perianth segments imparting a flattish and strongly pentagonal shape. Fl. Jun–Sep; Fr. Jul–Oct.

Kochia scoparia f. *trichophylla* (Voss) Stapf ex Schinz & Thell. (扫帚菜子 Sao Zhou Cai Zi) is a popular non-medicinal form widely cultivated worldwide for broom-making and as an ornamental.

HARVESTING, SOURCING AND NATURAL RANGE

The species is not native to China but naturalised throughout; ripe fruits are harvested in autumn from these naturalised plants which grow in valleys, riversides, field margins and disturbed places. The species is native in other parts of Asia and Europe and widely naturalised in Africa, Australia, North and South America.

CONSERVATION STATUS

China: Least Concern (MEPCAS, 2013, as *Kochia scoparia* var. *sieversiana* (Pall.) Ulbr. ex Asch. & Graebn.).
Global: Not Evaluated (IUCN Red List, accessed 8/10/2016).

Mature plants of *Kochia scoparia* with typical untidy habit

UNOFFICIAL SUBSTITUTES

The following species with similar-looking fruits have been traded as 'Di Fu Zi' in local markets in China. Their similar occurrence in international markets is unclear.

1. ***Chenopodium*** species (family Chenopodiaceae), including *C. album* L.
2. ***Dysphania ambrosioides*** (L.) Mosyakin & Clemants — Tu Jing Jie 土荆芥 (syn. *Chenopodium ambrosioides* L.) (family Chenopodiaceae); a taxonomically complex entity, widely naturalised as a weed in many parts of China but also cultivated as a medicine in northern China. A strong-smelling plant, with flowers in groups of 3—5 and with fruit more square compared to the Pharmacopoeia species.

Upper stem leaves of *K. scoparia* with pentagonal fruits (utricles) showing wavy perianth margins

TCM PROPERTIES AND USE

properties: **'sweet, bitter and cold'**
channels: **'Bladder and Kidney'**
actions: **'promotes urination, clears damp heat'**

indications: **dysuria; leucorrhoea; rashes, itching; (T)**

DRUG MORPHOLOGY

CRUDE DRUG/DECOCTING PIECES (DRIED FRUIT — UTRICLES): overall shape squat-spheroidal (1–3 mm diam.) with 5 membranous wings **[1]** (perianth segments) sometimes broken off forming a loose pentagonal shape. Externally hard, greyish-green, pale brown or pink; ventral surface with a shortly prominent fruit-stalk scar **[2]** and 5–10 radiating ridges **[3]**; membranous fruit-wall (pericarp) visible when perianth removed. Internally 1–2 seeds (c. 1 mm long), blackish-brown with a swollen horseshoe-shaped embryo [hl]. *Odour:* indistinct. *Taste:* slightly bitter. ***Good quality*:** plump and greyish-green.

PROCESSING

Fruits are sun-dried (= crude drug/decocting pieces).

OTHER COMMON NAMES

地肤 Di Fu (plant name); 扫帚菜子 Sao Zhou Cai Zi (refers to *K. scoparia f. trichophylla* — see *Plant Description*); belvedere fruit (CP2010); Fructus Kochiae.

Crude drug/decocting pieces of Di Fu Zi (*Kochia scoparia*) showing 2 colour forms; pale brown (left) and pale brownish-pink (right)

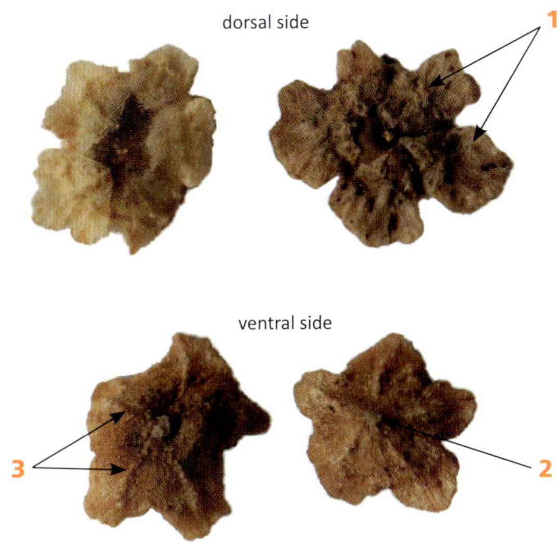

Fruits of Di Fu Zi (*Kochia scoparia*)

CONTAMINANTS

The small size of 'Di Fu Zi' fruit (1–3 mm diam) makes it vulnerable to accidental or intentional contamination. Unacceptably high levels (up to 40%) have been detected in UK-traded samples (Leon, pers. obs., 2014). Suspected contaminants include:

1. ***Cnidium monnieri*** fruits (family Apiaceae); this is the source of the separate drug 'She Chuang Zi' (Cnidii Fructus), page 716.
2. Fruits of other species of ***Kochia***.
3. Fruits of species of ***Leonurus*** (possibly *L. japonicus*), source of the separate drug 'Chong Wei Zi' (Leonuri Fructus), page 618.

Di Fu Zi (*Kochia scoparia* fruit)

contaminant *Kochia* species (possibly *K. scoparia* f. *trichophylla*)

Trade sample of Di Fu Zi (*Kochia scoparia* fruit) heavily contaminated with She Chuang Zi (*Cnidium monnieri* fruit) and other material

Dou Kou 豆蔻

Amomi Fructus Rotundus

Chinese and Indonesian cardomom

Drug part: FRUIT

PHARMACOPOEIA SPECIES 1

Amomum verum Blackw.

Zingiberaceae

synonym: *A. krervanh* Pierre ex Gagnep. (an illegitimate name, used in CP2015 as '*A. kravanh* Pierre ex Gagnep.').

Species of *Amomum* wild in China: 38

PLANT DESCRIPTION

A clump-forming, perennial herb with **rhizomes creeping along the soil surface.** *'Stems'* (pseudostems — densely overlapping leaf bases) 3–4 m tall. *Leaves* alternate, c. 18 per 'stem'; ligule (at base of each leaf blade) **0.4–0.8 cm long, leathery, outer surface and margin with long stiff hairs, apex straight or notched;** leaf blades simple, large and oblong (25–70 cm long; 6–18 cm wide), margins entire with long stiff hairs towards tips. *Flowers* bisexual, in dense cylindrical spikes (7–14 cm long; 3–4.5 cm diam.) **at base of plant and growing directly from the rhizomes; each flower usually hidden by a triangular and membranous bract** (3–4 cm long; 1–1.5 cm wide) overlapping those of adjacent flowers; calyx tubular, often inflated; petals white; **lip (labellum) yellow-tipped** (2.5–3 cm long). *Fruits* in groups of c. 16 or 17; each fruit a dry, globose capsule (1–1.2 cm diam.) enclosed when dry, by a persistent, membranous yellow bract; fruit surface finely ridged, yellowish or reddish when fresh, internally with 3 or 4 seed compartments. *Seeds* many and multi-angled. *Based on Lamxay & Newman, 2012*. See also *Crude Drug Morphology*.

HARVESTING, SOURCING AND NATURAL RANGE

Not native to China but locally cultivated in southern China (Guangxi, Hainan and southern Yunnan) as a culinary spice, healthfood and medicine. Ripe fruits are harvested in July or August or imported from Thailand, Cambodia and Indonesia (Sumatra), from where the species 'originates' in lowland evergreen forest.

CONSERVATION STATUS

China: not native.

Global: Data Deficient (IUCN Red List, accessed 28/8/2016). Although widely cultivated for food and medicine, the taxonomy of this spieces is complex and awaiting review before its conservation status can be assessed.

PHARMACOPOEIA SPECIES 2

A. compactum Sol. ex Maton

Zingiberaceae

PLANT DESCRIPTION

Similar to PS1 but smaller in most parts (except leaves). *'Stems'* up to 1–1.5 m tall. **Rhizomes growing below** (not along) **the ground.** *Leaves* alternate and simple; **ligule orbicular and very short (usually 1–2 mm long)**, 2-cleft and later, the margin becoming fringed with hairs; leaf blade lance-shaped to ovate (41–42 cm long; 10–17 cm wide), base narrowly extending down leaf stem, margins fringed with hairs and tip narrowed into a twisted point. Bracts harder and often with a dark margin. *Flower- and fruit-heads* **shorter and narrower (5.5–8 cm long** (excl. stalks), **c. 2.5 cm wide**); petals white or yellowish; **lip (labellum) yellowish with an orange mid-vein and purple margin.** *Fruits* **squat, sub-globose** (1–1.5 cm diam.), yellowish, c. 9-furrowed when dry and covered in thin, short hairs (pilose). *Seeds* many, multi-angular (c. 2–3 mm diam.). Fl. Feb–May, fr. Jun–Aug. *Based on Flora of China and Droop and Newman (2014).*

HARVESTING, SOURCING AND NATURAL RANGE

Not wild in China [native to Indonesia (Java)]. Ripe fruit are harvested from cultivated crop plants grown in the Chinese provinces of Hainan and Yunnan.

CONSERVATION STATUS

China: not native.

Global: Not Evaluated (IUCN Red List, accessed 20/8/2016).

TCM PROPERTIES AND USE

properties: **'pungent and warm'**
channels: **'Lung, Spleen and Stomach'**
action: **'binds the intestines, warms the interior, regulates qi, stops pain'**

indications: **abdominal distension, poor appetite, abdominal pain; nausea, vomiting**

Amomum verum medicinal crop, Yunnan province

A. verum freshly uprooted plant with flower spikes at base

A. verum with flower spikes growing directly from the rhizome (young flowers hidden inside membranous bracts)

RIGHT Flowering spikes of *A. verum* (longer head) and *A. compactum* (shorter head); note 1 or 2 white flowers emerging from apex of each

Freshly uprooted plants of *Amomum compactum* with flower spikes at base

A. compactum at different stages of maturity (one flower on left-hand flower-head; two fruits exposed between membranous bracts of fruit-head on right-hand side)

Amomum verum

Zingiberaceae

DRUG MORPHOLOGY

CRUDE DRUG (DRIED FRUIT): **globose [1] (1.2–1.8 cm diam.).** Externally fruit-wall yellowish-white or pale yellowish-brown with 3 or 4 main longitudinal furrows [2] (denoting 3 or 4 valves within) and many longitudinal ribs [3]; occasionally with scattered papillae [4]; apex with prominent short collar-like flower remnant [5]; base with sunken scar of fruit-stalk [6]; both ends softly hairy. Fruit-wall light and brittle, revealing 3 or 4 compartments [7] each with 7–10 seeds [8]. Seeds densely packed within the pale grey membranous aril [9]; seeds many angled, dark brown, wrinkled and aromatic (2–3 mm. diam.). *Texture (fruit-wall):* easily broken longitudinally. *Odour:* aromatic. *Taste:* pungent, cool and camphor-like. ***Good quality*:** large and plump, with a thin, intact fruit-wall; intensely aromatic.

PROCESSING

Fruit-stalks and persistent bracts are removed from the fruits and discarded then fruits are sun-dried (= crude drug); they are then sometimes sulphured before pulverising prior to decocting.

OTHER COMMON NAMES

白豆蔻 Bai Dou Kou 'white cardamom'; Cambodian cardomum; Chinese cardamom (SCN) (as *A. Kravanh* Pierre ex Gagnep.); Amomi Rotundus Fructus; Fructus Amomi Rotundus; round cardamon fruit (CP2010).

SAFETY

An official healthfood in China (Anon., 2002a,b).

A. compactum

Zingiberaceae

DRUG MORPHOLOGY

CRUDE DRUG (DRIED FRUIT): as for PS1 but *fruits* **slightly smaller (1–1.5 cm diam.),** slightly squat (oblate) or rounded [10], *fruit-wall* **thinner** [11], yellowish-white, occasionally purplish-brown and c. 9-furrowed when dry; seeds slightly larger (c. 4 mm diam.). *Odour:* indistinct.

PROCESSING

As for PS1.

OTHER COMMON NAMES

白豆蔻 Bai Dou Kou 'white cardamom'; 印尼白蔻 Yin Ni Bai Kou (Indonesian round cardamon fruit, CP2010); Fructus Amomi Rotundus (CP2010); round cardamom (SCN); round cardomon fruit (CP2010); Siam cardamom.

SAFETY

An official healthfood in China (Anon., 2002a,b).

Crude drug of Dou Kou (sourced from *Amomum verum*)

Crude drug of Dou Kou (sourced from *A. compactum)*

Fo Shou 佛手

Citri Sarcodactylis Fructus

Drug part: FRUIT

Citrus medica L. 'Fingered'

Rutaceae

synonym: *C. medica* var. *sarcodactylis* (Hoola van Nooten) Swingle (name used in CP2015, as '*C. medica* L. var. *sarcodactylis* Swingle').

Species and hybrids of *Citrus* wild in China: 11 (+ 5 hybrid cultivated species)

PLANT DESCRIPTION

A shrub or small evergreen tree with branches, leaf buds, and flowerbuds usually purplish when young. *Branches* with spines (c. 4 cm long). *Leaves* alternate and simple with stalks short, not winged; blade elliptic to ovate-elliptic (6–12 cm long; 3–6 cm wide or larger; margin shallowly saw-toothed; tip rounded, broad or rarely shortly pointed (mucronate); blade scattered with translucent and fragrant oil ducts. *Flowers* distinct in axillary, 12-flowered clusters or solitary; bisexual or occasionally male; fragrant. Each flower with calyx cup-shaped and 3–5-lobed; petals 5, white, spreading, equal and lance-shaped (1.5–2 cm long); stamens 30–50; style long and thick; stigma large and club-shaped; fragrant. *Fruit* large and distinctly fist-shaped or divided apically into long 'fingers' (up to c. 15 cm long; c. 10 cm wide), pale yellow, surface rough and leathery; fruit wall (pith) white to pale yellow and forming the bulk of the fruit; pulp (flesh) minimal, sometimes absent, when present divided into 10–15 colourless or milky yellow translucent vesicles (pulp); latter is acidic, slightly sweet and fragrant. *Seeds* small and smooth. Fl. Apr–May, fr. Oct–Nov.

HARVESTING, SOURCING AND NATURAL RANGE

Fruits are harvested in autumn when ripe or virtually ripe (yellow or yellowish-green) entirely from cultivated crop plants. This cultivar is mostly grown in China's SC provinces of Guangxi, Guangdong, Sichuan and Chongqing. The species is native to NE India and possibly Myanmar.

CONSERVATION STATUS

China/Global: this cultivar only known from cultivation.

Flower and flower-buds

Branch with mature fruit

Medicinal crop, Guangdong Province

Fruits of varying shape

TCM PROPERTIES AND USE

properties: **'pungent, bitter, sour and warm'**
channels: **'Liver, Spleen, Stomach and Lung'**
actions: **'regulates qi, resolves phlegm'**

indications: **epigastric pain, abdominal distension, loss of appetite; vomiting, nausea; cough, dyspnoea**

DRUG MORPHOLOGY

CRUD DRUG/DECOCTING PIECES (SLICED DRIED FRUIT): thinly sliced, irregularly oblong to ovate pieces (6–10 cm long; 3–7 cm wide; 2–4 mm thick) usually curled and undulating in outline **[1]** and on cut surfaces **[2]**; basal end sometimes distinctly truncate **[3]** with a central slightly raised or sunken fruit-stalk scar **[4]**; apical end usually divided into 2–5 round-tipped, curled lobes or 'fingers' **[5]**. Peel, if still present in places, yellowish-green (more typically from Sichuan) **[6]** or orangish-yellow (more typically from Guangdong), densely packed with aromatic oil ducts **[7]**. Cut surface (mesocarp) creamish-white or yellowish-white, strongly undulating and scattered with slightly raised, often pointed, vascular bundles **[8]**. *Texture*: hard and brittle (pliable when moistened). *Odour*: aromatic. *Taste*: slightly sweet then bitter.
Good quality: uniform, unbroken slices, fruit wall white; intensely aromatic.

PROCESSING

Fresh fruits are cut into longitudinal slices then dried at low temperature or in the sun (= crude drug/decocting pieces). Aged slices ('Chen Fo Shou') are traditionally considered clinically superior (Bensky *et al.*, 2004).

OTHER COMMON NAMES

川手柑 Chuan Shou Gan (Sichuan-sourced Fo Shou); 手片 Shou Pian; 佛手柑 Fo Shou Gan; 佛手片 Fo Shou Pian; 广手柑 Guang Shou Gan (Guangdong or Guangxi-sourced Fo Shou); 香橼 Xiang Yuan (plant name); Buddha's hand; Buddha-hand citron; flesh-finger citron (SCN); finger citron; Fructus Citri Sarcodactylis.

SAFETY

An official Chinese food and healthfood (Anon., 2002a,b).

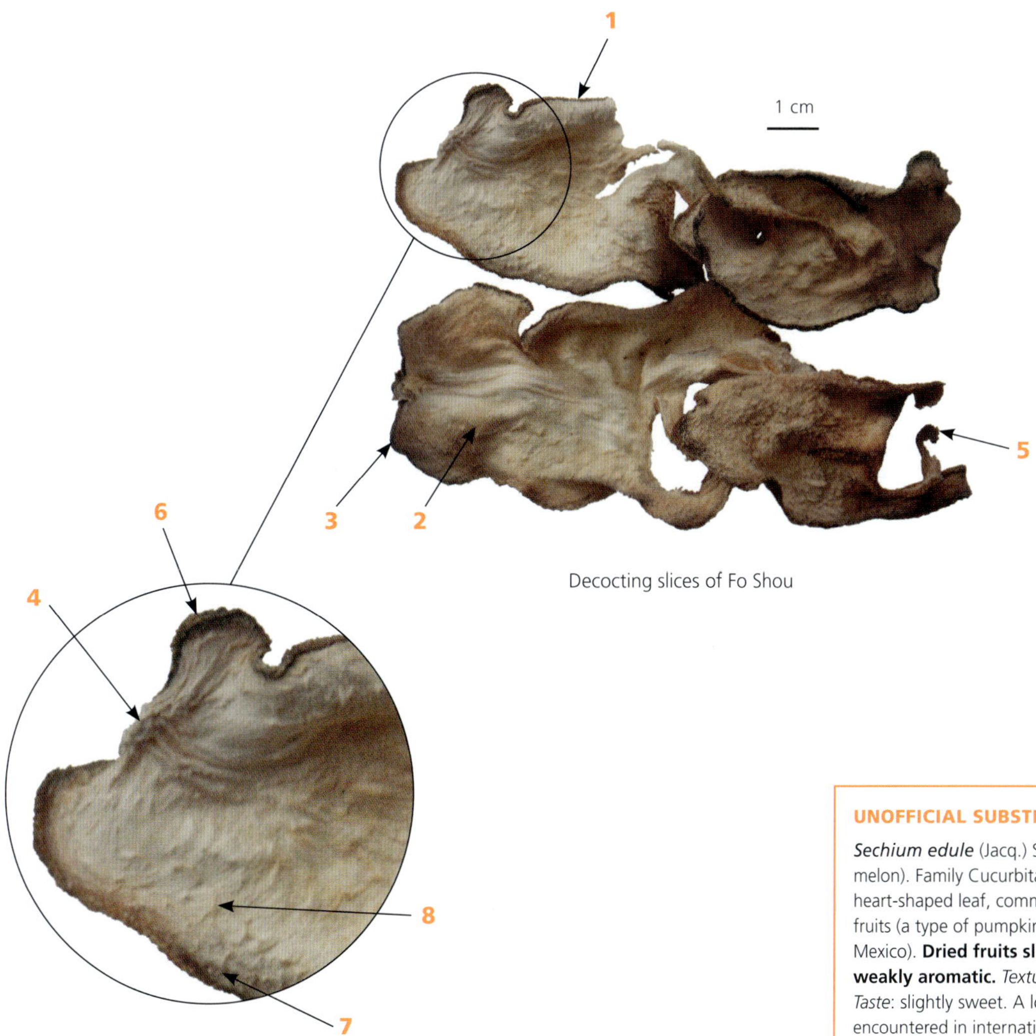

Decocting slices of Fo Shou

UNOFFICIAL SUBSTITUTE

Sechium edule (Jacq.) Sw. 佛手瓜 Fo Shou Gua (Fo Shu melon). Family Cucurbitaceae. A climbing perennial with a heart-shaped leaf, commonly grown in south China and whose fruits (a type of pumpkin) are eaten as a vegetable (native to Mexico). **Dried fruits slices are devoid of oil ducts and are weakly aromatic.** *Texture*: hard and brittle, somewhat powdery. *Taste*: slightly sweet. A local substitute in China unlikely to be encountered in international trade.

Fu Pen Zi 覆盆子 Rubi Fructus

Drug part: FRUIT

Rubus chingii Hu

Rosaceae

Species of *Rubus* wild in China: 208

synonym: *Rubus officinalis* Koidz.

taxonomic note: the *FoC* recognises 2 varieties, the type variety (var. *chingii*) and variety *suavissimus* (S. K. Lee) L. T. Lu; it is unclear if both are used as the source of 'Fu Pen Zi'.

PLANT DESCRIPTION

A rampant, deciduous perennial herb (1.5–3 m tall), sparsely covered in prickles (5–6 mm long). *Stems* arching, often with a **bluish-green** 'bloom', ageing to reddish-brown. *Leaves* alternate, **simple** and stalked; stipules green, linear- to lance-shaped (7–10 mm long) and flattened along base of leaf stalks; leaf blade in outline suborbicular (c. 5–11 cm long; c. 5–13 cm wide), **palmately-lobed** (**usually with 5 lobes, but varying from 3–8**); terminal lobe slightly longer; margin double saw-toothed. *Flowers* solitary (usually 2.5–4 cm diam.); radially symmetrical and bisexual; sepals 5, green or purplish-red, outer surface densely hairy, **with sepal apex shortly pointed and with a tapering tip**; petals 5, white, usually elliptic or ovate-oblong, each with shortly clawed base; stamens many and of unequal length; pistils many (shorter than longest stamens). *Fruit* a squat-conical aggregate (1.5–2 cm diam.), red when ripe (resembling a raspberry) and densely covered in very short, soft, grey hairs. Fl. Mar–Apr; fr. May–Jun.

HARVESTING, SOURCING AND NATURAL RANGE

Fruits are mostly wild harvested in early summer when immature (green to greenish-yellow). The species grows in forests, scrub and disturbed places (500–1,000 m altitude) in S and SE China: Anhui, Fujian, Guangxi, Jiangsu, Jiangxi and Zhejiang provinces [also wild in Japan]. The plant is also cultivated on small-holdings. In addition to the medicinal use of the immature fruits, the very sweet ripe fruits are eaten fresh and used to make jams and various drinks, including wine. Leaves of variety *suavissimus* are sugar-rich and used to make a sweet tea in China and, more recently, in the West.

CONSERVATION STATUS

China: Least Concern (var. *chingii*); Vulnerable (var. *suavissimus*) (MEPCAS, 2013).
Global: Not Evaluated (IUCN Red List; accessed 25/8/2016).

Rubus chingii with immature fruit, Anhui province

R. chingii, immature fruit

R. chingii fruits drying in the sun

TCM PROPERTIES AND USE

***properties*: 'sweet, sour and slightly warm'**
***channels*: 'Liver and Kidney'**
***action*: 'tonifies the Kidney, consolidates jing, tonifies the Liver'**

***indications*: spermatorrhoea, nocturnal emissions, impotence; enuresis; blurred vision**

DRUG MORPHOLOGY

CRUDE DRUG/DECOCTING PIECES (DRIED FRUIT): externally yellowish-green or pale brown. A squat, conical aggregate fruit (0.6–1.3 cm long; 0.5–1.2 cm diam.), with a broadly rounded tip **[1]** and sunken base **[2]**, with 10–60 densely packed single-seeded drupelets **[3]** and, at the base, a persistent 5-sepalled (sometimes broken), brown, scaly calyx **[4]**; **apex to each sepal shortly, pointed and with a tapering tip [5]**; fruit stalks sometimes present **[6]**. Individual drupelets easily removed; shortly crescent-shaped **[7]** (c. 1 mm long; 0.5 mm diam.) with outer (dorsal) surface densely clothed in very short soft hairs **[8]**; lateral surfaces netted [hl] **[9]**; and inner (ventral) surface with longitudinal ridges [hl] **[10]**. *Texture:* hard and light. *Odour:* indistinct. *Taste:* astringent and somewhat sour. ***Good quality***: fruit yellowish-green and unbroken.

PROCESSING

After removal of the fruit stalks, fruits are blanched and dried (= crude drug/decocting pieces).

OTHER COMMON NAMES

覆盆 Fu Pen; 盆子 Pen Zi; 甜茶 Tian Cha (plant name for variety *suavissimus*); 掌叶复盆子 Zhang Ye Fu Pen Zi (plant name for variety *chingii*); Fructus Rubi; palm-leaf raspberry (SCN).

SAFETY

Safety Class: 1; Interaction Class: A (Gardner & McGuffin, 2013). An official food and healthfood in China (Anon., 2002a,b).

Fu Pen Zi (dried fruits of *R. chingii*)

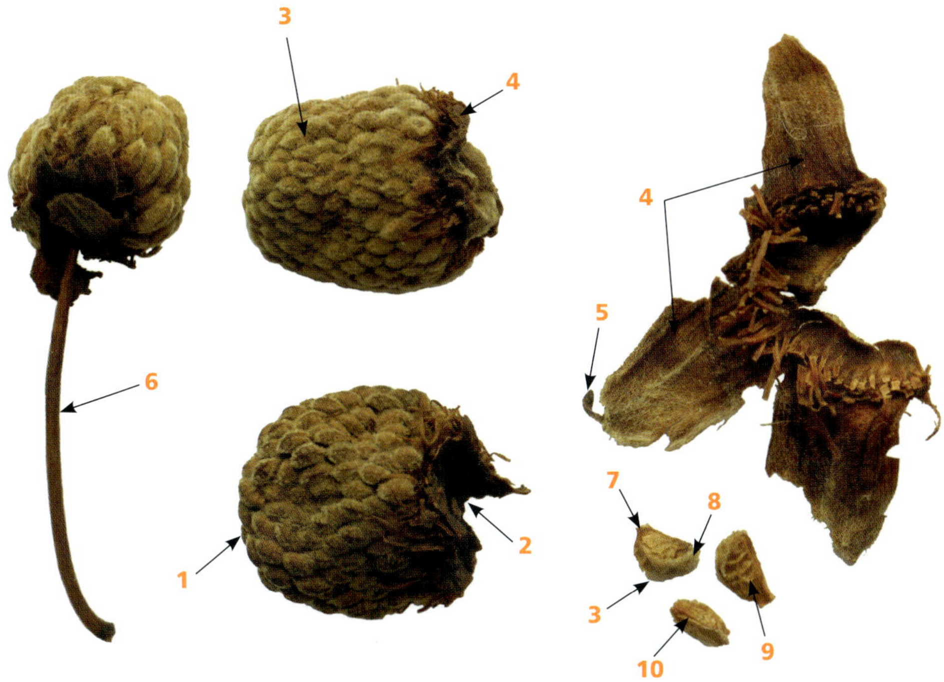

UNOFFICIAL SUBSTITUTES

Despite potential confusion with the other 200 or so species of *Rubus* growing wild in China, *R. chingii* is quite distinctive owing to its **palmately lobed simple leaves; most other *Rubus* species have pinnate leaves.** Substitution due to mis-identification is therefore rare but, in times of high demand, the use of look-alike fruits from other species of *Rubus* have been reported (examples given below) Xiao *et al.*, 2001–2002; NICPBP 1997–2002; Bensky *et al.*, 2004; Lu Lingdi & Boufford, 2003; Namba *et al.*,1986; laboratory-based authentication methods are advised to differentiate trade samples of the official drug from all its possible substitutes.

1. ***Rubus corchorifolius*** L. f. (山莓 Shan Mei); leaves simple, ovate (not lobed); fruits 1–1.2 cm diam.; drupelets wrinkled; fresh fruit widely used for jams, drinks, and wine; fruit, seeds, and roots used medicinally.
2. ***R. eucalyptus*** Focke (桉叶悬钩子 An Ye Xuan Gou Zi); leaves pinnately compound with 3–7 leaflets; **fruits larger than those of the official drug (1.2–2 cm diam.) and densely covered in long grey hairs**.
3. ***R. idaeus*** L. (覆盆子 Fu Pen Zi); leaves pinnately compound with 5–7 leaflets; ripe fruits red or orange, 1–1.4 cm diam., **drupelets prominently pitted**.
4. ***R. idaeopsis*** Focke (拟覆盆子 Ni Fu Pen Zi); leaves pinnately compound with 5–7 leaflets; fruits c. 1 cm diam., hairless or sparsely hairy; **drupelets strongly wrinkled and minutely punctate** [hl].
5. ***R. coreanus*** Miq. (插田泡 Cha Tian Pao); leaves pinnately compound with 5–7 leaflets; fruits 0.6–0.8 mm diam., hairless or almost hairless; endemic to Gansu, Henan and Shaanxi provinces. Included in the Korean Pharmacopoeia.
6. ***R. lasiostylus*** Focke (绵果悬钩子 Mian Guo Xuan Gou Zi); leaves pinnately compound with 3, rarely 7, leaflets; **fruits larger 1.5–2 cm diam., densely covered in long hairs with persistent styles**.

Gou Qi Zi 枸杞子 Lycii Fructus

Drug part: FRUIT

Lycium barbarum L.

Solanaceae

taxonomic/nomenclatural note: *FoC* and *Kew-MPNS* recognise the varieties 'var. *auranticarpum* K. F. Ching' and the type variety 'var. *barbarum*'; *Kew-MPNS* also recognises a third variety (published after the *FoC* account) 'var. *implicatum* T.Y. Chen & Xu L. Jiang'

Species of *Lycium* wild in China: 7

PLANT DESCRIPTION

A deciduous, many-branched perennial shrub (0.8–2 m tall). *Stems* thorny. *Leaves* solitary or clustered; leaf blades simple, lance-shaped or long elliptic (2–3 cm long; 3–6 mm wide); margins entire. *Inflorescences* solitary or clustered, arising from leaf axils. *Each flower* stalked, bisexual and radially symmetrical; **calyx usually 2-lobed** (4–5 mm long), lobes minutely 2- or 3-lobed at tips; flower purple, united basally into a funnel-shaped corolla tube (8–10 mm long) and obviously longer than the corolla lobes; corolla lobes 5, equal and spreading (5–6 mm long), corolla lobe margins virtually hairless; stamens 5; style slender; stamens and style slightly extending beyond corolla lobes. *Fruit* a fleshy, red or orangish-yellow berry, oblong or ovoid (0.4–2 cm long; 5–10 mm diam.), often glossy; calyx persistent. *Seeds* usually 4–20, brownish-yellow (c. 2 mm long) and kidney-shaped. *Many cultivars are grown and seed numbers per fruit are thought to vary accordingly.*

HARVESTING, SOURCING AND NATURAL RANGE

Fruits are harvested from medicinal crops grown extensively in NW China (provinces of Ningxia, Xinjiang, Gansu and Qinghai) with the bulk of production from Ningxia and where *Di Dao*-quality 'Gou Qi Zi' is sourced (Zhongning and Zhongwei districts). 'Xinjiang Gou Qi Zi', sourced from the same species grown in Xinjiang province, is considered 'superior' with local experts distinguishing the fruits by their rounder shape and sweeter taste (Zhao and Chen, 2014; K. Sulaiman pers. comm., 2012). (*FoC* uses the name 'Xinjiang Gou Qi Zi' for a different species (*L. dasystemum* Pojark.), see *Unofficial Substitutes*). Wild *L. barbarum* is endemic to China and grows in disturbed places, often near habitation; the 'type' variety (var. *barbarum*) is wild in Gansu, Hebei, Inner Mongolia, Ningxia, Qinghai, Shanxi, Sichuan and Xinjiang while varieties *auranticarpum* and *implicatum* are confined to Ningxia province. The species is naturalised in Asia and Europe.

From the beginning of the 21st century 'Gou Qi Zi' has become a popular Western 'superfood' traded as 'Goji berry' (Potterat, 2010)

CONSERVATION STATUS

Endemic to China: Least Concern (var. *barbarum*); Data Deficient (var. *aurantiirpum*) (MEPCAS, 2013); Not Evaluated (IUCN Red List, accessed 10/9/2016).

Fruiting branch of *Lycium barbarum*

Flowering branch of *L. barbarum*

EHTPA 4/14
GAP

TCM PROPERTIES AND USE

***properties*:** **'sweet and neutral'**
***channels*:** **'Liver, Kidney and Lung'**
***actions*:** **'tonifies Liver and Kidney, tonifies yin, moistens Lungs'**

***indications*:** **dizziness, tinnitus, blurred vision; infertility, impotence, spermatorrhoea; cough, haemoptysis**

DRUG MORPHOLOGY

CRUDE DRUG/DECOCTING PIECES (DRIED FRUIT): somewhat ellipsoid and usually tapering at both ends **[1]** (6–20 mm long; 3–10 mm diam.). Externally orangish-red and densely wrinkled **[2]**; apex with a tiny protuberant style scar **[3]**; base with a small round, white stalk scar **[4]**. Internally, fruit flesh viscous, enclosing 20–50, pale yellow or brownish-yellow, flattened, somewhat kidney-shaped seeds **[5]** (1.5–1.9 mm long; 1–1.7 mm diam.). Texture: pliable. *Odour*: indistinct. *Taste:* sweet. Fruits sourced from Ningxia province (Ningxia Gou Qi Zi) are often distinguished from those sourced from Xinjiang (Xinjiang Gou Qi Zi); the latter are usually described as rounder (less fusiform) and sweater. ***Good quality:*** large, sweet and fleshy.

PROCESSING

Ripe fruits are harvested when orangish-red in summer and autumn, stalks removed, then shade- or sun-dried (= crude drug and decocting pieces).

OTHER COMMON NAMES

杞子Qi Zi; 西杞子Xi Qi Zi; 新疆枸杞子 Xinjiang Gou Qi Zi; barberry wolfberry fruit (CP2010); Duke of Argyll's teaplant; Fructus Lycii; goji; goji berry; lycium (SCN); matrimony vine; wolfberry.

SAFETY

Safety Class: 1; Interaction Class: A (Gardner & McGuffin, 2013). An official food and healthfood in China (Anon., 2002a,b).

Crude drug/decocting pieces of Gou Qi Zi (dried fruits of *L. barbarum*)

UNOFFICIAL SUBSTITUTES

Fruits of other species (examples below) are reported to have been traded in China under the name 'Gou Qi Zi' as local or historical substitutes. Their occurrence as such in international trade is very unlikely given the extensive cultivation of the Pharmacopoeia species both as a medicinal and healthfood crop. Differentiating the fruits of some of these species from the Pharmacopoeia species requires laboratory-based techniques (Potterat, 2010).

1. *L. chinense* Mill. **Calyces 3–5-lobed.** Fruits used locally in China as a tonic; root-bark is one source of the separate TCM drug 'Di Gu Pi' (Lycii Cortex) see page 496.
2. *L. dasystemum* Pojark. *FoC* name: 新疆枸杞 Xinjiang Gou Qi. Native in Gansu, Qinghai and Xinjiang (NW China), 200–3,600 m altitude. Old stems with formidable thorns up to 6 cm long; leaves often larger than those of the Pharmacopoeia species (1.5–6 cm long; 0.5–1.5 cm wide); calyx 2- or 3-lobed; flower tube twice as long as flower lobes; seeds >20 (1.5–2 mm long).
3. *L. ruthenicum* Murray 黑果枸杞 Hei Guo Gou Ki. Gansu, Inner Mongolia, Ningxia, Qinghai, N Shaanxi, Tibet and Xinjiang. Fruits purplish-black.

ripe fruit of *Lycium chinense*

Gua Lou 瓜蒌

Trichosanthis Fructus

Drug part: FRUIT

Trichosanthes kirilowii Maxim.

Cucurbitaceae

synonym: *T. obtusiloba* C. Y. Wu ex C. Y. Cheng & Yueh

Species of *Trichosanthes* wild in China: 32

PLANT DESCRIPTION

Climbing annual or perennial herb with a long powdery root tuber. *Leaves* simple and alternate (5–20 cm long; 5–20 cm wide), **usually 3–5 (occasionally 7) -lobed, rarely deeply lobed or unlobed, lobes oblong or broadest near the tip;** leaf sometimes compound (with 3–5 leaflets), margin toothed; tendrils 2–5-parted. *Male and female flowers* on separate plants. Male flowers usually several along a stalk (in a raceme) each with a green leaf-like bract at base (1.5–2.5 (–3) cm long; 1–2 cm wide); female flowers solitary. Flowers usually white, sometimes pink or red, with flower tube (corolla) <7 cm long, extending into 5 spreading 'petal' lobes with long thread-like (filiform) tips. *Fruit* with long stalks 4–11 cm long; ovoid, gourd-like (7–10.5 cm diam.), brownish-yellow or orangish-yellow, smooth, hairless and many seeded; flesh of fresh fruit yellow. *Seed* flattened-ovoid (11–16 mm long; 7–12 mm wide), with continuous marginal ridge. Fl. May–Aug, fr. Aug–Oct.

HARVESTING, SOURCING AND NATURAL RANGE

Fruits, with stalks, are harvested in autumn when almost ripe (still slightly green) usually from medicinally farmed crops mainly grown in Shandong (E China) but also the neighbouring provinces of Henan, Hebei, Anhui, Shanxi and Shaanxi. *Di Dao* material is traditionally sourced form Henan province. In the wild, the species grows in NC, NE and SE China (Gansu, Hebei, Henan, Jiangsu, Jiangxi, Shandong, Shanxi, Zhejiang in open forest, scrub, grassland and field margins (200–1,800 m altitude); the species is also native to Japan and the Korean Peninsula.

CONSERVATION STATUS

China: Least Concern (MEPCAS, 2013).
Global: Not Evaluated (IUCN Red List, accessed 10/3/2016).

T. rosthornii Harms

Cucurbitaceae

PLANT DESCRIPTION

Similar to PS1 but leaves **deeply** 3–7-lobed, **usually 5-lobed**, **lobes linear to lance-shaped, lance-shaped or broadest at tips**; fruit-stalks sometimes shorter (4.5–8 cm long). Seed slightly longer and flatter (15–19 mm long; 8–10 mm wide; c. 2.5 mm thick) with broadest end **more obviously truncate**. Fl. Jun–Aug, fr. Aug–Oct.

HARVESTING, SOURCING AND NATURAL RANGE

Similar to PS1, however, the wild species is endemic to China and has a more southerly distribution: Anhui, Guangdong, Guangxi, Guizhou, Jiangxi, Sichuan and Yunnan provinces) where it grows in forests, scrub or grassland on mountainsides and in valleys (400–1,900 m altitude).

CONSERVATION STATUS

Endemic to China: Least Concern (MEPCAS, 2013); Not Evaluated (IUCN Red List, accessed 10/3/2016).

Note: For other drugs (sourced from these 2 species see 'Gua Lou Pi' (page 638) and 'Gua Lou Zi' (page 642).

TCM PROPERTIES AND USE

properties: 'sweet, slightly bitter and cold'
channels: 'Lung, Stomach and Large Intestine'
actions: 'clears lungs, resolves phlegm, regulates qi, disperses nodules'
indications: cough, phlegm; abscesses

Cultivated medicinal crop of *Trichosanthes kirilowii*

Male flowers of *T. kirilowii*

T. kirilowii — ripe fruit

Male flowering plant of *Trichosanthes rosthornii* with leaf-like bracts at base of flower tube

Female flowers (without leaf-like bracts) of *T. rosthornii*

Fruiting plant of *T. rosthornii* in the wild

Trichosanthes kirilowii

Cucurbitaceae

DRUG MORPHOLOGY

CRUDE DRUG (DRIED WHOLE FRUIT): ovoid or subspherical fruit (7–15 cm long; 6–10 cm diam.). Externally yellow, yellowish-orange or orangish-red; strongly crumpled from drying (with deep irregular grooves) or smooth in places; the more acute end (base) with remains of the hard fruit-stalk [1] (3–11 cm long when intact), the other end (apex) rounded and with a small round flower (style) scar [2]. Fruit-wall thin and brittle, c. 0.3 cm thick, inner surface yellowish-white and veined reddish-yellow, containing minimal orangish-yellow viscous flesh adhering to the many seeds. Seeds flattened-ellipsoid **(1.2–1.5 cm long; 0.6–0.7 cm wide; c. 3.5 mm thick)**, brown or brownish-cream, one end (apex) obtusely pointed or rounded, the other **somewhat angular or truncate;** an obvious fissure creates a continuous ring inside the perimeter of each flat surface (see seed photos under 'Gua Lou Zi' (Trichosanthis Semen) page 642). *Texture*: fruit wall fragile and easily broken; fruit pulp sticky. *Odour*: caramel-like. *Taste*: sweet and somewhat sour. Trade samples often cut open. ***Good quality*:** large, unbroken fruit, with a thick, orange and often wrinkled fruit wall.

DECOCTING PIECES (FLATTENED AND SLICED FRUIT LENGTHS): longitudinal strips with a corrugated appearance; white contents (cotyledons) of sliced seeds distinctly sandwiched between the fruit walls; short and seedless pieces of fruit wall yellowish-white on inner surface with reddish veining, orangish-brown or orangish-yellow on the outer surface; *texture*, *odour* and *taste* as for the crude drug.

PROCESSING

Fruits are dried in well-ventilated shade conditions (= crude drug) then stalks removed and fruits flattened before being sliced into decocting pieces.

OTHER COMMON NAMES

黄熟瓜蒌 Huang Shu Gua Luo (yellow mature Gua Lou); 全瓜蒌 Quan Gua Lou (whole Gua Lou); Chinese cucumber; Chinese snakegourd; Fructus Trichosanthis; Mongolian snakegourd; trichosanthes (SCN).

SAFETY

Safety Class 1; Interaction Class: A. Gardner & McGuffin (2013).

T. rosthornii

Cucurbitaceae

DRUG MORPHOLOGY

CRUDE DRUG (DRIED WHOLE FRUIT): as for PS1 but **seeds slightly larger and thinner (1.5–1.9 cm long; 0.8–1 cm wide; c. 0.2 mm thick)**; **the broader (basal) end more obviously truncate** (see seed photos under 'Gua Lou Zi' (Trichosanthis Semen) page 642).

DECOCTING PIECES (FLATTENED AND SLICED FRUIT LENGTHS): indistinguishable from PS1 unless seeds intact (see crude drug description of PS2).

PROCESSING

As for PS1.

OTHER COMMON NAMES

As for PS1 (excluding Mongolian snakegourd) but also 中华栝楼 Zhong Hua Gua Lou (plant name) and rosthorn snakegourd.

Gua Lou crude drug
(dried whole fruits sourced from *Trichosanthes kirilowii*)

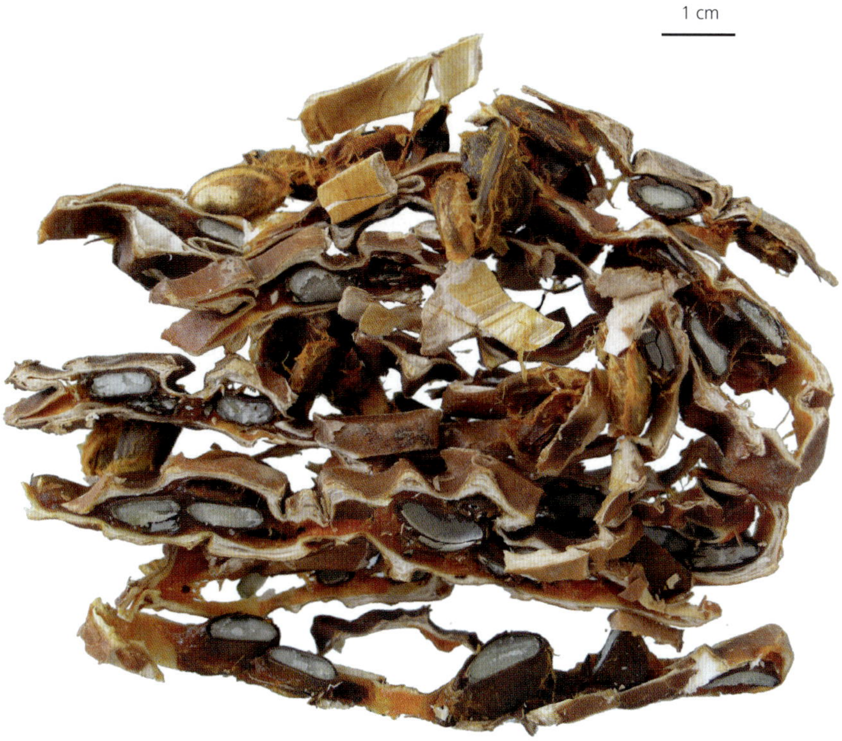

Gua Lou decocting pieces
(dried and sliced fruits sourced from *T. kirilowii*)

Comparison of the seeds of PS1 and PS2
left: seed of *T. rosthornii*;
right: seed of *T. kirilowii*

Fresh fruit (*T. kirilowii*) with seed mass and flesh within

UNOFFICIAL SUBSTITUTES

The fruits of *Trichosanthes* species have been reported as local or historical unofficial substitutes in China for reasons of clinical substitution and/or confusion (NICPBP 1997–2002; Bensky *et al.*, 2004). Their occurrence as substitutes in international trade is uncertain:

1. ***T. cucumeroides*** (Ser.) Maxim. 王瓜 Wang Gua. Cultivated for a variety of purposes: root (abortifacient), young fruit (vegetable) and root (industrial starch). Fruit-stalk shorter (0.5–0.2 cm) than official species when fresh and intact; fruits smaller (6–7 cm long; 4–5.5 cm diam.) when fresh, sometimes narrowly oblong and yellow: seeds sometimes triangular.
2. ***T. dunniana*** H. Lév. 糙点栝楼 Cao Dian Gua Lou. Fruit indistinguishable from that of the official species.

Gua Lou Pi 瓜蒌皮

Trichosanthis Pericarpium

Drug part: FRUIT PEEL

PHARMACOPOEIA SPECIES 1

Trichosanthes kirilowii Maxim.

Cucurbitaceae

synonym: *T. obtusiloba* C. Y. Wu ex C. Y. Cheng & Yueh

Species of *Trichosanthes* wild in China: 32

PLANT DESCRIPTION

A climbing annual or perennial herb with a long, powdery root tuber. *Leaves* simple and alternate (5–20 cm long; 5–20 cm wide), usually 3–5-lobed (occasionally 7-lobed), lobing seldom very deep, lobes oblong or broadest near the tip; leaf sometimes compound (with 3–5 leaflets), margin toothed; tendrils 2–5-parted. *Male and female flowers* on separate plants. **Male flowers** usually several along a stalk (in a raceme) each with a **green leaf-like bract at base (1.5–2.5(–3) long; 1–2 cm wide);** female flowers solitary. Flowers usually white, sometimes pink or red, with flower tube (corolla) <7 cm long, extending into 5 spreading 'petal' lobes with long thread-like (filiform) tips. *Fruit* on a long stalk (4–11 cm long), ovoid, gourd-like (7–10.5 cm diam.); brownish-yellow or orangish-yellow, smooth, hairless and many seeded; **flesh of fresh ripe fruit yellow.** *Seed* flattened-ovoid (11–16 mm long; 7–12 mm wide), with continuous marginal ridge. Fl. May–Aug, fr. Aug–Oct.

HARVESTING, SOURCING AND NATURAL RANGE

See 'Gua Lou' (Trichosanthis Fructus) page 634.

CONSERVATION STATUS

See 'Gua Lou' (Trichosanthis Fructus) page 634.

PHARMACOPOEIA SPECIES 2

T. rosthornii Harms

Cucurbitaceae

PLANT DESCRIPTION

Similar to PS1. *Leaves* usually 5-lobed (sometimes 3- or 7-lobed), lobing very deep, lobes linear to lance-shaped, lance-shaped near the tip. **Leaf-like bracts of male flowers small (6–14 mm long; 5–11 mm diam.). Flesh of fresh ripe *fruit* yellow.** *Seed* slightly longer and flatter (15–19 mm long; 8–10 mm wide; c. 2.5 mm thick) with broadest end obviously truncate. Fl. Jun–Aug, fr. Aug–Oct.

For other TCM drugs sourced from these 2 species see 'Gua Lou' (page 634) and 'Gua Lou Zi' (page 642).

HARVESTING, SOURCING AND NATURAL RANGE

See 'Gua Lou' (Trichosanthis Fructus) page 634.

CONSERVATION STATUS

See 'Gua Lou' (Trichosanthis Fructus) page 634.

UNOFFICIAL SUBSTITUTE

T. laceribractea Hayata

Cucurbitaceae

synonyms: *T. hupehensis* C. Y. Cheng & Yueh; *T. sinopunctata* C. Y. Cheng & Yueh

PLANT DESCRIPTION

Similar to both Pharmacopoeia species. *Leaves* simple or 3–7-lobed, lobes shallow or deep, **upper leaf surfaces rough with tiny raised (often white) glands [hl]; leaf-like bracts of male flowers much larger (5–8 cm diam.)** than those of both Pharmacopoeia species; **flesh of fresh ripe *fruit* dark greenish-blue;** *seeds* rounded or truncate at both ends and marginal ridge irregular.

HARVESTING, SOURCING AND NATURAL RANGE

Widely cultivated for its edible seed.

The wild species is endemic to S and C China (Guangdong, Guangxi, Hubei, Jiangxi, Sichuan and Taiwan) where it grows in forested valleys (200–1,100 m altitude).

CONSERVATION STATUS

Endemic to China: Least Concern (MEPCAS, 2013); Not Evaluated (IUCN Red List, accessed, 20/6/16).

TCM PROPERTIES AND USE

properties: 'sweet and cold'
channels: 'Large Intestine and Lung, Stomach'
actions: 'clears lungs heat, resolves phlegm, regulates qi'

Indications: cough, phlegm, dyspnoea

Male flowers of *Trichosanthes kirilowii*

Trichosanthes kirilowii — cultivated plant in fruit

Ripe fruit with yellow flesh; seeds scattered

Female flowers of *Trichosanthes rosthornii*

T. rosthornii with ripe fruit

Male flowers of *T. laceribractea* with large green bracts irregularly toothed

T. laceribractea with unripe fruit

T. laceribractea — fresh ripe fruits (seeds removed for food) showing remains of dark greenish-blue flesh inside

PHARMACOPOEIA SPECIES 1

Trichosanthes kirilowii

Cucurbitaceae

DRUG MORPHOLOGY

CRUDE DRUG (DRIED FRUIT PEEL): variously sized and shaped pieces (6–12 cm long), usually inwardly curved (involute) **[1]**. Externally smooth, orange to greenish-orange or brown, sometimes slightly dented **[2]** or shrivelled **[3]**, fruit-stalks sometimes persistent (3–11 cm long) at fruit base **[4]**; small, round and raised flower (style) scar at fruit apex **[5]**; pericarp thin. *Texture:* tough but brittle. Fruit-wall (pericarp) thin **[6]** and brittle, inner surface **yellowish-white [7], sometimes veined reddish-yellow**; sometimes 2 or more compartments divided by septa visible. *Odour:* caramel-like. *Taste:* slight or somewhat sour.

DECOCTING PIECES: as for crude drug but sliced into narrow strips c. 5 mm wide, 3–7 cm long **[8]**; folded flat , curved or with a few corrugations **[9]**.

PROCESSING

Contents of ripe fruits are removed and discarded, then peel is shade dried (= crude drug). Later it is washed before cutting into narrow strips (= decocting pieces).

OTHER COMMON NAMES

栝蒌皮 Gua Lou Pi; 药 瓜 Yao Gua; 药瓜皮 Yao Gua Pi; Pericarpium Trichosanthis; Mongolian snakegourd peel; snakegourd peel (CP2010); trichosanthes (SCN).

SAFETY

Safety Class 1; Interaction Class: A. (Gardner & McGuffin, 2013).

PHARMACOPOEIA SPECIES 2

T. rosthornii

Cucurbitaceae

DRUG MORPHOLOGY

Not macroscopically distinct from the drug of PS1; any colour variation as illustrated in the photos shown here is not a reliable differentiating character.

PROCESSING

As for PS1.

OTHER COMMON NAMES

栝蒌皮 Gua Lou Pi; 西蒌皮 Xi Lou Pi; 药瓜皮 Yao Gua Pi; Pericarpium Trichosanthis; rosthorn snakegourd peel; snakegourd peel (CP2010); trichosanthes (SCN).

UNOFFICIAL SUBSTITUTE

T. laceribractea

Cucurbitaceae

'DRUG SUBSTITUTE' KEY CHARACTERS

Indistinguishable macroscopically from that of PS1 and 2 except when fresh when **occasional remnants of dark greenish-blue veined flesh** are visible **[10]**.

OTHER COMMON NAMES

长萼栝楼 Chang E Gua Lou.

SUBSTITUTION COMMENT. This species is widely cultivated in China for its expensive, sweet and edible seeds; the fruit peel is a waste product and therefore readily available as a cheap substitute for official 'Gua Lou Pi' .

Crude drug of Gua Lou Pi (*Trichosanthes kirilowii* — fruit peel)

Decocting pieces of Gua Lou Pi (*T. kirilowii* — sliced peel)

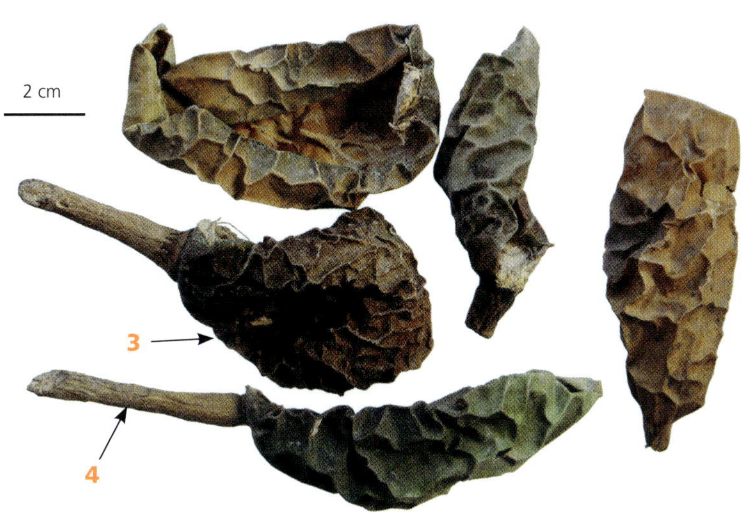

Crude drug of Gua Lou Pi (*T. rosthornii* — fruit peel)

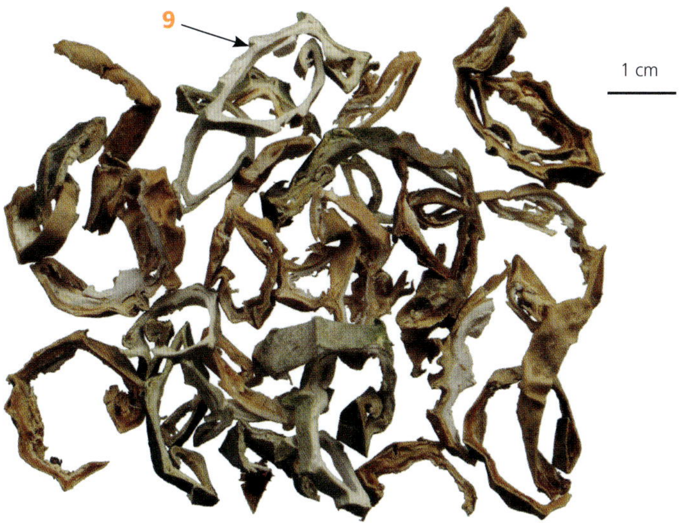

Decocting pieces of Gua Lou Pi (*T. rosthornii* — sliced fruit peel)

T. laceribractea — dried fruit peel (pericarps)

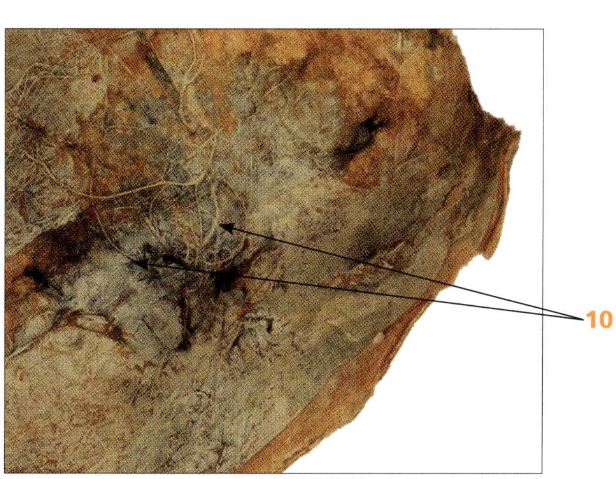

T. laceribractea — inner surface of fruit peel

Gua Lou Zi 瓜蒌子 Trichosanthis Semen

Drug part: SEED

PHARMACOPOEIA SPECIES 1

Trichosanthes kirilowii Maxim.

Cucurbitaceae

synonym: *T. obtusiloba* C. Y. Wu ex C. Y. Cheng & Yueh

Species of *Trichosanthes* wild in China: 32

PLANT DESCRIPTION

See 'Gua Lou' (Trichosanthis Fructus) page 634.

HARVESTING, SOURCING AND NATURAL RANGE

See 'Gua Lou' (Trichosanthis Fructus) page 634.

CONSERVATION STATUS

See 'Gua Lou' (Trichosanthis Fructus) page 634.

PHARMACOPOEIA SPECIES 2

T. rosthornii Harms

Cucurbitaceae

PLANT DESCRIPTION

See 'Gua Lou' (Trichosanthis Fructus) page 634.

HARVESTING, SOURCING AND NATURAL RANGE

See 'Gua Lou' (Trichosanthis Fructus) page 634.

CONSERVATION STATUS

See 'Gua Lou' (Trichosanthis Fructus) page 634.

For other TCM drugs sourced from these two species see 'Gua Lou' (page 634) and 'Gua Lou Pi' (page 638).

TCM PROPERTIES AND USE

properties: 'sweet and cold'
channels: 'Lung, Stomach and Large Intestine'
actions: 'resolves phlegm, moistens dryness, moistens intestines, disspiates nodules'
indications: cough, phlegm; constipation

Flowering plants of *Trichosanthes kirilowii*

Opened fresh fruit of *T. kirilowii*

Freshly harvested ripe fruits of *T. kirilowii*, one with seeds removed

T. rosthornii (female plant in flower)

T. rosthornii with ripe fruit

Trichosanthes kirilowii

Cucurbitaceae

DRUG MORPHOLOGY

CRUDE DRUG (DRIED SEED): flattened, ellipsoid seeds (12–15 mm long; 6–10 mm wide; c. 3.5 mm thick), one end (apex) obtusely pointed or rounded **[1]**, the other more angular **[2]**, sometimes slightly truncate with small, round hilum **[3]**. Seed coat brown or pale brown, smooth, with obvious fissure (marginal ridge) forming smooth ring **[4]** inside the perimeter of each flat surface. Seed coat (testa) texture hard; inner seed coat (tegumen) membranous and greyish-green. Cotyledons 2 yellowish-white and oily. *Odour:* indistinct. *Taste:* indistinct. ***Good quality:*** large, plump and oily.

PROCESSED DRUG (DRY-FRIED SEED 'CHAO GUA LOU ZI'): as for crude drug but slightly inflated and charred black or brown in places **[5]**.

PROCESSING

Seeds are removed from the fresh fruit and sun-dried (= crude drug) then pulverised prior to decocting. Sun-dried seeds are sometimes dry-fried until slightly charred (= Chao Gua Lou Zi 炒瓜蒌子).

OTHER COMMON NAMES

炒栝蒌仁 Chao Gua Lou Ren (dry-fried Gua Lou Ren); 炒瓜蒌仁 ChaoGua Lou Ren (dry-fried Gua Lou Ren); 栝楼 Gua Lou (plant name); 瓜蒌仁 Gua Lou Ren; Gua Lou Ren 栝蒌仁; Semen Trichosanthis; Semen Trichosanthis Tostum; trichosanthes (SCN); Trichosanthis Tostum Semen.

SAFETY

Safety Class 1; Interaction Class: A. (Gardner & McGuffin, 2013). Dry-frying the seeds is a common form of processing which claims to reduce any minor adverse effects such as nausea and vomiting (Bensky *et al.*, 2004; Chen & Chen, 2004).

T. rosthornii

Cucurbitaceae

DRUG MORPHOLOGY

CRUDE DRUG (DRIED SEED): similar to drug of PS1 but usually slightly longer and flatter (15–19 mm long; 8–10 mm wide; c. 2.5 mm thick). Seed coat brown and broadest end, with hilum, **more obviously truncate** **[6]**.

PROCESSED DRUG (DRY-FRIED SEED 'CHAO GUA LOU ZI'): as for crude drug but slightly inflated and charred black or brown in places **[7]**.

PROCESSING

As for PS1.

OTHER COMMON NAMES

炒栝蒌仁 Chao Gua Lou Ren (dry-fried Gua Lou Ren); 炒瓜蒌仁 Chao Gua Lou Ren (dry-fried Gua Lou Ren); 栝楼 Gua Lou (plant name); 瓜蒌仁 Gua Lou Ren; 栝蒌仁 Gua Lou Ren; Semen Trichosanthis; Semen Trichosanthis Tostum; trichosanthes (SCN); Trichosanthis Tostum Semen; Xi Lou Zi.

UNOFFICIAL SUBSTITUTES

The seed of many other species in *Trichosanthes* have been reported as local or historical unofficial substitutes (Bensky *et al.* 2004; NICPBP 1997–2002). Their occurrence as substitutes in international trade is uncertain:

1. ***T. cucumeroides*** (Ser.) Maxim. — 王瓜 Wang Gua
2. ***T. dunniana*** H. Lév. — 糙点栝楼 Cao Dian Gua Lou
3. ***T. fissibracteata*** C. Y. Wu ex C. Y. Cheng & C. H. Yueh — 裂苞栝楼 Lie Bao Gua Lou
4. ***T. lepiniana*** (Naudin) Cogn. — 马干铃栝楼 Ma Gan Ling Gua Lou
5. ***T. pilosa*** Lour. — 全缘栝楼 Quan Yuan Gua Lou (synonym: *T. ovigera* Blume)
6. ***T. rubriflos*** Thorel ex Cayla — 红花栝楼 Hong Hua Gua Lou
7. ***T. truncata*** C. B. Clarke — 大子栝楼 Da Zi Gua Lou

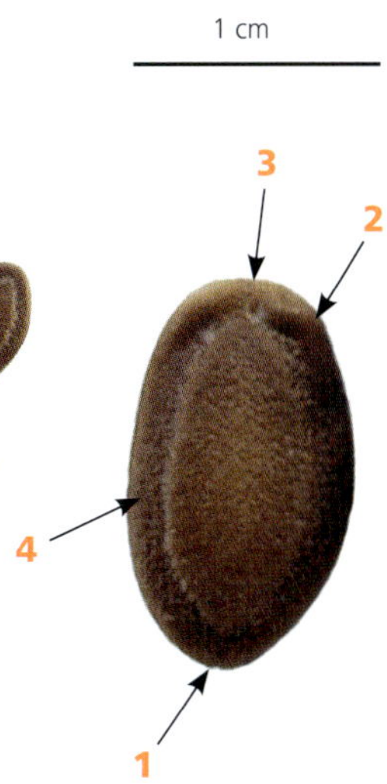

Crude drug of Gua Lou Zi (*Trichosanthes kirilowii*)

Processed (dry-fried) drug of Gua Lou Zi (*T. kirilowii*)

Crude drug of Gua Lou Zi (*T. rosthornii*)

Processed (dry-fried) drug of Gua Lou Zi (*T. rosthornii*)

Hei Zhi Ma 黑芝麻

Sesami Semen Nigrum

black sesame seed

Drug part: SEED

PHARMACOPOEIA SPECIES

Sesamum indicum L. (*black-seeded cultivar*)

Pedaliaceae

synonym: *Sesamum orientale* L.

Species of *Sesamum* wild in China: 1

PLANT DESCRIPTION

An upright, branched or unbranched, annual herb (up to 1.2 m tall). *Stems* 4-angled and covered in fine hairs. *Leaves* simple, opposite or alternate, becoming smaller up the stem; stipules absent; leaf blades lance-shaped to ovate, sometimes 3-parted (4–20 cm long; 2–10 cm wide), densely covered in mucilaginous glands; margins entire. *Flowers* in leaf axils, solitary, often closely spaced, large and conspicuous; flower-stalk short and with nectaries; calyx (4 or) 5-parted; petals white, pink or violet-pink with darker markings and fused to form an oblique trumpet (1.5–3.3 cm long) opening to 5 short, overlapping lobes; lowermost lobe longest; stamens 4 (sometimes 2); ovary superior (rarely inferior) 2- or 4-locular; style slender and thread-like; stigma 2-lobed. *Fruit* a dry, narrowly oblong capsule, rounded at the base (1.5–3 cm long; 6–7 mm diam.); 4-parted and dehiscing longitudinally. *Seeds* many, compressed, winged, fringed; rarely smooth. Fl. and ftg. dependent on planting time.

HARVESTING, SOURCING AND NATURAL RANGE

Ripe fruits are harvested entirely from cultivated plants in autumn, sun-dried and naturally dehisced seeds collected. The species is only known from cultivation in China where it is widely grown for food and medicine (native origin uncertain).

CONSERVATION STATUS

China: Only known from cultivation.
Global: Not Evaluated (IUCN Red List accessed 25/6/2016). Native origin uncertain.

UNOFFICIAL SUBSTITUTE

Linum usitatissimum L.

Linaceae

synonym: *Linum humile* Mill.

PLANT DESCRIPTION

Easily distinguished from *Sesamum indicum*. *Main differences*: a delicate-looking herbaceous annual, branching apically; *leaves* alternate, linear to lance-shaped and **much smaller (2–4 cm long; 1–5 mm wide);** flowers **with 5 spreading equal-sized petals (not fused into a trumpet),** (0.8–1.2 cm diam.), **petals usually blue**, rarely white or red, *fruit* **globose (6–9 mm diam.)**, hairless, **base enclosed in a 5-parted calyx.** Fl. Jun–Aug, fr. Jul–Oct.

HARVESTING, SOURCING AND NATURAL RANGE

Fruits are harvested entirely from crop plants in autumn, when fruits are ripe (brown), then threshed to release the seed. The species has been cultivated in China for millennia and is grown throughout except in Hainan and Taiwan. Many cultivars exists. The species originates from the Mediterranean.

CONSERVATION STATUS

China: not native; cultivated and occasionally naturalised.
Global: Not Evaluated (IUCN Red List, accessed 8/9/2016).

TCM PROPERTIES AND USE

properties: **'sweet and neutral'**
channels: **'Liver and Kidney'**
actions: **'tonifies yin, jing and Blood, moistens dryness'**

indications: **dizziness, blurred vision, tinnitus; problem with lactation; constipation**

Sesamum indicum — flowering plant

S. indicum — immature fruiting spike

S. indicum with ripe fruits ready for harvesting

Linum usitatissimum — flowering plants

L. usitatissimum — flowering and fruiting plants

L. usitatissimum — close-up of flower

L. usitatissimum — with developing fruit

Sesamum indicum — Pedaliaceae

DRUG MORPHOLOGY

CRUDE DRUG/DECOCTING PIECES-1 (DRIED SEED): laterally compressed and oval in outline (c. 3 mm long; 2 mm wide; 1 mm thick). **Seed coat black**, thin and smooth or with netted (reticulate) wrinkles [hl] **[1]**, often with a thin raised ridge around the longitudinal perimeter of one or both surfaces **[2]**; a pale, ovoid, dot-like hilum visible at the shortly pointed end **[3]**. Internally 2 white, oily cotyledons attached to a short embryonic stalk. *Odour*: indistinct; aromatic when crushed. *Taste:* indistinct; fragrant when chewed. Only the black-seeded cultivar is used medicinally in China; the white-seeded cultivar Bai Zhi Ma (白芝麻) is used for culinary purposes. ***Good quality*:** plump, black and aromatic.

DECOCTING PIECES-2 (DRY-FRIED SEED — CHAO HEI ZHI MA): as for crude drug/decocting pieces-1 but slightly swollen and cracked **[4]**; hilum blackened; and aromatic.

PROCESSING

Seeds are sun-dried (= crude drug/decocting pieces-1). Subsequently dry-frying may be performed until the seed cracks open 'Chao Hei Zhi Ma' (decocting pieces-2).

OTHER COMMON NAMES

胡麻 Hu Ma; 胡麻仁 Hu Ma Ren; 胡麻子 Hu Ma Zi; Semen Sesami Nigrum; Sesame (SCN, CP2010); 芝麻 Zhi Ma (plant name).

SAFETY

Safety Class A; Interaction Class: A (Gardner & McGuffin, 2013).

Linum usitatissimum — Linaceae

'DRUG SUBSTITUTE' MORPHOLOGY

Seed of this species is easily distinguished from that of the Pharmacopoeia species. *Main differences*: seed often slightly longer (4–6 mm long; 2–3 mm wide; 0.75–1.5 mm thick); seed coat **yellow** or **brown to reddish-brown** and **glossy [5]; narrow end with a laterally bent beak [6]** and from which a pale linear raphe extends towards the other end of the seed **[7]**.

OTHER NAMES

胡麻仁 Hu Ma Ren; 亚麻子 Ya Ma Zi; flax (SCN); Lini Semen; linseed; Semen Lini.

SAFETY

Safety Class A; Interaction Class: A.(Gardner & McGuffin, 2013). As a bulk-forming laxative, adequate fluids need to be taken at the same time (Gardner & McGuffin, 2013).

SUBSTITUTION COMMENT. Owing to a shared common name (Hu Ma Ren 胡麻仁), seed of this unofficial substitute (more commonly known as Ya Ma Zi 亚麻子) is occasionally traded in error. 'Ya Ma Zi' is linseed (or flax seed) (*Linum usitatissimum* L., family Linaceae) and is the source of a separate TCM drug (Lini Semen), see page 750. The two species are unrelated and their seed are easily distinguished macroscopically.

dried seed (decocting pieces-1)

dry-fried seed (decocting pieces-2)

Hei Zhi Ma crude drug/decocting pieces-1 (dried seeds of *Sesamum indicum*)

S. indicum

L. usitatissimum

golden cultivar

bronze cultivar

Ya Ma Zi, seed of *Linum usitatissimum*

Hua Ju Hong 化橘红

Citri Grandis Exocarpium

pomelo peel

Drug part: PEEL

PHARMACOPOEIA SPECIES 1

Citrus maxima (Burm.) Merr. **'Tomentosa'**

Rutaceae

synonym: *C. grandis* 'Tomentosa' (name used in CP2015).

Species and hybrids of *Citrus* wild in China: 11 (+ 5 hybrid cultivated species)

PLANT DESCRIPTION

A small evergreen tree whose young branches, lower leaf surfaces and flower stalks are covered in a fine mat of short, thin hairs. *Branches* often purplish. *Leaves* alternate and simple with a short and often conspicuously winged stalk (2–4 cm long; 0.5–3 cm wide or less) and articulated at point of attachment to the leaf blade; leaf blade broadly ovate or elliptic (9–16 cm long; 4–8 cm wide or larger), thick, dark green when mature, base rounded, tip rounded to broad (sometimes shortly pointed). *Flowers* axillary, solitary or in groups (racemes). Flower-buds often purplish; calyx 3–5-lobed; petals 4–5, white, spreading, equal and lance-shaped (1.5–2 cm long); stamens 25–35; style long and thick. *Fruit* (pomelo) fleshy, globose, squat or pear-shaped (usually 10 cm diam. or more); peel yellow or yellowish–green, **densely covered in a short mat of hairs**, dotted with sunken oil ducts; fruit wall (pith) broad, creamish-white and spongy; pulp (flesh) usually with 10–15 segments, white, pink, reddish or pale yellow. *Seeds* up to 200 or seedless; seeds irregularly shaped and ridged. Fl. Apr–May, fr. Sep–Dec.

HARVESTING, SOURCING AND NATURAL RANGE

Unripe or almost ripe fruits are harvested in summer from widely cultivated and sometimes naturalised trees in south China; the traditional *Di Dao* source is the Huazhou district of Guangdong province. The species is not native to China but is thought to originate from SE Asia.

CONSERVATION STATUS

China: not native.

Global: Not Evaluated (IUCN Red List, accessed 30/6/2015).

PHARMACOPOEIA SPECIES 2

C. maxima (Burm.) Merr.

Rutaceae

synonyms: *C. grandis* (L.) Osbeck (name used in CP2015); *Aurantium maximum* Burm.

PLANT DESCRIPTION

As for PS1 but **fruit surface hairless** and pith somewhat thinner.

HARVESTING, SOURCING AND NATURAL RANGE

As for PS1.

CONSERVATION STATUS

As for PS1.

TCM PROPERTIES AND USE

properties: **'pungent, bitter and warm'**
channels: **'Lung and Spleen'**
actions: **'invigorates qi, resolves damp, transforms phlegm**

indications: **nausea and vomiting; cough with profuse sputum**

Young leaves of *Citrus maxima* 'Tomentosa' showing winged stalks; inset: fruiting branch

Fruiting tree of *C. maxima* 'Tomentosa'

Unripe fruits of *C. maxima* 'Tomentosa' (left) and *C. maxima* (right) ready for processing

Leaves of *C. maxima*; inset: fruit ready for harvest

Crop of *C. maxima*, Guangxi province; inset: flower

Citrus maxima 'Tomentosa'

Rutaceae

DRUG MORPHOLOGY

CRUDE DRUG (DRIED PEEL OF MAO JU HONG): peel either folded in half towards the inner surface with lobes arranged alternately **[1]** or flattened to form a star-shape (the latter 15–28 cm diam.; 2–5 mm thick). Each lobe is traditionally described as 'willow-shaped'. External surface yellowish-green **[2]**, densely covered in a short mat of hairs **[3]** and dotted throughout with oil ducts **[4]**; internal surface yellowish-white or pale yellowish-brown, dotted with oil ducts and veined **[5]**. *Texture*: easily broken. *Fracture*: uneven with a row of oil ducts at the outermost edge **[6]**; pith slightly soft and elastic. *Odour*: aromatic. *Taste*: bitter and somewhat pungent. ***Good quality*:** peel with a uniform thickness and intensely aromatic.

DECOCTING PIECES (DRIED SLICED PEEL OF MAO JU HONG): curled slivers of peel (up to c. 6 cm long; 0.2–0.5 cm wide; up to 0.7 cm thick). External peel surface yellowish-green, densely covered in a short mat of hairs and dotted throughout with oil ducts **[2]**. Cut surface showing one row of oil ducts at the outermost edge **[6]** with pith yellowish-white or pale yellowish-brown **[5]**. *Texture*: slightly soft and elastic. *Odour*: aromatic. *Taste*: bitter and somewhat pungent.

PROCESSING

Fresh fruits are blanched, then the peel (exocarp) is scored into five or seven equal-sized lobes, the flesh (pulp) removed and the peel flattened and folded in a traditional style then dried (= crude drug). Peel sourced from this drug's *Di Dao* location may be tied in a traditional style into bundles for cosmetic purposes to reflect its higher value. Decocting pieces are made by washing and thoroughly softening the peel with water, then cutting into slivers or small pieces before drying again.

OTHER COMMON NAMES

副毛化橘红 Fu Mao Hua Ju Hong; 化州毛红 Hua Zhou Mao Hong; 化州橘红 Hua Zhou Ju Hong; 毛橘红 Mao Ju Hong (hairy Ju Hong); 毛七爪 Mao Qi Zhua; 正毛化橘红 Zheng Mao Hua Ju Hong; Exocarpium Citri Grandis; pomelo; pummelo peel (CP2010); shaddock.

SAFETY

An official food and healthfood in China (Anon., 2002a,b).

C. maxima

Rutaceae

DRUG MORPHOLOGY

CRUDE DRUG (DRIED PEEL OF GUANG JU HONG): as for PS1 above but outer surface **yellowish-green to yellowish-brown [7]** and **hairless [8].**

DECOCTING PIECES (DRIED SLICED PEEL OF GUANG JU HONG): as for PS1 but **peel surface yellowish-green to yellowish-brown and hairless [7]**.

Macroscopic characters are normally considered sufficent to differentiate this species from its cultivar above although for decocting slices this may prove more challenging.

PROCESSING

As for PS1.

OTHER COMMON NAMES

光橘红 Guang Ju Hong; 光七爪 Guang Qi Zhua; 光五爪 Guang Wu Zhua; 青光橘红 Qing Guang Ju Hong; 柚皮橘红 You Pi Ju Hong; Exocarpium Citri Grandis; pummelo peel (CP2010); shaddock.

SAFETY

As for PS1.

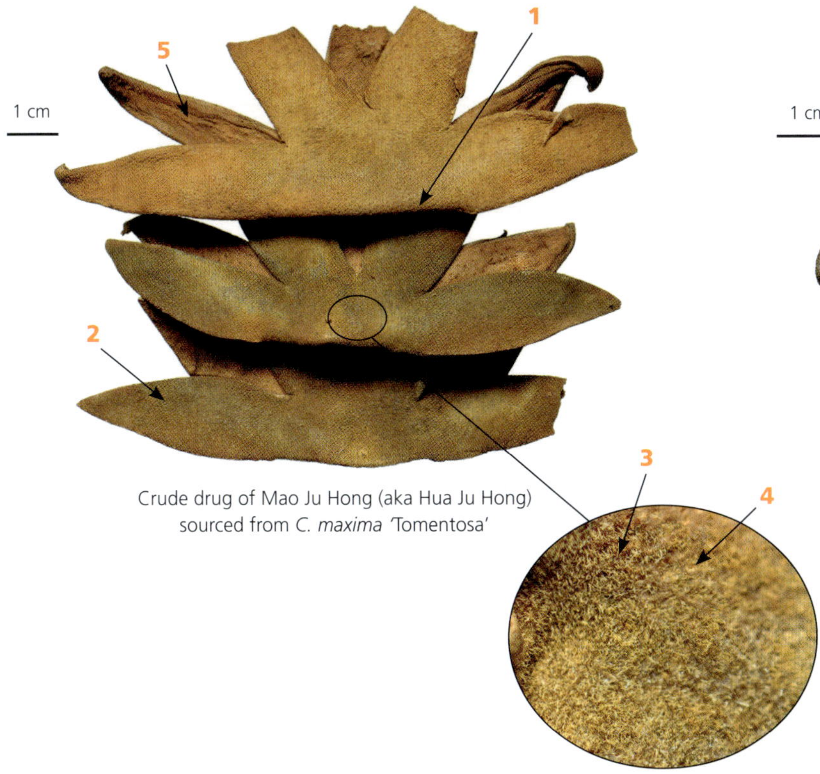

Crude drug of Mao Ju Hong (aka Hua Ju Hong)
sourced from *C. maxima* 'Tomentosa'

Decocting slivers of Hua Ju Hong (aka Mao Ju Hong)
sourced from *C. maxima* 'Tomentosa'

Crude drug of Hua Ju Hong (aka Guang Ju Hong)
sourced from *C. maxima*

Decocting pieces of Hua Ju Hong (aka Guang Ju Hong)
sourced from *C. maxima*

Hu Lu Ba 胡卢巴

Trigonellae Semen

fenugreek seed

Drug part: SEED

Trigonella foenum-graecum L.

Fabaceae

Species of *Trigonella* wild in China: 6

PLANT DESCRIPTION

An upright and highly aromatic annual (30–80 cm tall) with a taproot. *Leaves* divided into 3 obovate to oblong leaflets (trifoliolate), veins obvious (in 5 or 6 lateral pairs) and margins shallowly toothed. *Flowers* single or in pairs and tucked between the leaves towards the top of each plant; individual flowers 5-petalled, bilaterally symmetrical (c. 1 cm long), creamish-yellow, flushed violet basally, stalkless and never fully open (the standard petal somewhat laterally compressed and enveloping the other four petals). *Fruit* a dry pod (c.11 cm long), straight or slightly curved with a long beak up to 2 cm long; seeds 10–20 per fruit. Fl. Apr–Jul, fr. Jul–Sep.

HARVESTING, SOURCING AND DISTRIBUTION

The species is not native in China but widely cultivated throughout for medicinal and many other uses; the main production areas are Anhui, Sichuan and Henan provinces. Aerial plant parts are harvested in summer, when the fruits have ripened.

The species is native to the Himalayas and south-west Asia. In Mediterranean countries it has been cultivated since ancient times as a fodder and, today, is widely grown in India and neighbouring countries, north Africa and west Asia as a fodder, spice and pot-herb.

CONSERVATION STATUS

China: not native; only known from cultivation.
Global: 'not threatened' (Euro+Med Plantbase, accessed 22/7/16); Not Evaluated (IUCN Red List, accessed, 22/7/2016).

Cultivated plot of *Trigonella foenum-graecum*, Hebei province

Flowering plants of *T. foenum-graecum*

Fruiting plants of *T. foenum-graecum*

TCM PROPERTIES AND USE

***properties*: 'bitter and warm'**
***channel*: 'Kidney'**
***actions*: 'tonifies yang, dispels damp cold, relieves pain'**

***indications*: abdominal pain**

DRUG MORPHOLOGY

CRUDE DRUG/DECOCTING PIECES-1 (DRIED SEED): rhomboid or rectangular (3–4 mm long; 2–3 mm wide; 2 mm thick). Externally pale yellowish-brown, yellowish-green or dark brown, and smooth. An obvious longitudinal furrow runs obliquely along either side **[1]**, joining at a dot-like hilum **[2]**. Internally two seed leaves (cotyledons) are visible along with a curved and plump root shoot (radicle); endosperm pale yellow, translucent and mucilaginous. *Texture*: hard and seed coat thin. *Odour*: aromatic. *Taste*: slightly bitter. ***Good quality***: large, plump, heavy seed.

DECOCTING PIECES-2 (SALTED AND DRY-FRIED SEED): as for above but darker (yellowish-brown to brown), sometimes split and charred in places. *Taste*: slightly salty.

Crude drug/decocting pieces-1
(dried seed of *Trigonella foenum-graecum*, dark brown cultivar)

PROCESSING

Aerial plant parts are sun-dried then threshed to release the seed before they are washed and dried (= crude drug/decocting pieces-1). Seeds may undergo further processing by briefly immersing in salt-water before being dry-fried until they split and become aromatic (Yan Chao Hu Lu Ba 盐炒葫卢巴 'salted and dry-fried fenugreek seed' (= decocting pieces-2).

OTHER COMMON NAMES

胡巴子 Hu Ba Zi; common fenugreek seed (CP2010); fenugreek (SCN); Semen Trigonellae. Many other trade names exist for this drug reflecting its use as a medicine and culinary herb in other countries (see *Kew-MPNS*).

SAFETY

Seeds are used worldwide as a natural source of food flavouring, occasionally with dose restrictions (e.g. USA recommends <0.05%) (Davidson, 1999; Barnes *et al.*, 2007).

Salt-processed seed (Yan Chao Hu Lu Ba) of *T. foenum-graecum*

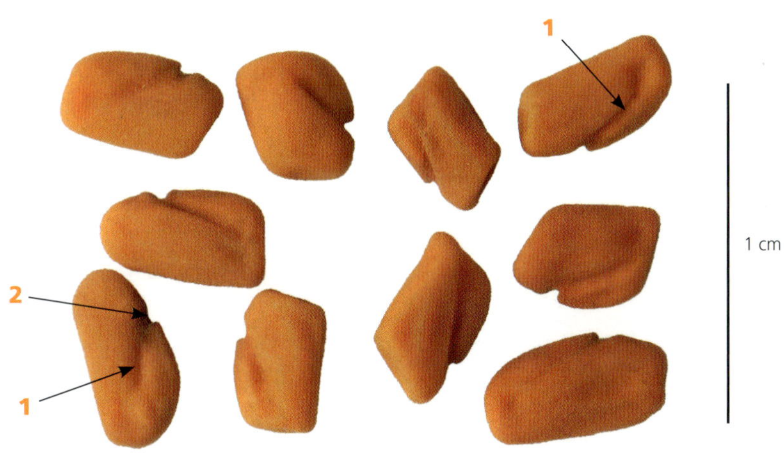

Hu Lu Ba close-up (yellowish-brown cultivar)

Huo Ma Ren 火麻仁

Cannabis Fructus

'hemp seed'

Drug part: FRUIT

Cannabis sativa L.

Cannabaceae

synonyms: *C. indica* Lam.; *C. sativa* var. *indica* (Lam.) Wehmer

Species of *Cannabis* wild in China: 1

PLANT DESCRIPTION

Owing to its long cultivation in China and worldwide many cultivars exist; there is much variation therefore in the morphological appearance of this plant.

A strong-smelling, upright, annual herb (1–3 m tall), usually with male and female flowers on separate plants. *Stems* furrowed. *Leaves* alternate or opposite, palmately compound (divided from a central point into 3–11 finger-like leaflets); leaflet margins saw-toothed. *Flowers* tiny and inconspicuous: male flowers yellowish-green and nodding, in loose clusters; female flowers green (without true petals) and crowded in upper leaf axils among leaf-like bracts. *Fruits* (colloquially 'seed') in small tight clusters; each fruit a dry single-seeded achene covered by a membranous calyx. Fl. May–Jun, fr. Jul.

HARVESTING, SOURCING AND NATURAL RANGE

Ripe fruit are harvested in autumn using cultivated plants or those naturalised throughout China. Long cultivation of this species in China and worldwide makes its original wild distribution uncertain; the *Flora of China* states that it probably originated in Central Asia.

CONSERVATION STATUS

China: native status uncertain.
Global: native status uncertain.

Leaves of *Cannabis sativa*

Nodding male flowers

Cultivated plants, Shaanxi province, China

Fruiting plant

TCM PROPERTIES AND USE

properties: 'sweet and neutral'
channels: 'Spleen, Stomach and Large Intestine'
actions: 'moistens intestines and dryness'

indications: constipation; topically for hair loss and dry hair

DRUG MORPHOLOGY

CRUDE DRUG/DECOCTING PIECES-1 (DRIED FRUIT – often erroniously called seeds): ovoid (egg-shaped) fruits (4–5.5 mm long; 2.5–4 mm diam.) with a perimeter keel **[1]**. Externally fruit wall (pericarp) slightly lustrous, green, brown, greyish-yellow or brownish-green, finely netted (reticulated) **[2]** [hl]; pericarp thin and brittle with a pale, rounded, slightly sunken hilum at the base **[3]** and a small pit (micropyle) at the shortly pointed apex **[4]**. Internally, one seed (virtually fused to fruit wall) with a large horseshoe-shaped embryo, large cotyledons and a long radical (all greyish-white, fleshy and oily). *Odour*: indistinct. *Taste*: mild. ***Good quality***: fruits large, yellowish-grey and plump.

DECOCTING PIECES-2 (DRY-FRIED FRUIT): as above but fruit contents yellowish and, when freshly processed, aromatic.

PROCESSING

Ripe fruit are sun-dried (= crude drug/decocting pieces); subsequent dry-frying produces 炒火麻仁 Chao Huo Ma Ren.

OTHER COMMON NAMES

大麻 Da Ma (plant name); 大麻仁 Da Ma Ren; 麻仁 Ma Ren; Fructus Cannabis; hemp (SCN, as subsp. *indica* and subsp. *sativa*); hemp fruit; hemp seed (CP2010).

SAFETY

Huo Ma Ren 'hemp seed' is an official food and healthfood in China (Anon., 2002a,b) and an increasingly popular healthfood in many other countries. A recent study (Sarmento *et al*., 2015) concluded that "*In general, the scientific community finds no health risks from the extremely low doses of THC* [the main psychoactive principle in this species] *present in hemp foods and body care products*" [i.e. those derived from hemp 'seed'].

Crude drug/decocting pieces of Huo Ma Ren (dried ripe fruit of *Cannabis sativa*)

Ji Li 蒺藜 (or Bai Ji Li 白蒺藜)

Tribuli Fructus

Drug part: FRUIT

Tribulus terrestris L.

Zygophyllaceae

Species of *Tribulus* wild in China: 2

PLANT DESCRIPTION

A prostrate, herbaceous annual, hairless or covered in stiff or soft hairs. *Branches* 20–60 cm long. *Stipule* (at base of leaf), green and distinct (4–6 mm long). *Leaves* opposite (1.5–5 cm long) and divided into 6–16 pairs of opposite leaflets; leaflet blades oblong to obliquely-oblong (5–10 mm long; 2–5 mm wide), margins entire, often with long white hairs; tips blunt or pointed. *Flowers* solitary, in leaf axils, with **flower stalk shorter than length of the pinnate leaf**; each flower bisexual (c. 1 cm diam.); sepals 5; petals 5, yellow, over-lapping, of equal size and shape; stamens 10, with glands; stigma 5-parted. *Fruit* dry, hard and distinctly spiny (4–6 mm diam.) with 5 carpels; each carpel with 2 pairs of symmetrically arranged hard spines, one large pair (4–6 mm long) and one shorter pair (c. 2 mm long) and arising either side of a low dorsal ridge densely covered in much shorter spines. Fl. May–Aug, fr. Jun–Sep.

HARVESTING, SOURCING AND NATURAL RANGE

Plants are wild-harvested in autumn when fruits are ripe. The species is a common and widespread native growing throughout much of China on sandy soils and waste ground up to 3,300 m altitude. It is also widely naturalised in many countries and a noxious weed in the USA.

CONSERVATION STATUS

China: Least Concern (MEPCAS, 2013).
Global: Not Evaluated (IUCN Red List, accessed 18/8/2016).

Tribulus terrestris — fruiting stem

T. terrestris — flower and leaves

T. terrestris — typical prostrate habit

UNOFFICIAL SUBSTITUTE

Tribulus cistoides L. (Da Hua Ji Li 大花蒺藜). A common and closely related species with a much smaller distribution than the Pharmacopoeia species. It grows wild in SC China (Hainan, Hunan, Taiwan and Yunnan). This species is easily confused with the Pharmacopoeia species and is likely to be used inadvertently. Whole plants distinguished by the length of their flower stalks (equal to or longer than each pinnate leaf).Fruits are morphologically indistinguishable.

TCM PROPERTIES AND USE

***properties*:** **'bitter, pungent and neutral'**
***channel*:** **'Liver'**
***actions*:** **'calms Liver, extinguishes wind, regulates qi, resolves stasis, dispels wind'**

***indications*:** **headaches, blurred vision, hypertension; abdominal distension; irregular menstruation; insufficient lactation; painful eyes; itchy skin**

DRUG DESCRIPTION

CRUDE DRUG AND DECOCTING PIECES-1 (DRIED FRUIT): ***unthreshed fruit* (*crude*):** externally greenish-white or yellowish-white (7–12 mm diam.), conspicuously covered in hard, straight very sharp spines of varying lengths. Fruit-stalk (if present) straight (c. 3–5 mm long) **[1]**. Intact fruits (schizocarps) **[2]** with 5 fused mericarps (carpels); each mericarp with 2 pairs of symmetrically arranged spines (larger pair 4–6 mm long **[3]**; shorter pair c. 2 mm long **[4]**) arising either side of a low dorsal ridge **[5]** covered in much shorter spines. *Fruit* sometimes splitting into separate mericarps (3–6 mm long) with the inner-facing mericarp surface white or yellowish-white with reticulations. *Texture:* hard and sharply spiny. *Odour:* indistinct. *Taste:* bitter and pungent.
***Threshed fruit* (*decocting pieces-1*):** slightly smaller and less spiny than unthreshed fruit **[6]** with blunted or broken spines **[7]**; yellowish-white. Intact fruit often split into individual mericarps **[8]**. *Texture:* hard and shortly spiny or strongly nodular (according to degree of threshing); overall surface roughly sculptured. *Odour:* indistinct. *Taste:* bitter and pungent. ***Good quality*:** hard and greyish-white.
DECOCTING PIECES-2 (DRY-FRIED): as for crude drug but pale yellow to brownish-yellow **[9]**. Fruits usually split into mericarps. *Odour:* slightly aromatic.

PROCESSING

Plants are sun-dried then beaten to release the fruits which are then usually threshed to remove the spines (= crude drug/ decocting pieces-1). Fruits may undergo subsequent dry-frying to produce 'Chao Ci Ji Li' (= decocting pieces-2).

OTHER COMMON NAMES

白蒺藜 Bai Ji Li 'white Ji Li'; 刺蒺藜 Ci Ji Li 'prickly Ji Li'; caltrop fruit; puncture vine; puncturevine caltrop fruit; tribulus (SCN).
The trade names 'Bai Ji Li' and 'Ci Ji Li 'are occasionally used to avoid confusion with the name 'Sha Yuan Ji Li' which is one of the names given a separate TCM drug sourced from the seed of *Astragalus complanatus* and more commonly known as 'Sha Yuan Zi'.

SAFETY

Slightly toxic (CP2015). Safety Class 2b; Interactions Class: A. Gardner & McGuffin (2013).

Crude drug of Ji Li — unthreshed fruit (schizocarps) of *Tribulus terrestris*

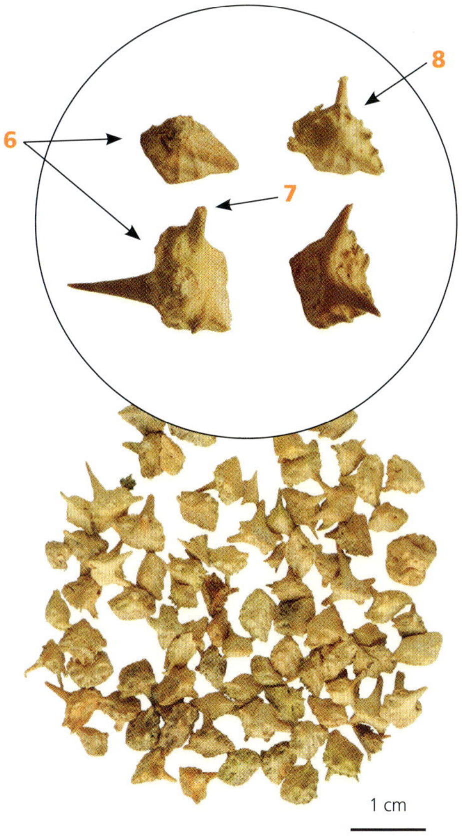

Decocting pieces-1 of Ji Li — threshed fruit (mericarps) of *T. terrestris*

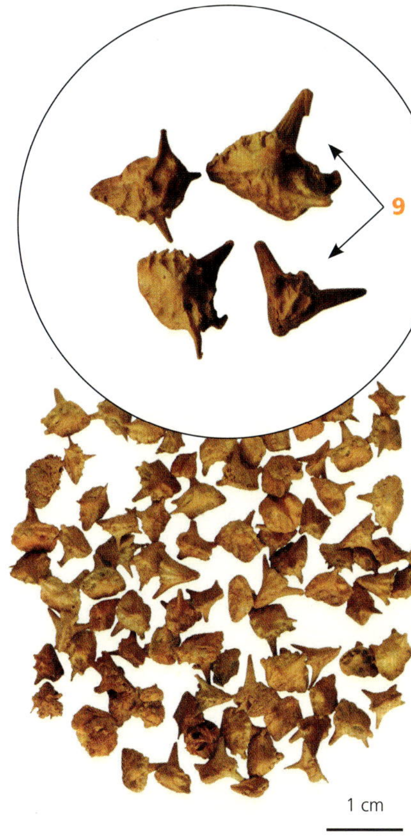

Decocting pieces-2 of Ji Li — dry-fried & threshed fruit (mericarps) of *T. terrestris*

Jin Deng Long 锦灯笼 Physalis Calyx seu Fructus

Drug parts: CALYX or CALYX WITH FRUIT

Physalis alkekengi L. var. ***franchetii*** (Mast.) Makino

Solanaceae

synonyms: *Alkekengi officinarum* Moench var. *franchetii* (Mast.) R. J. Wang (*Kew-MPNS* accepted name,11/3/2015); *Physalis franchetii* Mast.

Species of *Physalis* wild in China: 6

PLANT DESCRIPTION

An erect and hairless, perennial herb (40–80 cm tall) with a rhizome. *Stems* angular, slightly woody at base. Stipules absent. *Leaves* alternate or opposite, simple; blade narrowly to broadly ovate (5–15 cm long; 2–8 cm wide), base oblique, margin entire or coarsely toothed, tip pointed; almost hairless. *Flowers* solitary and bisexual, in leaf axils; calyx 5-lobed; petals fused to form a shallowly bell-shaped flower, slightly 5-lobed, white with distinct greenish or yellowish basal throat spots; stamens 5 and yellow; style 1. *Fruit* with persistent and conspicuously inflated ovate or rounded calyx or 'lantern' (3–4.5 cm long; 3–4 cm diam.), sparsely hairy, orange or orangish-red and membranous, with 10 (5 prominent) longitudinal ribs; fruit stalk persistent (2–3 cm long). 'Lantern' entirely enclosing a globose berry (1–1.5 cm diam.), orange, fleshy and many-seeded. Fl. May–Oct, fr. Jun–Nov.

HARVESTING, SOURCING AND NATURAL RANGE

Calyces ('lanterns') or calyces plus fruits are harvested in autumn when ripe (orange or reddish-orange), mostly from widely grown cultivated plants (often naturalised). This variety is wild in many parts of China and has become naturalised throughout (except Tibet). It grows in disturbed places often along forest margins between 800–2,500 m altitude; also wild in the Korean Peninsula. Widely grown in the West as an ornamental garden plant.

CONSERVATION STATUS

China: Least Concern (authors' pers. obs., 2010).
Global: Not Evaluated (IUCN Red List; accessed 19/8/2016).

Physalis alkekengi var. *franchetii* with solitary white flower and developing 'lanterns' below

Wild plant with ripe fruits

Freshly harvested Jin Deng Long with fleshy fruit removed, Hubei province

TCM PROPERTIES AND USE

properties: **'bitter and cold'**
channel: **'Lung'**
actions: **'clears heat, eliminate toxin, resolves phlegm'**

indications: **sore throat; dysuria; stranguria; (T)**

DRUG MORPHOLOGY

CRUDE DRUG/DECOCTING PIECES (DRIED CALYX sometimes including fleshy berry): calyx ('lantern') either intact **[1]** and somewhat crumpled or broken **[2]**, membranous, bright orange or orangish-red (3–4.5 cm long; 3–4 cm diam.), with 5 prominent longitudinal ribs **[3]**, the membranous surface between these finely reticulated **[4]**; apex shortly pointed **[5]**, often splitting into 5 lobes **[6]**; base rounded and dented **[7]**, often with persistent fruit stalk **[8]** (2–3 cm long). Fruit, when not removed, entirely enclosed by the 'lantern' (unless broken). Fruit a succulent globose berry (1–1.5 cm diam. unless squashed), orangish-red or brownish-red. *Seeds* many and reniform (1–2 mm diam.). *Texture (calyx)*: light and delicate. *Odour*: indistinct. *Taste*: bitter (calyx); sweet and sour (berry). ***Good quality*:** calyces large and orange or orangish-red.

PROCESSING

Calices with or without fruit are dried (= crude drug/decocting pieces).

OTHER COMMON NAMES

挂金灯 Gua Jin Deng; 红姑娘 Hong Gu Niang; 酸浆 Suan Jiang; Calyx Seu Fructus Physalis; Chinese lantern; Franchet groundcherry fruit (CP2010).

Crude drug/decocting pieces of Jin Deng Long

Jing Jie Sui 荆芥穗

incl. Jing Jie Sui Tan 荆芥穗炭

Schizonepetae Spica

incl. Schizonepetae Spica Carbonista

Drug part: FRUITING HEAD

Nepeta tenuifolia Benth.

Lamiaceae

synonym: *Schizonepeta tenuifolia* (Benth.) Briq. (name used in CP2015 as '*S. tenuisfolia* Briq.').

Species of *Nepeta* wild in China: 42

PLANT DESCRIPTION

An aromatic, annual herb (30–100 cm tall). *Stems* 4-sided, with many opposite branches; softly, greyish-white hairy. *Leaves* opposite, stalked and deeply 3- or 5-lobed with middle lobe longest; lobes lance-shaped, margin entire, tip pointed (entire leaf in outline c. 1–3.5 cm long; 1.5–2.5 cm wide); leaf blade glandular, lower surface greyish-green. *Flowers* many, in terminal spikes (2–13 cm long), in dense whorls (verticillasters), each formed from 2 virtually opposite clusters each cluster with a leaf-like bract and minute linear bracteoles; clusters towards apex closely abutting. Calyx tubular or bell-shaped, with 5 narrow triangular teeth; flowers tiny, pale violet, softly hairy and 2-lipped (upper lip 2-lobed, flat or concave; lower lip large 3-lobed with middle lobe larger); stamens 4; ovary superior and 4-lobed; style with 2-cleft tip. *Fruit* enclosed within persistent calyx, with 4 nutlets, these dry brown, 3-angled and oblong (c. 1.5 mm long; 0.7 mm diam.). Fl. Jul–Sep, fr. Aug–Oct.

NOTE — not to be confused with *Nepeta tenuiflora* Diels, a **perennial** herb up to 160 cm high, with opposite, oblong-lanceolate, serrated leaves (**not divided into lobes**), and **larger (1.3–1.5 cm) purplish-blue** flowers.

For other TCM drugs sourced from this species see 'Jing Jie' (Schizonepetae Herba) page 356.

HARVESTING, SOURCING AND NATURAL RANGE

Fruiting heads harvested when green in summer and autumn. Mostly sourced from medicinal crops (some GAP-Certified) grown in the Chinese provinces of Fujian, Henan, Hebei, Jiangsu, Yunnan and Zhejiang).

The species is wild in N, SW and C China (Gansu, Guizhou, Hebei, Heilongjiang, Henan, Liaoning, Qinghai, Shaanxi, Shanxi and Sichuan) and grows in forest margins and valleys (500–2,700 m altitude); [also native in Korean Peninsula].

CONSERVATION

China: Least Concern (MEPCAS, 2013); Least Concern (authors' pers. comm. 2011).

Global: Not Evaluated (IUCN Red List; accessed 24/3/2015).

Medicinal crop of *Nepeta tenuifolia*

N. tenuifolia flower-heads in bud

N. tenuifolia flower-heads in bud

Mature lower leaves of *N. tenuifolia*

TCM PROPERTIES AND USE

properties: **'pungent and slightly warm'**
[Jing Jie Sui Tan: 'pungent, sour and slightly warm']
channels: **'Lung and Liver'**
action: **'dispels wind cold'**
[Jing Jie Sui Tan: 'stops bleeding']

indications: **respiratory tract infection; headache; sores and abscesses**
[Jing Jie Sui Tan: haematochezia; metrorrhagia; postpartum faintness]

DRUG MORPHOLOGY

CRUDE DRUG/DECOCTING PIECES (FRUITING HEADS): heads cylindrical (3–15 cm long; ca. 7 mm diam.) with fruits in dense whorls (verticillasters) **[1]**; short sections of stem **[2]** often visible between whorls in lower part of spike; stems 4-sided (2–4 mm diam. with soft greyish-white hairs; each whorl with 2 indistinctly opposite fruit clusters **[3]** each with a green leaf-like bract **[4]** often still attached basally; bracts becoming smaller towards top of spike. Occasional flower remnants **[5]** may persist, tiny, pale violet and softly hairy (c. 4.5 mm long). Each fruit **[6]** with a persistent, papery and hairy, bell-shaped calyx (c. 3 mm long; 1.2 mm diam.), greenish-grey or yellowish-green, fragile in texture and with 5 sepal teeth [hl] **[7]**; calyx totally enclosing a group of 4 long-ovoid nutlets (c. 1.5 mm long; 0.7 mm diam.), dry, brown and 3-angled **[8]**. *Odour*: strongly aromatic. *Taste*: somewhat astringent, pungent and cool. ***Good quality*:** pale green with densely packed fruiting heads; strongly aromatic.

PROCESSED DECOCTING PIECES-2 (CARBONISED FRUITING HEADS): as for above but externally charred black **[9]** and internally yellow. *Odour*: charred. *Taste*: bitter and pungent.

PROCESSING

Fruiting heads are sun-dried (= crude drug/decocting pieces-1). These may subsequently be carbonised by dry-frying until charred black externally and yellow internally ('Jing Jie Sui Tan' = processed (carbonised) decocting pieces-2).

OTHER COMMON NAMES

裂叶荆芥 Lie Ye Jing Jie; fineleaf Schizonepeta spike (CP2010); Spica Schizonepetae.

Jing Jie Sui crude drug/decocting pieces

Nutlets

Processed (carbonised) fruit-heads (*N. tenuifolia*)

Ju He 橘核

Citri Reticulatae Semen

tangerine seed

Drug part: SEED

Citrus × reticulata Blanco

Rutaceae

taxonomic/nomenclatural note: in CP2015 as '*C. reticulata* Blanco'; *Kew-MPNS* accepted name 7/9/2016, *C.* × *aurantium* L. but *FoC* considers it to be a separate species. This drug is sourced from the same cultivars as those used for 'Chen Pi' (Citri Reticulatae Pericarpium) page 612.

Species and hybrids of *Citrus* wild in China: 11 (+ 5 hybrid cultivated species)

PLANT DESCRIPTION

See 'Chen Pi' (Citri Reticulatae Pericarpium) page 612.

HARVESTING, SOURCING AND NATURAL RANGE

Seeds are gathered as a by-product of ripe fruits harvested from cultivated trees used as the source of 'Chen Pi' (Citri Reticulatae Pericarpium) page 612.

See also other drugs sourced from this hybrid: 'Chen Pi' (page 612), 'Ju Hong' (page 666) and 'Qing Pi' (page 696).

CONSERVATION STATUS

As for 'Chen Pi' (Citri Reticulatae Pericarpium) page 612.

Crop of *Citrus* × *reticulata* 'Cha Chi', Guangdong province (source of Guang Chen Pi)

Ripe fruit of *C.* × *reticulata* 'Cha Chi' (a source of Chen Pi)

Ripe fruit of *C.* × *reticulata* 'Un Shui' (a source of Chen Pi)

TCM PROPERTIES AND USE

properties: 'bitter and neutral'
channels: 'Liver and Kidney'
actions: 'invigorates qi, resolves phlegm, disperses nodules'
indications: hernia, testicular, scrotal and mammary pain

DRUG MORPHOLOGY

CRUDE DRUG (DRIED SEED): sub-ovoid seed (8–12 mm long; 4–6 mm diam.), rounded at one end **[1]**, somewhat pointed or narrowing at the other with remnants of point of seed attachment **[2]**. External surface yellowish-white or pale greyish-white, widely reticulated to form an angled surface **[3]** with a thickened linear raised raphe running longitudinally **[4]**. Seed coat thin but strong; cotyledons 2, yellowish-green and oily. *Odour*: indistinct. *Taste*: bitter.

PROCESSED DRUG (SALT-PROCESSED DRIED SEED): as for the crude drug but externally darker and salty.

PROCESSING

Seeds are taken from ripe fruits, washed, then sun-dried (= crude drug). Optional processing may be performed: dried seeds are moistened thoroughly with salt-water then dry-fried (= 盐橘核 'salt-processed Ju He'). Seeds, whether salt-processed or not) are crushed prior to decocting .

OTHER COMMON NAMES

Semen Citri Reticulatae; tangerine seed (CP2010).

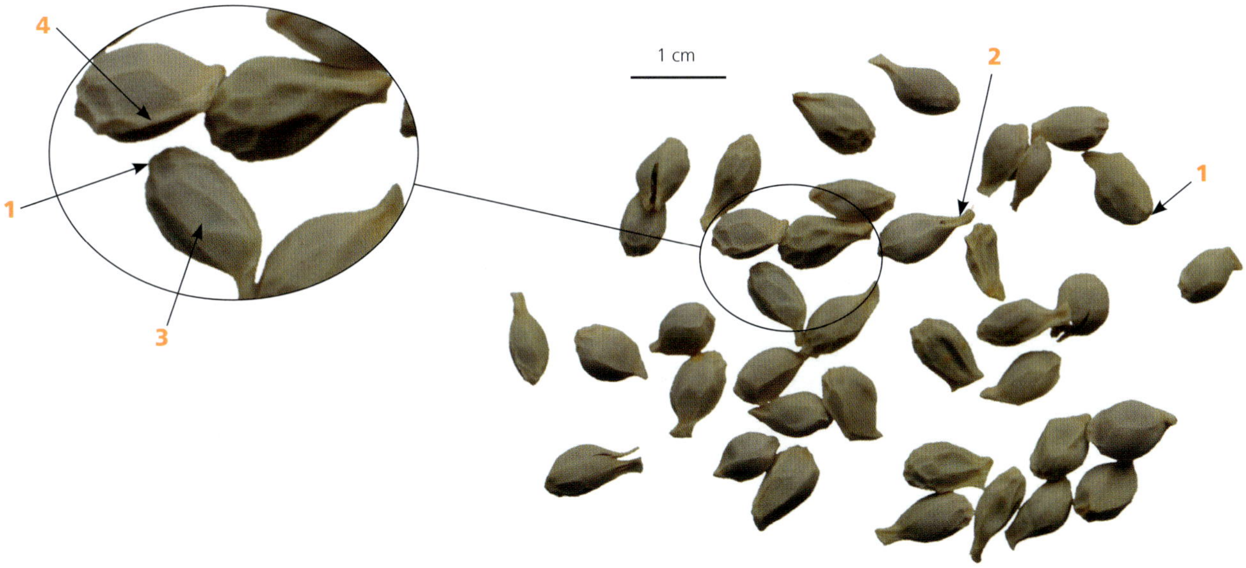

Crude drug of Ju He (*C.* × *reticulata* seed)

Salt-processed drug of Ju He (*C.* × *reticulata* seed)

Ju Hong 橘红

Citri Exocarpium Rubrum

red tangerine peel

Drug part: PEEL OF RIPE FRUIT

Citrus × reticulata Blanco

Rutaceae

taxonomic/nomenclatural note: in CP2015 as '*C. reticulata* Blanco'; *Kew-MPNS* accepted name 7/9/2016, *C. × aurantium*, but *FoC* considers it to be a separate species.

Species and hybrids of *Citrus* wild in China: 11 (+ 5 hybrid cultivated species)

PLANT DESCRIPTION

A small evergreen tree with many branchlets, occasional spines and fleshy fruits (tangerines/mandarin oranges). For full description see 'Chen Pi' (Citri Reticulatae Pericarpium) page 612.

See also other drugs sourced from this hybrid: 'Chen Pi' (page 612), 'Ju He' (page 664) and 'Qing Pi' (page 696)

HARVESTING, SOURCING AND NATURAL RANGE

Fruits are harvested when ripe (orange, yellow or reddish-orange) in late autumn and early winter from a wide variety of cultivars grown primarily for food. *Di Dao* sources are not known to exist for this drug.

CONSERVATION STATUS

As for 'Chen Pi' (Citri Reticulatae Pericarpium) page 612.

Grove of *Citrus × reticulata*, Guangdong province (S China)

Ripe crop of *C. × reticulata*

C. × reticulata fruits freshly harvested for medicinal use, Guangdong province

TCM PROPERTIES AND USE

***properties*:** **'pungent', 'bitter' and 'warm'**
***channels*:** **'Lung and Spleen'**
***actions*:** **'regulates qi, dries damp'**

***indications*:** **nausea, vomiting; cough with profuse sputum**

DRUG MORPHOLOGY

CRUDE DRUG (DRIED PEEL): irregularly shaped pieces of fruit peel (exocarp) (up to c. 8 cm long; c. 1 mm thick), margins often curled inwards **[1]**. External surface of peel orangish-red or yellowish-brown, darkening after storage; densely densely packed throughout with sunken oil ducts **[2]**. Inner surface yellowish-white with oil ducts many and conspicuous **[3]**. *Texture:* easily broken, fragile. *Odour:* aromatic. *Taste:* slightly bitter and numbing.

DECOCTING PIECES (DRIED, SLICED PEEL): as for the crude drug but in narrower strips (c. 1 cm wide).

PROCESSING

Fruits are peeled and the peel dried in the sun or shade (= crude drug) then cut into slices for decocting.

OTHER COMMON NAMES

川芸皮 Chuan Yun Pi; 芸皮 Yun Pi; Exocarpium Citri Rubrum; red tangerine peel (CP2010); tangerine (SCN).

SAFETY

An official food and healthfood in China (Anon., 2002a,b).

Crude drug of Ju Hong (sourced from *C.* × *reticulata*)

Decocting pieces of Ju Hong (sourced from *C.* × *reticulata*)

Ku Xing Ren 苦杏仁

Armeniacae Semen Amarum

Drug part: SEED

PHARMACOPOEIA SPECIES 1

Prunus armeniaca L.

Rosaceae

synonyms: *Armeniaca vulgaris* Lam. var. *vulgaris* (*FoC* accepted name); *P. armeniaca* var. *ansu* Maxim.

taxonomic/nomenclatural note: *Kew-MPNS* accepted name, 24/9/2016. The CP2015 recognises '*Prunus armeniaca* L. var. *ansu* Maxim.' and '*Prunus armeniaca* L.' as two separate sources of this drug; this present account follows the taxonomy in *Kew-MPNS* which does not treat var. *ansu* as distinct from the species.

PLANT DESCRIPTION

A deciduous tree (5–8 (rarely to 12) m tall) with g reyish bark; flowers appearing before the leaves. *Leaves* alternate and stalked; stipules present; leaf blades ovate to orbicular-ovate, bases rounded, wedge-shaped or somewhat heart-shaped with narrowed pointed tips and margins notched with round, symmetrical teeth (crenate). *Flowers* solitary (appearing to be clustered) and short stalked (1–3 mm long); bisexual; sepals 5, deep pink and saucer-shaped; petals 5, equal, white, pink or tinged red; stamens many. *Fruit* ovoid (1.5–2.5 cm diam. in wild (larger in cultivation), often succulent and edible (apricot) with a distinct longitudinal groove and ripening from green to white, yellow or orange and often red-tinged; not splitting when ripe; stone globose with a smooth or slightly rough surface, sometimes netted. Fl. Mar–Apr, fr. Jun–Jul.

Species of *Prunus* wild in China: >7

HARVESTING, SOURCING AND NATURAL RANGE

Fruits are harvested in summer mostly from wild and cultivated trees (the latter grown mainly for their edible fruit — apricots). Traditional *Di Dao* sites for 'Ku Xing Ren' are located in N and NE China (provinces of Hebei, Inner Mongolia, Liaoning and Shaanxi). 'Ku Xing Ren' is typically sourced from a 'bitter' form often called 北杏仁 Bei Xing Ren 'northern apricot seed' although a 'sweet' form (甜杏仁 Tian Xing Ren 'sweet apricot seed' or Nan Xing Ren 南杏仁 'southern apricot seed') can also be used.

In the wild, the species grows in open forests on mountain slopes in N and NW China (700–3,000 m altitude) but its populations are seriously declining due to habitat loss. Also native to Japan, Korean Peninsula and central Asia.

CONSERVATION STATUS

China: Near Threatened (as *Armeniaca vulgaris* var. *ansu*); Least Concern (as *A. vulgaris* var. *vulgaris*) (MEPCAS, 2013).
Global: Endangered (IUCN Red List, accessed 23/1/2016, as *A. vulgaris*).

PHARMACOPOEIA SPECIES 2

P. sibirica L.

Rosaceae

synonym: *Armeniaca sibirica* (L.) Lam. (*FoC* accepted name)

taxonomic/nomenclatural note: *Kew-MPNS* accepted name, 24/9/2016.

PLANT DESCRIPTION

Easily confused with the other Pharmacopoeia species. A shrub or small tree (2–5 m tall). Leaf bases rounded to somewhat heart-shaped; leaf margins minutely toothed. Flower- and fruit-stalks 1–2 mm long. Petals white with pink veins or pinkish. **Fruit flesh dry, compact and inedible and easily separated from the stone; splitting when ripe while still attached to the tree.** Stone **longer than wide** (1.2–2.2 cm long, nearly as broad), (those of PS1 and PS3 are wider than long). Fl. Mar–Apr, fr. Jun–Jul.

HARVESTING, SOURCING AND NATURAL RANGE

Fruits are harvested from wild populations which grow in dry sunny places, in montane forests, scrub and grassland (400–2,500 m altitude) in N, NE and W China (Gansu, Hebei, Heilongjiang, Henan, Inner Mongolia, Jilin, Liaoning, Ningxia, Shaanxi and Shanxi provinces). The species is also native to the Korean Peninsula, Mongolia and Russia.

CONSERVATION STATUS

China: Least Concern (MEPCAS, 2013).
Global: Not Evaluated (IUCN Red List, accessed 23/1/2016).

PHARMACOPOEIA SPECIES 3

P. mandshurica (Maxim.) Koehne

Rosaceae

synonyms: *Armeniaca mandshurica* (Maxim.) Skvortsov (*FoC* accepted name); *P. armeniaca* var. *mandshurica* Maxim.

taxonomic/nomenclatural note: *Kew-MPNS* accepted name, 24/9/2016.

PLANT DESCRIPTION

Easily confused with the other Pharmacopoeia species. A tree 5–15 m tall. Leaf bases variable; **leaf margins with irregular-sized teeth; flower- and fruit-stalks 0.7–1 cm long.** Petals pink or white. **Fruits slightly fleshy to dry, fragrant, not splitting when ripe**; stone subglobose (1.3–1.8 cm long; 1.1–1.8 cm wide), surface slightly rough. Fl. Apr–May, fr. May–Jul.

HARVESTING, SOURCING AND NATURAL RANGE

Fruits are harvested from wild populations which grow on sunny, scrubby slopes in mixed montane forests (200–1,000 m altitude) in NE China (Heilongjiang, Jilin and Liaoning). The species is also native to the Korean Peninsula and E Russia.

CONSERVATION STATUS

China: Least Concern (MEPCAS, 2013).
Global: Not Evaluated (IUCN Red List, accessed 23/1/2016).

TCM PROPERTIES AND USE

properties: **'bitter and slightly warm'**
channels: 'Lung and Large Intestine'
actions: 'relieves cough, moistens intestines'

indications: **coughs, dyspnoea; constipation**

Flowering branch of *Prunus armeniaca*

Immature fruiting branches of *P. armeniaca*

Mature fruit of *P. armeniaca*

Flowering branches of *P. sibirica*

Fruits of *P. sibirica* typically splitting open when ripe while still on the tree

Leaves and young fruit of *P. mandshurica*, Liaoning province

Leaves of *P. mandshurica*

Mature fruit of *P. mandshurica*, Jilin province

Prunus armeniaca

Rosaceae

DRUG MORPHOLOGY

CRUDE DRUG/DECOCTING PIECES-1 (DRIED SEED): laterally compressed-ovoid, almost heart-shaped [1] (1–1.9 cm long; 0.8–1.5 cm wide; 5–8 mm thick); one end shortly acute [2] and with a short linear hilum [3]; the other end thicker and broadly obtuse or truncate [4] and often asymmetrical with a vascular attachment scar (chalaza) [5]. Seed coat thin and yellowish-brown to deep brown, softly downy (tomentose) [6], and distinctly veined longitudinally [7]. Internally 2 milky-white, oily, cotyledons. *Texture*: hard and brittle. *Fracture*: smooth. *Odour*: indistinct. *Taste*: bitter. ***Good quality*:** large, plump and unbroken seeds.

DECOCTING PIECES-2 (SEED COAT REMOVED): smooth, milky-white or yellowish-white cotyledons either intact or separated [8]; innermost surface of separated ones flat or concave [9]. *Odour:* aromatic. *Taste:* bitter.

DECOCTING PIECES-3 (DRY-FRIED SEED): as above but externally yellow to brownish-yellow with occasional charred areas [10].

PROCESSING

After removal of the flesh, the hard endocarps are broken open to remove the seed which is sun-dried (= crude drug/decocting pieces-1); seeds often blanched and seed coats removed (see *Safety*) (= decocting pieces-2); subsequent dry-frying produces Chao Ku Xing Ren (炒苦杏仁) (= decocting pieces-3).

OTHER COMMON NAMES

北杏仁 Bei Xing Ren 'northern apricot'; 山杏 Shan Xing; 山杏仁 Shan Xing Ren; 野杏Ye Xing; 杏 Xing (plant name); 杏仁 Xing Ren; apricot (SCN); apricot kernel; bitter apricot seed (CP2010); Chinese bitter almond; Semen Armeniacae Amarum.

SAFETY

Slightly toxic (CP2015); Safety Class: 3; Interaction class: A (Gardner & McGuffin, 2013). Removal of seed coats and dry-frying is reported to reduce levels of toxic cyanogenic glycosides (Bensky *et al*., 2004) with the 'bitter' form containing higher levels than the 'sweet' form (Nan Xing Ren) (see *Harvesting*). Both are used as medicine, food and healthfood in China (Anon., 2002a,b; Gardner & McGuffin, 2013; Hu, 2005; Zhao & Xiao, 2009–2010).

P. sibirica

Rosaceae

DRUG MORPHOLOGY

CRUDE DRUG/DECOCTING PIECES-1 (DRIED SEED): similar to that of the other Pharmacopoeia species but **slightly smaller and broader than long (c. 0.9 cm long; 1.1 cm diam.) [11].**

DECOCTING PIECES-2 (SEED COAT REMOVED): as for crude drug.

DECOCTING PIECES-3 (DRY-FRIED SEED): as for crude drug.

PROCESSING

As for PS1.

OTHER COMMON NAMES

山杏 Shan Xing (plant name); 山杏仁 Shan Xing Ren; 西伯利亚杏 Xi Bo Li Ya Xing; 杏仁 Xing Ren; apricot kernel; bitter apricot seed (CP2010); Semen Armeniacae Amarum.

SAFETY

As for PS1.

P. mandshurica

Rosaceae

DRUG MORPHOLOGY

Neither the crude nor processed seeds can be reliably distinguished from those of PS1 using macroscopic characters alone. Compared to those of PS2, seeds are slightly bigger and narrower [12]. Laboratory-based methods are advised for robust seed differentiation of the three Pharmacopoeia species.

PROCESSING

As for PS1.

OTHER COMMON NAMES

东北杏 Dong Bei Xing; 山杏仁 Shan Xing Ren; 杏仁 Xing Ren; bitter apricot seed (CP2010).

SAFETY

As for PS1.

Crude drug/decocting pieces-1 (dried)
P. armeniaca

Decocting pieces-2 (blanched)
P. armeniaca

Decocting pieces-3 (dry-fried)
Chao Ku Xing Ren, *P. armeniaca*

Crude drug/decocting pieces-1 (dried), *P. sibirica*

Decocting pieces-2 (blanched), *P. sibirica*

Crude drug/decocting pieces-1 (dried), *P. mandshurica*

Decocting pieces-2 (blanched), *P. mandshurica*

Decocting pieces-3 (dry-fried)
Chao Ku Xing Ren, *P. mandshurica*

Lai Fu Zi 莱菔子

Raphani Semen

radish

Drug part: SEED

Raphanus sativus L.

Brassicaceae

synonyms: *R. chinensis* Mill.; *R. niger* Mill.; *R. raphanistrum* L. subsp. *sativus* (L.) Domin (*Kew-MPNS* accepted name, 18/5/2015).

taxonomic/nomenclatural note: many infraspecific taxa (varieties and forms) have been described and their taxonomy is highly confused.

Species of *Raphanus* wild in China: 2

PLANT DESCRIPTION

A highly variable species. An annual or biennial herb (10–130 cm tall), branched or unbranched; *roots* often fleshy (radish). *Leaves* alternate, simple, of various shapes and covered in stiff non-glandular hairs (simple trichomes); lower stem leaves stalked, shallowly or deeply lobed, margins toothed; upper stem leaves unstalked, lance-shaped, margins toothed or entire. *Inflorescences* terminal and lateral, with several, short-stalked flowers along a central axis and with lowermost flowers opening first (raceme). *Flowers* usually conspicuous owing to their bright petal colour: purple, pink or white (often with dark veins); sepals 4, oblong or linear; petals 4, equal size and shape, widest at tips (1.2–2.2 cm long; 3–8 mm wide); stamens 6 (4 long, 2 short); nectar glands 4. *Fruit* a smooth, dry, spindle- or lance-shaped, 2-valved capsule (occasionally ovoid or cylindric) with a single row of seeds becoming visible through the fruit wall, the latter sometimes constricted between seeds; seedless segment 1–3.5 mm long; seed-bearing part usually 3–15 cm long; 0.7–1.3 cm diam.; base rounded and corky; apex conical. *Seeds* globose or ovoid (2.5–4 mm diam.). Fl. and Fr. dependent on cultivation regime.

HARVESTING, SOURCING AND NATURAL RANGE

Ripe seeds are harvested in summer from cultivated plants long-grown in China as a vegetable crop (radish) and for its medicinal seed. The species is native to the Mediterranean and is cultivated worldwide.

CONSERVATION STATUS

China: not native.

Global: Not Evaluated (IUCN Red List, accessed 25/8/2016).

Flowering crop of *Raphanus sativus*, China

TOP RIGHT Fruiting crop of *R. sativus*, China

BOTTOM RIGHT Whole plant with fleshy root (radish)

TCM PROPERTIES AND USE

properties: **'pungent, sweet and neutral'**
channels: **'Spleen and Stomach'**
action: **'resolves food stagnation, descends qi, resolves phlegm'**

indications: **indigestion, abdominal distension, poor appetite, diarrhoea; chronic cough**

DRUG MORPHOLOGY

CRUDE DRUG (DRIED SEED): sub-ovoid or ellipsoid seeds, slightly flattened (2.5–4 mm long; 2–3 mm wide) **[1]**; often with distinct longitudinal furrows outlining the cotyledons and radical within. Seed coat (testa) thin and brittle, yellowish-, greyish-, or reddish-brown, and finely reticulate [hl] **[2]**. At one end, a small, rounded, dark brown or blackish hilum with rough, white border [hl] **[3]** sits adjacent to a smaller, rounded blackish micropyle [hl]. Internally: cotyledons 2, yellow and oily. *Odour*: indistinct. *Taste*: mild, somewhat bitter and pungent. ***Good quality***: seeds plump, oily and reddish-brown.

DECOCTING PIECES-1 (CRUSHED SEED): as for crude drug but partly crushed, some intact; some with seed coat (testa) removed and others broken into small pieces. Seed contents (cotyledons and radical) golden yellow throughout.

DECOCTING PIECES-2 (PROCESSED: DRY-FRIED SEED): (炒莱菔子 Chao Lai Fu Zi). As for crude drug but slightly enlarged, colour darker; texture crisp and fragile; faintly aromatic.

PROCESSING

Seed is sun-dried (= crude drug) then crushed immediately prior to decocting (= decocting pieces-1). Alternatively dried seeds may be dry-fried prior to crushing (= decocting pieces-2; 炒莱菔子 Chao Lai Fu Zi).

OTHER COMMON NAMES

萝卜 Luo Bo (plant name); 萝卜子 Luo Bo Zi; Semen Raphani; radish (SCN); radish seed (CP2010).

SAFETY

Not for long-term TCM use; for contraindications see Bensky *et al.*, 2004. An official food and healthfood in China (Anon., 2002).

Crude drug of Lai Fu Zi (seed of *R. sativus*)

Dry-fried Lai Fu Zi (processed seed of *R. sativus*) dp-2

Crushed decocting material (dp-1)

Lian Qiao 连翘 Forsythiae Fructus

Drug part: FRUIT

Forsythia suspensa (Thunb.) Vahl

Oleaceae

synonym: *Ligustrum suspensum* Thunb.

Species of *Forsythia* wild in China: 6

PLANT DESCRIPTION

A deciduous spreading or arching shrub, branches hollow, branchlets somewhat 4-angled. *Leaves* opposite, simple or 3-parted to trifoliate, ovate (2–10 cm long; 1.5–5 cm wide), margin saw-toothed. *Flowers* emerging before the leaves, bright yellow and conspicuous; solitary or in tight, axillary clusters. Flower calyx green, deeply 4-lobed; petals 4, yellow, fused at base into short tube, lobes oblong and much longer (1.2–2 cm) than basal tube; stamens 2; ovary superior, 2-valved. *Fruit* dry, brown and woody; ovoid to long ellipsoid (1.2–2.5 cm long; 6–12 mm diam.) with scattered raised lenticels; stalked and beaked. *Seeds* many, slender and membranous, with one lateral wing. Fl. Mar–Apr, fr. Jul–Sep.

HARVESTING, SOURCING AND NATURAL RANGE

Fruits are harvested in autumn mostly from wild plants. Two products are traded: 1. Qing Qiao 'green forsythia' which are unripe fruits that are still closed and greenish and 2. Lao Qiao 'old forsythia' fruits which are harvested when fully ripe (when yellowish-brown, opened or partly opened).

The species is endemic to China (Anhui, Hebei, Henan, Hubei, Jiangsu, Shaanxi, Shandong, Shanxi and Sichuan provinces) and grows wild in scrub or open grassland on slopes and in valleys (300–2,200 m altitude). Traditional *Di Dao* material is sourced from Shanxi and Henan provinces.

The species is also widely cultivated as an ornamental in China and many other countries.

CONSERVATION STATUS

Endemic to China: Least Concern (MEPCAS, 2013); Not Evaluated (IUCN Red List, accessed 9/4/2016).

Mature plants of *Forsythia suspensa* typically flowering before leaves appear

F. suspensa, flowering branch

F. suspensa, branch with immature green fruit ready to harvest as Qing Qiao

TCM PROPERTIES AND USE

***properties*: 'bitter and cool'**
***channels*: 'Lung, Heart and Gall Bladder'**
***actions*: 'clears heat, eliminates toxins'**

***indications*: respiratory tract infection, fever, sore throat; abscesses; dysuria**

DRUG MORPHOLOGY

CRUDE DRUG/DECOCTING PIECES (DRIED/STEAMED FRUIT):

1. Lao Qiao (old fruit): *fruit* halved (dehisced) **[1]**,or intact and partly open **[2]**; woody throughout; elongate to ovoid (1.5–2.5 cm long; 0.5–1.3 cm diam.), slightly compressed. Externally yellowish-brown or reddish-brown, faintly striated longitudinally **[3]**, or irregularly wrinkled **[4]**, lenticels many, scattered and raised **[5]**; a shallow longitudinal furrow **[6]** (midway between the raised valve suture lines and with a slightly curved, woody fruit-stalk **[7]**,or, if shed, a small fruit-stalk scar **[8]**; fruit tip narrowly pointed **[9]**. Internally with 2 large seed compartments **[10]** **(seeds mostly dispersed)** and divided by a hard, straight partition; internal surface pale yellowish-brown throughout, smooth and often lustrous. *Seeds*, when present, brown, slender and membranous, with one lateral wing. *Texture:* **hard but brittle.** *Odour:* slightly aromatic. *Taste:* bitter.

2. Qing Qiao (immature fruit): as for 'Lao Qiao' above but *fruit* **closed [11],** externally **greenish-brown or greenish-black,** lenticels greyish-white **[12]**. ***Seeds* many** (not yet dispersed), **yellowish-green** and slender, with one lateral wing. *Texture:* hard; **not brittle.**

PROCESSING

Lao Qiao (old forsythia) are sun-dried directly after harvesting; Qing Qiao (green forsythia) are steamed thoroughly, then sun-dried (= decocting pieces).

OTHER COMMON NAMES

带心连翘 Dai Xin Lian Qiao (forsythia with seeds); 黄连翘 Huang Lian Qiao (yellow forsythia fruit); 空壳 Kong Qiao; 落翘 Luo Qiao; forsythia; 青连翘 Qing Lian Qiao (green forsythia fruit); forsythia (SCN); Fructus Forsythiae; weeping forsythia capsule (CP2010).

SAFETY

Safety Class: 1; Interaction Class: A (Gardner & McGuffin, 2013).

Crude drug/decocting pieces of Lian Qiao (trade form 'Lao Qiao') sourced from ripe fruit of *Forsythia suspensa*)

Crude drug/decocting pieces of Lian Qiao (trade form 'Qing Qiao') sourced from immature fruit of *F. suspensa*)

Li Zhi He 荔枝核

Litchi Semen

lychee

Drug part: SEED

Litchi chinensis Sonn.

Sapindaceae

PLANT DESCRIPTION

A small, evergreen tree (up to 10 (rarely 15) m tall). *Bark* greyish-black; *branches* brownish-red with white lenticels. *Leaves* alternate and pinnate with 2, 3 or rarely 4 pairs of shiny, ovate and somewhat leathery leaflets (6–15 cm long; 2–4 cm wide); margins entire; tips pointed. Leaf (not leaflet) stalks basally swollen (10–25 mm long). *Flowers* in a branched and ascending, terminal head (thyrse), the latter clothed in short, golden hairs. Flowers small, white and single-sexed; bracts and bracteoles small; sepals cup-shaped and 4- or 5-lobed; petals absent. *Male flowers* (more prominent than female flowers) with 6 or 7 (rarely 8) white, thread-like stamens (c. 4 mm long); *female flowers* short stalked and heart-shaped, stigma 2- or 3-lobed. *Fruits* dark red, matte and ovoid (2–3.5 cm long) with a leathery fruit wall (often warty; crust-like when dry) encasing a white and intensely succulent flesh (arillode); *seed* single, brown and covered by fleshy arillode. Fl. spring, fr. summer.

Species of *Litchi* wild in China: 1

HARVESTING, SOURCING AND NATURAL RANGE

Fruits are harvested when ripe (dark red) in summer from plants widely cultivated for their edible fruit (litchi, lichee, lychee) in subtropical parts of S China, notably Fujian and Guangdong provinces. Approximately 10 cultivars are grown.

The species is native to SW Guangdong and Hainan and typically grows along river banks and by ponds and ditches [also native to Laos, Malaysia, Myanmar, New Guinea, Philippines, Thailand and Vietnam].

CONSERVATION STATUS

China: Not Evaluated.

Global: Not Evaluated (IUCN Red List, accessed 13/6/2015).

River bank in Fujian province (SE China) lined with *Litchi chinensis* trees

Fruiting branch of *L. chinensis*

TCM PROPERTIES AND USE

***properties*: 'sweet, slightly bitter and warm'**
***channels*: 'Liver and Kidney'**
***actions*: 'invigorates qi, relieves pain'**
***indications*: hernia pain; swollen and painful testes**

DRUG MORPHOLOGY

CRUDE DRUG (SEED): seed long-ovoid, slightly flattened laterally (1.5–2.2 cm long; 1–1.5 cm wide), sometimes dented **[1]**. External surface brown or purplish-brown, smooth and slightly lustrous **[2]**; truncate end with a pale, yellowish-brown scar (raphe) **[3]** (c. 7 mm diam.); other end rounded **[4]**. *Texture:* hard. *Fracture:* brownish-yellow with 2 seed leaves (cotyledons). *Odour*: indistinct. *Taste*: slightly sweet, bitter and astringent. ***Good quality*:** large and plump.

PROCESSING

After removal of the pulp and fleshy aril, seeds are washed and sun-dried (= crude drug) then pulverised prior to decocting. Dried seed may be further processed by thoroughly moistening with salt water before dry-frying (= Yan Li Zhi He 盐荔枝核).

OTHER COMMON NAMES

荔核 Li He; 荔仁 Li Ren; lychee seed (CP2010); Semen Litchi.

Crude drug

Pulverised

Salt-processed

Li Zhi He (dried seed of *Litchi sinensis*)

Man Jing Zi 蔓荆子 — Viticis Fructus

Drug part: FRUIT

PHARMACOPOEIA SPECIES 1

Vitex rotundifolia L. f.

Lamiaceae

synonyms: *V. trifolia* L. subsp. *litoralis* Steenis (*Kew-MPNS* accepted name 29/7/2016); *V. trifolia* L. var. *simplicifolia* Cham. (name used in CP2015).

Species of *Vitex* wild in China: 14

PLANT DESCRIPTION

A prostrate shrub with stems rooting at nodes and young stem branches silkily hairy. *Leaves* opposite, **usually 'simple' (rarely divided into 3 leaflets),** stalkless or shortly stalked; leaf blades ovate, broadest towards the tip or rounded; lower surfaces velvety, margins entire, with silky grey hairs. *Flowers* in terminal heads (3–10 cm long; 1–2.5 cm diam.); each flower bisexual and laterally symmetrical; calyx cup-shaped (4–5 mm long); corolla purplish-mauve to lilac blue, silkily hairy externally; petals fused to form a short tube (abruptly opening into 5 unequal-sized lobes (2 smallest lobes forming the upper lip and 3 larger ones the lower lip); middle lobe of lower lip largest and densely hairy along innermost surface; ovary superior; stamens and style fully exposed; latter thread-like with a 2-cleft stigma. *Fruit* dry, globose and dark brown when mature; mostly enclosed by a membranous 5-toothed persistent calyx. Fl. Jul.–Sep.; fr. Sep–Nov.

HARVESTING, SOURCING AND NATURAL RANGE

Mostly sourced from medicinal crops widely grown near lakes and along the eastern seaboard of China in the provinces of Shandong, Jiangxi and Zhejiang. Fruits are harvested when ripe in autumn.

The wild species grows in sandy soils often near the sea and is native in NE, SC and SE China: Anhui, Fujian, Guangdong, Hebei, Jiangsu, Jiangxi, Liaoning, Shandong, Taiwan and Zhejiang. Also native in Japan; SE Asia and Pacific Islands.

CONSERVATION STATUS

China: Least Concern (MEPCAS, 2013); 'rapidly decreasing' (Leung & Cheng, 2008). Red List, accessed 29/9/2016).
Global: Not Evaluated (IUCN Red List, accessed 12/8/2016).

PHARMACOPOEIA SPECIES 2

V. trifolia L.

Lamiaceae

PLANT DESCRIPTION

Easily distinguished from PS1 and the *Unofficial Substitute* below. An **erect shrub (1.5–5 m tall).** Leaves usually divided into **3–5 leaflets with middle leaflet always stalkless;** leaflet margins **entire.**

HARVESTING, SOURCING AND NATURAL RANGE

Harvesting and sourcing as for PS1. The wild species grows in open sites and along river banks (100–400 m altitude) in SC, SE and WC China: Fujian, Guangdong, Guangxi, Taiwan and Yunnan. Also native in south-east Asia, Australia and the Pacific Islands.

CONSERVATION STATUS

China: Not Evaluated.
Global: Not Evaluated (IUCN Red List, accessed 12/8/2016).

UNOFFICIAL SUBSTITUTE

V. negundo L. var. *cannabifolia* (Siebold & Zucc.) Hand.-Mazz.

Lamiaceae

synonym: *V. cannabifolia* Siebold & Zucc.

PLANT DESCRIPTION

Easily distinguished from the two Pharmacopoeia species. An **erect shrub**. Leaves usually divided into **3–5 leaflets with middle leaflet distinctly stalked**; **leaflet margins saw-toothed**.

CONSERVATION STATUS

China: Least Concern (MEPCAS, 2013).
Global: Not Evaluated (IUCN Red List, accessed 12/8/2016).

SUBSTITUTION COMMENT. Fruits of this common and widespread Chinese native species are superficially similar to those of PS1 and PS2 and are reported to be traded as 'Man Jing Zi' as a local substitute in China. Fruits are readily available as a cheap by-product of this wild-harvested species whose leaves are extracted on an industrial scale as the source of a TCM oil called 'Mu Jing You' (Vitex Oil) (CP2015).

The plant grows in scrub on mountain slopes and along roadsides (100–1,100 m altitude) in SC, SE and WC China. It is also native in India, Nepal and SE Asia.

TCM PROPERTIES AND USE

properties: **'acrid, bitter and cool'**
channels: **'Liver, Stomach and Bladder'**
action: **'dispels wind heat'**

indications: **headache; painful eyes; dizziness, vertigo; musculoskeletal pain**

Prostrate habit of *Vitex rotundifolia* (medicinal crop in Fujian province)

V. rotundifolia in flower with leaf margins entire

V. rotundifolia with immature fruit

Upright habit of *V. trifolia* in the wild (Guangdong province)

Flowering branch of *V. trifolia* with leaf margins entire

Fruiting branch of *V. trifolia*

V. negundo var. *cannabifolia* in flower

Inflorescence of *V. negundo* var. *cannabifolia*

V. negundo var. *cannabifolia*; leaves divided into 3–5 leaflets with margins saw-toothed

Vitex rotundifolia

Lamiaceae

DRUG MORPHOLOGY

CRUDE DRUG/DECOCTING PIECES-1 (DRIED FRUIT): globose (4–6 mm diam.) [1]. Externally, fruit wall greyish-black or dark brown, covered in a greyish-white pubescence [2], four indistinct longitudinal furrows and with an apex slightly concave [3]. **One to two thirds** of each fruit usually enclosed by a persistent, membranous calyx [4], latter soft, dense and greyish-white hairy and five-lobed (**two lobes deeply cut almost to calyx base [5]**). A densely pubescent fruit stalk is sometimes attached [6]. Fruit in transverse section, with four loculi (c. 1.5 by 1 mm) [7] **each locular usually with a single seed [8]**. *Texture of calyx:* hard, light, difficult to break. *Odour*: aromatic. *Taste*: indistinct. ***Good quality*:** large, plump and fragrant.

DECOCTING PIECES-2 (DRY-FRIED FRUIT): as for crude/dp-1 but slightly darker (black or brownish-black) and calyces often partly singed brown.

PROCESSING

Fruits are sun-dried and, sometimes, calyces are removed (Man Jing Zi = crude drug/decocting pieces-1); calyces are sometimes removed. Subsequent dry-frying produces Chao Man Jing Zi (= decocting pieces-2).

OTHER COMMON NAMES

单叶蔓荆 Dan Ye Man Jing (plant name); 京子 Jing Zi; 蔓荆实 Man Jing Shi; shrub chaste tree fruit (CP2010); simple-leaf chaste tree (SCN); Fructus Viticis.

V. trifolia

Lamiaceae

DRUG MORPHOLOGY

CRUDE DRUG/DECOCTING PIECES-1 (DRIED FRUIT): almost indistinguishable from those of PS1 but **locules slightly smaller** (1.5 mm by 0.2–0.5 mm) [9] **and normally devoid of seed**. The removal of calces is typical In Hong Kong [10].

DECOCTING PIECES-2 (DRY-FRIED FRUIT): as for crude/dp-1 but slightly darker and charred in places [11].

PROCESSING

As for PS1.

OTHER COMMON NAMES

蔓荆 Man Jing (plant name); 蔓荆实 Man Jing Shi; 京子 Jing Zi; 3-leaved chaste tree (plant name).

V. negundo var. *cannabifolia*

Lamiaceae

'DRUG SUBSTITUTE' MORPHOLOGY

DRIED FRUIT: fruits smaller (3–4mm long; 2–3 mm diam.) [12] then those of PS1 and PS2; The persistent **calyx encloses at least half of the fruit [13]** imparting a **somewhat conical shape**; **calyx very shallowly toothed [14]**; **locules large relative to fruit size [15]**.

OTHER COMMON NAMES

黄荆 Huang Jing (plant name); 黄荆子 Huang Jing Zi (seed); 牡荆 Mu Jing (plant name); Chinese chaste tree (plant name).

LESSER KNOWN UNOFFICIAL SUBSTITUTE

Cardiospermum halicacabum L. (倒地铃 Dao Di Ling), family Sapindaceae, is a common weed of tropical and subtropical China. Reports of its ***seed*** being used as a substitute or as an adulterant of official 'Man Jing Zi' locally in China may stem from its superficial resemblance having a **whitish-grey hilum [16] easily mistaken for the whitish-grey calyces of the fruit of PS1 and PS2; internally the endosperm is yellowish-white and oily [17].** The seed (and its roots) are used in Ayurvedic and Unani medicine (Anon., 1999–2011); Anon., (2007–2008).

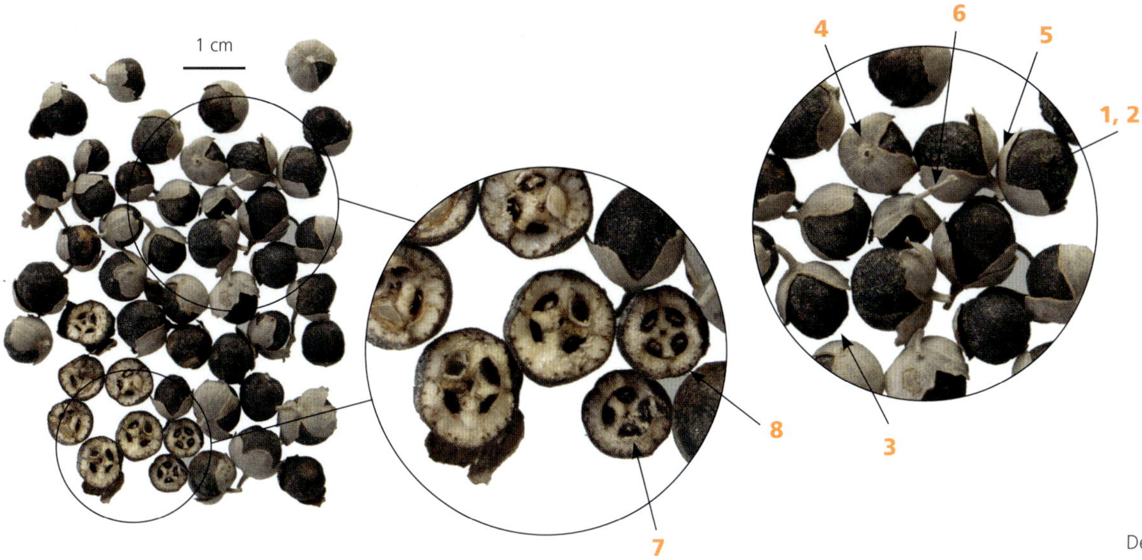

Crude drug/decocting pieces (dried) of Man Jing Zi sourced from style 1 fruits of *V. rotundifolia*

Decocting piece (dry-fried: Chao Man Jing Zi) sourced from fruits of *V. rotundifolia*

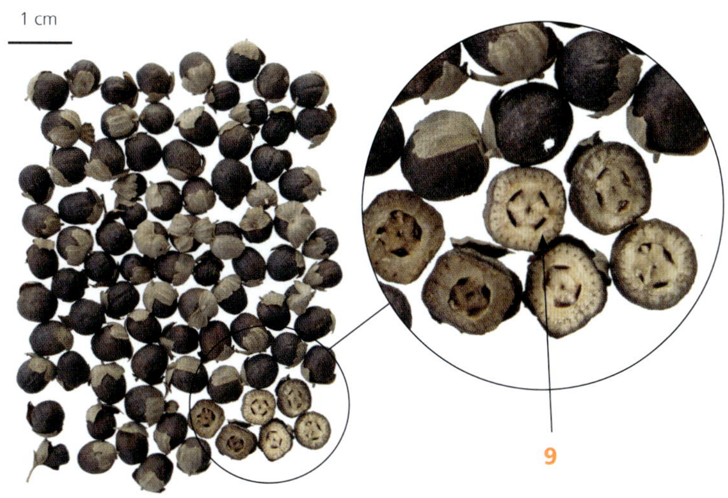

Crude drug/decocting pieces (dried) of Man Jing Zi sourced from fruits of *V. trifolia*

Decocting piece (dry-fried: Chao Man Jing Zi) sourced from fruits of *V. trifolia*

Hong Kong style of Man Jin Zi crude drug/decocting pieces of *V. trifolia* with calyces removed

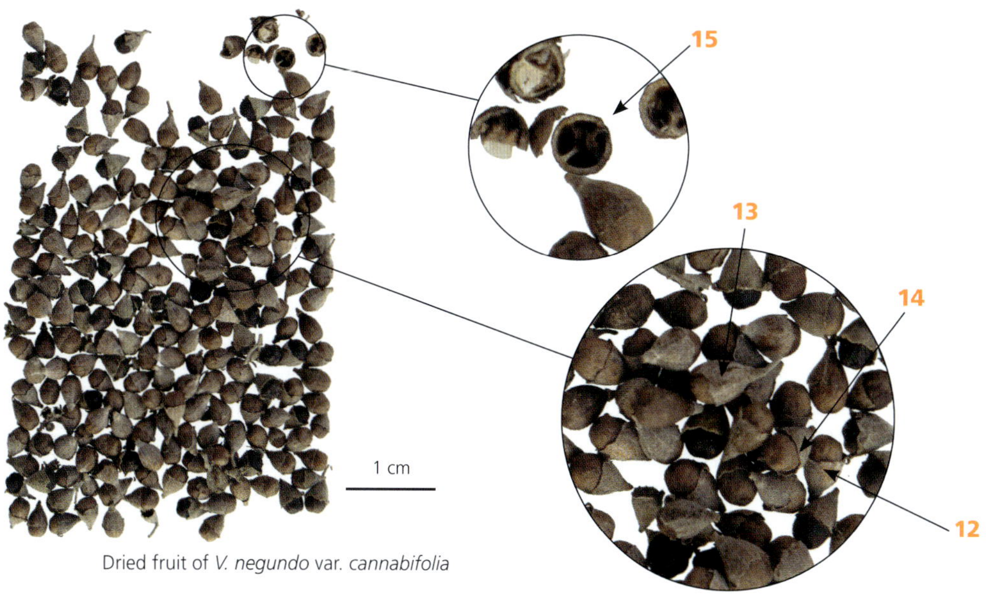

Dried fruit of *V. negundo* var. *cannabifolia*

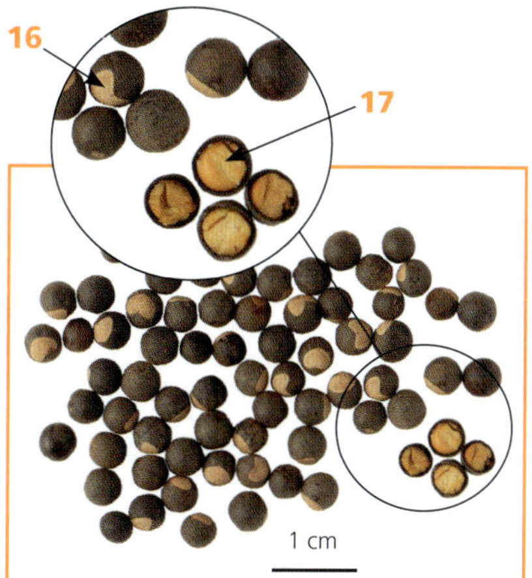

Seeds of *Cardiospermum halicacabum* with some (top left) in transverse section

Mu Gua 木瓜

Chaenomelis Fructus

Chinese quince

Drug part: FRUIT

PHARMACOPOEIA SPECIES

Chaenomeles speciosa (Sweet) Nakai

Rosaceae

synonyms: *C. lagenaria* (Loisel.) Koidz. (*Kew-MPNS* accepted name, 10/10/2016); *Cydonia speciosa* Sweet

PLANT DESCRIPTION

A deciduous, spreading, **thorny shrub (up to 2 m tall)**. *Leaves* alternate and simple, blades ovate to narrowly elliptic (3–9 cm long; 1.5–5 cm wide), margins shortly saw-toothed; mature leaves **hairless on lower surfaces**. *Flowers* showy (3–5 cm diam.), emerging before the leaves, in **3-5-flowered clusters on older branches**; sepals 5, **erect**, hairless on lower (outside) surface, and softly hairy on upper; petals 5, equal-sized, red, pink or white (3–5 cm diam.); stamens many (40–50); styles 5. *Fruit* a large ovoid, fleshy pome (4–6 cm diam.), yellow or yellowish-green; short-stalked or stalkless; many-seeded and fragrant. Fl. Mar–May, fr. Sep–Oct.

Note: Fruits of this species have been reported to adulterate those of the closely related edible quince (*Cydonia oblonga* Mill.). *C. oblonga* has entire leaf margins, flowers with 15–25 stamens and fruits pear-shaped.

HARVESTING, SOURCING AND NATURAL RANGE

Fruits are harvested in summer and autumn when almost ripe (greenish-yellow) mostly from plants grown on medicinal plant farms in the Chinese provinces of Anhui, Hubei, Sichuan and Zhejiang (*Di Dao*-quality 'Mu Gua' is sourced from Anhui, Hubei and Zhejiang). The tree is also widely grown in China as an ornamental. The wild species grows in parts of NC, SE and WC China (Fujian, Gansu, Guangdong, Guizhou, Hubei, Jiangsu, Shaanxi, Sichuan, Tibet and Yunnan) [also native to Myanmar].

CONSERVATION STATUS

China: Not Evaluated

Global: Not Evaluated (IUCN Red List, accessed 18/8/2016).

UNOFFICIAL SUBSTITUTE 1

C. cathayensis (Hemsl.) Schneid.

Rosaceae

synonyms: *C. lagenaria* Koidz. var. *cathayensis* (Hemsl.) Rehder; *Cydonia cathayensis* Hemsl.

PLANT DESCRIPTION

Easily confused with the Pharmacopoeia species. *Main differences*: **shrub or small tree (2–6 m tall)**; leaves elliptic or lance-shaped (5–11 cm long; 2–4 cm wide), **lower surface of young leaves densely covered in soft, brown hairs (tomentose)**; flowers 2–4 cm diam.; petals pink or white; fruit yellowish-red; wild fruits slightly larger (6–7 cm diam.) but cultivated fruits may be much larger (up to 15 cm long; 8 cm diam.) (Rumpunen, 2002). Fl. Mar–May; Fr. Sep–Oct.

CONSERVATION STATUS

Endemic to China: Least Concern (MEPCAS, 2013); Not Evaluated (IUCN Red List, accessed 18/8/2016).

UNOFFICIAL SUBSTITUTE 2

C. sinensis (Thouin) Koehne

Rosaceae

synonym: *Cydonia sinensis* Thouin

PLANT DESCRIPTION

Easily confused with the Pharmacopoeia species. *Main differences*: a **thornless shrub or small tree (5–10 m tall)**; **lower surface of young leaves densely covered in yellowish-white hairs**; **flowers solitary**; **sepals reflexed** and deciduous; fruit dark yellow, narrowly ellipsoid (10–15 cm long) and **woody**. Fl. Apr.; Fr. Sep–Oct.

CONSERVATION STATUS

Endemic to China: Least Concern (MEPCAS, 2013); Not Evaluated (IUCN Red List, accessed 18/8/2016).

LESSER KNOWN SUBSTITUTES

Fruit of the following Rosaceae species have also been reported locally in China as accidental or intentional substitutes of 'Mu Gua' (NICBP, Bensky *et al.* 1997–2002, 2004).

1. *Chaenomeles thibetica* T. T. Yu. 西藏木瓜 Xi Zang Mu Gua, 'Tibetan Mu Gua'. Some taxonomists consider this a hybrid between *C. cathayensis* and *Docynia delavayi* (see below) (*FoC*). Thorny; leaf margins entire; sepals reflexed; when fresh, **fruits oblong or pear-shaped (5–9 cm diam.)**. Wild in Tibet and W Sichuan.
2. *Docynia delavayi* (Franch.) C. K. Schneid. 云南移木衣 Yun Nan Duo Yi; 小木瓜 Xiao Mu Gua. Evergreen tree; leaves leathery with lower surfaces densely clothed in white hairs; flowers white; when fresh, **fruit smaller (2–3 cm diam.)**, yellow and, when young, densely hairy.
3. *D. indica* (Wall.) Decne. 移木衣 Duo Yi; 小木瓜 Xiao Mu Gua. Semi-evergreen or deciduous; leaves papery; sepals erect and softly hairy on both surfaces; petals white; when fresh, **fruit smaller (2–3 cm diam.), sub-spheroidal** and yellow. **Dried fruit and their decocting pieces shorter than the official drug**. Wild in Sichuan and Yunnan.

TCM PROPERTIES AND USE

***properties*: 'sour and warm'**
***channels*: 'Liver and Spleen'**
***actions*: 'dispels wind damp, opens collaterals'**

***indications*: musculoskeletal pain; vomiting, diarrhoea and cramps; indigestion**

Chaenomeles speciosa — flowering branch with young leaves

C. speciosa with ripe fruit

Chaenomeles cathayensis — branch with unopened flowers

C. cathayensis young leaves

Immature fruit of *C. cathayensis*

C. sinensis — branch with emerging flowers

Immature fruit of *C. sinensis*

Chaenomeles speciosa

Rosaceae

DRUG MORPHOLOGY

CRUDE DRUG (DRIED FRUIT): oblong fruit halves (4–9 cm long, 2–5 cm wide, 1–2.5 cm thick). External uncut surface dull, dark purplish-brown or purplish-red, deeply convoluted **[1]** and with an inrolled outer margin **[2]**; internal (cut) surface paler and reddish-brown or brownish-yellow, with longitudinal or narrowly ovoid furrows and one or more central locules (ovary chambers) **[3]**; latter usually devoid of seeds **[4]**; seed, if present, long, triangular and wrinkled. *Texture*: hard, solid and somewhat woody. *Odour*: faintly aromatic. *Taste*: sour and slightly astringent. ***Good quality*:** large, fleshy and purplish-red.

DECOCTING PIECES (SLICED, STEAMED FRUIT): size and shape variable but usually curved or crescent–shaped **[5]** (4–9 cm long; c. 1–1.5 cm wide, c. 2–5 mm thick); somewhat woody **[6]**. Outer uncut surface (fruit wall) purplish-brown or purplish-red, often convoluted or furrowed **[7]**; cut surface (flesh) with a granular structure and yellowish, brown or blackish-brown **[8]**; inner surface (lining of the seed chambers or locules) smooth and somewhat lustrous **[9]**. *Odour*: slightly aromatic. *Taste*: sour and slightly astringent.

PROCESSING

Fruits are boiled until skins turn greyish-white; they are then sliced in half longitudinally and sun-dried (= crude drug). Decocting pieces are made by softening (often using steam) before being thinly sliced.

OTHER COMMON NAMES

川木瓜 Chuan Mu Gua (Sichuan Mu Gua); 贴梗海棠 Tie Geng Hai Tang (short-stalked apple); 皱皮木瓜 Zhou Pi Mu Gua (wrinkly skinned woody melon); Cathay quince; Chinese quince; common flowering quince fruit (CP2010); Fructus Chaenomelis.

SAFETY

An official food and healthfood in China (Anon., 2002a). Gardner & McGuffin (2013) classify this herbal drug as Safety Class: 1, Interaction Class: A.

C. cathayensis

Rosaceae

'DRUG SUBSTITUTE' MORPHOLOGY

Dried fruits (crude drug and decocting pieces) cannot be reliably distinguished macroscopically from those of the Pharmacopoeia species. The use of laboratory-based authentication methods are required such as the microscopic and chemical ones described in Xiao *et al.* (2001–2).

OTHER COMMON NAMES

毛叶木瓜 Mao Ye Mu Gua (hairy leaved woody melon); Mua Gua (plant name).

SUBSTITUTION COMMENT. A species of quince closely related and easily mistaken for the Pharmacopoeia species (see *Plant Descriptions*) and widespread in the wild in S, C and W China (Fujian, Gansu, Guangxi, Guizhou, Hubei, Hunan, Jiangsu, Jiangxi, Shaanxi, Sichuan, Tibet, Yunnan and Zhejiang). Widely cultivated as an ornamental and for its edible fruits (when cooked) while its dried fruits are sometimes used in local medicine in China.

C. sinensis

Rosaceae

'DRUG SUBSTITUTE' MORPHOLOGY

Crude drug and decocting pieces easily distinguished macroscopically from the fruits of the Pharmacopoeia species: external surface smooth **[10]**, cut edges of fruit walls not obviously inrolled **[11]** and flesh somewhat woody **[12]**. *Taste*: sour and astringent.

OTHER COMMON NAMES

光皮木瓜 Guang Pi Mu Gua (smooth-surfaced woody melon); Mu Gua (plant name); Chaenomelis Fructus (Anon., 2002c).

SUBSTITUTION COMMENT. Closely related to the Pharmacopoeia species. The fruits have long been used in local Chinese medicine for a wide range of conditions including musculoskeletal ones similar to those of official 'Mu Gua' (*C. speciosa*) (Oku *et al.*, 2003; Sawai *et al.*, 2008). Its shared and similar names may be a cause of confusion in trade and but its characteristically woody fruits and, when dried, its smooth skin easily distinguishes it from official 'Mu Gua'. For other identification features see the *Plant Descriptions* and '*Drug Substitute' Morphology*). The fruits are the official source of 'Chaenomelis Fructus' in the Korean Herbal Pharmacopoeia (Anon., 2002c).

The species is endemic to China where it is common and widely distributed in the wild in the provinces of Anhui, Fujian, Guangdong, Guangxi, Guizhou, Hebei, Hubei, Jiangsu, Jiangxi, Shaanxi, Shandong and Zhejiang; it is also cultivated in China and elsewhere.

Crude drug of Mu Gua sliced lengthwise (source: *C. speciosa*)

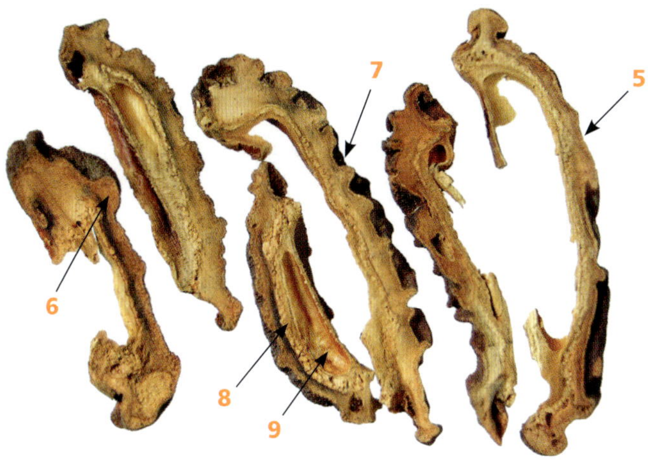

Decocting pieces of Mu Gua (source: *C. speciosa*)

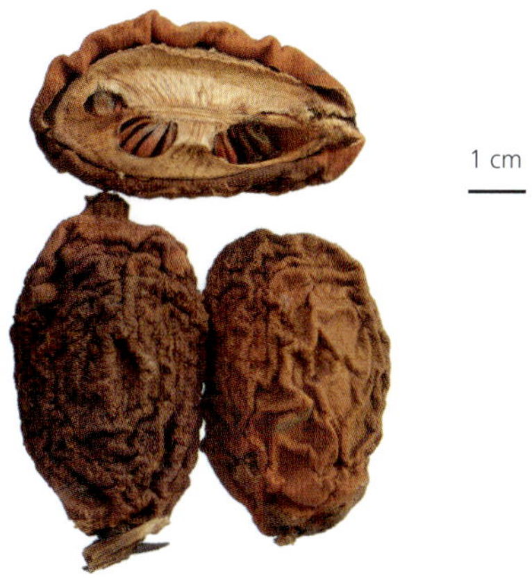

Mu Gua substitute: dried fruit of *C. cathayensis* showing convoluted fruit wall and interior

Sliced fruit

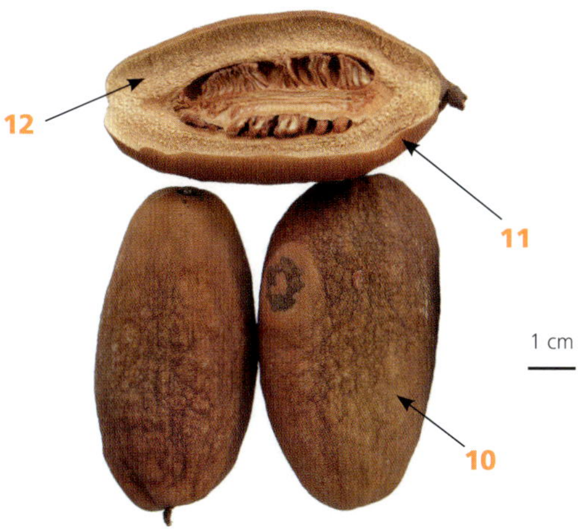

Mu Gua substitute: dried fruit of *C. sinensis* showing smooth fruit wall and interior

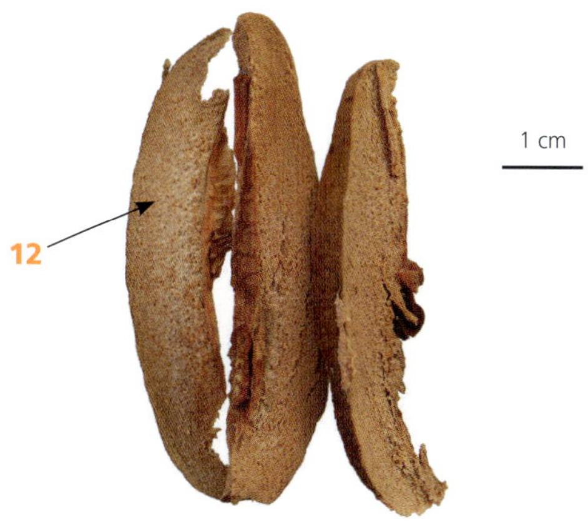

Sliced fruit

Nan Wu Wei Zi 南五味子 Schisandrae Sphenantherae Fructus

Drug part: FRUIT

Schisandra sphenanthera Rehder & E. H. Wilson

Schisandraceae

Species of *Schisandra* wild in China: 19

PLANT DESCRIPTION

A deciduous, woody vine. *Leaves* simple, somewhat clustered and papery; leaf blade, elliptic or ovate (rarely widest above the middle) 4–11(–13.5) cm long; 2–6(–8) cm wide, base wedge-shaped or truncate, margins finely toothed or (rarely) entire, tip short- to long-pointed. *Flowers* solitary, pendent and single-sexed (c. 1.5 cm diam.) with male and female flowers on different plants; stalk (peduncle) of male flowers 0.8–5.4 cm long; stalk of female flowers 1.8–6.4 cm long; 'petals' 5–9, yellow, orange, or red, largest 6–11.5 cm long; 3.5–8.5 mm wide; male flowers with 11–25 stamens; female flowers with 25–45 carpels. *Fruit* a linear and pendent aggregate of ovoid, pinkish-red to red carpels (latter 5.5–9 mm long; 4–8 mm diam.); stalk of fruit aggregate (peduncle) 2–10 cm long; individual carpels attached to an elongated fruit axis (torus) 2.5–14 (rarely up to 22) cm long. *Seed* 1–2 per fruit, almost kidney-shaped. Fl. Apr–Jul, fr. Jun–Oct.

See also 'Wu Wei Zi' (Schisandrae Chinensis Fructus) page 742

HARVESTING, SOURCING AND NATURAL RANGE

Sourcing is from wild-harvested plants mostly in the Chinese provinces of Henan, Shaanxi and Gansu (Zhao & Chen, 2014). Fruits are harvested in autumn when almost ripe (just before they turn bright red) to prevent subsequent rotting. The species is endemic to China and grows in forests and thickets usually between 700–2000 m altitude (sometimes higher) in Anhui, Gansu, Guizhou, Henan, Hubei, Hunan, Jiangsu, Shaanxi, Shanxi, Sichuan, NE Yunnan and Zhejiang.

In addition to their medicinal use in TCM, since 2007, the fruits have been the focus of an EU-China Biodiversity Programme (ECBP) project, a community-based livelihoods project, involving their sustainable wild harvest as an ingredient in growing world markets of functional foods and health drinks (Cunningham & Brinckmann, 2010).

CONSERVATION STATUS

Endemic to China: Data Deficient (MEPCAS, 2013); Not Evaluated (IUCN Red List, accessed 8/6/2016).

S. sphenanthera (orange-flowered form)

UNOFFICIAL SUBSTITUTES

The name 'Wu Wei Zi' has also been used to refer to many of China's 18 other species of *Schisandra* and their close relatives (Applequist, 2006; Ho *et al.*, 2006; Xiao *et al.*, 2001–2). See Lesser Known Unofficial Substitutes of 'Wu Wei Zi' (Schisandrae Chinensis Fructus) page 742.

Kadsura longipedunculata

Male flowers of *S. sphenanthera*

Male flowers of *S. sphenanthera* (left-hand flower with 'petals' partly removed to show stamens)

Young pendent fruiting head of *S. sphenanthera*

TCM PROPERTIES AND USE

properties: **'sour, sweet and warm'**
channels: **'Kidney, Lung and Heart'**
actions: **'astringes, tonifies qi, Lung and Kidney, generates fluids, secures essence'**

indications: **chronic cough, dyspnoea; spermatorrhoea; leucorrhoea; enuresis; spontaneous sweating; irritability, palpitations**

DRUG MORPHOLOGY

CRUDE DRUG (DRIED FRUIT): irregularly globose, somewhat angular where fruit wall has shrunk (4–6 mm diam.) **[1]**. External surface often dull, brownish-red to dark brown **[2]**, shrivelled with 1 or 2 seeds usually visible where flesh thin **[3]**; a brown oval stem scar visible at one end **[4]**. Pulp slightly oily and closely adhering to seed, making it difficult to remove cleanly. Seed somewhat ovoid to kidney-shaped **[5]** (3–3.5 mm long; 3.5–4 mm diam.), seed coat dark greyish-brown, matt **[6]**, minutely verrucose (warty) [hl] **[7]**, thin and fragile. *Taste*: pulp slightly sour. *Odour*: pulp indistinct; seed aromatic when crushed. ***Good quality***: large, purplish-red with fleshy pulp; intensely aromatic.

CRUDE DRUG-2 (VINEGAR-PROCESSED FRUIT): as for dried fruit but brownish-black or, where flesh is thin, deep red and somewhat oily. *Taste*: vinegar-tainted. *Odour*: vinegar-scented.

PROCESSING

Almost ripe fruits are sun or oven-dried or steamed, then stalks removed (= crude drug). Alternative processing (after drying) involves steaming with rice vinegar until fruits turn dark brownish-black (= Cu Nan Wu Wei Zi 醋南五味子, 'vinegar-processed Nan Wu Wei Zi'; crude drug-2). Fruits are pounded prior to decocting.

OTHER COMMON NAMES

华中五味子 Hua Zhong Wu Wei Zi (plant name); 五味子 Wu Wei Zi; five flavour fruit; Fructus Schisandrae Sphenantherae; schisandra (SCN); southern magnolia vine fruit (CP2010).

SAFETY

Safety Class: 1; Interaction Class: C (Gardner & McGuffin, 2013). An official healthfood in China (Anon., 2002a,b).

Crude drug Nan Wu Wei Zi (dried fruit of *S. sphenanthera*)

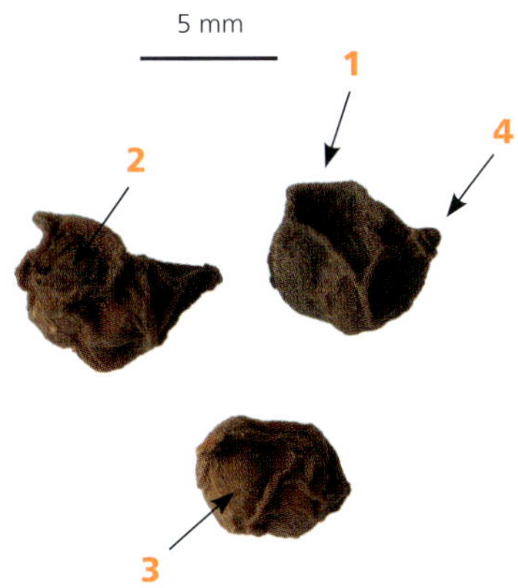

Fruit

Seed

Vinegar-processed Nan Wu Wei Zi (Ci Wu Wei Zi)

Niu Bang Zi 牛蒡子

Arctii Fructus

burdock fruit

Drug part: FRUIT

Arctium lappa L.

Asteraceae

PLANT DESCRIPTION

A branching, biennnial herb (up to 2 m) with a fleshy edible root, purplish stems and an untidy habit. *Leaves* alternate and simple; basal ones large and heart-shaped; upper ones smaller, varying in shape; leaf blade lower surface greyish-white or pale green and downy; upper surface dotted with yellow glands; **margins entire**. *Flowers* tiny and densely packed into globose thistle-like heads (15–42 mm diam.); heads solitary or more **usually in groups of 2–4**. Each *flower-head* with many linear to lance-shaped hook-tipped outer bracts (6–21 mm long; 0.7–2.6 mm wide) surrounding purplish-red florets. *Fruit-heads* with hook-tipped bracts persistent and surrounding the densely packed bristle-topped fruits (achenes), entire head often breaking free and attaching to clothing when mature; *fruits* (5–7 mm long) mottled pale brown, with short apical bristles (pappus) (1–3.5 mm long). Fl. and fr. Jun–Sep.

Species of *Arctium* wild in China: 2

HARVESTING, SOURCING AND NATURAL RANGE

Ripe fruit-heads are harvested in autumn from the wild or from plants extensively cultivated for their edible roots.

Wild plants grow in disturbed places, often near rivers, roads and villages (700–3,500 m altitude) throughout China except Hainan, Taiwan, and Tibet [also wild in Afghanistan, Bhutan, India, Japan, Nepal, Pakistan; SW Asia and Europe]. Traditional *Di Dao* quality Niu Bang Zi is sourced from NE China (Heilongjiang, Jilin and Liaoning provinces).

CONSERVATION STATUS

China: Least Concern (authors' pers. obs. 2014)

Global: Not Evaluated (IUCN Red List, accessed 12/3/2015).

Silybum marianum (L.) Gaertn.

Asteraceae

synonym: *Carduus marianus* L.

PLANT DESCRIPTION

Easily distinguished from the Pharmacopoeia species. Stout annual or biennial (40–150 cm tall); **leaves conspicuously white-marbled**, **margins spiny and irregularly lobed**; **flower-heads solitary** (25–50 mm diam.) with **bracts spiny,** broader (8–15 mm long; 6–10 mm wide) not hooked but with a stout apical spine (25–70 mm long); **bristles (pappus) of fruits longer (up to 15 mm) and denser.** (Sell & Murrell, 2006).

SUBSTITUTION COMMENT. Widely cultivated in China for its fruits (often incorrectly called seeds) and which are the source of the separate TCM drug 'Shui Fei Ji' (Silybi Fructus) and also widely used in Western herbal medicine and as an extract marketed as a dietary supplement to protect the liver. The superficial macroscopic similarity of its fruit occasionally causes confusion in trade; this is to be avoided since their clinical actions are quite different.

CONSERVATION STATUS

China: not native.

Global: Not Evaluated (IUCN Red List, accessed 13/7/2016).

TCM PROPERTIES AND USE

properties: **'pungent, bitter and cold'**
channels: **'Lung and Stomach'**
action: **'dispels wind heat'**

indications: **common cold or 'flu; measles, rubella and mumps; boils and sores**

Arctium lappa in fruit

A. lappa in flower with (inset) typically in groups of 2–4 flower-heads

A. lappa immature fruit-heads

A. lappa fruit-heads from above showing short bristles (pappus) atop each dry, sub-cylindrical fruit (achene)

Silybum marianum flowering plants

S. marianum with typical solitary flower-head

S. marianum fruiting heads with long dense feathery bristles atop each dry somewhat cylindrical fruit

Arctium lappa

Asteraceae

DRUG MORPHOLOGY

CRUDE DRUG (DRIED FRUIT): fruit oblong, somewhat laterally flattened and angular, often slightly curved (5–7 mm long; 2–3 mm wide), both ends rather truncated. Externally fruit coat dull, greyish-brown, purplish or greenish-black, usually speckled darker [hl]; with numerous longitudinal ridges **[1]**, 1 or 2 more prominent and often wrinkled near the base. Fruit apex (broadest end) with sunken and roundish flower scar **[2]**, often with small central and pointed persistent style base [hl]; fruit base usually with a creamish point of attachment scar [hl] **[3]**. Internally, 1 seed with 2 oily yellowish-white cotyledons. *Texture*: hard and difficult to break. *Odour*: slight. *Taste*: bitter, slightly pungent, astringent and numbing. ***Good quality*:** large, plump, greyish-brown fruits.

PROCESSED DRUG (DRY-FRIED FRUIT): as for crude drug but inflated, slightly darker and slightly aromatic.

DECOCTING PIECES (PULVERISED): amorphous mass of broken fruit walls and whitish oily seed contents.

PROCESSING

Fruit-heads are dried and threshed to release the fruit (= crude drug); they are pulverised before being decocted. The crude drug may also be dry-fried (= 'Chao Niu Bang Zi').

OTHER COMMON NAMES

牛蒡 Niu Bang; 牛子 Niu Zi; bardana; burdock (SCN); Fructus Arctii.

SAFETY

Safety Class 1; Interaction Class A (Gardner & McGuffin, 2013).

Silybum marianum

Asteraceae

'DRUG SUBSTITUTE' MORPHOLOGY

Fruits easily distinguished from those of the Pharmacopoeia species. Main differences: externally pale grey, greyish-brown or, **more usually, dark brown [4], fewer longitudinal ridges; apical scar much larger, slightly lustrous, creamish and collar-like [5] and typically encircling a large and often protruding, rounded oil body (elaiosome) [6]**.

Despite these differences, mixed batches containing both species may easily be overlooked because of the considerable colour variability of the fruits of both species. Furthermore, *S. marianum* fruits may be mistaken for the dry-fried fruits of Pharmacopoeia species since their colour and size are similar (see photo opposite of a contaminated sample).

OTHER COMMON NAMES

水飞蓟 Shui Fei Ji; milk thistle seed; holy thistle seed; Fructus Silybi, lady's thistle seed; Silybi Mariani Fructus.

SAFETY

Safety Class 1; Interaction Class A (Gardner & McGuffin, 2013; and where incorrectly referred to as 'seed'. An official healthfood in China (Anon., 2002a,b).

LESSER KNOWN UNOFFICIAL SUBSTITUTES

The fruits of other species have been reported as local or historical unofficial substitutes for reasons of clinical substitution and/or confusion (Bensky *et al.*, 2004; NICPBP 1997–2002).

1. ***Amorpha fruticosa*** L. (紫穗槐 Zi Sui Huai); family Fabaceae; fruits curved, slightly longer (6–10 mm) than those of the Pharmacopoeia species; markedly gland-dotted and apex beaked.
2. ***Arctium tomentosum*** Mill. (毛头牛蒡 Mao Tou Niu Bang or 绒毛牛蒡 Rong Mao Niu Bang); family Asteraceae; confined to Xinjiang province; fruits multi-striate.
3. ***Aucklandia costus*** Falc. (云木香 Yun Mu Xiang); family Asteraceae; fruits 6–8 mm long, apex wrinkled and with toothed crown. The root is the source of Aucklandiae Radix (木香 Mu Xiang) see page 210 CITES App. 1.
4. ***Onopordum acanthium*** L. (大翅蓟 Da Chi Ji); family Asteraceae; confined to Xinjiang province; fruits 3-ribbed, transversely wrinkled; when attached to plant, apical bristles (pappus) up to 1.2 cm long, reddish.

Crude drug of Niu Bang Zi
(*Arctium lappa* dried fruits)

Processed drug Chao Niu Bang Zi
(*A. lappa* fruits dry-fried)

Arctium lappa fruit (Niu Bang Zi)

Fruits of unofficial substitute Shui Fei Ji
(*Silybum marianum*)

Dry-fried fruits of unofficial substitute
Shui Fei Ji (*S. marianum*)

S. marianum fruit (Shui Fei Ji)

Dry-fried *A. lappa* fruit (Niu Bang Zi),
heavily contaminated with dry-fried
S. marianum fruit (Shui Fei Ji)

Nu Zhen Zi 女贞子

Ligustri Lucidi Fructus

Drug part: FRUIT

Ligustrum lucidum W. T. Aiton

Oleaceae

Species of *Ligustrum* wild in China: 27

PLANT DESCRIPTION

An evergreen or deciduous tree or shrub (up to 25 m tall). *Leaves* opposite, short-stalked, leaf blade ovate, broadly elliptic or lance-shaped (6–17 cm long; 3–8 cm wide), dark green (sometimes glossy), usually leathery, margins entire. *Stipules* absent. *Flowers* in conspicuous creamish-white, many-branched terminal panicles (8–25 cm long, 8–25 cm diam.). *Flowers* tiny and stalkless, calyx 4-parted, petal lobes 4, creamish-white, as long as petal tube, recurved to expose the 2 stamens; ovary superior, 2-valved; style 1 (shorter than stamens); stigma 2-cleft. *Fruits* dry, often crescent-shaped (7–10 mm long; 4–6 mm diam.), ripening in winter to bluish-black or reddish-black with a membranous outer skin and persistent calyx. *Seeds* 1 or 2. Fl. May–Jul, fr. Jul–May in southern China; in more northerly latitudes flowering and fruiting occurs some months later.

HARVESTING, SOURCING AND NATURAL RANGE

Mature or almost mature fruits are harvested in winter mainly from cultivated trees. In China the species is widely grown as a street tree in urban areas but also to culture wax insects used in the production of white wax candles etc.

The species is endemic to many parts of China except in the north (Anhui, Fujian, Gansu, Guangdong, Guangxi, Guizhou, Hainan, Henan, Hubei, Hunan, Jiangsu, Jiangxi, Shaanxi, Sichuan, Tibet, Yunnan and Zhejiang) and grows in woodland (up to 2,900 m altitude).

CONSERVATION STATUS

Endemic to China: Least Concern (MEPCAS, 2013); Not Evaluated (IUCN Red List, accessed 8/4/2015).

Ligustrum lucidum flowering in cultivation

Inflorescence of *L. lucidum*

Branch of *L. lucidum* with immature fruit-head

UNOFFICIAL SUBSTITUTES/CONTAMINANTS

Fruits of other species in the same or different families have occasionally been documented as local or historical unofficial substitutes in China. Their occurrence as substitutes in the international TCM trade is uncertain.

1. ***Ligustrum expansum*** Rehder — fruit (syns. *L. robustum* subsp. *chinense* P. S. Green and *L. thibeticum* Decne. — 粗壮女贞 Cu Zhuang Nu Zhen. A very local species whose wild populations are confined to western Hubei province. Fruit oblong to ellipsoid (c. 10 mm long; 5 mm diam.). Fruiting in September.
2. ***L. japonicum*** Thunb. — fruit (Japanese privet). Widely cultivated in China and the West as a popular hedging and ornamental plant. Fruits ovoid, not curved; reported as a substitute of the Pharmacopoeia species in Taiwan (Ho *et al.*, 2006).
3. ***L. sinense*** Lour. — fruit (小蜡 Xiao La). Widely distributed in the wild in temperate and tropical China. Fruits subglobose, 5–8 mm diam.
4. ***Viburnum schensianum*** Maxim. — fruit (陕西荚蒾 Shan Xi Jia Mi) (family Adoxaceae). Fruits superficially resemble those of the official drug: ovoid to globose (6–8 mm long; 4–5 mm diam.), maturing from red to black, base rounded, dorsally ridged and ventrally 3-grooved.

TCM PROPERTIES AND USE

***properties*: 'sweet, bitter and cool'**
***channels*: 'Liver and Kidney'**
***action*: 'tonifies Liver and Kidney, clears heat'**

***indications*: tinnitus, dizziness, vertigo; menopausal symptoms; constipation; thirst; weakness of back and knees**

DRUG MORPHOLOGY

CRUDE DRUG/DECOCTING PIECES (DRIED FRUIT): light, ovoid, elliptical or frequently crescent-shaped **[1]** (6–8.5 m long; 3.5–5.5 mm wide); both ends rounded **[2]**. Externally purplish-black, purple, greyish- or brownish-black; when almost mature, fruits have a shrunken surface with a conspicuous network of ridges **[3]**; mature fruits (less common) almost smooth or with occasional depressions; basal end with remains of greenish fruit stalk **[4]** and/or persistent calyx **[5]** present. Outer and inner fruit walls (epicarp and mesocarp respectively) thin and easily removed to reveal 1 (rarely 2) seeds **[6]** purplish-black or yellowish-brown, longitudinally furrowed [hl] and curved. *Texture*: dry and hard; sometimes pliable. *Fracture*: white (endosperm) **[7]**. *Odour*: indistinct. *Taste*: sweet, slightly bitter and astringent. ***Good quality*:** plump, solid and greyish-black.

DECOCTING PIECES (WINE-PROCESSED FRUIT — JIU NU ZHEN ZI): as for the dried fruits but slightly larger (swollen) **[8]**, externally bluish-black or black with a few large shrunken depressions **[9]**. *Odour*: wine-scented.

PROCESSING

After removal and discarding of leaves and stems, fruits are dried, or blanched then dried (= crude drug/decocting pieces). Dried fruits may subsequently be steamed or stewed with yellow rice wine to produce 酒女贞子 Jiu Nu Zhen Zi (wine-processed Nu Zhen Zi).

OTHER COMMON NAMES

Chinese privet; Fructus Ligustri Lucidi; glossy privet fruit (CP2010); ligustrum (SCN); white wax tree.

SAFETY

Safety Class: 1; Interaction Class: A (Gardner & McGuffin, 2013). An official healthfood in China (Anon., 2002a,b).

Crude drug

Wine-processed Nu Zhen Zi (= Jin Nu Zhen Zi)

Qian Shi 芡实

Euryales Semen

Drug part: RIPE SEED (kernel)

Euryale ferox Salisb.

Nymphaeaceae

Species of *Euryale* wild in China: 1

PLANT DESCRIPTION

An annual or short-lived aquatic perennial herb with rhizomes. *Leaves* simple, long-stalked, blades submerged or floating and fixed to stem at its centre (peltate). Submerged leaves not prickly, arrow-shaped or elliptic (4–10 cm diam.) with a deeply heart-shaped base. Floating leaves broadly elliptic to orbicular (1.3 (rarely up to 2.7) m diam.) sometimes with a distinctive notch on one side; leaf blades and leaf stems distinctly spiny; blade undersides dark purple with distinct netted ribbing, margin entire. *Flowers* solitary, floating or partly submerged (up to 5 cm diam.), long-stalked, sepals 4, greenish and densely prickly; petals many, in 5 concentric whorls; oblong to lance-shaped; purplish-violet fading to white centrally, stamens many, shorter than petals; ovary inferior and multilocular. *Fruit* dark purple, globose (5–10 cm diam.), spongy and densely spiny. *Seeds* black, 8 to many, globose (6–10 mm diam.). Fl. Jun–Aug.; ft. late autumn to early winter.

HARVESTING, SOURCING AND NATURAL RANGE

Ripe fruit is harvested in late autumn and early winter mostly from cultivated plants grown on small-holdings and medicinal plant farms. The main production areas are in the provinces of Jiangsu, Shandong, Anhui, Hunan and Hubei.

The species grows wild in lakes and ponds throughout China (except the north-west): Anhui, Fujian, Guangdong, Guangxi, Guizhou, Hainan, Hebei, Heilongjiang, Henan, Hubei, Hunan, Inner Mongolia, Jiangsu, Jiangxi, Jilin, Liaoning, Shaanxi, Shandong, Shanxi, Sichuan, Taiwan, Yunnan and Zhejiang [also wild in Bangladesh, India, Japan, Kashmir, the Korean Peninsula and Russia (Far East)].

In China the seeds are also eaten as a source of starch and as a healthfood.

CONSERVATION STATUS

China: Least Concern (MEPCAS, 2013).

Global: Least Concern (IUCN Red List, accessed 25/8/2016).

Euryale ferox, flowering plant in the wild

E. ferox — bursting flower bud

E. ferox, floating leaves with prickly upper surface

E. ferox, fruit interior showing developing seeds

TCM PROPERTIES AND USE

***properties*: 'sweet, sour and neutral'**
***channels*: 'Kidney and Spleen'**
***actions*: 'tonifies Spleen, dispels damp, tonifies the Kidneys, secures essence'**

***indications*: chronic diarrhoea; spermatorrhoea; enuresis, incontinence; leucorrhoea**

DRUG MORPHOLOGY

CRUDE DRUG/DECOCTING PIECES-1 (SEED): ovoid or sub-spherical, often broken; when whole (5–8 mm diam.). External surface (tegmen — inner seed coat) smooth, dull, reddish-brown for $^{2}/_{3}$ length of seed **[1]**, remaining $^{1}/_{3}$ pale brown grading into white or cream **[2]**; surface white when tegmen removed; apex with a large, circular and sunken hilum (2–3 mm diam.) **[3]**. Internally pure white and starchy, with 2 cotyledons. *Texture*: quite hard. *Fracture*: white and starchy **[4]**. *Odour*: none. *Taste*: bland. ***Good quality*:** unbroken, externally purplish-red or yellowish-white; powdery internally.

DECOCTING PIECES-2 (BRAN-FRIED SEED — 'Fu Chao Qian Shi'): as for dried seed but speckled with pale bran flakes **[5]** and broken surfaces tinged pale yellowish-brown **[6]**.

PROCESSING

The spiny fruit coat and thick, rigid, black seed coat (testa) is removed and discarded, then the seed is sun-dried (= crude drug/decocting pieces); further optional processing by dry-frying with wheat bran produces a Fu Chao Qian Shi (麸炒芡实 — bran-fried Qian Shi).

OTHER COMMON NAMES

刺莲蓬实 Ci Lian Peng Shi; 鸡头米 Ji Tou Mi; 苏芡实 Su Qian Shi; fox nut; euryale (SCN); Gordon Euryale Seed (CP2010); prickly water lily; Semen Euryales.

SAFETY

Safety Class: 1; Interaction Class: A (Gardner & McGuffin, 2013). An official Chinese food and healthfood (Anon., 2002a,b).

Crude drug/decocting pieces-1 Qian Shi (dried seed of *Euryale ferox*)

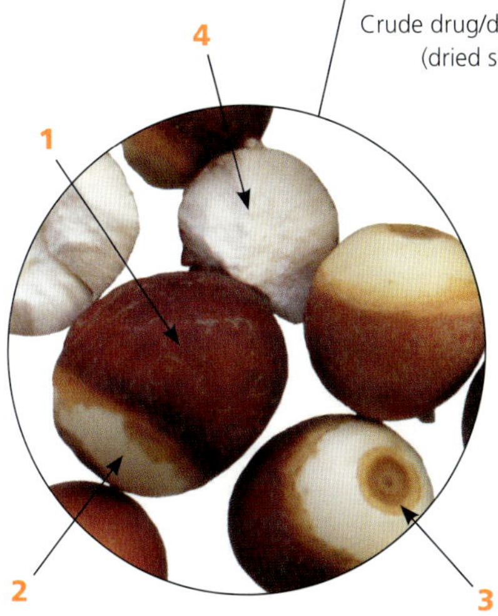

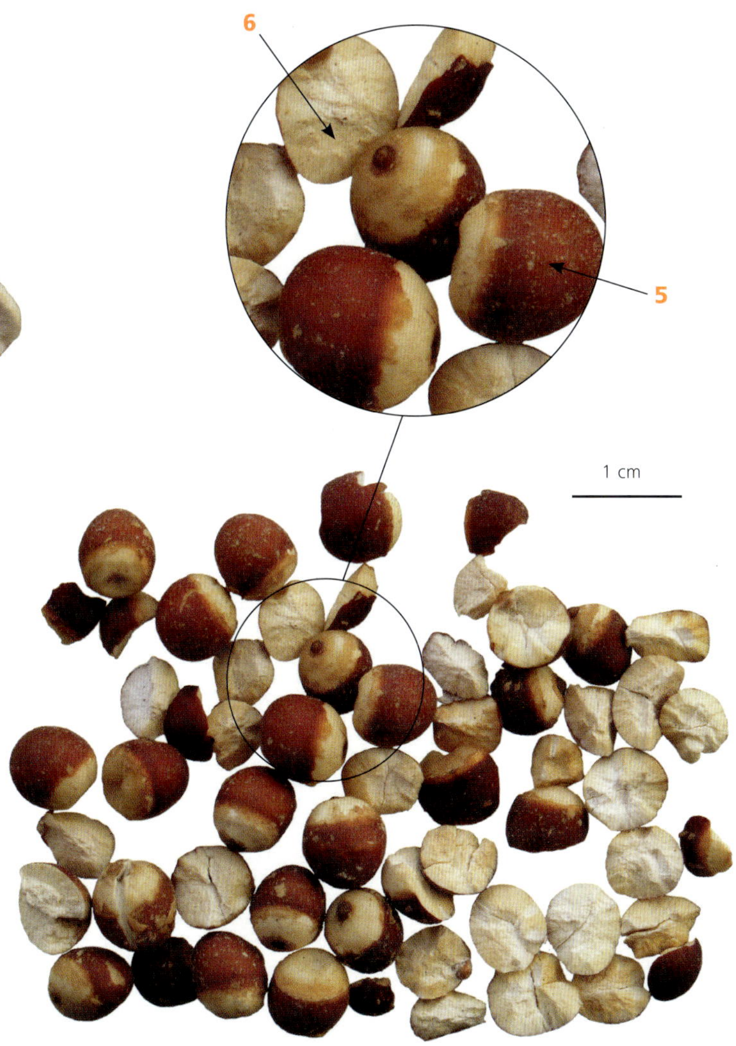

Decocting pieces-2, bran-fried Qian Shi (Fu Chao Qian Shi)

Qing Pi 青皮

Citri Reticulatae Pericarpium Viride

green tangerine peel

Drug part: IMMATURE FRUIT (PEEL AND WHOLE)

Citrus × reticulata Blanco

Rutaceae

taxonomic/nomenclatural note: in CP2015 as '*C. reticulata* Blanco'; *Kew-MPNS* accepted name 7/9/2016, *C.* × *aurantium* L., but *FoC* considers it to be a separate species. 4 main medicinal varieties are used (see 'Chen Pi' page 612)

Species and hybrids of *Citrus* wild in China: 11 (+ 5 hybrid cultivated species)

PLANT DESCRIPTION

(see 'Chen Pi' (Citri Reticulatae Pericarpium, page 612).

See also other drugs sourced from this hybrid: ('Chen Pi' (page 612), 'Ju He'(page 664) and 'Ju Hong' (page 666).

HARVESTING, SOURCING AND NATURAL RANGE

Using the very young fruit, two kinds of drug are created: the small young windfall fruit are collected in May and June to be traded as Ge Qing Pi (whole fruit) and the larger but still immature fruit on the tree are harvested in July and August, the peel removed and dried and traded as Si Hua Qing Pi (4-parted Qing Pi).

CONSERVATION STATUS

As for 'Chen Pi' (Citri Reticulatae Pericarpium) page 612.

Immature fruit of *C. reticulata* 'Un Shui' (a source of Qing Pi called Si Hua Qing Pi)

Young windfall fruit of *C. reticulata* 'Da Hong Pao' (a source of Qing Pi called Ge Qing Pi)

Ge Qing Pi

1 cm

Crude drug

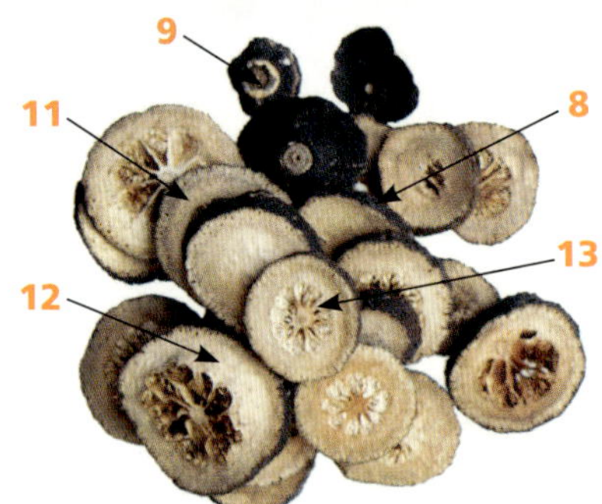

Decocting pieces – dried

Decocting pieces (vinegar processed

TCM PROPERTIES AND USE

***properties*: 'pungent, bitter and warm'**
***channels*: 'Liver, Gall Bladder and Stomach'**
***actions*: 'invigorates qi, courses Liver, dispels stasis'**

***indications*: chest and abdominal pain, distension; indigestion; hernia; irritability; mastitis; nausea and vomiting**

DRUG MORPHOLOGY

CRUDE DRUG (DRIED PEEL):

Si Hua Qing Pi (4-parted Qing Pi): intact peel of one fruit (c. 4–6 cm long; 0.1–0.2 cm thick), typically with 4 equal sized lobes **[1]** indicating cuts made when peel removed from fresh fruit; lobes elliptical and shortly pointed; central point of attachment with round central scar of fruit-stalk **[2]**. External surface greyish-green or blackish-green, pitted throughout with many oil ducts **[3]**. Inner surface almost white or yellowish-white, rough, with occasional yellowish-white or yellowish-brown veins **[4]**. *Texture*: somewhat hard, easily broken. *Fracture*: reveals 1–2 layers of oil ducts. *Odour*: aromatic. *Taste*: pungent and bitter. ***Good quality*:** externally blackish-green and hard; strongly aromatic.

Ge Qing Pi (whole Qing Pi): fruits sub-globose (0.5–2 cm diam.). Externally greyish-green or blackish-green, somewhat rough and densely pitted with oil ducts **[8]**; basal end with distinctive pale brown or cream, oval fruit-stalk scar **[9]**; apex with slightly raised style remnants **[10]**. *Texture*: hard. *Fracture*: fruit wall pale yellowish-brown or yellowish-white (1–2 mm thick), 1–2 layers of oil ducts visible at edge of fracture; pulp in transverse fracture with 8–10 pale brown vesicles. *Odour*: aromatic. *Taste*: bitter, sour and pungent. ***Good quality*:** externally blackish-green; internal pith yellowish-white; strongly aromatic.

DECOCTING PIECES (DRIED SLICED PEEL):

Si Hua Qing Pi (4-parted Qing Pi): sub-rounded or irregularly-shaped slivers (up to 3 cm long; 0.1–0.2 cm thick); often curved **[5]**. External and internal surfaces as for the crude drug of Si Hua Qing Pi (above). *Fracture*: reveals 1–2 layers of oil ducts **[6]**. *Texture*: somewhat hard, easily broken. *Odour*: aromatic. *Taste*: pungent and bitter.

Si Hua Qing Pi (Cu Qing Pi) *vinegar-processed*: as for sliced decocting pieces above but cut-surfaces darker (mid- to dark-brown) **[7]**. *Odour* and *taste*: vinegar-like.

Ge Qing Pi (Whole Qing Pi): round or sub-rounded slices (0.5–2 cm diam.; 2–3 mm thick). Uncut external surface as for crude drug of Ge Qing Pi (above). Cut surface pale yellowish- or greyish-white with many oil ducts confined to the outermost edge **[11]**); a continuous zone of undifferentiated pith **[12]** surrounds 8–10 pulp vesicles **[13]**. The appearance of the latter varies in longitudinal and transverse section as does the ratio in the cut surface of pith to pulp. *Odour*: aromatic. *Taste*: sour, bitter and pungent.

Ge Qing Pi (Cu Qing Pi) *vinegar-processed*: as for sliced Ge Qing Pi above but cut-surface darker (mid- to dark-brown) **[14]**. *Odour* and *taste*: vinegar-like.

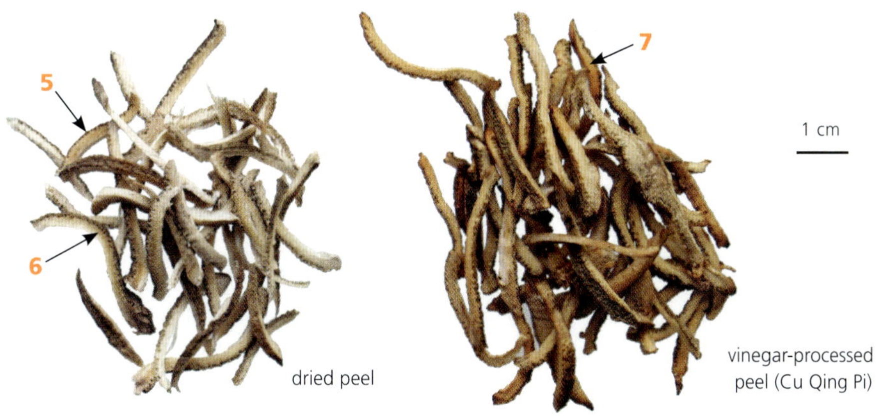

Decocting pieces of Sui Hua Qung Pi

PROCESSING

Fruits to be traded as Ge Qing Pi (i.e. from windfalls) are sun-dried as whole fruit, then later washed and moistened in a sealed container before being sliced into decocting pieces and dried again; further optional processing by moistening with rice-wine vinegar then dry-frying is also traditionally performed.

The larger tree-picked fruits, traded as Si Hua Qing Pi, are prepared when fresh by removing the peel in one 4-lobed longitudinal piece: the basal part (with the fruit-stalk scar) forms a central point which joins the 4 lobes; the flesh is discarded. The lobed peel is sun-dried and later cut into decocting slivers; the latter may be further processed with rice-wine vinegar to produce Cu Qing Pi.

OTHER COMMON NAMES

柑橘 Gan Ju (plant name); 个青皮 Ge Qing Pi (whole Qing Pi); 四化青皮 Si Hua Qing Pi (4-parted Qing Pi); green tangerine peel (CP2010); Pericarpium Citri Reticulatae Viride; tangerine (SCN).

SAFETY

Safety Class: 1; Interaction Class: A (Gardner & McGuffin, 2013). An official healthfood in China (Anon., 2002a,b)

Crude drug of Si Hua Qing Pi

Rou Dou Kou 肉豆蔻

Myristicae Semen

nutmeg

Drug part: SEED

Myristica fragrans Houtt.

Myristicaceae

Species of *Myristica* wild in China: 1 (+ 1 introduced)

PLANT DESCRIPTION

A small, bushy, evergreen tree (up to 10 m tall). *Leaves* alternate and simple, blades elliptic or elliptic to lance-shaped (4–8 cm long; 1.4–4 cm wide), somewhat leathery; hairless; upper surface often glossy; lateral veins 16 pinnate pairs; margin entire and tip shortly pointed. *Flowers* in separate inflorescences on the same tree. Male inflorescence slender (2.5–5 cm long), unbranched or forked, often with 4–8 flowers per branch. Male flowers 3- (rarely 4-) lobed, stamens 9–12 with filaments fused into a column. Female inflorescence with flowers solitary or few. Female flowers 3-lobed, yellowish-white (c. 6 mm long, 4 mm diam.); ovary superior; style 1; stigmas 2 and minute. Fruit solitary or in pairs, dry, sub-rounded to ovoid (3.5–5 cm diam.), pendant ripening to orange or yellow; fruit wall thick and leathery or fleshy, breaking into 2 halves when ripe. *Seed* solitary and surrounded by a fleshy, lustrous, reddish-brown and finely divided aril (source of mace); seed (nutmeg) large and ellipsoid (2–3 cm long; c. 2 cm diam.), woody, with endosperm (food store) characteristically marbled and rich in volatile oils.

HARVESTING, SOURCING AND NATURAL RANGE

Mature fruits are harvested from cultivated trees twice yearly (during Apr–Jun and Nov–Dec). The species is not native to China but is cultivated in tropical parts of S China (Guangdong, Taiwan and Yunnan) both as a source of 'Rou Dou Kou' and the popular culinary spices of mace (the aril) and nutmeg (the seed). The species is native to Indonesia (Moluccas) and widely cultivated as a spice throughout the tropics.

CONSERVATION STATUS

China: not native.

Global: Data Deficient (IUCN Red List, accessed 8/8/2016).

Myristica fragrans with unripe fruit

Leaves and ripe fruit of *M. fragrans* split to reveal red aril-covered seed

TCM PROPERTIES AND USE

properties: 'pungent and warm'
channels: 'Large Intestine, Spleen and Stomach'
actions: 'binds the intestines, warms the interior, regulates qi, stops pain'

indications: chronic diarrhoea; abdominal bloating

DRUG MORPHOLOGY

CRUDE DRUG (DRIED OR LIMED SEED — 'NUTMEG'): seed ovate to sub-spherical (2–3 cm long; 1.5–2.5 cm diam.). Externally pale or light to dark brown or greyish-brown, shallowly but densely furrowed throughout **[1]**; a broad and shallow longitudinal furrow (raphe) **[2]** connects a pale round, slightly raised hilum **[3]** at the basal (wider) end with a slightly sunken chalaza (growing point) at the opposite end **[4]**. Remnants of processing with 'lime milk' or flour sometimes present **[5]**. *Texture*: very hard and difficult to crush. *Fracture*: strongly ruminated (marbled) **[6]** with dark and light brown tissues where the pale endosperm has become 'laced' with the infoldings of the inner layer of the seed coat. *Odour*: intensely aromatic. *Taste*: warm and pungent. ***Good quality*:** large, heavy and solid seeds; oily and aromatic.

DECOCTING PIECES (BRAN-FRIED SLICED SEED — 'NUTMEG'): as for the crude drug but cracked and externally brown. *Odour*: intensely aromatic. *Taste*: pungent.

Rou Dou Kou crude drug

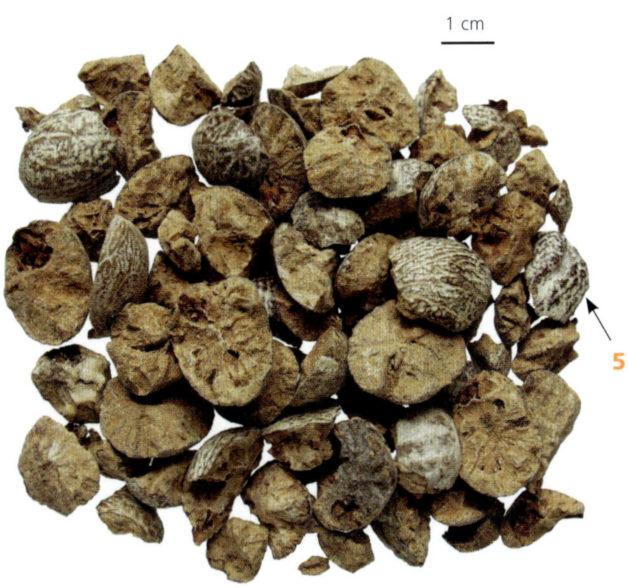

Rou Dou Kou pounded into decocting pieces

PROCESSING

The thick leathery fruit wall (pericarp), red aril and outer seed coat are removed. The remaining seed material is then either soaked in a lime solution for a day then dried, or dried directly at low temperature (= crude drug). The dried seed is pounded into small pieces prior to decocting. Alternative processing involves roasting the seed with wheat bran until the bran turns yellow (= Wei Rou Dou Kou 煨肉豆蔻 (roasted Rou Dou Kou); see also *Safety*.

OTHER COMMON NAMES

肉果 Rou Guo; 玉果 Yu Guo; nutmeg (SCN); Semen Myristicae. Many other trade names exist for this drug reflecting its use as a medicine and culinary spice in other countries.

SAFETY

An official food and healthfood in China (Anon., 2002a,b). Safety Class: 2b and 2d; Interaction Class: A (Gardner & McGuffin, 2013). Unroasted seeds (kernels) are **toxic** in doses above 5g (Barceloux, 2008). TCM normally uses roasted seeds reported to have lower toxicity (Bensky *et al.*, 2004; Chen & Chen, 2004).

Rou Dou Kou (one piece (seed) halved) plus dried and separated aril (mace) **[7]**

Sang Shen 桑椹

Mori Fructus

white mulberry fruit

Drug part: FRUIT

Morus alba L.

Moraceae

taxonomic note: two varieties and many cultivars exist and all are thought to be used as the source of this drug.

Species of *Morus* wild in China: 10

PLANT DESCRIPTION

A deciduous shrub or small tree (3–10 m tall). *Bark* grey and shallowly furrowed. *Stipules* (at base of leaf stalks) lance-shaped (2–3.5 cm long, usually falling). *Leaves* alternate, simple, stalked and large (5–30 cm long; 5–12 cm wide), ovate or irregularly lobed; base rounded or heart-shaped, margin coarsely toothed or shallowly scalloped, tip sharply pointed or broad; **leaf undersides sparsely hairy along mid-veins or in tufts in vein axils.** *Male and female flowers* in short cylindrical catkins arising from the leaf axils and usually on separate plants. Male catkins pendent and densely white hairy (2–3.5 cm long) with tiny male flowers each with 4 pale green, overlapping calyx lobes and 4 stamens; petals are absent. Female catkins 1–2 cm long (shorter than male), softly hairy and packed with tiny female flowers; each flower with 4 green ovoid and overlapping calyx lobes (becoming fleshy in fruit), style absent and stigma with 2 distinctively white-edged and diverging branches [hl]; petals absent. *Fruit* a cylindrical or ovoid multiple fruit (syncarp), resembling a blackberry and becoming very succulent; depending on the cultivar, ripening from red to either blackish-purple, purple or greenish-white. Fl. Apr–May, fr. May–Aug.

For other drugs sourced from this species see 'Sang Ye' (page 554), 'Sang Bai Pi' (page 522) and 'Sang Zhi' (page 476).

UNOFFICIAL SUBSTITUTES

Many species and cultivars of *Morus* have long been cultivated in China for a variety of uses such as food, medicine, silk-worm food and paper-making. Their ready availability may result in accidental or intentional confusion of their medicinal parts when traded in local markets in China; 'Sang Shen', for example, may be prone to substitution at a local level and identification is only likely if fruits can be traced to their source plants (*see below*). It is unlikely these issues relate to 'Sang Shen' destined for international markets.

1. ***Morus mongolica*** (Bureau) C. K. Schneid.. Mongolian mulberry. Leaves hairless; teeth of leaf margins with short, needle-like points.
2. ***M. australis*** Poir. Korean mulberry. Lower leaf surface sparsely clothed with thick hairs; leaf margin saw-toothed or entire.
3. ***M. cathayana*** Hemsl. Leaf undersides with a dense covering of white or yellowish-grey hairs. Female flowers with a very short style.

HARVESTING, SOURCING AND NATURAL RANGE

Fruits are harvested when almost ripe (red) from cultivated plants during April–June. Plants are grown throughout China primarily for their leaves for the rearing of silkworms in the silk industry. The main source of the medicinal fruits are from crops grown in Anhui, Hunan, Jiangsu and Zhejiang provinces (source Zhao & Chen, 2009–10).

CONSERVATION STATUS

China: Not Evaluated.

Global: Not Evaluated (IUCN Red List, accessed 25/8/2016).

Leaf underside of *M. alba* showing white hair tufts in axils of midvein and primary lateral veins

Crop of *Morus alba*, Anhui province

M. alba with ripening fruits

TCM PROPERTIES AND USE

properties: 'sweet, sour and cold'
channels: 'Heart, Liver and Kidney'
actions: 'tonifies yin and blood, generates fluids, moistens intestines'

indications: dizziness; tinnitus; palpitations; insomnia; constipation

DRUG MORPHOLOGY

CRUDE DRUG/DECOCTING PIECES (DRIED FRUIT): a multiple fruit (syncarp) (1–2 cm long, 0.5–0.8 cm wide) **[1]** with many succulent and densely packed yellowish-brown, brownish-red or purplish-black calices (fused sepals) ovoid, ellipsoid or cylindric (c. 2 mm long, 1 mm wide) **[2]**; each succulent calyx encloses a hard achene (single seeded fruit). Curved fruit stalks occasionally attached. *Texture*: hard. *Odour*: indistinct. *Taste*: slightly sour and sweet. ***Good quality*:** large purplish-red fruit with thick, sweet flesh.

PROCESSING

Fruits are sun-dried or briefly steamed then sun-dried (= crude drug and decocting pieces).

OTHER COMMON NAMES

桑 Sang (plant name); 桑果 Sang Guo; 桑椹子 Sang Shen Zi; Fructus Mori; white mulberry (SCN).

SAFETY

An official food and healthfood in China (Anon., 2002a,b). Safety Class: 1; Interaction Class: A (Gardner & McGuffin, 2013). Rare cases of adverse reaction in children and other allergic skin reactions have been reported (Bensky *et al.,* 2004).

greenish-white cultivar

purplish-black cultivar

Crude drug/decocting pieces

Shan Zha 山楂

Crataegi Fructus

Drug part: FRUIT

PHARMACOPOEIA SPECIES 1

Crataegus pinnatifida Bunge var. *major* N. E. Br.

Rosaceae

PLANT DESCRIPTION

Deciduous, usually thorny, tree (up to 6 m tall) with sickle-shaped stipules. **Thorns 1–2 cm long**. *Leaves* alternate and simple, **leaf-stalks 2–6 cm long**; blades ovate (5–10 cm long; 4–7.5 cm wide), leaf base truncate to broadly wedge-shaped, irregularly lobed, margins bi-serrate. *Flowers* in dense, white, conspicuous heads (4–6 cm diam.); flower-stalks softly hairy (pubescent) when young. Flowers bisexual, c. 1.5 cm diam., petals 5, white, equal-sized, obovate or sub-orbicular; stamens 20; ovary inferior, styles 3–5. *Fruits* in clusters; fruit-stalks softly hairy; each fruit sub-spherical, conspicuously red or dark red when ripe (**2–2.5 cm diam.**), conspicuously covered in grey macular dots, sepals persistent at apex, flesh dry and pulpy; seeds many. Fl. May–June , fr. Aug–Sep.

Species of *Crataegus* wild in China: 18

HARVESTING, SOURCING AND NATURAL RANGE

This variety is only known from cultivation and has long been grown in China for its fruit which are used for medicine, food and healthfood. The main production areas are in the NC Chinese provinces of Shandong, Henan, Hebei and Shanxi. Fruits are harvested when ripe (red) in autumn.

CONSERVATION STATUS

China/Global: only from cultivation.

PHARMACOPOEIA SPECIES 2

C. pinnatifida var. *pinnatifida*

Rosaceae

taxonomic/nomenclatural note: in CP2015 as '*C. pinnatifida* Bge.' but the type variety, i.e. var. *pinnatifida* is used here to differentiate it from *C. pinnatifida* var. *major* (PS1).

PLANT DESCRIPTION

As for PS1 but **fruits smaller (1–1.5 cm diam.)**, and macular dots less dense.

Leaves of PS1 and PS2 are the source of the separate TCM drug 'Shan Zha Ye' (Crataegi Folium), not included in this book.

HARVESTING, SOURCING AND DISTRIBUTION

Ripe fruits are harvested in autumn mostly from wild populations which grow in shrubby places on slopes (100–1,500 m altitude) in NC, NE and SE China (Hebei, Heilongjiang, Henan, Inner Mongolia, Jiangsu, Jilin, Liaoning, Shaanxi, Shandong, Shanxi and Zhejiang) [also native to the Korean Pensinsula].

CONSERVATION STATUS

China: Least Concern (IUCN vers. 3.1 (2000) (and authors' pers. obs.).

Global: Not Evaluated (IUCN Red List, accessed 20/8/2016).

UNOFFICIAL SUBSTITUTE

C. cuneata Siebold & Zucc.

Rosaceae

PLANT DESCRIPTION

Similar to PS1 and PS2 but thorns shorter **(5–8 mm long), leaf-stalks shorter (4–5 mm long)**, leaf bases **wedge-shaped or tapering, flower- heads smaller (2–2.5 cm diam.)** and **fruits red or yellow (1–2 cm diam.)** and often totally devoid of greyish macular dots.

CONSERVATION STATUS

China: Least Concern (MEPCAS, 2013).

Global: Not Evaluated (IUCN Red List, accessed 11/8/2016).

TCM PROPERTIES AND USE

properties: **'sweet, sour and slightly warm'**
channels: **'Spleen, Stomach and Liver'**
actions: **'resolves food stagnation, invigorates Blood'**

indications: **indigestion, abdominal distention; heart and chest pain; postpartum pain; diarrhoea**

Crataegus pinnatifida var. *major*

Leaves and fruits of *C. pinnatifida* var. *major*

Crataegus pinnatifida var. *pinnatifida*

Leaves and fruits of *C. pinnatifida* var. *pinnatifida*

Crataegus cuneata

Fresh ripe fruits of *C. cuneata*

Crataegus pinnatifida var. *major*

Rosaceae

DRUG MORPHOLOGY

DECOCTING PIECES-1 (SLICED DRIED FRUIT): transverse or longitudinal fruit slices, somewhat rounded **[1]** or shrunken or sometimes hemispherical **[2]** (2–2.5 cm diam.; 2–4 mm thick); margins often inrolled **[3]**. Uncut outer surface usually convex, dark red, dull and **densely covered in greyish dots [4]**. Cut surfaces usually concave, with dark yellow or pale brown pulp and five seed-bearing locules **[5]**, sometimes empty. Basal end of fruit often with persistent fruit-stalk **[6]** or calyx remnants; other end (apex) sometimes with tiny flower remnants **[7]**. *Texture*: tough. *Odour*: indistinct. *Taste*: sour and slightly sweet. ***Good quality*:** fruits red or reddish-brown externally with fleshy pulp.

DECOCTING PIECES-2 (SLICED DRY-FRIED FRUIT): as for the crude drug but with a smaller diameter, pulp darker (yellowish-brown) **[8]**; charred spots occasional **[9]**; slightly aromatic.

DECOCTING PIECES-3 (SLICED CHARRED FRUIT): as for DP-1 (above) but charred brownish-black **[10]**; loculi often seedless (fallen) **[11]**.

Processed fruits (dp-2 and dp-3) of this variety are significantly smaller than those of dp-1.

PROCESSING

Ripe fruits are sliced then dried (= decocting pieces-1); subsequent dry-frying produces Chao Shan Zha 炒山楂 (= decocting pieces-2) or charring Jiao Shan Zha 焦山楂 (= decocting pieces-3).

OTHER COMMON NAMES

北山楂 Bei Shan Zha 'northern hawthorn fruit'; 大山楂 Da Shan Zha 'big northern hawthorn fruit'; 红果 Hong Guo 'red fruit'; 山里红 Shan Li Hong; 酸楂 Suan Zha; Fructus Crataegi; hawthorn fruit (CP2010).

SAFETY

Safety Class: 1; Interaction Class: A (Gardner & McGuffin, 2013). An official Chinese food and healthfood (Anon., 2002a,b).

C. pinnatifida var. *pinnatifida*

Rosaceae

DRUG MORPHOLOGY

As for the decocting pieces of PS1 but **fruit slice diameter smaller [12] (1–1.5 cm)**.

PROCESSING

As for PS1.

OTHER COMMON NAMES

北山楂 Bei Shan Zha 'northern hawthorn fruit'; 红果 Hong Guo; 酸楂 Suan Zha; Fructus Crataegi; hawthorn fruit (CP2010); Wu Mao Shan Zha.

SAFETY

As for PS1.

C. cuneata

Rosaceae

'DRUG SUBSTITUTE' MORPHOLOGY

Easily distinguished macroscopically from fruits of PS1 since *C. cuneata* fruits are **much smaller [13] (1–2 cm diam.)**; they are a **similar size to those of PS2**.

C. cuneata fruits are usually traded as whole **[14]**) or squashed **[15]** fruits (rather than halved or sliced ones) (see Processing), the peel is sometimes yellowish rather than red and **often totally devoid of greyish macular dots** or with very few or indistinct ones **[16]**.

SUBSTITUTION COMMENT. A common species in C and S China, very similar and closely related to PS1 and PS2. Earlier editions of the CP (1963, 1977 and 1985) recognised this species as one of the official wild-harvested sources of Shan Zha and, although today it is likely to be considered as inferior, some local Chinese markets may continue to trade it as a readily available local substitute.

Fruits are totally wild harvested from plants which grow in shrubby places (200–2,200 m altitude) in NE, SC and SE China (Anhui, Fujian, Guangdong, Guangxi, Guizhou, Henan, Hubei, Hunan, Jiangsu, Jiangxi, S Shaanxi, Yunnan and Zhejiang [also native to Japan].

PROCESSING

Since fruits are smaller than those of PS1, they are more typically squashed when fresh rather than sliced to facilitate drying and perhaps also to disguise their smaller size when traded as a Shan Zha substitute.

OTHER COMMON NAMES

南山楂 Nan Shan Zha 'southern Chinese hawthorn fruit'; 野山楂 Ye Shan Zha 'wild Chinese hawthorn fruit'.

SAFETY

Not known but likely to be similar to official 'Shan Zha' species.

Shan Zha sourced from *C. pinnatifida* var. *major*

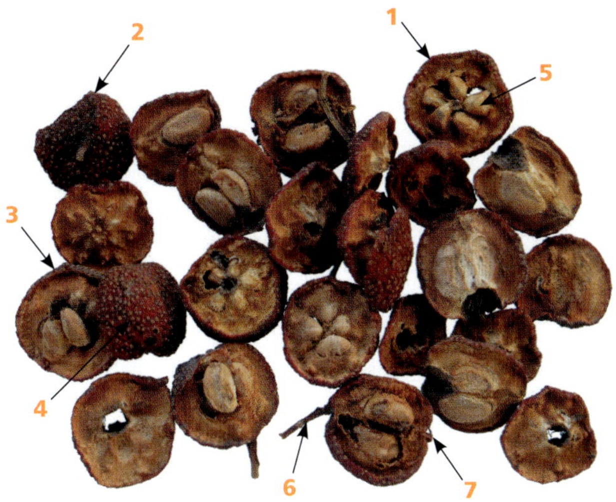

Decocting pieces-1 (sliced)

Decocting pieces-2
(sliced and dry-fried Chao Shan Zha)

Decocting pieces-3
(sliced and charred Jiao Shan Zha)

Shan Zha sourced from *C. pinnatifida* var. *pinnatifida*

Decocting pieces-1 (sliced)

Decocting pieces-2
(sliced and dry-fried Chao Shan Zha)

Decocting pieces-3
(sliced and charred Jiao Shan Zha)

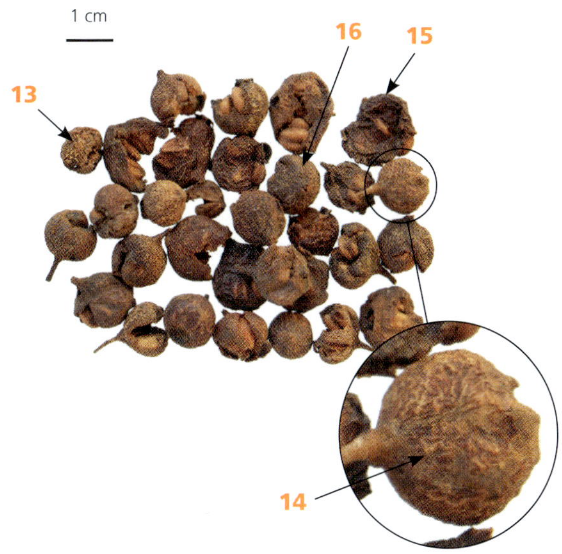

Dried ripe fruits of *C. cuneata*

LESSER KNOWN SUBSTITUTES

1. *Crataegus hupehensis* Sarg. (湖北山楂 Hu Bei Shan Zha). Sparsely thorny; leaf-stalk 4–5 mm long; **leaves with 2–4 pairs of shallow lobes at apical part**; flower-heads 3–4 cm diam., fruit sub-spherical, c. 2.5 cm diam., red, grey-dotted.
2. *C. scabrifolia* (Franch.) Rehder (云南山楂 Yun Nan Shan Zha). **Usually thornless**; **leaves rarely lobed**; leaf-stalk 1.5–4 cm; fruits red, rarely yellow, sub-spherical, 1.5–2.5 cm diam.
3. *Docynia delavayi* (Franch.) C. K. Schneid. (云南移木衣 Yun Nan Duo Yi). **Leaves not lobed but pinnately compound (divided into leaflets)**; petiole and lower leaf surfaces densely covered in yellow hairs; **fruits yellow**, ovoid or oblong, 2–3 cm diam.
4. *D. indica* (Wall.) Decne. (移木衣 Duo Yi). **Leaves not lobed**; **fruits yellow**, subspherical or ellipsoid (2–3 cm diam.).
5. *Malus doumeri* (Bois) A. Chev. (台湾海棠 Tai Wan Hai Tang). **Leaves not lobed**; leaf-stalk 1.5–3 cm long; flower-stalk densely covered in short white hairs; petals yellowish-white; **fruits yellowish-red**, spherical (2.5–5.5.cm diam.).
6. *M. prunifolia* (Willd.) Borkh. (秋子 Qiu Zi). Fruit 2–2.5 cm diam., fruit-stalk 3.2–3.5 cm long; flowers 4–5 cm diam.; fruit yellow or red, ovoid (2–2.5 cm diam.), with cavity at base.
7. *Pyrus calleryana* Decne. (豆梨 Dou Li). Leaves not lobed; fruit-stalk 1.5–3 cm long; fruit spherical, **blackish-brown** (c. 1 cm diam.).
8. *Sorbus pohuashanensis* (Hance) Hedl. (花楸树 Hua Qiu Shu). Fruit red or orangish-red (6–8 mm diam.).

Shan Zhu Yu 山茱萸 Corni Fructus

Drug part: FRUIT

Cornus officinalis Siebold & Zucc.

Cornaceae

synonym: *Macrocarpium officinale* (Siebold & Zucc.) Nakai

Species of *Cornus* wild in China: 25

PLANT DESCRIPTION

A deciduous shrub or tree (4–10 m tall) with a much branched and rounded crown. *Leaves* opposite and simple; leaf blades ovate-elliptic (5.5–10 cm long; 2.5–4.5 cm wide), undersides with long and short hairs (trichomes); veins pinnate and margin entire. *Flowers* appear before leaves in early spring in many-flowered umbel-like hemispherical heads on short, thick stalks (c. 2 mm long) at intervals along stems; bracts 4, conspicuous and ovate to valve-shaped (5–8 mm long), papery to leathery. Each flower tiny, on a slender, vivid-yellow stalk (0.5–1 cm long); sepals 4, fused and minute [hl]; petals 4, tiny, spreading, reflexed and vivid yellow; stamens 4; ovary inferior; style 1 (minute). *Fruit* narrowly ellipsoid (1.2–1.8 cm long; 0.5–0.7 cm wide), fleshy, bright red and 1-seeded. Fl. Mar–Apr, fr. Sep–Oct.

HARVESTING, SOURCING AND NATURAL RANGE

Ripe (red) fruits are harvested in late autumn or early winter, mostly from cultivated crop plants. Traditional *Di Dao* quality is sourced from Zhejiang province.

The wild species grows in mountain forests and forest edges, 400–2,100 m altitude, in SE and WC China: Anhui, Gansu, Henan, Hunan, Jiangsu, Jiangxi, Shaanxi, Shandong, Shanxi and Zhejiang [also native to Japan, Korean Peninsula].

CONSERVATION STATUS

China: Near Threatened (MEPCAS, 2013).
Global: Not Evaluated (IUCN Red List, accessed 11/8/2016).

Fruiting bushes of medicinal crop, Zhejiang province

Flowering branch

Fruiting branch

Shan Zhu Yu sun-drying, Shaanxi province

UNOFFICIAL SUBSTITUTES

Fruits of many other species in the same and other families have been reported as local or historical substitutes (Bensky *et al.*, 2004; NICPBP, 1997–2002); their occurrence in international trade very unlikely:

1. ***Berberis amurensis*** Rupr. (黄芦木 Huang Lu Mu); family Berberidaceae
2. ***B. poiretii*** C.K. Schneid. (细叶小檗 Xi Ye Xiao Bo); family Berberidaceae
3. ***Cerasus pleiocerasus*** (Koehne) T. T. Yu & C. L. Li (syn. *Prunus pleiocerasus* Koehne) (雕核樱桃 Diao He Ying Tao); family Rosaceae
4. ***Cornus oblonga*** Wall. (长圆叶梾木 Chang Yuan Ye Lai Mu); family Cornaceae
5. ***Crataegus pinnatifida*** Bunge; family Rosaceae. Source of 山楂 Shan Zha (Crataegi Fructus) see page 702.
6. ***Malus baccata*** (L.) Borkh. (山荆子 Shan Jing Zi); family Rosaceae
7. ***Viburnum schensianum*** Maxim. (陕西荚蒾 Shan Xi Jia Mi); family Caprifoliaceae
8. ***Vitis vinifera*** L. (葡萄 Pu Tao); family Vitaceae
9. ***V. amurensis*** Rupr. (山葡萄 Shan Pu Tao); family Vitaceae
10. ***Ziziphus mauritiana*** Lam. (滇刺枣 Dian Ci Zao); family Rhamnaceae. See account for 'Suan Zao Ren' (page 720).
11. ***Z. jujuba*** var. ***spinosa*** (Bunge) Hu ex H. F. Chow (酸枣 Suan Zao); family Rhamnaceae (source of 'Suan Zao Ren' (Ziziphi Spinosae Semen) page 720).

TCM PROPERTIES AND USE

***properties*: 'sour and slightly warm'**
***channels*: 'Kidney and Liver'**
***actions*: 'tonifies Kidney and Liver, secures the essence'**
***indications*: dizziness, vertigo; back and knee pain; impotence, spermatorrhoea; metrorrhagia; enuresis**

DRUG MORPHOLOGY

CRUDE DRUG/DECOCTING PIECES-1 (DRIED DE-SEEDED FRUIT): ovoid, de-seeded and somewhat deflated or bladder-like fruits **[1]** (1–1.5 cm long; 0.5–1 cm wide). Externally lustrous and red or purplish-red **[2]** with a tiny and slightly sunken calyx scar at one end **[3]** and a slightly raised fruit-stalk scar **[4]** at the other. *Texture*: slightly sticky and pliable. *Odour*: indistinct. *Taste*: sour, astringent and somewhat bitter. ***Good quality*:** fruits purplish-red and pliable or, if processed with yellow rice wine, purplish-black or black.

DECOCTING PIECES-2 (WINE-PROCESSED FRUIT): as for the above but purplish-black or black, and more shrivelled and lustreless **[5]**. *Texture*: moist and soft. *Odour*: slightly wine-scented.

PROCESSING

Ripe (red) fruits are blanched or part-dried at low temperature, seed discarded then sun-dried (= crude drug/decocting pieces-1). Subsequent steaming or simmering with yellow rice wine produces Jiu Yu Rou or Jiu Shan Yu (= decocting pieces -2).

OTHER COMMON NAMES

酒山萸肉 Jiu Shan Yu Rou (wine processed Shan Zhu Yu); 酒萸肉 Jiu Yu Ro (wine processed Shan Zhu Yu); 山萸 Shan Yu; 山茱萸肉 Shan Zhu Yu Rou (pulp of Shan Zhu Yu); 药枣 Yao Zao; 枣皮 Zao Pi; Asiatic cornel; Asiatic cornelian cherry; Asiatic dogwood; Fructus Corni.

SAFETY

Safety Class 1c; Interaction Class: C (Gardner & McGuffin, 2013).

Crude drug/decocting pieces-1 of Shan Zhu Yu

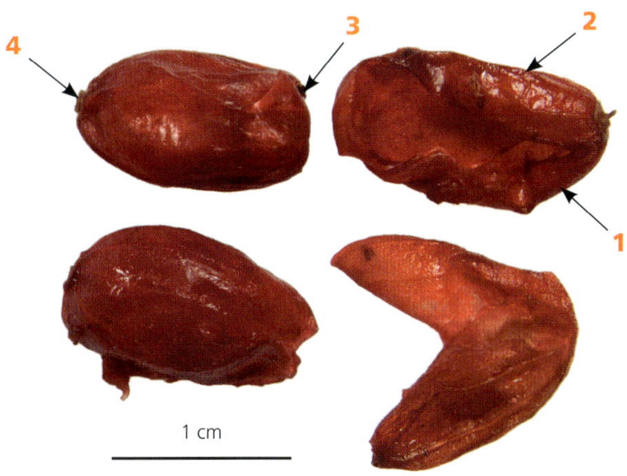

Wine-processed Shan Zhu Wu (decocting pieces-2)

Sha Ren 砂仁

Amomi Fructus

Drug part: FRUIT

Amomum villosum Lour. var. *villosum*

Zingiberaceae

Species of *Amomum* wild in China: 38

taxonomic/nomenclatural note: in CP2015 as '*Amomum villosum* Lour.'; the type variety name (i.e. var. *villosum*) is used here to differentiate this entity from *A. villosum* var. *xanthioides* below (PS2). A third variety, var. *zeylanicus* Karun. & Yakand., is restricted to Sri Lanka.

PLANT DESCRIPTION

An erect, perennial herb. Rhizomes scented, covered in brown scale-like sheaths and creeping along the soil surface. *'Stems'* (pseudostems — densely overlapping leaf bases) usually 1.5–3 m tall, with c. 20–40 leaves per 'stem'. *Leaves* alternate, almost stalkless and becoming rapidly smaller at top of each 'stem'; leaf sheaths with netted appearance; ligule (at base of each leaf blade) greenish-brown to dark brown, **semi-orbicular (0.2–1 cm long)**, leathery, hairless, tip notched or pointed and margin fringed with hairs; leaf blades simple, large, lance-shaped to linear (25–35 cm long; 3–7 cm wide), margins entire. *Flowers* bisexual, in dense ellipsoid spikes **at base of plant and growing directly from the rhizomes** on a 4–8 cm long stalk; calyx tubular; petal tube (c.1.8 cm long) with white lobes (1.6–2 cm long); **lip (labellum) white with yellow and purplish-red convex mid-vein, two purple spots at base and a yellow reflexed tip.** *Fruits* in very short-stalked clusters; each fruit a dry, ellipsoid, prickly capsule (1.5–2 cm long; 1–2 cm diam.), **purple when fresh and ripe, drying to brown** and covered in soft branched or unbranched spines. *Seeds* many, multi-angled and strongly aromatic. Fl. May–Jun, fr. Aug–Sep (based on *FoC* and *eMonocot*).

HARVESTING, SOURCING AND NATURAL RANGE

Ripe fruit (purple) are harvested during summer and autumn from cultivated crop plants grown in C and S China (Chongqing, Fujian, Guangdong, Guangxi, Sichuan and Yunnan provinces). The traditional *Di Dao* source is located in Guangdong.

The wild variety grows in wet, shady places in sparse forests between 100–600 m altitude in Fujian, Guangdong, Guangxi and Yunnan provinces (but see *Conservation Status*); [also wild in Cambodia, Laos, Thailand and Vietnam].

CONSERVATION STATUS

China: Not Evaluated. Possibly only known from cultivation in China (IUCN Red List, accessed 20/8/2016).
Global: Least Concern (IUCN Red List, accessed 20/8/2016).

A. villosum var. *xanthioides* (Wall. ex Baker) T. L. Wu & S. J. Chen

Zingiberaceae

synonym: *A. xanthioides* Wall. ex Baker
taxonomic/nomenclatural note: in CP2015 as '*A. villosum* Lour. var. *xanthioides* T. L. Wu et Senjen'.

PLANT DESCRIPTION

As for PS1 but fruits green or brownish when fresh and ripe.

HARVESTING, SOURCING AND NATURAL RANGE

Ripe fruit (green or brownish) are harvested during summer and autumn mostly from cultivated crop plants.

In the wild, this variety is native to the Chinese provinces of Guangxi and Yunnan where it grows in forests between 600–800 m altitude; it is also native to Cambodia, India, Laos, Myanmar, Thailand and Vietnam.

CONSERVATION STATUS

China: Not Evaluated.
Global: Data Deficient (conservation assessment is difficult since cultivated crop plants have become widely naturalised) (IUCN Red List, accessed 20/8/2016).

TCM PROPERTIES AND USE

properties: **'pungent and warm'**
channels: **'Spleen and Stomach'**
actions: **'transforms damp, tonifies Spleen, regulates qi'**

indications: **bloating, indigestion; nausea and vomiting; diarrhoea; morning sickness and to prevent miscarriage**

Cultivated crop of *Amomum villosum* var. *villosum*, Guangdong province

A. villosum var. *villosum* fresh fruit-heads

A. villosum var. *xanthioides* in the wild, China

A. villosum var. *xanthioides*

A. *villosum* var. *xanthioides* uprooted plant with rhizomes and fruiting heads (**at base**)

Amomum longiligulare T. L. Wu

Zingiberaceae

PLANT DESCRIPTION

As for PS1 but 1–1.5 m tall, ligule **lance-shaped (2–4.5 cm long)**, membranous and hairless; stalk of fruit spike shorter (1–3 cm long) and each fruit ovoid (1.5–2.2 cm long; 0.8–1.2 cm diam.) and covered with branched soft flaky spines (1 mm long). Fl. Apr–May, fr. Jun–Sep.

HARVESTING, SOURCING AND NATURAL RANGE

Ripe fruit (pinkish) are harvested in summer and autumn from cultivated crops grown in more southerly localities compared to PS1 and PS2 (Hainan and Guangdong provinces).

In the wild the species grows in tropical moist forests in Hainan province [also native in Laos, Thailand and Vietnam (Lamxay & Newman, 2012)].

CONSERVATION STATUS

China: Least Concern (MEPCAS, 2013); Vulnerable (Fu, 2011); Viulnerable A2cd (Wang & Xie, 2004)

Global: Least Concern (IUCN Red List, accessed 20/8/2016).

Alpinia zerumbet (Pers.) B. L. Burtt & R. M. Sm.

Zingiberaceae

synonym: *Costus zerumbet* Pers.

PLANT DESCRIPTION

Vegetative parts similar to those of PS1, 2 and 3 but leaf blades often longer and slightly wider (30–60 cm long; 5–10 cm wide); ligule 0.5–1 cm long and hairy. *Main differences*: **flowering and fruiting spikes emerging from the top of the leafy stems** (i.e not growing directly from the rhizomes at the base of the plant); flower spikes long (up to 30 cm) and usually pendent; rachis red and velvety. **Fruits scarlet or orangish-red and distinctly ribbed** (not covered in spines). Fl. Apr–Jun, fr. Jul–Oct.

See SUBSTITUTION COMMENT, page 714

CONSERVATION STATUS

China: Not Evaluated.

Global: Not Evaluated (IUCN Red List, accessed 20/8/2016).

Amomum longiligulare

Fresh purple fruit of *A. longiligulare* developing at base of plant

Alpinia zerumbet crop grown for ornamental purposes

Pendent flower spikes of *A. zerumbet* crop grown for ornamental purposes

Fresh scarlet fruits of *A. zerumbet* developing at top of leafy stems

PHARMACOPOEIA SPECIES 1

Amomum villosum var. *villosum*

Zingiberaceae

DRUG MORPHOLOGY

CRUDE DRUG (RIPE FRUIT): ovoid or ellipsoidal (1.5–2 cm long, 1–2 cm diam.). Externally fruit-wall (pericarp) with **3 indistinct longitudinal ridges [1]** (reflecting the 3 equal compartments within); dull brown, **covered throughout with soft, curving branched or unbranched spines creating a rugged surface [2]**; flower remnants visible at one end (apex) **[3]** and remains of the pale brown fruit-stalk (peduncle) at the other **[4]**. Fruit wall thin **[5]**, enclosing a pale brown membrane surrounding many, densely packed seeds arranged within 3 compartments and separated by white septa; with **5–26 seeds per compartment**. Seeds multi-angled **[6]** (2–3 mm diam.), **mostly brownish-red**, finely wrinkled [hl] and strongly aromatic; endosperm greyish-white. *Texture* (*fruit-wall*): brittle. *Odour*: strongly aromatic. *Taste*: cool, slightly bitter and pungent. ***Good quality*:** seeds oily; intensely aromatic.

PROCESSING

Ripe fruit are dried in the sun or at low temperature (= crude drug) then pulverised prior to decocting.

OTHER COMMON NAMES

春砂仁 Chun Sha Ren; 阳春砂 Yang Chun Sha; Mak Naeng Khieo (green fruit) — Lao name; Chinese amomum (SCN); Fructus Amomi; villous amomum fruit (CP2010).

SAFETY

An official food and healthfood in China (Anon., 2002a,b).

PHARMACOPOEIA SPECIES 2

A. villosum var. *xanthioides*

Zingiberaceae

DRUG MORPHOLOGY

As for PS1. Cannot be distinguished from PS1 using macroscopic characters. For numbered annotations in the photo see PS1 above; characters un-numbered in PS1 are numbered here as follows:
Fruit wall enclosing a pale brown membrane **[7]** surrounding many densely packed seeds **[8]** separated by white septa **[9]**.

PROCESSING

As for PS1.

OTHER COMMON NAMES

绿壳砂 Lu Ke Sha; Suo Sha; Xi Sha; Chinese amomum (SCN); Fructus Amomi; villous amomum fruit (CP2010).

SAFETY

An official food and healthfood in China (Anon., 2002a,b).

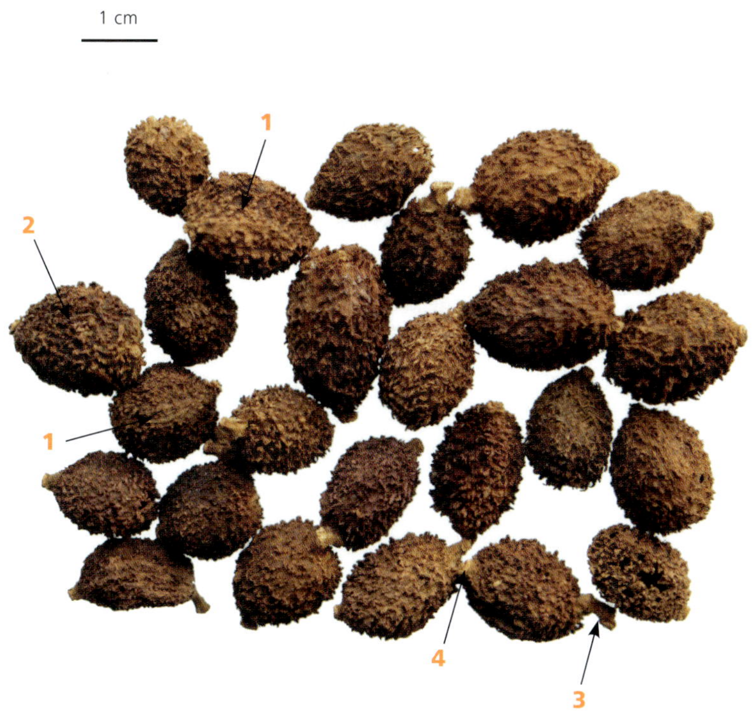

Crude drug of Sha Ren
(sourced from *Amomum villosum* var. *villosum*)

Decocting pieces (pulverised dried fruits) of Sha Ren
(sourced from *A. villosum* var. *villosum*)

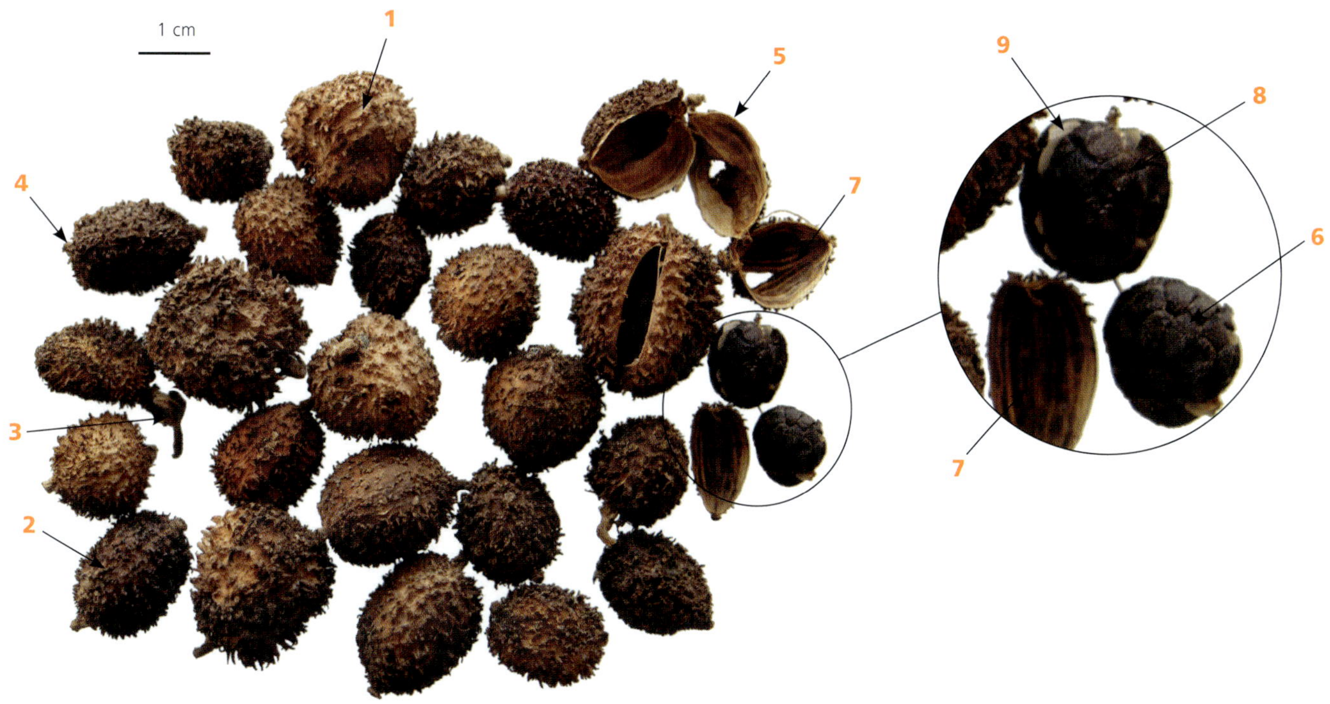

Crude drug of Sha Ren
(sourced from *A. villosum* var. *xanthioides*)

Amomum longiligulare

Zingiberaceae

DRUG MORPHOLOGY

CRUDE DRUG (RIPE FRUIT): similar to PS1 and 2; long ellipsoidal or ovoid (1.5–2 cm long; 0.8–1.2 cm diam.), **more obviously 3-ribbed [10], externally greyish-brown**, with spines easily removed, fruit wall slightly thicker [11]; **4–15 seeds per compartment [12].** *Odour*: **indistinct**. *Taste*: **indistinct**.

PROCESSING

As for PS1.

OTHER COMMON NAMES

海南壳砂 Hainan Qiao Sha; 海南砂 Hainan Sha; 海南砂仁 Hainan Sha Ren (Hainan sand seeds); Chinese amomum (SCN); Fructus Amomi; Villous amomum fruit (CP2020).

SAFETY

An official food and healthfood in China (Anon., 2002a,b).

Alpinia zerumbet

Zingiberaceae

'DRUG SUBSTITUTE' MORPHOLOGY

Fruits are easily distinguished from those of PS1, 2 and 3 by their **many and prominent longitudinal ribs [13]** and absence of spines. Seeds are more difficult to distinguish: 3–4 mm diam., **greyish-black or greyish-brown,** covered with a white membranous aril. *Odour*: strongly aromatic. *Taste*: pungent.

OTHER COMMON NAMES

川砂仁 Chuan Sha Ren; 土砂仁Tu Sha Ren; 艳山姜 Yan Shan Jiang (plant name).

SAFETY

The leaves are used as a flavouring for cooking in Fujian province.

SUBSTITUTION COMMENT. Fruits of *Alpinia zerumbet* are thought to be one of the commonest substitutes of Sha Ren among the many others often traded in China. Although they are easily distinguishable from official 'Sha Ren' they are readily available locally in China as a cheap and probably inferior local substitute. The species is widely cultivated in south China, either as an ornamental (usually those with variegated leaves), as a food flavouring (in Fujian province, using the leaves) or as a local medicine (fruits) in Guizhou province (Lin Yu-Lin, pers. obs.). Their appearance as a substitute in international trade is uncertain but is unlikely given the widespread cultivation of the Pharmacopoeia species (PS1, 2 and 3).

The species grows wild in the southern Chinese provinces of Guangdong, Guangxi, Hainan, Taiwan and Yunnan [and in Bangladesh, Cambodia, India, Indonesia, Laos, Malaysia, Myanmar, Philippines, Sri Lanka, Thailand and Vietnam].

Crude drug of Sha Ren (sourced from *Amomum longiligulare*)

Alpinia zerumbet dried fruits

LESSER KNOWN SUBSTITUTES

The fruits of many other species in the same family as the Pharmacopoeia species (Zingiberaceae) have been reported as local or historical substitutes in China for reasons of clinical substitution or confusion (NICBP, 1997–2002; Bensky *et al.*, 2004):

1. ***Alpinia blepharocalyx*** K. Schum. 云南草蔻 Yun Nan Cao Kou. Fruits slightly larger (c. 3 cm long; 2 cm diam.) and **softly hairy but without soft spines**; seeds greyish-yellow to dull brown.
2. ***A. japonica*** (Thunb.) Miq. 福建土砂仁 Fu Jian Tu Sha Ren, 山姜 Shan Jiang. Fruits salmon red when fresh, globose or ellipsoid **(1–1.5 cm in diam.), softly hairy but without soft spines; seeds smelling of camphor.**
3. ***Amomum neoaurantiacum*** T. L. Wu *et al.* 红壳砂仁 Hong Ke Sha Ren; (syn. *A. aurantiacum* H. T. Tsai & S. W. Zhao); endemic to Yunnan. Fruits orange when fresh, with **rust-coloured hairs and sparse soft spines**; seeds brown.
4. ***A. microcarpum*** C. F. Liang & D. Fang 细砂仁 Xi Sha Ren; wild in Guangxi only. Fruits purplish when fresh, **sparse appressed soft spines; seeds black**.
5. ***A. chinense*** Chun 海南假砂仁 Hai Nan Jia Sha Ren; endemic to Hainan. Fruits with soft spines 2–3 mm long.
6. ***A. muricarpum*** Elmer 疣果豆蔻 You Guo Dou Kou; wild in Guangdong and Guangxi. Fruits red when fresh (c. 2.5 cm diam.), covered in soft yellow hairs when fresh, with **branched fleshy spines 3–6 mm long**.
7. ***A. subulatum*** Roxb. 香豆蔻 Xiang Dou Kou; 印度砂仁 Yin Du Sha Ren; wild in Guangxi, Tibet and Yunnan. Fruits purple or reddish-brown when fresh, **2–2.5 cm diam., with 10 undulating wings**.
8. ***A. gagnepainii*** T. L. Wu *et al.* (syn. *A. thyrsoideum* Gagnep.) 长序砂仁 Chang Xu Sha Ren.

She Chuang Zi 蛇床子 Cnidii Fructus

Drug part: RIPE FRUIT (often called 'seed' in TCM texts)

Cnidium monnieri (L.) Cusson

Apiaceae

synonym: *Selinum monnieri* L.

Species of *Cnidium* wild in China: 5

PLANT DESCRIPTION

An upright, branching annual herb (10–80 cm tall) with a slender tap root (c. 2–3 mm thick). *Stem* furrowed. *Leaves* alternate and finely divided; basal and lower leaves 1–3 pinnate; upper leaves reduced. *Inflorescences* compound umbels (c. 2–3(–5) cm across); main umbels with bracts 6–10, linear to linear-lance-shaped and persistent; rays 8–30 (5–20 mm long); umbellets c. 15–20 flowered; bracteoles 5–9, linear, margin very finely ciliate; *flowers* tiny, calyx teeth minute or absent; petals 5, white with notched tips. *Fruit* dry, ovoid-ellipsoid schizocarp (1.5–3 mm long; 1–2 mm diam.) with 10 thin, cork-like winged ribs with lateral ribs slightly broader than dorsal ones; styles 2 and persistent, reflexed and 3–4 times longer than the swollen style base (stylopodium). Fl. Apr–Jul; fr. Jul–Oct.

HARVESTING, SOURCING AND NATURAL RANGE

Fruits are wild-harvested when ripe in summer or autumn. The species grows in grassy places near rivers and along field margins throughout much of China [also native to India, Korean Peninsula, Laos, Mongolia, Russia, Vietnam; Europe and naturalised in N America].

CONSERVATION STATUS

China: Least Concern (MEPCAS, 2013 as var. *formosanum*); Least Concern (authors' pers. comm., 1999).
Global: Not Evaluated (IUCN Red List, accessed 28/4/2015).

Cnidium monnieri, with ripening fruit-heads

C. monnieri; inset: notched petals

C. monnieri fruiting head with furrowed stem

Underside of *C. monnieri* compound umbel

CONTAMINANTS

'She Chuang Zi ' is prone to contamination (accidental or otherwise) especially with fruits of other Apiaceae species. This suggests problems of identification at the point of harvest in this notoriously difficult family. 'She Chuang Zi', recently traded on the UK market, was regularly found to contain contamination levels between 2–15% (van der Valk & Leon, pers obs. 2012 (see opposite).

Fruit contaminants reported to date include:

1. ***Apiaceae*** spp. — *Apium graveolens* L. (celery), *Daucus carota* L. (carrot), *Heracleum scabridum* Franch. and *Torilis* spp.
2. ***Bassia scoparia*** (L.) A. J. Scott (syn. *Kochia scoparia* (L.) Schrad.); family Amaranthaceae. Source of 'Di Fu Zi' (Kochiae Fructus) see page 622.
3. ***Leonurus japonicus*** Houtt.; family Lamiaceae. Source of 'Chong Wei Zi' (Leonuri Fructus) see page 618.

TCM PROPERTIES AND USE

***properties*: 'pungent, bitter and warm'**
***channel*: 'Kidney'**
***actions*: 'dispels wind cold damp, warms Kidney'**

***indications*: vaginal discharge; impotence, infertility; topically for lesions and itching (especially genital); (T)**

DRUG MORPHOLOGY

CRUDE DRUG/DECOCTING PIECES (DRIED FRUIT): fruits dry and, when whole (schizocarps) ellipsoid **[1]** (1.5–3 mm long; 1–2 mm diam.). Outer surface greyish-yellow or greyish-brown with (when whole) 10 distinctive longitudinal cork-like wings **[2]** with a persistent swollen style base (stylopodium) at apex **[3]**, occasionally with 2 recurved styles still intact **[4]**; basal end occasionally with remains of fruit-stalk. Most fruits broken longitudinally into 2 halves (mericarps), each with 5 longitudinal wings on the outer (dorsal) side **[5]** and a flat (commissural) surface on the inner side **[6]**; the latter with 2 brown longitudinal oil ducts (vittae) **[7]**; seed small, greyish-brown and oily. *Texture*: corky and fragile. *Odour*: aromatic. *Taste*: pungent, warm and numbing. ***Good quality*:** plump, yellowish-green and pungent when crushed.

PROCESSING

Fruits are sun-dried (= crude drug/decocting pieces).

OTHER COMMON NAMES

cnidium (SCN); common Cnidium fruit (CP2010); Fructus Cnidii.

SAFETY

Slightly toxic (CP2015). Safety Class: 1; Interaction Class: A (Gardner & McGuffin, 2013).

She Chuang Zi (dried fruit of *C. monnieri* (= crude drug/decocting pieces)

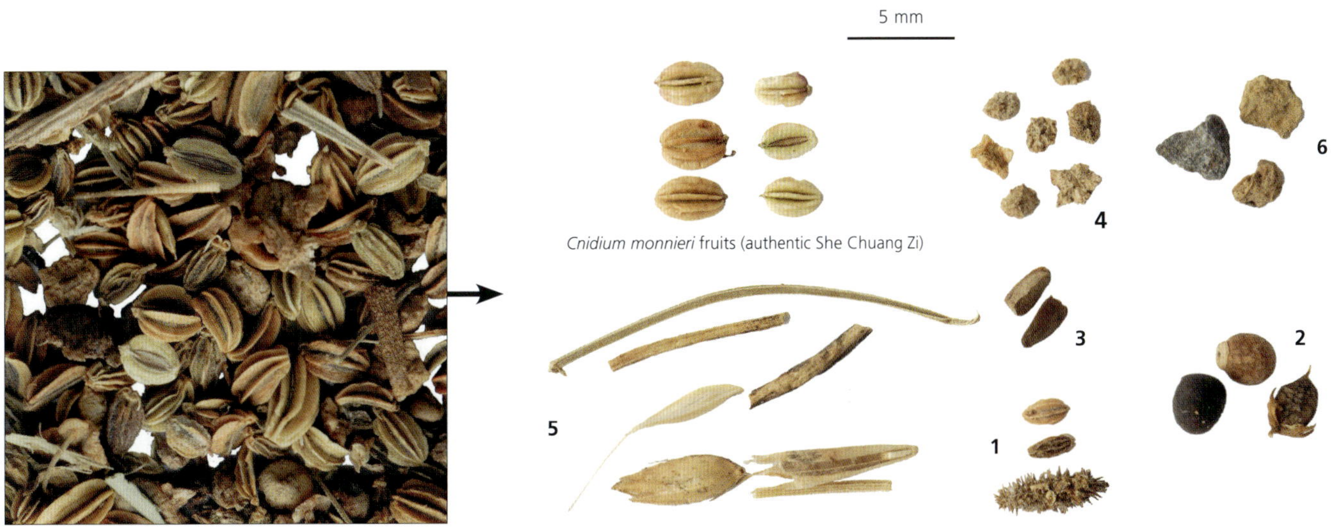

Heavily contaminated sample of She Chuang Zi on UK market. Contaminants include: Apiaceae fruit (?*Apium*) (1); Lamiaceae fruit (?*Perilla* (2) and ?*Leonurus* (3); Amaranthaceae fruit (4); miscellaneous plant material (5) and small stones (6)

Shi Liu Pi 石榴皮

Granati Pericarpium

pomegranate peel

Drug part: RIND (fruit wall or pericarp)

Punica granatum L.

Lythraceae

Species of *Punica* wild in China: 0

PLANT DESCRIPTION

A small, dense, deciduous and hairless shrub (2–3 m tall); young branches and branchlets 4-angled, often terminating in a hard spine. *Leaves* opposite (sometimes crowded on short lateral shoots), simple and short-stalked; leaf blade lance-shaped, elliptic or oblong (2–9 cm long; 1–2 cm wide), base gradually narrowing, tip broad or ending abruptly in a short, stiff point; upper leaf surface surface shiny; margin entire. *Flowers* bright and showy, solitary, terminal or 1–5 in axillary or terminal clusters; radially symmetrical and bisexual. Each flower with a smooth, thick and waxy calyx cup (same colour as the petals); petals 5–9, reddish-orange (occasionally white), strongly crumpled; stamens many; ovary inferior, many-loculed; style 1, with a pin-shaped head. '*Fruit*' (pomegranate) large (5–12 cm diam.) and somewhat flask-shaped with a crown of persistent sepals; 'fruit-wall' (rind) hard and leathery, often lustrous, ranging from red to reddish-brown to yellow-green. *Seeds* many, encased in a bright red translucent flesh. Fl. Mar–Jul.

HARVESTING, SOURCING AND NATURAL RANGE

Fruits are harvested in autumn, when ripe, from cultivated plants widely grown in China for food, medicine and as an ornamental. The species is not native to China but is naturalised in Western China [native from the Balkans to the Himalayas].

CONSERVATION STATUS

China: not native.

Global: Least Concern (IUCN Red List, accessed 25/8/2016).

Cultivated plants of *Punica granatum* in fruit, China

Flowering branch of *P. granatum*

Ripening fruit of *P. granatum*

TCM PROPERTIES AND USE

properties: **'sour and warm'**
channel: **'Stomach and Large Intestine'**
action: **'binds the intestines, stops bleeding'**

indications: **chronic diarrhoea; intestinal parasites; metrorrhagia; leucorrhoea**

DRUG MORPHOLOGY

CRUDE DRUG/DECOCTING PIECES-1 (RIND): varying in shape and size but usually 1.5–3 mm thick; often curled inwards. External surfaces somewhat convex, slightly warty throughout and sometimes lustrous; reddish-orange, pinkish-yellow or dark brown **[1]**. Internal surface pale dull yellow or mottled brown, with raised irregularly-shaped remains of inner fruit walls **[2]**. *Texture:* hard and brittle. *Odour:* none. *Taste:* bitter and astringent. ***Good quality***: large, clean pieces.

DECOCTING PIECES-2 (CARBONISED RIND — SHI LIU PI TAN): as for above externally blackened.

PROCESSING

After removal of the fleshy interior, the rind is sun-dried and cut into pieces (= crude drug/decocting pieces). The rind may undergo further processing by carbonising to produce Shi Liu Pi Tan 石榴皮炭 (carbonsied pomegranate rind; decocting pieces-2).

OTHER COMMON NAMES

石榴 Shi Liu (plant name); 石榴壳 Shi Liu Ke; 酸榴皮 Suan Liu Pi; Pericarpium Granati; pomegranate (SCN).

SAFETY

Slightly toxic (Bensky *et al.,* 2004); side effects reported within the normal dose range: nausea, vomiting, diarrhoea and abdominal pain. Safety Class: 2b and 2d (other specific use restrictions exist); Interaction Class: A (Gardner & McGuffin, 2013).

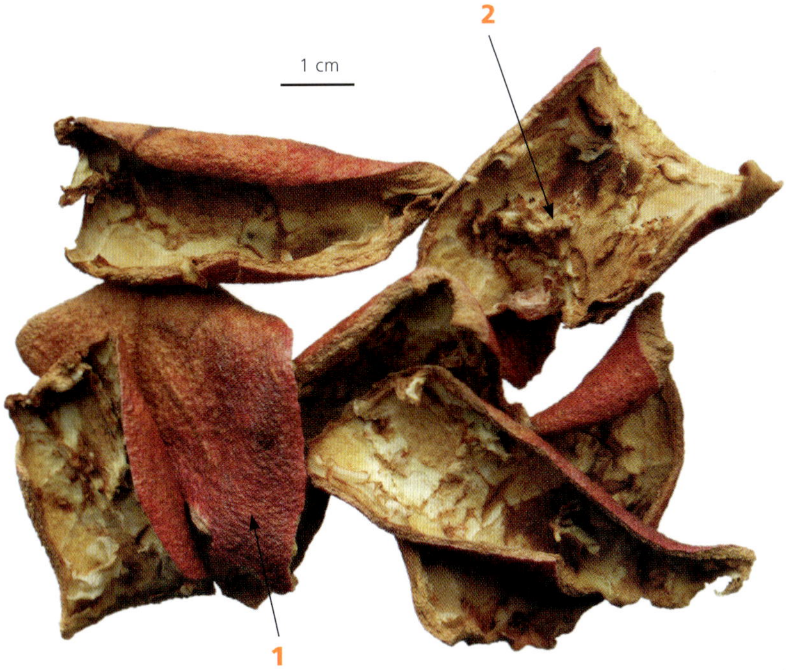

Crude drug/decocting pieces-1 of Shi Liu Pi
(dried and broken fruit wall of *Punica granatum*)

Decocting pieces-2 (Shi Liu Pi Tan; carbonised rind of *P. granatum*)

Suan Zao Ren 酸枣仁

Ziziphi Spinosae Semen

Drug part: SEED

PHARMACOPOEIA SPECIES

Ziziphus jujuba Mill. var. *spinosa* (Bunge) Hu ex H. F. Chow

Rhamnaceae

synonyms: *Z. vulgaris* Lam. var. *spinosa* Bunge; *Z. sativa* Gaertn.var. *spinosa* (Bunge) C. K. Schneid.; *Z. spinosa* (Bunge) Hu ex F. H. Chen

Species of *Ziziphus* wild in China: 12

PLANT DESCRIPTION

A **spiny**, deciduous **shrub** (to 10 m) with sinuous, smooth purplish-red or greyish-brown branches. *Leaves* simple, alternate, small and **shortly lance-shaped (3–7 cm long; 1.5–4 cm wide)** with rounded tips and scarcely toothed margins; blades hairless with upper surface somewhat waxy. *Flowers* inconspicuous, arising singly or in groups in leaf axils; fragrant. Each flower bisexual, sepals 5, triangular and yellowish-green; petals 5, tiny (smaller than sepals), obovate; ovary superior. Fruit reddish-purple (when ripe), sub-spherical or broadly long (7–12 mm diam.); base with persistent calyx tube; flesh spongy and sour-tasting, containing a single broadly rounded stone enclosing 1 or 2 flattened, orb-like seeds (**5–9 mm long; 5–7 mm diam.**). Fl. Jun–Jul, fr. Aug–Sep.

HARVESTING, SOURCING AND NATURAL RANGE

Fruits are harvested from wild plants in late autumn and early winter. The species grows on dry, sunny slopes and plains and is endemic to China (Anhui, Gansu, Hebei, Henan, Inner Mongolia, Jiangsu, Liaoning, Ningxia, Shaanxi, Shanxi, and Xinjiang).

CONSERVATION STATUS

China: Least Concern (MEPCAS, 2013).
Global: Least concern (IUCN Red List, accessed 28/8/2015, as '*Z. jujuba*').

UNOFFICIAL SUBSTITUTE 1

Z. mauritiana Lam.

Rhamnaceae

PLANT DESCRIPTION

Similar to *Z. jujuba* var. *spinosa* **but (up to 15 m), not spiny, young branches and leaf undersides densely covered with short, soft yellowish-white or greyish-white hairs, somewhat matted; leaves more rounded** (2.5–6 cm long; 1.5–4.5 cm wide). Fl. Aug–Nov, fr. Sep–Dec.

CONSERVATION STATUS

China: Least Concern (MEPCAS, 2013).
Global: Not Evaluated (IUCN Red List, accessed 26/8/2016).

SUBSTITUTION COMMENT. Seed of this species closely resembles that of the Pharmacopoeia species but it is unclear if its occurrence as an unofficial substitute is due to mis-identification of the plant at the point of harvest (when in shrub form the species is very similar to the Pharmacopoeia species) or whether it is due to intentional adulteration.
The species is native to southern and western China (Guangdong, Guangxi, Sichuan, and Yunnan) and grows in humid forests, thickets, slopes and riversides up to 1,800 m altitude. It is also cultivated in China (Fujian and Taiwan) for its edible fruit such that its otherwise discarded seed may provide a cheap and readily available look-alike to official 'Suan Zao Ren'. In recent years it has been traded on the UK TCM market as 'Suan Zao Ren' (i.e. the official drug) (van der Valk and Leon, pers. obs., 2012) and its occurrence as such in international TCM trade merits further investigation. The species has many other uses and is widely distributed in SE Asia, India, Africa and Australia.

UNOFFICIAL SUBSTITUTE 2

Hovenia dulcis Thunb.

Rhamnaceae

PLANT DESCRIPTION

Easily distinguished from the official species. A deciduous tree (rarely a shrub) to 10 m tall; **not spiny**. ***Leaves* larger** than those of the Pharmacopoeia species (**7–17 cm long; 4–11 cm wide**); **ovate or broadly oblong**; leaf blade margins irregularly serrate or coarsely serrate; **rachis fleshy** (central stem of the fruit-head); fruits smaller, black and leathery (6.5–7.5 mm diam.); **seed deep brown or blackish-purple (5–5.5 mm diam.)**. Fl. May–Jul, fr. Aug–Oct.

CONSERVATION STATUS

China: Least Concern (MEPCAS, 2013).
Global: Not Evaluated (IUCN Red List, accessed 25/8/2016).

SUBSTITUTION COMMENT. Seeds are only likely to be encountered as an unofficial substitute at a local level in China, Japan and the Korean Peninsula where the tree is widely cultivated on small-holdings for its sweet, fleshy rachis (central axis of the fruit-head) long used as a food supplement, folk medicine and in wine- and sweet-making (Hyun *et al.*, 2010); see also *Safety*.
The species grows wild in secondary forests in China (Anhui, Gansu, Hebei, Henan, NW Hubei, Jiangsu, Jiangxi, Shaanxi, Shandong, Shanxi and N Sichuan) between 200–1,400 m altitude [also wild in Japan, the Korean Peninsula and Thailand].

TCM PROPERTIES AND USE

properties: **'sweet, sour and neutral'**
channels: **'Heart and Liver'**
actions: **'calms the shen, tonifies Heart, nourishes yin'**

indications: **insomnia; anxiety, palpitations, irritability; spontaneous and night sweats**

Ziziphus jujuba var. *spinosa* in the wild

Z. jujuba var. *spinosa*, flowering branches

Z. jujuba var. *spinosa*, fruiting branches

Z. mauritiana – wild plant with ripening fruit

Z. mauritiana – immature fruit

Z. mauritiana – hairy leaf underside

Flowering branch of *Hovenia dulcis*

H. dulcis

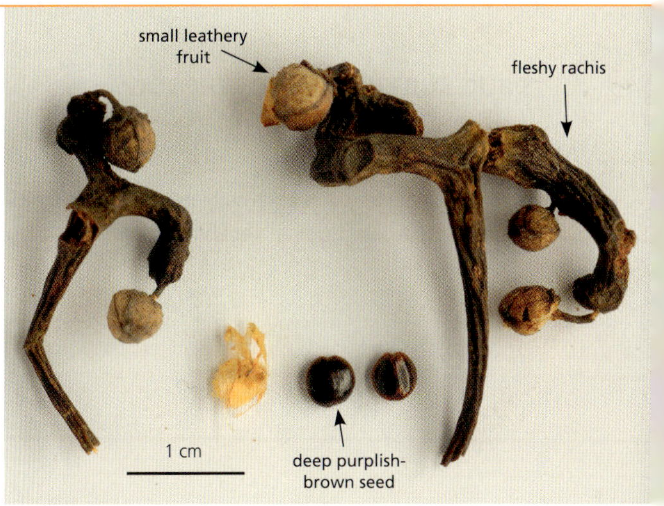

H. dulcis

Ziziphus jujuba var. *spinosa*

Rhamnaceae

DRUG MORPHOLOGY

CRUDE DRUG (DRIED SEED): **ellipsoidal or oblate [1] (5–9 mm long; 5–7 mm diam.**; c. 3 mm thick), and when one surface is flattened the perimeter is keeled. Externally smooth, slightly lustrous, reddish- or purplish-brown [2], occasionally ridged or furrowed longitudinally [3] or randomly fissured [4]; a linear and sometimes sunken hilum present at one end [5] and a small, protruding chalaza at the other [6]. Seed coat (testa) thin and brittle. Endosperm white; cotyledons 2, pale yellow and oily [7]. *Odour:* indistinct. *Taste:* weak. ***Good quality*:** plump, glossy and reddish-purple; internally yellowish-white.

DECOCTING PIECES (DRY-FRIED SEED): as for the crude drug but darker colour throughout and seed coat inflated.

PROCESSING

After discarding the spongy flesh, seeds are removed from the hard yellowish-brown endocarp then sun-dried (= crude drug). Seeds are pulverised prior to decocting. Sometimes they are dry-fried until they pop (= 'Chao Suan Zao Ren').

OTHER COMMON NAMES

酸枣 Suan Zao (plant name); 枣仁 Zao Ren; sour date (plant name); sour jujube (plant name); Semen Ziziphi Spinosae; spine date seed (CP2010); Ziziphus (SCN).

SAFETY

Safety Class: 2b; Interaction Class: A (Gardner & McGuffin, 2013). An official food and healthfood in China (Anon., 2002a,b).

Z. mauritiana

Rhamnaceae

'DRUG SUBSTITUTE' MORPHOLOGY

DRIED SEED: **Very similar to the official drug. Usually more rounded [8] (6–7 mm long; 5–6 mm wide**; 2 mm thick) **and often more glossy and yellowish- or reddish-brown [9]**.

While some trade batches may be easy to distinguish from the official drug using macroscopic characters alone, others (depending on their age and natural variability) may require laboratory-based methods for a reliable identification.

OTHER COMMON NAMES

滇刺枣 Dian Ci Zao; 缅甸枣仁 Mian Dian Zao Ren.

See also *Substitution Comment* for this species (page 720)

Hovenia dulcis

Rhamnaceae

'DRUG SUBSTITUTE' MORPHOLOGY

DRIED SEED: *Easily distinguished from the official drug*. **Slightly smaller** (4–5.5 mm long; 4–5.5 mm wide; 1–2 mm thick) [10] and more lustrous [11] and one end rounded-truncate [12]. Externally **deep brown or blackish-purple [13]**, sometimes with a shallow longitudinal keel on the lower surface.

OTHER COMMON NAMES

北枳椇 Bei Zhi Ju (plant name); 枳椇子 Zhi Ju Zi; Japanese raisin tree.

SAFETY

Fruit-stalks are an official food and healthfood in China (Anon., 2002a,b); safety of the seeds is unknown.

See also *Substitution Comment* for this species (page 720)

Crude drug of Suan Zao Ren (dried seed of *Ziziphus jujuba* var. *spinosa*)

Dry-fried seeds of Suan Zao Ren

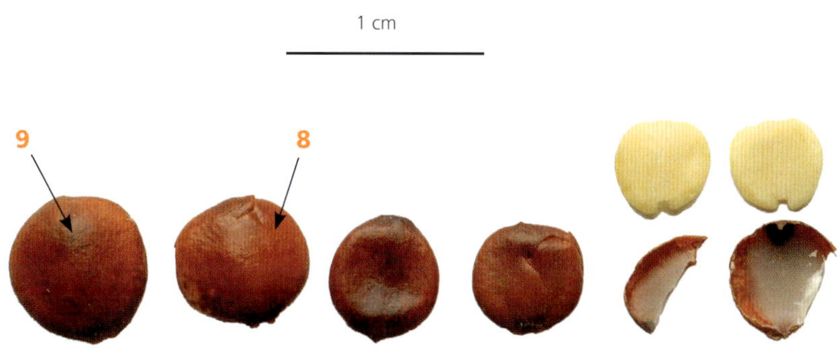

Dried seed of *Z. mauritiana*

Dry-fried seeds of *Z. mauritiana*

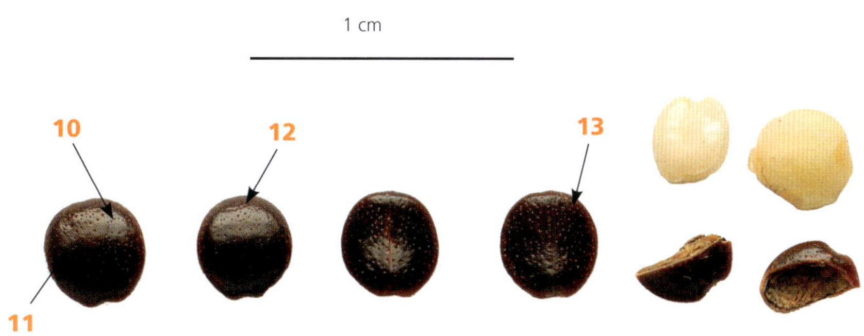

Dried seed of *H. dulcis*

Dry-fried seeds of *H. dulcis*

Tao Ren 桃仁

Persicae Semen

peach seed

Drug part: SEED

PHARMACOPOEIA SPECIES 1

Prunus persica (L.) Batsch

Rosaceae

synonym: *Amygdalus persica* L. (*FoC* accepted name)
taxonomic/nomenclatural note: *Kew-MPNS* accepted name, 20/2/2016

Species of *Prunus* wild in China: >7

PLANT DESCRIPTION

A deciduous tree (3–8 m tall) with a spreading crown, dark reddish-brown bark; flowers appearing before the leaves. *Leaves* alternate and stalked; leaf blades simple, variable in shape but generally lance-shaped (7–15 cm long; 2–3.5 cm wide), base broadly wedged-shaped, tip narrowly pointed, margins finely saw-toothed. *Flowers* solitary, stalkless or with short stalks (1–3 mm long); **sepals** 5, **softly hairy;** petals 5, equal, pink or white; stamens many; anthers purplish-red. ***Fruit* edible** (peach), ovoid to **laterally compressed** (usually 5–7 cm. diam.), with a ventral groove, ripening to greenish-white or orangish-yellow, usually with a red tinge on the exposed side; clothed in short, dense hairs; **flesh succulent, sweet or sour-sweet and fragrant**; **stone** (endocarp) ellipsoid to suborbicular, **laterally compressed** surface longitudinally and transversely furrowed and pitted, tip sharply pointed. *Seed* bitter. Fl. Mar–Apr, fr. Aug–Sep.

HARVESTING, SOURCING AND NATURAL RANGE

Fruits are harvested when ripe, entirely from cultivated trees which are grown throughout China as an ornamental as well as for their edible fruit (peaches) and medicinal seed (Tao Ren). The species is native to N China but is now extinct in the wild. Cultivated trees have often become naturalised in China in disturbed places (1,500–2,200 m. altitude).

CONSERVATION STATUS

China: 'truly wild peaches no longer exist' (*FoC*).
Global: Not Evaluated (IUCN Red List, accessed 8/10/2016)

PHARMACOPOEIA SPECIES 2

Prunus davidiana (Carrière) Franch.

Rosaceae

synonyms: *Amygdalus davidiana* (Carrière) de Vos ex L. Henry (*FoC* accepted name); *Persica davidiana* Carr.; *Prunus persica* var. *davidiana* (Carrière) Maxim.
taxonomic/nomenclatural note: *Kew-MPNS* accepted name, 20/2/2016

PLANT DESCRIPTION

As for PS1 but **sepals hairless, fruit flesh** (mesocarp) **thin, dry and inedible, stone** (endocarp) spherical, subspherical, ellipsoid or oblong but **not laterally compressed.** Fl. Mar–Apr, fr. Jul–Aug.

HARVESTING, SOURCING AND NATURAL RANGE

Although this species is widely cultivated in China as an ornamental, the fruits (source of Tao Ren) are completely wild-harvested. The tree grows in montane forests, scrub and abandoned fields (800–3,200 m altitude) in N, NE, C and W China (provinces of Gansu, Hebei, Heilongjiang, Henan, Inner Mongolia, Ningxia, Qinghai, Shaanxi, Shandong, Shanxi, Sichuan and Yunnan).

CONSERVATION STATUS

Endemic to China: Least Concern (Lin Yu-Lin' pers. obs. 2001). Not Evaluated (IUCN Red List, accessed 8/10/2016).

TCM PROPERTIES AND USE

properties: 'bitter, sweet and neutral '
channels: 'Heart, Liver and Large Intestine'
actions: 'invigorates Blood, dispels stasis, moistens intestines'

main indications: dysmenorrhoea, amenorrhoea; traumatic injury; constipation; dyspnoea

Prunus persica in cultivation

Fruiting branch of *P. persica*

LEFT
P. persica showing softly hairy sepals

Flowering branches of *Prunus davidiana*; (inset) *P. davidiana* showing hairless sepals

P. davidiana in fruit

Prunus persica

Rosaceae

DRUG MORPHOLOGY

CRUDE DRUG (DRIED SEED): **strongly laterally compressed-ovoid [1]** (1.2–1.8 cm long; 0.8–1.2 cm wide; **0.2–0.4 cm thick**); one end shortly acute **[2]**, with a short linear (long and narrow) hilum **[3]**; expanded in the middle; the other end often asymmetrical, broadly obtuse to rounded with a thin edge and a slightly darker chalaza **[4]**. Seed coat yellowish-brown to reddish-brown, with a rough granular surface and distinct longitudinal furrows/veins **[5]** extending from the chalaza. Seed coat thin, easily peeled away, enclosing 2 oily, white cotyledons **[6]**. *Texture:* hard and brittle. *Fracture:* smooth. *Odour:* indistinct. *Taste:* slightly bitter. ***Good quality:*** large and plump, with white, oily cotyledons.

DECOCTING PIECES-1 (PULVERISED SEED): sometimes blanched.

DECOCTING PIECES-2 (BLANCHED SEED): as for crude drug but cotyledons externally smooth **[7]**, pale yellowish-white, intact or separated **[8]**; inner surface of separated cotyledons flat or concave **[9]**. *Odour:* slightly aromatic. *Taste:* slightly bitter.

DECOCTING PIECES-3 (BLANCHED AND DRY-FRIED SEED): as for dp-2 but unevenly coloured pale to dark yellow or brown **[10]**, occasionally charred dark brown **[11]**. *Odour:* slightly aromatic. *Taste:* slightly bitter.

PROCESSING

After removal of the succulent flesh the stony endocarps are broken open to release the seed. The latter are sun-dried (= crude drug) and pounded into pieces or made into a paste (decocting pieces-1); removal of the seed coats (testa) by blanching reduces their toxicity and produces Chan Tao Ren 燀桃仁 (decocting pieces-2); subsequent dry-frying produces Chao Tao Ren 炒桃仁 (decocting pieces-3).

OTHER COMMON NAMES

桃 Tao 'peach' (plant name); Semen Persicae; peach (SCN); peach seed (CP2010).

SAFETY

SAFETY CLASS: 3; Interaction class: A (Gardner & McGuffin, 2013). Removal of seed coats and dry-frying is reported to reduce levels of toxic cyanogenic glycosides (Bensky *et al.*, 2004). Tao Ren is an official food and healthfood in China (Anon., 2002a,b).

P. davidiana

Rosaceae

DRUG MORPHOLOGY

CRUDE DRUG (SEED): slightly shorter and **thicker [12]** (**not strongly compressed laterally**) (c. 0.9 cm long; c. 0.7 cm wide; **c. 0.5 cm thick**) than that of *P. persica* (PS1). Cannot be reliably differentiated from that of PS1 using macroscopic characters alone.

DECOCTING PIECES-1, 2 AND 3: as above.

PROCESSING

As for PS1

OTHER COMMON NAMES

山桃 Shan Tao 'mountain peach' (plant name); 山桃仁 Shan Tao Ren 'mountain peach seed'; Semen Persicae; peach seed (CP2010).

UNOFFICIAL SUBSTITUTES

Substitutes are unlikely to be found on the market. Those reported historically are seed of *Prunus mira* Koehne, *P. tangutica* (Batalin) Koehne and *P. kansuensis* Rehd. (Xu & Xu, 1997).

1 cm

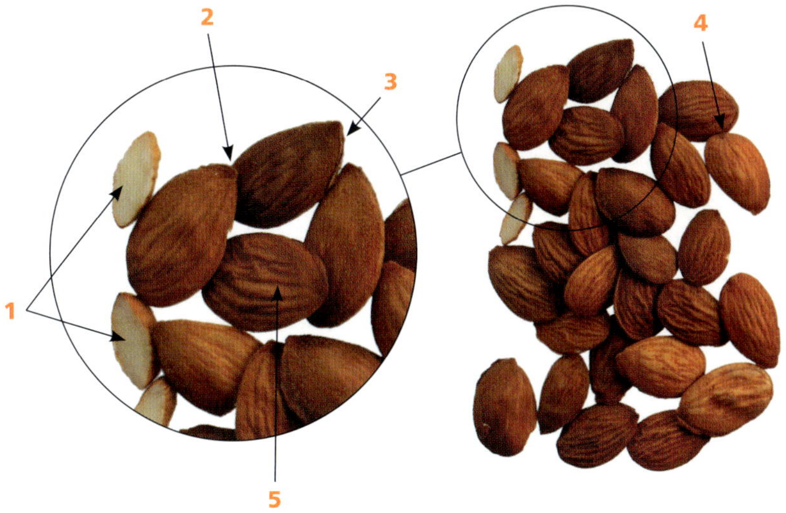

Crude drug of Tao Ren (sourced from *Prunus persica*)

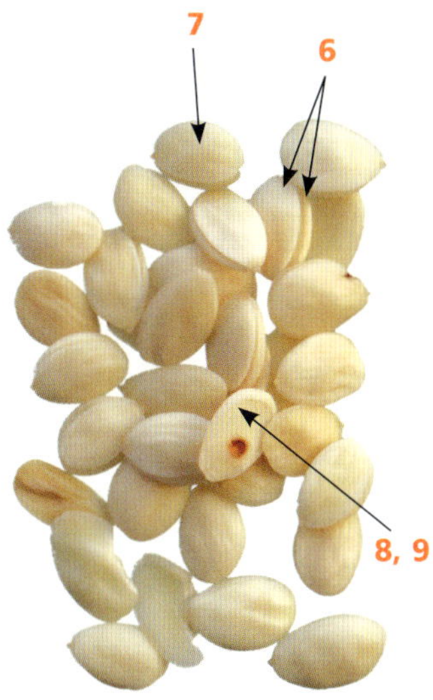

Blanched seed of Tao Ren (Chan Tao Ren) decocting pieces-2 sourced from *P. persica*

Blanched and dry-fried seed of Tao Ren (Chao Tao Ren) decocting pieces-3 sourced from *P. persica*

1 cm

Crude drug of Tao Ren (sourced from *Prunus davidiana*)

Blanched seed of Tao Ren (Chan Tao Ren) decocting pieces-2 sourced from *P. davidiana*

Blanched and dry-fried seed of Tao Ren (Chao Tao Ren) decocting pieces-3 sourced from *P. davidiana*

Tu Si Zi 菟丝子

Cuscutae Semen

Drug part: SEED

PHARMACOPOEIA SPECIES 1

Cuscuta australis R. Br.

Convolvulaceae

PLANT DESCRIPTION

A trailing parasitic plant with a distinctive habit of slender, stems (1 mm diam.) spreading over other herbs and woody plants (see *Harvesting* etc.), often forming a tangled mass. Stems often orange (**not streaked purple**). *Leaves* reduced to minute scales [hl]; not green. *Flowers* in compact, lateral, few to many-flowered heads; each flower tiny (c. 2 mm wide), with 3–5 sepals, corolla white or creamish and cup-shaped, corolla lobes not keeled; **styles 2**. *Fruit* a dry, globose capsule (3–4 mm. diam.); persistent corolla lobes (not keeled) **partly enclosing the 4-seeded fruit.**

Species of *Cuscuta* wild in China: 11

HARVESTING, SOURCING AND NATURAL RANGE

Plants are harvested in autumn, when fruits are ripe, mostly from medicinal crop plants parasitic on beans.

Host plants of the wild species are usually herbaceous or woody members of the families Fabaceae, Asteraceae and Verbenaceae, and the species is widely distributed in most provinces of China between 100–2,000 m altitude.

C. australis is also native in Asia, Australia and Europe.

CONSERVATION STATUS

China: Least Concern (authors' pers. obs. 2001).
Global: Not Evaluated (IUCN Red List, accessed 20/8/2016).

PHARMACOPOEIA SPECIES 2

C. chinensis Lam.

Convolvulaceae

PLANT DESCRIPTION

Very easily confused with PS1. *Main differences*: **corolla lobes keeled** and **persisting to enclose** most of the fruit.

HARVESTING, SOURCING AND NATURAL RANGE

Seeds are thought to be harvested from wild populations only, which grow on host species in the families Fabaceae, Asteraceae, and Zygophyllaceae. Plants grow in sunny places in mountains, thickets and on sandy beaches (200–3,000 m altitude) throughout much of China [also native to Afghanistan, Indonesia, Japan, Kazakhstan, the Korean Peninsula, Mongolia, Russia, Sri Lanka; Africa, SW Asia and Australia].

CONSERVATION STATUS

China: Least Concern (authors' pers. obs. 2002).
Global: Not Evaluated (IUCN Red List, accessed 20/8/2016).

UNOFFICIAL SYBSTITUTE

C. japonica Choisy

Convolvulaceae

PLANT DESCRIPTION

Very easily confused with PS1 and PS2. *Main differences*: stems often **purple-streaked**, flowers with **1 style** and **persistent corolla lobes not enclosing the fruit**.

SUBSTITUTION COMMENT. Although the living plant is easily confused with the two Pharmacopoeia species at point of harvest, the seed of *C. japonica* is significantly larger (see Drug Substitute Morphology). This suggests that its occurrence in trade as 'Tu Si Zi' is intentional rather than due to mis-identification. This view is supported by Hsu et al. (1986) and Zhao & Li (2007) who describe the seed as a frequent local substitute of 'Tu Si Zi' in China because of its similar clinical functions.

C. japonica is a common and widespread native in China with a similar distribution to that of PS1 and PS2.

CONSERVATION STATUS

China: Least Concern (Lin Yu-Lin pers. obs. 2001).
Global: Not Evaluated (IUCN Red List, accessed 20/8/2016).

TCM PROPERTIES AND USE

properties: 'pungent, sweet and neutral'
channels: 'Liver, Kidney and Spleen'
action: 'tonifies Kidney yang and yin, consolidates jing'

indications: **impotence; threatened miscarriage; leucorrhoea; incontinence; eye disorders; diarrhoea; (T)**

Cuscuta australis trailing over host plants

C. australis with flowers and fruit

Flowers and fruits of *C. australis*

C. chinensis trailing over host plant

Flowers and fruits of *C. chinensis*

Cuscuta japonica

C. japonica in flower

Fruiting plant of *C. japonica*

Cuscuta australis

Convolvulaceae

DRUG MORPHOLOGY

CRUDE DRUG/DECOCTING PIECES-1 (DRIED SEED): sub-spherical **(1–2 mm diam)**. Seed coat yellowish-brown or greyish-brown and **covered in small dot-like protuberances [1]**; **raphe** (usually a longitudinal ridge) **not obvious [2]** [hl]. *Texture*: hard and difficult to brerak. *Odour*: slight. *Taste:* indistinct.

CRUDE DRUG/DECOCTING PIECES-2 (SALT-PROCESSED SEED): as for the crude drug but seeds slightly convex, externally brownish-yellow, often split and slightly aromatic.

PROCESSING

After plants are sun-dried the seeds are removed (= crude drug/decocting pieces-1); subsequent and optional dry-frying with salt-water produces Yan Tu Si Zi 盐菟丝子 (salted Tu Si Zi) (= decocting pieces-2).

OTHER COMMON NAMES

南方菟丝子 Nan Fang Tu Si Zi; (plant name); 南菟丝子 Nan Tu Si Zi 'southern spitting silk seed'; Chinese dodder (SCN); dodder seed (CP2010); Semen Cuscutae.

C. chinensis

Convolvulaceae

DRUG MORPHOLOGY

CRUDE DRUG/DECOCTING PIECES-1 (DRIED SEED): very similar to the drug of PS1 but often distinguishable by the more **distinctive linear raphe** along one side **[3]** [hl].

CRUDE DRUG/DECOCTING PIECES-2 (SALT-PROCESSED SEED): difficult to distinguish from its counterpart sourced from PS1 owing to the effects of salt-processing on the shape of the seed (i.e. becoming rounder) **[4]**.

PROCESSING

As for PS1.

OTHER COMMON NAMES

Chinese dodder (SCN); dodder seed (CP2010); Semen Cuscutae.

SAFETY

Safety Class: 1; Interaction Class: A (Gardner & McGuffin, 2013). An official Chinese healthfood (Anon., 2002a,b).

C. japonica

Convolvulaceae

'DRUG SUBSTITUTE' MORPHOLOGY

DRIED SEED: similar to those of PS1 and PS2 but **c. 4 times larger [5] (2–2.5 mm diam.)**, **seed-coat smooth [6]** (not covered in small raised protuberances) and **raphe not distinctive [7]**. *Note*: colour of the seed is not a reliable identification character.

OTHER COMMON NAMES

大菟丝子 Da Tu Si Zi 'big spitting silk seed' (plant name); 金灯藤 Jin Deng Teng (plant name); Japanese dodder (SCN).

SAFETY

Safety Class: 1; Interaction Class: A (Gardner & McGuffin, 2013).

LESSER KNOWN UNOFFICIAL SUBSTITUTES

Seed of other species in the same genus (*Cuscuta*) have been reported as local historical substitutes of official 'Tu Su Zi' in China. Their occurrence as such in international trade is unlikely given that the main source of 'Tu Su Zi' (*C. australis*) has been established in medicinal cultivation and is therefore readily available. (Xiao *et al*., 2001–2002: vol. 2 p. 567; Zhao & Li, 2007).

Crude drug/decocting pieces-1:
Tu Su Zi sourced from *C. australis*

Crude drug/decocting pieces-1:
Tu Su Zi sourced from *C. chinensis*

Crude drug/decocting pieces-2:
salt-processed Tu Su Zi sourced from
C. chinensis (Yan Tu Si Zi)

Seed of *C. japonica*

Wang Bu Liu Xing 王不留行 Vaccariae Semen

Drug part: SEED

PHARMACOPOEIA SPECIES

Vaccaria hispanica (Mill.) Rauschert

Caryophyllaceae

synonyms: *V. segetalis* (Neck.) Garcke ex Asch. (name used in CP2015 as '*V. segetalis* (Neck.) Garcke'); *Saponaria hispanica* Mill.; *S. vaccaria* L.

Species of *Vaccaria* wild in China: 1

PLANT DESCRIPTION

An erect **annual herb** (30–70 cm tall). *Stems* greyish-green, hairless, apically branched into opposite pairs of lateral stems. *Leaves* opposite and simple; leaf-blade ovate to lance-shaped (3–9 cm long; 1.5–4 cm wide), stalkless, with base sometimes slightly clasping, **leaf undersides smooth**, tip pointed. *Flowers* at top of stems in a loose flat-topped inflorescence (corymb); leaf-like bracts present; each flower radially symmetrical, calyx 5-parted with 5 wings joined by pale grey membranes and becoming inflated in fruit; petals 5, each petal with a lower greenish 'claw' and an upper pink, apically notched 'limb'; stamens 10; styles 2. *Fruit* a dry capsule (8–10 mm long). *Seeds* many, black to reddish-brown (see *Drug Description*). Fl. Apr–Jul, fr. May–Aug.

HARVESTING, SOURCING AND NATURAL RANGE

Not native to China but a weed of wheat fields across many provinces: Anhui, Gansu, Guizhou, Hebei, Henan, Hubei, Hunan, Inner Mongolia, Jiangsu, Jiangxi, Ningxia, Qinghai, Shaanxi, Shandong, Shanxi, Xinjiang, Tibet and Yunnan [native in Asia and Europe]. Ripe seeds are harvested in summer.

CONSERVATION STATUS

China: not native.
Global: Not Evaluated (IUCN Red List, accessed 25/8/2016).

UNOFFICIAL SUBSTITUTE

Ficus pumila L.

Moraceae

PLANT DESCRIPTION

Easily distinguished from the official species. A **shrub or climber with milky latex.** *Leaves* alternate and simple, **much larger (5–12 cm long; 2–5 cm wide)** and of 2 different shapes (ovate-cordate and ovate-elliptic); **leaf undersides with conspicuous and honeycomb-like veins**. Flowers minute, without petals and completely enclosed within a yellowish-green to pale red, pear-shaped or obovate fleshy receptacle (a fig) (4–8 cm long; 3–5 cm wide) and **only visible by breaking open the fig**: the flowers develop on the inside wall and link to the outside by a single apical pore. The mature fig is densely packed with tiny single-seeded fruits (achenes). Fl. and fr. May–Aug.

SUBSTITUTION COMMENT. The fruits of this *Ficus* are recognised as a TCM drug in Guangdong province and Hong Kong (Lin, 1990–1996) and are occasionally confused with the Pharmacopoeia species since they share an identical Chinese name: 王不留行 Wang Bu Liu Xing. There is no mistaking the two drugs or their source plants.

Fruits are harvested from wild plants which grow in forests and along walls at low elevations in NC, SC and SE China: Anhui, Fujian, Guangdong, Guangxi, Guizhou, Henan, Hubei, Hunan, Jiangsu, Jiangxi, S Shaanxi, Sichuan, Taiwan, Yunnan and Zhejiang [also native in Japan and Vietnam].

CONSERVATION STATUS

China: Least Concern (MEPCAS, 2013).
Global: Not Evaluated (IUCN Red List, accessed 25/8/2016).

TCM PROPERTIES AND USE

properties: **'bitter and neutral'**
channels: **'Liver and Stomach'**
action: **'invigorates Blood, opens channels'**

indications: **dysmenorrhoea, amenorrhoea; musculoskeletal pain; sores and abscesses; insufficient lactation**

Vaccaria hispanica with developing fruit

V. hispanica coming into flower

V. hispanica in fruit

Climbing habit of *Ficus pumila*, in fruit

F. pumila with fig cut transversely to reveal milky latex and developing seeds within

Branch of *F. pumila* with ripening fruit (fig)

Vaccaria hispanica

Caryophyllaceae

DRUG MORPHOLOGY

CRUDE DRUG/DECOCTING PIECES-1 (DRIED SEED): spherical or sub-spherical (c. 2 mm diam.). Externally dull, black or reddish-brown [1], covered throughout with fine granular protuberances [2] and with one shallow longitudinal furrow [3] and a tiny white hilum dot at one end [4]. Internally, endosperm white, sometimes shrunken [5]; cotyledons two, cream and curved around the inner perimeter of the seed [6]. *Texture:* hard. *Odour:* indistinct. *Taste:* bitter and slightly astringent. ***Good quality*:** plump and black.

DECOCTING PIECES-2 (DRY-FRIED SEED): light and often dumb-bell shaped caused by the expansion of the white, endosperm into a spongy mass [7] with irregular-shaped pieces of torn dark seed coat persisting at either end [8]; cream cotyledons often remain intact forming tongue-like strips often around the seed middle [9].

PROCESSING

Ripe seeds are sun-dried (= crude drug/decocting pieces-1). Seeds are subsequently dry-fried until they burst and are called 炒王不留行 Chao [fried] Wang Bu Liu Xing (= decocting pieces-2).

OTHER COMMON NAMES

不留子 Bu Liu Zi, 留行子 Liu Xing Zi; 王不留 Wang Bu Liu; cow cockle (SCN); cowherb seed; Semen Vaccariae; Vaccaria seeds.

Ficus pumila

Moraceae

'DRUG SUBSTITUTE' MORPHOLOGY

DRIED FRUIT: **pear-shaped or obovate** receptacle (fig) **4–6 cm long and 4 cm diam.**; usually truncate at the broad apical end [10] where a central and prominent pore [11] is surrounded by a sunken zone and which in turn is bounded by a rounded fruit wall shoulder; the narrow basal end has a persistent short and woody fruit stalk [12] or fruit-stalk scar [13]. **Externally dull, yellowish-brown yellow or dark brown** with occasional faint longitudinal striations [14]. Internally a thick wall [15] surrounds a spongy mass of tiny single-sexed flowers or yellowish-brown achenes [16]. *Texture:* hard and light.

OTHER COMMON NAMES

薜荔果 Pi Li Guo; 凉粉果 Liang Fen Guo; 凉粉子 Liang Fen Zi; 木馒头 Mu Man Tou; 王不留行 Wang Bu Liu Xing.

LESSER KNOWN SUBSTITUTES

Seeds of other weedy and widespread species in the genus *Vicia* (family Fabaceae) have been reported as local and/or historical unofficial substitutes (NICBP, 1997–2002; Bensky *et al.*, 2004), however, their occurrence as substitutes in international trade today is uncertain:

1. ***Vicia sativa*** L. (救荒野豌豆 Jiu Huang Ye Wan Dou). Seeds circular-rectangular to circular-ovate, compressed laterally (4–7.5 mm), variously coloured from pale brown to dark green to almost black.
2. ***V. tetrasperma*** (L.) Schreb. (四籽野豌豆 Si Zi Ye Wan Dou). Seeds spherical (c. 1.6 mm diam.); hilum elliptical, surface dull, black or greyish-green with irregular dark purplish spots.
3. ***V. hirsuta*** (L.) Gray (小巢菜 Xiao Chao Cai). Seeds oblate-spheroid (1.5–3 mm), sometimes laterally flattened; hilum linear, raised and extending around ¼ seed circumference.

Crude drug/decocting pieces-1 (*Vaccaria hispanica* seed)

Decocting pieces-2 (*Vaccaria hispanica* dry-fried seed)

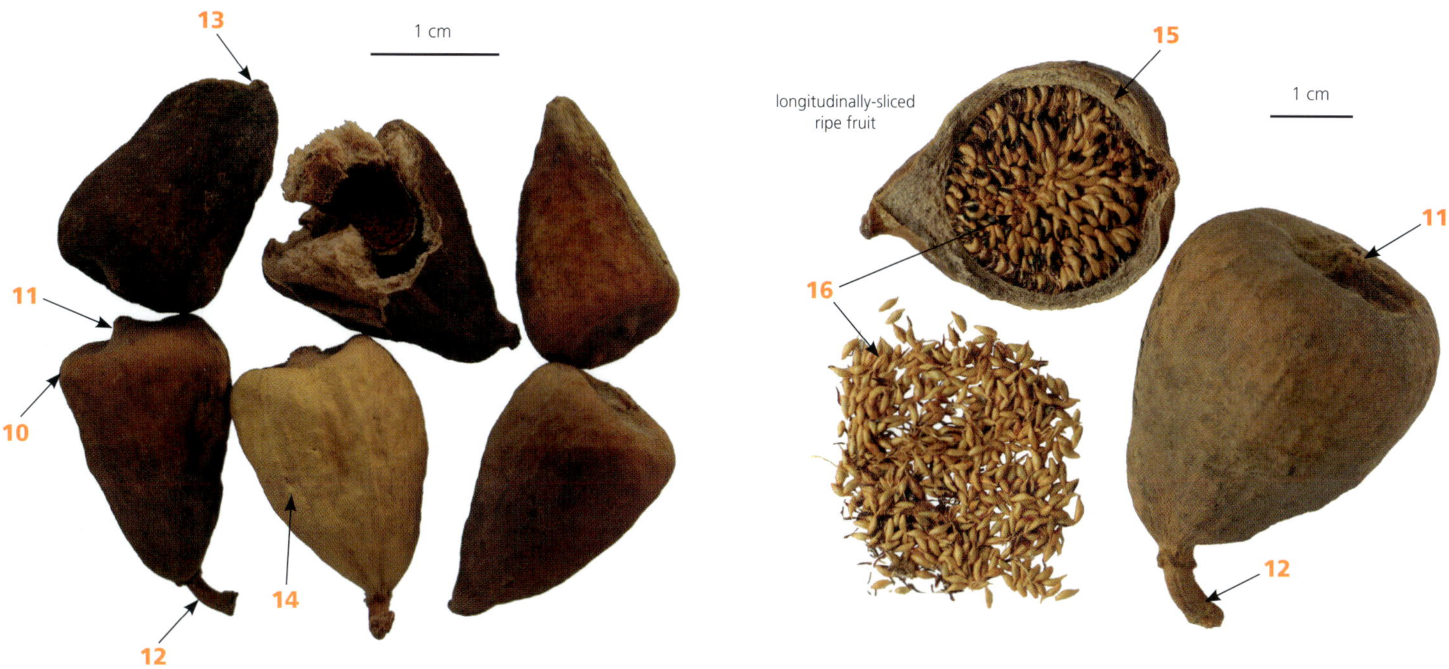

Dried receptacles (figs) of *Ficus pumila*

Wu Mei 乌梅 Mume Fructus

Drug part: FRUIT

Prunus mume (Siebold) Siebold & Zucc.

Rosaceae

Species of *Prunus* wild in China: >7

synonym: *Armeniaca mume* Siebold var. *mume* (*FoC* accepted name)

taxonomic/nomenclatural note: *Kew-MPNS* accepted name, 24/9/2016; there are a number of forms of *Prunus mume* recognised by some as varieties (e.g. *FoC*). The latter describes the type variety (i.e. var. *mume*) as the source of 'Wu Mei' and which is only known from cultivation in China; it is native to Japan and the Korean Peninsula. The emboldened characters in the Plant Description below aim to distinguish this cultivated form from the wild forms which are not used as a source of this drug.

PLANT DESCRIPTION

A deciduous tree, rarely a shrub (4–10 m tall). *Bark* smooth, greyish (or tinged green); **first year branchlets green**, hairless. *Leaves* alternate and single, stalks softly hairy when young and usually with nectaries; stipules present; leaf blades **thin** and greyish-green, hairless when mature, **oval-elliptic (4–8 cm long; 2.5–5 cm wide)**, bases tapering or rounded, margin minutely toothed, tips long and pointed. *Flowers* opening before leaves, solitary (2–2.5 cm diam.) or in pairs, on **short stalks (1–3 mm long),** conspicuous, saucer-shaped and radially symmetrical; fragrant; sepals 5, reddish-brown or green to greenish-purple (2.5–4 mm long); petals 5, equal, wedge-shaped, white or pink; stamens 15–45, spreading; ovary superior. *Fruit* rounded or near ellipsoid **(2–3 cm diam.)** and fleshy, with a distinct longitudinal groove; apex with stiff point; fleshy, yellow to greenish-white and softly hairy; **flesh sour; stone ellipsoid, base wedge-shaped and surface pitted.** Fl. winter to spring, fr. May–Jun (or Jul–Aug in N China).

Note: the ornamental blossoms of *P. mume* have inspired a long cultural history both in China and Japan and today over 300 cultivated varieties are grown. The flowers are also used as the source of the separate TCM drug 'Mei Hua' (Mume Flos) (not included in this book).

HARVESTING, SOURCING AND NATURAL RANGE

Fruits are harvested in summer, when nearly ripe (green), exclusively from cultivated plants which are widely grown throughout the country, especially south of the Chang Jiang (Yangtze river). Traditional *Di Dao* localities occur in the south-eastern provinces of Fujian and Zhejiang. See also ***taxonomic note***.

CONSERVATION STATUS

China: not native.

Global: Not Evaluated (IUCN Red List, accessed 17/8/2016).

Wu Mei plantation (*Prunus mume*), Anhui province

TCM PROPERTIES AND USE

properties: 'sour and neutral'
channels: 'Liver, Spleen, Lung and Large Intestine'
actions: 'astringes, tonifies Lung qi, binds the intestines, generates fluids'

indications: chronic cough; diarrhoea; rectal prolapse, haematochezia; metrorrhagia; abdominal parasites

Prunus mume with ripe fruit

P. mume with very short-stalked flowers

P. mume

Prunus mume Rosaceae

DRUG MORPHOLOGY

CRUDE DRUG/DECOCTING PIECES-1 (SMOKED FRUIT WITH STONE — WU MEI or MEI ZI): fruit sub-spherical or flattened (1.5–3 cm diam.). Externally black or brownish-black, mostly dull, irregularly wrinkled and uneven [1]; longitudinal groove sometimes visible [2]; basal end with roundish stalk scar [3]. Flesh closely adhering to the large central stone, the latter brownish-yellow, ellipsoid [4], base wedge-shaped and surface pitted [5]. Seed ovoid, flattened and pale yellow [6]. *Texture*: flesh pliable; stone very hard. *Odour*: indistinct. *Taste*: extremely sour. ***Good quality*:** large fruit, externally black, seeds small; extremely sour-tasting.

DECOCTING PIECES-2 (FRUIT WITHOUT STONES — WU MEI ROU): irregularly torn flesh or varying sizes [7] (most c. 0.5–2.5 cm long), black or brownish-black; externally mostly dull, irregularly wrinkled and uneven.

DECOCTING PIECES-3 (CARBONISED FRUIT — WU MEI TAN): as for Mei Zi above but charred black and swollen [8]. *Taste*: sour and somewhat bitter.

PROCESSING

Fruits are smoke-dried in a low temperature oven (c. 40°) then covered until they turn black (= 梅子 Mei Zi; crude drug/decocting pieces-1). The stones (endocarps) are sometimes removed later by softening the fruits with water or by steaming (= 乌梅肉 Wu Mei Rou [plum flesh]; decocting pieces-2). Subsequent carbonisation may be performed by dry-frying the smoked fruit over a high heat until the flesh bubbles, becomes dark brown throughout and loses its sticky texture; it is then baked dry. (= 乌梅炭 Wu Mei Tan [charred plum] decocting pieces-3).

OTHER COMMON NAMES

青梅 Qing Mei (green plum); Fructus Mume; Japanese apricot (SCN); smoked plum (2010).

SAFETY

Safety Class: 1; Interaction Class: A (Gardner & McGuffin, 2013). An official food and healthfood in China (Anon., 2002a,b).

UNOFFICIAL SUBSTITUTES

Fruits of closely related Rosaceae species (such as apricots and peaches) have been reported as local or historical substitutes in China (Bensky *et al.*, 2004; NICPBP, 1997–2002). Their occurrence as substitutes of 'Wu Mei' in international TCM trade is unlikely.

1. ***Prunus armeniaca*** L. (fruits are apricots). Young branchlets reddish-brown; leaf bases wedged-shaped or broadly wedge-shaped; leaf margins with rounded teeth; petals pink; fruit white, yellow or orange (often tinged red); stone globose, ovoid, or ellipsoid, **with keel-like ribs on ventral side [9], surface rough and netted [10].** Seeds are the source of the TCM drug 'Ku Xing Ren' (Armeniacae Semen Amarum) page 668.
2. ***P. persica*** (L.) Batsch (fruits are peaches). Tree often taller (3–8 m); branchlets green but reddish on exposed side; leaf bases broadly wedge-shaped; **fruits much larger (usually 5–7 cm diam.), densely hairy,** flesh edible (peach), succulent, **sweet to sour-sweet** and fragrant; stone ellipsoid to suborbicular, **surface longitudinally and transversely furrowed and pitted [11], tip sharply pointed [12].** Seeds are the source of the TCM drug 'Tao Ren' (Persicae Semen) page 742.
3. ***P. salicina*** Lindl. var. *salicina*; widely cultivated in China for its edible fruit. Leaves oblong to oblong-ovate or narrowly elliptic (6–8 (rarely 12) cm long; 3–5 cm wide), upper surface dark green and lustrous, base wedge-shaped, margin with 2 sizes of rounded teeth; flowers usually in 3s; **flower stalks longer (1–1.5 cm); fruit larger (3.5–5 cm diam. [up to 7 cm diam. in cultivated forms],** yellow or red, sometimes green or purple, **hairless;** stone **ovoid to oblong, wrinkled not pitted.**

Prunus armeniaca (apricot)

Prunus persica (peach)

Prunus mume stone

Wu Mei stone and kernel

Crude drug/decocting pieces-1 (Wu Mei or Mei Zi) — smoked fruit, stone present

Decocting pieces-2 (Wu Mei Rou) — smoked fruit, de-stoned

Decocting pieces-3 (Wu Mei Tan) — smoked and carbonised fruit

Wu Wei Zi 五味子 Schisandrae Chinensis Fructus

Drug part: FRUIT

Schisandra chinensis (Turcz.) Baill.

Schisandraceae

Species of *Schisandra* wild in China: 19

PLANT DESCRIPTION

A deciduous, woody climber (up to 8 m tall). *Leaves* simple, somewhat clustered and papery; leaf blade elliptic to widest above middle (4.5–8 (rarely up to 10.5) cm long; 2.5–6.5 cm wide), veins hairless or upper surface sparsely hairy; base wedge-shaped or extending into leaf stalk; margins finely toothed, tip abruptly pointed. *Flowers* solitary, pendent and single-sexed (c. 1.5 cm diam.), with male and female flowers on different plants; stalk (peduncle) of male flowers **0.6–2.8 cm long;** stalk of female flowers **0.7–3.5 cm long; 'petals' 5–9, white to yellow;** male flowers **usually with 5 stamens** (rarely 4 or up to 7) and fused into a short column; female flowers with 14–40 carpels. *Fruit* a linear, pendent and conspicuous aggregate of ovoid, pinkish-red to red carpels (latter 5–7.5 mm long; 4–6 mm diam.); stalk (peduncle) of fruit aggregate 2–7.5 cm long; **fruit aggregate 1.5–9.5 cm long.** *Seed* 1–2 per fruit, kidney-shaped, orangish-brown and lustrous. Fl. May–Jul, fr. Jul–Sep.

HARVESTING, SOURCING AND NATURAL RANGE

Until recently, fruits were totally wild-harvested in N and NE China (provinces of Hebei, Heilongjiang, Jilin, Liaoning, Inner Mongolia and Shanxi). Since c. 2010, medicinal plant farms have become established in Jilin and Liaoning provinces to provide a sustainable source (some GAP-certified) of this long-prized and increasingly popular herb of TCM and the healthfood industries worldwide (Lin Yu-Lin, pers. obs. 2013; Upton, 1999b). Fruits are harvested in autumn when ripe or almost ripe; *Di Dao* production sites are also located in Jilin and Liaoning provinces (Zhang *et al.*, 2010).

In the wild in China, the species grows in deciduous and mixed forests in mountain valleys and along river-sides (1,200-1,700 m altitude) and is also native in the Korean peninsula, Japan and Russia (Far East). In Europe it is cultivated as an ornamental along with other species in the genus.

CONSERVATION STATUS

China: Least Concern (MEPCAS, 2013).
Global: Not Evaluated (IUCN RED List, accessed 8/6/2016).

S. sphenanthera Rehder & E. H. Wilson

Schisandraceae

PLANT DESCRIPTION

Very similar to *S. chinensis* (the official source of 'Wu Wei Zi'). *Main differences*: **stalk (peduncle) of male flower longer (0.8–5.4 cm**); stalk of female flower sometimes longer **(1.8–6.4 cm long)**. **'Petals' yellow, orange or red**, largest 6–11.5 mm long; 3.5–8.5 mm wide; male flowers with **about twice as many stamens (11–25).** Fruit stalk (pendunclе) 2–10 cm long; **fruit aggregate often longer 2.5–14 (rarely up to 22) cm long;** fruits (carpels) red, 5.5–9 mm long; 4–8 mm diam. Seed somewhat ovoid to kidney-shaped, yellowish-grey to greyish-brown and matt. Fl. Apr–Jul, fr. Jun–Oct.

SUBSTITUTION COMMENT. There is a long history of similar clinical use of the fruits of *S. sphenanthera* and *S. chinensis* with the prefixes Bei 'north' and Nan 'south' used to differentiate them respectively. To reflect their significant chemical differences (Upton 1999b), editions of the CP from 2000 onwards (along with the Japanese Pharmacopoeia) have created separate drug entries for these species (see 'Nan Wu Wei Zi' (Schisandrae Sphenantherae Fructus) page 686). The latter has long been regarded as clinically inferior and hence cheaper, with the name 'Cinderella Schisandra' coined by some to reflect this (Cunningham & Brinckmann, 2010). *S. sphenanthera* fruit are often traded as 'Bei Wu Wei Zi' in an attempt to fetch a higher price and, in some instances, fruits traded on the UK market have been dyed a brighter red presumably to mimic those of *S. chinensis* (van der Valk & Leon, pers. obs.).

CONSERVATION STATUS

China: Data Deficient (MEPCAS, 2013).
Global: Not Evaluated (IUCN RED List, accessed 8/6/2016).

TCM PROPERTIES AND USE

properties: 'sour, sweet and warm'
channels: 'Kidney, Lung and Heart'
actions: 'astringes, tonifies qi, Lung and Kidney, calms the shen'

indications: chronic cough, dyspnoea; spermatorrhoea; leucorrhoea; enuresis; spontaneous sweating; irritability, palpitations

Female plant of *Schisandra chinensis*

Medicinal crop of *S. chinensis* with ripe fruit, Jilin province

S. chinensis with ripe fruit

S. sphenanthera (orange-flowered form)

Typical colour form of *S. sphenanthera* flowers, left-hand flower with 'petals' partly removed to show 11–25 stamens, Shaanxi province

Schisandra sphenanthera with developing fruit

Schisandra chinensis

Schisandraceae

DRUG MORPHOLOGY

CRUDE DRUG (DRIED FRUIT): irregularly globose, angular in outline where the flesh has shrunk **[1]**; sometimes partially flattened **(5–8 mm diam.)**. **Fruit wall often lustrous [2]**, **deep reddish, purplish-red, blackish-red** or dull 'white-frosted'; 1 or 2 seeds usually visible where flesh thin **[3]**; one brown oval stem scar visible at one end **[4]**. Seed relatively easy to remove from the flesh; seed kidney-shaped (3.2–3.8 mm long; (4.4–4.8) mm diam. **[5]**, **seed coat orangish-brown, smooth, lustrous [6]**, thin and fragile. *Taste:* pulp sour, salty and astringent; seed pungent and slightly bitter; aftertaste salty. *Texture:* flesh viscous, oily and pliable. *Odour:* pulp indistinct; seed aromatic when crushed. ***Good quality*:** large, purplish-red with thick, fleshy, pulp; intensely aromatic.

PROCESSED DRUG (VINEGARED FRUIT — CU WU WEI ZI): as for the crude drug but brownish-black **[7]** or, where flesh is thin, deep red; oily and often lustrous **[8]**. *Taste:* vinegar- tainted. *Odour:* vinegar-scented.

PROCESSING

Fruits are sun-dried or steamed, then stalks removed (= crude drug). Further processing may be undertaken by steaming with rice vinegar until fruits become a dark brownish-black (醋五味子 Cu Wu Wei Zi 'vinegared Wu Wei Zi'). Fruits are pounded prior to decocting.

OTHER COMMON NAMES

北五味子 Bei Wu Wei Zi 'northern five-flavoured seed'; Chinese magnolia vine fruit (CP2010); five flavor fruit; Fructus Schisandrae Chinensis; schisandra (SCN).

SAFETY

Safety Class: 1; Interaction Class: C (Gardner & McGuffin, 2013). An official healthfood in China (Anon., 2002a,b).

S. sphenanthera

Schisandraceae

'DRUG SUBSTITUTE' MORPHOLOGY

Very similar and easily confused with official 'Wu Wei Zi.' *Main differences*: **fruits less fleshy and often slightly smaller (c. 5–5.5 mm diam.) [9]**; **fruit wall dull, brownish-red to dark brown [10]** with seed difficult to remove cleanly. Seeds slightly smaller (3–3.5 mm long; 3.5–4 mm diam.) and **more rounded** (somewhat ovoid to kidney-shaped) **[11]**; seed coat **dark yellowish-grey or greyish-brown, dull (not lustrous) [12]**, minutely verrucose [hl] **[13]** (Saunders, 2000; Upton ed. 1999b); while these macroscopic characters are often useful, laboratory based methods are recommended for more reliable authentication especially where mixed batches are suspected and to eliminate some of the less common substitutes (see *Lesser Known Unofficial Substitutes*).

OTHER COMMON NAMES

华中五味子 Hua Zhong Wu Wei Zi 'middle China Wu Wei Zi'; 南五味子 Nan Wu Wei Zi 'southern Wu Wei Zi'; five flavour fruit; Fructus Schisandrae Sphenantherae; schisandra (SCN); Schisandrae Sphenantherae Fructus; southern magnolia vine fruit (CP2010).

SAFETY

Safety Class: 1; Interaction Class: C (Gardner & McGuffin, 2013). An official healthfood in China (Anon., 2002a,b).

LESSER KNOWN UNOFFICIAL SUBSTITUTES

While fruits of *S. sphenanthera* are generally considered the commonest substitute (Bensky *et al.*, 2004; Upton 1999b), the name 'Wu Wei Zi' has been used for many of the other 17 *Schisandra* species growing wild in China and species in closely related genera (Ho *et al.*, 2006; Xiao *et al.*, 2001–2). Their occurrence as substitutes of 'Wu Wei Zi' in international trade is uncertain.

1. other ***Schisandra*** species
2. ***Kadsura*** spp. (notably *K. japonica* (L.) Dunal (日本南五味子 Ri Ben Nan Wu Wei Zi) and *K. longipedunculata* Finet & Gagnep. (南五味子 Nan Wu Wei Zi) (Schisandraceae family)
3. ***Euonymus*** spp. (Celastraceae family)
4. ***Vitis*** spp. (notably *V. amurensis* Rupr. (山葡萄 Shan Pu Tao) (Vitaceae family)

Crude drug of Wu Wei Zi
(sourced from of *Schisandra chinensis*)

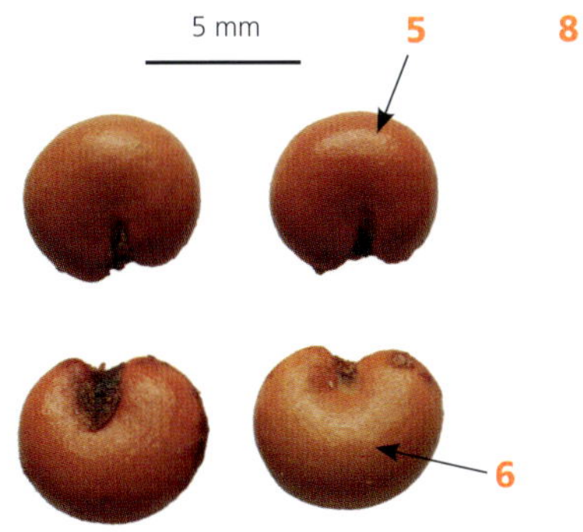

S. chinensis seed

Vinegared Wu Wei Zi (Ci Wu Wei Zi)
(sourced from *S. chinensis*)

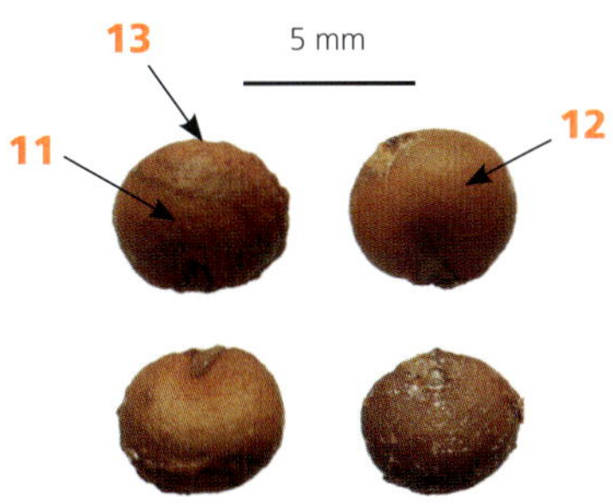

S. sphenanthera seed

Dried fruit of *S. sphenanthera*
(unofficial substitute 'Nan Wu Wei Zi')

Vinegared fruit of of *S. sphenanthera*
(unofficial substitute 'Nan Wu Wei Zi')

Wu Zhu Yu 吴茱萸

Euodiae Fructus

Drug part: FRUIT

Tetradium ruticarpum (A. Juss.) T. G. Hartley

Rutaceae

synonyms: *Evodia ruticarpa* (A. Juss.) Hook. f. & Thomson (in CP2015 as '*Euodia rutaecarpa* (Juss.) Benth.'); *E. ruticarpa* var. *bodinieri* (Dode) C. C. Huang (in CP2015 as '*Euodia rutaecarpa* (Juss.) Benth. var. *bodinieri* (Dode) Huang'); *E. ruticarpa* var. *officinalis* (Dode) C. C. Huang (in CP2015 as '*Euodia rutaecarpa* (Juss.) Benth. var. *officinalis* (Dode) Huang')

taxonomic/nomenclatural note: The source taxa listed in the CP2015 are treated here as synonyms of one species: *T. ruticarpum*; this follows the taxonomy of *FoC* and *Kew-MPNS*, 15/8/16.

PLANT DESCRIPTION

A deciduous shrub or tree (up to 9 m tall) packed with oil glands in most parts. Leaf buds in axils distinct. *Leaves* opposite and pinnate (15–40 cm long); usually with **2–6 pairs of ovate to elliptical leaflets**; leaflet margin entire or with tiny rounded teeth (crenulate). Each *inflorescence* a dense, many-flowered hemi-spherical head **(2.5–18 cm diam.)**. *Flowers* single-sexed, tiny, petals 5, green, white or yellow (3–5 mm long). *Fruit* sub-globular, densely glandular, with **5 sub-globose beakless carpels** (3.5–6 mm long) surrounding a central apical depression; **carpels** easily separating on maturity; seed per carpel 1, black plus an aborted ovule. Fl. Apr–Jun, fr. Aug–Nov.

Species of *Tetradium* wild in China: 7

HARVESTING, SOURCING AND NATURAL RANGE

Indehiscent fruiting branches (i.e. before fruits split open and release seeds) are harvested between August and November mostly from cultivated crop plants widely grown in Anhui, Fujian, Guangdong, Guangxi, Guizhou, Hebei, Henan, Hubei, Hunan, Jiangsu, Jiangxi, Shaanxi, Sichuan, Yunnan and Zhejiang.

The wild species grows in forest, scrub and open places (100–3,000 m altitude) in NC, C, S and SE China (Gansu, Guangdong, Guangxi, Guizhou, Hebei, Henan, Hubei, Hunan, Jiangsu, Jiangxi, Shaanxi, Sichuan, Yunnan and Zhejiang (also wild in Bhutan, NE India, Myanmar and Nepal).

CONSERVATION STATUS

China: Least Concern (MEPCAS, 2013).
Global: Not Evaluated (IUCN Red List, accessed 18/5/2015).

Medicinal crop of *T. ruticarpum*, Anhui province

Flowering branch of *Tetradium ruticarpum*

Young fruiting branch of *T. ruticarpum*

UNOFFICIAL SUBSTITUTES

The fruits of other species in the same genus have been reported as local substitutes in China (NICPBP, 1997–2002, Bensky *et al.*, 2004); their trade internationally under the name Wu Zhu Yu is unlikely given that the Pharmacopoeia species is widely available from commercial medicinal plant farms.

1. ***Tetradium glabrifolium*** (Champ. ex Benth.) T. G. Hartley (as the syn. *Evodia fargesii* in Bensky *et al.*, 2004); 楝叶吴萸 Lian Ye Wu Yu (plant name) or 臭辣树 Chou La Shu with 'Chou' meaning foul-smelling. **Carpels obtusely 3-angled**, 3.5–5 mm long; **foetid**; 2 seeds per carpel (1 abortive).
2. *T. austrosinense* (Hand.-Mazz.) T. G. Hartley (as the syn. *Euodia austrosinensis* in Bensky *et al.* 2004); 华南吴萸 Hua Nan Wu Yu). Carpels somewhat 3-sided; 1 seed per carpel (abortive seeds absent).
3. *T. daniellii* (Benn.) T. G. Hartley (as *Evodia daniella* Hemsl.; NICPBP, 1997–2002); 臭檀吴萸 Chou Tan Wu Yu. **Carpels longer and narrowly pear-shaped, 5–11 mm long**, apex beaked; 2 seeds per carpel (1 abortive).

TCM PROPERTIES AND USE

properties: **'pungent, bitter and hot'**
channels: **'Liver, Spleen and Stomach'**
actions: **'dispels cold damp, regulates qi, warms interior'**

indications: **abdominal pain; dysmenorrhoea; reflux, vomiting; diarrhoea; headache; ulcers, sores of head and mouth; (T)**

DRUG MORPHOLOGY

CRUDE DRUG/DECOCTING PIECES-1 (DRIED FRUIT): fruit squat and somewhat club-shaped (2–5 mm diam.) **[1]** with 5 (usually) fused carpels (3.5–6 mm long) often separating apically to form a 5-parted depression **[2]**. External surface of fruit dark, matt yellowish-brown, green or greyish-green and regularly covered with many oil glands visible as round protuberances or pits **[3]**; glistening beads of oil sometimes visible. Five brown pointed sepals (c. 1 mm long) persist below the fruit base **[4]**. Woody, sometimes branching, fruit-stalks often still attached and densely covered in yellow or cream bristle-like hairs **[5]**. Transverse fracture reveals 5 loculi **[6]**, each containing a single brown or yellow seed **[7]** (plus sometimes an additional aborted seed); oil glands scattered throughout. *Texture*: hard and brittle. *Odour*: intensely aromatic. *Taste*: pungent and bitter. ***Good quality***: fruit indehiscent (seeds present), greenish-black or dark green, firm and intensely aromatic.

DECOCTING PIECES-2 (LIQUORICE-PROCESSED FRUIT — ZHI WU ZHU YU: as for crude drug/unprocessed decocoting pieces-1 but externally brown or dark brown **[8]**.

PROCESSING

Fruiting branches are sun-dried or dried at low temperature, then fruits removed (= crude drug/decocting pieces-1). Further optional processing involves soaking the dried fruits in a decoction made from Gan Cao (Glycyrrhizae Radix et Rhizoma) not included in this book, then dry-frying to partial dryness then sun-drying (= 制吴茱萸 Zhi Wu Zhu Yu liquorice-prepared Evodia — decocting pieces-2).

OTHER COMMON NAMES

茶辣 Cha La; 淡吴茱萸 Dan Wu Zhu Yu; 石虎 Shi Hu; 吴芋 Wu Yu; 制吴茱萸 Zhi Wu Zhu Yu; Evodia (SCN); medicinal Euodia fruit (CP2010); Fructus Evodiae.

SAFETY

Slightly toxic (CP2015; Bensky *et al*., 2004). Liquorice-processed fruits are widely used to minimise the potential for adverse reactions. **Safety Class: 2d; Interaction Class: B** (Gardner & McGuffin, 2013). An official Chinese healthfood (Anon., 2002a,b).

Crude drug/unprocessed decocting pieces of Wu Zhu Yu

Liiquorice-processed decocting pieces of Zhi Wu Zhu Yu

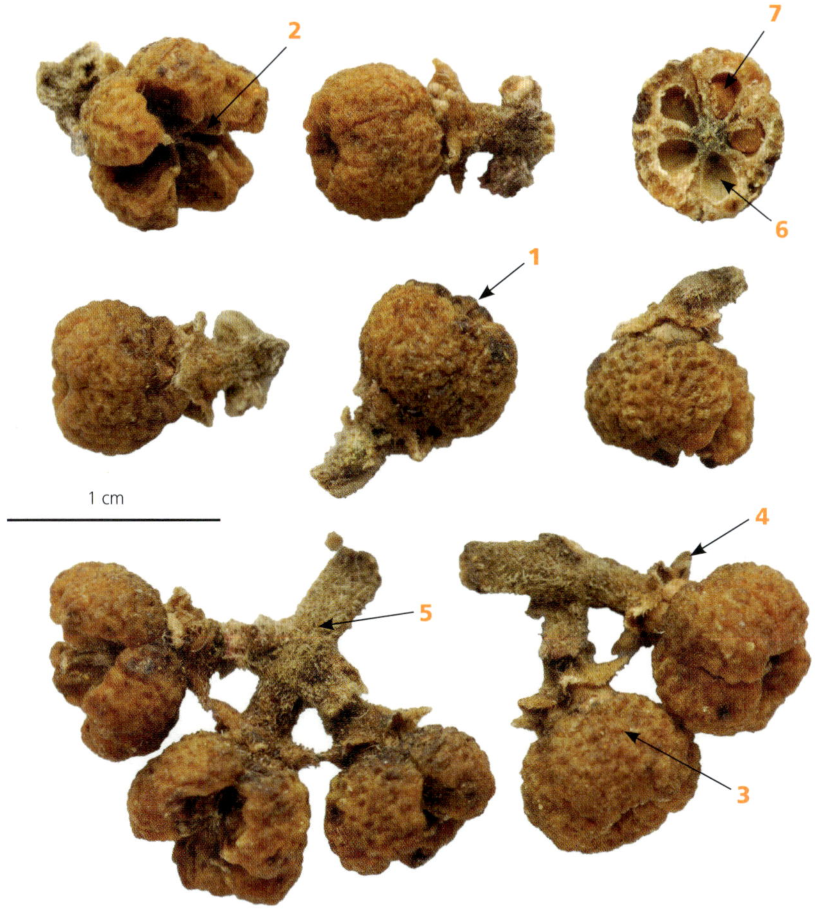

Xia Ku Cao 夏枯草 Prunellae Spica

Drug part: FRUIT-HEAD

Prunella vulgaris L.

Lamiaceae

PLANT DESCRIPTION

A low-growing perennial herb (5–30 cm tall), somewhat creeping and often rooting at the nodes. *Stems* 4-sided, multi-branched, purplish-red, with a sparse covering of coarse, white hairs. *Leaves* opposite; leaf stalks 0.7–2.5 cm long, becoming shorter in upper parts; leaf blades simple, ovate to rhomboid-ovate or lance-shaped (1.5–6 cm long; 0.6–2.5 cm wide), bases wedge-shaped or extending down the stem, tips broad to rounded, upper and lower surfaces almost hairless and margins shallowly scalloped or entire. *Flowers* in short but conspicuous and dense, straight-sided spikes (2–8 cm long) at the top of the plant; flowers in groups of 3 subtended by broad, membranous bracts (7 mm long; 11 mm wide) turning purplish-violet, tips abruptly pointed and margins membranous and distinctly fringed with hairs. Each flower bisexual, laterally symmetrical and short-stalked; calyx cup-shaped (c. 4 mm long) with 5 sharply tapering teeth; corolla purplish or white (c. 1.3 cm long) with a short corolla tube opening into a 2-lipped mouth (upper lip hollow and domed and c. ½ as long as lower lip; latter 3-lobed with middle lobe largest, concave and with a toothed margin; lateral lobes minute; stamens 4, free; ovary superior; style 2-cleft. *Fruit* with 4 dry, brown, sub-globose nutlets (c. 1.8 mm long; 0.9 mm diam.). Fl. Apr–Jun; fr. Jul–Oct.

Species of *Prunella* wild in China: 4

HARVESTING, SOURCING AND NATURAL RANGE

Fruiting heads are harvested in summer when brownish-red from wild and medicinal crop plants. The main cultivated production areas are in the Chinese provinces of Anhui , Jiangsu and Henan.

In the wild the species is widespread and common and grows in grassland, forest, margins of scrub, especially in damp places (0–3,200 m altitude). It has been recorded as native in Fujian, Gansu, Guangdong, Guangxi, Guizhou, Henan, Hubei, Hunan, Jiangxi, Shaanxi, Sichuan, Taiwan, Tibet, Xinjiang, Yunnan and Zhejiang [also wild in Japan, the Korean Peninsula, S. Central Asia, India, Africa, Europe and N America].

CONSERVATION STATUS

China: Least Concern (MEPCAS, 2013 (var. *lanceolata* only)
Global: Least Concern (IUCN Red List, accessed 29/9/2016).

Prunella vulgaris, in flower, Henan province

Young flowering heads of *P. vulgaris* with bracts yet to turn purple

Medicinal crop of *P. vulgaris*, Henan province

TCM PROPERTIES AND USE

properties: **'bitter, pungent and cold'**
channels: **'Liver and Gall Bladder'**
actions: **'clears heat and fire, resolves phlegm, disperses nodules'**

indications: **headaches, dizziness, hypertension, painful eyes; emotional instability; nodules, swollen glands; hypertension**

DRUG MORPHOLOGY

CRUDE DRUG/DECOCTING PIECES (DRIED FRUIT-HEADS): fruit-heads cylindrical, slightly flattened, pale brown to reddish-brown (1.5–8 cm long; 0.8–1.5 cm diam.) with several dense whorls of bracts and calices **[1]** (petals mostly fallen). Bracts broad (7 mm long; 11 mm wide) and membranous **[2]** with an abruptly pointed tip **[3]** and margins distinctly fringed with hairs **[4]**. Each bract surrounds 3 tiny flowers of which only the calyces remain **[5]**. Calyces c. same length (c. 4 mm), colour and texture as the bracts with 5 narrow, veined teeth, each ending in a long, narrow tip **[6]**. At the base of each calyx, 4 small, brown ovoid nutlets sometimes present. *Texture:* light. *Odour:* indistinct. *Taste:* indistinct. ***Good quality:*** long, reddish-brown fruit-heads with short stalks; leaves absent.

Note: Early editions of the Chinese Pharmacopoeia defined this drug as being sourced from all aerial parts of this species (not just the fruit-heads). Occasionally Xia Ku Cao continues to be traded according to this definition.

PROCESSING

Fruiting heads are harvested in summer when brownish-red, then sun-dried (= crude drug and decocting pieces).

OTHER COMMON NAMES

夏枯花 Xia Ku Hua; 夏枯球 Xia Ku Qiu; common self-heal fruit-spike (CP2010); heal all (SCN); Spica Prunellae. Many other trade names exist for this drug reflecting its medicinal use in other countries (see *Kew-MPNS*).

SAFETY

Safety Class: 1; Interaction Class: A (Gardner & McGuffin, 2013).

Crude drug of Xia Ku Cao 夏枯草
(dried fruit-heads of *P. vulgaris*).

Xiang Yuan 香橼

Citri Fructus

citron

Drug part: FRUIT

Citrus medica L.

Rutaceae

taxonomic/nomenclatural note: the CP2015 also lists *Citrus* × *wilsonii* Tanaka as an alternative source of this drug but its taxonomic status is currently uncertain. *FoC* includes it as a possible synonym of *C.* × *junos* Siebold ex Tanaka, and *Kew-MPNS* (8/9/2016) notes it is awaiting taxonomic review. It is therefore omitted until resolved.

Species and hybrids of *Citrus* wild in China: 11 (plus 5 hybrid cultivated species)

PLANT DESCRIPTION

A shrub or small evergreen tree (usually c. 2 m tall) with branches, leaf buds, and flower-buds usually purplish when young. Branches with spines (c. 4 cm long). *Leaves* alternate and simple, **stalks short, not winged and the articulated joint at leaf blade base inconspicuous or absent**; blades elliptic to ovate-elliptic (6–15 cm long; 3–6 cm wide or larger) margins shallowly saw-toothed, tip rounded, broad or (rarely) shortly pointed (mucronate); blade scattered with translucent and fragrant oil ducts. *Flowers* in axillary, 12-flowered clusters (racemes) or solitary; bisexual or occasionally male. Flower with calyx cup-shaped and 3–5-lobed; petals 5 (1.5–2 cm long), inner surface white, outer surface purplish; stamens 30–60; style long, thick and persistent; stigma large and club-shaped; fragrant. *Fruit* elliptic to sub-globose (1–25 cm long; 5–10 cm diam.; weighing up to 2 kg), yellow when mature; fragrant; surface rough and leathery; pith (pericarp) white to pale yellow, sometimes thicker than the fleshy pulp uneasily removed; pulp (flesh) divided into 10–17 segments, juice yellow; very sour to slightly sweet. *Seeds* small and smooth. Fl. Apr–May, fr. Oct–Nov (based on Lin Yu-Lin, 2010).

HARVESTING, SOURCING AND NATURAL RANGE

Fruits are harvested in autumn when ripe (yellow), totally from cultivated crop plants grown on hill-slopes in S, C and W China (Fujian, Guangxi, Guizhou, Hainan, Sichuan, Tibet and Yunnan). The species is native to NE India and possibly Myanmar.

CONSERVATION STATUS

China: Not Native; sometimes naturalised (*FoC*).
Global: Not Evaluated (IUCN Red List, accessed 24/6/2015).

Citrus medica cultivated on a smallholding, China

Immature fruit of *C. medica*

TCM PROPERTIES AND USE

***properties*: 'pungent, bitter and warm'**
***channels*: 'Liver, Spleen, Stomach and Lung'**
***actions*: 'invigorates qi, courses Liver, resolves phlegm'**

***indications*: chest and abdominal pain, distension; vomiting, nausea; cough**

DRUG MORPHOLOGY

CRUDE DRUG (DRIED SLICED FRUIT): round or oblong transverse fruit slices, often slightly sinuous **[1]** in outline (4–10 cm diam.; 0.2–0.5 cm thick). Peel orangish-brown, yellow or yellowish-green, rough, leathery and thin and densely pitted throughout with oil ducts **[2]**. Fruit wall (pith) 1–3 cm wide **[3]**, yellowish-white and scattered with irregularly shaped vascular tissues. Pulp (flesh) separated into 10–17 segments **[4]**. Central stele substantial (c. 3–6 mm diam.) **[5]**. *Texture*: soft and pliable. *Odour*: aromatic. *Taste*: bitter, pungent and slightly sweet. ***Good quality*:** fruit slices with a thick fruit wall (pericarp); intensely aromatic.

DECOCTING PIECES (DRIED FRUIT): small strips of whole fruit (up to c. 6 cm long; c. 0.5 cm wide; 0.1–0.5 cm thick) including outer peel, pith, pulp (flesh) and sometimes seed. Peel orangish-brown, yellow or yellowish-green, rough, leathery and thin and densely pitted throughout with oil ducts **[6]**. Fruit wall (pith) thicker than peel **[7]**, yellowish-white and scattered with irregularly shaped vascular tissues. Pulp (flesh) with walls of cut segments sometimes visible **[8]**. When present and intact, central stele substantial **[9]** (c. 3–6 mm diam.). *Texture*: soft and pliable. *Odour*: aromatic. *Taste*: bitter, pungent and slightly sweet.

PROCESSING

Fruits are transversely sliced when fresh then sun-dried or dried at low temperature (= crude drug); aged slices are preferred (Chen Xiang Yuan or Chen Xiang Yuan Pi). Decocting pieces are made by softening the slices before cutting them into slivers and drying again.

OTHER COMMON NAMES

枸橼 Ju Yuan; 香橼 Xiang Yuan (plant name); citron (SCN); citron fruit (CP2010); Fructus Citri.

SAFETY

An official food and healthfood in China (Anon., 2002a,b).

1 cm

Crude drug
(possibly low quality since fruit wall (pith) quite narrow)

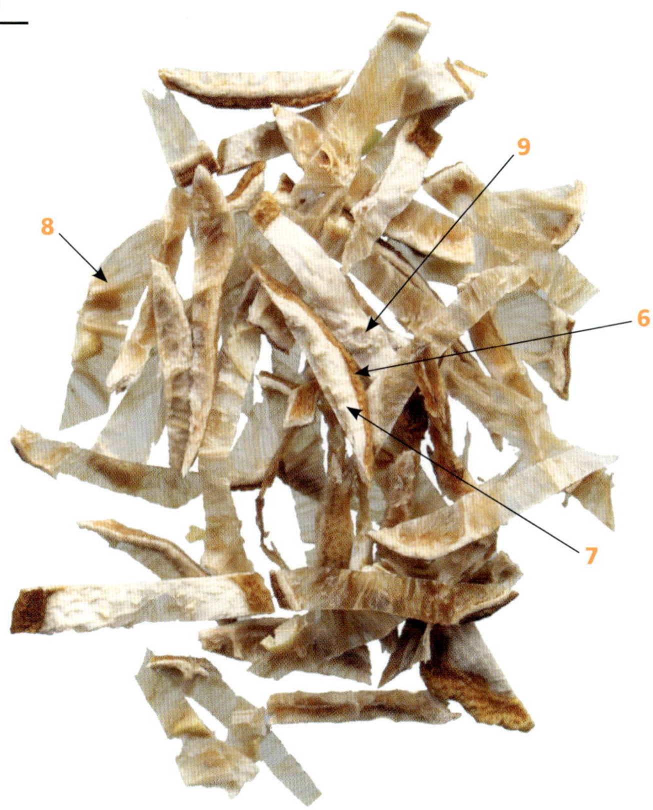

Decocting pieces

Ya Ma Zi 亚麻子

Lini Semen

flax seed; linseed

Drug part: SEED

Linum usitatissimum L.

synonym: *Linum humile* Mill.

Linaceae

Species of *Linum* wild in China: 8

PLANT DESCRIPTION

Over 20 cultivars have been bred worldwide for fibre, edible seed (mostly for linseed oil) or for both. Some variation in morphology is therefore to be expected.

A small, delicate-looking, herbaceous annual (30–120 cm tall), branching apically. *Leaves* alternate, simple and stalkless; leaf blade linear to lance-shaped (2–4 cm long; 1–5 mm wide), 3(or 5)-veined; margin usually entire. *Flowers* solitary (15–20 mm diam.) or in loose heads (cymes); flower-stalks 1–3 cm long; flower radially symmetrical and bisexual; sepals 4 or 5, ovate to lance-shaped and pointed; petals 5, usually blue (rarely white or red), of equal size and shape (obovate) and longer than sepals; stamens 5; ovary superior; styles 5, of equal length; stigmas linear or club-shaped. *Fruit* (when ripe) a dry, brown, hairless and globose capsule (6–9 mm in diam.) splitting into 10 segments; base with 5 persistent sepals. *Seeds* many, ovate in outline and flattened (4–6 mm long; 2–3 mm wide); see also *Drug Morphology.* Fl. Jun–Aug, *fr.* Jul–Oct.

HARVESTING, SOURCING AND NATURAL RANGE

Fruits are harvested entirely from crop plants in autumn, when fruits are ripe (brown), then threshed to release the seed. The species has been cultivated in China for millennia and is grown throughout except in Hainan and Taiwan. Many cultivars exist. The species originates from the Mediterranean.

China: not native; cultivated and occasionally naturalised.
Global: Not Evaluated (IUCN Red List, accessed 10/9/2015).

Linum usitatissimum — flowering plants

L. usitatissimum, with immature fruits

TCM PROPERTIES AND USE

properties: **'sweet and neutral'**
channels: **'Lung, Kidney and Large Intestine'**
action: **'moistens intestines, tonifies Blood, dispels wind'**

indications: **constipation; itchy dry skin; hair loss**

DRUG MORPHOLOGY

CRUDE DRUG (SEED): oval in outline and laterally compressed and (4–6 mm long; 2–3 mm wide; 0.75–1.5 mm thick). Externally smooth and glossy **[1]**, reddish-brown or yellow; broadly rounded at one end, narrowing at the other. Narrow end with a laterally bent beak **[2]** with, on the concave side, a tiny [hl] pale sunken hilum and micropyle **[3]**. A pale linear raphe [hl] **[4]** extends from the beak towards the other end of the seed. Seed coat thin and brittle, enclosing membranous endosperm and 2 yellowish-white and oily cotyledons. *Odour:* indistinct. *Taste:* mild and mucilaginous on chewing. ***Good quality***: large, full, reddish-brown.

PROCESSING

Seeds are sun-dried (= crude drug). They are ground prior to decocting or dry-fried then ground.

OTHER COMMON NAMES

胡麻仁 Hu Ma Ren; flax (SCN); linseed (CP2010); Semen Lini.

SAFETY

Safety Class A; Interaction Class: A. (Gardner & McGuffin, 2013). As a bulk-forming laxative, adequate fluids need to be taken at the same time (Gardner & McGuffin, 2013).

Crude drug (left, bronze cultivar; right golden cultivar

bronze cultivar

golden cultivar

UNOFFICIAL SUBSTITUTE

Owing to a shared common name (Hu Ma Ren 胡麻仁), seed of Hei Zhi Ma (黑芝麻) is occasionally traded in error (and *vice versa*). 'Hei Zhi Ma' is sesame seed (*Sesamum indicum* L., family Pedaliaceae) and is the source of a separate TCM drug (Sesami Semen Nigrum) see page 646. The two species are unrelated and their seed are easily distinguished macroscopically.

Yi Yi Ren 薏苡仁

Coicis Semen

Drug part: SEED

Coix lacryma-jobi L. var. *ma-yuen* (Rom. Caill.) Stapf

Poaceae

synonyms: *C. lacryma-jobi* subsp. *ma-yuen* (Rom. Caill.) T. Koyama; *C. ma-yuen* Rom. Caill.

taxonomic/nomenclatural note: in CP2015 as *'Coix lacryma-jobi* L. var. *mayuen* (Roman.) Stapf'. The seed (strictly a fruit) of the very closely related type variety (*C. lacryma-jobi* var. *lacryma-jobi*) is not used as a source of 'Yi Yi Ren' because its husk (utricle) is too hard to remove easily.

Species of *Coix* wild in China: 2

PLANT DESCRIPTION

A tall (1–3 m), robust, annual grass with a somewhat arching habit. *Ligule* (structure at base of leaf blade) membranous (0.6–1.2 mm long). *Leaves* simple, large, flat and linear (10–40 cm long; **1.5–7 cm wide**), usually hairless; margins entire and rough. *Flowers* (without petals) in clusters in leaf axils; female flower enclosed in a smooth, soft, elliptical to sub-globose husk (utricle) (8–12 mm long; 4–9 mm wide) longitudinally striate **and maturing from green to pale or dark brown; husk brittle and easily broken;** greenish-yellow male flowers emerge as a stalked cluster from the beaked apex of the husk. *Fruit* ovoid (5–8 mm long; 4–6 mm diam.) contains a single white or yellow true seed, rich in starch. Fl. and fr. Jul–Dec.

HARVESTING, SOURCING AND NATURAL RANGE

Husks (utricles) harvested in autumn, when ripe, from cultivated crop plants, primarily grown in the Chinese provinces of Fujian, Hebei, Liaoning and Zhejiang.

The wild species grows in Southern, central and parts of northern China up to 2,000 m altitude: Anhui, Fujian, Guangdong, Guangxi, Hebei, Henan, Hubei, Jiangsu, Jiangxi, Liaoning, Shaanxi, Sichuan, Taiwan, Yunnan and Zhejiang [also wild in Bhutan, India, Indonesia, Laos, Malaysia, Myanmar, Philippines, Thailand and Vietnam].

CONSERVATION STATUS

China: Least Concern (MEPCAS, 2013).

Global: Not Evaluated (IUCN Red List, accessed 25/8/2016).

Coix lacryma-jobi var. *ma-yuen*, flowering crop

Soft thin walled husk of Yi Yi Ren (left) and hard, thick walled husks of var. *lacryma-jobi* (right)

C. lacryma-jobi var. *ma-yuen*, flower-head

TCM PROPERTIES AND USE

EHTPA 4/14
GAP

properties: 'sweet, bland and cool'
channels: 'Spleen, Stomach and Lung'
actions: 'resolves damp, tonifies qi, clears heat'

indications: musculoskeletal pain, sores and abscesses

DRUG MORPHOLOGY

CRUDE DRUG/DECOCTING PIECES-1 (DRIED 'SEED'): 'seed' broadly ovoid (4–8 mm long; 3–6 mm diam.), externally smooth and milky white **[1]**, sometimes with remnants of yellowish-brown skin **[2]**; one end broadly rounded **[3]**, the other slightly concave **[4]** and with a pale dotted hilum **[5]**. A deep, wide longitudinal furrow, usually brown and rough, extends the length of the 'seed' **[6]**; the opposite side of the seed is broadly convex **[7]** *Texture*: hard and difficult to break. *Fracture*: white and starchy. *Odour*: indistinct. *Taste*: slightly sweet. ***Good quality*:** large, white 'seed'.

DECOCTING PIECES-2 (BRAN-PROCESSED 'SEED' — FU CHAO YI YI REN): as for crude drug/decocting pieces-1 but externally pale yellow or light brown.

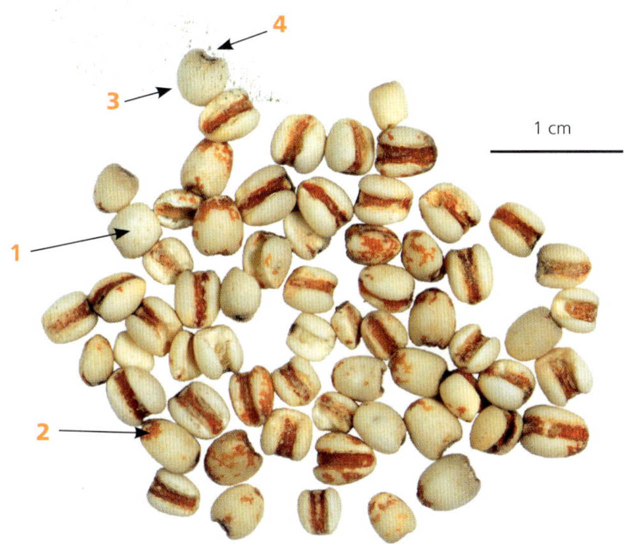

Yi Yi Ren (dried seed; crude drug/decocting pieces-1)

Fu Chao Yi Yi Ren (bran-processed seed)

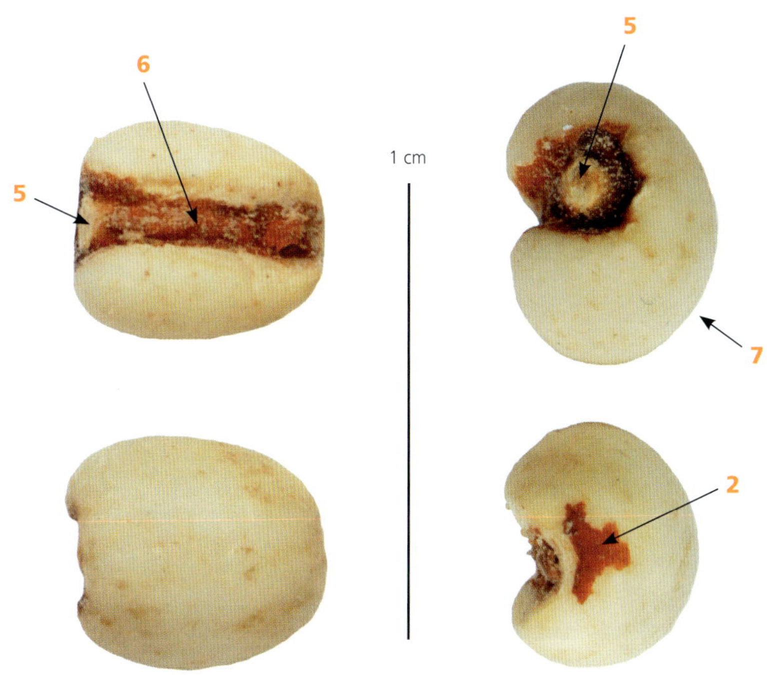

Lateral view

Medial view

PROCESSING

The soft and fragile husks (utricles) are sun-dried, then seed removed (= crude drug/decocting pieces-1); further optional dry-frying with wheat bran produces 'Fu Chao Yi Yi Ren' (= decocting pieces-2).

OTHER COMMON NAMES

川谷 Chuan Gu; 薏米 Yi Mi; Chinese pearl barley; Job's tears (SCN); Semen Coicis.

SAFETY

Safety Class: 2b; Interaction Class: A (Gardner & McGuffin, 2013). An official food and healthfood in China (Anon., 2002a,b).

Yi Zhi 益智 Alpiniae Oxyphyllae Fructus

Drug part: FRUIT

Alpinia oxyphylla Miq.

Zingiberaceae

PLANT DESCRIPTION

An erect, perennial herb with creeping rhizomes. *'Stems'* (pseudostems — densely overlapping leaf bases) 1–3 m tall. *Leaves* many, in 2 rows on opposite sides of the 'stem'; ligule (at base of each leaf blade) membranous, brownish, 2-cleft (1–2 cm long) and minutely hairy [hl];leaf stalks short; leaf blades simple, lance-shaped (25–35 cm long; 3–6 cm wide), base sub-rounded, margins entire and tips long and tail-like. *Flowers* bisexual, densely packed in large and conspicuous, long-pyramidal spikes arising from tips of leafy stems; before opening, flower spike enclosed by a long, pointed and conical bract; each flower with a tubular calyx (c. 1.2 cm long), split along one side, tip 3-toothed; petals fused to form a very short pinkish-white 3-lobed tube (8–10 mm long); petal lobes oblong (c. 1.8 cm long), central one white and wider than lateral ones; the much larger petal-like lip or labellum (modified stamens) white with red or pinkish stripes and a distinctly wavy margin. *Fruit* a globose capsule, becoming spindle-shaped when dry (1.5–2 cm long; c. 1 cm diam.); 3-valved internally; seeds many. Fl. Mar–May, fr. Apr–Sep.

Species of *Alpinia* wild in China: 51

HARVESTING, SOURCING AND NATURAL RANGE

Fruits are harvested when ripe (red) in summer and autumn mostly from wild populations which grow in shady forests. The species is endemic to China and confined to the SC and SE provinces of Fujian, Guangdong, Guangxi, Hainan and Yunnan; Hainan is the source of traditional *Di Dao* quality. Cultivated as a crop on small-holdings in Hainan.

CONSERVATION STATUS

China: Not Evaluated.

Global: Not Evaluated (IUCN Red List, accessed 20/8/2016).

A. oxyphylla coming into flower

Ripening fruits

TCM PROPERTIES AND USE

properties: **'pungent and warm'**
channels: **'Kidney and Spleen'**
action: **'warms the Spleen and Kidney'**

indications**: diarrhoea, abdominal pain, poor appetite; enuresis, spermatorrhoea**

DRUG MORPHOLOGY

CRUDE DRUG-1 (DRIED FRUIT): fruit ellipsoidal, narrowing acutely at both ends **[1]** (1.2–2 cm long; 1–1.3 cm diam.); apical end usually more pronounced with persistent remnants of flower bracts long, membranous and often strongly bent **[2]**; sometimes broken off; basal end with short scar of fruit-stalk **[3]**. Externally mottled brown or greyish-brown usually with 13–20 distinct, pale yellow, interrupted longitudinal ridges **[4]**. Seeds many, closely adhering to one another within a greyish-brown web-like aril **[5]**; seeds in groups of 6–11 within each of the 3 compartments defined by membranous septa **[6]**. Individual seeds irregularly globose, flattened at each end **[7]** (c. 3 mm diam.); externally greyish-brown with a reddish-brown seed attachment scar on one lateral face **[8]**; endosperm white. *Texture*: fruit wall rough and slightly pliable; seeds hard. *Fracture*: fruit-wall thin **[9]**, closely adhering to seeds which contain a bright white endosperm. *Odour*: aromatic. *Taste*: acrid and slightly bitter. ***Good quality*****:** large, plump and intensely aromatic.

CRUDE DRUG-2 (SEED MASSES — YI ZHI REN): as for crude drug-1 but without the fruit-wall.

PROCESSED DRUG (SALT-PROCESSED SEED MASSES — YAN YI ZHI REN): as for crude drug-2 but tasting of salt.

PROCESSING

Fruits are dried at low temperature (= crude drug-1), then the fruit-wall (pericarp) is removed and discarded leaving the seed masses ('Yi Zhi') (= crude drug-2); optional processing by subsequently moistening with salt-water then stir-frying produces 'Yan Yi Zhi Ren' (salt-processed 'Yi Zhi Ren') (= processed drug). All drug forms are pulverised before decocting.

OTHER COMMON NAMES

盐益智仁 Yan Yi Zhi Ren 'salt-processed Yi Zhi seed'; 盐炒益智仁 Yan Chao Yi Zhi Ren 'salt-processed and dry-fried Yi Zhi seed'; 盐水益智仁 Yan Shui Yi Zhi Ren 'salt-watered Yi Zhi seed'; 益智子Yi Zhi Zi ; 益智仁 Yi Zhi Ren; Alpiniae Fructus; bitter cardamom; bitter cardamon; black cardamom; Fructus Alpiniae Oxyphyllae; sharp-leaf galangal (SCN).

SAFETY

An official Chinese food and healthfood (Anon., 2002a,b).

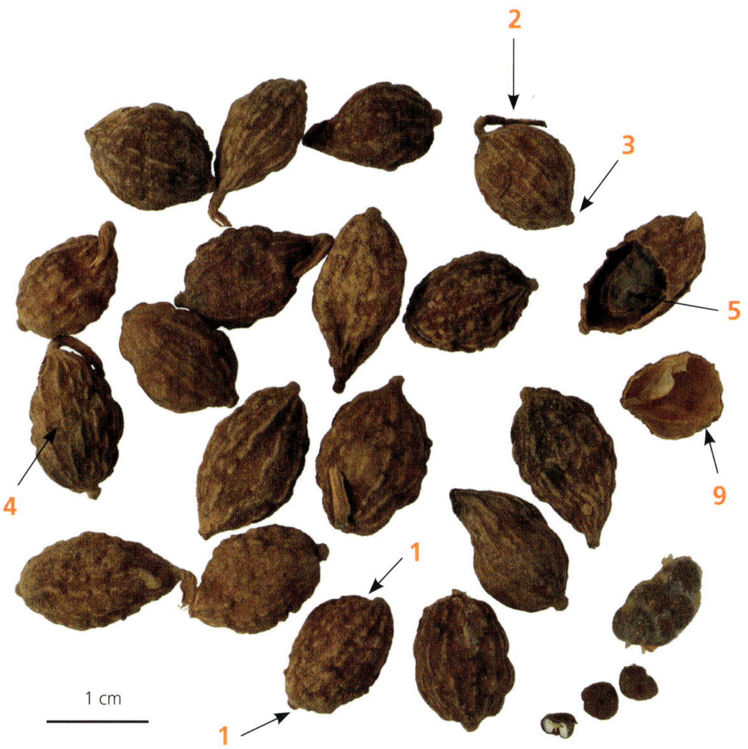

Yi Zhi (dried fruit with fruit walls intact) = crude-drug-1

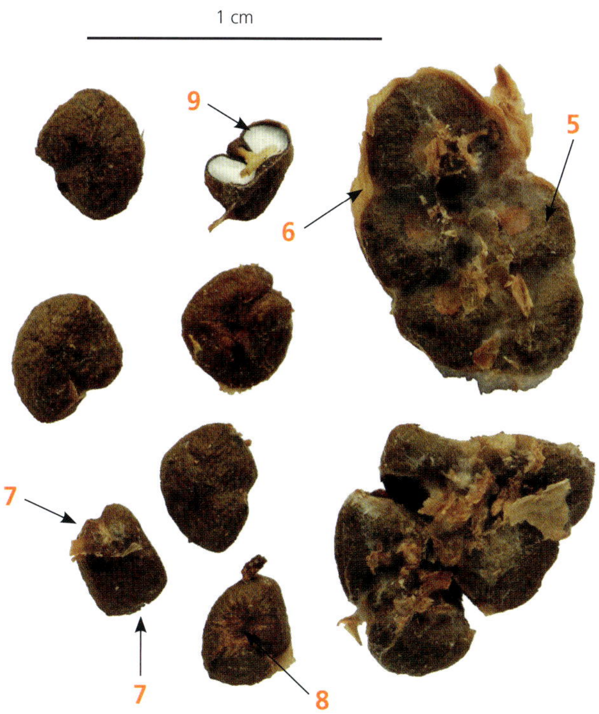

Yi Zhi (or Yi Zhi Ren) (seed masses still joined by web-like membrane or separated) = crude drug-2

Yu Li Ren 郁李仁 Pruni Semen

Drug part: SEED

PHARMACOPOEIA SPECIES 1

Prunus humilis Bunge

Rosaceae

Species of *Prunus* wild in China: >7

synonyms: *Cerasus humilis* (Bunge) S. Ya. Sokolov (*FoC* accepted name); *Prunus japonica* Thunb. var. *salicifolia* Kom.

taxonomic/nomenclatural note: *Kew-MPNS* accepted name, 24/9/2016

PLANT DESCRIPTION

A small, deciduous shrub (0.4–1.5 m tall) with slender greyish-brown branches. *Leaves* alternate or clustered, simple and short-stalked; leaf blades oval and broadest above the middle (2.5–5 cm long; 1–2 cm wide), leaf bases wedge-shaped; margins toothed (sometimes teeth of 2 different sizes). *Flowers* conspicuous, emerging just before or with the leaves; solitary or in groups of 3 in the leaf axils; radially symmetrical with petals of equal size; sepals 5; petals 5, free, pink or white, oval and spreading; stamens 30–35; ovary superior. *Fruit* a sub-globose, hairless, succulent, red to purplish-red drupe (1.5–1.8 cm diam.) without a longitudinal groove and, when ripe, not splitting; stone globose to ovoid and sparsely ribbed. Fl. Apr–May, fr. Jun–Oct.

HARVESTING, SOURCING AND NATURAL RANGE

Ripe fruits are harvested in summer and autumn mostly from wild plants which grow in thickets on sunny mountain slopes (400–1,800 m altitude). The species is native in N, NE and W China: Hebei, Heilongjiang, Henan, Inner Mongolia, Jiangsu, Jilin, Liaoning, Shandong, Shaanxi, Shanxi and Sichuan. It is usually cultivated for its edible fruit.

CONSERVATION STATUS

China: Least Concern (MEPCAS, 2013 as *C. humilis*).
Global: Not Evaluated (IUCN Red List, accessed 2/9/2015).

PHARMACOPOEIA SPECIES 2

P. japonica Thunb.

Rosaceae

synonym: *Cerasus japonica* (Thunb.) Loisel. (*FoC* accepted name)

taxonomic/nomenclatural note: *Kew-MPNS* accepted name, 24/9/2016

PLANT DESCRIPTION

Very similar to *P. humilis* (PS1) but **leaves broadest below the middle** and **leaf bases rounded**; **fruits slightly smaller (c. 1 cm diam.)**. Fl. May, fr. Jun–Aug.

HARVESTING, SOURCING AND NATURAL RANGE

Harvesting as for PS1. The species is native in N, C and NE China (Hebei, Heilongjiang, Henan, Jilin, Liaoning, Shandong and Zhejiang provinces) and is sometimes cultivated as a medicinal crop; [also native in Japan and Korean Peninsula].

CONSERVATION STATUS

China: Least Concern (MEPCAS, 2013 as *C. japonica* var. *nakai* H. Lév.).
Global: Not Evaluated (IUCN Red List, accessed 2/9/2015).

PHARMACOPOEIA SPECIES 3

P. pedunculata (Pall.) Maxim.

Rosaceae

synonym: *Amygdalus pedunculata* Pall. (*FoC* accepted name)

taxonomic/nomenclatural note: *Kew-MPNS* accepted name, 24/9/2016; in CP2015 as '*P. pendiculata* Maxim.'

PLANT DESCRIPTION

Very similar to *P. humilis* (PS1) but **fruits softly hairy** (tomentose), **flesh dry** (not succulent), splitting when ripe. Fl. May–Jun, fr. Jul–Aug.

HARVESTING, SOURCING AND NATURAL RANGE

Harvesting as for PS1. The natural distribution of the species is confined to N China where it typically grows on rocky slopes in Inner Mongolia, Ningxia and Shaanxi provinces; [also native in Mongolia and Russia].

CONSERVATION STATUS

China: Near Threatened (MEPCAS, 2013).
Global: Not Evaluated (IUCN Red List, accessed 2/9/2015).

TCM PROPERTIES AND USE

properties: 'pungent, bitter, sweet and neutral'
channels: 'Spleen, Large and Small Intestines'
actions: 'moistens intestines, regulates water'

indications: constipation; abdominal distension, oedema

Flowering branch of *Prunus humilis*

Wild plant of *P. humilis* with immature fruit, Shaanxi province

Ripe fruit of *P. humilis*

Flowering branch of *P. japonica*

Branch of *P. japonica* with ripening fruit

Wild plant of *P. pedunculata* (Inner Mongolia)

Fruiting branch of *P. pedunculata* with ripening fruit

Prunus humilis

Rosaceae

DRUG MORPHOLOGY

CRUDE DRUG/DECOCTING PIECES (DRIED SEED): externally with a thin, yellowish-white or pale brown membranous seed coat (testa). Internally, 2 oily, creamish-white cotyledons. *Fracture:* creamish-white. Seed **ovoid (5–8 mm long; 3–5 mm diam.)** **[1]**, the acute end with a linear hilum on one side **[2]**; the rounded end with a slightly darker vascular scar of attachment (chalaza) **[3]** and from which several veined furrows extend longitudinally **[4]**. *Odour*: indistinct. *Taste*: slightly bitter. ***Good quality*:** smooth and plump.

PROCESSING

After removal of the flesh and hard stone wall, seeds are dried (= crude drug and decocting pieces). They are pulverised prior to decocting.

OTHER COMMON NAMES

欧李 Ou Li (plant name); 小李仁 Xiao Li Ren 'small Chinese dwarf cherry seed'; Chinese dwarf cherry seed (CP2010); Semen Pruni.

SAFETY

An official food and healthfood in China (Anon., 2002a,b).

P. japonica

Rosaceae

DRUG MORPHOLOGY

CRUDE DRUG/DECOCTING PIECES (DRIED SEED): cannot be reliably distinguished from the drug of PS1 using macroscopic characters. Smaller than that of PS3 (see below).

PROCESSING

As for PS1.

OTHER COMMON NAMES

小李仁 Xiao Li Ren 'small Chinese dwarf cherry seed'; 郁李 Yu Li (plant name); Chinese dwarf cherry seed (CP2010); Japanese bush cherry (plant name); Semen Pruni.

SAFETY

An official food and healthfood in China (Anon., 2002a,b).

P. pedunculata

Rosaceae

DRUG MORPHOLOGY

CRUDE DRUG/DECOCTING PIECES (DRIED SEED): only distinguishable from that of PS1 and PS2 by its size which is **proportionally larger (6–10 mm long; 5–7 mm diam.)** **[5]**; colour of seed coat often brighter golden yellowish-brown **[6]**.

UNOFFICIAL SUBSTITUTES

Seeds of other species of Rosaceae have been reported as local or historical substitutes; their occurrence as substitutes in international trade is uncertain.

1. ***Cerasus dictyoneura*** (Diels) Holub 毛叶欧李 Mao Ye Ou Li. Cultivated in China for its edible fruit;
2. ***Prinsepia uniflora*** Batalin. Seeds are the source of the TCM drug 蕤仁 Rui Ren (Prinsepiae Nux). **Seed slightly larger** (7–10 mm long, 6–8 mm wide) and externally **pale yellowish-brown or dark brown with distinct reticulated furrows;**
3. ***Prunus tomentosa*** Thunb. 毛樱桃 Mao Ying Tao. Widely cultivated in China as an ornamental and for its edible fruit;
4. ***P. triloba*** Lindl. 榆叶梅 Yu Ye Mei;
5. ***P. salicina*** Lindl. var. ***salicina*** 山李子 Shan Li Zhi. Widely cultivated in China for its edible fruit; fruits larger (3.5–5 cm diam.; up to 7 cm diam. in cultivated forms);
6. ***P. simonii*** Carrière 杏李 Xing Li.

Seed of *Prinsepia uniflora*

PROCESSING

As for PS1.

OTHER COMMON NAMES

长柄扁桃 Chang Bing Bian Tao; 长梗扁桃 Chang Geng Bian Tao (plant name); 大李仁 Da Li Ren 'large Chinese dwarf cherry seed'; Semen Pruni.

SAFETY

An official food and healthfood in China (Anon., 2002a,b).

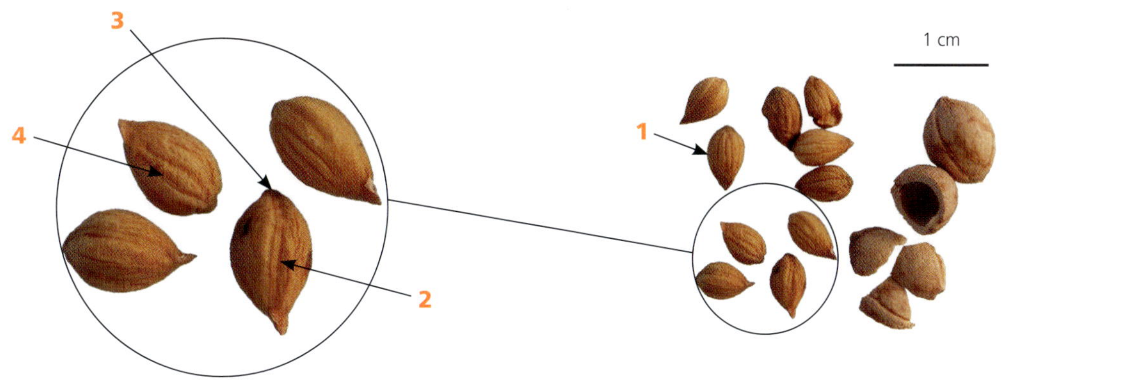

Crude drug/decocting pieces (dried seed) of Yu Li Ren (aka 'Xiao Li Ren') sourced from *P. humilis* (thick shells of broken endocarps (stones): right)

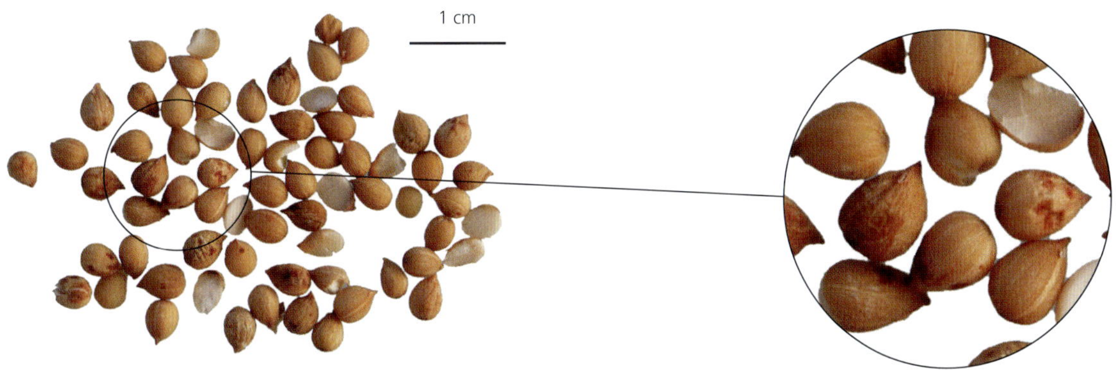

Crude drug/decocting pieces (dried seed) of Yu Li Ren (aka 'Xiao Li Ren') sourced from *P. japonica*

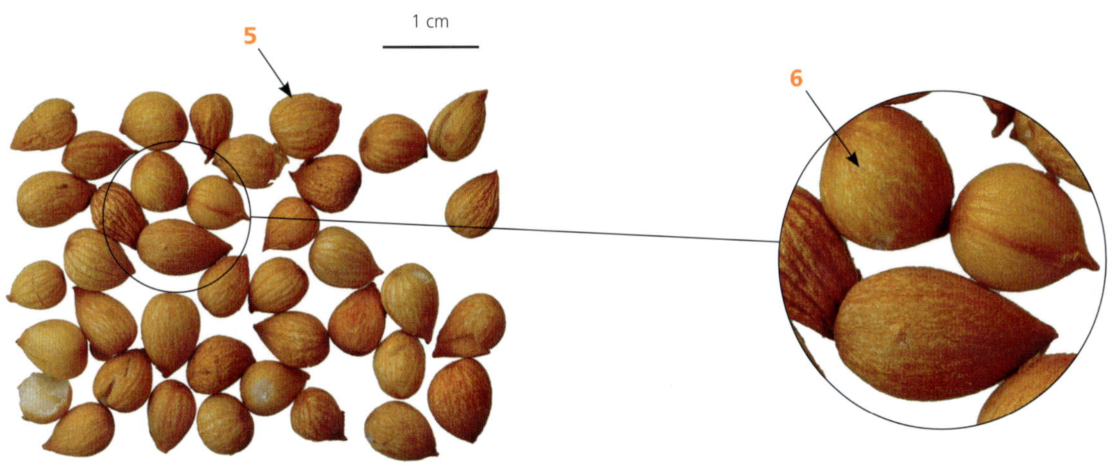

Crude drug/decocting pieces (dried seed) of Yu Li Ren (aka 'Da Li Ren') sourced from *P. pedunculata*

Zhi Qiao or Zhi Ke 枳壳

Aurantii Fructus

bitter orange

Drug part: IMMATURE FRUIT

Citrus × aurantium L.

Rutaceae

synonyms: too many to list (*Kew-MPNS*, 1/7/2016), including *C.* × *sinensis* (L.) Osbeck.
taxonomic nomenclatural note: in CP2015 as '*C. aurantium* L.'; includes certain cultivars, notably 'Chulan', 'Huang Pi', and 'Tangcheng' (Bensky *et al.*, 2004).

Species and hybrids of *Citrus* wild in China: 11 (plus 5 hybrid cultivated species)

PLANT DESCRIPTION

A small evergreen tree. *Branches* ribbed and with spines (up to 8 cm long). *Leaves* alternate and simple, stalks all clearly articulated at the base of the leaf blade but vary in length and width of wing according to the cultivar (1–3 cm long; 0.6–1.5 cm wide); leaf blade elliptic or broadly ovate (7–12 cm long; 4–7 cm wide), thick and dark green, base rounded or broadly wedge-shaped, margin almost entire, tip often 1-notched. Each flower in few-flowered axillary groups (racemes) or solitary; bisexual or male. Flower with calyx cup-shaped and 4- or 5-lobed; petals 5, white, spreading, equal and oblong/ lance-shaped (2–3.5 mm diam.); stamens 20–25, usually fused basally. *Fruit* large and conspicuous when ripe, globose to squat (up to 10 cm diam.), orange to reddish, peel surface coarse, fruit wall (pith) thick and often difficult to remove; pulp (flesh) divided into 7–13 segments. Cultivars used are bitter and slightly sour. *Seeds* many, large and ridged. Fl. Apr–May, fr. Sep–Dec.

HARVESTING, SOURCING AND NATURAL RANGE

This hybrid is only known from cultivation but is sometimes naturalised. Immature fruits are harvested when still green in July from cultivated trees grown south of the Qin Ling mountains of NC China. Five main production areas are recognised and their corresponding trade items are named as follows: Sichuan (producing Chuan Zhi Qiao); Jiangxi (Jiang Zhi Qiao); Hunan (Xiang Zhi Qiao) and Jiangsu and Zhejiang (Su Zhi Qiao or Dai Dai Ha Zhi Qiao) Bensky *et al.* (2004) . Traditional *Di Dao* quality is sourced from Hunan (Yuanjiang), Jiangxi (Qingjiang) and Sichuan (Wangxian).

CONSERVATION STATUS

China/Global: only known from cultivation.

See 'Zhi Shi' (Aurantii Fructus Immaturus) page 762, also sourced from this hybrid.

Fruiting branches of *C.* × *aurantium* showing variation of leaf stalks between different cultivars

Medicinal crop of *Citrus* × *aurantium*, Jiangxi province

Immature fruits of *C.* × *aurantium* sun-drying, Jiangxi province

TCM PROPERTIES AND USE

properties: **'pungent, bitter and cool'**
channels: **'Spleen and Stomach'**
action: **'regulates qi'**

indications: **abdominal distension, constipation**

DRUG MORPHOLOGY

CRUDE DRUG (DRIED HALVED FRUIT): hemispherical transverse fruit halves (3–5 cm diam.). Peel dark brown or brown, scattered throughout with paler dot-like oil ducts **[1]**; uncut basal end with conspicuous pale, round scar of fruit stalk **[2]**. Cut end reveals a broad, yellowish-white, smooth and slightly prominent pith **[3]** (0.4–1.3 cm wide) with 1–2 rows of oil ducts towards the outer edge. *Pith texture*: hard and difficult to break. The central pulp is divided clearly into 7–13 (rarely up to 15) shrunken juice segments **[4]**, brown to dark brown and sometimes containing large, ridged seeds **[5]**. *Odour*: aromatic. *Taste*: bitter and slightly sour. ***Good quality*:** hard, solid and aromatic fruit; externally greenish-brown (exocarp); pulp (flesh) more than half the diameter of the fruit.

DECOCTING PIECES-1 (SLICED FRUIT): straight or often irregularly curved transverse slices of pith **[6]** with thin outer peel still attached (up to 7 cm long; usually c. 1 cm wide); cut pieces of fruit ends occasional, more angular **[7]** and variously sized. Peel thin dark brown or brown, scattered throughout with dot-like oil ducts **[8]**. Pith yellowish-white to yellowish-brown and smooth (0.4–1 cm wide, sometimes wider) with rows of oil ducts towards the outer edge **[9]**. Purplish-brown pulp segments or their remnants occasionally still attached to inner surface **[10]**. *Odour*: aromatic. *Taste*: bitter and slightly sour.

DECOCTING PIECES-2 (BRAN-PROCESSED SLICED FRUIT): as for decocting pieces above but darker and occasionally charred.

PROCESSING

Immature fruits are sliced in half transversely then dried at low temperature, sometimes in the sun (= crude drug); decoction pieces are made by washing and thoroughly soaking the fruit halves with water before cutting into thin slices. Further optional processing with wheat bran (Mai Fu 麦麸) involves heating the bran until it smokes then adding the fruit slices until they turn yellow or dark; the bran is removed by sifting and the slices cooled (= Fu Chao Zhi Qiao 麸炒枳壳 bran-fried Zhi Qiao).

OTHER COMMON NAMES

川枳壳 Chuan Zhi Qiao; 麸炒枳壳 Fu Chao Zhi Qiao; 江枳壳 Jiang Zhi Qiao; 酸橙 Suan Cheng (plant name); 湘枳壳 Xiang Zhi Qiao; 枳壳 Zhi Ke; bitter orange (SCN); Fructus Aurantii; orange fruit (CP2010).

SAFETY

Safety Class: 1; Interaction Class: C (latter applies to the juice only which is reported to inhibit the action of certain pharmaceuticals) (Gardner & McGuffin, 2013). An official healthfood in China (Anon., 2002a,b).

Crude drug of Zhi Qiao

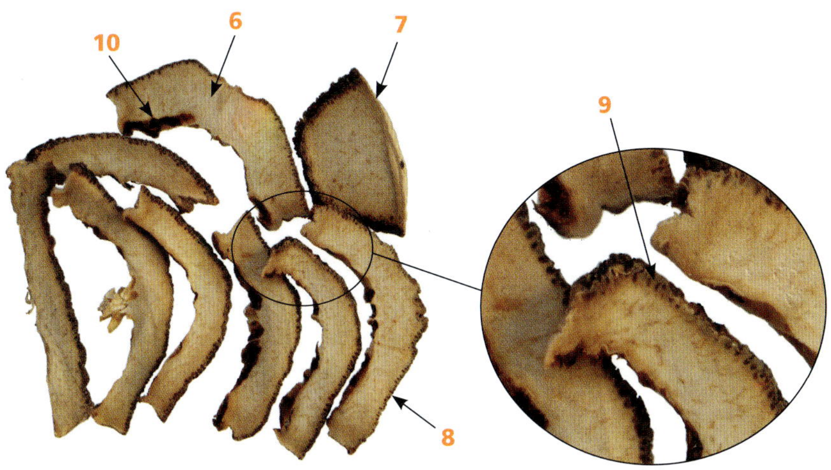

Decocting pieces-1 (Zhi Qiao)

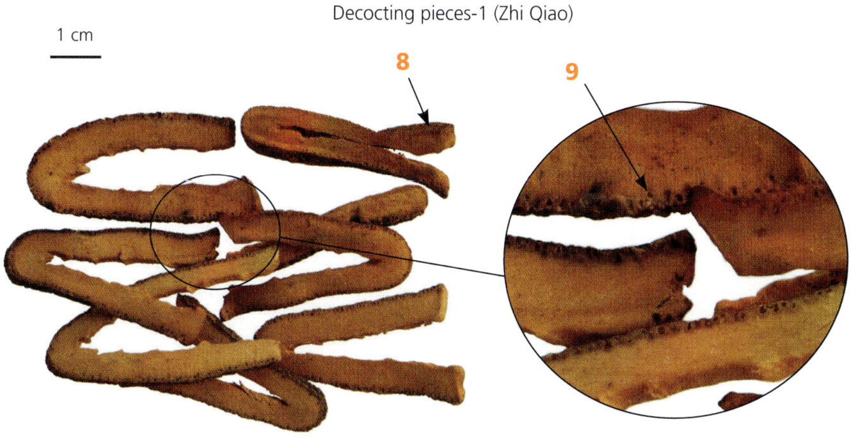

Decocting pieces-2 (Fu Chao Zhi Qiao)

UNOFFICIAL SUBSTITUTES

Other medicinal *Citrus* reported as local substitutes in China, their occurrence as substitutes in international trade is unlikely:

1. ***C. trifoliata*** L. (as *Poncirus trifoliata*).
2. ***C. maxima*** (Burm.) Merr. (synonym *C. grandis*).
3. ***C. medica*** L. (source of Xiang Yuan (Citri Fructus) see page 748).

Zhi Shi 枳实

Aurantii Fructus Immaturus

unripe bitter orange

Drug part: YOUNG FRUIT

Citrus × aurantium L.

Rutaceae

synonyms: too many to list (*Kew-MPNS*, 1/7/2016) including *C.* ×*sinensis* (L.) Osbeck (alternative source of Zhi Shi in CP2015 as '*C. sinensis* Osbeck').

taxonomic/nomenclatural note: in CP2015 as '*C. aurantium* L.'; includes certain named cultivars, notably 'Chulan', 'Huang Pi', and 'Tangcheng' (Bensky *et al.*, 2004).

Species and hybrids of *Citrus* wild in China: 11 (plus 5 hybrid cultivated species)

PLANT DESCRIPTION

A small evergreen tree. For full description see 'Zhi Qiao'/'Zhi Ke' (Aurantii Fructus) page 760.

UNOFFICIAL SUBSTITUTES

As for 'Zhi Qiao' (Aurantii Fructus) page 760 but excluding *C. medica*.

HARVESTING, SOURCING AND NATURAL RANGE

This hybrid is only known from cultivation but it is sometimes naturalised. Young fruit (younger than those of Zhi Qiao (page 760) sourced from the same plant) are harvested as windfalls during May and June from cultivated trees grown south of the Qin Ling mountains of north-central China. Main production areas are Sichuan (producing 'Chuan Zhi Shi'), Jiangxi (producing 'Jiang Xi Zhi Shi' and regarded as top quality) and Hunan (producing 'Xiang Zhi Shi'), Bensky *et al.* (2004).

CONSERVATION STATUS

China/Global: only known from cultivation.

Medicinal crop of *Citrus* × *aurantium*, Jiangxi province

Immature fruit of *C.* × *aurantium* ready to harvest for Zhi Shi

Flowering branches of *C.* × *aurantium*

Young fruit of *C.* × *aurantium* (too mature to harvest as the source of Zhi Shi)

TCM PROPERTIES AND USE

***properties*: 'pungent, bitter sour and cool'**
***channels*: 'Spleen and Stomach'**
***actions*: 'regulates qi, resolves phlegm'**

***indications*: abdominal distension, pain, constipation, diarrhoea**

DRUG MORPHOLOGY

CRUDE DRUG (DRIED HALVED OR WHOLE FRUIT): hemispherical (transversely halved) **[1]** or smaller spherical (whole) fruit **[2]** (0.5–2.5 cm diam.). Peel dark green or dark brownish-green **[3]**, scattered throughout with raised oil ducts; uncut basal end with conspicuous pale, round scar of fruit stalk **[4]**. Cut surface reveals a broad, yellowish-white or yellowish-brown, smooth and usually prominent pith **[5]** (0.3–1.2 cm wide) with 1–2 rows of oil ducts towards the outer edge **[6]**. *Pith texture*: hard and difficult to break. The central pulp is divided clearly into 7–13 (rarely up to 15) brown, shrunken juice segments **[7]**, often containing ridged seeds. *Odour*: aromatic. *Taste*: bitter and slightly sour. ***Good quality*:** small, hard, solid fruit with a fresh aroma; outer surface (peel) greenish-black; pith (fruit-wall) white; small central pulp (flesh).

DECOCTING PIECES (DRIED FRUIT SLICES): fruit slices of varying shape according to local processing tradition: rounded (up to 2.5 cm diam.), curved or strip-like. Peel thin, dark green or dark brownish-green and scattered with raised oil-ducts **[8]**. Pith broad (0.3–1.2 cm wide), yellowish-white to yellowish-brown **[9]** with 1–2 rows of oil ducts towards the outer edge. Pulp segments brown and shrivelled; when intact 7–13 (rarely up to 15), often damaged and fragmented with large cavities **[10]**. Fruit ends **[11]** usually devoid of pulp segments; with pith continuous throughout the cut surface. *Odour*: aromatic. *Taste*: bitter and slightly sour.

DECOCTING PIECES (BRAN-FRIED FRUIT SLICES): as for decocting pieces above but darker and occasionally charred.

PROCESSING

Fruits are sliced in half then dried in the sun or at low temperature (very small fruits are left whole) (= crude drug). Decocting pieces are made by thoroughly soaking the fruit in water then cutting into 1–2 mm slices. Subsequent optional processing with wheat bran (Mai Fu 麦麸) involves heating the bran until it smokes then adding the fruit slices until they turn yellow or dark; the bran is removed by sifting and the slices cooled (= Fu Chao Zhi Shi 麸炒枳实 'bran-fried Zhi Shi').

OTHER COMMON NAMES

川枳实 Chuan Zhi Shi; 鹅眼枳实 E Yan Zhi Shi; 江枳实 Jiang Zhi Shi; 湘枳实 Xiang Zhi Shi; bitter orange (SCN): Fructus Aurantii Immaturus; immature orange fruit (CP2010).

SAFETY

See Zhi Qiao (Aurantii Fructus) page 760.

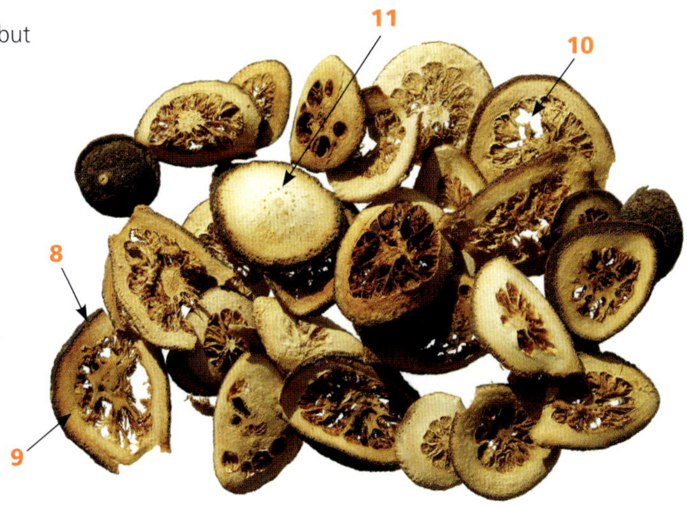

Decocting slices of Zhi Shi

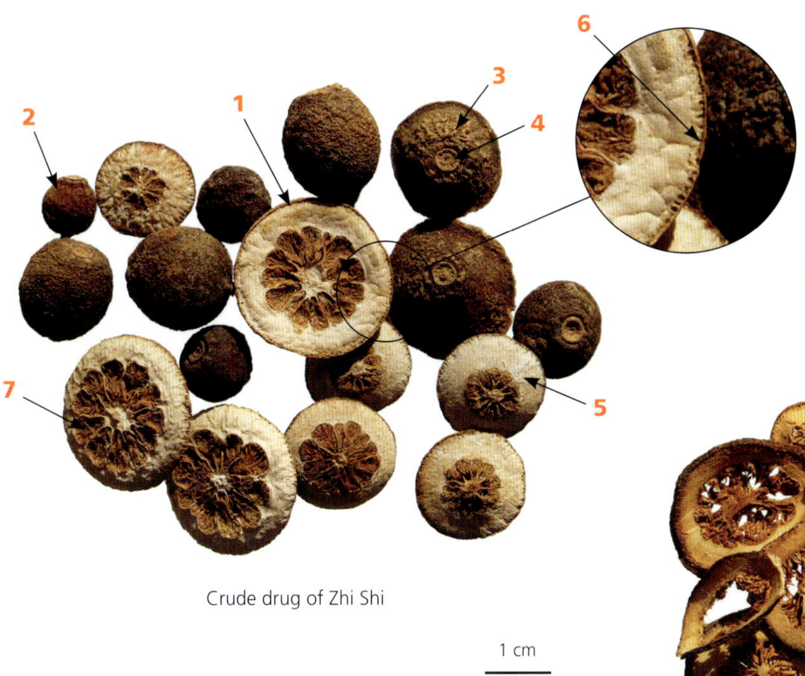

Crude drug of Zhi Shi

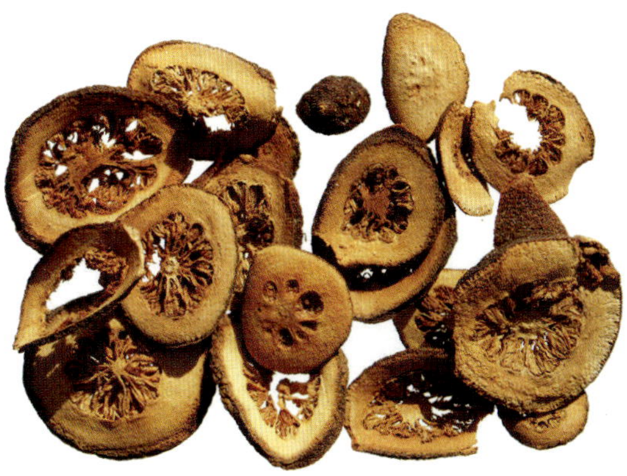

Bran-fried decocting slices (Fu Chao Zhi Shi)

Zhi Zi 栀子 Gardeniae Fructus

Drug part: FRUIT

Gardenia jasminoides J. Ellis

Rubiaceae

synonyms: *G. grandiflora* Lour.; *G. jasminoides* var. *grandiflora* (Lour.) Nakai

Species of *Gardenia* wild in China: 5

PLANT DESCRIPTION

A small dense, evergreen shrub (up to 3 m tall) and rather variable in respect of the size of its leaves and flowers. *Leaves* opposite and simple, ovate to elliptic (6–12 cm long; 1.5–4 cm wide); upper surface glossy; margin entire. *Stipules* persistent or deciduous. *Flowers* solitary, terminal, stalked and bisexual, conspicuous and fragrant (c. 2.8 cm diam.); petals white (ageing to yellow), 5–8, spreading, wedge shaped or rounded, basally fused into a tube (3–5 cm long; 6–28 mm wide); stamens 5–12; ovary inferior; stigma 2-lobed. *Fruits* dry and quite leathery, ovoid or ellipsoid (1.5–7 cm long; 1.2–2 cm diam.); calyx of fused sepals, deep orange-yellow or yellow, with 5–9 longitudinal wings and 5–8 persistent long narrow calyx lobes at the tip. *Seeds* many. Fl. Mar–Jul, fr. May–Feb.

HARVESTING, SOURCING AND NATURAL RANGE

Fruits are harvested when reddish-yellow during September to November from medicinal crop plants grown in the C and E Chinese provinces of Fujian, Hubei, Hunan, Jiangxi and Zhejiang. The wild species in China grows in forests, scrub, fields and along stream-sides (0–1,500 m altitude) in Anhui, Fujian, Guangdong, Guangxi, Guizhou, Hainan, Hubei, Hunan, Jiangsu, Jiangxi, Shandong, Sichuan, Taiwan, Yunnan and Zhejiang [also native to Cambodia, India, Japan, N Korea, Laos, Nepal, Pakistan and Vietnam; cultivated in Europe, North America and the Pacific islands].

CONSERVATION STATUS

China: Least Concern (MEPCAS, 2013).
Global: Not Evaluated (IUCN Red List, accessed 11/8/2016).

Gardenia jasminoides, flowering plant medicinally farmed, Jiangxi province, China

G. jasminoides in full flower

G. jasminoides, branches with immature fruits

UNOFFICIAL SUBSTITUTES

The below have been reported, sometimes in international trade:

1. A cultivated ***Gardenia*** called Shui Zhi Zi (水栀子 water gardenia) or Da Zhi Zi (大栀子 big gardenia). Probably a cultivar of *G. jasminoides* long cultivated in China for its larger fruit, the source of a commercial dye. Fruit also used as a topical medicine but **not recommended for internal use** (Bensky *et al.* (2004); NICPBP, 1997–2002; Xiao *et al.* (2001–2). Fruits larger (3–7 cm long); longitudinal ribs broader. Often called *G. jasminoides* Ellis var. *grandiflora* Nakai in TCM texts (Bensky *et al.*, 2004; Xiao *et al.*, 2001–2002) but the *FoC* treats this name as a synonym of the Pharmacopoeia species.
2. ***G. sootepensis*** Hutch. (sometimes misspelt *G. soontepense*); 大黄栀子Da Huang Zhi Zi. Fruits larger (2.5–5.5 cm long; 1.5–3.5 cm diam.), longitudinal ribs 5–6 and very narrow. In China, confined to Yunnan province.

TCM PROPERTIES AND USE

***properties*: 'bitter and cold'**
***channels*: 'Heart, Lung, Stomach and san jiao'**
***actions*: 'clears hear and fire, drains damp'**

***indications*: fever, irritability; insomnia; dysuria; traumatic injury; (T)**

DRUG MORPHOLOGY

CRUDE DRUG (FRUIT STEAMED OR BLANCHED): almost globose or ellipsoid **[1]** (1.5–3.5 cm long; 1–1.5 cm diam.). Externally reddish-yellow or orange-yellow, with 5–9 longitudinal winged ribs **[2]** and a conspicuous longitudinal vein, often branching, between ribs **[3]**. Base often tapering **[4]**; apex often with calyx remnants sometimes visible at one end **[5]**. Fruit wall (pericarp) thin and brittle, slightly lustrous **[6]**; inner surface with 2–3 raised septa. Seeds many, flattened (c. 3.5 mm long; 3 mm diam.); clustered in a mass **[7]**; seed coat deep red or orange-yellow and densely warty. *Odour*: indistinct. *Taste*: bitter and somewhat sour. ***Good quality*:** plump, reddish-orange fruits with a thin fruit wall.

PROCESSED DRUG (DRY-FRIED FRUIT): as for the crude drug but fruit wall darker and yellowish-brown **[8]**.

PROCESSED DRUG (CHARRED): as for the crude drug but fruit wall charred brown or black **[9]**; contents yellowish-brown or dark brown **[10]**. *Odour*: indistinct. *Taste*: slightly bitter and sour.

PROCESSING

After removal of stalks, fruits are steamed or blanched then dried (= crude drug). Fruits may be further processed by dry-frying until yellowish-brown (= 炒栀子 Chao Zhi Zi 'dry-fried gardenia') or charred (焦栀子 Jiao Zhi Zi). Fruits are crushed prior to decocting.

OTHER COMMON NAMES

炒山 栀子 Chao Shan Zhi Zi 'dry-fried mountain gardenia'; Hei Shan Zhi Zi 'black gardenia'; 江山栀子 Jiang Shan Zhi Zi; 山栀子 Shan Zhi Zi 'mountain gardenia'; Shan Zhi Tan 'charred mountain gardenia'; 小山栀子 Xiao Shan Zhi Zi 'small mountain gardenia'; cape jasmine fruit; Fructus Gardeniae; gardenia (SCN); Gardenia Fructus Praeparatus (refers to Charred Gardenia Fruit in CP2015).

SAFETY

Safety Class: 1; Interaction Class: A (Gardner & McGuffin, 2013). An official food and healthfood in China (Anon., 2002a,b).

Zhi Zi (left) crude drug; (right) pulverised for decocting

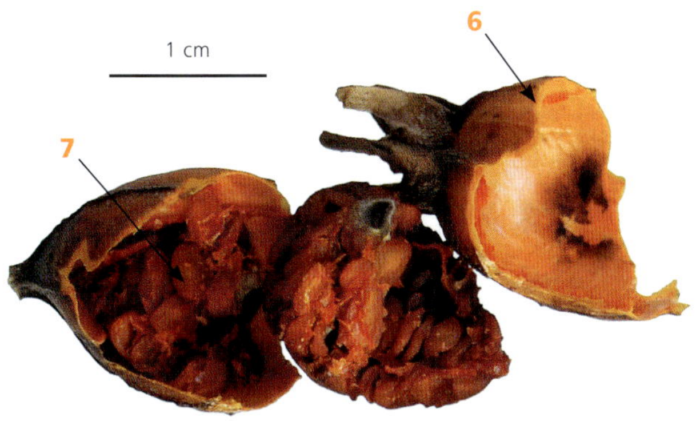

Broken fruits of Zhi Zi showing interior and seeds

Chao Zhi Zi (dry-fried Zhi Zi); (left) whole fruit; (right) pulverised for decocting

Jiao Zhi Zi (charred Zhi Zi; charred Fructus Gardeniae); (left) whole fruit; (right) pulverised for decocting

Zi Su Zi 紫苏子 (or Su Zi 苏子) Perillae Fructus

Drug parts: FRUIT (NUTLET)

Perilla frutescens (L.) Britton

Lamiaceae

synonym: *Ocimum frutescens* L.
taxonomic/nomenclatural note: 3 varieties are recognised by *FoC* and *Kew-MPNS*, 20/8/2016.

Species of *Perilla* wild in China: 1

PLANT DESCRIPTION

An erect, herbaceous annual (0.3–2 m tall). *Stems* 4-sided, green or purple, sparsely or densely covered in fine hairs. *Leaves* opposite with each pair at right angles to the next; stalks 3–5 cm long; **leaf blades** simple, **broadly ovate to circular** (4–13 cm long; 2.8–10 cm wide), green, purplish, or purplish-black, bases rounded to broadly wedge-shaped, tips pointed and margins narrowly to coarsely saw-toothed, **undersides without conspicuous sunken glands.** *Flowers* tiny, in densely hairy, sub-cylindrical many-flowered, 1-sided spikes (1.5–15 cm long) arising from the leaf axils; each flower subtended by an ovate bract (c. 4 mm long; 4 mm wide), green or purple, membranous, pointed and of similar length to the flower. Each flower bisexual, laterally symmetrical and short-stalked; calyx cup-shaped (c. 3 mm long), green to yellowish, glandular, densely hairy at its base and 2-lipped at apex with 3 pointed teeth in the upper lip and 2 pointed teeth in the lower lip; corolla white to purplish-red, bell-shaped and opening into a 2-lipped mouth (lower lip 3-lobed and slightly longer than upper lip); **stamens 4, shorter than corolla;** ovary superior; style 1. *Fruit* with 4 dry, sub-globose nutlets; each nutlet **(c.1.5 mm diam.)**, greyish-brown, greyish-yellow or purplish and **with a netted surface**. Fl. Aug–Nov, fr. Aug–Dec.

For other drugs sourced from this species see 'Zi Su Geng' (Perillae Caulis) page 488 and 'Zi Su Ye' (Perillae Folium) page 556.

HARVESTING, SOURCING AND NATURAL RANGE

Fruits are harvested, when ripe, in autumn. Sourcing is mainly from cultivated crop plants well-established in China with many cultivars bred for medicinal or other purposes; most are thought to be used in TCM. The main production areas for Zi Su Zi are the central and eastern Chinese provinces of Jiangsu, Zhejiang, Hebei, Hubei and Henan. The purple-leaved cultivars are widely used as a condiment and salad vegetable.

In the wild, the species is common on disturbed ground and has a wide distribution in China (Fujian, Guangdong, Guangxi, Guizhou, Hebei, Hubei, Jiangsu, Jiangxi, Shanxi, Sichuan, Taiwan, Tibet, Yunnan and Zhejiang [also native in Bhutan, Cambodia, India, Indonesia (Java), Japan, Korean Peninsula, Laos and Vietnam].

CONSERVATION STATUS

China: Least Concern (MEPCAS, 2013, as '*P. frutescens* var. *purpurascens*').
Global: Not Evaluated (IUCN Red List, accessed 18/8/2016).

Mosla species

Lamiaceae

PLANT DESCRIPTION

Leaves usually much narrower and lance-shaped (1–4 cm; 0.5–2 cm wide) **and undersides with conspicuous sunken glands.** Nutlets c. 0.7–2.1 mm diam., basal areola often absent; surface regularly netted or with shallow pits.

CONSERVATION STATUS

China: no *Mosla* species recorded as threatened.
Global: Not Evaluated (IUCN Red List, accessed 18/8/2016).

TCM PROPERTIES AND USE

properties: 'pungent and warm'
channels: 'Lung and Large Intestines'
actions: 'relieves cough, moistens intestines'

Indications: cough, dyspnoea, constipation

Medicinal crop of *Perilla frutescens*

Inflorescence of *P. frutescens*

P. frutescens (form with purple leaf undersides)

Flowering heads of *Mosla chinensis*

Leaves of *M. chinensis* showing glandular leaf surface

Close-up of flower-head of *M. chinensis*

Perilla frutescens

Lamiaceae

DRUG MORPHOLOGY

CRUDE DRUG/DECOCTING PIECES-1 (DRIED NUTLET): nutlet ovoid or sub-spherical **(c. 1.5 mm diam.)**. Externally **greyish-brown, greyish-yellow or purplish, mottled with fine, reticulated veining** [hl] **[1]**. Due to the large number of varieties and cultivars grown, the colour and degree of reticulation on the fruit is highly variable; Basal end with a **broad and distinct greyish-white areola [2]** surrounding a slightly raised nutlet stalk scar **[3]**. Nutlet wall thin and membranous **[4]**, easily broken, enclosing a single yellowish-white seed with membranous seed coat (testa); cotyledons 2, whitish and oily; endosperm absent or minimal. *Odour*: aromatic when crushed. *Taste*: slightly acrid.

***Good quality*:** plump fruits (usually greyish-brown); aromatic when crushed.

DECOCTING PIECES-2 (DRY-FRIED): as for crude drug/decocting pieces-1 but slightly darker brown **[5]** and sometimes cracked **[4]**. Odour: toasted.

PROCESSING

Fruits are sun-dried, then nutlets removed to form the crude drug/decocting pieces-1); they may be subsequently dry-fried until aromatic or when they begin to pop (= Chao Zi Su Zi 炒紫苏子; = decocting pieces-2).

OTHER COMMON NAMES

白苏子 Bai Su Zi; Fructus Perillae; perilla (SCN); 苏子 Su Zi.

Mosla species

Lamiaceae

'DRUG SUBSTITUTE' MORPHOLOGY

DRIED NUTLET: distinguishing nutlets of *Mosla* species from those of *P. frutescens* (Zi Su Zi) is challenging morphologically, especially given the many cultivars of *P. frutescens* grown in China and hence morphological variation. Furthermore there are 12 species of *Mosla* native to China (with varying nutlet characters) with one (*M. chinensis*) medicinally cultivated as the source of Xiang Ru (Moslae Herba) see page 386. The nutlets of 3 *Mosla* species are illustrated here (including *M. chinensis*) but their smaller size **[6]** (compared to *P. frutescens*) is not necessarily indicative of other Chinese *Mosla* species whose nutlets may be up to 2.1 mm in diameter (larger than those of *P. frutescens*). Other, possibly more reliable characters are an **indistinct or absent basal areola [7]** and a **nutlet surface regularly pitted [8]** rather than reticulated. Further research is needed into the robustness of these macroscopic characters. In the meantime, laboratory-based authentication and quality assurance methods are advised when substitution or contamination is suspected.

OTHER COMMON NAMES

石荠苎属 Shi Qi Zhu Shu (plant name for the genus *Mosla*).

SUBSTITUTION COMMENT. 'Zi Su Zi' (Perillae Fructus) in international trade may occasionally be substituted or contaminated with nutlets from species in the closely related genus *Mosla* (NICBP, 1997–2002; Bensky *et al.* 2014; Van der Valk and Leon *pers. obs.*; Xiao *et al.*, 2001–2). The cause is not known (see also *'Drug Substitute' Morphology*).

Many species of *Mosla* are common and widespread in the wild in N and NW China and occupy a variety of habitats such as forests, streams and roadsides. See also 'Xiang Ru' (Moslae Herba) page 386 sourced from medicinal crops of *M. chinensis*.

Zi Su Zi: sun-dried (left); dry-fried (right)

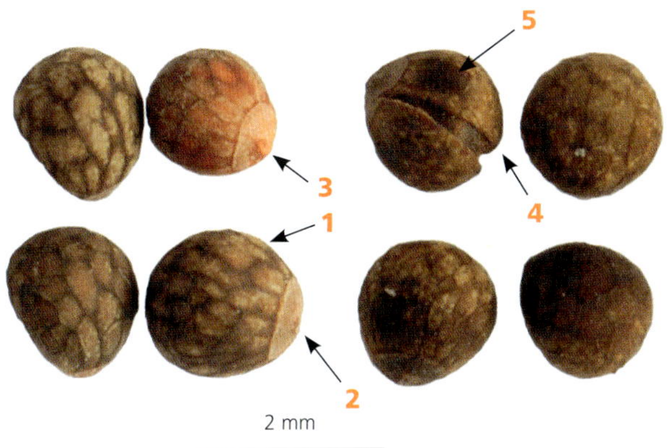

Zi Su Zi: close-up of sun-dried nutlets of *P. frutescens* (crude drug/decocting pieces-1)

Comparison of nutlets of 3 species of *Mosla* with those of *P. frutescens* (Zi Su Zi).

Commercial sample of 'Zi Su Zi' bought in TCM dispensary in the UK (2013) showing contamination (probably nutlets of *Mosla* spp.)

Bibliography

Aguilar, N.O. (2001). *Bacopa monnieri* (L.) Pennell. In: J.L.C.H. van Valkenburg & N. Bunyapraphatsara (eds), *Plant Resources of South-East Asia. Medicinal Plants* 12 (2): 93-96. Backhuys, Leiden.

Anon. (1999–2011). *The Ayurvedic Pharmacopoeia of India*, Part I, Vol. 1–7. First edition. Government of India, Ministry of Health and Family Welfare, New Delhi. www.plimism.nic.in/publications.html

Anon. (2002a). 按照传统既是食品又是中药材的物质目录（2013版）（征求意见稿）*List of materials traditionally recognized as both food and Chinese herbal medicine*. Ministry of Health, People's Republic of China.

Anon. (2002b). *A Notice for the Re-inforcement of the Regulation of Healthcare Product Ingredients*. HMC Directive 2002–51. Ministry of Health, People's Republic of China.

Anon. (2002c). *The Korean Herbal Pharmacopoeia.* English edition (Eighth edition). Korea Food and Drug Administration.

Anon. (2003a). *Restricted Herb List, January 2003.* UK Register of Chinese Herbal Medicine, London.

Anon. (2007–2008). *The Unani Pharmacopoeia of India*, Vol. 1–5. Government of India, Ministry of Health and Family Welfare, New Delhi. www.ccrum.net/research/upc/

Applequist, W. (2006). *The Identification of Medicinal Plants. Handbook of the Morphology of Botanicals in Commerce.* Missouri Botanical Garden Press, St. Louis.

Barden, A. (2000). *Heart of the Matter: Agarwood Use and Trade and CITES Implementation for* Aquilaria malaccensis TRAFFIC International, Cambridge.

Barceloux, D. G. (2008). *Medical Toxicology of Natural Substances: Foods, Fungi, Medicinal Herbs, Plants and Venomous Animals*. Wiley, Hoboken.

Barnes, J., Anderson, L. A. & Phillipson, J. D. (2007). *Herbal Medicines.* Third edition. Pharmaceutical Press, London.

Barnes, J., Veitch, N., Anderson, L. A. & Phillipson, J. D. (2013). *Herbal Medicines.* Fourth edition. Pharmaceutical Press, London.

Bauer, R. & Franz, G. (2010). Modern European monographs for quality control of Chinese herbs. *Planta Medica* 76: 2004–2011.

Beentje, H. (2016). *The Kew Plant Glossary: an illustrated dictionary of plant terms.* Second edition. Kew Publishing, Royal Botanic Gardens, Kew.

Bensky, D., Clavey, S. & Stoeger, E. (2004). *Chinese Herbal Medicine: Materia Medica.* Third edition. Eastland Press, Seattle.

Bing, Q. Z, Zhang, B. G, Zhang, Z. & Chen, Z. H. (2008). Study on good agricultural practice for *Tulipa edulis*; planting density and sowing depth tests. [In Chinese]. *Zhong Guo Zhong Yao Za Zhi* [China Journal of Chinese Materia Medica] 33: 2463–2466.

Bisset, N. & Wichtl, M. (1984). *Herbal Drugs and Phytopharmaceuticals: a handbook for practice on a scientific basis.* Medpharm, Stuttgart.

Blatter, A. & Reich, E. (2004a). High performance thin-layer chromatographic analysis of aristolochic acids in Chinese drugs. *Journal of Planar Chromatography — Modern TLC* 17: 355–359.

Blatter, A. & Reich, E. (2004b). Qualitative and quantitative HPTLC methods for quality control of *Stephania tetrandra. Journal of Liquid Chromatography & Related Technologies* 27: 2087–2100.

Bone, K. & Mills, S. (2013). *Principles and Practice of Phytotherapy: modern herbal medicine*. Second edition. Elsevier.

Booker, A., Jalil, B., Frommenwiler, D., Reich, E., Zhai, L., Kulic, Z. & Heinrich, M. (2016a). The authenticity and quality of *Rhodiola rosea* products. *Phytomedicine* 23: 754–762

Booker, A., Zhai, L. X., Gkouva, Ch., Li, S. Y. & Heinrich M. (2016b). From traditional resource to global commodities: a comparison of *Rhodiola* species using NMR spectroscopy–metabolomics and HPTLC. *Frontiers in Pharmacology* 7: 254.

Boumba, V. A., Mitselou, A. & Vougiouklakis, T. (2004). Fatal poisoning from ingestion of *Datura stramonium* seeds. *Veterinary and Human Toxicology* 46: 81–82.

Brand, E. (2016). Changes in herbal medicine from ancient times to present. *The Mayway Mailer*. www.mayway.com/mayway-mailers/

Buriani, A., Garcia-Bermejo, M. L., Bosisio, E., Xu, Q., Li, H., Dong, X., Simmonds, M. S., Carrara, M., Tejedor, N., Lucio-Cazana, J. & Hylands, P. J. (2012). Omic techniques in systems biology approaches to traditional Chinese medicine research: present and future. *Journal of Ethnopharmacology* 140: 535–544.

Burrows, G. E. & Tyrl, R. J. (2013). *Toxic Plants of North America*. Wiley, Ames.

But, P. P. (1994). Herbal poisoning caused by adulterants or erroneous substitutes. *Journal of Tropical Medicine and Hygiene* 97: 1–4.

Cha, J. D., Jeong, M. R., Jeong, S. I., Moon, S. E., Kim, J. Y., Kil, B. S. & Song, Y. H. (2005). Chemical composition and antimicrobial activity of the essential oils of *Artemisia scoparia* and *A. capillaris*. *Planta Medica* 71: 186–190.

Chan, K., Shaw, D., Simmonds, M. S. J., Leon, C. J., Xu, Q., Lu, A., Sutherland, I., Ignatova, S., Zhu, Y.–P., Verpoorte, R., Williamson, E. M. & Duez, P. (2012). Good practice in reviewing and publishing studies on herbal medicine, with special emphasis on traditional Chinese medicine and Chinese *materia medica*. *Journal of Ethnopharmacology* 140: 469–475.

Chang, H. M. & But, P. P. (1986). *Pharmacology and Applications of Chinese Materia Medica*. 2 vols. World Scientific, Singapore.

Chang, Y. S. & Ho, Y. L. (2013). *Color Illustrations of Taiwan Herbal Pharmacopeia.* [In Chinese]. Ministry of Health and Welfare, Taipei.

Chen, J. K. & Chen, T. T. (2004). *Chinese Medical Herbology and Pharmacology*. Art of Medicine Press, City of Industry.

Chen, S.L., Jia, M. R., Wang, Y., Xue, G. & Xiao, P. G. (2003). Study on the plant community of *Fritillaria cirrhosa* in China. [In Chinese]. *Zhong Guo Zhong Yao Za Zhi* [China Journal of Chinese Materia Medica] 28: 398–402.

Chen, S. L. & Lin, Y. L. (2008). *Illustrated Handbook of Chinese Medicinal Herbs — Distinguishing Common Chinese Medicinal Herbs and their Counterfeits.* 3 vols. [In Chinese]. Zhong Yi Eu Ji Publishing House, Beijing.

Chen, S. L. & Lin, Y. L. (2013). *Zhong yao yin pian biao zhun tu jian*

[Chinese Herbal Medicine Standards Illustrations. In Chinese]. Fujian Science and Technology Press, Fuzhou.

Chen, S. L., Lin, Y. L., Qian, Z. Z. & Leon, C. J. (eds) (2010a). *A Colored Identification Atlas of Chinese Materia Medica and Plants as Specified in the Pharmacopoeia of the People's Republic of China*. 2 vols. [In Chinese]. Peoples Medical Publishing, Beijing.

Chen, S., Pang, X., Song, J., Shi, L., Yao, H., Han, J. & Leon, C. (2014). A renaissance in herbal medicine identification; from morphology to DNA. *Biotechnology Advances*, Jul 31. pii: S0734–9750(14)00109–8.

Chen, S. L., Yao, H., Han, J., Liu, C., Song, J., Shi, L., Zhu, Y., Ma, X., Gao, T., Pang, X., Luo, K., Li, Y., Li, X., Jia, X., Lin, Y. & Leon, C. J. (2010b). Validation of the ITS2 region as a novel DNA barcode for identifying medicinal plant species. *PloS one* 5: e8613.

Cheung, F. (2011). TCM: made in China. *Nature* 480: S82–S83.

China State Food and Drug Administration. *GAP Inspection Announcements*. www.sda.gov.cn/WS01/CL0082

Choi, Y. E., Kim, Y. S., Yi, M. J., Park, W. G., Yi, J. S., Chun, S. R., Han, S. S. & Lee, S. J. (2007). Physiological and chemical characteristics of field- and mountain-cultivated ginseng roots. *Journal of Plant Biology* 50: 198–205.

CITES (Convention on International Trade in Endangered Species of Wild Fauna and Flora). *Appendices*. www.cites.org/eng/app/appendices.shtml

CITES-EU. *The European Union and Trade in Wild Fauna and Flora*. ec.europa.eu/environment/cites/legislation_en.htm

Court, W. E. (2000). *Ginseng; the Genus* Panax. Harwood Academic, Amsterdam.

CP2010. (2011). *Pharmacopoeia of the People's Republic of China 2010* [English edition]. China Medical Science Press, Beijing.

CP2015. (2015). *Pharmacopoeia of the People's Republic of China* [Chinese edition]. Medical Science and Technology Press, Beijing.

Cronquist, A. (1994). *Intermountain Flora: vascular plants of the intermountain west, USA. Volume 5. Asterales*. New York Botanical Garden, New York.

Crook, V. (2013). *Arisaema heterophyllum. IUCN Red List of Threatened Species*. Version 2014.1.: www.iucnredlist.org

Cunningham, A. & Brinckmann, J. (2010). "Cinderella" *Schisandra*. A project linking conservation and local livelihoods in the Upper Yangtze ecoregion of China. *Herbalgram* 85: 28–29.

Dang, H., Wang, Q., Wang, H., Yan, M. & Liu, X. (2016). The integration of Chinese material medica into the Chinese health care delivery system, an update. *Phytotherapy Research* 30: 292–297.

Davidson, A. (1999). *The Oxford Companion to Food*. Oxford University Press, Oxford.

Debelle, F. D., Vanherweghem, J. L & Nortier, J. (2008). Aristolochic acid nephropathy: a worldwide problem. *Kidney International* 74: 158–169.

Deng, H., Ye, Z. H. & Wong, M. H. (2004). Accumulation of lead, zinc, copper and cadmium by 12 wetland plant species thriving in metal-contaminated sites in China. *Environmental Pollution* 132: 29–40.

Department of Health. Hong Kong (2005–). *Hong Kong Chinese Materia Medica Standards*. Chinese Medicine Division, Department of Health, Hong Kong.

Ding, G., Zhang, D. Z., Ding, X. Y., Zhou, Q., Zhang, W. C. & Li X. X. (2008). Genetic variation and conservation of the endangered Chinese endemic herb *Dendrobium officinale* based on SRAP analysis. *Plant Systematics and Evolution* 276: 149–156.

Ding, X. Y., Xu, L. S., Wang, Z. T., Zhou, K. Y., Xu, H. & Wang, Y. Q. (2002). Authentication of stems of *Dendrobium officinale* by rDNA ITS region sequences. *Planta Medica* 68: 191–192.

Edwards, S. E., Heinrich, M. H., Rocha, I. & Williamson, E. M. (2015). *Phytopharmacy: an evidence-based guide to herbal medicinal products*. Wiley Blackwell, Oxford.

EHTPA. European Herbal & Traditional Medicine Practitioners Association (2014). *The Core Curriculum for Herbal & Traditional Medicine*. Fourth edition. www.ehtpa.eu/pdf/Core%20Curriculum%20 2014%20V4%20-%20SK.pdf

EMA (2014). Committee on Herbal Medicinal Products (HMPC). Public statement on the use of herbal medicinal products containing toxic, unsaturated pyrrolizidine alkaloids (PAs). EMA/HMPC/893108/2011; 24 November 2014. www.ema.europa.eu/docs/en_GB/document_library/Public_statement/2014/12/WC500179559.pdf.

eMonocot portal. A global online biodiversity information resource for monocotyledons. www.emonocot.org

Euro+Med (2006–): Euro+Med PlantBase — the information resource for Euro-Mediterranean plant diversity. ww2.bgbm.org/EuroPlusMed/

Fang, L. Q. & Guo, J. X. (1992). Morphological and histological studies on the Chinese drug shan-dou-gen. [In Chinese]. *Yao Xue Xue Bao* (*Acta Pharmaceutica Sinica*) 27: 308–316.

Farah, M. H., Edwards, R., Lindquist, M., Leon, C. & Shaw, D. (2000). International monitoring of adverse health effects associated with herbal medicines. *Pharmacoepidemiology & Drug Safety* 9: 105–112

Farah, M.H., Olsson, S., Bate, J., Lindquist, M., Edwards, R., Simmonds, M. S. J., Leon, C., de Boer, H. J. & Thulin, M. (2006) Botanical nomenclature in pharmacovigilance and a recommendation for standardisation. *Drug Safety* 29: 1023–1029.

FDA (U.S. Food and Drug Administration). *Poisonous Plants Database*. www.accessdata.fda.gov/scripts/plantox/textResults.cfm

Federal Register (2004). Final rule declaring Dietary Supplements containing ephedrine alkaloids adulterated because they present an unreasonable risk. *Federal Register* 69 (28). 11 February, 2004.

Flower, A., Witt, C., Liu, J. P., Ulrich-Merzenich, G., Muir, K., Yu, H., Prude, M. & Lewith, G. (2014). GP-TCM Unabridged guidelines for randomised controlled trials investigating Chinese herbal medicine (CHM). *European Journal of Integrative Medicine* 6: 186–210.

Flurer, R. A., Jones, M. B., Vela, N., Ciolino, L. A. & Wolnik, K. A. (2001). *Determination of aristolochic acids in traditional Chinese medicines and dietary supplements*. DFS/ORO/ORA(4212) (updated 2001/11/06): web.archive.org/web/20150908070642/http://ahpa.org/portals/0/pdfs/FDA_AnalyticalMethod_AA.pdf

FNA. *Flora of North America North of Mexico*. www.fna.org/FNA

FoC. *Flora of China*. (1994–2013). www.efloras.org/search_page.aspx?flora_id=2

Foster, S. (2012). Exploring the peripatetic maze of black cohosh adulteration: a review of the nomenclature, distribution, chemistry, market status, analytical methods and safety. *Herbalgram* 98: 32–51.

Foster, S. & Yue C. X. (1992). *Herbal Emissaries. Bringing Chinese Herbs to the West*. Healing Arts, Rochester.

FRPS (1959–2004). *Flora Reipublicae Popularis Sinicae*. 80 vols. Academica Sinica, Beijing.

Fu, D. Z. (2011). *List of Threatened Plants of China*. Lanzhou.

Fu, L. K. (1992). *China Plant Red Data Book; Rare and Endangered Plants*. Vol. 1. Science Press, Beijing.

Fu, L. K. (ed.) (1999–). *Higher Plants of China*. Qingdao Publishing, Quingdao.

Gardner, Z. E. & McGuffin, M. (2013). *American Herbal Products Association's Botanical Safety Handbook*. Second edition. CRC Press, Boca Raton.

Gasson, P. (2011). How precise can wood identification be? Wood anatomy's role in support of the legal timber trade, especially CITES. *IAWA Journal* 32: 137–154.

Gasson, P., Baas, P. & Wheeler, E. (2011). Wood anatomy of CITES-listed tree species. *IAWA Journal* 32: 155–197.

Government of Canada. *Species at Risk Public Registry, SCHEDULE 1 (Subsections 2(1), 42(2) and 68(2)) List of Wildlife Species at Risk*. www.registrelep-sararegistry.gc.ca

Gordaliza, M., Garcıa, P. A., Del Corral, J. M., Castro, M. A., & Gómez-Zurita, M. A. (2004). Podophyllotoxin: distribution, sources, applications and new cytotoxic derivatives. *Toxicon* 44: 441-459.

Grey-Wilson, C. & Cribb, P. (2011). *Guide to the Flowers of Western China*. Kew Publishing, Royal Botanic Gardens, Kew.

GRIN. USDA Agricultural Research Service. *Germplasm Resources Information Network*. www.ars-grin.gov/~sbmljw/cgi-bin/tax_search.pl?

Guo, H. B., Cui, X. M., An, N., & Cai, G. P. (2010a). Sanchi ginseng (*Panax notoginseng* (Burkill) F. H. Chen) in China: distribution, cultivation and variations. *Genetic Resources and Crop Evolution* 57: 453-460.

Guo, H., Wang, W., Yang, N., Guo, B., Zhang, S., Yang, R., Yuan, Y., Yu, J., Hu, S., Sun, Q. & Yu, J. (2010c). DNA barcoding provides distinction between *Radix Astragali* and its adulterants. *Science China: Life Sciences* 53: 992–999.

Guo, J. L., Ren, Y., Chen, L., Pei, J. & Wan, D. G. (2010b). Authentication of *Caulis clematidis armandii* ("Chuanmutong") and differentiation of its common adulterants using RAPD and SCAR markers. *Journal of Medicinal Plants Research* 4: 697–701.

Hao D. C., Gu X. J., Xiao, P. G. & Peng, Y. (2013). Chemical and biological research of *Clematis* medicinal resources–review paper. *Chinese Science Bulletin* 58: 1120–1129.

Harley, R. & Paton, A. (2001). *Leonurus japonicus* Houtt. (Labiatae): the correct name for a common tropical weed. *Kew Bulletin* 56: 243–244.

Harrington, R. A., Kujawski, R., Dennis, H. & Ryan, P. (2003). Invasive plants and green industry. *Journal of Arboriculture* 29: 42–48.

Hashimoto, K., Higuchi, M., Makino, B., Sakakibara, I., Kubo, M., Komatsu, Y., Maruno, M. & Okada, M. (1999). Quantitative analysis of aristolochic acids, toxic compounds, contained in some medicinal plants. *Journal of Ethnopharmacology* 64: 185–189.

Haw, S. G. (2001). Tree peonies: a review of their history and taxonomy. *New Plantsman* 8: 156–171.

Heinrich, M., Chan, J., Wanke, S., Neinhuis, C. & Simmonds, M. S. J. (2009). Local uses of *Aristolochia* species and contents of nephrotoxic aristolochic acids 1 and 2 — a global assessment based on bibliographic sources. *Journal of Ethnopharmacology* 125: 108–144.

Hind, D. J. N. (2011). *An Annotated Preliminary Checklist of the Compositae of Bolivia* (version 2). www.kew.org/science/tropamerica/boliviacompositae

HMPC (2016). *Public statement on contamination of herbal medicinal products/traditional herbal medicinal products with pyrrolizidine alkaloids. Transitional recommendations for risk management and quality control*, s.l.: EMA, Committee on Herbal Medicinal Products (HMPC). www.ehtpa.eu/pdf/PA%20Public%20Statement%20HMPC%202016.pdf

Ho, Y. L., Lin, I. & Chang, Y. S. (2006). *Illustrations of Commonly Misused Chinese Crude Drug Species in Taiwan*. [In Chinese]. Taiwan's Chinese Medicine Medication Safety, Environment Regulations Series (VI), Committee on Chinese Medicine and Pharmacy, Department of Health, Executive Yuan, Taiwan.

Hong, B. G., Xiu, M. C., Na, A. & Ge, P. C. (2010). Sanchi ginseng (*Panax notoginseng* (Burkill) F. H. Chen) in China: distribution, cultivation and variations. *Genetic Resources and Crop Evolution* 57: 453–460.

Hong, D. Y. (2010). *Peonies of the World. Taxonomy and Phytogeography*. Kew Publishing, Royal Botanic Gardens, Kew.

Hong, D. Y. & Pan, K. Y. (1999). A revision of the *Paeonia suffruticosa* complex (Paeoniaceae). *Nordic Journal of Botany* 19: 289–300.

Hong Kong (1999). *Chinese Medicine Ordinance (1999) Schedules I and II*: www.legislation.gov.hk/blis_pdf.nsf/6799165D2FEE3FA94825755E0033E532/6B6574BC51B2EB7D482575EF00194933/$FILE/CAP_549_e_b5.pdf

Hsu, H. Y., Chen, Y. P., Shen, S. J., Hsu, C. S., Chen, C. C. & Chang, H. C. (1986). *Oriental Materia Medica: a Concise Guide*. Oriental Healing Arts Institute, Long Beach.

Hu, S. L. (1989). *The Authentic and Superior Medicinal Herb in China*. [In Chinese]. Heilongjiang Publishing, Heirbin.

Hu, S. Y. (2005). *Food Plants of China*. Chinese University Press, Hong Kong.

Huai, H. Y., Dong, Q. Q. & Liu, A. H. (2010). Ethnomedicinal analysis of toxic plants from five ethnic groups in China. *Ethnobotany Research and Applications* 8: 169–179. dx.doi.org/10.17348/era.8.0.169–179

Huang, H. Q., Zhang, X., Shen, Y. H., Su, J., Liu, X. H., Tian, J. M., Lin, S., Shan, L. & Zhang, W. D. (2009a). Polyacetylenes from *Bupleurum longiradiatum*. *Journal of Natural Products* 72: 2153–2157.

Huang, K. C. (1999). *The Pharmacology of Chinese Herbs*. Second edition. CRC Press, Boca Raton.

Huang, T. C., Huang, L. H., Chang, Y. S. & Lin, I. H. (2009b). *The Illustration of Common Medicinal Plants in Taiwan*. Committee on Chinese Medicine and Pharmacy, Department of Health, Taipei.

Hyun, T. K., Eom, S. H., Yu, C. Y. & Roitsch, T. (2010). *Hovenia dulcis* — an Asian traditional herb. *Planta Medica* 76: 943–949.

Inside Wood (2004–). insidewood.lib.ncsu.edu/search

Iranshahy, M. & Iranshahi, M. (2011). Traditional uses, phytochemistry and pharmacology of asafoetida (*Ferula assa-foetida* oleo-gum-resin) — a review. *Journal of Ethnopharmacology* 134: 1–10.

IUCN (2012a). IUCN SSC China Plant Specialists Group. *List of Threatened Plants in China*. Unpublished.

IUCN (2012b). *IUCN Red List Categories and Criteria: Version 3.1.* Second edition. International Union for Conservation of Nature, Gland and Cambridge.

IUCN (2012c). *IUCN Red List of Threatened Species*. Version 2012.2. www.iucnredlist.org

Iwatsuki, K., Yamazaki, T., Boufford, D. E. & Ohba, H. (1995). *Flora of Japan, Vol. IIIb, Angiospermae, Dictotyledoneae, Sympetalae (b).* Kodansha, Tokyo.

Japanese Pharmacopoeia. (2001): pdb.nihs.go.jp/jp15e/JP15.pdf

JP (2012). *Japanese Pharmacopoeia.* Sixteenth edition. [English edition]. www.mhlw.go.jp/topics/bukyoku/iyaku/yakkyoku/english.html]

Jellin, J. M., *et al.* (2012). *Natural Medicines Comprehensive Database*. Therapeutic Research Faculty, Stockton.

Joshi, V. C., Avla, B. & Khan, I. A. (2008). Authentication of *Stephania tetrandra* S. Moore (fang ji) and differentiation of its common adulterants using microscopy and HPLC analysis. *Journal of Natural Medicine* 62: 117–121.

Kala, C. P. (2005). Indigenous uses, population density, and conservation of threatened medicinal plants in protected areas of the Indian Himalayas. *Conservation Biology* 19: 368-378.

Kamboj, A. & Saluja, A. K. (2010). Phytopharmacological review of *Xanthium strumarium* L. (Cocklebur). *International Journal of Green Pharmacy* 4: 129–139.

Kaur, R., Kashyap, A., Majeed, S., Chauhan, N. S., & Bhardwaj, S. B. (2010). *In vitro* propagation and conservation of *Inula racemosa* Hook. f. An endangered medicinal plant of temperate origin. *Journal of Advanced Laboratory Research in Biology* 1: 88–91.

Kasimu, R., Tanmaka, K., Tezuka, Y., Gong, Z.-N., Li J.-X., Basnet, P. & Namba, T. (1998). Comparative study of seventeen *Salvia* plants: aldose reductase inhibitory activity of water and MeOH extracts and liquid chromatography-mass spectrometry (LC-MS) analysis of water extracts. *Chemical and Pharmaceutical Bulletin* 46: 500–504.

Keung, W. M. (2002). *Pueraria*. Taylor and Francis, London.

Kew-MPNS. *Medicinal Plant Names Services*. Royal Botanic Gardens, Kew. mpns.kew.org/mpns-portal

Khatoon, S. & Mehrotra, S. (2009). *Bark drugs.* National Institute of Science Communication and Information Resources CSIR, New Delhi.

Kharkwal, A. C., Kushwaha, R., Prakash, O., Ogra, R. K., Bhattacharya, A., Nagar, P. K. & Ahuja, P. S. (2008). An efficient method of propagation of *Podophyllum hexandrum*: an endangered medicinal plant of the Western Himalayas under ex situ conditions. *Journal of Natural Medicines* 62: 211–216.

Kite, G. C., Howes, M. J. R., Leon, C. J. & Simmonds, M. S. R. (2003). Liquid chromatography/mass spectrometry of malonyl-ginsenosides in the authentication of ginseng. *Rapid Communications in Mass Spectrometry* 17: 238–244.

Kite, G. C., Yule, M. A., Leon, C. J. & Simmonds, M. S. J. (2002). Detecting aristolochic acids in herbal remedies by liquid chromatography/serial mass spectrometry. *Rapid Communications in Mass Spectrometry* 16: 585–590.

Klein-Schwartz, W. & Oderda, G. M. (1984). Jimsonweed intoxication in adolescents and young adults. *JAMA Paediatrics* 138: 737–739.

Koh, H. L., Wang, H., Zhou, S., Chan, E. & Woo, S. O. (2006). Detection of aristolochic acid I, tetrandine and fangchinoline in medicinal plants by high performance liquid chromatography and liquid chromatography/mass spectrometry. *Journal of Pharmaceutical and Biomedical Analysis* 40: 653–661.

KP (2007). *Korean Pharmacopoeia*. Ninth edition. Korea Food and Drug Administration. www.mfds.go.kr/eng/index.do?nMenuCode=50

Kum, K. Y., Booker, A., Leon, C., Garcia, J.P & Heinrich, M. (2016). *Salvia miltiorrhiza* (Danshen); an NMR-metabolomic and HPTLC-based analysis of products' quality. *Planta Medica* 81(S 01) S1-S381.

Kuniyal, C. P., Rawat, D. S. & Sundriya, R. C. (2015). Cultivation of *Saussurea costus* cannot be treated as 'artificially propagated'. *Current Science* 108: 1587–1589.

Lamxay, V. & Newman, M. F. (2012). A revision of *Amomum* (Zingiberaceae) in Cambodia, Laos and Vietnam. *Edinburgh Journal of Botany* 69: 99–206.

Lau, T. W., Shaw, P. C., Wang, J. & But, P. H. (2001). Authentication of medicinal *Dendrobium* species by the Internal Transcribed Spacer of ribosomal DNA. *Planta Medica* 67: 456–60.

Leon C. J., Fay, M. & Rix, M. (2009). *Fritillaria yuminensis*. *Curtis's Botanical Magazine* 26: 21–32.

Leung, P. C. & Cheng, K. F. (2008). Good Agricultural Practice (GAP): does it ensure a perfect supply of medicinal herbs for research and drug development? *International Journal of Applied Research in Natural Products*, 1(2): 1–8.

Li, M., Qin, X. Q., Chen, S. B. & Feng, Y. X. (1994). Morphological and microscopical identification of genus *Astragalus* on mainly used commercial Huang qi. *Acta Pharmaceutica Sinica* 29: 862–871.

Li, M. H., Chen, J. M., Peng, Y., Wu, Q. L. & Xiao, P. G. (2008). Investigation of Danshen and related medicinal plants in China. *Journal of Ethnopharmacoogy* 120: 419–426.

Li, S. Z. & Luo, X. (2003). *Ben Cao Gang Mu* [Compendium of Materia Medica]. 6 vols. [In Chinese]. Foreign Languages Press, Beijing.

Li, T. & Zhang, H. (2008). Identification and comparative determination of rhodionin in traditional Tibetan medicinal plants of fourteen *Rhodiola* species by high-performance liquid chromatography-photodiode array detection and electrospray ionization-mass spectrometry. *Chemical and Pharmaceutical Bulletin* 56: 807–814.

Li, X. B., Liu, H. T., Qi, Y. D., Zhang, J., Zhu Y. X., Zhang Z. & Zhang, B. (2012a). [Advances in the research of chemical constituents and bioactiviites of *Conioselinum*. *Lishizhen Medicine and Materia Medica Research* 23(12): 3097–3100.

Li, X. B. Qi, Y. D Liu, H. T., Zhang, J., Zhang, Z. & Zhang, B. (2013). Chemical constituents of *Conioselinum vaginatum*. *Zhong Guo Zhong Yao Za Zhi* [China Journal of Chinese Materia Medica] 38: 1543–1547.

Li, X. W., Song, J. Y., Wei, J. H., Hu, Z. G., Xie, C. X. *et al.* (2012). Natural fostering in *Fritillaria cirrhosa*: integrating herbal medicine production with biodiversity conservation. *Acta Pharmaceutica Sinica B* 2(1): 77–82.

Lin, B. (1990–96). *Guang dong zhong yao zhi* [In Chinese Traditional Chinese Medicines in Guangdong]. 2 vols. Guangdong Science and Technology, Guangdong.

Lin, R. H., Li, Z. F., Lin, J. M., Ye, J. X, Cai, Q. Y., Chen, L. D. & Jun, P. (2015). Ethanolic extract of *Tulipa edulis Bak* induces apoptosis in SGC-7901 human gastric carcinoma cells via the mitochondrial signaling pathway. *Oncology Letters* 10: 2371–2377.

Lin, Y. P. *et al.* (2003). *Regional Standard of Chinese Medicinal and Nationality Materials in Gui Zhou*. [In Chinese]. Gui Zhou Science & Technology Press, Guiyang.

Liu, Y. *et al.* (2013). Whole-tree agarwood-Inducing technique: an efficient novel technique for producing hi-quality agarwood in cultivated *Aquilaria sinensis* trees. *Molecules* 18(3): 3086–3106.

Lou, Z. C. (ed.) (1995–6). *Species Systematization and Quality Evaluation of Commonly Used Chinese Traditional Drugs*. North China. 6 vols. [In Chinese]. Beijing yi ke da xue chu ban she, Beijing.

Lou, Z. C., Wang, X., Masayuki, M. & Tsuneo, N. (1988). Pharmacognostical studies on the Chinese crude drug "Da-huang" (大黄 Rhubarb)-1-Botanical Origin of the Official Da-huang. *Japanese Journal of Pharmacognosy* 42: 291–301.

Lord, G. M., Tagore, R., Cook, T., Gower, P. & Pusey, C. D. (1999). Nephropathy caused by Chinese herbs in the UK. *Lancet* 354: 481–482.

Ma, H. L., Zhu, Z. B., Zhang, X. M., Miao, Y. Y. & Guo, Q. S. (2014). Species identification of the medicinal plant *Tulipa edulis* (Liliaceae) by DNA barcode marker. *Biochemical Systematics and Ecology* 55:362–368.

Ma, J. S., Wei, C., Liu, Q. R., Ming, Y. & Han, L. J. (2006). A revision of *Phellodendron* (Rutaceae). *Edinburgh Journal of Botany* 63: 131–151.

Marieschi, M., Torelli, A. & Brunui, R. (2012). Quality control of saffron (*Crocus sativus* L). Development of SCAR markers for the detection of plant adulterants used as bulking agents. *Journal of Agricultural and Food Chemistry* 60: 10998–11004.

Martindale, W. & Reynolds, J. E. F. (1996). *The Extra Pharmacopoeia*. Royal Pharmaceutical Society, London.

Martindale (2011). *Martindale: The Complete Drug Reference*. Thirty-seventh edition. Volume A. Pharmaceutical Press, London.

Mazzanti, G., Battinelli, L., Daniele, C., Mastroianni, C. M., Lichtner, M., Coletta, S. & Costantini, S. (2004). New case of acute hepatitis following the consumption of Shou Wu Pian, a Chinese herbal product derived from *Polygonum multiflorum*. *Annals of Internal Medicine* 140: W30.

McGraw, J. B., Lubbers, A. E., Van der Voort, M., Mooney, E. H., Furedi, M. A., Souther, S., Turner, J. B. & Chandler, J. (2013). Ecology and conservation of ginseng (*Panax quinquefolius*) in a changing world. *Annals of the New York Academy of Sciences* 1286: 62–91.

McGuffin, M., Kartesz, J. T., Leung, A.Y. & Tucker, A. O. (eds) (2000). *Herbs of Commerce*. American Herbal Products Association, Silver Spring.

MEPCAS. Ministry of Environmental Protection and the Chinese Academy of Sciences. (2013). *China Biodiversity Red List — higher plant volumes*. www.mep.gov.cn/gkml/hbb/bgg/201309/W020130917614244055331.pdf

MHRA (2012). *Safety of Herbal Medicinal Products*. Medicines and Healthcare products Regulatory Agency. July 2002. www.mhra.gov.uk/home/groups/es-herbal/documents/websiteresources/con009293.pdf.

MHRA (2014). *Banned and Restricted Herbal Ingredients*. UK Medicines and Healthcare products Regulatory Agency, 18 December, 2014. www.gov.uk/government/publications/list-of-banned-or-restricted-herbal-ingredients-for-medicinal-use/banned-and-restricted-herbal-ingredients.

Mei, W. L., *et al.* (2013). Advances in the mechanism, artificial agarwood-introduction techniques and chemical constituents of artificial agarwood production. *Chinese Journal of Tropical Crops* 34: 2513–2520.

Miao, Y. Y., Zhu, Z. B., Guo, Q. S., Ma, H. L., Yang, Y., & Zhu, L. F. (2014). Effects of flower bud removal and artificial pollination on growth and yield of *Tulipa edulis*. [In Chinese]. *Zhong Guo Zhong Yao Za Zhi* [China Journal of Chinese Materia Medica] 39: 2016-2018.

Michl, J. M., Jennings, H. M., Kite, G., Ingrouille, M., Simmonds, M. S. J. & Heinrich, M. (2013). Is aristolochic acid nephropathy a widespread problem in developing countries? A case study of *Aristolochia indica* L. in Bangladesh using an ethnobotanical-phytochemical approach. *Journal of Ethnopharmacology* 149: 235–244.

Milne, S. T. & Meek, P. D. (1998). Fatal colchicine overdose: report of a case and review of the literature. *American Journal of Emergency Medicine* 16: 603–608.

Mitch, L. W. (1987). Cockleburs. *Weed Technology* 1: 359–360.

Molur, S. & Walker, S. (eds) (1998). *Conservation Assessment and Management Plan. (CAMP) Workshop Report. Selected Medicinal Plants of Northern, Northeastern and Central India*. Lucknow 21–25 January 1997. ZOO Outreach Organisation, CBSG India, Coimbatore.

Mukhopadhyay, M. J., Mukhopadhyay, S. & Sen, S. (2002). *In vitro* propagation of *Iphigenia indica*, an alternative source of colchicine. *Plant Cell, Tissue and Organ Culture* 69: 101–104.

Nadeem, M., Palni, L. M. S., Purohit, A. N., Pandey, H. & Nandi, S. K. (2011). Propagation & conservation of *Podophyllum hexandrum* Royle: an important medicinal herb. *Biological Conservation* 92: 121–129.

Nagata, T., DuVal, A., Schmull, M., Tchernaja, T. A. & Crane, P. R. (2013). *Paulownia tomentosa*: a Chinese plant in Japan. *Curtis's Botanical Magazine* 30: 261–274.

Namba, T., Komatus, K. & Mikage, M. (1986). Pharmacognostical studies on the Chinese crude drug "Fu-pen-zi", botanical origin of "Fu-pen-zi" from Korea (II). *Journal of Natural Medicines* 40: 95–102.

Nayar, M. P. & Sastry, A. R. K. (eds) (1998). *Red Data Book of Indian Plants 2*. Botanical Survey of India, Calcutta.

Ngyen Tien Ban (ed.) (2007). *Sach Do Viet Nam, Phan II. Thuc Vat*. [Vietnam Red Data Book, Part II. Plants]. Vietnamese Academy of Science and Technology, Hanoi.

NICPBP. (1997–2002). National Institute for the Control of Pharmaceutical and Biological Products (NICPBP) and Guangdong Provincial Institute for Drug Control (eds). *Zhong Guo Zhong Yao Cai Zhen Wei Jian Bie Tu Dian* [Illustrated Dictionary of Chinese Herbal Medicines for Identification and Authentication. In Chinese]. 4 vols. Guangdong Science and Technology Press, Guangzhou.

Nortier, J. L., Schmeiser, H. H., Muniz-Martinez, M. C., Arlt, V. M., Vervaet, C., Garbar C. H., Daelemans, P. & Vanherweghem, J. L. (2003). Invasive urothelial carcinoma after exposure to Chinese herbal medicine containing aristolochic acid may occur without severe renal failure. *Nephrology Dialysis Transplantation* 18: 426–428.

Oku, H. Ueda, Y. & Ishiguro, K. (2003). Antipruritic effects of the fruits of *Chaenomeles sinensis*. *Biological and Pharmaceutical Bulletin* 26: 1031–1034.

Okuda, T., Yoshida, T., Chen, X. M., Xie, J. X. & Fukushima, M. (1987). Corianin from *Coriaria japonica* A. Gray and sesquiterpene lactones from *Loranthus parasiticus* Merr. used for treatment of schizophrenia. *Chemical and Pharmacutical Bulletin* 35: 182–187.

Pan, Z. H., Li, Y., Liu, J. L., Ning, D. S., Li, D. P., Wu, X. D. & Wen, Y. X. (2012). A cytotoxic cardenolide and a saponin from the rhizomes of *Tupistra chinensis. Fitoterapia* 83: 1489–1493.

Perry, L. M. (1980). *Medicinal Plants of East and Southeast Asia: attributed properties and uses.* MIT Press, Cambridge.

Potterat, O. (2010). Goji (*Lycium barbarum* and *L. chinense*): phytochemistry, pharmacology and safety in the perspective of traditional uses and recent popularity. *Planta Medica*, 76: 7–19.

Qin, P., Chen, Q. P., Wang, B. R. & Shi, L. W. (1995). Studies of Zi Hua Di Ding. In: Z. Q. Lou & P. Qin (eds), *Species Systematization and Quality Evaluation of Commonly Used Chinese Traditional Drugs*, pp. 565–670. Beijing Medical University and Beijing Union Medical University, Beijing.

Qiu, S. H., Zeng, Y. Q., Pan, K. T., Tang, Y. C. & Xu, J. M. (1979). On the nomenclature of the Chinese plant drug "Chuanxiong". *Acta Phytotaxonomica Sinica* 17: 101–103.

Roeder, E. (2000). Medicinal plants in China containing pyrrolizidine alkaloids. *Pharmazie* 55: 711–726.

Sahebkar, A. & Iranshahi, M. (2011). Volatile constituents of the genus *Ferula* (Apiaceae): a review. *Journal of Essential Oil Bearing Plants* 14: 504–531.

Sarmento, L., Carus, M., Grotenhermen, F. & Kruse, D. (2015). *Scientifically Sound Guidelines for THC in Food in Europe*. Nova-Institute, 75 pages. eiha.org/media/2015/08/15-07-24-Report-Scientifically-Safe-Guidelines-THC-Food-nova-EIHA.pdf

Saunders, R. K. (2000). *Monograph of* Schisandra *(Schisandraceae).* American Society of Plant Taxonomists, Ann Arbor.

Sawai, R., Kuroda, K., Shibata, T., Gomyou, R., Osawa, K., & Shimizu, K. (2008). Anti-influenza virus activity of *Chaenomeles sinensis. Journal of Ethnopharmacology* 118: 108-112.

Schippmann, U. (2001). *Medicinal Plants Significant Trade Study*. CITES Project S-109; Plants Committee Document PC9 9.1.3. (rev.) Bundesamt für Naturschutz (BfN), German Federal Agency for Nature Conservation, Bonn.

Schischkin, B. K. (2000). *Flora of the USSR.* Vol. 23. *Bignoniaceae–Valerianaceae*. Smithsonian Institution, Washington.

Seca, A. M., Grigore, A., Pinto, D. C., & Silva, A. M. (2014). The genus *Inula* and their metabolites: from ethnopharmacological to medicinal uses. *Journal of Ethnopharmacology* 154: 286-310.

Sell, P. & Murrell, G. (2006). *Flora of Great Britain and Ireland.* Vol. 4. *Campanulaceae-Asteraceae*. Cambridge University Press, Cambridge.

SEPA (1999). *Regulations of the Peoples Republic of China on Wild Plants Protection*; Article 16. First version. Provisional second version. www.plant.csdb.cn/protectlist

SFDA (2003). Notice on the Termination of the Medicinal Use of Guan Mutong. SFDA-REGISTRATION No. 121.

Shang, X. F., He, X. R, He, X. Y., Li, M. X., Zhang, R. X., Fan, P. C., Zhang, Q. L. & Jia Z. P. (2010). The genus *Scutellaria* an ethnopharmacological and phytochemical review. *Journal of Ethnopharmacology* 128: 279–313.

Shang Z. J. (ed.) (1991). *Lei gong Pao Zhi Lun.* [In Chinese]. Anhui Science and Technology Press, Hefei.

Shaw, D. (2010). Toxicological risks of Chinese herbs. *Planta Medica* 76: 2012–2018.

Shaw, D., Ladds, G., Duez, P., Williamson, E. & Chan, K. (2012). Pharmacovigilance of herbal medicine. *Journal of Ethnopharmacology* 140: 513–518.

Shaw, D., Leon, C. J., Kolev S. & Murray V. (1997). Traditional remedies and food supplements; a 5-year toxicological study (1991–1995). *Drug Safety* 17: 342–356.

Sionneau, P. (1995). *Pao Zhi: an Introduction to the Use of Processed Chinese Medicinals*. Blue Poppy Press, Boulder.

Sorenson, W. R. & Sullivan, D. (2007). Determination of aristolochic acid I in botanicals and dietary supplements potentially contaminated with aristolochic acid I using LC-UV with confirmation by LC/MS: Collaborative study. *Journal of AOAC INTERNATIONAL* 90: 925–933.

Stace, C.A. (2010). *New Flora of the British Isles*. Cambridge University Press, Cambridge.

Stevens, P. F. (2001–). *Angiosperm Phylogeny Website. Version 13,* 2013. www.mobot.org/MOBOT/research/APweb

Stewart, M. & Steenkamp, V. (2001). Pyrrolizidine poisoning: a neglected area in human toxicology. *Therapeutic Drug Monitoring* 23: 698–708.

Sun, Z., Gao, T., Yao, H., Shi, L., Zhu, Y. & Chen, S. (2011). Identification of *Lonicera japonica* and its related species using the DNA barcoding method. *Planta Medica* 77: 301–306.

Tang, W. & Eisenbrand, G. (2011). *Handbook of Chinese Medicinal Plants: Chemistry. Pharmacology, Toxicology*. 2 vols. Wiley, Weinheim.

The Plant List. www.theplantlist.org

Timoshyna A., Li, C. Y., Ke, Z., Morgan, B. & Tsipidis, V. (2015) *Engaging China's private sector in sustainable management of medicinal plants (EGP MAPs) — the multiplier effect*. TRAFFIC. Cambridge. www.traffic.org/species-reports/traffic_species_plants30.pdf

TOXNET. *US National Library of Medicine Toxnet Database*. toxnet.nlm.nih.gov/cgi-bin/sis/htmlgen?TOXLINE

Tu, P. F., Xu, G. J., Xu, L. S. & Jin, R. L. (1991). Bencaological studies on Shashen and Qiui (*Adenophora* spp.). *Zhong Guo Zhong Yao Za Zhi* [China Journal of Materia Medica] 16: 200–201.

Tu, P. F., Xu, G. J., Xu, L. S. & Jin, R. L. (1994). A survey of commercial Chinese crude drug Sha Shen (Radix Adenophorae and Glehniae). *Zhong Guo Zhong Yao Za Zhi* [China Journal of Chinese Materia Medica] 19: 646–7, 701.

UK Government (2001). Statutory Instrument no. 1841. *The Medicines (Aristolochia and Mu tong etc.) (prohibitions) order 2001*. The Stationery Office, London. www.legislation.gov.uk/uksi/2001/1841/pdfs/uksi_20011841_en.pdf

Upton, R. (ed.) (1999a). Astragalus *Root;* Astragalus membranaceus *&* Astragalus membranaceus *var.* mongholicus. *Analytical, Quality Control and Therapeutic Monograph.* American Herbal Pharmacopoeia and Therapeutic Compendium, American Herbal Pharmacopoeia, Scotts Valley.

Upton, R. (ed.) (1999b). *Schisandra berry;* Schisandra chinensis*; Analytical, Quality Control and Therapeutic Monograph.* American Herbal Pharmacopoeia and Therapeutic Compendium, American Herbal Pharmacopoeia, Scotts Valley.

Upton, R. (ed.) (2003). *Dang Gui Root.* Angelica sinensis *(Oliv.) Diels; Standards of Analysis, Quality Control and Therapeutics*. American Herbal Pharmacopoeia and Therapeutic Compendium, American Herbal Pharmacopoeia, Scotts Valley.

Upton, R. (2012). *American Ginseng Root:* Panax quinquefolius. *Standards of Analysis, Quality Control, and Therapeutics*. American Herbal Pharmacopoeia, Scott's Valley.

Upton, R. & Graff, A. (eds). (2006). *Characterization of Selected Plants that may Contain or be Adulterated with Aristolochic Acid*. American Herbal Pharmacopoeia, Scotts Valley.

Upton, R., Graff, A., Jolliffe, G., Langer, R. & Williamson, E. (eds). (2011). *Microscopic Characterization of Botanical Medicines.* American Herbal Pharmacopoeia, CRC Press, Boca Raton.

USFDA. US Food and Drug Administration. (2000). *Dietary Supplements: aristolochic acid*. www.cfsan.fda.gov/dms/ds-bot.html

USFDA. US Food and Drug Administration. (2001a). *Listing of Botanical Ingredients of Concern. Attachment A — Botanicals known or suspected to contain aristolochic acid*. www.fda.gov/Food/RecallsOutbreaksEmergencies/SafetyAlertsAdvisories/ucm095283.htm

USFDA. US Food and Drug Administration (2001b). *Listing of Botanical Ingredients of Concern. Attachment D — Botanical products determined by FDA to contain aristolochic acids*. www.fda.gov/Food/RecallsOutbreaksEmergencies/SafetyAlertsAdvisories/ucm095297.htm

Vanherweghem, J. L., Tielemans, C., Abramowicz, D., Depierreux, M., Vanhaelen-Fastre, R., Vanhaelen, M., Dratwa, M., Richard, C., Vandervelde, D., Verbeelen, D. & Jadoul, M. (1993). Rapidly progressive interstitial renal fibrosis in young women: association with slimming regimen including Chinese herbs. *The Lancet* 341: 387–391.

Verheij, E.W.M. (1999). *Syzygium aromaticum* (L.) Merrill & Perry. In: C.C. de Guzman & J.J. Siemonsma (eds), *Plant Resources of South-East Asia. Spices* 13: 211-218. Backhuys, Leiden.

Waddick, J. W. & Zhao, Y. T. (1992). *Iris of China*. Timber Press, Portland.

Walter, K. & Gillett, H. J. (eds) (1998). *1997 IUCN Red List of Threatened Plants*. IUCN World Conservation Union, Gland.

Wang, A. L., Zhang, P., Liu, X., Liang, J. Q. & Li, W. W. (2016). Genetic structure and diversity of *Glehnia littoralis*, an endangered medicinal plant in China. *Biochemical Systematics and Ecology* 66: 265–271.

Wang, G. Q. (ed.) (2014). *National Survey of Chinese Medicine Resources*. Third edition. 4 vols. [In Chinese]. People's Medical Publishing House, Beijing.

Wang, S. & Xie, Y. (eds) (2004). *China Species Red List. Volume 1. Red List.* Higher Education Press, Beijing.

Wang, T., Zhang, X., & Xie, W. (2012). *Cistanche deserticola* Y. C. Ma,"Desert ginseng": a review. *American Journal of Chinese Medicine* 40: 1123-1141.

Wang, X. M. & Ren, Y. (2009). *Rheum tanguticum*, an endangered medicinal plant endemic to China. *Journal of Medicinal Plants Research* 3: 1195–1203.

Wang, Z. T., Xie, P. S. & Hu, Z. (2013). *Monographs for Quality Evaluation of Chinese Crude Drugs*. In English & Chinese. Shanghai Scientific & Technical Publishers, Shanghai.

WCS. *World Checklist of Selected Plant Families*. Royal Botanic Gardens, Kew. www.kew.org/wcsp

Wei, K. H., Li, L. X., Huang, Y. C., Wang, M. Y., Li, C. & Miao, J. H. (2013). Tissue culture of *Sophora tonkinensis* Gapnep. and its quality evaluation. *Pharmacognosy Magazine* 9: 323–30.

WHO. World Health Organisation (2006). *WHO Monograph on Good Agricultural and Collection Practices (GACP) for* Artemisia annua *L.* www.who.int/medicines/publications/traditional/ArtemisiaMonograph.pdf

Wichtl, M. (2004). *Herbal Drugs and Phytopharmaceuticals. A Handbook for Practice on a Scientific Basis.* Medpharm Scientific, Stuttgart.

Williamson, E. M., Lorenc, A., Booker, A. & Robinson, N. (2013). The rise of traditional Chinese medicine and its materia medica: a comparison of the frequency and safety of materials and species used in Europe and China. *Journal of Ethnopharmacology* 149: 453–462.

Wink, M. & van Wyk, B. E. (2008). *Mind-altering and Poisonous Plants of the World*. Timber Press, Portland.

Wiseman, N. & Feng, Y. (1998). *A Practical Dictionary of Chinese Medicine*. Second edition. Paradigm Publications.

Wong, A. & Chan, C. (2005). Review of adverse events related to Chinese medicines in Hong Kong, July 2004–June 2005. *Public Health Epidemiology Bulletin* 14: 45–51.

Wong, K. H., Razmovski-Naumovski, V. R., Li, K. M., Li, G. Q. & Chan, K. (2015). The quality control of two *Pueraria* species using Raman spectroscopy coupled with partial least squares analysis. *Journal of Raman Spectroscopy* 46: 361–368.

Wright, C. W. (2002). *Artemisia.* Taylor & Francis, London.

Wright, M. (2004). *An Introduction to Chinese Herbal Medicine with Particular Reference to the Chinese Umbelliferae.* Greenbank, Edinburgh.

Xiao, P. G. & Chen, S. L. (2003). Establishment of the national macro-manage system on Chinese medicinal resource. *Zhong Guo Zhong Yao Za Zhi* [China Journal of Chinese Materia Medica] 28: 4–6.

Xiao, P. G., Li, D. P. & Yang, S. L. (eds) (2001–2). *Modern Chinese Materia Medica*. 5 vols. [In Chinese]. Chemical Industry Publishing House, Beijing.

Xiao, X. H., Chen, S. L., Huang, L. Q. & Xiao, P. G. (2009). Survey of investigations on Daodi Chinese medicinal materials in China since 1980. *Zhong Guo Zhong Yao Za Zhi* [China Journal of Chinese Materia Medica] 34: 519–523.

Xiao X. H., Fang, Q. M., Xia W. J., Yin, G. P., Shu Z. W. & Qiao C. Z. (1997). Numerical taxonomy of medicinal *Salvia* L. and the genuineness of Danshen. *Journal of Plant Resources and Environment* 6: 17–21.

Xie P. S. & Leung, A. Y. (2009). Understanding the traditional aspect of Chinese medicine in order to achieve meaningful quality control of Chinese *materia medica*. *Journal of Chromatography A* 1216: 1933–1940.

Xin, T. Y., Li, X. J., Yao, H., Lin, Y. L., Ma, X. C., Cheng, R. Y., Song, J. Y., Ni, L. H., Fan, C. Z., & Chen, S. L. (2015). Survey of commercial *Rhodiola* products revealed species diversity and potential safety issues. *Scientific Reports* 5: 8337.

Xu, G. B. (1997). The authentication of several *Viola* species used in herbal medicine. [In Chinese]. *Zhong Yao Cai* 20: 371–373.

Xu, G. J. & Xu, L. S. (eds) (1997). *Species Systematization and Quality Evaluation of Commonly Used Chinese Traditional Drugs*. Southern set. 4 vols. [In Chinese]. Fujian Science and Techology Press, Fuzhou.

Xu, R., Chen, J., Chen, S. L., Liu, T. N., Zhu, W. C., & Xu, J. (2009). *Cistanche deserticola* Ma cultivated as a new crop in China. *Genetic Resources and Crop Evolution* 56: 137-142.

Yang, D. Y., Hirotoshi, F., Cai, S. Q. & Komatsu, K. (2004). Molecular analysis of *Rheum* species used as Rhei Rhizoma based on chloroplast *mat*K gene sequence and its application for identification. *Biological and Pharmaceutical Bulletin* 27: 375–83.

Yen, K. Y. (1992). *The Illustrated Chinese Materia Medica. Crude and Prepared*. SMC Publishing, Taipei.

Yin, Y. L. & Guo, L. (2009). Nutrient Analysis of *Tulipa edulis*. *Forest By-product and Speciality in China*. 2009 (3). doi: 10.13268/j .cnki .fbsic.2009.03.013

Yu, H., Xie, C. X., Song, J. Y., Zhou, Y. Q. & Chen, S. L. (2010). TCMGIS-II based prediction of medicinal plant distribution for conservation planning: a case study of *Rheum tanguticum*. *Chinese Medicine* 5: 31. doi: 10.1186/1749-8546-5-31

Zeng, L. H. (2008). *The Study and Demonstration of the Management of Secondary Forests in Tropical Regions for the Purpose of Enhancing Economic and Ecological Benefits (Final Technical Report PD 294/04 Rev.4 (F), 1 August 2009),* Guangdong Academy of Forest, Guangzhou, ITTO. tree-breeding-gaf.com/upload/200810061 65831_6e9df84ce3a9b1ff176ea3717ea452c75885.pdf

Zeng, W. Z., Xu, G. B. & Wan, D. G. (1996). The investigation and authentication of Zi Hua Di Ding traded in Sichuan. [In Chinese]. *Zhong Yao Cai* 19: 127–8.

Zhang, B., Peng, Y., Zhang, Z., Liu, H., Qi, Y., Liu, S. & Xiao, P. G. (2010). GAP production of TCM herbs in China. *Planta Medica* 76: 1948–1955.

Zhang, H. Y., Bi, W. G., Yu, Y. & Liao, W. B. (2012a). *Angelica sinensis* (Oliv.) Diels in China: distribution, cultivation, utilization and variation. *Genetic Resources and Crop Evolution* 59: 607–13.

Zhang, L., Yan, J. B., Liu, X .M., Ye, Z. G., Yang, X. H., Meyboome, R., Chan, K., Shaw, D. & Duez, P. (2012b). Pharmacovigilance practice and risk control of Traditional Chinese Medicine drugs in China: Current status and future perspective. *Journal of Ethnopharmacology* 140: 519–525.

Zhang, Q. X. (1995–96). *Colour Atlas of Chinese Medicinal Herbs*. 3 vols. [In Chinese]. People's Health Publishing House, Beijing.

Zhang, X. Z. (2008). *Di Dao Yao Cai Tu Jian* Illustrations of Good Quality Chinese Medicines. 7 vols. [In Chinese]. China Medical University, Taichun.

Zhao, G. P., Dai, S. & Chen, R. S. (2006). *Zhong Yao Da Ci Dian.* 2 vols. [In Chinese]. Shanghai Scientific & Technical Publishers, Shanghai.

Zhao, Z. Z. (2004). *An Illustrated Chinese Materia Medica in Hong Kong*. School of Chinese Medicine, Hong Kong Baptist University, Hong Kong.

Zhao, Z. Z. & Chen, H. (eds) (2014). *Chinese Medicinal Identification: an Illustrated Approach*. Paradigm, New Mexico.

Zhao, Z., Guo, P. & Brand, E. (2012). The formation of daodi medicinal materials. *Journal of Ethnopharmacology* 140: 476–481.

Zhao, Z. Z., Hu, Y., Liang, Z. T., Yuen, J. P. S., Jiang, Z. H. & Leung, K. S. Y. (2006a). Authentication is fundamental for standardization of Chinese Medicines. *Planta Medica* 72: 865–874.

Zhao, Z. Z. & Li, Y. S. (2007). *Easily Confused Chinese Medicines in Hong Kong*. Chinese Medicine Merchants Association, Hong Kong.

Zhao, Z., Liang, Z., Chan, K., Lu, G., Lee, E. L., Chen, H. & Li, L. (2010). A unique issue in the standardization of Chinese materia medica: processing. *Planta Medica* 76: 1975–1986.

Zhao, Z. Z., Liang, Z. T., Jiang, Z. H., Leung, K. S. Y., Chan, C. L., Chan, H. Y., Sin, J., Man, T. O. & Law, K. W. (2008). Comparative study on the aristolochic acid I content of Herba Asari for safe use. *Phytomedicine* 15: 741–748.

Zhao, Z. Z. & Xiao, P. G. (eds) (2009–10). *Encyclopedia of Medicinal Plants*. 4 vols. Hong Kong Jockey Club Institute of Chinese Medicine, Hong Kong.

Zhao, Z. Z., Yuen, J. P. S., Wu, J. L., Yu, T. & Huang, W. H. (2006b). A systematic study on confused species of Chinese materia medica in the Hong Kong market. *Annals, Academy of Medicine, Singapore* 35: 764–769.

Zhou, S. L., Xiong, G. M., Li, Z. Y. & Wen, J. (2005). Loss of genetic diversity of domesticated *Panax notoginseng* F. H. Chen as evidenced by ITS sequence and AFLP polymorphism: a comparative study with *P. stipuleanatus* H. T. Tsai et K. M. Feng. *Journal of Integrative Plant Biology* (formerly *Acta Botanica Sinica)* 47: 107–115.

Zhou, Y. (2006). Preliminary study of the evaluation system of precious and endangered wild officinal plants in Changbai Mountains. *Acta Botanica Boreali-Occidentali Sinica* 26: 599–605.

Zhu, Y. P. (1998). *Chinese Materia Medica: chemistry, pharmacology and applications*. Harwood Academic, Amsterdam.

Zhu, Y. P. (2002). Toxicity of the Chinese Herb Mu Tong (*Aristolochia manshuriensis*). What history tells us. *Adverse Drug Reactions and Toxicological Reviews* 21: 171–177.

Zhu, S. H., Sunnassee, A., Yuan R. X., Ren L., Chen X. R. & Liu, L. (2010). Fatal renal failure due to the Chinese herb "GuanMu Tong" (*Aristolochia manshuriensis*): Autopsy findings and review of literature. *Forensic Science International* 199: e5–e7.

Zhu, X. Z., Zhang, Y. X., Liu, X. & Gao, T. (2015). Authentication of commercial processed *Glehniae Radix* (*Beishashen*) by DNA barcodes. *Chinese Medicine* 10: 35.

Zhuravlev, Y., Koren, O. & Khrolenko, Y. A. (2008). *Panax ginseng* natural populations; their past, current state and perspectives. *Acta Pharmacologica Sinica* 29: 1127–1136.

Zou, Z. C. & Qin, B. (1996). *Species Systematization and Quality Evaluation of Commonly used Chinese Traditonal Drugs*. North-China edition. [In Chinese]. Beijing Medical University Press, Beijing.

Index to plant and drug names

Key: Accepted plant names **bold black;** plant name synonyms *italic;* plant family names, common names and Pin Yin names roman black; accepted Pharmaceutical names **bold red** (i.e. those used as headers for accounts); other Pharmaceutical names roman red. Names in 'single quotes' are directly cited from earlier texts. Page numbers in **bold** indicate the first page of drug accounts.

Index to Chinese and Pin Yin names

Key: Page numbers in **bold** indicate the first page of drug accounts.

Index of indications (medical conditions treated)

Elevation

- 4,000 metres and higher
- 2,000 – 4,000 metres
- 500 – 2,000 metres
- 200 – 500 metres
- 0 – 200 metres

Expedition areas

Anhui: 2002, 2004, 2007

Chongging: 2006

Fujian: 2001

Gansu: 1999, 2007

Guangdong: 2001

Guangxi: 1998, 2001, 2007, 2009

Guizhou: 2006

Hainan:1998

Hebei: 1998, 1999, 2003, 2008, 2009, 2006, 2011

Henan: 2002

Hubei: 2007

Inner Mongolia: 1999, 2013

Jiangsu: 2004

Jiangxi: 2001

Jilin: 1999, 2001

Ningxia: 1999

Shaanxi: 1999, 2000, 2004, 2006

Sichuan: 1999, 2002, 2006, 2007

Xinjiang: 2012, 2013

Yunnan: 1998, 2007

Zhejiang: 2001, 2004

0 Miles 500

0 Kilometres 500 1000

Joint Kew-IMPLAD medicinal plant field expeditions in China (1998 – 2014)

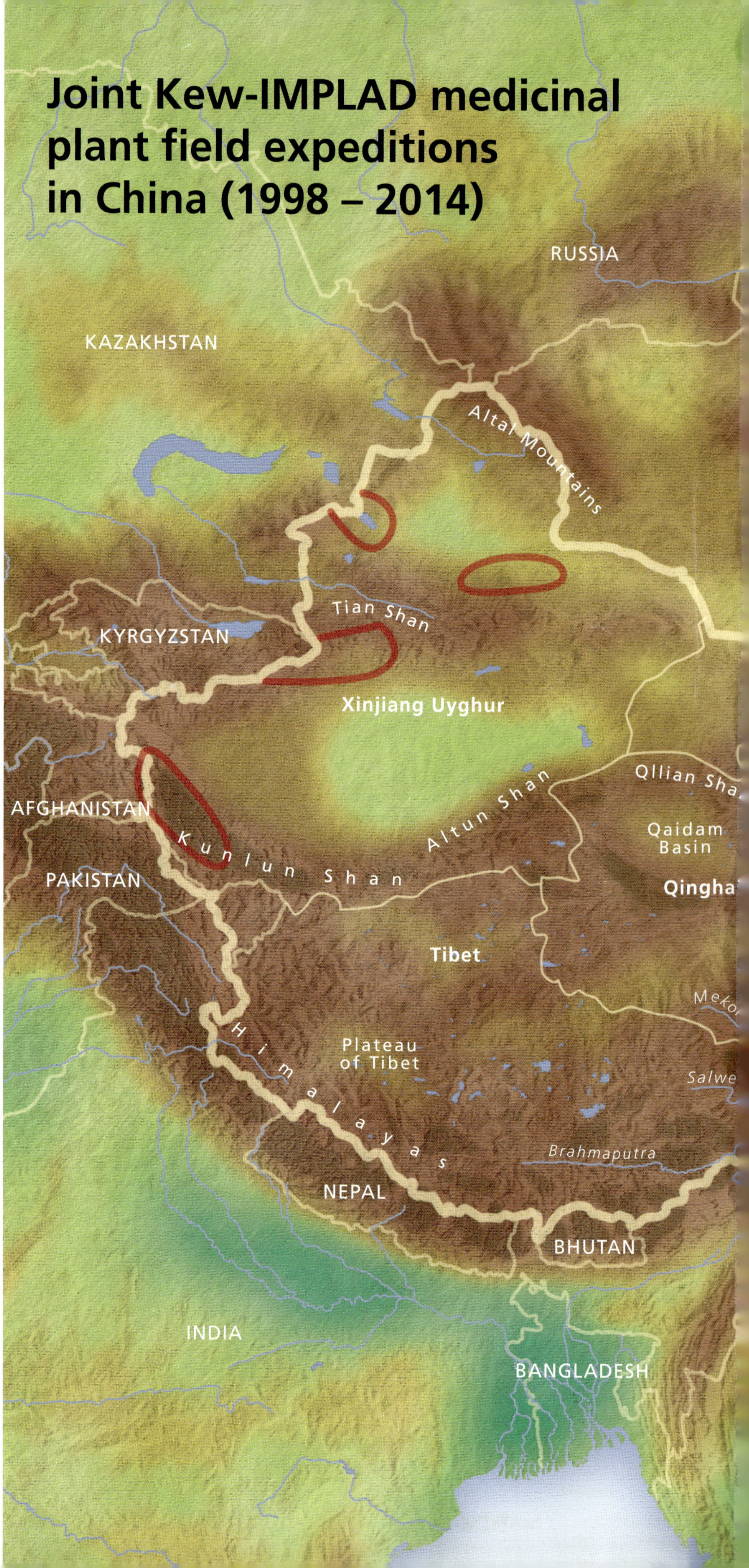